Fourth Edition

Microbiology
FUNDAMENTALS

A Clinical Approach

CDC/ Hannah A Bullock; Azaibi Tamin

Marjorie Kelly Cowan
Miami University Middletown

Heidi Smith
Front Range Community College

Jennifer Lusk
BSN RN CCRN, Clinical Advisor

Mc
Graw
Hill

MICROBIOLOGY FUNDAMENTALS: A CLINICAL APPROACH, FOURTH EDITION

1 2 3 4 5 6 7 8 9 LWI 24 23 22 21

ISBN 978-1-260-70243-9 (bound edition)
MHID 1-260-70243-X (bound edition)
ISBN 978-1-260-78605-7 (loose-leaf edition)
MHID 1-260-78605-6 (loose-leaf edition)

Portfolio Manager: *Lauren Vondra*
Product Developer: *Darlene M. Schueller*
Marketing Manager: *Tami Hodge*
Content Project Managers: *Jessica Portz and Tammy Juran*
Buyer: *Laura Fuller*
Designer: *David W. Hash*
Content Licensing Specialist: *Lori Hancock*
Cover Image: *(Earth Map): McGraw-Hill Education/Mark O. Martin, PhD, and Jennifer J. Quinn, PhD/photographers; (Micrograph): CDC/Hannah A. Bullock; Azaibi Tamin; (Stethoscope): ©JethroT/Shutterstock*
Compositor: *MPS Limited*

All credits appearing on page or at the end of the book are considered to be an extension of the copyright page.

Library of Congress Cataloging-in-Publication Data

Names: Cowan, M. Kelly, author. | Smith, Heidi (College teacher), author. | Lusk, Jennifer, author.
Title: Microbiology fundamentals : a clinical approach / Marjorie Kelly Cowan, Heidi Smith, Jennifer Lusk.
Description: Fourth edition. | New York, NY : McGraw-Hill Education, 2022. | Includes bibliographical references and index.
Identifiers: LCCN 2020008517 (print) | LCCN 2020008518 (ebook) | ISBN 9781260702439 (bound edition; alk. paper) | ISBN 126070243X (bound edition; alk. paper) | ISBN 9781260786057 (loose-leaf edition; alk. paper) | ISBN 1260786056 (loose-leaf edition; alk. paper) | ISBN 9781260786071 (epub)
Subjects: MESH: Microbiological Phenomena | Communicable Diseases
Classification: LCC QR41.2 (print) | LCC QR41.2 (ebook) | NLM QW 4 | DDC 579–dc23
LC record available at https://lccn.loc.gov/2020008517
LC ebook record available at https://lccn.loc.gov/2020008518

mheducation.com/highered

Brief Contents

Contributions by Ronald M. Atlas, University of Louisville

About the Authors

Marjorie Kelly Cowan, PhD, started teaching microbiology at Miami University in 1993. Her specialty is teaching microbiology for pre-nursing/allied health students at the university's Middletown campus, a regional open-admissions campus. She started life as a dental hygienist. She then went on to attain her PhD at the University of Louisville, and later worked at the University of Maryland's Center of Marine Biotechnology and the University of Groningen in The Netherlands. Kelly has published (with her students) 24 research articles stemming from her work on bacterial adhesion mechanisms and plant-derived antimicrobial compounds. But her first love is teaching—both doing it and studying how to do it better. She is past chair of the Undergraduate Education Committee of the American Society for Microbiology (ASM). Her current research focuses on the student achievement gap associated with economic disparities, as well as literacy in the science classroom. In her spare time, Kelly hikes, reads, and still tries to (s)mother her three grown kids.

©Kelly Cowan

Heidi Smith, MS, leads the microbiology department at Front Range Community College in Fort Collins, Colorado. Collaboration with other faculty across the nation, the development and implementation of new digital learning tools, and her focus on student learning outcomes have revolutionized Heidi's face-to-face and online teaching approaches and student performance in her classes. The use of digital technology has given Heidi the ability to teach courses driven by real-time student data and with a focus on active learning and critical thinking activities.

Heidi is an active member of the American Society for Microbiology and participated as a task force member for the development of their Curriculum Guidelines for Undergraduate Microbiology Education. At FRCC, Heidi directs a federal grant program designed to increase student success in transfer and completion of STEM degrees at the local university as well as facilitate undergraduate research opportunities for underrepresented students.

Off campus, Heidi spends as much time as she can enjoying the beautiful Colorado outdoors with her husband and four children.

©Heidi Smith

Jennifer Lusk, BSN, RN, CCRN, is a registered nurse at a large academic children's hospital in Denver, Colorado. She has practiced in pediatric intensive care for 10 years in large inner-city pediatric hospitals. Jennifer has spent her nursing career caring for critically ill children as a bedside nurse, charge nurse, and Continuous Renal Replacement Therapy (CRRT) specialist. She is the CRRT Clinical Program Coordinator, providing oversight and program development for the critical care dialysis therapy. She enjoys her diverse clinical role, which involves educating nurses and physicians, mentoring, researching, writing policies, and quality improvement work. In her time away from work, Jennifer enjoys spending time outdoors with her husband, son, and dog, especially hiking and exploring national parks.

©Tia Brayman

Preface

Students:

Welcome! I am so glad you are here. I am very excited for you to try this book. I wrote it after years of frustration, teaching from books that didn't focus on the right things that my students needed. My students (and, I think, you) need a solid but not overwhelming introduction to microbiology and infectious diseases. I asked myself: What are the major concepts I want my students to remember five years from now? And then I worked backward from there, making sure everything pointed to the big picture. And of course, the COVID-19 pandemic has made it clear how important this subject matter is to all of us.

While this book has enough detail to give you context, there is not so much detail that you will lose sight of the major principles. Biological processes are described right next to the illustrations that illustrate them. The format is easier to read than most books because there is only one column of text on a page and wider margins. The margins gave me space to add interesting illustrations and clinical content. A working nurse, Jennifer Lusk, brings her experience to life on the pages and shows you how this information will matter to you when you are working as a health care provider. We have interesting and up-to-the-moment Case Files, Medical Moments, Microbiome selections, and NCLEX® questions in every chapter. COVID-19 content is woven in from beginning to end. My coauthor, Heidi Smith, writes all of the online content specifically for this book. I don't think you'll find a better online set of learning tools anywhere.

I really wanted this to be a different kind of book. I use it in my own classes and my students love it! Well, maybe they *have* to say that, but I hope you truly do enjoy it and find it to be a refreshing kind of science book.

—Kelly Cowan

I dedicate this book to every front-line health care worker. During the COVID-19 pandemic your heroism has become obvious to all.–Kelly

I dedicate this book to the newest addition to our family, Kume, and all of the other diligent students who one day want to help people prevent and heal from disease.–Heidi

Instructors: Student Success Starts with You

Tools to enhance your unique voice

Want to build your own course? No problem. Prefer to use our turnkey, prebuilt course? Easy. Want to make changes throughout the semester? Sure. And you'll save time with Connect's auto-grading too.

65%
Less Time
Grading

Laptop: McGraw Hill; Woman/dog: George Doyle/Getty Images

Study made personal

Incorporate adaptive study resources like SmartBook® 2.0 into your course and help your students be better prepared in less time. Learn more about the powerful personalized learning experience available in SmartBook 2.0 at **www.mheducation.com/highered/connect/smartbook**

Affordable solutions, added value

Make technology work for you with LMS integration for single sign-on access, mobile access to the digital textbook, and reports to quickly show you how each of your students is doing. And with our Inclusive Access program you can provide all these tools at a discount to your students. Ask your McGraw Hill representative for more information.

Padlock: Jobalou/Getty Images

Solutions for your challenges

A product isn't a solution. Real solutions are affordable, reliable, and come with training and ongoing support when you need it and how you want it. Visit **www .supportateverystep.com** for videos and resources both you and your students can use throughout the semester.

Checkmark: Jobalou/Getty Images

SUPPORT AT
every step

Students: Get Learning that Fits You

Effective tools for efficient studying

Connect is designed to make you more productive with simple, flexible, intuitive tools that maximize your study time and meet your individual learning needs. Get learning that works for you with Connect.

Study anytime, anywhere

Download the free ReadAnywhere app and access your online eBook or SmartBook 2.0 assignments when it's convenient, even if you're offline. And since the app automatically syncs with your eBook and SmartBook 2.0 assignments in Connect, all of your work is available every time you open it. Find out more at
www.mheducation.com/readanywhere

> *"I really liked this app—it made it easy to study when you don't have your textbook in front of you."*
>
> - Jordan Cunningham,
> Eastern Washington University

Everything you need in one place

Your Connect course has everything you need—whether reading on your digital eBook or completing assignments for class, Connect makes it easy to get your work done.

Calendar: owattaphotos/Getty Images

Learning for everyone

McGraw Hill works directly with Accessibility Services Departments and faculty to meet the learning needs of all students. Please contact your Accessibility Services Office and ask them to email accessibility@mheducation.com, or visit
www.mheducation.com/about/accessibility
for more information.

UNIQUE INTERACTIVE QUESTION TYPES

Unique Interactive Question Types in Connect, Tagged to ASM's Curriculum Guidelines for Undergraduate Microbiology

1 **Case Study:** Case studies come to life in a learning activity that is interactive, self-grading, and assessable. The integration of the cases with videos and animations adds depth to the content, and the use of integrated questions forces students to stop, think, and evaluate their understanding. Pre- and post-testing allow instructors and students to assess their overall comprehension of the activity.

2 **Concept Maps:** Concept maps allow students to manipulate terms in a hands-on manner in order to assess their understanding of chapter-wide topics. Students become actively engaged and are given immediate feedback, enhancing their understanding of important concepts within each chapter.

3 **What's the Diagnosis:** Specifically designed for the disease chapters of the text, this is an integrated learning experience designed to assess the student's ability to utilize information learned in the preceding chapters to successfully culture, identify, and treat a disease-causing microbe in a simulated patient scenario. This question type is true experiential learning and allows the students to think critically through a real-life clinical situation.

4 **SmartGrid Questions:** SmartGrid questions replace the traditional end-of-chapter questions, and all of these questions are available for assignment in Connect. These questions were carefully constructed to assess chapter material as it relates to all six concepts outlined in the American Society of Microbiology curriculum guidelines plus the competency of "Scientific Thinking." The questions are cross-referenced with Bloom's taxonomy of learning level. Seven concepts × three increasing Bloom's levels = a robust assessment tool of 21 questions.

5 **Animations:** Animation quizzes pair our high-quality animations with questions designed to probe student understanding of the illustrated concepts.

6 **Animation Learning Modules:** Animations, videos, audio, and text all combine to help students understand complex processes. These tutorials take a stand-alone, static animation and turn it into an interactive learning experience for your students and include real-time remediation. Key topics have an Animated Learning Module assignable through Connect.

7 **Labeling:** Using the high-quality art from the textbook, check your students' visual understanding as they practice interpreting figures and learning structures and relationships.

8 **Classification:** Ask students to organize concepts or structures into categories by placing them in the correct "bucket."

9 **Sequencing:** Challenge students to place the steps of a complex process in the correct order.

10 **Composition:** Fill in the blanks to practice vocabulary, and then reorder the sentences to form a logical paragraph (these exercises may qualify as "writing across the curriculum" activities).

All McGraw Hill Connect content is tagged to Learning Outcomes for each chapter as well as topic, section, Bloom's Level, ASM topic, and ASM Curriculum Guidelines to assist you in customizing assignments and in reporting on your students' performance against these points. This will enhance your ability to assess student learning in your courses by allowing you to align your learning activities to peer-reviewed standards from an international organization.

NCLEX®

NCLEX® Prep Questions: Sample questions are available in Connect to assign to students, and there are questions throughout the book as well.

McGraw Hill Create is a self-service website that allows you to create custom course materials using McGraw Hill's comprehensive, cross-disciplinary content and digital products.

Writing Assignment

Available within McGraw Hill Connect® and McGraw Hill Connect® Master, the Writing Assignment tool delivers a learning experience to help students improve their written communication skills and conceptual understanding. As an instructor you can assign, monitor, grade, and provide feedback on writing more efficiently and effectively.

Tegrity in Connect is a tool that makes class time available 24/7 by automatically capturing every lecture. With a simple one-click start-and-stop process, you capture all computer screens and corresponding audio in a format that is easy to search, frame by frame. Students can replay any part of any class with easy-to-use, browser-based viewing on a PC, Mac, or other mobile device.

Educators know that the more students can see, hear, and experience class resources, the better they learn. Tegrity's unique search feature helps students efficiently find what they need, when they need it, across an entire semester of class recordings. Help turn your students' study time into learning moments immediately supported by your lecture.

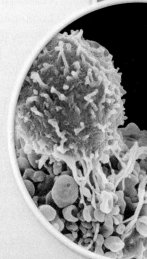

Steve Gschmeissner/
Getty Images

LearnSmart® Prep is designed to get students ready for a forthcoming course by quickly and effectively addressing prerequisite knowledge gaps that may cause problems down the road. This question bank highlights a series of questions, including Fundamentals of Science, Fundamentals of Math and Statistics, Fundamental Skills for the Scientific Laboratory, and Student Success, to give students a refresher on the skills needed to enter and be successful in their course! LearnSmart Prep maintains a continuously adapting learning path individualized for each student, and tailors content to focus on what the student needs to master in order to have a successful start in the new class.

Remote Proctoring & Browser-Locking Capabilities

 connect + proctorio

New remote proctoring and browser-locking capabilities, hosted by Proctorio within Connect, provide control of the assessment environment by enabling security options and verifying the identity of the student.

Seamlessly integrated within Connect, these services allow instructors to control students' assessment experience by restricting browser activity, recording students' activity, and verifying students are doing their own work.

Instant and detailed reporting gives instructors an at-a-glance view of potential academic integrity concerns, thereby avoiding personal bias and supporting evidence-based claims.

Steve Gschmeissner/
Science Source

Microbiology Fundamentals Laboratory Manual, Fourth Edition

Steven Obenauf, Broward College
Susan Finazzo, Perimeter College, Georgia State University

Written specifically for pre-nursing and allied health microbiology students, this manual features brief, visual exercises with a clinical emphasis.

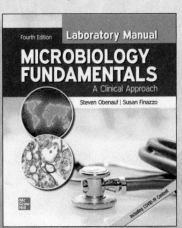

Fourth Edition   Laboratory Manual

MICROBIOLOGY
FUNDAMENTALS
A Clinical Approach

Steven Obenauf | Susan Finazzo

CLINICAL

Clinical applications help students see the relevance of microbiology.

COVID-19 Content   Throughout the text, small COVID-19 boxes appear in the margins next to content that can be connected to the pandemic virus. Since COVID is immediately relevant to students, it helps to pique their interest and imbues many fundamental concepts with a real life application.

Case File   Each chapter begins with a case written from the perspective of a former microbiology student who is working in health care now.

These high-interest introductions provide a specific example of how the chapter content is relevant to real life and future health care careers.

CASE FILE

Wound Care

I was an RN working in a large city hospital on a medical floor. A lot of our patients had diabetes and were suffering various complications of the disease, particularly diabetic wounds caused by poor circulation. Wound care was a large part of my job. After 2 years on the unit, I decided to pursue wound care certification. Once I became a wound care specialist, I continued to work in the same hospital and saw patients with complicated and/or chronic wounds.

Mr. Jones was one of the first patients I consulted about after I became certified. He was an elderly gentleman who had lost his sight due to diabetes

Shutterstock/GagliardiImages

SmartGrid: From Knowledge to Critical Thinking

SmartGrid

In place of traditional end-of-chapter questions, Kelly Cowan has created a grid made up of three columns and seven rows, for a total of 21 questions. The rows contain the six major curricular guidelines (and the competency of *scientific thinking*) from the American Society for Microbiology. The columns represent increasing levels of Bloom's taxonomy of learning. Each question is carefully constructed of material from the chapter that meets both the ASM guideline and the Bloom's level indicated. Instructors can assign a row (to emphasize a curriculum guideline) or a column (asking a variety of questions at a particular Bloom's level). The questions in column 3 (Bloom's level 5 and 6) can easily be used for group problem solving and other higher-order learning activities.

Medical Moment

Plastic Bottles for Clean Water

Every week around the world, 30,000 people die from lack of clean water. Ninety percent of these are children under 5 years old. Clean water—taken for granted in the developed world—is a resource more precious than gold on the rest of the planet. Even though we take it for granted, the processes and infrastructure used to deliver it to us are complex and expensive. How can we export those to other settings? Maybe we don't have to. Solar water disinfection is a method of safely disinfecting drinking water by simply placing contaminated water in a transparent plastic bottle and leaving it in the sun for 6 hours. Ultraviolet light kills bacteria and parasites and inactivates viruses, making the water safe. This technique has been used all over the world in impoverished nations where citizens have no access to clean drinking water, and it has proven to be an effective way of preventing diarrheal disease.

Q. Would you suspect that the water treated this way becomes sterile?

Answer in Appendix B.

©Khalil Senosi/AP Images

Medical Moment   These boxes give students a more detailed clinical application of a nearby concept in the chapter. Each Medical Moment ends with a question. Answers appear in Appendix B.

NCLEX® PREP

1. Which of the following factors would promote progression of an infection? Select all that apply.
   a. *low microbial virulence*
   b. *proper portal of entry*
   c. *genetic profile of host resistance to microbe*
   d. *no previous exposure to this infection*
   e. *host immunosuppression*

NCLEX® Prep Questions   Found throughout the chapter, these multiple-choice questions are application-oriented and designed to help students learn the microbiology information they will eventually need to pass the NCLEX® examination. Students will begin learning to think critically, apply information, and, over time, prep themselves for the examination.

Additional questions are available in Connect for homework and assessment.

The Microbiome   Each chapter ends with a reading about a microbiome discovery or story that is relevant to that chapter.

The Microbiome — The Gut and the Brain

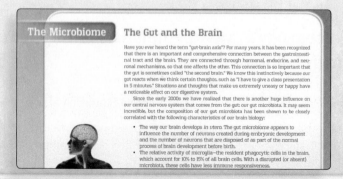

Have you ever heard the term "gut-brain axis"? For many years, it has been recognized that there is an important and comprehensive connection between the gastrointestinal tract and the brain. They are connected through hormonal, endocrine, and neuronal mechanisms, so that one affects the other. This connection is so important that the gut is sometimes called "the second brain." We know this instinctively because our gut reacts when we think certain thoughts, such as "I have to give a class presentation in 5 minutes." Situations and thoughts that make us extremely uneasy or happy have a noticeable effect on our digestive system.

Since the early 2000s we have realized that there is another huge influence on our central nervous system that comes from the gut: our gut microbiota. It may seem incredible, but the composition of our gut microbiota has been shown to be closely correlated with the following characteristics of our brain biology:

- The way our brain develops *in utero*. The gut microbiome appears to influence the number of neurons created during embryonic development and the number of neurons that are disposed of as part of the normal process of brain development before birth.
- The relative activity of microglia—the resident phagocytic cells in the brain, which account for 10% to 15% of all brain cells. With a disrupted (or absent) microbiota, these cells have less immune responsiveness.

VISUAL

Visually appealing layouts and vivid art closely linked to narrative for easier comprehension.

Engaging, Accurate, and Educational Art   Single column of text is easier to read and leaves space for eye-catching art to keep students engaged.

Infographics   New infographic-style visual summaries that students can relate to.

a pseudohypha, a chain of yeast cells formed when buds remain attached in a row (**figure 4.13**). Because of its manner of formation, it is not a true hypha like that of molds. While some fungal cells exist only in a yeast form and others occur primarily as hyphae, a few are classified as **dimorphic**. This means they can take either form, depending on growth conditions, such as changing temperature. Several fungi that cause human disease are dimorphic.

Many fungi make their home on the human body, as part of the normal human microbiome. Yet nearly 300 species of fungi can also cause human disease. The Centers for Disease Control and Prevention currently identifies three types of fungal disease in humans: (1) community-acquired infections in the general population caused by environmental pathogens, (2) hospital-associated infections caused by fungal pathogens in clinical settings, and (3) opportunistic infections caused by low-virulence species infecting already-weakened individuals (**table 4.3**).

Mycoses (the term for fungal infections) vary in the way the pathogen enters the body and the degree of tissue involvement they display. Even so-called

voriconazole?
a. *cell membrane*
b. *nucleus*
c. *ribosomes*
d. *mitochondria*

©BSIP/UIG/Getty Images

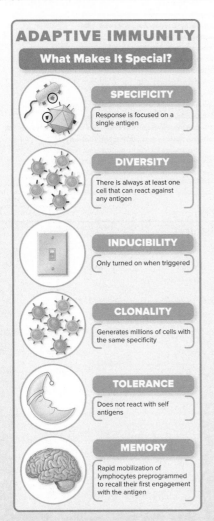

Visual Tables   The most important points explaining a concept are distilled into table format and paired with the relevant art.

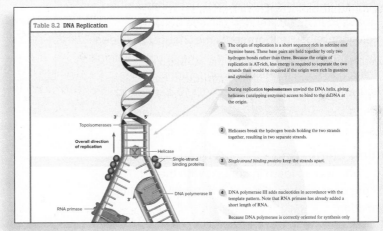

Table 8.2 DNA Replication

1   The origin of replication is a short sequence rich in adenine and thymine bases. These base pairs are held together by only two hydrogen bonds rather than three. Because the origin of replication is AT-rich, less energy is required to separate the two strands than would be required if the origin were rich in guanine and cytosine.

During replication **topoisomerases** unwind the DNA helix, giving helicases (unzipping enzymes) access to bind to the dsDNA at the origin.

2   Helicases break the hydrogen bonds holding the two strands together, resulting in two separate strands.

3   *Single-strand binding proteins keep the strands apart.*

4   DNA polymerase III adds nucleotides in accordance with the template pattern. Note that RNA primase has already added a short length of RNA.

Because DNA polymerase is correctly oriented for synthesis only

Topoisomerases
Overall direction of replication
Helicase
Single-strand binding proteins
DNA polymerase III
RNA primase

Process Figures   Complex processes are broken into easy-to-follow steps. Numbered steps help students walk through the figure.

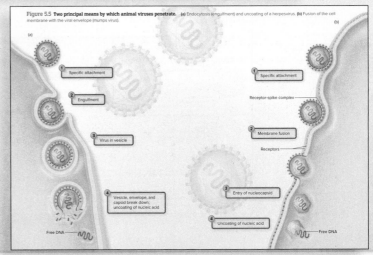

Figure 5.5 **Two principal means by which animal viruses penetrate.**   (a) Endocytosis (engulfment) and uncoating of a herpesvirus. (b) Fusion of the cell membrane with the viral envelope (mumps virus).

(a)
1 Specific attachment
2 Engulfment
Virus in vesicle
Vesicle, envelope, and capsid break down; uncoating of nucleic acid
Free DNA

(b)
1 Specific attachment
Receptor-spike complex
2 Membrane fusion
Receptors
3 Entry of nucleocapsid
4 Uncoating of nucleic acid
Free DNA

BRIEF

Streamlined coverage of core concepts helps students retain the information they will need for advanced courses.

Martin Oeggerli/
Science Source

Chemistry topics required for understanding microbiology are combined with the foundational content found in chapter 1.

Basic genetics and genetic engineering are synthesized into one chapter covering the concepts that are key to microbiology students.

A chapter in microbiology textbooks that is often not used in health-related classes becomes relevant because it presents the 21st-century idea of "One Health"—that the environment and animals influence human health and infections. This is extremely relevant to the COVID-19 pandemic.

Duplication Eliminated   Detail is incorporated into figures so students can learn in context with the art. This allows a more concise narrative flow while still retaining core information.

Contributions by Ronald M. Atlas, University of Louisville

iii

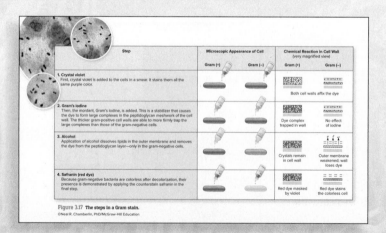

Figure 3.17 The steps in a Gram stain.
O'Neal R. Chamberlin, PhD/McGraw-Hill Education

Changes to the Fourth Edition

Significant Changes

COVID-19 content (in addition to being featured in the respiratory disease chapter and the One Health chapter) is included throughout the text, in small COVID-19 boxes. Since this is immediately relevant to students, it helps to pique their interest and embed many different concepts in a real life scenario.

New chapter summaries The chapter summaries have been converted into succinct, visual infographics, a format students are accustomed to and that provides the high points in a format that is reproduced chapter after chapter.

Short author commentaries Several figures in each chapter are accompanied by a word bubble from the authors, emphasizing key parts of the figure or interesting aspects.

Chapter Highlights

Chapter 1 New infographic about the types of microorganisms we will study in the book. Added section on the cautions we should take about research on the microbiome: that it *is* big but many results are still preliminary.

Chapter 3 I describe my own recent experience with *C. diff.* Throughout the book, the genus *Clostridium* has been updated to *Clostridioides* in the case of *C. diff.*

Chapter 4 Emphasis on the rise of fungal infections in immunocompromised persons. "The progress of these infections in immunosuppressed people was vividly described: They'll just rot you down quick as a flash," said one of the authors."

Chapter 5 New infographic explaining cytopathic effects.

Chapter 6 Made illustrations of levels of oxygen growth clearer. Clarified the illustration of serial dilution counting of growth.

Chapter 7 The commentaries on the (often difficult) figures are particularly helpful in this chapter.

Chapter 8 A comment on the flow-of-genetic-information figure, emphasizing how the knowledge of the regulatory RNAs is new, and how science works. Same type of commentary about mutations. More epigenetic discussion. A commentary about the similarity between binary fission and PCR processes. Discussion of using DNA for information storage.

Chapter 10 New discussion about how the microbiome affects efficacy of antimicrobial drugs. FDA ban of antimicrobials in agriculture.

Chapter 11 New sites of the microbiome: placenta, etc. Changes in epidemiology of hospital-acquired infections (HAIs). Important information about the emergence of *Candida auris*.

Chapter 12 Comment bubble points out the discovery of lymphatic system in brain. Made figures of phagocytosis and of interferon much clearer, with commentary bubble on the latter.

Chapter 13 More explanation on the figure of genetic rearrangements for antibody diversity.

Chapter 14 Added the newly appreciated influence of T cells on allergies.

Chapter 15 Explanation of the NAAT (nucleic acid amplification techniques) and how they relate to the techniques we have already described. The new practice of PCR'ing tissue samples.

Chapter 16 Measles epidemics due to lack of vaccinations. *Malassezia* association with pancreatic cancer.

Chapter 17 Demoted Zika virus disease from a Highlight disease; latest data about acute flaccid myelitis. Highlighted the Global Polio Eradication Initiative. Updated the arbovirus epidemiology.

Chapter 18 Did some clarification of the size of insects involved in various diseases. Ebola updates.

Chapter 19 COVID-19 featured as a highlight disease. Added *Candida auris* otitis media; expanded and updated discussion of influenza vaccines; new epidemiology of whooping cough; use of gene amplification and antibiotic susceptibility testing for tuberculosis; new section on novel coronaviruses.

Chapter 20 Thirty-state *Campylobacter* outbreak from handling puppies; rise of antibiotic resistance in *Helicobacter;* new infographic about foodborne disease outbreaks.

Chapter 21 A commentary that helps students understand data reported as "per 100,000 people."

Chapter 22 Critical information and graphics describing how the COVID-19 pandemic illustrates the principles of One Health. Added discussion of the emergence of *Candida auris* and the influence of climate change; the creep northward of mosquito and tick-borne diseases; hepatitis A epidemic in the United States; wildfires in California and Australia; longer duration of mosquito-borne disease season.

Acknowledgments

I am always most grateful to the students in my classes. They teach me every darned day how to do a better job helping them understand these concepts that are familiar to me but new to them. All the instructors who reviewed the manuscript were also great allies. I thank them for lending me some of their microbiological excellence. Heidi Smith improves everything she touches, especially this book. Jennifer Lusk added very meaningful medical insights and clinical content. My handlers at McGraw Hill make the wheels go around. They include Lauren Vondra, Tami Hodge, Jessica Portz, David Hash, Tammy Juran, Laura Fuller, and Lori Hancock. Darlene Schueller, my day-to-day editor, is a wonderful human being and taskmaster, in that order. In short, I'm just a lucky girl surrounded by talented people.

—Kelly Cowan

I, too, am thankful for the many students that enter my classroom every year. My work is greatly inspired by their questions, their ideas, and their unique way of looking at complex things. I am filled with gratitude to Kelly Cowan for her willingness to partner with me; I constantly learn from her about better ways to reach all students. I appreciate the entire microbiology team at McGraw Hill who truly believes in producing materials that lead to student success.

—Heidi Smith

Reviewers

Janet M. Dowding
Miami Dade College

Tod R. Fairbanks
Palm Beach State College

Shonteria L. Johnson
Southeastern Community College

Mustapha Lahrach
Hillsborough Community College

Christopher Lupfer
Missouri State University

Laurie Shannon Meadows
Rowan College at Burlington County

Dick Wells
Ozarks Technical Community College

Erin Windsor
College of Southern Nevada

Elizabeth Yelverton
Pensacola State College

We are very pleased to have been able to incorporate real student data points and input, derived from thousands of our SmartBook users, to help guide our revision. SmartBook heat maps provided a quick visual snapshot of usage of portions of the text and the relative difficulty students experienced in mastering the content. With these data, we were able to hone not only our text content but also the SmartBook questions.

Contents

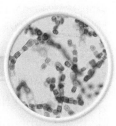

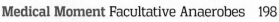

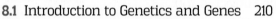

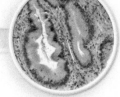

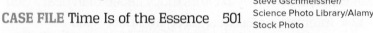

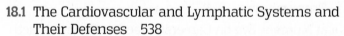

1

Introduction to Microbes and Their Building Blocks

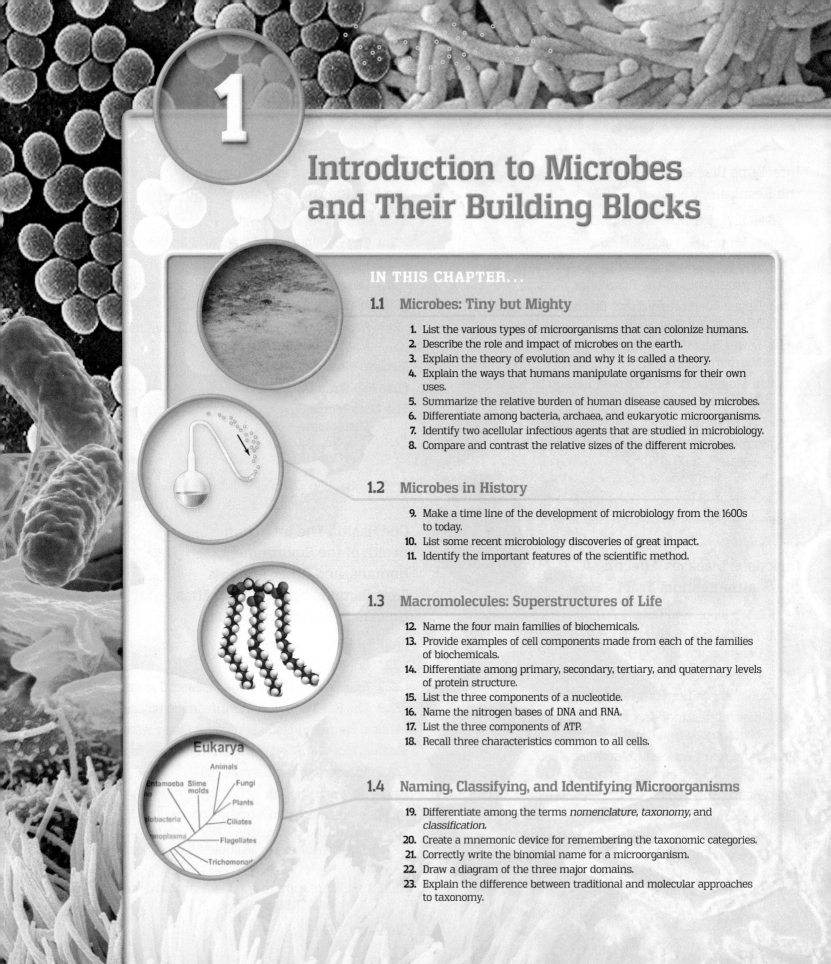

Eukarya

Animals

Entamoeba Slime molds

Fungi

Plants

Ciliates

obacteria

noplasma

Flagellates

Trichomona

CASE FILE

The Subject Is You!

At the beginning of every chapter in this book, a different health care worker will tell you a story about something "microbiological" that happened to him or her in the line of duty. For this first chapter, though, I am claiming "dibs" as author and am going to introduce myself to you by telling you about the first day of class in my course.

Long ago I noticed that students have a lot of anxiety about their microbiology course. I know that starts you out with one strike against you because attitudes are powerful determinants of our success. So on the first day of class I often spend some time talking with students about how much they already know about microbiology.

Sometimes I start with "How many of you have taken your kids for vaccinations?" since in the classes I teach very many students are parents. Right away students will tell me why they or friends they know have not vaccinated their children, and I can tell them there's a sophisticated microbiological concept they are referencing, even if they aren't naming it: *herd immunity*, discussed in chapter 11 of this book.

Of course, nowadays, everyone has some type of experience with COVID-19, and the virus that causes it, SARS-CoV-2. The whole world has undergone a crash course in infection, and in epidemiology. This course will help you sort out all of the information that has been pouring out about COVID-19.

- Think about how many times you have taken antibiotics in the past few years. What is special about antibiotics that they are only given to treat infections?

- What is the most unusual infection you have ever encountered among family or friends or patients you have cared for?

Case File Wrap-Up appears at the end of the chapter.

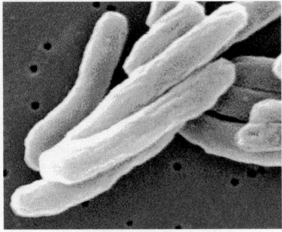

Mycobacterium tuberculosis bacteria

Source: CDC/Janice Carr

1.1 Microbes: Tiny but Mighty

Microbiology is a specialized area of biology that deals with living things ordinarily too small to be seen without magnification. Such **microscopic** organisms are collectively referred to as **microorganisms** (my″-kroh-or′-gun-izms), **microbes,** or several other terms depending on the kind of microbe or the purpose. There are several major groups of microorganisms that we'll be studying. They can be either cellular or noncellular. The cellular microorganisms we will study are **bacteria, archaea, fungi, and protozoa.** Another cellular organism that causes human infections is not technically a microorganism. **Helminths** are multicellular animals whose mature form is visible to the naked eye. Acellular microorganisms causing human disease are the **viruses** and **prions. Table 1.1** gives you a first glimpse at these microorganisms. There is another very important group of organisms called algae. They are critical to the health of the biosphere but do not directly infect humans, so we will not consider them in this book. Each of the other seven groups contains members that colonize humans, so we will focus on them.

The nature of microorganisms makes them both very easy and very difficult to study—easy because they reproduce so rapidly and we can quickly grow large populations in the laboratory, and difficult because we usually can't see them directly. We rely on a variety of indirect means of analyzing them in addition to using microscopes.

Table 1.1 The Types of Microorganisms We Will Study in this Book

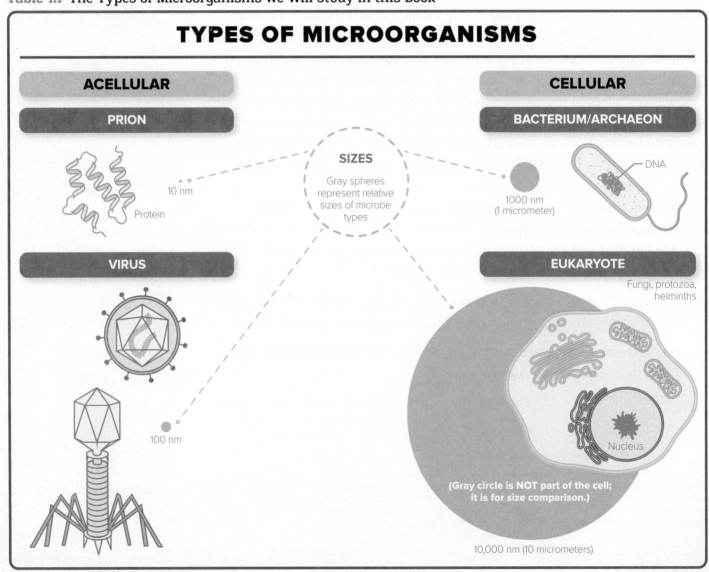

Microbes and the Planet

For billions of years, microbes have extensively shaped the development of the earth's habitats and the evolution of other life forms. It is understandable that scientists searching for life on other planets first look for signs of microorganisms.

Single-celled organisms appeared on this planet about 3.8 billion years ago according to the fossil record. One of these organisms–referred to as LCA, or the Last Common Ancestor–eventually led to the appearance of two newer single cell types, called bacteria and archaea. A little bit later this single-celled ancestor gave rise to **eukaryotic** (yoo-kar'-ee-ot-ic) cells. The type of cell known as LCA no longer exists. Only its "offspring"–bacteria, archaea, and eukaryotes–remain. *Eu-kary* means "true nucleus," and these were the only cells containing a nucleus. Bacteria and archaea have no true nucleus. For that reason, they have traditionally been called **prokaryotes** (pro-kar'-ee-otes), meaning "prenucleus." But researchers are suggesting we no longer use the term *prokaryote* to lump them together because archaea and bacteria are so distinct genetically. Some scientists have started calling them **akaryotes,** meaning "no nucleus." If you consider the seven types of microorganisms we will be dealing with in this book, you will recognize bacteria and archaea as each having their own domain. The protozoa, fungi, and helminths are all in the domain Eukarya. Viruses and prions do not appear on the tree of life because they are not cells, and not considered living. That sounds strange, but we will delve into that in the virus chapter, which comes later.

Figure 1.1 depicts the resulting tree of life–a diagram of all organisms on the planet. There are two important things to note about this figure. First, all of biologic life falls into these three categories, known as domains. Most of the organisms you are familiar with (animals, plants, etc.) are in one category, Eukarya. Second, these three domains all emerged from a single common cell type (the "stem" at the bottom).

Bacteria and archaea are predominantly single-celled organisms. Many eukaryotic organisms are also single-celled, but the eukaryotic cell type also developed into highly complex multicellular organisms such as worms and humans. In terms of numbers, eukaryotic cells are a small minority compared to the bacteria and archaea, but their larger size (and our own status as eukaryotes!) makes us perceive them as dominant to–and more important than–bacteria and archaea.

 Important Note to Students!

This is your author here. I wanted to alert you right up front that you should look at the figures and read the tables in the chapters. I know that it is human nature to skip these when you see the reference in the main text (like "**figure 1.4**") and just move on with the next sentence. But in this book I made a real point to put a lot of information in the figures and tables because it is easier to digest things such as processes and categories when they are presented in a more visual format. And there are a lot of "processes" and "categories" in biology! So I opted for a bit less text, and a bit more pictures and tables. So be sure to make it a point to stop by and examine these visual features. Thanks! Kelly

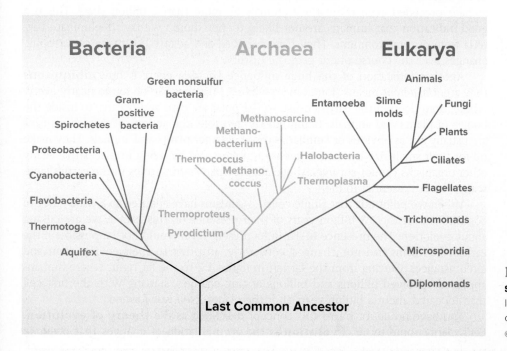

Figure 1.1 The tree of life: A phylogenetic system.   A system for representing the origins of cell lines and major taxonomic groups. There are three distinct cell lines placed in superkingdoms called domains.

©McGraw-Hill Education

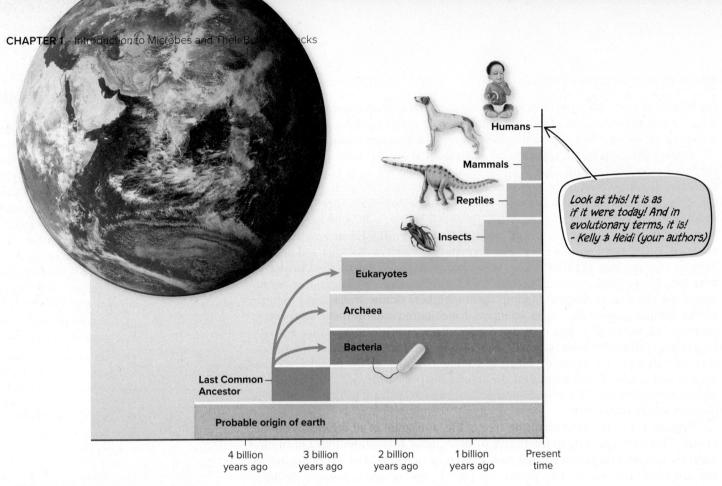

Figure 1.2 Evolutionary time line.
(photo): NASA/Goddard Space Flight Center

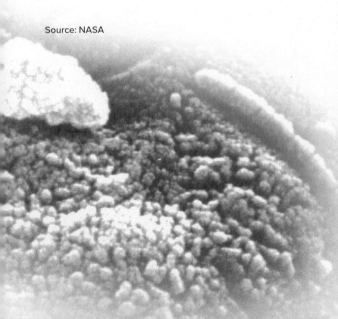

Source: NASA

Figure 1.2 depicts the time line of appearances of different types of organisms on earth. Starting on the left, you see that the ancestor cell type was here alone for quite a while before giving rise to the three domains of life. Eukaryotes came along last, and it took a very long time for single-celled eukaryotes to develop into more complex eukaryotic organisms (insects, reptiles, and mammals). On the scale pictured in the figure, humans just barely appeared in very recent earth history. Bacteria and archaea preceded even the earliest animals by more than 2 billion years. This is a good indication that humans are not likely to–nor should we try to–eliminate bacteria from our environment. They have survived and adapted to many catastrophic changes over the course of our geologic history.

Another indication of the huge influence bacteria exert is how **ubiquitous** they are. *Ubiquitous* means "found everywhere." Microbes can be found nearly everywhere, from deep in the earth's crust, to the polar ice caps and oceans, to inside the bodies of plants and animals. Being mostly invisible, the actions of microorganisms are usually not as obvious or familiar as those of larger plants and animals. They make up for their small size by their immense numbers and by living in places that many other organisms cannot survive. Above all, they play central roles in the earth's landscape that are essential to life.

When we point out that single-celled organisms have adapted to a wide range of conditions over the 3.5 billion years of their presence on this planet, we are talking about evolution. The presence of life in its present form would not be possible if the earliest life forms had not changed constantly, adapting to their environment and circumstances. Getting from the far left in figure 1.2 to the far right, where humans appeared, involved billions and billions of tiny changes, starting with the first cell that appeared about a billion years after the planet itself was formed.

You have no doubt heard this concept described as the **theory of evolution.** Let's clarify some terms. **Evolution** is the accumulation of changes that occur in organisms as they adapt to their environments. It is documented every day in all corners of the planet, an observable phenomenon testable by science. Scientists use the term *theory* in a different way than the general public does, which often leads to great confusion. In science, a theory begins as a hypothesis, or an educated guess to explain an observation. By the time a hypothesis has been labeled a *theory* in science, it has undergone years and years of testing and not been disproved. It is taken as fact. This is much different from the common usage, as in "My theory is that he overslept and that's why he was late." The theory of evolution, like the germ theory and many other scientific theories, refers to a well-studied and well-established natural phenomenon, not just a random guess.

How Microbes Shape Our Planet

Microbes are deeply involved in the flow of energy and food through the earth's ecosystems. Most people are aware that plants carry out **photosynthesis,** which is the light-fueled conversion of carbon dioxide to organic material, accompanied by the formation of oxygen (called oxygenic photosynthesis). However, bacteria invented photosynthesis long before the first plants appeared, first as a process that did not produce oxygen (*anoxygenic photosynthesis*). This anoxygenic photosynthesis later evolved into oxygenic photosynthesis, which not only produced oxygen but also was much more efficient in extracting energy from sunlight. Hence, these ancient, single-celled microbes were responsible for changing the atmosphere of the earth from one without oxygen to one with oxygen. The production of oxygen also led to the use of oxygen for aerobic respiration and the formation of ozone, both of which set off an explosion in species diversification. Today, photosynthetic microorganisms (mainly bacteria and algae) account for more than 70% of the earth's photosynthesis, contributing the majority of the oxygen to the atmosphere **(figure 1.3).**

In the long-term scheme of things, microorganisms are the main forces that drive the structure and content of the soil, water, and atmosphere. For example:

- The temperature of the earth is regulated by gases emitted by living organisms. These gases include carbon dioxide, nitrous oxide, and methane, which create an insulation layer in the atmosphere and help retain heat. Many of these gases are produced by microbes living in the environment and the digestive tracts of animals.
- The most abundant cellular organisms in the oceans are not fish but bacteria. Think of a 2-liter soda bottle. Two liters of surface ocean water contains approximately 1,000,000,000 (1 billion) bacteria. Each of these bacteria likely harbors thousands of viruses inside of it, making viruses the most abundant inhabitants of the oceans. The bacteria and their viruses are major contributors to photosynthesis and other important processes that create our environment. (Be careful here. The first sentence in this paragraph said that bacteria are the most abundant *cellular* organisms in oceans. But viruses, which are not cellular, far outnumber them.)
- Bacteria and fungi live in complex associations with plants that assist the plants in obtaining nutrients and water and may protect them against disease. Microbes form similar interrelationships with animals, notably, in the stomach of cattle, where a rich assortment of bacteria digests the complex carbohydrates of the animals' diets and causes the animals to release large amounts of methane into the atmosphere.

Microbes and Humans

Microorganisms clearly have monumental importance to the earth's operation. Their diversity and versatility make them excellent candidates for being used by humans

Medical Moment

Medications from Microbes

Penicillin is a worthy example of how microorganisms can be used to improve human life. Alexander Fleming, a Scottish bacteriologist, discovered penicillin quite by accident in 1928. While growing several bacterial cultures in Petri dishes, he accidentally forgot to cover them. They remained uncovered for several days. When Fleming checked the Petri dishes, he found them covered with mold. Just before Fleming went to discard the Petri dishes, he happened to notice that there were no bacteria to be seen around the mold—in other words, the mold was killing all of the bacteria in its vicinity.

Recognizing the importance of this discovery, Fleming experimented with the mold (of the genus *Penicillium*) and discovered that it effectively stopped or slowed the growth of several bacteria. The chemical that was eventually isolated from the mold—penicillin—became widely used during the Second World War and saved many soldiers' lives, in addition to cementing Fleming's reputation.

Q. Can you think of a logical reason that a microbe (the fungus) would produce a chemical that harms another microbe (the bacteria)?

Answer in Appendix B.

Figure 1.3 **A rich photosynthetic community.**
Summer pond with a thick mat of algae.
Jerome Wexler/Science Source

Figure 1.4 The 2011 Gulf oil spill.   There is evidence that ocean bacteria metabolized ("chewed up") a lot of the spilled oil.

Chief Petty Officer John Kepsimelis/US Coast Guard

COVID-19

The top causes of death in table 1.2 do not include data for COVID-19. As of mid-2020, COVID-19 deaths would not have registered on the graph for worldwide deaths, but the numbers of deaths it caused in the United States would register in the top five.

for our own needs, and for them to "use" humans for their needs, sometimes causing disease along the way. We'll look at both of these kinds of microbial interactions with humans in this section.

By accident or choice, humans have been using microorganisms for thousands of years to improve life and even to shape civilizations. Baker's and brewer's yeasts are types of single-celled fungi that cause bread to rise and ferment sugar into alcohol to make wine and beers. Other fungi are used to make special cheeses such as Roquefort or Camembert. Historical records show that households in ancient Egypt kept moldy loaves of bread to apply directly to wounds and lesions. When humans manipulate microorganisms to make products in an industrial setting, it is called **biotechnology.** For example, some specialized bacteria have unique capacities to mine precious metals or to clean up human-created contamination.

Genetic engineering is an area of biotechnology that manipulates the genetics of microbes, plants, and animals for the purpose of creating new products and genetically modified organisms (GMOs). One powerful technique for designing GMOs is called **recombinant DNA technology.** This technology makes it possible to transfer genetic material from one organism to another and to deliberately alter DNA. Bacteria and fungi were some of the first organisms to be genetically engineered. This was possible because they are single-celled organisms and they are so adaptable to changes in their genetic makeup. Recombinant DNA technology has unlimited potential in terms of medical, industrial, and agricultural uses. Microbes can be engineered to synthesize desirable products such as drugs, hormones, and enzymes. It has become popular to dislike GMOs. As with any technological advance, the capacity to create GMOs can have both positive and negative aspects. Your job is to learn about them, so that you can have an informed opinion.

Another way of tapping into the unlimited potential of microorganisms is the science of **bioremediation** (by'-oh-ree-mee-dee-ay"-shun). This term refers to the ability of microorganisms—ones already present or those introduced intentionally—to restore the stability of an ecosystem or to clean up toxic pollutants. Microbes have a surprising capacity to break down chemicals that would be harmful to other organisms **(figure 1.4).** This includes even human-made chemicals that scientists have developed and for which there are no natural counterparts.

Microbes Harming Humans

One of the most fascinating aspects of the microorganisms with which we share the earth is that, despite all of the benefits they provide, they also contribute significantly to human misery as **pathogens** (path'-oh-jenz). The vast majority of microorganisms that associate with humans cause no harm. In fact, they provide many benefits to their human hosts. Note that a diverse microbial biota living in and on humans is an important part of human well-being. However, humankind is also plagued by nearly 2,000 different microbes that can cause various types of disease. Any disease caused by a microorganism is termed an **infectious disease.** Many diseases are not caused by microorganisms, but by genetic defects, imbalances in body systems, exposure to chemicals in the environment, among others. Infectious diseases still devastate human populations worldwide, despite significant strides in understanding and treating them. The World Health Organization (WHO) estimates there are a total of 10 billion new infections across the world every year. Infectious diseases are important common causes of death in much of humankind, and they still kill a significant percentage of the U.S. population. **Table 1.2** depicts the 10 top causes of death per year (by all causes, infectious and noninfectious) in the United States and worldwide.

We are also witnessing an increase in the number of new (emerging) and older (reemerging) diseases. AIDS, hepatitis C, Zika virus, West Nile virus, and tuberculosis are examples. It is becoming clear that human actions in the form of deforestation, industrial farming techniques, and chemical and antibiotic usage can foster the emergence or reemergence of particular infectious diseases. These patterns will be discussed in chapter 22.

Table 1.2 Top Causes of Death—All Causes

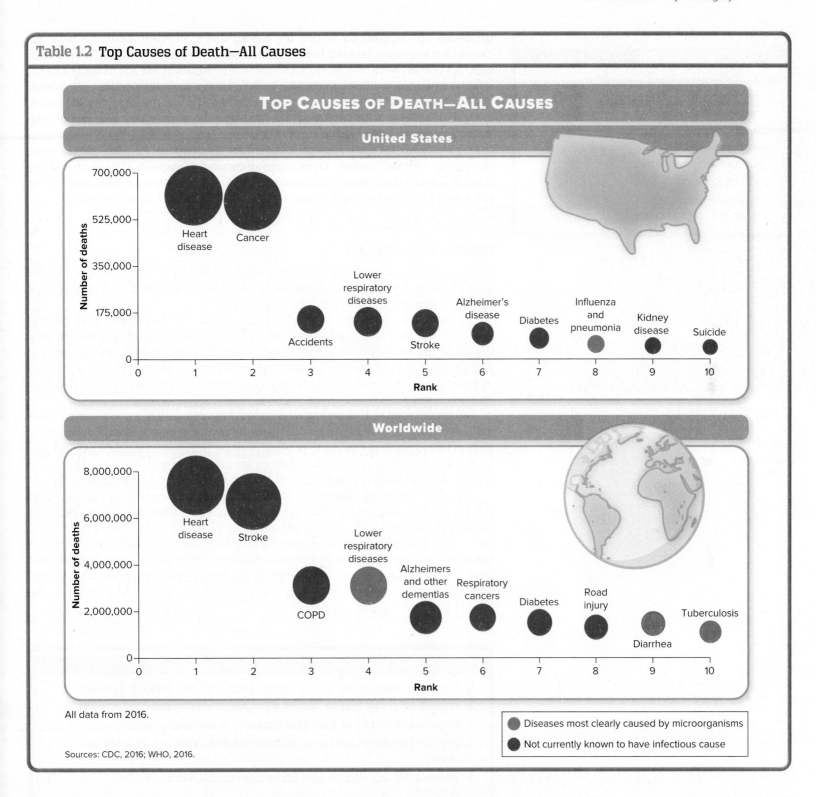

TOP CAUSES OF DEATH—ALL CAUSES

United States

Worldwide

All data from 2016.

Sources: CDC, 2016; WHO, 2016.

● Diseases most clearly caused by microorganisms
● Not currently known to have infectious cause

One of the most eye-opening discoveries in recent years is that many diseases that used to be considered noninfectious probably do involve microbial infection. One well-known example is that of gastric ulcers, now known to be caused by a bacterium called *Helicobacter*. But there are more. Diseases as different as multiple sclerosis, obsessive compulsive disorder, coronary artery disease, and even obesity have been linked to chronic infections with microbes. It seems that the golden age of microbiological discovery, during which all of the "obvious" diseases were characterized and cures or preventions were devised for them, should more accurately be referred to as the *first* golden age. We're now discovering the subtler side of microorganisms. Later in this chapter we will introduce the human microbiome–the microbes that call the human body home from birth onward. We will see that variations in the microbiome also determine a person's tendency to develop both infectious and noninfectious conditions.

Another important development in infectious disease trends is the increasing number of patients with weakened defenses, who, because of welcome medical advances, are living active lives instead of enduring long-term disability or death from their conditions. They are subject to infections by common microbes that are not pathogenic to healthy people. There is also an increase in microbes that are resistant to drugs. It appears that even with the most modern technology available to us, microbes still have the "last word," as the great French scientist Louis Pasteur observed.

What Are They Exactly?

Cellular Organization

As discussed earlier, two basic cell types appeared during evolutionary history. The bacteria and archaea, along with eukaryotic cells, differ not only in the complexity of their cell structure but also in contents and function.

In general, bacterial and archaeal cells are about 10 times smaller than eukaryotic cells, and they lack many of the eukaryotic cell structures such as **organelles.** Organelles are small, membrane-bound structures in the eukaryotic cell that perform specific functions and include the nucleus, mitochondria, and chloroplasts. Examples of bacteria, archaea, and eukaryotic microorganisms are covered in more detail in chapters 3 and 4.

All bacteria and archaea are microorganisms, but only some eukaryotes are microorganisms **(figure 1.5)**. Also, of course, humans are eukaryotes. Certain small eukaryotes–such as helminths (worms), many of which can be seen with the naked eye–are also included in the study of infectious diseases because of the way they are transmitted and the way the body responds to them, though they are not microorganisms.

As stated previously, viruses are not independently living cellular organisms. Instead, they are small particles that are at a level of complexity somewhere between large molecules and cells. Viruses are much simpler than cells. Outside their host, they are composed of a small amount of hereditary material (either DNA or RNA but never both) wrapped up in a protein covering. Some viruses have an additional layer, a lipid membrane that is exterior to the protein part. Then we have prions, which are even simpler than viruses. They contain no nucleic acid, only protein, but act like infectious microorganisms.

1.1 LEARNING OUTCOMES—Assess Your Progress

1. List the various types of microorganisms that can colonize humans.
2. Describe the role and impact of microbes on the earth.
3. Explain the theory of evolution and why it is called a theory.
4. Explain the ways that humans manipulate organisms for their own uses.
5. Summarize the relative burden of human disease caused by microbes.
6. Differentiate among bacteria, archaea, and eukaryotic microorganisms.
7. Identify two acellular infectious agents that are studied in microbiology.
8. Compare and contrast the relative sizes of the different microbes.

1.2 Microbes in History

If not for the extensive interest, curiosity, and devotion of thousands of microbiologists over the last 300 years, we would know little about the microscopic realm that surrounds us. Each additional insight, whether large or small, has added to our current knowledge of living things and processes. And the discoveries continue. Every day brings new surprises and insights into the microbial world. This section summarizes the prominent discoveries made in the past 300 years.

Spontaneous Generation

From very earliest history, humans noticed that when certain foods spoiled, they became inedible or caused illness, and yet other "spoiled" foods did no harm and even had enhanced flavor. Indeed, several centuries ago, there was already a sense that diseases such as the Black Plague and smallpox were caused by some sort of transmissible matter. But the causes of such phenomena were vague and obscure because, frankly, we couldn't *see* anything amiss. Consequently, they remained cloaked in mystery and regarded with superstition–a trend that led even well-educated scientists to believe in a concept called **spontaneous generation.** This was the belief that invisible vital forces present in matter led to the creation of life. The belief was continually reinforced as people observed that meat left out in the open soon "produced" maggots, that mushrooms appeared on rotting wood, seemingly out of nowhere, that rats and mice emerged from piles of litter, and other similar phenomena. Though some of these early ideas seem quaint and ridiculous in light of modern knowledge, we must remember that, at the time, mysteries in life were accepted and the scientific method was not widely practiced. Even after single-celled organisms

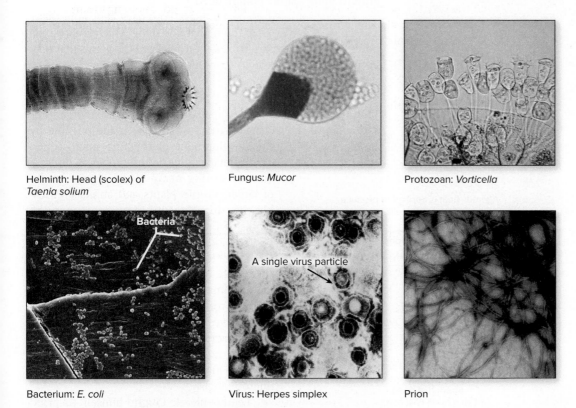

Helminth: Head (scolex) of *Taenia solium*

Fungus: *Mucor*

Protozoan: *Vorticella*

Bacterium: *E. coli*

Virus: Herpes simplex

Prion

Figure 1.5 Six types of microorganisms. Archaea are not pictured here.

Source: CDC/Dr. Mae Melvin (*Taenia solium*); CDC/Dr. Lucille K. Georg (*Mucor*); Nancy Nehring/E+/Getty Images (*Vorticella*); Janice Haney Carr/CDC (*E. coli*); Dr. Erskine Palmer/CDC (Herpes simplex); Cultura/Shutterstock (Prion)

Wine, cheese, and bread are each made using bacteria and fungi.

©lynx/iconotec.com/ Glow Images

were discovered during the mid-1600s, the idea of spontaneous generation continued to exist. Some scientists assumed that microscopic beings were an early stage in the development of more complex ones.

Over the subsequent 200 years, scientists waged an experimental battle over the two hypotheses that could explain the origin of simple life forms. Some tenaciously clung to the idea of **abiogenesis** (*a* = "without"; *bio* = "life"; *genesis* = "beginning"–"beginning in absence of life"), which embraced spontaneous generation. On the other side were advocates of **biogenesis** ("beginning with life") saying that living things arise only from others of their same kind. There were serious advocates on both sides, and each side put forth what appeared on the surface to be plausible explanations of why its evidence was more correct. Finally in the mid-1800s, the acclaimed chemist and microbiologist Louis Pasteur entered the arena. He had recently been studying the roles of microorganisms in the fermentation of beer and wine, and it was clear to him that these processes were brought about by the activities of microbes introduced into the beverage from air, fruits, and grains. The methods he used to discount spontaneous generation were simple yet brilliant.

To demonstrate that air and dust were the source of microbes, Pasteur filled flasks with broth and shaped their openings into long, swan-neck-shaped tubes **(figure 1.6).** The flasks' openings were freely open to the air but were curved so that gravity would cause any airborne dust particles to deposit in the lower part of the necks. He heated the flasks to sterilize the broth and then incubated them. As long as the flask remained intact, the broth remained sterile; but if the neck was broken off so that dust fell directly down into the container, microbial growth immediately commenced.

Pasteur summed up his findings, "For I have kept from them, and am still keeping from them, that one thing which is above the power of man to make; I have kept from them the germs that float in the air, I have kept from them life."

The Role of the Microscope

True awareness of the widespread distribution of microorganisms and some of their characteristics was finally made possible by the development of the first microscopes. These devices revealed microbes as discrete entities sharing many of the characteristics of larger, visible plants and animals. Probably the earliest record of microbes is in the works of Englishman Robert Hooke. In the 1660s, Hooke studied many different materials, including household objects, plants, and trees. He described cellular structures for the first time (in tree bark) and drew sketches of "little structures" that seemed to be alive. Hooke paved the way for even more exacting observations of microbes by Antonie van Leeuwenhoek (lay′-oo-wun-hook), a Dutch linen merchant and self-made microbiologist.

Leeuwenhoek taught himself to grind glass lenses to ever-finer specifications so he could see the threads in his fabrics with better clarity. Eventually, he became interested

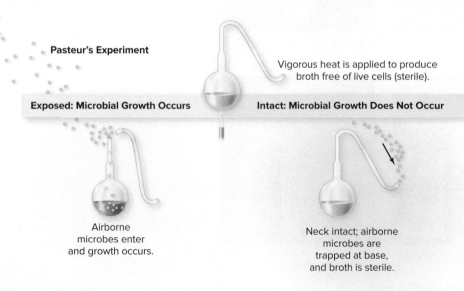

Pasteur's Experiment

Vigorous heat is applied to produce broth free of live cells (sterile).

Exposed: Microbial Growth Occurs   **Intact: Microbial Growth Does Not Occur**

Airborne microbes enter and growth occurs.

Neck intact; airborne microbes are trapped at base, and broth is sterile.

Figure 1.6 Pasteur's swan-neck flask experiment disproving spontaneous generation.
He left the flask open to air but bent the neck so that gravity would trap any airborne microbes.

in things other than thread counts. He took rainwater from a clay pot, smeared it on his specimen holder, and peered at it through his finest lens. He found "animals appearing to me ten thousand times less than those which may be perceived in the water with the naked eye."

He didn't stop there. He scraped the plaque from his teeth and from the teeth of some volunteers who had never cleaned their teeth in their lives and took a good close look at that. He recorded: "In the said matter there were many very little living animalcules, very prettily a-moving. . . . Moreover, the other animalcules were in such enormous numbers, that all the water . . . seemed to be alive." Leeuwenhoek started sending his observations to the Royal Society of London, and eventually he was recognized as a scientist of great merit.

Leeuwenhoek constructed more than 250 small, powerful microscopes that could magnify objects up to 300 times **(figure 1.7).** Considering that he had no formal training in science, his descriptions of bacteria and protozoa (which he called "animalcules") were accurate and precise.

These events marked the beginning of our understanding of microbes and the diseases they can cause. Discoveries continue at a breakneck pace, however. In fact, the 2000s are being widely called the Century of Biology, fueled by our new abilities to study genomes and harness biological processes. To give you a feel for what has happened most recently, **table 1.3** provides a glimpse of some recent discoveries that have had huge impacts on our understanding of microbiology.

The changes to our view of the role of RNAs that you see in table 1.3 highlight a feature of biology–and all of science–that is perhaps underappreciated. Because we have thick textbooks containing all kinds of assertions and "facts," many people think science is an ironclad collection of facts. Wrong! Science is an ever-evolving collection of new information, gleaned from observable phenomena and combined with old information to come up with the current understandings of nature. Some of the hypotheses explaining these observations have been confirmed so many times over such a long period of time that they are, if not "fact," very close to fact. Many other hypotheses will be altered over and over again as new findings emerge. And that is the beauty of science.

It is important to understand that modern science is conducted according to a set of widely accepted "rules," termed the *scientific method* **(figure 1.8).** Researchers form a hypothesis, and then perform experiments or other types of studies that allow them to either reject or accept the hypothesis. In real life it can also get much more complicated than that. Note the blue arrows in the figure indicating that the process can stop and revisit earlier steps. After the results are communicated (last step in the figure), other scientists review and repeat the studies in order to verify them or bring them into question. This is a process that distinguishes true research from even the most educated guesses or opinions.

The Beginnings of Medical Microbiology

Early experiments on the sources of microorganisms led to the profound realization that microbes are everywhere: Not only are air and dust full of them, but the entire surface of the earth, its waters, and all objects are inhabited by them. This discovery led to immediate applications in medicine. So you see that the seeds of medical microbiology were sown in the mid to latter half of the 19th century (the 1800s) with the introduction of the germ theory of disease and the resulting use of sterile, aseptic, and pure culture techniques.

Figure 1.7 Leeuwenhoek's microscope. A brass replica of a Leeuwenhoek microscope. The lens is held in front of one eye with the specimen holder facing outward.
Tetra Images/Alamy Stock Photo

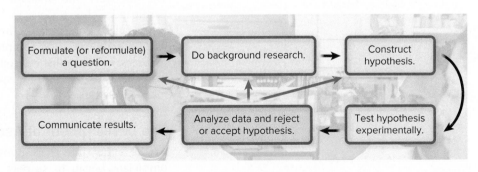

Figure 1.8 An overview of the scientific method.
Ryan McVay/Getty Images

Table 1.3 Recent Advances in Microbiology

Discovery of restriction enzymes—1970s.   Three scientists, Daniel Nathans, Werner Arber, and Hamilton Smith, discovered these little molecular "scissors" inside bacteria. They chop up DNA in specific ways. This was a huge moment that enabled scientists to use these enzymes to cut DNA in tailor-made ways. This opened the floodgates to genetic engineering—and all that has meant for the treatment of diseases, the investigation into biological processes, and the biological "revolution" of the 21st century.

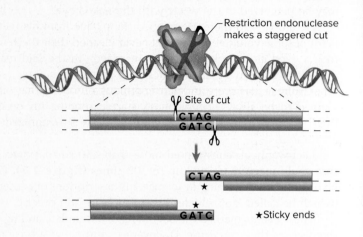

The invention of the PCR technique—1980s.   The polymerase chain reaction (PCR) was a breakthrough in our ability to detect tiny amounts of DNA and then amplify them into quantities sufficient for studying. It has provided a new and powerful method for discovering new organisms, diagnosing infectious diseases, and doing forensic work such as crime scene investigation. Its inventor is Kary Mullis, a scientist working at a company in California at the time. He won the Nobel Prize for this invention in 1993.

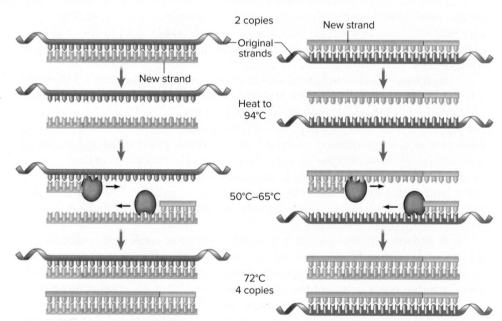

The Discovery of Spores and Sterilization

The discovery and detailed description of heat-resistant bacterial endospores by Ferdinand Cohn, a German botanist, clarified the reason that heat would sometimes fail to completely eliminate all microorganisms. The modern sense of the word **sterile,** meaning completely free of all life forms (including spores) and virus particles, was established from that point on. The capacity to sterilize objects and materials is an absolutely essential part of microbiology, medicine, dentistry, and many industries.

The Development of Aseptic Techniques

At the same time that spontaneous generation was being hotly debated, a few physicians began to suspect that microorganisms could cause not only spoilage and decay but also human diseases. It occurred to these rugged individualists that even the human body itself was a source of infection. In 1843, Dr. Oliver Wendell

The importance of small RNAs—2000s. Once we were able to sequence entire genomes (another big move forward), scientists discovered something that turned a concept we literally used to call "dogma" on its head. The previously held "Central Dogma of Biology" was that DNA makes RNA, which leads to the creation of proteins. Genome sequencing has revealed that perhaps only 2% of DNA actually codes for a protein. Much RNA doesn't end up with a protein counterpart. These pieces of RNA are usually small. It now appears that they have critical roles in regulating what happens in the cell. It has led to new approaches to how diseases are treated. For example, if some small RNAs are important in bacteria that infect humans, they can be new targets for antimicrobial therapy.

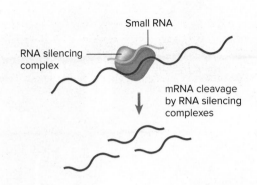

Genetic identification of the human microbiome—2010s and beyond. The first detailed information produced by the Human Microbiome Project (HMP) was astounding: Even though the exact types of microbes found in and on different people are highly diverse, the overall set of metabolic capabilities the bacterial communities possess is remarkably similar among people. This and other groundbreaking discoveries have set the stage for new knowledge of our microbial guests and their role in our overall health and disease.

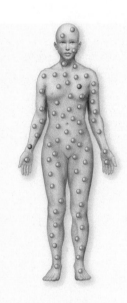

Holmes, an American physician, published an article in which he observed that mothers who gave birth at home experienced fewer infections than did mothers who gave birth in the hospital. A few years later, the Hungarian Dr. Ignaz Semmelweis showed quite clearly that women became infected in the maternity ward after being examined by physicians coming directly from the autopsy room—without washing their hands.

In the 1860s, the English surgeon Joseph Lister took notice of these observations and was the first to introduce **aseptic** (ay-sep′-tik) **techniques** aimed at reducing microbes in a medical setting. Lister's concept of asepsis was much more limited than our modern precautions. It mainly involved disinfecting the hands and the air with strong antiseptic chemicals, such as phenol, prior to surgery. It is hard for us to believe, but as recently as the late 1800s surgeons wore street clothes in the operating room and had little idea that hand washing was important **(figure 1.9).** Lister's techniques and the application of heat for sterilization became the foundations for microbial control by physical and chemical methods, which are still in use today.

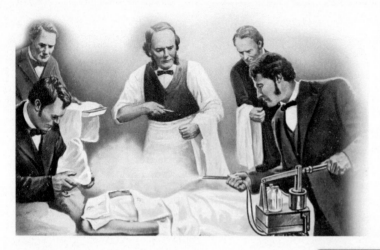

Figure 1.9 **An artist's depiction of Joseph Lister's operating theater in the mid-1800s.**

Bettmann/Getty Images

The Germ Theory of Disease

Louis Pasteur made enormous contributions to our understanding of the microbial role in wine and beer formation. He invented pasteurization and conducted some of the first studies showing that human diseases could arise from infection. These studies, supported by the work of other scientists, became known as the **germ theory of disease.** Pasteur's contemporary, Robert Koch, established *Koch's postulates*, a series of logical steps that verified the germ theory and could establish whether an organism was pathogenic and which disease it caused (see chapter 11). About 1875, Koch used this experimental system to show that anthrax was caused by a bacterium called *Bacillus anthracis*. So useful were his postulates that the causative agents of 20 other diseases were discovered between 1875 and 1900, and even today, they are the standard for identifying pathogens of plants and animals.

1.2 LEARNING OUTCOMES—Assess Your Progress

9. Make a time line of the development of microbiology from the 1600s to today.
10. List some recent microbiology discoveries of great impact.
11. Identify the important features of the scientific method.

1.3 Macromolecules: Superstructures of Life

In this book, we won't be presenting the basics of chemistry, though of course it is important to understand chemical concepts to understand all of biology. But that is what chemistry textbooks are for! However, there will be so much emphasis on some important *bio*chemicals in this book and in your course that we want to present a concise description of cellular macromolecules.

All microorganisms—indeed, all organisms—are constructed from just a few major types of biological molecules, called **macromolecules**—"macro" because they are often very large. They include four main families: carbohydrates, lipids, proteins, and nucleic acids **(table 1.4).** All macromolecules except lipids are formed by polymerization, a process in which repeating subunits termed **monomers** are bound into chains of various lengths termed **polymers.** For example, proteins (polymers) are composed of a chain of amino acids (monomers). In the following section and in later chapters, we consider numerous concepts relating to the roles of macromolecules in cells. **Table 1.5** presents the important structural features of the four main macromolecules.

Carbohydrates: Sugars and Polysaccharides

The term **carbohydrate** originates from the composition of members of this class: They are combinations of carbon (*carbo-*) and water. Carbohydrates can be generally represented by the formula $(CH_2O)_n$, in which *n* indicates the number of units of this combination of atoms. Some carbohydrates also contain additional atoms of sulfur or nitrogen **(figure 1.10).**

Monosaccharides and disaccharides are specified by combining a prefix that describes some characteristic of the sugar with the suffix *-ose*. For example, **hexoses** are composed of 6 carbons, and **pentoses** contain 5 carbons. **Glucose** (Gr. *glyko*, "sweet") is the most common and universally important hexose; **fructose** is named for fruit (one place where it is found); and xylose, a pentose, derives its name from the Greek word for wood. Disaccharides are named similarly: **lactose** (L. *lacteus*, "milk") is an important component of milk; **maltose** means malt sugar; and **sucrose** (Fr. *sucre*, "sugar") is common table sugar or cane sugar.

The green specks are microorganisms in the stomach of a tube worm.

©Stephen Durr

Table 1.4 Macromolecules and Their Functions

Macromolecule	Basic Structure	Examples	Notes About the Examples
Carbohydrates			
Monosaccharides	3- to 7-carbon sugars	Glucose, fructose	Sugars involved in metabolic reactions; building block of disaccharides and polysaccharides
Disaccharide	Two monosaccharides	Maltose (malt sugar)	Composed of two glucoses; an important breakdown product of starch
		Lactose (milk sugar)	Composed of glucose and galactose
		Sucrose (table sugar)	Composed of glucose and fructose
Polysaccharides	Chains of monosaccharides	Starch, cellulose, glycogen	Cell wall, food storage
Lipids			
Triglycerides	Fatty acids + glycerol	Fats, oils	Major component of cell membranes; storage
Phospholipids	Fatty acids + glycerol + phosphate	Membrane components	
Waxes	Fatty acids, alcohols	Mycolic acid	Cell wall of mycobacteria
Steroids	Ringed structure	Cholesterol, ergosterol	In membranes of eukaryotes and some bacteria
Proteins			
	Chains of amino acids	Enzymes; part of cell membrane, cell wall, ribosomes, antibodies	Serve as structural components and perform metabolic reactions
Nucleic acids	Nucleotides (pentose sugar + phosphate + nitrogen base) Nitrogen bases Purines: adenine (A), guanine (G) Pyrimidines: cytosine (C), thymine (T), uracil (U)		
Deoxyribonucleic acid (DNA)	Contains deoxyribose sugar and thymine, not uracil	Chromosomes; genetic material of viruses	Mediate inheritance
Ribonucleic acid (RNA)	Contains ribose sugar and uracil, not thymine	Ribosomes; mRNA, tRNA, small RNAs, genetic material of viruses	Facilitate expression of genetic traits

The Functions of Polysaccharides

Polysaccharides contribute to structural support and protection and also serve as nutrient and energy stores. The cell walls in plants and many microscopic algae derive their strength and rigidity from **cellulose,** a long, fibrous polysaccharide. Because of this role, cellulose is probably one of the most common organic substances on the earth. Interestingly, it is digestible only by certain bacteria, fungi, and protozoa. These microbes, called decomposers, play an essential role in breaking down and recycling plant materials.

Structural polysaccharides can be conjugated (chemically bonded) to amino acids, nitrogen bases, lipids, or proteins. **Agar,** an indispensable polysaccharide in preparing solid culture media, is a natural component of certain seaweeds. It is a complex polymer of galactose and sulfur-containing carbohydrates. The exoskeletons of certain fungi contain **chitin** (ky'-tun), a polymer of glucosamine (a sugar with an amino functional group). **Peptidoglycan** (pep-tih-doh-gly'-kan) is one special class of compounds in which polysaccharides (glycans) are linked to peptide fragments (a short chain of amino acids). This molecule provides the main source of structural support to the bacterial cell wall. The outer covering of gram-negative bacteria also contains **lipopolysaccharide,** a complex of lipid and polysaccharide responsible for symptoms such as fever and shock (see chapters 3 and 11).

The outer surface of many cells has a "sugar coating" composed of polysaccharides bound in various ways to proteins (this combination is termed a glycoprotein). This structure, called the **glycocalyx,** serves as a protective outer layer, and also can play a role in attachment of the cells to other cells or surfaces. Small sugar

Medical Moment

Delivering Essential Nutrients

It is important to maintain homeostasis in ill patients. Supplemental nutrition is often necessary. For some patients, intake of calories by mouth or feeding tube is not possible. In order to supply essential minerals, vitamins, electrolytes, amino acids, glucose, and fluid, nutrition via intravenous (IV) line may be initiated. Total parenteral nutrition (TPN) is prescribed to meet the entirety of the patient's daily nutritional needs, accounting for their specific disease process and organ function. In addition, fatty acids might be prescribed alongside TPN because they provide essential energy stores and help with the absorption of some vitamins. This intravenous nutrition provides the body with energy patients need for maintenance of cellular processes and promotion of healing.

Q. Use context in the paragraph above to deduce what the word "parenteral" means.

Answer in Appendix B.

Table 1.5 Macromolecules in the Cell

Carbohydrates. Another word for *sugar* is **saccharide.** A **monosaccharide** is a simple sugar containing from 3 to 7 carbons; a **disaccharide** is a combination of two monosaccharides; and a **polysaccharide** is a polymer of five or more monosaccharides bound in linear or branched chain patterns.

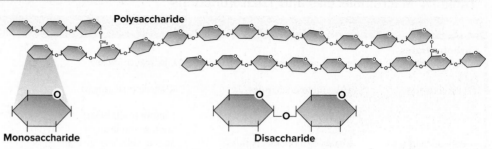

Polysaccharide

Monosaccharide

Disaccharide

Lipids. The term **lipid,** derived from the Greek word *lipos,* meaning fat, is not a chemical designation but an operational term for a variety of substances that are not soluble in polar solvents such as water but will dissolve in nonpolar solvents such as benzene and chloroform. Here we see a model of a single molecule of a phospholipid. The phosphate-alcohol head leads a charge to one end of the molecule; its long, trailing hydrocarbon chain is uncharged.

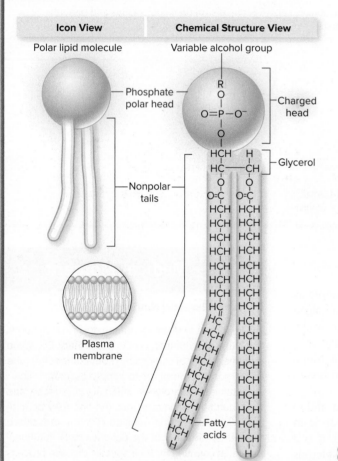

Icon View	Chemical Structure View

Polar lipid molecule — Variable alcohol group

Phosphate polar head — Charged head

Nonpolar tails — Glycerol

Plasma membrane — Fatty acids

Proteins. Proteins are chains of amino acids. Amino acids have a basic skeleton consisting of a carbon (called the α carbon) linked to an amino group (NH_2), a carboxyl group (COOH), a hydrogen atom (H), and a variable R group. The variations among the amino acids occur at the R group, which is different in each amino acid and confers the unique characteristics to the molecule and to the proteins that contain it. A covalent bond called a **peptide bond** forms between the amino group on one amino acid and the carboxyl group on another amino acid.

- R group
- peptide backbone
- — peptide bond

Nucleic acids. Both DNA and RNA are polymers of repeating units called **nucleotides,** each of which is composed of three smaller units: a **nitrogen base,** a **pentose** (5-carbon) sugar, and a **phosphate.** The nitrogen base is a cyclic compound that comes in two forms: *purines* (two rings) and *pyrimidines* (one ring). There are two types of purines–**adenine (A)** and **guanine (G)**–and three types of pyrimidines–**thymine (T), cytosine (C),** and **uracil (U).** The nitrogen base is covalently bonded to the sugar *ribose* in RNA and *deoxyribose* (because it has one less oxygen than ribose) in DNA. The backbone of a nucleic acid strand is a chain of alternating phosphate-sugar-phosphate-sugar molecules, and the nitrogen bases branch off the side of this backbone.

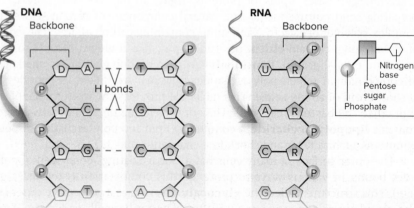

DNA Backbone

H bonds

RNA Backbone

Nitrogen base
Pentose sugar
Phosphate

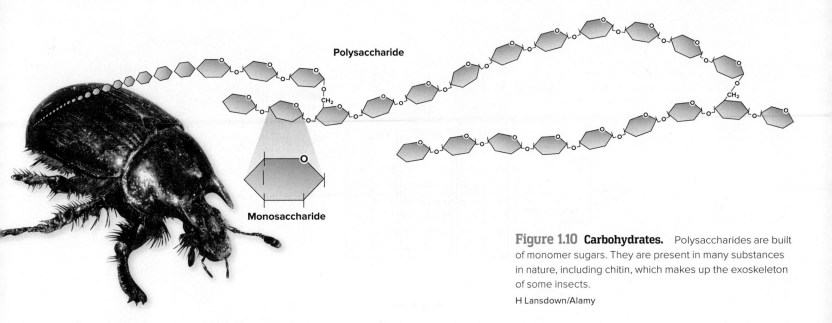

Polysaccharide

Monosaccharide

Figure 1.10 Carbohydrates. Polysaccharides are built of monomer sugars. They are present in many substances in nature, including chitin, which makes up the exoskeleton of some insects.

H Lansdown/Alamy

molecules on cell surfaces also account for the differences in human blood types. Viruses also have glycoproteins on their surface with which they bind to and invade their host cells.

Lipids: Fats, Phospholipids, Steroids, and Waxes

There are four main types of compounds classified as lipids: triglycerides, phospholipids, steroids, and waxes.

The **triglycerides** are an important storage lipid. This category includes fats and oils. Triglycerides are composed of a single molecule of glycerol bound to three fatty acids **(figure 1.11). Glycerol** is a 3-carbon alcohol with three OH groups, and fatty acids are long-chain hydrocarbon molecules with a carboxyl group (COOH) at one end that is free to bind to the glycerol. The hydrocarbon portion of a fatty acid can vary in length from 4 to 24 carbons–and, depending on the fat, it may be saturated or unsaturated. If all carbons in the chain are single-bonded to 2 other carbons and 2 hydrogens, the fat is saturated. If there is at least one C=C double bond in the chain, it is unsaturated. The structure of fatty acids is what gives fats and oils (liquid fats) their greasy, insoluble nature. In general, solid fats (such as butter) are more saturated, and liquid fats (such as oils) are more unsaturated.

In most cells, triglycerides are stored in long-term concentrated form as droplets or globules. When they are acted on by digestive enzymes called lipases, the fatty acids and glycerol are freed to be used in metabolism. Fatty acids are a superior source of energy, yielding twice as much per gram as other storage molecules (carbohydrates). Soaps are K^+ or Na^+ salts of fatty acids whose qualities make them excellent grease removers and cleaners (see chapter 9).

Membrane Lipids

Phospholipids in membranes have a hydrophilic ("water-loving") region and a hydrophobic ("water-fearing") region. The hydrophilic region carries a negative charge due to a phosphate group attached to an alcohol group. The long fatty acid chains are uncharged and make that portion of the molecule hydrophobic **(figure 1.12a).** When exposed to an aqueous solution, the charged heads are attracted to the water phase, and the nonpolar tails are repelled from the water phase **(figure 1.12b).** This property causes lipids to naturally form into single and double layers (bilayers). When two single layers of polar lipids come together to form a double layer, the outer

Oils on duck feathers keep these two canvasback ducks insulated and dry, no matter how much time they spend in the water.

©Guy Crittenden/Getty Images

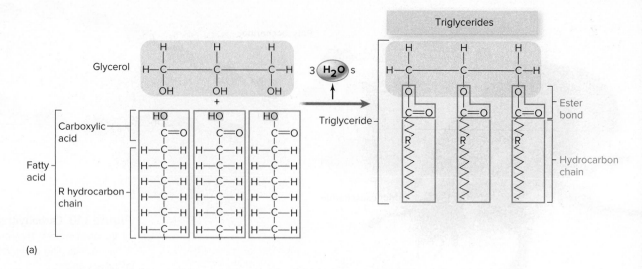

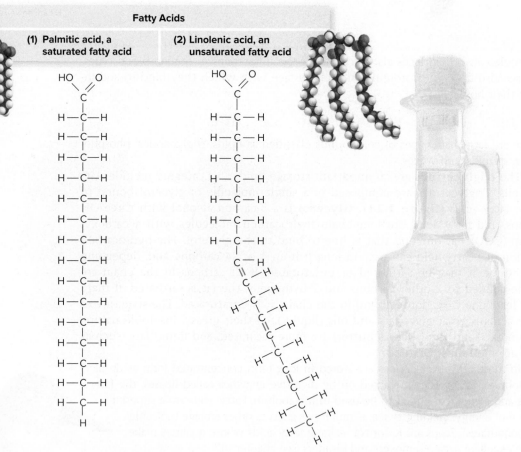

Figure 1.11 Synthesis and structure of a triglyceride.
(a) Because a water molecule is released at each ester bond, this is an example of dehydration synthesis. The jagged lines and R symbol represent the hydrocarbon chains of the fatty acids, which are commonly very long. (b) Structural and three-dimensional models of fatty acids and triglycerides.
(1) A saturated fatty acid has long, straight chains that readily pack together and form solid fats.
(2) An unsaturated fatty acid—here a polyunsaturated one with 3 double bonds—has a bend in the chain that prevents packing and produces oils (right).
Stockbyte/PunchStock

hydrophilic face of each single layer will orient itself toward the solution, and the hydrophobic portions will become immersed in the core of the bilayer. This behavior allows phospholipids to be the main constituent of all cell membranes.

Steroids and Waxes

Steroids are complex ringed compounds commonly found in cell membranes and as animal hormones. The best known of these is the sterol called **cholesterol (figure 1.13).** (A sterol is a steroid that has an OH group.) Cholesterol reinforces the structure of the

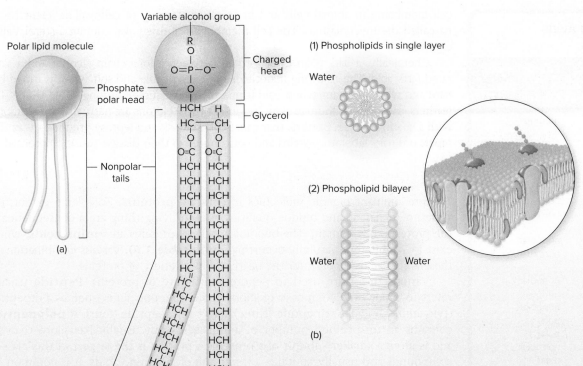

Figure 1.12 Phospholipids—membrane molecules.
(a) A model of a single molecule of a phospholipid. The phosphate-alcohol head lends a charge to one end of the molecule; its long, trailing hydrocarbon chain is uncharged. **(b)** The behavior of phospholipids in water-based solutions causes them to become arranged (1) in single layers called micelles, with the charged head oriented toward the water phase and the hydrophobic nonpolar tail buried away from the water phase, or (2) in double-layered phospholipid systems with the hydrophobic tails sandwiched between two hydrophilic layers.

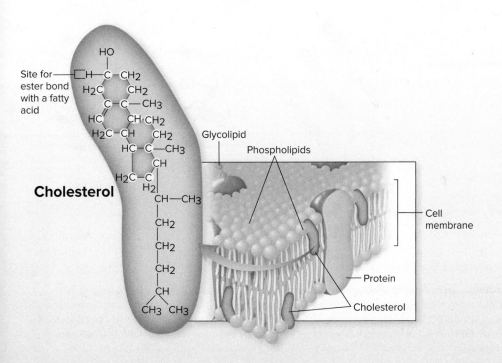

Figure 1.13 Cutaway view of a membrane with its bilayer of lipids. The primary lipid is phospholipid—however, cholesterol is inserted in some membranes.

Table 1.6 Twenty Amino Acids and Their Abbreviations

Acid	Abbreviation	Characteristic of Their R Groups
Alanine	Ala	nonpolar
Arginine	Arg	+
Asparagine	Asn	polar
Aspartic acid	Asp	−
Cysteine	Cys	polar
Glutamic acid	Glu	−
Glutamine	Gln	polar
Glycine	Gly	polar
Histidine	His	+
Isoleucine	Ile	nonpolar
Leucine	Leu	nonpolar
Lysine	Lys	+
Methionine	Met	nonpolar
Phenylalanine	Phe	nonpolar
Proline	Pro	nonpolar
Serine	Ser	polar
Threonine	Thr	polar
Tryptophan	Trp	nonpolar
Tyrosine	Tyr	polar
Valine	Val	nonpolar

+ = positively charged; − = negatively charged.

cell membrane in animal cells and in an unusual group of cell-wall-deficient bacteria called the mycoplasmas. The cell membranes of fungi also contain a sterol, called ergosterol.

Chemically, a *wax* is an ester formed between a long-chain alcohol and a saturated fatty acid. The resulting material is typically pliable and soft when warmed but hard and water resistant when cold (think of a wax candle). Among living things, fur, feathers, fruits, leaves, human skin, and insect exoskeletons are naturally waterproofed with a coating of wax. Bacteria that cause tuberculosis and leprosy produce a wax that repels ordinary laboratory stains and contributes to their disease-causing potential.

Proteins: Shapers of Life

The predominant organic molecules in cells are **proteins.** To a large extent, the structure, behavior, and unique qualities of each living thing are a consequence of the proteins they contain. The building blocks of proteins are **amino acids,** which exist in 20 different naturally occurring forms **(table 1.6).** Various combinations of these amino acids account for the nearly infinite variety of proteins.

Various terms are used to denote the nature of proteins. **Peptide** usually refers to a molecule composed of short chains of amino acids, such as a dipeptide (two amino acids), a tripeptide (three), and a tetrapeptide (four). A **polypeptide** contains an unspecified number of amino acids but usually has more than 20 and is often a smaller subunit of a protein. A protein is the largest of this class of compounds and usually contains a minimum of 50 amino acids. It is common for the term *protein* to be used to describe all of these molecules. In chapter 8, we see that protein synthesis is not just a random connection of amino acids; it is directed by information provided in DNA.

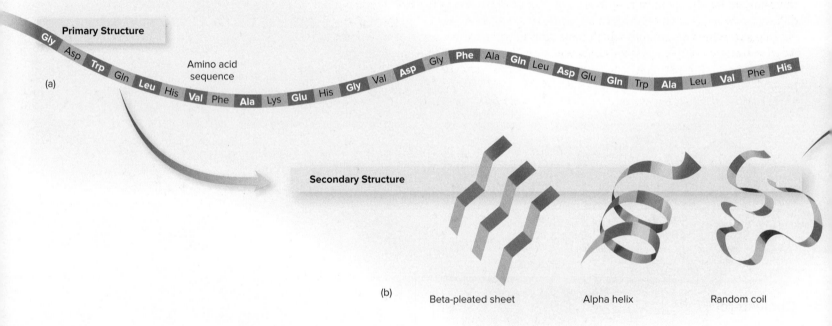

Primary Structure

Gly Asp Trp Gln Leu His Val Phe Ala Lys Glu His Gly Val Asp Gly Phe Ala Gln Leu Asp Glu Gln Trp Ala Leu Val Phe His

Amino acid sequence

(a)

Secondary Structure

(b)

Beta-pleated sheet   Alpha helix   Random coil

Figure 1.14 Stages in the formation of a functioning protein. (a) Its primary structure is a series of amino acids bound in a chain. (b) Its secondary structure develops when the chain forms hydrogen bonds that fold it into one of several configurations such as an α helix or β-pleated sheet. Some proteins have several configurations in the same molecule. (c) A protein's tertiary structure is due to further folding of the molecule into a three-dimensional mass that is stabilized by hydrogen, ionic, and disulfide bonds between functional groups. (d) The quaternary structure exists only in proteins that consist of more than one polypeptide chain. Each of the separate chains in this protein has a different color.

Protein Structure and Diversity

The reason that proteins are so varied and specific is that the way they are folded makes all the difference with respect to their function. A protein has a natural tendency to assume more complex levels of organization, called the secondary, tertiary, and quaternary structures **(figure 1.14).** The **primary (1°) structure** is the type, number, and order of amino acids in the chain, which varies extensively from protein to protein. The **secondary (2°) structure** arises when various functional groups (called R groups) exposed on the outer surface of the molecule interact by forming hydrogen bonds. This interaction causes the amino acid chain to twist into a coiled configuration called the *alpha helix* (α helix) or to fold into an accordion pattern called a *beta-pleated sheet* (β-pleated sheet). Many proteins contain both types of secondary configurations. Proteins at the secondary level undergo a third degree of torsion called the **tertiary (3°) structure** created by additional bonds between functional groups (figure 1.14*c*). In proteins with the sulfur-containing amino acid **cysteine,** further tertiary stability is achieved through covalent disulfide bonds between sulfur atoms on two different parts of the molecule. Some complex proteins also participate in a **quaternary (4°) structure,** in which more than one polypeptide forms a large, multiunit protein. This is typical of antibodies and some enzymes that act in cell synthesis.

The most important outcome of the various forms of bonding and folding is that each different type of protein develops a unique shape, and its surface displays a distinctive pattern of pockets and bulges. As a result, a protein can react only with molecules that complement or fit its particular surface features like a lock and key. Such a degree of specificity can provide the functional diversity required for many thousands of different cellular activities. For example, **enzymes** serve as the catalysts for all chemical reactions in cells, and nearly every reaction requires a different enzyme (see chapter 7). This specificity comes from the architecture of the binding site, which determines which molecules fit it. The same is true of antibodies: **Antibodies** are complex glycoproteins with specific regions of attachment for bacteria, viruses, and other microorganisms. The functional three-dimensional form of a protein is termed the *native state*. If it is disrupted by some means, the protein is said to be *denatured*. Agents such as heat, acid, alcohol, and some disinfectants disrupt

Protein is a major component of meats, eggs, and nuts.
©Comstock Images/Getty Images

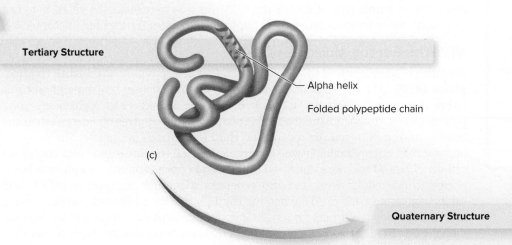

Tertiary Structure

Alpha helix

Folded polypeptide chain

(c)

Quaternary Structure

Two or more polypeptide chains

(d)

Curly hair is the result of particular protein folding patterns as described in figure 1.14.

©Ingram Publishing

(i.e., denature) the stabilizing bonds within the chains and cause the molecule to become nonfunctional, as described in chapter 9.

The Nucleic Acids: A Cell Computer and Its Programs

The fourth type of macromolecule is **nucleic acid.** DNA and RNA are two major representatives of this group. DNA contains a special coded genetic program with detailed and specific instructions for each organism's heredity. It transfers the details of its program to RNA, "helper" molecules responsible for carrying out DNA's instructions and translating the DNA program into proteins that can perform life functions. For now, let us briefly consider the structure and some functions of DNA, RNA, and a close relative, adenosine triphosphate (ATP).

The Double Helix of DNA

DNA is a huge molecule formed by two long nucleotide strands linked along their length by hydrogen bonds between nitrogen bases. The pairing of the nitrogen bases occurs according to a predictable pattern: Adenine always pairs with thymine, and cytosine with guanine. The bases are attracted in this way because each pair shares oxygen, nitrogen, and hydrogen atoms exactly positioned to align perfectly for hydrogen bonds **(figure 1.15).**

Owing to the manner of nucleotide pairing and stacking of the bases, the actual configuration of DNA is a *double helix* that looks somewhat like a spiral staircase. Just as with proteins, the structure of DNA is intimately related to its function. DNA molecules are usually extremely long.

RNA: Organizers of Protein Synthesis

Like DNA, RNA consists of a long chain of nucleotides. However, RNA is usually a single strand, except in some viruses. It contains ribose sugar instead of deoxyribose and uracil instead of thymine (see table 1.5). Several functional types of RNA are formed using the DNA template through a process called transcription. Three major types of RNA are directly used for protein synthesis. Messenger RNA (mRNA) is a copy of a gene (a single functional part of the DNA) that provides the order and type of amino acids in a protein; transfer RNA (tRNA) is a carrier that delivers the correct amino acids for protein assembly; and ribosomal RNA (rRNA) is a major component of ribosomes (described in chapter 3). A fourth type of RNA is the RNA that acts to regulate the genes and gene expression. More information on these important processes is presented in chapter 8.

ATP: The Energy Molecule of Cells

A relative of RNA involved in an entirely different cell activity is **adenosine triphosphate (ATP).** ATP is a nucleotide containing adenine, ribose, and three phosphates rather than just one **(figure 1.16).** It belongs to a category of high-energy compounds (also including guanosine triphosphate [GTP]) that gives off energy when the bond is broken between the second and third (outermost) phosphate. The presence of these high-energy bonds makes it possible for ATP to release and store energy for cellular chemical reactions. Breakage of the bond of the terminal phosphate releases energy to do cellular work and also generates adenosine diphosphate (ADP). ADP can be converted back to ATP when the third phosphate is restored, thereby serving as an energy depot. Carriers for oxidation-reduction activities (nicotinamide adenine dinucleotide [NAD], for instance) are also derivatives of nucleotides (see chapter 8).

Cells: Where Chemicals Come to Life

As we proceed in this chemical survey from the level of simple molecules to increasingly complex levels of macromolecules, at some point we cross a line from the realm of lifeless molecules and arrive at the fundamental unit of life called a **cell.** A cell is indeed a huge aggregate of carbon, hydrogen, oxygen, nitrogen, and many other

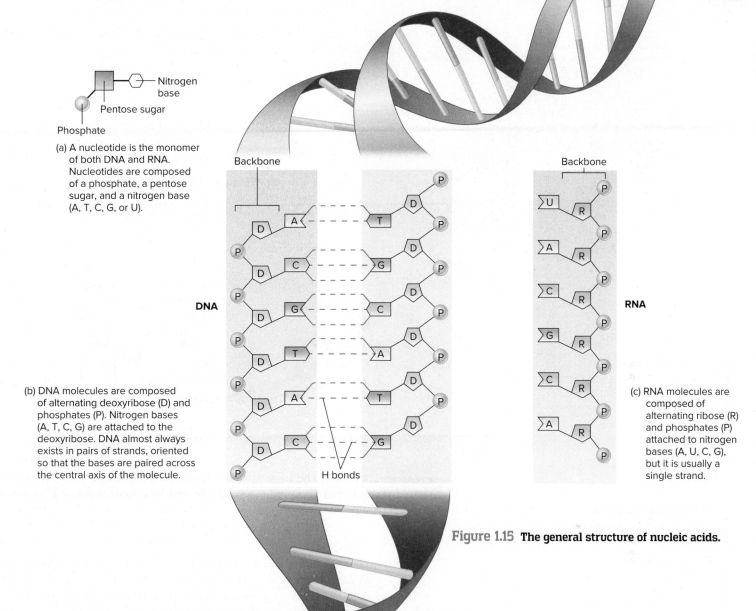

(a) A nucleotide is the monomer of both DNA and RNA. Nucleotides are composed of a phosphate, a pentose sugar, and a nitrogen base (A, T, C, G, or U).

(b) DNA molecules are composed of alternating deoxyribose (D) and phosphates (P). Nitrogen bases (A, T, C, G) are attached to the deoxyribose. DNA almost always exists in pairs of strands, oriented so that the bases are paired across the central axis of the molecule.

(c) RNA molecules are composed of alternating ribose (R) and phosphates (P) attached to nitrogen bases (A, U, C, G), but it is usually a single strand.

Figure 1.15 **The general structure of nucleic acids.**

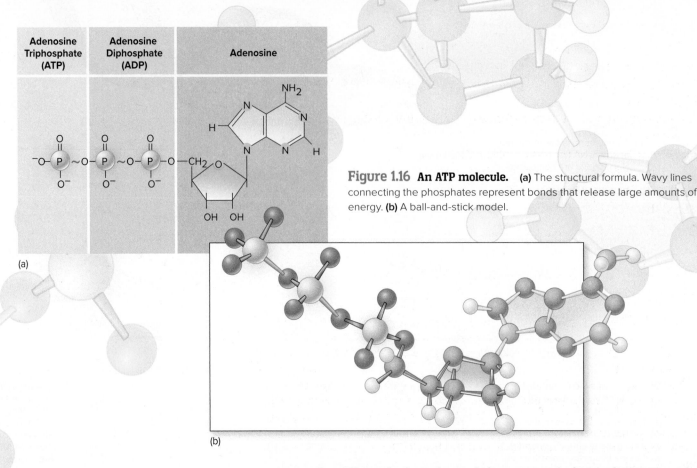

Adenosine Triphosphate (ATP)	Adenosine Diphosphate (ADP)	Adenosine

(a)

Figure 1.16 An ATP molecule.   **(a)** The structural formula. Wavy lines connecting the phosphates represent bonds that release large amounts of energy. **(b)** A ball-and-stick model.

(b)

atoms, and it follows the basic laws of chemistry and physics, but it is much more. The combination of these atoms produces characteristics, reactions, and products that can only be described as *living*.

Fundamental Characteristics of Cells

The bodies of some living things, such as bacteria and protozoa, consist of only a single cell, whereas those of animals and plants contain trillions of cells. Regardless of the organism, all cells have a few common characteristics. They tend to be spherical, polygonal, cubical, or cylindrical; and their protoplasm (internal cell contents) is encased in a cell or cytoplasmic membrane. They have chromosomes containing DNA, and ribosomes for protein synthesis, and they are exceedingly complex in function. Aside from these few similarities, the contents and structure of the three different cell types–bacterial, archaeal, and eukaryotic–differ significantly.

Animals, plants, fungi, and protozoa are all made up of eukaryotic cells. Such cells contain a number of complex internal parts called organelles that perform useful functions for the cell involving growth, nutrition, or metabolism. Organelles are distinct cell components that perform specific functions and are enclosed by membranes. Organelles also partition the eukaryotic cell into smaller compartments. The most visible organelle is the nucleus, a roughly ball-shaped mass surrounded by a double membrane that contains the DNA of the cell. Other organelles include the Golgi apparatus, endoplasmic reticulum, vacuoles, and mitochondria.

Bacterial and archaeal cells may seem to be the cellular "have nots" because, for the sake of comparison, they are described by what they lack. They have no nucleus and generally no other organelles. This apparent simplicity is misleading, however, because the fine structure of these cells is complex. Overall, bacterial and archaeal cells can engage in nearly every activity that eukaryotic cells can, and many can function in ways that eukaryotes cannot. Chapters 3 and 4 delve deeply into the properties of bacterial, archaeal, and eukaryotic cells.

A chicken egg is a single large cell.

©Ingram Publishing

1.4 Naming, Classifying, and Identifying Microorganisms

The science of classifying living beings is **taxonomy.** It originated more than 250 years ago when Carl von Linné (also known as Linnaeus; 1701–1778), a Swedish botanist, laid down the basic rules for *classification* and established taxonomic categories, or **taxa** (singular, *taxon*).

Von Linné realized early on that a system for recognizing and defining the properties of living beings would prevent chaos in scientific studies by providing each organism with a unique name and an exact "slot" in which to catalog it. This classification would then serve as a means for future identification of that same organism and permit people working in many biological fields to know if they were indeed discussing the same organism.

The primary concerns of modern taxonomy are still naming, classifying, and identifying. These three areas are interrelated and play a vital role in keeping a dynamic inventory of the extensive array of living and extinct beings. In general,

Nomenclature (naming) is the assignment of scientific names to the various taxonomic categories and to individual organisms.

Classification is the orderly arrangement of organisms into a hierarchy.

Identification is the process of discovering and recording the traits of organisms so that they may be recognized or named and then classified.

Nomenclature

Many macroorganisms are known by a common name suggested by their dominant features. For example, a bird species may be called a "red-headed blackbird" or a flowering plant species a "black-eyed Susan." Some species of microorganisms are also called by informal names, including human pathogens such as "gonococcus" (*Neisseria gonorrhoeae*) or fermenters such as "brewer's yeast" (*Saccharomyces cerevisiae*), or the recent "Iraqiabacter" (*Acinetobacter baumannii*), but this is not the usual practice. If we were to adopt common names such as the "little yellow coccus," the terminology would become even more cumbersome and challenging than scientific names.

The method of assigning a scientific or specific name is called the **binomial** (two-name) **system** of nomenclature. The scientific name is always a combination of the genus name followed by the species name. The genus part of the scientific name is capitalized, and the species part begins with a lowercase letter. Both should be italicized (or underlined if using handwriting), as follows:

Escherichia coli

The treehopper and its discoverer, Sylvie (below.). Sylvie, at the age of 5, with her mother the biologist Dr. Laura Sullivan-Beckers (above).

©Robyn Pizzo (top); ©Dr. Laura Sullivan-Beckers (bottom)

The two-part name of an organism is sometimes abbreviated to save space, as in *E. coli,* but only if the genus name has already been stated. The inspiration for names is extremely varied and often rather imaginative. A biology professor from Murray State University in Kentucky was planting flowers in her garden with her 2-year-old daughter Sylvie, when Sylvie discovered something. It was a species of treehopper insect that her mother was not familiar with. It turns out that it was an as-yet-undiscovered species, and Sylvie's mom, Dr. Laura Sullivan-Beckers, named it in honor of its discoverer, her daughter. Its binomial name is now *Hebetica sylviae.* Some species of microbes have also been named in honor of a microbiologist who originally discovered the microbe or who has made outstanding contributions to the field. Other names may designate a characteristic of the microbe (shape, color), a location where it was found, or a disease it causes. Two examples of specific names, their pronunciations, and their origins are

- *Staphylococcus aureus* (staf'-i-lo-kok'-us ah'-ree-us) Gr. *staphule,* "bunch of grapes," *kokkus,* "berry," and Gr. *aureus,* "golden." The species looks like a bunch of grapes under a microscope, and its colonies are golden yellow on agar. It is a common bacterial pathogen of humans.
- *Lactobacillus sanfrancisco* (lak'-toh-bass-ill'-us san-fran-siss'-koh) L. *lacto,* "milk," and *bacillus,* "little rod." A bacterial species used to make sourdough bread, for which San Francisco is known.

Here's a helpful hint: These names may seem difficult to pronounce and the temptation is to simply "slur over them." But when you encounter the names of microorganisms in the chapters ahead, it will be extremely useful to take the time to sound them out and repeat them until they seem familiar. Even experienced scientists stumble the first few times through new names. Stumbling out loud is a great way to figure them out, and you are much more likely to remember them that way–they are less likely to end up in a tangled heap with all of the new language you will be learning.

Classification

Classification schemes are organized into several descending ranks, beginning with the most general all-inclusive taxonomic category and ending with the smallest and most specific category. This means that all members of the highest category share only one or a few general characteristics, whereas members of the lowest category are essentially the same kind of organism–that is, they share the majority of their characteristics. The taxonomic categories from top to bottom are **domain, kingdom, phylum** or **division, class, order, family, genus,** and **species.** That means that each kingdom can be subdivided into a series of phyla or divisions, each phylum is made up of several classes, each class contains several orders, and so on. Because taxonomic schemes are to some extent artificial, certain groups of organisms may not exactly fit into the main categories. In such a case, additional taxonomic levels can be imposed above (super) or below (sub) a taxon, giving us such categories as "superphylum" and "subclass."

Let's compare the taxonomic breakdowns of a human and a protozoan (pro-tuh-zoh'-un) to illustrate the fine points of this system **(figure 1.17).** Humans and protozoa are both organisms with nucleated cells (eukaryotes); therefore, they are in the same domain (Eukarya), but they are in different kingdoms. Humans are multicellular animals (kingdom Animalia), whereas protozoa are single-cellular organisms that, together with algae, belong to the kingdom Protista. To emphasize just how broad the category "kingdom" is, ponder the fact that we humans belong to the same kingdom as jellyfish. Of the several phyla within this kingdom, humans belong to the phylum Chordata, but even a phylum is rather all-inclusive, considering that humans share it with other vertebrates as well as with creatures called sea squirts. The next level, class Mammalia, narrows the field considerably by grouping only those vertebrates that have hair and suckle their young. Humans belong to the order Primates, a group that also includes apes, monkeys, and lemurs. Next comes the family Hominoidea, containing only humans and apes. The final levels are our genus, *Homo*

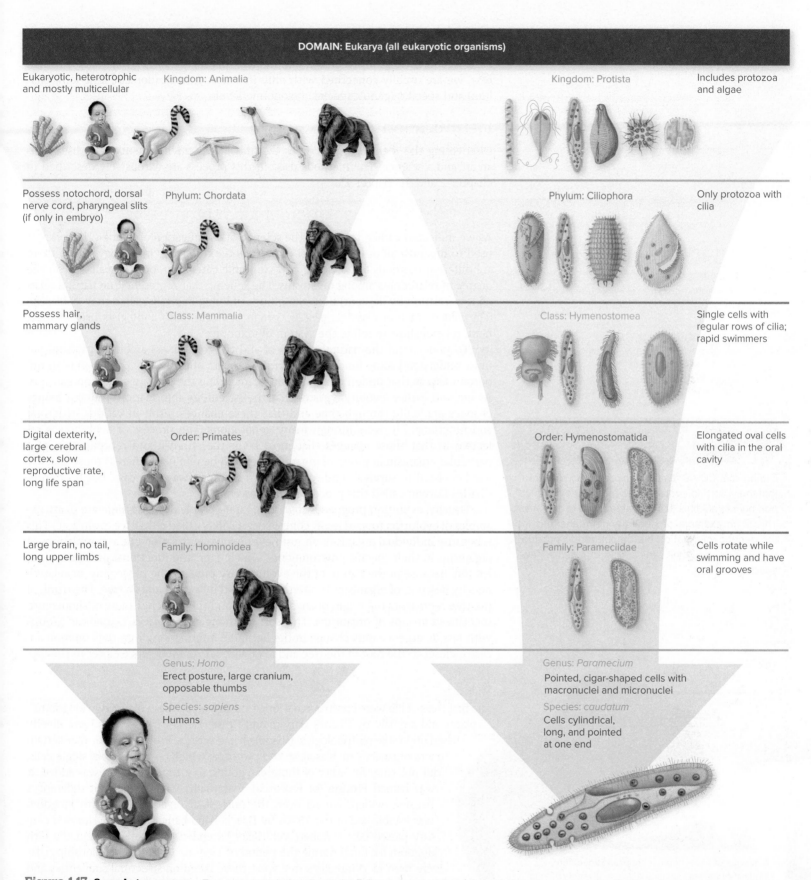

DOMAIN: Eukarya (all eukaryotic organisms)

Eukaryotic, heterotrophic and mostly multicellular — Kingdom: Animalia

Kingdom: Protista — Includes protozoa and algae

Possess notochord, dorsal nerve cord, pharyngeal slits (if only in embryo) — Phylum: Chordata

Phylum: Ciliophora — Only protozoa with cilia

Possess hair, mammary glands — Class: Mammalia

Class: Hymenostomea — Single cells with regular rows of cilia; rapid swimmers

Digital dexterity, large cerebral cortex, slow reproductive rate, long life span — Order: Primates

Order: Hymenostomatida — Elongated oval cells with cilia in the oral cavity

Large brain, no tail, long upper limbs — Family: Hominoidea

Family: Parameciidae — Cells rotate while swimming and have oral grooves

Genus: *Homo*
Erect posture, large cranium, opposable thumbs

Species: *sapiens*
Humans

Genus: *Paramecium*
Pointed, cigar-shaped cells with macronuclei and micronuclei

Species: *caudatum*
Cells cylindrical, long, and pointed at one end

Figure 1.17 Sample taxonomy. Two organisms belonging to the Eukarya domain, traced through their taxonomic series. On the left, modern humans, *Homo sapiens*. On the right, a common protozoan, *Paramecium caudatum*.

(all modern and ancient humans), and our species, *sapiens* (meaning "wise"). Notice that for the human as well as the protozoan, the taxonomic categories in descending order become less inclusive and the individual members more closely related. In this text, we are usually concerned with only the most general (domain, kingdom, phylum) and specific (genus, species) taxonomic levels.

Identification

Discovering the identity of microbes we find in the environment or in diseases is an art and a science. The methods used in this process are extensively described in chapter 2 and in chapter 15.

The Origin and Evolution of Microorganisms

As we indicated earlier, *taxonomy*, the science of classification of biological species, is used to organize all of the forms of modern and extinct life. In biology today, there are different methods for deciding on taxonomic categories, but they all rely on the degree of relatedness among organisms. The scheme that represents the natural relatedness (relation by descent) between groups of living beings is called their *phylogeny* (Gr. *phylon*, "race or class"; L. *genesis*, "origin or beginning"). Biologists use phylogenetic relationships to refine the system of taxonomy.

To understand the natural history of and the relatedness among organisms, we must understand some fundamentals of the process of evolution. Evolution is an important theme that underlies all of biology, including the biology of microorganisms. As we said earlier, evolution states that the hereditary information in living beings changes gradually through time and that these changes result in various structural and functional changes through many generations. The process of evolution is selective in that those changes that most favor the survival and reproduction of a particular organism or group of organisms tend to be retained, whereas those that are less beneficial to survival tend to be lost. This is not always the case, but it often is. Charles Darwin called this process *natural selection*.

Usually, evolution progresses toward greater complexity, but there are many examples of evolution toward lesser complexity. (This is called reductive evolution.) This is because individual organisms almost never evolve in isolation but as populations of organisms in their specific environments, which exert the functional pressures of selection. Because of the nature of the evolutionary process, the phylogeny, or relatedness by descent, of organisms is often represented by a diagram of a tree. The trunk of the tree represents the origin of ancestral lines, and the branches show offshoots into specialized groups of organisms. This sort of arrangement places taxonomic groups with less divergence (less change in the heritable information) from the common ancestor closer to the root of the tree and taxa with lots of divergence closer to the top.

A Universal Web of Life

The first trees of life were constructed a long time ago on the basis of just two kingdoms—plants and animals—by Charles Darwin and Ernst Haeckel. These trees were chiefly based on visible morphological (shape) characteristics. It became clear that certain (micro)organisms such as algae and protozoa, which only existed as single cells, did not truly fit either of those categories, so a third kingdom was added. It was named Protista (or Protozoa). Eventually, when significant differences became evident among even the unicellular organisms, a fourth kingdom was established in the 1870s by Haeckel and named Monera. Almost a century passed before Robert Whittaker extended this work and added a fifth kingdom for fungi during the period of 1959 to 1969. The relationships that were used in Whittaker's tree were those based on structural similarities and differences, and the way these organisms obtained their nutrition. These criteria indicated that there were five major taxonomic units, or kingdoms: the monera, protists, plants, fungi, and animals. Each organism in these five categories consisted

COVID-19

During the COVID-19 pandemic, it was declared that the virus had evolved from being able to infect non-human animals to being able to infect humans. This is an example of small (random) changes in the genetic information that happened to provide the virus a new "skill": the ability to infect a new species.

A handful of soil is home to thousands of different kinds of organisms, including a wide diversity of fungi, bacteria, viruses, and protozoa.
©Pixtal/age fotostock

of one of the two cell types, those cells lacking a nucleus and the eukaryotic cells. Whittaker's five-kingdom system quickly became the standard.

With the rise of genetics as a molecular science, newer methods for determining phylogeny have led to the development of a differently shaped tree–with important implications for our understanding of evolutionary relatedness. Molecular genetics allowed an in-depth study of the structure and function of the genetic material at the molecular level. In 1975, Carl Woese discovered that one particular macromolecule, the *ribonucleic acid in the small subunit of the ribosome* (abbreviated "ssu rRNA, or "16S rRNA"") was highly conserved–meaning that it was nearly identical in organisms within the smallest taxonomic category, the species. Based on a vast amount of experimental data, Woese hypothesized that 16S rRNA provides a "biological chronometer" or a "living record" of the evolutionary history of a given organism. Extended analysis of this molecule in prokaryotic and eukaryotic cells indicated that all members in a certain group of bacteria, then known as archaebacteria, had 16S rRNA with a sequence that was significantly different from the 16S rRNA found in other bacteria and in eukaryotes. This discovery led Carl Woese and collaborator George Fox to propose a separate taxonomic unit for the archaebacteria, which they named **Archaea.** Under the microscope, they resembled the structure of bacteria, but molecular biology has revealed that the archaea, though seemingly bacterial in nature, were actually more closely related to eukaryotic cells. To reflect these relationships, Woese and Fox proposed an entirely new system that assigned all known organisms to one of the three major taxonomic units, the domains, each being a different type of cell. Turn to the beginning of this chapter and reexamine figure 1.1, which is a depiction of the three-domain system.

The domains are the highest level in hierarchy and can contain many kingdoms and superkingdoms. Cell types lacking a nucleus are represented by the domains Archaea and **Bacteria,** whereas eukaryotes are all placed in the domain **Eukarya.** Analysis of the 16S rRNAs from all organisms in these three domains suggests that all modern and extinct organisms on earth arose from a common ancestor. Therefore, eukaryotes did not emerge from bacteria and archaea. Instead, it appears that bacteria, archaea, and eukaryotes all emerged separately from a different, now extinct, cell type.

To add another level of complexity, the most current data suggest that "trees" of life do not truly represent the relatedness of organisms in their totality. It has become obvious that genes travel horizontally–meaning from one species to another in nonreproductive ways–and that the tidy generation-to-generation changes are complicated by neighbor-to-neighbor exchanges of DNA. For example, it is estimated that 40% to 50% of human DNA has been carried to humans from other species (by viruses). For these reasons, most scientists like to think of a *web* as the proper representation of life these days. Nevertheless, this new scheme does not greatly affect our presentation of most microbes because we will discuss them at the genus or species level. Keep in mind that our methods of classification or evolutionary schemes reflect our current understanding and will change as new information is uncovered.

Please note that viruses and prions are not included in any of the classification or evolutionary schemes because they are not cells or organisms. Their special taxonomy is discussed in chapter 5.

©Tony Sweet/Digital Vision/Getty Images

1.4 LEARNING OUTCOMES—Assess Your Progress

19. Differentiate among the terms *nomenclature*, *taxonomy*, and *classification*.
20. Create a mnemonic device for remembering the taxonomic categories.
21. Correctly write the binomial name for a microorganism.
22. Draw a diagram of the three major domains.
23. Explain the difference between traditional and molecular approaches to taxonomy.

CASE FILE WRAP-UP

©Michael Williams (photo of Kelly Cowan with student); Science Photo Library RF/Getty Images (white blood cell)

If you have a bacterial infection, your doctor is likely (but not in all cases) to prescribe an antibiotic. Antibiotics are drugs that are designed to harm microbes but not harm the human host. That is their specific job—to target the microorganism. So if you have an illness that is not caused by a microorganism, you should not take antibiotics.

If you decide to go into health care as a profession, you will see a few common infections very frequently, but there will also be a wide variety of infections that you will likely only encounter once or a few times in your career. No one expects you to remember everything about every possible infection you study here. What's important is that you become familiar with important patterns of disease and the ways that our body—and the treatments we apply—affect them.

Stephen Durr

Meet Your Microbiome

In the past few years, a new word has popped up on newsfeeds and websites: *microbiome*. It refers to the sum total of all the microbes in a certain environment. The human microbiome consists of all the viruses, bacteria, fungi, and protozoa that call the human body home. Trillions of these microorganisms live on our body, as part of our natural biology. Unless something goes wrong, they do not cause disease, and, in fact, are necessary parts of human development and ongoing life.

The Human Microbiome Project (HMP) began in 2008, using techniques to identify body microbes that did not require growing the microbes separately in the lab (a technique scientists have relied on since the mid-1800s), but instead identified them on the basis of their genetic material. The HMP produced a staggering array of results, and they keep coming at breakneck pace.

We have learned that the microbiome may differ whether you were delivered via cesarean or vaginal birth. We have learned that the gut microbiome—the microbes living in your intestinal tract—may influence not just your intestinal health but also your likelihood to experience autoimmune disease, your weight, and even your mood! We know how the composition of the microbiome of different body systems (your skin, your eyes, your lungs) differs in health and in disease.

In short, we have learned that the characteristics of your microbiome determine your own, human, biology—and what types of experiences you will have as an organism. But there is a big note of caution here: The explosion of information about the human microbiome has led to some very preliminary science, and a lot of careless journalism. One leading microbiome researcher even calls it "Microbiomania." Many journalists gloss over the difference between correlation (these two things were observed at the same time) and causation (this thing *led to* this other thing). So we get reporting that suggests that elite cyclists should be "poop-doping" (getting transplanted fecal material that contains the donor's gut microbiome) to improve their performance, when there is no real evidence for it. Right here at the end of every chapter in this book, we will tell a short story about the microbiome as it pertains to the subject matter in the chapter, and try to address the quality of the science - and the reporting - behind the story.

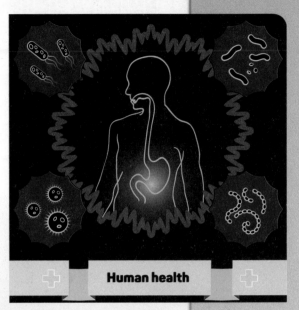

Human health

©Anna Smirnova/Alamy Stock Vector

Source: Shutterstock/Barbo

INTRODUCTION TO MICROBES AND THEIR BUILDING BLOCKS

1.1 MICROBES: TINY BUT MIGHTY

Microorganisms in this book are: archaea, bacteria, fungi, protozoa, helminths, viruses, and prions. Some are cellular and some are not cells. Microbes are everywhere on the planet and are critical for nutrient and energy cycling.

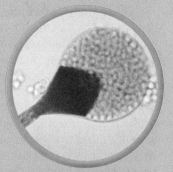

CDC/Dr. Lucille K. Georg

The first golden age of microbiology was in the mid-1800s, when scientists developed culturing methods and the germ theory of disease.

MICROBES IN HISTORY 1.2

lynx/ iconotec.com/Glow Images

1.3 MACROMOLECULES: SUPERSTRUCTURES OF LIFE

Cells and microbes are constructed of four main macromolecules: carbohydrates, lipids, proteins, and nucleic acids. Macromolecules are generally built up by polymerization of smaller molecular subunits (monomers).

Naming, classifying, and identifying microorganisms is called taxonomy. All cellular organisms belong to one of three domains: Eukarya, Bacteria, and Archaea.

NAMING, CLASSIFYING, AND IDENTIFYING 1.4 MICROORGANISMS

Order: Primates

Family: Hominoidea

SmartGrid: From Knowledge to Critical Thinking

This *21 Question Grid* takes the topics from this chapter and arranges them with respect to the American Society for Microbiology's Undergraduate Curriculum guidelines—all six of the important "Concepts" as well as the important "Competency" of scientific literacy. Three questions are supplied about chapter content that refer to the Concept or Competency, in increasing levels of Bloom's taxonomy for learning.

ASM Concept/ Competency	A. Bloom's Level 1, 2—Remember and Understand (Choose one.)	B. Bloom's Level 3, 4—Apply and Analyze	C. Bloom's Level 5, 6—Evaluate and Create

Evolution	1. Which of the following is an acellular microorganism lacking a nucleus?
	  a. bacterium
	  b. helminth
	  c. protozoan
	  d. virus
		2. Name seven types of microorganisms that we are studying in this book and use each one in a sentence.	3. Defend the argument that a web of life is a more accurate representation of evolutionary relatedness than a tree of life.

Cell Structure and Function	4. Which of the following is a macromolecule that assembles into bilayers?
	  a. protein
	  b. phospholipid
	  c. nucleic acid
	  d. carbohydrate
		5. Often when there is a local water main break, a town will post an advisory to boil water before ingesting it. Identify the biological basis behind the effectiveness of this procedure in minimizing illness.	6. Imagine a way you might design a drug to destroy microbes that will not harm human cells.

Metabolic Pathways	7. Identify the process or environment in this list that is not affected by microorganisms.
	  a. oxygen cycles
	  b. global temperatures
	  c. human health
	  d. all of the above have microbial involvement
		8. Provide an argument about why metabolic capabilities are so much more diverse in single-celled organisms like bacteria and archaea than they are in multicellular eukaryotes.	9. Provide a possible interpretation of the finding that the identity of microbes found in different people may differ, but the metabolic pathways that they bring are similar in different people.

Information Flow and Genetics	10. DNA leads to RNA, which can lead to the creation of
	  a. proteins.
	  b. lipids.
	  c. cells.
	  d. oxygen.
		11. Compare and contrast the RNA molecule with the DNA molecule.	12. Suggest an argument for why eukaryotic cells have developed an enclosed nucleus unlike bacteria and archaea.

Microbial Systems	13. Microbes are found in which habitat?
	  a. human body
	  b. earth's crust
	  c. oceans
	  d. all of the above
		14. Defend or refute this statement: *Microbes intend to cause human disease.*	15. *Coevolution* is a term describing the influence that two organisms occupying the same niche have on each other. Sketch a scenario for coevolution between a bacterium and a human living in the same environment.

ASM Concept/Competency	A. Bloom's Level 1, 2—Remember and Understand (Choose one.)	B. Bloom's Level 3, 4—Apply and Analyze	C. Bloom's Level 5, 6—Evaluate and Create

Impact of Microorganisms	16. Which of the following processes can be the result of human manipulation of microbial genes?
	  a. the central dogma
	  b. natural selection
	  c. bioremediation
	  d. abiogenesis
		17. Speculate about why scientists believe there are more microbial species that we do not yet know about than those that we do know about.	18. Imagine you are a guest speaker in a middle school science class. Explain to the students why human life would not be possible without microorganisms.

Scientific Thinking	19. When a hypothesis has been thoroughly supported by long-term study and data, it is considered
	  a. a law.
	  b. a speculation.
	  c. a theory.
	  d. proven.
		20. Defend the use of complicated-sounding names for identifying microorganisms.	21. Identify the most important component of the scientific method and defend your answer.

Answers to the multiple-choice questions appear in Appendix A.

Visual Connections

This question connects content within and between chapters.

Figure 1.2. Look at the red bar (the time that bacteria have been on earth) and at the time that humans appeared. Speculate on the probability that we will be able to completely eliminate all bacteria from our planet, and discuss whether or not this would even be a beneficial action.

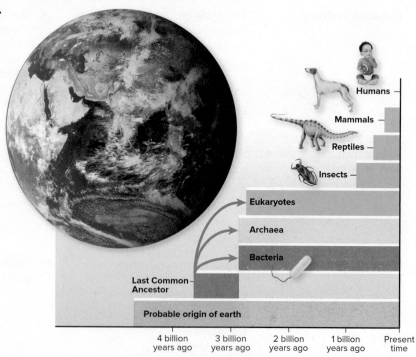

Source: NASA/Goddard Space Flight Center (Earth)

2

Tools of the Laboratory

Methods for the Culturing and Microscopic Analysis of Microorganisms

IN THIS CHAPTER...

2.1 How to Culture Microorganisms

1. Explain what the Five I's are and what each step entails.
2. Discuss three physical states of media and when each is used.
3. Compare and contrast selective and differential media, and give an example of each.
4. Provide brief definitions for *defined media* and *complex media*.

2.2 The Microscope

5. Convert among the different units of the metric system.
6. List and describe the three elements of good microscopy.
7. Differentiate between the principles of light microscopy and the principles of electron microscopy.
8. Give examples of simple, differential, and special stains.

CASE FILE

Treating the Unknown

As a nurse in a hospital inpatient surgical unit, an important part of my role is to assess patients for any complications related to their procedure. Most of my patients have incisions from their surgery. Some have implanted equipment, drains, and intravenous (IV) lines. Others are predisposed to developing infection because of the injury that caused them to need surgery. Because of these risks, one of the most important things I look for in my patients is any sign of infection.

I was caring for a 24-year-old male after his orthopedic surgery. He had been involved in a car accident and had a complex fracture in his lower leg. The surgery team had repaired the injury in the operating room. The patient now had implanted rods and screws to stabilize the bones. He had a large incision from the procedure.

Two days after his surgery, the patient developed a high fever. He was reporting increased pain in his leg. I called the provider and we discussed my concern that the patient may have an infection. Blood work was sent to the lab to look at the patient's complete blood count (CBC) with white blood cell differential, electrolyte panel, and a sample to be cultured for microorganism growth. I collected the blood culture in a sterile manner to ensure there was no contamination from the patient or environment that would alter the results.

While I waited for the lab results, I carefully assessed the patient for other signs of infection. His heart rate was high, but his blood pressure was normal. The surgical incision appeared red and was warm to the touch. His leg was swollen. There was a small amount of greenish drainage from the surgical incision.

The patient's white blood cell count was elevated and the differential white blood cell count indicated infection, so the physician ordered the start of IV antibiotics and, to take a closer look at the surgical site itself, an X ray of the leg.

- What are this patient's signs of infection?

- How would the physician decide what type of antibiotics to prescribe?

Case File Wrap-Up appears at the end of the chapter.

Medical Moment

The Making of the Flu Vaccine: An Example of a Live Growth Medium

Have you ever wondered why health care workers ask about allergic reactions to eggs prior to immunizing patients? Live attenuated vaccines are sometimes created by culturing a virus, such as the influenza virus, in live animals, often chick embryos. The virus is inoculated into fertilized eggs, which are then incubated to encourage the replication of large numbers of virus particles. The contents of the eggs are then collected and purified to create the vaccine.

Today, influenza vaccine preparations contain such low levels of egg protein that they can be safely administered even in most individuals with allergies, though it is recommended that they be medically monitored after receiving the dose. Vaccines in which the virus has been grown in cell culture, rather than in chicken eggs, are also available.

Q. Based on the information here, do you think that influenza viruses can infect (adult) birds? Why or why not?

Answer in Appendix B.

2.1 How to Culture Microorganisms

The Five I's

When you're trying to study microorganisms, you are confronted by some unique problems. First, most habitats (such as the soil, or the human mouth) contain microbes in complex associations, so it is often necessary to separate the species from one another. Second, to maintain and keep track of such small research subjects, microbiologists usually have to grow them under artificial (and thus distorting) conditions. A third difficulty in working with microbes is that they are not visible to the eye. Fourth, microbes are everywhere, and undesirable ones can be introduced into your experiment, causing misleading results.

Microbiologists use five basic techniques to manipulate, grow, examine, and characterize microorganisms in the laboratory **(figure 2.1):**

1. inoculation,
2. incubation,
3. isolation,
4. inspection, and
5. identification.

These procedures make it possible to handle and maintain microorganisms as discrete entities whose detailed biology can be studied and recorded. Having said that, keep in mind as we move through this chapter: It is not necessary to grow a microorganism to identify it anymore, though it still remains a very common method. You will read about noncultivation methods of identifying microbes in chapter 15. Sometimes growing microbes in isolated cultures can tell you very little about how they act in a mixed-species environment, but being able to isolate them and study them is also valuable, as long as you keep in mind that it is an unnatural state for them.

① Inoculation

To grow, or **culture,** microorganisms, one introduces a tiny sample (the inoculum) into a container of nutrient **medium** (plural, *media*). The medium provides an environment in which they multiply. This process is called **inoculation.** To avoid introducing unwanted microorganisms to the medium, any instrument used for picking up the sample and transferring it must be *sterile*. The inoculated medium is then incubated under appropriate conditions (next step) and the resulting growth is called a culture. (Note that we use "culture" both as a verb and as a noun.) Clinical specimens are obtained from body fluids (blood, cerebrospinal fluid, peritoneal fluid), discharges (sputum, urine, feces), anatomical sites (throat, nose, ear, eye, genital tract), or diseased tissue (such as an abscess or wound). Then the specimens are inoculated into medium in order to identify the microorganisms in them. Other samples subject to microbiological analysis are soil, water, sewage, foods, air, and inanimate objects. Procedures for proper specimen collection are discussed in chapter 15.

② Incubation

Once a container of medium has been inoculated, it is **incubated,** which means it is placed in a temperature-controlled chamber (incubator) to encourage multiplication. Although there are microbes that can grow at temperatures ranging from freezing to boiling, the usual temperatures used in laboratories–especially in medical facilities–fall between 20°C and 45°C. Incubators can also control the content of atmospheric gases such as oxygen and carbon dioxide that may be required for the growth of certain microbes. During the incubation period (ranging from a day to several weeks), the microbe multiplies and produces growth that is observable macroscopically. Microbial growth in a liquid medium

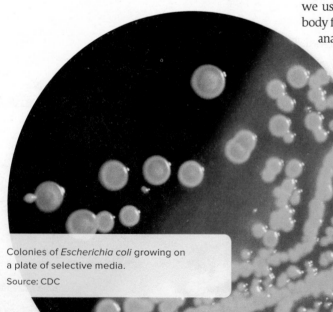

Colonies of *Escherichia coli* growing on a plate of selective media.
Source: CDC

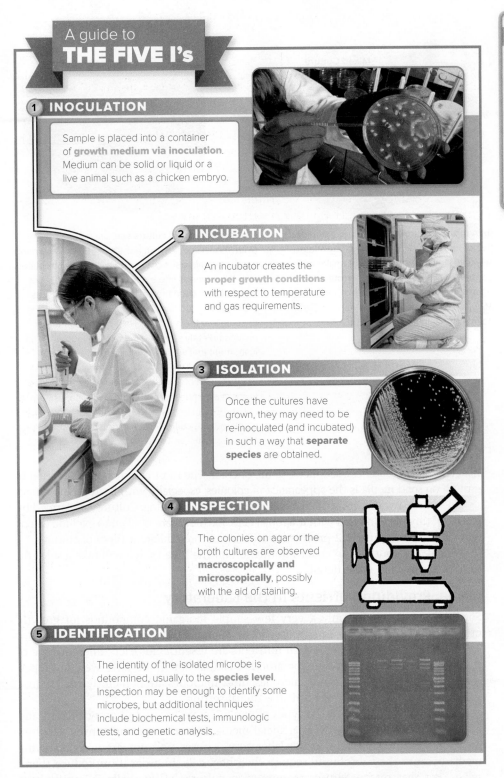

A guide to
THE FIVE I's

① INOCULATION

Sample is placed into a container of **growth medium via inoculation**. Medium can be solid or liquid or a live animal such as a chicken embryo.

② INCUBATION

An incubator creates the **proper growth conditions** with respect to temperature and gas requirements.

③ ISOLATION

Once the cultures have grown, they may need to be re-inoculated (and incubated) in such a way that **separate species** are obtained.

④ INSPECTION

The colonies on agar or the broth cultures are observed **macroscopically and microscopically**, possibly with the aid of staining.

⑤ IDENTIFICATION

The identity of the isolated microbe is determined, usually to the **species level**. Inspection may be enough to identify some microbes, but additional techniques include biochemical tests, immunologic tests, and genetic analysis.

Figure 2.1 A summary of the general laboratory techniques carried out by microbiologists.   It is not necessary to perform all the steps shown or to perform them exactly in this order, but all microbiologists participate in at least some of these activities. In some cases, one may proceed right from the sample to inspection, and in others, only inoculation and incubation on special media are required.

Source: CDC/James Gathany (lab worker holding petri dish); Monty Rakusen/Getty Images (scientist in lab); Pixtal/age fotostock (kneeling scientist); Centers for Disease Control and Prevention (*Vibrio cholerae* petri dish); Lisa Burgess/McGraw-Hill Education (DNA gel electrophoresis)

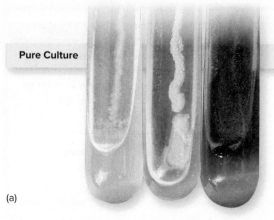

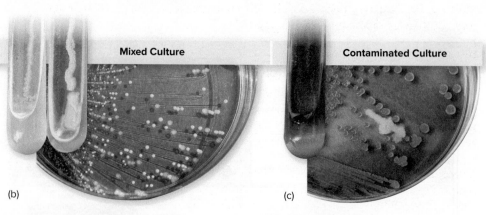

(a)

(b)

(c)

Figure 2.2 Various conditions of cultures.
(a) Three tubes containing pure cultures of *Escherichia coli* (white), *Micrococcus luteus* (yellow), and *Serratia marcescens* (red). A **pure culture** is a growth medium that contains only a single known species or type of microorganism. This type of culture is most frequently used for laboratory study because it allows the systematic examination and control of one microorganism by itself.

(b) A **mixed culture** is a container that holds two or more *identified*, easily differentiated species of microorganisms, not unlike a garden plot containing both carrots and onions. Pictured here is a mixed culture of *M. luteus* (bright yellow colonies) and *E. coli* (faint white colonies).

(c) A **contaminated culture** was once pure or mixed (with known species) but has since had **contaminants** (unwanted microbes of uncertain identity) introduced into it, like weeds into a garden. Contaminants get into cultures when the lids of tubes or Petri dishes are left off for too long, allowing airborne microbes to settle into the medium. They can also enter on an incompletely sterilized inoculating loop or on an instrument that you have inadvertently reused or touched to the table or your skin.

This plate of *S. marcescens* was overexposed to room air, and it has developed a large, white colony. Because this intruder is not desirable and not identified, the culture is now contaminated.

All photos: Kathleen Talaro

materializes as cloudiness, sediment, scum, or color. The most common manifestation of growth on solid media is the appearance of colonies, especially with bacteria and fungi.

In some ways, culturing microbes is analogous to gardening. Cultures are formed by "seeding" tiny plots (media) with microbial cells. Extreme care is taken to exclude weeds (contaminants). **Figure 2.2** provides a summary of three different types of cultures.

Before we continue to cover information on the Five I's, we will take a side trip to look at media in more detail.

Media: Providing Nutrients in the Laboratory

Some microbes require only a very few simple inorganic compounds for growth; others need a complex list of specific inorganic and organic compounds.

The trick is figuring out which nutrients are essential for each different type of microorganism to grow in the laboratory. In fact, although we have well-developed methods of growing very many microbes, most microbes are non-cultivable because we don't know what they need in an artificial setting.

This chapter focuses on the types of nonliving media used to grow bacteria and fungi. (We also refer to these as artificial media.) Protozoa and helminths are often more complex to cultivate in the laboratory. Viruses will not grow on nonliving media because they absolutely require their host cells in order to reproduce themselves. For that reason, viruses are grown in cultures of live cells. Scientists have developed a method to amplify the numbers of prions in the laboratory, but it is not considered cultivation in the true sense.

Types of Media

Media can be classified according to three properties **(table 2.1):**

1. physical state,
2. chemical composition, and
3. functional type (purpose).

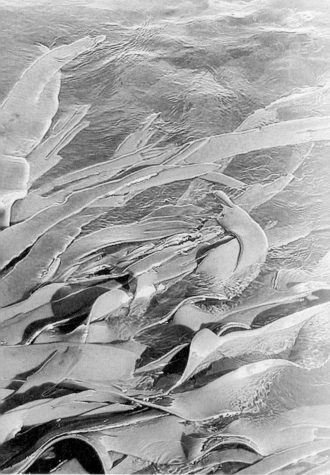

Agar, the main component of media, is commonly harvested from seaweed.
©Dan Ippolito

Most media discussed here are designed for bacteria and fungi, though algae and some protozoa can also be grown in media.

Physical States of Media

Figure 2.3 provides a good summary of three physical types of media: liquid, semisolid, and solid.

Agar, a complex polysaccharide isolated from the alga *Gelidium*, is a critical tool in the microbiology lab. The benefits of agar are numerous. It is solid at room temperature, and it melts (liquefies) at the boiling temperature of water (100°C). Once it is liquefied, agar does not resolidify until it cools to 42°C, so it can be inoculated and poured in liquid form at temperatures (45°C to 50°C) that will not harm the microbes or the handler. Agar is flexible and moldable, and it provides a basic framework to hold moisture and nutrients. Importantly, it is not itself a digestible nutrient for most microorganisms.

Chemical Content of Media

Media whose compositions are precisely chemically defined are termed *defined* (also known as *synthetic*). Such media contain pure organic and inorganic compounds that vary little from one source to another and have a molecular content specified by means of an exact formula. Defined media may contain nothing more than a few essential compounds such as salts and amino acids dissolved in water or may be composed of a variety of defined organic and inorganic chemicals. Such standardized and reproducible media are most useful for research applications when the exact concentration of components in the media is controlled so that metabolic processes of the microbe can be precisely monitored.

If even one component of a given medium is not precisely defined, the medium belongs in the *complex* category. Complex media contain extracts of animals, plants, or yeasts, including such materials as ground-up cells, tissues, and secretions. Examples are blood, serum, and meat extracts or infusions. These materials are all sure to contain a rich supply of nutrients, although the type and amounts will vary from

Table 2.1 Three Categories of Media Classification

Physical State	Chemical Composition	Functional Type
1. Liquid 2. Semisolid 3. Solid (that can be converted to liquid) 4. Solid (that cannot be liquefied)	1. Chemically defined 2. Complex; not chemically defined	1. General purpose 2. Enriched 3. Selective 4. Differential 5. Anaerobic growth 6. Specimen transport 7. Assay 8. Enumeration

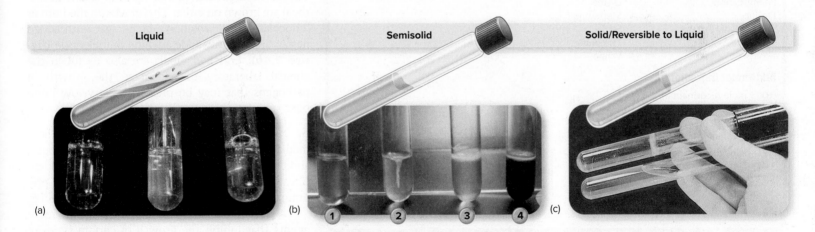

Liquid	Semisolid	Solid/Reversible to Liquid

(a)   (b)   1   2   3   4   (c)

Figure 2.3 Media in different physical forms.

(a) Liquid media are water-based solutions that do not solidify at temperatures above freezing and that tend to flow freely when the container is tilted. Growth occurs throughout the container and will appear cloudy, or can appear as flakes or settle to the bottom of the vessel. The tubes here hold urea broth. Urea is added to the broth in order to see if the microbe being cultured contains an enzyme that digests urea and releases ammonium. If this happens, the pH of the broth is raised and the dye becomes increasingly pink. Left: uninoculated broth, pH 7; middle: weak positive, pH 7.5; right: strong positive, pH 8.0.

(b) Semisolid media are firmer than liquid media but not as firm as solid media. They do not flow freely and have a soft, clotlike consistency at room temperature. Semisolid media are used to examine the motility of bacteria and to provide a backdrop for visible reactions to occur. Here, sulfur indole motility (SIM) medium is pictured. The (1) medium is stabbed with an inoculum and incubated. Location of growth indicates nonmotility (2) or motility (3). Additionally, if H_2S gas is released, a black precipitate forms (4).

(c) Media containing 1%–5% agar are solid enough to remain in place when containers are tilted or inverted. They are reversibly solid and can be liquefied with heat, poured into a different container, and resolidified. Solid media provide a firm surface on which cells can form discrete colonies. Nutrient gelatin contains enough gelatin (12%) to take on a solid consistency. The top tube shows it as a solid. The bottom tube indicates what happens when it is warmed or when microbial enzymes digest the gelatin and liquefy it.

All photos: Kathleen Talaro

batch to batch—which is not a problem for many situations. Other possible ingredients are milk, yeast extract, soybean digests, and peptone. Nutrient broth, blood agar, and MacConkey agar are all complex media.

Tables 2.2A and **2.2B** provide an illustration of the difference between defined and complex media, in this case, for growing *Staphylococcus aureus*.

Media for Different Purposes

Microbiologists have many types of media at their disposal. For that reason, until recently, microbiologists knew of only a few species of bacteria or fungi that could not be cultivated artificially. However, newer DNA detection technologies have shown us that there are many more microbes that we don't know how to cultivate in the lab than those that we do. Although we can now study some vital traits of bacteria without actually growing the bacteria, developing new media is still important for growing the bacteria that we are discovering using those genomic methods.

General-purpose media are those that will allow the growth of as broad a spectrum of microbes as possible. As a rule, they are of the complex variety. Examples include nutrient agar and broth, brain-heart infusion, and trypticase soy agar (TSA). An **enriched medium** contains complex organic substances such as blood, serum, hemoglobin, or special **growth factors** (specific vitamins, amino acids) that certain species must have in order to grow. Bacteria that require growth factors and complex nutrients are termed **fastidious.** Blood agar is made by adding sterile sheep, horse, or rabbit blood to a sterile agar base **(figure 2.4*a*).** It is widely used to grow fastidious streptococci and other pathogens. Disease-causing *Neisseria* (one species causes gonorrhea) are grown on either Thayer-Martin medium or "chocolate" agar, which is a blood agar with added hemin and nicotinamide adenine dinucleotide **(figure 2.4*b*).** Enriched media are also useful in the clinical laboratory to encourage the growth of pathogens that may be present in very low numbers, such as in urine or blood specimens.

Selective and Differential Media   These media are designed for special microbial groups, and they are extremely useful in isolation and identification. They can permit, in a single step, the preliminary identification of a genus or even a species.

A **selective medium** contains one or more agents that inhibit the growth of certain types of microbe or microbes (call them A, B, and C) but not others (D). This *selects* microbe D, and only microbe D is able to grow. Selective media are very important in the initial stages of isolating a specific type of microorganism from samples containing dozens of different species—for example, feces, saliva, skin, water, and soil. They speed up isolation by suppressing the unwanted background organisms and favoring growth of the desired ones.

Media for isolating intestinal pathogens (MacConkey agar, Hektoen enteric [HE] agar) contain

©Image Source/Corbis

Table 2.2A Defined Medium for Growth and Maintenance of Pathogenic *Staphylococcus aureus*

0.25 Gram Each of These Amino Acids	0.5 Gram Each of These Amino Acids	0.12 Gram Each of These Amino Acids
Cystine	Arginine	Aspartic acid
Histidine	Glycine	Glutamic acid
Leucine	Isoleucine	
Phenylalanine	Lysine	
Proline	Methionine	
Tryptophan	Serine	
Tyrosine	Threonine	
	Valine	

Additional ingredients

0.005 mole nicotinamide ⎤
0.005 mole thiamine ⎥ — Vitamins
0.005 mole pyridoxine ⎥
0.5 microgram biotin ⎦

1.25 grams magnesium sulfate ⎤
1.25 grams dipotassium hydrogen phosphate ⎥ — Salts
1.25 grams sodium chloride ⎥
0.125 gram iron chloride ⎦

Ingredients dissolved in 1,000 milliliters of distilled water and buffered to a final pH of 7.0.

Table 2.2B Brain-Heart Infusion Broth: A Complex Medium for Growth and Maintenance of Pathogenic *Staphylococcus aureus*

27.5 grams brain, heart extract, peptone extract
2   grams glucose
5   grams sodium chloride
2.5 grams disodium hydrogen phosphate

Ingredients dissolved in 1,000 milliliters of distilled water and buffered to a final pH of 7.0.

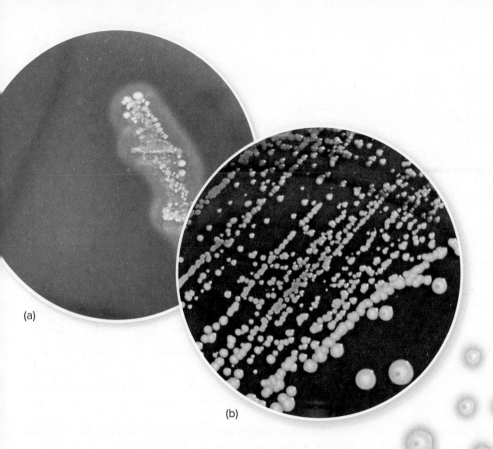

Figure 2.4 Examples of enriched media. **(a)** Blood agar plate growing bacteria from the human throat. Enzymes from the bacterial colonies break down the red blood cells in the agar, leaving a clear "halo." **(b)** Culture of *Neisseria* sp. on chocolate agar. Chocolate agar gets its brownish color from cooked blood (not chocolate) and does not produce hemolysis.
Lisa Burgess/McGraw-Hill Education (a); Kathleen Talaro (b)

bile salts as a selective agent. Other agents that have selective properties are dyes, such as methylene blue and crystal violet, and antimicrobial drugs. **Table 2.3** gives multiple examples of selective media and what they do.

Differential media do not inhibit the growth of any particular microorganisms but are designed to display visible differences in how they grow. Differentiation shows up as variations in colony size or color **(figure 2.5),** in media color changes, or in the formation of gas bubbles and precipitates. These variations often come from the type of chemicals these media contain and the ways that microbes react to them. For example, when microbe X metabolizes a certain substance not used by organism Y, then X will cause a visible change in the medium and Y will not **(figure 2.6).** The simplest differential media show just two reaction types, leading to a "yes" or "no" situation. For example, the medium will cause one type of bacterial colony to change color, while the other types of bacteria do not. Some media are sufficiently complex to allow for three or four different reactions. Importantly, you should know that a single medium can be both selective and differential, owing to its different ingredients. MacConkey agar, for example, appears in table 2.3 (selective media) and **table 2.4** (differential media) due to its ability to suppress the growth of some organisms while producing a visual distinction among the ones

Table 2.3 Selective Media, Agents, and Functions

Medium	Selective Agent	Used for
Enterococcus faecalis broth	Sodium azide, tetrazolium	Isolation of fecal enterococci
Tomato juice agar	Tomato juice, acid	Isolation of lactobacilli from saliva
MacConkey agar	Bile, crystal violet	Isolation of gram-negative intestinal bacteria
Salmonella/ Shigella (SS) agar	Bile, citrate, brilliant green	Isolation of *Salmonella* and *Shigella*
Lowenstein-Jensen	Malachite green dye	Isolation and maintenance of *Mycobacterium*
Sabouraud's agar	pH of 5.6 (acid)	Isolation of fungi (inhibits bacteria)

Figure 2.5 A medium that is both selective and differential. MacConkey agar selects against gram-positive bacteria. Therefore, you will not see them here! It also differentiates between lactose-fermenting bacteria (indicated by a pink-red reaction in the center of the colony) and lactose-negative bacteria (indicated by an off-white colony with no dye reaction).
Kathleen Talaro

Figure 2.6 Comparison of selective and differential media with general-purpose media. **(a)** A mixed sample containing three different species is streaked onto plates of general-purpose nonselective medium and selective medium. **(b)** Another mixed sample containing three different species is streaked onto plates of general-purpose nondifferential medium and differential medium.

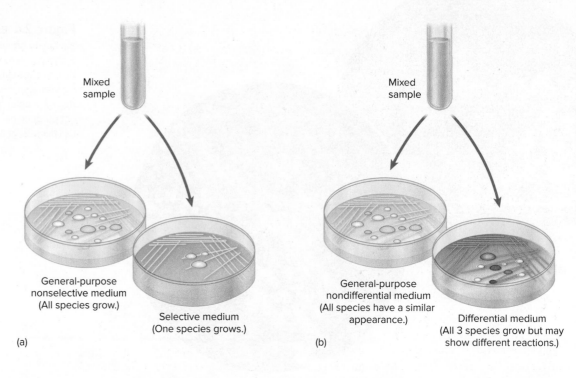

Mixed sample

Mixed sample

General-purpose nonselective medium (All species grow.)

Selective medium (One species grows.)

(a)

General-purpose nondifferential medium (All species have a similar appearance.)

Differential medium (All 3 species grow but may show different reactions.)

(b)

that do grow. The agar in figure 2.5 illustrates this activity; you just can't see the colonies that were suppressed. Media that are both selective and differential allow for microbial isolation and identification to occur at the same time, which can be very useful in the screening of patient specimens as well as food and water samples.

Dyes are frequently used as differential agents because many of them are pH indicators that change color in response to the production of an acid or a base. For example, MacConkey agar contains neutral red, a dye that is yellow when neutral and pink or red when acidic. A common intestinal bacterium such as *Escherichia coli* that gives off acid when it metabolizes the lactose in the medium develops red to pink colonies, and one like *Salmonella* that does not give off acid remains its natural color (off-white).

Miscellaneous Media   A **reducing medium** contains a substance (sodium thioglycollate or cystine) that absorbs oxygen or slows the penetration of oxygen in a medium, thus reducing its availability. Reducing media are important for growing bacteria that don't require oxygen (termed anaerobic) or for determining the oxygen requirements of isolates. **Carbohydrate fermentation media** contain sugars that can be fermented (converted to acids) and a pH indicator to show this reaction **(figure 2.7)**.

NCLEX® PREP

2. A patient with presumed urinary tract infection (UTI) has a urine sample collected and sent to the laboratory for culture. To encourage the growth of scant pathogens in the sample, an enriched medium may be used. Which of the following are substances that may be present in enriched media? Choose all that apply.

   a. *serum*

   b. *hemoglobin*

   c. *vitamins*

   d. *growth factors*

Table 2.4 Differential Media

Medium	Substances That Result in Differentiation	Differentiates Between or Among
Blood agar	Intact red blood cells	Types of hemolysis displayed by different species of *Streptococcus*
Mannitol salt agar	Mannitol, phenol red	Species of *Staphylococcus*
MacConkey agar	Lactose, neutral red	Bacteria that ferment lactose (lowering the pH) from those that do not
Urea broth	Urea, phenol red	Bacteria that hydrolyze urea to ammonia from those that do not
Sulfur indole motility (SIM)	Thiosulfate, iron	H_2S gas producers from nonproducers
Triple-sugar iron agar (TSIA)	Triple sugars, iron, and phenol red dye	Fermentation of sugars, H_2S production
Birdseed agar	Seeds from thistle plant	*Cryptococcus neoformans* and other fungi

Control   **Alkaline**   **Acid: No gas**   **Acid with gas**

Figure 2.7 Carbohydrate fermentation in broth.
This medium is designed to show fermentation (acid production) and gas formation. The medium changes color in the presence of acid (from red to yellow). Gas formation is monitored by placing a small tube upside-down in the medium. If gas bubbles form, they are captured inside the small tube.
Lisa Burgess/McGraw-Hill Education

Transport media are used to maintain and preserve specimens that have to be held for a period of time before clinical analysis or to sustain delicate species that die rapidly if not held under stable conditions. **Assay media** are used by technologists to test the effectiveness of antimicrobial drugs (see chapter 10) and by drug manufacturers to assess the effect of disinfectants, antiseptics, cosmetics, and preservatives on the growth of microorganisms. **Enumeration media** are used by industrial and environmental microbiologists to count the numbers of organisms in milk, water, food, soil, and other samples.

③ Isolation: Separating One Species from Another

Certain **isolation** techniques are based on the concept that if an individual bacterial cell is separated from other cells and provided adequate space on a solid nutrient surface, it will grow into a discrete mound of cells called a **colony (figure 2.8).** If it was formed from a single cell, a colony consists of just that one species and no other. Proper isolation requires that a small number of cells be inoculated into a relatively large volume or over a large area of medium. It generally requires the following materials: a medium that has a relatively firm surface, a Petri dish (a clear, flat dish with a cover), and inoculating tools.

There are three main ways to accomplish isolation:

- the streak plate technique
- the pour plate technique
- the spread plate technique

With one method, called the streak plate method, a small droplet of culture or sample is spread over the surface of the medium with an *inoculating loop* in a pattern that gradually thins out the sample and separates the cells spatially over several sections of the plate **(figure 2.9a).** The goal here is to allow a single cell to grow into an isolated colony.

In another method, the loop dilution, or pour plate, technique, the sample is inoculated serially into a series of cooled but still liquid agar tubes so as to dilute the number of cells in each successive tube in the series **(figure 2.9b).** Inoculated tubes are then plated out (poured) into sterile Petri dishes and allowed to solidify (harden). The end result (usually in the second or third plate) is that the number of cells per volume is so decreased that cells have ample space to grow into separate colonies. One difference between this and the streak plate method is that in this technique some of the colonies will develop deep in the medium itself. In the streak plate method, the agar is already solidified in the Petri dish before the bacteria are streaked on it.

With the spread plate technique, a small volume of liquid sample is pipetted onto the surface of the medium and spread around evenly by a sterile spreading

Figure 2.8 Isolation technique.   Stages in the formation of an isolated colony, showing the microscopic events and the macroscopic result. Separation techniques such as streaking can be used to isolate single cells. After numerous cell divisions, a macroscopic mound of cells, or a colony, will be formed. This is a relatively simple yet successful way to separate different types of bacteria in a mixed sample.

Lisa Burgess/McGraw-Hill Education (test tube); Richard Hutchings/McGraw-Hill Education (petri dish); Lisa Burgess/McGraw-Hill Education (colonies growing on agar medium)

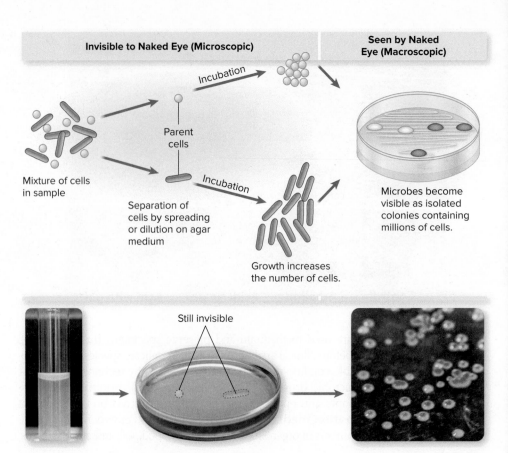

tool (sometimes called a "hockey stick" because of its shape). Like the streak plate, cells are pushed onto separate areas on the surface so that they can form individual colonies **(figure 2.9c).**

④ ⑤ Inspection and Identification

How does one determine (i.e., identify) what sorts of microorganisms have been isolated in cultures? Generally, their microscopic appearance is of limited value because many bacteria have similar shapes. The microscope *can* be useful in differentiating the smaller, simpler bacterial cells from the larger, more complex eukaryotic cells. Appearance can be useful in identifying eukaryotic microorganisms to the level of genus or species because of their distinctive morphological features. Unfortunately, bacteria are generally not identifiable by these methods because very different species may appear quite similar. For them, we have to include other techniques, some of which characterize their cellular metabolism. These methods, called biochemical tests, can determine fundamental chemical characteristics such as nutrient requirements, products given off during growth, presence of enzymes, and mechanisms for deriving energy. Their genetic and immunologic characteristics are also used for identification. In chapter 15, we present more detailed examples of the most current genotypic and immunologic identification methods.

> **2.1 LEARNING OUTCOMES—Assess Your Progress**
>
> 1. Explain what the Five I's are and what each step entails.
> 2. Discuss three physical states of media and when each is used.
> 3. Compare and contrast selective and differential media, and give an example of each.
> 4. Provide brief definitions for *defined media* and *complex media*.

©Ryan McVay/Getty Images

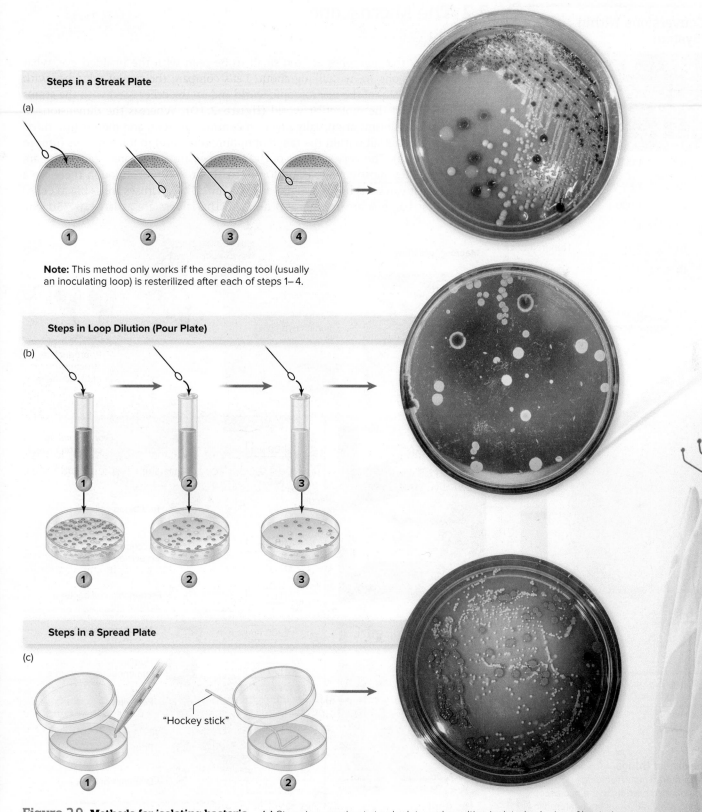

Steps in a Streak Plate

(a)

Note: This method only works if the spreading tool (usually an inoculating loop) is resterilized after each of steps 1–4.

Steps in Loop Dilution (Pour Plate)

(b)

Steps in a Spread Plate

(c)

"Hockey stick"

Figure 2.9 Methods for isolating bacteria. **(a)** Steps in a quadrant streak plate and resulting isolated colonies of bacteria. **(b)** Steps in the loop dilution method and the appearance of plate 3. **(c)** Spread plate and its result.
Kathleen Talaro (a), (b); Kathleen Talaro and Harold Benson (c)

Table 2.5 Conversions Within the Metric System

	Symbol	Factor (Meter multiplied by —)	Numerically	
Giga	Gm	10^9	1 000 000 000	Billion
Mega	Mm	10^6	1 000 000	Million
Kilo	km	10^3	1 000	Thousand
Meter	m		1	
Centi	c	10^{-2}	0.01	Hundredth
Milli	mm	10^{-3}	0.001	Thousandth
Micro	μ	10^{-6}	0.000 001	Millionth
Nano	n	10^{-9}	0.000 000 001	Billionth

2.2 The Microscope

Microbial Size

When we say that microbes are too small to be seen with the unaided eye, what sorts of dimensions are we talking about? Let's compare the size of microbes with the larger organisms of the macroscopic world and on the other hand with the atoms and molecules of the molecular world **(figure 2.10)**. Whereas the dimensions of macroscopic organisms are usually given in centimeters (cm) and meters (m), those of microorganisms fall within the range of millimeters (mm) to micrometers (μm) to nanometers (nm). The very smallest of the microbes are the prions. Then there are the viruses. They mostly range from 20 nm to about 400 nm, although there are a few types that can be as big as 800 nm or 1500 nm. (Those viruses are as big as cells.) The smallest bacteria are around 200 nm, and the largest are as large as 750 μm.

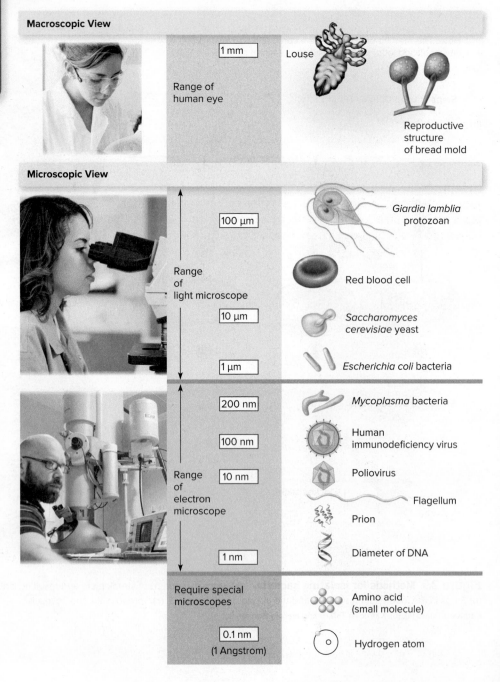

Figure 2.10 The size of things. Common measurements encountered in microbiology and a scale of comparison from the macroscopic to the microscopic, molecular, and atomic. Most microbes encountered in our studies will fall between 100 μm and 10 nm in overall dimensions. The microbes shown are more or less to scale within size zone but not between size zones.

Cultura Creative/Alamy Stock Photo (woman looking at plate); Corbis/SuperStock (woman looking in microscope); McGraw-Hill Education (a graduate student works with a TEM electron microscope)

Yeasts (a single-celled form of fungus) are generally 3-4 μm, though some can be much larger. Protozoa are generally around 100-300 μm. It is much easier to get a feel for the relative sizes by looking at a visual (figure 2.10). Also consult **table 2.5** for a reminder of metric measurements and relative size.

Magnification and Microscope Design

The microbial world is of obvious importance, but it would remain largely uncharted without an essential tool: the microscope. The fundamental parts of a modern compound light microscope are illustrated in **figure 2.11.**

Principles of Light Microscopy

Microscopes provide three important qualities:

- magnification,
- resolution, and
- contrast.

Magnification

Magnification occurs in two phases. The first lens in this system (the one closest to the specimen) is the objective lens, and the second (the one closest to the eye) is the ocular lens, or eyepiece **(figure 2.12).** The objective lens forms the initial image of the specimen, called the **real image.** When this image is projected up through the microscope body to the plane of the eyepiece, the ocular lens forms a second image, the **virtual image.** The virtual image is the one that will be received by the eye and converted to a retinal and visual image. The magnifying power of the objective lens

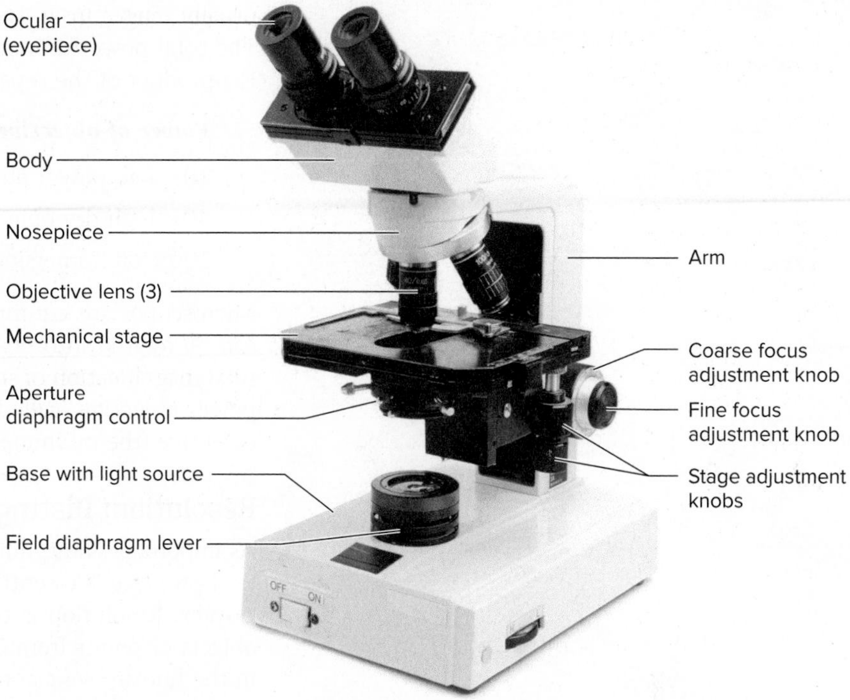

Figure 2.11 The parts of a student laboratory microscope. This microscope is a compound light microscope with two oculars (called binocular). It has three objective lenses.
Peter Skinner/Science Source

Ocular (eyepiece) · Body · Nosepiece · Objective lens (3) · Mechanical stage · Aperture diaphragm control · Base with light source · Field diaphragm lever · Arm · Coarse focus adjustment knob · Fine focus adjustment knob · Stage adjustment knobs · OFF ON

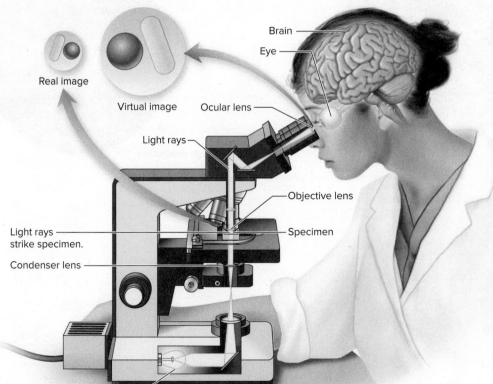

Brain · Eye · Real image · Virtual image · Ocular lens · Light rays · Objective lens · Specimen · Light rays strike specimen. · Condenser lens · Light source

Figure 2.12 The pathway of light and the two stages in magnification of a compound microscope. As light passes through the condenser, it forms a solid beam that is focused on the specimen. Light leaving the specimen that enters the objective lens is refracted so that an enlarged primary image, the real image, is formed. One does not see this image, but its degree of magnification is represented by the smaller circle. The real image is projected through the ocular, and a second image, the virtual image, is formed by a similar process. The virtual image is the final magnified image that is received by the retina and perceived by the brain. Notice that the lens systems cause the image to be reversed.

usually ranges from 4× to 100×, and the power of the ocular lens is usually 10×. The total power of magnification of the final image formed by the combined lenses is a product of the separate powers of the two lenses:

Power of objective	Power of ocular	Total magnification
10× low power objective	× 10×	= 100×
40× high dry objective	× 10×	= 400×
100× oil immersion objective	× 10×	= 1,000×

Microscopes are equipped with a nosepiece holding three or more objectives that can be rotated into position as needed. Depending on the power of the ocular, the total magnification of standard light microscopes can vary from 40× with the lowest power objective (called the scanning objective) to 1,000× with the highest power objective (the oil immersion objective).

Resolution: Distinguishing Magnified Objects Clearly

As important as magnification is for visualizing tiny objects or cells, an additional optical property is essential for seeing clearly. That property is resolution, or **resolving power.** Resolution is the capacity of an optical system to distinguish two adjacent objects or points from one another. For example, at a certain fixed distance, the lens in the human eye can resolve two small objects as separate points as long as the two objects are no closer than 0.2 mm apart. The eye examination given by optometrists is in fact a test of the resolution of the human eye for various-size letters read at a particular distance. **Figure 2.13** should help you understand the concept of resolution.

The oil immersion lens (100× magnification) uses oil to capture some of the light that would otherwise be lost to scatter **(figure 2.14)**. Reducing this scatter

Figure 2.13 **Effect of wavelength on resolution.**
A simple model demonstrates how the wavelength of light influences the resolving power of a microscope. The size of the balls illustrates the relative size of the wave. Here, a human cell (fibroblast) is illuminated with long wavelength light **(a)** and short wavelength light **(b)**. In (a), the waves are too large to penetrate the tighter spaces and produce a fuzzy, undetailed image.
Courtesy Nikon Instruments Inc.

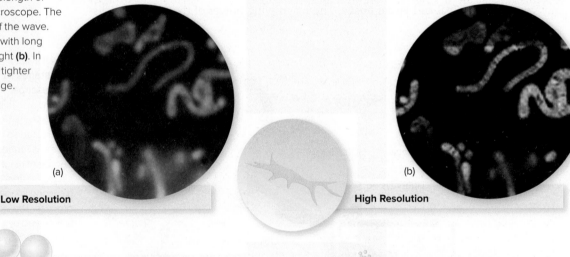

(a) Low Resolution

(b) High Resolution

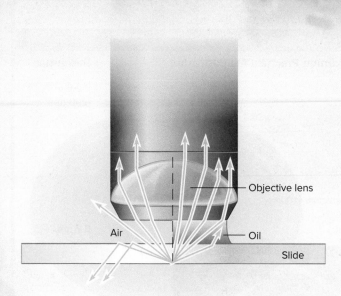

Appearance in Microscope	Appearance in Reality

Small bacterial cells

Eukaryotic cells

0.2 µm   2 µm   0.2 µm   2 µm

Figure 2.14 Workings of an oil immersion lens. Without oil, some of the peripheral light that passes through the specimen is scattered into the air or onto the glass slide; this scattering decreases resolution.

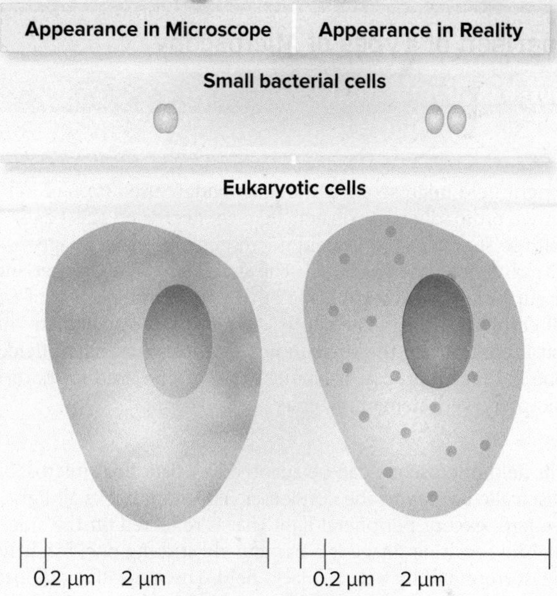

Figure 2.15 The importance of resolution. If a microscope has a resolving power of 0.2 µm, then the bacterial cells would not be resolvable as two separate cells. Likewise, the small specks inside the eukaryotic cell will not be visible.

increases resolution. In practical terms, the oil immersion lens can resolve any cell or cell part as long as it is at least 0.2 µm in diameter, and it can resolve two adjacent objects as long as they are at least 0.2 µm apart **(figure 2.15)**. In general, organisms that are 0.5 µm or more in diameter are easily seen. This includes fungi and protozoa, some of their internal structures, and most bacteria. However, a few bacteria and most viruses are far too small to be resolved by the optical microscope and require electron microscopy (discussed later in this chapter). In summary, then, the factor that most limits the clarity of a microscope's image is its resolving power. Even if a light microscope were designed to magnify several thousand times, its resolving power could not be increased, and the image it produced would simply be enlarged and fuzzy.

Contrast

The third quality of a well-magnified image is its degree of contrast from its surroundings. The contrast is measured by a quality called the **refractive index.** Refractive index refers to the degree of bending that light undergoes as it passes from one medium, such as water or glass, to another medium, such as bacterial cells. The higher the difference in refractive indexes (the more bending of light), the sharper the contrast that is registered by the microscope and the eye. Because too much light can reduce contrast and burn out the image, an adjustable iris diaphragm on most microscopes controls the amount of light entering the condenser. The lack of contrast in cell components can be compensated for by using special lenses (the phase-contrast microscope) and by adding dyes.

Different Types of Light Microscopes

Optical microscopes that use visible light can be described by the nature of their *field*, meaning the circular area viewed through the ocular lens. We will discuss three types of visible-light microscopes: bright-field, dark-field, and phase-contrast. A fourth type of optical microscope, the fluorescence microscope, uses ultraviolet radiation as the illuminating source; another, the confocal microscope, uses a laser beam. Each of these microscopes is adapted for viewing specimens in a particular way, as described in **table 2.6.**

NCLEX® PREP

3. The capacity of an optical system to distinguish or separate two adjacent objects or points from one another is known as
   a. *the real image.*
   b. *the virtual image.*
   c. *resolving power.*
   d. *numerical aperture.*
   e. *power.*

Table 2.6 Comparison of Types of Microscopy

Visible light as source of illumination

Microscope	Maximum Practical Magnification	Resolution
Bright-Field The bright-field microscope is the most widely used type of light microscope. Although we ordinarily view objects (like the words on this page) with light reflected off the surface, a bright-field microscope forms its image when light is transmitted *through* the specimen. The specimen, being denser and more opaque than its surroundings, absorbs some of this light, and the rest of the light is transmitted directly up through the ocular. As a result, the specimen will produce an image that is darker than the surrounding brightly illuminated field. The bright-field microscope is a multipurpose instrument that can be used for both live, unstained material and preserved, stained material.	2,000×  *Paramecium (400×)*	0.2 μm (200 nm)
Dark-Field A bright-field microscope can be adapted as a dark-field microscope by adding a special disc called a *stop* to the condenser. The stop blocks all light from entering the objective lens—except peripheral light that is reflected off the sides of the specimen itself. The resulting image is a particularly striking one: brightly illuminated specimens surrounded by a dark (black) field. The most effective use of dark-field microscopy is to visualize living cells that would be distorted by drying or heat or that cannot be stained with the usual methods.	2,000×  *Paramecium (400×)*	0.2 μm
Phase-Contrast If similar objects made of clear glass, ice, cellophane, or plastic are immersed in the same container of water, an observer would have difficulty telling them apart because they have similar optical properties. Internal components of a live, unstained cell also lack contrast and can be difficult to distinguish. But cell structures do differ slightly in density, enough that they can alter the light that passes through them in subtle ways. The phase-contrast microscope has been constructed to take advantage of this characteristic. This microscope contains devices that transform the subtle changes in light waves passing through the specimen into differences in light intensity. For example, denser cell parts such as organelles alter the pathway of light more than less dense regions (the cytoplasm). Light patterns coming from these regions will vary in contrast. The amount of internal detail visible by this method is greater than by either bright-field or dark-field methods. The phase-contrast microscope is most useful for observing intracellular structures such as bacterial endospores, granules, and organelles, as well as the locomotor structures of eukaryotic cells such as cilia.	2,000× 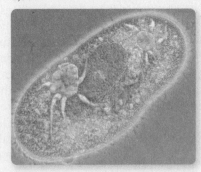 *Paramecium (400×)*	0.2 μm

Ultraviolet rays as source of illumination

Microscope	Maximum Practical Magnification	Resolution
Fluorescence The fluorescence microscope is a specially modified compound microscope furnished with an ultraviolet (UV) radiation source. The name of this type of microscopy comes from the use of certain dyes (acridine, fluorescein) and minerals that are **fluorescent.** The dyes emit visible light when bombarded by short ultraviolet rays. For an image to be formed, the specimen must first be coated or placed in contact with a source of fluorescence. Shining ultraviolet radiation on the specimen causes it to give off light that will form its own image, usually an intense red, blue, or green against a black field. Fluorescence microscopy has its most useful applications in diagnosing infections and pinpointing particular cellular structures.	2,000× 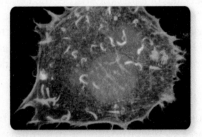 Fluorescence image of a eukaryotic cell	0.2 μm

(Background Image): Science Photo Library/Getty Images

Table 2.6 (continued)

Ultraviolet rays as source of illumination (continued)

Microscope	Maximum Practical Magnification	Resolution
Confocal The scanning confocal microscope overcomes the problem of cells or structures being too thick. This is a problem for other types of microscopes, as they are unable to focus on all of their levels. This microscope uses a laser beam of light to scan various depths in the specimen and deliver a sharp image focusing on just a single plane. It is thus able to capture a highly focused view at any level, ranging from the surface to the middle of the cell. It is most often used on fluorescently stained specimens, but it can also be used to visualize live unstained cells and tissues.	2,000×  Myofibroblasts, cells involved in tissue repair (400×)	0.2 μm

Electron beam forms image of specimen

Microscope	Maximum Practical Magnification	Resolution
Transmission Electron Microscope (TEM) Transmission electron microscopes are the method of choice for viewing the detailed structure of cells and viruses. This microscope produces its image by transmitting electrons through the specimen. Because electrons cannot easily penetrate thick preparations, the specimen must be sectioned into extremely thin slices (20-100 nm thick) and stained or coated with metals that will increase image contrast. The darker areas of TEM micrographs represent the thicker (denser) parts, and the lighter areas indicate the more transparent and less dense parts.	100,000,000× 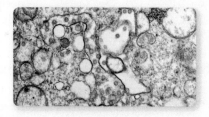 This is SARS-CoV-2 (also on the cover of this book). It is the cause of the COVID-19 pandemic. (100,000×)	0.5 nm
Scanning Electron Microscope (SEM) The scanning electron microscope provides some of the most dramatic and realistic images in existence. This instrument is designed to create an extremely detailed three-dimensional view of all kinds of objects–from plaque on teeth to tapeworm heads. To produce its images, the SEM bombards the surface of a metal-coated specimen with electrons while scanning back and forth over it. A shower of electrons deflected from the surface is picked up with great accuracy by a sophisticated detector, and the electron pattern is displayed as an image on a television screen. You will often see these images in vivid colors. The color is always added afterward; the actual microscopic image is black and white.	100,000,000×  Algae showing cell walls made of calcium discs (10,000×)	10 nm

M. I. Walker/Science Source (light microscope image); Kent Wood/Science Source (dark-field image); Roland Birke/Photolibrary/Getty Images (phase-contrast); Dr. Dan Kalman/Science Source (immunofluorescence image); Dr. Gopal MurtScience Source (confocal light image micrograph); Source: CDC/ Dr. Erskine Palmer (transmission electron micrograph); Science Photo Library/Getty Images (scanning electron micrograph)

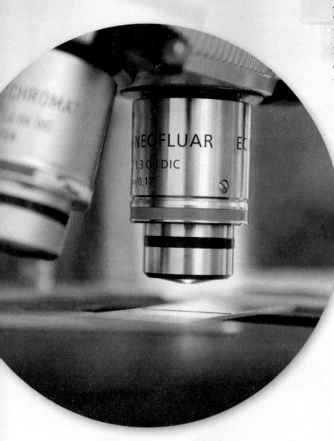

Preparing Specimens for the Microscope

A specimen for optical microscopy is generally prepared by placing a sample on a glass slide. This is then placed on the microscope stage between the condenser and the objective lens. The manner in which a slide specimen (also called a mount) is prepared depends upon (1) the condition of the specimen, either in a living or preserved state; (2) the aims of the examiner, whether to observe overall structure, identify the microorganisms, or see movement; and (3) the type of microscopy available, whether it is bright-field, dark-field, phase-contrast, or fluorescence.

Fresh, Living Preparations

Live samples of microorganisms are examined using wet mounts or in hanging drop mounts so that they can be observed as close to their natural state as possible. The cells are suspended in a suitable fluid (water, broth, saline) that temporarily maintains viability and provides space and a medium for locomotion. A wet mount consists of a drop or two of the culture placed on a slide and overlaid with a coverslip. The hanging drop preparation is made with a special concave (depression) slide, a Vaseline adhesive or sealant, and a coverslip from which a tiny drop of sample is suspended **(figure 2.16)**. These short-term mounts provide a true assessment of the size, shape, arrangement, color, and motility of cells. However, if you need to visualize greater cellular detail, you will have to use phase-contrast microscopy, which requires that the specimens are "fixed" and therefore no longer alive.

Fixed, Stained Smears

A more permanent slide for long-term study can be obtained by preparing fixed, stained specimens. "Fixed" in this context means that the organisms are dried and attached to the glass slide. The so-called smear technique, developed by Robert Koch more than 150 years ago, consists of spreading a thin film made from a liquid suspension of cells on a slide and air-drying it. Next, the air-dried smear is usually heated gently by a process called heat fixation that simultaneously kills the specimen and secures it to the slide.

Stains

Like images on undeveloped photographic film, the unstained cells of a fixed smear are very difficult to see, no matter how great the magnification or how fine the resolving power of the microscope. To solve this problem, stains can be added to the fixed smear, either coloring the objects (microbes) or coloring the background so that the unstained microbes stand out. Staining is any procedure that applies colored chemicals called *dyes* to specimens. Dyes give a color to cells or cell parts by becoming affixed to them through a chemical reaction. Dyes can be classified as basic (cationic) dyes, which have a positive charge, or acidic (anionic) dyes, which have a negative charge. Because chemicals of opposite charge are attracted to each other, cell parts that are negatively charged will attract basic dyes, and those that are

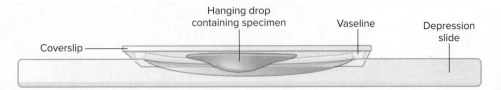

Figure 2.16 Hanging drop technique. Cross-section view of slide and coverslip. (Vaseline actually surrounds entire well of slide.)

Table 2.7 Comparison of Positive and Negative Stains

	Positive Staining	Negative Staining
Appearance of cell	Colored by dye	Clear and colorless
Background	Not stained (generally white)	Stained (dark gray or black)

Dyes employed	Basic dyes:
	  Crystal violet
	  Methylene blue
	  Safranin Malachite green
		Acidic dyes:
		  Nigrosin
		  India ink

Subtypes of stains	Simple stain
	
	Differential stains:
	  Gram stain
	  Acid-fast stain
	  Spore stain
	
	Special stains:
	  Capsule
	  Flagella
	  Endospore
	  Granules
	  Nucleic acid
		Capsule
		
		Endospore

Photos: Lisa Burgess/McGraw-Hill Education

positively charged will attract acidic dyes. Many cells, especially those of bacteria, have numerous negatively charged acidic substances on their surfaces and thus stain more readily with basic dyes. Acidic dyes, on the other hand, tend to be repelled by cells, so they are good for negative staining (staining the background and leaving cells unstained).

Negative versus Positive Staining   Two basic types of staining technique are used, depending upon how a dye reacts with the specimen (summarized in **table 2.7).** Most procedures involve a **positive stain,** in which the dye actually sticks to the specimen and gives it color. A **negative stain,** on the other hand, is just the reverse (like a photographic negative). The dye does not stick to the specimen but settles some distance from its outer boundary, forming a silhouette. Nigrosin (blue-black) and India ink (a black suspension of carbon particles) are the dyes most commonly used for negative staining. The cells themselves do not stain because these dyes are negatively charged and are repelled by the negatively charged surface of the cells. The value of negative staining is its simplicity and the reduced shrinkage or distortion of cells because the smear is not heat fixed. Negative staining is also used to accentuate the capsule that surrounds certain bacteria and yeasts.

Simple versus Differential Staining   Positive staining methods fall into three categories: simple, differential, or special. **Simple stains** require only a single dye. **Differential stains** use two differently colored dyes, called the *primary dye* and the *counterstain,* to distinguish between cell types or parts. Differential staining techniques tend to be more complex and sometimes require additional chemical reagents to produce the desired reaction.

   Simple stains cause all cells in a smear to appear more or less the same color, regardless of type, but they can still reveal bacterial characteristics such as shape, size, and arrangement **(figure 2.17).**

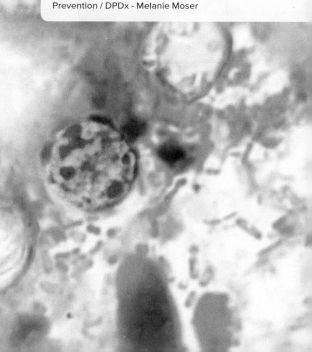

Photomicrograph of stool sample stained with acid-fast stain revealing *Cyclospora.*

Source: Centers for Disease Control and Prevention / DPDx - Melanie Moser

Medical Moment

Gram-Positive versus Gram-Negative Bacteria

The Gram stain is one type of differential stain that can help to identify bacterial species and guide treatment decisions. Differentiating between gram-positive and gram-negative organisms is important. One of the main differences between gram-positive and gram-negative bacteria is that gram-negative bacteria have an outer membrane containing LPS (lipopolysaccharide). This molecule is also called endotoxin, which can cause a severe reaction if it enters the circulatory system, causing symptoms of shock (high fever, dangerously low blood pressure, and elevated respiratory rate). This is known as endotoxic shock.

Q. This molecule has two names: lipopolysaccharide and endotoxin. Which of these is most often used by medical doctors and which is more likely to be used by microbiology researchers?

Answer in Appendix B.

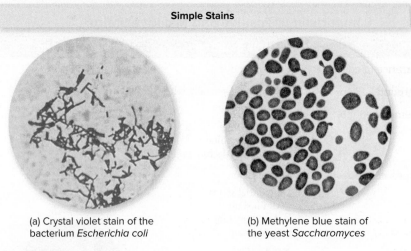

Simple Stains

(a) Crystal violet stain of the bacterium *Escherichia coli*

(b) Methylene blue stain of the yeast *Saccharomyces*

Figure 2.17 Simple stains.

Kathleen Talaro (a); Source: Centers for Disease Control and Prevention (b)

Types of Differential Stains

A typical differential stain uses differently colored dyes to clearly contrast two cell types or cell parts. Common combinations are red and purple, red and green, or pink and blue **(figure 2.18)**. Typical examples include the Gram stain, the acid-fast stain, and the endospore stain. Some staining techniques (endospore, capsule), which are differential, are also in the "special" category.

The Gram Stain

In 1884, Hans Christian Gram discovered a staining technique that could be used to make bacteria in infectious specimens more visible. His technique consisted of sequential applications of crystal violet (the primary dye), Gram's iodine (the mordant), an alcohol rinse (decolorizer), and a contrasting counterstain. Bacteria that stain purple are called gram-positive, and those that stain red are called gram-negative. Gram-variable organisms produce both pink- and purple-staining cells.

The different results in the Gram stain are due to differences in the structure of the cell envelope and how it reacts to the series of reagents applied to the cells. We will study it in more detail in chapter 3.

This century-old staining method remains a universal basis for bacterial classification and identification. The Gram stain can also be a practical aid in diagnosing infection and in guiding drug treatment. For example, Gram staining a fresh sputum or spinal fluid specimen can help pinpoint the possible cause of infection, and in some cases it is possible to begin drug therapy on the basis of this stain. Even in this day of elaborate and expensive medical technology, the Gram stain remains an important first tool in diagnosis.

Figure 2.18
Differential stains.

ASM/Science Source (a); Richard J. Green/Science Source (b); CDC/Courtesy Larry Stauffer, Oregon State Public Health Laboratory (c)

Differential Stains

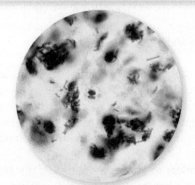

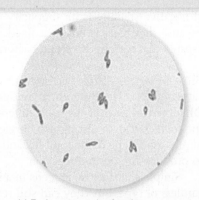

(a) Gram stain. Here both gram-negative (pink) rods and gram-positive (purple) cocci are visible.

(b) Acid-fast stain. Reddish-purple cells are acid-fast. Blue cells are non-acid-fast.

(c) Endospore stain showing endospores (green) and vegetative cells (pink).

Special Stains

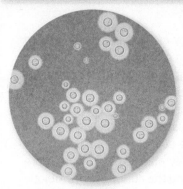

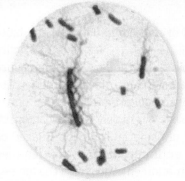

(a) India ink capsule stain of *Cryptococcus neoformans*

(b) Flagellar stain of *Proteus vulgaris*

Figure 2.19 Special stains.
Dr. Leanor Haley/Centers for Disease Control and Prevention (a); David Fankhauser (b)

Other Differential Stains The **acid-fast stain,** like the Gram stain, is an important diagnostic stain that distinguishes acid-fast bacteria (pink) from non-acid-fast bacteria (blue). (In this context *fast* means "resistant to.") This stain originated as a specific method to detect *Mycobacterium tuberculosis* in specimens. It was determined that these bacterial cells have a particularly impervious outer wall that holds fast (tightly or tenaciously) to the dye (carbol fuchsin) even when washed with a solution containing acid or acid alcohol. This stain is used for other medically important bacteria, fungi, and protozoa. Often it is performed when a gram-variable result is seen in a specimen.

The **endospore** stain (spore stain) is similar to the acid-fast method in that a dye is forced by heat into resistant bodies called endospores (their characteristics and significance are discussed in chapter 3). This stain is designed to distinguish between endospores and the cells that they come from (so-called **vegetative** cells). Of significance in medical microbiology are the gram-positive, endospore-forming members of the genus *Bacillus* (the cause of anthrax) and *Clostridium* (the cause of botulism and tetanus)—dramatic diseases that we consider in later chapters.

Special Stains Special stains are used to emphasize certain cell parts that are not revealed by conventional staining methods **(figure 2.19). Capsular staining** is a method of observing the microbial capsule, an unstructured protective layer surrounding the cells of some bacteria and fungi. Because the capsule repels most stains, it is often negatively stained with India ink, or it may be demonstrated by special positive stains. One example is *Cryptococcus*, which causes a serious form of fungal meningitis in AIDS patients (see chapter 17).

Flagellar staining is a method of revealing **flagella** (singular, *flagellum*), the tiny, slender filaments used by bacteria for locomotion. Because the width of bacterial flagella lies beyond the resolving power of the light microscope, in order to be seen, they must be enlarged by depositing a coating on the outside of the filament and then staining it. Their presence, number, and arrangement on a cell are useful for identification of the bacteria.

©Flying Colours Ltd/Getty Images

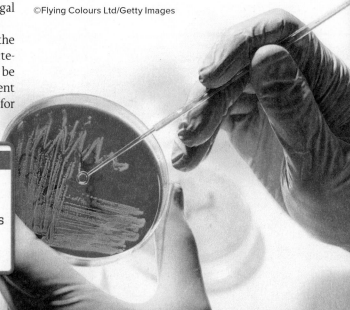

2.2 LEARNING OUTCOMES—Assess Your Progress

5. Convert among the different units of the metric system.
6. List and describe the three elements of good microscopy.
7. Differentiate between the principles of light microscopy and the principles of electron microscopy.
8. Give examples of simple, differential, and special stains.

CASE FILE WRAP-UP

Staphylococcus epidermidis growing on blood agar.

©Terry Vine/Blend Images, LLC (health care worker); James Redfearn/McGraw-Hill Education (agar dish)

Identification of microorganisms in the laboratory can take several days to allow for culture growth. Patients with clinical signs of infection, such as fever, rapid heart rate, low blood pressure, high white blood cell count, or localized signs of infection (tenderness, redness, swelling, heat, purulent drainage) are often started on treatment to curb infection. Some antibiotics, referred to as broad-spectrum antibiotics, treat a wide variety of bacterial infections. Based on the suspected source of infection, antifungal or antiviral treatment may also be started. As more information emerges from the lab regarding the culture, the antibiotic treatment is modified to treat more specific types of microorganisms, and eventually, the exact organism may be identified from the culture.

The patient's fever did not initially improve with broad-spectrum antibiotics. The X ray of the patient's leg revealed an abscess, or pocket of pus. The patient returned to the operating room to have the surgical site washed out. A sample of fluid from the abscess was collected and sent to the lab for culture. In just a few hours, the wound culture had already been identified as gram-positive bacteria. The provider changed the antibiotic prescription to one known to specifically treat gram-positive organisms.

After 3 days, the laboratory confirmed that there were no bacteria seen in the patient's blood culture. The specimen in the abscess fluid was identified as *Staphylococcus epidermidis*. Through further testing, the lab was able to recommend the antibiotics that would best treat the organism. The source of infection had been recognized and effective treatment initiated.

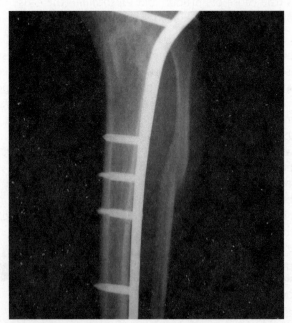

X ray of lower leg with implanted rods and screws.

©Thinkstock Images/Stockbyte/Getty Images

The Back Story

My goal in this Microbiome box is to impress upon you the revolution that took place when scientists were able, for the first time, to detect the presence of microbes *without culturing them*—a little odd, I know, because this chapter is about culturing microbes, and then seeing them under a microscope.

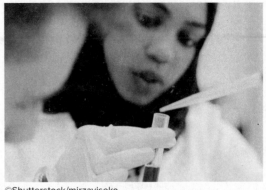
©Shutterstock/mirzavisoko

These methods were, indeed, the only ways available to scientists until about the beginning of the 2000s. And make no mistake, the pioneers who figured out how to culture, and how to see, microbes were indeed groundbreaking!

But as science advanced, as it always does, we figured out that the vast majority of microbes can still not be cultured with human-invented media, and for the most part cannot be seen because they cannot be cultured in numbers large enough to do so. But then DNA technology brought us methods that allow us to detect the presence of microbes without culturing or microscopy.

Using techniques to "read" sequences of nucleic acids (both DNA and RNA), scientists were able to sample human sites, separate the human from the microbial sequences, and then identify thousands and thousands of previously unknown microbes associated with humans.

This ability has brought us to the point where we truly are experiencing a new quantum leap in our understanding of how microbes influence all kinds of "environments." The environment we are concerned with in this book is the human body. Scientists expect that everything we now accept as standard practice in health maintenance and disease treatment will be changed by the information the microbiome provides us, given enough rigorous investigation.

TOOLS OF THE LABORATORY: Methods for the Culturing and Microscopic Analysis of Microorganisms

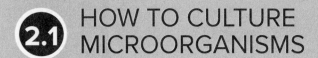

2.1 HOW TO CULTURE MICROORGANISMS

Scientists have spent more than 150 years devising methods to grow microorganisms in the laboratory. There are many effective methods that are useful in medicine. But it is now clear that there are many times more microbes in the world that we cannot yet cultivate artificially.

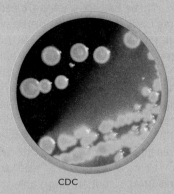

CDC

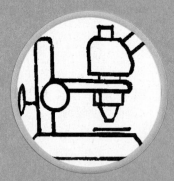

Because most microorganisms are too small to be seen with the naked eye, microscopes and special visualizing techniques (such as staining) help us identify and understand them. One particularly important staining technique is the Gram stain, which is the basis for an extremely useful way of describing bacteria.

THE MICROSCOPE 2.2

SmartGrid: From Knowledge to Critical Thinking

This *21 Question Grid* takes the topics from this chapter and arranges them with respect to the American Society for Microbiology's Undergraduate Curriculum guidelines—all six of the important "Concepts" as well as the important "Competency" of scientific literacy. Three questions are supplied about chapter content that refer to the Concept or Competency, in increasing levels of Bloom's taxonomy for learning.

ASM Concept/ Competency	A. Bloom's Level 1, 2—Remember and Understand (Choose one.)	B. Bloom's Level 3, 4—Apply and Analyze	C. Bloom's Level 5, 6—Evaluate and Create

Evolution	1. The identities of microorganisms on our planet
	  a. are mostly known.
	  b. have nearly all been identified via microscopy.
	  c. have nearly all been identified via culturing techniques.
	  d. are still mostly unknown.
		2. Before the availability of molecular techniques, archaea and bacteria were considered to be very closely related. Explain why that is understandable, based on what you learned about microscopy.	3. Often bacteria that are freshly isolated from a patient or the environment have thick sugary outer layers (capsules). After they are transferred many times on laboratory medium, they tend to lose their capsule. Speculate about why that might happen.

Cell Structure and Function	4. Which of these types of organisms is least likely to be identified to the genus level with light microscopy?
	  a. bacteria
	  b. protozoa
	  c. fungus
	  d. helminth
		5. Obviously, *structures* of microorganisms can be revealed by various types of microscopy. Can you think of any *functions* that might also be revealed? Which ones, and by which microscopic technique?	6. Some bacteria can produce a structure called an endospore, which is used as a survival structure. It is surrounded by a very tough and impervious outer layer that keeps it protected in adverse environmental conditions. What kinds of things would you need to know about this structure if you wanted to devise a way to stain it and see it under a microscope?

Metabolic Pathways	7. A fastidious organism must be grown on what type of medium?
	  a. general-purpose medium
	  b. differential medium
	  c. defined medium
	  d. enriched medium
		8. Write a short paragraph to differentiate among the following: *selective medium, differential medium, enriched medium.*	9. Alexander Fleming discovered penicillin in 1928. The first part of the story is that he noticed that agar plates upon which he was growing a bacterium were contaminated with obvious fungal colonies. Instead of throwing the plates out because they were contaminated, he examined them and realized that around the areas where the fungi were growing, there were no bacteria growing. Finish this story.

Information Flow and Genetics	10. Viruses are commonly grown in/on
	  a. animal cells or tissues.
	  b. agar plates.
	  c. broth cultures.
	  d. all of the above.
		11. Can you devise a growth medium with ingredients that would detect whether a culture of MRSA (antibiotic-resistant *Staphylococcus aureus)* that you have been storing in your lab is still antibiotic resistant?	12. There is a type of differential medium that can reveal whether bacteria growing on it produce and secrete an enzyme that breaks down DNA. Why would a bacterium secrete an enzyme that destroys DNA?

ASM Concept/ Competency	A. Bloom's Level 1, 2—Remember and Understand (Choose one.)	B. Bloom's Level 3, 4—Apply and Analyze	C. Bloom's Level 5, 6—Evaluate and Create
Microbial Systems	13. Most of the time, microbes in natural circumstances exist a. as single cells relationship. b. in relationship with other species. c. as single species. d. as colonies on agar.	14. Several bacteria live naturally in a material on your teeth called *plaque* that contains many different species, which interact with each other in significant ways. Identify some of the problems of studying one of these bacterial species after isolating it through a streak plate procedure and examining its behavior.	15. Archaea often grow naturally in extreme environments. Those called *extreme thermophiles* thrive in hot springs and in high-temperature hydrothermal vents in the ocean. Invent a culturing technique you could use that would be selective for these organisms.
Impact of Microorganisms	16. Diagnosis of infections in a hospitalized person is often accomplished via a. microscopy. b. culture of samples from patient. c. palpitation of infected area. d. two of the above.	17. After performing the streak plate procedure on a bacterial specimen, the culture was incubated. When you viewed the plate, there was heavy growth with no isolated colonies in the first quadrant. There was no other growth on the plate (in the second quadrant and beyond). Discuss possible errors that could have led to this result.	18. You are a scientist studying a marsh area contaminated with PCBs, toxic chemical compounds found in industrial waste. You look at a water sample using light microscopy and discover motile single cells that are about 1 micrometer in circumference. You hypothesize that these microbes might be useful in cleaning up the toxic waste. Because there is no instruction book for how to culture them, how might you go about creating a growth medium for them?
Scientific Thinking	19. If a Gram stain result is unclear, which of the following should be considered? a. getting a new sample b. restaining the same sample c. identity of microbe may actually be gram-variable d. all of the above	20. Explain the difference between multiplying a meter by 10^9 and multiplying it by 10^{-9} to get a gigameter and a nanometer, respectively.	21. You perform the special stain for bacterial flagella on a specimen. You don't know if the bacteria you are staining possess flagella or not. When you examine the specimen, you see no flagella on the bacteria. What are all the possibilities for why you didn't see them?

Answers to the multiple-choice questions appear in Appendix A.

Visual Connections

This question connects content within and between chapters.

Figure 2.9a. If you were using the quadrant streak plate method to plate a very dilute broth culture (with many fewer bacteria than the broth used for the plate pictured here), would you expect to see single, isolated colonies in quadrant 4 or quadrant 3? Explain your answer.

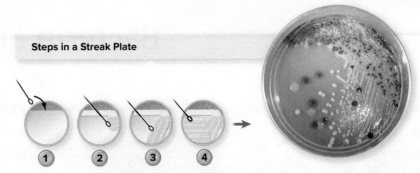

Steps in a Streak Plate

1 2 3 4

Note: This method only works if the spreading tool (usually an inoculating loop) is resterilized after each of steps 1–4.

(photo): Kathleen Talaro and Harold Benson

3

Bacteria and Archaea

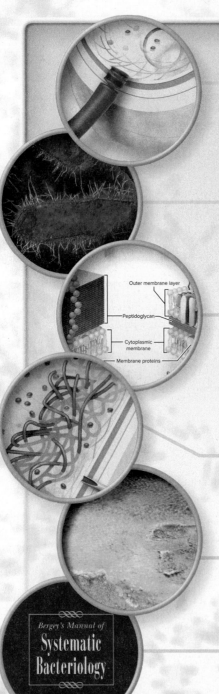

IN THIS CHAPTER...

3.1 Bacteria and Archaea: An Overview

1. List the structures all bacteria possess.
2. Identify three structures some but not all bacteria possess.
3. Describe three major shapes of bacteria.
4. Provide at least four terms to describe bacterial arrangements.

3.2 External Structures

5. Describe the structure and function of six different structures found on the exterior of a bacterial cell.
6. Explain how a flagellum works in the presence of an attractant.

3.3 The Cell Envelope: The Wall and Membrane(s)

7. Differentiate between the two main types of bacterial envelope structure.
8. Discuss why gram-positive cell walls are stronger than gram-negative cell walls.
9. Name a substance in the envelope structure of some bacteria that can cause severe symptoms in humans.

3.4 Bacterial Internal Structure

10. Identify seven structures that may be contained in bacterial cytoplasm.
11. Detail the causes and mechanisms of sporulation and germination.

3.5 The Archaea

12. Compare and contrast the major features of archaea, bacteria, and eukaryotes.

3.6 Classification Systems for Bacteria and Archaea

13. Differentiate between *Bergey's Manual of Systematic Bacteriology* and *Bergey's Manual of Determinative Bacteriology*.
14. Name four divisions ending in *–cutes* and describe their characteristics.
15. Define a *species* in terms of bacteria.

CASE FILE

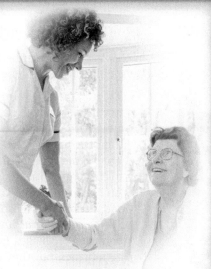

Dave and Les Jacobs/Blend Images LLC

Extreme Endospores

While working as a newly graduated nurse, I was caring for an elderly female patient from a local nursing home who had been admitted for a hip replacement. The patient seemed to be recovering well until she developed redness, increased swelling, and a pus discharge at the surgical site. The wound was cultured and the patient was started on a cephalosporin antibiotic. The results from microbiological testing revealed that the infection was caused by *Staphylococcus aureus*, and it tested sensitive to the cephalosporin drug she was already taking.

The patient successfully completed the course of antibiotic therapy, and within a few days all signs of infection had subsided. The patient was progressing well with physiotherapy, and we were beginning to plan for discharge back to the nursing home when the patient suddenly began to experience diarrhea. At first I assumed that the diarrhea was because of an expected side effect from the antibiotic, but it soon became clear that this was something more than a general side effect. On the first day, the patient had two loose bowel movements. By the second day, the episodes of diarrhea were occurring every 2 to 3 hours. The stools were watery and foul-smelling and contained large amounts of mucus. The patient complained of abdominal pain and cramping, and she subsequently developed a fever. The physician was notified, and a stool specimen was collected for laboratory testing.

I was surprised when the stool culture came back showing that the patient's diarrhea was actually caused by the bacterium *Clostridioides difficile*. The patient was placed on contact isolation and was started on intravenous metronidazole (Flagyl). With this treatment, the diarrhea gradually slowed and finally stopped. Repeat cultures, performed after the metronidazole therapy was completed, showed that the infection had been successfully cleared.

- How is *C. difficile* spread?
- What risk factors made this patient particularly vulnerable to infection with *C. difficile*?

Case File Wrap-Up appears at the end of the chapter.

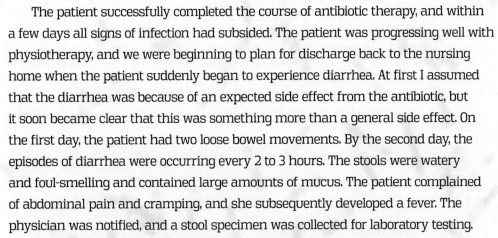

3.1 Bacteria and Archaea: An Overview

In chapter 1, we described bacteria and archaea as being cells with no true nucleus. Let's look at how these organisms are different from eukaryotes:

- *The way their DNA is packaged:* Bacteria and archaea have nuclear material that is free inside the cell (i.e., they do not have a nucleus). Eukaryotes have a membrane around their DNA (making up a nucleus). Eukaryotes and some archaea wrap their DNA around **histones;** bacteria do not.
- *The makeup of their cell wall:* Bacteria have a wall structure that is unique compared to eukaryotes. Bacteria have sturdy walls made of a chemical called peptidoglycan. Archaeal walls are also tough and made of other chemicals, distinct from bacteria and distinct from eukaryotic cells.
- *Their internal structures:* Bacteria and archaea don't have complex, membrane-bounded organelles (eukaryotes do). A few bacteria and archaea have internal membranes, but they don't surround organelles.

Both non-eukaryotic and eukaryotic microbes are ubiquitous in the world today. Although both can cause infectious diseases, treating them with drugs requires different types of approaches. In this chapter and coming chapters, you'll discover why that is.

The evolutionary history of non-eukaryotic cells extends back almost 4 billion years. The fact that these organisms have endured for so long in such a variety of habitats can be attributed to a cellular structure and function that are amazingly versatile and adaptable.

The Structure of the Bacterial Cell

In this chapter, the descriptions, except where otherwise noted, refer to *bacterial cells.* We will focus on the features of bacteria because you will encounter them more often than archaea in a clinical environment. We will analyze the significant ways in which archaea are unique later in the chapter.

The general cellular organization of a bacterial cell can be represented with this infographic:

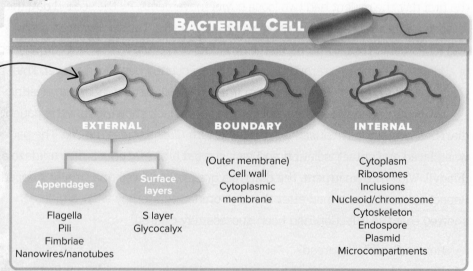

Escherichia coli
©Steve Gschmeissner/Science Photo Library/Getty Images

All bacterial cells invariably have a cytoplasmic membrane, cytoplasm, ribosomes, a cytoskeleton, and one (or a few) chromosome(s). The majority have a cell wall and a surface coating called a glycocalyx. Specific structures that are found in some but not all bacteria are flagella, an outer membrane, pili, fimbriae, nanowires/nanotubes, plasmids, inclusions, endospores, and microcompartments. Most of these structures are observed in archaea as well.

Figure 3.1 presents a three-dimensional anatomical view of a generalized, rod-shaped bacterial cell. As we survey the principal anatomical features of this cell, we

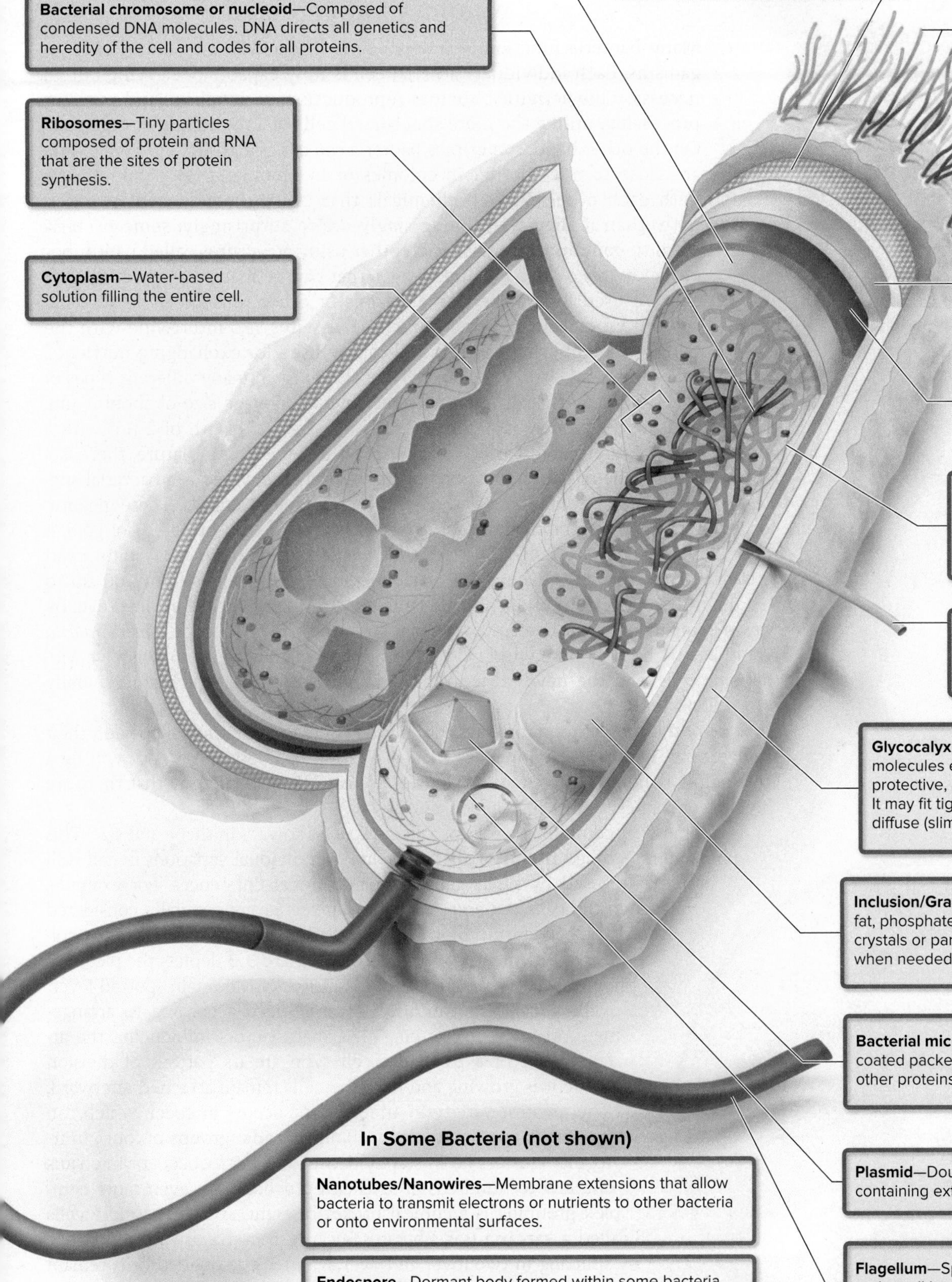

Figure 3.1 Structure of a bacterial cell. Cutaway view of a typical rod-shaped bacterium, showing major structural features.

In All Bacteria

Cell (cytoplasmic) membrane—A thin sheet of lipid and protein that surrounds the cytoplasm and controls the flow of materials into and out of the cell pool.

Bacterial chromosome or nucleoid—Composed of condensed DNA molecules. DNA directs all genetics and heredity of the cell and codes for all proteins.

Ribosomes—Tiny particles composed of protein and RNA that are the sites of protein synthesis.

Cytoplasm—Water-based solution filling the entire cell.

In Some Bacteria

S layer—Monolayer of protein used for protection and/or attachment.

Fimbriae—Fine, hairlike bristles extending from the cell surface that help in adhesion to other cells and surfaces.

Outer membrane—Extra membrane similar to cytoplasmic membrane but also containing lipopolysaccharide. Controls flow of materials, and portions of it are toxic to mammals when released.

Cell wall—A semirigid casing that provides structural support and shape for the cell.

Cytoskeleton—Long fibers of proteins that encircle the cell just inside the cytoplasmic membrane and contribute to the shape of the cell.

Pilus—An appendage used for drawing another bacterium close in order to transfer DNA to it.

Glycocalyx (tan coating)—A coating or layer of molecules external to the cell wall. It serves protective, adhesive, and receptor functions. It may fit tightly (capsule) or be very loose and diffuse (slime layer).

Inclusion/Granule—Stored nutrients such as fat, phosphate, or glycogen deposited in dense crystals or particles that can be tapped into when needed.

Bacterial microcompartments—Protein-coated packets used to localize enzymes and other proteins in the cytoplasm.

Plasmid—Double-stranded DNA circle containing extra genes.

Flagellum—Specialized appendage attached to the cell by a basal body that holds a long, rotating filament. The movement pushes the cell forward and provides motility.

In Some Bacteria (not shown)

Nanotubes/Nanowires—Membrane extensions that allow bacteria to transmit electrons or nutrients to other bacteria or onto environmental surfaces.

Endospore—Dormant body formed within some bacteria that allows for their survival in adverse conditions.

Intracellular membranes

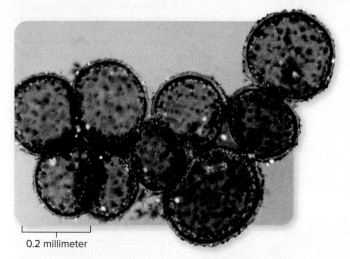

0.2 millimeter

Figure 3.2 *Thiomargarita namibiensis.* The bacteria are each about half the width of a common paperclip wire—in other words, quite large!

©Max Planck Institute/AFP/Newscom

Figure 3.3 **Pleomorphic bacteria.** If you look closely at this micrograph of violet-stained *Rickettsia rickettsii* bacteria, you will see some coccoid cells, rod-shaped cells, and hybrid forms.

Source: Billie Ruth Bird/CDC

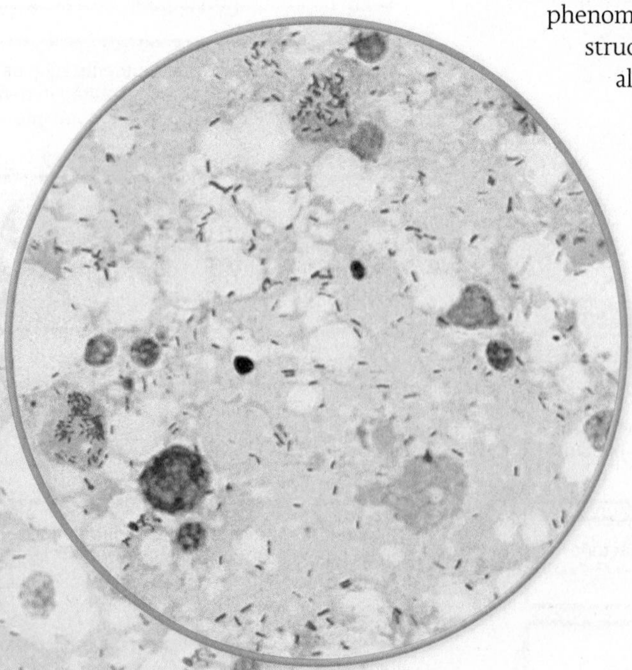

will perform a microscopic dissection of sorts, beginning with the outer cell structures and proceeding to the internal contents.

Bacterial Shapes and Arrangements

Many bacteria function as independent single-celled, or unicellular, organisms. Each individual bacterial cell is fully capable of carrying out all necessary life activities, such as reproduction, metabolism, and nutrient processing, unlike the more specialized cells of a multicellular organism. On the other hand, sometimes bacteria *can* act as a group. When bacteria are close to one another in colonies or in biofilms, they communicate with each other through chemicals that cause them to behave differently than if they were living singly. More surprisingly, some bacteria seem to communicate with each other using structures called nanotubes or nanowires, which are appendages that can be many micrometers long and are used for transferring electrons or other substances outside the cell onto metals in the environment. The wires also intertwine with the wires of neighboring bacteria and can be used for exchanging nutrients.

Bacteria come in all shapes and sizes. They also exist in many different types of groupings. Let's start with size. Bacterial cells have an average size of about 1 μm. Cocci have a circumference of 1 μm, and rods may have a length of 2 μm with a width of 1 μm. But that's just the average. As with everything in nature, there is a lot of variation. One of the largest non-eukaryotes yet discovered is a bacterial species living in ocean sediments near the African country of Namibia. These gigantic cocci are arranged in strands that look like pearls and contain hundreds of golden sulfur granules. They are called *Thiomargarita namibiensis*, which means "sulfur pearl of Namibia" **(figure 3.2).** The size of the individual cells ranges from 100 up to 750 μm (0.1 to 0.75 mm), and many are large enough to see with the naked eye. By way of comparison, if the average bacterium were the size of a mouse, *Thiomargarita* would be as large as a blue whale! On the other end of the spectrum, we have *Mycoplasma* cells, and the newly discovered ultra-small bacteria, which are generally 0.15 to 0.30 μm.

One of the most important ways to describe bacteria is by their shape and their arrangement. **Table 3.1** summarizes shapes. Getting to know these now will be a great help for the rest of your studies in this course. We generally say that there are three major shapes: round, rod-shaped, and spiral.

It is somewhat common for cells of a single species to vary in shape and size. This phenomenon, called **pleomorphism,** is due to individual variations in cell wall structure caused by slight genetic or nutritional differences. For example, although the cells of *Corynebacterium diphtheriae* are generally considered rod-shaped, in culture they display variations such as club-shaped, swollen, curved, filamentous, and coccoid. **Figure 3.3** depicts the pleomorphic *Rickettsia ricketsii*, which causes Rocky Mountain spotted fever.

Bacterial cells can also be categorized according to arrangement, or style of grouping. The main factors influencing the arrangement of a particular cell type are its pattern of division while it is growing and how the cells remain attached afterward. The greatest variety in arrangement occurs in cocci, which can be single, in pairs (diplococci), in **tetrads** (groups of four), in irregular clusters (as in staphylococci and micrococci), or in chains of a few to hundreds of cells (streptococci). An even more complex grouping is a cubical packet of eight, sixteen, or more cells called a **sarcina** (sar′-sih-nah). Because bacteria usually reproduce by splitting in two (over and over again), if the divided cells remain attached, they lead to these diverse arrangements. After division, the resultant daughter cells remain attached **(figure 3.4).**

Table 3.1 Bacterial Shapes

(a) Coccus

If the cell is spherical or ball-shaped, the bacterium is described as a **coccus** (kok′-us). Cocci (kok′-sī) can be perfect spheres, but they also can exist as oval, bean-shaped, or even pointed variants. The photo is of *Staphylococcus* (10,000×).

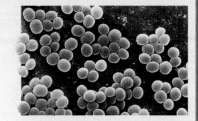

(b) Rod/Bacillus

A cell that is cylindrical is termed a rod, or **bacillus** (bah-sil′-lus). There is also a genus named *Bacillus*. Rods are quite varied in their actual form. Depending on the species, they can be blocky, spindle-shaped, round-ended, long and threadlike (filamentous), or even club-shaped or drumstick-shaped. Note: When a rod is short and plump, it is called a **coccobacillus.** The photo is of *Legionella pneumophila*.

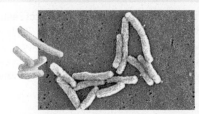

(c) Vibrio

Singly occurring rods that are gently curved are called **vibrio** (vib′-ree-oh). The photo is of *Vibrio vulnificus* (13,000×).

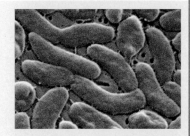

(d) Spirillum

A bacterium having a slightly curled or spiral-shaped body is called a **spirillum** (spy-ril′-em)—a rigid helix, twisted twice or more along its axis (like a corkscrew). The photo is of *Campylobacter jejuni*.

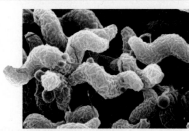

(e) Spirochete

Another spiral cell (which contains periplasmic flagella) is the **spirochete** (spy′-roh-keet), a more flexible form that resembles a spring. The photo is of spirochetes (7,500×).

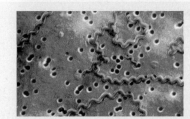

(f) Branching filaments

A few bacteria produce multiple branches off of a basic rod structure, a form called branching **filaments.** The photo is of *Streptomyces* (1,500×).

Source: CDC/Janice Haney Carr (a); Source: Janice Haney Carr/CDC (b); Source: Janice Haney Carr/CDC (c); Source: Photo by De Wood, digital colorization by Chris Pooley, USDA-ARS (d); VEM/Science Source (e); Eye of Science/Science Source (f)

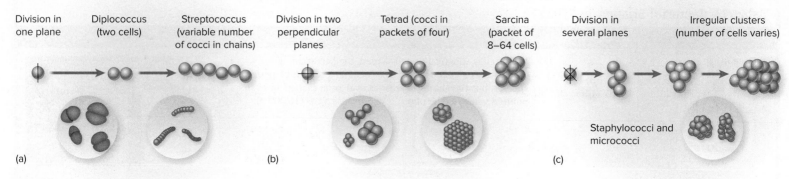

Figure 3.4 Arrangement of cocci resulting from different planes of cell division. **(a)** Division in one plane produces diplococci and streptococci. **(b)** Division in two or three planes at right angles produces tetrads and packets. **(c)** Division in several planes produces irregular clusters.

Rods are less varied in arrangement because they divide only in one plane. They occur either as single cells, as a pair of cells with their ends attached (diplobacilli), or as a chain of several cells (streptobacilli). A **palisades** (pal'-ih-saydz) arrangement, typical of the *Corynebacterium* species, is formed when the cells of a chain remain partially attached by a small hinge region at the ends. The cells tend to fold (snap) back upon each other, forming a row of cells oriented side by side **(figure 3.5).** Spirilla are occasionally found in short chains, but spirochetes rarely remain attached after division.

3.1 LEARNING OUTCOMES—Assess Your Progress

1. List the structures all bacteria possess.
2. Identify three structures some but not all bacteria possess.
3. Describe three major shapes of bacteria.
4. Provide at least four terms to describe bacterial arrangements.

3.2 External Structures

Appendages: Cell Extensions

Several different types of accessory structures sprout from the surface of bacteria. These long **appendages** are common but are not present on all species. Appendages can be divided into two major groups: those that provide motility (flagella and axial filaments) and those that provide attachment points or channels (fimbriae, pili, and nanotubes/nanowires).

Flagella—Bacterial Propellers

The bacterial flagellum (flah-jel'-em), an appendage of truly amazing construction, is unique in the biological world. The primary function of flagella is to confer **motility,** or self-propulsion—that is, the capacity of a cell to swim freely through an aqueous habitat. The flagellum has three distinct parts: the filament, the hook (sheath), and the basal body **(figure 3.6).** The **filament,** a helical structure composed of proteins, is approximately 20 nm in diameter and varies from 1 to 70 µm in length. It is inserted into a curved, tubular hook. The hook is anchored to the cell by the basal body, a stack of rings firmly anchored through the cell wall, to the cytoplasmic membrane and the outer membrane. This arrangement permits the hook with its filament to rotate 360°, rather than undulating back and forth like a whip as was once thought. Although many archaea possess flagella, recent studies have shown that the structure is quite different than the bacterial flagellum. It is called *archaellum* by some scientists.

Figure 3.5 *Corynebacterium* **cells illustrating the palisades (stacking) tendency.**

©De Agostini/Getty Images

Palisades

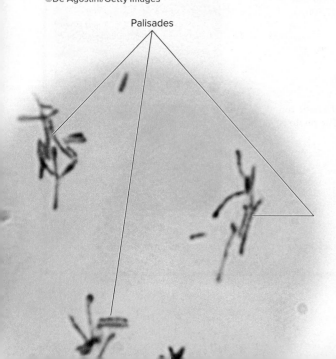

All spirilla, about half of the rods, and a small number of cocci are flagellated. Bacteria can have single or multiple flagella. They can be found either at the ends of a rod-shaped cell or dispersed across a round- or rod-shaped cell. If they are found on the ends of a cell, they are termed *polar* because the ends of rod-shaped bacteria are called poles.

1. In a *polar* arrangement, the flagella are attached at one or both ends of the cell. Three subtypes of this pattern are:
   - **monotrichous** (mah′-noh-trik′-us), with a single flagellum;
   - **lophotrichous** (lo′-foh-), with small bunches or tufts of flagella emerging from the same site; and
   - **amphitrichous** (am′-fee-), with flagella at both poles of the cell.
2. In a **peritrichous** (per′-ee-) arrangement, flagella are dispersed randomly over the surface of the cell **(figure 3.7).**

Motility is one piece of information used in the laboratory identification or diagnosis of pathogens. Flagella are hard to visualize via light microscopy, but often it is sufficient to know simply whether a bacterial species is motile. One way to detect motility is to stab a tiny mass of cells into a soft (semisolid) medium in a test tube. Growth spreading rapidly through the entire medium is indicative of motility. Alternatively, cells can be observed microscopically with a hanging drop slide. A truly motile cell will flit, dart, or wobble around the field, making some progress, whereas one that is nonmotile jiggles about in one place but makes no progress.

Fine Points of Flagellar Function

Flagellated bacteria can perform some rather sophisticated feats. They can move in response to chemical signals—a type of behavior called **chemotaxis** (ke′-moh-tak′-sis). Positive chemotaxis is movement of a cell in the direction of a favorable chemical stimulus (usually a nutrient). Negative chemotaxis is movement away from a potentially harmful compound.

The flagellum is effective in guiding bacteria through the environment primarily because the system for detecting chemicals is linked to the mechanisms that drive the flagellum. There are clusters of receptors located in the cytoplasmic membrane that bind specific molecules coming from the immediate environment. The attachment of sufficient numbers of these molecules transmits signals to the flagellum and sets it into rotary motion. The actual "fuel" for the flagellum to turn is a gradient of protons (hydrogen ions) that are generated by the metabolism of the bacterium and that bind

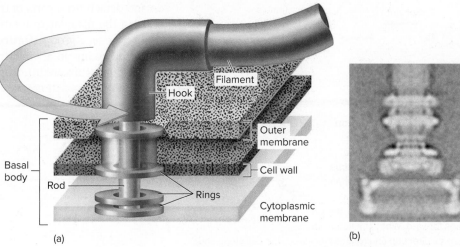

(a)

(b)

Figure 3.6 Details of the basal body of a flagellum in a gram-negative cell. **(a)** The hook, rings, and rod function together as a tiny device that rotates the filament 360°. **(b)** An electron micrograph of the basal body of a bacterial flagellum.

©Sarkar MK1, Paul K, Blair D., "Chemotaxis signaling protein CheY binds to the rotor protein FliN to control the direction of flagellar rotation in *Escherichia coli*," PNAS May 18, 2010 vol. 107 no. 20 9370-9375

Figure 3.7 Four types of flagellar arrangements.
(a) Monotrichous flagellum on the bacterium *Clostridioides phytofermentans*. **(b)** Peritrichous flagella on *Salmonella* (the shorter hairs are fimbriae). **(c)** Lophotrichous flagella on *Helicobacter*. **(d)** Amphitrichous flagella on *Alcaligenes*.
©Science Photo Library/Alamy Stock Photo (a); CDC/James Archer (b); Heather Davies/Science Source (c); Smith Collection/Gado/ Getty Images (d)

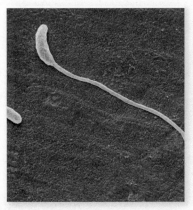

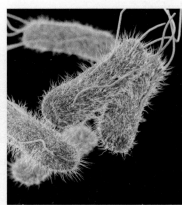

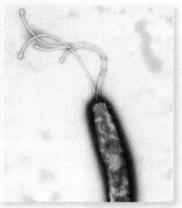

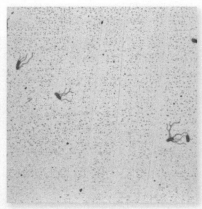

(a)

(b)

(c)

(d)

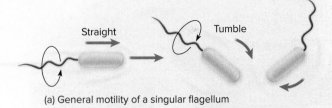

(a) General motility of a singular flagellum

Figure 3.8 The operation of flagella and the mode of locomotion in bacteria with polar and peritrichous flagella.   (a) In general, when a polar flagellum rotates in a counterclockwise direction, the cell swims forward. When the flagellum reverses direction and rotates clockwise, the cell stops and tumbles. **(b)** In peritrichous forms, all flagella sweep toward one end of the cell and rotate as a single group. During tumbles, the flagella lose coordination.

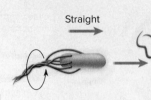

(b) Peritrichous motility

Key

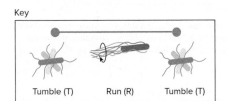

Tumble (T)   Run (R)   Tumble (T)

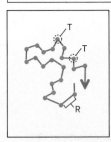
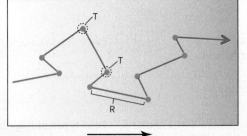

(a) No attractant or repellent

(b) Gradient of attractant concentration

Figure 3.9 Chemotaxis in bacteria.
(a) A bacterium moves via a random series of short runs and tumbles when there is no attractant or repellent. **(b)** The cell spends more time on runs as it gets closer to the attractant.

to and detach from parts of the flagellar motor within the cytoplasmic membrane, causing the filament to rotate. If several flagella are present, they become aligned and rotate as a group **(figure 3.8)**. As a flagellum rotates counterclockwise, the cell itself swims in a smooth linear direction toward the stimulus. This action is called a *run*. Runs are interrupted at various intervals by *tumbles*, during which the flagellum reverses direction and causes the cell to stop and change its course. Alternation between runs and tumbles generates what is termed a *random walk* form of motility in these bacteria. However, in response to a concentration gradient of an attractant molecule, the bacterium will begin to slow down its tumbles, permitting longer runs and overall progress toward the stimulus **(figure 3.9)**. The movement now becomes a *biased* random walk in which movement is favored (biased) in the direction of the attractant. But what happens when a flagellated bacterium wants to run away from a toxic environment? In this case, the random walk then favors movement away from the concentration of repellent molecules. By delaying tumbles, the bacterium increases the length of its runs, allowing it to redirect itself away from the negative stimulus.

Periplasmic Flagella   Corkscrew-shaped bacteria called *spirochetes* show an unusual, wriggly mode of locomotion caused by two or more long, coiled threads, the periplasmic flagella or *axial filaments*. A periplasmic flagellum is a type of internal flagellum that is enclosed in the space between the cell wall and the cytoplasmic membrane. Bacteria that cause the diseases syphilis and Lyme disease utilize their periplasmic flagella to penetrate tissues.

Appendages for Attachment or Channel Formation

The structures termed **pilus** (pil-us) and **fimbria** (fim′-bree-ah) are both bacterial surface appendages that provide some type of attachment. **Nanotubes** or nanowires are used as channels for nutrient or energy exchange.

Fimbriae are small, bristlelike fibers sprouting off the surface of many bacterial cells **(figure 3.10)**. Their exact composition varies, but most of them contain protein. Fimbriae have an inherent tendency to stick to each other and to surfaces. They are partially responsible for the mutual clinging of cells that leads to biofilms and other thick aggregates of cells on the surface of liquids and for the microbial colonization of inanimate solids such as rocks and glass. Some pathogens can colonize

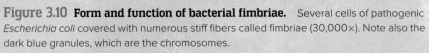

Figure 3.10 Form and function of bacterial fimbriae.   Several cells of pathogenic *Escherichia coli* covered with numerous stiff fibers called fimbriae (30,000×). Note also the dark blue granules, which are the chromosomes.
©Eye of Science/Science Source

and infect host tissues because of a tight adhesion between their fimbriae and epithelial cells. For example, the gonococcus (agent of gonorrhea) colonizes the genitourinary tract, and *Escherichia coli* colonizes the intestine by this means. Mutant forms of these pathogens that lack fimbriae are unable to cause infections.

A pilus is a long, rigid tubular structure made of a special protein, *pilin.* Conjugation pili are utilized in a "mating" process between cells called **conjugation,** which involves partial transfer of DNA from one cell to another **(figure 3.11).** A conjugation pilus from the donor cell unites with a recipient cell, thereby providing a cytoplasmic connection for making the transfer. Production of these pili is controlled genetically. Conjugation using a pilus takes place only between compatible gram-negative cells. Although conjugation does occur in gram-positive bacteria, it does not involve a conjugation pilus. The roles of pili and conjugation are further explored in chapter 8. There is a special type of structure in some bacteria called a type IV pilus. Like the pili described here, it can transfer genetic material. In addition, it can act like fimbriae and assist in attachment, and act like flagella and make a bacterium motile.

Nanotubes (also called nanowires) have only recently been sufficiently understood. They are very thin, long, tubular extensions of the cytoplasmic membrane that bacteria use as channels either to transfer amino acids (i.e., food) among one another or to harvest energy by shuttling electrons from an electron-rich surface in the environment. In this latter scenario, bacteria generate energy in the absence of oxygen. When we study metabolism, we will see that cells generate the necessary energy to grow by consuming nutrients and transferring electrons from the nutrients to oxygen. But many bacteria can generate their energy in the absence of oxygen by sharing electrons with substances in the environment containing iron, such as iron-rich rocks. They use long nanowires to do this, transferring electrons up and down the tubular extensions of the membrane. Scientists call this activity "breathing rock instead of oxygen."

Surface Coatings: The S Layer and the Glycocalyx

The bacterial cell surface is frequently exposed to severe environmental conditions. Bacterial cells protect themselves with either an S layer or a glycocalyx, or both. **S layers** are single layers of thousands of copies of a single protein linked together like tiny chain link fences. They are often called "the armor" of a bacterial cell **(figure 3.12).** It took scientists a long time to discover them because bacteria only produce them when they are in a hostile environment. The nonthreatening conditions of growing in a lab in a nutritious broth with no competitors around ensured that bacteria usually did not produce the layer. We now know that many different species have the ability to produce an S layer, including pathogens such as *Clostridioides difficile* and *Bacillus anthracis.* Some bacteria use S layers to aid in attachment as well.

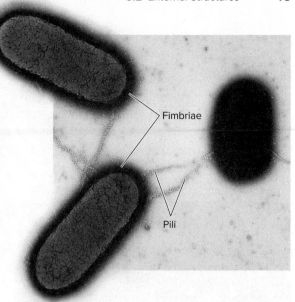

Fimbriae

Pili

Figure 3.11 Three bacteria in the process of conjugating. Clearly evident are the pili forming mutual conjugation bridges between a donor (middle cell) and two recipients (cells on the left side). Fimbriae can also be seen on the two left-hand cells.
©L. Caro/SPL/Science Source

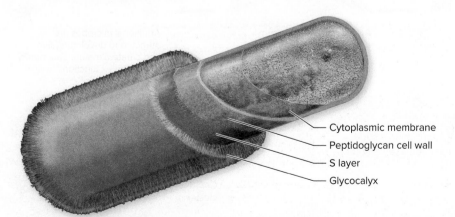

Cytoplasmic membrane
Peptidoglycan cell wall
S layer
Glycocalyx

Figure 3.12 Position of the bacterial S layer, shown in purple.
©Russell Kightley/Science Source

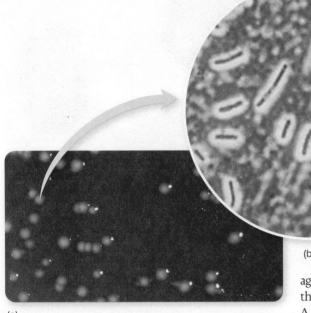

(a)

The glycocalyx develops as a coating of repeating polysaccharide or glycoprotein units. This protects the cell and, in some cases, helps it adhere to its environment. Glycocalyces differ among bacteria in thickness, structure, and chemical composition. Some bacteria are covered with a loose shield called a **slime layer** that can protect them from loss of water and nutrients. A glycocalyx is called a **capsule** when it is bound more tightly to the cell than a slime layer is and it is denser and thicker. Capsules are often visible in negatively stained preparations **(figure 3.13b)** and lead to a prominently sticky (mucoid) character to colonies on agar **(figure 3.13a)**.

(b)

Specialized Functions of the Glycocalyx

Capsules are formed by many pathogenic bacteria, such as *Streptococcus pneumoniae* (a cause of pneumonia, an infection of the lung), *Haemophilus influenzae* (one cause of meningitis), and *Bacillus anthracis* (the cause of anthrax). Encapsulated bacterial cells generally have greater disease-causing abilities because capsules protect the bacteria against host white blood cells called phagocytes. Phagocytes are a natural body defense that can engulf and destroy foreign cells through phagocytosis, thus preventing infection. A capsular coating blocks the mechanisms that phagocytes use to attach to and engulf bacteria. By escaping phagocytosis, the bacteria are free to multiply and infect body tissues. Encapsulated bacteria that mutate to nonencapsulated forms usually lose their ability to cause disease.

Glycocalyces can be important in formation of biofilms **(figure 3.14a)**. The thick, white plaque that forms on teeth comes in part from the surface slimes produced by certain streptococci in the oral cavity. This slime protects them from being dislodged from the teeth and provides a niche for other oral bacteria that, in time, can lead to dental disease. The glycocalyx of some bacteria is so highly adherent that it has a significant role in the persistent colonization of nonliving materials such as plastic catheters, intrauterine devices, and metal pacemakers that are in common medical use **(figure 3.14b)**.

Figure 3.13 Encapsulated bacteria.
(a) Colonies of *Citrobacter* bacteria that are encapsulated. The characteristic of interest here is the relative *glossiness* of the colonies. Bacteria that produce capsules appear glossier on agar. They are called "smooth" by microbiologists, whereas those without capsules are called "rough." **(b)** Special stain of encapsulated bacteria.
Source: CDC (a); Michael Abbey/Science Source (b)

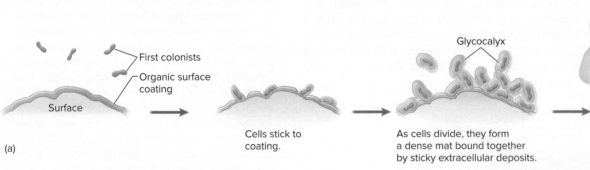

Glycocalyx

First colonists

Organic surface coating

Surface

Cells stick to coating.

As cells divide, they form a dense mat bound together by sticky extracellular deposits.

Additional microbes are attracted to developing film and create a mature community with complex function.

(a)

Figure 3.14 Biofilm formation. **(a)** The stepwise formation of a biofilm on a surface. **(b)** Scanning electron micrograph of *Staphylococcus aureus* cells attached to a catheter by a slime secretion.
©Scimat/Science Source (b)

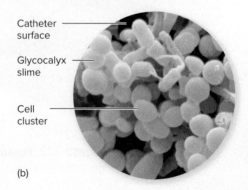

Catheter surface

Glycocalyx slime

Cell cluster

(b)

3.3 The Cell Envelope: The Wall and Membrane(s)

The boundary of a bacterium is marked by a multilayer structure that we will call the *cell envelope*. It is composed of two or three basic layers: the cell wall, the cytoplasmic membrane, and, in some bacteria, an outer membrane. Although each envelope layer performs a distinct function, together they act as a single protective unit.

Differences in Cell Envelope Structure

Bacteria can be divided into two distinct groups, called gram-positive and gram-negative, based on whether they stain a certain way (-positive) or not (-negative). We will describe the structural differences, and later will show you how that stain, called the Gram stain, works.

If you made a cross section through a gram-positive cell envelope **(figure 3.15)**, it would resemble an open-faced sandwich: The thick cell wall would represent a tall stack of roast beef. Underneath the meat is the "bread," the thin cytoplasmic membrane. A similar cross section of a gram-negative cell envelope shows a complete sandwich with three layers: an outer membrane, a thin cell wall (less meat on this sandwich), and the cytoplasmic membrane. The cell wall layers in both gram-positive and gram-negative

Figure 3.15 A comparison of the detailed structure of gram-positive and gram-negative cell envelopes. The images at the top are electron micrographs of actual gram-positive and gram-negative cells.

Dr. Kari Lounatmaa/Science Source (gram-positive cell); Dennis Kunkel Microscopy, Inc./Science Source (gram-negative cell)

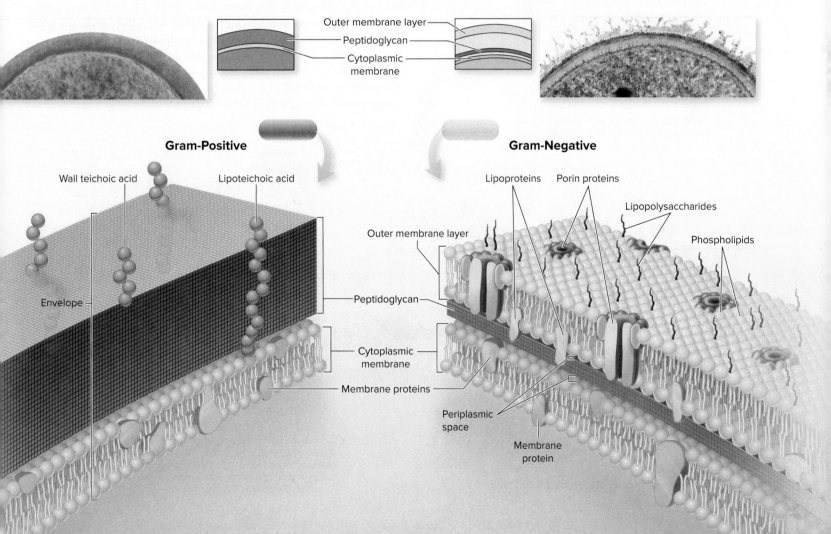

bacteria are made of a unique molecule called peptidoglycan. In gram-positive cells there are several layers of it, and in gram-negative cells there is only one layer.

Moving from outside to in (see figure 3.1), the outer membrane (if present) lies just under the glycocalyx. Next comes the cell wall. Finally, the innermost layer is always the cytoplasmic membrane. Because only some bacteria have an outer membrane, we discuss the cell wall first.

The Cell Wall

The **cell wall** accounts for a number of important bacterial characteristics. In general, it helps determine the shape of a bacterium, and it also provides the kind of strong structural support necessary to keep a bacterium from bursting or collapsing because of changes in osmotic pressure.

The cell walls of most bacteria are relatively rigid because of a unique macromolecule called peptidoglycan (PG). The only place peptidoglycan is found on earth is here in bacterial cell walls. This compound is composed of a repeating framework of long *glycan* (*sugar*) chains cross-linked by short peptide (protein) fragments to provide a strong but flexible support framework **(figure 3.16).** The amount and exact composition of peptidoglycan vary among the major bacterial groups.

Bacteria live and grow in environments that have water in them. (Water-based environments are termed "aqueous.") Because of this, they are constantly absorbing excess water by osmosis. Were it not for the strength and rigidity of the peptidoglycan in the cell wall, they would burst from internal pressure. This feature of the cell wall provides us with an obvious target in designing antibacterial drugs. Several types of drugs used to treat infection (penicillin, cephalosporins) are effective because they target the peptide crosslinks in the peptidoglycan, causing it to disintegrate. With their cell walls incomplete or missing, such cells have very little protection from **lysis** (ly'-sis), which is the disintegration or rupture of the cell. Our bodies themselves contain a powerful antibacterial substance: Lysozyme, an enzyme contained in tears and saliva, breaks the bonds in the glycan chains and causes the wall to break down.

More than a hundred years ago, long before the detailed anatomy of bacteria was even remotely known, a Danish physician named Hans Christian Gram developed a staining technique, the **Gram stain,** that distinguishes two generally different groups of bacteria. The two major

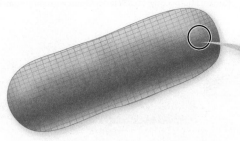

(a) The peptidoglycan can be seen as a crisscross network pattern similar to a chain-link fence.

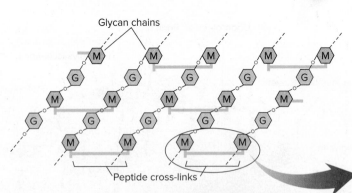

(b) It contains alternating sugar glycans (G and M) bound together in long strands. The G stands for *N*-acetyl glucosamine, and the M stands for *N*-acetyl muramic acid.

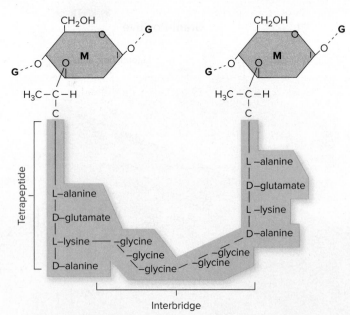

(c) A detailed view of the links between the muramic acids. Tetrapeptide chains branching off the muramic acids connect by interbridges also composed of amino acids. It is this linkage that provides rigid yet flexible support to the cell and that may be targeted by drugs like penicillin.

Figure 3.16 Structure of peptidoglycan in the cell wall.

groups shown by this technique are the **gram-positive** bacteria and the **gram-negative** bacteria. The designations *gram-positive* and *gram-negative* are based largely on differences in the peptidoglycan layer of the cell envelope, as you will see next.

The Gram-Positive Cell Wall

The bulk of the gram-positive cell wall is a thick, homogeneous sheath of peptidoglycan ranging from 20 to 80 nm in thickness. It also contains tightly bound acidic polysaccharides, including **teichoic acid** and **lipoteichoic acid** (see figure 3.15). Teichoic acid is a polymer of ribitol or glycerol (alcohols) and phosphate that is embedded in the peptidoglycan matrix. Lipoteichoic acid is similar in structure but is attached to the lipids in the cytoplasmic membrane. These molecules appear to function in cell wall maintenance and enlargement during cell division, and they also contribute to the acidic charge on the cell surface.

The Gram-Negative Cell Wall

The gram-negative cell wall is a single, thin (1–3 nm) sheet of peptidoglycan. Although it acts as a somewhat rigid protective structure as previously described, its thinness makes gram-negative bacteria more susceptible to lysis.

The Gram Stain

The technique of Hans Christian Gram consisted of timed, sequential applications of crystal violet (the primary dye), Gram's iodine (the mordant), an alcohol rinse (decolorizer), and safranin (a contrasting counterstain). Bacteria that stain purple are called gram-positive, and those that stain red are called gram-negative. The different results in the Gram stain are due to differences in the structure of the cell wall and how it reacts to the series of reagents applied to the cells **(figure 3.17).**

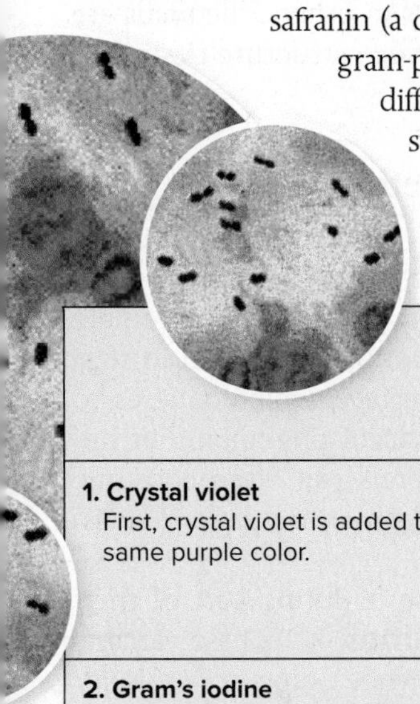

Step	Microscopic Appearance of Cell		Chemical Reaction in Cell Wall (very magnified view)	
	Gram (+)	Gram (−)	Gram (+)	Gram (−)
1. Crystal violet First, crystal violet is added to the cells in a smear. It stains them all the same purple color.				Both cell walls affix the dye
2. Gram's iodine Then, the mordant, Gram's iodine, is added. This is a stabilizer that causes the dye to form large complexes in the peptidoglycan meshwork of the cell wall. The thicker gram-positive cell walls are able to more firmly trap the large complexes than those of the gram-negative cells.			Dye complex trapped in wall	No effect of iodine
3. Alcohol Application of alcohol dissolves lipids in the outer membrane and removes the dye from the peptidoglycan layer—only in the gram-negative cells.			Crystals remain in cell wall	Outer membrane weakened; wall loses dye
4. Safranin (red dye) Because gram-negative bacteria are colorless after decolorization, their presence is demonstrated by applying the counterstain safranin in the final step.			Red dye masked by violet	Red dye stains the colorless cell

Figure 3.17   The steps in a Gram stain.

©Neal R. Chamberlin, PhD/McGraw-Hill Education

Medical Moment

Collecting Sputum

The nurse is often responsible for collecting sputum samples for acid-fast staining when a patient with a cough is suspected of having tuberculosis. A sterile container must be provided, and the patient should be instructed that early morning specimens are best, usually collected upon first awakening. This is due to the fact that sputum often "pools" in the bronchi when the patient is sleeping at night; therefore, it is easier to collect a larger sample in the morning after the patient has been lying down all night.

If the patient is unable to produce any sputum, giving him or her an aerosolized dose of saline inhaled by mask may help moisten secretions, making it easier for the patient to produce the sample. Samples are also sometimes collected by suctioning the patient. Doctors may order acid-fast sputum samples for tuberculosis to be collected on three consecutive mornings. This helps to increase the likelihood of identifying the bacteria if they are present.

Q. *Mycobacterium tuberculosis* colonizes the lungs. What technical problem would you anticipate if you were to plate sputum on agar to look for it?

Answer in Appendix B.

This century-old staining method is still the universal basis for bacterial classification and identification. The Gram stain can also be a practical aid in diagnosing infection and in guiding drug treatment. For example, Gram staining a fresh urine or throat specimen can help pinpoint the possible cause of infection, and in some cases it is possible to begin drug therapy on the basis of this stain. Even in this day of elaborate and expensive medical technology, the Gram stain remains an important and unbeatable first tool in diagnosis.

Nontypical Cell Walls

Several bacterial groups have a different cell wall structure than gram-positive or gram-negative bacteria, and some bacteria have no cell wall at all. Although these exceptional forms can stain positive or negative in the Gram stain, examination of their fine structure shows that they do not really fit the descriptions for typical gram-negative or gram-positive cells. For example, the cells of *Mycobacterium* and *Nocardia* contain peptidoglycan and stain gram-positive, but the bulk of their cell wall is composed of unique types of lipids. One of these is a very-long-chain fatty acid called *mycolic acid,* or cord factor, that contributes to the pathogenicity of this group (see chapter 19). The thick, waxy nature bestowed on the cell wall by these lipids also means that these bacteria are highly resistant to certain chemicals and dyes. Such resistance is the basis for the acid-fast stain used to diagnose tuberculosis and leprosy.

The archaea exhibit unusual and chemically distinct cell walls. In some, the walls are composed almost entirely of polysaccharides, and in others, the walls are pure protein. As a group, they all lack the true peptidoglycan structure described previously.

Mycoplasmas and Other Cell-Wall-Deficient Bacteria

Mycoplasmas are bacteria that naturally lack a cell wall. Although other bacteria require an intact cell wall to prevent the bursting of the cell, the mycoplasma cytoplasmic membrane is stabilized by sterols and is resistant to lysis. These extremely tiny, pleomorphic cells are very small bacteria, ranging from 0.1 to 0.5 μm in size. The most important medical species is *Mycoplasma pneumoniae*, which adheres to the epithelial cells in the lung and causes an atypical form of pneumonia in humans (often called "walking pneumonia" because its sufferers can often continue their daily activities, and the illness can often be treated on an outpatient basis) (described in chapter 19).

Some bacteria that ordinarily have a cell wall can lose it during part of their life cycle. These wall-deficient forms are referred to as **L forms** or L-phase variants (for the Lister Institute, where they were discovered). Evidence points to a role for L forms in persistent infections that can be resistant to antibiotic treatment.

The Gram-Negative Outer Membrane

The **outer membrane (OM)** (see figure 3.15) is similar in composition to most membranes, except that it contains specialized types of polysaccharides and proteins. Remember that each membrane is a bilayer of phospholipids. The uppermost layer of the OM is different because some of the phospholipid molecules are substituted with a macromolecule called *lipopolysaccharide* (LPS) **(figure 3.18)**. The polysaccharide chains extending off the surface function as signaling molecules and receptors. The lipid portion of LPS has been referred to as *endotoxin* because it stimulates fever and shock reactions in gram-negative infections such as meningitis and typhoid fever. The innermost layer of the OM is a phospholipid layer anchored by means of lipoproteins to the peptidoglycan layer below. The outer membrane serves as a sieve by allowing only relatively small molecules to penetrate. These molecules pass through the membrane via special channels formed by *porin proteins* that completely span the outer membrane.

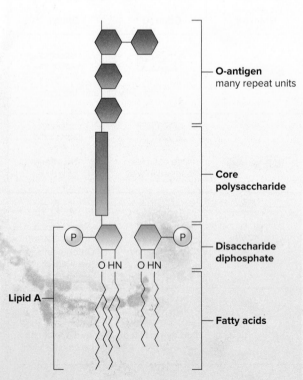

Figure 3.18 The lipopolysaccharide molecule.

O-antigen
many repeat units

Core
polysaccharide

Disaccharide
diphosphate

Fatty acids

Lipid A

Cytoplasmic Membrane Structure

Appearing just beneath the cell wall is the cytoplasmic membrane, often also called the cell membrane. We use "cytoplasmic membrane" in this book because the term makes it immediately clear that we are talking about the membrane in direct contact with the cytoplasm. This membrane is a very thin (5-10 nm), flexible sheet molded completely around the cytoplasm. It is a lipid bilayer with proteins embedded to varying degrees. Bacterial cytoplasmic membranes have this typical structure, containing primarily phospholipids (making up about 30%-40% of the membrane mass) and proteins (contributing 60%-70%). Major exceptions to this description are the membranes of mycoplasmas, which contain high amounts of sterols–rigid lipids that stabilize and reinforce the membrane–and the membranes of archaea, which contain unique branched hydrocarbons rather than fatty acids.

Some environmental bacteria, including photosynthesizers and ammonia oxidizers, contain dense stacks of internal membranes. In many cases, they derive from the cytoplasmic membrane, and they are studded with enzymes or photosynthetic pigments. The inner membranes allow for a higher concentration of these enzymes and pigments and also create a compartmentalization that allows for higher energy production.

Functions of the Cytoplasmic Membrane

Because bacteria have none of the typical eukaryotic organelles, the cytoplasmic membrane provides a site for functions such as energy reactions, nutrient processing, and synthesis. A major action of the cytoplasmic membrane is to regulate *transport*, that is, the passage of nutrients into the cell and the discharge of wastes. Although water and small uncharged molecules can diffuse freely across the membrane, the membrane is a *selectively permeable* structure with special carrier mechanisms for the passage of most molecules (see chapter 6). The glycocalyx and cell wall can inhibit the passage of large molecules, but they are not the primary transport apparatuses.

In eukaryotic cells, important functions such as respiration and ATP synthesis take place on organelles called mitochondria. Because bacteria do not have mitochondria, they carry out these and other important activities using proteins embedded in their membranes.

Practical Considerations of Differences in Cell Envelope Structure

The different envelopes in gram-positive and gram-negative bacteria have practical implications. The outer membrane contributes an extra barrier in gram-negative bacteria that makes them resistant to some antimicrobial chemicals such as dyes and disinfectants. For this reason, they are generally more difficult to inhibit or kill than are gram-positive bacteria. One exception is for alcohol-based compounds, which can dissolve the lipids in the outer membrane and therefore damage the cell. This is why alcohol swabs are often used to cleanse the skin prior to certain medical procedures, such as venipuncture. Treating infections caused by gram-negative bacteria often calls for different drugs from gram-positive infections, especially drugs that can cross the outer membrane.

Lipopolysaccharide molecules in an outer membrane of a gram-negative bacterium.
- Kelly & Heidi

3.3 LEARNING OUTCOMES—Assess Your Progress

7. Differentiate between the two main types of bacterial envelope structure.
8. Discuss why gram-positive cell walls are stronger than gram-negative cell walls.
9. Name a substance in the envelope structure of some bacteria that can cause severe symptoms in humans.

3.4 Bacterial Internal Structure

Contents of the Cell Cytoplasm

The **cytoplasm** is a gelatinous solution encased by the cytoplasmic membrane. Its major component is water (70%-80%), which serves as a solvent for the contents of the cell, which is a complex mixture of nutrients including sugars, amino acids, and salts. Some people distinguish the cytoplasm from the *cytosol*. In this situation, cytoplasm refers to everything inside the cytoplasmic membrane, and cytosol is only the fluid portion. We won't use that distinction.

Bacterial Chromosomes and Plasmids

The hereditary material of most bacteria exists in the form of a single circular strand of DNA designated as the **bacterial chromosome.** Some bacteria have multiple chromosomes. By definition, bacteria do not have a nucleus; that is, their DNA is not enclosed by a nuclear membrane but instead is aggregated in a dense area of the cell called the **nucleoid.** The chromosome is actually an extremely long molecule of double-stranded DNA that is tightly coiled around special basic protein molecules so as to fit inside the cell. You have already learned that DNA exists as a double helix. The bacterial chromosome, then, is a long double helix formed into a circle. There are no loose ends. Arranged along its length are genetic units (genes) that carry information required for bacterial maintenance and growth.

Although the chromosome is the minimal genetic requirement for bacterial survival, many bacteria contain other, nonessential pieces of DNA called **plasmids** (refer to figure 3.1). These tiny strands exist as separate double-stranded circles of DNA, although at times they can become integrated into the chromosome. In certain circumstances, they may be duplicated and passed on to related nearby bacteria. During bacterial reproduction, they are duplicated and passed on to offspring. They are not essential to bacterial growth and metabolism, but they do provide protective traits such as resisting drugs and producing toxins and enzymes (see chapter 8). Because they can be readily manipulated in the laboratory and transferred from one bacterial cell to another, plasmids are an important agent in genetic engineering techniques.

Ribosomes: Sites of Protein Synthesis

A bacterial cell contains thousands of tiny **ribosomes,** which are the site of protein synthesis. When viewed even by very high magnification, ribosomes show up as fine, spherical specks dispersed throughout the cytoplasm that often occur in chains called polysomes. Many are also attached to the cytoplasmic membrane. Chemically, a ribosome is a combination of a special type of RNA called ribosomal RNA, or rRNA (about 60%), and protein (40%). Ribosomes are characterized by their density, designated by "S units." Ribosomes consist of a small subunit and a large subunit **(figure 3.19).** Both of these are made of a mixture of rRNA and protein. The small subunit has an S value of 30, and the large subunit has an S value of 50. Overall, the bacterial ribosome has a density of 70S. (It is not simply an additive property; that is why the total S value is not a sum of the small and large subunits.) The two subunits fit together to form a miniature platform upon which protein synthesis is performed. Note that eukaryotic ribosomes are similar but still different. Because of this, we can design drugs to target bacterial ribosomes that do not harm our own. Eukaryotic ribosomes are designated 80S. Although archaea possess 70S ribosomes, they are more similar in structure to that of 80S eukaryotic ribosomes.

Inclusion Bodies and Microcompartments

Bacteria manufacture structures referred to as inclusion bodies to respond to their environmental conditions. They can store nutrients in inclusion bodies to respond to periods of low food availability. They can pack gas into inclusion vesicles to provide buoyancy in an aquatic environment. They can even store crystals of iron oxide with magnetic properties in inclusion bodies. These magnetotactic bacteria use the granules to orient themselves

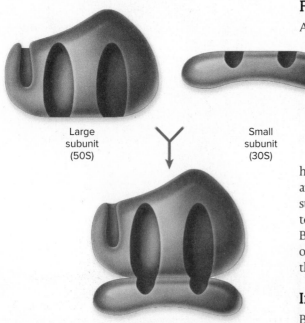

Large subunit (50S)

Small subunit (30S)

Ribosome (70S)

Figure 3.19 A model of a bacterial ribosome, showing the small (30S) and large (50S) subunits, both separate and joined.

in polar and gravitational fields to bring them to environments with the proper oxygen content. **Figure 3.20** illustrates a bacterium with these iron oxide crystals.

In the early 2000s, new compartments inside bacterial cells were discovered. These were named *bacterial microcompartments* (BMCs). Their outer shells are made of protein, arranged geometrically, and are packed full of enzymes that are designed to work together in biochemical pathways, thereby ensuring that they are in close proximity to one another.

The Cytoskeleton

Until recently, scientists thought that the shape of all bacteria was completely determined by the peptidoglycan layer (cell wall). Although this is true of many bacteria, particularly the cocci, other bacteria produce long polymers of proteins that are very similar to structures in eukaryotic cells for the cytoskeleton. (Cytoskeleton means "cell skeleton.") In eukaryotes, the cytoskeleton is made of three main types of proteins: actin, tubulin, and intermediate filament proteins. Bacteria have proteins that are similar to each of these three, and contain additional types of proteins in their cytoskeletons. These proteins are arranged in helical ribbons around the cell just under the cytoplasmic membrane. These fibers contribute to cell shape, perhaps by influencing the way peptidoglycan is manufactured, and also to cell division. Cytoskeletal proteins have also been identified in archaea. Because these proteins are sufficiently different from those in eukaryotic cells, they are a potentially powerful target for future antibiotic development.

Bacterial Endospores

The anatomy of bacteria helps them adjust rather well to challenging habitats. But of all microbial structures, nothing can compare to the bacterial endospore for withstanding hostile conditions and facilitating survival.

Endospores are dormant bodies produced by bacteria such as *Bacillus, Clostridium,* and *Sporosarcina.* These bacteria can exist in two different forms–a vegetative cell and an endospore **(figure 3.21).** The vegetative cell is a metabolically active and growing cell. When environmental conditions become challenging, these bacteria will form endospores in a process called **sporulation.** The endospore exists initially inside the vegetative cell, but eventually the cell disintegrates and the endospore is on its own. Both gram-positive and gram-negative bacteria can form endospores, but the medically relevant ones are all gram-positive. Most bacteria form only one endospore; therefore, this is not a reproductive function for them.

Bacterial endospores are the hardiest of all life forms, capable of withstanding extremes in heat, drying, freezing, radiation, and chemicals that would readily kill vegetative cells. Their survival under such harsh conditions is due to several factors. The heat resistance of endospores is due to their high content of calcium and dipicolinic acid. We know, for instance, that heat destroys cells by inactivating proteins and DNA. This process requires a certain amount of water in the protoplasm. There is a chemical in endospores called calcium dipicolinate that removes that water. This leaves the endospore very dehydrated, making it less vulnerable to the effects of heat. The thick, impervious cortex and endospore coats also protect against radiation and chemicals. Bacterial endospores are capable of surviving indefinitely. Microbiologists have even unearthed a viable endospore from a 250-million-year-old salt crystal. Initial analysis of this ancient microbe indicates it is a species of *Bacillus.*

Endospore Formation: Sporulation

What causes a vegetative cell to make the switch to begin endospore formation? It is usually the fact that nutrients have started to become scarce. Carbon depletion and nitrogen depletion are common triggers. Once this stimulus has been received by the vegetative cell, it undergoes a conversion to become a sporulating cell called a **sporangium.** Complete

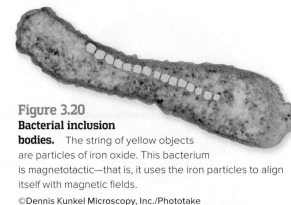

Figure 3.20
Bacterial inclusion bodies. The string of yellow objects are particles of iron oxide. This bacterium is magnetotactic—that is, it uses the iron particles to align itself with magnetic fields.
©Dennis Kunkel Microscopy, Inc./Phototake

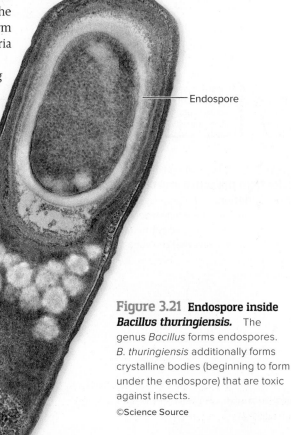

Endospore

Figure 3.21 Endospore inside
Bacillus thuringiensis. The genus *Bacillus* forms endospores. *B. thuringiensis* additionally forms crystalline bodies (beginning to form under the endospore) that are toxic against insects.
©Science Source

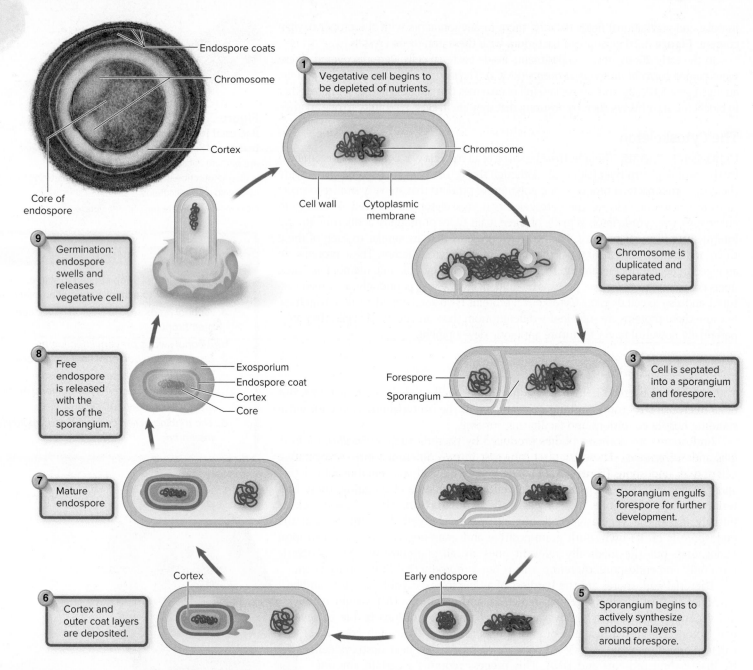

Endospore coats

Chromosome

Cortex

Core of endospore

1 Vegetative cell begins to be depleted of nutrients.

Chromosome

Cell wall   Cytoplasmic membrane

2 Chromosome is duplicated and separated.

Forespore

Sporangium

3 Cell is septated into a sporangium and forespore.

4 Sporangium engulfs forespore for further development.

Early endospore

5 Sporangium begins to actively synthesize endospore layers around forespore.

6 Cortex and outer coat layers are deposited.

Cortex

7 Mature endospore

8 Free endospore is released with the loss of the sporangium.

Exosporium
Endospore coat
Cortex
Core

9 Germination: endospore swells and releases vegetative cell.

Figure 3.22 A typical sporulation cycle in *Bacillus* species from the active vegetative cell to release and germination. The process takes, on average, from 6–10 hours. Inset is a high magnification (10,000×) cross section of a single endospore showing the dense protective layers that surround the core with its chromosome.

©Science Source

transformation of a vegetative cell into a sporangium and then into an endospore requires 6 to 10 hours in most endospore-forming species. **Figure 3.22** illustrates the major physical and chemical events in this process.

Return to the Vegetative State: Germination

After lying in a state of inactivity for an indefinite time, endospores can be revitalized when favorable conditions arise. Germination–the reversal of dormancy–happens when water and a specific chemical or environmental stimulus (germination agent) are present. Once it begins, it proceeds to completion quite rapidly (1½ hours). Although the specific germination agent varies among species, it is generally a small organic molecule such as an amino acid or an inorganic salt. This agent stimulates the formation of hydrolytic (digestive) enzymes by the endospore membranes. These enzymes digest the cortex and allow water into the core. As the core rehydrates and takes up nutrients, it begins to grow out of the endospore coats. In time, it reverts to a fully active vegetative cell, resuming the vegetative cycle.

Medical Significance of Bacterial Endospores

Although the majority of endospore-forming bacteria are relatively harmless to humans, several bacterial pathogens are endospore formers. In fact, some aspects of the diseases they cause are related to the persistence and resistance of their spores. *Bacillus anthracis* is the agent of anthrax. Its ability to form endospores makes it an ideal candidate for bioterrorism. The genus *Clostridium* includes even more pathogens, such as *C. tetani*, the cause of tetanus (lockjaw); *C. perfringens*, the cause of gas gangrene; and *C. botulinum*, the cause of botulism. (Each of these disease conditions is discussed in the infectious disease chapters, according to the organ systems it affects.)

Because they inhabit the soil and dust, endospores are constant intruders where sterility and cleanliness are important. They resist ordinary cleaning methods that use boiling water, soaps, and disinfectants; and they frequently contaminate cultures and media. Hospitals and clinics must take precautions to guard against the potential harmful effects of endospores, especially those of *Clostridioides difficile*, the cause of a severe gastrointestinal disease known as "*C. diff.*" Endospore destruction is a particular concern of the food-canning industry. Several endospore-forming species cause food spoilage or poisoning. Ordinary boiling (100°C) will usually not destroy such endospores, so canning is carried out in pressurized steam at 120°C for 20 to 30 minutes. Such rigorous conditions ensure that the food is sterile and free from viable bacteria.

3.4 LEARNING OUTCOMES—Assess Your Progress

10. Identify seven structures that may be contained in bacterial cytoplasm.

11. Detail the causes and mechanisms of sporogenesis and germination.

3.5 The Archaea

The discovery and characterization of novel cells resembling bacteria but with unusual anatomy, physiology, and genetics changed our views of microbial taxonomy and classification (see chapter 1). These single-celled, simple organisms, called archaea, are a third cell type in a separate superkingdom (the domain Archaea). We include them in this chapter because they share many bacterial characteristics. But it has become clear that they are actually more closely related to domain Eukarya than to bacteria. For example, archaea and eukaryotes share a number of ribosomal RNA sequences that are not found in bacteria, and their protein synthesis and ribosomal subunit structures are similar. **Table 3.2** outlines selected points of comparison of the three domains.

Among the ways that the archaea differ significantly from other cell types are that they have entirely unique sequences in their rRNA. They exhibit a novel method of DNA compaction, and they contain unique membrane lipids, cell wall components, and pilin proteins. It is clear that the archaea are the most primitive of all life forms and are most closely related to the first cells that originated on the earth 3.8 billion years ago. The early earth is thought to have contained a hot, anaerobic "soup" with sulfuric gases and salts in abundance. The modern archaea still live in the habitats on the earth that have these same ancient conditions—the most extreme habitats in nature. It is for this reason that they are often called *extremophiles*, meaning that they "love" extreme conditions in the environment.

Some archaea thrive in extremely high temperatures. Others need extremely high concentrations of salt or acid to survive. Some archaea live on sulfur, reducing it to hydrogen sulfide to get their energy. Members of the group called *methanogens*

Table 3.2 Comparison of Three Cellular Domains

Characteristic	Bacteria	Archaea	Eukarya
Chromosomes	Single or a few, circular	Single, circular	Multiple, linear
Types of ribosomes	70S	70S but structure is similar to 80S	80S
Contains unique ribosomal RNA signature sequences	+	+	+
Eukarya(-like) protein synthesis	−	+	+
Cell wall made of peptidoglycan	+	−	−
Cytoplasmic membrane lipids	Fatty acids with ester linkages	Long-chain, branched hydrocarbons with ether linkages	Fatty acids with ester linkages
Sterols in membrane	− (some exceptions)	−	+
Nucleus and membrane-bound organelles	No	No	Yes
Flagellum	Bacterial flagellum	Archaellum	Eukaryotic flagellum

can convert CO_2 and H_2 into methane gas (CH_4) through unusual and complex pathways. Archaea that are adapted to growth at very low temperatures are called *psychrophilic* (loving cold temperatures). Those growing at very high temperatures are *hyperthermophilic* (loving high temperatures). Hyperthermophiles flourish at temperatures between 80°C and 113°C and find 50°C (which is 122°F) too cold to grow. They live in volcanic waters and soils and submarine vents and are often salt- and acid-tolerant as well.

Archaea are not just environmental microbes. They have been isolated from human tissues such as the colon, the mouth, and the vagina. Recently, an association was found between the degree of severity of periodontal disease and the presence of archaeal RNA sequences in the gingiva, suggesting that archaea may be capable of causing human disease.

3.5 LEARNING OUTCOMES—Assess Your Progress

12. Compare and contrast the major features of archaea, bacteria, and eukaryotes.

3.6 Classification Systems for Bacteria and Archaea

Scientists use classification systems to differentiate and identify species in medical and applied microbiology. They are also useful in organizing microorganisms and as a means of studying their relationships and origins. Since classification began around 200 years ago, several thousand species of bacteria and archaea have been identified, named, and cataloged.

There are two comprehensive databases that help scientists classify bacteria and archaea. One, called *Bergey's Manual of Systematic Bacteriology*, presents a comprehensive view of bacterial and archaeal relatedness, based largely on rRNA sequencing information. (We need to remember that all bacteria and archaea classification systems are in a state of constant flux; no system is ever finished.)

A separate database, called *Bergey's Manual of Determinative Bacteriology*, is based entirely on phenotypic characteristics (not genetics), which are the observable

Thermophilic archaea and cyanobacteria colonizing a thermal pool in Yellowstone National Park.

©Robert Glusic/Getty Images

characteristics of an organism such as its shape (seen by microscopy) and its metabolic capabilities (such as what it requires to grow). It is utilitarian in focus, categorizing bacteria by traits commonly tested for in clinical, teaching, and research labs. It is widely used by microbiologists who need to identify bacteria but do not need to know their evolutionary backgrounds. This type of classification is more useful for students of medical microbiology as well.

Taxonomic Scheme

Bergey's Manual of Determinative Bacteriology organizes the bacteria and archaea into four major divisions. These divisions are based on the nature of the cell wall. The **Gracilicutes** (gras'-ih-lik'-yoo-teez) have gram-negative cell walls and so they are thin-skinned; the **Firmicutes** have gram-positive cell walls that are thick and strong; the **Tenericutes** (ten'-er-ik'-yoo-teez) lack a cell wall and thus are soft; and the **Mendosicutes** (men-doh-sik'-yoo-teez) are the archaea (also called archaebacteria), primitive cells with unusual cell walls and nutritional habits. The system used in *Bergey's Manual* further organizes bacteria and archaea into subcategories such as classes, orders, and families, but these are not available for all groups.

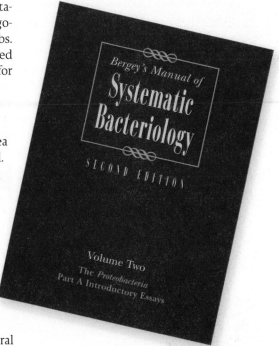

Courtesy Bergey's Manual Trust

Species and Subspecies in Bacteria and Archaea

Among most organisms, the species level is a distinct, readily defined, and natural taxonomic category. In animals, for instance, a species is a distinct type of organism that can produce viable offspring only when it mates with others of its own kind. This definition does not work for bacteria and archaea primarily because they do not use a typical mode of sexual reproduction. Also, they can accept genetic information from unrelated forms, and they can alter their genetic makeup by a variety of mechanisms. For that reason, it is necessary to hedge a bit when we define a *bacterial species*. Theoretically, it is a collection of bacterial cells, all of which share an overall similar pattern of traits, in contrast to other groups whose patterns differ significantly. Members of a bacterial species should also share approximately 95% average matches

Purestock/Getty Images

in their gene sequences. Although the boundaries that separate two closely related species in a genus are in some cases arbitrary, this definition serves as a method to separate the bacteria and archaea into various kinds that can be cultured and studied.

Individual members of given species can show variations, as well. Therefore, more categories within species exist, but they are not well defined. Microbiologists use terms like *subspecies, strain,* or *type* to designate bacteria of the same species that have differing characteristics. *Serotype* refers to representatives of a species that stimulate a distinct pattern of antibody (serum) responses in their hosts because of distinct surface molecules.

3.6 LEARNING OUTCOMES—Assess Your Progress

13. Differentiate between *Bergey's Manual of Systematic Bacteriology* and *Bergey's Manual of Determinative Bacteriology.*

14. Name four divisions ending in –cutes and describe their characteristics.

15. Define a species in terms of bacteria.

CASE FILE WRAP-UP

The circle contains an electron micrograph of *Clostridioides difficile,* the endospore-forming bacterium that causes a common healthcare-associated intestinal infection.

©Dave and Les Jacobs/Blend Images LLC (nurse); Source: CDC/Dr. Gilda Jones (*Clostridioides difficile* bacteria)

Clostridioides difficile is an endospore-forming bacterium that has gained attention over the last few years as the causative agent of a common (and potentially deadly) healthcare-associated infection. Often called "*C. diff,*" this disease is spread by direct contact with an infected individual or the pathogen itself. In the hospital, the bacterium and more often its endospores can be present on bedrails, bedside tables, sinks, and even surfaces such as stethoscopes and blood pressure cuffs. Endospores are most often the source of infection because they are extremely resistant to many cleaning agents.

Individuals at higher risk of contracting the disease include the elderly, individuals with weakened immune systems, people with intestinal disorders, and people who have recently taken antibiotics. The patient in the opening case file was elderly, had recently had major surgery, and was already battling another infection that was being treated with antibiotics, all risk factors for the development of *C. diff.* The disease can range from a mild infection to a life-threatening illness causing severe diarrhea up to 15 times a day. Note that some people are asymptomatic carriers of this pathogen, which makes controlling the disease that much more difficult, especially in health care settings.

Dear students: In the last year, I (one of the authors, Marjorie Kelly Cowan) had an infection with *C. diff.* I kept ignoring the symptoms because I am a healthy person, but one night I ended up doubled over and had to go to the emergency room. The pain was unbearable. I got treatment, after many tears and many tests. I apparently contracted the infection from many hours spent with my mother in her rehab facility after she broke her hip!

A Sticky Situation

The human microbiome generally exists in a peaceful relationship with its host. However, when circumstances change, the microbiome can become more sinister. One way that circumstances can change is when some type of foreign material is implanted in the body. Artificial heart valves, an intrauterine device, and an artificial hip are all susceptible to being colonized by bacteria from the microbiome. Nonhuman surfaces in the body are particularly susceptible to attachment by bacteria. When the microorganisms continue to accumulate, they form a structure called a **biofilm.**

A study published in the *Proceedings of the National Academy of Sciences* in 2013 revealed just how quickly biofilms can clog commonly used medical devices, such as cardiovascular stents. Researchers from Princeton utilized narrow tubes closely resembling those found in certain medical devices. Specific materials were chosen to replicate the surface of the equipment, and the tubes were then exposed to fluid under pressure in order to closely mimic conditions within the human body.

The researchers used microbes that are known to contaminate medical devices and engineered them to produce a pigment that could be observed microscopically. After forcing a stream of these microbes through the experimental tubes for approximately 40 hours, microscopic analysis revealed the formation of a biofilm on the inside walls of the device.

Over the next few hours, the researchers then forced a stream of different microbes into the experimental tubes. Within a short period of time, these cells were noted adhering to the biofilm-coated inner walls of the tubes. Further analysis revealed that the flow within the narrow tubes nudged the trapped cells into threadlike "streamers" that rippled along with the moving fluid.

Initially, the formation of these microbial threads only slightly decreased the rate of fluid flow within the experimental tubes. However, after 55 hours, the streamers began to weave together, creating a net similar to a spider's web. This newly formed structure spanned the diameter of the narrow tube and trapped even more cells, triggering a total blockage of the experimental tube within an hour. This experiment revealed an important phenomenon that may explain why devices such as stents often fail.

A study in 2019 offered hints about why nonhuman surfaces might be more susceptible to biofilm formation than are human surfaces. Their experiments showed that a natural human enzyme called thrombin broke up biofilms as they built on human mucosal surfaces.

Source: *Proceedings of the National Academy of Sciences.* February 11, 2013.
Source: *Nature Communications.* July 19, 2019.

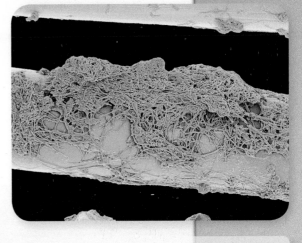

An accumulation of bacteria on a single fiber of a gauze bandage.

©Steve Gschmeissner/Science Source

Source: Shutterstock/Barbo

BACTERIA AND ARCHAEA

3.1 BACTERIA AND ARCHAEA: An Overview

Bacteria and archaea are single-celled and seem to be simple organisms. They are very different from one another, and both are very sophisticated.

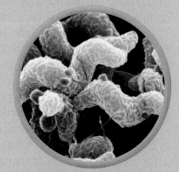

Photo by De Wood, digital colorization by Chris Pooley, USDA-ARS

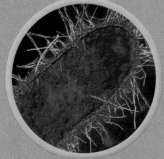

Bacteria have a wide variety of appendages: flagella, fimbriae, pili, and nanotubes. Many also have two external layers: a sticky capsule (slime layer) and a layer called the S layer.

EXTERNAL STRUCTURES 3.2

Eye of Science/Science Source

3.3 THE CELL ENVELOPE: The Wall and Membrane(s)

The bacterial cell envelope performs many functions. It consists of a cell wall made of peptidoglycan and either one or two membranes, depending on whether the cell is gram-positive or gram-negative.

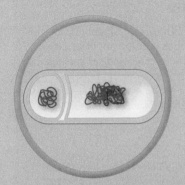

Compared to eukaryotic organisms, bacteria have relatively few structures in their cytoplasm. There is a DNA chromosome, and possibly plasmids. Like all cells, bacteria have ribosomes. Some bacteria have inclusion bodies, cytoskeletons, and microcompartments. Endospores are a survival structure in some bacteria.

BACTERIAL INTERNAL STRUCTURE

3.5 THE ARCHAEA

The archaea are one of three domains of life—and are more closely related to Eukarya, even though they physically resemble Domain Bacteria. They often live in extreme environments, although they are also found in humans.

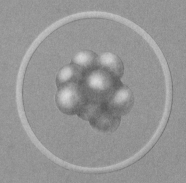

Classification systems help scientists and clinicians speak a common language when referring to organisms. Bacteria and archaea are divided into four classes: Gracilicutes, Firmicutes, Tenericutes, and Mendosicutes.

CLASSIFICATION SYSTEMS FOR BACTERIA AND ARCHAEA

SmartGrid: From Knowledge to Critical Thinking

This *21 Question Grid* takes the topics from this chapter and arranges them with respect to the American Society for Microbiology's Undergraduate Curriculum guidelines—all six of the important "Concepts" as well as the important "Competency" of scientific literacy. Three questions are supplied about chapter content that refer to the Concept or Competency, in increasing levels of Bloom's taxonomy for learning.

ASM Concept/ Competency	A. Bloom's Level 1, 2—Remember and Understand (Choose one.)	B. Bloom's Level 3, 4—Apply and Analyze	C. Bloom's Level 5, 6—Evaluate and Create

Evolution	1. Archaea
	  a. are most genetically related to bacteria.
	  b. contain a nucleus.
	  c. cannot cause disease in humans.
	  d. lack peptidoglycan in their cell walls.
		2. Recall from section 1.1 that both bacteria and archaea appear to have evolved from the Last Common Ancestor (LCA). Based on what you know about their habitats and structures, speculate on *why* there are differences between the two.	3. Suppose an argument in your city has erupted about whether bacteria can be called species. Write a letter to the editor for your local newspaper arguing one side or the other.

Cell Structure and Function	4. Which of the following is present in both gram-positive and gram-negative cell walls?
	  a. an outer membrane
	  b. peptidoglycan
	  c. teichoic acid
	  d. lipopolysaccharides
		5. As a supervisor in the infection control unit, you hire a local microbiologist to analyze samples from your hospital's hot-water tank for microbial contamination. Although she was unable to culture any microbes, she reports that basic microscopic analysis revealed the presence of cells 0.8 µm in diameter that lacked a nucleus. Transmission electron microscopy showed that the cells lacked membrane-bound organelles but did contain ribosomes. Which domain of life do you hypothesize the cells represent? What further analysis would you perform?	6. Clearly bacteria are very different from eukaryotic cells. Yet they have similar processes and structures. Their DNA is very similar. Their ribosomes are very similar. The way they make proteins is similar. Speculate as to what led to this similarity in such radically different organisms.

Metabolic Pathways	7. Bacterial endospores usually function in
	  a. reproduction.
	  b. metabolism of nutrients.
	  c. survival.
	  d. storage.
		8. Creating extremely long tubes (nanotubes) out of existing cell membrane material seems like a lot of hard work. Why would this be justified for an organism in an oxygen-poor environment?	9. Bacteria and archaea have a much greater diversity in their metabolic capabilities than do eukaryotes. They can live at 113°C (235°F) and at incredibly high atmospheric pressures. Construct an argument for why these simple cells have developed the ability to do this.

Information Flow and Genetics	10. Which structure plays a direct role in the exchange of genetic material between bacterial cells?
	  a. flagellum
	  b. pilus
	  c. capsule
	  d. fimbria
		11. Bacteria have been found to change the structures they produce when they are placed in a situation of high flow rates, especially in a narrow space, like a tube. Often these new structures allow them to attach themselves or trap other bacteria in a biofilm. Identify some advantages to this strategy from the bacterium's perspective.	12. Bacterial and archaeal chromosomes are not enclosed in a membrane-bound nucleus. You will learn about some advantages of this arrangement in another chapter, but can you think of some disadvantages?

ASM Concept/Competency	A. Bloom's Level 1, 2—Remember and Understand (Choose one.)	B. Bloom's Level 3, 4—Apply and Analyze	C. Bloom's Level 5, 6—Evaluate and Create
Microbial Systems	**13.** Nanotubes are extensions of the _____ that can function in _____. **a.** membrane, genetic exchange **b.** pilus, genetic exchange **c.** flagellum, motility **d.** membrane, nutrient transfer	**14.** The results of your patient's wound culture just arrived, and Gram staining revealed the presence of pink, rod-shaped bacterial cells organized in pairs. You quickly realize that this patient could be at risk for developing fever and shock. Explain how the culture results indicated this potential risk.	**15.** We know that bacteria/archaea and their genetics are clearly influenced by their environment because they have developed membranes and structures that allow them to live in a wide variety of conditions. Do you think the environment in which they live has likewise experienced change due to *their* presence? Can you cite any examples?
Impact of Microorganisms	**16.** Find the true statement about biofilms. **a.** They are found only in outdoor environments. **b.** They are found only on artificial medical implants. **c.** They consist of many representatives of a single bacterial species. **d.** They complicate the treatment of some infections.	**17.** Suggest more than one reason why bacteria may express glycocalyces in natural circumstances but not in the laboratory.	**18.** Construct arguments agreeing with and refuting this statement: *Human infections may have originated as accidental encounters between humans and microbes that were actually meant to interact with other parts of the natural environment.*
Scientific Thinking	**19.** Which of the following would be used to identify an unknown bacterial culture that came from a patient in the intensive care unit? **a.** *Gray's Anatomy* **b.** *Bergey's Manual of Determinative Bacteriology* **c.** *Bergey's Manual of Systematic Bacteriology* **d.** *The Physician's Desk Reference*	**20.** During the cold war between the Soviet Union and the United States, both countries were interested in "weaponizing" *Bacillus anthracis* (anthrax) endospores to deliver via missiles and other weapons. Why did they choose this particular organism?	**21.** During the cold war between the Soviet Union and the United States, both countries were interested in "weaponizing" *Bacillus anthracis* (anthrax) endospores to deliver via missiles and other weapons. Do you believe there is any ethical difference between using infectious agents as weapons compared to traditional weapons? Explain.

Answers to the multiple-choice questions appear in Appendix A.

Visual Connections

This question connects content within and between chapters.

From chapter 2, figure 2.18. Explain why some cells are pink and others are purple in this image of a Gram-stained bacterial smear.

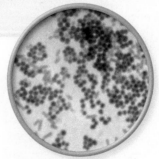

ASM/Science Source

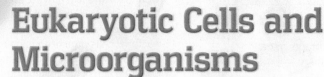

Eukaryotic Cells and Microorganisms

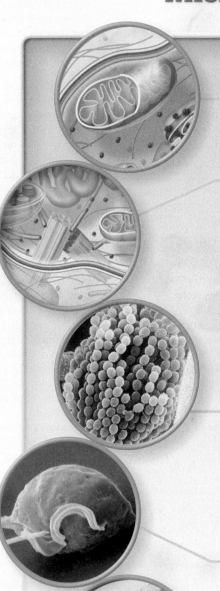

IN THIS CHAPTER...

4.1 Overview of the Eukaryotes

1. Relate bacterial, archaeal, and eukaryotic cells to the *Last Common Ancestor*.
2. List the types of eukaryotic microorganisms, and identify which are unicellular and which are multicellular.

4.2 Structures of the Eukaryotic Cell

3. Differentiate among the flagellar structures of bacteria, eukaryotes, and archaea.
4. List similarities and differences between eukaryotic and bacterial cytoplasmic membranes.
5. Describe the main structural components of a nucleus.
6. Diagram how the nucleus, endoplasmic reticulum, and Golgi apparatus act together with vesicles during the transport process.
7. Explain the function of the mitochondrion.
8. Explain the importance of ribosomes, and differentiate between eukaryotic and bacterial types.
9. List and describe the three main fibers of the cytoskeleton.
10. Explain how endosymbiosis contributed to the development of eukaryotic cells.

4.3 The Fungi

11. List two detrimental and two beneficial activities of fungi (from the viewpoint of humans).
12. List three general features of fungal anatomy.
13. Differentiate among the terms *heterotroph, saprobe,* and *parasite.*
14. Explain the relationship between fungal hyphae and the production of a mycelium.
15. Describe two ways in which fungal spores arise.

4.4 The Protozoa

16. Describe the protozoan characteristics that illustrate why protozoa are informally placed into a single group.
17. List three means of locomotion exhibited by protozoa.
18. Explain why a cyst stage may be useful to a protozoan.
19. Give an example of a disease caused by each of the four types of protozoa.

4.5 The Helminths

20. List the two major groups of helminths, and provide examples representing each body type.
21. Summarize the stages of a typical helminth life cycle.

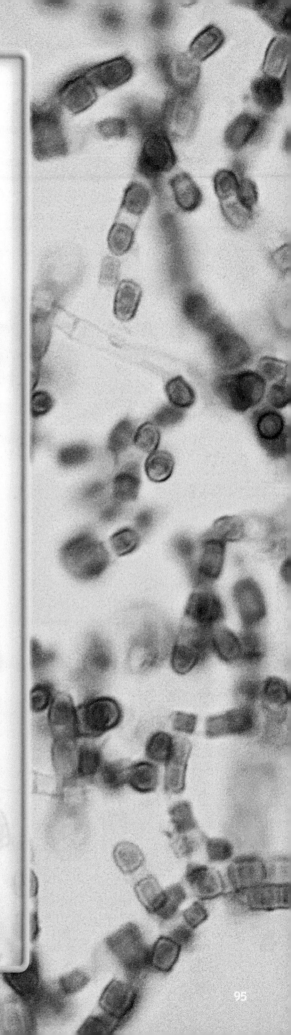

CASE FILE

Puzzle in the Valley

Working as a newly graduated radiology technologist in a rural hospital in California, I encountered a case that would prove to be a challenge for everyone involved. The patient was a male migrant farm worker in his mid-30s who presented to the ER with common flulike symptoms: fever, chills, weakness, cough, muscular aches and pains, and headache. He also had a painful red rash on his lower legs.

The emergency room physician believed that the patient likely had pneumonia, but she found the rash puzzling. She asked me to obtain a chest X ray. I performed anterior-posterior and lateral views of the chest, which revealed two nodules approximately 2 cm in size in the patient's left upper lobe. The physician stated that the nodules were consistent with pneumonia, but the possibility of cancer could not be ruled out. The patient's age and the fact that he was a nonsmoker, however, made a diagnosis of lung cancer less likely than pneumonia.

The patient was admitted to the hospital for IV antibiotic treatment. Before the antibiotic therapy was started, a sputum sample was collected and sent to a larger center for culture and sensitivity (C&S) testing. Despite IV fluids, rest, and broad-spectrum antibiotics targeting both gram-positive and gram-negative bacteria, the patient showed no improvement. After receiving the C&S report, it became clear why the intravenous antibiotics were not working: The patient had a fungal infection, not a bacterial infection as first suspected. The physician immediately started the patient on amphotericin B, a potent antifungal medication that would properly treat the patient's case of coccidioidomycosis.

- How might the patient have contracted this infection?

- Why did the initial antibiotic therapy fail to improve the patient's symptoms?

Case File Wrap-Up appears at the end of the chapter.

Table 4.1 Eukaryotic Organisms Studied in Microbiology

Always Unicellular	May Be Unicellular or Multicellular	Always Multicellular
Protozoa	Fungi Algae	Helminths (have unicellular egg or larval forms)

4.1 Overview of the Eukaryotes

Evidence from **paleontology** indicates that the first eukaryotic cells appeared on the earth approximately 2 to 3 billion years ago. While it used to be thought that eukaryotic cells evolved directly from ancient prokaryotic cells, we now believe that bacteria, archaea, and eukaryotes evolved from a different kind of cell, a precursor to both prokaryotes and eukaryotes that biologists call the *Last Common Ancestor*. This ancestor was neither prokaryotic nor eukaryotic but gave rise to all three current cell types.

The first primitive eukaryotes were probably single-celled and independent, but, over time, some cells began to aggregate, forming colonies. With further evolution, some of the cells within colonies became *specialized*, or adapted to perform a particular function advantageous to the whole colony, such as movement, feeding, or reproduction. Complex multicellular organisms evolved as individual cells in the organism lost the ability to survive apart from the intact colony.

Only certain eukaryotes are traditionally studied by microbiologists–primarily the protozoa, the microscopic algae and fungi, and helminths **(table 4.1)**. Because the vast majority of algae do not cause infections of humans, we will discuss only the other three eukaryotic microbes in this chapter.

4.1 LEARNING OUTCOMES—Assess Your Progress

1. Relate bacterial, archaeal, and eukaryotic cells to the *Last Common Ancestor*.
2. List the types of eukaryotic microorganisms, and identify which are unicellular and which are multicellular.

4.2 Structures of the Eukaryotic Cell

Nearly all eukaryotic microbial cells have a cell membrane, nucleus, mitochondria, endoplasmic reticulum, Golgi apparatus, vacuoles, cytoskeleton, cytoplasm, and glycocalyx. Only some types of eukaryotes have a cell wall, appendages for moving, and chloroplasts **(figure 4.1)**. In the following sections, we cover the microscopic structure and functions of the eukaryotic cell. As with the bacteria, we begin on the outside and proceed inward through the cell.

©EM Research Services, Newcastle University

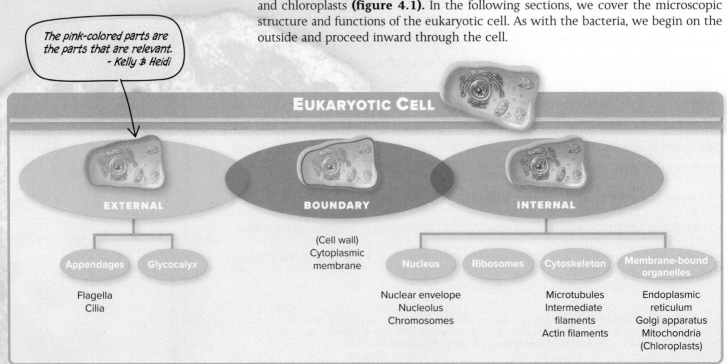

The pink-colored parts are the parts that are relevant.
- Kelly & Heidi

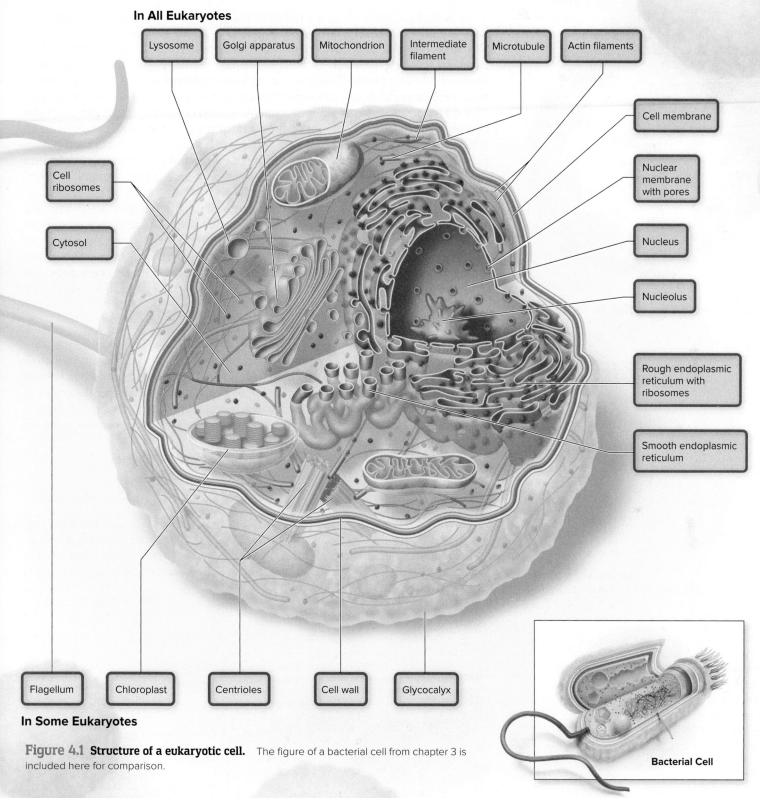

Figure 4.1 Structure of a eukaryotic cell. The figure of a bacterial cell from chapter 3 is included here for comparison.

External Structures

Appendages for Moving: Cilia and Flagella

Motility allows microorganisms to locate nutrients and to migrate toward positive stimuli such as sunlight; it also enables them to avoid harmful substances and stimuli. Most eukaryotic microbes move by using flagella or cilia. This type of motility is common in protozoa, many algae, and a few fungal and animal cells.

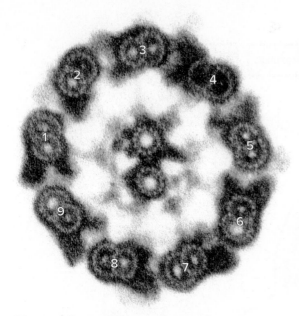

Figure 4.2 Microtubules in flagella.   A cross section that reveals the typical 9 + 2 arrangement found in both flagella and cilia.

Aaron J. Bell/Science Source

Eukaryotic flagella are much different from those of bacteria, even though they have the same name. The eukaryotic flagellum is about 10 times thicker, structurally more complex, and covered by an extension of the cell membrane. A single flagellum is a long, sheathed cylinder containing regularly spaced hollow tubules–microtubules–that extend along its entire length **(figure 4.2).** A cross section reveals nine pairs of closely attached microtubules surrounding a single central pair. This scheme, called the 9 + 2 arrangement, is found in eukaryotic flagella and also cilia (figure 4.2). During locomotion, the adjacent microtubules slide past each other, whipping the flagellum back and forth. Although details of this process are too complex to discuss here, it involves expenditure of energy and a coordinating mechanism in the cell membrane. The placement and number of flagella can be useful in identifying flagellated protozoa and certain algae.

Cilia are very similar in overall architecture to flagella, but they are shorter and more numerous (some cells have several thousand). They are found only on a single group of protozoa and in certain animal cells. In the ciliated protozoa, the cilia occur in rows over the cell surface, where they beat back and forth in regular oar-like strokes. These protozoa are among the fastest of all motile cells. On some cells, cilia also function as feeding and filtering structures.

The Glycocalyx

Many eukaryotic cells have a glycocalyx, an outermost layer that comes into direct contact with the environment (see figure 4.1). This structure, which is sometimes called an extracellular matrix, is usually composed of polysaccharides. It appears as a network of fibers, a slime layer, or a capsule much like the glycocalyx of bacteria and archaea. Because of its position on the outside of the cell, the glycocalyx contributes to protection, adherence of cells to surfaces, and reception of signals from other cells and from the environment.

Boundary Structures

The Cell Wall

Protozoa and helminths do not have cell walls, but fungi do. The cell walls of fungi are rigid and provide structural support and shape, but they are different in chemical composition from bacterial and archaeal cell walls. They have an inner layer of polysaccharide fibers composed of chitin or cellulose, and an outer layer of mixed glycans **(figure 4.3).**

Figure 4.3 Cross-sectional views of fungal cell walls.   **(a)** An electron micrograph of fungal cells. **(b)** A drawing of the section of the wall inside the square in **(a)**.

Thomas Deerinck, NCMIR/Science Source

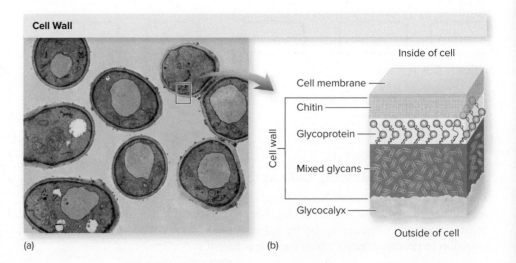

The Cell Membrane

The cell (or cytoplasmic) membrane of eukaryotic cells is a typical bilayer of phospholipids in which protein molecules are embedded. In addition to phospholipids, eukaryotic membranes also contain *sterols* of various kinds. Sterols are a form of lipid, but they are different from phospholipids in both structure and behavior. They are fairly rigid molecules, and this makes eukaryotic membranes more stable than those of non-eukaryotic cells. This strengthening feature is extremely important in those cells that don't have a cell wall. Cytoplasmic membranes of eukaryotes have a similar function as those in bacteria and archaea, serving as selectively permeable barriers.

Internal Structures

Unlike bacteria and archaea, eukaryotic cells contain a number of individual membrane-bound organelles that are extensive enough to account for 60% to 80% of their volume. They are contained in a fluid called the cytosol. The entire contents of the cell (outside of the nucleus) is called the cytoplasm. The terms cytosol and cytoplasm are sometimes used interchangeably.

The Nucleus

The nucleus is a compact sphere that is the most prominent organelle of eukaryotic cells. It is separated from the cell cytoplasm by an external boundary called a nuclear envelope. The envelope has a unique architecture. It is composed of two parallel membranes–that means two lipid bilayers–separated by a narrow space. It is perforated with small, regularly spaced openings, or pores. The pores are formed by the connection of the inner and outer nuclear membranes **(figure 4.4).** The nuclear pores are passageways. Macromolecules use these passages to move from the nucleus to the cytoplasm and vice versa. The nucleus is filled with a matrix called the nucleoplasm and a granular mass, the **nucleolus,** that stains more intensely than the immediate surroundings because of its RNA content. The nucleolus is the site for ribosomal RNA synthesis and a collection area for ribosomal subunits. The subunits are transported through the nuclear pores into the cytoplasm for final assembly into ribosomes.

A prominent feature of the nucleoplasm in stained preparations is a network of dark fibers known as **chromatin.** Chromatin is made of linear DNA, which, of course,

The cell in this corner shows the parts described in the figure colored in purple.
- Kelly & Heidi

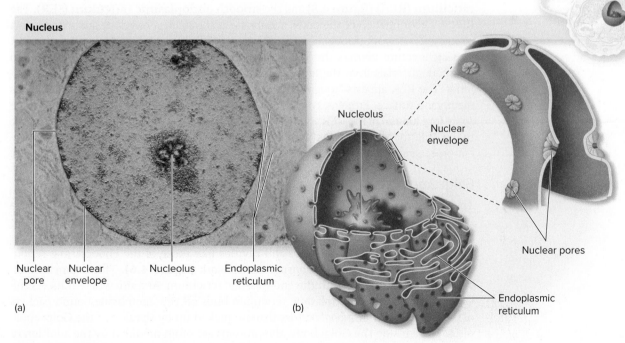

Nucleus

Nucleolus

Nuclear envelope

Nuclear pores

Endoplasmic reticulum

Nuclear pore   Nuclear envelope   Nucleolus   Endoplasmic reticulum

(a)   (b)

Figure 4.4 The nucleus.
(a) Electron micrograph section of a nucleus, showing its most prominent features. **(b)** Cutaway three-dimensional view of the relationships of the nuclear envelope and pores.
D Spector/Photolibrary/Getty Images

Endoplasmic Reticulum

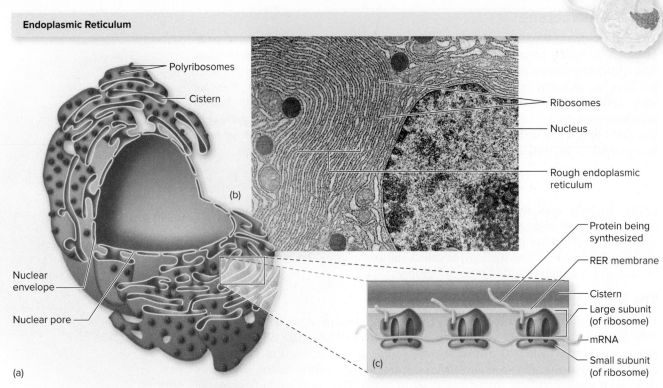

Figure 4.5 **The origin and detailed structure of the rough endoplasmic reticulum (RER).** **(a)** Schematic view of the origin of the RER from the outer membrane of the nuclear envelope. **(b)** TEM of the RER. **(c)** Detail of the orientation of a ribosome on the RER membrane. ©Don W. Fawcett/Science Source

is the genetic material of the cell. When it is wound around histone proteins, chromatin forms structures called chromosomes.

Endoplasmic Reticulum

The **endoplasmic reticulum (ER)** is a series of membrane tunnels used in transport and storage. There are two kinds of endoplasmic reticulum: the **rough endoplasmic reticulum (RER) (figure 4.5)** and the **smooth endoplasmic reticulum (SER).** The RER is a continuation of the outer membrane of the nuclear envelope and extends in a continuous network through the cytoplasm, even all the way out to the cell membrane. This architecture permits the spaces in the RER, called cisternae (singular, *cistern*), to transport materials from the nucleus to the cytoplasm and ultimately to the cell's exterior. The RER appears "rough" because of large numbers of ribosomes attached to its membrane surface. Proteins synthesized on the ribosomes are shunted into the inside space (the lumen) of the RER and held there for later packaging and transport. In contrast to the RER, the SER is a closed tubular network without ribosomes. It functions in nutrient processing and in synthesis and storage of nonprotein macromolecules such as lipids.

Golgi Apparatus

The **Golgi apparatus,** also called the Golgi complex or Golgi body, is the site in the cell in which proteins are modified and then sent to their final destinations. It is an organelle consisting of a stack of several flattened, disc-shaped sacs called cisternae. These sacs have both membranes and cavities like those of the endoplasmic reticulum, but they do not form a continuous network **(figure 4.6).** This organelle is always closely associated with the endoplasmic reticulum. At a site where it meets the Golgi apparatus, the endoplasmic reticulum buds off tiny membrane-bound packets of protein called *transitional vesicles* that are picked up by the face of the Golgi apparatus. Once inside the Golgi body, the proteins are often modified by the addition of polysaccharides and lipids. The final action of this apparatus is to pinch off finished

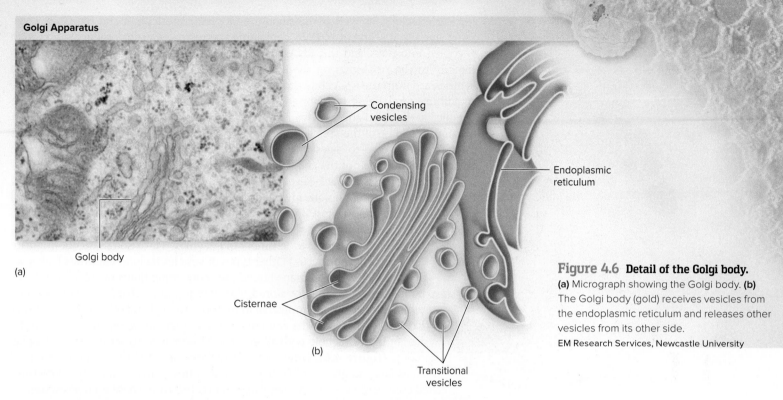

Golgi Apparatus

(a)

Golgi body

Condensing vesicles

Cisternae

(b)

Transitional vesicles

Endoplasmic reticulum

Figure 4.6 Detail of the Golgi body.
(a) Micrograph showing the Golgi body. **(b)** The Golgi body (gold) receives vesicles from the endoplasmic reticulum and releases other vesicles from its other side.
EM Research Services, Newcastle University

condensing vesicles that will carry the modified proteins to organelles such as lysosomes or outside the cell as secretory vesicles.

Nucleus, Endoplasmic Reticulum, and Golgi Apparatus: Nature's Assembly Line

As the keeper of the eukaryotic genetic code, the nucleus ultimately governs and regulates all cell activities. But, because the nucleus is located in a specific cellular site, it has to direct these activities using a structural and chemical network. This network includes ribosomes, which originate in the nucleus, and the rough endoplasmic reticulum **(figure 4.7).** Remember that the RER is continuously connected with the

Figure 4.7 The transport process. The cooperation of organelles in protein synthesis and transport: Nucleus → RER → Golgi apparatus → vesicles → secretion.

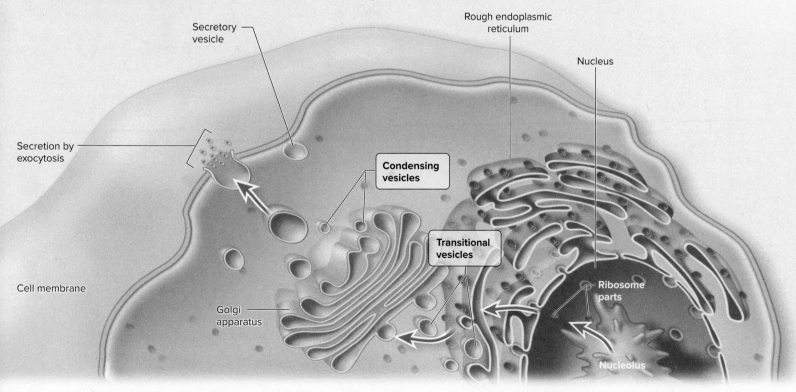

Secretory vesicle

Rough endoplasmic reticulum

Nucleus

Secretion by exocytosis

Condensing vesicles

Transitional vesicles

Ribosome parts

Cell membrane

Golgi apparatus

Nucleolus

nuclear envelope, as well as the smooth endoplasmic reticulum and the Golgi apparatus. Initially, a segment of DNA containing the instructions for producing a protein is copied into RNA, and this RNA transcript is passed out through the nuclear pores directly to the ribosomes on the endoplasmic reticulum. Here, specific proteins are synthesized from the RNA code and deposited in the lumen (space) of the endoplasmic reticulum. After being transported to the Golgi apparatus, the protein products are chemically modified and packaged into vesicles that can be used by the cell in a variety of ways. Some of the vesicles contain enzymes to digest food inside the cell. Other vesicles are secreted to digest materials outside the cell, and others are important in the growth and repair of the cell wall and membrane.

A **lysosome** is a vesicle that buds off of the Golgi apparatus. It contains a variety of enzymes. Lysosomes are involved in digestion of food particles inside the cell and in protection against invading microorganisms. They also participate in digestion and removal of cell debris in damaged tissue. Another type of vesicle, the peroxisome, contains a wide variety of enzymes. (Peroxisomes do not originate from the Golgi apparatus.) Other types of vesicles include **vacuoles** (vak'-yoo-ohlz), which are membrane-bound sacs containing fluids or solid particles to be digested, excreted, or stored. They are found in phagocytic cells (certain white blood cells and protozoa) in response to food and other substances that have been engulfed. The contents of a food vacuole are digested through the merger of the vacuole with a lysosome. This merged structure is called a phagosome **(figure 4.8).** Other types of vacuoles are used in storing reserve food such as fats and glycogen. Protozoa living in freshwater habitats use structures called contractile vacuoles to regulate osmotic pressure. These vacuoles systematically expel excess water that has diffused into the cell (described later).

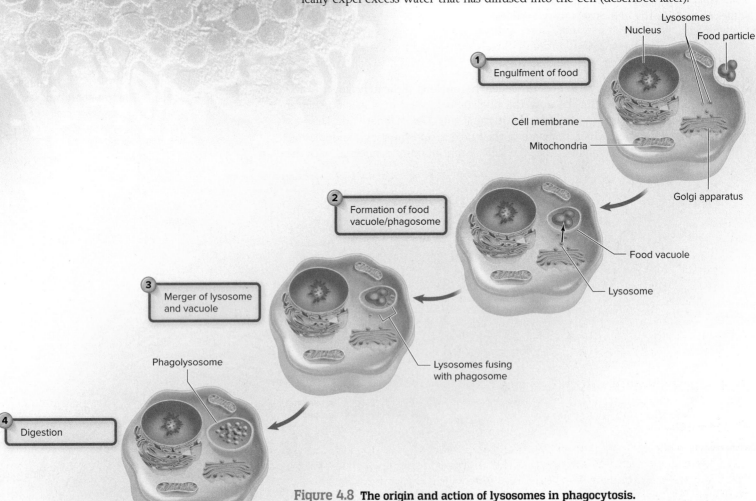

1 Engulfment of food

Lysosomes
Nucleus
Food particle
Cell membrane
Mitochondria
Golgi apparatus

2 Formation of food vacuole/phagosome

Food vacuole
Lysosome

3 Merger of lysosome and vacuole

Lysosomes fusing with phagosome

4 Digestion

Phagolysosome

Figure 4.8 The origin and action of lysosomes in phagocytosis.

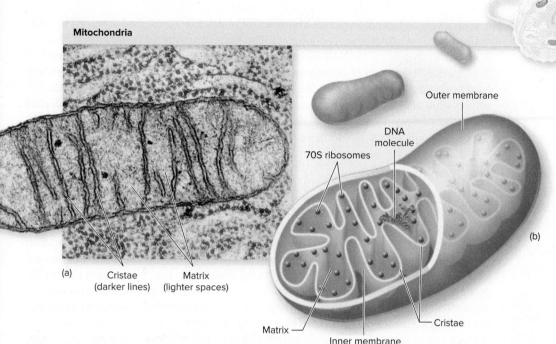

Mitochondria

(a) Cristae (darker lines) Matrix (lighter spaces)

Outer membrane
DNA molecule
70S ribosomes
(b)
Matrix
Inner membrane
Cristae

Figure 4.9 General structure of a mitochondrion.
CNRI/Science Photo Library/Getty Images

Mitochondria

Although the nucleus is the cell's control center, none of the cellular activities it commands could proceed without a constant supply of energy. The bulk of a cell's energy is generated in most eukaryotes by **mitochondria** (my″-toh-kon′-dree-uh). When viewed with light microscopy, mitochondria appear as round or elongated particles scattered throughout the cytoplasm. The internal ultrastructure reveals that a single mitochondrion consists of a smooth, continuous outer membrane that forms the external contour, and an inner, folded membrane nestled neatly within the outer membrane **(figure 4.9).** The folds on the inner membrane are called **cristae** (kris′-te).

The cristae membranes hold the enzymes and electron carriers needed in aerobic respiration. This is an oxygen-using process that extracts chemical energy contained in nutrient molecules and stores it in the form of high-energy molecules, or ATP. Mitochondria (along with chloroplasts) are unique among organelles in that they divide independently of the cell. They also contain circular molecules of DNA and have bacteria-sized 70S ribosomes. These characteristics have caused scientists to suggest that mitochondria were once bacterial cells that developed into eukaryotic organelles over time. While it was previously thought that all eukaryotic organisms must have mitochondria, scientists have discovered that some protozoa have pared-down versions of mitochondria (called mitosomes) and one species has even been found that contains neither mitochondria nor mitosomes. You have probably noticed that in biology, almost every "rule" gives way to one or two exceptions in the natural world.

Chloroplasts

Chloroplasts are remarkable organelles found in algae and plant cells that are capable of converting the energy of sunlight into chemical energy through photosynthesis. Another important product of the photosynthesis process in chloroplasts is oxygen gas. Although chloroplasts resemble mitochondria, chloroplasts are larger, contain special pigments, and are much more varied in shape.

Ribosomes

In an electron micrograph of a eukaryotic cell, ribosomes are numerous, tiny particles that give a "dotted" appearance to the cytoplasm. Ribosomes are distributed throughout the cell: Some are scattered freely in the cytoplasm and cytoskeleton; others are attached to the rough endoplasmic reticulum as previously described. Still others appear inside the

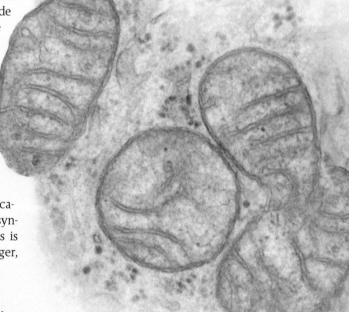

Mitochondria.
©EM Research Services, Newcastle University

mitochondria and in chloroplasts. Multiple ribosomes are often found arranged in short chains called polyribosomes (polysomes). The basic structure of eukaryotic ribosomes is similar to that of bacterial ribosomes, described in chapter 3. Both are composed of large and small subunits of ribonucleoprotein (see figure 4.5). By contrast, however, the eukaryotic ribosome (except in the mitochondrion) is the larger 80S variety that is a combination of 60S and 40S subunits. This difference means that we can use antibiotics that target prokaryotic ribosomes without harming our own eukaryotic ribosomes. As in the bacteria, eukaryotic ribosomes are the staging areas for protein synthesis.

The Cytoskeleton

The cytoplasm of a eukaryotic cell is criss-crossed by a flexible framework called the cytoskeleton. This framework appears to have several functions, such as anchoring organelles, moving RNA and vesicles, and permitting shape changes and movement in some cells **(figure 4.10).** The three main types of cytoskeletal elements are *actin filaments, intermediate filaments,* and *microtubules.* **Actin** filaments are long, thin protein strands about 7 nm in diameter. They are found throughout the cell but are most highly concentrated just inside the cell membrane. Actin filaments are responsible for cellular movements such as contraction, crawling, pinching during cell division, and formation of cellular extensions. **Microtubules** are long, hollow tubes that maintain the shape of those eukaryotic cells that don't have walls, and they transport substances from one part of a cell to another. The spindle fibers that play an essential role in mitosis are actually microtubules that attach to chromosomes and separate them into daughter cells. As indicated earlier, microtubules are also responsible for the movement of cilia

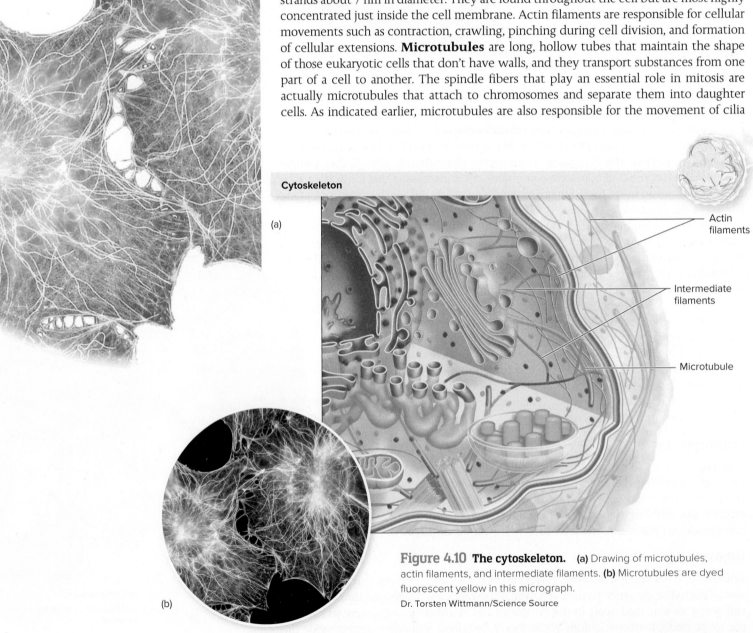

Cytoskeleton

(a)

Actin filaments

Intermediate filaments

Microtubule

(b)

Figure 4.10 The cytoskeleton.   (a) Drawing of microtubules, actin filaments, and intermediate filaments. **(b)** Microtubules are dyed fluorescent yellow in this micrograph.
Dr. Torsten Wittmann/Science Source

and flagella. **Intermediate filaments** are ropelike structures that are about 10 nm in diameter. (Their name comes from their intermediate size, between actin filaments and microtubules.) Their main role is in structural reinforcement to the cell and to organelles. For example, they support the structure of the nuclear envelope.

Table 4.2 summarizes the differences between eukaryotic and bacterial and archaeal cells. Viruses (discussed in chapter 5) are included as well.

Becoming Eukaryotic

Now is a good time to introduce the concept of **endosymbiosis.** As you recall, many scientists believe that both bacterial and eukaryotic cells emerged from an earlier, now-extinct, cell type, called the Last Common Ancestor (LCA). You remember that eukaryotes are thought to have evolved later than bacteria and archaea. It is believed that a bacterial cell parasitized another descendant cell of the LCA **(figure 4.11)** and eventually became a permanent part of that cell as its

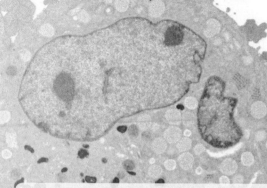

This human epithelial cell has turned cancerous. It has an irregular surface and an enlarged nucleus.
©Science Photo Library/Getty Images

Table 4.2 Components and Characteristics of Each Organism Type*

Function or Structure	Characteristic	Bacterial/Archaeal Cells	Eukaryotic Cells	Viruses**
Genetics	Nucleic acids	+	+	+
	Chromosomes	+	+	−
	True nucleus	−	+	−
	Nuclear envelope	−	+	−
Reproduction	Mitosis	−	+	−
	Production of sex cells	+/−	+	−
	Binary fission	+	+	−
Biosynthesis	Independent	+	+	−
	Golgi apparatus	−	+	−
	Endoplasmic reticulum	−	+	−
	Ribosomes	+***	+	−
Respiration	Mitochondria	−	+	−
Photosynthesis	Pigments	+/−	+/−	−
	Chloroplasts	−	+/−	−
Motility/locomotor structures	Flagella	+/−***	+/−	−
	Cilia	−	+/−	−
Shape/protection	Membrane	+	+	+/− (called "envelope" when present)
	Cell wall	+***	+/−	− (have capsids instead)
	Glycocalyx	+/−	+/−	−
Complexity of function		+	+	+/−
Size (in general)		0.5–3 μm****	2–300 μm	0.2 μm

*+ Means most members of the group exhibit this characteristic; − means most lack it; +/− means some members have it and some do not.

**Viruses cannot participate in metabolic or genetic activity outside their host cells.

***The bacterial/archaeal type is functionally similar to the eukaryotic but structurally unique.

****Much smaller and much larger bacteria exist.

Figure 4.11 Endosymbiosis. Many scientists believe that bacteria and archaea—and a pre-eukaryotic cell type—developed from the Last Common Ancestor, possibly under the influence of DNA viruses. Eukaryotes then evolved from an interaction (endosymbiosis) between an archaea-like cell (also originating from the LCA) and bacteria.

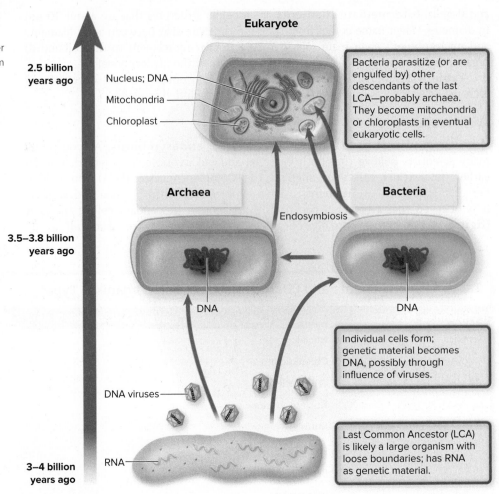

mitochondrion. Similarly, photosynthetic bacteria are thought to have become part of the precursor cell, eventually becoming the chloroplast in eukaryotic plant cells. That could explain why mitochondria and chloroplasts have their own (circular) DNA, 70S ribosomes, and own two-layer membranes. Figure 4.11 depicts a widely held view of how this happened.

4.2 LEARNING OUTCOMES—Assess Your Progress

3. Differentiate among the flagellar structures of bacteria, eukaryotes, and archaea.
4. List similarities and differences between eukaryotic and bacterial cytoplasmic membranes.
5. Describe the main structural components of a nucleus.
6. Diagram how the nucleus, endoplasmic reticulum, and Golgi apparatus act together with vesicles during the transport process.
7. Explain the function of the mitochondrion.
8. Explain the importance of ribosomes, and differentiate between eukaryotic and bacterial types.
9. List and describe the three main fibers of the cytoskeleton.
10. Explain how endosymbiosis contributed to the development of eukaryotic cells.

4.3 The Fungi

The kingdom Fungi is large and filled with a great variety and complexity of forms. In medical microbiology we are most concerned with the fungi known as yeasts and molds. Other fungi include mushrooms and puffballs. Although the majority of fungi are either unicellular or colonial (i.e., they form colonies), a few complex forms such as mushrooms and puffballs are truly multicellular. Cells of the microscopic fungi exist in two basic forms: yeasts and hyphae. A yeast cell is distinguished by its round to oval shape and by its mode of asexual reproduction. It grows swellings on its surface called buds, which then become separate cells. **Hyphae** (hy′-fee) are long, threadlike cells found in the bodies of fungi of the filamentous type. These are called molds **(figure 4.12).** Some species form a **pseudohypha,** a chain of yeast cells formed when buds remain attached in a row **(figure 4.13).** Because of its manner of formation, it is not a true hypha like that of molds. While some fungal cells exist only in a yeast form and others occur primarily as hyphae, a few are classified as **dimorphic.** This means they can take either form, depending on growth conditions, such as changing temperature. Several fungi that cause human disease are dimorphic.

Many fungi make their home on the human body, as part of the normal human microbiome. Yet nearly 300 species of fungi can also cause human disease. The Centers for Disease Control and Prevention currently identifies three types of fungal disease in humans: (1) community-acquired infections in the general population caused by environmental pathogens, (2) hospital-associated infections caused by fungal pathogens in clinical settings, and (3) opportunistic infections caused by low-virulence species infecting already-weakened individuals **(table 4.3).**

Mycoses (the term for fungal infections) vary in the way the pathogen enters the body and the degree of tissue involvement they display. Even so-called

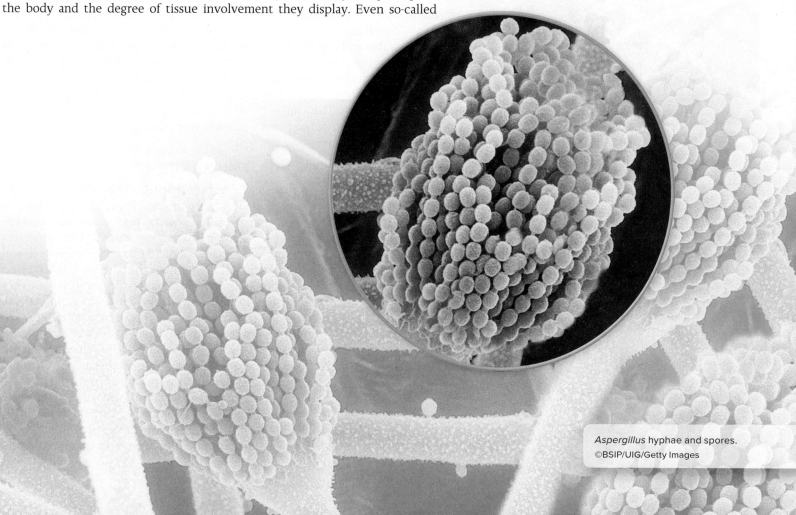

Aspergillus hyphae and spores.
©BSIP/UIG/Getty Images

Figure 4.12 *Diplodia maydis,* **a pathogenic fungus of corn plants.** **(a)** Scanning electron micrograph of a single colony showing its filamentous texture (24×). **(b)** Close-up of hyphal structure (1,200×). **(c)** Basic structural types of hyphae.

Courtesy Dr. Judy A. Murphy (a-b)

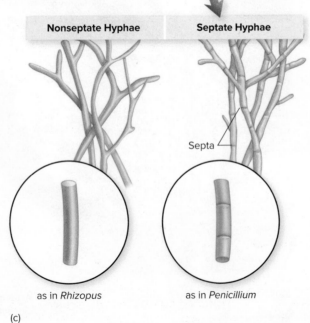

Septum

Nonseptate Hyphae

Septate Hyphae

Septa

as in *Rhizopus*

as in *Penicillium*

(a)

(b)

(c)

harmless species found in the air and dust around us may be able to cause infections, especially in individuals who already have cancer, diabetes, or AIDS.

This last category, opportunistic infections, has become very troubling in recent years. In 2018 an important review was published that revealed that deaths from fungal infections have surpassed deaths due to either malaria or breast cancer in the world. Transplant patients, cancer patients, and HIV-positive patients are particularly susceptible to opportunistic fungi. The progress of these infections in immunosuppressed people was vividly described in the review. "They'll just rot you down quick as a flash," said one of the authors.

Fungi can cause other dangerous medical conditions without establishing an actual infection. Fungal cell walls give off chemical substances that can trigger allergies. The toxins produced by poisonous mushrooms can induce neurological disturbances and even death. The mold *Aspergillus flavus* synthesizes a potentially lethal poison called aflatoxin. The consumption of grain contaminated with this mold has led to increased cases of liver cancer in developing nations.

Fungi pose problems to the agricultural industry. A number of species can be damaging to field plants such as corn and grain. This reduces crop production and can also cause disease in domestic animals that eat the contaminated feed crops. Fungi can also rot fresh produce during shipping and storage. It has been estimated that as much as 40% of the yearly fruit crop is consumed not by humans but by fungi. On the beneficial side, however, fungi play an essential role in decomposing organic matter and returning essential minerals to the soil. They form stable associations with plant roots, forming structures called mycorrhizae, that increase the ability of the roots to absorb water and nutrients. Industry has tapped the biochemical potential of fungi to produce large quantities of antibiotics, alcohol, organic acids, and vitamins. Some fungi are eaten or used to provide flavorings to food. The yeast *Saccharomyces* produces the alcohol in beer and wine and the gas that causes bread to rise. Blue cheese, soy sauce, and cured meats derive their unique flavors from the actions of fungi.

Fungal Nutrition

All fungi are **heterotrophic.** They acquire nutrients from a wide variety of organic materials. (Sources of nutrition are called **substrates.**) Most fungi are **saprobes,** meaning that they obtain these substrates from the remnants of dead plants and animals in soil or aquatic habitats. Fungi can also be **parasites** on the bodies of living animals or plants, although very few fungi absolutely require a living host. In general, the fungus penetrates the substrate and secretes enzymes that reduce it to small molecules that can be absorbed by the cells. Fungi have enzymes for digesting an incredible array of substances, including feathers, hair, cellulose, petroleum products, wood, and rubber. Fungi are often found in nutritionally poor or adverse environments. Various fungi thrive in substrates with high salt or sugar content, at relatively high temperatures, and even in snow and glaciers.

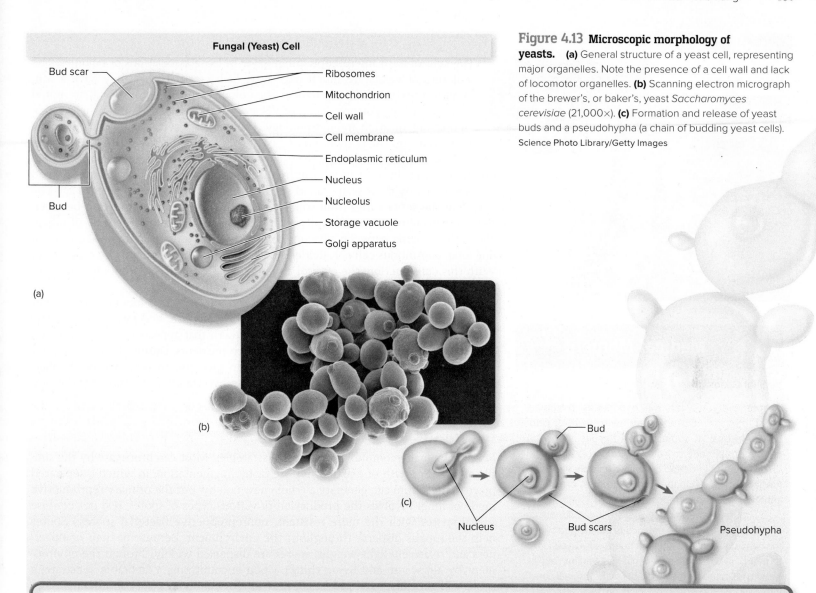

Fungal (Yeast) Cell

Bud scar
Ribosomes
Mitochondrion
Cell wall
Cell membrane
Endoplasmic reticulum
Nucleus
Nucleolus
Storage vacuole
Golgi apparatus

Bud

(a)

(b)

(c)

Bud

Nucleus   Bud scars   Pseudohypha

Figure 4.13 Microscopic morphology of yeasts. **(a)** General structure of a yeast cell, representing major organelles. Note the presence of a cell wall and lack of locomotor organelles. **(b)** Scanning electron micrograph of the brewer's, or baker's, yeast *Saccharomyces cerevisiae* (21,000×). **(c)** Formation and release of yeast buds and a pseudohypha (a chain of budding yeast cells). Science Photo Library/Getty Images

Table 4.3 Three Categories of Fungal Infections of Humans

Source of Fungus	Who Is Affected?	Examples	Site of Symptoms
Community-Acquired Infections			
Environmental species	General population	Ringworm, a fungal infection	Affects skin in various parts of the body, including scalp, groin, and feet.
		Coccidioidomycosis (also called Valley fever), caused by a fungus that lives in dust and soil, particularly in the SW United States	Flulike symptoms in respiratory tract and rest of body; includes headaches, night sweats, and muscle and joint pain; possible rash
Hospital-Associated Infections			
Pathogens in clinical settings	People in hospital or long-term care	Various fungi that contaminate health care facilities	Could be anywhere
Opportunistic Infections			
Environmental species or normal fungal biota	People who have compromised immune systems, or disrupted microbiota	Mucormycosis, caused by a fungus found in soil, leaves, compost, and rotting wood	Affects lungs and sinuses most often; skin can also be the target
		Candidiasis, caused by a yeast that is normal biota on human mucosal surfaces	The mucosal site where the microbe normally lives in small numbers; when the normal balance of microbiota is disrupted, the yeast proliferates and causes inflammation

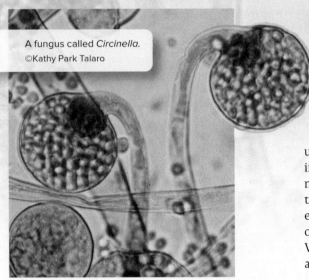

A fungus called *Circinella.*
©Kathy Park Talaro

Medical Moment

Vaginal Candidiasis

Almost every woman will experience a vaginal yeast infection, caused by an overgrowth of *Candida albicans,* at some time in her life. Although uncomfortable due to irritation, itching, and vaginal discharge, the infection is easily treatable with antifungal medication in the form of creams, oral medications, or vaginal suppositories.

The female reproductive system is quite amazing, and a very delicate balance is maintained within this environment. A small amount of *C. albicans* is nearly always present within the vagina, but its growth is limited by the acidic pH of the vaginal canal. Interestingly, the acid-producing bacteria also living within the vagina help to tightly control this pH level. When something disrupts the vaginal pH, *C. albicans* takes the opportunity to proliferate, leading to an overgrowth of this microbe, which can become a "yeast infection."

But what causes disruption of the normal vaginal pH? Although pregnancy, diabetes, obesity, and monthly hormonal changes can cause pH levels to fluctuate, by far the most common cause that leads to the development of a yeast infection is antibiotic therapy. This is due to the fact that the drug kills the protective bacteria (normal biota) of the vagina along with whichever pathogen is being targeted by the antibiotic. Because these are critical for keeping *C. albicans* in check, the yeast can then overpopulate the vagina.

Q. Yeast infection is one common side effect of antibiotic treatment. Diarrhea is another common side effect. Suggest an explanation for this.

Answer in Appendix B.

Morphology of Fungi

The cells of most microscopic fungi grow in loose associations or colonies. The colonies of yeasts are much like those of bacteria in that they have a soft, uniform texture and appearance. The colonies of filamentous fungi are noted for the striking cottony, hairy, or velvety textures that arise from their microscopic organization and morphology. The woven, intertwining mass of hyphae that makes up the body or colony of a mold is called a **mycelium.**

Although hyphae contain the usual eukaryotic organelles, they also have some unique features. In most fungi, the hyphae are septate, meaning they are divided into segments by cross walls called **septa** (singular, *septum;* see figure 4.12c). The nature of the septa varies from solid partitions with no communication between the compartments to partial walls with small pores that allow the flow of organelles and nutrients between adjacent compartments. Nonseptate hyphae consist of one long, continuous cell *not* divided into individual compartments by cross walls. With this construction, the cytosol and organelles move freely from one region to another, and each hyphal element can have several nuclei.

Hyphae can also be classified according to their particular function. Two types are vegetative hyphae and reproductive, or fertile, hyphae. Vegetative hyphae (mycelia) are responsible for the visible mass of growth that appears on the surface of a substrate and penetrates it to digest and absorb nutrients. During the development of a fungal colony, the vegetative hyphae can give rise to reproductive hyphae, which branch off a vegetative mycelium. These hyphae are responsible for the production of fungal reproductive bodies called **spores.**

Reproductive Strategies and Spore Formation

Fungi have many complex reproductive strategies. Most can propagate by the simple outward growth of existing hyphae or by fragmentation, in which a separated piece of mycelium can generate a whole new colony. But the primary reproductive mode of fungi involves the production of various types of spores. (Do not confuse fungal spores with the more resistant, nonreproductive bacterial spores.) Spores help the fungus disperse throughout the environment. Because of their compactness and relatively light weight, spores are dispersed widely through the environment by air, water, and living things. Upon encountering a favorable substrate, a spore will germinate and produce a new fungus colony in a very short time.

Fungal spores are explicitly responsible for multiplication. Different fungi have such a wide variety of different spores that they are largely classified and identified by their spores and spore-forming structures, but we won't cover this information. Instead, we will focus on the most general subdivision, which is based on the way the spores arise. Asexual spores are the products of mitotic division of a single parent cell, and sexual spores are formed by the fusing of two parental nuclei followed by meiosis. An important consequence of meiosis and sexual reproduction is that it serves to increase the genetic variation among spores.

Asexual Spore Formation

There are two types of asexual spores, **sporangiospores** and **conidiospores,** which are also called conidia **(figure 4.14):**

1. Sporangiospores **(figure 4.14a)** are formed by successive cleavages within a saclike head called a sporangium, which is attached to a stalk, the sporangiophore.
2. Conidiospores, or **conidia,** are free spores not enclosed by a spore-bearing sac. They develop either by the pinching off of the tip of a special fertile hypha or by the segmentation of a preexisting vegetative hypha. There are many different forms of conidia, illustrated in **figure 4.14b.**

Figure 4.14 Types of asexual mold spores.
(a) Sporangiospores: (1) *Absidia,* (2) *Syncephalastrum.*
(b) Conidial variations: (1) arthrospores (e.g., *Coccidioides*),
(2) chlamydospores and blastospores (e.g., *Candida
albicans*), (3) phialospores (e.g., *Aspergillus*),
(4) macroconidia and microconidia (e.g., *Microsporum*), and
(5) porospores (e.g., *Alternaria*).

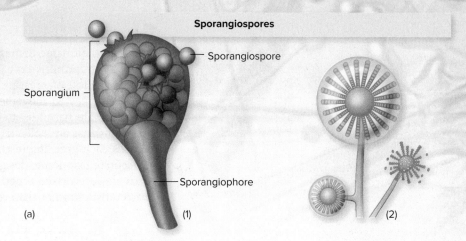

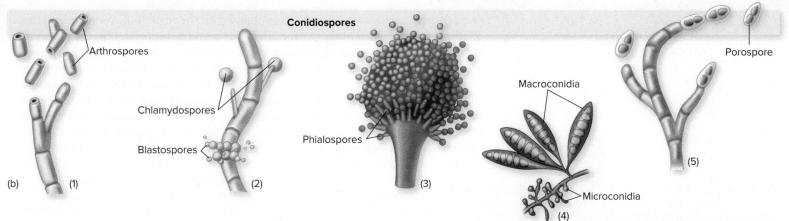

Sexual Spore Formation

Fungi are hugely successful at reproduction with their millions of asexual spores. That being the case, why is the production of sexual spores necessary? The answer lies in important variations that occur when fungi of different genetic makeup combine their genetic material. Just as in plants and animals, this mixing of DNA from two parents creates offspring with combinations of genes different from that of either parent. The offspring from such a union can have slight variations in form and function that are potentially advantageous in the adaptation and survival of their species.

The majority of fungi produce sexual spores at some point. The details of this process vary greatly. It could be as simple as the fusion of fertile hyphae of two different strains, or as complicated as a complex union of differentiated male and female structures or the development of special fruiting structures. It may be a surprise to discover that the fleshy part of a mushroom is actually a fruiting body designed to protect and help disseminate its sexual spores.

4.3 LEARNING OUTCOMES—Assess Your Progress

11. List two detrimental and two beneficial activities of fungi (from the viewpoint of humans).

12. List three general features of fungal anatomy.

13. Differentiate among the terms *heterotroph, saprobe,* and *parasite.*

14. Explain the relationship between fungal hyphae and the production of a mycelium.

15. Describe two ways in which fungal spores arise.

Medical Moment

Opportunistic Fungal Infection

Ordinarily harmless organisms that cause disease in a weakened host are described as opportunistic pathogens. One example is mucormycosis, a rare but serious fungal infection. *Mucormycetes* are fungi commonly found in soil or decaying matter. The fungal conidia may be inhaled from the air or introduced by skin trauma. Most humans are not affected by these spores, but those with impaired immune systems or poorly controlled diabetes are a perfect host for the fungus. In inhaled mucormycosis, hyphae invade the tissues and blood vessels of the sinuses, lungs, eyes, and face. The infection is treated with intravenous antifungal medications. The fungus spreads rapidly, often requiring extensive, repeated surgeries to restore affected facial bones and tissues. If not aggressively treated, the fungal infection may spread into the brain and is frequently fatal.

Q. It is difficult to treat fungal infections without doing some harm to the host. Drugs that treat bacterial infections are less likely to harm the human host. Can you think of why that might be?

Answer in Appendix B.

4.4 The Protozoa

Although their name comes from the Greek for "first animals," protozoa are far from being simple, primitive organisms. The protozoa constitute a very large group (about 12,000 species) of creatures that, although single-celled, have startling properties when it comes to movement, feeding, and behavior. Although most members of this group are harmless, free-living inhabitants of water and soil, a few species are pathogens that are collectively responsible for hundreds of millions of infections of humans each year. Interestingly, the term *protozoan* is more of a convenience than an accurate taxonomic designation. As we next describe them, you will see why protozoa are categorized together. As it turns out, it is because of their similar physical characteristics rather than their genetic relatedness.

Protozoan Form and Function

Most protozoan cells are single cells containing all the major eukaryotic organelles. Their organelles can be highly specialized for feeding, reproduction, and locomotion. The cytoplasm is usually divided into a clear outer layer called the **ectoplasm** and a more granular inner region called the endoplasm. Ectoplasm is involved in locomotion, feeding, and protection. Endoplasm houses the nucleus, mitochondria, and food and contractile vacuoles. Some protozoa even have organelles that work somewhat like a primitive nervous system to coordinate movement. Protozoa can move through fluids by means of *pseudopods* ("false feet"), flagella, or cilia. Because protozoa lack a cell wall, they have a certain amount of flexibility. Their outer boundary is a cell membrane that regulates the movement of food, wastes, and secretions. Their cell shape can remain constant (as in most ciliates) or can change constantly (as in amoebas). Certain amoebas encase themselves in hard shells made of calcium carbonate. The size of most protozoan cells falls within the range of 3 to 300 µm. Some notable exceptions are giant amoebas and ciliates that are large enough (3 to 4 mm in length) to be seen swimming in pond water.

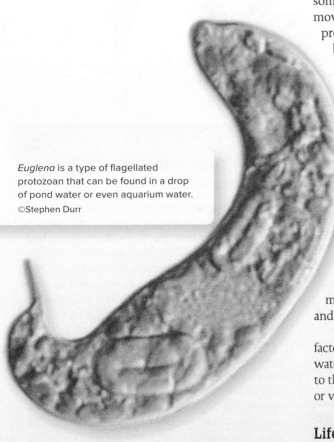

Euglena is a type of flagellated protozoan that can be found in a drop of pond water or even aquarium water.
©Stephen Durr

Nutritional and Habitat Range

The protozoa we will be interested in are typically heterotrophic and usually require their food in a complex organic form. Free-living species scavenge dead plant or animal debris and even graze on live bacteria and algae. Some species have special feeding structures, such as oral grooves, that carry food particles into a passageway or gullet that packages the captured food into vacuoles for digestion. Some protozoa absorb food directly through the cell membrane. Pathogenic species may live on the fluids of their host, such as plasma and digestive juices, or they can actively feed on tissues.

Although protozoa have adapted to a wide range of habitats, their main limiting factor is the availability of moisture. Their predominant habitats are fresh and marine water, soil, plants, and animals. Even extremes in temperature and pH are not a barrier to their existence. Hardy species are found in hot springs, ice, and habitats with very low or very high pH.

Life Cycles and Reproduction

Protozoa are called **trophozoites** when they are in their motile feeding stage. This is a stage that requires ample food and moisture to remain active. A large number of species are also capable of entering into a dormant, resting stage called a **cyst** when conditions in the environment become unfavorable for growth and feeding. During *encystment*, the trophozoite cell rounds up into a sphere, and its ectoplasm secretes a tough, thick cuticle around the cell membrane **(figure 4.15)**. Because cysts are more resistant than ordinary cells to heat, drying, and chemicals, they can survive adverse periods. They can also be dispersed by air currents and may even be an important factor in the spread of diseases such as amoebic dysentery. If provided with moisture and nutrients, a cyst breaks open and releases the active trophozoite.

The life cycles of protozoans vary from simple to complex. Several protozoan groups exist only in the trophozoite state. Many alternate between a trophozoite and a cyst stage, depending on the conditions of the habitat. The life cycle of a parasitic protozoan dictates its mode of transmission to other hosts. For example, the flagellate *Trichomonas vaginalis* causes a common sexually transmitted infection. Because it does not form cysts, it is more delicate and must be transmitted by intimate contact between sexual partners. In contrast, intestinal pathogens such as *Entamoeba histolytica* and *Giardia lamblia* form cysts and are readily transmitted in contaminated water and foods.

All protozoa reproduce by relatively simple, asexual methods, usually mitotic cell division. Several pathogenic species, including the causative agents of malaria and toxoplasmosis, reproduce asexually by multiple rounds of division inside a host cell. Sexual reproduction also occurs during the life cycle of most protozoa. Ciliates participate in conjugation, a form of genetic exchange in which two cells fuse temporarily and exchange micronuclei. This process of sexual recombination yields new and different genetic combinations that can be advantageous in evolution.

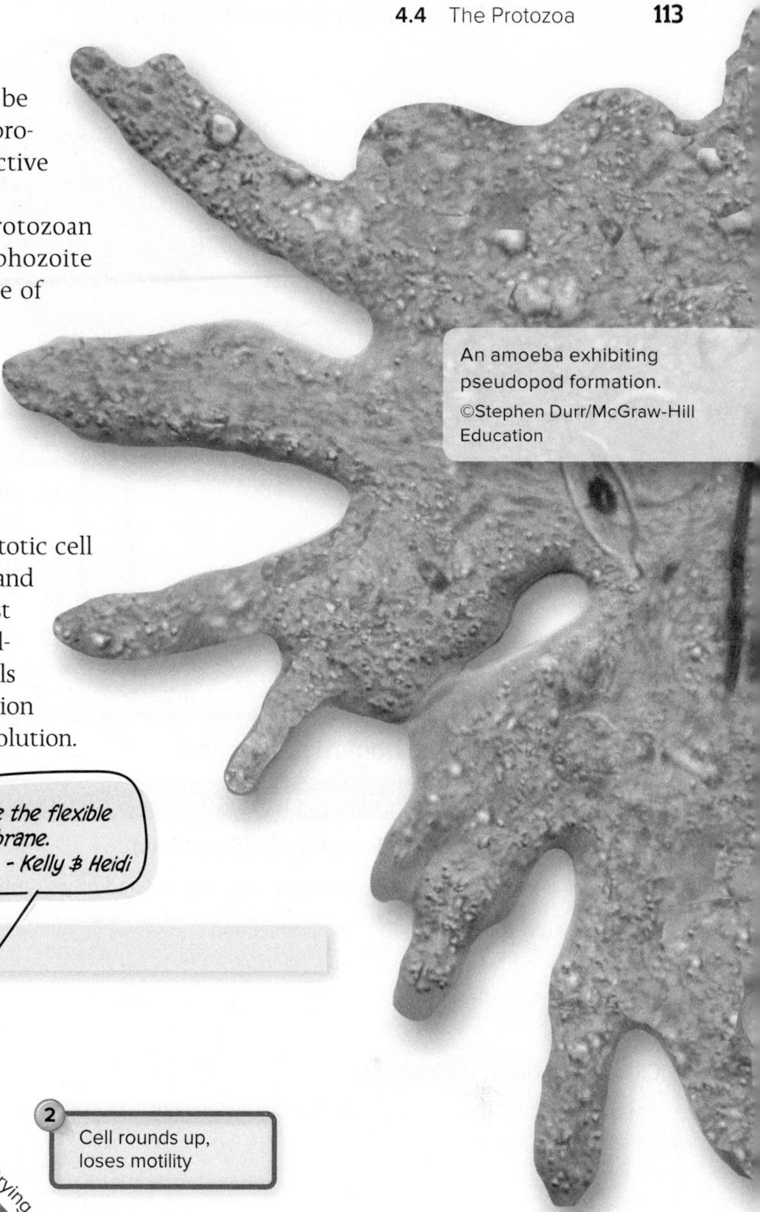

An amoeba exhibiting pseudopod formation.
©Stephen Durr/McGraw-Hill Education

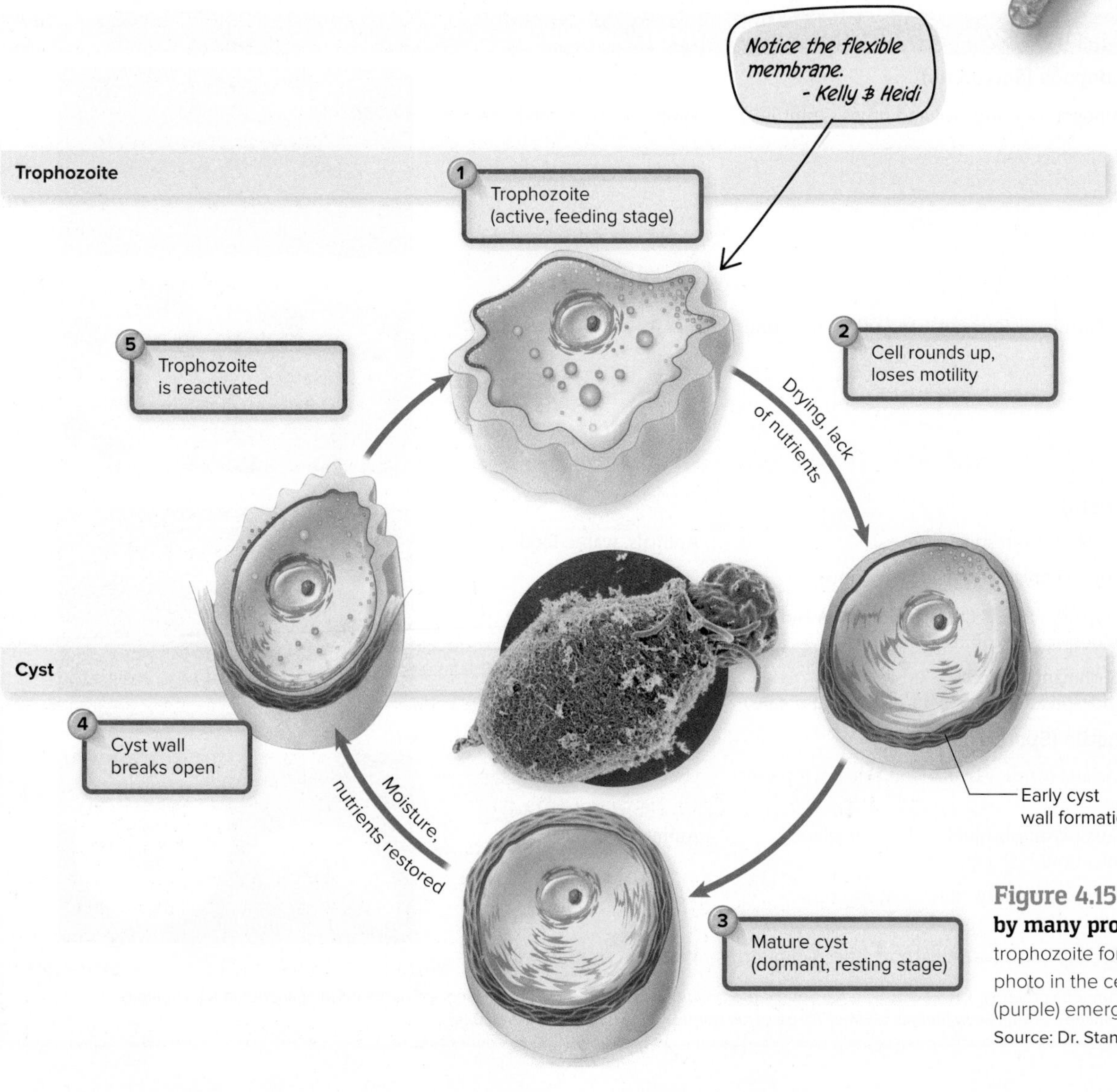

Notice the flexible membrane.
- Kelly & Heidi

Trophozoite

1 Trophozoite (active, feeding stage)

5 Trophozoite is reactivated

2 Cell rounds up, loses motility

Drying, lack of nutrients

Cyst

4 Cyst wall breaks open

Moisture, nutrients restored

Early cyst wall formation

3 Mature cyst (dormant, resting stage)

Figure 4.15 The general life cycle exhibited by many protozoa. All protozoa have a trophozoite form, but not all produce cysts. The photo in the center shows a *Giardia* trophozoite (purple) emerging from its cyst form (orange).
Source: Dr. Stan Erlandsen/CDC

Giardia lamblia.
©Dr. Tony Brain/Science Source

Classification of Selected Medically Important Protozoa

As has been stated, taxonomists have problems classifying protozoa. They are very diverse and frequently frustrate attempts to generalize or place them in neat groupings. We will use a common and simple system of four groups, based on their method of motility: Sarcodina (pseudopods), Ciliophora (cilia), Mastigophora (flagella), and Sporozoa (gliding motility) **(table 4.4).**

4.4 LEARNING OUTCOMES—Assess Your Progress

16. Describe the protozoan characteristics that illustrate why protozoa are informally placed into a single group.

17. List three means of locomotion exhibited by protozoa.

18. Explain why a cyst stage may be useful to a protozoan.

19. Give an example of a disease caused by each of the four types of protozoa.

Table 4.4 Major Pathogenic Protozoa

Protozoan	Disease	Reservoir/Source	
Amoeboid Protozoa Using Pseudopods (Sarcodina)			
Entamoeba histolytica	Amoebiasis (intestinal and other symptoms)	Humans, water, food	
Naegleria, Acanthamoeba	Brain infection	Water	
Ciliated Protozoa (Ciliophora)			
Balantidium coli	Balantidiosis (intestinal and other symptoms)	Pigs, cattle, primates	
Flagellated Protozoa (Mastigophora)			
Giardia lamblia	Giardiasis (intestinal distress)	Animals, water, food	
Trichomonas vaginalis	Trichomoniasis (vaginal symptoms)	Human	
Trypanosoma brucei, T. cruzi	Trypanosomiasis (intestinal distress and widespread organ damage)	Animals, vector-borne	
Leishmania donovani, L. tropica, L. brasiliensis	Leishmaniasis (either skin lesions or widespread involvement of internal organs)	Animals, vector-borne	
Apicomplexan Protozoa—Nonmotile (Sporozoa)			
Plasmodium vivax, P. falciparum, P. malariae	Malaria (cardiovascular and other symptoms)	Human, vector-borne	
Toxoplasma gondii	Toxoplasmosis (flulike illness or silent infection)	Animals, vector-borne	
Cryptosporidium	Cryptosporidiosis (intestinal and other symptoms)	Water, food	
Cyclospora cayetanensis	Cyclosporiasis (intestinal and other symptoms)	Water, fresh produce	

4.5 The Helminths

Tapeworms, flukes, and roundworms are collectively called *helminths*, from the Greek word meaning "worm." Adult specimens are usually large enough to be seen with the naked eye, and they range from the longest tapeworms, measuring up to about 25 m in length, to roundworms less than 1 mm in length. Helminths are animals. They are included in the study of microbes mainly due to their infective abilities and the fact that they produce microscopic eggs and larvae.

On the basis of body type, the two major groups of pathogenic helminths are the flatworms (phylum Platyhelminthes) and the roundworms (phylum Aschelminthes, also called **nematodes**). Flatworms have a very thin, often segmented body plan **(figure 4.16),** and roundworms have an elongated, cylindrical, unsegmented body **(figure 4.17).** The flatworm group is subdivided into the **cestodes,** or tapeworms, named for their long, ribbonlike arrangement, and the **trematodes,** or flukes, characterized by flat, ovoid bodies. Not all flatworms and roundworms are parasites by nature; many live free in soil and water. Because most disease-causing

Figure 4.16 Pathogenic flatworms. **(a)** A cestode (tapeworm), showing the scolex; long, tapelike body; and magnified views of immature and mature proglottids (body segments). The photo shows an actual tapeworm. **(b)** The structure of a trematode (liver fluke). Note the suckers that attach to host tissue and the dominance of reproductive and digestive organs.

Geoff Brightling/Dorling Kindersley/Getty Images (a); Eye of Science/Science Source (b)

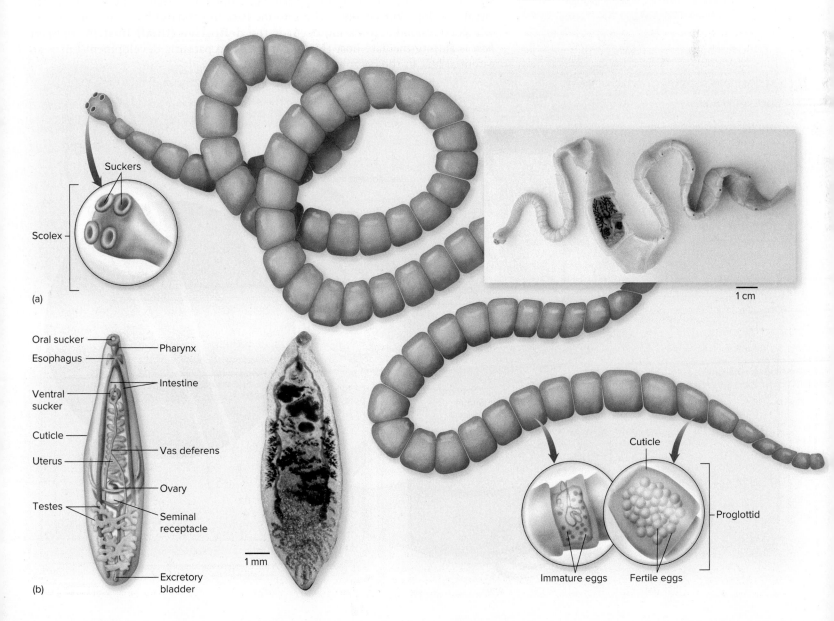

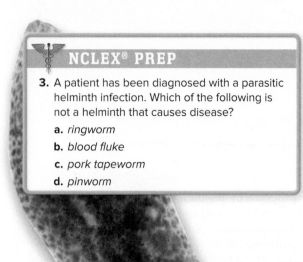

Planaria is a flatworm that is often studied in high school and college biology labs.
Source: NHPA/M. I. Walker/Photoshot

helminths spend part of their lives in the gastrointestinal tract, they are discussed in chapter 20.

General Worm Morphology

All helminths are multicellular animals equipped to some degree with organs and organ systems. In pathogenic helminths, the most developed organs are those of the reproductive tract, with some degree of reduction in the digestive, excretory, nervous, and muscular systems.

Life Cycles and Reproduction

The complete life cycle of helminths includes the fertilized egg (embryo), larval, and adult stages. In the majority of helminths, adults derive nutrients and reproduce sexually in a host's body. In nematodes, the sexes are separate and usually different in appearance. In trematodes, the sexes can be either separate or **hermaphroditic,** meaning that male and female sex organs are in the same individual worm. Cestodes are generally hermaphroditic. Helminths must complete their life cycle by transmitting an infective form, usually an egg or larva, to the body of another host, either of the same or a different species. The host in which larval development occurs is the intermediate (secondary) host, and the host in which adulthood and mating occur is the **definitive (final) host.** A *transport host* is an intermediate host that experiences no parasitic development but is an essential link in the completion of the cycle.

Figure 4.17 Pathogenic roundworm. **(a)** A male *Ascaris* nematode (roundworm). **(b)** Female (left) and male (right) *Ascaris* worms.
Source: CDC

Mouth
Pseudocoelom
Cuticle
Pharynx
Brain
Dorsal nerve cord
Lateral nerve cord
Gut
Sperm duct
Ventral nerve cord
Excretory pore
Testis
Seminal vesicle
Cloaca
Spicules
Anus
(a)
(b)

Table 4.5 Examples of Helminths and How They Are Transmitted

Common Name	Disease or Worm	Host Requirement	Spread to Humans By
Roundworms			
Nematodes			
Intestinal Nematodes			
Infective in egg (embryo) stage			
Ascaris lumbricoides	Ascariasis	Humans	Fecal pollution of soil with eggs
Enterobius vermicularis	Pinworm	Humans	Close contact
Infective in larval stage			
Trichinella spiralis	Trichina worm	Pigs, wild mammals	Consumption of meat containing larvae
			Burrowing of larva into tissue
Tissue Nematodes			
Onchocerca volvulus	River blindness	Humans, black flies	Fly bite
Dracunculus medinensis	Guinea worm	Humans and *Cyclops* (an aquatic invertebrate)	Ingestion of water containing *Cyclops*
Flatworms			
Trematodes			
Schistosoma japonicum	Blood fluke	Humans and snails	Skin penetration of larval stage
Cestodes			
Taenia solium	Pork tapeworm	Humans, swine	Consumption of undercooked or raw pork
Diphyllobothrium latum	Fish tapeworm	Humans, fish	Consumption of undercooked or raw fish

Fertilized eggs are usually released to the environment and are provided with a protective shell and extra food to aid their development into larvae. Even so, most eggs and larvae are vulnerable to heat, cold, drying, and predators and are destroyed or unable to reach a new host. To counteract this, certain worms have adapted a reproductive capacity that borders on the incredible: A single female *Ascaris* can lay 200,000 eggs a day, and a large female can contain over 25 million eggs at varying stages of development! If only a tiny number of these eggs makes it to another host, the parasite will have been successful in completing its life cycle.

In general, humans become infected by ingesting the worm, or by the worm penetrating tissues, such as the feet. The sources of the infective stage may be contaminated food, soil, or water, or other infected animals. Humans are the definitive hosts for many of the parasites listed in **table 4.5.** In about half the diseases, they are also the sole biological reservoir. In other cases, animals or insect vectors serve as reservoirs or are required to complete worm development.

A Helminth Cycle: The Pinworm

To illustrate a helminth cycle in humans, we use the example of a roundworm, *Enterobius vermicularis,* the pinworm or seatworm. This worm causes a very common infestation of the large intestine. Worms range from 2 to 12 mm long and have a tapered, curved cylindrical shape **(figure 4.18).** The condition they cause, enterobiasis, is usually a simple, uncomplicated infection that does not spread beyond the intestine.

A cycle starts when a person swallows microscopic eggs picked up from another infected person by direct contact or by touching articles that person has touched. The eggs hatch in the intestine and then release larvae that mature into adult worms within about 1 month. Male and female worms mate, and the female migrates out to the anus to deposit eggs, which cause intense itchiness that the infected human relieves by scratching. Scratching contaminates the fingers, which, in turn, transfers eggs to bedclothes and other inanimate objects. This

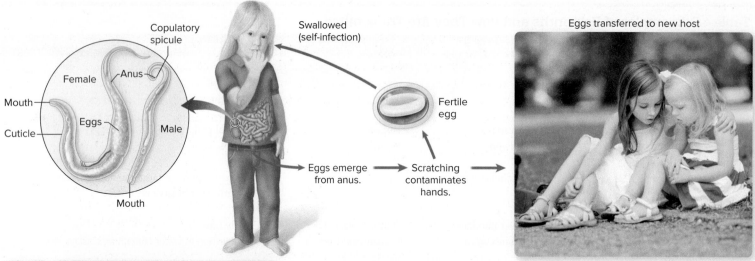

Figure 4.18 The life cycle of the pinworm, a roundworm. Eggs are the infective stage and are transmitted by contaminated hands. Children frequently reinfect themselves and also pass the parasite on to others.
©MNStudio/Shutterstock

Medical Moment

Neglected Parasitic Infections

It might come as a surprise to you that one-fourth of the world's population is infected with intestinal roundworms. Or maybe it is easier to stomach because we are confident that protozoan and helminthic infections are relatively rare in the United States.

Think again. The CDC has begun a campaign against what they are calling "neglected parasitic infections (NPIs)" in the United States. Three of them are protozoan infections, and two are caused by helminths. They are:

- Chagas disease–a protozoan disease caused by *Trypanosoma cruzi*
- Neurocysticercosis–caused by the tapeworm *Taenia sodium*
- Toxocariasis–caused by worms that travel through tissues and can cause blindness
- Toxoplasmosis–60 million people in the United States are infected by the protozoan causing this disease.
- Trichomoniasis–a protozoan infection of the genital tracts

Neurocysticercosis is the single most common infectious cause of seizures in some areas of the United States. Up to 300,000 people in the United States are infected with the protozoan that causes Chagas disease.

It is time to stop thinking of these infections as "other people's diseases."

Q. There is a protozoan disease that is not considered "neglected." It causes new infections in over 200 million people per year, and millions of dollars are poured into vaccine and treatment research every year. What protozoan disease is this?

Answer in Appendix B.

person becomes a host and a source of eggs, and can spread them to others in addition to reinfecting himself or herself. Enterobiasis occurs most often among families and in other close living situations. Its distribution is worldwide among all socioeconomic groups, but it seems to attack younger people more frequently than older ones.

Distribution and Importance of Pathogenic Worms

About 50 species of helminths cause disease in humans. The pathogenic helminths are frequently called *parasites*, though we will encounter a more specific definition of this word in chapter 11. They are distributed in all areas of the world. Some worms are restricted to a given geographic region, and many have a higher incidence in tropical areas. This knowledge must be tempered with the realization that jet-age travel, along with human migration, is gradually changing the patterns of helminth infections, especially of those species that do not require alternate hosts or special climatic conditions for development. The yearly estimate of worldwide cases numbers in the billions, and these are not confined to developing countries. A conservative estimate places 50 million helminth infections in North America alone.

4.5 LEARNING OUTCOMES–Assess Your Progress

20. List the two major groups of helminths, and provide examples representing each body type.

21. Summarize the stages of a typical helminth life cycle.

CASE FILE WRAP-UP

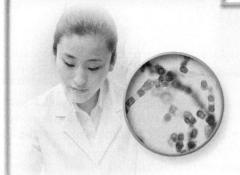

Jade LLC/Blend Images LLC (health care worker); Source: Dr. Lucille K. Georg/CDC (*Coccidioides immitis* arthroconidia)

Coccidioidomycosis develops when an individual inhales spores produced by the fungus *Coccidioides immitis*. This disease, which is often called *Valley fever,* is endemic to the desert regions of the southwestern United States, and cases are commonly seen in both South and Central America. Most people who become exposed to the fungus never exhibit any signs or symptoms of illness. Others develop flulike symptoms or pneumonia that may persist for months. Individuals with weakened immune systems tend to experience the most severe forms of Valley fever, and in some cases the disease is fatal.

The patient in the opening case file was a farm worker and likely inhaled spores while working outdoors because California is one of the states where *Coccidioides immitis* can be found. His chest X ray revealed the lung nodules typical of this disease (which may be mistaken for cancer), and the patient's blood tested positive for the fungus. Microscopic evidence of the fungus was identified in his sputum sample. Once the patient was started on an antifungal medication, he began to gradually improve and fully recovered from the disease. Due to misinterpretation of the initial symptoms, the medication the patient was initially started on was an antibiotic drug. Although effective against bacteria, antibiotics will not lead to targeted destruction of this fungus.

Robert Glusic/Exactostock/Superstock

Are Eukaryotic Microorganisms Part of Our Microbiome?

Malassezia furfur at 100× magnification.
©Creatas/PunchStock

Yes! Most definitely. Let's focus on fungi. Fungi are well known to be normal inhabitants—and sometimes pathogens—for humans. One example is the fungus that causes yeast infections of the gut, mouth, and vagina, called *Candida albicans*.

Fungal infections of the skin (and hair and nails) are quite common as well. Twenty-nine million people in the United States experience this every year. The Human Microbiome Project has finally made it possible to look at the fungal species that live normally on our skin and other surfaces.

A research team from the National Institutes of Health (NIH) documented the fungi they found on human surfaces. They sampled 14 body sites from each of 10 healthy adults. Their DNA analysis identified more than 80 fungal genera. Previously, when these studies were conducted using culture techniques, only 18 different genera were found. They discovered that one species, *Malassezia*, is commonly found as a normal inhabitant of skin on the head and on most of the body (the trunk). *Malassezia* has been recognized before when it causes very superficial skin infections, and it has been associated with the condition of dandruff. This study tells us that most of the time it is a normal inhabitant of our surfaces.

For some reason, the body part displaying the most diverse fungal population was the heel, containing about 80 fungal genera. Toenails had about 60 genera, and the webs of the toes had about 40. Skin surfaces like the head and trunk displayed just 2 to 10 genera each.

EUKARYOTIC CELLS AND MICROORGANISMS

4.1 OVERVIEW OF THE EUKARYOTES

Eukaryotes are one of the three domains of life; humans are eukaryotes. There are several microbes that colonize humans that are eukaryotic. Three of the major ones are fungi, protozoa, and helminths.

BSIP/UIG/Getty Images

The eukaryotic cell has many more intercellular structures than bacteria and archaea do. One of the major distinctive structures for eukaryotic cells is the membrane-bound nucleus.

STRUCTURES OF THE EUKARYOTIC CELL 4.2

4.3 THE FUNGI

The fungi can be multicellular or unicellular. They can reproduce asexually as well as sexually. They cause three categories of disease: community-acquired disease, hospital-associated infections, and opportunistic infections.

Protozoa are not a taxonomically cohesive group, but they are informally placed in a single group by medical professionals. They are classified by how they move, as either ciliates, flagellates, amoebas, or nonmotiles. The malaria parasite is a protozoan.

THE PROTOZOA 4.4

4.5 THE HELMINTHS

There are two major groups of helminths that infect humans, the flatworms (Platyhelminthes) and the roundworms (Aschelminthes). Flatworms have two categories: cestodes and trematodes. Helminths have complex life cycles that can involve multiple hosts.

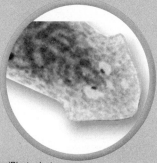

NHPA/M. I. Walker/Photoshot

SmartGrid: From Knowledge to Critical Thinking

This *21 Question Grid* takes the topics from this chapter and arranges them with respect to the American Society for Microbiology's Undergraduate Curriculum guidelines—all six of the important "Concepts" as well as the important "Competency" of scientific literacy. Three questions are supplied about chapter content that refer to the Concept or Competency, in increasing levels of Bloom's taxonomy for learning.

ASM Concept/ Competency	A. Bloom's Level 1, 2—Remember and Understand (Choose one.)	B. Bloom's Level 3, 4—Apply and Analyze	C. Bloom's Level 5, 6—Evaluate and Create

Evolution	1. Mitochondria likely originated from
	  a. archaea.
	  b. invagination of the cell membrane.
	  c. bacteria.
	  d. chloroplasts.
		2. Summarize the endosymbiotic theory and explain how it accounts for major structural similarities and differences between bacterial and eukaryotic cells.	3. The best theory of the evolution of the three domains of life (figure 4.11) envisions cellular organisms switching to a DNA vs. an RNA lifestyle very early. Create a scenario in which it would be advantageous for the Last Common Ancestor to use RNA, while it would be advantageous for its descendants to use DNA.

Cell Structure and Function	4. Yeasts are _____ fungi, and molds are _____ fungi.
	  a. macroscopic; microscopic
	  b. unicellular; filamentous
	  c. motile; nonmotile
	  d. water; terrestrial
		5. Compare and contrast the structure and function of the following among bacteria and eukaryotes.
		  a. ribosome
		  b. flagellum
		  c. glycocalyx
			6. Using components from all three types of eukaryotic organisms in this chapter, design an ideal organism, and explain your choices.

Metabolic Pathways	7. The Golgi apparatus
	  a. receives vesicles from the mitochondria.
	  b. packages products into transitional vesicles.
	  c. modifies proteins.
	  d. synthesizes proteins and sterols.
		8. Considering the role of fungi in nature, speculate on why they have such a wide array of metabolic capabilities.	9. Speculate on why fungi evolved to produce products such as alcohol, antibiotics, and vitamins.

Information Flow and Genetics	10. Fungi produce which structures for reproduction and multiplication?
	  a. endospores
	  b. cysts
	  c. spores
	  d. eggs
		11. Write a paragraph illustrating the life of a protein, from DNA to mature polypeptide, and the course of its travels within a cell throughout its synthesis.	12. Investigate whether there are other organisms, besides fungi, that have both sexual and asexual forms of reproduction and devise a hypothesis about why that might be advantageous.

Microbial Systems	13. Which of these organisms has the best potential to survive in extreme environments?
	  a. fungi
	  b. trophozoite form of protozoa
	  c. helminth
	  d. yeast
		14. Why do you think the incidence of opportunistic fungal diseases in humans has increased dramatically in the last 50 years?	15. Do you suppose any of the eukaryotic microbes populate biofilms in nature? Defend your answer.

ASM Concept/ Competency	A. Bloom's Level 1, 2—Remember and Understand (Choose one.)	B. Bloom's Level 3, 4—Apply and Analyze	C. Bloom's Level 5, 6—Evaluate and Create
Impact of Microorganisms	**16.** Which of these groups causes the most casualties on an annual basis globally? **a.** helminths **b.** protozoa **c.** fungi **d.** algae	**17.** Provide at least three examples illustrating both beneficial and detrimental aspects of fungi in the modern world today.	**18.** Do you suspect that the fact that humans use microbes such as fungi to manufacture massive amounts of metabolic products for our use will change the evolution of these organisms? Defend your answer.
Scientific Thinking	**19.** Which of the following is not useful to determine whether a clinical isolate is a bacterium, fungus, or protozoan? **a.** its size under a light microscope **b.** whether it has a cell wall **c.** whether it can form protective structures under stress **d.** all of the above are reliable	**20.** Why were protozoa originally considered a single group, and why is that no longer the case?	**21.** Write a paragraph that would explain the difference between heterotroph, saprobe, and parasite to a middle school class.

Answers to the multiple-choice questions appear in Appendix A.

Visual Connections

This question connects content within and between chapters.

From chapter 2, figure 2.1. Discuss how the techniques of the Five I's of microbiology would be completed if your patient's infection was due to a protozoan, a eukaryotic microbe.

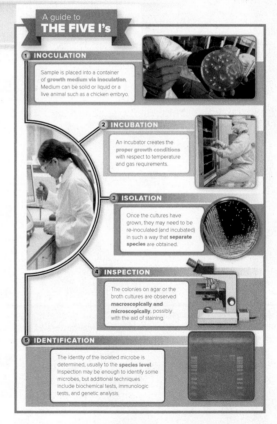

Monty Rakusen/Getty Images (scientist in lab); Source: CDC/James Gathany (lab worker holding petri dish); Pixtal/AGE Fotostock (kneeling scientist); Source: CDC (*Vibrio cholerae* petri dish); Centers for Disease Control and Prevention (drawing of microscope); Lisa Burgess/McGraw-Hill Education (DNA gel electrophoresis)

5

Viruses and Prions

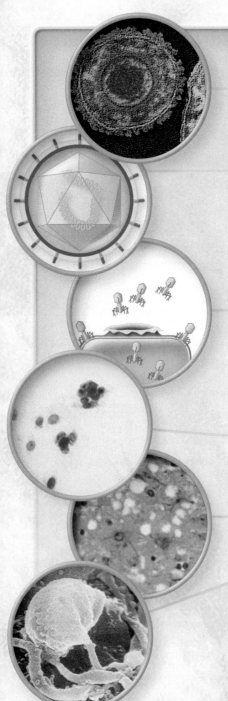

IN THIS CHAPTER...

CASE FILE

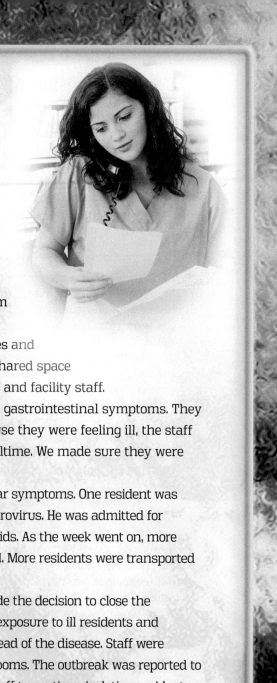

Outbreak in Assisted Living

I was an LPN working in an assisted living facility and caring for a population of elderly residents. In this facility, residents shared a room and there was a communal dining area. During the day, the seniors would participate in activities and therapy. I could never have imagined that this shared space would one day become a danger to the residents and facility staff.

One day last March, two residents developed gastrointestinal symptoms. They were experiencing vomiting and diarrhea. Because they were feeling ill, the staff kept them in their respective rooms during mealtime. We made sure they were drinking fluids, as they were able.

The next day, more residents fell ill with similar symptoms. One resident was transported to the hospital and diagnosed with norovirus. He was admitted for treatment of dehydration with intravenous (IV) fluids. As the week went on, more residents were affected, and even some staff fell ill. More residents were transported to the local hospital needing IV fluids.

The medical director of the nursing home made the decision to close the communal areas of the facility. He explained that exposure to ill residents and contaminated surfaces may be facilitating the spread of the disease. Staff were required to wear gowns and gloves in all patient rooms. The outbreak was reported to the county health department, who advised the staff to continue isolating residents as much as possible. When residents fell ill, special cleaning techniques were used to disinfect the area.

- Why was norovirus spreading so quickly?

- How should viral illness be treated?

Case File Wrap-Up appears at the end of the chapter.

This variegated tulip gets its beautiful colors from a viral infection.
©Anna Yu/Getty Images

5.1 The Position of Viruses in the Biological Spectrum

Viruses are a unique group of biological entities known to infect every type of cellular, such as bacteria, algae, fungi, protozoa, plants, and animals. Viruses are extremely abundant on our planet. For example, it is documented that seawater can contain 10 million viruses per milliliter, and human feces probably contain 100 times that many. It is estimated that the sum of viruses in the ocean represents 270 million metric tons of organic matter. We are just beginning to understand the impact of these huge numbers of viruses on our environment.

For many years, the cause of viral infections such as smallpox and polio was unknown, even though it was clear that the diseases were transmitted from person to person. The French scientist Louis Pasteur was certainly on the right track when he hypothesized that rabies was caused by a "living thing" smaller than bacteria. In 1884 he was able to develop the first vaccine for rabies. Pasteur also proposed the term *virus* (which is Latin for poison) to name this special group of infectious agents.

The first important hints about the unique characteristics of viruses occurred in the 1890s. First, D. Ivanovski and M. Beijerinck showed that a disease in tobacco was caused by a virus (tobacco mosaic virus). Then, Friedrich Loeffler and Paul Frosch discovered an animal virus that causes foot-and-mouth disease in cattle. These early researchers found that when infectious fluids from the victims were passed through porcelain filters designed to trap bacteria, the fluid that came through the filter remained infectious. This result proved that an infection could be caused by a fluid containing agents smaller than bacteria and thus first introduced the concept of a *filterable virus.*

Over the succeeding decades, a remarkable picture of the physical, chemical, and biological nature of viruses has taken form. Viruses are noncellular particles with a definite size, shape, and chemical composition. The development of special techniques meant that many of them could be cultured in the laboratory. Then, thanks to new genomic techniques, including DNA arrays and "next-generation" nucleic acid sequencing techniques, we developed a much clearer picture of the number and variety of viruses on earth. Studies of the human virome (a part of the human microbiome) and of the world's oceans are showing us that there are vast multitudes of viruses that have roles we cannot even guess about.

The exceptional and curious nature of viruses raises many questions, including the following:

1. Are they organisms; that is, are they alive?
2. What role did viruses play in the evolution of life?
3. How can they jump from other species to suddenly cause severe disease in humans?
4. How can particles so small and simple be capable of causing disease and death?
5. What is the connection between viruses and cancer?

In this chapter, we address these questions and many others.

The unusual structure and behavior of viruses have led to debates about their connection to the rest of the microbial world. One viewpoint holds that since viruses are unable to multiply independently from the host cell, they are not living things but should instead be called infectious molecules. Another viewpoint proposes that even though viruses do not exhibit most of the life processes of cells, they can *direct* them, and, for that reason, they are more than inert and lifeless molecules. In many ways, this debate doesn't matter for our purposes. Viruses are agents of disease and must be dealt with through control, therapy, and prevention, whether we regard them as living or not. In keeping with all of this, it is best to describe viruses as either *active* or *inactive* (rather than alive or dead).

Table 5.1 Properties of Viruses

- Are obligate intracellular parasites of bacteria, protozoa, fungi, algae, plants, and animals
- Estimated 10^{31} virus particles on earth, approximately 10 times the number of bacteria and archaea combined
- Are ubiquitous in nature and have had major impact on development of biological life
- Are ultramicroscopic in size, ranging from 20 nm up to 1,000 nm (diameter)
- Are not cells; structure is very compact and economical
- Do not independently fulfill the characteristics of life
- Basic structure consists of protein shell (capsid) surrounding nucleic acid core
- Nucleic acid can be either DNA or RNA, but not both
- Nucleic acid can be double-stranded DNA, single-stranded DNA, single-stranded RNA, or double-stranded RNA
- Molecules on virus surface give them high specificity for attachment to host cell
- Multiply by taking control of host cell's genetic material and regulating the synthesis and assembly of new viruses
- Lack enzymes for most metabolic processes
- Lack machinery for synthesizing proteins

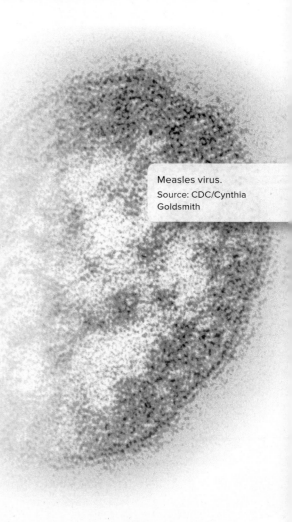

Measles virus.
Source: CDC/Cynthia Goldsmith

Recent discoveries suggest that viruses have been absolutely vital in forming cells and other life forms as they are today. By infecting other cells, and sometimes influencing their genetic makeup, they have shaped the way cells, tissues, bacteria, plants, and animals have evolved to their present forms. For example, scientists think that approximately 8% of the human genome consists of sequences that come from viruses that have inserted their genetic material permanently into human DNA. Bacterial DNA also contains 10% to 20% viral sequences. As you learn more about how viruses work, you will see how this could happen.

Viruses are different from their host cells in size, structure, behavior, and physiology. They are *obligate intracellular parasites* that cannot multiply unless they invade a specific host cell and instruct its genetic and metabolic machinery to make and release quantities of new viruses. Other unique properties of viruses are summarized in **table 5.1.**

How Viruses Are Classified and Named

For many years, the animal viruses were classified mainly on the basis of their hosts and the kind of diseases they caused. Newer systems for naming viruses also take into account the actual nature of the virus particles themselves, with only partial emphasis on host and disease. The main criteria currently used to group viruses are structure, chemical composition, and similarities in genetic makeup.

A group called the International Committee on the Taxonomy of Viruses (ICTV) is in charge of classifying and categorizing viruses. It is very difficult, even for professional taxonomists, to classify viruses, and their classifications are constantly changing. In this book, we will focus on obvious characteristics of the viruses and the host cells they infect.

5.1 LEARNING OUTCOMES—Assess Your Progress

1. Explain what it means when viruses are described as *filterable.*
2. Identify better terms for viruses than *alive* or *dead.*

NCLEX® PREP

1. Which statements are accurate regarding properties of viruses? Select all that apply.
   a. *Viruses are cell structures that are compact and economical.*
   b. *Viruses can be living molecules outside the host cell.*
   c. *The basic structure of a virus is composed simply of a protein shell.*
   d. *Viruses do not contain enzymes for most metabolic processes.*
   e. *Viruses contain either DNA or RNA.*

5.2 The General Structure of Viruses

Size Range

As a group, viruses represent the smallest infectious agents (with some unusual exceptions to be discussed later in this chapter). They are dwarfed by their host cells: More than 2,000 bacterial viruses could fit into an average bacterial cell, and more than 50 million polioviruses could be accommodated by an average human cell. Common animal viruses range in size from the small parvoviruses (around 20 nm [0.02 μm] in diameter) to the herpes simplex virus (around 150 nm) **(figure 5.1).** Unusual viruses have been discovered recently that are huge, in virus terms. *Pandoravirus,* pictured in figure 5.1, is about the same size as a coccus-shaped bacterial cell. Some cylindrical viruses are relatively long (800 nm [0.8 μm] in length) but so narrow in diameter (15 nm [0.015 μm]) that even with the high magnification and resolution of an electron microscope, they are very difficult to see. Figure 5.1 compares the sizes of several viruses with bacterial and eukaryotic cells and molecules.

Viral architecture is most readily observed using special stains in combination with electron microscopy **(figure 5.2).**

Viral Components: Capsids, Nucleic Acids, and Envelopes

It is important to realize that viruses bear no real resemblance to cells and that they do not have the protein-synthesizing machinery found in even the simplest cells. Their outer surface is composed of regular, repeating subunits that give rise to their crystalline appearance. The general plan of virus organization is the utmost in simplicity and compactness. Most viruses contain only those parts needed to invade and control a host cell: an external coating and a core containing one or more nucleic acid strands of either DNA or RNA, and sometimes one or two enzymes. The graphic titled "Virus Particle" summarizes viral structure with a diagram similar to the ones we have seen for bacteria and eukaryotes:

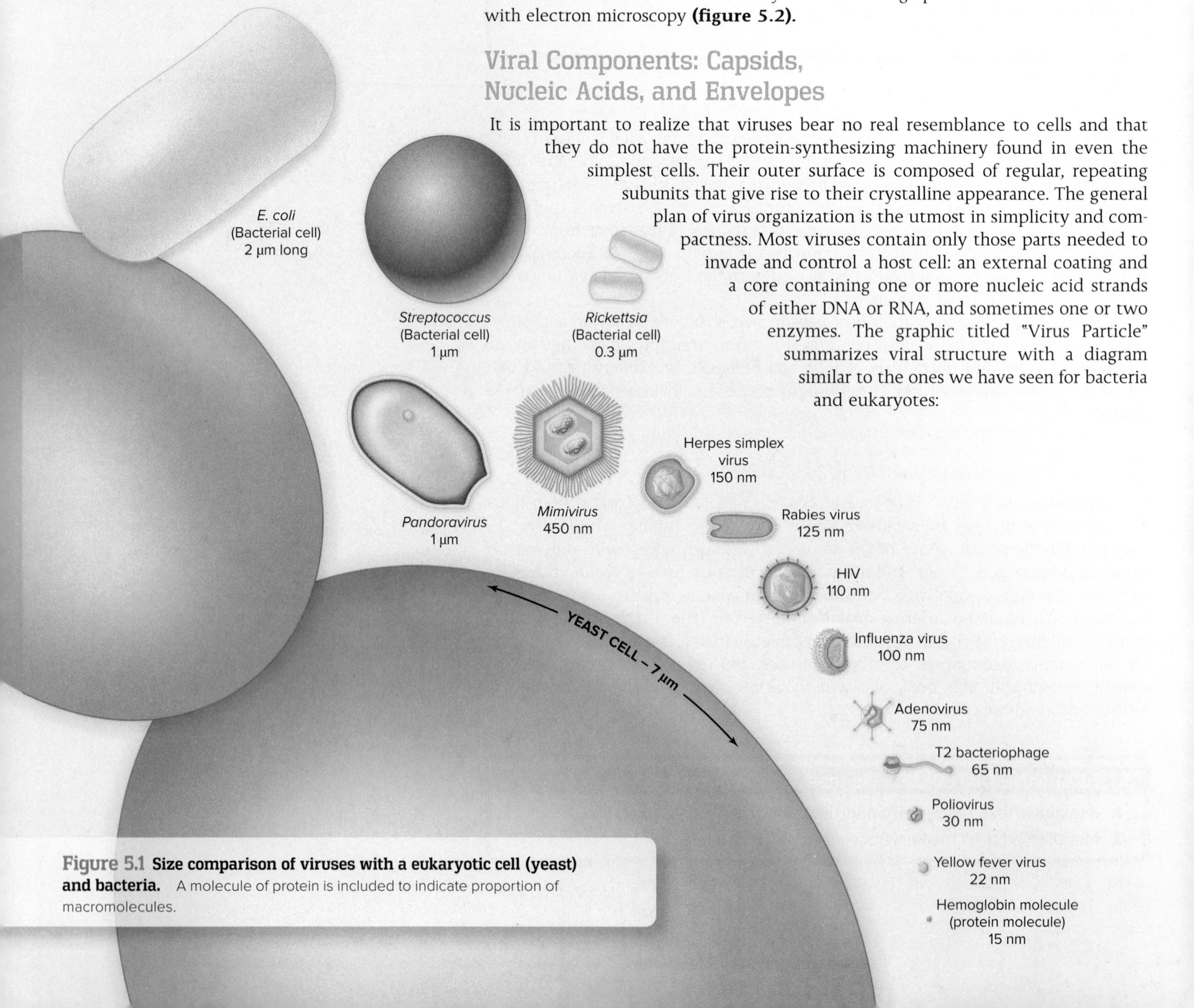
E. coli
(Bacterial cell)
2 μm long
Streptococcus
(Bacterial cell)
1 μm
Rickettsia
(Bacterial cell)
0.3 μm
Pandoravirus
1 μm
Mimivirus
450 nm
Herpes simplex virus
150 nm
Rabies virus
125 nm
HIV
110 nm
Influenza virus
100 nm
Adenovirus
75 nm
T2 bacteriophage
65 nm
Poliovirus
30 nm
Yellow fever virus
22 nm
Hemoglobin molecule
(protein molecule)
15 nm
YEAST CELL – 7 μm

Figure 5.1 Size comparison of viruses with a eukaryotic cell (yeast) and bacteria. A molecule of protein is included to indicate proportion of macromolecules.

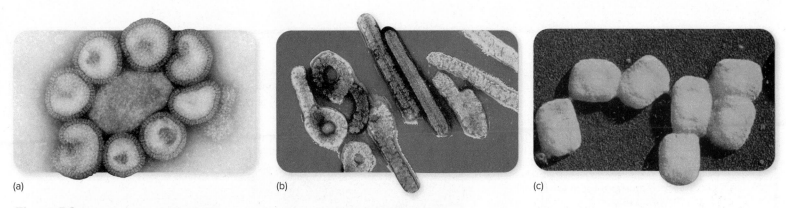

Figure 5.2 Methods of viewing viruses.   (a) Negative staining of influenza viruses. (b) Positive stain of the Ebola virus. Note the textured capsid. (c) Shadowcasting image of a vaccinia virus.
Source: Centers for Disease Control and Prevention/Dr. F. A. Murphy; Phototake; A. Barry Dowsett/Science Source

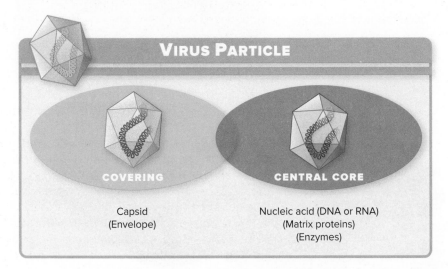

All viruses have a protein **capsid,** or shell, that surrounds the nucleic acid in the central core. Together the capsid and the nucleic acid are referred to as the **nucleocapsid (figure 5.3).** Many animal viruses also possess an additional covering external to the capsid called an *envelope,* which is usually a modified piece of the host's cell membrane **(figure 5.3*b*).** Most viruses that infect humans have envelopes. Viruses that consist of only a nucleocapsid are considered *naked viruses* **(figure 5.3*a*).** Both naked and enveloped viruses possess proteins on their outer surfaces that project from either the nucleocapsid or the envelope. They are the molecules that allow viruses to dock with their host cells and are called *spikes.* As we shall see later, the enveloped viruses differ from the naked viruses in the way that they enter and leave a host cell. A fully formed virus that is able to establish an infection in a host cell is often called a **virion.**

The Viral Capsid and Envelope

When a virus particle is magnified several hundred thousand times, the capsid appears as the most prominent geometric feature. In general, each capsid is constructed from identical subunits called **capsomeres** that are constructed from protein molecules. The capsomeres spontaneously self-assemble into the finished capsid. Depending on how the capsomeres are shaped and arranged, this assembly results in two different types: helical and icosahedral. **Tables 5.2** through **5.4** depict the variations on these two themes.

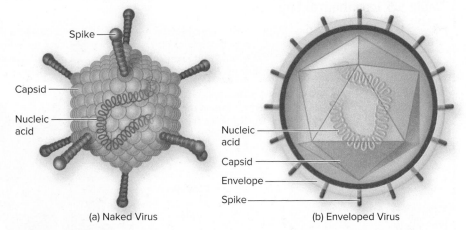

Figure 5.3 Generalized structure of viruses.   (a) The simplest virus is a naked virus (nucleocapsid), consisting of a geometric capsid assembled around a nucleic acid strand or strands. (b) An enveloped virus is composed of a nucleocapsid surrounded by a flexible membrane called an envelope.

Table 5.2 Helical Capsids

The simpler **helical capsids** have rod-shaped capsomeres that bond together to form a series of hollow discs resembling a bracelet. During the formation of the nucleocapsid, these discs link with other discs to form a continuous helix into which the nucleic acid strand is coiled.

Naked	The nucleocapsids of naked helical viruses are very rigid and tightly wound into a cylinder-shaped package. An example is the *tobacco mosaic virus*, which attacks tobacco leaves (right).	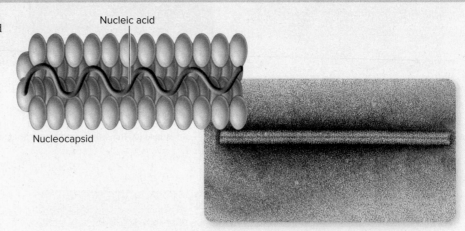

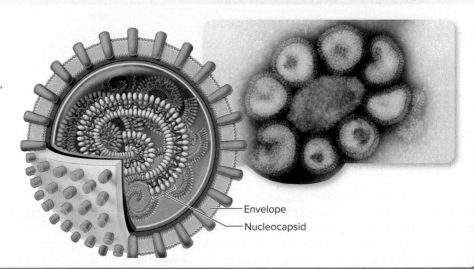

Enveloped	Enveloped helical nucleocapsids are more flexible and tend to be arranged as a looser helix within the envelope. This type of morphology is found in several enveloped human viruses, including influenza, measles, SARS–CoV–2, and rabies.	

Omikron/Science Source (tobacco mosaic virus. TEM Magnification: 260,000x); Source: Centers for Disease Control and Prevention/Dr. F. A. Murphy negative-stained (TEM) of influenza virus particles. Magnification not available

Table 5.3 Icosahedral Capsids

These capsids form an **icosahedron** (eye″-koh-suh-hee′-drun)—a three-dimensional, 20-sided figure with 12 evenly spaced corners. The arrangements of the capsomeres vary from one virus to another. Some viruses construct the capsid from a single type of capsomere, while others may contain several types of capsomeres. There are major variations in the number of capsomeres; for example, a poliovirus has 32, and an adenovirus has 252 capsomeres.

Naked	Adenovirus is an example of a naked icosahedral virus. In the photo you can clearly see the spikes, some of which have broken off.	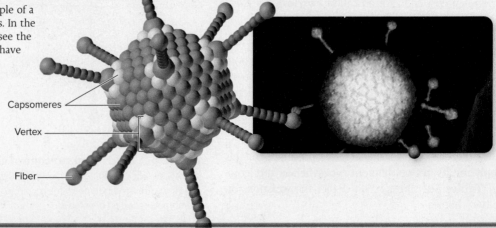

Table 5.3 Icosahedral Capsids (continued)

Enveloped   Two very common viruses, hepatitis B virus (left) and the herpes simplex virus (right), possess enveloped icosahedrons.

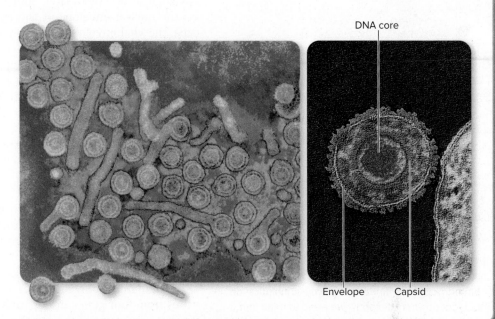

Dr. Linda M. Stannard, University of Cape Town/SPL/Science Source (adenovirus); Dr. Linda M. Stannard, University of Cape Town/SPL/Science Source (hepatitis B virus); Eye of Science/ Science Source (TEM—Herpes simplex viruses)

Table 5.4 Complex Capsids

Complex capsids, only found in the viruses that infect bacteria, may have multiple types of proteins and take shapes that are not symmetrical. They are never enveloped. The one pictured on the right is a T4 bacteriophage.

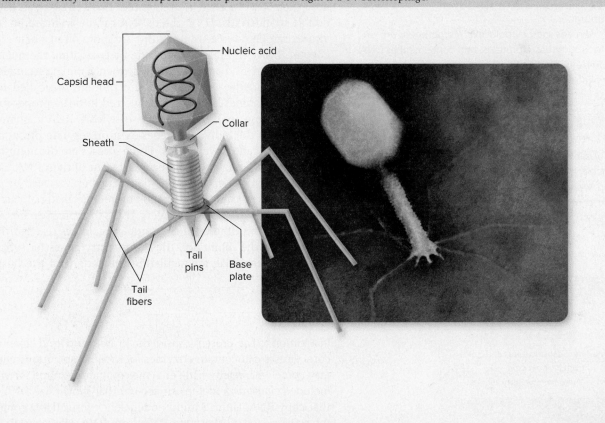

©Ami Images/Science Source

When **enveloped viruses** (mostly animal viruses) are released from the host cell, they take with them a bit of the cell's membrane system in the form of an envelope, as described later. Some viruses bud off the cell membrane; others leave via the nuclear envelope or the endoplasmic reticulum. Whichever avenue of escape, the viral envelope differs significantly from the host's membranes. In the envelope, some or all of the regular membrane proteins are replaced with special viral proteins. Some of the envelope proteins attach to the capsid of the virus, and glycoproteins (proteins bound to a carbohydrate) remain exposed on the outside of the envelope. These protruding molecules, called spikes, are essential for the attachment of viruses to the next host cell. Because the envelope is more flexible than the capsid, enveloped viruses are pleomorphic (of variable shape) and range from spherical to filamentous in shape.

Nucleic Acids: At the Core of a Virus

The sum total of the genetic information carried by an organism is called its **genome.** We know that the genetic information of living cells is carried by nucleic acids (DNA, RNA). Viruses—even though they are not alive, neither are they cells—are no exception to this rule, but there is a significant difference. Unlike cells, which contain both DNA and RNA, viruses contain either DNA or RNA, *but not both.*

The number of viral genes is quite small compared with that of a cell. It varies from four genes in hepatitis B virus to hundreds of genes in some herpesviruses. Viruses possess only the genes needed to invade host cells and redirect their activity. By comparison, the bacterium *Escherichia coli* has approximately 4,000 genes, and a human cell has approximately 23,000 genes. These additional genes allow cells to carry out the complex metabolic activity necessary for independent life.

In chapter 1, you learned that DNA usually exists as a double-stranded molecule and that RNA is single-stranded. Viruses are different. They exhibit wide variety in how their RNA or DNA is configured. DNA viruses can have single-stranded (ss) or double-stranded (ds) DNA; the dsDNA can be arranged linearly or in ds circles. RNA viruses can be double-stranded but are more often single-stranded. You will learn in chapter 8 that all proteins are made by translating the nucleic acid code on a single strand of RNA into an amino acid sequence. Single-stranded RNA genomes that are ready for immediate translation into proteins are called *positive-sense* RNA. Other viral RNA genomes have to be converted into the proper form to be made into proteins, and these are called *negative-sense* RNA. RNA genomes may also be *segmented,* meaning that the individual genes exist on separate pieces of RNA. A special type of RNA virus is called a *retrovirus.* These viruses are distinguished by the fact that they carry their own enzymes to create DNA out of their RNA. **Table 5.5** gives examples of each configuration of viral nucleic acid.

In all cases, these tiny strands of genetic material carry the blueprint for viral structure and functions. In a very real sense, viruses are genetic parasites because they cannot multiply until their nucleic acid has reached the internal habitat of the host cell. At the minimum, they must carry genes for synthesizing the viral capsid and genetic material, for regulating the actions of the host, and for packaging the mature virus.

Other Substances in the Virus Particle

In addition to the protein capsid, the protein and lipid envelopes, and the nucleic acid core, viruses can contain enzymes for specific operations within their host cell. They may come with ready-made enzymes that are required for viral replication. Examples include *polymerases* (pol-im'-ur-ace-uz) that synthesize DNA and RNA, and replicases that copy RNA. Human immunodeficiency virus (HIV) comes equipped with *reverse transcriptase* for synthesizing DNA from RNA. However, the vast majority of viruses

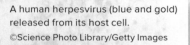

A human herpesvirus (blue and gold) released from its host cell.
©Science Photo Library/Getty Images

Table 5.5 Viral Nucleic Acid

	Diagram	Virus Name	Disease It Causes
DNA Viruses–Examples			
Double-stranded DNA		Variola virus	Smallpox
		Herpes simplex 2	Genital herpes
Single-stranded DNA		Parvovirus	Erythema infectiosum (skin condition)
RNA Viruses–Examples			
Single-stranded (+) sense		Poliovirus	Poliomyelitis
Single-stranded (−) sense		Influenza virus	Influenza
Double-stranded RNA		Rotavirus	Gastroenteritis
Single-stranded RNA (carries reverse transcriptase)		HIV	AIDS

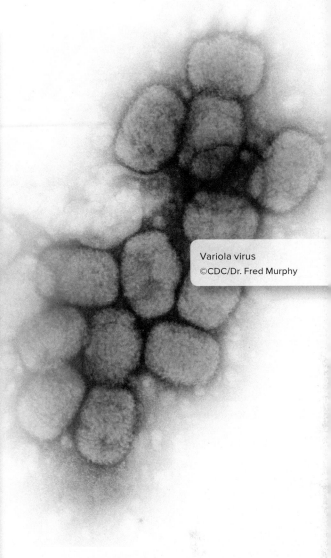

Variola virus
©CDC/Dr. Fred Murphy

COVID-19

SARS-CoV-2, like all coronaviruses, is a single-stranded (+) sense RNA virus.

completely lack the genes for synthesis of metabolic enzymes. As we shall see, this deficiency is not an obstacle because viruses have adapted to completely take over their hosts' metabolic resources. Some viruses can actually carry away substances from their host cell. For instance, arenaviruses pack along host ribosomes, and retroviruses "borrow" the host's tRNA molecules.

5.2 LEARNING OUTCOMES—Assess Your Progress

3. Discuss the size of viruses relative to other microorganisms.
4. Describe the function and structure(s) of viral capsids.
5. Distinguish between enveloped and naked viruses.
6. Explain the importance of viral surface proteins, or spikes.
7. Diagram the possible nucleic acid configurations that viruses may possess.

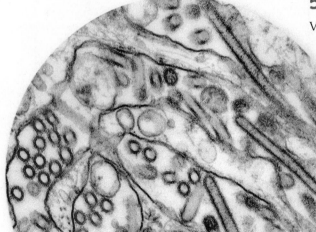

Avian flu viruses (gold) inside host cells (green).
Source: CDC/Cynthia Goldsmith

5.3 How Viruses Multiply

Viruses are minute parasites that seize control of the synthetic and genetic machinery of cells. The way this cycle works dictates the way the virus is transmitted and what it does to its host, the responses of the immune defenses, and human measures to control viral infections.

Multiplication Cycles in Animal Viruses

The general phases in the life cycle of animal viruses are **adsorption, penetration and uncoating, synthesis, assembly,** and **release** from the host cell. The length of the entire multiplication cycle varies from 8 hours in polioviruses to 36 hours in some herpesviruses. **Table 5.6** walks through the major phases of the viral life cycle, using a + strand RNA virus (of which rubella virus is an example) as a model.

Phases of the Multiplication Cycle

Adsorption   Because a virus can invade its host cell only through making an exact fit with a specific host molecule, the range of hosts it can infect is limited **(figure 5.4).** This limitation, known as the **host range,** may be highly restricted, as in the case of hepatitis B, which infects only liver cells of humans. Other viruses are considered moderately restrictive, like the poliovirus, which infects intestinal and nerve cells of primates (humans, apes, and monkeys), or broad, like the rabies virus, which can infect various cells of all mammals. Cells that lack compatible virus receptors are resistant to adsorption and invasion by that virus. This explains why, for example, human liver cells are not infected by the canine hepatitis virus and dog liver cells cannot host the human hepatitis A virus. It also explains why viruses usually have tissue specificities called *tropisms* (troh'-pizmz) for certain cells in the body. The hepatitis B virus targets the liver, and the mumps virus targets salivary glands.

Figure 5.4 The viral attachment process.   An enveloped coronavirus with spikes. The configuration of the spike has a complementary fit for cell receptors. The process in which the virus lands on the cell and plugs into receptors is termed *docking.*

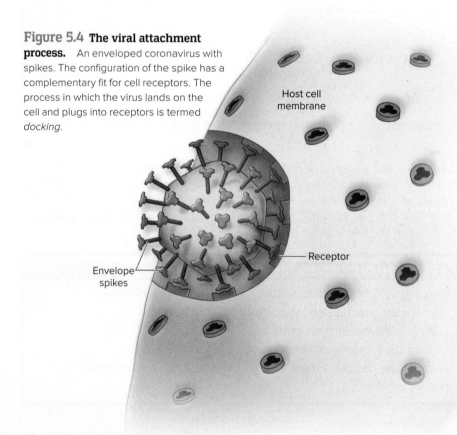

Host cell membrane

Receptor

Envelope spikes

Table 5.6 Life Cycle of Animal Viruses

COVID-19

SARS-CoV-2 uses this process to infect human cells.

1 **Adsorption.** The virus encounters a susceptible host cell and adsorbs specifically to receptor sites on the cell membrane. The membrane receptors that viruses attach to are usually proteins that the cell requires for its normal function. Glycoprotein spikes on the envelope (or on the capsid of naked viruses) bind to the cell membrane receptors.

2 **Penetration and Uncoating.** In this example, the entire virus is engulfed (endocytosed) by the cell and enclosed in a vacuole or vesicle. When enzymes in the vacuole dissolve the envelope and capsid, the virus is said to be uncoated, a process that releases the viral nucleic acid into the cytoplasm.

3 **Synthesis: Replication and Protein Production.** Almost immediately, the viral nucleic acid begins to synthesize the building blocks for new viruses. First, the + ssRNA, which is ready to serve as mRNA, starts being translated into viral proteins, especially those useful for further viral replication. The + strand is then replicated into − ssRNA. This RNA becomes the template for the creation of many new + ssRNAs, used as the viral genomes for new viruses. Additional + ssRNAs are synthesized and used for late-stage mRNAs. Some viruses come equipped with the necessary enzymes for synthesis of viral components; others utilize those of the host. Proteins for the capsid, spikes, and viral enzymes are synthesized on the host's ribosomes using its amino acids.

4 **Assembly.** Toward the end of the cycle, mature virus particles are constructed from the growing pool of parts. In most instances, the capsid is first laid down as an empty shell that will serve as a receptacle for the nucleic acid strand. One important event leading to the release of enveloped viruses is the insertion of viral spikes into the host's cell membrane so they can be picked up as the virus buds off with its envelope.

5 **Release.** Assembled viruses leave their host in one of two ways. Nonenveloped and complex viruses that reach maturation in the cell nucleus or cytoplasm are released when the cell lyses or ruptures. Enveloped viruses are liberated by budding from the membranes of the cytoplasm, nucleus, endoplasmic reticulum, or vesicles. During this process, the nucleocapsid binds to the membrane, which curves completely around it and forms a small pouch. Pinching off the pouch releases the virus with its envelope.

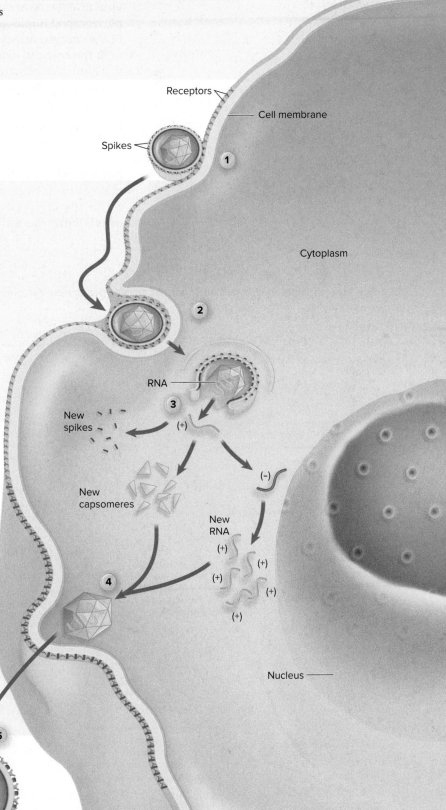

Penetration and Uncoating Animal viruses have some impressive mechanisms for entering a host cell. The flexible cell membrane of the host is penetrated either by the whole virus or by its nucleic acid **(figure 5.5).** In penetration by **endocytosis (figure 5.5a),** the entire virus is engulfed by the cell and enclosed in a vacuole or vesicle. Enzymes inside the vesicle may break down the viral capsid. This leaves the virus in an *uncoated* state, in which the nucleic acid is freed. In other cases, the whole nucleocapsid is released into the cytoplasm, and the virus is not uncoated until later in the process. Another means of entry involves direct fusion of the viral envelope with the host cell membrane (as in influenza and mumps viruses) **(figure 5.5b).** In this form of penetration, the envelope merges directly with the cell membrane, and by doing this the nucleocapsid is freed into the cell's interior.

Synthesis: Replication and Protein Production Most DNA viruses enter the host cell's nucleus and are replicated and assembled there. With a few exceptions (such as retroviruses), RNA viruses are replicated and assembled in the cytoplasm.

In chapter 8 you will learn that cellular organisms make new copies of their new genomes by duplicating their DNA. They also use DNA to make mRNA that directs the creation of proteins. These processes can be very different in viruses.

In the life cycle of dsDNA viruses, the synthesis phase is divided into two parts. During the early phase, viral DNA enters the nucleus, where several genes–

Figure 5.5 Two principal means by which animal viruses penetrate. (a) Endocytosis (engulfment) and uncoating of a herpesvirus. (b) Fusion of the cell membrane with the viral envelope (mumps virus).

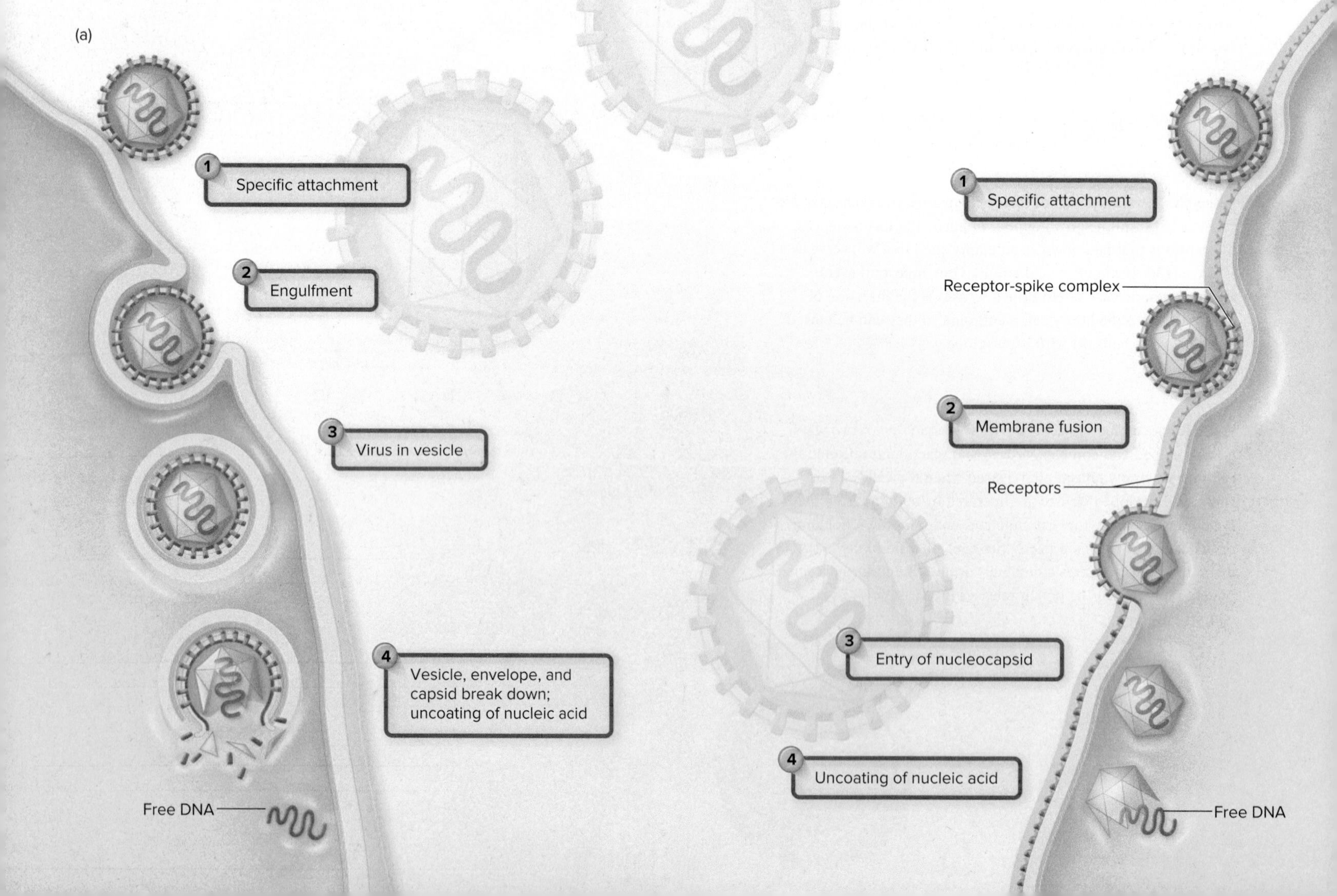

(a)

1 Specific attachment

2 Engulfment

3 Virus in vesicle

4 Vesicle, envelope, and capsid break down; uncoating of nucleic acid

Free DNA

(b)

1 Specific attachment

Receptor-spike complex

2 Membrane fusion

Receptors

3 Entry of nucleocapsid

4 Uncoating of nucleic acid

Free DNA

usually the ones that make proteins needed to make new viral DNA—are transcribed into a messenger RNA. That newly synthesized mRNA then moves into the cytoplasm to be translated into viral proteins (enzymes) needed to replicate the viral DNA. This DNA replication occurs in the nucleus. The host cell's own DNA polymerase is often involved, though some viruses (herpes, for example) have their own. During the late phase, other parts of the viral genome are transcribed and translated into proteins required to form the capsid and other structures. The new viral genomes and capsids are assembled, and the mature viruses are released by budding or cell disintegration. In some viruses, the viral DNA becomes silently *integrated* into the host's genome by insertion at a particular site on the host genome. This integration may later lead to the transformation of the host cell into a cancer cell and the production of a tumor.

Assembly   As illustrated in table 5.6, this step actually puts together the new viruses using the "parts" manufactured in the synthesis process: new capsids and new nucleic acids.

Release   **Figure 5.6** illustrates the mechanics of viral release from host cells. The number of viruses released by infected cells is variable, controlled by factors such as the size of the virus and the health of the host cell. About 3,000 to 4,000 virions are released from a single cell infected with poxviruses, whereas a poliovirus-infected cell can release over 100,000 virions. If even a small number of these virions happen to meet another susceptible cell and infect it, the potential for rapid viral proliferation is immense.

Damage to the Host Cell

Cytopathic effects (CPEs) are defined as virus-induced damage to the cell that changes its microscopic appearance. Individual cells can undergo obvious changes in shape or size, or develop intracellular changes **(figure 5.7a)**. It is common to find *inclusion bodies,* or compacted masses of viruses or damaged cell organelles, in the nucleus and cytoplasm **(figure 5.7b)**. Examination of cells and tissues for cytopathic

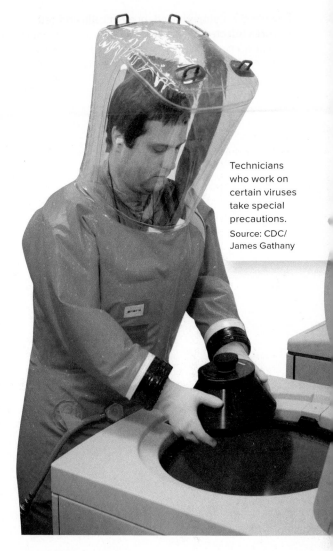

Technicians who work on certain viruses take special precautions.
Source: CDC/ James Gathany

Figure 5.6 Maturation and release of enveloped viruses.   **(a)** As the virus is budded off the membrane, it simultaneously picks up an envelope and spikes. **(b)** A micrograph of HIV leaving its host T cell by budding off its surface.
Chris Bjornberg/Science Source (b)

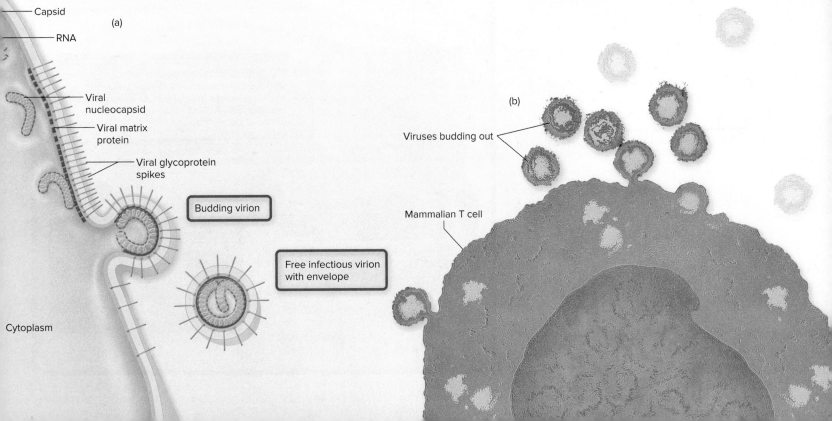

Capsid

(a)

RNA

Viral nucleocapsid

Viral matrix protein

Viral glycoprotein spikes

Budding virion

Free infectious virion with envelope

Cytoplasm

(b)

Viruses budding out

Mammalian T cell

Figure 5.7 Cytopathic changes in cells and cell cultures infected by viruses. **(a)** Overview of types of cytopathic effects. **(b)** Human epithelial cells infected by herpes simplex virus demonstrate giant cells with multiple nuclei. **(c)** Fluorescent-stained human cells infected with cytomegalovirus. Note the inclusion bodies. Note also that both viruses disrupt the cohesive junctions between cells, which would ordinarily be arranged side by side in neat patterns.

Centers for Disease Control and Prevention (b); Courtesy Massimo Battaglia, INeMM CNR, Rome, Italy (c)

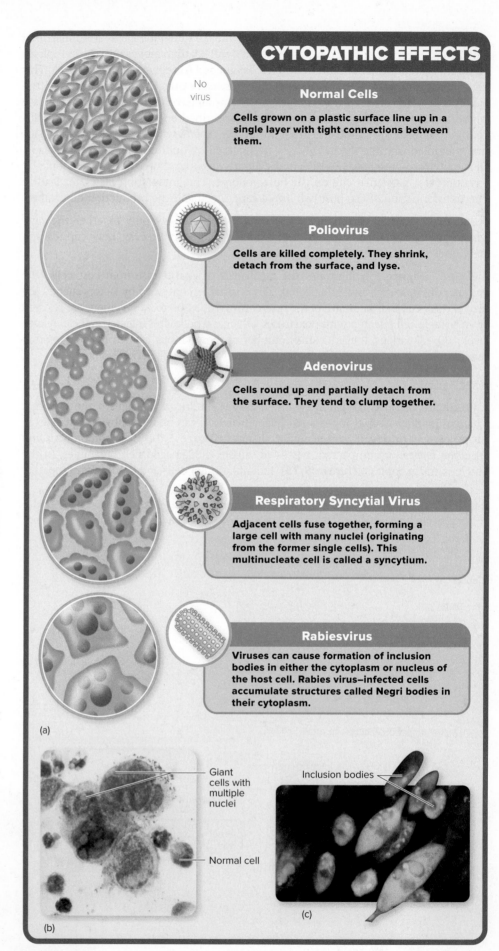

CYTOPATHIC EFFECTS

Normal Cells

Cells grown on a plastic surface line up in a single layer with tight connections between them.

No virus

Poliovirus

Cells are killed completely. They shrink, detach from the surface, and lyse.

Adenovirus

Cells round up and partially detach from the surface. They tend to clump together.

Respiratory Syncytial Virus

Adjacent cells fuse together, forming a large cell with many nuclei (originating from the former single cells). This multinucleate cell is called a syncytium.

Rabiesvirus

Viruses can cause formation of inclusion bodies in either the cytoplasm or nucleus of the host cell. Rabies virus–infected cells accumulate structures called Negri bodies in their cytoplasm.

(a)

Giant cells with multiple nuclei

Normal cell

(b)

Inclusion bodies

(c)

effects can be an important part of the diagnosis of viral infections. One very common CPE is the fusion of multiple damaged host cells into single large cells containing multiple nuclei. These are called **syncytia** (singular, *syncytium*). Sometimes they are called giant cells. They come about as a result of some viruses' ability to fuse membranes. One virus (respiratory syncytial virus) is even named for this effect.

Persistent Infections

Although accumulated damage from a virus infection kills most host cells, some cells maintain a carrier relationship, in which the cell harbors the virus and is not immediately lysed. These so-called *persistent infections* can last from a few weeks to the remainder of the host's life. Viruses can remain latent in the cytoplasm of a host cell, or can incorporate into the DNA of the host. When viral DNA is incorporated into the DNA of the host, it is called a **provirus.** One of the more serious complications occurs with the measles virus. It may remain hidden in brain cells for many years, causing progressive damage and loss of function. Several types of viruses remain in a *chronic latent state*, periodically becoming reactivated. Examples of this are herpes simplex virus (cold sores and genital herpes) and herpes zoster virus (chickenpox and shingles). Both viruses can go into latency in nerve cells and later emerge under the influence of various stimuli to cause recurrent symptoms.

Viruses and Cancer

Some animal viruses enter a host cell and permanently alter its genetic material, leading to cancer. Experts estimate that about 13% of human cancers are caused by viruses. (The percentage is higher in developing countries.) These viruses are termed *oncogenic*, and their effect on the cell is called **transformation.** Viruses that cause cancer in animals act in several different ways, illustrated in **figure 5.8.** In some cases, the virus carries genes that directly cause the cancer. In other cases, the virus produces proteins that induce a loss of growth regulation in the cell, leading to cancer. Transformed cells have an increased rate of growth. They also have changes in their chromosomes. These include changes in the cell's surface molecules and the capacity to divide for an indefinite period, unlike normal animal cells. Mammalian viruses capable of initiating tumors are called **oncoviruses.** Some of these are DNA viruses such as papillomavirus (genital warts are associated with cervical cancer), herpesviruses (one herpesvirus, Epstein-Barr virus, causes a cancer called Burkitt's lymphoma), and hepatitis B virus (liver cancer). A virus related to HIV, called HTLV-I, is also involved in human cancers. These findings have spurred a great deal of speculation on the possible involvement of viruses in cancers and other diseases such as multiple sclerosis.

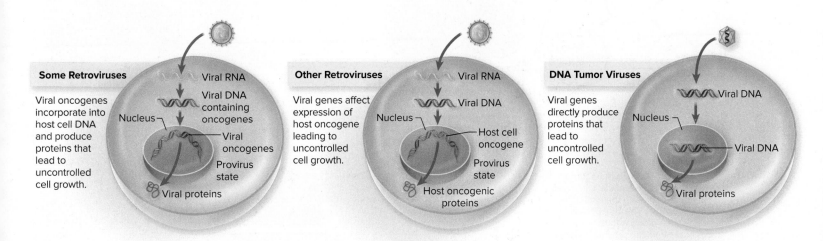

Figure 5.8 Three mechanisms for viral induction of cancer.

Viruses That Infect Bacteria

We now turn to the life cycle of another type of virus called **bacteriophage.** When Frederick Twort and Felix d'Herelle discovered bacterial viruses in 1915, it first appeared that the bacterial host cells were being eaten by some unseen parasite. For that reason, the name *bacteriophage* was used (*phage* coming from the Greek word for "eating"). These organisms are often referred to as "phages." Most bacteriophages contain double-stranded DNA, although single-stranded DNA and RNA types exist as well. So far as is known, every bacterial species is parasitized by one or more specific bacteriophages. Bacteriophages are of great interest to medical microbiologists because they often make the bacteria they infect more pathogenic for humans (more about this later). Probably the most widely studied bacteriophages are those of the intestinal bacterium *Escherichia coli*–especially the ones known as the T-even phages such as T2 and T4. They have an icosahedral capsid head containing DNA, a central tube (surrounded by a sheath), collar, base plate, tail pins, and fibers, which in combination make an efficient package for infecting a bacterial cell.

T-even bacteriophages go through similar stages as the animal viruses described earlier **(figure 5.9).** They *adsorb* to host bacteria using specific receptors on the

Figure 5.9 Events in the lytic cycle of T-even bacteriophages.   The lytic cycle (1–7) involves full completion of viral infection through lysis and release of virions. Occasionally, the virus enters a reversible state of lysogeny (left) and its genetic material is incorporated into the host's genetic material.

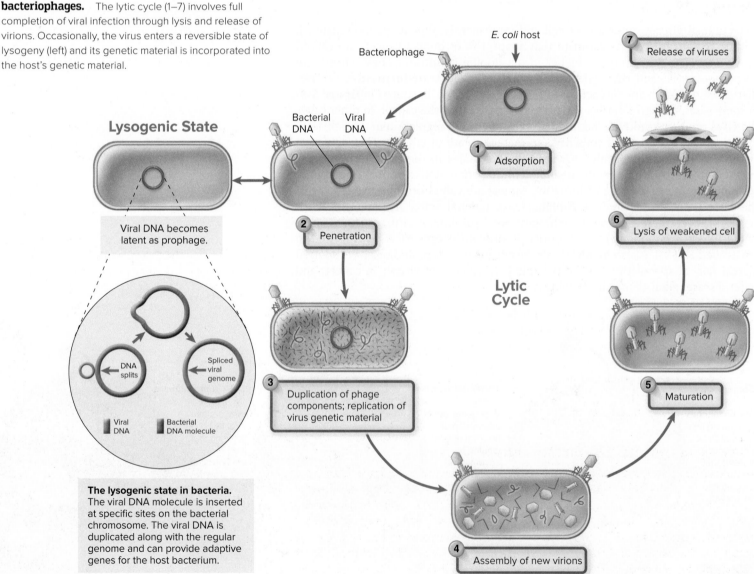

Lysogenic State

Viral DNA becomes latent as prophage.

DNA splits

Spliced viral genome

Viral DNA

Bacterial DNA molecule

The lysogenic state in bacteria. The viral DNA molecule is inserted at specific sites on the bacterial chromosome. The viral DNA is duplicated along with the regular genome and can provide adaptive genes for the host bacterium.

E. coli host

Bacteriophage

Bacterial DNA   Viral DNA

1 Adsorption

2 Penetration

Lytic Cycle

3 Duplication of phage components; replication of virus genetic material

4 Assembly of new virions

5 Maturation

6 Lysis of weakened cell

7 Release of viruses

bacterial surface. Although the entire phage does not enter the host cell, the nucleic acid *penetrates* the host after being injected through a rigid tube the phage inserts through the bacterial membrane and wall **(figure 5.10).** This eliminates the need for *uncoating.* Entry of the nucleic acid causes host cell DNA replication and protein synthesis to stop. Soon the host cell machinery is used for viral *replication* and synthesis of viral proteins. As the host cell produces new phage parts, the parts spontaneously *assemble* into bacteriophages.

An average-size *Escherichia coli* cell can contain up to 200 new phage units at the end of this period. Eventually, the host cell becomes so packed with viruses that it **lyses**–splits open–and releases the mature virions **(figure 5.11).** This process is hastened by viral enzymes produced late in the infection cycle that digest the cell envelope, thereby weakening it. Upon release, the virulent phages can spread to other susceptible bacterial cells and begin a new cycle of infection.

Bacteriophage infection may result in lysis of the cell, as just described. When this happens, the phage is said to have been in the *lytic* phase or cycle. Alternatively, phages can be less obviously damaging in a cycle called the *lysogenic cycle.*

Recently a new type of virus was discovered. These have been named *virophages.* They parasitize other viruses that are infecting the same host cell they infect, using genes from other (usually larger) viruses for their own replication and production. Even though these are parasites of viruses, note that they must be in a host cell, along with their "host" virus.

Lysogeny: The Silent Virus Infection

One category of DNA phages, called **temperate phages,** can participate in a lytic phase or in the very different *lysogenic cycle.* In this cycle they undergo adsorption and penetration into the bacterial host but then do not undergo replication or release immediately. Instead, the viral DNA enters an inactive **prophage** state, reminiscent of the provirus state in animal viruses, during which it is inserted into the bacterial chromosome. This viral DNA will be retained by the bacterial cell and copied during its normal cell division so that the cell's progeny will also have the temperate phage DNA (see figure 5.9). This condition, in which the host chromosome carries bacteriophage DNA, is called **lysogeny** (ly-soj'-uhn-ee). Because viral particles are not produced, the bacterial cells carrying temperate phages do not lyse, and they appear entirely normal. On occasion, in a process called **induction,** the prophage in a lysogenic cell will be activated and progress directly into viral replication and the lytic cycle. Lysogeny is a less deadly form of parasitism than the full lytic cycle and is thought to be an advancement that allows the virus to spread without killing the host.

Bacteriophages are just now receiving their due as important shapers of biological life. Scientists believe that there are more bacteriophages than all other forms of life in the biosphere combined. As we mentioned in the opening paragraphs of this chapter, viral genes linger in human, animal, plant, and bacterial genomes in huge numbers. As such, viruses can contribute what are essentially permanent traits to the host cells, so much so that it could be said that all organisms are really hybrids of themselves and the viruses that infect them.

The Role of Lysogeny in Human Disease

Many bacteria that infect humans are lysogenized by phages. Sometimes that is very bad news for the human: Occasionally phage genes in the bacterial chromosome cause the production of toxins or enzymes that the bacterium would not otherwise have. When a bacterium acquires a new trait from its temperate phage, it is called **lysogenic conversion.** The phenomenon was first discovered in the 1950s in the bacterium that causes diphtheria, *Corynebacterium diphtheriae.* The diphtheria toxin responsible for the deadly nature of the disease is actually a bacteriophage product.

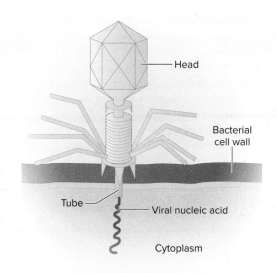

Figure 5.10 Penetration of a bacterial cell by a T-even bacteriophage. After adsorption, the phage plate becomes embedded in the cell wall and the sheath contracts, pushing the tube through the cell wall and releasing the nucleic acid into the interior of the cell.

Figure 5.11 A weakened bacterial cell, crowded with viruses. The cell has ruptured and released numerous virions that can then attack nearby susceptible host cells. Note the empty heads of "spent" phages lined up around the ruptured wall.

Lee D. Simon/Science Source

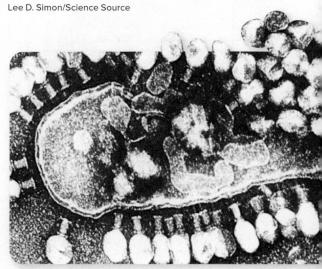

C. diphtheriae without the phage are harmless. Other bacteria that are made virulent by their prophages are *Vibrio cholerae*, the agent of cholera, and *Clostridium botulinum*, the cause of botulism.

5.3 LEARNING OUTCOMES—Assess Your Progress

8. Diagram the five-step life cycle of animal viruses.
9. Define the term *cytopathic effect* and provide one example.
10. Discuss both persistent and transforming infections.
11. Provide thorough descriptions of both lysogenic and lytic bacteriophage infections.

5.4 Techniques to Cultivate and Identify Animal Viruses

In order to study viruses, it is necessary to cultivate them. This presents many problems with organisms that require living cells as their "medium." Scientists have developed methods, which include inoculation of animals and embryonic bird tissues. Methods using living embryos or animals are called *in vivo*. Another strategy is to use cells or tissues that are cultivated in the lab. These are called *in vitro* methods.

The primary purposes of viral cultivation are to

1. isolate and identify viruses in clinical specimens;
2. prepare viruses for vaccines; and
3. do detailed research on viral structure, multiplication cycles, genetics, and effects on host cells.

Using Live Animal Inoculation

Specially bred strains of white mice, rats, hamsters, guinea pigs, and rabbits are the usual choices for animal cultivation of viruses. Invertebrates (insects) or nonhuman primates are occasionally used as well. Because viruses can exhibit host specificity, certain animals can allow a given virus to grow more readily than others.

Using Bird Embryos

A bird egg containing an embryo provides an intact and self-supporting unit, complete with its own sterile environment and nourishment. Furthermore, it furnishes several embryonic tissues that readily support viral multiplication. Chicken, duck,

Avian flus often originate in parts of Southeast Asia where contact between avians and humans is commonplace.
©Shutterstock / Tong_stocker

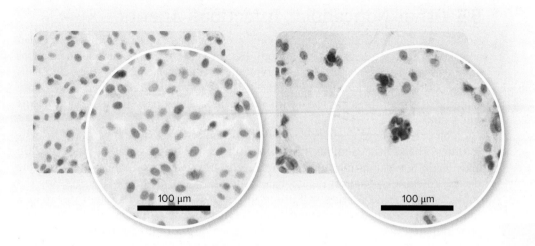

Figure 5.12 Appearance of normal and infected cell culture. Microscopic view of a layer of animal cells before infection with the appropriate virus (left), and after (right).
Source: Bakonyi T, Lussy H, Weissenböck H, Hornyák A, Nowotny N /CDC

and turkey eggs are the most common choices for inoculation. The virus must be injected through the egg shell, usually by drilling a hole or making a small window in the shell.

Using Cell (Tissue) Culture Techniques

The most important early discovery that led to easier cultivation of viruses in the laboratory was the development of a simple and effective way to grow populations of isolated animal cells in culture. These types of *in vitro* cultivation systems are called *cell culture*, or *tissue culture*. Animal cell cultures are grown in sterile chambers with special media that contain the correct nutrients required by animal cells to survive. The cultured cells grow in the form of a *monolayer*, a single, confluent sheet of cells that supports viral multiplication and allows researchers to closely inspect the culture for signs of infection (contamination) **(figure 5.12).**

One way to detect the growth of a virus in culture is to observe degeneration and lysis of infected cells in the monolayer of cells. The areas where virus-infected cells have been destroyed show up as clear, well-defined patches in the cell sheet called **plaques** (figure 5.12). Plaques are essentially the visible manifestation of CPEs, discussed earlier. This same technique is used to detect and count bacteriophages because they also produce plaques when grown in soft agar cultures of their host cells (bacteria). A plaque develops when the viruses released by an infected host cell radiate out to adjacent host cells. As new cells become infected, they die and release more viruses, and so on. As this process continues, the infection spreads gradually and symmetrically from the original point of infection, causing the macroscopic appearance of round, clear spaces that correspond to areas of dead cells.

Because of ongoing fears of the sudden appearance of pandemic strains of influenza virus in humans, scientists have long tried to find faster and more efficient ways to grow the vaccine strains of influenza virus, which has been grown in chicken eggs since the 1950s. Growing the vaccine strains in eggs takes a long time. Now the vaccine strains can also be grown in cell culture. Cell-culture-grown influenza virus is now one of the available vaccine options.

5.4 LEARNING OUTCOMES—Assess Your Progress

12. List the three principal purposes of cultivating viruses.
13. Describe three ways in which viruses are cultivated.

NCLEX® PREP

4. Which of the following is a known association between viruses and cancers?
   a. *Papillomavirus causes brain cancer.*
   b. *Infection with herpesvirus leads to AIDS.*
   c. *Hepatitis B is associated with liver cancer.*
   d. *Papillomavirus is associated with gastric cancer.*

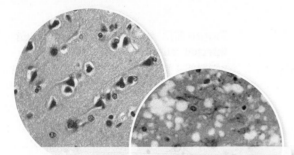

The damage inflicted on brain tissue by Creutzfeldt-Jakob disease. Diseased tissue (*right*) shows spongelike holes not seen in healthy brains (*left*).

©Michael Abbey/Science Source (left); Source: CDC/Dr. Al Jenny (right)

Medical Moment

Differentiating Between Bacterial and Viral Infections

Viral and bacterial diseases can share many of the same symptoms. How can physicians tell the difference? This is an important question, as treatment for bacterial infections often requires antibiotic therapy, whereas treatment for viral illnesses is often supportive—antibiotic therapy is ineffective against viruses.

Many viral illnesses will cause milder symptoms than their bacterial counterpart. For example, viral meningitis is typically a less serious disease than bacterial meningitis, and viral pharyngitis is likewise a less serious condition than bacterial pharyngitis. Therefore, doctors will take into account how sick the patient is when trying to determine whether a patient is suffering from a viral or bacterial infection. They will also look at duration of symptoms, time of year, and known illnesses circulating in the community.

However, none of the above are hard and fast rules. Doctors become very proficient at recognizing various illnesses, but sometimes even they cannot tell the difference. When in doubt, blood work, cultures, and other diagnostic tests can help them make the diagnosis.

Q. Would a light microscope be useful for the diagnosis of viral infections? Why or why not?

Answer in Appendix B.

NCLEX® PREP

5. Which of the following diseases is not caused by a noncellular infectious agent?
   a. *acquired immunodeficiency syndrome*
   b. *tuberculosis*
   c. *smallpox*
   d. *Creutzfeldt-Jakob disease*
   e. *measles*

5.5 Other Noncellular Infectious Agents

Not all noncellular infectious agents are viruses. One group of unusual forms, even smaller and simpler than viruses, can cause serious diseases in humans and animals. The diseases are progressive and universally fatal. A common feature of each of these conditions is the deposition of distinct protein fibrils in the brain tissue. Researchers have determined that these fibrils are the agents of the disease and have named them **prions** (pree′-onz).

The first of these diseases discovered in humans was *Creutzfeldt-Jakob* disease. It afflicts the central nervous system of humans and causes gradual degeneration and death. Several animals (sheep, mink, elk) are victims of similar transmissible diseases. Bovine spongiform encephalopathy (BSE), or "mad cow disease," was the subject of fears and a crisis in Europe in the 1980s and 1990s when researchers found evidence that the disease could be acquired by humans who consumed contaminated beef. This was the first incidence of prion disease transmission from animals to humans. In 2015, a new prion disease of humans was recognized. Shy-Drager syndrome (SDS) or multiple system atrophy (MSA) resembles Parkinson's disease. It is characterized by the accumulation of a protein called alpha-synuclein in the brain, and leads to the symptoms throughout the body. Interestingly, both Parkinson's and Alzheimer's display the accumulation of protein fibrils in the brain. Researchers are investigating whether these conditions might also possibly be caused by prion infection.

The exact mode of prion infection is currently being investigated. The fact that prions are composed primarily of protein (no nucleic acid) has certainly revolutionized our ideas of what can constitute an infectious agent. One of the most compelling questions is just how a prion could be replicated because all other infectious agents require some nucleic acid.

There are other fascinating viruslike agents in human disease. Satellite viruses are viruslike particles that are dependent on other viruses for replication. One remarkable example is the adeno-associated virus (AAV). It was named that because it was originally thought that it could replicate only in cells infected with an adenovirus. But it can also infect cells that are infected with other viruses. Another satellite virus, called the delta agent, is a naked circle of RNA that is expressed only in the presence of the hepatitis B virus and can worsen the severity of liver damage.

Plants are also parasitized by viruslike agents called **viroids** that differ from ordinary viruses by being very small (about one-tenth the size of an average virus) and being composed of only naked strands of RNA. They lack a capsid or any other type of coating. Viroids are significant pathogens in several economically important plants, including tomatoes, potatoes, cucumbers, citrus trees, and chrysanthemums.

5.5 LEARNING OUTCOMES—Assess Your Progress
14. Name three noncellular infectious agents besides viruses.

5.6 Viruses and Human Health

About 260 different viruses are known to infect humans. The number of viral infections that occur on a worldwide basis is nearly impossible to measure accurately. Certainly, viruses are extremely common causes of acute infections such as colds, hepatitis, chickenpox, influenza, herpes, and warts. If one also takes into account prominent viral infections found only in certain regions of the world, such as Dengue fever, Rift Valley fever, and yellow fever, the total could easily exceed several billion cases each year. Although most viral infections do not result in death, some, such as COVID-19, rabies, AIDS, and Ebola, have high mortality rates, and others can lead to long-term consequences (polio, neonatal rubella). Current research is

Table 5.7 Some Important Human Virus Families, Genera

Family	Genus of Virus	Common Name of Genus Members	Name of Disease
DNA Viruses			
Poxviridae	Orthopoxvirus	Variola and vaccinia	Smallpox, cowpox
Herpesviridae	Simplexvirus	Herpes simplex 1 virus (HSV)	Fever blister, cold sores
		Herpes simplex 2 virus (HSV)	Genital herpes
	Varicellovirus	Varicella zoster virus (VZV)	Chickenpox, shingles
	Cytomegalovirus	Human cytomegalovirus (CMV)	CMV infections
Adenoviridae	Mastadenovirus	Human adenoviruses	Adenovirus infection
Papovaviridae	Papillomavirus	Human papillomavirus (HPV)	Several types of warts
	Polyomavirus	JC virus (JCV)	Progressive multifocal leukoencephalopathy (PML)
Hepadnaviridae	Orthohepadnavirus	Hepatitis B virus (HBV or Dane particle)	Serum hepatitis
RNA Viruses			
Coronaviridae	Coronavirus	SARS-CoV-2	COVID-19
Picornaviridae	Enterovirus	Poliovirus	Poliomyelitis
		Coxsackie virus	Hand-foot-mouth disease
	Hepatovirus	Hepatitis A virus (HAV)	Short-term hepatitis
	Rhinovirus	Human rhinovirus	Common cold, bronchitis
Caliciviridae	Norovirus	Norwalk virus	Viral diarrhea, Norwalk virus syndrome
Flaviviridae	Flavivirus	Dengue fever virus	Dengue fever
		West Nile fever virus	West Nile fever
		Zika virus	Zika virus disease
Filoviridae	Ebolavirus	Ebola, Marburg virus	Ebola fever
Orthomyxoviridae	Influenza A virus	Influenza virus, type A (Asian, Hong Kong, and swine influenza viruses)	Influenza or "flu"
Paramyxoviridae	Morbillivirus	Measles virus	Measles
	Pneumovirus	Respiratory syncytial virus (RSV)	Common cold syndrome
Rhabdoviridae	Lyssavirus	Rabies virus	Rabies
Retroviridae	Deltaretrovirus	Human T-lymphotropic virus I (HTLV-I)	T-cell leukemia
	Lentivirus	HIV (human immunodeficiency viruses 1 and 2)	Acquired immunodeficiency syndrome (AIDS)

Source: CDC/C. S. Goldsmith and A. Balish

focused on the possible connection of viruses to chronic afflictions of unknown cause, such as type 1 diabetes, multiple sclerosis, various cancers, Alzheimer's, and even obesity. Additionally, as mentioned earlier, several cancers have their origins in viral infection.

Table 5.7 provides a list of the most common viruses causing diseases in humans.

Treatment of Animal Viral Infections

The nature of viruses makes it difficult to design effective therapies against them. Because viruses are not bacteria, antibiotics aimed at disrupting bacterial cells do not work on them. Because viruses "borrow" host proteins and functions to propagate themselves, it is difficult to find drugs that will affect the virus without damaging host cells. Almost all antiviral drugs so far licensed have been designed to target one of the steps in the viral life cycle you learned about earlier in this chapter. The

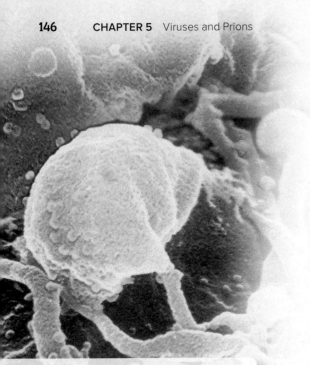

HIV virions (green) being released from the surface of an infected cell (pink).
Source: CDC/Cynthia Goldsmith

integrase inhibitor class of HIV drugs interrupts the ability of HIV genetic information to incorporate into the host cell DNA.

Because antiviral drugs are so much trickier to design than antibacterial drugs, scientists in the 20th century turned to vaccine development to prevent the viral diseases. This is why nearly all of the vaccines available today are targeted at viral diseases. The 21st century will certainly see advances in antiviral drug and vaccine development.

5.6 LEARNING OUTCOMES—Assess Your Progress

15. Analyze the relative importance of viruses in human infection and disease.

16. Discuss the primary reason that antiviral drugs are more difficult to design than antibacterial drugs.

CASE FILE WRAP-UP

©DreamPictures/Pam Ostrow/Blend Images LLC (health care worker); CDC/Charles D. Humphrey (Norovirus)

Norovirus causes inflammation of the gastrointestinal tract. Infected persons experience symptoms such as body aches, fever, nausea, vomiting, and diarrhea. The illness occurs commonly in the community and spreads quickly through contaminated food, liquid, or surfaces. An infected individual may shed billions of viral particles, especially when symptoms are active. It takes only a few (5–20) of these norovirus particles to infect another individual.

The best methods to prevent transmission are handwashing, safe food handling, and cleaning of infected surfaces. In the assisted living facility, isolating residents from shared spaces prevents the spread of disease on common surfaces. Using thorough disinfection techniques to clean up after a resident experiences diarrhea or vomiting helps eliminate viral particles from surfaces.

No treatment exists to eliminate norovirus or many other viruses. While norovirus runs its course, supportive treatment is focused on preventing dehydration for patients with gastrointestinal fluid losses. Drinking liquids, especially those intended for rehydration, are most helpful in replacing fluid and nutrients the body has lost. For those with severe dehydration, hospitalization may be required for intravenous fluid administration.

Are Viruses Part of the Microbiome?

Yes. The sum total of the viruses associated with your body is called the **virome**. As you are learning, viruses must occupy cells as their homes. So the cells of your body are capable of hosting viruses. In many cases, they have little effect on your physiology. But we still don't understand the influence these "quiet" human viruses have on human health. And then there are all the viruses, called *bacteriophages*, that are infecting the bacteria that are part of your microbiome. This adds up to perhaps 10^{15} viruses as part of your microbiome. (Written out, that's 1,000,000,000,000,000.) Scientists estimate that there are between 10^{13} and 10^{14} cells of the human body and even more bacterial cells in and on the average human. So the viruses win in terms of numbers—because there are hundreds to thousands in bacterial cells.

Most people are not aware of this teeming landscape in and on their body. Until rather recently, medical research was concerned only with the viruses that cause diseases—such as rabies, polio, and influenza. But sophisticated molecular techniques have been able to identify the "quiet" viruses—both those that are seemingly always quiet and those that are intermittently quiet but do become pathogenic from time to time. For example, we know that almost every adult is infected with the Epstein-Barr (EB) virus. EB virus is a herpesvirus, and virtually all people are infected with one or more of the herpesviruses. It is estimated that more than 50% of people are infected with the virus causing genital herpes, though the majority—yes, the majority—don't even know it. Most adults have been infected with cytomegalovirus (CMV), which does not seem to affect adults, except to perhaps worsen illnesses with other microbes, but it can cause serious birth defects if infection occurs during pregnancy.

Having said all this, the vast majority of viruses in the human body are not pathogenic. We know that the development of the mammalian placenta was influenced by viruses. Recent research even suggests that the presence of bacteriophages in our intestinal mucosa may play a role in protecting us against bacterial infection. Because of their relatively easy movement between hosts, they have probably contributed genes from other organisms in the biosphere that have affected the development of our own physiology in significant ways.

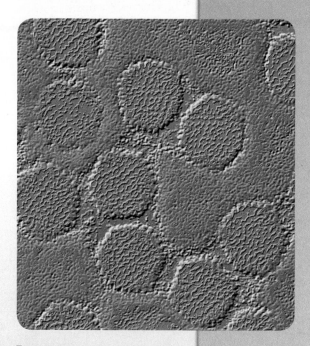

Transmission electron micrograph of the adenovirus. It causes a wide range of human infections, but is so common that it is probably also a part of the human microbiome.
©Science Photo Library RF/Getty Images

Source: Shutterstock/Baloo

VIRUSES AND PRIONS

THE POSITION OF VIRUSES IN THE BIOLOGICAL SPECTRUM
5.1

Viruses are not independent life forms. But they are extremely adept at manipulating their host cells for their own purposes. They have an incredible influence on our planet and on human health.

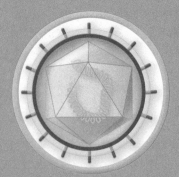

Viruses are very diverse. Unlike other organisms, they have either RNA or DNA, but not both. But they have evolved mechanisms to recreate themselves in varied ways. All viruses have nucleic acid and a capsid. Some of them also have envelopes, which they acquire from the host cell they infect.

THE GENERAL STRUCTURE OF VIRUSES
5.2

HOW VIRUSES MULTIPLY
5.3

Viruses infect all cells. When they infect animal cells, there is a 5-step process they use. When they infect bacteria, there is a similar process. In both hosts, they can become dormant and then reactivate under certain stimuli.

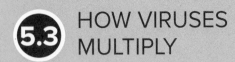

Chris Bjornberg/Science Source

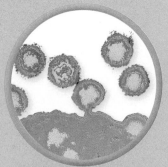

Because viruses require living cells to propagate, it can be complicated to "grow" them. Live bacterial and animal cell cultures can support viral growth in the lab.

TECHNIQUES TO CULTIVATE AND IDENTIFY ANIMAL VIRUSES
5.4

5.5 OTHER NONCELLULAR INFECTIOUS AGENTS

Satellite viruses, viroids, and prions are less complex than viruses. Prions are made entirely of proteins—with no nucleic acid—but they can still replicate themselves and have been found to cause severe human disease.

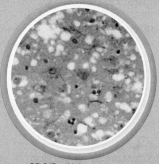

CDC/Dr. Al Jenny

Viruses are easily responsible for several billion infections each year. It is conceivable that many chronic diseases of unknown cause will eventually be connected to viral agents.

5.6 VIRUSES AND HUMAN HEALTH

SmartGrid: From Knowledge to Critical Thinking

This *21 Question Grid* takes the topics from this chapter and arranges them with respect to the American Society for Microbiology's Undergraduate Curriculum guidelines—all six of the important "Concepts" as well as the important "Competency" of scientific literacy. Three questions are supplied about chapter content that refer to the Concept or Competency, in increasing levels of Bloom's taxonomy for learning.

ASM Concept/ Competency	A. Bloom's Level 1, 2—Remember and Understand (Choose one.)	B. Bloom's Level 3, 4—Apply and Analyze	C. Bloom's Level 5, 6—Evaluate and Create
Evolution	1. ___% of human DNA is thought to consist of viral DNA sequences. a. 0.08 b. 8 c. 1 d. 45	2. Discuss the influence that viruses have on the idea of "species" in bacteria.	3. Construct a scenario in which viral latency and lysogeny provide an evolutionary advantage to viruses.
Cell Structure and Function	4. The host cells that viruses can infect are determined by the a. receptors on the host cells. b. DNA in host cells. c. proximity of host cells. d. concentration of host cells in vicinity.	5. If viruses that normally form envelopes were prevented from budding, would they still be infectious? Why or why not?	6. Viruses use the host cell cytoplasmic space as their "factories" and their enzymes and other macromolecules as their "tools." Does that make them more sophisticated or less sophisticated than cells? Justify your answer.
Metabolic Pathways	7. The general steps in a viral multiplication cycle are a. adsorption, penetration, synthesis, assembly, and release. b. endocytosis, uncoating, replication, assembly, and budding. c. adsorption, uncoating, duplication, assembly, and penetration. d. endocytosis, penetration, replication, maturation, and exocytosis.	8. Compare and contrast the processes of latency and lysogeny in viruses.	9. Pathogenic bacteria lysogenized by phages can cause more serious disease than their counterparts that are not lysogenized. Speculate on whether it is to the bacterium's advantage to cause more serious disease or to cause less serious disease.

ASM Concept/ Competency	A. Bloom's Level 1, 2—Remember and Understand (Choose one.)	B. Bloom's Level 3, 4—Apply and Analyze	C. Bloom's Level 5, 6—Evaluate and Create
Information Flow and Genetics	**10.** When phage nucleic acid is incorporated into the nucleic acid of its host cell and is replicated when the host DNA is replicated, this is considered part of which cycle? a. lytic cycle b. virulence cycle c. lysogenic cycle d. cell cycle e. multiplication cycle	**11.** Describe the cell locations in which RNA and DNA viruses multiply.	**12.** RNA viruses tend to mutate—that is, to accumulate changes in their nucleic acid sequences—more frequently than any other organism. Speculate on why that might be and how that can be a positive trait if the virus is a human pathogen.
Microbial Systems	**13.** A virus that undergoes lysogeny is a/an a. temperate phage. b. intemperate phage. c. T-even phage. d. animal virus. e. DNA virus.	**14.** In figure 4.11, viruses are suggested to have contributed to cellular nucleic acid becoming DNA instead of RNA. What stages in the viral life cycle could have led to this phenomenon?	**15.** Imagine and describe a scenario in which viruses in the human microbiome could exert a positive influence on human health.
Impact of Microorganisms	**16.** Clear patches in cell cultures that indicate sites of virus infection are called a. plaques. b. pocks. c. colonies. d. prions.	**17.** Name and discuss one or more medical procedures or treatments that use viruses to benefit human health.	**18.** Construct an argument for whether humans or viruses have had more of an impact on the planet earth as it exists today.
Scientific Thinking	**19.** What is the source of enveloped viruses' envelopes? a. the capsid b. the host cell membrane c. the viral enzymes d. the DNA inside the virus	**20.** Write a paragraph targeted at middle-school students about why antibiotics do not work on viruses, and why it is difficult to design antiviral agents.	**21.** The earliest drugs developed to treat HIV disease were "anti-reverse-transcriptase" agents. Explain why these would be effective and relatively safe.

Answers to the multiple-choice questions appear in Appendix A.

Visual Connections

This question connects content within and between chapters.

From chapter 1, table 1.2. This chart from chapter 1 identified diseases most clearly caused by microorganisms. Considering what you have learned in this chapter, are there more deaths caused by microorganisms than might be accounted for by the diseases in pink? Can you make a rough guess of how many total deaths might be caused by viruses?

TOP CAUSES OF DEATH—ALL CAUSES

United States

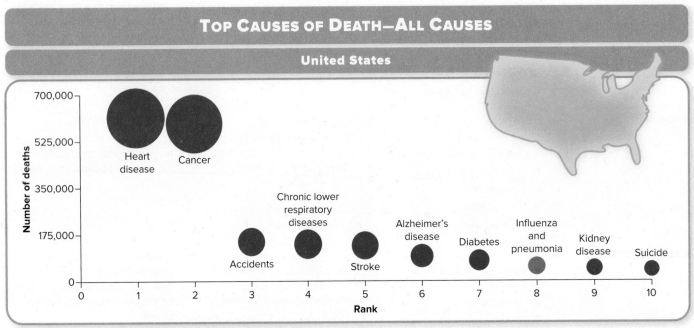

Worldwide

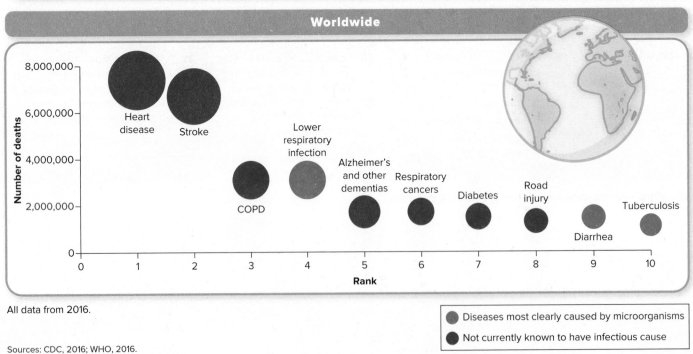

All data from 2016.

Sources: CDC, 2016; WHO, 2016.

- Diseases most clearly caused by microorganisms
- Not currently known to have infectious cause

Chapter design elements: (Covid): CDC/Alissa Eckert, MS; Dan Higgins, MAMS; (Note): McGraw-Hill Education; (NCLEX): Shutterstock/Abert; (Doctor): David Gould/Getty Images

6

Microbial Nutrition and Growth

Terese M. Barta, Ph.D.

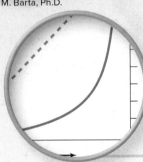

IN THIS CHAPTER...

6.1 Microbial Nutrition

1. List the essential nutrients of a bacterial cell.
2. Differentiate between macronutrients and micronutrients.
3. List and define four different terms that describe an organism's sources of carbon and energy.
4. Define *saprobe* and *parasite*, and explain why these terms can be an oversimplification.
5. Compare and contrast the processes of diffusion and osmosis.
6. Identify the effects of isotonic, hypotonic, and hypertonic conditions on a cell.
7. Name two types of passive transport and one type of active transport.

6.2 Environmental Factors That Influence Microbes

8. List and define five terms used to express the temperature-related growth capabilities of microbes.
9. Summarize three ways in which microorganisms function in the presence of differing oxygen conditions.
10. Identify three important environmental factors (other than temperature and oxygen) with which organisms must cope.
11. List and describe the five types of associations microbes can have with their hosts.
12. Discuss characteristics of biofilms that differentiate them from planktonic bacteria.

6.3 The Study of Bacterial Growth

13. Summarize the steps of bacterial binary fission.
14. Define *doubling time*, and describe how it leads to exponential growth.
15. Compare and contrast the four phases of growth in a bacterial growth curve.
16. Identify one culture-based and one non-culture-based method used for analyzing bacterial growth.

CASE FILE

Wound Care

I was an RN working in a large city hospital on a
medical floor. A lot of our patients had diabetes and were
suffering various complications of the disease, particularly
diabetic wounds caused by poor circulation. Wound care was a
large part of my job. After 2 years on the unit, I decided to pursue wound care
certification. Once I became a wound care specialist, I continued to work in the
same hospital and saw patients with complicated and/or chronic wounds.

Mr. Jones was one of the first patients I consulted about after I became
certified. He was an elderly gentleman who had lost his sight due to diabetes.
When I met Mr. Jones, he had a chronic wound on his lower leg that had been
present for months. The wound was circumferential, taking up half of his lower
leg. It was also grossly infected. Mr. Jones had been admitted to the hospital
for antibiotics to treat his infection. It was clear that if the antibiotics failed to
improve his wound, Mr. Jones was in danger of losing his leg.

Within a day of admission, we realized that antibiotic therapy alone was not
going to be enough. Mr. Jones developed signs of gas gangrene. Wound cultures
were positive for *Clostridium perfringens*, which produces toxins that destroy
muscle tissue and results in sepsis and death if untreated.

Mr. Jones was taken immediately to surgery, where his wound was debrided,
meaning that dead or devitalized tissue was removed. Following surgery, he was
given large doses of penicillin in an effort to stop the spread of the infection.

The next day, Mr. Jones was started on daily hyperbaric oxygen therapy,
with sessions lasting for 45 minutes. Slowly, his wound began to improve. The
wound was debrided twice more under anesthesia, and the patient remained on
antibiotics until wound cultures came back free of *C. perfringens*. Although the
wound took several months to heal, Mr. Jones kept his leg.

- What is hyperbaric oxygen therapy, and why is it used to treat wounds
  infected with *C. perfringens*?

- Is *C. perfringens* an aerobe or an anaerobe?

Case File Wrap-Up appears at the end of the chapter.

Source: Don Stalons/Centers for Disease Control and Prevention (*C. perfringens*); Terese M. Barta, Ph.D. (*tubes*);
Shutterstock/GagliardiImages (*health care worker*)

6.1 Microbial Nutrition

With respect to nutrition, microbes are not really so different from humans. All organisms require a constant influx of certain substances from their habitat. Even bacteria living in mud on a diet of inorganic sulfur, or protozoa digesting wood in a termite's intestine, have some basic nutritional needs. In general, all living things require a source of elements such as carbon, hydrogen, oxygen, phosphorus, potassium, nitrogen, sulfur, calcium, iron, sodium, chlorine, and magnesium. But the ultimate source of a particular element, its chemical form, and how much of it the microbe needs are all points of variation between different types of organisms.

Any substance that must be provided to an organism is called an **essential nutrient.** Two categories of essential nutrients are **macronutrients** and **micronutrients.** Macronutrients are required in relatively large quantities and play principal roles in cell structure and metabolism. Examples of macronutrients are carbon, hydrogen, and oxygen. Micronutrients, or **trace elements,** such as manganese, zinc, and nickel, are present in much smaller amounts and are involved in enzyme function and maintenance of protein structure.

Another way to categorize nutrients is according to their carbon content. An inorganic nutrient is an atom or simple molecule that contains a combination of atoms other than carbon and hydrogen. The natural reservoirs of inorganic compounds are mineral deposits in the crust of the earth, bodies of water, and the atmosphere. Examples include metals and their salts (magnesium sulfate, ferric nitrate, sodium phosphate), gases (oxygen, carbon dioxide), and water. In contrast, the molecules of organic nutrients contain carbon and hydrogen atoms and are usually the products of living things. They range from the simplest organic molecule, methane (CH_4), to large polymers (carbohydrates, lipids, proteins, and nucleic acids). The source of nutrients is extremely varied: Some microbes obtain their nutrients entirely from inorganic sources, and others require a combination of organic and inorganic sources.

Chemical Analysis of Microbial Cytoplasm

Table 6.1 lists the major contents of the bacterium *Escherichia coli.* Some of these components are absorbed in a ready-to-use form, and others must be synthesized by the cell from simple nutrients. The important features of cell composition can be summarized as follows:

- Water is the most abundant of all the components (70%).
- Proteins are the next most prevalent chemical.
- About 97% of the dry cell weight is composed of organic compounds.
- About 96% of the dry cell weight is composed of just six elements (represented by CHONPS and shown later in table 6.3).
- Chemical elements are needed in the overall scheme of cell growth, but most of them are available to the cell as compounds and not as pure elements.
- A cell as "simple" as *E. coli* contains on the order of 5,000 different compounds, yet it needs to absorb only a few types of nutrients to synthesize this great diversity.

What Microbes Eat

The earth's limitless habitats and microbial adaptations make for an elaborate menu of microbial nutritional schemes. Fortunately, most organisms show consistent trends and can be described by a few general categories **(table 6.2)** and a few selected terms. To keep these straight you should always remember that the main determinants of a microbe's nutritional type are its source of carbon and its source of energy.

We'll start with an organism's *carbon source:* In this regard, microbes are either **heterotrophs** or **autotrophs.** A heterotroph is an organism that must obtain its carbon

Table 6.1 The Chemical Composition of an *Escherichia coli* Cell

	% Dry Weight		% Dry Weight
Organic Compounds		**Elements**	
Proteins	50	Carbon (C)	50
Nucleic Acids		Oxygen (O)	20
RNA	20	Nitrogen (N)	14
DNA	3	Hydrogen (H)	8
Carbohydrates	10	Phosphorus (P)	3
Lipids	10	Sulfur (S)	1
Miscellaneous	4	Potassium (K)	1
Inorganic Compounds		Sodium (Na)	1
Water	(none in dry weight)	Calcium (Ca)	0.5
All others	3	Magnesium (Mg)	0.5
		Chlorine (Cl)	0.5
		Iron (Fe)	0.2
		Trace metals	0.3

in an organic form. An autotroph ("self-feeder") is an organism that uses inorganic CO_2 as its carbon source. Because autotrophs have the special capacity to convert CO_2 into organic compounds, they are not nutritionally dependent on other living things.

The next way that microbes are categorized is via their *energy source*. They are either **phototrophs** or **chemotrophs.** Microbes that photosynthesize (create energy from sunlight) are phototrophs, and those that get their energy from chemical compounds are chemotrophs.

Escherichia coli
Source: CDC/ Peggy S. Hayes & Elizabeth H. White, M.S.

Table 6.2 Nutritional Categories of Microbes by Energy and Carbon Source

Category	Energy Source	Carbon Source	Example
Autotroph			
Photoautotroph	Sunlight	CO_2	Photosynthetic organisms, such as algae, plants, cyanobacteria
Chemoautotroph			
Chemoorganic autotrophs	Organic compounds	CO_2	Methanogens
Chemolithoautotrophs	Inorganic compounds (minerals)	CO_2	*Thiobacillus*, "rock-eating" bacteria
Heterotroph			
Photoheterotroph	Sunlight	Organic	Purple and green photosynthetic bacteria
Chemoheterotroph	Metabolic conversion of the nutrients from other organisms	Organic	Protozoa, fungi, many bacteria, animals
Saprobe	Metabolizing the organic matter of dead organisms	Organic	Fungi, bacteria (decomposers)
Parasite	Utilizing the tissues, fluids of a live host	Organic	Various parasites and pathogens; can be bacteria, fungi, protozoa, animals

The terms for carbon and energy source are often merged into a single word for convenience. The categories described here are meant to describe only the major nutritional groups and do not include unusual exceptions. **Figure 6.1** illustrates two examples.

Autotrophs and Their Energy Sources

Autotrophs—that is, organisms that get their carbon from CO_2—derive energy from one of two possible nonliving sources: sunlight and chemical reactions involving simple chemicals. When they get their energy from sunlight, they are called photo-autotrophs, and when they get it from chemicals, they are called chemoautotrophs. **Photoautotrophs** are photosynthetic—that is, they capture the energy of light rays and transform it into chemical energy that can be used in cell metabolism. Because photosynthetic organisms (algae, plants, some bacteria) produce organic molecules that can be used by themselves and by heterotrophs, they form the basis for most food webs.

Chemoautotrophs are of two types: One of these is the group called chemoorganic autotrophs. These use organic compounds for energy and inorganic compounds as a carbon source. The second type of chemoautotroph is a group called **lithoautotrophs,** which require neither sunlight nor organic nutrients, relying totally on inorganic minerals. These bacteria derive energy in diverse and rather amazing ways. In very simple terms, they remove electrons from inorganic substrates—such as hydrogen gas, hydrogen sulfide, sulfur, or iron—and combine them with carbon dioxide and hydrogen.

Heterotrophs and Their Energy Sources

The majority of heterotrophic microorganisms are **chemoheterotrophs** that derive both carbon and energy from organic compounds. Processing these organic molecules by cellular respiration or fermentation releases energy in the form of ATP. Chemoheterotrophic microorganisms belong to one of two main categories that differ in how they obtain their organic nutrients: Saprobes are free-living microorganisms that feed primarily on organic detritus from dead organisms, and parasites ordinarily derive nutrients from the cells or tissues of a living host.

Saprobes occupy a niche as decomposers of plant litter, animal matter, and dead microbes. If not for the work of these decomposers, the earth would gradually fill up with organic material, and the nutrients it contains would not be recycled.

Parasites live in or on the body of a host, which they harm to some degree. Because many parasites cause damage to tissues (disease) or even death, they are also called *pathogens.* Parasites range from viruses to helminths (worms), and they can live on the body (ectoparasites), in the organs and tissues (endoparasites), or even within cells (intracellular parasites, the most extreme type). *Obligate parasites* (for example, the leprosy bacillus and the syphilis spirochete) are unable to grow outside of a living host. Parasites that have less strict requirements can be cultured artificially if provided with the correct nutrients and environmental conditions. Bacteria such as *Streptococcus pyogenes* (the cause of strep throat) and *Staphylococcus aureus* can grow on artificial media.

The binary "saprobe" versus "parasite" distinction is a bit of a simplification, which we will address in a later section. For now, just note that the vast majority of microbes causing human disease are chemoheterotrophs.

Essential Nutrients

Chemicals that are necessary for particular organisms, which they cannot manufacture by themselves, are called *essential.* For microbes, the essential nutrients are carbon, hydrogen, oxygen, nitrogen, phosphate, and sulfur. Biologists remember these elements with the acronym CHONPS **(table 6.3).**

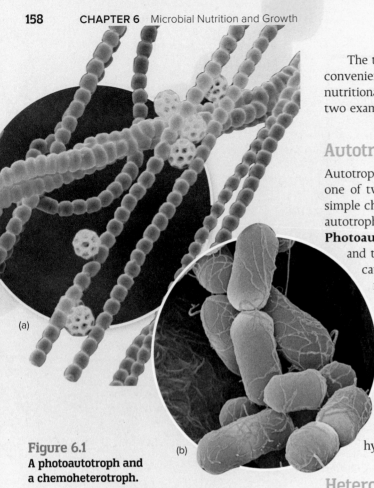

(a)

(b)

Figure 6.1

A photoautotroph and a chemoheterotroph.

(a) *Cyanobacterium,* in green, a photosynthetic autotroph.

(b) *Escherichia coli,* a chemoheterotroph.

(a) STEVE GSCHMEISSNER/Getty Images; (b) Martin Oeggerli/Science Source

 NCLEX® PREP

1. A saprobe derives its energy from
   a. *sunlight.*
   b. *conversion of nutrients from other organisms.*
   c. *utilizing the tissues/fluids of a living host.*
   d. *metabolizing the organic matter of dead organisms.*

 NCLEX® PREP

2. Mineral ions used in microbial nutrition include
   a. *sodium.*
   b. *potassium.*
   c. *calcium.*
   d. *magnesium.*
   e. *all of these.*

Table 6.3 Essential Nutrients—Where They Can Be Obtained

Carbon	Among the common organic molecules that can satisfy this requirement are proteins, carbohydrates, lipids, and nucleic acids. In most cases, these molecules provide several other nutrients as well.
Hydrogen	Hydrogen is a major element in all organic and several inorganic compounds, including water (H_2O), salts ($Ca[OH]_2$), and certain naturally occurring gases (H_2S, CH_4, and H_2). These gases are both used and produced by microbes. Hydrogen helps cells maintain their pH, is useful for forming hydrogen bonds between molecules, and also serves as a source of free energy in respiration.
Oxygen	Because oxygen is a major component of organic compounds such as carbohydrates, lipids, nucleic acids, and proteins, it plays an important role in the structural and enzymatic functions of the cell. Oxygen is likewise a common component of inorganic salts such as sulfates, phosphates, nitrates, and water. Free gaseous oxygen (O_2) makes up 20% of the atmosphere.
Nitrogen	The main reservoir of nitrogen is nitrogen gas (N_2), which makes up 79% of the earth's atmosphere. This element is indispensable to the structure of proteins, DNA, RNA, and ATP. Such compounds are the primary nitrogen source for heterotrophs, but to be useful, they must first be degraded into their basic building blocks (proteins into amino acids; nucleic acids into nucleotides). Some bacteria and algae utilize inorganic nitrogenous nutrients (NO_3^-, NO_2^-, or NH_3). A small number of bacteria and archaea can transform N_2 into compounds usable by other organisms through the process of nitrogen fixation. Regardless of the initial form in which the inorganic nitrogen enters the cell, it must first be converted to NH_3, the only form that can be directly combined with carbon to synthesize amino acids and other compounds.
Phosphate	The main inorganic source of phosphorus is phosphate (PO_4^{3-}), derived from phosphoric acid (H_3PO_4) and found in rocks and oceanic mineral deposits. Phosphate is a key component of nucleic acids and is therefore essential to the genetics of cells and viruses. Because it is also found in ATP, it serves in cellular energy transfers. Other phosphate-containing compounds are phospholipids in cytoplasmic membranes and coenzymes such as NAD^+.
Sulfur	Sulfur is widely distributed throughout the environment in mineral form. Rocks and sediments (such as gypsum) can contain sulfate (SO_4^{2-}), sulfides (FeS), hydrogen sulfide gas (H_2S), and elemental sulfur (S). Sulfur is an essential component of some vitamins (vitamin B_1) and the amino acids methionine and cysteine; the latter help determine shape and structural stability of proteins by forming unique linkages called disulfide bonds.

Other Important Nutrients

Mineral ions are also important components in microbial metabolism. Potassium is essential to protein synthesis and membrane function. Sodium is important for certain types of cell transport. Calcium is a stabilizer of the cell wall and endospores of bacteria. Magnesium is a component of chlorophyll and a stabilizer of membranes and ribosomes. Iron is an important component of the cytochrome proteins of cell respiration. Zinc is an essential regulatory element for eukaryotic genetics. It is a major component of "zinc fingers"–binding factors that help enzymes adhere to specific sites on DNA. Copper, cobalt, nickel, molybdenum, manganese, silicon, iodine, and boron are needed in small amounts by some microbes but not others. On the other hand, in chapter 9 you will see that metals can also be very toxic to microbes. The concentration of metal ions can even influence the diseases microbes cause. For example, the bacteria that cause gonorrhea and meningitis grow more rapidly in the presence of iron.

How Microbes Eat: Transport Mechanisms

A microorganism's habitat provides necessary nutrients—some abundant, others scarce. But they must still be taken into the cell. Survival also requires that cells transport waste materials out of the cell and into the environment. Whatever the direction, transport occurs across the cytoplasmic membrane, the structure specialized for this role. This is true even in organisms with cell walls (bacteria, algae, and fungi) because the cell wall is usually too nonselective to screen the entrance or exit of molecules.

The driving force of transport is atomic and molecular movement—the natural tendency of atoms and molecules to be in constant random motion. This phenomenon of molecular movement, in which atoms or molecules move in a gradient from an area of higher density or concentration to an area of lower density or concentration, is **diffusion.**

Medical Moment

Osmosis and IV Fluids

Administering intravenous (IV) solutions is a very common practice in medicine. The osmotic movement of water occurs as the body attempts to create a balance between the different solute concentrations that exist on either side of a semipermeable membrane. Keeping that in mind, let's look at different types of IV solutions commonly used in medicine.

Isotonic solutions have a tonicity that is the same as the body's plasma. When isotonic solutions are administered, there will be very little movement, if any, between the body tissues and the blood vessels.

Hypertonic solutions have a tonicity that is higher than the body plasma. Administering hypertonic solutions will cause water to shift from the extravascular spaces into the bloodstream to increase the intravascular volume. This is how the body attempts to dilute the higher concentration of electrolytes in the IV fluid.

Hypotonic solutions have a tonicity that is lower than the body plasma, causing water to shift from the intravascular to the extravascular space, and eventually into the cells of the tissues. In this case, the body moves water from the intravascular space to the cells in order to dilute the electrolytes in the cells.

Q. In this situation, what structure is acting as the semipermeable membrane?

Answer in Appendix B.

NCLEX® PREP

3. A physician has ordered hypotonic parenteral therapy for a postoperative client over a 24-hour period. Based on this order, what is the *priority* nursing action?

   a. *Assess the patient's labs and verify the order with the pharmacy prior to hanging the ordered fluid.*

   b. *Contact the physician and request an order for hypertonic solution instead.*

   c. *Place the client on intake and output measures.*

   d. *Begin intravenous hypotonic fluids immediately.*

Figure 6.2 **Model system to demonstrate osmosis.** Here we have a solution enclosed in a sack-shaped membrane and attached to a hollow tube. The membrane is permeable to water (solvent) but not to solute. The sack is immersed in a container of pure water. In the inset, you see that the net direction of water diffusion is into the sack.

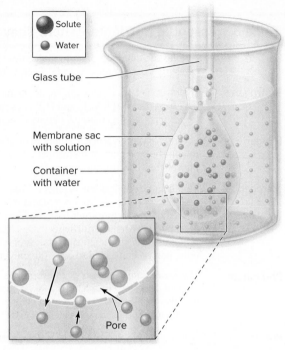

The Movement of Water: Osmosis

Diffusion of water through a selectively permeable membrane, a process called **osmosis,** is a physical phenomenon that is easily demonstrated in the laboratory with nonliving materials. It provides a model of how cells deal with various solute concentrations in aqueous solutions. In an osmotic system, the membrane is *selectively,* or *differentially, permeable.* This means that they have passageways that allow free diffusion of water but that block other dissolved molecules **(figure 6.2).** When this membrane is placed between solutions of differing concentrations and the solute cannot pass through the membrane, then under the laws of diffusion, water will diffuse at a faster rate from the side that has more water to the side that has less water. As long as the concentrations of the solutions differ, one side will experience a net loss of water and the other a net gain of water, until equilibrium is reached and the rate of diffusion is equalized.

Osmosis in living systems is similar to the model shown in **figure 6.3.** Living membranes generally block the entrance and exit of larger molecules and permit free diffusion of water. Because most cells are surrounded by some free water, the amount of water entering or leaving has a far-reaching impact on cellular activities and survival. This osmotic relationship between cells and their environment is determined by the relative concentrations of the solutions on either side of the cytoplasmic membrane (figure 6.3). Such systems can be compared using the terms *isotonic,* *hypotonic,* and *hypertonic.* (The root *-tonic* means "tension." *Iso-* means "the same," *hypo-* means "less," and *hyper-* means "more" or "greater.")

Under **isotonic** conditions, the environment is equal in solute concentration to the cell's internal environment, and because diffusion of water proceeds at the same rate in both directions, there is no net change in cell volume. Isotonic solutions are generally the most stable environments for cells because they are already in an osmotic steady state with the cell. Microorganisms living in host tissues are most likely to be living in isotonic habitats.

Under **hypotonic** conditions, the solute concentration of the external environment is lower than that of the cell's internal environment. Pure water provides the most hypotonic environment for cells because it has no solute. The net direction of osmosis is from the hypotonic solution into the cell, and cells without walls swell and can burst.

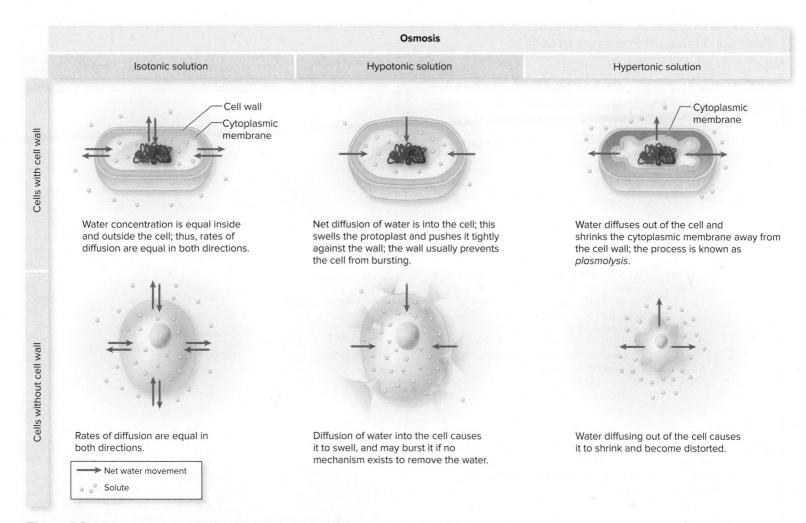

Figure 6.3 Cell responses to solutions of differing osmotic content. Note that, unlike in figure 6.2, there is no tube into which the extra fluid can rise.

A slightly hypotonic environment can be a good situation for bacterial cells. The constant slight tendency for water to flow into the cell keeps the cytoplasmic membrane fully extended and the cytoplasm full. This is the optimum condition for the many processes occurring in and on the membrane. Slight hypotonicity is tolerated quite well by most bacteria because of their rigid cell walls.

Hypertonic conditions are also out of balance with the tonicity of the cell's cytoplasm, but in this case, the environment has a higher solute concentration than the cytoplasm. Because a hypertonic environment will force water to diffuse out of a cell, it is said to have high *osmotic pressure*, or potential. Microbes in hypertonic solutions often cannot grow or metabolize because their membranes are so wrinkled up that many cell processes are impossible. This is the principle behind using concentrated salt and sugar solutions as preservatives for food, such as in salted hams, because the microbes in them will be stopped from growing by the hypertonic state.

Movement of Solutes

Simple diffusion is limited to small nonpolar molecules like oxygen or lipid-soluble molecules that may pass through the membranes. But it is imperative that a cell be able to move polar molecules and ions across the plasma membrane as well, and this is impossible via simple diffusion. So microbes have developed multiple mechanisms to move substances across membranes. We look at these (and simple diffusion) in **table 6.4.**

Table 6.4 Transport Processes in Cells

	Examples	Description	Energy Requirements	
Passive	Simple diffusion	A fundamental property of atoms and molecules that exist in a state of random motion	None. Substances move on a gradient from higher concentration to lower concentration.	
	Facilitated diffusion	Molecule binds to a specific receptor in membrane and is carried to other side. Molecule-specific. Goes both directions. Rate of transport is limited by the number of binding sites on transport proteins.	None. Substances move on a gradient from higher concentration to lower concentration.	Membrane
Active	Carrier-mediated active transport	Atoms or molecules are pumped into or out of the cell by specialized receptors.	Driven by ATP or the proton motive force	Membrane

As you see in table 6.4, very often energy is required to move molecules into or out of cells. In that case, the process is more accurately called transport as opposed to diffusion, and is seen as "active." Features inherent in **active transport** systems are

1. the transport of nutrients against the diffusion gradient or in the same direction as the natural gradient but at a rate faster than by diffusion alone;
2. the presence of specific membrane proteins (permeases and pumps); and
3. the expenditure of energy.

Examples of substances transported actively are monosaccharides, amino acids, organic acids, phosphates, and metal ions.

Endocytosis: Eating and Drinking by Cells

Some eukaryotic cells transport large molecules, particles, liquids, or even other cells across the cell membrane. Because the cell usually expends energy to carry out this transport, it is also a form of active transport. The substances transported do not pass physically through the membrane but are carried into the cell by endocytosis. First the cell encloses the substance in its membrane, simultaneously forming a vacuole and engulfing it. Amoebas and certain white blood cells ingest whole cells or large solid

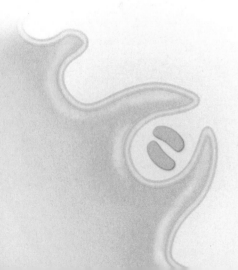

matter by a type of endocytosis called **phagocytosis.** Liquids, such as oils or molecules in solution, enter the cell through a type of endocytosis called **pinocytosis.**

6.1 LEARNING OUTCOMES—Assess Your Progress

1. List the essential nutrients of a bacterial cell.
2. Differentiate between macronutrients and micronutrients.
3. List and define four different terms that describe an organism's sources of carbon and energy.
4. Define *saprobe* and *parasite*, and explain why these terms can be an oversimplification.
5. Compare and contrast the processes of diffusion and osmosis.
6. Identify the effects of isotonic, hypotonic, and hypertonic conditions on a cell.
7. Name two types of passive transport and one type of active transport.

6.2 Environmental Factors That Influence Microbes

Microbes are exposed to a wide variety of environmental factors in addition to nutrients. These include such factors as heat, cold, gases, pH, radiation, osmotic and hydrostatic pressures, and even the effects of other microbes. For most microbes, environmental factors fundamentally affect the function of metabolic enzymes. Survival in a changing environment is largely a matter of whether the enzyme systems of microorganisms can adapt to alterations in their habitat.

Temperature

Microbial cells are unable to control their own temperature and therefore take on the ambient temperature of their natural habitats. Their survival is dependent on adapting to whatever temperature variations are encountered in that habitat. The range of temperatures for the growth of a given microbial species can be expressed as three *cardinal temperatures*. The **minimum temperature** is the lowest temperature that permits a microbe's continued growth and metabolism. Below this temperature, its activities stop. The **maximum temperature** is the highest temperature at which growth and metabolism can proceed. If the temperature rises slightly above maximum, growth will temporarily stop, but if it continues to rise beyond that point, the enzymes and nucleic acids will eventually become permanently inactivated (otherwise known as denaturation), and the cell will die. This is why heat works so well as an agent in microbial control. The **optimum temperature** covers a small range, intermediate between the minimum and maximum, which promotes the fastest rate of growth and metabolism (rarely is the optimum a single point).

Depending on their natural habitats, some microbes have a narrow cardinal range, others a broad one. Some strict parasites will not grow if the temperature varies more than a few degrees below or above the host's body temperature. For instance, the typhus bacterium multiplies only in the range of 32°C to 38°C, and rhinoviruses (one cause of the common cold) multiply most successfully in tissues that are slightly below normal body temperature (33°C to 35°C). Other organisms are not so limited. Strains of *Staphylococcus aureus* grow within the range of 6°C to 46°C, and the intestinal bacterium *Enterococcus faecalis* grows within the range of 0°C to 44°C.

Another way to express temperature adaptation is to describe whether an organism grows optimally in a cold, moderate, or hot temperature range. The terms used for these ecological groups are *psychrophile, mesophile,* and *thermophile* **(figure 6.4),** respectively.

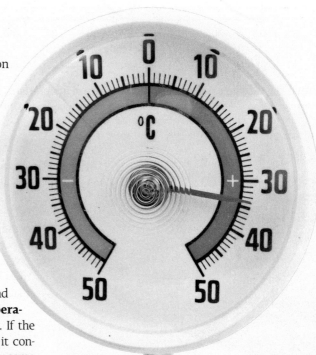

©Boris Sosnovyy/Shutterstock.com

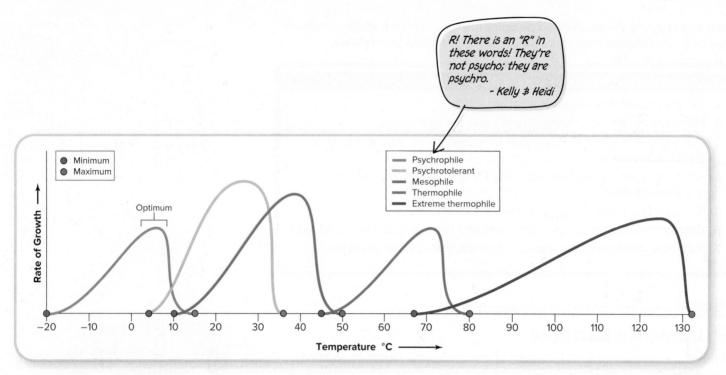

Figure 6.4 Ecological groups by temperature of adaptation. Psychrophiles can grow at or below 0°C and have an optimum below 15°C. Psychrotolerants have an optimum of from 15°C to 30°C. As a group, mesophiles can grow between 10°C and 50°C, but their optima usually fall between 20°C and 40°C. Generally speaking, thermophiles require temperatures above 45°C and grow optimally between this temperature and 80°C. Extreme thermophiles have optima above 80°C. Note that the ranges can overlap to an extent.

(a)

(b)

Figure 6.5 Red snow. (a) An early summer snowbank provides a perfect habitat for psychrophilic photosynthetic organisms like *Chlamydomonas nivalis*.
(b) Microscopic view of this snow alga (actually classified as a green alga, although a red pigment dominates at this stage of its life cycle).
(a) Francois Gohier/Science Source; (b) Courtesy Nozomu Takeuchi

A **psychrophile** (sy´-kroh-fyl)–the blue line in figure 6.4–is a microorganism that has an optimum temperature below 15°C and is still capable of growth at 0°C. It is obligate with respect to cold (i.e., it can only grow in cold temperatures) and generally cannot grow above 20°C. Unlike most laboratory cultures, storage in the refrigerator causes them to grow, rather than inhibiting them. As one might predict, the habitats of psychrophilic bacteria, fungi, and algae are lakes and rivers, snowfields **(figure 6.5)**, polar ice, and the deep ocean. Rarely, if ever, are they pathogenic. True psychrophiles must be distinguished from the less extreme *psychrotolerant* (the gold line in figure 6.4) that grow slowly in cold but have an optimum temperature between 15°C and 30°C.

The majority of medically significant microorganisms are **mesophiles** (mez´-oh-fylz; the green line in figure 6.4), organisms that grow at intermediate temperatures. The optimum growth temperatures (optima) of most mesophiles fall into the range of 20°C to 40°C. Organisms in this group inhabit animals and plants as well as soil and water in temperate, subtropical, and tropical regions. Most human pathogens grow optimally somewhere between 30°C and 40°C (human body temperature is 37°C). Some mesophilic bacteria, such as *Staphylococcus aureus*, grow optimally at body temperature but are also facultatively *psychrotolerant* meaning they can survive and multiply slowly at refrigerator temperatures, causing concern for food storage. (**Facultative** is a term used in biology that designates an organism as capable of growing under differing sets of conditions. We will see the term again when we discuss oxygen requirements.) *Listeria monocytogenes* is a human pathogen that is truly psychrotolerant, meaning its optimum growth is between 30°C and 40°C. But it will grow slowly at temperatures as low as 1°C, and often grows in ice cream and refrigerated meat. *Thermoduric* microbes, which can survive short exposure to high temperatures but are normally mesophiles, are common contaminants of heated or pasteurized foods. Examples include heat-resistant endospore formers such as *Bacillus* and *Clostridium*.

A **thermophile** (thur´-moh-fyl; the pink line in figure 6.4) is a microbe that grows optimally at temperatures greater than 45°C. Such heat-loving microbes live in soil and water associated with volcanic activity, in compost piles, and in habitats directly exposed to the sun. Thermophiles vary in heat requirements, with a general range of growth of 45°C to 80°C. Most eukaryotic forms cannot survive above 60°C, but a few thermophilic bacteria, called extreme thermophiles (the brown line in figure 6.4), grow between 80°C and 121°C.

Gases

The atmospheric gases that most influence microbial growth are O_2 and CO_2. Of these, oxygen gas has the greatest impact on microbial growth. Not only is it an important respiratory gas, but it is also a powerful oxidizing agent that exists in many toxic forms. In general, microbes fall into one of three categories:

- those that use oxygen and can detoxify it,
- those that can neither use oxygen nor detoxify it, and
- those that do not use oxygen but can detoxify it.

How Microbes Process Oxygen

As oxygen is acted on by cellular enzymes, it is transformed into several toxic products. *Singlet oxygen* (O) is an extremely reactive molecule. Notably, it is one of the substances produced by phagocytes to kill invading bacteria. The buildup of singlet oxygen and the oxidation of membrane lipids and other molecules can damage and destroy a cell. The highly reactive *superoxide ion* (O_2^-), *hydrogen peroxide* (H_2O_2), and *hydroxyl radicals* (OH^-) are other destructive metabolic by-products of oxygen. To protect themselves against damage, most cells have developed enzymes that go about the business of scavenging and neutralizing these chemicals. The complete conversion of superoxide ion into harmless oxygen requires a two-step process and at least two enzymes:

$$\text{Step 1.} \quad 2O_2^- + 2H^+ \xrightarrow{\substack{\text{Superoxide} \\ \text{dismutase}}} H_2O_2 \text{ (hydrogen peroxide)} + O_2$$

$$\text{Step 2.} \quad 2H_2O_2 \xrightarrow{\text{Catalase}} 2H_2O + O_2$$

In this series of reactions (essential for aerobic organisms), the superoxide ion is first converted to hydrogen peroxide and normal oxygen by the action of an enzyme called superoxide dismutase. Because hydrogen peroxide is also toxic to cells (after all, it is used as a disinfectant and antiseptic), it must be degraded by the enzyme catalase into water and oxygen. With this system, toxic superoxide ion is completely broken down into harmless water and oxygen. If a microbe is not capable of dealing with toxic oxygen by these or similar mechanisms, it is forced to live in habitats free of oxygen.

Because oxygen requirements of microorganisms differ so dramatically and are so important clinically, microbes are grouped into several general categories **(table 6.5)**. **Figure 6.6** depicts the growth of different microbes in tubes of fluid thioglycollate. The location of growth indicates the oxygen requirements of the microbes.

Carbon Dioxide

Although all microbes require some carbon dioxide in their metabolism, *capnophiles* grow best at a higher CO_2 tension than is normally present in the atmosphere. Some notable capnophiles are *Neisseria* (a genus causing gonorrhea and meningitis), *Brucella* (undulant fever), and *Streptococcus pneumoniae*. Growing these from clinical specimens requires providing a higher CO_2 tension than normal.

pH

The term **pH** is defined as the degree of acidity or alkalinity (basicity) of a solution. It is expressed by the pH scale, a series of numbers ranging from 0 to 14. The pH of fresh pure water (7.0) is neutral, neither acidic nor basic. As the pH value decreases toward 0, the acidity increases, and as the pH increases toward 14, the alkalinity increases. The majority of organisms are *neutrophiles* and live or grow in habitats between pH 6 and 8 because strong acids and bases can be highly damaging to enzymes and other cellular substances.

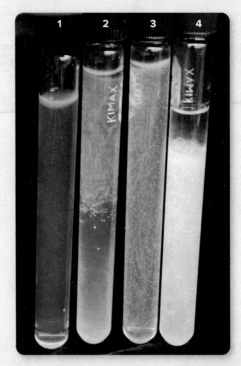

Figure 6.6 Four tubes showing three different patterns of oxygen utilization. These tubes use thioglycollate, which reduces oxygen to water, to restrict oxygen diffusion through the agar. So, whereas there is an oxygen-rich layer at the top of the agar, the oxygen concentration rapidly decreases deeper in the agar. In tube 1, the obligately aerobic *Pseudomonas aeruginosa* grows only at the very top of the agar. Tubes 2 and 3 contain two different examples of facultatively anaerobic bacteria. Many facultatives, though able to grow both aerobically and anaerobically, grow more efficiently in the aerobic mode. This is more obvious in tube 2 (*Staphylococcus aureus*) and less obvious in tube 3 (*Escherichia coli*). Tube 4 contains *Clostridium butyricum,* an obligate anaerobe.
Terese M. Barta, Ph.D.

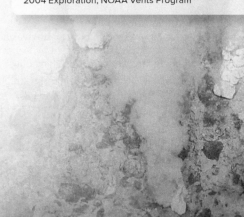

Communities of extreme thermophilic microorganisms live around hot deep-sea ocean vents.

Image courtesy of Submarine Ring of Fire 2004 Exploration, NOAA Vents Program

Table 6.5 Oxygen Usage and Tolerance Patterns in Microbes

Culture Appearance

In all cases, bacteria below are grown in a medium called thioglycollate, which allows anaerobic bacteria to grow in tubes exposed to air. Oxygen concentration is highest at the top of the tube, and lowest at the bottom.

Aerobes

Can use gaseous oxygen in their metabolism and possess the enzymes needed to process toxic oxygen products. An organism that cannot grow without oxygen is an *obligate aerobe.*

Examples: Most fungi, protozoa, and many bacteria, such as *Bacillus* species and *Mycobacterium tuberculosis*

(Obligate aerobe)

Growth

Microaerophiles

Are harmed by normal atmospheric concentrations of oxygen but require a small amount of it in metabolism.

Examples: Organisms that live in soil or water or in mammalian hosts, not directly exposed to atmosphere; *Helicobacter pylori*, *Borrelia burgdorferi*

Growth

Facultative anaerobes

Do not require oxygen for metabolism but use it when it is present. Can also perform anaerobic metabolism. In the tube to the right, the bacteria are growing throughout, but there is heavier growth in the aerobic portion of the tube (upper) because aerobic growth can proceed more quickly in some facultative anaerobes.

Examples: Many gram-negative intestinal bacteria, staphylococci

Growth

Anaerobes

Lack the metabolic enzyme systems for using oxygen in respiration. *Obligate anaerobes* also lack the enzymes for processing toxic oxygen and die in its presence.

Examples: Many oral bacteria, intestinal bacteria

(Obligate anaerobe)

Growth

Aerotolerant anaerobes

Do not utilize oxygen but can survive and grow to a limited extent in its presence. They are not harmed by oxygen, mainly because they possess alternate mechanisms for breaking down peroxides and superoxide.

Examples: Certain lactobacilli and streptococci, clostridial species

Growth

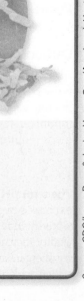

Source: CDC/Laura Rose & Janice Haney Carr (Aerobess); Heather Davies/Science Source (Microaerophiles); Janice Haney Carr/CDC (Facultative anaerobes); Source: CDC/Janice Carr (Anaerobes); Janice Haney Carr/CDC (Aerotolerant anaerobes)

A few microorganisms live at pH extremes. Obligate *acidophiles* include *Euglena mutabilis*, an alga that grows in acid pools between 0 and 1.0 pH, and *Thermoplasma*, an archaea that lacks a cell wall and lives in hot coal piles at a pH of 1 to 2, and will die if exposed to pH 7. *Picrophilus* thrives at a pH of 0.7, and can grow at a pH of 0. Because many molds and yeasts tolerate moderate acid, they are the most common spoilage agents of pickled (acidic) foods. *Alkalinophiles*, such as *Natronomonas* species, live in hot pools and soils that contain high levels of basic minerals (up to pH 12.0).

Some neutrophiles can grow in environments outside of their pH range due to compounds they produce during metabolism. *Helicobacter* is an example of this. Bacteria that decompose urine create alkaline conditions because ammonium (NH_4^+) can be produced when urea (a component of urine) is digested. (Ammonium raises the pH of solutions.) Metabolism of urea is one way that *Proteus* spp. can neutralize the acidity of the urine to colonize and infect the urinary system.

Osmotic Pressure

Although most microbes exist under hypotonic or isotonic conditions, a few, called **osmophiles,** live in habitats with a high solute concentration. One common type of osmophile prefers high concentrations of salt; these organisms are called **halophiles** (hay´-loh-fylz). Obligate halophiles such as *Halobacterium* and *Halococcus* inhabit salt lakes, ponds, and other hypersaline habitats. They grow optimally in solutions of 25% NaCl but require at least 9% NaCl (combined with other salts) for growth. These archaea have significant modifications in their cell walls and membranes and will lyse in hypotonic habitats. Facultative halophiles are remarkably resistant to salt, even though they do not normally reside in high-salt environments. For example, *Staphylococcus aureus* can grow on NaCl media ranging from 0.1% up to 20%.

Radiation and Hydrostatic/Atmospheric Pressure

Radiation

Some microbes (phototrophs) can use visible light rays as an energy source, but nonphotosynthetic microbes tend to be damaged by the toxic oxygen products produced by contact with light. Some microbial species produce yellow carotenoid pigments to protect against the damaging effects of light by absorbing and dismantling toxic oxygen. Other types of radiation that can damage microbes are ultraviolet and ionizing rays (X rays and cosmic rays). In chapter 9, you will see just how these types of energy are used to control the growth of microbes.

Pressure

The ocean depths subject organisms to increasing hydrostatic pressure. Deep-sea microbes called **barophiles** (referring to barometric pressure) exist under pressures that range from a few times to over 1,000 times the pressure of the atmosphere. These bacteria are so strictly adapted to high pressures that they will rupture when exposed to normal atmospheric pressure.

Other Organisms

Up to now, we have considered the importance of nonliving environmental influences on the growth of microorganisms. Another profound influence comes from other organisms that share (or sometimes *are*) their habitats. In all but the rarest instances, microbes live in shared habitats, which give rise to complex and fascinating associations. Some associations are between similar or dissimilar types of microbes; others involve multicellular organisms such as animals or plants. Interactions can have beneficial, harmful, or no particular effects on the organisms involved. They can be obligatory or nonobligatory to the members; and they often involve nutritional interactions. This outline provides an overview of the major types of microbial associations:

Cucumbers placed in a hypertonic solution turn into the smaller, denser pickles.
©Photodisc/Alamy

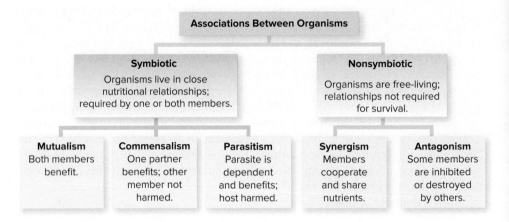

Strong Partnerships: Symbioses

A general term used to denote a situation in which two organisms live together in a close partnership is **symbiosis**, and the members are termed *symbionts*. In everyday language, the term "symbiosis" is almost always used in a positive sense, as in, "Those roommates have a symbiotic relationship." In biology, symbiotic relationships can be either positive or negative. The term simply denotes a close partnership. Three main types of symbiosis occur:

- **Mutualism** exists when organisms live in an obligatory and mutually beneficial relationship. Many microbes that live in or on humans fall in this category. The microbes receive necessary nutrients, and the host gets a variety of benefits, ranging from protection from pathogenic microbes to the healthy development of the immune system that is only possible when a robust resident microbiome is present.
- In a relationship known as **commensalism**, the member called the commensal receives benefits, while its partner is neither harmed nor benefited. Most "normal biota" microbes used to be considered commensals, with the assumption that they provided neither harm nor benefit to their hosts. We now know that the normal biota contribute greatly to the health of their human host, so they are more accurately classified as mutualists.
- **Parasitism** is a relationship in which the host organism provides the parasitic microbe with nutrients and a habitat, and the host suffers from the relationship. Microbes that make humans sick fall in this category.

There is an important point to be made here about the use of the word *parasite* in this context. Earlier in this chapter we put organisms into the categories of "saprobe" or "parasite," meaning that they live either off of dead material or a live host, respectively. Scientists have come to learn that not all organisms that require a living host necessarily *harm* the host. So a more nuanced view of the nutritional requirement categories might be "saprobe" and "nonsaprobe"–with the "nonsaprobe" category having both parasitic and commensal members.

Associations but Not Partnerships: Nonsymbiotic Interactions

Even when organisms are not engaged in symbiotic relationships, they are interacting. Relationships between free-living species can have either negative or positive results. **Antagonism** is an association between free-living species that arises when members of a community compete. In this interaction, one microbe secretes chemical substances into the surrounding environment that inhibit or destroy another microbe in the same habitat. The first microbe may gain a competitive advantage by increasing the space and nutrients available to it. Interactions of this type are common in the soil, where mixed communities often compete for space and food. *Antibiosis*–the production of inhibitory compounds such as antibiotics–is actually a form of antagonism. Hundreds of naturally occurring antibiotics have been isolated from bacteria and fungi and used as drugs to control diseases.

Synergism is an interrelationship between two or more free-living organisms that benefits them but is not necessary for their survival. Together, the participants cooperate to produce a result that none of them could do alone. Gum disease, dental caries, and some bloodstream infections involve mixed infections by bacteria interacting synergistically.

Biofilms: The Epitome of Synergy

Biofilms are communities of bacteria and/or other microbes that are attached to a surface and to each other, forming a multilayer conglomerate of cells and intracellular material. Usually there is a "pioneer" colonizer, a bacterium that initially attaches to a surface, such as a tooth or the lung tissue **(figure 6.7).** Other microbes then attach either to those bacteria or to the complex sugar and protein substance that inevitably is secreted by microbial colonizers of surfaces. In many cases, once the cells are attached, they are stimulated to release chemicals that accumulate as the cell population grows. By this means, they can monitor the size of their own population. This is a process called **quorum sensing.** Bacteria can use quorum sensing to interact with other members of the same species, as well as members of other species that are close by. Eventually large complex communities are formed, which have different physical and biological characteristics in different locations of the community. The bottom of a biofilm may have very different pH and oxygen conditions than the surface of a biofilm, for example. It is now clearly established that microbes in a biofilm, as opposed to those in a planktonic (free-floating) state, behave and respond very differently to their environments. Different genes are even activated in the two situations. At any rate, a single biofilm is usually a partnership among multiple microbial inhabitants and for that reason cannot be eradicated by traditional methods targeting individual infections. This kind of synergism has led to the necessity of rethinking treatment of a great many different conditions.

Figure 6.7 Steps in the formation of a biofilm.
The photograph is an SEM of a biofilm formed on a gauze bandage. The blue bacteria are methicillin-resistant *Staphylococcus aureus* (MRSA). The orange substance is the extracellular matrix; and the green tubes are fibers of the gauze. This is probably a single-species biofilm.
Ellen Swogger and Garth James, Center for Biofilm Engineering, Montana State University

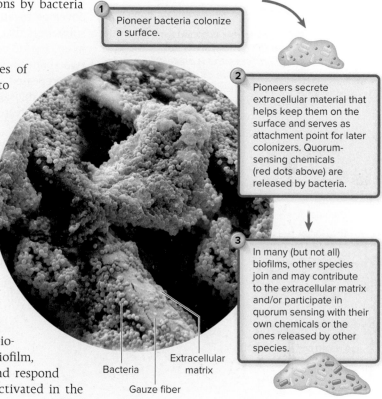

1. Pioneer bacteria colonize a surface.

2. Pioneers secrete extracellular material that helps keep them on the surface and serves as attachment point for later colonizers. Quorum-sensing chemicals (red dots above) are released by bacteria.

3. In many (but not all) biofilms, other species join and may contribute to the extracellular matrix and/or participate in quorum sensing with their own chemicals or the ones released by other species.

Bacteria

Extracellular matrix

Gauze fiber

4. Biofilms serve as a constant source of bacteria that can "escape" and become free-living again.

6.2 LEARNING OUTCOMES—Assess Your Progress

8. List and define five terms used to express the temperature-related growth capabilities of microbes.

9. Summarize three ways in which microorganisms function in the presence of differing oxygen conditions.

10. Identify three important environmental factors (other than temperature and oxygen) with which organisms must cope.

11. List and describe the five types of associations microbes can have with their hosts.

12. Discuss characteristics of biofilms that differentiate them from planktonic bacteria.

6.3 The Study of Bacterial Growth

Microorganisms can "grow" in many different ways. In chapter 5 you learned how eukaryotic microbes grow. Here we focus on bacterial growth. While bacteria can be found that grow through budding and hyphal formation (similar to fungi), the majority of bacteria grow by a process called **binary fission.**

Binary Fission

Binary fission refers to the fact that one cell becomes two. During binary fission, the parent cell enlarges, duplicates its chromosome, and then starts to pull its cell envelope together in the center of the cell using a band of protein that is made of substances that resemble actin and tubulin–the protein component of microtubules in eukaryotic cells. The cell wall eventually forms a complete septum between the two about-to-be cells. This process divides the cell into two daughter cells, and is then repeated at intervals by each new daughter cell in turn. With each successive round of division, the population increases. The stages in this continuous process are shown in greater detail in **figure 6.8.**

Keep in mind that even though we use the terms "parent" and "daughter" cells, this is asexual reproduction. Both daughter cells are identical to each other and the parent cell no longer exists.

Kelly & Heidi

The Rate of Population Growth

The time required for a complete fission cycle–from one parent cell to two new daughter cells–is called the **generation, or doubling, time.** The term *generation* has a similar meaning as it does in humans. It is the period between an individual's birth and the time it produces offspring. In bacteria, each new fission cycle, or generation, increases the population by a factor of 2, or doubles it. The initial parent stage consists of 1 cell, the first generation consists of 2 cells, the second 4, the third 8, then 16, 32, 64, and so on. As long as the environment remains favorable, this doubling effect can continue at a constant rate. With the passing of each generation, the population will double, over and over again.

The length of the generation time is a measure of the growth rate of an organism. Compared with the growth rates of most other living things, bacteria are notoriously rapid. The average generation time is 30 to 60 minutes under optimum conditions. The shortest generation times can be 10 to 12 minutes, although some bacteria have generation times of days. For example, *Mycobacterium leprae,* the cause of Hansen's disease, has a generation time of 10 to 30 days–as long as that of some animals. Environmental bacteria commonly have generation times measured in months. Most pathogens have relatively short doubling times. *Salmonella enteritidis* and *Staphylococcus aureus,* bacteria that cause foodborne illness, double in 20 to 30 minutes, which is why leaving food at room temperature even for a short period has caused many cases of foodborne disease. In a few hours, a population of these bacteria can easily grow from a small number of cells to several million.

Figure 6.9 shows several quantitative characteristics of growth: The cell population size can be represented by the number 2 with an exponent (2^1, 2^2, 2^3, 2^4); the exponent increases by one in each generation; and the number of the exponent is also the number of the generation. This growth pattern is termed **exponential.** Because these populations often contain very large numbers of cells, it is useful to express them by means of exponents or logarithms. The data from a growing bacterial population are graphed by plotting the number of cells as a function of time. Plotting the logarithm number (or exponent) over time provides a straight line indicative of exponential growth. Plotting the data arithmetically gives a constantly curved slope. In general, logarithmic graphs

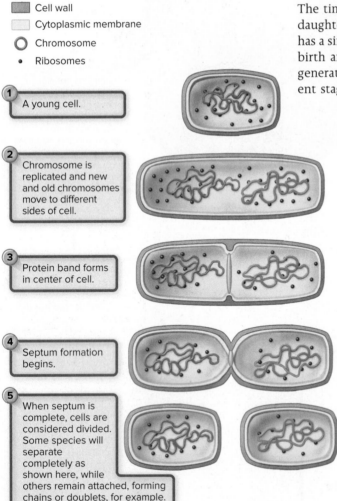

Cell wall
Cytoplasmic membrane
Chromosome
Ribosomes

1 A young cell.

2 Chromosome is replicated and new and old chromosomes move to different sides of cell.

3 Protein band forms in center of cell.

4 Septum formation begins.

5 When septum is complete, cells are considered divided. Some species will separate completely as shown here, while others remain attached, forming chains or doublets, for example.

Figure 6.8 Steps in binary fission of a rod-shaped bacterium.

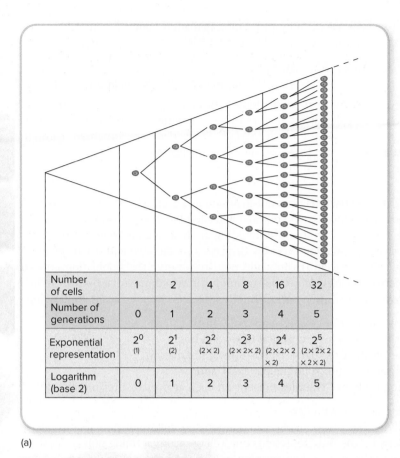

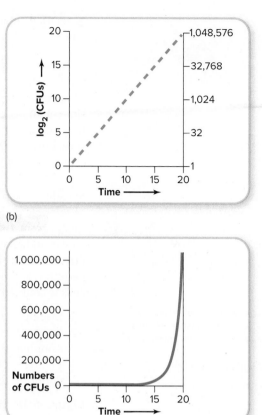

(b)

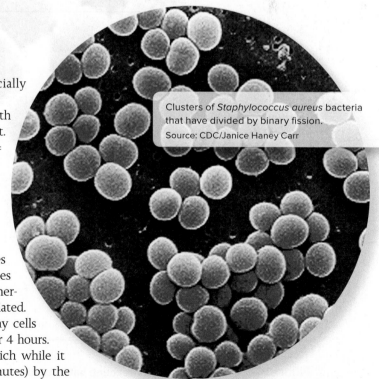

(c)

(a)

Figure 6.9 **The mathematics of population growth.** **(a)** Starting with a single cell, if each product of reproduction goes on to divide by binary fission, the population doubles with each new cell division or generation. This process can be represented by logarithms using exponents or by simple numbers. **(b)** Plotting the logarithm of the cells produces a straight line indicative of exponential growth, whereas **(c)** plotting the cell numbers arithmetically gives a curved slope.

Clusters of *Staphylococcus aureus* bacteria that have divided by binary fission.

Source: CDC/Janice Haney Carr

are preferred because an accurate cell number is easier to read, especially during early growth phases.

Predicting the number of cells that will arise during a long growth period (yielding millions of cells) is based on a relatively simple concept. One could use the method of addition ($2 + 2 = 4$; $4 + 4 = 8$; $8 + 8 = 16$; $16 + 16 = 32$; and so on) or a method of multiplication (for example, $2^5 = 2 \times 2 \times 2 \times 2 \times 2$), but it is easy to see that for 20 or 30 generations, this calculation could be very tedious. An easier way to calculate the size of a population over successive generations is to use this equation:

$$N_t = (N)2^n$$

Here, N_t is the total number of cells in the population (the "t" denotes "at some point in time t"). N is the starting number, the exponent n denotes the generation number, and 2^n represents the number of cells in that generation. If we know any two of the values, the other values can be calculated. We will use the example of *Staphylococcus aureus* to calculate how many cells (N_t) will be present in an egg salad sandwich after it sits in a warm car for 4 hours. We will assume that N is 10 (number of cells deposited in the sandwich while it was being prepared). To derive n, we need to divide 4 hours (240 minutes) by the generation time (we will use 20 minutes). This calculation comes out to 12, so 2^n is equal to 2^{12}. Using a calculator, we find that 2^{12} is 4,096.

$$\text{Final number } (N_t) = 10 \times 4,096$$
$$= 40,960 \text{ bacterial cells in the sandwich}$$

This same equation, with modifications, is used to determine the generation time, a more complex calculation that requires knowing the number of cells at the beginning and end of a growth period. Such data are obtained through actual testing by a method discussed in the following section.

The Population Growth Curve

In reality, a population of bacteria does not maintain its potential growth rate and does not double endlessly because in closed systems (called batch cultures) numerous factors prevent the cells from continuously dividing at their maximum rate. Laboratory studies indicate that a population typically displays a predictable pattern, or **growth curve,** over time. The method traditionally used to observe the population growth pattern is a *standard* (or *viable*) *plate count* technique, in which the total number of live cells is counted over a given time period. Briefly, this method involves the following:

1. placing a tiny number of cells into a sterile broth,
2. incubating this culture over a period of several hours,
3. sampling the broth at regular intervals during incubation,
4. plating each sample onto solid media (agar), and
5. counting the number of colonies present on each agar plate after incubation.

Figure 6.10 illustrates this process.

Evaluating these agar plates involves a common and important principle in microbiology: One colony on the plate represents one cell or colony-forming unit (CFU) from the original sample. (The term "colony-forming unit" is used sometimes, acknowledging

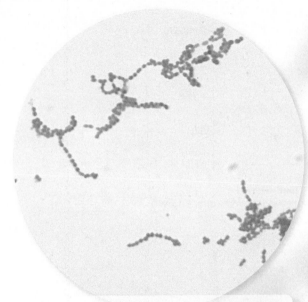

Chains of *Streptococcus* bacteria that have divided by binary fission. When mature cells stay attached like this, it means they stop their binary fission before completely separating.

Source: Centers for Disease Control and Prevention

Equally spaced time intervals	60 min	120 min	180 min	240 min	300 min	360 min	420 min	480 min	540 min	600 min
0.1 mL sample added to tube										
Sample is diluted in liquid agar medium and poured or spread over surface of solidified medium										
Plates are incubated, colonies are counted	None									
Number of colonies (CFUs) per 0.1 mL	<1*	2	4	7	13	23	45	80	135	230
Number of colonies per mL	<1*	20	40	70	130	230	450	800	1,350	2,300
	<1*	(2.0×10^1)	(4.0×10^1)	(7.0×10^1)	(1.3×10^2)	(2.3×10^2)	(4.5×10^2)	(8.0×10^2)	(1.35×10^3)	(2.3×10^3)

500 mL inoculated flask

*Only means that too few cells are present to be assayed.

Figure 6.10 Steps in a viable plate count: batch culture method. The calculation is number of colonies × 10 (to convert the 0.1 mL sample to a 1.0 mL standard) × total volume of flask (500 mL).

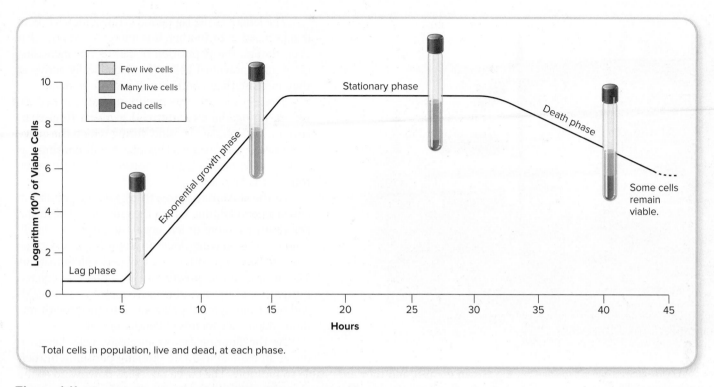

Figure 6.11 The growth curve in a bacterial culture. On this graph, the number of viable cells expressed as a logarithm (log) is plotted against time. See text for discussion of the various phases. Note that with a generation time of 35 minutes, the population has risen from 10 (10^1) cells to over 1,000,000,000 (10^9) cells in only 16 hours.

the fact that a group of cells–such as a chain or a pair of connected bacteria–may be the one "unit" placed by itself on that area of the agar.) Multiplication of the number of colonies in a single sample by the container's volume gives a fair estimate of the total population size (number of cells) at any given point. In other words, you multiply the number of colonies arising from the fluid you removed, by the dilution factor. For example, if you want to express the results as "per milliliter" but you sampled only 1/10 of a milliliter, you just multiply the number of visible CFUs × 10. The growth curve is determined by graphing the number for each sample in sequence for the whole incubation period.

Because there are few cells in the early stages of growth, very early samples can give a zero reading even if there are viable cells in the culture. Also, the sampling itself can remove enough viable cells to alter the tabulations, but because the purpose is to compare relative trends in growth, these factors do not significantly change the overall pattern.

Stages in the Normal Growth Curve

The system of batch culturing just described is *closed,* meaning that nutrients and space are finite and there is no mechanism for the removal of waste products. Data from an entire batch growth period typically produce a curve with a series of phases termed the *lag phase,* the *exponential growth (log) phase,* the *stationary phase,* and the *death phase* **(figure 6.11).**

The **lag phase** is a relatively "flat" period on the graph when the population appears not to be growing or is growing at less than the exponential rate. Growth lags primarily because

1. the newly inoculated cells require a period of adjustment, enlargement, and synthesis;
2. the cells are not yet multiplying at their maximum rate; and
3. the population of cells is so sparse or dilute that the sampling misses them.

The bacteria in cold yogurt are probably in stationary phase. Once inside a warm body, the bacteria can reenter lag and then exponential phases and increase their numbers in the gut.
©McGraw-Hill Education

The cloudiness in fish tanks is due in large part to the growth of bacteria in the water.
©Getty Images/WIN-Initiative RM

The length of the lag period varies somewhat from one population to another. It is important to note that even though the population of cells is not increasing (growing), individual cells are metabolically active as they increase their contents and prepare to divide.

The cells reach the maximum rate of cell division during the **exponential growth (logarithmic or log) phase,** a period during which the curve increases geometrically. This phase will continue as long as cells have adequate nutrients and the environment is favorable.

At the **stationary growth phase,** the population enters a period during which the rates of cell birth and cell death are more or less equal. At this time, the division rate is slowing down (making it easier for cell death to catch up with the rate of new cell formation). The decline in the growth rate is caused by depleted nutrients and oxygen plus excretion of organic acids and other biochemical pollutants into the growth medium, due to the increased density of cells.

As the limiting factors intensify, cells begin to die at an exponential rate (literally perishing in their own wastes), and they are unable to multiply. The curve now dips downward as the **death phase** begins. The speed with which death occurs depends on the relative resistance of the species and how toxic the conditions are, but it is usually slower than the exponential growth phase. It is now clear that many cells in a culture stay alive, but more or less dormant, for long periods of time. They are so dormant that, although they are alive, they won't grow on culture medium and therefore are missed in colony counts. The name for this state is the **viable nonculturable (VNC)** state.

Practical Importance of the Growth Curve

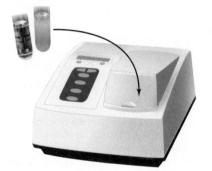

The tendency for populations to exhibit phases of rapid growth, slow growth, and death has important implications in controlling microbes in the environment, in managing microbial diseases, in industrial microbiology, and more. Antimicrobial agents such as heat and disinfectants rapidly accelerate the death phase in all populations, but microbes in the exponential growth phase are more vulnerable to these agents than are those that have entered the stationary phase.

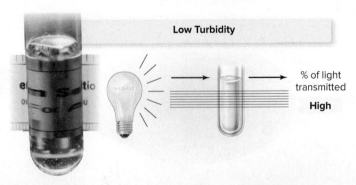

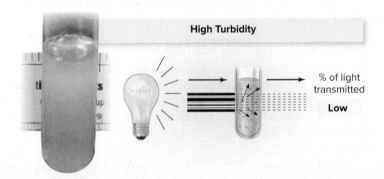

Figure 6.12 Turbidity measurements as indicators of growth. Holding a broth to the light is one method of checking for gross differences in cloudiness (turbidity). The broth on the left is transparent, indicating little or no growth; the broth on the right is cloudy and opaque, indicating heavy growth. The eye is not sensitive enough to pick up fine degrees in turbidity; more sensitive measurements can be made with a spectrophotometer. On the left you will see that a tube with no growth will allow light to easily pass. Therefore more light will reach the photodetector and give a higher transmittance value. In a tube with growth (on the right), the cells scatter the light, resulting in less light reaching the photodetector and, therefore, giving a lower transmittance value.

Kathleen Talaro

In general, actively growing cells are more vulnerable to growth inhibition and destruction.

In recent years, scientists have begun to consider experiments with microbes that are growing at a slower rate than the "optimal" rate. They point out that in most natural situations (even in the human body), microbes are in competition for substances that are important for their growth. To understand what happens in the natural state, these scientists argue, we should see what happens with them when we apply a little stress to them, in the form of competitors or lower amounts of nutrients or other growth factors.

Analyzing Population Size Without Culturing

Turbidity

Microbiologists have developed several alternative ways of analyzing bacterial growth without actually having to grow them. One of the simplest methods for estimating the size of a population is through turbidometry. This technique relies on the simple observation that a tube of clear nutrient solution becomes cloudy, or **turbid,** as microbes grow in it. In general, the greater the turbidity, the larger the population size, which can be measured by means of sensitive instruments **(figure 6.12).**

Counting

If you know how many cells are present in a specified amount of fluid, you can multiply that to determine the number in the total volume of the fluid. It is possible to conduct a **direct cell count** microscopically **(figure 6.13).** This technique, very similar to that used in blood cell counts, employs a special microscope slide (cytometer) calibrated to accept a tiny sample that is spread over a premeasured grid. One inherent inaccuracy in this method, as well as in spectrophotometry, is that no distinction can be made between dead and live cells, both of which are included in the count.

Counting can be automated by sensitive devices such as the *Coulter counter*, which electronically scans a fluid as it passes through a tiny pipette. As each cell flows by, it is detected and registered on an electronic sensor **(figure 6.14).** A *flow cytometer* works on a similar principle, but in addition to counting, it can measure cell size and even differentiate between live and dead cells. When used in conjunction with fluorescent dyes and antibodies to tag cells, it has been used to differentiate between gram-positive and gram-negative bacteria. It has been adapted for use as a rapid method to identify pathogens in patient specimens and to differentiate blood cells. More sophisticated forms of the flow cytometer can also sort cells of

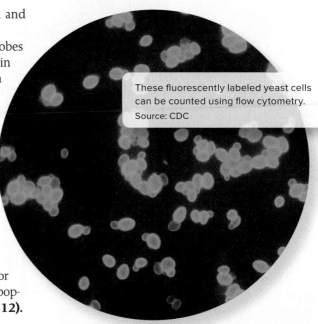

These fluorescently labeled yeast cells can be counted using flow cytometry.
Source: CDC

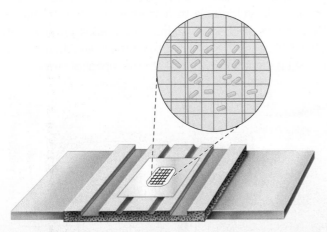

Figure 6.13 Direct microscopic count of bacteria. A small sample is placed on the grid under a cover glass. Individual cells, both living and dead, are counted. This number can be used to calculate the total cell count of a sample.

Medical Moment

MRSA: PCR Over Culture

Methicillin-resistant *Staphylococcus aureus* (MRSA) is a term for bacterial infection caused by an antibiotic-resistant strain of *S. aureus*. Many people are exposed in the community, becoming carriers without even knowing they are infected. This has driven hospitals to proactively identify carriers.

In many facilities, patients are routinely screened for MRSA upon admission. This surveillance testing allows for appropriate antibiotic selection and prevents spread to other patients. For patients with known MRSA, contact isolation (gown and gloves for patient contact) is necessary.

The MRSA carrier state is typically diagnosed using PCR testing by nasal swab, rather than culture. Polymerase chain reaction (PCR) testing allows for more rapid identification of MRSA by detecting the bacterial DNA in the sample. Conventional culture can take 3 days to return results, providing opportunity for transmission to other patients during the time results are pending.

To collect the PCR sample, the patient's nostrils are swabbed. The cotton swab is then sent to the lab for testing. Studies have shown PCR accuracy to be comparable to culture in detecting the presence of MRSA.

Q. Two contact precautions are mentioned in the story. Can you name some other precautions that should be considered?

Answer in Appendix B.

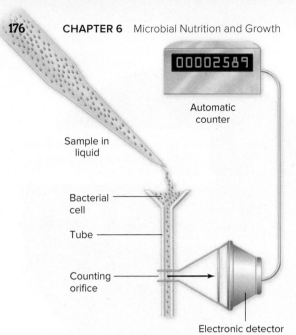

Figure 6.14 Coulter counter. As cells pass through this device, they trigger an electronic sensor that tallies their numbers.

different types into separate compartments of a collecting device. Although flow cytometry can be used to count bacteria in natural samples without the need for culturing them, it requires fluorescent labeling of the cells you are interested in detecting, which is not always possible.

Genetic Probing

A variation of the **polymerase chain reaction (PCR),** called real-time PCR, allows scientists to quantify bacteria and other microorganisms that are present in environmental or tissue samples without isolating them and without culturing them. We will learn more about these methods in later sections.

6.3 LEARNING OUTCOMES—Assess Your Progress

13. Summarize the steps of bacterial binary fission.
14. Define *doubling time,* and describe how it leads to exponential growth.
15. Compare and contrast the four phases of growth in a bacterial growth curve.
16. Identify one culture-based and one non-culture-based method used for analyzing bacterial growth.

CASE FILE WRAP-UP

The bacteria pictured are Clostridium perfringens.
©Shutterstock/GagliardiImages; Source: Don Stalons/Centers for Disease Control and Prevention

C. perfringens is commonly found in soil but is also found in the normal biota of the intestines. It produces a toxin known as alpha toxin, which can enter into muscle tissue through a wound. The toxin destroys tissue and produces gas, hence the term *gas gangrene.* Massive infection often leads to sepsis and death.

Treatment of gas gangrene includes wound debridement and penicillin; however, penicillin alone is unable to penetrate infected muscle tissue deeply enough to kill all of the bacteria, so penicillin is used in combination with surgical treatment. Amputation is often required to definitively treat *C. perfringens.*

Hyperbaric oxygen therapy (HBOT) has been used to successfully treat wound infections, including infections caused by *C. perfringens.* Anaerobic bacteria such as *C. perfringens* are susceptible to increased concentrations of oxygen. With HBOT, large quantities of oxygen-free radicals are generated, which renders bacteria vulnerable to oxidative death. HBOT has also been found to inhibit the release of alpha toxins. Lastly, HBOT enhances the effects of antibiotics. HBOT in cases of *C. perfringens* has resulted in avoidance of amputation, increased patient survival, and shorter hospital stays.

Hyperbaric chamber
©ERproductions Ltd/Blend Images LLC

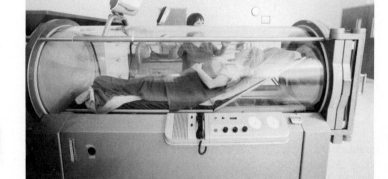

The Great Oxidation Event and Earth's Microbiome

There has been a lot of debate about when oxygen-producing microbes appeared on the earth. Until recently, scientists have estimated that only anaerobic cells existed until about 2 or 3 billion years ago. It was not until what is called the *great oxidation event* (GOE) occurred that life, as we know it, was possible on the earth. What is the GOE, and why is it significant? Until that time, atmospheric oxygen in the form of O_2 was very low–probably less than 1/10,000 of present levels. Previous to the GOE, the atmosphere contained significant amounts of methane, a greenhouse gas, which kept the earth's temperatures at high levels. The production of oxygen was a significant event in the evolution of life on earth. Until O_2 first accumulated in the atmosphere, the earth was too hot, too full of methane, and too acidic for life to exist.

Scientists led by geomicrobiologist Kirt Konhauser at the University of Alberta have studied oxidation of iron pyrite (FeS_2, also known as fool's gold) into iron oxide. This process released acid that dissolved rocks into chromium and other metals in ancient sea beds. Their data show that chromium levels in the sea beds increased significantly about 2.48 billion years ago, about 100 million years earlier than was previously thought. These **acidophilic** microbes thrived in the extremely low pH of the ocean waters of ancient earth. Konhauser has found evidence of these same bacterial life forms living off of pyrite in the highly acidic wastewater of mining sites (pictured). These organisms are the most ancient oxygen producers known and may be the driving force behind the atmosphere in which we live today.

Source: NOAA Restoration Center & Damage Assessment and Restoration Program

Source: Shutterstock/Barbo

MICROBIAL NUTRITION AND GROWTH

6.1 MICROBIAL NUTRITION

Bacteria have a wide array of nutritional strategies. They need macronutrients and micronutrients, and are deeply dependent on the characteristics of the fluids around them.

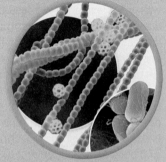

STEVE GSCHMEISSNER/Getty Images; Martin Oeggerli/Science Source

Important factors for microbial growth and survival include temperature, oxygen conditions, pH, radiation, etc. Microorganisms live in symbiotic or nonsymbiotic associations with other species.

ENVIRONMENTAL FACTORS THAT INFLUENCE MICROBES **6.2**

6.3 THE STUDY OF BACTERIAL GROWTH

Most bacteria reproduce by binary fission. Microbial growth refers to both increase in cell size and increase in the number of cells in a population. There are four distinct stages of bacterial growth in a batch environment.

SmartGrid: From Knowledge to Critical Thinking

This *21 Question Grid* takes the topics from this chapter and arranges them with respect to the American Society for Microbiology's Undergraduate Curriculum guidelines—all six of the important "Concepts" as well as the important "Competency" of scientific literacy. Three questions are supplied about chapter content that refer to the Concept or Competency, in increasing levels of Bloom's taxonomy for learning.

ASM Concept/ Competency	A. Bloom's Level 1, 2—Remember and Understand (Choose one.)	B. Bloom's Level 3, 4—Apply and Analyze	C. Bloom's Level 5, 6—Evaluate and Create
Evolution	1. Which descriptors are likely to have applied to the earliest microbes on the planet? a. chemoautotrophic b. thermophilic c. chemoheterotrophic d. two of the above	2. Can you think of any advantages gained when eukaryotes and multicellular organisms evolved to use sexual reproduction instead of binary fission? Discuss.	3. Speculate about how earth's atmosphere came to contain oxygen, and the role early bacteria and archaea played.
Cell Structure and Function	4. Which of the following is true of passive transport? a. It requires a gradient. b. It uses the cell wall. c. It includes endocytosis. d. It only moves water.	5. Compare the effects of a hypertonic, hypotonic, and isotonic solution on an amoeba versus a bacterium.	6. Usually scientists looking for life on other planets try to determine first if the planet had or has evidence of water. Develop a rationale for this based on cellular chemistry.
Metabolic Pathways	7. An organism that can synthesize all its required organic components from CO_2 using energy from the sun is a a. photoautotroph. b. photoheterotroph. c. chemoautotroph. d. chemoheterotroph.	8. Provide evidence in support of or refuting this statement: Microbial life can exist in the complete absence of sunlight and organic compounds.	9. Develop an explanation for why biofilm bacteria use quorum sensing to regulate the size of their community.
Information Flow and Genetics	10. Most bacteria increase their numbers by a. sexual reproduction. b. hyphae formation. c. binary fission. d. endocytosis.	11. Looking at figure 6.3, explain how a cell in a hypertonic solution might have trouble dividing.	12. In binary fission, the parent chromosome is duplicated so that each of the two daughter cells has the same copy of the original chromosome. If this is the case, speculate on what happened to allow bacteria to become diversified into so many types?

ASM Concept/ Competency	A. Bloom's Level 1, 2—Remember and Understand (Choose one.)	B. Bloom's Level 3, 4—Apply and Analyze	C. Bloom's Level 5, 6—Evaluate and Create
Microbial Systems	**13.** A cell exposed to a hypertonic environment will _____ by osmosis. a. gain water b. lose water c. neither gain nor lose water d. burst	**14.** Considering figure 6.6, discuss whether tube 2 or tube 3 is more likely to grow in diverse niches or conditions. Defend your answer.	**15.** Bacteria and archaea are everywhere on the planet. Would it be to our advantage (if it were possible) to eliminate all of them, in an effort to protect the human population from current and emerging diseases? Why or why not?
Impact of Microorganisms	**16.** A pathogen would most accurately be described as a a. parasite. b. commensal. c. symbiont. d. saprobe.	**17.** How can you explain the fact that an unopened carton of milk in the refrigerator will eventually spoil even though it has never been opened?	**18.** Describe a scenario in which it would be to a pathogen's advantage to have a very long generation time.
Scientific Thinking	**19.** In a viable count, each _____ represents a _____ from the sample population. a. CFU, colony b. colony, CFU c. hour, generation d. cell, generation	**20.** If an egg salad sandwich sitting in a car on a warm day contains 40,960 bacteria after 4 hours, how many bacteria will it contain after another generation time for the bacteria?	**21.** Scientists now believe that most bacteria in nature exist in biofilms. Can you suggest some technical reasons that this was not understood for the first 150 years of microbiology?

Answers to the multiple-choice questions appear in Appendix A.

Visual Connections

This question connects content within and between chapters.

From chapter 5, figure 5.9. What type of symbiotic relationship is pictured here?

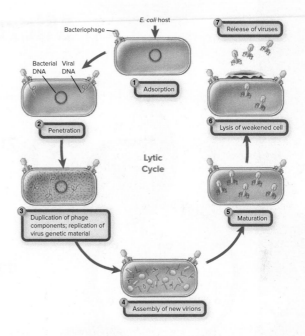

7

Microbial Metabolism

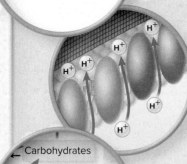

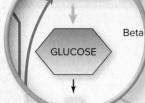

CASE FILE

Surviving Sepsis

I was working as a nurse in an emergency department. A medical transport team brought in a patient from a long-term care facility. The elderly man was lethargic and required a lot of stimulation before he could respond to my questions. The attendant who accompanied the patient reported he had become drowsier in the past 2 days, causing concern for his caregivers. That morning, his blood pressure was low so the patient was transported to the hospital for further assessment.

In the emergency department, his blood pressure and oxygen saturation levels were low. The patient had a high fever. His pulses were weak and heart rate was rapid. I applied an oxygen mask to the patient with 10 liters of flow. The provider ordered intravenous (IV) fluids to be given quickly as a bolus. Another nurse obtained IV access and started the fluids. I called the pharmacy to notify them that we suspected the patient had septicemia and needed STAT antibiotics.

Next, I prepared to draw labs from the patient. The provider requested a full laboratory workup, including blood and urine cultures. The provider also requested a serum lactate level. Knowing this required special precautions for collection, I obtained supplies for a free-flowing sample and prepared to send it immediately to the laboratory for processing. I obtained blood cultures immediately before the antibiotic doses arrived.

- Why was a serum lactate test ordered for this patient?

- How is sepsis treated?

Case File Wrap-Up appears at the end of the chapter.

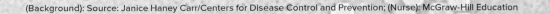

(Background): Source: Janice Haney Carr/Centers for Disease Control and Prevention; (Nurse): McGraw-Hill Education

7.1 Metabolism and the Role of Enzymes

Metabolism, from the Greek term *metaballein,* meaning "change," pertains to all chemical reactions and physical workings of the cell. Although metabolism entails thousands of different reactions, most of them fall into one of two general categories. **Anabolism,** sometimes also called *biosynthesis,* is any process that results in synthesis of cell molecules and structures. It is a building and bond-making process that forms larger macromolecules from smaller ones, and it usually requires the input of energy. **Catabolism** is the opposite of anabolism. Catabolic reactions break the bonds of larger molecules into smaller molecules and often release energy. In a cell, linking anabolism to catabolism ensures the efficient completion of many thousands of processes. Another fundamental fact about metabolism is that electrons are critical to the process.

In summary, a cell collects and spends energy by transferring electrons from an external source to internal carriers that eventually shuttle it into a series of proteins that make the energy usable for a cell. Electron flow is the key. Along the way, metabolism accomplishes the following **(figure 7.1):**

1. assembles smaller molecules into larger macromolecules needed for the cell; in this process, ATP (energy) is utilized to form bonds (anabolism);
2. degrades macromolecules into smaller molecules, a process that yields energy (catabolism); and
3. collects and spends energy in the form of ATP (adenosine triphosphate).

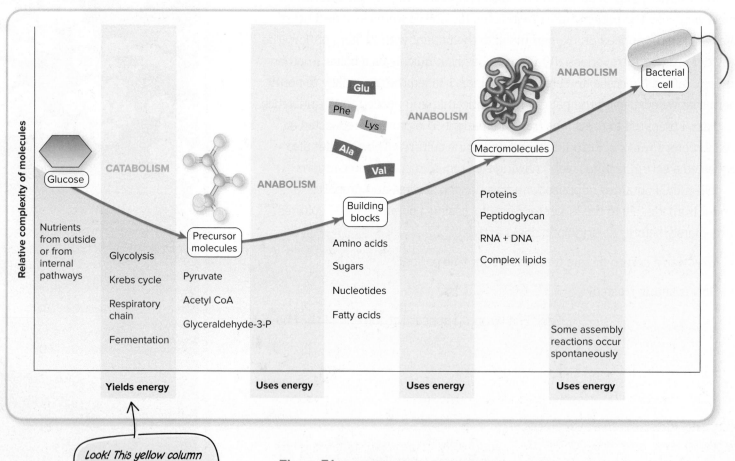

Look! This yellow column takes up much of our discussion in this chapter.
- Kelly & Heidi

Figure 7.1 Simplified model of metabolism. Cellular reactions fall into two major categories. Catabolism (yellow) involves the breakdown of complex organic molecules to extract energy and form simpler end products. Anabolism (blue) uses the energy to synthesize necessary macromolecules and cell structures from precursors.

Table 7.1 Checklist of Enzyme Characteristics
• Are made of protein or RNA; may require cofactors
• Act as organic catalysts to speed up the rate of cellular reactions
• Have unique characteristics such as shape, specificity, and function
• Enable metabolic reactions to proceed at a speed compatible with life
• Have an active site for target substrates
• Are much larger than their substrates
• Associate closely with substrates but do not become integrated into the reaction products
• Are not used up or permanently changed by the reaction
• Can be recycled, so they function in extremely low concentrations
• Are greatly affected by temperature and pH
• Can be regulated by feedback and genetic mechanisms

Enzymes: Catalyzing the Chemical Reactions of Life

The chemical reactions of life cannot proceed without a special class of macromolecules called *enzymes.* Enzymes are a remarkable example of **catalysts,** chemicals that increase the rate of a chemical reaction without becoming part of the products or being consumed in the reaction. It is easy to think that an enzyme creates a reaction, but that is not true. The major characteristics of enzymes are summarized in **table 7.1.**

How Do Enzymes Work?

An enzyme speeds up the rate of a metabolic reaction, but just how does it do this? During a chemical reaction, reactants are converted to products either by bond formation or by bond breakage. A certain amount of energy is required to initiate every such reaction, which limits the rate at which reactions can happen. While the rate could be sped up by increased heat or other means, biological cells use enzymes to vastly increase the speed of important reactions.

The molecules that are acted upon by enzymes are called *substrates*. At the molecular level, an enzyme promotes a reaction by serving as a physical site upon which the substrates can be positioned. The enzyme is almost always much larger in size than its substrate, and it presents a unique active site that matches only that particular substrate. Although an enzyme binds to the substrate and participates directly in changes to the substrate, it does not become a part of the products, is not used up by the reaction, and can function over and over again. Enzyme speeds are very rapid. Speeds range from a thousand substrate molecules converted per second for lactate dehydrogenase to several million converted per second by the enzyme catalase.

Enzyme Structure

Most enzymes are proteins–although there is a special class that are made of RNA–and they are classified as simple or conjugated. Simple enzymes are made of protein alone, whereas conjugated enzymes **(figure 7.2)** contain protein plus nonprotein molecules. A conjugated enzyme, sometimes referred to as a **holoenzyme,** is a combination of a protein and one or more **cofactors.** In this situation, the actual protein portion is called the **apoenzyme.** Cofactors are either organic molecules, called **coenzymes,** or inorganic elements (metal ions). For example, catalase, an enzyme that we studied in chapter 6, breaks down hydrogen peroxide and requires iron as a metallic cofactor.

Enzyme-Substrate Interactions

For a reaction to take place, a temporary enzyme-substrate union must occur at the active site **(figure 7.3).** The fit is so specific that it is often described as a "lock-and-key" fit in which the substrate is inserted into the active site's pocket.

The bonds formed between the substrate and enzyme are weak and easily reversible. Once the enzyme-substrate complex has formed, the designated reaction(s)

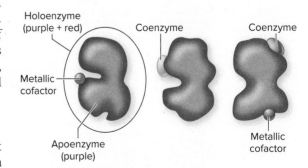

Figure 7.2 Conjugated enzyme structure.
Conjugated enzymes have an apoenzyme (polypeptide or protein) component and one or more cofactors.

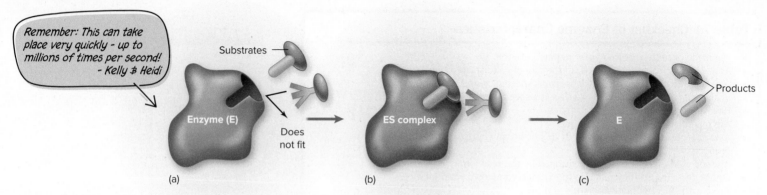

Figure 7.3 Enzyme-substrate reactions.   **(a)** When the enzyme and substrate come together, the substrate (S) must show the correct fit and position with respect to the enzyme (E). **(b)** When the ES complex is formed, it enters a transition state. During this temporary but tight interlocking union, the enzyme participates directly in breaking or making bonds. **(c)** Once the reaction is complete, the enzyme releases the products.

occur on the substrate, often with the aid of a cofactor, and a product is formed and released. The enzyme can then attach to another substrate molecule and repeat this action. Although enzymes can potentially catalyze reactions in both directions, most examples in this chapter depict them working in one direction only.

Cofactors: Supporting the Work of Enzymes

In chapter 6, you learned that microorganisms require specific metal ions called trace elements and also certain organic growth factors. In many cases, the need for these substances arises from their roles as cofactors for enzymes. The metallic cofactors, including iron, copper, magnesium, manganese, zinc, cobalt, selenium, and many others, have precise functions between the enzyme and its substrate. In general, metals activate enzymes, help bring the active site and substrate close together, and participate directly in chemical reactions with the enzyme-substrate complex.

Coenzymes are a type of cofactor. They are organic compounds that work in conjunction with an apoenzyme to perform the alteration of a substrate. The general function of a coenzyme is to remove a chemical group from one substrate molecule and add it to another substrate, thereby serving as a transient carrier of this group. In a later section of this chapter, we shall see that coenzymes carry and transfer hydrogen atoms, electrons, carbon dioxide, and amino groups. Many coenzymes are derived from **vitamins,** which explains why vitamins are important to nutrition and may be required as growth factors for living things. Vitamin deficiencies can prevent the complete holoenzyme from forming. Consequently, both the chemical reaction and the structure or function dependent upon that reaction are compromised.

Naming Enzymes

Enzymes are classified and named according to characteristics such as site of action, type of action, and substrate. In general, an enzyme name is composed of two parts: a prefix or stem word derived from a certain characteristic—usually the substrate acted upon, the type of reaction catalyzed, or both—followed by the ending *-ase.*

Each enzyme is also assigned a common name that indicates the specific reaction it catalyzes. With this system, an enzyme that digests a carbohydrate substrate is a *carbohydrase.* A specific carbohydrase, *amylase,* acts on starch (amylose is a major component of starch). The enzyme *maltase* digests the sugar maltose. An enzyme that breaks peptide bonds of a protein is a *proteinase, protease,* or *peptidase.* Some fats and other lipids are digested by *lipases.* DNA is broken down by *deoxyribonuclease,* generally shortened to *DNase.* A *synthetase* or *polymerase* bonds together many small molecules into large molecules.

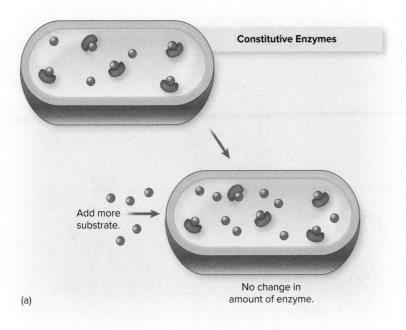

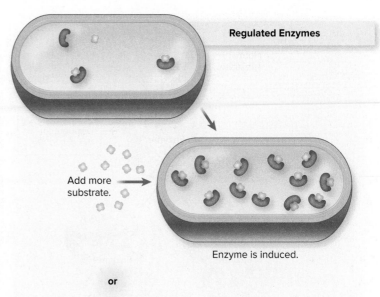

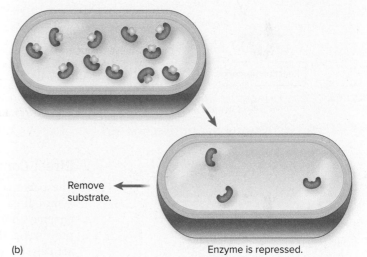

Figure 7.4 Constitutive and regulated enzymes.
(a) Constitutive enzymes are present in constant amounts in a cell. The addition of more substrate does not increase the numbers of these enzymes. **(b)** The concentration of regulated enzymes in a cell increases or decreases in response to substrate levels.

Regulation of Enzyme Action

Enzymes are not all produced in equal amounts or at equal rates. Some, called **constitutive enzymes (figure 7.4*a*),** are always present and in relatively constant amounts, regardless of the amount of substrate in the vicinity. The enzymes involved in utilizing glucose, for example, are very important in metabolism and so they are classified as constitutive. Other enzymes are **regulated enzymes (figure 7.4*b*),** meaning they are either turned on (induced) or turned off (repressed) in response to changes in concentration of the substrate.

The activity of an enzyme is highly influenced by the cell's environment. In general, enzymes operate only under the natural temperature, pH, and osmotic pressure of an organism's habitat. When enzymes are subjected to changes in these normal conditions, they tend to be chemically unstable, or **labile.** Low temperatures inhibit enzyme reactions, and high temperatures denature the apoenzyme. **Denaturation** is a process by which the weak bonds that collectively maintain the native shape of the apoenzyme are broken. This disruption causes extreme distortion of the enzyme's shape and prevents the substrate from attaching to the active site. When enzymes are nonfunctional, metabolic reactions fail to happen and cell death can follow. Low or high pH or certain chemicals (heavy metals, alcohol) are also denaturing agents.

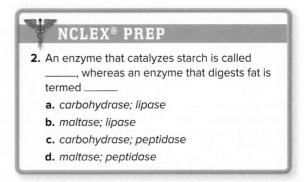

NCLEX® PREP

2. An enzyme that catalyzes starch is called _____, whereas an enzyme that digests fat is termed _____.
   a. *carbohydrase; lipase*
   b. *maltase; lipase*
   c. *carbohydrase; peptidase*
   d. *maltase; peptidase*

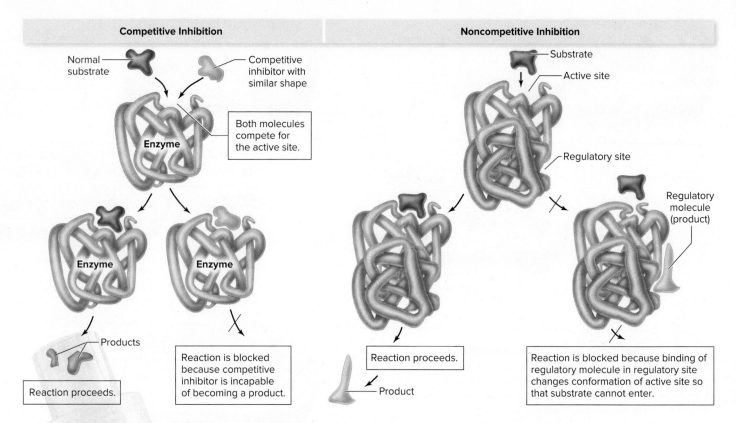

| Competitive Inhibition | Noncompetitive Inhibition |

Normal substrate

Competitive inhibitor with similar shape

Both molecules compete for the active site.

Enzyme

Enzyme   **Enzyme**

Products

Reaction proceeds.

Reaction is blocked because competitive inhibitor is incapable of becoming a product.

Substrate

Active site

Regulatory site

Regulatory molecule (product)

Reaction proceeds.

Product

Reaction is blocked because binding of regulatory molecule in regulatory site changes conformation of active site so that substrate cannot enter.

Figure 7.5 **Examples of two common control mechanisms for enzymes.**

Direct Controls on the Action of Enzymes

The bacterial cell has many ways of directly influencing the activity of its enzymes. It can inhibit enzyme activity by supplying a molecule that resembles the enzyme's normal substrate. The "mimic" can then occupy the enzyme's active site, preventing the actual substrate from binding there. Because the mimic cannot actually be acted on by the enzyme or function in the way the product would have, the enzyme is effectively shut down. This form of inhibition is called **competitive inhibition** because the mimic is competing with the substrate for the binding site **(figure 7.5).** (In chapter 10, you will see that some antibiotics use the same strategy of competing with enzymatic active sites to shut down metabolic processes.)

Another form of inhibition can occur with special types of enzymes that have two binding sites—the active site and another area called the regulatory site (figure 7.5). These enzymes are regulated by the binding of molecules other than the substrate in a different site—a regulatory site. Often the regulatory molecule is the product of the enzymatic reaction itself. This provides a negative feedback mechanism that can slow down enzymatic activity once a certain concentration of product is produced. This is **noncompetitive inhibition** because the regulator molecule does not bind in the same site as the substrate.

Controls on Enzyme Synthesis

Controlling enzymes by controlling their synthesis is another effective mechanism because enzymes do not last indefinitely. Some wear out, some are deliberately degraded, and others are diluted with each cell division. For catalysis to continue, enzymes eventually must be replaced. This cycle works into the scheme of the cell, where replacement of enzymes can be regulated according to cell demand.

Many laundry detergents contain enzymes that target proteins, lipids, and carbohydrates, which are the chemical constituents of stains on clothing.
©Jill Braaten/McGraw-Hill Education

Figure 7.6 One type of genetic control of enzyme synthesis: enzyme repression.   (1)–(5) The enzyme is synthesized continuously via uninhibited transcription and translation until enough product has been made. (6)–(7) Excess product reacts with a site on DNA that regulates the enzyme's synthesis, thereby inhibiting further enzyme production.

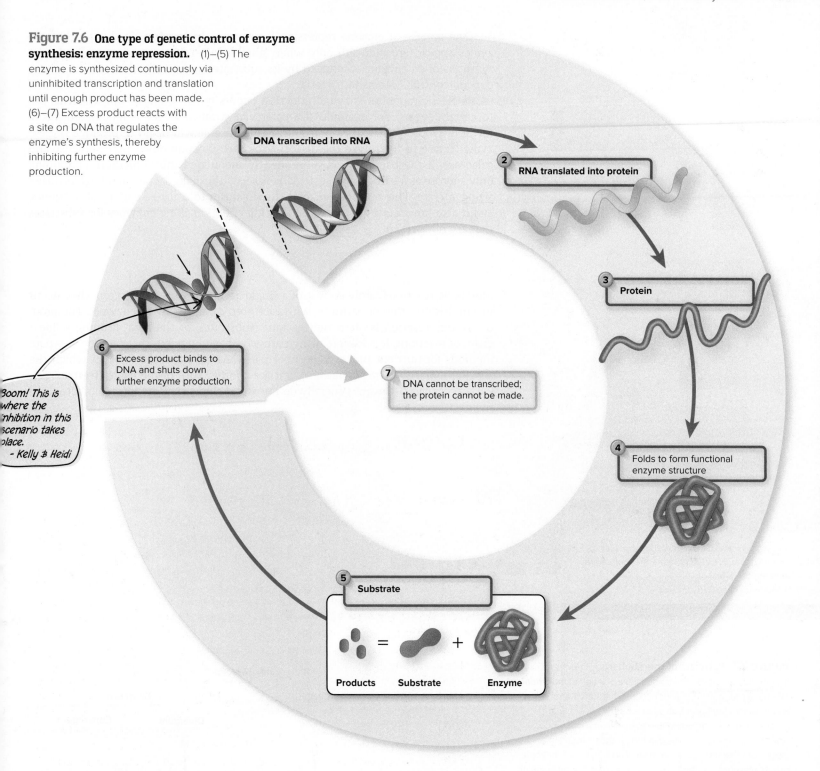

1 DNA transcribed into RNA

2 RNA translated into protein

3 Protein

4 Folds to form functional enzyme structure

5 Substrate

Products   Substrate   Enzyme

6 Excess product binds to DNA and shuts down further enzyme production.

7 DNA cannot be transcribed; the protein cannot be made.

Boom! This is where the inhibition in this scenario takes place.
— Kelly & Heidi

The mechanisms of this system are genetic in nature; that is, they require regulation of DNA and the protein synthesis machinery–topics we shall encounter once again in chapter 8.

   Enzyme repression is a means to stop further synthesis of an enzyme somewhere along its pathway. As the level of the end product from a given enzymatic reaction has built to excess, the genetic apparatus responsible for replacing these enzymes is automatically suppressed **(figure 7.6).**

The inverse of enzyme repression is **enzyme induction.** In this process, enzymes appear (are induced) only when suitable substrates are present—that is, the synthesis of an enzyme is induced by its substrate. Both mechanisms are important genetic control systems in bacteria.

A classic model of enzyme induction occurs in the response of *Escherichia coli* to certain sugars. For example, if a particular strain of *E. coli* is inoculated into a medium whose principal carbon source is lactose, it will begin to produce the enzyme lactase to hydrolyze the disaccharide into its component parts, glucose and galactose. If the bacterium is subsequently inoculated into a medium containing only sucrose as a carbon source, it will cease synthesizing lactase and begin synthesizing sucrase. This response enables the organism to utilize a variety of nutrients, and it also prevents a microbe from wasting energy making enzymes for substrates that are not present.

Metabolic Pathways

Metabolic reactions rarely consist of a single action or step. More often, they occur in a multistep series or pathway, with each step catalyzed by an enzyme. The product of one reaction is often the reactant (substrate) for the next, forming a linear chain of reactions. In addition, many pathways have branches that provide alternate methods for nutrient processing. Other pathways take a cyclic form, in which the starting molecule is regenerated to initiate another turn of the cycle **(figure 7.7).** On top of that, pathways generally do not stand alone; they are interconnected and merge at many sites.

Enzymes isolated from microbes are used by denim manufacturers to make the fabric softer and impart different colors to it.
©Frederick Bass/Getty Images

Figure 7.7 Patterns of metabolism.   In general, metabolic pathways consist of a linked series of individual chemical reactions that produce intermediary metabolites and lead to a final product. These pathways occur in several patterns, including linear, cyclic, and branched. Anabolic pathways involved in biosynthesis result in a more complex molecule, each step adding on a functional group, whereas catabolic pathways involve the dismantling of molecules and can generate energy. Virtually every reaction in a series—represented by an arrow—involves a specific enzyme.

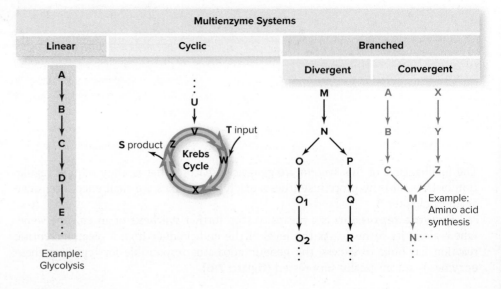

7.2 The Pursuit and Utilization of Energy

In order to carry out the work of all of their metabolic processes, cells require constant input of some form of usable energy. The energy can come directly from sunlight (in photosynthesizers), or from free electrons (in electricity-harvesting bacteria). In most bacteria we examine in this book, the energy comes from organic substances (like sugars) when their bonds are broken, releasing and transferring electrons. The energy is mostly stored in ATP.

Plant cells, such as in this energy-rich sugar cane, create ATP from photosynthesis.
©Digital Vision/PunchStock

Energy in Cells

Cells manage energy in the form of chemical reactions that change molecules. This often involves activities such as the making or breaking of bonds and the transfer of electrons. Not all cellular reactions are equal with respect to energy. Some release energy, and others require it to proceed. For example, a reaction that proceeds as follows:

$$X + Y \xrightarrow{\text{Enzyme}} Z + \text{Energy}$$

releases energy as it goes forward. This type of reaction is an **exergonic** (ex-er-gon′-ik) **reaction.** Energy of this type is available for doing cellular work. Energy transactions such as the following:

$$\text{Energy} + A + B \xrightarrow{\text{Enzyme}} C$$

are called **endergonic** (en-der-gon′-ik) **reactions** because they require the input of energy. In cells, exergonic and endergonic reactions are often coupled, so that released energy is immediately put to use.

Summaries of metabolism may make it seem that cells "create" energy from nutrients, but they do not. What they actually do is extract chemical energy already present in nutrient fuels and apply that energy toward useful work in the cell, much like a gasoline engine releases energy as it burns fuel. The engine does not actually produce energy, but it converts some of the potential energy to do work.

At the simplest level, cells possess specialized enzyme systems that trap the energy present in the bonds of nutrients as they are progressively broken. During exergonic reactions, energy released by bonds is stored in certain high-energy phosphate bonds, such as in ATP. The ability of ATP to temporarily store and release the energy of chemical bonds provides the fuel for endergonic cell reactions. Before discussing ATP, we examine the process behind electron transfer: redox reactions.

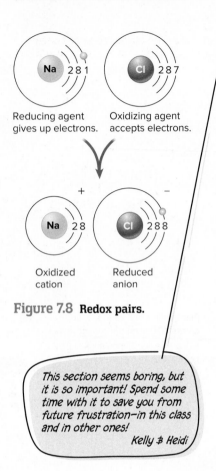

Reducing agent gives up electrons.

Oxidizing agent accepts electrons.

Oxidized cation

Reduced anion

Figure 7.8 Redox pairs.

This section seems boring, but it is so important! Spend some time with it to save you from future frustration—in this class and in other ones!

Kelly & Heidi

Oxidation and Reduction

Some atoms and compounds readily give or receive electrons and participate in oxidation (the loss of electrons) or reduction (the gain of electrons). The compound that loses the electrons is **oxidized,** and the compound that receives the electrons is **reduced (figure 7.8).** Such **oxidation-reduction (redox)** reactions are common in the cell and indispensable to the required energy transformations. Important components of cellular redox reactions are oxidoreductases, which remove electrons from one substrate and add them to another. The enzymes carry coenzymes that are vitally important to the transfer of electrons. These coenzymes are nicotinamide adenine dinucleotide (NAD) **(figure 7.9)** and flavin adenine dinucleotide (FAD). As coenzymes, they sit in a groove on the enzymes and accept and donate the electrons as the enzymes do their work. (Take note: Even if by now your eyes are glazing over at all the terms and details, this paragraph is a valuable one! If you remember the statements in this paragraph, then the rest of metabolism will be much easier to understand. A handy mnemonic device for remembering this is OIL RIG: Oxidation is Losing, Reduction is Gaining.)

Redox reactions always occur in pairs, with an electron donor and an electron acceptor, which constitute a *redox pair*. Oxidation-reduction reactions salvage electrons along with the energy they contain.

This changes the energy balance, leaving the just-reduced compound with more energy than the now-oxidized one. The energy now present in the electron acceptor can be captured to **phosphorylate** (add an inorganic phosphate) to ADP or to some other compound. This process stores the energy in a high-energy molecule (ATP, for example). In many cases, the cell does not handle electrons as separate entities but rather as parts of an atom such as hydrogen, which contains a proton and an electron. For simplicity's sake, we will continue to use the term *electron transfer*, but keep in mind that hydrogens are often involved in the transfer process. The removal of hydrogens from a compound during a redox reaction is called dehydrogenation. The job of handling these protons and electrons falls to one or more carriers (the NAD and FAD mentioned earlier), which function as short-term repositories for the electrons until they can be transferred.

Electron Carriers: Molecular Shuttles

The electron carriers NAD and FAD resemble shuttles that are alternately loaded and unloaded, repeatedly accepting and releasing electrons and hydrogens to facilitate the

Figure 7.9 Details of NAD reduction. The coenzyme NAD contains the vitamin nicotinamide (niacin) and the purine adenine attached to double ribose phosphate molecules (a dinucleotide). The principal site of action is on the nicotinamide (boxed areas). Hydrogens and electrons donated by a substrate interact with a carbon on the top of the ring. One hydrogen bonds there, carrying two electrons, and the other hydrogen is carried in solution as H^+ (a proton).

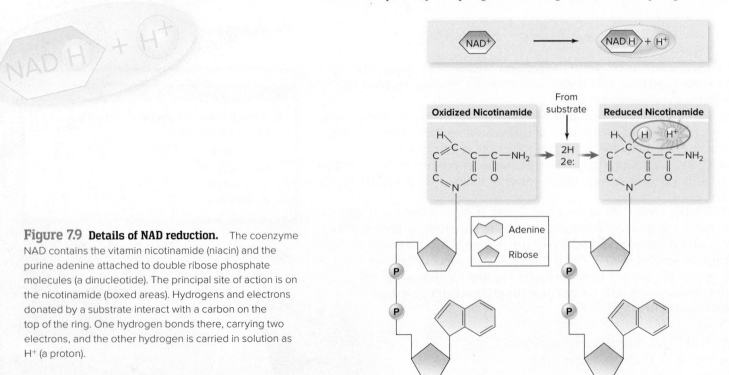

transfer of redox energy. In catabolic pathways, electrons are extracted and carried through a series of redox reactions until the final electron acceptor at the end of a particular pathway is reached. In aerobic metabolism, this acceptor is molecular oxygen. In anaerobic metabolism, it is some other inorganic or organic compound.

Adenosine Triphosphate: Metabolic Money

Let's look more closely at the powerhouse molecule adenosine triphosphate. ATP has also been described as metabolic money because it can be earned, banked, saved, spent, and exchanged. As an energy repository, ATP provides a connection between energy-yielding catabolism and the other cellular activities that require energy. Some clues to its energy-storing properties lie in its unique molecular structure.

The Molecular Structure of ATP

ATP is a three-part molecule consisting of a nitrogen base (adenine) linked to a 5-carbon sugar (ribose), with a chain of three phosphate groups bonded to the ribose **(figure 7.10)**. The high energy of ATP comes from the orientation of the phosphate groups, which are relatively bulky and carry negative charges. The proximity of these repelling electrostatic charges imposes a strain–especially between the last two phosphate groups. The strain on the phosphate bonds accounts for the energetic quality of ATP because removal of the terminal phosphates releases free energy.

Breaking the bonds between the two outermost phosphates of ATP yields adenosine diphosphate (ADP), which can then be further converted to adenosine monophosphate (AMP). AMP derivatives help form the backbone of RNA and are also a major component of certain coenzymes (NAD, FAD, and coenzyme A).

The Metabolic Role of ATP

ATP is the primary energy currency of the cell, and when its energy is used in a chemical reaction, it must then be replaced. Therefore, ATP utilization and replenishment make up an ongoing cycle. Often, the energy released during ATP hydrolysis drives biosynthesis by providing an activating phosphate to an individual substrate before it is enzymatically linked to another molecule. ATP is also used to prepare molecules for catabolism, such as when a 6-carbon sugar is phosphorylated during the early stages of glycolysis:

$$\text{Glucose} \xrightarrow[\quad\quad\quad]{\text{ATP} \quad \text{ADP}} \text{Glucose-6-phosphate}$$

When ATP is utilized by the removal of the terminal phosphate to release energy plus ADP, ATP then needs to be re-created. The reversal of this process–that is, adding the terminal phosphate to ADP–will replenish ATP, but it requires an input of energy:

Substrate — PO$_4$ + ADP ⟶ ○ + ATP

In heterotrophs, the energy infusion that regenerates a high-energy phosphate comes from certain steps of catabolic pathways, in which nutrients such as carbohydrates are degraded and yield energy. ATP is formed when substrates or electron carriers provide a high-energy phosphate that becomes bonded to ADP.

7.2 LEARNING OUTCOMES—Assess Your Progress

6. Name the chemical in which energy is stored in cells.
7. Create a general diagram of a redox reaction.
8. Identify electron carriers used by cells.

Figure 7.10 The structure of adenosine triphosphate (ATP). Removing the left-most phosphate group yields ADP; removing the next one yields AMP.

Adenosine Triphosphate (ATP)	Adenosine Diphosphate (ADP)	Adenosine

Adenine

~ Bond that releases energy when broken

Ribose

ATP is used for energy in all cells, including human cells.
©Michael Svoboda/Getty Images

7.3 Catabolism

Now you have an understanding of all the tools a cell needs to *metabolize*. Metabolism uses *enzymes* to drive reactions that break down (*catabolize*) organic molecules to materials (*precursor molecules*) that cells can then use to build (*anabolize*) larger, more complex molecules that they need. This process was presented symbolically in figure 7.1. Another very important point about metabolism is that *reducing power* (the electrons available when NAD and FAD are in their reduced forms–NADH and FADH$_2$) and *energy* (stored in the bonds of ATP) are needed in large quantities for the anabolic parts of metabolism (the blue bars in that figure). They are produced during the catabolic part of metabolism (the yellow bar).

Metabolism starts with "nutrients" from the environment, usually discarded molecules from other organisms. Cells have to get the nutrients inside; to do this, they use the transport mechanisms discussed in chapter 6. Some of these require energy, which is available from catabolism already occurring in the cell. In the next step, intracellular nutrients have to be broken down to the appropriate precursor molecules. These catabolic pathways are discussed next.

Getting Materials and Energy

Nutrient processing is extremely varied, especially in bacteria, yet in most cases it is based on three basic catabolic pathways. Frequently, the nutrient is glucose. In previous discussions, microorganisms were categorized according to their requirement for oxygen gas. This designation is related directly to these nutrient processing pathways. **Figure 7.11** provides an overview of the three major pathways for producing the needed precursors and energy (i.e., catabolism).

Figure 7.11 Overview of the three main pathways of catabolism.

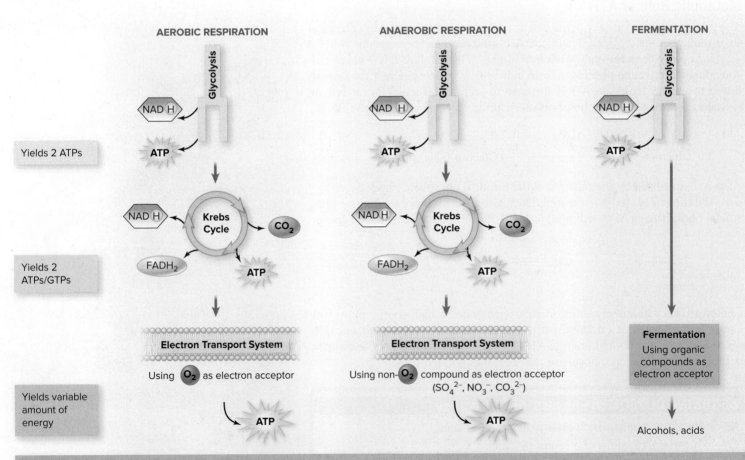

Aerobic respiration is a series of reactions (glycolysis, the Krebs cycle, and the respiratory chain) that converts glucose to CO_2 and allows the cell to recover significant amounts of energy. Aerobic respiration relies on free oxygen as the final acceptor for electrons and hydrogens and produces a relatively large amount of ATP. Aerobic respiration is characteristic of many bacteria, fungi, protozoa, and animals.

Anaerobic respiration is a metabolic strategy used by many microorganisms, some strictly anaerobic and others who are able to metabolize with or without oxygen. This system involves the same three pathways as aerobic respiration, but it does not use oxygen as the terminal electron acceptor; instead, NO_3^-, SO_4^{2-}, CO_3^{3-}, and other oxidized compounds are utilized.

Fermentation is an adaptation used by facultative and aerotolerant anaerobes to incompletely oxidize (ferment) glucose. In this case, oxygen is not required, organic compounds are the terminal electron acceptors, and a relatively small amount of ATP is produced.

Glycolysis

All three of the metabolic pathways begin with **glycolysis,** which turns glucose into two copies of pyruvic acid–a chemical uniquely capable of yielding energy in the pathways that follow. **Table 7.2** illustrates glycolysis.

Pyruvic Acid—Central to All Three Metabolic Strategies

In strictly aerobic organisms and some anaerobes, pyruvic acid enters the Krebs cycle for further processing and energy release. Facultative anaerobes can use fermentation, in which pyruvic acid is re-reduced into acids or other products.

After Pyruvic Acid, Aerobic and Anaerobic Respiration Strategies

Bacterial respiration, whether done aerobically or anaerobically, utilizes the Krebs cycle and the electron transport system to harvest the energy and products needed to build cell parts.

Microbes often obtain nutrients from dead plants and organisms.
©imageshop - Zefa Visual Media UK Ltd/Alamy

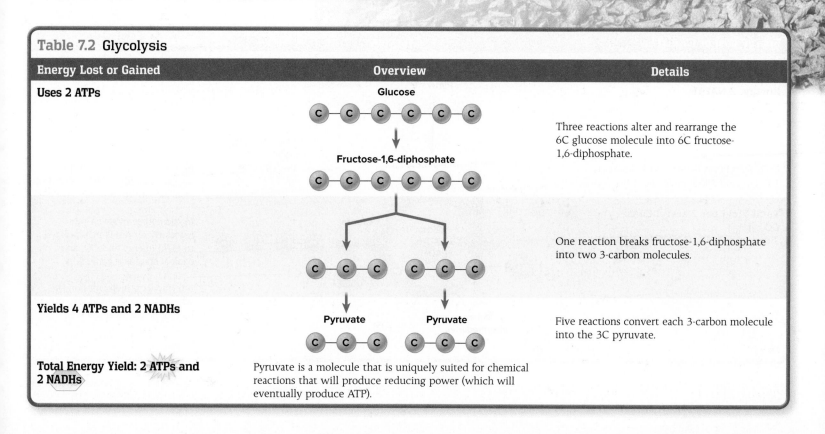

Table 7.2 Glycolysis		
Energy Lost or Gained	**Overview**	**Details**
Uses 2 ATPs	Glucose	Three reactions alter and rearrange the 6C glucose molecule into 6C fructose-1,6-diphosphate.
	Fructose-1,6-diphosphate	One reaction breaks fructose-1,6-diphosphate into two 3-carbon molecules.
Yields 4 ATPs and 2 NADHs	Pyruvate Pyruvate	Five reactions convert each 3-carbon molecule into the 3C pyruvate.
Total Energy Yield: 2 ATPs and 2 NADHs	Pyruvate is a molecule that is uniquely suited for chemical reactions that will produce reducing power (which will eventually produce ATP).	

The Krebs cycle operates similarly in aerobic and anaerobic respiration and is covered next.

The Krebs Cycle—A Carbon and Energy Wheel

As you have seen, the cleaving and oxidation of glucose during glycolysis yield a comparatively small amount of energy and gives off pyruvic acid. Pyruvic acid is still energy-rich, which means it contains a number of extractable hydrogens and electrons to power ATP synthesis. The hydrogens and electrons are only made available during the work of the second and third phases of respiration, in which pyruvic acid's hydrogens are transferred to oxygen. In the following section, we examine the second phase of catabolism, which takes place in the cytoplasm of bacteria and in the mitochondrial matrix in eukaryotes. **Table 7.3** summarizes the Krebs cycle.

The **Krebs cycle** (also known as the tricarboxylic acid [TCA] cycle) transfers the energy stored in acetyl CoA to NAD^+ and FAD by reducing them (transferring electrons to them). Thus, the main products of the Krebs cycle are these reduced molecules (as well as 2 ATPs for each glucose molecule). The reduced coenzymes NADH and $FADH_2$ are vital to the energy production that will occur in electron transport. Along the way, the 2-carbon acetyl CoA joins with a 4-carbon compound, oxaloacetic acid, and then participates in seven additional chemical transformations while "spinning off" the NADH and $FADH_2$. That's why we sometimes call the Krebs cycle the "carbon and energy wheel."

The "H's" indicate that the carriers are fully loaded and carrying energy.
Kelly & Heidi

The Respiratory Chain: Electron Transport

We now come to the energy chain, which is the final "processing mill" for electrons and hydrogen ions and the major generator of ATP. It is the final step in both aerobic and anaerobic respiration. Overall, the electron transport system (ETS) consists of a chain of special redox carriers that receives electrons from reduced carriers (NADH, $FADH_2$) generated by glycolysis and the Krebs cycle and passes

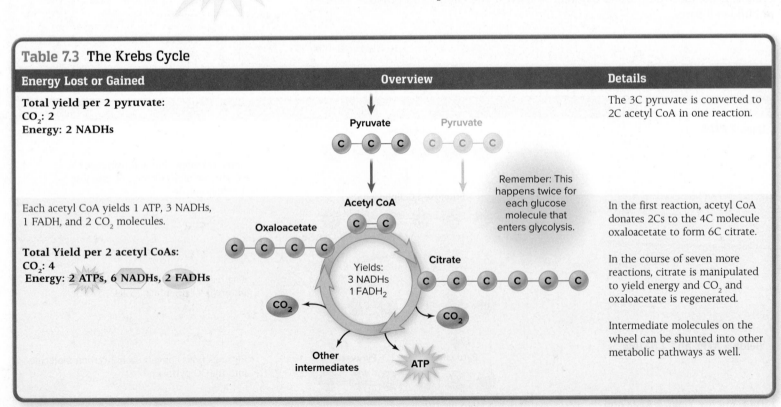

Table 7.3 The Krebs Cycle

Energy Lost or Gained	Overview	Details
Total yield per 2 pyruvate: CO_2: 2 **Energy: 2 NADHs**	Pyruvate / Pyruvate — C C C C C C **Acetyl CoA** — C C **Oxaloacetate** — C C C C **Citrate** — C C C C C C Yields: 3 NADHs, 1 $FADH_2$ CO_2 CO_2 Remember: This happens twice for each glucose molecule that enters glycolysis. Other intermediates ATP	The 3C pyruvate is converted to 2C acetyl CoA in one reaction.
Each acetyl CoA yields 1 ATP, 3 NADHs, 1 FADH, and 2 CO_2 molecules. **Total Yield per 2 acetyl CoAs:** CO_2: 4 **Energy: 2 ATPs, 6 NADHs, 2 FADHs**		In the first reaction, acetyl CoA donates 2Cs to the 4C molecule oxaloacetate to form 6C citrate. In the course of seven more reactions, citrate is manipulated to yield energy and CO_2 and oxaloacetate is regenerated. Intermediate molecules on the wheel can be shunted into other metabolic pathways as well.

them in a sequential and orderly fashion from one redox molecule to the next. The flow of electrons down this chain is full of energy and allows the active transport of hydrogen ions to the outside of the membrane where the respiratory chain is located. In aerobic respiration, the step that finalizes the transport process is the acceptance of electrons and hydrogen by oxygen, producing water. This process consumes oxygen. Some variability exists from one organism to another, but the principal compounds that carry out these complex reactions are NADH dehydrogenase, flavoproteins, coenzyme Q (ubiquinone), and **cytochromes** (sy′-toh-krohmz). The cytochromes contain a tightly bound metal atom at their center that is actively involved in accepting electrons and donating them to the next carrier in the series. The highly compartmentalized structure of the respiratory chain is an important factor in its function. Note in **table 7.4** that the electron transport carriers and enzymes are embedded in the cytoplasmic membrane in bacteria. The equivalent structure for housing them in eukaryotes is the inner mitochondrial membranes pictured in **figure 7.12.**

Table 7.4 The Respiratory (Electron Transport) Chain

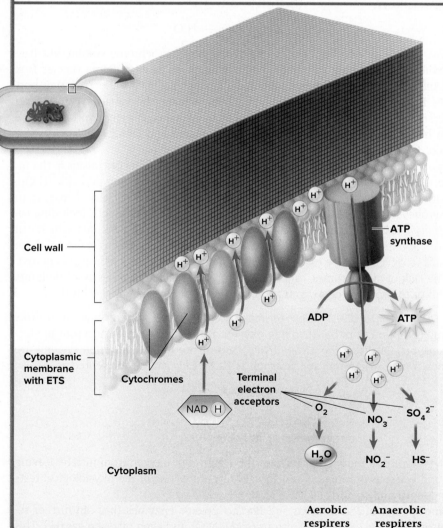

Reduced carriers (NADH, FADH$_2$) transfer electrons and H$^+$ to first electron carrier in chain: NADH dehydrogenase.

These are then sequentially transferred to the next four to six carriers with progressively more positive reduction potentials. The carriers are called cytochromes. The number of carriers varies, depending on the bacterium.

Simultaneous with the reduction of the electron carriers, protons are moved to the outside of the membrane, creating a concentration gradient (more protons outside than inside the cell). The extracellular space becomes more positively charged and more acidic than the intracellular space. This condition creates the *proton motive force*, by which protons flow down the concentration gradient through the ATP synthase embedded in the membrane. This results in the conversion of ADP to ATP.

Once inside the cytoplasm, protons combine with O$_2$ to form water (in aerobic respirers [left]), or with a variety of O-containing compounds (in anaerobic respirers [right]) to produce more reduced compounds.

Aerobic respiration yields a maximum of 3 ATPs per oxidized NADH and 2 ATPs per oxidized FADH.

Anaerobic respiration yields less per NADH and FADH$_2$.

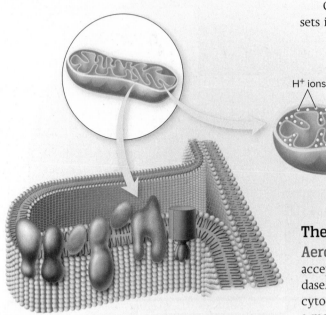

Figure 7.12 **The electron transport system on the inner membrane of the mitochondrial cristae in eukaryotes.**

Medical Moment

Facultative Anaerobes

Some organisms are capable of utilizing either aerobic or anaerobic respiration, depending on what is available in their environment. As discussed in chapter 6, these are called facultative anaerobes. If oxygen is available, aerobic respiration is utilized as it is most efficient and yields more ATP than other metabolic pathways. In the absence of oxygen, these organisms use either anaerobic respiration or fermentation.

One example is *Listeria*. Foods ranging from dairy products and uncooked meats to fruits and vegetables have been contaminated with *Listeria monocytogenes*. Exposure to this bacterium causes foodborne illness and in those at high risk, may cause death. Being a facultative anaerobe contributes to the resiliency of *Listeria*, allowing it to survive and multiply in diverse environments. Most people are regularly exposed to this bacterium in their daily settings. Following safe food preparation and handling techniques can prevent foodborne illness.

Q. From this information, can you identify another way in which *Listeria* adapts to environmental conditions, in addition to oxygen utilization?

Answer in Appendix B.

Conveyance of the NADHs from glycolysis and the Krebs cycle to the first carrier sets in motion the remaining steps. The hydrogen ions sequestered outside the membrane travel back into the cell using the **ATP synthase** complex as a channel. When they do so, they provide the energy to add a phosphate to ADP, creating ATP in the process. Each NADH that enters electron transport can give rise to a maximum of 3 ATPs (though actual numbers are probably lower due to inefficiencies in the pathways). This coupling of ATP synthesis to electron transport is termed **oxidative phosphorylation.** Because the electrons from $FADH_2$ from the Krebs cycle enter the respiratory chain at a later point than the NAD reactions, there is less energy to release, and only 2 ATPs are the result.

The Final Step

Aerobic Respiration   The final step in aerobic respiration, during which oxygen accepts the electrons, is catalyzed by cytochrome aa_3, also called cytochrome oxidase. This large enzyme complex is specifically adapted to receive electrons from cytochrome c, pick up hydrogens from the solution, and react with oxygen to form a molecule of water. This reaction, though in actuality more complex, can be summarized as follows:

$$2H^+ + 2e^- + \tfrac{1}{2}O_2 \rightarrow H_2O$$

Most eukaryotic aerobes have a fully functioning cytochrome system, but bacteria exhibit wide-ranging variations in this part of the system. Some species lack one or more of the redox steps; others have several alternative electron transport schemes. Because many bacteria lack cytochrome oxidase, this variation can be used to differentiate among certain genera of bacteria. An oxidase detection test can be used to help identify members of the genera *Neisseria* and *Pseudomonas* and some species of *Bacillus*.

A potential side reaction of the respiratory chain in aerobic organisms is the incomplete reduction of oxygen to superoxide ion (O_2^-) and hydrogen peroxide (H_2O_2). As mentioned in chapter 6, these toxic oxygen products can be very damaging to cells. Aerobes have neutralizing enzymes to deal with these products, including *superoxide dismutase* and *catalase*. One exception is the genus *Streptococcus*, which can grow well in oxygen yet lacks both cytochromes and catalase. The tolerance of these organisms to oxygen can be explained by the neutralizing effects of a special peroxidase. The lack of cytochromes, catalase, and peroxidases in anaerobes as a rule limits their ability to process free oxygen and contributes to its toxic effects on them.

Anaerobic Respiration   The terminal step in anaerobic respiration utilizes inorganic compounds, rather than free oxygen, as the final electron acceptor in electron transport. Of these, the nitrate (NO_3^-) and nitrite (NO_2^-) reduction systems are best known. The reaction in species such as *Escherichia coli* is represented as:

$$\begin{array}{c} \text{Nitrate reductase} \\ \downarrow \\ NO_3^- + NADH \rightarrow NO_2^- + H_2O + NAD^+ \\ \text{nitrate} \qquad\qquad \text{nitrite} \end{array}$$

The enzyme nitrate reductase catalyzes the removal of oxygen from nitrate, leaving nitrite and water as products. A test for this reaction is one of the physiological tests used in identifying bacteria.

Some species of *Pseudomonas* and *Bacillus* possess enzymes that can further reduce nitrite to nitric oxide (NO), nitrous oxide (N_2O), and even nitrogen gas (N_2). This process, called **denitrification,** is a very important step in recycling nitrogen in the biosphere. Other oxygen-containing nutrients reduced anaerobically by various bacteria are carbonates and sulfates. None of the anaerobic pathways produce as much ATP as aerobic respiration. They produce a variable amount of ATP, from 2 to 36 ATPs.

After Pyruvic Acid, the Fermentation Strategy

The definition of *fermentation* is the *incomplete oxidation of glucose or other carbohydrates in the absence of oxygen*. This process uses organic compounds as the terminal electron acceptors and yields a small amount of ATP (see figure 7.11). This pathway is used by organisms that do not have an electron transport chain and therefore cannot respire. Other organisms can repress the production of electron transport chain proteins when oxygen is lacking in their environment. They can then revert to fermentation.

Without an electron transport chain to churn out large quantities of ATP from reduced carriers, it may seem that fermentation would yield only small amounts of energy (2 ATPs maximum per glucose), and that would slow down growth. What actually happens, however, is that many bacteria can grow as fast as they would in the presence of oxygen. This rapid growth is made possible by an increase in the rate of glycolysis. From another standpoint, fermentation permits independence from molecular oxygen and allows colonization of anaerobic environments. It also enables microorganisms with a versatile metabolism to adapt to variations in the availability of oxygen. For them, fermentation provides a means to grow even when oxygen levels are too low for aerobic respiration.

Bacteria that digest cellulose in the rumens of cattle are largely fermentative. After initially hydrolyzing cellulose to glucose, they ferment the glucose to organic acids, which are then absorbed as the bovine's principal energy source. Even human muscle cells can undergo a form of fermentation that permits short periods of activity after the oxygen supply in the muscle has been exhausted. Muscle cells convert pyruvic acid into lactic acid, which allows anaerobic production of ATP to proceed for a time. But this cannot go on indefinitely, and after a few minutes, the accumulated lactic acid causes muscle fatigue.

Table 7.5 gives an overview of fermentation.

Yeasts turn the sugar in grapes into alcohol through fermentation.
©Ingram Publishing/SuperStock

drew Hagen/
tterstock.com

Table 7.5 Fermentation

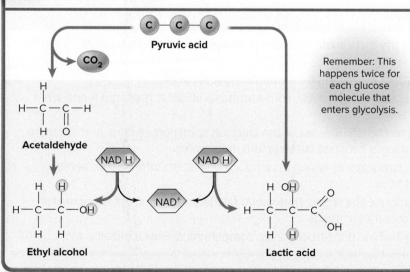

Pyruvic acid

CO_2

Acetaldehyde

NAD H   NAD H

NAD$^+$

Ethyl alcohol   Lactic acid

Remember: This happens twice for each glucose molecule that enters glycolysis.

Pyruvic acid from glycolysis can itself become the electron acceptor.

It can continue through alcoholic fermentation (on the left) or acidic fermentation (on the right).

Pyruvic acid can also be enzymatically altered and then serve as the electron acceptor.

The NADs are recycled to reenter glycolysis.

The organic molecules that became reduced in their role as electron acceptors are extremely varied, and often yield useful products such as ethyl alcohol, lactic acid, propionic acid, butanol, and others.

Products of Fermentation in Microorganisms

Alcoholic beverages (wine, beer, whiskey) are perhaps the most well-known fermentation products. Note that the products of alcoholic fermentation are not only ethanol but also CO_2, a gas that accounts for the bubbles in champagne and beer and the rising of bread dough.

Other fermentation products are solvents (acetone, butanol), organic acids (lactic, acetic acids), dairy products, and many other foods. Derivatives of proteins, nucleic acids, and other organic compounds are fermented to produce vitamins, antibiotics, and even hormones such as hydrocortisone.

We have provided only a brief survey of fermentation products, but it is worth noting that microbes can be harnessed to synthesize a variety of other substances simply by varying the raw materials provided them. Large-scale industrial processes using microorganisms often utilize entirely different mechanisms from those described here, and they even occur aerobically, particularly in antibiotic, hormone, vitamin, and amino acid production. Even so, they are also referred to as *fermentation*.

Catabolism of Noncarbohydrate Compounds

We have given you one version of events for catabolism, using glucose, a carbohydrate, as our example. Other compounds serve as fuel, as well. The more complex polysaccharides are easily broken down into their component sugars, which can enter glycolysis at various points. Microbes also break down other molecules for their own use, of course. Two other major sources of energy and building blocks for microbes are lipids (fats) and proteins. Both of these must be broken down to their component parts to produce precursor metabolites and energy.

Recall from chapter 1 that fats are fatty acids joined to glycerol. Enzymes called **lipases** break these apart. The glycerol is then converted to dihydroxyacetone phosphate (DHAP), which can enter a step midway through glycolysis. The fatty acid component goes through a process called **beta oxidation.** Fatty acids have a variable number of carbons; in beta oxidation, 2-carbon units are successively transferred to coenzyme A, creating acetyl CoA, which enters the Krebs cycle. This process can yield a large amount of energy. Oxidation of a 6-carbon fatty acid yields 50 ATPs, compared with 38 for a 6-carbon sugar.

Proteins are chains of amino acids. Enzymes called **proteases** break proteins down to their amino acid components, after which the amino groups are removed by a reaction called **deamination.** This leaves a carbon compound, which is easily converted to one of several Krebs cycle intermediates.

7.3 LEARNING OUTCOMES—Assess Your Progress

9. Name the three main catabolic pathways and the estimated ATP yield for each.
10. Construct a paragraph summarizing glycolysis.
11. Describe the Krebs cycle, with emphasis on what goes into it and what comes out of it.
12. Discuss the significance of the electron transport system, and compare the process between bacteria and eukaryotes.
13. State two ways in which anaerobic respiration differs from aerobic respiration.
14. Summarize the steps of microbial fermentation, and list three useful products it can create.
15. Describe how noncarbohydrate compounds are catabolized.

7.4 Anabolism and the Crossing Pathways of Metabolism

Our discussion now turns from catabolism and energy extraction to anabolic functions and biosynthesis. In this section, we present aspects of intermediary metabolism, including amphibolic pathways, the synthesis of simple molecules, and the synthesis of macromolecules.

The Efficiency of the Cell

It must be obvious by now that cells have mechanisms for careful management of carbon compounds. Rather than being dead ends, most catabolic pathways contain strategic molecular intermediates (metabolites) that can be diverted into anabolic pathways. In this way, a given molecule can serve multiple purposes, and the maximum benefit can be derived from all nutrients and metabolites of the cell pool. The ability of a system to integrate catabolic and anabolic pathways to improve cell efficiency is termed **amphibolism** (am-fee-bol'-izm).

At this point in the chapter, you can appreciate a more complex view of metabolism than the one we presented at the beginning in figure 7.1. **Table 7.6** demonstrates the amphibolic nature of intermediary metabolism.

Are you seeing that glucose is the center of this whole thing?
- Kelly & Heidi

Table 7.6 Amphibolic Pathways of Glucose Metabolism

Anabolic Pathways

Intermediates from glycolysis are fed into the amino acid synthesis pathway. From there, the compounds are formed into proteins. Amino acids can then contribute nitrogenous groups to nucleotides to form nucleic acids.

Glucose and related simple sugars are made into additional sugars and polymerized to form complex carbohydrates.

The glycolysis product acetyl CoA can be oxidized to form fatty acids, critical components of lipids.

Catabolic Pathways

In addition to the respiration and fermentation pathways already described, bacteria can deaminate amino acids, which leads to the formation of a variety of metabolic intermediates, including pyruvate and acetyl CoA.

Also, fatty acids can be oxidized to form acetyl CoA.

Anabolism: Formation of Macromolecules

Monosaccharides, amino acids, fatty acids, nitrogen bases, and vitamins—the building blocks that make up the various macromolecules and organelles of the cell—come from two possible sources. They can enter the cell from the outside "ready to use," or they can be synthesized through various cellular pathways. The degree to which an organism can synthesize its own building blocks (simple molecules) is determined by its genetic makeup, a factor that varies tremendously from group to group. In chapter 6, you learned that autotrophs require only CO_2 as a carbon source, a few minerals to synthesize all cell substances, and no organic nutrients. Some heterotrophic organisms (*E. coli*, yeasts) are also very efficient in that they can synthesize all cellular substances from minerals and a single organic carbon source such as glucose. Compare this with a strict parasite that has few synthetic abilities of its own and derives most precursor molecules from the host.

Whatever their source, once these building blocks are added to the metabolic pool, they are available for the synthesis of polymers by the cell.

Carbohydrate Biosynthesis

Glucose has a central role in metabolism and energy utilization. For that reason, there are multiple pathways for manufacturing it in cells. Certain structures in the cell depend on an adequate supply of glucose as well. It is the major component of the cellulose cell walls of some eukaryotes and of certain storage granules (starch, glycogen). One of the intermediaries in glycolysis, glucose-6-P, is used to form glycogen. Monosaccharides other than glucose are important in the synthesis of bacterial cell walls. Peptidoglycan contains a linked polymer of muramic acid and glucosamine. Fructose-6-P from glycolysis is used to form these two sugars. Carbohydrates (deoxyribose, ribose) are also essential building blocks in nucleic acids. Polysaccharides are the predominant components of cell surface structures such as capsules and the glycocalyx, and they are commonly found in slime layers.

Amino Acids, Protein Synthesis, and Nucleic Acid Synthesis

Proteins account for a large proportion of a cell's constituents. They are essential components of enzymes, the cytoplasmic membrane, the cell wall, and cell appendages. As a general rule, 20 amino acids are needed to make these proteins. Although some organisms (*E. coli*, for example) have pathways that will synthesize all 20 amino acids, others, including animals, lack some or all of the pathways for amino acid synthesis and must acquire the essential ones from their diets. Protein synthesis itself is a complex process that requires a genetic blueprint and the operation of intricate cellular machinery, as you will see in chapter 8.

DNA and RNA are responsible for the hereditary continuity of cells and the overall direction of protein synthesis. Because nucleic acid synthesis is a major topic of genetics and is closely allied to protein synthesis, it will also be covered in chapter 8.

Assembly of the Cell

The component parts of a bacteria cell are synthesized on a continuous basis, and catabolism is also taking place as long as nutrients are present and the cell is in a nondormant state. When anabolism produces enough macromolecules to serve two cells, and when DNA replication produces duplicate copies of the cell's

---

Medical Moment

Amino Acids: Essential, Nonessential, and Conditionally Essential Amino Acids

Essential amino acids are those amino acids that must be obtained from our diet, as our bodies lack the ability to synthesize them, while nonessential amino acids are those that can be synthesized by our bodies.

However, as is often the case, there are exceptions to every rule just to keep things interesting! *Conditionally essential* amino acids are not normally required in the diet, except in specific populations that are unable to synthesize them in adequate amounts. For example, individuals with PKU (phenylketonuria), a genetic condition, must strictly restrict the levels of phenylalanine in their bodies to prevent serious complications such as developmental disabilities; however, because they cannot synthesize enough tyrosine from the trace amounts of phenylalanine in their bodies, tyrosine becomes an essential amino acid for these individuals.

Q. From the above, infer how tyrosine—nonessential in healthy people—is normally produced in the body.

Answer in Appendix B.

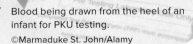

Blood being drawn from the heel of an infant for PKU testing.
©Marmaduke St. John/Alamy

genetic material, the cell undergoes binary fission, which results in two cells from one parent cell. The two cells will need twice as many ribosomes, twice as many enzymes, and so on. The cell has created these during the anabolic phases we have described. Before cell division, the membrane(s) and the cell wall will have increased in size to create a cell that is almost twice as big as a "newborn" cell. Once synthesized, the phospholipid bilayer components of the membranes assemble themselves spontaneously with no energy input. Other assembly reactions require the input of energy. Proteins and other components must be added to the membranes. Growth of the cell wall, accomplished by the addition and coupling of sugars and peptides, requires energy input. The energy gained during catabolic processes provides all the energy for these complex building reactions.

7.4 LEARNING OUTCOMES—Assess Your Progress

16. Provide an overview of the anabolic stages of metabolism.

17. Define *amphibolism*.

CASE FILE WRAP-UP

(Nurse): McGraw-Hill Education; (Inset): Source: Janice Haney Carr/Centers for Disease Control and Prevention

In low oxygen states, the body uses other sources of energy. When organs are unable to extract the oxygen they need for energy production, the body converts to anaerobic metabolism to produce ATP. The by-product of this process is lactic acid. High levels of lactic acid can alter the patient's blood pH—a state called lactic acidosis. Generally, a high serum lactate level indicates poor organ function.

Prompt identification and treatment of sepsis are associated with better outcomes and decreased mortality. The priority for sepsis management is early fluid resuscitation and antibiotic administration. It is essential to support hemodynamics (blood pressure and perfusion) to ensure adequate oxygen delivery to body tissues and prevent further organ damage.

In this case, the nurse correctly prioritized collection of body fluid samples for culture while waiting for the arrival of antibiotics. Collecting these samples after antibiotics have been given may reflect altered growth of bacteria, clouding the picture for treatment of a specific organism. Patients with septicemia are treated with broad-spectrum antibiotics until the specific causative organism is identified.

Electricity Eaters

This chapter describes metabolism, and it states right up front that metabolism has three goals:

- break down large molecules (usually sugars) to get building blocks,
- build molecules the cell needs, and
- harvest energy to do the building work.

Put another way, electron flow is necessary for energy to be gained. The flow goes like this:

an energy source → electron carriers → the electron transport chain

Well, nature continues to surprise us with its many different variations on central themes. Recently, scientists discovered microbes that accomplish their metabolic goals using nothing but electrons. Recall from table 6.2 that many autotrophs run on CO_2 and sunlight or minerals. But these newly discovered bacteria are capable of metabolizing and growing using only electrons from electricity. We know that some bacteria can run the TCA cycle in reverse, creating acetyl CoA from carbon dioxide. However, the very newest bacteria discovered can live with no carbon input at all—not even CO_2.

Dr. Ken Nealson, at the University of Southern California, whose lab conducted these studies, states: "This is huge. What it means is there's a whole part of the microbial world that we don't know about." In addition there are a slew of practical applications, including waste recycling and creating biological fuel cells. Dr. Nealson is also guessing that this might be the predominant mode of life on other planets.

Electricity-eating bacteria have been found in many places with little oxygen but lots of minerals. One group was isolated from rocks on the sea bed near Catalina Island in California.
©incamerastock/Alamy Stock Photo

MICROBIAL METABOLISM

7.1 METABOLISM AND THE ROLE OF ENZYMES

Enzymes are the workhouses of all cells. They catalyze reactions that turn big things into small things, and small things into big things. These reactions enable metabolism, which is the combination of catabolism (breaking compounds down) and anabolism (building compounds up).

Products

Converting the energy contained in nutrients to energy that is usable requires storing it in molecules in the cell. These energy carriers store the energy in the form of electrons. Electron carriers pass off energy (electrons) in a chain called a redox chain.

THE PURSUIT AND UTILIZATION OF ENERGY **7.2**

7.3 CATABOLISM

Catabolism is the "feeding phase" of metabolism. The chemical structure of carbohydrates yields a lot of electrons (energy). Breaking down the carbohydrates often leads to the intermediate pyruvic acid, which can go into several different pathways, producing products and energy.

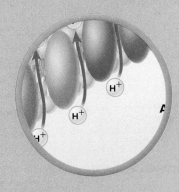

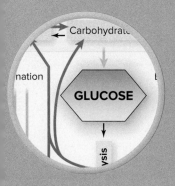

Cells and systems can integrate catabolic and anabolic pathways to improve efficiency. The ability to use both is called amphibolism.

ANABOLISM AND THE CROSSING PATHWAYS OF METABOLISM **7.4**

SmartGrid: From Knowledge to Critical Thinking

This *21 Question Grid* takes the topics from this chapter and arranges them with respect to the American Society for Microbiology's Undergraduate Curriculum guidelines—all six of the important "Concepts" as well as the important "Competency" of scientific literacy. Three questions are supplied about chapter content that refer to the Concept or Competency, in increasing levels of Bloom's taxonomy for learning.

ASM Concept/ Competency	A. Bloom's Level 1, 2–Remember and Understand (Choose one.)	B. Bloom's Level 3, 4–Apply and Analyze	C. Bloom's Level 5, 6–Evaluate and Create
Evolution	1. The electron transport system in bacteria is located on the _____ and in eukaryotic cells on the _____. a. plastid, chloroplast b. cytoplasmic membrane, mitochondrion c. cell wall, mitochondrion d. mitochondrion, cytoplasmic membrane	2. Speculate on why glycolysis, which is the beginning point of many catabolic pathways, does not require oxygen.	3. Provide evidence in support of or refuting the following statement: The evolution of aerobic respiration was driven by the success of photosynthetic microbes.
Cell Structure and Function	4. Many coenzymes are formed from a. metals. b. vitamins. c. proteins. d. substrates.	5. Describe the roles played by ATP and NAD in metabolism.	6. Explain why electron transport systems are always found in a membrane.
Metabolic Pathways	7. Energy is carried from catabolic to anabolic reactions in the form of a. ADP. b. high-energy ATP bonds. c. coenzymes. d. inorganic phosphate.	8. What is meant by the concept of the "final electron acceptor"?	9. Investigate the creation of lactic acid in human cells via fermentation, and write a paragraph contrasting it with microbial fermentation.
Information Flow and Genetics	10. Enzyme action can be blocked by competitive molecules binding in the active site, by repressors binding in a distant site, and by a. product binding to the DNA used to make enzymes. b. substrates being in high concentration. c. incorrect temperature conditions. d. two of the above.	11. Compare and contrast the molecular structure of ATP with the molecular structure of an RNA nucleotide.	12. Suggest some reasons that pyruvate is central to so many metabolic strategies.
Microbial Systems	13. Enzymes that can be shut down or activated based on the presence of chemicals in their environment are called a. constitutive. b. repressed. c. holoenzymes. d. regulated.	14. Which of the three major catabolic strategies do you think the Last Common Ancestor possessed? Defend your answer.	15. Write a paragraph explaining why metabolic pathways that are amphibolic are advantageous for cells.

ASM Concept/ Competency	A. Bloom's Level 1, 2–Remember and Understand	B. Bloom's Level 3, 4–Apply and Analyze	C. Bloom's Level 5, 6–Evaluate and Create
Impact of Microorganisms	**16.** The type of microbial metabolic pathway that is most often exploited to make acids and alcohols industrially is a. aerobic respiration. b. anaerobic respiration. c. fermentation. d. none of the above.	**17.** Defend this statement: Microbes (and their metabolic strategies) are absolutely essential to plant life on earth.	**18.** Construct an argument for the possibility that scientists will continue to discover novel metabolic strategies that are not yet known.
Scientific Thinking	**19.** Which of the following is true? a. The suffix "-ase" indicates an enzyme. b. Often enzymes are named for the substrates they act upon. c. Enzymes are larger than their substrates. d. All of the above are true.	**20.** Find three enzymatic capabilities that are commonly used to identify microbial species in a clinical lab.	**21.** Polymerase chain reaction (PCR) is a lab technology that requires high temperatures to make copies of DNA. The copies are made using an enzyme called a polymerase. PCR would not have been possible before thermophilic archaea were discovered. Explain why.

Answers to the multiple-choice questions appear in Appendix A.

Visual Connections

This question connects previous images to a new concept.

From chapter 3, figure 3.15. On these depictions of the gram-positive and gram-negative envelopes, draw protons in the proper compartment in such a way that creates a proton motive force.

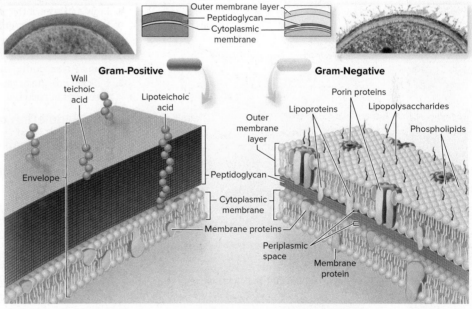

Dr. Kari Lounatmaa/Science Source; Dennis Kunkel Microscopy, Inc./Science Source

8

Microbial Genetics and Genetic Engineering

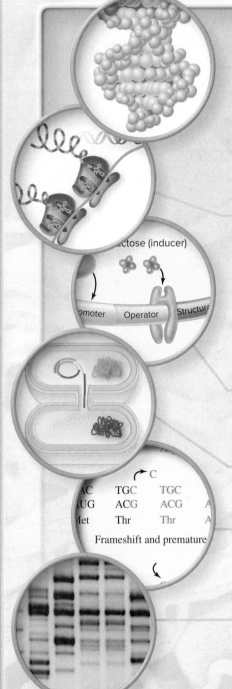

IN THIS CHAPTER...

8.1 Introduction to Genetics and Genes

1. Define the terms *genome* and *gene*.
2. Differentiate between genotype and phenotype.
3. Draw a segment of DNA, labeling all important chemical groups within the molecule.
4. Summarize the steps of bacterial DNA replication, and identify the enzymes used in this process.
5. Compare and contrast the synthesis of leading and lagging strands during DNA replication.

8.2 Transcription and Translation

6. Provide an overview of the relationship among DNA, RNA, and proteins.
7. Identify important structural and functional differences between RNA and DNA.
8. Draw a picture of the process of transcription.
9. List the three types of RNA directly involved in translation.
10. Define the terms *codon* and *anticodon*, and list three start and stop codons.
11. Identify the locations of the promoter, the start codon, and the A and P sites during translation.
12. Indicate how eukaryotic transcription and translation differ from these processes in bacteria.

8.3 Genetic Regulation of Protein Synthesis

13. Define the term *operon*, and explain one advantage it provides to a bacterial cell.
14. Highlight the main points of *lac* operon operation.

8.4 DNA Recombination Events

15. Explain the defining characteristics of a recombinant organism.
16. Describe three forms of horizontal gene transfer used in bacteria.

8.5 Mutations: Changes in the Genetic Code

17. Define the term *mutation*, and discuss one positive and one negative example of it in microorganisms.
18. Differentiate among frameshift, nonsense, silent, and missense mutations.
19. Explain the influence of single-nucleotide polymorphisms on an organism.

8.6 Studying DNA in the Laboratory and Genetic Engineering

20. Explain the importance of restriction endonucleases to genetic engineering.
21. List the steps in the polymerase chain reaction.
22. Describe how you can clone a gene into a bacterium.
23. Differentiate between DNA profiling and DNA sequencing.
24. Provide a definition of *synthetic biology*.
25. Name two genetic techniques that are designed to treat human diseases.

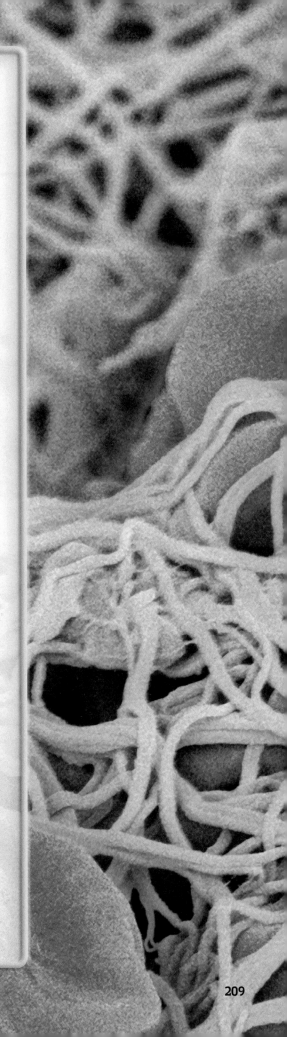

CASE FILE

Factors to Consider

As a nurse for an elementary school, my
responsibilities range from wellness education
to caring for acute health needs of students and
staff. I also help administer medications and manage
medical needs for students with chronic conditions.
One student whom I care for at the elementary school has
hemophilia. I work closely with the patient, his parents, and school staff to ensure
his medical needs are met while still allowing him to be at school.

The 7-year-old male student, Rayshawn, was diagnosed with Hemophilia A
as a toddler. The condition is often inherited, but neither of his parents have
hemophilia. After further genetic testing, it was determined that the student's
hemophilia was caused by a spontaneous genetic mutation.

Having Hemophilia A means that Rayshawn is predisposed to frequent
bleeding. He is not able to produce the blood clotting proteins necessary for normal
clot formation. Small traumatic events could cause bleeding that is difficult to stop.
Of greater concern is internal bleeding that could go unnoticed for some time and
ultimately be life-threatening.

To manage his hemophilia, Rayshawn receives injections of clotting factor VIII
three times per week. The infusions are administered at home, but I have vials
of factor VIII at school in case it is needed during the school day. If the student
has active bleeding, I bring him to the clinic to slowly infuse the intravenous (IV)
factor replacement. He is restricted from high-contact activities but is allowed to
participate in physical activities that have minimal risk of injury.

Another important part of my assessment of the patient is looking for joint
swelling and tenderness. I have educated the school staff to be aware that
spontaneous joint bleeds often occur with hemophilia and require immediate
attention.

- Why does hemophilia cause uncontrollable bleeding?

- How is factor VIII produced?

Case File Wrap-Up appears at the end of the chapter.

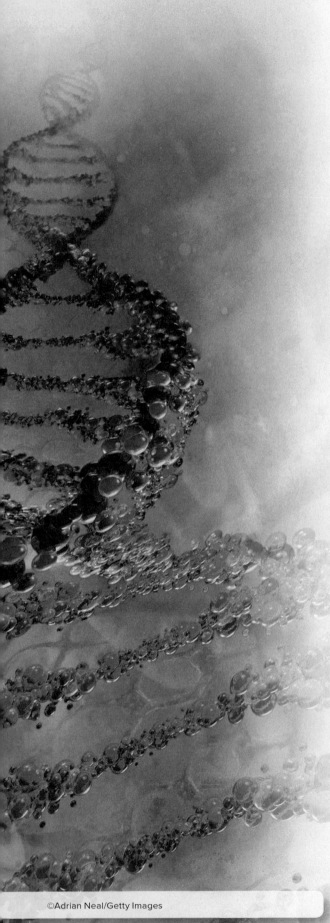

8.1 Introduction to Genetics and Genes

Genetics is the study of the inheritance, or **heredity,** of living things. It is a wide-ranging science that explores

- the transmission of biological traits from parent to offspring,
- how those traits are expressed in an organism,
- the structure and function of the genetic material, and
- how this material changes.

This chapter will explore DNA, which is the genetic material, and the proteins and other products that it gives rise to in a cell. Coming out of chapter 7, we should point out that the production of new DNA, RNA, and proteins is an example of an anabolic process.

The Nature of the Genetic Material

The genome is the sum total of genetic material of an organism. Although most of the genome exists in the form of chromosomes, genetic material can appear as nonchromosomal material as well **(figure 8.1).** For example, bacteria and some fungi contain tiny extra pieces of DNA (plasmids). Also certain organelles of eukaryotes (the mitochondria and chloroplasts) are equipped with their own DNA. Genomes of cells are composed exclusively of DNA, but the viruses contain either DNA or RNA as the principal genetic material. Although the specific genome of an individual organism is unique, the general pattern of nucleic acid structure and function is similar among all organisms.

A **chromosome** is a distinct cellular structure composed of a neatly packaged DNA molecule. The chromosomes of eukaryotes and bacterial cells differ in several respects. The structure of eukaryotic chromosomes consists of a DNA molecule tightly wound around histone proteins, whereas a bacterial chromosome is condensed into a packet by proteins that are histone-like, but not actually histones. Eukaryotic chromosomes are located in the nucleus. They vary in number from a few to hundreds. They can occur in pairs (diploid) or singles (haploid), and they are double-stranded DNA in a more-or-less linear arrangement. In contrast, most bacterial chromosomes are circular double strands of DNA. Bacteria generally have one, two, or sometimes several chromosomes.

The chromosomes of all cells are subdivided into basic informational packets called genes. A **gene** can be defined from more than one perspective. In classical genetics, the term refers to the fundamental unit of heredity responsible for a given trait in an organism. In the molecular and biochemical sense, it is a site on the chromosome that provides information for a certain cell function. More specifically still, it has traditionally been characterized as a certain segment of DNA that contains the necessary code to make a protein or RNA molecule. Genes fall into three basic categories: (1) **structural genes** that code for proteins, (2) genes that code for the RNA machinery used in protein production, and (3) regulatory genes that control gene expression. The sum of all of these types of genes constitutes an organism's distinctive genetic makeup, or **genotype** (jee'-noh-typ). The expression of the genotype creates traits (certain structures or functions) referred to as the **phenotype** (fee'-noh-typ). Just as a person inherits a combination of genes (genotype) that gives a certain eye color or height (phenotypes), a bacterium inherits genes that direct the formation of a flagellum, and a virus contains genes for its capsid structure. All organisms contain more genes in their genotypes than are manifested as a phenotype at any given time. In other words, the phenotype can change depending on which genes are "turned on" (expressed).

The Size and Packaging of Genomes

Genomes vary greatly in size. The smallest viruses have four or five genes; the bacterium *Escherichia coli* has a single chromosome containing 4,288 genes, and a human cell has about 19,000 to 20,000 genes on 23 chromosome pairs (46 chromosomes).

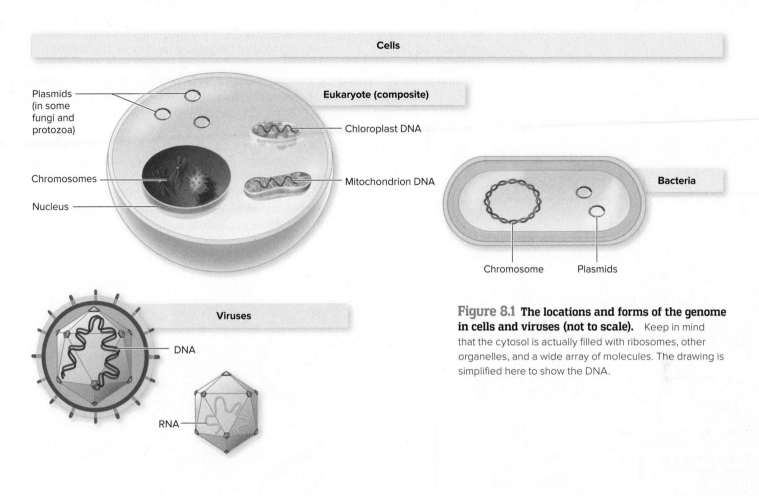

Cells

Plasmids (in some fungi and protozoa)

Eukaryote (composite)

Chloroplast DNA

Chromosomes

Mitochondrion DNA

Nucleus

Bacteria

Chromosome   Plasmids

Viruses

DNA

RNA

Figure 8.1 The locations and forms of the genome in cells and viruses (not to scale). Keep in mind that the cytosol is actually filled with ribosomes, other organelles, and a wide array of molecules. The drawing is simplified here to show the DNA.

The chromosome of *E. coli* would measure about 1 mm if unwound and stretched out linearly, and yet this fits within a cell that measures just over 1 μm across, making the stretched-out DNA 1,000 times longer than the cell. This is possible because the DNA is very tightly wound around protein in the cytoplasm **(figure 8.2).** Still, the bacterial chromosome takes up only about one-third to one-half of the cell's volume. Likewise, if the sum of all DNA making up the 46 human chromosomes in a cell were unraveled and laid end to end, it would measure about 6 feet.

The DNA Code

The general structure of DNA is universal, except in some viruses that contain single-stranded DNA. The basic unit of DNA structure is a nucleotide, and a chromosome in a typical bacterium consists of several million nucleotides linked end to end. Each nucleotide is composed of a phosphate, a **deoxyribose,** and a **nitrogenous base.** The nucleotides covalently bond to each other in a sugar-phosphate linkage that becomes the backbone of each strand. Each sugar attaches in a repetitive pattern to two phosphates. One of the bonds is to the number 5′ (read "five prime") carbon on deoxyribose, and the other is to the 3′ carbon, which creates a certain order and direction on each strand **(figure 8.3).** (In cyclical carbon molecules, like sugars, the carbons are numbered so we can keep track of them. Deoxyribose has 5 carbons numbered 1–5.)

The nitrogenous bases, **purines** and **pyrimidines,** attach along a strand by covalent bonds at the 1′ position of the sugar **(figure 8.3a).** They join with complementary bases from the other strand using hydrogen bonds. Such weak bonds

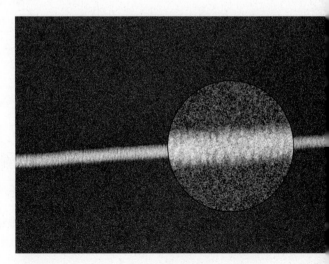

Figure 8.2 Transmission electron micrograph of DNA. The helical structure is clearly visible in the magnified area.
©Professor Enzo Di Fabrizio, IIT/Science Source

are easily broken, allowing the molecule to be "unzipped" into its complementary strands. This feature is of great importance in gaining access to the information encoded in the nitrogenous base sequence. Pairing of purines and pyrimidines is not random; it is dictated by the formation of hydrogen bonds between certain bases. So, in DNA, the purine adenine (A) always pairs with the pyrimidine thymine (T), and the purine guanine (G) always pairs with the pyrimidine cytosine (C). The bases are attracted to each other in this pattern because each has a complementary three-dimensional shape that matches its pair. Although the base-pairing partners generally do not vary, the sequence of base pairs along the DNA molecule can assume any order, resulting in a nearly infinite number of possible nucleotide sequences.

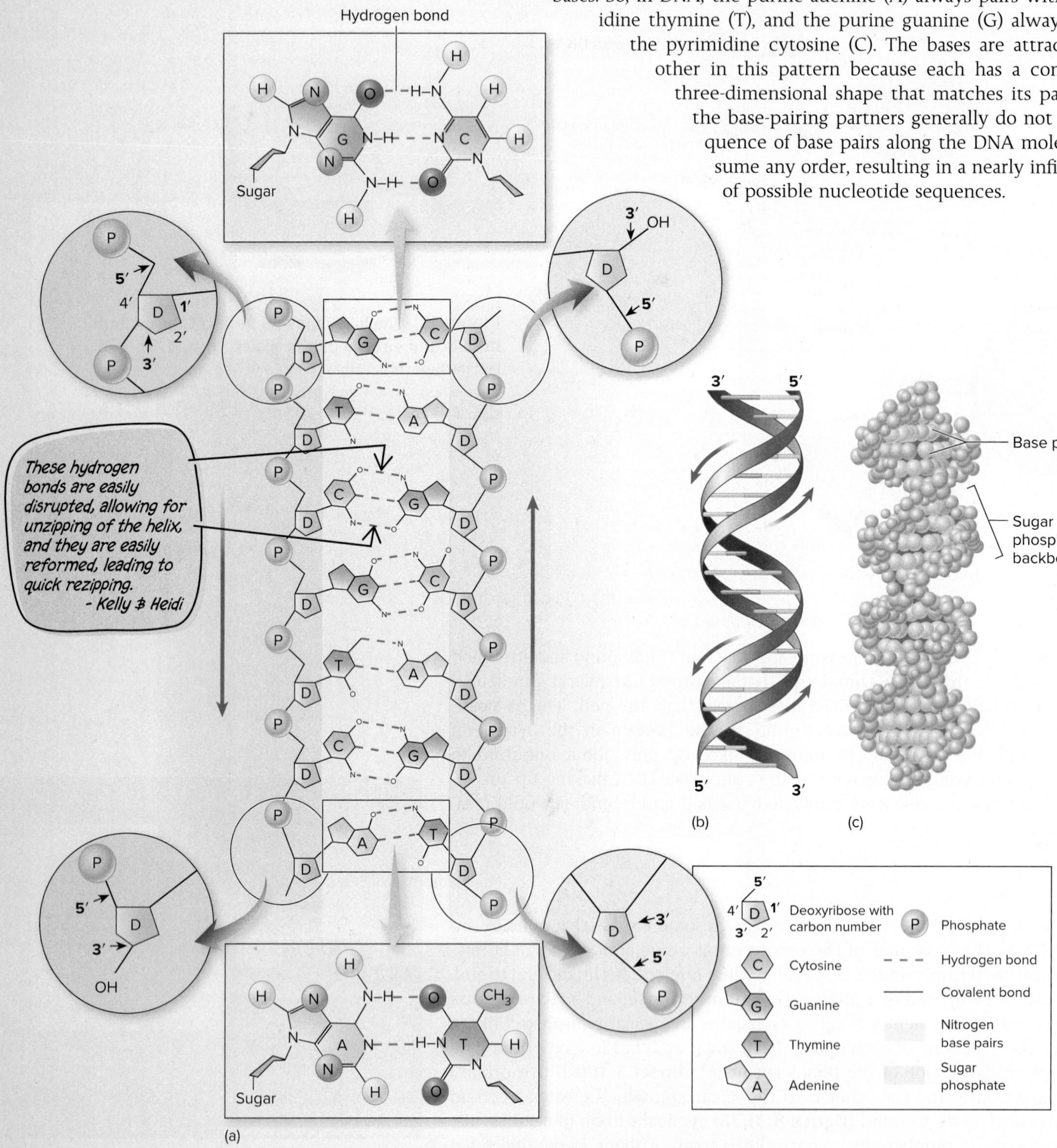

These hydrogen bonds are easily disrupted, allowing for unzipping of the helix, and they are easily reformed, leading to quick rezipping.
   - Kelly & Heidi

Figure 8.3 Three views of DNA structure.   **(a)** A schematic nonhelical model, to show the arrangement of the molecules it is made of. Note that the order of phosphate and sugar bonds differs between the two strands, going from the 5′ carbon to the 3′ carbon on one strand, and from the 3′ carbon to the 5′ carbon on the other strand. Insets show details of the nitrogen *bases*. **(b)** Simplified model that highlights the antiparallel arrangement. **(c)** Space-filling model that more accurately depicts the three-dimensional structure of DNA.

Other important considerations of DNA structure concern the nature of the double helix itself. The halves are not oriented in the same direction. One side of the helix runs in the opposite direction of the other, in what is called an **antiparallel** arrangement (**figure 8.3b**). The order of the bond between the carbon on deoxyribose and the phosphates is used to keep track of the direction of the two sides of the helix. This means that one helix runs from the 5′ to 3′ direction, and the other runs from the 3′ to 5′ direction. This characteristic is a significant factor in DNA synthesis and protein production.

The Significance of DNA Structure

The English language, based on 26 letters, can create an infinite variety of words, but how can an apparently complex genetic language such as DNA be based on just four nitrogen base "letters"? A mathematical example can explain the possibilities. For a segment of DNA that is 1,000 nucleotides long, there are $4^{1,000}$ different sequences possible. Calculated out, this number would approximate 1.5×10^{602}, a number so huge that it provides nearly endless degrees of variation.

In this chapter, we will address two separate reactions DNA enters into: its own replication (facilitating cell division) and its role in producing proteins.

DNA Replication

The process of duplicating DNA is called *DNA replication*. In the following example, we will show replication in bacteria; but with some exceptions, it also applies to the process as it works in eukaryotes and some viruses.

The Overall Replication Process

DNA replication requires a careful orchestration of the actions of 30 different enzymes (partial list in **table 8.1**), which separate the strands of the existing DNA molecule, copy one strand, and produce two complete daughter molecules. A critical feature of DNA replication is that each daughter molecule will be identical to the parent in composition, but neither one is completely new; one of the strands in each double-stranded daughter molecule comes from the parent DNA. The preservation of the parent molecule in this way, termed **semiconservative replication**—*semi*-meaning "half," as in *semicircle*—helps explain how the replication process maintains accuracy and fidelity during successive cycles.

Refinements and Details of Replication

The specific step of synthesizing a new daughter strand of DNA using the parental strand as a template is carried out by the enzyme DNA polymerase III. The process of replication can be most easily understood by keeping in mind a couple of points concerning both the structure of the DNA molecule and the limitations of DNA polymerase III:

1. DNA polymerase III is unable to *begin* synthesizing a chain of nucleotides but can only continue to add nucleotides to an already-existing chain.
2. DNA polymerase III can add nucleotides in only one direction, so new strands are always synthesized 5′ to 3′.

Interestingly, the enzyme gyrase is a good target for antibacterial drugs, since it is different than the gyrase used in eukaryotes. The process of DNA replication is outlined in **table 8.2.**

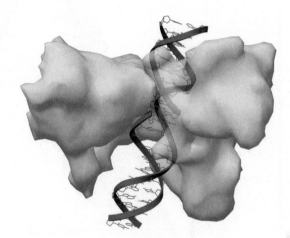

DNA nestled in the active site of a topoisomerase.
©McGraw-Hill Education/Jeramia Ory, Ph.D., photographer

Table 8.1 Some Enzymes Involved in DNA Replication and Their Functions

Enzyme	Function
Helicase	Unzipping the DNA helix
Gyrase	Helping to untangle the DNA supercoils
Primase	Synthesizing an RNA primer
DNA polymerase III	Adding bases to the new DNA chain; proofreading the chain for mistakes
DNA polymerase I	Removing primer, closing gaps, repairing mismatches
Ligase	Final binding of nicks in DNA during synthesis and repair
Topoisomerases I and II	Supercoiling and untangling

Table 8.2 DNA Replication

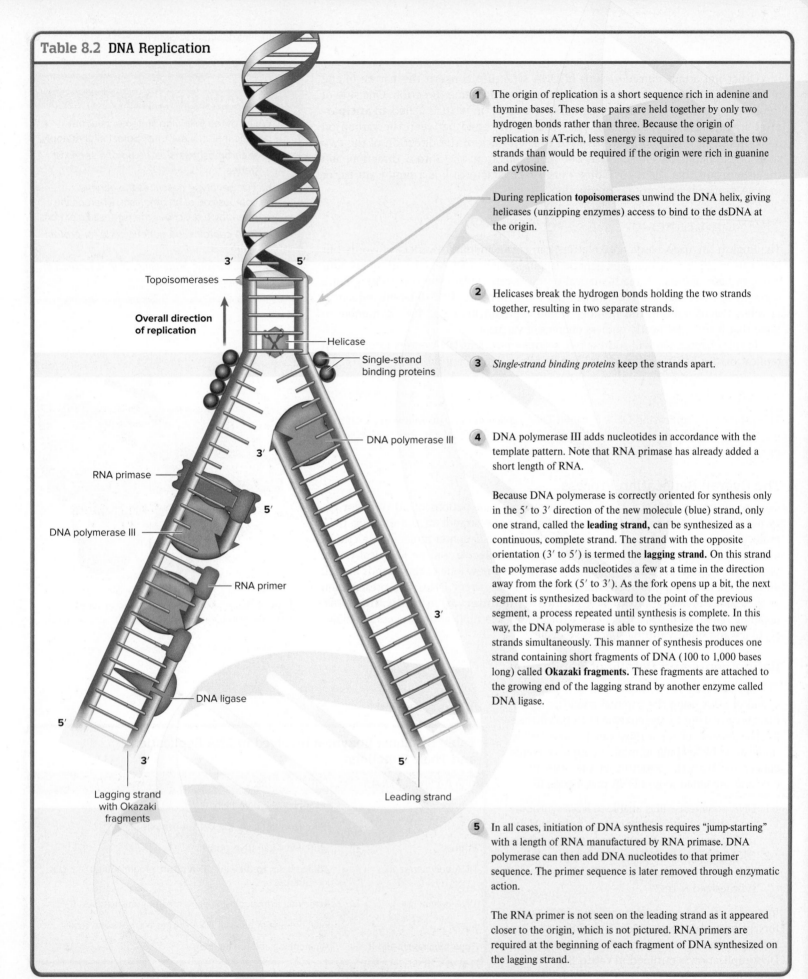

Topoisomerases

Overall direction of replication

Helicase

Single-strand binding proteins

DNA polymerase III

RNA primase

DNA polymerase III

RNA primer

DNA ligase

Lagging strand with Okazaki fragments

Leading strand

1. The origin of replication is a short sequence rich in adenine and thymine bases. These base pairs are held together by only two hydrogen bonds rather than three. Because the origin of replication is AT-rich, less energy is required to separate the two strands than would be required if the origin were rich in guanine and cytosine.

   During replication **topoisomerases** unwind the DNA helix, giving helicases (unzipping enzymes) access to bind to the dsDNA at the origin.

2. Helicases break the hydrogen bonds holding the two strands together, resulting in two separate strands.

3. *Single-strand binding proteins* keep the strands apart.

4. DNA polymerase III adds nucleotides in accordance with the template pattern. Note that RNA primase has already added a short length of RNA.

   Because DNA polymerase is correctly oriented for synthesis only in the 5′ to 3′ direction of the new molecule (blue) strand, only one strand, called the **leading strand,** can be synthesized as a continuous, complete strand. The strand with the opposite orientation (3′ to 5′) is termed the **lagging strand.** On this strand the polymerase adds nucleotides a few at a time in the direction away from the fork (5′ to 3′). As the fork opens up a bit, the next segment is synthesized backward to the point of the previous segment, a process repeated until synthesis is complete. In this way, the DNA polymerase is able to synthesize the two new strands simultaneously. This manner of synthesis produces one strand containing short fragments of DNA (100 to 1,000 bases long) called **Okazaki fragments.** These fragments are attached to the growing end of the lagging strand by another enzyme called DNA ligase.

5. In all cases, initiation of DNA synthesis requires "jump-starting" with a length of RNA manufactured by RNA primase. DNA polymerase can then add DNA nucleotides to that primer sequence. The primer sequence is later removed through enzymatic action.

   The RNA primer is not seen on the leading strand as it appeared closer to the origin, which is not pictured. RNA primers are required at the beginning of each fragment of DNA synthesized on the lagging strand.

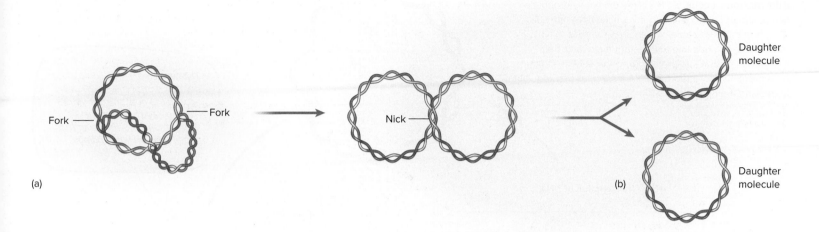

(a)

(b)

Daughter molecule

Daughter molecule

Fork

Fork

Nick

Elongation and Termination of the Daughter Molecules The addition of nucleotides proceeds at an astonishing pace, estimated in some bacteria to be 750 bases per second at each fork! As replication proceeds, the newly produced double strand loops away from the original molecule **(figure 8.4a)**. DNA polymerase I removes the RNA primers used to initiate DNA synthesis and replaces them with DNA. When the forks come full circle and meet, ligases move along the lagging strand to begin the initial linking of the fragments and to complete synthesis and separation of the two circular daughter molecules **(figure 8.4b)**.

As with any language, DNA is occasionally "misspelled" when an incorrect base is added to the growing chain. In bacteria such mistakes are made once in approximately 10^9 to 10^{10} bases. Think of it this way: An overnight test tube culture of *E. coli* can be expected to have around 10^9 to 10^{10} bacteria per mL in it. That means a random mutation in every single base in the *E. coli* genome will be represented in at least one cell in the culture. Because continued cellular integrity is very dependent on accurate replication, cells have evolved their own proofreading function for DNA. DNA polymerase III, the enzyme that elongates the molecule, can detect incorrect, unmatching bases; excise them; and replace them with the correct base. DNA polymerase I can also proofread the molecule and repair damaged DNA.

Figure 8.4 Completion of chromosome replication in bacteria. **(a)** As replication proceeds, one double strand loops away from the other. **(b)** Final separation is achieved through repair and the release of two completed molecules. The daughter cells receive these during binary fission.

Reminder: Be Sure to Read the Tables

If you skip the tables, you will miss much of the most important information about genetic processes. Follow the drawings and the accompanying text to save yourself time later catching up with important information.

8.1 LEARNING OUTCOMES—Assess Your Progress

1. Define the terms *genome* and *gene*.
2. Differentiate between genotype and phenotype.
3. Draw a segment of DNA, labeling all important chemical groups within the molecule.
4. Summarize the steps of bacterial DNA replication, and identify the enzymes used in this process.
5. Compare and contrast the synthesis of leading and lagging strands during DNA replication.

8.2 Transcription and Translation

Although the genome is full of critical information, the DNA itself does not perform cell processes directly. Its stored information is conveyed to RNA molecules, which carry out the instructions. The concept that genetic information flows from DNA to RNA to protein is a central theme of biology **(figure 8.5a)**. More precisely, it states that the master code of DNA is first used to synthesize an RNA molecule via

Figure 8.5 Summary of the flow of genetic information in cells. DNA is the ultimate storehouse and distributor of genetic information. **(a)** DNA must be deciphered into a usable cell language. It does this by transcribing its code into RNA helper molecules that translate that code into protein. **(b)** Other sections of the DNA produce very important RNA molecules that regulate genes and their products.

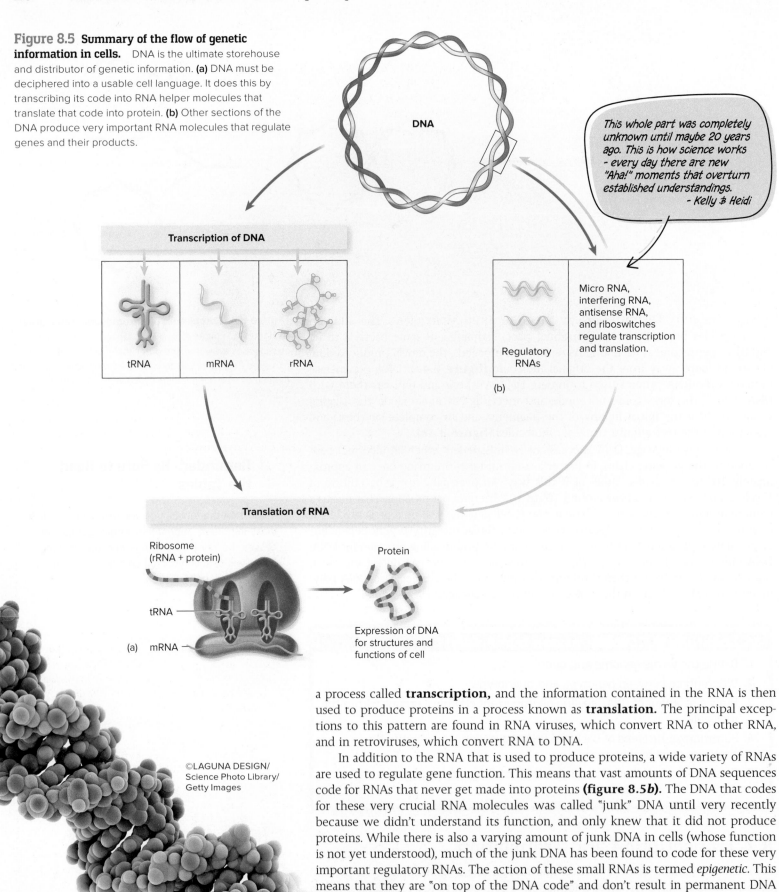

DNA

This whole part was completely unknown until maybe 20 years ago. This is how science works - every day there are new "Aha!" moments that overturn established understandings.
— Kelly & Heidi

Transcription of DNA

tRNA   mRNA   rRNA

Regulatory RNAs — Micro RNA, interfering RNA, antisense RNA, and riboswitches regulate transcription and translation.

(b)

Translation of RNA

Ribosome (rRNA + protein)

Protein

tRNA

(a)   mRNA

Expression of DNA for structures and functions of cell

©LAGUNA DESIGN/ Science Photo Library/ Getty Images

a process called **transcription,** and the information contained in the RNA is then used to produce proteins in a process known as **translation.** The principal exceptions to this pattern are found in RNA viruses, which convert RNA to other RNA, and in retroviruses, which convert RNA to DNA.

In addition to the RNA that is used to produce proteins, a wide variety of RNAs are used to regulate gene function. This means that vast amounts of DNA sequences code for RNAs that never get made into proteins **(figure 8.5b).** The DNA that codes for these very crucial RNA molecules was called "junk" DNA until very recently because we didn't understand its function, and only knew that it did not produce proteins. While there is also a varying amount of junk DNA in cells (whose function is not yet understood), much of the junk DNA has been found to code for these very important regulatory RNAs. The action of these small RNAs is termed *epigenetic.* This means that they are "on top of the DNA code" and don't result in permanent DNA changes (although sometimes epigenetic changes can be passed down to subsequent generations).

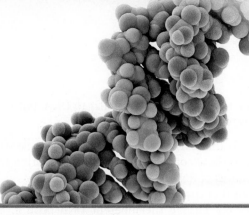

Transcription: The First Stage of Gene Expression

During transcription, the DNA code is converted to RNA in a stepwise fashion, directed by a large and very complex enzyme system, **RNA polymerase.** Only one strand of the DNA–the **template strand**–is used as the instructions for synthesis of a functioning polypeptide.

   Table 8.3 describes transcription.

Table 8.3 Transcription

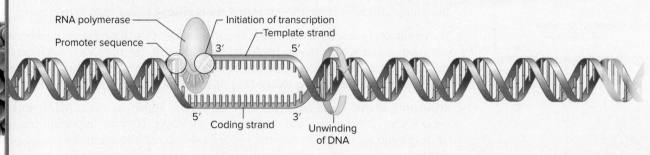

1 *Initiation.* Transcription is initiated when RNA polymerase recognizes a segment of the DNA called the **promoter** region. This region consists of two sequences of DNA just prior to the beginning of the gene to be transcribed. These promoter sequences provide the signal for RNA polymerase to bind to the DNA. Then there is a special sequence, which is where the RNA polymerase begins its transcription. As the DNA helix unwinds, the polymerase first pulls the early parts of the DNA into itself, a process called "DNA scrunching," and then, having acquired energy from the scrunching process, begins to advance down the DNA strand to continue synthesizing an RNA molecule complementary to the template strand of DNA. The nucleotide sequence of promoters differs only slightly from gene to gene, with all promoters being rich in adenine and thymine.

Only one strand of DNA, called the template strand, is copied by RNA polymerase.

2 *Elongation.* During elongation, which proceeds in the 5′ to 3′ direction (with regard to the growing RNA molecule), the mRNA is assembled by the addition of nucleotides that are complementary to the DNA template. Remember that uracil (U) is placed as adenine's complement. As elongation continues, the part of DNA already transcribed is rewound into its original helical form.

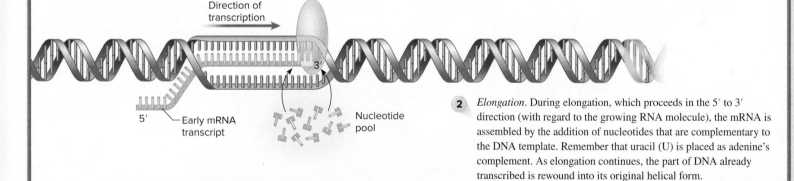

3 *Termination.* At termination, the polymerases recognize another code that signals the separation and release of the mRNA strand, or transcript. The smallest mRNA might consist of 100 bases; an average-size mRNA might consist of 1,200 bases; and a large one might consist of several thousand.

NCLEX® PREP

2. The following are all true of RNA, *except*
    a. *it is a single-stranded molecule existing in helical form.*
    b. *it contains uracil rather than thymine as the complementary base-pairing mate for adenine.*
    c. *uracil does not follow the pairing rules; therefore, the inherent DNA code is changed.*
    d. *the sugar in RNA is ribose rather than deoxyribose.*

The RNAs

In terms of its general properties, ribonucleic acid is similar to DNA, but its general structure is different in several ways:

1. It is a single-stranded molecule that exists in helical form. This single strand can form secondary and tertiary levels of complexity due to bonds within the molecule, leading to specialized forms of RNA (tRNA and rRNA).
2. RNA contains *uracil (U)*, instead of thymine, as the complementary base-pairing mate for adenine. This does not change the inherent DNA code in any way because the uracil still follows the pairing rules.
3. Although RNA, like DNA, contains a backbone that consists of alternating sugar and phosphate molecules, the sugar in RNA is **ribose** rather than deoxyribose.

The products of transcription belong in two major categories: those that are necessary for translation and those that have other functions in the cell. The translation machinery includes messenger RNA, transfer RNA, and ribosomal RNA. We will spend a lot of time on these molecules in the coming section, but first we will list the other RNA varieties:

- Regulatory RNAs: These are small RNA molecules, known as micro RNAs, antisense RNAs, riboswitches, and small interfering RNAs. They are important regulators of gene expression in bacteria and eukaryotes and also act in the coiling of chromatin in eukaryotic cells.
- Primer RNA: This RNA is laid down in DNA replication, as a template of sorts for the DNA sequence. Primer RNAs are operative in both bacterial and eukaryotic cells.
- Ribozymes: These enzymes are made of RNA and, in eukaryotes, remove unneeded sequences from other RNAs.

In eukaryotes, transcription occurs in the nucleus (pictured here) and translation occurs in the cytoplasm. In bacteria and archaea, both processes occur in the cytoplasm.
©EM Research Services, Newcastle University

Table 8.4 The Three RNAs Involved in Translation

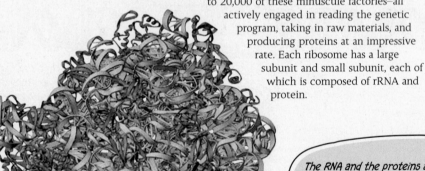

The bacterial (70S) ribosome is a macromolecule composed of tightly packaged **ribosomal RNA (rRNA)** (large gray and blue areas) and protein (smaller lavender and purple areas). A metabolically active bacterial cell can contain up to 20,000 of these minuscule factories–all actively engaged in reading the genetic program, taking in raw materials, and producing proteins at an impressive rate. Each ribosome has a large subunit and small subunit, each of which is composed of rRNA and protein.

The RNA and the proteins are both highly folded. The folding serves to put the right amino acids and the right nucleotides in the right places, and makes their location less susceptible to single changes that could move them far away from each other.

– Kelly & Heidi

©Center for Molecular Biology of RNA, UC-Santa Cruz

All of these–and tRNA, mRNA, and rRNA–are products of the transcription of distinct genes in the chromosome.

Now, on to translation.

Transcription Products Used in Translation

Three different RNA products of transcription are needed for the final step of protein expression. They are described in **table 8.4.**

The ribosomes of bacteria and eukaryotes are of different sizes. Ribosomes in bacteria, as well as the ribosomes in chloroplasts and mitochondria of eukaryotes, are of a 70S size, made up of a 50S (large) subunit and a 30S (small) subunit. The "S" is a measurement of sedimentation rates, which is how ribosomes are characterized. It is a nonlinear measure; therefore, 30S and 50S add up to 70S. Eukaryotic ribosomes are 80S (a large subunit of 60S and a 40S small subunit). The small subunit binds to the 5′ end of the mRNA, and the large subunit supplies enzymes for making peptide bonds on the protein.

Translation: The Second Stage of Gene Expression

In translation, all of the elements needed to synthesize a protein, from the mRNA to the amino acids, are brought together on the ribosomes **(figure 8.6).**

The Master Genetic Code: The Message in Messenger RNA

Translation relies on a central principle: The mRNA nucleotides are read in groups of three. Three nucleotides are called a **codon,** and it is the codon that dictates which amino acid is added to the growing peptide chain. In **figure 8.7,** the mRNA codons

Background image: ©Center for Molecular Biology of RNA, UC-Santa Cruz

Amino acids

Exit site

Large subunit

Small subunit

5′

Ribosomal proteins

tRNAs

mRNA transcript

Figure 8.6 The "players" in translation. A ribosome serves as the stage for protein synthesis. Assembly of the small and large subunits results in specific sites for holding the mRNA and two tRNAs with their amino acids. This depiction of the ribosome matches the depiction on the left-hand page so you can see the connection between the molecular image and the image we will use in the book.

Transfer RNA (tRNA) is also a copy of a specific region of DNA; however, it differs from mRNA. Each one is only 75 to 95 nucleotides long, and contains sequences of bases that form hydrogen bonds with complementary sections within the same tRNA strand. At these points, the molecule bends back upon itself into several hairpin loops, giving the molecule a secondary cloverleaf structure that folds even further into a complex, three-dimensional helix. This compact molecule is an adaptor that converts RNA language into protein language. The bottom loop of the cloverleaf exposes a triplet, the **anticodon,** that both designates the specificity of the tRNA and complements mRNA's codons. At the opposite end of the molecule is a binding site for the amino acid that is specific for that tRNA's anticodon. For each of the 20 amino acids, there is at least one specialized type of tRNA to carry it.

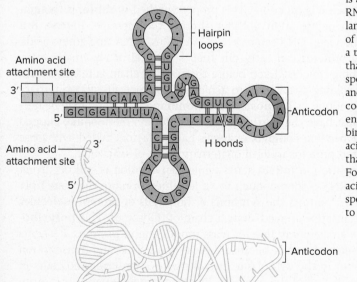

Hairpin loops

Amino acid attachment site

3′

5′

Amino acid attachment site

3′

5′

H bonds

Anticodon

Anticodon

Messenger RNA (mRNA) is a transcript (copy) of a structural gene or genes in the DNA. The complementary base-pairing rules ensure that the code will be faithfully copied in the mRNA transcript. The message of this transcribed strand is later read as a series of triplets called codons. The length of the mRNA molecule varies from about 100 nucleotides to several thousand. It carries the sequence that will dictate the eventual amino acid sequence of the protein.

AUG CUC GGG AGA UUU GUG CGC

Codon

AGU CAC CGG UAA

Figure 8.7 The genetic code: codons of mRNA that specify a given amino acid. The master code for translation is found in the mRNA codons.

		Second Base Position					
		U	**C**	**A**	**G**		
First Base Position	**U**	UUU } Phenylalanine UUC UUA } Leucine UUG	UCU UCC UCA Serine UCG	UAU } Tyrosine UAC UAA } STOP** UAG	UGU } Cysteine UGC UGA STOP** UGG Tryptophan	U C A G	Third Base Position
	C	CUU CUC CUA Leucine CUG	CCU CCC CCA Proline CCG	CAU } Histidine CAC CAA } Glutamine CAG	CGU CGC CGA Arginine CGG	U C A G	
	A	AUU AUC Isoleucine AUA AUG START f-Methionine*	ACU ACC ACA Threonine ACG	AAU } Asparagine AAC AAA } Lysine AAG	AGU } Serine AGC AGA } Arginine AGG	U C A G	
	G	GUU GUC GUA Valine GUG	GCU GCC GCA Alanine GCG	GAU } Aspartic acid GAC GAA } Glutamic acid GAG	GGU GGC GGA Glycine GGG	U C A G	

* This codon initiates translation.

**For these codons, which give the orders to stop translation, there are no corresponding tRNAs and no amino acids.

and their corresponding amino acid specificities are given. Because there are 64 different triplet codes and only 20 different amino acids, it is not surprising that some amino acids are represented by several codons. For example, leucine and serine can each be represented by any of six different triplets, and only tryptophan and methionine are represented by a single codon. This property is called **redundancy** and allows for the insertion of correct amino acids (sometimes) even when mistakes occur in the DNA sequence, as they do with regularity. Also, in codons such as those for leucine, only the first two nucleotides are required to encode the correct amino acid. Any of the four nucleotides can appear in the third position of the codon without changing the fact that leucine is called for. This property, called **wobble,** is thought to permit some variation or mutation without altering the message. **Figure 8.8** shows the relationship between DNA sequence, RNA codons, tRNA, and amino acids.

Before newly made proteins can carry out their structural or enzymatic roles, they often require finishing touches. Even before the peptide chain is released from the ribosome, it begins folding upon itself to achieve its biologically active tertiary conformation. Other alterations, called **posttranslational** modifications, may be necessary. Some proteins must have the starting amino acid (formyl methionine) clipped off. Proteins destined to become complex enzymes have cofactors added; and some join with other completed proteins to form quaternary levels of structure.

In bacteria, the translation of mRNA starts while transcription is still occurring **(figure 8.9).** A single mRNA is long enough to be fed through more than one ribosome simultaneously. This permits the synthesis of hundreds of protein molecules from the same mRNA transcript arrayed along a chain of ribosomes. This **polyribosomal complex** is like an assembly line for mass production of proteins. It occurs in bacteria, but not in eukaryotic cells, because there is no nucleus; and transcription and translation both occur in the cytoplasm. (In eukaryotes, transcription occurs in the nucleus.) Remember that all of the processes involved in gene expression are anabolic processes. Protein synthesis consumes an enormous amount of energy. Nearly 1,200 ATPs are required just for the synthesis of an average-size protein. **Table 8.5** contains the details of translation.

NCLEX® PREP

3. The function of a codon in protein synthesis is best described as a three-nucleotide sequence
   a. *that specifies the initiation of translation.*
   b. *that specifies the termination of translation.*
   c. *on tRNA that complements the mRNA code.*
   d. *that dictates a specific amino acid.*

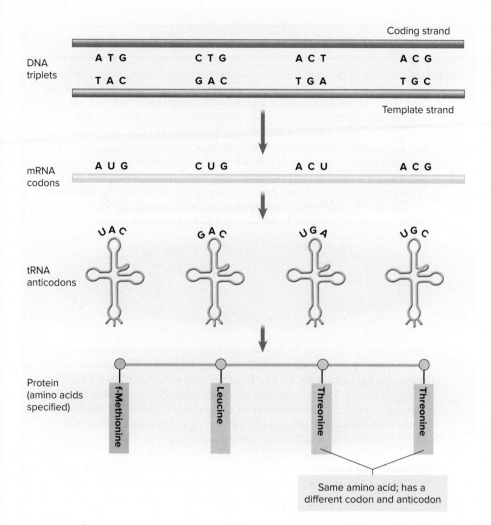

Coding strand

DNA triplets	A T G	C T G	A C T	A C G
	T A C	G A C	T G A	T G C

Template strand

mRNA codons	A U G	C U G	A C U	A C G

tRNA anticodons: UAC   GAC   UGA   UGC

Protein (amino acids specified): f-Methionine   Leucine   Threonine   Threonine

Same amino acid; has a different codon and anticodon

Figure 8.8 Interpreting the DNA code. If the DNA sequence is known, the mRNA codon can be surmised. If a codon is known, the anticodon and, finally, the amino acid sequence can be determined. The reverse is not as straightforward (determining the exact codon or anticodon from amino acid sequence) due to the redundancy of the code.

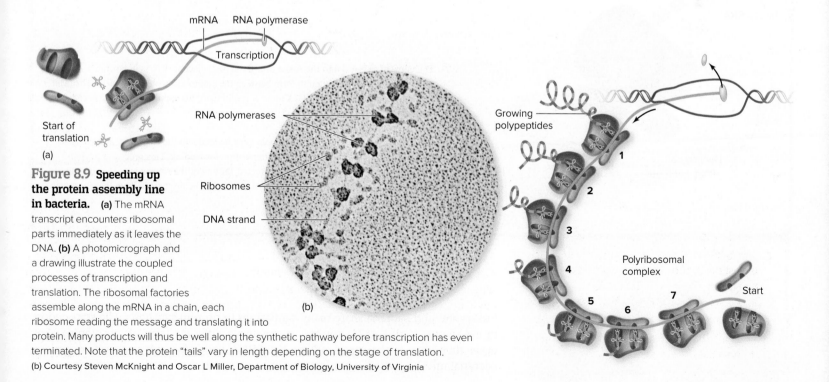

Figure 8.9 Speeding up the protein assembly line in bacteria. (a) The mRNA transcript encounters ribosomal parts immediately as it leaves the DNA. (b) A photomicrograph and a drawing illustrate the coupled processes of transcription and translation. The ribosomal factories assemble along the mRNA in a chain, each ribosome reading the message and translating it into protein. Many products will thus be well along the synthetic pathway before transcription has even terminated. Note that the protein "tails" vary in length depending on the stage of translation.

(b) Courtesy Steven McKnight and Oscar L Miller, Department of Biology, University of Virginia

mRNA   RNA polymerase   Transcription

Start of translation   (a)

RNA polymerases

Ribosomes

DNA strand

(b)

Growing polypeptides

Polyribosomal complex

Start

Table 8.5 Translation

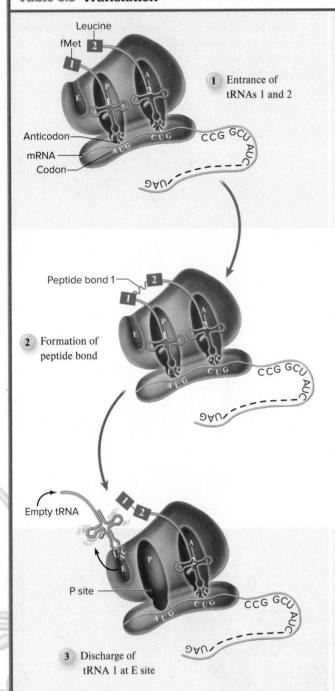

1 Entrance of tRNAs 1 and 2

2 Formation of peptide bond

3 Discharge of tRNA 1 at E site

1 The mRNA molecule leaves the DNA transcription site and is transported to ribosomes in the cytoplasm. Ribosomal subunits come together and form sites to hold the mRNA and tRNAs. The ribosome begins to scan the mRNA by moving in the 5′ to 3′ direction along the mRNA. The first codon it encounters is called the START codon, which is almost always AUG. With the mRNA message in place on the assembled ribosome, the next step in translation involves entrance of tRNAs with their amino acids. The pool of cytoplasm contains a complete array of tRNAs, already charged by having the correct amino acid attached. The step in which the complementary tRNA meets with the mRNA code is guided by the two sites on the large subunit of the ribosome called the P site (left) and the A site (right). The ribosome also has an exit or E site where used tRNAs are released. (P stands for peptide site; A stands for aminoacyl (amino acid) site; E stands for exit site.)

2 Rules of pairing dictate that the anticodon of this tRNA must be complementary to the mRNA codon AUG; thus, the tRNA with anticodon UAC will be the first tRNA to occupy site P. It happens that the amino acid carried by the initiator tRNA in bacteria is formyl methionine. The ribosome shifts its "reading frame" to the right along the mRNA from one codon to the next. This brings the next codon into place on the ribosome and makes a space for the next tRNA to enter the A position. A peptide bond is formed between the amino acids on the adjacent tRNAs, and the polypeptide grows in length.

Elongation then begins with the filling of the A site by a second tRNA. The identity of this tRNA and its amino acid is dictated by the second mRNA codon.

3 The entry of tRNA 2 into the A site brings the two adjacent tRNAs in favorable proximity for a peptide bond to form between the amino acids (aa) they carry. The fMet is transferred from the first tRNA to aa 2, resulting in two coupled amino acids called a dipeptide.

For the next step to proceed, some room must be made on the ribosome, and the next codon in sequence must be brought into position for reading. This process is accomplished by translocation, the enzyme-directed shifting of the ribosome to the right along the mRNA strand, which causes the blank tRNA 1 to be discharged from the ribosome at the E site.

Differences Between Eukaryotic and Bacterial Transcription and Translation

Eukaryotes and bacteria share many similarities in protein synthesis. The start codon in eukaryotes is also AUG, but eukaryotes use a different form of methionine. Another difference is that each eukaryotic mRNA codes for just one protein, unlike bacterial mRNAs, which often contain information from several genes in series.

Table 8.5 Translation (continued)

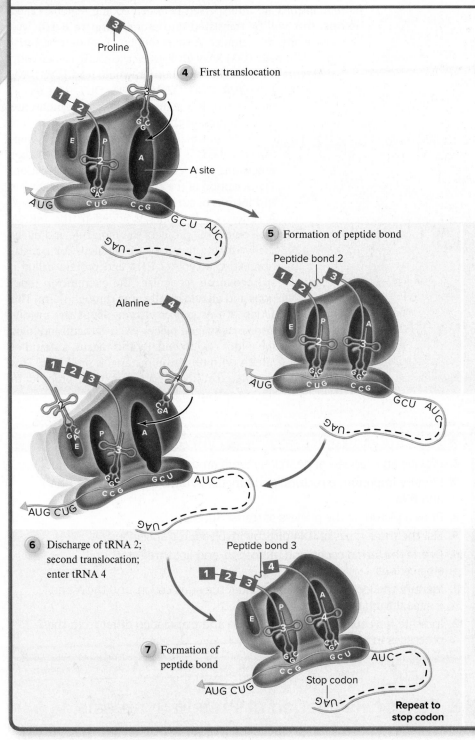

4 First translocation

Proline

A site

5 Formation of peptide bond

Peptide bond 2

Alanine **4**

6 Discharge of tRNA 2; second translocation; enter tRNA 4

Peptide bond 3

7 Formation of peptide bond

Stop codon

Repeat to stop codon

4 This also shifts the tRNA holding the dipeptide into P position. Site A is temporarily left empty. The tRNA that has been released is now free to drift off into the cytoplasm and become recharged with an amino acid for later additions to this or another protein.

The stage is now set for the insertion of tRNA 3 at site A as directed by the third mRNA codon. This insertion is followed once again by peptide bond formation between the dipeptide and amino acid 3 (making a tripeptide), splitting of the peptide from tRNA 2, and translocation.

5 This releases tRNA 2, shifts mRNA to the next position, moves tRNA 3 to position P, and opens position A for the next tRNA (which will be called tRNA 4).

6 From this point on, peptide elongation proceeds repetitively by this same series of actions out to the end of the mRNA.

7 The termination of protein synthesis is brought about by the presence of at least one special codon occurring just after the codon for the last amino acid. Termination codons—UAA, UAG, and UGA—are codons for which there is no corresponding tRNA. Although they are often called nonsense codons, they carry a necessary and useful message: Stop here. When this codon is reached, a special enzyme breaks the bond between the final tRNA and the finished polypeptide chain, releasing it from the ribosome.

As just mentioned, the presence of the DNA in a separate compartment (the nucleus) means that eukaryotic transcription and translation cannot be simultaneous. The mRNA transcript must pass through pores in the nuclear membrane and be carried to the ribosomes in the cytoplasm for translation.

We have given the simplified definition of a gene that works well for bacteria, but most eukaryotic genes (and, surprisingly, archaeal genes) do *not* exist as an uninterrupted series of triplets coding for a protein. A eukaryotic gene contains the code for a protein, but located along the gene are one to several intervening sequences of bases, called **introns,** that do not code for protein. Introns are interspersed between coding regions, called **exons,** that will be translated into protein **(figure 8.10).** We can use words as examples. A short section of colinear bacterial gene might read TOM SAW OUR DOG DIG OUT; a eukaryotic gene that codes for the same portion would read TOM SAW XZKP FPL OUR DOG QZWVP DIG OUT. The recognizable words are the exons, and the nonsense letters represent the introns.

This unusual genetic architecture, sometimes called a split gene, requires further processing for eukaryotes before translation. Transcription of the entire gene with both exons and introns occurs first, producing a pre-mRNA. A series of adenosines is added to the mRNA molecule. This protects the molecule and eventually directs it out of the nucleus for translation. Next, a type of RNA and protein called a **spliceosome** recognizes the exon-intron junctions and enzymatically cuts through them. The action of this splicer enzyme loops the introns into lariat-shaped pieces, excises them, and joins the exons end to end. By this means, a strand of mRNA with no intron material is produced. This completed mRNA strand can then proceed to the cytoplasm to be translated.

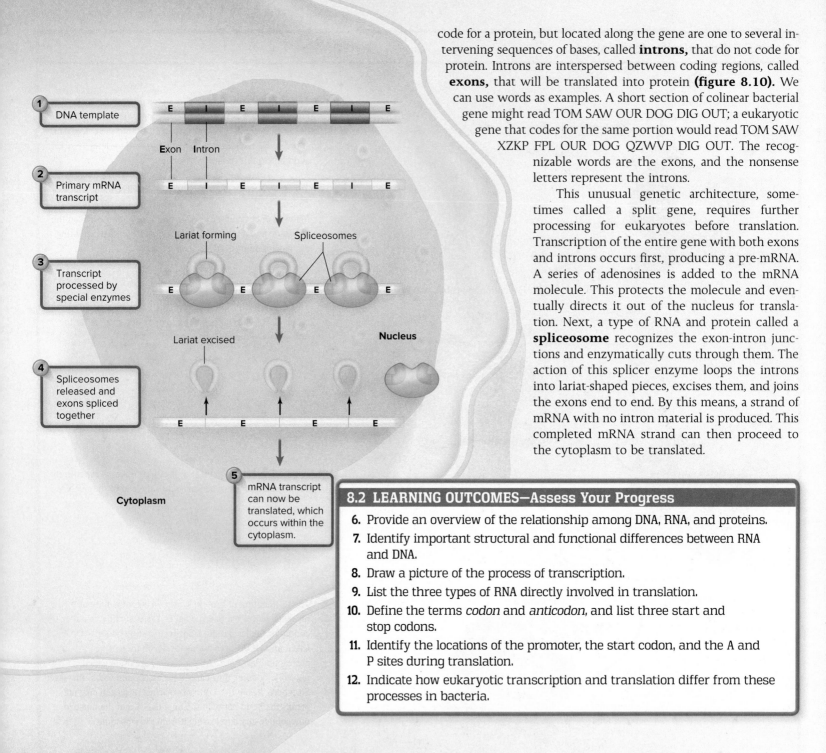

Figure 8.10 The split gene of eukaryotes.
Eukaryotic genes have an additional complicating factor in their translation. Their coding sequences, or exons (E), are interrupted at intervals by segments called introns (I) that are not part of that protein's code. Introns are transcribed but not translated, which necessitates their removal by RNA splicing enzymes before translation.

8.2 LEARNING OUTCOMES—Assess Your Progress

6. Provide an overview of the relationship among DNA, RNA, and proteins.
7. Identify important structural and functional differences between RNA and DNA.
8. Draw a picture of the process of transcription.
9. List the three types of RNA directly involved in translation.
10. Define the terms *codon* and *anticodon,* and list three start and stop codons.
11. Identify the locations of the promoter, the start codon, and the A and P sites during translation.
12. Indicate how eukaryotic transcription and translation differ from these processes in bacteria.

8.3 Genetic Regulation of Protein Synthesis

In chapter 7, we surveyed the metabolic reactions in cells and the enzymes involved in those reactions. At that time, we mentioned that some enzymes are regulated and that one form of regulation occurs at the genetic level. Control mechanisms ensure that genes are active only when their products are required. In this way, enzymes will be produced as they are needed and prevent the waste of energy and materials in dead-end synthesis. Antisense RNAs, micro RNAs, and riboswitches provide regulation in bacteria, archaea, and eukaryotes. Bacteria and archaea have an additional strategy: They organize collections of genes into **operons.** Operons consist of a coordinated set of genes, all of which are regulated as a single unit. Operons are described as either inducible or repressible. Many catabolic operons, or operons encoding enzymes that act in catabolism, are inducible, meaning that the operon is turned on (induced) by the substrate of the enzyme(s) for which the structural genes code. In this way, the enzymes needed to metabolize a nutrient (lactose, for example) are produced only when that nutrient is present in the environment. Repressible operons are normally in the "on" position, but are shut down in certain circumstances. Repressible operons often contain genes coding for anabolic enzymes, such as those used to synthesize amino acids. In the case of these operons, several genes in series are turned off (repressed) by the product synthesized by the enzyme.

The Lactose Operon: A Model for Inducible Gene Regulation in Bacteria

The best-understood system for control through genetic induction is the **lactose (*lac*) operon.** This system, first described in 1961 by François Jacob and Jacques Monod, accounts for the regulation of lactose metabolism in *Escherichia coli.* Many other operons with similar modes of action have since been identified, and together they show us that the environment of a cell can have great impact on gene expression.

The lactose operon has three important features **(table 8.6):**

1. the **regulator,** the gene that codes for a protein capable of repressing the operon (a **repressor**);
2. the control locus, composed of two areas, the **promoter** (recognized by RNA polymerase) and the **operator,** a sequence that acts as an on/off switch for transcription; and
3. the structural locus, made up of three genes, each coding for a different enzyme needed to catabolize lactose. One of the enzymes, β-galactosidase, hydrolyzes the lactose into its monosaccharides; another, permease, brings lactose across the cytoplasmic membrane.

The operon provides an efficient strategy that permits genes for a particular metabolic pathway to be induced or repressed all together by a single regulatory element. The promoter, operator, and structural components usually lie adjacent to one another, but the regulator can be at a distant site.

Table 8.6 supplies the details of how the *lac* operon works.

A fine but important point about the *lac* operon is that it functions only in the absence of glucose or if the cell's energy needs are not being met by the available glucose. Glucose is the preferred carbon source because it can be used immediately in growth and does not require induction of an operon. When glucose is present, a second regulatory system ensures that the *lac* operon is inactive, regardless of lactose levels in the environment.

Phase Variation

When bacteria turn on or off a complement of genes that leads to obvious phenotypic changes, it is sometimes called **phase variation.** Phase variation is a type of phenotypic variation, but it has its own name because it has some special characteristics, the most important of which is that the phenotype is heritable, meaning it is passed down to subsequent generations. The process of turning on genes is often mediated by regulatory proteins, as described with operons. The term *phase variation* is most often applied to traits affecting the bacterial cell surface and was originally coined to describe the ability of bacteria to change components of their surface that marked them for targeting by the host's immune system. Because these surface molecules also influenced the bacterium's ability to attach to surfaces, the ability to undergo phase variation allowed the microbes to adapt to—and stick in—different environments. Examples of phase variation include the ability of *Neisseria gonorrhoeae* strains to produce attachment fimbriae and the ability of *Streptococcus pneumoniae* to produce a capsule.

Lactose used by bacteria is the same disaccharide found in the milk we drink.
©John Fedele/Getty Images

Table 8.6 The *lac* Operon

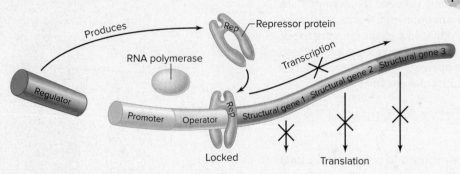

1 This operon is normally in "off" mode and does not initiate transcription when the correct substrate is absent. The operon is maintained in the off position by the repressor protein that is coded by the regulatory gene. This relatively large molecule is allosteric, meaning it has two binding sites, one for the operator sequence on the DNA and another for lactose. In the absence of lactose, this repressor binds to the operator locus. This blocks the transcription of the structural genes lying downstream. Think of the repressor as a lock on the operator, and if the operator is locked, the structural genes cannot be transcribed. Importantly, the regulator gene lies upstream (to the left) of the operator region and is transcribed constitutively because it is not controlled in tandem with the operon.

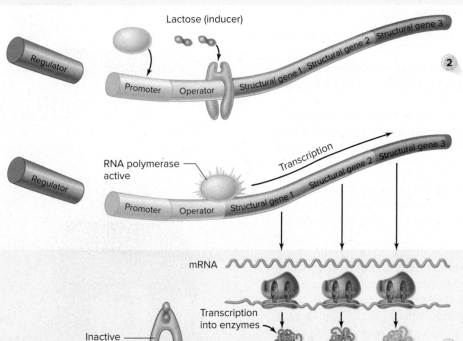

2 If lactose is added to the cell's environment, it triggers events that turn the operon on. The binding of lactose to the repressor protein causes a conformational change in the repressor that dislodges it from the operator segment of the DNA. With the operator opened up, RNA polymerase can now bind to the promoter, and proceed.

3 The structural genes are transcribed in a single unbroken transcript coding for all three enzymes. (During translation, however, each protein is synthesized separately.)

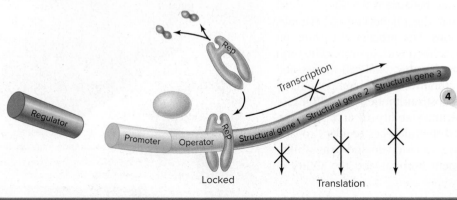

4 As lactose is depleted, further enzyme synthesis is not necessary, so the order of events reverses. At this point, there is no longer sufficient lactose to inhibit the repressor; hence, the repressor is again free to attach to the operator. The operator is locked, and transcription of the structural genes and enzyme synthesis related to lactose both stop.

8.4 DNA Recombination Events

Genetic recombination through sexual reproduction is an important means of genetic variation in eukaryotes. Although bacteria have no exact equivalent to sexual reproduction, they exhibit a primitive means for sharing or recombining parts of their genome. An event in which one bacterium donates DNA to another bacterium is a type of genetic transfer termed **recombination,** the end result of which is a new strain different from both the donor and the original recipient strain. Recombination in bacteria depends in part on the fact that bacteria contain extrachromosomal DNA—that is, plasmids—and are adept at interchanging genes. Genetic exchanges have tremendous effects on the genetic diversity of bacteria. They provide additional genes for resistance to drugs and metabolic poisons, new nutritional and metabolic capabilities, and increased virulence and adaptation to the environment.

In general, any organism that contains genes that originated in another organism is called a **recombinant.**

Horizontal Gene Transfer in Bacteria

Any transfer of DNA that results in organisms acquiring new genes that did not come directly from parent organisms is called **horizontal gene transfer.** (Acquiring genes from parent organisms during reproduction would be vertical gene transfer.) Bacteria have been known to engage in horizontal gene transfer for decades. It is now becoming clear that eukaryotic organisms—including humans—also engage in horizontal gene transfer, often aided and abetted by microbes such as viruses. Earlier in this book we discussed the concept of "pangenomes," made necessary by the discovery that different representatives of the same species had dozens of genes that the reference organisms did not. This comes about mainly from horizontal gene transfer.

DNA transfer between bacterial cells typically involves small pieces of DNA in the form of plasmids or chromosomal fragments. Plasmids are small, circular pieces of DNA that contain their own origin of replication and therefore can replicate independently of the bacterial chromosome. Plasmids are found in many bacteria (as well as some fungi) and typically contain, at most, only a few dozen genes. Although plasmids are not necessary for bacterial survival, they often carry useful traits, such as antibiotic resistance. Chromosomal fragments that have escaped from a lysed bacterial cell are also commonly involved in the transfer of genetic information between cells. An important difference between plasmids and fragments is that while a plasmid has its own origin of replication and is stably replicated and inherited, chromosomal fragments must integrate themselves into the bacterial chromosome in order to be replicated and eventually passed to progeny cells.

Depending on the mode of transmission, the means of genetic recombination in bacteria is called conjugation, transformation, or transduction. *Conjugation* requires the attachment of two related species and the formation of a bridge that can transport DNA. *Transformation* entails the transfer of naked DNA and requires no special vehicle. *Transduction* is the transfer of DNA from one bacterium to another via a bacterial virus **(table 8.7).**

Conjugation: Exchanging Genes

Conjugation is a mode of genetic exchange in which a plasmid or other genetic material is transferred by a donor to a recipient cell via a direct connection. Both gram-negative and

Researchers have found DNA that originated in snakes in many different animals, including cows. They suspect it was transferred there by ticks or other parasites.
©Imagestate Media (John Foxx)/Imagestate (cow); bennymarty/123RF (snake)

Table 8.7 Types of Horizontal Gene Transfer in Bacteria

Examples of Mode	Factors Involved	Direct or Indirect*	Genes Commonly Transferred in Nature**
Conjugation	Donor cell with pilus Fertility plasmid in donor Both donor and recipient alive Bridge forms between cells to transfer DNA	Direct	Drug resistance; resistance to metals; toxin production; enzymes; adherence molecules
Transformation	Free donor DNA (fragment) Live, competent recipient cell	Indirect	Polysaccharide capsule
Transduction	Donor is lysed bacterial cell Defective bacteriophage is carrier of donor DNA Live recipient cell of same species as donor	Indirect	Toxins; enzymes for sugar fermentation; drug resistance

*Direct means the donor and recipient are in contact during exchange; indirect means they are not.
**In the lab almost any gene can be transferred.

F⁺ and F⁻ bacteria share the same space in many settings, including the human gut.

Source: Centers for Disease Control and Prevention/Janice Haney Carr

gram-positive cells can conjugate. In gram-negative cells, the donor's plasmid (called a **fertility,** or **F factor**) has genes that direct the synthesis of a conjugative pilus. The recipient cell has a recognition site on its surface. A cell's role in conjugation is denoted by F⁺ for the cell that has the F plasmid and by F⁻ for the cell that lacks it. Contact is made when a pilus grows out from the F⁺ cell, attaches to the surface of the F⁻ cell, contracts, and draws the two cells together. In gram-positive cells, an opening is created between two adjacent cells, and the replicated DNA passes across from one cell to the other. Conjugation is a conservative process, in that the donor bacterium generally retains ("conserves") a copy of the genetic material being transferred.

There are hundreds of conjugative plasmids with some variations in their properties. One of the best understood plasmids is the F factor in *E. coli*, which can do either of two things **(figure 8.11):**

1. The donor (F⁺) cell makes a copy of its F factor and transmits this to a recipient (F⁻) cell. The F⁻ cell is thereby changed into an F⁺ cell capable of producing a pilus and conjugating with other cells. No additional donor genes are transferred.
2. In high-frequency recombination (Hfr) donors, the plasmid becomes integrated into the F⁺ donor chromosome, which, when replicated, begins to transfer to the recipient cell. This means that some chromosomal genes get transferred to the recipient. Plasmid genes may or may not be transferred.

Figure 8.11
Conjugation: genetic transmission through direct contact between two cells.

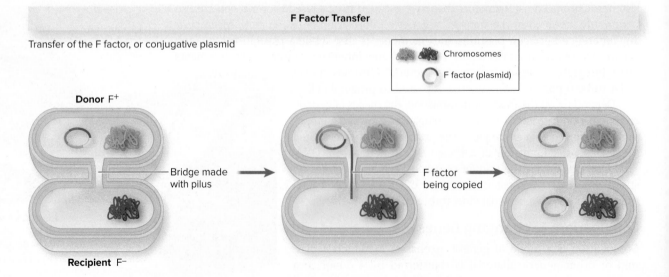

F Factor Transfer

Transfer of the F factor, or conjugative plasmid

Chromosomes
F factor (plasmid)

Donor F⁺

Bridge made with pilus

F factor being copied

Recipient F⁻

Conjugation has great biomedical importance. Special **resistance (R) plasmids,** or **factors,** that bear genes for resisting antibiotics and other drugs are commonly shared among bacteria through conjugation. Transfer of R factors can pass on resistance to antibiotics such as tetracycline, chloramphenicol, streptomycin, sulfonamides, and penicillin. Other types of R factors carry genetic codes for resistance to heavy metals (nickel and mercury) or for synthesizing virulence factors (toxins, enzymes, and adhesion molecules) that increase the pathogenicity of the bacterial strain. One important example of this is *E. coli* strains that have acquired the dangerous toxin from *Shigella* bacteria via conjugation, called shiga toxin.

Transformation: Capturing DNA from Solution

We now know that a chromosome released by a lysed cell breaks into fragments small enough to be accepted by a recipient cell. This nonspecific acceptance by a bacterial cell of small fragments of soluble DNA from the surrounding environment is termed *transformation.* Transformation is facilitated by special DNA-binding proteins on the cell wall that capture DNA from the surrounding medium. Cells that are capable of accepting genetic material through this means are termed **competent.** The new DNA is transported into the cytoplasm, where some of it is inserted into the bacterial chromosome. Transformation is a natural event found in several groups of gram-positive and gram-negative bacterial species.

Because transformation requires no special appendages, and the donor and recipient cells do not have to be in direct contact, the process is useful for certain types of recombinant DNA technology. With this technique, foreign genes from a completely unrelated organism are inserted into a plasmid, which is then introduced into a competent bacterial cell through transformation in the same way that small pieces are taken up naturally. These recombinations can be carried out in a test tube, and human genes can be experimented upon and even expressed outside the human body by placing them in a microbial cell. This same phenomenon in eukaryotic cells, termed **transfection,** is an essential aspect of genetically engineered yeasts, plants, and mice.

Transduction: The Case of the Piggyback DNA

Earlier we described bacteriophages as bacterial parasites. Viruses can in fact serve as genetic vectors (an entity that can bring foreign DNA into a cell). The process by which a bacteriophage serves as the carrier of DNA from a donor cell to a recipient cell is *transduction.* It occurs naturally in a broad spectrum of bacteria. The participating bacteria in a single transduction event must generally be the same species because of the specificity of viruses for host cells.

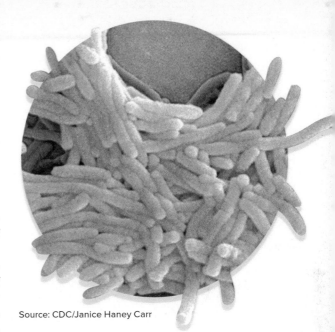

Source: CDC/Janice Haney Carr

Hfr Transfer

High-frequency (Hfr) transfer involves transmission of chromosomal genes from a donor cell to a recipient cell. The plasmid jumps into the chromosome, and when the chromosome is duplicated, the plasmid and part of the chromosome are transmitted to a new cell through conjugation. This plasmid/chromosome hybrid then incorporates into the recipient chromosome.

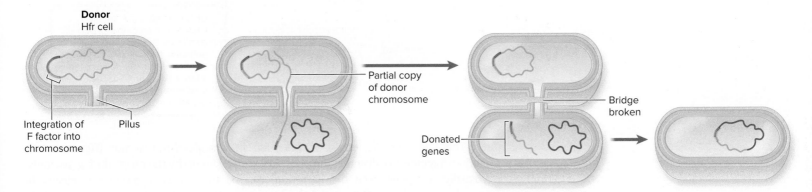

Donor
Hfr cell

Integration of F factor into chromosome

Pilus

Partial copy of donor chromosome

Donated genes

Bridge broken

Figure 8.12 Generalized transduction: genetic transfer by means of a virus carrier.

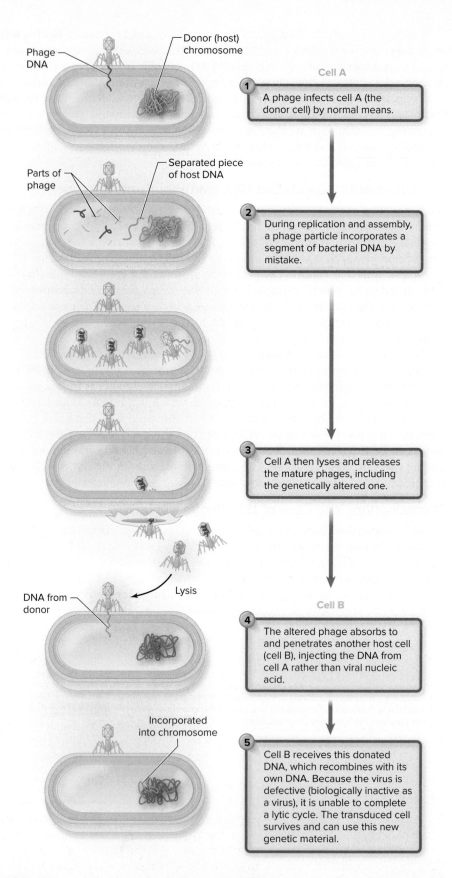

Phage DNA

Donor (host) chromosome

Cell A

1 A phage infects cell A (the donor cell) by normal means.

Parts of phage

Separated piece of host DNA

2 During replication and assembly, a phage particle incorporates a segment of bacterial DNA by mistake.

3 Cell A then lyses and releases the mature phages, including the genetically altered one.

Lysis

DNA from donor

Cell B

4 The altered phage absorbs to and penetrates another host cell (cell B), injecting the DNA from cell A rather than viral nucleic acid.

Incorporated into chromosome

5 Cell B receives this donated DNA, which recombines with its own DNA. Because the virus is defective (biologically inactive as a virus), it is unable to complete a lytic cycle. The transduced cell survives and can use this new genetic material.

There are two versions of transduction. In *generalized transduction* **(figure 8.12),** random fragments of disintegrating host DNA are taken up by the phage during assembly. Virtually any gene from the bacterium can be transmitted through this means. In *specialized transduction* **(figure 8.13),** a highly specific part of the host genome is regularly

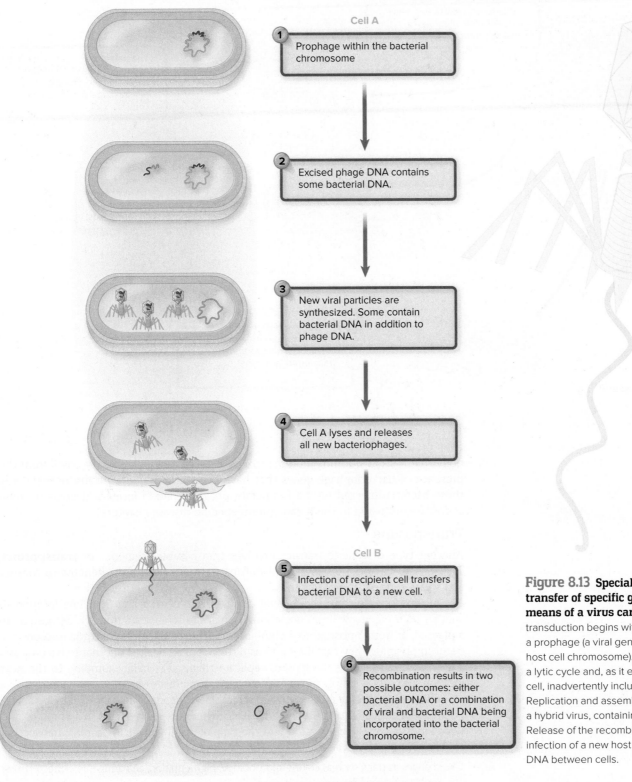

Cell A

1. Prophage within the bacterial chromosome

2. Excised phage DNA contains some bacterial DNA.

3. New viral particles are synthesized. Some contain bacterial DNA in addition to phage DNA.

4. Cell A lyses and releases all new bacteriophages.

Cell B

5. Infection of recipient cell transfers bacterial DNA to a new cell.

6. Recombination results in two possible outcomes: either bacterial DNA or a combination of viral and bacterial DNA being incorporated into the bacterial chromosome.

Figure 8.13 Specialized transduction: transfer of specific genetic material by means of a virus carrier. Specialized transduction begins with a cell that contains a prophage (a viral genome integrated into the host cell chromosome). Rarely, the virus enters a lytic cycle and, as it excises itself from its host cell, inadvertently includes some bacterial DNA. Replication and assembly result in production of a hybrid virus, containing some bacterial DNA. Release of the recombinant virus and subsequent infection of a new host result in transfer of bacterial DNA between cells.

incorporated into the virus. This specificity is explained by the prior existence of a temperate prophage inserted in a fixed site on the bacterial chromosome. When activated, the prophage DNA separates from the bacterial chromosome, carrying a small segment of host genes with it. During a lytic cycle, these specific viral-host gene combinations are incorporated into the viral particles and carried to another bacterial cell.

Several cases of specialized transduction have biomedical importance. The virulent strains of bacteria such as *Corynebacterium diphtheriae*, *Clostridium* spp., and *Streptococcus pyogenes* all produce toxins with profound physiological effects. The

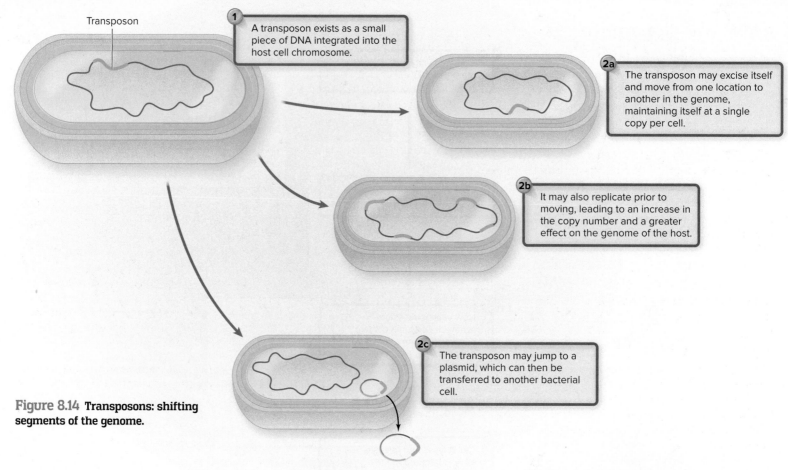

Figure 8.14 Transposons: shifting segments of the genome.

nonvirulent strains do not produce toxins. It turns out that toxicity arises from the presence of bacteriophage genes that have been introduced by transduction. Only those bacteria infected with a temperate phage are toxin formers. (Details of toxin action are discussed in the organ system-specific disease chapters.)

Transposons

Another type of genetic transfer involves transposable elements, or **transposons.** Transposons have the distinction of shifting from one part of the genome to another and so are termed "jumping genes."

All transposons share the general characteristic of traveling from one location to another on the genome—from one chromosomal site to another, from a chromosome to a plasmid, or from a plasmid to a chromosome **(figure 8.14).** Because transposons can occur in plasmids, they can also be transmitted from one cell to another in bacteria and a few eukaryotes. Some transposons replicate themselves before jumping to the next location, and others simply move without replicating first.

The overall effect of transposons—to scramble the genetic language—can be beneficial or adverse, depending upon such variables as where insertion occurs in a chromosome, what kinds of genes are relocated, and the type of cell involved. In bacteria, transposons are known to be involved in

- changes in traits such as colony morphology, pigmentation, and antigenic characteristics;
- replacement of damaged DNA; and
- the intermicrobial transfer of drug resistance (in bacteria).

8.4 LEARNING OUTCOMES—Assess Your Progress

15. Explain the defining characteristics of a recombinant organism.

16. Describe three forms of horizontal gene transfer used in bacteria.

8.5 Mutations: Changes in the Genetic Code

As precise and predictable as the rules of genetic expression seem, permanent changes do occur in the genetic code. Indeed, genetic change is the driving force of evolution. Any change to the nucleotide sequence in the genome is called a mutation. Mutations are most noticeable when the genotypic change leads to a change in phenotype. Mutations can involve the loss of base pairs, the addition of base pairs, or a rearrangement in the order of base pairs. Do not confuse this with genetic recombination, in which microbes transfer whole segments of genetic information among themselves.

A microorganism that exhibits a natural, nonmutated characteristic is known as a **wild type,** or wild strain with respect to that trait. If a microorganism bears a mutation, it is called a **mutant strain.** Mutant strains can show variance in morphology, nutritional characteristics, genetic control mechanisms, resistance to chemicals, temperature preference, and nearly any type of enzymatic function.

Causes of Mutations

Mutations can be spontaneous or induced, depending upon their origin. A **spontaneous mutation** is a random change in the DNA arising from errors in replication that occur randomly. The frequency of spontaneous mutations has been measured for a number of organisms. Mutation rates vary tremendously, from one mutation in 10^3 replications (a high rate) to one mutation in 10^{10} replications (a low rate). The rapid rate of bacterial reproduction allows these mutations to be observed more readily in bacteria than in most eukaryotes.

Induced mutations result from exposure to known **mutagens,** which are physical or chemical agents that interact with DNA in a disruptive manner. Examples of mutagens are some types of radiation (UV light, X rays) and certain chemicals such as nitrous acid.

Categories of Mutations

Mutations range from large mutations, in which large genetic sequences are gained or lost, to small ones that affect only a single base on a gene. These latter mutations, which involve addition, deletion, or substitution of single bases, are called **point mutations.**

To understand how a change in DNA influences the cell, remember that the DNA code appears in a particular order of triplets (three bases) that is transcribed into mRNA codons, each of which specifies an amino acid. A permanent alteration in the DNA that is copied faithfully into mRNA and translated can change the structure of the protein. A change in a protein can likewise change the morphology and physiology of a cell. Some mutations have a harmful effect on the cell, leading to cell dysfunction or death; these are called lethal mutations. Neutral mutations produce neither adverse nor helpful changes. A small number of mutations are beneficial in that they provide the cell with a useful change in structure or physiology.

Any change in the code that leads to placement of a different amino acid is called a **missense mutation.** A missense mutation can do one of the following:

- create a faulty, nonfunctional (or less functional) protein;
- produce a protein that functions in a different manner; or
- cause no significant alteration in protein function (see **table 8.8** to see what missense mutations look like).

A **nonsense mutation,** on the other hand, changes a normal codon into a stop codon that does not code for an amino acid and stops the production of the protein wherever it occurs. A nonsense mutation almost always results in a nonfunctional protein. (Table 8.8, row d, shows a nonsense mutation resulting from a frameshift [described in next paragraph].) A **silent mutation** (table 8.8, row c) alters a base but does not change the amino acid and thus has no effect. For example, because of the redundancy of the code, ACU, ACC, ACG, and ACA all code for threonine, so a mutation that

Barbara McClintock won the Nobel Prize for Physiology or Medicine in 1983 for the discovery of mobile genetic elements, or transposons, in the corn plant.
©Science Source

Medical Moment

Mutations Caused by Life-Saving Radiation

One effective treatment for some cancers is radiation therapy. A high dose of ionizing radiation is delivered to the cancer site, causing damage to DNA and cancerous cell death. However, cells surrounding the cancer site are also exposed to the radiation. DNA mutations may occur in these healthy cells.

DNA mutations in healthy cells can lead to development of a secondary cancer in an individual treated with radiation therapy due to the changes in the genetic code. The mutation may alter cellular reproductive processes, causing uncontrolled duplication. Secondary cancer may not appear for many years but is a known risk for individuals with cancer as a child. Commonly these secondary cancers are skin or bone marrow cancers, or leukemia.

Q: Do mutations from radiation therapy cause changes in the DNA code that are passed down to the next generation?

Answer in Appendix B.

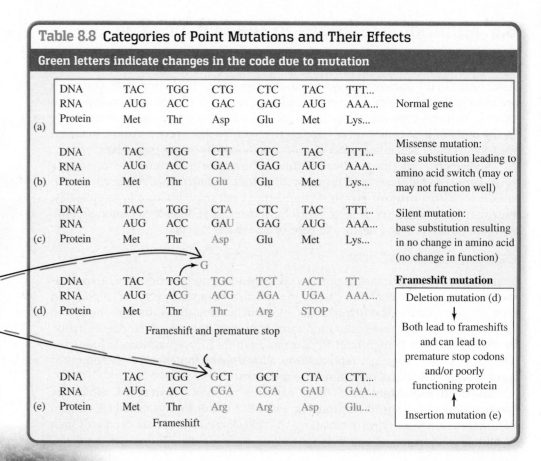

Table 8.8 Categories of Point Mutations and Their Effects

Green letters indicate changes in the code due to mutation

(a)	DNA	TAC	TGG	CTG	CTC	TAC	TTT...	Normal gene
	RNA	AUG	ACC	GAC	GAG	AUG	AAA...	
	Protein	Met	Thr	Asp	Glu	Met	Lys...	

								Missense mutation: base substitution leading to amino acid switch (may or may not function well)
(b)	DNA	TAC	TGG	CTT	CTC	TAC	TTT...	
	RNA	AUG	ACC	GAA	GAG	AUG	AAA...	
	Protein	Met	Thr	Glu	Glu	Met	Lys...	

								Silent mutation: base substitution resulting in no change in amino acid (no change in function)
(c)	DNA	TAC	TGG	CTA	CTC	TAC	TTT...	
	RNA	AUG	ACC	GAU	GAG	AUG	AAA...	
	Protein	Met	Thr	Asp	Glu	Met	Lys...	

G

								Frameshift mutation
(d)	DNA	TAC	TGC	TGC	TCT	ACT	TT	
	RNA	AUG	ACG	ACG	AGA	UGA	AAA...	
	Protein	Met	Thr	Thr	Arg	STOP		

Frameshift and premature stop

Deletion mutation (d)
↓
Both lead to frameshifts and can lead to premature stop codons and/or poorly functioning protein
↑
Insertion mutation (e)

(e)	DNA	TAC	TGG	GCT	GCT	CTA	CTT...
	RNA	AUG	ACC	CGA	CGA	GAU	GAA...
	Protein	Met	Thr	Arg	Arg	Asp	Glu...

Frameshift

Does this make sense to you? The arrow here means that the G nucleotide has been skipped during DNA replication and no longer appears. The arrow here means it was accidentally put in the code where it doesn't belong.
— Kelly & Heidi

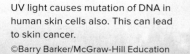

UV light causes mutation of DNA in human skin cells also. This can lead to skin cancer.
©Barry Barker/McGraw-Hill Education

changes only the last base will not alter the sense of the message in any way. A **back-mutation** occurs when a gene that has undergone mutation reverses (mutates back) to its original base composition.

Mutations also occur when one or more bases are inserted into or deleted from a newly synthesized DNA strand. This type of mutation, known as a **frameshift** (table 8.8, rows d and e), is so named because the reading frame of the mRNA has been changed. Frameshift mutations nearly always result in a nonfunctional protein because most amino acids after the mutation are different from what was coded for in the original DNA. Also note that insertion or deletion of bases in multiples of three (3, 6, 9, etc.) results in the addition or deletion of amino acids but does not disturb the reading frame.

Single Nucleotide Polymorphism

One type of mutation recently found to be important in individual traits is called **single nucleotide polymorphism (SNP)** because only a single nucleotide is altered. This is a result of a point mutation at some point in the organism's ancestry, and it is passed on genetically. Tens of thousands of these differences at a single locus (when two different individuals are compared) are known to exist throughout the genome. The human genome contains 10 million SNPs. These variations are currently a hot area of research and commerce.

The ability to identify SNPs has proven critical to the new field of personalized medicine, which is customized to a person's genetic makeup. One example is when patients' genomes are examined for SNPs that have been found to be associated with a particular disease, to determine their risk. For example, in a condition called thrombophilia (a blood-clotting disorder), a point mutation in the gene for a clotting factor (factor V) causes an arginine to become a glutamine **(figure 8.15)**.

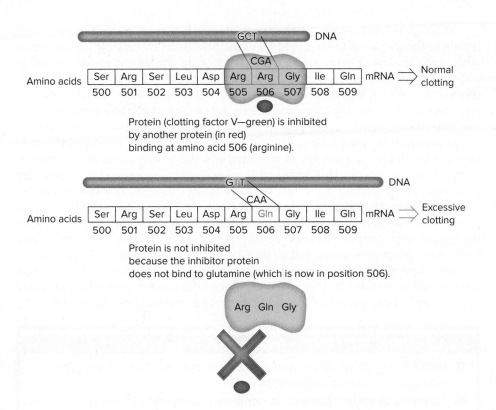

This leads to increased clotting in the patient. Also, SNPs can determine whether a patient will respond favorably to a particular treatment. The new field of pharmacogenomics tailors drug treatments using this knowledge of SNPs.

Repair of Mutations

Earlier we indicated that DNA has a proofreading mechanism to repair mistakes in replication that might otherwise become permanent. Because mutations are potentially life-threatening, the cell has systems for finding and repairing DNA that has been damaged by various mutagenic agents and processes. Most ordinary DNA damage is resolved by enzymatic systems specialized for finding and fixing such defects.

DNA that has been damaged by ultraviolet radiation can be restored by photoactivation—or light repair. This repair mechanism requires visible light and a light-sensitive enzyme, DNA photolyase, an enzyme that can detect and attach to the damaged areas (sites of abnormal pyrimidine binding). Ultraviolet repair mechanisms are successful only for a relatively small number of UV mutations. Cells cannot repair severe, widespread damage and will die.

Mutations can be excised, or cut out, by a series of enzymes that remove the incorrect bases and add the correct ones. This process is known as excision repair. First, enzymes break the bonds between the bases and the sugar-phosphate strand at the site of the error. A different enzyme subsequently removes the defective bases one at a time, leaving a gap that will be filled in by DNA polymerase I and ligase. A repair system can also locate mismatched bases that were missed during proofreading—for example, C mistakenly paired with A, or G with T.

Positive and Negative Effects of Mutations

Many mutations are not repaired. How the cell copes with them depends on the nature of the mutation and the strategies available to that organism. Mutations are permanent and heritable, meaning these changes will be passed on to the offspring of organisms and become a long-term part of the gene pool. Many mutations are harmful to organisms; others provide adaptive advantages.

Medical Moment

Gene Mutations

In human cellular function, even small changes in the genetic code can have devastating consequences. Consider the following examples of mutations, which are caused by an alteration in just one or two amino acids:

- *Alpers-Huttenlocher syndrome*: A point mutation causes the insertion of one incorrect nucleotide, resulting in the placement of amino acid threonine in place of alanine. This mutation inhibits DNA synthesis and causes decreased cellular energy. Patients with this syndrome present between ages 2 and 4 with developmental regression, liver problems, and recurrent seizures. The condition is progressive and eventually fatal.

- *Cystic fibrosis*: This disease affects cellular chloride channels, causing thick and sticky mucus production by secretory glands. Thick mucus causes problems for the airways, pancreas, reproductive organs, and digestive system. The mutation in the *CFTR* gene is most commonly caused by the deletion of a single amino acid. Though symptoms may be managed, patients with cystic fibrosis are vulnerable to infections and shorter life expectancy.

- *Epidermolysis bullosa*: A missense mutation is responsible for this inherited condition, which causes skin fragility. Those affected develop blisters and lesions in response to even minor skin contact. The condition affects two nucleotides that code for amino acids that build keratin, a protein that gives skin stability and strength.

Q. Without the redundancy of the genetic code, would there be more or fewer cases of these diseases in the population?

Answer in Appendix B.

Although most spontaneous mutations are not beneficial, a small number contribute to the success of the individual and the population by creating variant strains with alternate ways of expressing a trait. Microbes are not "aware" of this advantage and do not direct these changes; they simply respond to the environment they encounter. A mutation is beneficial if it allows the organism to more readily survive or reproduce. In the long-range view, mutations and the variations they produce are the raw materials for change in the population and, thus, for adaptation and evolution.

Mutations that create variants occur frequently enough that any population contains mutant strains for a number of characteristics, but as long as the environment is stable, these mutants (for that particular trait) will never comprise more than a tiny percentage of the population. When the environment changes, however, it can become hostile for the survival of certain individuals, and only those microbes bearing protective mutations will be equipped to survive in the new environment. In this way, the environment naturally "selects" certain mutant strains that will reproduce, give rise to subsequent generations, and, in time, be the dominant strain in the population. Through these means, any change that confers an advantage due to this selection pressure will be retained by the population. One of the clearest models for this sort of selection and adaptation is acquired drug resistance in bacteria (see chapter 10).

8.5 LEARNING OUTCOMES—Assess Your Progress

17. Define the term *mutation*, and discuss one positive and one negative example of it in microorganisms.
18. Differentiate among frameshift, nonsense, silent, and missense mutations.
19. Explain the influence of single-nucleotide polymorphisms on an organism.

8.6 Studying DNA in the Laboratory and Genetic Engineering

Knowing how DNA works within the cell to carry out the goals of a microbe allows scientists to utilize these processes to accomplish goals more to the liking of human beings. Since the 1970s, discoveries and advances have led to an explosion of new capabilities and, as a result, an explosion of new knowledge about microbes and about biology in general. In this section, we will highlight a few techniques that have relevance for microbiology and in particular for infectious diseases.

Working with DNA in the Laboratory

Two very important techniques were developed in the 1970s and 1990s that made massive advances in DNA study and manipulation possible. These are, first, the discovery of **restriction endonucleases,** and, second, the invention of the polymerase chain reaction.

The groundbreaking discovery in 1971 of restriction endonucleases made almost everything we discuss in this section possible. These enzymes come from bacterial cells. They recognize foreign DNA and are capable of breaking the phosphodiester bonds between adjacent nucleotides on both strands of DNA, leading to a break in the DNA strand. In the bacterial cell, this ability protects against the incompatible DNA of bacteriophages or plasmids. In the biotechnologist's lab, the enzymes can be used to cleave DNA at desired sites and are necessary for the techniques of recombinant DNA technology.

Thousands of restriction endonucleases have been discovered in bacteria. Each one has a known sequence of 4 to 40 or more base pairs as its target, so sites of cutting can be finely controlled. Many of these enzymes have the unique property of recognizing and clipping at base sequences called **palindromes (figure 8.16).**

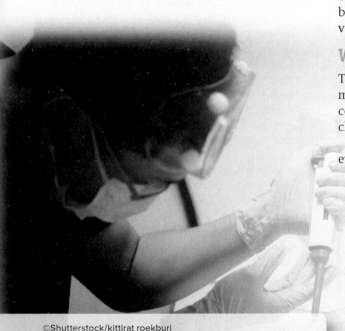

Figure 8.16 Some useful properties of DNA.

DNA Heating and Cooling

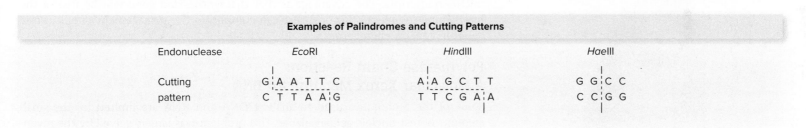

DNA responds to heat by denaturing—losing its hydrogen bonding and then separating into its two strands. When cooled, the two strands rejoin at complementary regions. The two strands need not be from the same organism as long as they have matching nucleotides.

Examples of Palindromes and Cutting Patterns

Endonuclease	*Eco*RI	*Hin*dIII	*Hae*III
Cutting pattern	G A A T T C C T T A A G	A A G C T T T T C G A A	G G C C C C G G

Action of Restriction Endonucleases

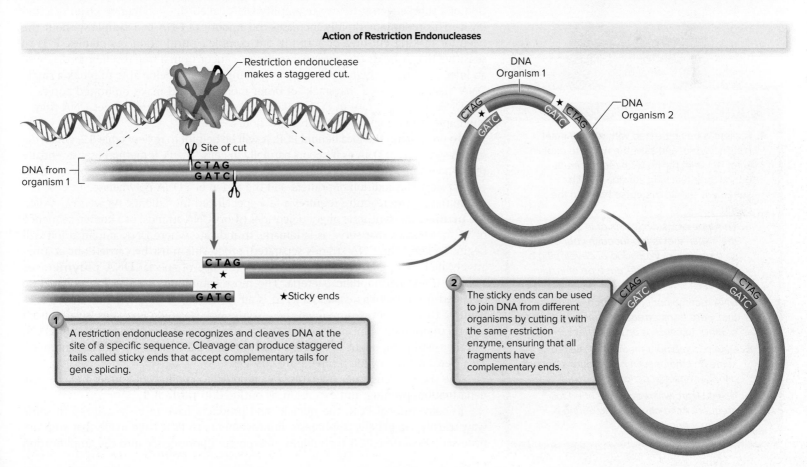

Restriction endonuclease makes a staggered cut.

Site of cut

DNA from organism 1

C T A G
G A T C

★Sticky ends

1 A restriction endonuclease recognizes and cleaves DNA at the site of a specific sequence. Cleavage can produce staggered tails called sticky ends that accept complementary tails for gene splicing.

DNA Organism 1

DNA Organism 2

2 The sticky ends can be used to join DNA from different organisms by cutting it with the same restriction enzyme, ensuring that all fragments have complementary ends.

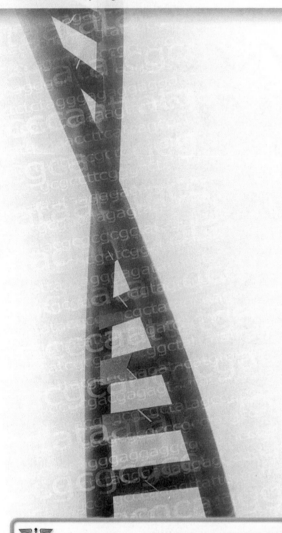

Palindromes are sequences of DNA that are identical when read from the 5′ to 3′ direction on one strand and the 5′ to 3′ direction on the other strand.

Endonucleases are usually named by combining the first letter of the bacterial genus, the first two letters of the species, and the endonuclease number. For example, *Eco*RI is the first endonuclease found in *Escherichia coli*, and *Hin*dIII is the third endonuclease discovered in *Haemophilus influenzae* type d (figure 8.16).

The pieces of DNA produced by restriction endonucleases are termed **restriction fragments.** Because DNA sequences vary, even among members of the same species, differences in the cutting pattern of specific restriction endonucleases give rise to restriction fragments of differing lengths, known as **restriction fragment length polymorphisms (RFLPs).** RFLPs allow the direct comparison of the DNA of two different organisms at a specific site.

Another enzyme, called a **ligase,** is necessary to seal the sticky ends together by rejoining the phosphate-sugar bonds cut by endonucleases. Its main application is in final splicing of genes into plasmids and chromosomes.

An enzyme called **reverse transcriptase (RT)** is best known for its role in the replication of the AIDS virus and other retroviruses. It also provides geneticists with a valuable tool for converting RNA into DNA. Copies called **complementary DNA,** or **cDNA,** can be made from messenger, transfer, ribosomal, and other forms of RNA. The technique provides a valuable means of synthesizing eukaryotic genes from mRNA transcripts. The advantage is that the synthesized gene will be free of the intervening sequences (introns) that can complicate the management of eukaryotic genes in genetic engineering.

Polymerase Chain Reaction: A Molecular Xerox Machine for DNA

Some of the techniques used to analyze DNA and RNA are limited by the small amounts of test nucleic acid available. This problem was largely solved by the invention of a simple, versatile way to amplify DNA called the polymerase chain reaction (PCR). This technique rapidly increases the amount of DNA in a sample without the need for making cultures or carrying out complex purification techniques. It is so sensitive that it holds the potential to detect cancer from a single cell or to diagnose an infection from a single gene copy. It is comparable to being able to pluck a single DNA "needle" out of a "haystack" of other molecules and make unlimited copies of the DNA. The rapid rate of PCR makes it possible to replicate a target DNA from a few copies to billions of copies in a few hours.

To understand the idea behind PCR, it will be helpful to review table 8.2, which describes synthesis of DNA as it occurs naturally in cells. The PCR method uses essentially the same events, with the opening up of the double strand, using the exposed strands as templates, the addition of primers, and the action of a DNA polymerase.

Initiating the reaction requires a few specialized ingredients. As we saw earlier, the **primers** are synthetic oligonucleotides (short DNA strands) of a known sequence of 15 to 30 bases that serve as landmarks to indicate where DNA amplification will begin. To keep the DNA strands separated, processing must be carried out at a relatively high temperature. This necessitates the use of special **DNA polymerases** isolated from thermophilic bacteria. The most commonly used is Taq polymerase obtained from *Thermus aquaticus.* (Taq is an abbreviation of the genus and species name of this microbe, from which the enzyme was isolated.) Enzymes isolated from this thermophilic organism remain active at the elevated temperatures used in PCR. Another vital component of PCR is a machine called a thermal cycler that automatically performs the cyclic temperature changes.

The PCR technique operates by repetitive cycling of three basic steps: denaturation, priming, and extension, as outlined in **table 8.9.**

In conventional PCR, the nucleic acid products have to be visualized in some way, usually via gel electrophoresis. Improvements to PCR have made that step unnecessary. Real-time PCR techniques incorporate fluorescence into the amplification

NCLEX® PREP

4. A client is being treated with recombinant insulin, *Lispro,* and asks the nurse to explain what is meant by the term *recombinant* during their teaching session. Which statement should the nurse make to the client?

   a. *This type of insulin is similar in structure to the insulin that is produced in your body, so there is less likelihood of developing a reaction than if a different type of insulin was used.*

   b. *DNA recombinant medication utilizes genes that were transferred from one organism to another to cause a desired effect.*

   c. *DNA recombinant medication, Lispro, is the same as other insulin preparations but has a longer onset of action.*

   d. *Using Lispro will reduce the need for frequent Accu-chek testing during a 24-hour period.*

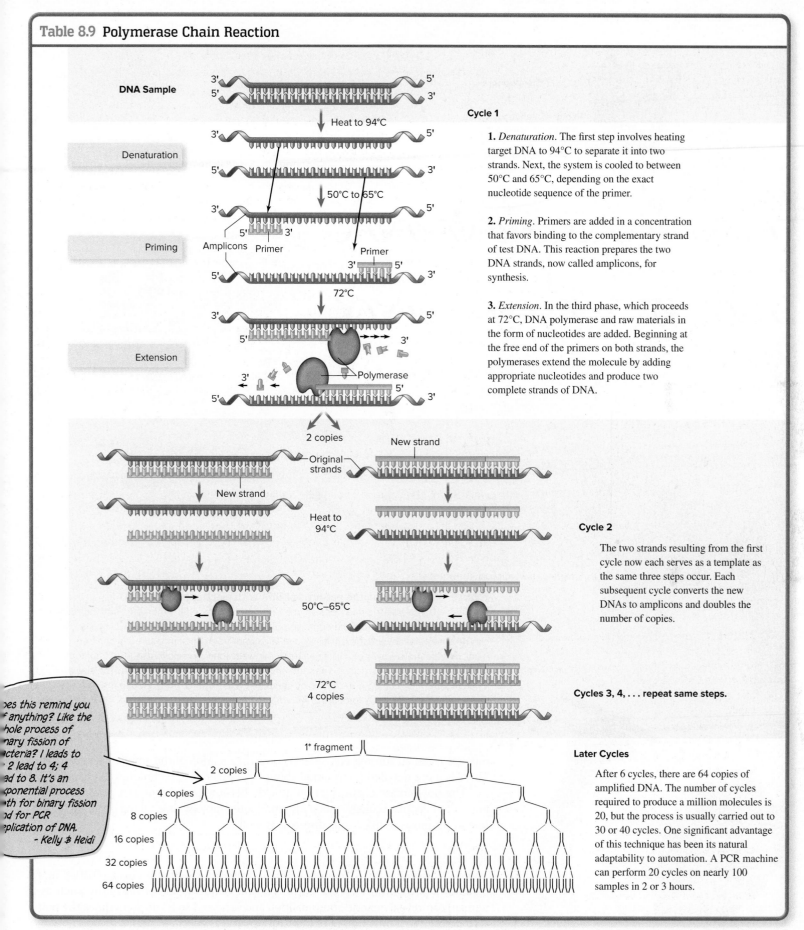

Table 8.9 Polymerase Chain Reaction

DNA Sample

Heat to 94°C

Denaturation

50°C to 65°C

Priming

Amplicons   Primer   Primer

72°C

Extension

Polymerase

2 copies

Original strands

New strand

New strand

Heat to 94°C

50°C–65°C

72°C
4 copies

Cycle 1

1. *Denaturation.* The first step involves heating target DNA to 94°C to separate it into two strands. Next, the system is cooled to between 50°C and 65°C, depending on the exact nucleotide sequence of the primer.

2. *Priming.* Primers are added in a concentration that favors binding to the complementary strand of test DNA. This reaction prepares the two DNA strands, now called amplicons, for synthesis.

3. *Extension.* In the third phase, which proceeds at 72°C, DNA polymerase and raw materials in the form of nucleotides are added. Beginning at the free end of the primers on both strands, the polymerases extend the molecule by adding appropriate nucleotides and produce two complete strands of DNA.

Cycle 2

The two strands resulting from the first cycle now each serves as a template as the same three steps occur. Each subsequent cycle converts the new DNAs to amplicons and doubles the number of copies.

Cycles 3, 4, . . . repeat same steps.

> es this remind you
> f anything? Like the
> hole process of
> nary fission of
> cteria? I leads to
> 2 lead to 4; 4
> ad to 8. It's an
> xponential process
> th for binary fission
> nd for PCR
> eplication of DNA.
> - Kelly & Heidi

1* fragment

2 copies

4 copies

8 copies

16 copies

32 copies

64 copies

Later Cycles

After 6 cycles, there are 64 copies of amplified DNA. The number of cycles required to produce a million molecules is 20, but the process is usually carried out to 30 or 40 cycles. One significant advantage of this technique has been its natural adaptability to automation. A PCR machine can perform 20 cycles on nearly 100 samples in 2 or 3 hours.

*For simplicity's sake, we have omitted the elongation of the complete original parent strand during the first cycles. Ultimately, templates that correspond only to the smaller fragments dominate and become the primary population of replicated DNA.

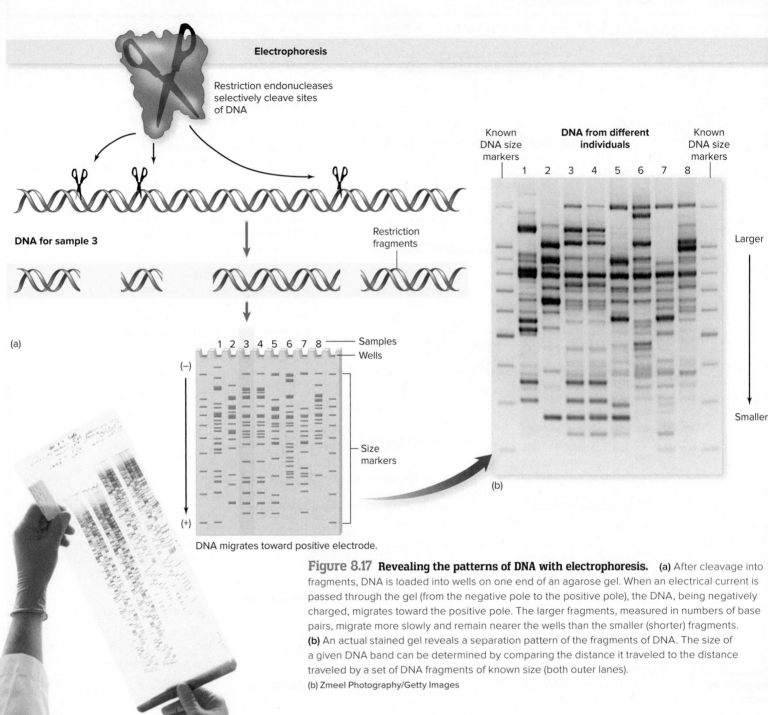

Electrophoresis

Restriction endonucleases selectively cleave sites of DNA

DNA for sample 3

Restriction fragments

(a)

1 2 3 4 5 6 7 8 — Samples

Wells

(−)

Size markers

(+)

DNA migrates toward positive electrode.

Known DNA size markers

DNA from different individuals

1 2 3 4 5 6 7 8

Known DNA size markers

Larger

Smaller

(b)

Figure 8.17 Revealing the patterns of DNA with electrophoresis. **(a)** After cleavage into fragments, DNA is loaded into wells on one end of an agarose gel. When an electrical current is passed through the gel (from the negative pole to the positive pole), the DNA, being negatively charged, migrates toward the positive pole. The larger fragments, measured in numbers of base pairs, migrate more slowly and remain nearer the wells than the smaller (shorter) fragments. **(b)** An actual stained gel reveals a separation pattern of the fragments of DNA. The size of a given DNA band can be determined by comparing the distance it traveled to the distance traveled by a set of DNA fragments of known size (both outer lanes).

(b) Zmeel Photography/Getty Images

©Rob Melnychuk/Getty Images

mix itself, and an automated reader records the creation of the amplicons in real time. There is more detailed information about the newer PCR techniques in chapter 15.

The polymerase chain reaction quickly became prominent as a powerful workhorse of molecular biology, medicine, and biotechnology. It is taking on an important role in diagnosis of infectious diseases, as well.

Analysis of DNA

There are many reasons scientists would want to know the characteristics of an organism's DNA. Knowing the entire genome sequence of an organism (such as a human) can reveal genetic abnormalities, ancestry, and so forth. Sometimes the point is to determine if one sample of DNA is the same as another sample. If you know the DNA patterns of a pathogen causing an outbreak, for example, you can look at the

DNA patterns of a microbe isolated from a patient to see if they were made ill by that same pathogen. Also, you can look at the pattern of human DNA left at a crime scene and match it against a suspect's DNA. There are multiple ways to analyze DNA, and we will highlight two of them: **DNA profiling** and **DNA sequencing.**

DNA Profiling

You've no doubt seen images like the one in **figure 8.17b,** with its ladderlike images, on TV crime shows. These are examples of DNA profiles. They rely on the fact that the DNA in different individuals–even of the same species–contain small differences. This is true for humans, and it is true for bacteria of the same species. When the individual's DNA is digested with a set of restriction enzymes, the small sequence differences will cause the DNA to be cut up at different places, resulting in fragments of different lengths in the digested samples of different individuals. To actually see the array of different fragments, and to compare the fragments of different individuals, scientists use a technique called **gel electrophoresis.**

In this technique, samples are placed in compartments (wells) in a soft agar gel and subjected to an electrical current. The phosphate groups in DNA give the entire molecule an overall negative charge, which causes the DNA to move toward the positive pole in the gel. The rate of movement is based primarily on the size of the fragments. The larger fragments move more slowly and remain nearer the top of the gel, whereas the smaller fragments migrate faster and end up farther from the wells. The positions of DNA fragments are determined by staining the DNA fragments in the gel **(figure 8.17a).**

DNA Sequencing

There are many instances in which a detailed picture of the entire sequence of a genome is needed.

By far the most detailed maps of a genome are **sequence maps,** which give an exact order of bases in a plasmid, a chromosome, or an entire genome. Genomes of thousands of organisms have been sequenced, including viruses, bacteria, and eukaryotic organisms (including humans). One of the remarkable discoveries in this huge enterprise has been how similar the genomes of relatively unrelated organisms are. Humans share approximately 80% of their DNA codes with mice, about 60% with rice, and even 30% with the worm *C. elegans.*

So how is this sequencing performed? We can illustrate the process by describing an older method, called shotgun sequencing **(figure 8.18).** Shotgun sequencing can be broken down into seven steps:

1. First, the genome of an organism is broken down into smaller, manageable fragments.

2. The fragments are separated through gel electrophoresis.

3. Each fragment is inserted into a plasmid and is cloned into an *E. coli* cell. This produces a complete library of fragments. The library exists in the bacterial cultures, which can be preserved indefinitely and sampled repeatedly.

4. The plasmids are purified and the DNA fragments are sequenced by automated sequencers, machines that add labeled primers to each fragment. The primers usually recognize the plasmid sequences that flank the genome insert, so that the machine can tell where the fragment begins and ends. Each section of the genome ends up being sequenced multiple times in an overlapping fashion.

5. A computer program takes all the sequence data and is able to find where the sequence overlaps. This automated process results in a larger, contiguous set (contigs) of nucleotide sequences.

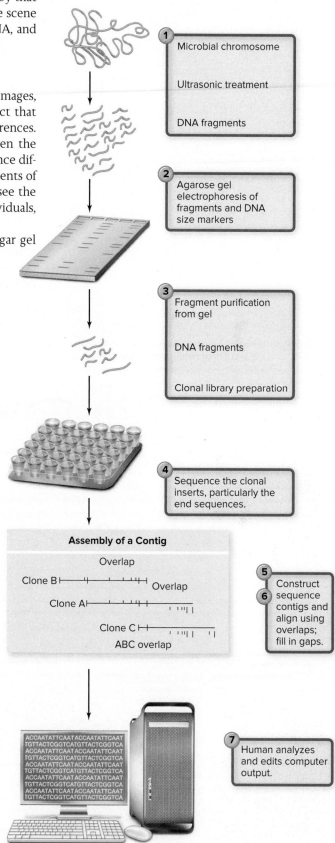

1 Microbial chromosome

Ultrasonic treatment

DNA fragments

2 Agarose gel electrophoresis of fragments and DNA size markers

3 Fragment purification from gel

DNA fragments

Clonal library preparation

4 Sequence the clonal inserts, particularly the end sequences.

Assembly of a Contig

Overlap

Clone B

Overlap

Clone A

Clone C

ABC overlap

5
6 Construct sequence contigs and align using overlaps; fill in gaps.

7 Human analyzes and edits computer output.

Figure 8.18 Whole-genome shotgun sequencing. See text for explanation of numbers.

6 The contigs are put in the proper order to determine the entire sequence. This step is tricky; there are often gaps between the contigs, but there are a variety of methods to resolve this issue.

7 An important last step is editing. A human examines the sequence, looking for irregularities, frameshifts, and ambiguities.

In modern sequencing, similar principles are used, but the whole process is scaled up in what is called "high throughput" genome sequencing. High-throughput sequencing (also called "deep sequencing" or "next generation sequencing") requires four steps **(figure 8.19)**: (1) A library of DNA (or RNA) is prepared by fragmenting the DNA and fitting each fragment with some common adaptors, or short sequences that are designed to match the probes in the next step. (2) The collection of sequences is placed in a flow cell, whose surface is coated with the probes that will bind with adaptor molecules on the DNA or RNA. Inside the flow cell, the conditions and reagents are provided to use one or another form of PCR to amplify each fragment many times. (3) Then all of the amplified fragments are sequenced so that each nucleotide is "read" thousands of times, each time it is present on a sequence of any length. The details of how this happens differ for each of the automated systems, but all rely on techniques that originated in the

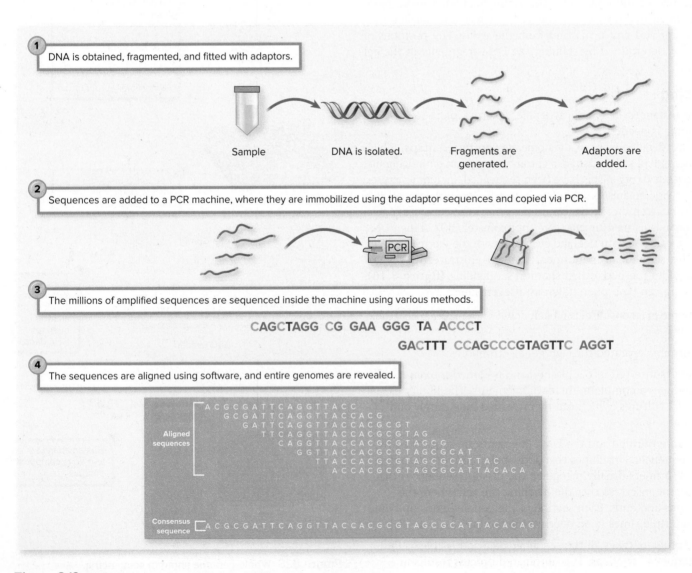

1 DNA is obtained, fragmented, and fitted with adaptors.

Sample   DNA is isolated.   Fragments are   Adaptors are
                                          generated.   added.

2 Sequences are added to a PCR machine, where they are immobilized using the adaptor sequences and copied via PCR.

PCR

3 The millions of amplified sequences are sequenced inside the machine using various methods.

CAGCTAGG CG GAA GGG TA ACCCT

GACTTT CCAGCCCGTAGTTC AGGT

4 The sequences are aligned using software, and entire genomes are revealed.

Aligned sequences

ACGCGATTCAGGTTACC
GCGATTCAGGTTACCACG
GATTCAGGTTACCACGCGT
TTCAGGTTACCACGCGTAG
CAGGTTACCACGCGTAGCG
GGTTACCACGCGTAGCGCAT
TTACCACGCGTAGCGCATTAC
ACCACGCGTAGCGCATTACACA

Consensus sequence

ACGCGATTCAGGTTACCACGCGTAGCGCATTACACAG

Figure 8.19 **High-throughput sequencing of DNA.**

Sanger method. (4) Finally, the millions of sequence lengths–also called "reads"–are aligned using bioinformatics software. Putting all the reads together, the entire sequence can be deduced.

This whole process is mostly automated, using machines developed over the course of the last decade. The analysis of 16s RNA to determine how different organisms are related is accomplished through high-throughput sequencing.

The New "-omics"

The ability to obtain the entire sequences of organisms has spawned new vocabulary that refers to the "total picture" of some aspect of a cell or organism.

genomics The systematic study of an organism's genes and their functions.

proteomics The study of an organism's complement of proteins (its "proteome") and functions mediated by the proteins.

metagenomics (also called "community genomics") The study of all the genomes in a particular ecological niche, as opposed to individual genomes from single species.

metabolomics The study of the complete complement of small chemicals present in a cell at any given time. Provides a snapshot of the physiological state of the cell and the end products of its metabolism.

Recombinant DNA Technology

The primary intent of recombinant DNA technology is to deliberately remove genetic material from one organism and combine it with that of a different organism. Its origins can be traced to 1970, when microbiologists first began to duplicate the clever tricks bacteria do naturally with bits of extra DNA such as plasmids, transposons, and proviruses. As mentioned earlier, humans have been trying to artificially influence genetic transmission of traits for centuries. The discovery that bacteria can readily accept, replicate, and express foreign DNA made them powerful agents for studying the genes of other organisms in isolation. The practical applications of this work were soon realized by scientists. Bacteria could be genetically engineered to mass-produce substances such as hormones, enzymes, and vaccines that were difficult to synthesize by the usual industrial methods.

This process is called molecular cloning, or gene cloning. (Don't confuse it with another use of the word "clone," namely, the cloning of whole organisms.) Gene cloning involves the removal of a selected gene from an animal, plant, or microorganism (the genetic donor), followed by its propagation in a different host organism. Cloning requires that the desired donor gene first be selected, excised by restriction endonucleases, and isolated. The gene is next inserted into a **vector** (usually a plasmid or a virus) that will insert the DNA into a **cloning host.** The cloning host is usually a bacterium or a yeast that can replicate the gene and translate it into the protein product for which it codes. In the next section, we examine the elements of gene isolation, vectors, and cloning hosts and show how they participate in a complete recombinant DNA procedure.

Construction of a Recombinant, Insertion into a Cloning Host, and Genetic Expression

Table 8.10 is a step-by-step guide to cloning a gene.

Synthetic Biology

In recent years, researchers have staked out entirely new territory in genetic manipulation: They are creating new biological molecules and organisms from scratch. This field is called *synthetic biology*. A pioneer in the field is one of the same men who sequenced the human genome, Craig Venter. In 2010, he successfully created

Table 8.10 Gene Cloning

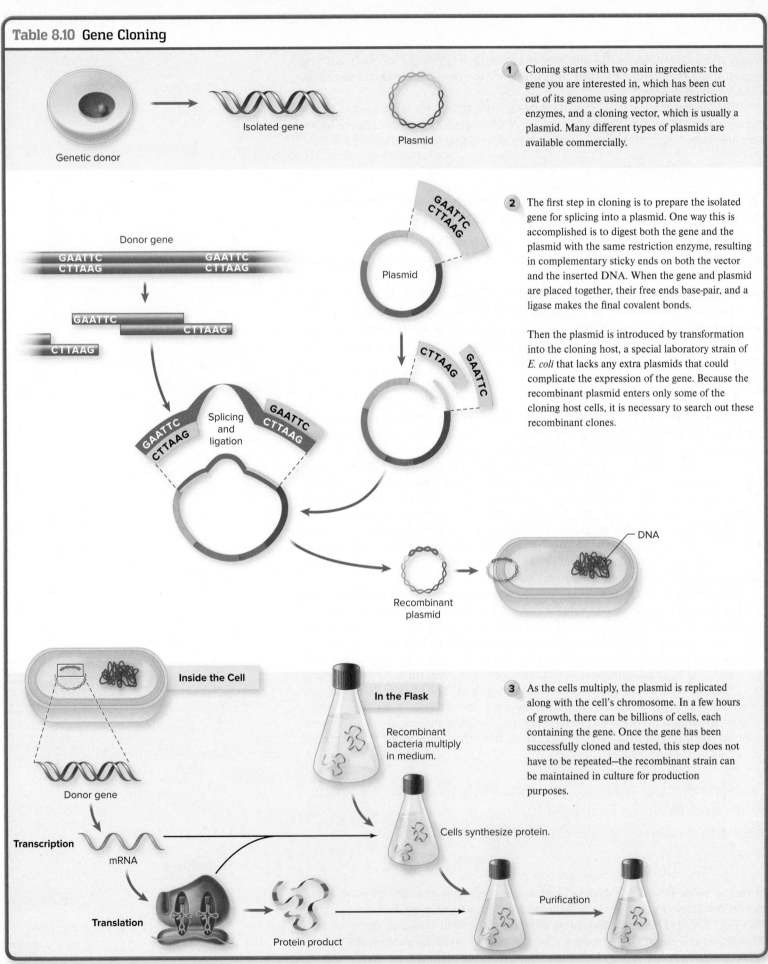

1 Cloning starts with two main ingredients: the gene you are interested in, which has been cut out of its genome using appropriate restriction enzymes, and a cloning vector, which is usually a plasmid. Many different types of plasmids are available commercially.

Genetic donor

Isolated gene

Plasmid

2 The first step in cloning is to prepare the isolated gene for splicing into a plasmid. One way this is accomplished is to digest both the gene and the plasmid with the same restriction enzyme, resulting in complementary sticky ends on both the vector and the inserted DNA. When the gene and plasmid are placed together, their free ends base-pair, and a ligase makes the final covalent bonds.

Then the plasmid is introduced by transformation into the cloning host, a special laboratory strain of *E. coli* that lacks any extra plasmids that could complicate the expression of the gene. Because the recombinant plasmid enters only some of the cloning host cells, it is necessary to search out these recombinant clones.

Donor gene

GAATTC
CTTAAG   GAATTC
                CTTAAG

GAATTC
CTTAAG

CTTAAG

GAATTC
CTTAAG

Plasmid

GAATTC
CTTAAG

CTTAAG
GAATTC

Splicing and ligation

GAATTC
CTTAAG

DNA

Recombinant plasmid

Inside the Cell

In the Flask

Recombinant bacteria multiply in medium.

3 As the cells multiply, the plasmid is replicated along with the cell's chromosome. In a few hours of growth, there can be billions of cells, each containing the gene. Once the gene has been successfully cloned and tested, this step does not have to be repeated—the recombinant strain can be maintained in culture for production purposes.

Donor gene

Transcription

mRNA

Cells synthesize protein.

Translation

Protein product

Purification

a self-replicating bacterial cell from four bottles of chemicals: the four nucleotides of DNA. This was a breakthrough of major proportions, as it was the first time a living, replicating cell had been synthesized from chemicals. Synthetic biology uses engineering-type methods to assemble molecules and cells. Medical science is poised to be revolutionized when scientists can create precise chemicals to replace those missing in disease, assemble customized immune components, or construct biological molecules that can precisely target cancerous cells or pathogenic microbes. Synthetic biology also holds promise for alternative energy production, for offering new and different manufacturing processes, and for data storage. For example, researchers say that using DNA to store information, they could fit the entire contents of the Internet in a shoebox. Of course, the ability of scientists to "create" life, in a sense, makes many people nervous. The scientific community, and those who monitor it, are engaged in intense conversations about the ethics—as well as security issues—of synthetic biology.

Using Genetic Techniques to Treat Disease

Advances in genetic manipulation have allowed the development of several new strategies to treating diseases such as cancer and other debilitating or life-threatening conditions. Two of these are **gene therapy** and **CRISPR.**

Gene therapy involves replacing a faulty gene that is responsible for disease with a gene from a healthy organism. For example, a blood condition called beta-thalassemia comes about when a gene for beta-globulin is defective. Patients require monthly blood transfusions for survival.

There are various strategies for this type of therapy. In general, the normal gene is cloned in vectors such as retroviruses or adenoviruses that are infectious but relatively harmless. In one technique, tissues can be removed from the patient and incubated with these genetically modified viruses to transfect them with the normal gene. The transfected cells are then reintroduced into the patient's body by transfusion **(figure 8.20)**. Alternatively, naked DNA or a virus vector is directly introduced into the patient's tissues. This is the basis of a successful immunotherapy treatment for melanoma today.

Experimentation with various types of gene therapy, or clinical testing, is performed on human volunteers with the particular genetic condition. Thousands of these trials have been and are being carried out in the United States and other countries. Most trials target cancer, single-gene defects, and infections; and most gene deliveries are carried out by virus vectors. Early therapeutic trials were hampered by several difficulties relating to effectiveness and safety. Some of the safety issues were related to the use of (seemingly safe) viruses as delivery vehicles, which then ended up causing malignancies.

The strategies described so far are called *somatic cell* gene therapy. This means that the changes are permanent in the individual who is treated, but they are not passed on to offspring. The ultimate sort of gene therapy is germline therapy, in which genes are inserted into an egg, sperm, or early embryo. In this type of therapy, the new gene will be present in all cells of the individual. The therapeutic gene is also heritable (that is, can be passed on to subsequent generations).

There is another system found in bacteria and archaea that can be exploited by scientists to alter genomes. The system is called *clustered regularly interspaced short palindromic repeats.* You will see it in the media as CRISPR. In bacteria and archaea, these are short lengths of DNA with repeating nucleotides. After the repeats, short segments of spacer DNA are found. These turn out to be the leftovers of DNA left behind by "invading" bacteriophages or plasmids. The CRISPR areas of the genome are capable of recognizing

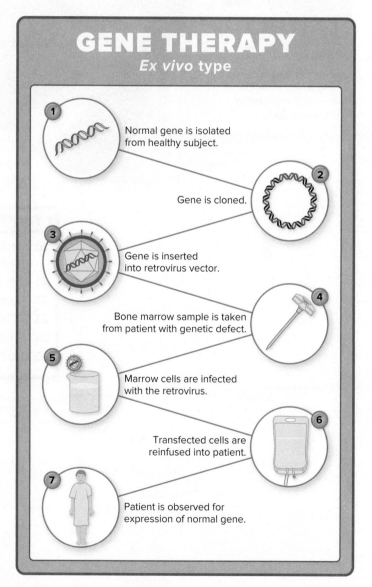

GENE THERAPY
Ex vivo type

1 Normal gene is isolated from healthy subject.

2 Gene is cloned.

3 Gene is inserted into retrovirus vector.

4 Bone marrow sample is taken from patient with genetic defect.

5 Marrow cells are infected with the retrovirus.

6 Transfected cells are reinfused into patient.

7 Patient is observed for expression of normal gene.

Figure 8.20 Protocol for the *ex vivo* type of gene therapy in humans.

and cutting this foreign DNA, keeping the bacterium or archaea from being invaded. It is thought to be an adaptive immune system used by bacteria. In other words, the bacteria "learn" the identity of an attacking phage by placing bits of its DNA in its own genome, and in the future can cut it up before it causes trouble.

The system turns out to be highly adaptable for laboratory use, and scientists have started using CRISPR in many genetic engineering applications. It is cheap, relatively easy to perform, and very powerful. Researchers need only design a correct **guide RNA** that targets specific gene sequences and mix it with nucleases associated with the CRISPR system, and they can cut DNA in just about any organism exactly where they want to. It has already been used to make changes in animals, and human trials began in 2019. Scientists hope to repair blood disorders and cancers by removing malfunctioning cells and applying CRISPR to fix the DNA defects. They are also injecting CRISPR molecules into the eyes of people with diseased retinas. The therapeutic and research potential for this technology is huge. Some scientists warn that we don't yet understand the wide-reaching implications of changing DNA, especially in the germline, and that more caution is needed. Another concern is that the CRISPR system can be used in a process called **gene drive.** In this scenario, CRISPR is used to artificially cause an organism's offspring to accrue a particular mutation at a much accelerated rate. This might be a good thing, for instance, if CRISPR can cause mosquitoes to no longer be susceptible to the malaria protozoan. The fear is that this speeded-up evolution could have many unintended consequences.

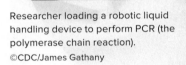

Researcher loading a robotic liquid handling device to perform PCR (the polymerase chain reaction).
©CDC/James Gathany

8.6 LEARNING OUTCOMES—Assess Your Progress

20. Explain the importance of restriction endonucleases to genetic engineering.

21. List the steps in the polymerase chain reaction.

22. Describe how you can clone a gene into a bacterium.

23. Differentiate between DNA profiling and DNA sequencing.

24. Provide a definition of *synthetic biology*.

25. Name two genetic techniques that are designed to treat human diseases.

CASE FILE WRAP-UP

©stockbroker/123RF (health care worker); Steve Gschmeissner/Science Source (blood cells and fibrin)

The clotting cascade is the complex process that leads to formation of a blood clot and prevents uncontrolled bleeding. It involves numerous steps and the activation of multiple clotting factors and other cofactors. Different types of hemophilia inhibit different proteins that form clotting factors. Hemophilia A is characterized by low factor VIII clotting activity. Diagnosis is made by laboratory tests measuring factor levels.

To replace the missing clotting factor, IV infusions are administered. Depending on the severity of hemophilia, patients may require regular scheduled infusions to prevent internal bleeding. Some patients with mild hemophilia only require factor infusions when there are signs of injury or bleeding.

The administered factor VIII is produced using recombinant DNA technology. The synthetic product is nearly identical to that encoded by the human factor VIII gene. The gene coding for the factor is cloned into a plasmid vector and then animal cells in a laboratory are injected with the vector. They then produce mass quantities of the clotting protein. This recombinant factor VIII has advantages over factor products derived from human blood sources, as it cannot transmit potentially dangerous microorganisms.

Host Genetics and the Microbiome

©Pete Collins/iStock/Getty Images

The composition of the human microbiota shows a lot of variability from person to person. Of course, we know that humans themselves show a lot of variation, which comes from their different genetic makeup. This led scientists to wonder whether the composition of the microbiota is influenced by the host's genetics.

One good way to test this is to look at two different types of pairs of people: monozygotic (identical) twins and dizygotic (fraternal) twins. Fraternal twins do not share the same genes, but identical twins do. If the microbiomes of identical twins were significantly more similar than the microbiomes of fraternal twins, it would suggest that the human genome influences what microbiome the person acquires.

To ask this question the way scientists do, you would construct a hypothesis: The degree of difference between the microbiota of fraternal twins will be no greater than the degree of difference between the microbiota of identical twins. (This is written as a null hypothesis, meaning it is a statement that there will be no difference between two groups.) Then you would set up an experiment to test the hypothesis, using a large number of pairs of both types of twins. In this study, 416 pairs of twins were examined.

In this study, the identical twins turned out to have more similar microbiomes than the fraternal twins. They had what they called "a hub of heritable taxa," chief among them a newly discovered bacterial group named *Christensenellaceae*.

So the hypothesis was disproven; there was a significant difference between the two groups. The paper's authors suggest that a person's microbiome is heritable, like having blue eyes. Only here it is a bit more indirect—a person's genotype is heritable, which determines his or her phenotype, which may determine his or her microbiome.

There is a saying in science, "Chance favors the prepared mind." In the case of this study, the scientists found something extra they were not counting on: The presence of *Christensenellaceae* was associated with low body mass index (BMI). Because this was just an association and the study could not prove causation, they did another experiment in which they deliberately exposed mice to *Christensenellaceae*. Those mice had reduced weight gain compared to mice not fed *Christensenellaceae*. So the studies continue. This is what many scientists love about their jobs: discovering surprises, and finding answers to questions that practically ask themselves!

Source: *Cell Host and Microbe*, vol. 19, pp. 731–743, 2016.

MICROBIAL GENETICS AND GENETIC ENGINEERING

8.1 INTRODUCTION TO GENETICS AND GENES

Nucleic acids contain the blueprints of life in the form of genes. DNA is the double-stranded blueprint molecule for all cellular organisms. The blueprints of viruses can be either DNA or RNA. In cells, DNA is replicated through a process that is called semiconservative because a new two-stranded molecule contains one parental strand and one daughter strand.

DNA is transcribed into RNA. Much of the RNA is used to regulate cell function. Other RNAs are in turn translated into proteins, which do much of the work of a cell. Eukaryotes transcribe DNA in the nucleus and then translate it in the cytoplasm. Bacteria perform transcription and translation simultaneously in the cytoplasm.

TRANSCRIPTION AND TRANSLATION 8.2

8.3 GENETIC REGULATION OF PROTEIN SYNTHESIS

If many bacterial genes are involved in a related function, they might be organized into operons. If these units are generally off but can be turned on, they are called inducible. If they are normally on but can be turned off, they are repressible. Other genes are constitutive, which means they are always on.

Recombination refers to DNA from one organism being permanently incorporated into the DNA of another organism. It provides bacteria, which are non-sexual cells, with a mechanism for genetic variation, through processes called horizontal gene transfer. Eukaryotes also use horizontal gene transfer to supplement their sexual reproduction.

DNA RECOMBINATION EVENTS 8.4

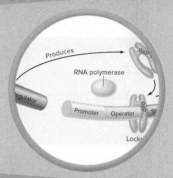

8.5 MUTATIONS: CHANGES IN THE GENETIC CODE

Mutations are changes in the nucleotide sequence of an organism's genome. They can be spontaneous or be caused by exposure to some external agent. All cells have enzymes that repair damaged DNA, but they don't fix all mutations. Mutations provide a mechanism for changes in an organism's properties.

Scientists can use nature's tools in intentional ways to analyze and to manipulate the genetic contents of organisms. DNA can be profiled and sequenced, and it can be cloned and put into new organisms. These methods have led to two disease treatment techniques: gene therapy and CRISPR.

STUDYING DNA IN THE LABORATORY AND GENETIC ENGINEERING 8.6

SmartGrid: From Knowledge to Critical Thinking

This *21 Question Grid* takes the topics from this chapter and arranges them with respect to the American Society for Microbiology's Undergraduate Curriculum guidelines—all six of the important "Concepts" as well as the important "Competency" of scientific literacy. Three questions are supplied about chapter content that refer to the Concept or Competency, in increasing levels of Bloom's taxonomy for learning.

ASM Concept/ Competency	A. Bloom's Level 1, 2—Remember and Understand (Choose one.)	B. Bloom's Level 3, 4—Apply and Analyze	C. Bloom's Level 5, 6—Evaluate and Create

Evolution	1. Single nucleotide polymorphisms are found in
	  a. DNA.
	  b. RNA.
	  c. plasmids.
	  d. siRNA.
		2. Using your knowledge of DNA from this chapter, imagine two different ways antibiotic resistance may develop in a bacterium.	3. Conduct research on CRISPR and explain in layperson's terms why there are ethical concerns around its use in genetic engineering.

Cell Structure and Function	4. Which of the following is a characteristic of RNA?
	  a. RNA is double-stranded.
	  b. RNA contains thymine, which pairs with adenine.
	  c. RNA contains deoxyribose.
	  d. RNA molecules are necessary for translation and gene regulation.
		5. List some advantages and disadvantages to a cell having DNA in the cytoplasm as opposed to in a nucleus.	6. Construct an argument for why tRNA contains a lot of secondary structure but mRNA does not.

Metabolic Pathways	7. The *lac* operon is usually in the _____ position and is activated by a/an _____ molecule.
	  a. on; repressor
	  b. off; inducer
	  c. on; inducer
	  d. off; repressor
		8. Discuss the intersection between the "metabolome" and phenotype.	9. Defend this statement: *All of biology is dependent on binding reactions.*

Information Flow and Genetics	10. DNA is semiconservative because the _____ strand will become half of the _____ molecule.
	  a. RNA; DNA
	  b. template; finished
	  c. sense; mRNA
	  d. codon; anticodon
		11. Examine the DNA triplets here and determine the amino acid sequence they code for. Then provide a different DNA sequence that will produce the same protein.
		TAC CAG ATA CAC TCC CCT
			12. Using the same piece of DNA in question 11 (TAC CAG ATA CAC TCC CCT), create the following:
			(i) a frameshift mutation; (ii) a silent mutation; and (iii) a nonsense mutation.

Microbial Systems	13. Which of the following is not a mechanism of horizontal gene transfer?
	  a. spontaneous mutations
	  b. transformation
	  c. transduction
	  d. conjugation
		14. Why is it advisable to use reverse transcriptase to create DNA from an mRNA transcript when preparing eukaryotic genes for use in bacterial cells? Why not use the eukaryotic DNA directly?	15. Metagenomics is providing insight into the possible causes of some difficult-to-understand diseases, such as irritable bowel syndrome. Develop a possible explanation for why metagenomic approaches might be better than traditional diagnostic approaches.

ASM Concept/ Competency	A. Bloom's Level 1, 2—Remember and Understand	B. Bloom's Level 3, 4—Apply and Analyze	C. Bloom's Level 5, 6—Evaluate and Create

Impact of Microorganisms	16. The creation of biological molecules and cells entirely from chemicals is called
	  a. recombination.
	  b. sequencing.
	  c. artificial biology.
	  d. synthetic biology.
		17. You are a public health official trying to determine if 20 different people in your city were sickened by the same microbe, or if they were unrelated. Which technique from this chapter would you use, and why?	18. Genetically modified organisms (GMOs)—especially in foods—are considered by popular culture to be "bad." Construct a counterargument.

Scientific Thinking	19. How many cycles of PCR would it take to create 16 copies of a DNA molecule?
	  a. two
	  b. three
	  c. four
	  d. eight
		20. Construct an analogy using your clothes closet to describe the difference between genotype and phenotype.	21. What experiment could you devise to determine if a particular phage engaged in general or specialized transduction?

Answers to the multiple-choice questions appear in Appendix A.

Visual Connections

This question connects previous images to a new concept.

From chapter 3, figure 3.10. Speculate on why these cells contain two chromosomes (shown in blue).

©Eye of Science/Science Source

9

Physical and Chemical Control of Microbes

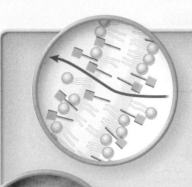

IN THIS CHAPTER...

9.1 Controlling Microorganisms

1. Clearly define the terms *sterilization, disinfection, decontamination, sanitization, antisepsis,* and *degermation.*
2. Identify the microorganisms that are most resistant and least resistant to control measures.
3. Compare the action of microbicidal and microbistatic agents.
4. Name four categories of cellular targets for physical and chemical agents.

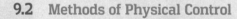

9.2 Methods of Physical Control

5. Name six methods of physical control of microorganisms.
6. Discuss both moist and dry heat methods, and identify multiple examples of each.
7. Define *thermal death time* and *thermal death point.*
8. Explain methods of moist heat control.
9. Explain two methods of dry heat control.
10. Identify advantages and disadvantages of cold and desiccation.
11. Differentiate between the two types of radiation control methods.
12. Explain how filtration and osmotic pressure function as control methods.

9.3 Methods of Chemical Control

13. Name the desirable characteristics of chemical control agents.
14. Discuss chlorine and iodine and their uses.
15. List advantages and disadvantages to phenolic compounds.
16. Explain the mode of action of chlorhexidine.
17. Explain the applications of oxidizing agents.
18. Identify some heavy metal control agents.
19. Discuss the disadvantages of aldehyde agents.
20. Identify applications for ethylene oxide sterilization.

ERproductions Ltd/Blend Images LLC (forceps); Charles D. Winters/McGraw-Hill Education (boiling water); Kathy Park Talaro (heavy metals); shapecharge/E+/Getty Images (health care worker)

CASE FILE

Control of Microorganisms in the Operating Room

As a surgical technician in the operating room, I assist with all phases of surgical procedures. Before the surgery begins, I prepare the operation suite. During the procedure, I assist the surgeons by ensuring they have the supplies needed. Following the procedure, I prepare surgical supplies to be appropriately cleaned or disposed.

One day, I was the surgical tech for a kidney transplant. The patient was a 55-year-old male with chronic kidney disease who was receiving a donor organ from a relative. Prior to the surgery, I met with the family and explained my role in the procedure. I made sure that their questions were answered and appropriate consent was obtained.

Prior to the procedure, I prepared the operating room with the supplies necessary for surgery. First, I cleaned large metal carts with a bleach solution. I spread large plastic sterile drapes over the carts, ensuring that I did not touch or contaminate them. I opened the packages containing trays of sterilized equipment and, using proper technique, transferred them to the tables. For this procedure, even ice made from sterile water was prepared.

When the patient was in the operating room and we were ready to begin the procedure, I donned surgical scrubs, a mask, hair covering, and shoe covers. I then scrubbed my hands, arms, and elbows with antimicrobial soap and water, followed by alcohol-based hand cleaner, and finally a scrub brush with antiseptic. Keeping my hands raised in the air, I was assisted by another nurse in putting on a sterile gown and gloves. I then used an iodine-impregnated sponge to scrub the patient's surgical site.

During the procedure, because my personal protective equipment was sterile, I only touched supplies that were on the sterile field and table. Throughout the case, I handed the necessary supplies to the surgeons and kept the equipment organized and counted. If supplies were needed that were not on the operating table, a circulating nurse obtained them and opened the package onto the sterile field in a manner that prevented contamination of the contents. I was then able to handle the supplies and give them to the surgeons as they were needed.

- Why is sterile technique in the operating room so important?
- How is equipment sterilized to be reused between cases?

Case File Wrap-Up appears at the end of the chapter.

9.1 Controlling Microorganisms

Much of the time in the developed world, we take for granted tap water that is drinkable, food that is not spoiled, shelves that are full of products to eradicate "germs," and drugs that are always available to treat infections. Controlling our degree of exposure to potentially harmful microbes is a monumental concern in our lives. The ancient Greeks learned to burn corpses and clothing during epidemics and the Egyptians embalmed the bodies of their dead, using strong salts and pungent oils. These methods may seem rather archaic by modern measures, but these examples illustrate that controlling microbes has been a concern for several centuries.

The methods of microbial control used outside of the body are designed to result in four possible outcomes:

1. sterilization,
2. disinfection,
3. decontamination (also called sanitization), or
4. antisepsis (also called degermation).

These terms are categorized in **table 9.1.** While it may seem clumsy to have more than one word for some of these processes, it is important that you recognize them when you hear them used so we include them here. To complicate matters, the everyday use of some of these terms can at times be vague and inexact. For example, occasionally one may be directed to "sterilize" or "disinfect" a patient's skin, even though this usage does not fit the technical definition of either term. We also provide a flowchart **(figure 9.1)** to summarize the major applications and aims in microbial control.

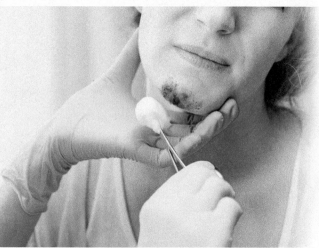
©Photographee.eu/Shutterstock

Table 9.1 Concepts in Antimicrobial Control

Techniques and chemicals that are capable of sterilizing are highlighted with a pink background.

Term	Definition	Key Points	Examples of Agents
Sterilization	Process that destroys or removes all viable microorganisms (including viruses)	The term *sterile* should be used only in the strictest sense to refer to materials that have been subjected to the process of sterilization (there is no such thing as slightly sterile). Generally reserved for inanimate objects as it would be impractical or dangerous to sterilize parts of the human body Common uses: surgical instruments, syringes, commercially packaged food	Heat (autoclave) Sterilants (chemical agents capable of destroying endospores)
Disinfection	Physical process or a chemical agent to destroy vegetative pathogens but not bacterial endospores Removes harmful products of microorganisms (toxins) from material	Normally used on inanimate objects because the concentration of disinfectants required to be effective is harmful to human tissue Common uses: boiling food utensils, applying 5% bleach solution to an examining table, immersing thermometers in an iodine solution between uses	Bleach Iodine Heat (boiling)
Decontamination/ Sanitization	Cleansing technique that mechanically removes microorganisms as well as other debris to reduce contamination to safe levels	Important to restaurants, dairies, breweries, and other commercial entities that handle large numbers of soiled utensils/containers Common uses: Cooking utensils, dishes, bottles, and cans must be sanitized for reuse.	Soaps Detergents Commercial dishwashers
Antisepsis/ Degermation	Reduces the number of microbes on the human skin A form of decontamination but on living tissues	Involves scrubbing the skin (mechanical friction) or immersing it in chemicals (or both)	Alcohol Surgical hand scrubs

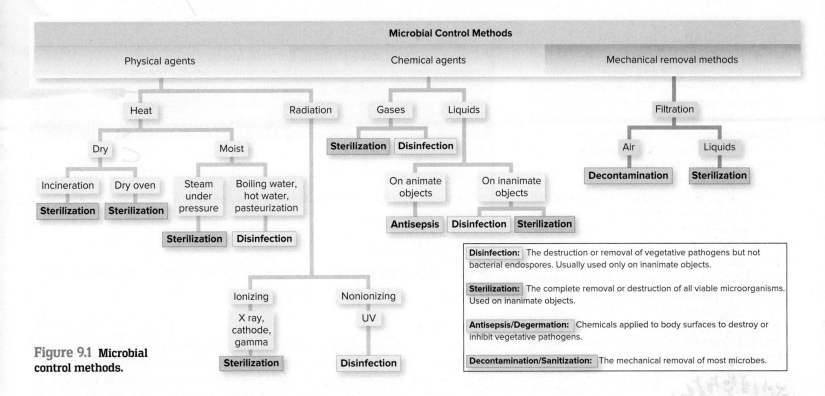

Figure 9.1 Microbial control methods.

Relative Resistance of Microbial Forms

The primary targets of microbial control are microorganisms capable of causing infection or spoilage that are present in the external environment and on the human body. This targeted population is rarely simple or uniform. It often contains mixtures of microbes with extreme differences in resistance and harmfulness. **Figure 9.2** compares the general resistance these forms have to physical and chemical methods of control.

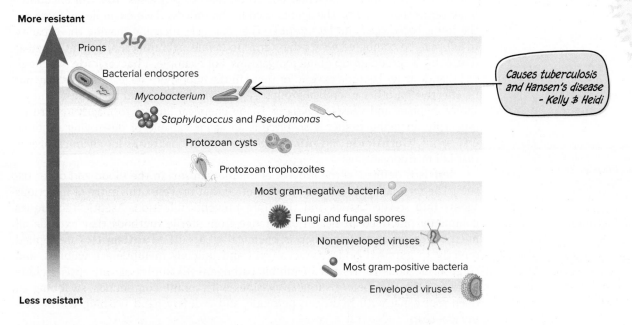

Figure 9.2 Relative resistance of different microbial types to microbial control agents. This is a very general hierarchy; different control agents are more or less effective against the various microbes.

A Note About Prions

Prions are in a class of their own when it comes to "sterilization" procedures. This chapter defines "sterile" as the absence of all viable microbial life—but none of the procedures described in this chapter are necessarily sufficient to destroy prions. Prions are extraordinarily resistant to heat and chemicals. If instruments or other objects become contaminated with these unique agents, they must either be discarded as biohazards or, if this is not possible, a combination of chemicals and heat must be applied in accordance with the Centers for Disease Control and Prevention (CDC) guidelines. The guidelines themselves are constantly evolving as new information becomes available. In the meantime, this chapter discusses sterilization using bacterial endospores as the toughest form of microbial life. When tissues, fluids, or instruments are suspected of containing prions, consultation with infection control experts and/or the CDC is recommended when determining effective sterilization conditions. Chapter 17 describes prions in detail.

Table 9.2 Comparative Resistance of Bacterial Endospores and Vegetative Cells to Control Agents

Method	Required to Destroy Endospores	Required to Destroy Vegetative Forms	Endospores Are ___× More Resistant
Heat (moist)	120°C	80°C	1.5
Radiation (X-ray) dosage	4,000 Grays	1,000 Grays	4
Sterilizing gas (ethylene oxide)	1,200 mg/L	700 mg/L	1.7
Sporicidal liquid (2% glutaraldehyde)	3 h	10 min	18

Actual comparative figures on the requirements for destroying various groups of microorganisms are shown in **table 9.2.** Bacterial endospores have traditionally been considered the most resistant microbial entities, being as much as 18 times harder to destroy than their counterpart vegetative cells. Because of their resistance to microbial control, their destruction is the goal of *sterilization* because any process that kills endospores will invariably kill all less-resistant microbial forms. (Prions are a special case; see note in the margin.) Other methods of control (disinfection, antisepsis) act primarily upon microbes that are less hardy than endospores.

Agents versus Processes

The terms *sterilization, disinfection,* and so on refer to processes. You will encounter other terms that describe the agents used in the process. Two examples of these are the terms *bactericidal* and *bacteristatic.* The root *-cide,* meaning "having the capacity to kill," can be combined with other terms to define an antimicrobial agent aimed at destroying a certain group of microorganisms. For example, a **bactericide** is a chemical that destroys bacteria except for those in the endospore stage. It may or may not be effective on other microbial groups. A *fungicide* is a chemical that can kill fungal spores, hyphae, and yeasts. A *virucide* is any chemical known to inactivate viruses, especially on living tissue. A *sporicide* is an agent capable of destroying bacterial endospores. **Germicide** and **microbicide** are additional terms for chemical agents that kill microorganisms.

Sepsis is defined as the growth of microorganisms in the blood and other tissues. The term **asepsis** refers to any practice that prevents the entry of infectious agents into sterile tissues and therefore prevents infection. Aseptic techniques commonly practiced in health care range from sterile methods that exclude all microbes to *antisepsis.* In antisepsis, chemical agents called **antiseptics** are applied directly to exposed body surfaces (skin and mucous membranes), wounds, and surgical incisions to destroy or inhibit pathogens. Examples of antisepsis include preparing the skin before surgical incisions with iodine compounds, swabbing an open root canal with hydrogen peroxide, and ordinary hand washing with a germicidal soap.

The Greek words *stasis* and *static* mean "to stand still." They can be used in combination with various prefixes to indicate a state in which microbes are temporarily prevented from multiplying but are not killed outright. Although killing or permanently inactivating microorganisms is the usual goal of microbial control, microbistasis does have meaningful applications. **Bacteristatic** agents prevent the growth of bacteria on tissues or on objects in the environment, and *fungistatic* chemicals inhibit fungal growth. Materials used to control microorganisms in the body (antiseptics and drugs) often have **microbistatic** effects because many microbicidal compounds can be highly toxic to human cells. Note that even a *-cidal* agent doesn't necessarily result in sterilization, depending on how it is used.

Practical Matters in Microbial Control

Numerous factors should be taken into consideration when selecting a workable method of microbial control. One useful framework for determining how devices that come in contact with patients should be handled is whether they are considered *critical, semicritical,* or *noncritical. Critical* medical devices are those that are expected to come into contact with sterile tissues. A good example of this would be a syringe needle or an artificial hip. These must be sterilized before use. *Semicritical* devices are those that come into contact with mucosal membranes. An endoscopy tube is an example. These must receive at least high-level disinfection and preferably should be sterilized. *Noncritical* items are those that do not touch the patient or are only expected to touch intact skin, such as blood pressure cuffs or crutches. They require only low-level disinfection unless they become contaminated with blood or body fluids.

A remarkable variety of substances can require sterilization. They range from durable solids such as rubber to sensitive liquids such as serum. Hundreds of situations requiring sterilization confront the network of persons involved in health care, whether technician, nurse, doctor, or manufacturer, and no universal method works well in every case.

Considerations such as cost, effectiveness, and method of disposal are all important. For example, disposable plastic items such as catheters and syringes that are used in invasive medical procedures have the potential for infecting the tissues. These must be sterilized during manufacture by a nonheating method (gas or radiation) because heat can damage plastics. After these items have been used, it is often necessary to destroy or decontaminate them before they are discarded because of the potential risk to the handler. Steam sterilization (autoclaving), which is quick and sure, is a sensible choice at this point because it does not matter if the plastic is destroyed. As we become more aware of the plastic pollution problem, it remains to be seen how these practices will change.

What Is Microbial Death?

Death is a phenomenon that involves the permanent termination of an organism's vital processes. Signs of life in complex organisms such as animals are self-evident, and death is made clear by loss of nervous function, respiration, or heartbeat. In contrast, death in microscopic organisms that are composed of just one or a few cells is often hard to detect because they reveal no conspicuous vital signs to begin with. The permanent loss of reproductive capability, even under optimum growth conditions, has become the accepted microbiological definition of death.

Factors Affecting Death Rate

The individual cells of a culture show marked variation in susceptibility to a given microbicidal agent. Death of the whole population is not instantaneous but begins

Crops, like these soybeans, are grown using fungicidal chemicals.
©Doug Sherman/Geofile

when a certain threshold of microbicidal agent (some combination of time and concentration) is met. Death continues in a logarithmic manner as the time or concentration of the agent is increased.

Because many microbicidal agents target the cell's metabolic processes, active cells (younger, rapidly dividing) tend to die more quickly than those that are less metabolically active (older, inactive). Eventually, a point is reached at which survival of any cells is highly unlikely. This point is equivalent to sterilization.

The effectiveness of a particular agent is governed by several factors besides time. These additional factors influence the action of antimicrobial agents:

1. The number of microorganisms. A higher load of contaminants requires more time to destroy.
2. The nature of the microorganisms in the population. In most actual circumstances of disinfection and sterilization, the target population is not a single species of microbe but a mixture of bacteria, fungi, endospores, and viruses, presenting a broad spectrum of microbial resistance.
3. The temperature and pH of the environment.
4. The concentration (dosage, intensity) of the agent. For example, ultraviolet (UV) radiation is most effective at 260 nanometers (nm), and most disinfectants are more active at higher concentrations.
5. The mode of action of the agent. How does it kill or inhibit the microorganism?
6. The presence of solvents, interfering organic matter, and inhibitors. Saliva, blood, and feces can inhibit the actions of disinfectants and even of heat.

The influence of these factors is discussed in greater detail in subsequent sections.

Modes of Action of Antimicrobial Agents

An antimicrobial agent's adverse effect on cells is known as its *mode* (or *mechanism) of action*. Agents affect one or more cellular targets, inflicting damage progressively until the cell is no longer able to survive. Antimicrobials have a range of cellular targets, with the agents that are least selective in their targeting tending to be effective against the widest range of microbes (examples include heat and radiation). More selective agents (drugs, for example) tend to target only a single cellular component and are much more restricted as to the microbes they are effective against.

The cellular targets of physical and chemical agents fall into four general categories:

1. the cell wall,
2. the cytoplasmic membrane,
3. cellular synthetic processes (DNA, RNA), and
4. proteins.

Table 9.3 depicts the effects of various agents on cellular structures and processes. For example, **figure 9.3** illustrates what happens to a membrane when it is exposed to surfactants.

Table 9.3 Actions of Various Physical and Chemical Agents upon the Cell

Cellular Target	Effects of Agents	Examples of Agents Used
Cell wall	Chemical agents can damage the cell wall by • blocking its synthesis or • digesting it	Chemicals Detergents Alcohol
Cytoplasmic membrane	Agents disrupt the lipid layer of the cytoplasmic membrane. This opens up the cytoplasmic membrane and allows damaging chemicals to enter the cell and important ions to exit the cell.	Detergents Alcohol
Cellular synthesis	Agents can interrupt the synthesis of proteins via the ribosomes, inhibiting proteins needed for growth and metabolism and preventing multiplication. Agents can change genetic codes (mutation).	Formaldehyde Radiation Ethylene oxide
Proteins	Some agents are capable of denaturing proteins (breaking of protein bonds, which results in breakdown of the protein structure). Agents may attach to the active site of a protein, preventing it from interacting with its chemical substrate.	Moist heat Alcohol Phenolics

COVID-19

Alcohol-based hand sanitizers are a mainstay of disinfection of surfaces and antisepsis of hands during the COVID-19 pandemic. SARS-CoV-2 is an enveloped virus, and its envelope is susceptible to both detergents and alcohol.

9.1 LEARNING OUTCOMES—Assess Your Progress

1. Clearly define the terms *sterilization, disinfection, decontamination, sanitization, antisepsis,* and *degermation.*
2. Identify the microorganisms that are most resistant and least resistant to control measures.
3. Compare the action of microbicidal and microbistatic agents.
4. Name four categories of cellular targets for physical and chemical agents.

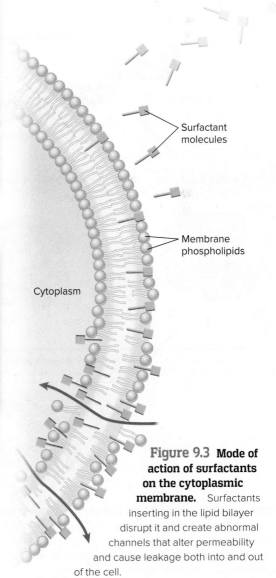

Figure 9.3 Mode of action of surfactants on the cytoplasmic membrane. Surfactants inserting in the lipid bilayer disrupt it and create abnormal channels that alter permeability and cause leakage both into and out of the cell.

Surfactant molecules

Membrane phospholipids

Cytoplasm

©Bruce Chambers, The Orange County Register/Newscom

9.2 Methods of Physical Control

We can divide our methods of controlling microorganisms into two broad categories: physical and chemical. We'll start with physical methods. Microorganisms have adapted to the tremendous diversity of habitats the earth provides, even severe conditions of temperature, moisture, pressure, and light. For microbes that normally withstand such extreme physical conditions, our attempts at control would probably have little effect. Fortunately for us, we are most interested in controlling microbes that flourish in the same environment in which humans live. The vast majority of these microbes are readily controlled by abrupt changes in their environment. Most prominent among antimicrobial physical agents is heat. Other agents include radiation, filtration, ultrasonic waves, and even cold. The following sections examine some of these methods and explore their practical applications in medicine, commerce, and the home.

Heat

As a rule, elevated temperatures (exceeding the maximum growth temperature) are microbicidal, whereas lower temperatures (below the minimum growth temperature) are microbistatic. We'll start with heat. Heat can be applied in either moist or dry forms. *Moist heat* occurs in the form of hot water, boiling water, or steam (vaporized water). In practice, the temperature of moist heat usually ranges from 60°C to 135°C. As we shall see, the temperature of steam can be regulated by adjusting its pressure in a closed container. *Dry heat* refers to hot air (such as in an oven) or an open flame. In practice, the temperature of dry heat ranges from 160°C to several thousand degrees Celsius.

Mode of Action and Relative Effectiveness of Heat

Moist heat and dry heat differ in their modes of action as well as in their efficiency. Moist heat can achieve the same effectiveness as dry heat but with lower temperatures and shorter exposure times **(table 9.4).** Although many cellular structures are damaged by moist heat, its most lethal effect is the coagulation and denaturation of proteins, which quickly and permanently halts cellular metabolism.

Dry heat dehydrates the cell, removing the water necessary for metabolic reactions, and it also denatures proteins **(figure 9.4).** However, the lack of water actually increases the stability of some protein conformations, necessitating the use of higher temperatures when dry heat is employed as a method of microbial control. At very high temperatures, dry heat oxidizes cells, burning them to ashes. This method is the one used in the laboratory when a loop is flamed or in industry when medical waste is incinerated.

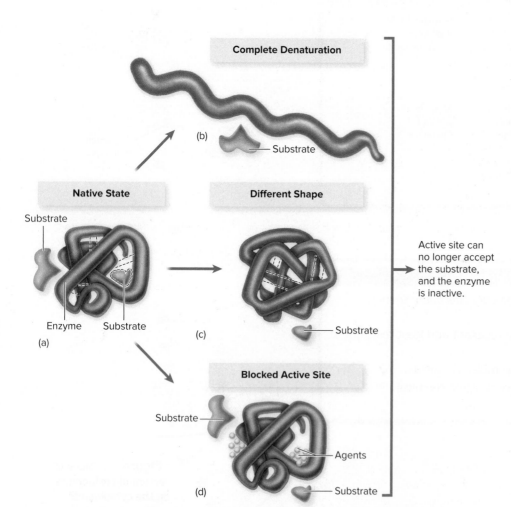

Figure 9.4 The action of heat and chemicals on proteins.   (a) The native (functional) state is maintained by bonds that create active sites to fit the substrate. Heat and agents denature the protein by breaking all or some secondary and tertiary bonds. Results are **(b)** complete unfolding or **(c)** random bonding and incorrect folding. **(d)** Some agents react with functional groups on the active site and interfere with bonding.

Table 9.4 Comparison of Times and Temperatures to Achieve Sterilization with Moist and Dry Heat

	Temperature (°C)	Time to Sterilize (Min)
Moist heat	121	15
	125	10
	134	3
Dry heat	121	600
	140	180
	160	120
	170	60

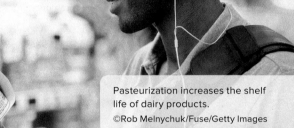

Pasteurization increases the shelf life of dairy products.
©Rob Melnychuk/Fuse/Getty Images

Heat Resistance and Thermal Death: Endospores and Vegetative Cells

Bacterial endospores exhibit the greatest resistance and the vegetative forms of bacteria and fungi are the least resistant to both moist and dry heat. Destruction of endospores requires temperatures above boiling.

Vegetative cells also vary in their sensitivity to heat. Among bacteria, the death times with moist heat range from 50°C for 3 minutes (*Neisseria gonorrhoeae*) to 60°C for 60 minutes (*Staphylococcus aureus*). It is worth noting that vegetative cells of endospore formers are just as susceptible as vegetative cells of non-endospore-formers, and that pathogens are neither more nor less susceptible than nonpathogens. Other microbes, including fungi, protozoa, and worms, are rather similar in their sensitivity to heat. Viruses are surprisingly resistant to heat, with a tolerance range extending from 55°C for 2 to 5 minutes (adenoviruses) to 60°C for 600 minutes (hepatitis A virus).

Susceptibility of Microbes to Heat: Thermal Death Measurements

A combination of the two variables, heat and time, constitutes the **thermal death time,** or **TDT,** defined as the shortest length of time required to kill all test microbes at a specified temperature. The TDT has been experimentally determined for the microbial species that are common or important contaminants in various heat-treated materials. Another way to compare the susceptibility of microbes to heat is the **thermal death point (TDP),** defined as the lowest temperature required to kill all microbes in a sample in 10 minutes.

Easy to confuse. Bear down on these!
Kelly & Heidi

Many perishable substances are processed with moist heat. Some of these products are intended to remain on the shelf at room temperature for several months or even years. The chosen heat treatment must render the product free of agents of spoilage or disease. At the same time, the quality of the product and the speed and cost of processing must be considered. For example, in the commercial preparation of canned green beans, one of the manufacturer's greatest concerns is to prevent growth of botulism bacteria. From several possible TDTs (i.e., combinations of time and temperature) for *Clostridium botulinum* endospores, the factory must choose one that kills all endospores but does not turn the beans to mush. Out of these many considerations emerges an optimal TDT for a given processing method. Commercial canneries heat low-acid foods at 121°C for 30 minutes, a treatment that sterilizes these foods. Because of such strict controls in canneries, cases of botulism due to commercially canned foods are rare.

 Note: In all of the tables in this chapter, agents that are capable of *sterilizing* will feature a red/pink background.

Moist Heat Control Methods

The three ways that moist heat is employed to control microbes are described in **table 9.5.**

Table 9.5 Moist Heat Methods

Techniques and chemicals that are capable of sterilizing are highlighted with a pink background.

Method	Applications

©Charles D. Winters/McGraw-Hill Education

Boiling Water: Disinfection   A simple boiling water bath can quickly decontaminate items in the clinic and home. Because a single processing at 100ºC will not kill all resistant cells, this method can be relied on only for disinfection and not for sterilization. Exposing materials to boiling water for 30 minutes will kill most non-endospore-forming pathogens, including resistant species such as the tubercle bacillus and staphylococci. Probably the greatest disadvantage with this method is that the items can be easily recontaminated when removed from the water.

Useful in the home for disinfection of water, materials for babies, food and utensils, bedding, and clothing from the sickroom

©James King-Holmes/Science Source

Pasteurization: Disinfection of Beverages   Fresh beverages such as milk, fruit juices, beer, and wine are easily contaminated during collection and processing. Because microbes have the potential for spoiling these foods or causing illness, heat is frequently used to reduce the microbial load and destroy pathogens. **Pasteurization** is a technique in which heat is applied to liquids to kill potential agents of infection and spoilage, while at the same time retaining the liquid's flavor and nutritional value.

Ordinary pasteurization techniques require special heat exchangers that expose the liquid to 71.6ºC for 15 seconds (flash method) or to 63ºC to 66ºC for 30 minutes (batch method). The first method is preferable because it is less likely to change flavor and nutrient content, and it is more effective against certain resistant pathogens such as *Coxiella* and *Mycobacterium*. Although these treatments inactivate most viruses and destroy the vegetative stages of 97% to 99% of bacteria and fungi, they do not kill endospores or particularly heat-resistant microbes. Milk is not sterile after regular pasteurization. In fact, it can contain 20,000 microbes per milliliter or more, which explains why even an unopened carton of milk will eventually spoil. (Newer techniques can also produce *sterile milk* that has a storage life of 3 months. This milk is processed with ultrahigh temperature [UHT]–134ºC– for 1 to 2 seconds.) This is not generally considered pasteurization, so we don't consider pasteurization a sterilization method.

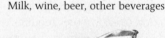
Milk, wine, beer, other beverages

©John A. Rizzo/Getty Images

Table 9.5 (continued)

Method	Applications
Steam Under Pressure: Autoclaving At sea level, normal atmospheric pressure is 15 pounds per square inch (psi), or 1 atmosphere. At this pressure, water will boil (change from a liquid to a gas) at 100°C, and the resultant steam will remain at exactly that temperature, which is unfortunately too low to reliably kill all microbes. In order to raise the temperature of steam, the pressure at which it is generated must be increased. As the pressure is increased, the temperature at which water boils and the temperature of the steam produced both rise. For example, at a pressure of 20 psi (5 psi above normal), the temperature of steam is 109°C. As the pressure is increased to 10 psi above normal, the steam's temperature rises to 115°C, and at 15 psi above normal (a total of 2 atmospheres), it will be 121°C. It is not the pressure by itself that is killing microbes but the increased temperature it produces.	Heat-resistant materials such as glassware, cloth (surgical dressings), metallic instruments, liquids, paper, some media, and some heat-resistant plastics. If items are heat-sensitive (plastic Petri dishes) but will be discarded, the autoclave is still a good choice. However, it is ineffective for sterilizing substances that repel moisture (oils, waxes), or for those that are harmed by it (powders).

Such pressure-temperature combinations can be achieved only with a special device that can subject pure steam to pressures greater than 1 atmosphere. Health and commercial industries use an **autoclave** for this purpose, and a comparable home appliance is the pressure cooker. The most efficient pressure-temperature combination for achieving sterilization is 15 psi, which yields 121°C. It is important to avoid overpacking materials or haphazardly loading the chamber, which prevents steam from circulating freely around the contents and prevents the full contact that is necessary. The duration of the process is adjusted according to the bulkiness of the items in the load (thick bundles of material or large flasks of liquid) and how full the chamber is. The range of autoclaving times varies from 10 minutes for light loads to 40 minutes for heavy or bulky ones; the average time is 20 minutes.

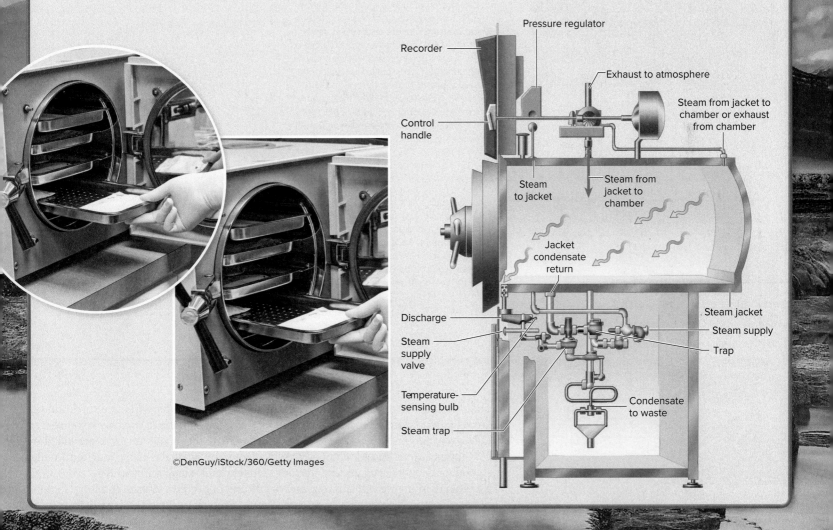

©DenGuy/iStock/360/Getty Images

Dry Heat: Hot Air and Incineration

Dry heat is not as versatile or as widely used as moist heat, but it has several important sterilization applications. The temperatures and times employed in dry heat vary according to the particular method, but in general, they are higher than with moist heat. **Table 9.6** describes the two methods.

Table 9.6 Dry Heat Methods
Techniques and chemicals that are capable of sterilizing are highlighted with a pink background.

Method		Applications
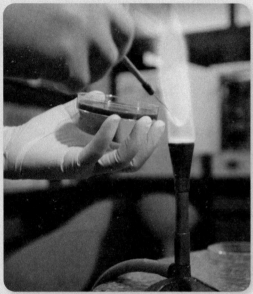 ©UIG/Getty Images	**Incineration** in a flame is perhaps the most rigorous of all heat treatments. The flame of a Bunsen burner reaches 1,870°C at its hottest point, and furnaces/incinerators operate at temperatures of 800°C to 6,500°C. Direct exposure to such intense heat ignites and reduces microbes and other substances to ashes and gas. Incineration of microbial samples on inoculating loops and needles using a Bunsen burner is a very common practice in the microbiology laboratory. This method is fast and effective, but it is also limited to metals and heat-resistant glass materials. This method also presents hazards to the operator (an open flame) and to the environment (contaminants on needle or loop often spatter when placed in flame). Tabletop infrared incinerators have replaced Bunsen burners in many labs for these reasons. Large incinerators are regularly employed in hospitals and research labs for complete destruction of infectious materials.	Bunsen burners/small incinerators: laboratory instruments such as inoculating loops. Large incinerators: syringes, needles, culture materials, dressings, bandages, bedding, animal carcasses, and pathology samples.
 ©RayArt Graphics/Alamy	**The hot-air oven** provides another means of dry-heat sterilization. The so-called *dry oven* is usually electric (occasionally gas) and has coils that radiate heat within an enclosed compartment. Heated, circulated air transfers its heat to the materials in the oven. Sterilization requires exposure to 150°C to 180°C for 2 to 4 hours, which ensures thorough heating of the objects and destruction of endospores.	Glassware, metallic instruments, powders, and oils that steam does not penetrate well. Not suitable for plastics, cotton, and paper, which may burn at the high temperatures, or for liquids, which will evaporate.

The Effects of Cold and Desiccation

The principal benefit of cold treatment is to slow down the growth of cultures and microbes in food during processing and storage. *Remember that cold can only be counted on to be microbistatic, not microbicidal.* Although it is true that some microbes are killed by cold temperatures, most are not adversely affected by gradual cooling, long-term refrigeration, or deep-freezing. In fact, freezing temperatures, ranging from −70°C to −135°C, are often used in research labs to preserve cultures of bacteria, viruses, and fungi for long periods. Some psychrophiles grow very slowly even at freezing temperatures and can continue to secrete toxic products. Ignorance of these facts is probably responsible for

©PhotoAlto/PunchStock

A wide variety of foods are irradiated to control microbial growth.
Source: USDA—Agricultural Research Service

numerous cases of food poisoning from frozen foods that have been defrosted at room temperature and then inadequately cooked. Pathogens able to survive several months in the refrigerator are *Staphylococcus aureus*, *Clostridium* species (endospore formers), *Streptococcus* species, and several types of yeasts, molds, and viruses. Outbreaks of *Salmonella* food infection traced back to refrigerated foods such as ice cream, eggs, and tiramisu are testimony to the inability of cold temperatures to reliably kill pathogens.

Vegetative cells directly exposed to normal room air gradually become dehydrated, or **desiccated.** Delicate pathogens such as *Streptococcus pneumoniae*, the spirochete of syphilis, and *Neisseria gonorrhoeae* can die after a few hours of air drying, but many others are not killed and some are even preserved. Endospores of *Bacillus* and *Clostridium* are viable for thousands of years under extremely dry conditions. Staphylococci and streptococci in dried secretions and the microbe that causes tuberculosis surrounded by sputum can remain viable in air and dust for lengthy periods. Many viruses (especially nonenveloped) and fungal spores can also withstand long periods of desiccation. Desiccation can be a valuable way to preserve foods because it greatly reduces the amount of water available to support microbial growth.

It is interesting to note that a combination of freezing and drying–called **lyophilization** (ly-off″-il-ih-za′-shun)–is a common method of *preserving* microorganisms and other cells in a viable state for many years. Pure cultures are frozen instantaneously and exposed to a vacuum that rapidly removes the water (it goes straight from the frozen state into the vapor state, skipping the liquid state altogether). This method avoids the formation of ice crystals that would damage the cells. Although not all cells survive this process, enough of them do to permit future reconstitution of that culture.

As a general rule, chilling, freezing, and desiccation should not be construed as methods of disinfection or sterilization because their antimicrobial effects are erratic and uncertain, and one cannot be sure that pathogens subjected to them have been killed.

Radiation

Energy, in the form of radiation, is a useful source of antimicrobial activity. **Radiation** is defined as energy emitted from atomic activities and dispersed at high velocity through matter or space. **Figure 9.5** illustrates the different wavelengths of radiation. In our discussion, we consider only those types suitable for microbial control: gamma rays, X rays, and ultraviolet radiation. These are the forms with shorter wavelengths than those of visible light. There are several especially useful practices that take advantage of these forms of radiation **(table 9.7).** Hospitals have begun using specialized UV-light machines to clean hospital rooms after patients have been released and the rooms have been cleaned with traditional methods **(figure 9.6)**. The UV apparatus must be moved to more than one position so that the radiation reaches each area of the room, but it can disinfect areas that are not usually reached by hand cleaning methods.

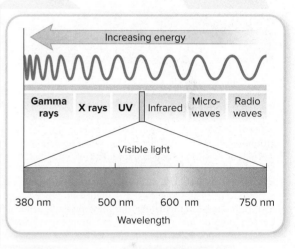

Figure 9.5 The electromagnetic spectrum, showing different types of radiation.

©tatadonets/123RF

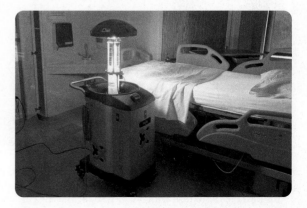

Figure 9.6 A UV light disinfection unit for hospital rooms.
©Essdras M Suarez/The Boston Globe/Getty Images

Table 9.7 Radiation Methods

Techniques and chemicals that are capable of sterilizing are highlighted with a pink background.

Method	Applications

©Adam Hart-Davis/Science Source

Ionizing Radiation: Gamma Rays and X Rays Ionizing radiation is a highly effective alternative for sterilizing materials that are sensitive to heat or chemicals. Devices that emit ionizing rays include gamma-ray machines containing radioactive cobalt, X-ray machines similar to those used in medical diagnosis, and cathode-ray machines. Items are placed in these machines and irradiated for a short time with a carefully chosen dosage. The dosage of radiation is measured in *Grays* (which has replaced the older term, *rads*). Depending on the application, exposure ranges from 5 to 50 kiloGrays (kGray; a kiloGray is equal to 1,000 Grays). Although all ionizing radiation can penetrate liquids and most solid materials, gamma rays are most penetrating, X rays are intermediate, and cathode rays are least penetrating.

Foods have been subject to **irradiation** in limited circumstances for more than 50 years. From flour to pork and ground beef, to fruits and vegetables, radiation is used to kill not only bacterial pathogens but also insects and worms and even to inhibit the sprouting of white potatoes.

Irradiation may lead to a small decrease in the amount of thiamine (vitamin B_1) in food, but this change is small enough to be inconsequential. The irradiation process does produce short-lived free radical oxidants, which disappear almost immediately (this same type of chemical intermediate is produced through cooking as well). Certain foods do not irradiate well and are not good candidates for this type of antimicrobial control. The white of eggs becomes milky and liquid, grapefruit gets mushy, and alfalfa seeds do not germinate properly. Lastly, it is important to remember that food is not made radioactive by the irradiation process, and many studies, in both animals and humans, have concluded that there are no ill effects from eating irradiated food. In fact, NASA relies on irradiated meat for its astronauts.

Drugs, vaccines, medical instruments (especially plastics), syringes, surgical gloves, tissues such as bone and skin, and heart valves for grafting.

After the anthrax attacks of 2001, mail delivered to certain Washington, D.C., ZIP codes was irradiated with ionizing radiation. Its main advantages include speed, high penetrating power (it can sterilize materials through outer packages and wrappings), and the absence of heat. Its main disadvantages are potential dangers to radiation machine operators from exposure to radiation and possible damage to some materials.

©Tom Pantages

Nonionizing Radiation: Ultraviolet Rays **Ultraviolet (UV) radiation** ranges in wavelength from approximately 100 to 400 nm. It is most lethal from 240 to 280 nm (with a peak at 260 nm). Owing to its lower energy state, UV radiation is not as penetrating as ionizing radiation. Because UV radiation passes readily through air, slightly through liquids, and only poorly through solids, the object to be disinfected must be directly exposed to it for full effect.

Ultraviolet rays are a powerful tool for destroying fungal cells and spores, bacterial vegetative cells, protozoa, and viruses. Bacterial spores are about 10 times more resistant to radiation than are vegetative cells, but they can be killed by increasing the time of exposure. Even though it is possible to sterilize with UV, it is so technically challenging that we don't regularly call it a sterilizing technology.

Usually directed at disinfection rather than sterilization. Germicidal lamps can cut down on the concentration of airborne microbes as much as 99%. They are used in hospital rooms, operating rooms, schools, food preparation areas, and dental offices. Ultraviolet disinfection of air has proved effective in reducing postoperative infections, preventing the transmission of infections by respiratory droplets, and curtailing the growth of microbes in food-processing plants and slaughterhouses.

Ultraviolet irradiation of liquids requires special equipment to spread the liquid into a thin, flowing film that is exposed directly to a lamp. This method can be used to treat drinking water and to purify other liquids (milk and fruit juices) as an alternative to heat. The photo shows a UV treatment system for the disinfection of water.

UV radiation damages cells by causing the inappropriate formation of bonds between two adjacent bases on a DNA strand **(figure 9.7).** It affects pyrimidines (Ts and Cs). If the new "dimers" are not repaired, they can prevent that segment of DNA from being correctly replicated or transcribed. Massive dimerization is lethal to cells.

Other Physical Methods: Filtration

Filtration is an effective method to remove microbes from air and liquids. In practice, a fluid, or the air, is strained through a filter with openings large enough for the fluid to pass through but too small for microorganisms to pass through **(figure 9.8a).**

Most modern microbiological filters are thin membranes of cellulose acetate, polycarbonate, and a variety of plastic materials (Teflon, nylon) whose pore size can be carefully controlled and standardized. Ordinary substances such as charcoal, diatomaceous earth, or unglazed porcelain are also used in some applications. Viewed microscopically, most filters are perforated by very precise, uniform pores **(figure 9.8b).** The pore diameters vary from coarse (8 μm) to ultrafine (0.02 μm), permitting selection of the minimum particle size to be trapped. Those with even smaller pore diameters permit true sterilization by removing viruses, and some will even remove large proteins. A sterile liquid filtrate is typically produced by suctioning the liquid through a sterile filter into a presterilized container. These filters are also used to separate mixtures of microorganisms and to count bacteria in water analysis.

Filtration is used to prepare liquids that cannot withstand heat, including serum and other blood products, vaccines, drugs, IV fluids, enzymes, and media. Filtration has been employed as an alternative method for decontaminating milk and beer without altering their flavor. It is also an important step in water purification. Its use extends to filtering out particulate impurities (crystals, fibers, and so on) that can cause severe reactions in the body.

Filtration is also an efficient means of removing airborne contaminants that are a common source of infection and spoilage. High-efficiency particulate air (HEPA) filters are widely used to provide a flow of decontaminated air to hospital rooms and sterile rooms.

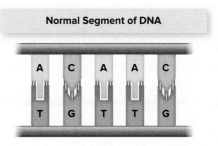

Normal Segment of DNA

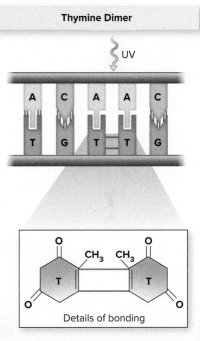

Thymine Dimer

UV

Details of bonding

Figure 9.7 Formation of pyrimidine dimers by the action of ultraviolet (UV) radiation.   This shows what occurs when two adjacent thymine bases on one strand of DNA are induced by UV rays to bond laterally with each other. The result is a thymine dimer (shown in greater detail). Dimers can also occur between adjacent cytosines, and thymine and cytosine bases.

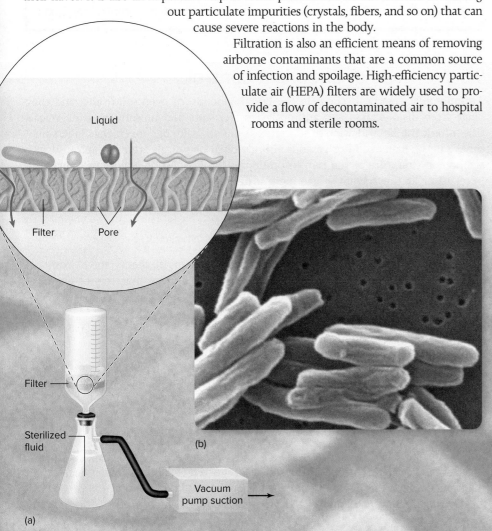

Liquid

Filter   Pore

Filter

Sterilized fluid

(b)

Vacuum pump suction

(a)

Figure 9.8 Membrane filtration.   **(a)** Vacuum assembly for achieving filtration of liquids through suction. The surface of the filter is shown magnified in the blown-up section, with tiny passageways (pores) too small for the microbial cells to enter but large enough for liquid to pass through. **(b)** Scanning electron micrograph (21,000×) of filter, showing relative size of pores and bacteria trapped on its surface.

(b) Source: CDC/Dr. Ray Butler; Janice Carr

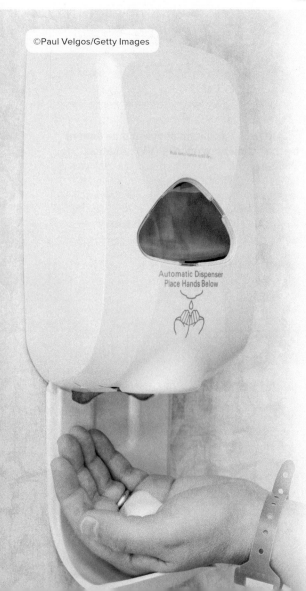

Canned sardines remain edible for 2–3 years, preserved by the osmotic pressure of high salt.
©Burke/Triolo Productions/Getty Images

©Paul Velgos/Getty Images

Automatic Dispenser
Place Hands Below

Osmotic Pressure

In chapter 6, you learned about the effects of osmotic pressure on cells. This fact has long been exploited as a means of preserving food. Adding large amounts of salt or sugar to foods creates a hypertonic environment for bacteria in the foods, causing plasmolysis and making it impossible for the bacteria to multiply (remember chapter 6?). People knew that these techniques worked long before they knew why. Even in ancient times, people used pickling, smoking, and drying of foods to control growth of microorganisms. This is why meats are "cured," or treated with high salt concentrations so they can be kept for long periods without refrigeration. High sugar concentrations in foods like jellies have the same effect. Osmotic pressure, however, is never a sterilizing technique.

9.2 LEARNING OUTCOMES—Assess Your Progress

5. Name six methods of physical control of microorganisms.
6. Discuss both moist and dry heat methods, and identify multiple examples of each.
7. Define *thermal death time* and *thermal death point.*
8. Explain methods of moist heat control.
9. Explain two methods of dry heat control.
10. Identify advantages and disadvantages of cold and desiccation.
11. Differentiate between the two types of radiation control methods.
12. Explain how filtration and osmotic pressure function as control methods.

9.3 Methods of Chemical Control

Antimicrobial chemicals occur in the liquid, gaseous, or even solid state, and they range from disinfectants and antiseptics to sterilants and preservatives (chemicals that inhibit the deterioration of substances). For the sake of convenience (and sometimes safety), many solid or gaseous antimicrobial chemicals are dissolved in water, alcohol, or a mixture of the two to produce a liquid solution. Solutions containing pure water as the solvent are termed **aqueous,** whereas those dissolved in pure alcohol or alcohol-water mixtures are termed **tinctures.**

Selecting a Microbicidal Chemical

The choice and appropriate use of antimicrobial chemical agents are of constant concern in medicine and dentistry. Although actual clinical practices of chemical decontamination vary widely, some desirable qualities in a germicide have been identified, including the following:

- rapid action even in low concentrations,
- solubility in water or alcohol and long-term stability,
- broad-spectrum microbicidal action without being toxic to human and animal tissues,
- penetration of inanimate surfaces to sustain a cumulative or persistent action,
- resistance to becoming inactivated by organic matter,
- not corrosive and nonstaining,
- sanitizing and deodorizing properties, and
- affordability and availability.

Germicides are evaluated in terms of their effectiveness in destroying microbes in medical and dental settings. The three levels of chemical decontamination procedures are *high, intermediate,* and *low.* High-level germicides kill endospores and, if properly

*Look back at figure 9.2.
Kelly & Heidi*

used, are sterilants. Materials that necessitate high-level control are medical devices–for example, catheters, heart-lung equipment, and implants–that are not heat-sterilizable and are intended to enter body tissues during medical procedures. Intermediate-level germicides kill fungal (but not bacterial) spores, resistant pathogens such as the bacterium that causes tuberculosis, and viruses. They are used to disinfect items (respiratory equipment, thermometers) that come into intimate contact with the mucous membranes but are noninvasive. Low levels of disinfection eliminate only vegetative bacteria, vegetative fungal cells, and some viruses. They are used to clean materials such as electrodes, straps, and pieces of furniture that touch the skin surfaces but not the mucous membranes.

Factors Affecting the Germicidal Activity of Chemicals

Factors that control the effect of a germicide include the nature of the microorganisms being treated, the nature of the material being treated, the degree of contamination, the time of exposure, and the strength and chemical action of the germicide. The variations in concentration and time needed can be quite wide **(table 9.8).**

The Chemical Categories

The modes of action of most germicides are to attack the cellular targets discussed earlier: proteins, nucleic acids, the cell wall, and the cytoplasmic membrane. **Table 9.9** on the following pages provides details about the most commonly used chemicals and their modes of action.

Table 9.8 Required Concentrations and Times for Chemical Destruction of Selected Microbes

Organism	Concentration	Time
Agent: Chlorine		
Mycobacterium tuberculosis	50 ppm	50 sec
Entamoeba cysts (protozoa)	0.1 ppm	150 min
Hepatitis A virus	3 ppm	30 min
Agent: Ethyl Alcohol		
Staphylococcus aureus	70%	10 min
Escherichia coli	70%	2 min
Poliovirus	70%	10 min
Agent: Hydrogen Peroxide		
Staphylococcus aureus	3%	12.5 sec
Neisseria gonorrhoeae	3%	0.3 sec
Herpes simplex virus	3%	12.8 sec
Agent: Quaternary Ammonium Compound		
Staphylococcus aureus	450 ppm	10 min
Salmonella typhi	300 ppm	10 min
Agent: Ethylene Oxide Gas		
Streptococcus faecalis	500 mg/L	2-4 min
Influenza virus	10,000 mg/L	25 h

The main antimicrobial ingredient in many mouthwashes is alcohol.
©Stockbyte/PunchStock

Table 9.9 Germicidal Categories According to Chemical Group

Techniques and chemicals that are capable of sterilizing are highlighted with a pink background.

Agent	Target Microbes	Form(s)	Mode of Action	Indications for Use	Limitations
Halogens: chlorine	Can kill endospores (slowly); all other microbes	Liquid/gaseous chlorine (Cl_2), hypochlorites (OCl), chloramines (NH_2Cl)	In solution, these compounds combine with water and release hypochlorous acid (HOCl); denature enzymes permanently and suspend metabolic reactions	Chlorine kills bacteria, endospores, fungi, and viruses; gaseous/liquid chlorine: used to disinfect drinking water, sewage and waste water; hypochlorites: used in health care to treat wounds, disinfect bedding and instruments, sanitize food equipment and in restaurants, pools, and spas; chloramines: alternative to pure chlorine in treating drinking water; also used to treat wounds and skin surfaces	Less effective if exposed to light, alkaline pH, and excess organic matter
Halogens: iodine	Can kill endospores (slowly); all other microbes	Free iodine in solution (I_2) Iodophors (complexes of iodine and alcohol)	Penetrates cells of microorganisms where it interferes with a variety of metabolic functions; interferes with the hydrogen and disulfide bonding of proteins	2% iodine, 2.4% sodium iodide (aqueous iodine) used as a topical antiseptic 5% iodine, 10% potassium iodide used as a disinfectant for plastic and rubber instruments, cutting blades, etc. Iodophor products contain 2% to 10% of available iodine, which is released slowly; used to prepare skin for surgery, in surgical scrubs, to treat burns, and as a disinfectant	Can be extremely irritating to the skin and is toxic when absorbed **Many iodophors banned in consumer products in 2017.**
Oxidizing agents	Kill endospores and all other microbes	Hydrogen peroxide, peracetic acid	Oxygen forms free radicals (−OH), which are highly toxic and reactive to cells	As an antiseptic, 3% hydrogen peroxide used for skin and wound cleansing, mouth washing, bedsore care Used to treat infections caused by anaerobic bacteria 35% hydrogen peroxide used in low-temperature sterilizing cabinets for delicate instruments	Sporicidal only in high concentrations
Aldehydes	Kill endospores and all other microbes	Organic substances bearing a −CHO functional group on the terminal carbon	Glutaraldehyde can irreversibly disrupt the activity of enzymes and other proteins within the cell Ortho-phthalaldehyde	Glutaraldehyde kills rapidly and is broad-spectrum; used to sterilize respiratory equipment, scopes, kidney dialysis machines, dental instruments Ortho-phthalaldehyde is safer than glutaraldehyde and just as effective	Glutaraldehyde is somewhat unstable, especially with increased pH and temperature Ortho-phthalaldehyde is much more expensive than glutaraldehyde
Gaseous sterilants/disinfectants	Ethylene oxide kills endospores; other gases less effective	Ethylene oxide is a colorless substance that exists as a gas at room temperature	Ethylene oxide reacts vigorously with functional groups of DNA and proteins, blocking both DNA replication and enzymatic actions Chlorine dioxide is a strong alkylating agent	Ethylene oxide is used to disinfect plastic materials and delicate instruments; can also be used to sterilize syringes, surgical supplies, and medical devices that are prepackaged	Ethylene oxide is explosive–it must be combined with a high percentage of carbon dioxide or fluorocarbon It can damage lungs, eyes, and mucous membranes if contacted directly Ethylene oxide is rated as a carcinogen by the government

Table 9.9 (continued)

Agent	Target Microbes	Form(s)	Mode of Action	Indications for Use	Limitations
Phenol (carbolic acid)	Some bacteria, viruses, fungi	Derived from the distillation of coal tar Phenols consist of one or more aromatic carbon rings with added functional groups	In high concentrations, they are cellular poisons, disrupting cell walls and membranes, proteins In lower concentrations, they inactivate certain critical enzyme systems	Phenol remains one standard against which other (less toxic) phenolic disinfectants are rated; the phenol coefficient quantitatively compares a chemical's antimicrobial properties to those of phenol Phenol is now used only in certain limited cases, such as in drains, cesspools, and animal quarters	Toxicity of many phenolics makes them dangerous to use as antiseptics **Many phenols banned in consumer products in 2017, including triclosan and triclocarban.**
Chlorhexidine	Most bacteria, viruses, fungi	Complex organic base containing chlorine and two phenolic rings	Targets bacterial membranes, where selective permeability is lost, bacterial cell walls, and proteins, resulting in denaturation	Mildness, low toxicity, and rapid action make chlorhexidine a popular choice of agents Used in hand scrubs, prepping skin for surgery, as an obstetric antiseptic, as a mucous membrane irrigant, etc.	Effects on viruses and fungi are variable
Alcohol	Most bacteria, viruses, fungi	Colorless hydrocarbons with one or more −OH functional groups Ethyl and isopropyl alcohol are suitable for antimicrobial control	Concentrations of 50% and higher dissolve membrane lipids, disrupt cell surface tension, and compromise membrane integrity	Germicidal, nonirritating, and inexpensive Routinely used as skin degerming agents (70% to 95% solutions)	Rate of evaporation decreases effectiveness Inhalation of vapors can affect the nervous system
Detergents	Some bacteria, viruses, fungi	Polar molecules that act as surfactants Anionic detergents have limited microbial power Cationic detergents, such as quaternary ammonium compounds ("quats"), are much more effective antimicrobials	Positively charged end of the molecule binds well with the predominantly negatively charged bacterial surface proteins Long, uncharged hydrocarbon chain allows the detergent to disrupt the cytoplasmic membrane Cytoplasmic membrane loses selective permeability, causing cell death	Effective against viruses, algae, fungi, and gram-positive bacteria Rated only for low-level disinfection in the clinical setting Used to clean restaurant utensils, dairy equipment, equipment surfaces, restrooms	Ineffective against tuberculosis bacterium, hepatitis virus, *Pseudomonas*, and endospores Activity is greatly reduced in presence of organic matter Detergents function best in alkaline solutions **Some quats banned in consumer products in 2017.**
Heavy metal compounds	Some bacteria, viruses, fungi	Heavy metal germicides contain either an inorganic or an organic metallic salt; may come in tinctures, soaps, ointment, or aqueous solution	Mercury, silver, and other metals exert microbial effects by binding onto functional groups of proteins and inactivating them	Organic mercury tinctures are fairly effective antiseptics Organic mercurials serve as preservatives in cosmetics, ophthalmic solutions, and other substances Silver nitrate solutions are used for topical germicides and ointments	Microbes can develop resistance to metals Not effective against endospores Can be toxic if inhaled, ingested, or absorbed May cause allergic reactions in susceptible individuals
Acids and alkalis	Some bacteria, viruses, fungi	Organic acids in food production	pH alteration	Food manufacture, deodorants, deodorizers	Large changes in pH can be corrosive

Charged Head

Medical Moment

Zap VAP: The Role of Chlorhexidine

Patients in the intensive care unit often require support for their breathing. A plastic tube is inserted in the nose or mouth so a ventilator can deliver oxygen and breaths. However, the breathing tube establishes a path by which microbes may easily invade the lower respiratory tract and lungs, as well as restricts the patient's natural ability to clear secretions. Therefore, patients with an artificial airway are at risk for hospital-acquired lung infection, or ventilator-associated pneumonia (VAP).

There are many recommendations for care that are implemented by nurses and the care team to prevent VAP. One of the interventions is routine oral care with an antiseptic, such as chlorhexidine. The surfaces of the oral cavity are rinsed with the antiseptic wash to kill organisms in the mouth that could migrate to the lungs.

Chlorhexidine has a low toxicity and is safe for use on mucous membranes. It is applied directly to the teeth and surfaces of the oral cavity by a soft sponge or irrigation. The use of chlorhexidine in patients on ventilator support has been shown to decrease the incidence of VAP.

Q. What is the mode of action of chlorhexidine?

Answer in Appendix B.

In 2016 the Food and Drug Administration (FDA) placed a ban (effective fall 2017) on 19 active ingredients that may no longer be used in consumer products such as soaps or cleaning solutions. They did this because many of the active ingredients can select for resistance in microbes, and some of them can even increase the likelihood that microbes will become resistant to antibiotics themselves. There has been a proliferation of antimicrobial chemicals in everyday products, such as hand lotions and detergents. Most scientists agree that soaps and detergents are sufficient on their own for the cleaning they are expected to do, and that "spiking" them with antimicrobial agents causes more harm than good in the home environment. Of course, the hospital environment is a different situation. But in order for the antimicrobial agents used to clean surfaces and skin in a hospital to remain effective, we need to decrease their overall presence in the (outpatient) environment. The agents banned in the FDA ruling are pointed out in table 9.9.

A chemical's strength or concentration is expressed in various ways, depending on accepted practice and the method of preparation. In dilutions, a small volume of the liquid chemical (solute) is diluted in a larger volume of solvent to achieve a certain ratio. For example, a common laboratory phenolic disinfectant such as Lysol is usually diluted 1:200; that is, one part of chemical has been added to 200 parts of water by volume. Solutions such as chlorine that are effective in very diluted concentrations are expressed in parts per million (ppm). In percentage solutions, the solute is added to water by weight or volume to achieve a certain percentage in the solution. Alcohol, for instance, is used in percentages ranging from 50% to 95%. Alcohol requires a bit of water to be most effective.

Another factor that contributes to germicidal effectiveness is the length of exposure. Most compounds require adequate contact time to allow the chemical to penetrate and to act on the microbes present. The composition of the material being treated must also be considered. Smooth, solid objects are more reliably disinfected than are those with pores or pockets that can trap soil. An item contaminated with common biological matter such as serum, blood, saliva, pus, fecal material, or urine presents a problem in disinfection. Large amounts of organic material can hinder the penetration of a disinfectant and, in some cases, can form bonds that inactivate the effectiveness of the disinfectant. Adequate cleaning of instruments and other reusable materials must occur before use of the germicide or sterilant. Otherwise, there is no way to predict whether your procedure will be effective.

For a look at the antimicrobial chemicals found in some common household products, see **table 9.10.**

Table 9.10 Active Ingredients of Various Commercial Antimicrobial Products

Product	Specific Chemical Agent	Antimicrobial Category
Lysol® Sanitizing Wipes	Dimethyl benzyl ammonium chloride	Detergent (quat)
Clorox® Disinfecting Wipes	Dimethyl benzyl ammonium chloride	Detergent (quat)
Tilex® Mildew Remover	Sodium hypochlorites	Halogen
Lysol® Mildew Remover	Sodium hypochlorites	Halogen
Lysol® Disinfecting Spray	Alkyl dimethyl benzyl ammonium saccharinate/ethanol	Detergent (quats)/alcohol
ReNu® Contact Lens Solution	Polyaminopropyl biguanide	Chlorhexidine
Scope® Mouthwash	Ethanol	Alcohol
Purell® Instant Hand Sanitizer	Some ethanol-based; some detergent-based	Alcohol or detergent
Allergan® Eye Drops	Sodium chlorite	Halogen

9.3 LEARNING OUTCOMES—Assess Your Progress

13. Name the desirable characteristics of chemical control agents.
14. Discuss chlorine and iodine and their uses.
15. List advantages and disadvantages to phenolic compounds.
16. Explain the mode of action of chlorhexidine.
17. Explain the applications of oxidizing agents.
18. Identify some heavy metal control agents.
19. Discuss the disadvantages of aldehyde agents.
20. Identify applications for ethylene oxide sterilization.

NCLEX® PREP

3. A nurse is preparing to place a peripheral IV. Prior to inserting the IV catheter, he scrubs the patient's skin with a 70% isopropyl alcohol swab per hospital policy. Which of the following statements best represents the rationale for this action?
   a. *Alcohol sterilizes the patient's skin.*
   b. *Alcohol is irritating to the skin but necessary to prevent infection.*
   c. *The patient's skin must be scrubbed for an adequate amount of time to allow the alcohol to kill the microbes that are present.*
   d. *Alcohol inhibits reproduction of microbes present on the skin, though it cannot kill what is already present.*

CASE FILE WRAP-UP

At the end of the case, the sterile equipment and surgical supplies were counted to ensure that none had been left in the patient. Human tissue waste and blood-saturated supplies were disposed of in biohazard waste containers. Waste that was potentially pathogenic was sent to be sterilized by an autoclave prior to disposal. Plastic supplies, syringes, gauze sponges, and other single-use equipment were discarded. The surgical tools were rinsed of visible debris and placed in large trays for sterilization.

Some surgical equipment is reused after appropriate sterilization. In the hospital sterile processing department, an autoclave is used to eliminate all viable microorganisms, including bacterial endospores and viruses. (Prions may not be destroyed by this process.) The surgical tools are packed loosely onto a tray covered with cloth and loaded into the autoclave. There it is exposed to high heat under pressure. To ensure the equipment within the packaging has been brought to the appropriate temperature, indicator tape is applied to the outside of the package. This autoclave tape changes color when a high enough temperature has been maintained for the necessary time for sterilization to be complete.

The prevention of surgical site infections (SSIs) is one of the major benchmarks of quality in a hospital. SSIs create complications for the postoperative patient with increases in morbidity and hospital length of stay, and can even be fatal.

The Microbiome

©O. Dimier/PhotoAlto

 COVID-19

Take note: This study was conducted before the onset of the COVID-19 pandemic. Soon there may be studies that show how health care workers and civilians changed their hand hygiene practices during the pandemic and how that impacted the skin microbiome.

Hand Hygiene

How many times have you heard that one of the best ways to prevent infections in hospitals is for health care workers to wash their hands? OK, but now that we know about how important our resident microbiota is, we should ask what the frequent hand washing does to those microorganisms. A recent study followed 34 health care workers in a surgical intensive care unit. Twenty-four of them were registered nurses, six were respiratory therapists, and four were nurse technologists.

During a typical 12-hour shift,

- 53% of them reported washing their hands with soap and water between 6 and 20 times,
- 41% used alcohol rubs more than 20 times, and
- 62% donned more than 40 pairs of gloves.

The researchers then tested for common healthcare-associated pathogens and found that approximately 45% of the health care workers' dominant hands were positive for *Staphylococcus aureus*. They detected methicillin-resistant *Staphylococcus aureus* (MRSA) on 3.9% of the dominant hands. About 4% of hands were positive for the fungus *Candida albicans*.

These rates look very similar to the rates found in the general population—those of us who are not washing our hands 20 times a day. This might be an illustration of the difference between *colonization*—when microbes become part of your "normal" microbiota—and *contamination*—when you pick up microbes from your environment and they stay only transiently. Contaminants are usually easily washed off, while colonizers are embedded deeper in your skin oils and the top layers of epithelial cells.

One question raised, but not answered, by this study is whether altering the normal microbiome of the hand by frequent hand hygiene practices might make it (a) easier or (b) more difficult for pathogens to contaminate and/or colonize caregivers' hands.

Source: Rosenthal, et al., "Healthcare Workers' Hand Microbiome May Mediate Carriage of Hospital Pathogens," *Pathogens*, vol. 3, no. 4, pp. 1–13. March 2014.

Source: Shutterstock/Barbo

PHYSICAL AND CHEMICAL CONTROL OF MICROORGANISMS

9.1 CONTROLLING MICROORGANISMS

Microorganisms in the environment might need to be controlled to prevent human infection. This can be accomplished with physical methods or chemical methods. Sterilization destroys all microbes, including viruses. Antisepsis, disinfection, decontamination/sanitization, and antisepsis/degermation reduce the numbers of microorganisms.

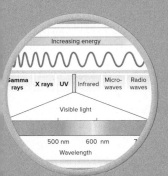

Heat, cold, radiation, drying, filtration, and osmotic pressure are all methods of physical control of microorganisms. Heat can be deployed dry (flames, ovens) or wet. Autoclaving (steam under pressure) achieves sterilization but is destructive to many materials.

METHODS OF PHYSICAL CONTROL 9.2

9.3 METHODS OF CHEMICAL CONTROL

Chemical agents can be either microbicidal or microbistatic. They are also classified as high-, medium-, or low-level germicides. They can be liquids or gases. Factors that determine the effectiveness of these agents are the types and numbers of microbes, the material involved, the strength of the agent, and the exposure time.

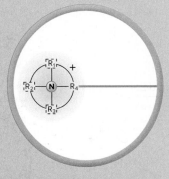

SmartGrid: From Knowledge to Critical Thinking

This *21 Question Grid* takes the topics from this chapter and arranges them with respect to the American Society for Microbiology's Undergraduate Curriculum guidelines—all six of the important "Concepts" as well as the important "Competency" of scientific literacy. Three questions are supplied about chapter content that refer to the Concept or Competency, in increasing levels of Bloom's taxonomy for learning.

ASM Concept/ Competency	A. Bloom's Level 1, 2—Remember and Understand (Choose one.)	B. Bloom's Level 3, 4—Apply and Analyze	C. Bloom's Level 5, 6—Evaluate and Create
Evolution	1. DNA repair mechanisms can help alleviate the effects of a. UV radiation. b. alcohol disinfection. c. autoclaving. d. dry heat.	2. Explain why all cells in a population do not die instantaneously when exposed to an antimicrobial agent, even when they are of a single species.	3. Triclosan and other antimicrobial compounds have been banned from consumer products. Explain why we should not be using every chemical available to us to combat microbes.
Cell Structure and Function	4. Microbial control methods that kill _____ are able to sterilize. a. viruses b. the tuberculosis bacteria c. endospores d. cysts	5. Why are enveloped viruses generally more susceptible to disinfection, when they have an extra "layer"?	6. Construct an argument about why some antimicrobial agents are toxic for skin and human tissue.
Metabolic Pathways	7. Cytoplasmic enzymes are most likely to be disrupted by a. UV light. b. low temperatures. c. high temperatures. d. detergents.	8. Most antimicrobials that target protein function are nonselective as to the microbes they affect. Explain why that might be so.	9. There is a bacterium that has a metabolic pathway that enables it to survive in the presence of acid. What kind of pathway or product would you expect in this case?
Information Flow and Genetics	10. Transcription is targeted most directly by a. quats. b. detergents. c. UV radiation. d. alcohol.	11. Do you think UV radiation is a good way to disinfect living tissue? Why or why not?	12. Suggest a possible mechanism for how a microbe that becomes resistant to an antimicrobial chemical can transfer that to another microorganism, and also transfer resistance to antibiotics in the process.
Microbial Systems	13. Disinfection procedures must take into account a. the mixture of microbes being targeted. b. whether the microbes in biofilms are dispersed. c. the amount of organic matter associated with the microbes. d. all of the above.	14. Provide a rationale for how a particular microbe could be considered a serious contaminant in one situation and completely harmless in another.	15. What additional considerations would you have to address if the device you were trying to disinfect—such as a water line in a dental office—was known to contain a microbial biofilm?

ASM Concept/ Competency	A. Bloom's Level 1, 2—Remember and Understand (Choose one.)	B. Bloom's Level 3, 4—Apply and Analyze	C. Bloom's Level 5, 6—Evaluate and Create

Impact of Microorganisms	16. Sanitization is a process by which
	  a. the microbial load on an object is reduced.
	  b. objects are made sterile with chemicals.
	  c. utensils are scrubbed.
	  d. skin is debrided.
		17. Tissue transplantation carries a risk of transferring microbes in the tissue to the recipient. What are the most appropriate ways of ensuring that does not happen?	18. Defend this statement: It is possible to make our environment too clean.

Scientific Thinking	19. The most versatile method for sterilizing heat-sensitive liquids is
	  a. UV radiation.
	  b. exposure to ozone.
	  c. peracetic acid.
	  d. filtration.
		20. Precisely what is microbial death?	21. Devise an experiment that will differentiate between bactericidal versus bacteriostatic effects.

Answers to the multiple-choice questions appear in Appendix A.

Visual Connections

This question connects content within and between chapters.

From chapter 3, figure 3.15. Study this illustration of a gram-negative cell envelope. In what ways could alcohol damage these two membranes? How would that harm the cell?

Gram-Negative

Outer membrane layer

Peptidoglycan

Cytoplasmic membrane

Membrane proteins

Antimicrobial Treatment

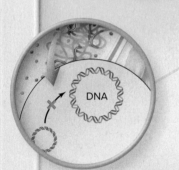

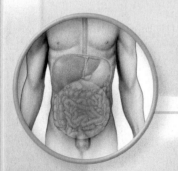
Glow Images (pills); Source: CDC/Dr. Richard Facklam (Etest); Ryan McVay/Getty Images (health care worker)

CASE FILE

Not What We Were Expecting

I was working in a pediatric hospital in the emergency room. A 2-year-old girl with an obvious rash was brought in by her mother, who wondered whether the rash might be chickenpox.

I took the child and her mother back to a cubicle and began to take the child's history. The child was healthy and had no major health problems. Her vaccinations were all current, including the chickenpox vaccine, which made a diagnosis of chickenpox unlikely. She had recently had otitis media (an ear infection) and was on her second-to-last day of antibiotic therapy with Ceclor, a cephalosporin antibiotic. The child had taken Ceclor on one other occasion for an ear infection.

The vital signs were normal. She did not have a fever. The rash was maculopapular, a flat red rash with tiny pimplelike eruptions in the center. The rash covered her face, chest, back, arms, and legs. In fact, it covered almost her entire body except the palms of her hands and soles of her feet. The child was clearly uncomfortable and was scratching exposed areas of skin.

The mother reported that she had taken the child for a haircut in the afternoon and noticed a few spots on the child's neck. She did not think much of the spots at the time. She dropped the little girl off at her mother-in-law's for child care while she went to work as a waitress. When she returned to her mother-in-law's house to pick the child up after her shift, the child was awake, irritable, and covered in the rash. Alarmed, the mother brought the child immediately to the emergency room.

After recording the history and the child's vital signs, I went to find the physician and reported my findings. He told me that he felt he knew what the problem was but would quickly examine the child first before telling me his diagnosis. The doctor looked at the patient's rash and told the mother, "Just as I thought. The rash is a reaction to the Ceclor your daughter has been taking. Stop the Ceclor and the rash will go away." The mother was surprised because she had always thought that an allergic reaction to a drug would start with the first dose. The physician told her that an allergic reaction could begin at any time, even after taking the same drug numerous times.

The child's ears were checked and there was no sign of infection. The girl and her mother were discharged after receiving a prescription for an antihistamine to help control the itching.

- What category of antibiotic does Ceclor fall under?

- What is the mechanism underlying the allergic response to an antibiotic?

Case File Wrap-Up appears at the end of the chapter.

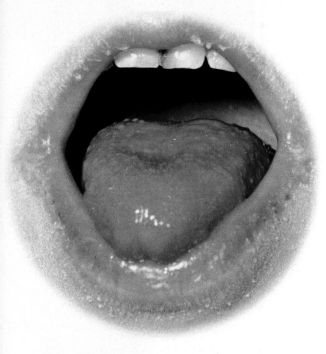

Scarlet fever is an uncommon sequel to strep throat. It is treatable with antibiotics. Strawberry tongue is one of its symptoms.
©BSIP/Universal Images Group/Getty Images

10.1 Principles of Antimicrobial Therapy

A hundred years ago in the United States, one out of three children was expected to die of an infectious disease before the age of 5. Early death or severe lifelong debilitation from scarlet fever, diphtheria, tuberculosis, meningitis, and many other bacterial diseases was a fearsome yet undeniable fact of life to most of the world's population. The introduction of modern drugs to control infections in the 1940s was a medical revolution that has added significantly to the life span and health of humans. It is no wonder that, for many years, antibiotics were regarded as miracle drugs. But even though antimicrobial drugs have greatly reduced the incidence of certain infections, they have definitely not eradicated infectious disease and probably never will. In fact, we are dangerously close to a postantibiotic era, where the drugs we have are no longer effective.

The goal of antimicrobial chemotherapy is deceptively simple: Administer a drug to an infected person, which destroys the infective agent without harming the host's cells. In actuality, this goal is rather difficult to achieve because many (often contradictory) factors must be taken into account. The ideal drug should be easy to administer, yet be able to reach the infectious agent anywhere in the body. It should also be toxic to the infectious agent, while being nontoxic to the host. It should remain active in the body as long as needed, yet be safely and easily broken down and excreted. Additionally, microbes in biofilms often require different drugs than when they are not in biofilms. In short, the perfect drug does not exist–but by balancing drug characteristics against one another, a satisfactory compromise can often be achieved **(table 10.1).**

Antimicrobial agents are characterized with regard to their origin, range of effectiveness, and whether they are naturally produced or chemically synthesized. A few of the more important terms you will encounter are found in **table 10.2.**

In this chapter, we describe different types of antimicrobial drugs, their mechanism of action, and the types of microbes on which they are effective. The organ system chapters 16 through 21 list specific disease agents and the drugs used to treat them.

Table 10.1 Characteristics of the Ideal Antimicrobial Drug

- Toxic to the microbe but nontoxic to host cells
- Microbicidal rather than microbistatic
- Relatively soluble; functions even when highly diluted in body fluids
- Remains potent long enough to act and is not broken down or excreted prematurely
- Does not lead to the development of antimicrobial resistance
- Complements or assists the activities of the host's defenses
- Remains active in tissues and body fluids
- Readily delivered to the site of infection
- Does not disrupt the host's health by causing allergies or predisposing the host to other infections

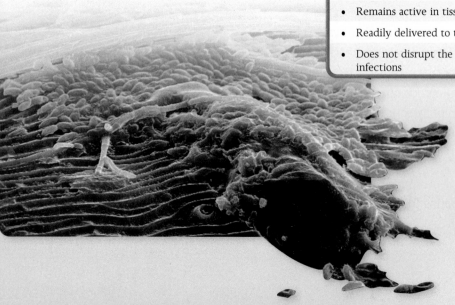

Bacterial biofilm formed on the surface of a spider.
Source: Centers for Disease Control and Prevention/Janice Haney Carr

Table 10.2 Terminology of Antimicrobials

Prophylaxis	Use of a drug to prevent infection of a person at risk
Antimicrobial Chemotherapy	The use of drugs to control infection
Antimicrobials	All-inclusive term for any antimicrobial drug, regardless of its origin
Antibiotics	Substances produced by the natural metabolic processes of some microorganisms that can inhibit or destroy other microorganisms; generally, the term is used for drugs targeting bacteria and not other types of microbes
Semisynthetic Drugs	Drugs that are chemically modified in the laboratory after being isolated from natural sources
Synthetic Drugs	Drugs produced entirely by chemical reactions
Narrow-Spectrum (Limited Spectrum)	Antimicrobials effective against a limited array of microbial types—for example, a drug effective mainly against gram-positive bacteria
Broad-Spectrum (Extended Spectrum)	Antimicrobials effective against a wide variety of microbial types—for example, a drug effective against both gram-positive and gram-negative bacteria

First mass-produced in 1944, penicillin saved many lives in WWII.
Source: Library of Congress Prints and Photographs Division

The Origins of Antimicrobial Drugs

Antibiotics are originally metabolic products of bacteria and fungi. They are produced by microbes in order to reduce competition for nutrients and space in their habitat. The greatest numbers of current antibiotics are derived from bacteria in the genera *Streptomyces* and *Bacillus* and from molds in the genera *Penicillium* and *Cephalosporium*. Also, chemists have created new drugs by altering the structure of naturally occurring antibiotics. More recent innovations include entirely new chemicals synthesized in the lab. Very recently, scientists have studied genes found in our own microbiome research to detect genes that could lead to new antimicrobials, but that have been previously overlooked.

Identifying the Microbe and Starting Treatment

Before actual antimicrobial therapy can begin, it is best that at least three factors be considered:

1. the identity of the microorganism causing the infection,
2. the degree of the microorganism's susceptibility (also called sensitivity) to various drugs, and
3. the overall medical condition of the patient.

Identification of infectious agents from body specimens should be attempted as soon as possible. It is especially important that such specimens be taken before any antimicrobial drug is given, before the drug reduces the numbers of the infectious agent. Direct examination of body fluids, sputum, or stool is a rapid initial method for detecting and perhaps even identifying bacteria or fungi. A doctor often begins the therapy on the basis of such immediate findings, or even on the basis of an informed best guess. For instance, if a sore throat appears to be caused by *Streptococcus pyogenes*, the physician might prescribe penicillin because this species seems to be almost universally sensitive to it so far. If the infectious agent is not or cannot be isolated, epidemiological statistics may be required to predict the most likely agent in a given infection. For example, *Streptococcus pneumoniae* accounts for the majority of cases of bacterial meningitis in children, followed by *Neisseria meningitidis* (discussed in detail in chapter 17).

Testing for the Drug Susceptibility of Microorganisms

Some bacteria require antimicrobial sensitivity testing and some do not. Testing is essential in those groups of bacteria commonly showing resistance, such as *Staphylococcus* species, *Neisseria gonorrhoeae*, *Streptococcus pneumoniae*, *Enterococcus faecalis*, and the aerobic gram-negative intestinal bacilli. Drug testing in fungal or protozoal infections used to be unwarranted because the agents used against them generally worked on all the representatives of that group. That is no longer true, and if initial treatment is ineffective, drug testing is essential. In general, these tests involve exposing a pure culture of the microbe to several different drugs and observing the effects of the drugs on growth.

The *Kirby-Bauer* technique is an agar diffusion test that provides useful data on antimicrobial susceptibility. In this test, the surface of a plate of special medium is spread with the test bacterium (for example), and small discs containing a premeasured amount of antimicrobial are dispensed onto the bacterial lawn. After incubation, the zone of inhibition surrounding the discs is measured and compared with a standard for each drug **(table 10.3** and **figure 10.1)**. An alternative diffusion system that provides additional information on drug effectiveness is the E-test **(figure 10.2)**.

More sensitive and quantitative results can be obtained with tube dilution tests. First, the antimicrobial is diluted serially in tubes of broth (often miniaturized tubes called wells in 96-well plates). Then each tube is inoculated with a small uniform sample of pure culture, incubated, and examined for growth (turbidity). The smallest concentration (highest dilution) of drug that visibly inhibits growth is called the **minimum inhibitory concentration,** or **MIC.** The MIC is useful in determining the smallest effective dosage of a drug and in providing a comparative index against other antimicrobials **(figure 10.3)**. In many clinical laboratories, these antimicrobial testing procedures are performed in automated machines that can test dozens of drugs simultaneously.

Penicillins damage bacterial cell walls.
Source: CDC/Janice Haney Carr

Source: Centers for Disease Control and Prevention/Janice Carr

Table 10.3 Results of a Sample Kirby-Bauer Test

Drug	Zone Sizes (mm) Required for Susceptibility (S)	Resistance (R)	Example Results (mm) for *Staphylococcus aureus*	Evaluation
Bacitracin	>13	<8	15	S
Chloramphenicol	>18	<12	20	S
Erythromycin	>18	<13	15	I
Gentamicin	>13	<12	16	S
Kanamycin	>18	<13	20	S
Neomycin	>17	<12	12	R
Penicillin G	>29	<20	10	R
Polymyxin B	>12	<8	10	I
Streptomycin	>15	<11	11	R
Vancomycin	>12	<9	15	S
Tetracycline	>19	<14	25	S

R = resistant, I = intermediate, S = sensitive.

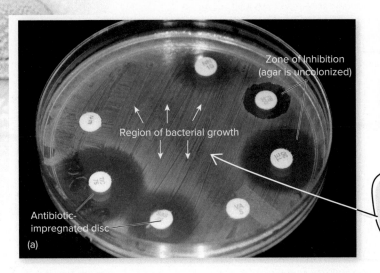

(a)

Figure 10.1 **Technique for preparation and interpretation of disc diffusion tests.** **(a)** Photograph of an agar dish with a bacterial isolate distributed evenly all over its surface. After inoculation, the antibiotic-containing discs are dropped onto the agar and it is incubated. The damp agar moistens the discs and antibiotic diffuses out of them in widening arcs. The concentration of the antibiotic decreases as it gets farther from the disc. If the test bacterium is sensitive to a drug, a zone of inhibition develops around its disc. Roughly speaking, the larger the size of this zone, the greater is the bacterium's sensitivity to the drug. The diameter of each zone is measured in millimeters and evaluated for susceptibility or resistance by means of a comparative standard (see table 10.3). **(b)** The principles illustrated in a drawing.

(a) Don Rubbelk/McGraw-Hill Education

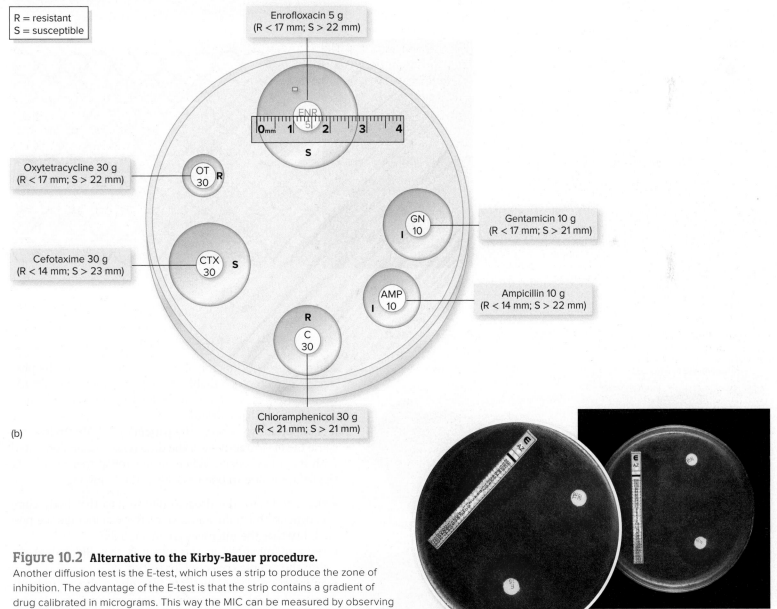

Figure 10.2 **Alternative to the Kirby-Bauer procedure.**
Another diffusion test is the E-test, which uses a strip to produce the zone of inhibition. The advantage of the E-test is that the strip contains a gradient of drug calibrated in micrograms. This way the MIC can be measured by observing the mark on the strip that corresponds to the edge of the zone of inhibition.
(photo): Source: CDC/Dr. Richard Facklam

Figure 10.3 Tube dilution test for determining the minimum inhibitory concentration (MIC).
(a) The antibiotic is diluted serially through tubes of liquid nutrient from right to left. All tubes are then inoculated with an identical amount of a test bacterium and then incubated. The first tube on the left is a control that lacks the drug and shows maximum growth. The dilution of the first tube in the series that shows no growth (no turbidity) is the MIC. **(b)** Microbroth dilution in a multiwell plate adapted for eukaryotic pathogens. Here, amphotericin B, flucytosine, and several azole drugs are tested on a pathogenic yeast. Pink indicates growth and blue, no growth. Numbers indicate the dilution of the MIC, and Xs show the first well without growth.

(b) Dr. David Ellis

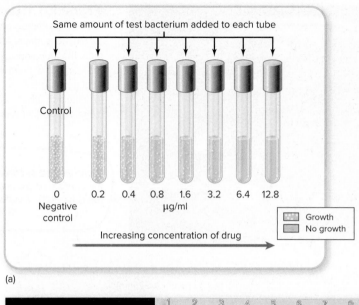

(a)

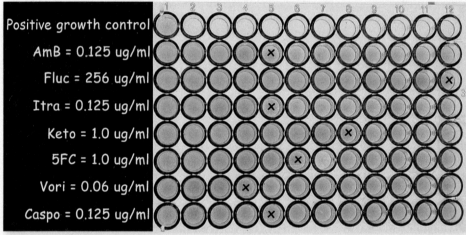

(b)

©Terry Vine/Blend Images LLC

The MIC and Therapeutic Index

The results of antimicrobial sensitivity tests guide the physician's choice of a suitable drug. If treatment has already been started, it is important to determine if the tests bear out the use of that particular drug. Once therapy has begun, it is important to observe the patient's clinical response because the *in vitro* activity of the drug is not always correlated with its *in vivo* effect. When antimicrobial treatment fails, the failure is due to one or more of the following:

- the inability of the drug to diffuse into that body compartment (the brain, joints, skin); this can include the possibility that the microbes are in a biofilm;
- resistant microbes in the infection that didn't make it into the sample collected for testing;
- an infection caused by more than one pathogen (mixed), some of which are resistant to the drug; or
- in outpatient situations you have to also consider the possibility that the patient did not take the antimicrobials correctly.

If therapy does fail, a different drug, combined therapy, or a different method of administration must be considered.

Because drug toxicity to the host is a concern, it is best to choose the one with high selective toxicity for the infectious agent and low human toxicity. The **therapeutic index (TI)** is defined as the ratio of the dose of the drug that is toxic to humans compared to its minimum effective (therapeutic) dose. The closer these two figures are (the smaller the ratio), the greater the potential for toxic drug reactions. For example, a drug that has a therapeutic index of

$$\frac{10 \ \mu g/mL \ (\text{toxic dose})}{9 \ \mu g/mL \ (\text{MIC})} \ TI = 1.1$$

is a riskier choice than one with a therapeutic index of

$$\frac{10 \ \mu g/mL}{1 \ \mu g/mL} \ TI = 10$$

When a series of drugs being considered for therapy have similar MICs, the drug with the highest therapeutic index usually has the widest margin of safety.

The physician must also take a careful history of the patient to discover any preexisting medical conditions that will influence the activity of the drug or the response of the patient. A history of allergy to a certain class of drugs precludes the use of that drug and any drugs related to it. Underlying liver or kidney disease will ordinarily require changing the drug therapy because these organs play such an important part in metabolizing or excreting the drug. Infants, the elderly, and pregnant women require special precautions. For example, age can diminish gastrointestinal absorption and organ function, and most antimicrobial drugs cross the placenta and could affect fetal development.

Patients must be asked about other drugs they are taking because incompatibilities can result in increased toxicity or failure of one or more of the drugs. This includes over-the-counter drugs and dietary supplements. For example, the combination of aminoglycosides and cephalosporins can be toxic to kidneys; antacids reduce the absorption of isoniazid; and the interaction of tetracycline or rifampin with oral contraceptives can abolish the contraceptive's effect. Some drug combinations (penicillin with certain aminoglycosides, or amphotericin B with flucytosine) act synergistically, so that reduced doses of each can be used in combined therapy.

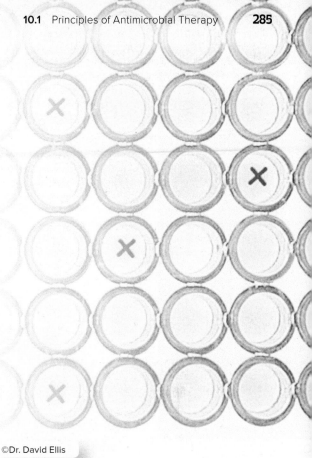

©Dr. David Ellis

The Art and Science of Choosing an Antimicrobial Drug

Even when all the information is in, the final choice of a drug is not always easy or straightforward. Consider the hypothetical case of an elderly alcoholic patient with pneumonia caused by *Klebsiella* and complicated by diminished liver and kidney function. All drugs must be given by injection because of prior damage to the gastrointestinal lining and poor absorption. Drug tests show that the infectious agent is sensitive to third-generation cephalosporins, gentamicin, imipenem, and azlocillin. The patient's history shows previous allergy to the penicillins, so these would be ruled out. Drug interactions occur between alcohol and the cephalosporins, which are also associated with serious bleeding in elderly patients, so this may not be a good choice. Aminoglycosides such as gentamicin are toxic to the kidneys and poorly cleared by damaged kidneys. Imipenem causes intestinal discomfort, but it has less toxicity and would be a viable choice.

In the case of a cancer patient with severe systemic *Candida* infection, there will be fewer criteria to weigh. Intravenous antifungals are the only possible choices, despite drug toxicity and other possible adverse side effects.

Adding to the complexity of choosing an antimicrobial is our new knowledge of the human microbiome, especially in the gut. Oral antimicrobials are chemically modified by the microbes in the gut, in different ways depending on which microbes are there. This accounts for substantial variability in how different humans respond to the drugs, and is an active area of research.

©Jupiterimages/Thinkstock/Getty Images

While choosing the right drug is an art and a science, requiring the consideration of many different things, the process has been made simpler–or at least more portable–with the advent of smartphones and applications ("apps"). Most doctors now have the information literally at their fingertips, when they pull their smartphones out of their pockets.

10.1 LEARNING OUTCOMES—Assess Your Progress

1. State the main goal of antimicrobial treatment.
2. Identify the sources for the most commonly used antimicrobials.
3. Describe two methods for testing antimicrobial susceptibility.
4. Define *therapeutic index*, and identify whether a high or a low index is preferable.

10.2 Interactions Between Drug and Microbe

The goal of antimicrobial drugs is either to disrupt the cell processes or structures of bacteria, fungi, and protozoa or to inhibit virus replication. Many antimicrobial drugs interfere with the function of enzymes required to synthesize or assemble macromolecules, or they destroy structures already formed in the cell. Above all, drugs should be **selectively toxic,** which means they should kill or inhibit microbial cells without simultaneously damaging host tissues. This concept of selective toxicity is central to antibiotic treatment, and the best drugs in current use are those that block the actions or synthesis of molecules in microorganisms but not in vertebrate cells. Examples of drugs with excellent selective toxicity are those that block the synthesis of the cell wall in bacteria (e.g., penicillins). They have low toxicity and few direct effects on human cells because human cells lack the chemical peptidoglycan and are thus unaffected by this action of the antibiotic. Among the most toxic to human cells are drugs that act upon a structure common to both the infective agent and the host cell, such as the cytoplasmic membrane (e.g., amphotericin B, used to treat fungal infections). As the characteristics of the infectious agent become more and more similar to those of the host cell, selective toxicity becomes more difficult to achieve, and undesirable side effects are more likely to occur.

Mechanisms of Drug Action

The goal of chemotherapy is to disrupt the structure or function of an organism to the point where it can no longer survive. For that reason, the first step toward this goal is to identify the structural and metabolic needs of a living cell. Once the requirements of a living cell have been determined, methods of removing, disrupting,

Coxiella burnetii multiplying inside a vacuole of a human cell.

Source: National Institute of Allergy and Infectious Diseases (NIAID)/NIH/USHHS

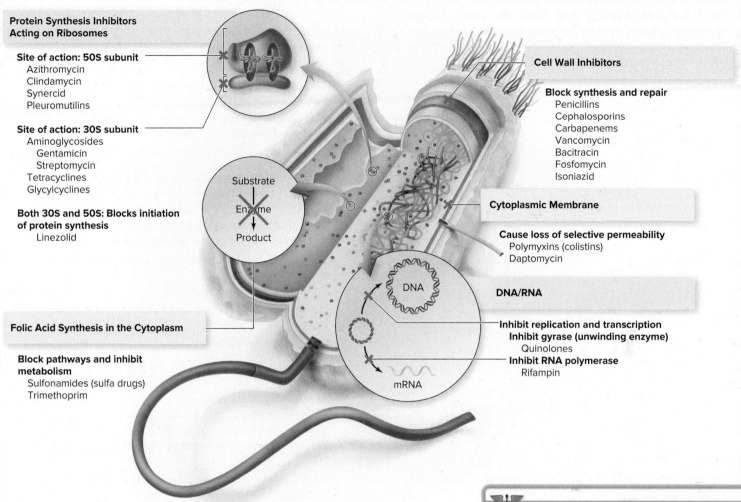

Protein Synthesis Inhibitors
Acting on Ribosomes

Site of action: 50S subunit
Azithromycin
Clindamycin
Synercid
Pleuromutilins

Site of action: 30S subunit
Aminoglycosides
Gentamicin
Streptomycin
Tetracyclines
Glycylcyclines

Both 30S and 50S: Blocks initiation
of protein synthesis
Linezolid

Folic Acid Synthesis in the Cytoplasm

Block pathways and inhibit
metabolism
Sulfonamides (sulfa drugs)
Trimethoprim

Substrate
Enzyme
Product

Cell Wall Inhibitors

Block synthesis and repair
Penicillins
Cephalosporins
Carbapenems
Vancomycin
Bacitracin
Fosfomycin
Isoniazid

Cytoplasmic Membrane

Cause loss of selective permeability
Polymyxins (colistins)
Daptomycin

DNA/RNA

Inhibit replication and transcription
Inhibit gyrase (unwinding enzyme)
Quinolones
Inhibit RNA polymerase
Rifampin

DNA
mRNA

Figure 10.4 **Primary sites of action of antimicrobial drugs on bacterial cells.**

or interfering with these requirements can be used as potential antimicrobial strategies. The metabolism of an actively dividing cell is marked by the production of new cell wall components (in most cells), DNA, RNA, proteins, and cytoplasmic membrane. For that reason, most currently used antimicrobial drugs are divided into categories based on which of these metabolic targets they affect. These categories are outlined in **figure 10.4** and include the following:

1. inhibition of cell wall synthesis,
2. inhibition of nucleic acid (RNA and DNA) structure and function,
3. inhibition of the ribosome in protein synthesis,
4. interference with cytoplasmic membrane structure or function, and
5. inhibition of folic acid synthesis.

As you will see, these categories are not completely discrete, and some effects can overlap. **Table 10.4** describes these categories, as well as common drugs comprising each of these categories. We will see later that new drugs being developed often have other modes of action.

NCLEX® PREP

2. A patient is admitted to the hospital with pneumonia. A chest x-ray is obtained and a specimen is sent to the laboratory for culture and sensitivity testing. The provider orders antibiotic therapy to begin. Which of the following statements by the RN is appropriate when providing patient education about antibiotic therapy?

a. *"We will be giving you antibiotics because respiratory viruses are common in the community right now and we suspect you have a viral pneumonia."*

b. *"We are starting broad-spectrum and narrow-spectrum antibiotics so we can be sure to treat all the organisms that could be causing your infection."*

c. *"Your chest x-ray results appear to show bacterial pneumonia. As results from the lab specimen return, your antibiotic therapy may be changed."*

d. *"It is important to notify me immediately if you develop nausea or diarrhea as it indicates the antibiotic is not working."*

Table 10.4 Specific Drugs and Their Metabolic Targets

Drug Class/Mechanism of Action	Subgroups	Uses/Characteristics
Drugs That Target the Cell Wall		
Penicillins	Penicillins G and V	Natural forms used to treat gram-positive cocci, some gram-negative bacteria (meningococci, syphilis, spirochetes)
	Ampicillin, carbenicillin, amoxicillin	Have a broader spectrum of action, are semisynthetic; used against gram-negative enteric rods
	Nafcillin, cloxacillin	Useful in treating infections caused by some penicillinase-producing bacteria (penicillinase is one type of beta-lactamase, a class of enzymes that destroy the beta-lactam ring in some antibiotics; some bacteria can produce these enzymes, making them resistant to these types of antibiotics)
©j4m3z/iStockphoto/Getty Images	Clavulanic acid	Inhibits beta-lactamase enzymes; added to penicillins to increase their effectiveness in the presence of penicillinase-producing bacteria
Cephalosporins	Cefazolin	First generation*; most effective against gram-positive cocci, few gram-negative bacteria
	Cefaclor	Second generation; more effective than first generation against gram-negative bacteria such as *Enterobacter, Proteus*, and *Haemophilus*
	Cephalexin, cefotaxime	Third generation; broad-spectrum, particularly against enteric bacteria that produce beta-lactamases
	Ceftriaxone	Third generation; semisynthetic broad-spectrum drug that treats wide variety of urinary, skin, respiratory, and nervous system infections
	Cefepime	Fourth generation
	Cegtaroline	Fifth generation; used against methicillin-resistant *Staphylococcus aureus* (MRSA) and also against penicillin-resistant gram-positive and gram-negative bacteria
Carbapenems	Doripenem, imipenem	Powerful but potentially toxic; reserved for use when other drugs are not effective
	Aztreonam	Narrow-spectrum; used to treat gram-negative aerobic bacilli causing pneumonia, septicemia, and urinary tract infections; effective for those who are allergic to penicillin
Miscellaneous Drugs That Target the Cell Wall	Bacitracin	Narrow-spectrum; used to combat superficial skin infections caused by streptococci and staphylococci; main ingredient in Neosporin
	Isoniazid	Used to treat *Mycobacterium tuberculosis*, but only against growing cells; used in combination with other drugs in active tuberculosis
	Vancomycin	Narrow-spectrum of action; used to treat staphylococcal infections in cases of penicillin and methicillin resistance or in patients with an allergy to penicillin
	Fosfomycin tromethamine	Phosphoric acid agent; effective as an alternative treatment for urinary tract infection caused by enteric bacteria
Drugs That Target Protein Synthesis		
Aminoglycosides Insert on sites on the 30S subunit and cause the misreading of the mRNA, leading to abnormal proteins	Streptomycin	Broad-spectrum; used to treat infections caused by gram-negative rods, certain gram-positive bacteria; used to treat bubonic plague, tularemia, and tuberculosis

*New improved versions of drugs are referred to as new "generations."

Table 10.4 (continued)

Drug Class/Mechanism of Action	Subgroups	Uses/Characteristics
Drugs That Target Protein Synthesis (*continued*)		
Tetracyclines Block the attachment of tRNA on the A acceptor site and stop further protein synthesis	Tetracycline	Effective against gram-positive and gram-negative rods and cocci, aerobic and anaerobic bacteria, mycoplasmas, rickettsias, and spirochetes
Glycylcyclines	Tigecycline	Newer derivative of tetracycline; effective against bacteria that have become resistant to tetracyclines
Macrolides Inhibit translocation of the subunit during translation (erythromycin)	Erythromycin, clarithromycin, azithromycin	Relatively broad-spectrum, semisynthetic; used in treating ear, respiratory, and skin infections, as well as *Mycobacterium* infections in AIDS patients
Miscellaneous Drugs That Target Protein Synthesis	Clindamycin	Broad-spectrum antibiotic used to treat penicillin-resistant staphylococci, and serious anaerobic infections of the stomach and intestines unresponsive to other antibiotics
	Quinupristin + dalfopristin (Synercid)	A combined antibiotic from the streptogramin group of drugs; effective against *Staphylococcus* and *Enterococcus* species causing endocarditis and surgical infections, including resistant strains
	Linezolid	Synthetic drug from the oxazolidinones; a novel drug that inhibits the initiation of protein synthesis; used to treat antibiotic-resistant organisms such as MRSA and VRE
Drugs That Target Folic Acid Synthesis		
Sulfonamides Interfere with folate metabolism by blocking enzymes required for the synthesis of tetrahydrofolate, which is needed by the cells for folic acid synthesis and eventual production of DNA, RNA, and amino acids	Sulfamethoxazole	Used to treat shigellosis, acute urinary tract infections, certain protozoal infections
	Silver sulfadiazine	Used to treat burns, eye infections (in ointment and solution forms)
	Trimethoprim	Inhibits the enzymatic step in an important metabolic pathway that comes just before the step inhibited by sulfonamides; trimethoprim often given in conjunction with sulfamethoxazole because of this synergistic effect; used to treat *Pneumocystis jiroveci* in AIDS patients
Drugs That Target DNA or RNA		
Fluoroquinolones Inhibit DNA unwinding enzymes or helicases, thereby stopping DNA transcription	Ciprofloxacin, ofloxacin	Second generation
	Levofloxacin	Third generation; used against gram-positive organisms, including some that are resistant to other drugs
Miscellaneous Drugs That Target DNA or RNA	Rifampin	Limited in spectrum because it cannot pass through the cell envelope of many gram-negative bacilli; mainly used to treat infections caused by gram-positive rods and cocci and a few gram-negative bacteria; used to treat leprosy and tuberculosis
Drugs That Target Cytoplasmic or Cell Membranes		
Polymyxins (colistins) Interact with membrane phospholipids; distort the cell surface and cause leakage of protein and nitrogen bases, particularly in gram-negative bacteria	Polymyxin B	Used to treat drug-resistant *Pseudomonas aeruginosa* and severe urinary tract infections caused by gram-negative rods. Colistins are toxic to humans. They are used as a last-resort antibiotic when bacteria are resistant to all other antibiotics.
	Daptomycin	Most active against gram-positive bacteria

©Ingram Publishing

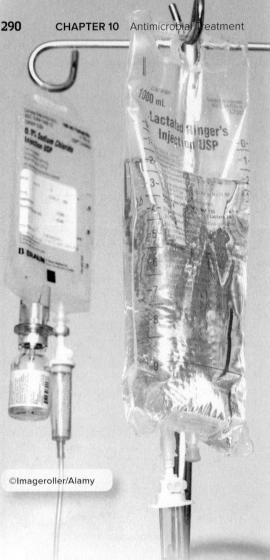

©Imageroller/Alamy

Spectrum of Activity

One of the most useful ways of categorizing antimicrobials, which you have already encountered in the previous section, is to designate them as either **broad-spectrum** or **narrow-spectrum.** Broad-spectrum drugs are effective against more than one group of bacteria, whereas narrow-spectrum drugs generally target a specific group. **Table 10.5** demonstrates that tetracyclines are broad-spectrum, whereas polymyxin and even penicillins are narrow-spectrum agents.

Because penicillin is such a familiar antibiotic, and because the alterations in the molecule over the years illustrate how antibiotics are developed and improved upon, we provide an overview in **table 10.6.** Here you will see that original penicillin was narrow-spectrum and susceptible to microbial counterattacks. Later penicillins were developed to overcome those two limitations.

Unfortunately, we don't seem to be getting it right often enough. A 2016 study by the Centers for Disease Control and Prevention (CDC) and the Pew Charitable Trust looked at the three most common infections in the United States: sinus infections, sore throats, and ear infections. The study found that 48% of these patients were not prescribed the antibiotics recommended by national standards for these infections. While we just discussed reasons the first-line antibiotic might not be prescribed, experts say that only about 20% of patients are likely to have reasons the first-line drug should not be used. Many times the nonrecommended antibiotics are more broad-spectrum than the recommended ones, and are therefore more likely to lead to microbiome disruption and increased antibiotic resistance.

Referring back to table 10.4, you can view details about various antimicrobial drugs based on which of the five major mechanisms they target.

Antibiotics and Biofilms

As you read in chapter 6, biofilm inhabitants behave differently than their free-living counterparts. One of the major ways they differ—at least from a medical perspective—is that they are often unaffected by the same antimicrobials that work against them

Table 10.5 Spectrum of Activity for Antibiotics

Bacteria	Mycobacteria	Gram-Negative Bacteria	Gram-Positive Bacteria	Chlamydias	Rickettsias
Examples of diseases	Tuberculosis	Salmonellosis, plague, gonorrhea	Strep throat, staph infections*	Chlamydia, trachoma	Rocky Mountain spotted fever
Spectrum of activity of various antibiotics	Isoniazid Streptomycin Tobramycin Polymyxin	Carbapenems Tetracyclines Sulfonamides Cephalosporins Penicillins			
Are there normal biota in this group?	Yes	Yes	Yes	Probably	None known

*Note that some members of a bacterial group may not be affected by the antibiotics indicated, due to acquired or natural resistance. In other words, exceptions do exist.

Table 10.6 Characteristics of Selected Penicillin Drugs

Name	Spectrum of Action	Uses, Advantages	Disadvantages
Penicillin G	Narrow	Best drug of choice when bacteria are sensitive; low cost; low toxicity	Can be hydrolyzed by penicillinase; allergies occur; requires injection
Penicillin V	Narrow	Good absorption from intestine; otherwise, similar to Penicillin G	Hydrolysis by penicillinase; allergies
Methicillin, nafcillin	Narrow	Not usually susceptible to penicillinase	Poor absorption; allergies; growing resistance; methicillin rarely used now
Ampicillin	Broad	Works on gram-negative bacilli	Can be hydrolyzed by penicillinase; allergies; only fair absorption
Amoxicillin	Broad	Gram-negative infections; good absorption	Hydrolysis by penicillinase; allergies
Azlocillin, mezlocillin, ticarcillin	Very broad	Effective against *Pseudomonas* species; low toxicity compared with aminoglycosides	Allergies; susceptible to many beta-lactamases

when they are free-living. When this was first recognized, it was assumed that it was a problem of penetration, that the (often ionically charged) antimicrobial drugs could not penetrate the sticky extracellular material surrounding biofilm organisms. While that is a factor, there is something more important contributing to biofilm resistance: the different phenotype expressed by biofilm bacteria. When secured to surfaces, they express different genes and therefore have different antibiotic susceptibility profiles.

Years of research have so far not yielded an obvious solution to this problem, though there are several partially successful strategies. One of these involves interrupting the quorum-sensing pathways that mediate communication between cells and may change phenotypic expression. Daptomycin (trade name: Cubicin), a lipopeptide that is effective in deep tissue infections with resistant bacteria, has also shown some success in biofilm infection treatment. Also, some researchers have found that adding DNase to their antibiotics can help with penetration of the antibiotic through the extracellular debris–apparently some of which is DNA from lysed cells.

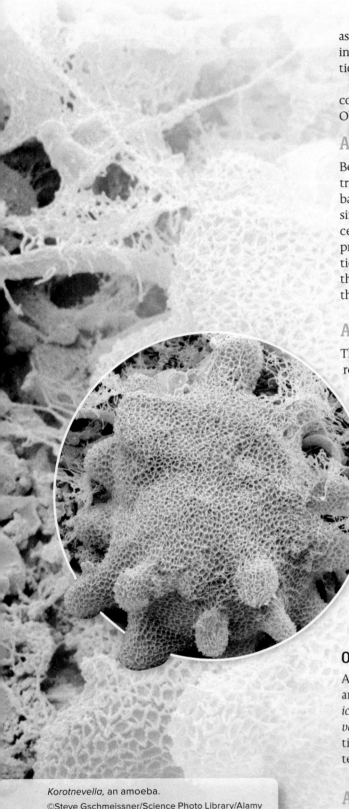

Korotnevella, an amoeba.
©Steve Gschmeissner/Science Photo Library/Alamy

Many biofilm infections can be found on biomaterials inserted in the body, such as cardiac or urinary catheters. These can be impregnated with antibiotics prior to insertion to prevent colonization. This, of course, cannot be done with biofilm infections of natural tissues, such as the prostate or middle ear.

Interestingly, it appears that chemotherapy with some antibiotics—notably aminoglycosides—can cause bacteria to form biofilms at a higher rate than they otherwise would. Obviously there is much more to come in understanding biofilms and their control.

Agents to Treat Fungal Infections

Because the cells of fungi are eukaryotic, they present special problems in drug treatment. For one, the great majority of antimicrobial drugs are designed to act on bacteria and are generally ineffective in combating fungal infections. For another, the similarities between fungal and human cells often mean that drugs toxic to fungal cells are also capable of harming human tissues. A few agents with special antifungal properties have been developed for treating systemic and superficial fungal infections. Four main drug groups currently in use are the macrolide polyene antibiotics, the azoles, the echinocandins, and allylamines. **Table 10.7** describes in further detail the antifungal drug groups and their actions.

Agents to Treat Protozoal Infections

The enormous diversity among protozoal and helminthic parasites and their corresponding therapies reach far beyond the scope of this textbook; however, a few of the more common drugs are surveyed here and described again for particular diseases in the organ systems chapters.

Antimalarial Drugs: Quinine and Its Relatives

Quinine, extracted from the bark of the cinchona tree, was the principal treatment for malaria for hundreds of years, but it has been replaced by the synthesized quinolones, mainly chloroquine and primaquine, which have less toxicity to humans. Because there are several species of *Plasmodium* (the malaria parasite) and many stages in its life cycle, no single drug is universally effective for every species and stage. Furthermore, the geographical region where the malaria infection was acquired must be taken into account, as different parts of the world harbor strains of the parasite with differing drug resistances. A drug called artemisinin, which comes originally from a plant called sweet wormwood, used for centuries in Chinese traditional medicine, is one of the staples for malaria treatment.

Other Anti-Protozoal Drugs

A widely used amoebicide, metronidazole (Flagyl), is effective in treating mild and severe intestinal infections and hepatic disease caused by *Entamoeba histolytica*. Given orally, it also can treat infections by *Giardia lamblia* and *Trichomonas vaginalis* (described in chapters 20 and 21, respectively). Other drugs with antiprotozoal activities are quinacrine (a quinine-based drug), sulfonamides, and tetracyclines.

Agents to Treat Helminthic Infections

Treating helminthic infections has been one of the most difficult and challenging of all chemotherapeutic tasks. Flukes, tapeworms, and roundworms are much larger parasites than other microorganisms and, being animals, have greater similarities to human physiology. Also, the usual strategy of using drugs to block their reproduction is usually not successful in eradicating the adult worms. The most effective drugs immobilize, disintegrate, or inhibit the metabolism in all stages of the life cycle.

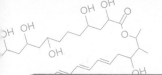

Table 10.7 Agents Used to Treat Fungal Infections

Drug Group	Drug Examples	Action
Macrolide polyenes	Amphotericin B (structure shown above in gray)	• Bind to fungal membranes, causing loss of selective permeability; extremely versatile • Can be used to treat skin, mucous membrane lesions caused by *Candida albicans* • Injectable form of the drug can be used to treat histoplasmosis and *Cryptococcus* meningitis
Azoles	Ketoconazole, fluconazole, miconazole, and clotrimazole	• Interfere with sterol synthesis in fungi • Ketoconazole–cutaneous mycoses, vaginal and oral candidiasis, systemic mycoses • Fluconazole–AIDS-related mycoses (aspergillosis, *Cryptococcus* meningitis) • Clotrimazole and miconazole–used to treat infections in the skin, mouth, and vagina
Echinocandins	Micafungin, caspofungin	• Inhibit fungal cell wall synthesis • Used against *Candida* strains and aspergillosis
Allylamines	Terbinafine, naftifine	• Inhibit enzyme critical for ergosterol synthesis • Used to treat ringworm and other cutaneous mycoses

Albendazole is a broad-spectrum antiparasitic drug used for several types of roundworm intestinal infestations. This drug works locally in the intestine to inhibit the function of the microtubules of worms, eggs, and larvae. This means the parasites can no longer utilize glucose, which leads to their demise. The compound pyrantel paralyzes the muscles of intestinal roundworms. Consequently, the worms are unable to maintain their grip on the intestinal wall and are expelled along with the feces by the normal peristaltic action of the bowel. Two newer antihelminthic drugs are praziquantel, a treatment for various tapeworm and fluke infections, and ivermectin, used for strongyloidiasis and onchocerciasis in humans. Helminthic diseases are described in chapter 20 because these organisms spend a large part of their life cycles in the digestive tract.

Agents to Treat Viral Infections

The treatment of viral infections presents unique problems. With a virus, we are dealing with an infectious agent that relies upon the host cell for the vast majority of its metabolic functions. With currently used drugs, disrupting viral metabolism requires that we disrupt the metabolism of the host cell to a much greater extent than is desirable. Put another way, selective toxicity with regard to viral infection is difficult to achieve because a single metabolic system is responsible for the well-being of both virus and host. Although viral diseases such as measles, mumps, and hepatitis are routinely prevented by the use of effective vaccinations, epidemics of AIDS, influenza, and even the "commonness" of the common cold attest to the need for more effective medications for the treatment of viral pathogens.

The currently used antiviral drugs were developed to target specific points in the infectious cycle of viruses. Three major modes of action are as follows:

1. barring penetration of the virus into the host cell,
2. blocking the transcription and translation of viral molecules, and
3. preventing the maturation of viral particles.

Table 10.8 presents an overview of antivirals from each of these categories.

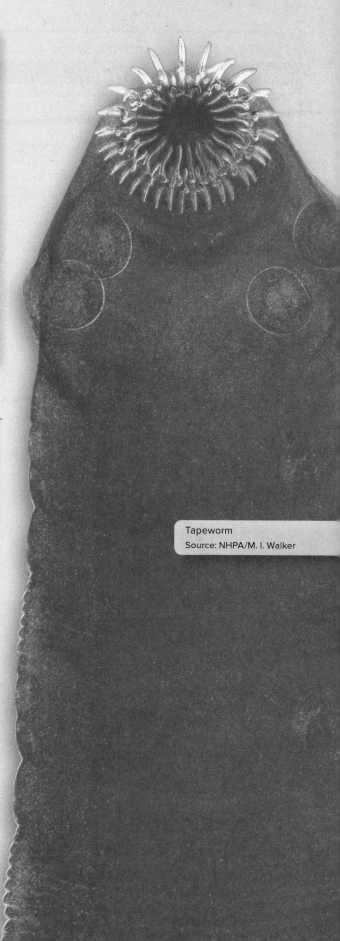

Tapeworm
Source: NHPA/M. I. Walker

Table 10.8 Actions of Antiviral Drugs

Mode of Action	Examples	Effects of Drug
Inhibition of Virus Entry Receptor/fusion/uncoating inhibitors	Enfuvirtide (Fuzeon®)	Blocks HIV infection by preventing the binding of viral proteins to cell receptor, thereby preventing fusion of virus with cell
	Amantadine (Symmetrel®)	Blocks entry of influenza virus by interfering with fusion of virus with cell membrane
Inhibition of Nucleic Acid Synthesis	Acyclovir (Zovirax®), other "cyclovirs," vidarabine	Nucleoside analogs; take the place of actual nucleosides and terminate DNA replication in **herpesviruses**
	Ribavirin, Remdesivir	Purine analogs, used for respiratory syncytial virus (RSV) and some hemorrhagic fever viruses, Remdesivir in trials for SARS-CoV-2
	Zidovudine (AZT), lamivudine (3TC), didanosine (ddI), zalcitabine (ddC), and stavudine (d4T)	Nucleotide analog reverse transcriptase (RT) inhibitors; stop the action of reverse transcriptase in HIV, blocking viral DNA production
	Nevirapine, efavirenz, delavirdine	Nonnucleotide analog reverse transcriptase inhibitors; attach to HIV RT binding site, stopping its action
Inhibition of Viral Assembly/Release	Indinavir, saquinavir	Protease inhibitors; insert into HIV protease, stopping its action and resulting in inactive noninfectious viruses
	Zanamivir (Relenza®) and oseltamivir (Tamiflu®)	Neuraminidase inhibitors; block exit of influenza viruses from host cells

The influenza virus.

Source: National Institute of Allergy and Infectious Diseases (NIAID)/NIH/USHHS

10.2 LEARNING OUTCOMES—Assess Your Progress

5. Explain the concept of selective toxicity.
6. List the five major targets of antimicrobial agents.
7. Identify which categories of drugs are most selectively toxic and why.
8. Distinguish between broad-spectrum and narrow-spectrum antimicrobials, and explain the significance of the distinction.
9. Identify the microbes against which the various penicillins are effective.
10. Explain the mode of action of penicillinases and their role in treatment decisions.
11. Identify two antimicrobials that act by inhibiting protein synthesis.
12. Explain how drugs targeting folic acid synthesis work.
13. Identify one example of a fluoroquinolone.
14. Describe the mode of action of drugs that target the cytoplasmic or cell membrane.
15. Discuss how treatments of biofilm and nonbiofilm infections differ.
16. Name the four main categories of antifungal agents.
17. Explain why antiprotozoal and antihelminthic drugs are likely to be more toxic than antibacterial drugs.
18. List the three major targets of action of antiviral drugs.

10.3 Antimicrobial Resistance

One unfortunate outcome of the use of antimicrobials is the development of microbial **drug resistance,** an adaptive response in which microorganisms begin to tolerate an amount of drug that would ordinarily be inhibitory. The ability to escape the effects of antimicrobial drugs is due to the genetic versatility and adaptability of microbial populations. The property of drug resistance can be intrinsic as well. Resistance is called intrinsic when it is a fixed trait of a microbe. For example, all microbes are intrinsically resistant to antibiotics they themselves produce. Of much greater importance is the acquisition of resistance to a drug by a microbe that was previously sensitive to the drug. In our context, the term *antibiotic resistance* will refer to this last type of acquired resistance.

How Does Resistance Develop?

Contrary to popular belief, antibiotic resistance is an ancient phenomenon. Because most of our oldest therapeutically used antibiotics are natural products from fungi and bacteria, resistance to them has been a survival strategy for *other* microbes for as long as the microbes have been around. The scope of the problem in terms of using the antibiotics as treatments for humans became apparent in the 1980s and 1990s, when scientists and physicians witnessed treatment failures on a large scale. The acquisition of drug resistance is not always a result of exposure to the drug. This adds another dimension to the efforts to prolong antibiotic effectiveness, which so far have focused on limiting the amount of antibiotic in the environment. We see now that this is important but not enough to prevent microorganisms from developing resistance altogether.

Whether antibiotics are present or not, microbes become newly resistant to a drug after one of the following two events occurs:

1. spontaneous mutations in critical chromosomal genes or
2. acquisition of entire new genes or sets of genes via horizontal transfer from another species.

There may be a third mechanism of acquiring resistance to a drug, which is a phenotypic, not a genotypic, adaptation. Recent studies suggest that bacteria can "go to sleep" when exposed to antibiotics, meaning they will slow or stop their metabolism so that they cannot be harmed by the antibiotic. They can then rev back up after the antibiotic concentration decreases. Sometimes these bacteria are called "persisters." (This is one reason biofilm bacteria are less susceptible to antibiotics than free-living bacteria are.)

Also, some fungi have an additional option for becoming antibiotic-resistant, as discovered in 2014. In these species of fungi, a small regulatory RNA known as interfering RNA–or RNAi–has been found to bind to a genetic sequence temporarily. When it is bound, the gene is silenced and the target of the antibiotic is not manufactured by the fungus, thus rendering it temporarily resistant to that drug. This provides the fungus more flexibility, allowing it to express the gene later when the antibiotic is no longer present. This reversible mechanism is called an **epigenetic** event, and the gene silencing is called an **epimutation.** In the next sections, we will focus on the two genetic changes that can result in acquired resistance.

Drug resistance that is found on chromosomes usually results from spontaneous random mutations in bacterial populations. The chance that such a mutation will be advantageous is minimal, and the chance that it will confer resistance to a specific drug is lower still. Nevertheless, given the huge numbers of microorganisms in any population and the constant rate of mutation, such mutations do occur. The end result varies from slight changes in microbial sensitivity, which can be overcome by larger doses of the drug, to complete loss of sensitivity.

Resistance occurring through horizontal transfer originates from plasmids called resistance (R) factors that are transferred through conjugation, transformation, or transduction. Such traits are "lying in wait" for an opportunity to be expressed and to confer adaptability on the species. Many bacteria also maintain transposable drug resistance sequences (transposons) that are duplicated and inserted from one plasmid to another or from a plasmid to the chromosome. Chromosomal genes and plasmids containing codes for drug resistance are often faithfully replicated and inherited by all subsequent progeny. This sharing of resistance genes accounts for the rapid proliferation of drug-resistant species. As you have read in earlier chapters, gene transfers are extremely frequent in nature, with genes coming from totally unrelated bacteria, viruses, and other organisms living in the body's normal biota and the environment.

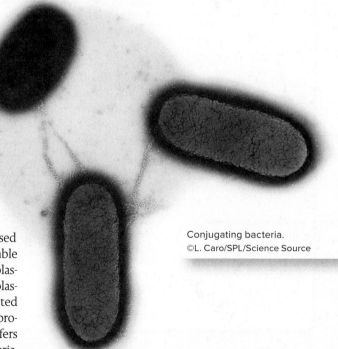

Conjugating bacteria.
©L. Caro/SPL/Science Source

Specific Mechanisms of Drug Resistance

Mutations and horizontal transfer, just described, result in mutants that have acquired one of several mechanisms of drug resistance. The actual changes that take place inside the cell as a result of these gene changes mostly fall into five categories **(table 10.9)**. Think of the two pathways to acquiring resistance as the "how" and these five mechanisms as the "what."

Table 10.9 Mechanisms of Drug Resistance

Mechanism	Example	
New enzymes are synthesized, inactivating the drug (occurs when new genes are acquired).	Bacterial exoenzymes called beta-lactamases or penicillinases hydrolyze the beta-lactam ring structure of some penicillins and cephalosporins, rendering the drugs inactive.	
Permeability or uptake of the drug into the bacterium is decreased (occurs via mutation).		
Drug is immediately eliminated (occurs through the acquisition of new genes).	Many bacteria possess multidrug-resistant (MDR) pumps that actively transport drugs out of cells, conferring drug resistance on many gram-positive and gram-negative pathogens.	
Binding sites for drugs are decreased in number and/or affinity (occurs via mutation or through the acquisition of new genes).	Erythromycin and clindamycin resistance is associated with an alteration on the 50S ribosomal binding site.	
An affected metabolic pathway is shut down, or an alternative pathway is used (occurs via mutation of original enzymes).	Sulfonamide and trimethoprim resistance develop when microbes deviate from the usual patterns of folic acid synthesis.	

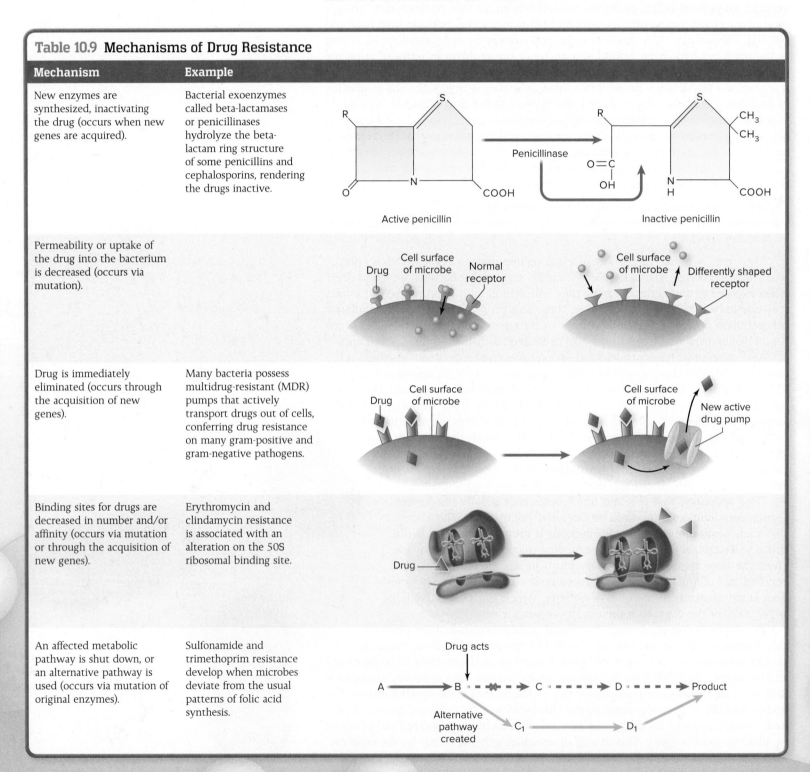

Natural Selection and Drug Resistance

So far, we have been considering drug resistance at the cellular and molecular levels, but its full impact is felt only if this resistance occurs throughout the cell population. Let us examine how this might happen and its long-term therapeutic consequences.

Any large population of microbes is likely to contain a few individual cells that are already drug resistant because of prior mutations or transfer of plasmids **(figure 10.5).** While we now know that many things can cause these "odd balls" to start overtaking the population, one of the most reliable ways to make this happen is for the correct antibiotic to be present. Individuals that are sensitive to the antibiotic are inhibited or destroyed, and resistant individuals survive and proliferate. During the population growth that follows, offspring of these resistant microbes will inherit this drug resistance. In time, the replacement population will have a preponderance of the drug-resistant forms and can eventually become completely resistant (figure 10.5). In ecological terms, the environmental factor (in this case, the drug) has put selection pressure on the population, allowing the more "fit" microbe (the drug-resistant one) to survive, and the whole population has evolved to a condition of drug resistance.

An Urgent Problem

Textbooks generally avoid using superlatives and exclamation points. But the danger of antibiotic resistance can hardly be overstated. The CDC issued a "Threat Report" about this issue for the first time in 2013, and they continue to monitor the situation, which they label "potentially catastrophic." Now it is such a large problem that an economist recently predicted that worldwide deaths from antibiotic-resistant microbes will surpass deaths from cancer by 2050. This crisis has the world's attention. In September 2016 a meeting of the General Assembly of the United Nations was held on this topic, only the fourth time in history a health issue was the topic of such a meeting. (The other issues were HIV, Ebola, and noncommunicable diseases.)

Even though the antibiotic era began less than 80 years ago, we became so confident it would be permanent that we may have forgotten what it was like before antibiotics were available. Certain types of pneumonia had a 50% fatality rate. Strep throat could turn deadly overnight. Infected wounds often required amputations or led to death. Yet the effectiveness of our currently available antibiotics is declining, in some cases very rapidly. There is a real possibility that we will enter a post-antibiotic era, in which some infections will be untreatable.

New and effective antibiotics have been slow to come to market. There are a variety of reasons for this, including the economic reality that antibiotics (taken in

The United Nations building in New York City.
©Arnaldo Jr/Shutterstock

Figure 10.5 **How antibiotic resistance happens.**

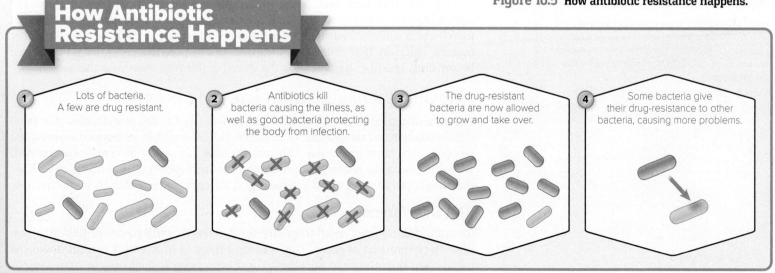

How Antibiotic Resistance Happens

1. Lots of bacteria. A few are drug resistant.
2. Antibiotics kill bacteria causing the illness, as well as good bacteria protecting the body from infection.
3. The drug-resistant bacteria are now allowed to grow and take over.
4. Some bacteria give their drug-resistance to other bacteria, causing more problems.

short courses) are not as lucrative for drug manufacturers as drugs for chronic diseases, which must often be taken for life, even though they are just as time-consuming and expensive to develop. Policy makers are starting to create incentives for the discovery and manufacture of new antibiotics.

The CDC has categorized resistant bacteria into three groups, termed "hazard levels." The three hazard levels are *concerning, serious,* and *urgent.* We will look at them individually in the disease chapters later in the book.

Urgent Threats

- *Clostridioides difficile* (*C. diff*)
- Carbapenem-resistant Enterobacteriaceae (CRE)
- Drug-resistant *Neisseria gonorrhoeae*

Serious Threats

- Multidrug-resistant *Acinetobacter*
- Drug-resistant *Campylobacter*
- Fluconazole-resistant *Candida* (a fungus)
- Extended spectrum β-lactamase-producing Enterobacteriaceae (ESBLs)
- Vancomycin-resistant *Enterococcus* (VRE)
- Multidrug-resistant *Pseudomonas aeruginosa*
- Drug-resistant non-typhoidal *Salmonella*
- Drug-resistant *Salmonella* typhi
- Drug-resistant *Shigella*
- Methicillin-resistant *Staphylococcus aureus* (MRSA)
- Drug-resistant *Streptococcus pneumoniae*
- Drug-resistant tuberculosis

Concerning Threats

- Vancomycin-resistant *Staphylococcus aureus* (VRSA)
- Erythromycin-resistant Group A *Streptococcus*
- Clindamycin-resistant Group B *Streptococcus*

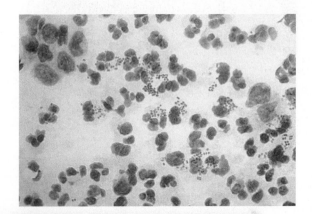

Gram stain of discharge from a person infected with *Neisseria gonorrhoeae*. The large pink objects are *neutrophils* and the small *diplococci* are the gram-negative *N. gonorrhoeae* inside the neutrophils.
©CDC/Dr. Norman Jacobs

In the United States alone, 2 million people a year become infected with resistant bacteria, and at least 23,000 deaths are attributed to them.

In the fall of 2015 bacteria containing the gene *mcr-1* were found in China. This gene makes them resistant to colistin (also known as polymyxin), which is considered a last-resort antibiotic for bacteria that are multiple-drug resistant, including the carbapenem-resistant Enterobacteriaceae (CRE) in the Urgent Threat category above. Subsequently they have been found in humans and animals in the United States and dozens of other countries. A recent study found that 8% of Enterobacteriaceae collected from hospital surfaces contained the *mcr-1* gene. The gene can easily be spread to other bacteria. This raises the possibility of the spread of infections that can no longer be treated by any drug. The CDC is monitoring the spread of this gene closely for that reason.

New Approaches to Antimicrobial Therapy

The pace of antibiotic discovery and creation was rapid in the few decades after penicillin revolutionized the practice of medicine. But new antibiotics stopped entering the pipeline in the latter part of the 20th century. Now that we are in a crisis situation with antibiotic resistance, researchers are taking novel approaches to developing new drugs that they hope can replace, and improve upon, the ones that are losing their effectiveness.

RNA Interference

RNA interference, you recall from chapter 8, refers to small pieces of RNA that regulate the expression of genes. This is being exploited in attempts to shut down the metabolism of pathogenic microbes. There have been several human trials of RNA interference, including trials to evaluate the effectiveness of synthetic RNAs in treating hepatitis C and respiratory syncytial virus.

Defense Peptides

Other researchers are looking into proteins called defense peptides. Host defense peptides are peptides of 20 to 50 amino acids that are secreted as part of the mammalian innate immune system. They have names such as defensin, magainins, and protegrins. Some bacteria produce similar peptides. These are called bacteriocins and lantibiotics. Both host and bacterial defense peptides have multiple activities against bacteria–inserting in their membranes and also targeting other structures in the cells. Other peptides target viruses. Fuzeon, a drug used in HIV infections, is a peptide that inhibits fusion of the virus to human cells.

Researchers have also had success with artificially engineered peptides. A molecule called SNAPP–structurally nanoengineered antimicrobial peptide polymer–has shown promise in killing multiple types of bacteria that have become resistant to other drugs.

CRISPR

The CRISPR system, which you read about in chapter 8, provides the first true hope that science may be capable of overcoming antibiotic resistance. CRISPR is a system found in bacteria that can cause very specific cuts in genes. Researchers turned this into a powerful genetic engineering tool and one possible way to use it is as a method to treat antibiotic-resistant infections, together with an antibiotic. Scientists in Israel delivered CRISPR packaged in a phage vector that only infected the offending bacteria. CRISPR then destroyed the genes that were causing the resistance, allowing the antibiotics to be effective.

Drugs from Noncultivable Bacteria

Our current antibiotics were either derived from or inspired by natural chemicals secreted by microorganisms themselves. But, considering the fact that more than 99% of all microbes are noncultivable, some researchers have reasoned that there could be a treasure chest of new antimicrobial substances in soil and elsewhere in the environment that will never be identified with traditional methods. Researchers in Massachusetts designed a unique chamber in which to allow soil bacteria to grow in their natural environment and then harvested an antibiotic they call teixobactin. It inhibits cell wall synthesis using a unique mechanism. Most strikingly, bacteria seem unable to–or at least very slow to–develop resistance to the drug.

©MedicalRF.com

Bacteriophages

Sometimes the low-tech solution can be the best one. Before antibiotics were discovered and developed in the 1930s and 1940s, scientists in the former Soviet Union and other regions were treating patients with mixtures of bacteriophages, whose host bacterium was presumed to be causing the illness. The antibiotic era put a temporary end to serious consideration of this strategy, but it is receiving renewed attention as we search for alternatives to antibiotics. One recent human trial used a mixture of bacteriophages specific for *Pseudomonas aeruginosa* to treat ear infections caused by the bacterium. These infections are found in the form of biofilms and have been extremely difficult to treat. The phage preparation called Biophage-PA successfully treated patients who had experienced long-term antibiotic-resistant infections. Other researchers are incorporating phages into wound dressings. One clear advantage to bacteriophage treatments is the extreme specificity of the phages–only one species of bacterium is affected, leaving the normal inhabitants of the body, and the body itself, alone. Phage treatments are currently licensed to decontaminate food and food-processing facilities, as well as to treat poultry for *Salmonella*.

Helping Nature Along

Other novel approaches to controlling infections include the use of **probiotics** and **prebiotics.** Probiotics are preparations of live microorganisms that are fed to animals and humans to improve the intestinal biota. This can serve to replace microbes lost

Medical Moment

Why Do Antibiotics Cause Diarrhea?

You are prescribed an antibiotic for strep throat. You take it as prescribed. Then, suddenly, you have diarrhea to go along with your fever and sore throat—just what you didn't need!

Why do we often get diarrhea when we take antibiotics? We have resident microbial biota in our intestines. We can refer to these helpful bacteria as "good" bacteria and the potentially illness-causing bacteria as "bad" bacteria. When we take antibiotics, we upset the delicate balance between numbers of good and bad bacteria so that the bad can begin to outnumber the good. This may result in diarrhea, an unpleasant side effect of many antibiotics.

Q. Is postantibiotic diarrhea considered drug toxicity, an allergic reaction, or neither?

Answer in Appendix B.

Figure 10.6 Examples of probiotic grocery items. The bacteria pictured at right top are *Lactobacillus casei,* a common live culture in probiotic foods.
Kathleen Talaro; Steve Gschmeissner/Science Photo Library/Alamy Stock Photo

during antimicrobial therapy or simply to augment the biota that is already there. This is a slightly more sophisticated application of methods that have long been used in an empiric fashion, for instance, by people who consume yogurt because of the beneficial microbes it contains. Recent years have seen a huge increase in the numbers of probiotic products sold in ordinary grocery stores **(figure 10.6).** Experts generally find these products safe, and in some cases they can be effective. Probiotics are thought to be useful for the management of food allergies. Their role in the stimulation of mucosal immunity is also being investigated.

Prebiotics are nutrients that encourage the growth of beneficial microbes in the intestine. For instance, certain sugars such as fructans are thought to encourage the growth of the beneficial *Bifidobacterium* in the large intestine and to discourage the growth of potential pathogens.

A technique that is gaining mainstream acceptance is the use of fecal transplants in the treatment of recurrent *Clostridioides difficile* infection and ulcerative colitis. This procedure involves the transfer of feces from healthy patients via colonoscopy. This is, in fact, just an adaptation of probiotics. But instead of a few beneficial bacterial species being given orally with the hope that they will establish themselves in the intestines, a rich microbiota is administered directly to the site it must colonize–the large intestine. Work is also underway to develop a pill containing the appropriate species, with a coating that will enable it to remain intact as it traverses the stomach and small intestine and releases the bacteria in the lower intestine. Clearly, the use of these agents is a different type of antimicrobial strategy than we are used to, but it may have its place in a future in which traditional antibiotics are more problematic.

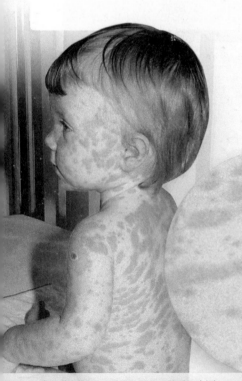

An allergic reaction to an antimicrobial medication.
Source: Centers for Disease Control and Prevention

10.3 LEARNING OUTCOMES—Assess Your Progress

19. Discuss two main ways that microbes acquire antimicrobial resistance.
20. List five cellular or structural mechanisms that microbes use to resist antimicrobials.
21. Discuss at least two novel antimicrobial strategies that are under investigation.

10.4 Interactions Between Drug and Host

Until now, this chapter has focused on the interaction between antimicrobials and the microorganisms they target. During an infection, the microbe is living in or on a host; therefore, the drug is administered to the host even though its target is the microbe. For that reason, the effect of the drug on the host must always be considered.

It is estimated that at least 5% of all persons taking an antimicrobial drug experience some type of serious adverse reaction to it. The major side effects of drugs fall into one of three categories: direct damage to tissues through toxicity, allergic reactions, and disruption in the balance of normal microbial biota. The damage incurred by antimicrobial drugs can be short term and reversible or permanent, and it ranges in severity from cosmetic to lethal.

Toxicity to Organs

Drugs most often adversely affect the following organs: the liver (hepatotoxic), kidneys (nephrotoxic), gastrointestinal tract, cardiovascular system and blood-forming tissue (hemotoxic), nervous system (neurotoxic), respiratory tract, skin, bones, and teeth. The potential toxic effects of drugs on the body, along with the responsible drugs, are detailed in **table 10.10.**

NCLEX® PREP

3. Which medication could be used against gram-negative bacteria, gram-positive bacteria, chlamydias, and rickettsias?
   a. *tobramycin*
   b. *penicillin*
   c. *tetracyclines*
   d. *cephalosporins and sulfonamides*

Table 10.10 Major Adverse Toxic Reactions to Common Drug Groups

Antimicrobial Drug	Primary Damage or Abnormality Produced
Antibacterials	
Penicillin G	Rash, hives, watery eyes
Carbenicillin	Abnormal bleeding
Ampicillin	Diarrhea and enterocolitis
Cephalosporins	Inhibition of platelet function
	Decreased circulation of white blood cells; nephritis
Sulfonamides	Formation of crystals in kidney; blockage of urine flow
	Hemolysis
	Reduction in number of red blood cells
Polymyxin (colistin)	Kidney damage
	Weakened muscular responses
Quinolones (ciprofloxacin, norfloxacin)	Headache, dizziness, tremors, GI distress
Rifampin	Damage to hepatic cells
	Dermatitis
Antifungals	
Amphotericin B	Disruption of kidney function
Flucytosine	Decreased number of white blood cells
Antiprotozoal Drugs	
Metronidazole	Nausea, vomiting
Antihelminthics	
Pyrantel	Intestinal irritation
	Headache, dizziness
Antivirals	
Acyclovir	Seizures, confusion
	Rash
Amantadine	Nervousness, light-headedness
	Nausea
AZT	Immunosuppression, anemia

NCLEX® PREP

4. A patient is prescribed an antibiotic known to be nephrotoxic. Which of the following are important metrics to follow in assessing if kidney function is impaired?
   a. *renal function labs*
   b. *urine output*
   c. *stool output*
   d. *a and b*
   e. *a, b, and c*

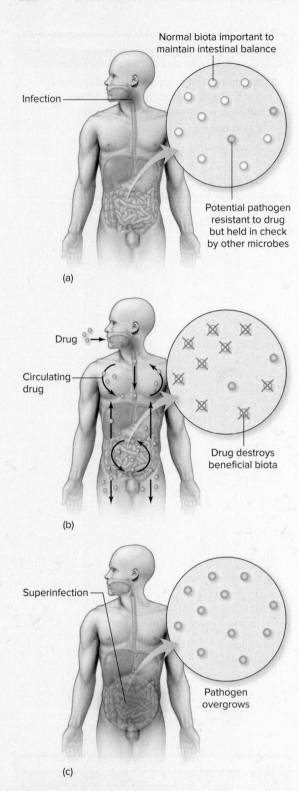

Normal biota important to maintain intestinal balance

Infection

Potential pathogen resistant to drug but held in check by other microbes

(a)

Drug

Circulating drug

Drug destroys beneficial biota

(b)

Superinfection

Pathogen overgrows

(c)

Figure 10.7 The role of antimicrobials in disrupting microbial biota and causing superinfections. **(a)** A primary infection in the throat is treated with an oral antibiotic. **(b)** The drug is carried to the intestine and is absorbed into the circulation. **(c)** The primary infection is cured, but drug-resistant pathogens have survived and create an intestinal superinfection.

Allergic Responses to Drugs

One of the most frequent drug reactions is **allergy.** This reaction occurs because the drug acts as an antigen (a foreign material capable of stimulating the immune system) and stimulates an allergic response. This response can be provoked by the intact drug molecule or by substances that develop from the body's metabolic alteration of the drug. In the case of penicillin, for instance, it is not the penicillin molecule itself that causes the allergic response but a product, *benzylpenicilloyl.* Allergic reactions have been reported for every major type of antimicrobial drug, but the penicillins account for the greatest number of antimicrobial allergies, followed by the sulfonamides.

People who are allergic to a drug become sensitized to it during the first contact, usually without symptoms. Once the immune system is sensitized, a second (or later) exposure to the drug can lead to a reaction such as a skin rash (hives), respiratory inflammation, and, rarely, anaphylaxis, an acute, overwhelming allergic response that develops rapidly and can be fatal. (This topic is discussed in greater detail in chapter 14.)

Suppression and Alteration of the Microbiota by Antimicrobials

Most normal, healthy body surfaces, such as the skin, large intestine, outer openings of the urogenital tract, and oral cavity, provide numerous habitats for a virtual "garden" of microorganisms. These normal colonists, or residents, called the **biota,** or microbiota, consist mostly of harmless or beneficial bacteria, but a small number can potentially be pathogens. Although we defer a more detailed discussion of this topic to chapter 11 and later chapters, here we focus on the general effects of drugs on this population.

If a broad-spectrum antimicrobial is introduced into a host to treat an infection, it will destroy "good" microbes as well as the disease-causing microbe. When this therapy destroys beneficial resident species, other microbes that were once in small numbers can begin to overgrow and cause disease. This complication is called a **superinfection (figure 10.7).** (This topic is also addressed in the Microbiome feature at the end of the chapter.)

Some common examples demonstrate how a disturbance in microbial biota leads to replacement biota and superinfection. A broad-spectrum cephalosporin used to treat a urinary tract infection by *Escherichia coli* will cure the infection, but it will also destroy the lactobacilli in the vagina that normally maintain a protective acidic environment there. The drug has no effect, however, on *Candida albicans,* a yeast that also resides in normal vaginas. Released from the inhibitory environment provided by lactobacilli, the yeasts proliferate and cause symptoms. *Candida* can cause similar superinfections of the oropharynx (thrush) and the large intestine.

Oral therapy with some antimicrobials is associated with a serious and potentially fatal condition known as *antibiotic-associated colitis* (pseudomembranous colitis). This condition is caused by the overgrowth of the endospore-forming bacterium *Clostridioides difficile.* It invades the intestinal lining and releases toxins that induce diarrhea, fever, and abdominal pain. It has become an urgent public health problem, as you read earlier.

An Antimicrobial Drug Dilemma

In the past, many physicians tended to use a "shotgun" antimicrobial therapy for minor infections, which involves administering a broad-spectrum drug instead of a more specific narrow-spectrum one. This practice led to superinfections and other adverse reactions. Importantly, it also caused the development of resistance in "bystander" microbes (normal biota) that were exposed to the drug as well. This helped to spread antibiotic resistance to pathogens. With growing awareness of the problems of antibiotic resistance, this practice is less frequent.

Tons of antimicrobial drugs produced in this country are exported to other countries, where controls are not as strict. Nearly 200 different antibiotics are sold over the counter in Latin America and Asian countries. It is common for people in these countries to self-medicate without understanding the correct medical indication. Drugs used in this way are not only likely to be ineffective, but, worse yet, they are known to be responsible for emergence of drug-resistant bacteria that subsequently cause epidemics. On top of that, about 140,000 tons of antibiotics are used at low levels in livestock around the world in order to enhance their growth. As you know, low levels of antibiotics used for the long term are the best conditions for creating antibiotic-resistant bacteria. They can then transfer into human beings. In 2017, the U.S. Food and Drug Administration banned the use of medically important antibiotics for the use of growth promotion in livestock.

10.4 LEARNING OUTCOMES—Assess Your Progress

22. Distinguish between drug toxicity and allergic reactions to drugs.

23. Explain what a superinfection is and how it occurs.

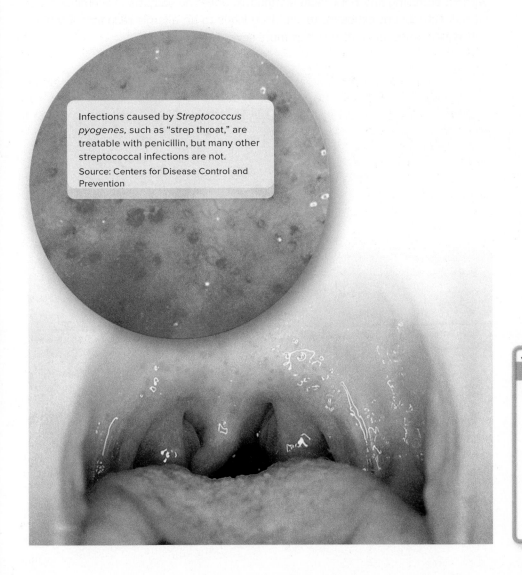

Infections caused by *Streptococcus pyogenes*, such as "strep throat," are treatable with penicillin, but many other streptococcal infections are not.

Source: Centers for Disease Control and Prevention

NCLEX® PREP

5. Mary has a urinary tract infection and is prescribed cephalexin for 10 days. Toward the end of her course of treatment, Mary develops a vaginal yeast infection. The yeast infection is an example of a/an

  a. *superinfection.*

  b. *expected complication.*

  c. *allergic reaction.*

  d. *toxic reaction.*

CASE FILE WRAP-UP

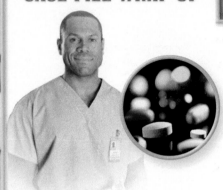

©Design Pics/Kristy-Anne Glubish (pills in boxes); ©Ryan McVay/Getty Images (health care worker); Glow Images (pills); Markos Dolopikos/Alamy (pills in package)

Cefaclor, which goes by a variety of brand names, including Ceclor, is a second-generation cephalosporin antibiotic used to treat gram-negative bacteria. When it first came out, it was popular among physicians for treating otitis media infections; however, cefaclor caused rash in a large number of patients. It has now fallen out of favor as newer cephalosporins have come along.

People with an allergy to penicillin may not be able to take cefaclor, as there is a possibility of a cross-reaction occurring. This is due to a similarity in the side chain structure of penicillins and some cephalosporins. The choice of whether to avoid the use of cephalosporins in individuals who are allergic to penicillin is often based on the allergic manifestations and the drug under consideration. Some people are able to take cephalosporins without suffering any adverse effects but should be aware of the possibility of reaction, however remote.

Allergic response to an antibiotic occurs because the drug acts as an antigen, a foreign agent that stimulates the immune response. People who are allergic to antibiotics usually become sensitized during the first contact, usually without suffering any noticeable symptoms. Once the body has become sensitized, subsequent exposure to the drug leads to an allergic response. Each subsequent exposure can result in more severe symptoms.

How Antibiotics Impact the Microbiome

While scientists have long had an appreciation of the fact that our "normal biota" can be beneficial, it is only in the past several years that we have discovered the absolutely integral role the microbiome plays in human health. Now consider this: For the last 80 years we have been dousing ourselves with antibiotics. Antibiotics were and are a modern medical miracle, one of the most important public health advances in history. But now we have to start considering what effects they have on the entire microbiome, not just the one causing the offending infection.

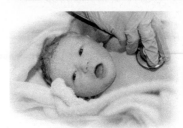

©Design Pics/LJM Photo/Getty Images

One study in 2016 tried to do just this. A group from Washington University (WU) in St. Louis examined the collateral effects of antibiotics on the microbiome. We have already discovered that a disrupted microbiome can lead to consequences such as altered nutrient metabolism, decreased vitamin production, decreased protection from pathogenic microbes, among other things. So the WU group collected information from many sources to provide an overview of the effects of antibiotic usage on the microbiome over the life of a human. This table summarizes some of their key findings.

Birth	The guts of babies born by cesarean section are colonized by skin bacteria as opposed to the vaginal biota that colonizes vaginal-born babies. This has been associated with higher rates of immunological disorders, asthma, and type 1 diabetes. Antibiotics are also routinely administered to the mother during a cesarean section, further disrupting the formation of a healthy gut microbiota in the infant. Premature birth results in aggressive administration of broad-spectrum antibiotics. This is associated with increased risk of sepsis and other serious consequences.
Childhood	The risk of developing type 2 diabetes has been found to be increased with the repeated use of some types of antibiotics, reflecting the role of the microbiome in metabolic diseases such as diabetes.
Adulthood	Some antibiotic-induced changes observed: (i) increased inappropriate immune activity leading to asthma, allergies, and autoimmune disease; (ii) increased susceptibility to infection and diarrhea due to decreased diversity of gut microbiome; (iii) vitamin K deficiency leading to uncontrolled bleeding due to depletion of vitamin-producing bacteria.

Source: *Genome Medicine*, vol. 8, p. 39, 2016.

Source: Shutterstock/Barbo

ANTIMICROBIAL TREATMENT

10.1 PRINCIPLES OF ANTIMICROBIAL TREATMENT

Humans are treated with antimicrobials when they are experiencing disease symptoms from microbial infection. Antimicrobials, which are naturally produced by bacteria and fungi, can be broad-spectrum or narrow-spectrum in their scope of effectiveness. The best antimicrobials are those that harm the microbe but not the human host. These are called "selectively toxic."

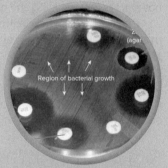

Region of bacterial growth

©Don Rubbelk/McGraw-Hill Education

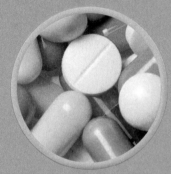

©Peter Dazeley/Photographer's Choice/Getty Images

Currently there are five main cellular targets for antibiotics in bacteria: cell wall synthesis, nucleic acid structure and function, protein synthesis, cytoplasmic membranes, and folic acid synthesis. Bacteria in biofilms respond differently to antibiotics than when they are free-floating. It is therefore difficult to eradicate biofilms in the human body. There are a great number of antibacterial drugs but a limited number that are effective against protozoa, helminths, fungi, and viruses.

INTERACTIONS BETWEEN DRUG AND MICROBE 10.2

10.3 ANTIMICROBIAL RESISTANCE

Microbes acquire genes that code for methods of inactivating or escaping the antimicrobial, or acquire mutations that affect the drug's impact. There are some microorganisms that have become resistant to nearly every antibiotic available. This situation constitutes an urgent public health problem.

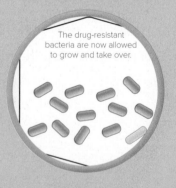

The drug-resistant bacteria are now allowed to grow and take over.

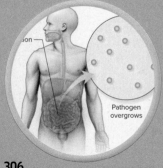

Pathogen overgrows

The three major side effects of antimicrobials are toxicity to host tissues, allergic reactions, and problems resulting from alteration of normal biota. Disruption of the gut microbiome can have wide-reaching effects on human health.

INTERACTIONS BETWEEN DRUG AND HOST 10.4

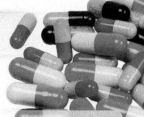

SmartGrid: From Knowledge to Critical Thinking

This *21 Question Grid* takes the topics from this chapter and arranges them with respect to the American Society for Microbiology's Undergraduate Curriculum guidelines—all six of the important "Concepts" as well as the important "Competency" of scientific literacy. Three questions are supplied about chapter content that refer to the Concept or Competency, in increasing levels of Bloom's taxonomy for learning.

ASM Concept/ Competency	A. Bloom's Level 1, 2—Remember and Understand (Choose one.)	B. Bloom's Level 3, 4—Apply and Analyze	C. Bloom's Level 5, 6—Evaluate and Create

Evolution	1. Microbial resistance to drugs is acquired through
	  a. conjugation.
	  b. transformation.
	  c. transduction.
	  d. all of these.
		2. Your friend was recently diagnosed with strep throat. One week after his treatment, his symptoms returned. Your friend tells you, "I must have become immune to the drug the doctor gave me." Discuss the validity of your friend's statement, providing evidence in support of or refuting his claim.	3. You are invited to participate in debate about the concept of evolution. Construct an opening statement supporting evolution, using antibiotic resistance in microorganisms as an example.

Cell Structure and Function	4. Drugs that prevent the formation of the bacterial cell wall are
	  a. quinolones.
	  b. penicillins.
	  c. tetracyclines.
	  d. aminoglycosides.
		5. Why does the penicillin group of antibiotics have milder toxicity than other antibiotics?	6. This chapter discussed five major cellular targets for antibiotic action. Can you imagine a different one? Which one and why?

Metabolic Pathways	7. A compound synthesized by bacteria or fungi that destroys or inhibits the growth of other microbes is a/an
	  a. synthetic drug.
	  b. antibiotic.
	  c. interferon.
	  d. competitive inhibitor.
		8. Conduct research to find out why drugs blocking folic acid synthesis are highly selective (not harmful to humans).	9. What kind of organisms do you suppose first possessed beta-lactamases, and for what reason?

Information Flow and Genetics	10. R factors are _____ that contain a code for _____.
	  a. genes, replication
	  b. plasmids, drug resistance
	  c. transposons, interferon
	  d. plasmids, conjugation
		11. You take a sample from a growth-free portion of the zone of inhibition in the Kirby-Bauer test and inoculate it onto a plate of nonselective medium. What does it mean if growth occurs on the new plate?	12. Develop an argument about why the CRISPR technique could potentially provide a permanent solution to antibiotic resistance, where other "fixes" have been temporary.

ASM Concept/ Competency	A. Bloom's Level 1, 2—Remember and Understand (Choose one.)	B. Bloom's Level 3, 4—Apply and Analyze	C. Bloom's Level 5, 6—Evaluate and Create
Microbial Systems	**13.** Treating malarial infections is theoretically difficult because **a.** the protozoal parasite is eukaryotic and therefore similar to human cells. **b.** there are several species of *Plasmodium*. **c.** no single drug can target all the life stages of *Plasmodium*. **d.** all of the above are true.	**14.** Can you think of a situation in which it would be better for a drug to be microbistatic rather than microbicidal? Discuss thoroughly.	**15.** Scientists have found evidence that exposing some bacteria to antibiotics makes them much more likely to form biofilms. Construct an argument for why this might be.
Impact of Microorganisms	**16.** Pick the true statement about antibiotics. **a.** Antibiotics were invented by chemists in the 1830s. **b.** Antibiotics are natural products of microorganisms. **c.** Antibiotics are inherently dangerous for the environment. **d.** None of the above are true.	**17.** Speculate on whether having large concentrations of antibiotics in water sources or farmland would be a positive or negative phenomenon. Defend your answer.	**18.** Imagine a postantibiotic era in which there are no drug treatments for microbial infections. What alternative strategies will make up our arsenal against morbidity (sickness) and mortality (death) due to microbes?
Scientific Thinking	**19.** An antimicrobial drug with a _____ therapeutic index is a better choice than one with a _____ therapeutic index. **a.** low; high **b.** high; low	**20.** A critically ill patient enters your emergency room, exhibiting signs and symptoms of severe septic shock. In this case, should you immediately begin treatment with a broad-spectrum drug or a narrow-spectrum drug? Explain your answer.	**21.** You are the director of the Centers for Disease Control and Prevention. What are your top three recommendations for halting the spread of *mcr-1*-containing bacteria?

Answers to the multiple-choice questions appear in Appendix A.

Visual Connections

This question connects content within and between chapters.

From chapter 8, table 8.5. Place Xs over this figure in places where bacterial protein synthesis might be inhibited by drugs.

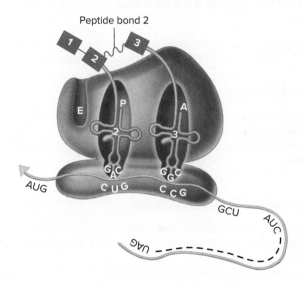

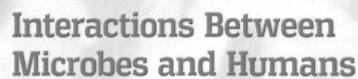

11

Interactions Between Microbes and Humans

IN THIS CHAPTER...

11.1 The Human Microbiome

1. Differentiate among the terms *colonization*, *infection*, and *disease*.
2. Identify the sites where normal biota are found in humans.
3. Discuss how microbiome research is changing our understanding of normal biota.

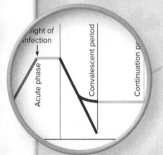

11.2 When Colonization Leads to Disease

4. Differentiate between a microbe's pathogenicity and its virulence.
5. List the steps a microbe has to take to get to the point where it can cause disease.
6. Explain the significance of polymicrobial infections.
7. List several portals of entry and exit.
8. Define *infectious dose*, and explain its role in establishing infection.
9. Describe three ways microbes cause tissue damage.
10. Compare and contrast major characteristics of endotoxin and exotoxins.
11. Provide a definition of *virulence factors*.
12. Draw a diagram of the stages of disease in a human.

11.3 Important Features of Infectious Disease Transmission

13. Differentiate among the various types of reservoirs, providing examples of each.
14. List several different modes of transmission of infectious agents.
15. Define *healthcare-associated infection*, and list two common types.
16. List Koch's postulates, and discuss when they might not be appropriate in establishing causation.

11.4 Epidemiology: The Study of Disease in Populations

17. Summarize the goals of epidemiology, and differentiate it from traditional medical practice.
18. Explain what is meant by a disease being "notifiable" or "reportable," and provide examples.
19. Define *incidence* and *prevalence*, and explain the difference between them.
20. Discuss the three major types of epidemics, and identify the epidemic curve associated with each.

©Science Photo Library/Alamy Stock Photo (secondary lung cancer); Source: NIH Roadmap for Medical Research website (Human Microbiome Project); ©silverlining56/Getty Images (opening door); Image Source/Monty Rakusen (surgeon and nurses during surgery)

CASE FILE

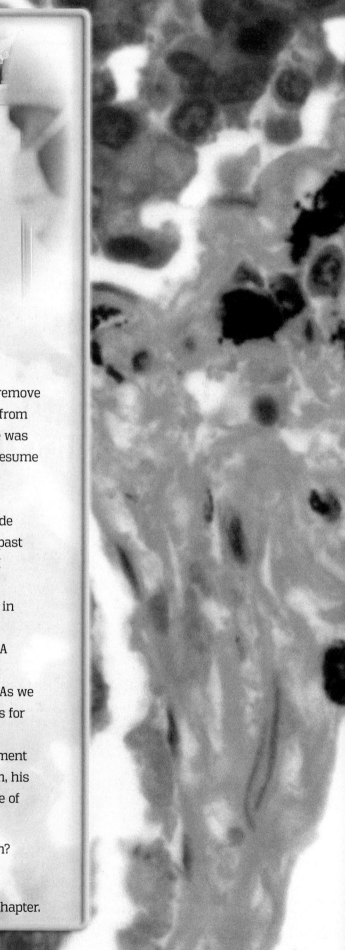

Surgical Site Infection

I was working as a nurse practitioner with a general surgery team in a large urban hospital. We consulted and performed surgery for patients with a wide variety of diagnoses. One day I was called to the emergency department to examine a 64-year-old male who had been released just three days prior. Ricardo had undergone a lobectomy procedure following a diagnosis of lung cancer. My team had performed that procedure, creating a six-inch incision in his right chest to remove a lobe of his lung. Ricardo had recovered as expected and been discharged from the hospital six days after surgery. As part of his discharge instructions, he was given specific instructions about how to care for the incision and when to resume normal activities.

When I arrived in the emergency department to examine Ricardo, he appeared uncomfortable. His heart rate was elevated and he had a low-grade fever. He reported that the pain in his chest had been worsening over the past day and he was having difficulty breathing. When I looked at his incision, I noticed that it appeared pink and scabbed. I was relieved that the incision appeared to be healing but was concerned about what could be happening in Ricardo's chest cavity.

Ricardo provided a sputum sample and I sent it to the lab for a culture. A blood sample revealed an elevated white blood cell count. A chest x-ray was obtained to assess the thoracic cavity and showed a right-side pneumonia. As we awaited the results from the sputum culture, I started a course of antibiotics for Ricardo.

Ricardo was eventually admitted to the inpatient surgical unit for treatment of his infection and oxygen support. After discussion with the surgical team, his pneumonia was determined to be a surgical site infection, which is one type of healthcare-associated infection.

- Why is Ricardo's infection classified as a healthcare-associated infection?

- What are some strategies for preventing surgical site infections?

Case File Wrap-Up appears at the end of the chapter.

11.1 The Human Microbiome

It is easy to think of humans and other mammals as discrete, stand-alone organisms that are also colonized by some nice, nonharmful microorganisms. In fact, that's what scientists thought for the last 150 years or so. But the truer picture is that humans and other mammals have the form and the physiology that they have due to *having been formed in intimate contact with their microbes.* Do you see the difference? The **human microbiome,** the sum total of all microbes found on and in a normal human, is critically important to the health and functioning of its host organism. This chapter describes the relationship between the human and microorganisms, both the good (the majority) and the bad.

Colonization, Infection, Disease—A Continuum

Our bodies consist of more microbial cells than human cells. Some microbes are colonists (normal biota), some are rapidly lost (transients), and others invade the tissues. Sometimes resident biota become invaders. For the most part our resident microbiota, the human microbiome, **colonize** us for the long term, and do not cause disease. We encounter transient microbes in our environment, most of which are harmless, although occasionally they lead to **infection,** a condition in which microbes get past the host defenses, enter the tissues, and multiply. We may or may not be aware that an infection is taking place. When the cumulative effects of the infection damage or disrupt tissues and organs, the pathologic state that results is **disease.**

A disease is defined as any deviation from health. There are hundreds of different human diseases, caused by such factors as infections, genetics, aging, and malfunctions of systems or organs. In this chapter we discuss only **infectious disease**—a pathologic state caused directly by microorganisms or their products.

When you consider the evolutionary time line (refer to figure 1.2) of bacteria and humans, it is quite clear that humans evolved in an environment that had long been populated by bacteria and single-celled eukaryotes. It should not be surprising, therefore, that humans actually require the microbes in and on their bodies for proper growth and development.

The extent to which this is true has been surprising even to the scientists studying it. Since 2007, a worldwide research effort has been underway that utilizes the powerful techniques of genome sequencing and "big data" tools. The American effort is called the **Human Microbiome Project (HMP),** and there are similar projects occurring around the world. The aim has been not only to characterize the microbes living on human bodies when they are healthy but also to determine how the microbiome differs in various diseases. Previous to this international project, scientists and clinicians mainly relied on culture techniques to determine what the "normal biota" consisted of. That meant we only knew about bacteria and fungi that we could grow in the laboratory, which vastly undercounts the actual number and variety because many—even the majority of—microbes cannot be cultured in the laboratory, though they grow quite happily on human tissues.

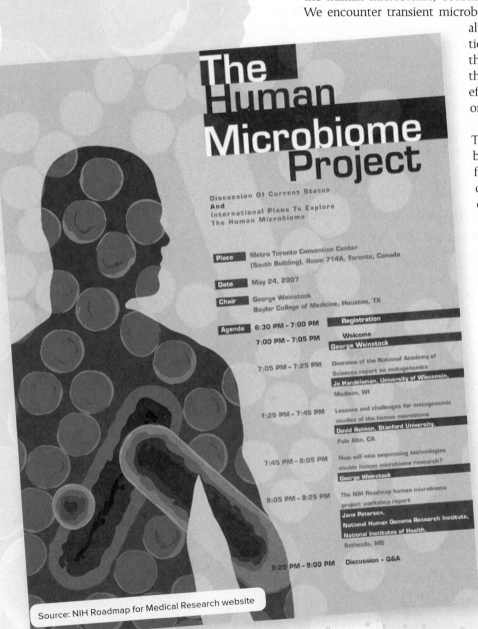

Source: NIH Roadmap for Medical Research website

Several important and surprising results have already emerged. Here are just a few examples.

- Human cells contain about 21,000 protein-encoding genes, but the microbes living in/on us contain millions. Many of the proteins produced by these genes are enzymes that help us digest our food and metabolize all kinds of substances that we could not otherwise use.
- We have a lot of microbes in places we used to think were sterile. One of the most striking examples is the lungs. They were previously thought to be sterile but actually seem to contain their own sparse but diverse microbiota.
- Viruses are not traditionally discussed in the context of normal biota. However, they are certainly present in healthy humans in vast quantities, both those that infect human cells and those infecting all the cells of our resident biota. For example, researchers think that there are 100 million viruses per gram of human feces (see The Microbiome feature at the end of this chapter). Throughout evolutionary history, viral infections (of cells of all types) have influenced the way cells and organisms and communities and, yes, earth's entire ecosystem have developed. The critical contributions of viruses are just now being rigorously studied.
- All healthy people seem to also harbor potentially dangerous pathogens in low numbers. This is a tribute to the normal biota that rarely cause disease. Their presence keeps the pathogens in check.
- The makeup of the intestinal biota can influence many facets of one's overall health. As an example, differences in the gut microbiome have preliminarily been associated with risk for Crohn's disease and obesity. Though this may not seem surprising, other research is finding associations between the composition of the gut microbiome and heart disease, asthma, autism, diabetes, and even moods. A recent study provided evidence that the composition of the gut microbiome influenced the body's response to cancer drugs, influencing how many toxic side effects occurred. As with all revolutionary breakthroughs, some of these early results will be overturned with further study. But one thing is absolutely clear: Our microbiome has a huge influence on our physiology.
- Clearly, incoming microbes that might cause disease have to encounter the microbiome as part of their path to establishing themselves in the host. Studies are showing that the relative success of viruses such as influenza and HIV in causing disease can be greatly influenced by the composition of the host microbiomes. This has implications for future treatment strategies.

Microbiome research has shown that among healthy adults, the normal microbiota varies significantly. For instance, the microbiota on a person's right hand was found to be significantly different than that on the same person's left hand. What seemed to be more important than the exact microbial profile of any given body site was the profile of proteins, especially the enzymatic capabilities. That profile remained stable across subjects, though the microbes that were supplying those enzymes could differ broadly. Scientists are in the process of cataloging other microorganisms besides bacteria via metagenomics—and just beginning to appreciate their numbers in the human microbiome.

The information about the human microbiome presented in this chapter reflects new findings. We will try to show you the differences between the old picture of normal biota in various organ systems and the new, emerging picture. At this point in medical history, it will be important to appreciate the transitioning view.

Acquiring the Microbiota

The human body offers a seemingly endless variety of environmental niches, with wide variations in temperature, pH, nutrients, and oxygen tension occurring from one area to another. Because the body provides such a range of habitats, it should not be surprising that the body supports a wide range of microbes. **Table 11.1** provides a breakdown of our current understanding of the microbiota living in and on a healthy

Table 11.1 A Shifting Understanding of Sites with Microbiota

Sites Previously Known to Harbor Normal Microbiota	
Skin and adjacent mucous membranes	External genitalia
	Vagina
Upper respiratory tract	External ear canal
Gastrointestinal tract, including mouth	External eye (lids, conjunctiva)
Outer portion of urethra	

Additional Sites Now Thought to Harbor at Least Some Normal Microbiota (or Their DNA)
Lungs (lower respiratory tract)
Bladder (and urine)
Breast and breast milk
Amniotic fluid and fetus

Sites in Which DNA from Microbiota Has Been Detected
Brain
Bloodstream

©Ian Hooton/Science Photo Library/Getty Images

host. The uppermost row contains the set of sites that microbiologists have long known to host a normal biota. The middle row presents some new sites recently found to harbor microbiota in a healthy human. The bottom row reports that two sites, the brain and the bloodstream, have both been found to contain DNA from multiple species of bacteria. Their exact role there is not entirely clear yet.

The generally antagonistic effect "good" microbes have against intruder microorganisms is called **microbial antagonism.** Normal biota exist in a steady established relationship with the host and are unlikely to be displaced by incoming microbes. This antagonistic protection is partly the result of a limited number of attachment sites in the host site, all of which are stably occupied by normal biota. This antagonism is also enabled by the chemical or physiological environment created by the resident biota, which is hostile to most other microbes. There are often members of the "normal" biota that would cause disease if they were allowed to multiply to larger numbers. Microbial antagonism is also responsible for keeping them in check.

It is important to note that hosts with compromised immune systems could very easily experience disease caused by their normal biota. Factors that weaken host defenses and increase susceptibility to infection include the following:

- old age and extreme youth (infancy, prematurity);
- genetic defects in immunity, and acquired defects in immunity (AIDS);
- surgery and organ transplants;
- underlying disease: cancer, liver malfunction, diabetes;
- chemotherapy/immunosuppressive drugs;
- physical and mental stress;
- pregnancy; and
- other infections.

Colonization of the Fetus and Newborn

Until rather recently, the uterus and its contents were thought to be sterile during embryonic and fetal development and remain essentially germ-free until just before birth **(figure 11.1).** We now know that the placenta harbors a small but significant array of bacteria.

The next important source of microbiota for a newborn is its trip through the vagina. The vaginas of healthy women of child-bearing age contain a variety of bacteria. They are especially rich in *Lactobacillus* bacteria, which are capable of digesting milk, and in other species that have been found useful in protecting the baby from skin disorders and other conditions. Of course, in the United States about 30% of babies are born by cesarean section (either through medical necessity or as elective procedures). Studies have shown that babies born this way have different gut microbiomes than those born vaginally, at least initially. There is a lot of research examining the long-term effects of this difference. Some people advocate for swabbing the newborn baby with gauze containing the mother's vaginal fluids, but there is still no evidence that this produces a long-term benefit. So far there are more questions than answers about the difference in microbiomes resulting from C-sections and vaginal births.

Where Babies Get a Microbiome

In Utero
Previously thought to be sterile, the womb has its own microbiota

Birth
Vaginal and C-section births contribute different initial microbiomes to baby

Milk
Breast milk has a microbiome, but sterilized baby formula has none

Caregivers
Family, siblings, and others share microbes with the baby

Environment
Baby can pick up microbes from anything she comes in contact with

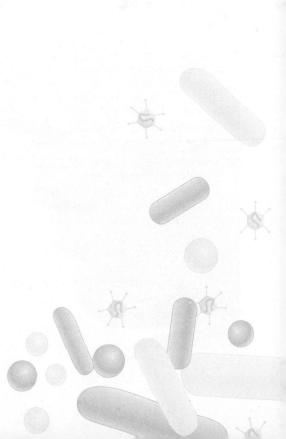

Figure 11.1 **The origins of microbiota in newborns.**

Jim Connely (ultrasound); Adam Gault/SPL/Getty Images (woman lying by doctor holding plastic spatula over a jar of water); alexmak72427/Getty Images (scar); Tetiana Mandziuk/Shutterstock (breast feeding); Pixtal/ SuperStock (bottle); Marc Romanelli/Blend Images (family); Kwame Zikomo/Purestock/SuperStock (with dog)

The baby continues to acquire resident microbiota from the environment, notably from its diet. Throughout most of evolutionary history, of course, that means human breast milk. Scientists have found that human milk contains around 600 species of bacteria and a lot of sugars that babies cannot digest. The sugars are used by healthy gut bacteria, suggesting a role for breast milk in maintaining a healthy gut microbiome in the baby. The skin, gastrointestinal tract, and portions of the respiratory and genitourinary tracts all continue to be colonized as contact continues with family members, health care personnel, the environment, and food.

11.1 LEARNING OUTCOMES—Assess Your Progress

1. Differentiate among the terms *colonization*, *infection*, and *disease*.
2. Identify the sites where normal biota are found in humans.
3. Discuss how microbiome research is changing our understanding of normal biota.

Source: NIAID, NIH, Rocky Mountain Laboratories

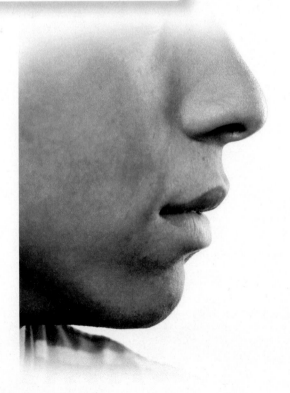

The respiratory tract is the most common portal of entry.
©Vladimir Gjorgiev/Shutterstock

11.2 When Colonization Leads to Disease

A microbe whose relationship with its host is parasitic and results in infection and disease is termed a pathogen. Various aspects of the host influence whether a microbe will have severe, mild, or no effects. Variation in the genes coding for components of the immune system—or even the anatomy of infection sites—is one of these factors. Gender, hormone levels, and overall health also play a role. **Pathogenicity,** then, is a broad concept that describes an organism's potential to cause disease and is used to divide pathogenic microbes into one of two groups. **True pathogens** (primary pathogens) are capable of causing disease in healthy persons with normal immune defenses. They are generally associated with a specific, recognizable disease, which may vary in severity from mild (colds) to severe (malarial) to fatal (rabies). Examples of true pathogens include the influenza virus, plague bacillus, and malarial protozoan.

Opportunistic pathogens cause disease when the host's defenses are compromised or when the pathogens become established in a part of the body that is not natural to them. Opportunists are not considered pathogenic to a normal healthy person and, unlike true pathogens, do not generally possess well-developed virulence properties. Examples of opportunistic pathogens include *Pseudomonas* species and *Candida albicans*. In practice, the distinction between true pathogens and opportunistic pathogens is becoming less useful as we gain more understanding of the many factors controlling the disease process. We'll discuss this more later.

The relative severity of the disease caused by a particular microorganism depends on the **virulence** of the microbe. Although the terms *pathogenicity* and *virulence* are often used interchangeably, *virulence* is the accurate term for describing the degree of pathogenicity. The virulence of a microbe is determined by its ability to

1. establish itself in the host and
2. cause damage.

There is much involved in both of these steps. To establish themselves in a host, microbes must enter the host, attach firmly to host tissues, negotiate the microbiome, and survive the host defenses. To cause damage, microbes produce toxins or induce a host response that is actually injurious to the host. Any characteristic or structure of the microbe that contributes to the preceding activities is called a **virulence factor.** Virulence can be due to single or multiple factors. In some microbes, the sources of virulence are clearly established, but in others they are not.

Polymicrobial Infections

From the very earliest days of infectious disease studies, in the 1800s, scientists isolated single microorganisms to study their effects on animals or humans. Later in this chapter you will learn about Koch's postulates, a set of rules for determining the cause of an unknown infectious condition. The postulates depend on obtaining microbes in pure culture. This procedure is extremely valuable because, as you know, most scientific experimentation requires isolating the variables and holding all but one constant.

This aspect of scientific rigor has probably kept us from understanding the roles that interacting microbes play in causing disease. Many scientists now believe that the majority of infections are **polymicrobial,** with contributions from more than one microbe. One classic set of infections is influenza (caused by a virus) and pneumonia (often caused by a bacterium). Influenza infection frequently leads to pneumonia. In another example, several types of skin infections are known to be caused by either *Staphylococcus* or *Streptococcus* species. In fact, researchers have found that when these two are cultivated together with another common skin resident, *Moraxella*, both staph and strep increase their transcription of virulence factors. Perhaps upon diagnosis one or the other is isolated from the skin, but it seems possible that the three of them together led to the disease symptoms.

In the next decade, many more polymicrobial causes will no doubt be discovered, which can help us look for unique prevention and treatment strategies.

Step One: Becoming Established—Portals of Entry

To initiate an infection, a microbe enters the tissues of the body by a characteristic route, the **portal of entry,** usually the skin or a mucous membrane. The source of the infectious agent can be exogenous or endogenous. Organisms coming from outside the body are exogenous. Organisms coming from somewhere in the same human host are considered endogenous.

The majority of pathogens have adapted to a specific portal of entry, one that provides a favorable habitat for further growth and spread. This adaptation can be so restrictive that if certain pathogens enter the "wrong" portal, they will not cause infections. For instance, inoculation of the nasal mucosa with the influenza virus is likely to give rise to the flu, but if this virus contacts only the skin, no infection will result. Occasionally, an infective agent can utilize more than one portal. For instance, *Mycobacterium tuberculosis* enters through both the respiratory and gastrointestinal tracts, and pathogens in the genera *Streptococcus* and *Staphylococcus* have adapted to invasion through several portals of entry such as the skin, urogenital tract, and respiratory tract. **Table 11.2** outlines common portals of entry, the organisms and diseases associated with these portals, and methods of entry.

Infectious Dose

Another factor crucial to the course of an infection is the quantity of microbes in the inoculating dose. For most agents, infection will proceed only if a minimum number, called the *infectious dose* (*ID*), is present. This number has been determined experimentally for many microbes. On the low end of the scale, the ID for *Coxiella burnetii*, the causative agent of Q fever, is only a single cell, and the ID is only about 10 infectious cells in tuberculosis, giardiasis, and coccidioidomycosis. The ID is 1,000 bacteria for gonorrhea and 10,000 bacteria for typhoid fever, in contrast to 1,000,000,000 bacteria in cholera. Numbers below an infectious dose will generally not result in an infection. But if the quantity is far in excess of the ID, the onset of disease can be extremely rapid.

Step Two: Becoming Established—Attaching to the Host and Interacting with the Microbiome

Adhesion is a process by which microbes gain a more stable foothold on host tissues. Because adhesion is dependent on binding between specific molecules on both the host and pathogen, a particular pathogen is limited to only those cells (and organisms)

Medical Moment

When the Portal of Entry Is Compromised

Different portals of entry have protective mechanisms to prevent infectious agents from gaining entry. For example, the eye produces tears, which not only rinse pathogens out of the eye but also contain pathogen-fighting chemicals. The skin acts as a physical barrier, providing it is intact.

What happens when there is a breach at a portal of entry? The respiratory tract is lined with cilia, fingerlike projections that protrude from cells that sweep back and forth to move particles toward the throat so that they can be swallowed rather than remain in the respiratory tract. In *primary ciliary dyskinesia,* affected individuals lack properly functioning cilia. These individuals have frequent respiratory tract infections beginning in early childhood. Chronic respiratory infections lead to bronchiectasis, which results from damage affecting the bronchial tubes leading to the lungs. This condition affects approximately one in 16,000 individuals and is usually passed down from two parents who have the defective gene but do not have the disease themselves (autosomal recessive pattern).

Q. Cigarette smokers often have a higher rate of respiratory tract infections. Can you suggest a reason for this?

Answer appears in Appendix B.

Table 11.2 Portals of Entry and Organisms Typically Involved

Portal of Entry	Organism/Disease	How Access Is Gained
Skin	*Staphylococcus aureus, Streptococcus pyogenes, Clostridium tetani*	Via nicks, abrasions, punctures, areas of broken skin
	Herpes simplex (type 1)	Via mucous membranes of the lips
	Helminth worms	Burrow through the skin
	Viruses, rickettsias, protozoa (i.e., malaria, West Nile virus)	Via insect bites
	Haemophilus aegyptius, Chlamydia trachomatis, Neisseria gonorrhoeae	Via the conjunctiva of the eye
Gastrointestinal tract	*Salmonella, Shigella, Vibrio, Escherichia coli,* poliovirus, hepatitis A, echovirus, rotavirus, enteric protozoans (*Giardia lamblia, Entamoeba histolytica*)	Through eating/drinking contaminated foods and fluids; Via fomites (inanimate objects contaminated with the infectious organism)
Respiratory tract	Bacteria causing meningitis, influenza, measles, mumps, rubella, chickenpox, common cold, *Streptococcus pneumoniae, Klebsiella, Mycoplasma, Cryptococcus, Pneumocystis, Mycobacterium tuberculosis, Histoplasma*	Via inhalation of offending organism
Urogenital tract	HIV, *Trichomonas,* hepatitis B, syphilis, *Treponema pallidum, Neisseria gonorrhoeae, Chlamydia trachomatis,* herpes, genital warts	Enter through the skin/mucosa of penis, external genitalia, vagina/cervix, urethra; may enter through an unbroken surface or through a cut or abrasion

Source: NIAID, NIH, Rocky Mountain Laboratories

to which it can bind. Once attached, the pathogen is poised advantageously to invade the body compartments. Bacterial, fungal, and protozoal pathogens attach most often by mechanisms such as fimbriae (pili), surface proteins, and adhesive slimes or capsules. Viruses attach by means of specialized receptors. In addition, parasitic worms are mechanically fastened to the portal of entry by suckers, hooks, and barbs. There are many different methods in which microbes can attach themselves to host tissues. Firm attachment to host tissues is almost always a prerequisite for causing disease because the body has so many mechanisms for flushing microbes and foreign materials from its tissues.

Of course when a new microbe enters a body site, it will also encounter the resident microbiota. The microbiome is generally very stable and "settled in" in body sites and its presence can prevent newcomers from attaching or becoming established (microbial antagonism). Recently some evidence has arisen of cases where the microbiome assisted disease-causing microbes. We know one thing for sure: The microbiome is part of the host environment that incoming microbes must deal with.

Step Three: Becoming Established—Surviving Host Defenses

Microbes that are not established in a normal biota relationship in a particular body site in a host are likely to encounter resistance from host defenses when first entering, especially from certain white blood cells called **phagocytes.** These cells ordinarily engulf and destroy pathogens by means of enzymes and antimicrobial chemicals (see chapter 12).

Antiphagocytic factors are used by some pathogens to avoid phagocytes. The antiphagocytic factors of microorganisms help them to circumvent some part of the phagocytic process (see figure 11.2*b*). The most aggressive strategy involves bacteria that kill phagocytes outright. Species of both *Streptococcus* and *Staphylococcus* produce **leukocidins,** substances that are toxic to white blood cells. Some microorganisms secrete an extracellular surface layer (slime or capsule) that makes it physically difficult for the phagocyte to engulf them. *Streptococcus pneumoniae, Salmonella typhi, Neisseria meningitidis,* and *Cryptococcus neoformans* are notable examples. Some bacteria are well adapted to survive inside phagocytes after ingestion. For instance, pathogenic species of *Legionella, Mycobacterium,* and many rickettsias are readily engulfed but are capable of avoiding destruction. The ability to survive intracellularly in phagocytes has special significance because it provides a place for the microbes to hide, grow, and be spread throughout the body.

Step Four: Causing Disease

How Virulence Factors Contribute to Tissue Damage

Virulence factors are structures or capabilities that allow a pathogen to cause infection in a host. From a microbe's perspective, they are simply adaptations it uses to invade and establish itself in the host. The effects of a pathogen's virulence factors on tissues vary greatly. Cold viruses, for example, invade and multiply but cause relatively little damage to their host. At the other end of the spectrum, pathogens such as *Clostridium tetani* or HIV severely damage or kill their host. There are three major ways that microorganisms damage their host:

1. directly through the action of enzymes or toxins (both endotoxin and exotoxins; **figure 11.2*a*),**
2. indirectly by inducing the host's defenses to respond excessively or inappropriately **(figure 11.2*b*),** and
3. epigenetic changes made to host cells by microbes **(figure 11.2*c*).**

It is obvious that enzymes, endotoxin, and exotoxins are virulence factors, but other characteristics of microbes that lead to host overreaction are also considered virulence

Bacteria attaching to one another and to a tooth in an accumulation of plaque.

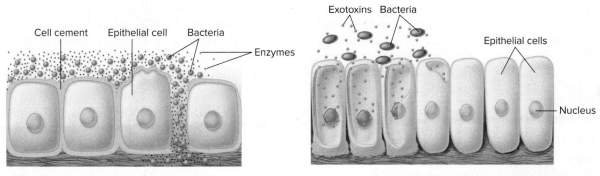

(a) **Microbes Secrete Enzymes and Toxins**

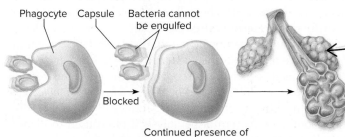

> When bacteria in the lungs have capsules, the immune system continues to pour fluids into the lungs to try to get defensive cells there. This creates the symptoms of pneumonia.
>
> – Kelly & Heidi

(b) **Host Defenses Do Damage**

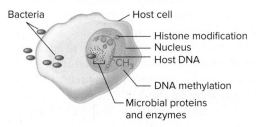

(c) **Microbes Induce Epigenetic Changes on Host DNA**

Figure 11.2   Three ways microbes damage the host.   (a) Microbial enzymes and exo- and endotoxins disrupt host cell structure or connections between host cells. **(b)** Microbes evade initial host defenses, and the host continues to react to the presence of the microbe, causing (host) damage with its response. **(c)** Microbial products make epigenetic changes to the DNA and/or supporting structures, such as histones, altering the host genes that are expressed.

factors. The capsule of *Streptococcus pneumoniae* is a good example. Its presence prevents the bacterium from being cleared from the lungs by phagocytic cells, leading to a continuous influx of fluids into the lung spaces, and the condition we know as pneumonia.

1. Enzymes and Toxins

Extracellular Enzymes   Many pathogenic bacteria, fungi, protozoa, and worms secrete **exoenzymes** that break down and inflict damage on tissues. Other enzymes dissolve the host's defense barriers and promote the spread of microbes to deeper tissues.
   Examples of enzymes are

1. mucinase, which digests the protective coating on mucous membranes and is a factor in amoebic dysentery; and
2. hyaluronidase, which digests hyaluronic acid, the substance that cements animal cells together. This enzyme is an important virulence factor in staphylococci, clostridia, streptococci, and pneumococci.

Many gastrointestinal (GI) diseases are caused by bacterial toxins that affect the GI tract. These are called enterotoxins.
©Colin Anderson/Brand X Pictures/Getty Images

Figure 11.3 Beta-hemolysis and alpha-hemolysis by different bacteria on blood agar. Beta-hemolysis, in the lower right, results in complete clearing of the red blood cells incorporated in the agar. Alpha-hemolysis, on the lower left, refers to incomplete lysis of the red blood cells, leaving a greenish tinge to the colonies and the area surrounding them.

Lisa Burgess/McGraw-Hill Education

Some enzymes react with components of the blood. Coagulase, an enzyme produced by pathogenic staphylococci, causes clotting of blood or plasma. By contrast, the bacterial kinases (streptokinase, staphylokinase) do just the opposite, dissolving fibrin clots and assisting in the invasion of damaged tissues. In fact, one form of streptokinase is used as a therapy to dissolve blood clots in patients who have problems with thrombi and embolisms.

Bacterial Toxins: A Potent Source of Cellular Damage   A **toxin** is a specific chemical product of microbes that is poisonous to other organisms. A toxin is named according to its specific target of action: Neurotoxins act on the nervous system; enterotoxins act on the intestine; hemotoxins lyse red blood cells; and nephrotoxins damage the kidneys.

There are two broad categories of bacterial toxins. **Exotoxins** are proteins with a strong specificity for a target cell and extremely powerful, sometimes deadly, effects. They generally affect cells by damaging the cell membrane and causing lysis or by disrupting intracellular function. **Hemolysins** (hee-mahl'-uh-sinz) are a class of bacterial exotoxin that disrupts the cell membrane of red blood cells (and some other cells, too). This damage causes the red blood cells to **hemolyze**—to burst and release hemoglobin pigment. Hemolysins that increase pathogenicity include the streptolysins of *Streptococcus pyogenes* and the alpha (α) and beta (β) toxins of *Staphylococcus aureus*. When colonies of bacteria growing on blood agar produce hemolysin, distinct zones appear around the colony. The pattern of hemolysis is often used to identify bacteria and determine their degree of virulence **(figure 11.3)**.

In contrast to the category *exotoxin*, which contains many different examples, the word *endotoxin* refers to a single substance. Endotoxin is actually a chemical called lipopolysaccharide (LPS), which is part of the outer membrane of gram-negative cell walls. Gram-negative bacteria shed these LPS molecules into tissues or into the circulation. Endotoxin differs from exotoxins because it has a variety of systemic effects on tissues and organs. Depending upon the amounts present, endotoxin can cause fever, inflammation, hemorrhage, and diarrhea. Blood infections by gram-negative bacteria such as *Salmonella*, *Shigella*, *Neisseria meningitidis*, and *Escherichia coli* are particularly dangerous, in that they can lead to fatal endotoxic shock. **Figure 11.4** contains important information about exotoxins and endotoxin.

2. Inducing a Damaging Host Response

Despite the extensive discussion on direct virulence factors, such as enzymes and toxins, it is probably the case that more microbial diseases are the result of indirect damage, or the host's excessive or inappropriate response to a microorganism. This is an extremely important point because it means that pathogenicity is not always a trait inherent in microorganisms but is really a consequence of the interplay between microbe and host.

3. Epigenetic Changes in Host Cells

This mechanism that microbes use to damage host cells is the most recently discovered. Microbes have been shown to shut down or activate regions of DNA in the host cell, via epigenetic processes. They can include binding to host cell histones, binding to the small RNAs used for the silencing of genes, binding to chromatin itself, and so forth. These changes can harm the host cell, or change its function in some way that favors persistence of the microbe in or on it. Some pathogens have been found to secrete proteins that interact with regions of host cell DNA that are responsible for cytoskeleton structure, and thereby cause disorganization of the cell. Sometimes these changes are passed on to new host cells, causing persistent symptoms. Some researchers speculate that this could be one source of unexplained illnesses or symptoms where no causative microbes are found.

Whether Disease Will Occur

Now that you have studied what it takes for a microbe to cause disease, we can put that in perspective with respect to a particular human host. In previous times, microbes were assigned to categories such as pathogen or nonpathogen. It is true that

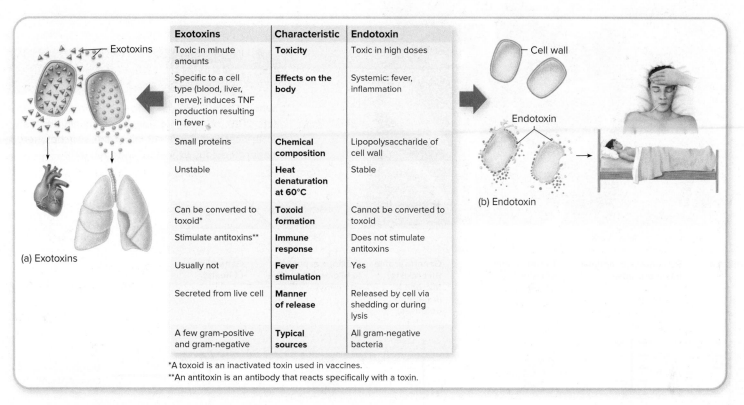

Exotoxins	Characteristic	Endotoxin
Toxic in minute amounts	**Toxicity**	Toxic in high doses
Specific to a cell type (blood, liver, nerve); induces TNF production resulting in fever	**Effects on the body**	Systemic: fever, inflammation
Small proteins	**Chemical composition**	Lipopolysaccharide of cell wall
Unstable	**Heat denaturation at 60°C**	Stable
Can be converted to toxoid*	**Toxoid formation**	Cannot be converted to toxoid
Stimulate antitoxins**	**Immune response**	Does not stimulate antitoxins
Usually not	**Fever stimulation**	Yes
Secreted from live cell	**Manner of release**	Released by cell via shedding or during lysis
A few gram-positive and gram-negative	**Typical sources**	All gram-negative bacteria

*A toxoid is an inactivated toxin used in vaccines.
**An antitoxin is an antibody that reacts specifically with a toxin.

Figure 11.4 **The origins and effects of circulating exotoxins and endotoxin.** **(a)** Exotoxins, given off by live cells, have highly specific targets and physiological effects. **(b)** Endotoxin, given off when the cell walls of gram-negative bacteria disintegrate, has more generalized physiological effects.

some microbes are very likely to cause disease (the rabies virus), and other microbes are unlikely to cause disease (*Lactobacillus* bacteria in our gut). But most microbes fall somewhere in between, meaning they can cause disease under the right circumstances. In addition, the other part of the disease equation is the human host. There are many variables among humans, both more or less permanent (genetics) and temporary (fatigue or pregnancy) characteristics. When there is a decrease in a host's ability to mount an immune defense, these patients are termed *immunocompromised*. This situation can come about due to an underlying condition (that is, pregnancy or an autoimmune disease) or due to drugs being taken for another disease, such as cancer. Because of advances in treatment of many chronic diseases and cancer, many more people who are functioning in daily society are, in fact, immunocompromised. This makes them more susceptible to infections.

Microbes that don't usually cause disease in healthy people are called opportunistic pathogens. These are microbes that take advantage of immunocompromised hosts to cause disease. It is more useful to assess each host-microbe encounter on its own terms. In every encounter between a microbe and a host, several factors determine whether disease will result. It is why you may be the only member of your family not sickened by that "bad" potato salad that everyone ate, for example.

Figure 11.5 lays out the factors on both sides—the microbe's and the host's. Several examples are illustrated, with the likely outcomes resulting from the varying "settings" on the host and microbe sides. On the microbe side (the blue column), three factors are (1) its native virulence (for example, does it produce exotoxins?), (2) how many organisms encountered the host (how close it is to its optimal infectious dose), and (3) whether it made contact with the correct portal of entry. On the host side, important factors are (1) natural genetic variability, which will impact how well components of our defense respond; (2) whether the host has seen the microbe before, either through infection or vaccination; and (3) the host's general level of health. Read each row from left to right to see how one particular interaction turns out.

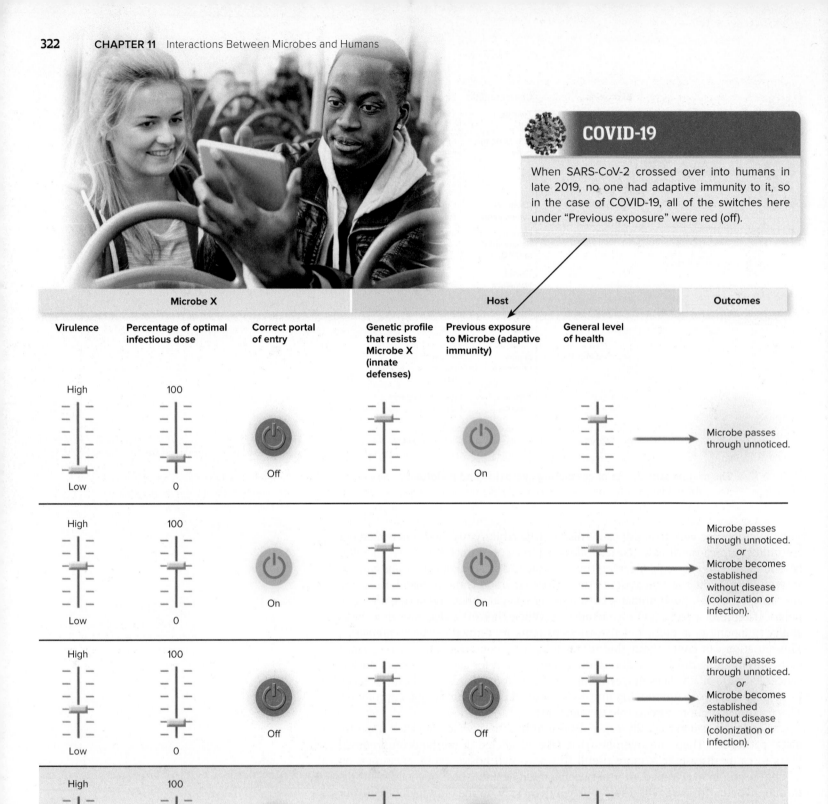

COVID-19

When SARS-CoV-2 crossed over into humans in late 2019, no one had adaptive immunity to it, so in the case of COVID-19, all of the switches here under "Previous exposure" were red (off).

Figure 11.5 Will disease result from an encounter between a (human) host and a microorganism? In most cases, all of the slider bars must be in the correct ranges and the microbe's toggle switch must be in the "on" position, while the host's toggle switch must be in the "off" position in order for disease to occur. These are just a few examples and not the only options. For instance, you can see from the third row that even when the host has no adaptive immunity, for example, the microbe does not have enough advantages to cause disease.

Dave and Les Jacobs/Kolostock/Blend Images

Table 11.3 Definitions of Infection Types

Type of Infection	Definition	Example
Localized infection	Microbes enter the body, remain confined to a specific tissue	Boils, warts, fungal skin infections
Systemic infection	Infection spreads to several sites and tissue fluids–usually via the bloodstream–but may travel by other means such as nerves (rabies) and cerebrospinal fluid (meningitis)	Mumps, rubella, chickenpox, AIDS, anthrax, typhoid, syphilis
Focal infection	Infectious agent spreads from a (usually asymptomatic) local site and is carried to other tissues	Periodontal infections leading to cardiovascular consequences
Mixed infection (polymicrobial infection)	Several agents establish themselves simultaneously at the infection site	Human bite infections, wound infections, gas gangrene
Primary infection	The initial infection	Can be any infection
Secondary infection	A second infection caused by a different microbe, which complicates a primary infection; often a result of lowered host immune defenses	Influenza complicated by pneumonia, common cold complicated by bacterial otitis media
Acute infection	Infection comes on rapidly, with severe but short-lived effects	Influenza
Chronic infection	Infection that progresses and persists over a long period of time	HIV

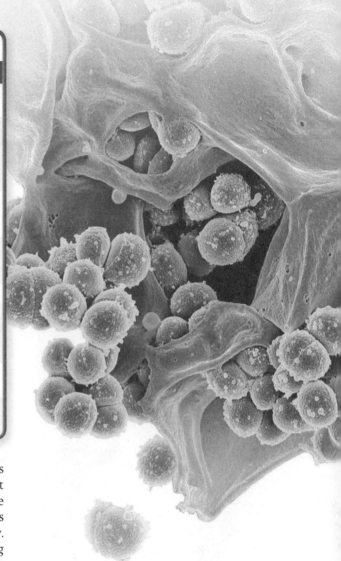

Methicillin-resistant *Staphylococcus aureus* (MRSA) (gold balls) being engulfed by a neutrophil (red).
©Callista Images/Image Source

Let's look at row 2 as an example from figure 11.5. In this scenario, the anonymous microbe has a moderate ability to be virulent, it has encountered the host with about half of its optimal infectious dose, but it did access the correct portal of entry. The host does have a genetic profile that provides high defenses for this microbe and has seen the microbe before, meaning that it will have some degree of adaptive immunity. Lastly, the host is generally healthy. So, even though the microbe has some things going for it, it will not be likely to cause disease but will simply be transient, or, if it does become established in the host, it will not cause disease. Spend some time on the other rows and you will gain an appreciation of when an infectious disease occurs or not.

Manifestations of Disease

Patterns of Infection

There is more than one way to have an infection. Clinicians use specific terms to talk about infections that are only in one location (localized infection) or in a location that is different than the initial infection (focal infection). See **table 11.3** for important terms that describe infections.

Signs and Symptoms: Warning Signals of Disease

When an infection causes pathologic changes that lead to disease, it is often accompanied by a variety of signs and symptoms. A **sign** is any objective evidence of disease as noted by an observer; a **symptom** is the subjective evidence of disease as sensed by the patient. In general, signs are more precise than symptoms, though both can have the same underlying cause. For example, an infection of the brain might present with the sign of bacteria in the spinal fluid and symptom of headache. When a disease can be identified or defined by a certain complex of signs and symptoms, it is termed a **syndrome.** Specific signs and symptoms for particular infectious diseases are covered in chapters 16 through 21.

Signs and Symptoms of Inflammation The earliest symptoms of disease usually come from the activation of the body defense process called **inflammation.** The inflammatory response includes cells and chemicals that respond innately to disruptions in the tissue. Some common symptoms of inflammation include fever, pain, soreness, and swelling. Signs of inflammation include **edema,** the accumulation of fluid in an afflicted tissue; **granulomas** and **abscesses,** walled-off collections of inflammatory cells and microbes in the tissues; and **lymphadenitis,** swollen lymph nodes.

Signs of Infection in the Blood   Changes in the number of circulating white blood cells, as determined by special counts, are considered to be signs of possible infection. **Leukocytosis** (loo″-koh′-sy-toh′-sis) is an increase in the level of white blood cells, whereas **leukopenia** (loo″-koh-pee′-nee-uh) is a decrease. Other signs of infection revolve around the occurrence of a microbe or its products in the blood. The clinical term for blood infection, **septicemia,** refers to a general state in which microorganisms are multiplying in the blood and are present in large numbers. When small numbers of bacteria or viruses are found in the blood, the correct terminology is **bacteremia,** or **viremia,** respectively which means that these microbes are present in the blood but are not necessarily multiplying.

During infection, a normal host will invariably show signs of an immune response in the form of antibodies in the serum. This fact is the basis for several serological tests used in diagnosing infectious diseases such as AIDS or syphilis. Such adaptive immune reactions indicate the body's attempt to develop specific immunities against pathogens. We concentrate on this role of the host defenses in chapters 12 and 13.

Infections That Go Unnoticed

It is rather common for an infection to produce no noticeable symptoms, even though the microbe is active in the host tissue. In other words, although infected, the host does not manifest the disease. An infection of this nature is known as **asymptomatic** or **subclinical** (inapparent) because the patient experiences no symptoms or disease and does not seek medical attention.

Step Five: Vacating the Host—Portals of Exit

Earlier, we introduced the idea that a pathogen is considered *unsuccessful* if it does not have a mechanism for leaving its host and moving to other susceptible hosts. With few exceptions, pathogens depart by a specific avenue called the **portal of exit (figure 11.6).** In most cases, the pathogen is shed or released from the body through secretion, excretion, discharge, or sloughed tissue. The usually very high number of infectious agents in these materials increases the likelihood that the pathogen will reach other hosts. In many cases, the portal of exit is the same as the portal of entry, but some pathogens use a different route. As we see in the next section, the portal of exit is a concern for epidemiologists because it greatly influences the dissemination of infection in a population.

Figure 11.7 illustrates the five steps of disease causation by microbes.

Long-Term Infections and Long-Term Effects

The apparent recovery of the host does not always mean that the microbe has been completely removed or destroyed by the host defenses. After the initial symptoms in certain chronic infectious diseases, the infectious agent retreats into a dormant state called **latency.** Throughout this latent state, the microbe can periodically become active and produce a recurrent disease. The viral agents of herpes simplex, herpes zoster, hepatitis B, AIDS, and Epstein-Barr can persist in the host for long periods. The agents of syphilis, typhoid fever, tuberculosis, and malaria can also enter into latent stages. The person harboring a persistent infectious agent may or may not shed it during the latent stage. If it is shed, such persons are chronic carriers who serve as sources of infection for the rest of the population.

Some diseases leave **sequelae** (the medical term for consequences) in the form of long-term or permanent damage to tissues or organs. For example, meningitis can result in deafness, strep throat can lead to rheumatic heart disease, Lyme disease can cause arthritis, and polio can produce paralysis.

Humans shed about 1 million skin cells—and the microbes on them—every day.
©Rubberball/Getty Images

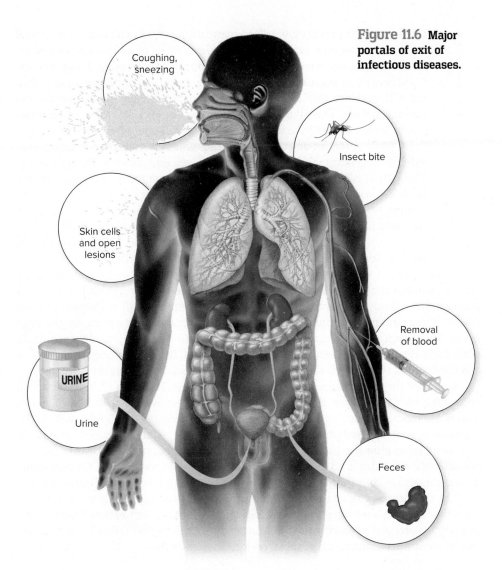

Figure 11.6 Major portals of exit of infectious diseases.

Coughing, sneezing

Insect bite

Skin cells and open lesions

Removal of blood

URINE

Urine

Feces

The Course of an Infection

There are four distinct phases of infection and disease: the incubation period, the prodrome, the acute period, and the convalescent period. The **incubation period** is the time from initial contact with the infectious agent (at the portal of entry) to the appearance of the first symptoms. During the incubation period, the agent is multiplying at the portal of entry but has not yet caused enough damage to elicit symptoms. Although this period is relatively well defined and predictable for each microorganism, it does vary according to host resistance, degree of virulence, and distance between the target organ and the portal of entry (the farther apart, the longer

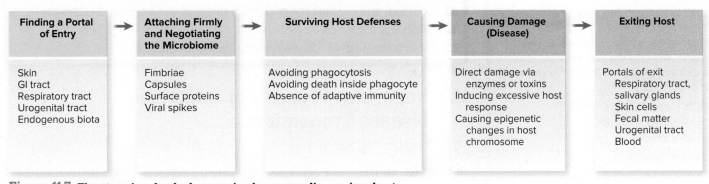

Finding a Portal of Entry	Attaching Firmly and Negotiating the Microbiome	Surviving Host Defenses	Causing Damage (Disease)	Exiting Host
Skin GI tract Respiratory tract Urogenital tract Endogenous biota	Fimbriae Capsules Surface proteins Viral spikes	Avoiding phagocytosis Avoiding death inside phagocyte Absence of adaptive immunity	Direct damage via enzymes or toxins Inducing excessive host response Causing epigenetic changes in host chromosome	Portals of exit Respiratory tract, salivary glands Skin cells Fecal matter Urogenital tract Blood

Figure 11.7 **The steps involved when a microbe causes disease in a host.**

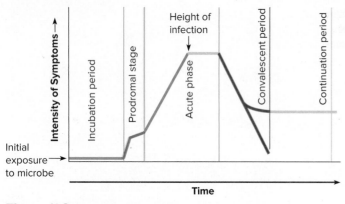

Figure 11.8 Stages in the course of infection and disease. The stages have different durations in different infections.

the incubation period). Overall, an incubation period can range from several hours in pneumonic plague to several years in leprosy. The majority of infections, however, have incubation periods ranging between 2 and 30 days.

The earliest notable symptoms of many infections appear as a vague feeling of discomfort, such as head and muscle aches, fatigue, upset stomach, and general malaise. This short period (1-2 days) is known as the **prodromal stage.** Some diseases have very specific prodromal symptoms. Next, the infectious agent enters an **acute** phase, during which it multiplies at high levels, exhibits its greatest virulence, and becomes well established in its target tissue. This period is often marked by fever and other prominent and more specific signs and symptoms, which can include cough, rashes, diarrhea, loss of muscle control, swelling, jaundice, discharge of exudates, or severe pain, depending on the particular infection. The length of this period is extremely variable.

As the patient begins to respond to the infection, the symptoms decline—sometimes dramatically, other times slowly. During the recovery that follows, called the **convalescence,** the patient's strength and health gradually return, owing to the healing nature of the immune response. During this period, many patients stop taking their antibiotics, even though there are still pathogens in their system. Think about it—the microbes still alive at this stage of treatment are the ones in the population with the most intrinsic resistance to the antibiotic. In most cases, continuing the antibiotic dosing will take care of them. But stop taking the drug now and the bacteria that are left to repopulate are the ones with the higher resistance.

These are the four phases all infectious diseases have. There is a fifth phase, which only some infections have: the **continuation** phase, in which *either* the organism lingers for months, years, or indefinitely after the patient is completely well *or* the organism is gone but symptoms continue. Typhoid fever is one disease in which the organism lingers after the patient has fully recovered. An example of a disease that lingers after the organism is no longer detectable is syphilis (in some cases). Chronic Lyme disease can also be put in that category. The stages are summarized in **figure 11.8.**

The transmissibility of the microbe during these five stages is different for each microorganism. A few agents are released mostly during incubation (measles, for example); many are released primarily during the acute period (*Shigella*); and others can be transmitted during all of these periods (hepatitis B).

11.2 LEARNING OUTCOMES—Assess Your Progress

4. Differentiate between a microbe's pathogenicity and its virulence.
5. List the steps a microbe has to take to get to the point where it can cause disease.
6. Explain the significance of polymicrobial infections.
7. List several portals of entry and exit.
8. Define *infectious dose,* and explain its role in establishing infection.
9. Describe three ways microbes cause tissue damage.
10. Compare and contrast major characteristics of endotoxin and exotoxins.
11. Provide a definition of *virulence factors.*
12. Draw a diagram of the stages of disease in a human.

11.3 Important Features of Infectious Disease Transmission

Reservoirs: Where Pathogens Come From

In order for an infectious agent to continue to exist and be spread, it must have a permanent place to reside. The **reservoir** is the primary habitat in the natural world where a potential pathogen makes its home. Often it is a human or animal carrier, although soil, water, and plants are also reservoirs. The reservoir can be distinguished from the infection *transmitter*, which is the individual or object from which an infection is actually acquired. In diseases such as syphilis, the reservoir and the transmitter are the same (the human body). In the case of hepatitis A, the reservoir (a human carrier) is usually different from the mode of transmission (contaminated food). **Table 11.4** shows how reservoirs and transmission are interrelated.

Table 11.4 Reservoirs and Transmitters

Reservoirs	Transmission Examples
Living Reservoirs	
Animals (Other than humans and arthropods) Mammals, birds, reptiles, etc.	Pathogens from animals • can be directly transmitted to humans, as in the example of bats transmitting rabies to humans • can be transmitted to humans via vectors, as with fleas passing the plague from rats to people • can be transmitted through vehicles such as water, as in the case of leptospirosis, which is often transmitted from animal urine to human skin via bodies of water
Humans Actively ill	©Thinkstock/Getty Images (writing); Ingram Micro/Media Bakery (pen in mouth)  A person suffering from a cold contaminates a pen, which is then picked up by a healthy person. That is indirect transmission. Alternatively, a sick person can transmit the pathogen directly by sneezing on a healthy person.
Carriers	A person who is fully recovered from his hepatitis but is still shedding hepatitis A virus in his feces may use suboptimal hand-washing technique. He contaminates food, which a healthy person ingests (indirect transmission). Carriers can also transmit through direct means, as when an incubating carrier of HIV, who does not know she is infected, transmits the virus through sexual contact.
Arthropods Biological vectors	 When an arthropod is the host (and reservoir) of the pathogen, it is also the mode of transmission. Source: CDC/James Gathany
Nonliving Reservoirs	
Soil **Water** **Air** **The built environment**	 Some pathogens, such as the TB bacterium, can survive for long periods in nonliving reservoirs. They are then directly transmitted to humans when they come in contact with the contaminated soil, water, or air. ©Christopher Kerrigan/McGraw-Hill Education

Living Reservoirs

The list of both living and nonliving reservoirs is presented in table 11.4. You can probably guess that a great number of infections that affect humans have their reservoirs in other humans. Persons or animals with obvious symptomatic infection are obvious sources of infection, but a **carrier** is, by definition, an individual who *inconspicuously* shelters a pathogen and spreads it to others without any notice. The duration of the carrier state can be short or long term, and it is important to remember that the carrier may or may not have experienced disease due to the microbe.

Several situations can produce the carrier state. **Table 11.5** describes the various carrier states and provides examples of each.

Animals as Reservoirs and Sources   Animals deserve special consideration as reservoirs of infections. The majority of animal reservoir agents are arthropods such as fleas, mosquitoes, flies, and ticks. Larger animals can also spread infection–for example, mammals (rabies), birds (psittacosis), or lizards (salmonellosis).

Many vectors and animal reservoirs spread their own infections to humans. An infection indigenous to animals but also transmissible to humans is a **zoonosis** (zoh''-uh-noh'-sis). Some zoonotic infections (rabies, for instance) can have multihost involvement, and others can have very complex cycles in the wild (see plague in chapter 18). Zoonotic spread of disease is promoted by close associations of humans with animals, and people in animal-oriented or outdoor professions are at greatest risk. At least 150 zoonoses exist worldwide; the most common ones are listed in **table 11.6.** Zoonoses make up a full 70% of all new emerging diseases worldwide. It is worth noting that zoonotic infections are impossible to completely eradicate without also eradicating the animal reservoirs.

Nonliving Reservoirs

Microorganisms have adapted to nearly every habitat in the biosphere. They thrive in soil and water and are in the air. They also colonize what is known as "the built environment," surfaces in homes, office buildings, and structures of all kinds. Although most of these microbes are saprobic and cause little harm and considerable benefit to humans, some are opportunists and a few are regular pathogens. Because human hosts are in regular contact with these environmental sources, acquisition of pathogens from nonliving reservoirs is always a possibility.

The Transmission of Infectious Agents

Infectious diseases can be categorized on the basis of how they are acquired. A disease is **communicable** when an infected host can transmit the infectious agent to another host and establish infection in that host. (Although this terminology is standard, one must realize that it is not the disease that is communicated but the microbe. Also be aware that the word *infectious* is sometimes used interchangeably with the word *communicable*, but this is not precise usage.) The transmission of the agent can be direct or indirect, and the ease with which the disease is transmitted varies considerably from one agent to another. If the agent is highly communicable, especially through direct contact, the disease is **contagious.** Influenza and measles move readily from host to host and, for that reason, are called contagious, whereas Hansen's disease (leprosy) is only weakly communicable. Because they can be spread through the population, communicable diseases are our main focus in the following sections.

In contrast, a **noncommunicable** infectious disease does *not* arise through transmission of the infectious agent from host to host. The infection and disease are acquired through some other special circumstance. Noncommunicable infections sometimes occur when a compromised person is invaded by his or her own microbiota. Ear infections are good examples of this. Other instances of this are when an individual has accidental contact with a microbe that exists in a nonliving reservoir such as soil. Some examples are certain mycoses, acquired through

COVID-19

SARS-CoV-2 belongs to a class of infections that has a zoonotic *origin*, but has "jumped species" to be transmissible between humans. In this book, zoonotic infections refer to infections that are transmitted from animals to human routinely.

Table 11.5 Carrier States

Carrier State	Explanation	Example
Asymptomatic carriers	Infected but show no symptoms of disease	Gonorrhea, genital herpes with no lesions, human papillomavirus
Incubating carriers	Spread the infectious agent during the incubation period	Infectious mono-nucleosis,
Convalescent carriers	Recuperating patients without symptoms; they continue to shed viable microbes and convey the infection to others	Hepatitis A
Chronic carriers	Individuals who shelter the infectious agent for a long period after recovery because of the latency of the infectious agent	Tuberculosis, typhoid fever
Passive carriers	Medical and dental personnel who must constantly handle patient materials that are heavily contaminated with patient secretions and blood risk picking up pathogens mechanically and accidentally transferring them to other patients	Various healthcare-associated infections

COVID-19 can be found in asymptomatic carriers and incubating carriers. As of publication of this book, we aren't yet sure about convalescent or chronic carriers.
– Kelly & Heidi

Microbes are multiplying.

Asymptomatic STD

Incubation

Convalescent

Chronic

©Vitalii Hulai/Shutterstock

Table 11.6 Common Zoonotic Infections

Disease	Primary Animal Reservoirs
Viruses	
Rabies	Mammals
Yellow fever	Wild birds, mammals, mosquitoes
Viral fevers	Wild mammals
Hantavirus	Rodents
Influenza	Chickens, birds, swine
West Nile virus	Wild birds, mosquitoes
Bacteria	
Rocky Mountain spotted fever	Dogs, ticks
Psittacosis	Birds
Leptospirosis	Domestic animals
Anthrax	Domestic animals
Brucellosis	Cattle, sheep, pigs
Plague	Rodents, fleas
Salmonellosis	Mammals, birds, reptiles, and rodents
Tularemia	Rodents, birds, arthropods
Other Microbes	
Ringworm	Domestic mammals
Toxoplasmosis	Cats, rodents, birds
Trypanosomiasis	Domestic and wild mammals
Trichinosis	Swine, bears
Tapeworm	Cattle, swine, fish

©imagebroker/Alamy (raccoons); McGraw-Hill Education (pig); Ingram Publishing/SuperStock (cat)

inhalation of fungal spores, and tetanus, in which *Clostridium tetani* spores from a soiled object enter a cut or wound. Persons with these infections do not become a source of disease to others.

Patterns of Transmission in Communicable Diseases

There are many different ways that microbes can be transmitted. They are summarized in **table 11.7**. First of all, we speak of horizontal versus vertical transmission. The term *horizontal* means the disease is spread through a population from one infected individual to another. *Vertical* signifies transmission from parent to offspring via the ovum, sperm, placenta, or milk.

Within the category of horizontal transmission, we can further subdivide the mechanisms into three broad groups: direct contact, indirect routes, and vector transmission. Direct transmission involves very close contact between people. Indirect transmission occurs when an object or substance carries the agent from one person to another. A special category of indirect transmission is parenteral transmission. It refers to a puncture, in which material from the environment is deposited directly in deeper tissues. This can be intentional, in the case of contaminated needles,

NCLEX® PREP

5. A patient is found to have a dental abscess. Antibiotic therapy is initiated to prevent spread of the infection to other tissues. If it does spread, what type of infection is it?

a. *focal infection*

b. *primary infection*

c. *secondary infection*

d. *mixed infection*

Table 11.7 Patterns of Transmission in Communicable Diseases

Mode of Transmission	Definition
Vertical	Transmission is from parent to offspring via the ovum, sperm, placenta, or milk
Horizontal	Disease is spread through a population from one infected individual to another

> The rest of this table is about horizontal transmission.
> — Kelly & Heidi

Direct (contact) transmission

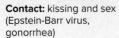

Involves very close proximity or actual physical contact between two hosts

Types:

- Touching, kissing, sex
- Droplet contact, in which fine droplets are sprayed directly upon a person during sneezing, coughing, or speaking

Droplets (colds, chickenpox)

Contact: kissing and sex (Epstein-Barr virus, gonorrhea)

Indirect transmission

Infectious agent must pass from an infected host to an intermediate conveyor (a vehicle) and from there to another host

Infected individuals contaminate objects, food, or air through their activities

Types:

- **Fomite**—inanimate object that harbors and transmits pathogens (doorknobs, phones, faucet handles)
- **Vehicle**—a natural, nonliving material that can transmit infectious agents

  - **Air**—smaller particles evaporate and remain in the air and can be encountered by a new host; **aerosols** are suspensions of fine dust or moisture particles in the air that contain live pathogens
  - **Water**—some pathogens survive for long periods in water and can infect humans long after they were deposited in the water
  - **Soil**—microbes resistant to drying live in and can be transmitted from soil
  - **Food**—meats may contain pathogens with which the animal was infected; foods can also be contaminated by food handlers
  - **Parenteral** transmission via intentional or unintentional injection into deeper tissues (needles, knives, branches, broken glass, etc.)

Special Category: oral-fecal route—using either vehicles or fomites. A fecal carrier with inadequate personal hygiene contaminates food during handling, and an unsuspecting person ingests it; alternatively, a person touches a surface that has been contaminated with fecal material and touches his or her mouth, leading to ingestion of fecal microbes

Vector transmission

Types:

- **Mechanical vector**—insect carries microbes to host on its body parts
- **Biological vector**—insect injects microbes into host; part of microbe life cycle completed in insect

or unintentional, as in the case of puncture injuries. Lastly, vectors are arthropods that harbor an infectious agent and transmit it to a human.

Vectors can be further subdivided into two categories, biological and mechanical. A vector is termed "biological" if the microorganism lives and multiplies within the insect. Mechanical vectors are not necessary to the life cycle of an infectious agent and merely transport it without being infected. The external body parts of these animals become contaminated when they come into physical contact with a source of pathogens. Houseflies are mechanical vectors. They feed on decaying garbage and feces, and while they are feeding, their feet and mouthparts easily become contaminated. Then they land on your burger at the cookout.

Healthcare-Associated Infections

Infectious diseases that are acquired or develop during a hospital or health care facility stay are known as **healthcare-associated infections** (HAIs), also called **nosocomial** (nohz″-oh-koh′-mee-al) **infections.** This concept seems strange at first thought because a hospital is regarded as a place to get treatment for a disease, not a place to acquire a disease. Yet it is not uncommon for a surgical patient's incision to become infected or a burn patient to develop a case of pneumonia in the clinical setting. The rate of healthcare-associated infections can be as low as 0.1% or as high as 20% of all admitted patients, depending on the clinical setting, with an average of about 4%. This adds up to about 750,000 cases in acute care hospitals every year, which result in nearly 75,000 deaths.

When you examine the circumstances, you see that a certain number of HAIs are virtually unavoidable. After all, the hospital both attracts and creates compromised patients, and it serves as a collection point for pathogens. Some patients become infected when surgical procedures or lowered defenses permit resident biota to invade their tissues. Other patients acquire infections directly or indirectly from fomites, medical equipment, other patients, medical personnel, visitors, air, and water.

The health care process itself increases the likelihood that infectious agents will be transferred from one patient to another. Indwelling devices such as catheters, prosthetic heart valves, grafts, drainage tubes, and tracheostomy tubes form ready portals of entry and habitats for infectious agents. Because such a high proportion of the hospital population receives antimicrobial drugs during their stay, drug-resistant microbes are selected for at a much greater rate than is the case outside the hospital.

The most common HAIs include pneumonia, gastrointestinal illness, urinary tract infections, bloodstream infections, and surgical site infections.

Five common hospital pathogens, and what they usually cause, are

Clostridioides difficile (*C. diff*): GI infections;

Staphylococcus aureus: pneumonia, surgical site infections, bloodstream infections;

Klebsiella species: surgical site infections, urinary tract infections, pneumonia;

Escherichia coli: urinary tract infections, surgical site infections, bloodstream infections;

Enterococcus species: surgical site infections, urinary tract infections, bloodstream infections.

The frequency of HAIs and the microbes responsible for them are dynamic situations. As hospitals target problem areas, the most prominent infections shift. For example, the number of HAIs in U.S. hospitals declined in 2015 as compared to 2011,

Airplanes can play a role in spreading diseases—but not in the way you might think. Studies have shown that airborne diseases are *not* more easily spread in airplane cabins. However, airplanes can move sick people from one continent to another quickly and thus widen an epidemic into a pandemic.

©Tony Cordoza/Alamy (airplane); Smith Collection/Gado/Archive Photos/Getty Images (aisle); ColorBlind Images/Getty Images (passengers)

from 4.0% to 3.2% of admitted patients. This was seen to be mostly due to the decrease in numbers of surgical site infections and urinary tract infections. Gastrointestinal infections (most due to *C. difficile*) and pneumonia are the top two continuing concerns.

A new organism has begun to worry U.S. health care officials. A multidrug-resistant yeast called *Candida auris* has been causing HAIs in several countries around the world, since 2009. It causes highly invasive infections with high mortality rates. Only one U.S. case was cited in the CDC alert issued in 2016, but due to its appearance over such a wide geographic area, the agency issued the alert to encourage hospitals to be on the lookout for it. It worries public health officials because it is resistant to antifungal agents with different modes of action, seems to have appeared simultaneously on different continents, and is extremely difficult to eradicate from contaminated surfaces. Scientists still do not know where the fungus resides in nature, which makes it difficult to design prevention strategies.

Hospitals generally employ an *infection control officer* who not only implements proper practices and procedures throughout the hospital but is also charged with tracking potential outbreaks, identifying breaches in asepsis, and training other health care workers in aseptic technique. Among those most in need of this training are nurses and other caregivers whose work, by its very nature, exposes them to needlesticks, infectious secretions, blood, and physical contact with the patient. The same practices that interrupt the routes of infection in the patient can also protect the health care worker. It is for this reason that hospitals have adopted universal precautions that recognize that all secretions from all persons in the clinical setting are potentially infectious and that transmission can occur in either direction.

©Paul Burns/Blend Images LLC

Which Agent Is the Cause? Using Koch's Postulates to Determine Etiology

An essential aim in the study of infection and disease is determining the precise **etiologic,** or causative, **agent** of a newly recognized condition. In our modern technological age, we take for granted that a certain infection is caused by a certain microbe, but such has not always been the case. More than a century ago, Robert Koch realized that in order to prove the germ theory of disease, he would have to develop a standard for determining causation that would stand the test of scientific scrutiny. Out of his experimental observations on the transmission of anthrax in cows came a series of proofs, called **Koch's postulates,** that established the principal criteria for etiologic studies. **Table 11.8** demonstrates the principles of Koch's postulates.

The general idea of Koch's postulates is that you must isolate what you think the cause is, then apply it to a naive population, and then produce the same effect. This general progression is still considered the gold standard for determining that any given factor is causing—not simply correlated with—some observed condition. Many of the media reports you read about the latest "findings" are based on studies reporting correlations, not causation. Consider the claim that Facebook usage leads to shorter attention spans. Finding a causative relationship between the two things would require taking large numbers of people who had never used Facebook, measuring their attention spans, and then forcing them to use Facebook for some allotted time period. Then you would need to remeasure their attention spans. This would approach the gold standard of Koch's postulates, although you would not be able to completely hold all other variables constant in your subjects, unless you kept them all in your laboratory for the whole duration of your experiment. So you begin to see how difficult true causation is to prove.

Koch's postulates are reliable for many infectious diseases, but they cannot be completely fulfilled in certain situations. For example, some infectious agents are not readily isolated or grown in the laboratory. If one cannot elicit an infection similar to that seen in humans by inoculating it into an animal, it is very difficult to

Medical Moment

Eye on Careers: Infection Control Practitioner

An infection control practitioner (ICP) holds an integral position in many hospitals and health care organizations. An ICP's biggest role is to reduce the spread of healthcare-associated infections, limiting the spread of infectious disease. Many different professionals may be designated as ICPs, including nurses, doctors, epidemiologists, or others who have taken specialized training to prepare them for this important role.

An ICP may be responsible for the following:

- tracking positive cultures to ensure treatment has been implemented;
- following up actual and potential exposures to communicable diseases;
- training and educating staff regarding infection control practices and protocols;
- communicating with government entities such as the CDC (the Centers for Disease Control and Prevention), state public health departments, workers' compensation boards, and OSHA (the Occupational Safety and Health Administration).

Q. Can you think of what type of incident might lead an ICP to get involved with the care of a health care worker in a hospital?

Answer in Appendix B.

Table 11.8 Koch's Postulates

Postulate #1	Postulate #2
Find evidence of a particular microbe in every case of a disease.	**Isolate** that microbe from an infected subject and **cultivate** it in pure culture in the laboratory; **characterize** it fully.

Antibody

Antigens

ID   C   T

Control Test   Sample well

NCLEX® PREP

6. A client reported onset of these symptoms a few days ago: low-grade fever, arthralgia, and fatigue. The client was seen in the outpatient clinic by a health care provider, diagnosed with a bacterial infection, and prescribed a course of antibiotic therapy. At present, the client has 3 days of antibiotic treatment remaining. The client would be in which stage of infection?

   a. *acute phase*
   b. *convalescent period*
   c. *incubation period*
   d. *prodromal stage*

prove the etiology. It is difficult to satisfy Koch's postulates for viral diseases because viruses usually have a very narrow host range. Human viruses may cause disease only in humans, or perhaps in primates, though the disease symptoms in apes will often be different. To address this, there are modified postulates for viral infections.

One very important reason Koch's postulates are increasingly viewed with caution is the idea, presented earlier, that perhaps the majority of human infections are polymicrobial. In those infections, Koch's postulates cannot be satisfied.

11.3 LEARNING OUTCOMES—Assess Your Progress

13. Differentiate among the various types of reservoirs, providing examples of each.
14. List several different modes of transmission of infectious agents.
15. Define *healthcare-associated infection*, and list two common types.
16. List Koch's postulates, and discuss when they might not be appropriate in establishing causation.

11.4 Epidemiology: The Study of Disease in Populations

So far, our discussion has revolved primarily around the impact of an infectious disease in a single individual. Let us now turn our attention to the effects of diseases on the community—the realm of **epidemiology.** By definition, this term involves the study of the frequency and distribution of disease and other health-related factors in defined populations. It involves many disciplines—not only microbiology but also anatomy, physiology, immunology, medicine, psychology, sociology, ecology, and statistics—and it considers all forms of disease, including heart disease, cancer, drug addiction, and mental illness.

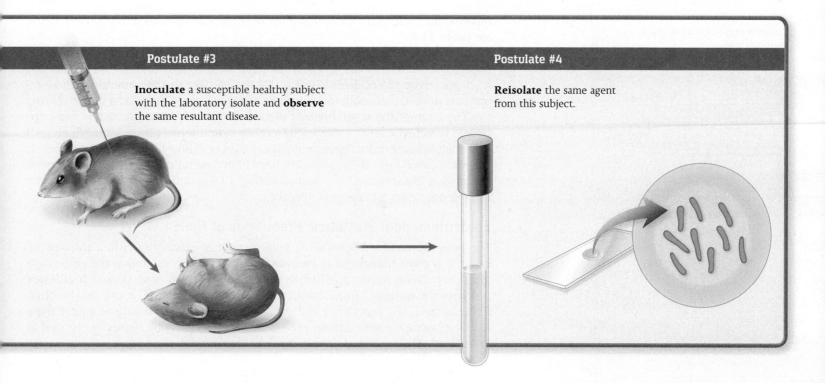

Postulate #3

Inoculate a susceptible healthy subject with the laboratory isolate and **observe** the same resultant disease.

Postulate #4

Reisolate the same agent from this subject.

A groundbreaking British nurse named Florence Nightingale helped to lay the foundations of modern epidemiology. She arrived in the Crimean war zone in Turkey in the mid-1850s, where the British were fighting and dying at an astonishing rate. Estimates suggest that 20% of the soldiers there died (by contrast, 2.6% of U.S. soldiers in the Vietnam war died). Even though this was some years before the discovery of the germ theory, Nightingale understood that filth contributed to disease and instituted methods that had never been seen in military field hospitals. She insisted that separate linens and towels be used for each patient, and that the floors be cleaned and the pipes of sewage unclogged. She kept meticulous notes of what was killing the patients and was able to demonstrate that many more men died of disease than of their traumatic injuries. She used statistical analysis to convince government officials that these patterns were real. This was indeed one of the earliest forays into epidemiology—trying to understand how diseases were being transmitted and using statistics to do so.

Epidemiologists try to identify causative agents, using adaptations of Koch's postulates when the disease is not an infectious disease, or using mostly nonexperimental types of analyses. The techniques of epidemiology are also used to track behaviors, such as exercise or smoking. The epidemiologist is a medical sleuth who collects clues on the causative agent, pathology, sources, and modes of transmission and tracks the numbers and distribution of cases of disease in the community. The outcomes of these studies help public health departments develop prevention and treatment programs and establish a basis for predictions.

Tracking Disease in the Population

Surveillance is an epidemiological term that basically means "keeping track of trends in nonemergency times." Weather patterns and human behaviors are two examples of things that are tracked as an aid to predicting disease patterns. Of course, an important task for surveillance involves keeping data for a large number of diseases seen by the medical community and reported to public health authorities. By law, certain **reportable,** or notifiable, **diseases** must be reported to authorities; others are

Epidemiology is the study of disease in populations.
©Donald Nausbaum/Getty Images

reported on a voluntary basis. (For a list of reportable diseases in the United States, see **table 11.9.**)

A well-developed network of individuals and agencies at the local, district, state, national, and international levels keeps track of infectious diseases. Physicians and hospitals report all notifiable diseases that are brought to their attention. These reports are made either about individuals or in the aggregate, depending on the disease.

The Internet has revolutionized disease tracking. For example, a few years ago Google launched a service that used trending search terms ("fever," "flu symptoms") to predict flu outbreaks, theoretically faster than traditional surveillance methods. It had some flaws, but other companies have improved on the concept. Most experts expect that this use of what is known as "big data" promises to speed up prediction and response time for disease outbreaks.

Epidemiological Statistics: Frequency of Cases

The **prevalence** of a disease is the total number of existing cases in a given population. It is often thought of as a snapshot and is usually reported as the percentage of the population having a particular disease at any given time. Disease **incidence** measures the number of *new* cases over a certain time period. It can also be of interest to track the prevalence and incidence of *infection*, in situations where there is a lot of asymptomatic transmission. In that case, the variable "cases" is changed to "infections". The equations used to figure these rates are

$$\text{Prevalence} = \frac{\text{Total number of cases in population}}{\text{Total number of persons in population}} \times 100 = \%$$

Example: The prevalence of smoking among adults in the United States is 14% currently.

$$\text{Incidence} = \frac{\begin{array}{c}\text{Number of new cases}\\\text{in a designated time period}\end{array}}{\text{Total number of susceptible persons}} \quad \begin{array}{l}\text{(Usually reported}\\\text{per 100,000 persons)}\end{array}$$

Example: The incidence of new tuberculosis cases in the United States in 2018 was 2.8 per 100,000.

Changes in incidence and prevalence are usually followed over a seasonal, yearly, and long-term basis and are helpful in predicting trends **(figure 11.9).** Also of concern to the epidemiologist are the rates of disease with regard to sex, race, or geographic region. **Figure 11.9a** displays incidence trends over time as well as within different age groups; **figure 11.9b** shows another way of depicting epidemiological data, in a map format. Also of importance is the **mortality rate,** which measures the total number of deaths in a population due to a certain disease. Over the past century, the overall death rate from infectious diseases in the developed world has dropped, although the number of persons afflicted with infectious diseases (the **morbidity rate**) has remained relatively high.

Clearly, different infections have differing degrees of communicability in populations. They also have varying abilities to cause death. When epidemiologists keep track of mortality rates, they can use these as predictive tools. As they keep track of incidence rates in epidemic and nonepidemic situations, they can calculate a factor called the **reproductive rate (R_0)** of a disease. That is also extremely useful in predicting and managing the spread of infectious disease. Please take a look at the Note near the end of the chapter About Epidemiology: Communicability and Deadliness in which we introduce these two important characteristics of infectious disease. In each of the disease chapters, we will circle back to the two triangles introduced here.

Table 11.9 Nationally Notifiable Infectious Diseases in the United States, 2019*

- Anthrax

- Arboviral neuroinvasive and non-neuroinvasive diseases

  - California serogroup virus disease
  - Chikungunya virus disease
  - Eastern equine encephalitis virus disease
  - Powassan virus disease

  - St. Louis encephalitis virus disease

  - West Nile virus disease
  - Western equine encephalitis disease

- Babesiosis
- Botulism
- Brucellosis
- Campylobacteriosis
- Carbapenemase-producing carbapenem-resistant *Enterobacteriaceae* (CP-CRE)
  - CP-CRE, *Enterobacter* species
  - CP-CRE, *E. coli*
  - CP-CRE, *Klebsiella* species
- Chancroid
- *Chlamydia trachomatis* infections
- Cholera
- Coccidioidomycosis
- Congenital syphilis/syphilitic stillbirth
- Cryptosporidiosis
- Cyclosporiasis
- Dengue (including Dengue fever, hemorrhagic fever, and shock syndrome)
- Diphtheria
- Ehrlichiosis/anaplasmosis
- Foodborne disease outbreak
- Giardiasis

- Gonorrhea

- *Haemophilus influenzae,* invasive disease

  - Hansen's disease (leprosy)
  - Hantavirus pulmonary syndrome
  - Hemolytic uremic syndrome

  - Hepatitis A, B, C

  - HIV infection

  - Influenza-associated pediatric mortality
  - Invasive pneumococcal disease (IPD)/ *Streptococcus pneumoniae,* invasive disease
  - Latent TB infection
- Legionellosis
- Leptospirosis
- Listeriosis
- Lyme disease

- Malaria
- Measles
- Meningococcal disease
- Mumps
- Novel influenza A infections
- Pertussis
- Plague
- Poliomyelitis, paralytic
- Poliovirus infection, nonparalytic
- Psittacosis
- Q fever
- Rabies, animal or human
- Rubella (German measles)
- Rubella, congenital syndrome
- Salmonellosis

- Severe acute respiratory syndrome–associated coronavirus (SARS-CoV) disease (This refers to the virus causing the SARS epidemic in 2003. SARS-CoV-2 is now also on this list.)

- Shiga toxin–producing *Escherichia coli* (STEC)
- Shigellosis
- Smallpox
- Spotted fever rickettsiosis
- Streptococcal toxic-shock syndrome
- *Streptococcus pneumoniae,* invasive disease
- Syphilis
- Tetanus

- Toxic shock syndrome (other than streptococcal)
- Trichinellosis (trichinosis)
- Tuberculosis
- Tularemia
- Typhoid fever
- Vancomycin-intermediate *Staphylococcus aureus* (VISA)
- Vancomycin-resistant *Staphylococcus aureus* (VRSA)
- Varicella
- Vibriosis
- Viral hemorrhagic fevers due to:
  - Crimean-Congo hemorrhagic fever virus
  - Ebola virus
  - Lassa virus
  - Lujo virus
  - Marburg virus
  - New world arenaviruses (Guanarito, Machupo, Junin, and Sabia viruses)
- Yellow fever
- Zika virus infection and disease

The bacterium that causes syphilis, *Treponema pallidum.*
Source: CDC

*Reportable to the CDC; other diseases may be reportable to state departments of health.

Source: Centers for Disease Control and Prevention, 2019 (latest year that it was updated).

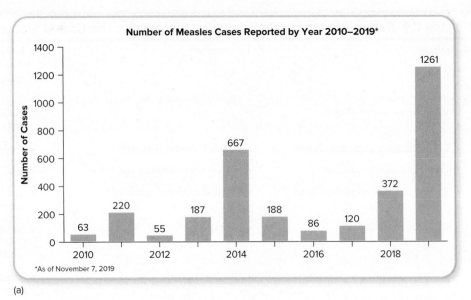

(a)

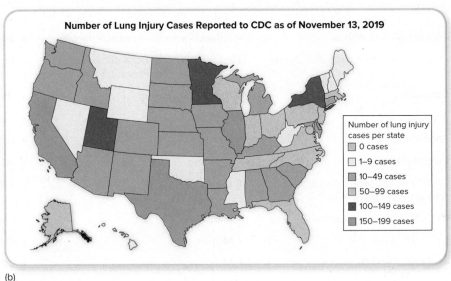

(b)

Figure 11.9 Depictions of epidemiological data.   **(a)** A bar graph that displays the incidence of measles cases in the United States from 2010 to 2019; **(b)** a geographical representation of where vaping lung injuries have occurred in the United States in 2019. Keep in mind that epidemiological methods are also used for noninfectious diseases.
Source: CDC

When there is an increase in disease in a particular geographic area, it can be helpful to examine the epidemic curve–which is incidence plotted over time–to determine if the infection is a **common-source** or **propagated epidemic.** A common-source epidemic, illustrated in **figure 11.10a,** is one in which the infectious agent was present in a single source that had widespread distribution, infecting people in a wide geographic area. When food is contaminated at a factory, it can result in common-source epidemics. A particular type of common-source epidemic is referred to as **point-source epidemic.** It is still a common source, but usually it is a single small batch of food or water that was contaminated, and everyone was infected at once, such as spoiled potato salad at a church picnic. That curve would have a much shorter time span on the horizontal axis. A propagated epidemic **(figure 11.10b)** results from an infectious agent that is communicable from person to person and therefore is sustained–propagated–over time in a population. Influenza is the classic example of this. The graph illustrates an entire year of influenza incidence in the United States. In propagated epidemics, the incidence

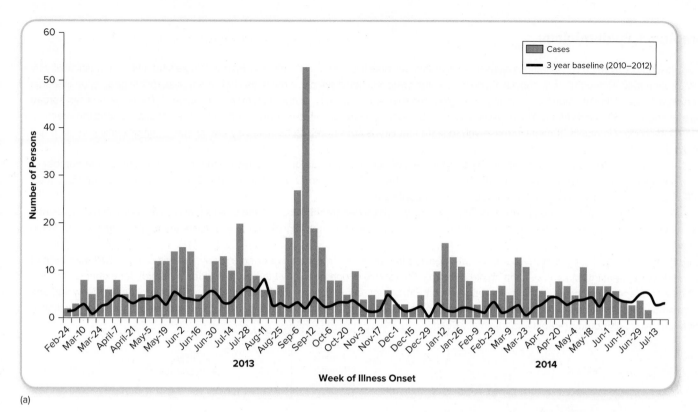

(a)

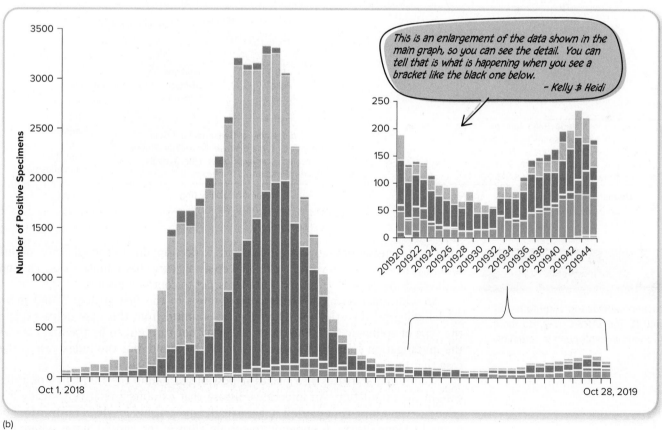

(b)

Figure 11.10 Epidemic curves. **(a)** A common-source epidemic curve. This illustrates a multistate outbreak of drug-resistant *Salmonella* caused by Foster Farms chicken that was contaminated. **(b)** A propagated epidemic curve of the incidence of influenza in the United States from fall of 2018 through fall of 2019.

*These are written in year-week format, as in YYYYWW. So "201920" means the 20th week of 2019.

Source: CDC

A Note About Epidemiology

There are two big descriptors of any given infectious disease—how communicable it is and how deadly it is. Epidemiologists quantify communicability by a factor called R_0 (pronounced "R-naught" or "R-sub-zero"), defined as the basic reproduction rate. It describes how many susceptible people, on average, one infected person will spread the infection to. The highly contagious measles virus has an R_0 of about 15, meaning that one infected person can spread the infection, on average, to 15 other individuals. The R_0 assumes those 15 people are unvaccinated for the microbe and have not experienced the infection, therefore having no secondary immunity. It might surprise you to learn that HIV is considered to be relatively low on the communicability scale. It has an R_0 of only about 3.4.

Deadliness is calculated via the case fatality rate (CFR): the numbers of persons who die of the disease within a specified time ÷ the number of persons infected. This calculation is based on persons who receive no treatment. The CFR for rabies is 100%. The CFR for cholera is about 1%. Understand that a CFR of even 1% is high—indicating that 1 of 100 infected people dies.

These measures of infectious disease are approximate and can vary based on geographic location and other situational factors. But a general idea of R_0 and CFR can guide a lot of health care decisions and policies. Diseases with an extremely high R_0, for example, are the ones for which vaccination is most needed.

When we get to the disease chapters of this book, we will highlight key diseases in each chapter on a graph like this one. We will take both of these measures and divide them into quarters—lumping all CFRs from 0% to 25% in one quarter (at the bottom of the pyramid), for example. The R_0s will also be divided into quarters. Practice reading the graph a little bit: Find an infection that is highly communicable but not very deadly. Now find one that is not very communicable but deadly. The good news is that none of these common infections are both highly deadly and highly communicable. And most of them are minimally communicable *and* minimally deadly.

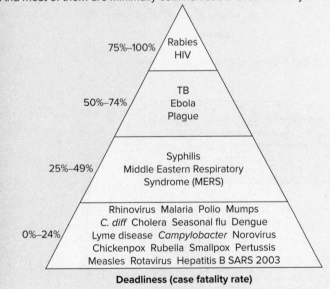

Deadliness (case fatality rate)

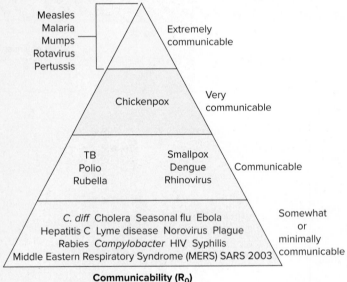

Communicability (R_0)

COVID-19

At the time of publication, it was too soon to report a definitive CFR and R_0 for SARS-CoV-2. So far, it seems to be very communicable with a *relatively* low case fatality rate.

constantly increases until it peaks, then begins to decrease due to natural factors or due to control measures. Each year the influenza epidemic curve has a similar shape. The end of the epidemic is due to features of the virus, not to control measures.

An additional term, the **index case,** refers to the first patient found in an epidemiological investigation. How the cases unfurl from this case helps explain the type of epidemic it is. The index case may not turn out to be the first case—as the investigation continues, earlier cases may be found—but the index case is the case that brought the epidemic to the attention of officials.

Monitoring statistics also makes it possible to define the frequency of a disease in the population. An infectious disease that exhibits a relatively steady frequency over a long time period in a particular geographic locale is **endemic.** For example, Lyme disease is endemic to certain areas of the United States where the tick vector is found. A certain number of new cases are expected in these areas every year. When a disease is **sporadic,** occasional cases are reported at irregular intervals in random locales. Tetanus and diphtheria are reported sporadically in the United States (fewer than 50 cases a year).

When statistics indicate that the prevalence of an endemic or sporadic disease is increasing beyond what is expected for that population, the pattern is described as an **epidemic.** The spread of an epidemic across continents is a **pandemic,** as exemplified by COVID-19, AIDS, and influenza.

11.4 LEARNING OUTCOMES—Assess Your Progress

17. Summarize the goals of epidemiology, and differentiate it from traditional medical practice.

18. Explain what is meant by a disease being "notifiable" or "reportable," and provide examples.

19. Define *incidence* and *prevalence,* and explain the difference between them.

20. Discuss the three major types of epidemics, and identify the epidemic curve associated with each.

Medical Moment

Typhoid Mary

One of the most famous index cases in medical literature is the case of Mary Mallon, nicknamed "Typhoid Mary." As a cook in New York in the early 1900s, she was identified as the source of exposing at least 120 people to typhoid. As she moved between families as a cook, small typhoid outbreaks would follow. Despite showing no symptoms of infection herself, Mary hosted the *Salmonella typhi* bacteria.

Mary Mallon was the first asymptomatic typhoid carrier in the United States. She was forced into confinement twice by the department of health but protested this treatment. After her release to the general public, she continued to spread typhoid throughout her life. Today the term "Typhoid Mary" is used as an idiom for a person who spreads harm, knowingly or innocently.

Q: What is the value in identifying an index case in an epidemic?

Answer in Appendix B.

CASE FILE WRAP-UP

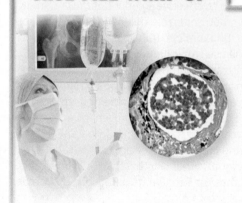

©Image Source/Monty Rakusen(health care worker); Science Photo Library/Alamy Stock Photo(micrograph)

Ricardo had a right-side pneumonia, most likely caused by his surgical procedure. The surgical incision provided a portal of entry for organisms to be introduced into his chest cavity. A healthcare-associated infection is an infectious disease that develops during hospitalization. Though Ricardo had gone home appearing healthy, and then returned with the infection, the organism was most likely introduced during his lobectomy. The Centers for Disease Control and Prevention maintains specific definitions, specifying criteria for when an infection should be defined as hospital-associated.

Hospital infection control policies are aimed at stopping the spread of infectious disease. Hand hygiene practices are the most important method of preventing the spread of pathogens, by either proper hand washing or use of an alcohol-based hand cleanser.

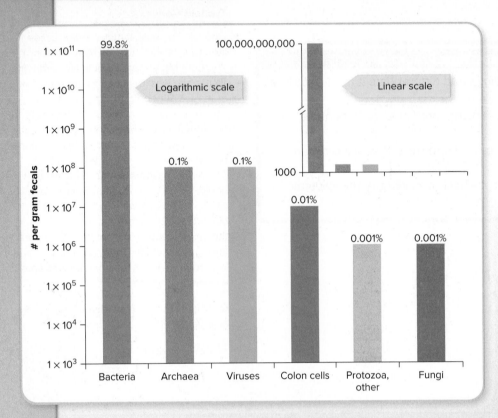

Graph depicts estimates of all components in a typical fecal sample.

Source: *PLOS Biology*, 2016

Clostridioides difficile infection is an extreme example of what happens when the normal microbiota of the bowel is destroyed. *C. difficile* is a gram-positive organism that forms endospores. It grows slowly and can be challenging to isolate.

C. difficile is one of the most common, and most serious, healthcare-associated infections. The microbe, in its worst manifestation, causes a condition known as pseudomembranous colitis, in which pseudomembranous plaques form on the colon mucosa. Antibiotic therapy is one of the most common causes of *C. difficile*. Antibiotics so drastically alter the normal microbiota of the colon that *C. difficile* is allowed to proliferate, releasing toxins that attack the mucosa of the bowel, resulting in cramping and diarrhea. A substantial number of patients develop bloodstream infections, and 50% of these die.

Treatment of *C. difficile* infection frequently entails potent drugs known to be effective against *C. difficile*, namely metronidazole (Flagyl), vancomycin, or fidaxomicin (Dificid). Dificid is used in more severe cases of the disease. However, relapse can still occur.

One therapy that is gaining attention is fecal transplant. The theory behind fecal transplant is to combat bacterial infection of the colon by infusing the colon with feces from a healthy person in an effort to displace the pathogenic microbes, restoring the normal balance in the colon. The treatment may sound distasteful but has proven to be effective.

The administration can be via enema, or sometimes by an epigastric tube, or even a pill. Donors are generally family members, and the donor feces are extensively tested to be sure that the donor is free of any bacterial, viral, or other harmful infections. Recipients may be administered the fecal enemas for 5 to 10 days, although sometimes one treatment is sufficient.

A study published in the *NEJM* (*New England Journal of Medicine*) determined a 94% cure rate in patients treated with fecal microbiota transplants, compared to only a 31% cure rate with vancomycin. The results were so overwhelmingly positive that the study was stopped early, as it was considered unethical not to offer fecal microbiota transplants to all of the study participants. In 2013, the FDA approved the use of fecal transplants in patients with *C. difficile* who do not respond to standard medical therapy.

Since 2011, when this became a more common practice, the assumption has been made that the cure is brought about because of the rich complement of bacteria in the healthy feces that restores health by colonizing the distressed colon. A paper from 2016 points out that bacteria are by far not the only components in transferred feces (see figure), and that scientists should pick apart the possible contributions of other components of the transplants.

INTERACTIONS BETWEEN MICROBES AND HUMANS

11.1 THE HUMAN MICROBIOME

Normal biota are easily cultured from the skin and in the respiratory tract, the gastrointestinal tract, the outer parts of the urethra, the vagina, the eye, and the external ear canal. Molecular techniques have also revealed normal biota to be present in the lungs, bladder, breast milk, and amniotic fluid.

Source: NIH Roadmap for Medical Research website

True pathogens cause infectious disease in healthy hosts; opportunistic pathogens cause damage only when the host immune system is compromised in some way. Every encounter between a susceptible host and a potential disease-causing organism does not result in disease. Multiple characteristics of the two partners (microbe and host) come into play.

WHEN COLONIZATION LEADS TO DISEASE 11.2

11.3 IMPORTANT FEATURES OF INFECTIOUS DISEASE TRANSMISSION

A communicable disease can be transmitted from an infected host to others, but not all infectious diseases are communicable.

The spread of infectious disease from person to person is called horizontal transmission.

Healthcare-associated infections are acquired in a hospital or health care facility from surgical procedures, equipment, personnel, and exposure to drug-resistant microorganisms.

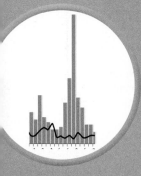

Epidemiology is the study of the determinants and distributions of infectious and noninfectious diseases in populations.

The prevalence of a disease is the percentage of existing cases in a given population. The disease incidence, or morbidity rate, is the number of newly infected members in a population during a specified time period.

EPIDEMIOLOGY: THE STUDY OF DISEASE IN POPULATIONS 11.4

SmartGrid: From Knowledge to Critical Thinking

This *21 Question Grid* takes the topics from this chapter and arranges them with respect to the American Society for Microbiology's Undergraduate Curriculum guidelines—all six of the important "Concepts" as well as the important "Competency" of scientific literacy. Three questions are supplied about chapter content that refer to the Concept or Competency, in increasing levels of Bloom's taxonomy for learning.

ASM Concept/ Competency	A. Bloom's Level 1, 2—Remember and Understand (Choose one.)	B. Bloom's Level 3, 4—Apply and Analyze	C. Bloom's Level 5, 6—Evaluate and Create
Evolution	1. The best descriptive term for the resident microbiota is 　a. commensal. 　b. parasitic. 　c. pathogenic. 　d. mutualistic.	2. In some circumstances, microbes can be quite virulent when they first infect a new species (such as in a zoonosis), but over decades of association with the new human host, cause milder and milder disease. Can you speculate about why this is evolutionarily advantageous to the pathogen?	3. Conduct research on germ-free mice. Use what you find to write a paragraph about the coevolution of microbes and humans.
Cell Structure and Function	4. Which of the following are virulence factors? 　a. toxins 　b. enzymes 　c. capsules 　d. all of the above	5. Why do you suppose specific adhesive structures, such as fimbriae, are critical to the disease-causing capabilities of many bacteria? Be thorough in your answer.	6. Discuss the role of endospores in ensuring the ongoing transmission of a bacterium in a population.
Metabolic Pathways	7. When the resident microbiota prevents the establishment of a pathogen, it is called 　a. disruption. 　b. a superinfection. 　c. a nonliving reservoir. 　d. microbial antagonism.	8. Correlate the stages in the course of an infectious disease with what you know about the growth of bacterial cultures.	9. *Helicobacter pylori,* a microbe that can cause disease in the stomach, produces an enzyme called urease that breaks down urea into ammonia and oxygen. What advantage does that confer on the bacterium?
Information Flow and Genetics	10. Most microbial exotoxins would be created using the process of 　a. DNA replication. 　b. protein synthesis. 　c. mutation. 　d. fatty acid synthesis.	11. Explain what you can about microbe-induced epigenetic changes to the host DNA.	12. Some of the most pathogenic *E. coli* possess an exotoxin gene originally found in *Shigella* bacteria. Construct a scenario for how and in what biological setting that gene was shared.
Microbial Systems	13. The _____ is the time between an encounter with a pathogen and the first symptoms. 　a. prodrome 　b. acute stage 　c. convalescence 　d. incubation period	14. A line of thinking called the "hygiene hypothesis" suggests that when children have more exposure to the environment (dirt, soil, animals), they have better health as adults. Why might this be?	15. At one time it was common practice to prescribe tetracycline for adolescents for months or years at a time to combat acne. Glance back at table 10.5 and predict what consequences tetracycline would have on the microbiome and the host.

ASM Concept/ Competency	A. Bloom's Level 1, 2—Remember and Understand (Choose one.)	B. Bloom's Level 3, 4—Apply and Analyze	C. Bloom's Level 5, 6—Evaluate and Create
Impact of Microorganisms	**16.** Which factors might determine if you get sick from an infection when someone else might not? **a.** your astrological sign **b.** your general health **c.** whether the microbe is gram-negative or gram-positive **d.** your height	**17.** In figure 11.5, it was noted that the switch under "adaptive immunity" was definitively red in the beginning of the COVID-19 pandemic. What does that say about the herd immunity in the population at that time?	**18.** In question 2, we discussed pathogens that became less virulent the longer they were associated with the human host. But there is another category of pathogen that never loses any virulence over hundreds of years of association with humans. Can you deduce what circumstances make lower virulence of no advantage to a pathogen?
Scientific Thinking	**19.** The number of cases, including new cases as well as already-existing cases, in a defined period of time is the **a.** incidence. **b.** prevalence. **c.** It could be either; need more information.	**20.** You are performing Koch's postulates on sputum specimens taken from patients with a new unexplained respiratory illness. Agar cultures of all of the sputum show an abundance of a particular kind of colony, to the exclusion even of the normal variety of respiratory biota. No healthy subjects' cultures have this particular kind of colony. But when you administer this bacterium in aerosol form to the proper animal model, you do not get disease. What are the possible reasons for this?	**21.** One important way to critically analyze a study that finds a correlation between two factors is the *biological plausibility* of the finding. This means whether there are already known biological facts or phenomena that might explain the correlation. Assess the biological plausibility of a correlation between the composition of the gut microbiome and one's likelihood of having irritable bowel syndrome. Do the same for a correlation between the gut microbiome and the likelihood of having the autoimmune disease rheumatoid arthritis.

Answers to the multiple-choice questions appear in Appendix A.

Visual Connections

This question connects content within and between chapters.

From chapter 2, figure 2.4a. What chemical is the organism in this illustration producing? How does this add to an organism's pathogenicity?

©Lisa Burgess/McGraw-Hill Education

12

Host Defenses I
Overview and Innate Defenses

IN THIS CHAPTER...

12.1 Defense Mechanisms of the Host: An Overview

1. Summarize the three lines of host defenses.
2. Define "marker" and discuss its importance in the second and third lines of defense.
3. Name the body systems that participate in immunity.
4. Describe the structure and function of the lymphatic system and its connection with the circulatory system.
5. Name three kinds of blood cells that function in innate immunity.
6. Connect the mononuclear phagocyte system to innate immunity.
7. Describe how T and B lymphocytes are involved in adaptive immunity.
8. Summarize the importance of cytokines, and list one pro-inflammatory and one anti-inflammatory cytokine.

12.2 The First Line of Defense

9. Identify the three components of the first line of defense.
10. Identify the four body systems that participate in the first line of defense.
11. Describe two examples of how the normal microbiota contribute to the first line of defense.

12.3 The Second Line of Defense

12. List the four major categories of the second line of defense.
13. Outline the steps of phagocytosis.
14. Outline the steps of inflammation.
15. Discuss the mechanism of fever and how it helps defend the body.
16. Name four types of antimicrobial host-derived products.
17. Compose one good overview sentence about the purpose and the mode of action of the complement system, and another about the purpose and mode of action of interferon.

CASE FILE

Bacteria Cause That?

I was working in an endoscopy unit as a registered nurse when I met Natasha, a 20-year-old college student. Natasha was having an endoscopy to determine the cause of her persistent stomach pain.

Natasha had been having episodes of epigastric pain intermittently for 5 years. Approximately every 4 to 6 weeks, she would experience intense upper abdominal pain that left her unable to engage in her usual activities, including her classes and work. She would experience what she described as intense hunger, but pain would worsen immediately after eating so she learned to avoid eating during the worst of her "attacks," which could last as long as 5 days.

She had been admitted to the emergency room after a particularly intense bout of pain. She was dehydrated and weak and her pain was constant. Lab studies revealed an elevated hemoglobin level, indicative of dehydration, and a normal white blood cell count. She had no fever. Her epigastric area was moderately tender on palpation. Her heart rate was elevated, likely as a response to the pain she was experiencing. The ER physician who examined her had her admitted to the hospital and arranged an urgent endoscopy.

Once in the endoscopy unit, Natasha was given medications to induce conscious sedation. The back of her throat was anesthetized and the endoscopy scope was passed down her throat. The esophagus was normal. The stomach, however, was inflamed, with evidence of prior healed ulcerations. One small unhealed lesion remained.

Just as the ER physician had expected, Natasha had a gastric ulcer. Biopsies of the inflamed area were obtained, as well as photographs of the lesion, and the scope was removed. Natasha was then given medication to reverse the effects of the sedatives. Once fully recovered, Natasha was released to go home.

A week later, Natasha had an appointment with the gastroenterologist, who told Natasha that biopsies had revealed the presence of *Helicobacter pylori* in her stomach, which had caused the ulcers. She was prescribed a combination of antibiotics and an acid reducer and was given an appointment to follow up after she had completed the therapy to confirm the efficacy of the treatment. She was also instructed to contact them again if she experienced any more episodes of stomach pain.

- How does *H. pylori* survive in the acidic environment of the stomach?
- What classic response to injury did Natasha display?

Case File Wrap-Up appears at the end of the chapter.

12.1 Defense Mechanisms of the Host: An Overview

The host defenses are a multilevel network of innate protections and adaptive protections that are commonly referred to as the first, second, and third lines of defense **(figure 12.1).** The interaction and cooperation of these three levels of defense normally provide comprehensive protection against infection. The *first line of defense* includes any barrier that blocks invasion at the portal of entry. This nonspecific line of defense limits access to the internal tissues of the body. It is very general in action. The *second line of defense,* called innate immunity, is a more internal system of protective cells, fluids, and processes that includes inflammation and phagocytosis. It acts rapidly at both the local and systemic levels once the first line of defense has been overcome. The highly specific *third line of defense* is acquired only as each foreign substance is encountered by white blood cells called lymphocytes. The goal of this response is to make lymphocytes that are specifically adapted to each individual invader. For that reason, it is called adaptive immunity. The reaction with each different foreign microbe produces unique protective substances and cells that can come into play if that microbe is encountered again. The third line of defense provides long-term immunity. It is discussed in detail in chapter 13. This chapter focuses on the first and second lines of defense.

Human systems are armed with various levels of defense that do not operate in a completely separate fashion. Most defenses overlap and are even redundant in some of their effects. This bombards microbial invaders with an entire assault force, making their survival unlikely. Figure 12.1 provides an overview of the three systems. The light blue box near the bottom that connects both the second and third lines of defense is a good example of how the systems overlap.

Immunology encompasses the study of all features of the body's second and third lines of defense. Although this chapter is concerned, not surprisingly, with infectious microbial agents, be aware that immunology is central to the study of fields as diverse as cancer and allergy.

The skin is the largest organ of your body.
©Erik Palmer/Fogstock Images/Media Bakery

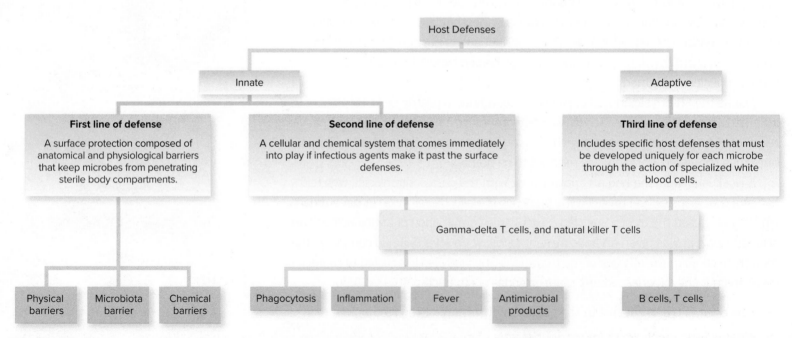

Figure 12.1 **Flowchart summarizing the major components of the host defenses.** Defenses are classified into one of two general categories: (1) innate or (2) adaptive. These can be further subdivided into the first, second, and third lines of defense, each being characterized by a different level and type of protection. There is also a set of cells that straddle the categories of innate and adaptive defenses (light blue box).

In the body, the mandate of the immune system can be easily stated. A healthy functioning immune system is responsible for the following:

1. surveillance of the body,
2. recognition of foreign material, and
3. destruction of entities deemed to be foreign.

All cells (microbial and otherwise), as well as some particles such as pollen, display a unique mix of macromolecules on their surfaces that the immune system "senses" to determine if they are foreign or not. These chemicals are called *antigens*. Because foreign cells or particles could potentially enter through any number of portals, the cells of the immune system constantly move about the body, searching for potential pathogens. This process is carried out primarily by white blood cells, which have been trained to recognize body cells (so-called **self**) and differentiate them from any foreign material in the body, such as an invading bacterial cell **(nonself).** The ability to evaluate cells and macromolecules as either self or nonself is central to the functioning of the immune system. While foreign substances must be recognized as a potential threat and dealt with appropriately, self cells and chemicals must not come under attack by the immune defenses. Many autoimmune disorders are a result of the immune system mistakenly attacking the body's own tissues and organs. For example, in rheumatoid arthritis, the body attacks its own joints and tissues, causing pain and loss of function.

Another word for antigens is **markers (figure 12.2).** Markers, which generally consist of proteins and/or sugars, can be thought of as the cellular equivalent of facial characteristics in humans and allow the cells of the immune system to identify whether or not a newly discovered cell poses a threat. While cells deemed to be self are left alone, cells and other objects designated as foreign are marked for destruction by a number of methods, the most common of which is phagocytosis. Markers that many different kinds of microbes have in common are called **pathogen-associated molecular patterns (PAMPs).** Host cells with important roles in the innate immunity of the second line of defense have **pattern recognition receptors (PRRs)** that they use to recognize PAMPs. There is a middle ground as well. Nonself proteins that are not harmful—such as those found in food we ingest and on commensal microorganisms—are generally recognized as such and the immune system is signaled not to react or to react differently.

Unlike many systems, the immune system does not exist in a single, well-defined site; rather, it is a large, complex, and diffuse network of cells and fluids that permeate every organ and tissue. It is this arrangement that promotes the surveillance and recognition processes that help screen the body for harmful substances.

Tissues, Organs, and Cells Participating in Immunity

The body is partitioned into several fluid-filled spaces called the intracellular, extracellular, lymphatic, cerebrospinal, and circulatory compartments. Although these compartments are physically separated, they have numerous connections. For effective immune responsiveness, the activities in one fluid compartment must be communicated to other compartments.

At the microscopic level, clusters of tissue cells are in direct contact with the **mononuclear phagocyte system (MPS),** which is described shortly, and the extracellular fluid (ECF). Blood and lymphatic capillaries penetrate into these tissues. This close association allows cells that originate in the MPS and ECF to diffuse or migrate into the blood and lymphatics. Any products of a lymphatic reaction can be transmitted directly into the blood through the connection between these two systems. Certain cells and chemicals originating in the blood can move through the vessel walls into the extracellular spaces and migrate into the lymphatic system.

Lymphatic System Vessels and Fluids

The **lymphatic system** is a compartmentalized network of vessels, cells, and specialized accessory organs **(figure 12.3).** It begins in the farthest reaches of the tissues

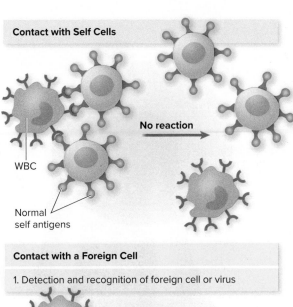

Contact with Self Cells

No reaction

WBC

Normal self antigens

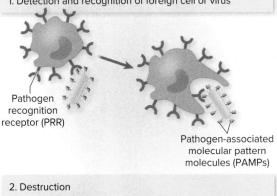

Contact with a Foreign Cell

1. Detection and recognition of foreign cell or virus

Pathogen recognition receptor (PRR)

Pathogen-associated molecular pattern molecules (PAMPs)

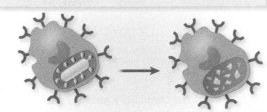

2. Destruction

Figure 12.2 *Search,* *recognize,* and *destroy* **is the mandate of the immune system.** White blood cells are equipped with a very sensitive sense of "touch." As they travel through the tissues, they feel surface markers that help them determine what is self and what is not. When self markers are recognized, no response occurs. However, when nonself is detected, a reaction to destroy it is mounted.

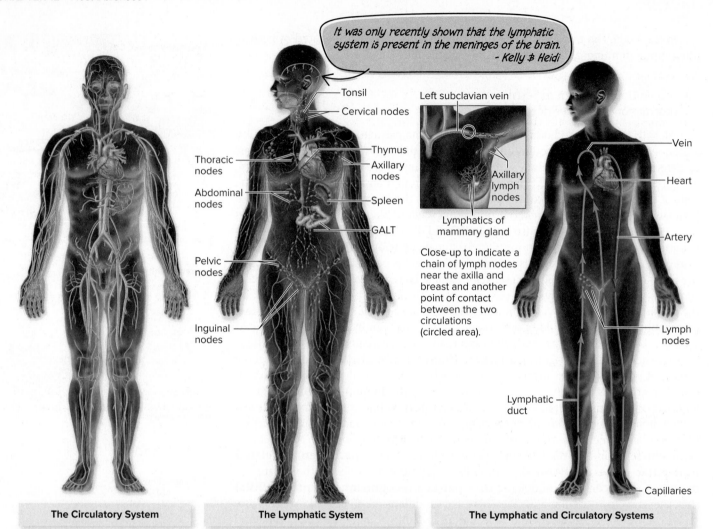

It was only recently shown that the lymphatic system is present in the meninges of the brain.
— Kelly & Heidi

Tonsil
Cervical nodes
Thoracic nodes
Thymus
Axillary nodes
Abdominal nodes
Spleen
GALT
Pelvic nodes
Inguinal nodes

Left subclavian vein
Axillary lymph nodes
Lymphatics of mammary gland

Close-up to indicate a chain of lymph nodes near the axilla and breast and another point of contact between the two circulations (circled area).

Vein
Heart
Artery
Lymph nodes
Lymphatic duct
Capillaries

The Circulatory System

Body compartments are screened by circulating WBCs in the cardiovascular system.

Figure 12.3 The circulatory and lymphatic systems.

The Lymphatic System

The lymphatic system consists of a branching network of vessels that extend into most body areas. Note the higher density of lymphatic vessels in the "dead-end" areas of the hands, feet, and breast, which are frequent contact points for infections. Other lymphatic organs include the lymph nodes, spleen, gut-associated lymphoid tissue (GALT), the thymus, and the tonsils.

The Lymphatic and Circulatory Systems

Comparison of the generalized circulation of the lymphatic system and the blood. Although the lymphatic vessels parallel the regular circulation, they transport in only one direction unlike the cyclic pattern of blood. Direct connection between the two circulations occurs at points near the heart where large lymph ducts empty their fluid into veins (circled area).

Red blood cells (red), T cells (white blood cells: blue), and platelets (gold).
©Science Photo Library/Alamy Stock Photo

as tiny capillaries that transport a special fluid (lymph) through an increasingly larger tributary system of vessels and filters (lymph nodes), and it leads to major vessels that drain back into the regular circulatory system. Some major functions of the lymphatic system are as follows:

1. to provide a route for the return of extracellular fluid to the circulatory system proper;
2. to act as a "drain-off" system for the inflammatory response; and
3. to render surveillance, recognition, and protection against foreign materials through a system of lymphocytes, phagocytes, and antibodies.

Lymphatic Fluid   Lymph is a plasmalike liquid carried by the lymphatic circulation. It is formed when certain blood components move out of the blood vessels into the extracellular spaces and diffuse or migrate into the lymphatic capillaries. Like blood, it transports numerous white blood cells (especially lymphocytes) and miscellaneous materials such as fats, cellular debris, and infectious agents that have gained access to the tissue spaces.

Lymphatic Vessels The system of vessels that transport lymph is constructed along the lines of blood vessels. As the lymph is never subjected to high pressure, the lymphatic vessels appear more similar to thin-walled veins than to thicker-walled arteries. The tiniest vessels, lymphatic capillaries, accompany the blood capillaries and extend into all parts of the body except parts of the central nervous system and certain organs such as bone, placenta, and thymus. Their thin walls are easily permeated by extracellular fluid that has escaped from the circulatory system. Lymphatic vessels are found in particularly high numbers in the hands and feet and around the areola of the breast.

Soon you will read about the bloodstream and blood vessels. Two overriding differences between the bloodstream and the lymphatic system should be mentioned. First, because one of the main functions of the lymphatic system is returning lymph to the circulation, the flow of lymph is in one direction only, with lymph moving from the extremities toward the heart. Eventually, lymph will be returned to the bloodstream through the thoracic duct or the right lymphatic duct to the subclavian vein near the heart. The second difference concerns how lymph travels through the vessels of the lymphatic system. While blood is transported through the body by means of a dedicated pump (the heart), lymph is moved only through the contraction of the skeletal muscles through which the lymphatic ducts wend their way. This dependence on muscle movement helps to explain the swelling of the hands and feet that sometimes occurs during the night (when muscles are inactive) that clears up soon after waking.

Primary Lymphatic Organs In the lymphatic system, the sites of immune cell birth and the locations where they mature are considered **primary lymphatic organs.** Locations in the body where immune cells become activated, reside, or carry out their functions are called **secondary lymphatic organs.**

Red Bone Marrow The red bone marrow is an important intersection between the circulatory system, skeletal system, and lymphatic system. It is typically found in flat bones and the ends of long bones and is the site of blood cell production. All blood cells originate in the bone marrow, including B- and T-lymphocyte precursors. After B lymphocytes begin to express markers that identify them as B cells to the rest of the body, they complete their maturation process while still in the bone marrow. Once this is complete, these B cells then migrate to secondary lymphatic organs, such as the spleen or lymph nodes, where they wait to encounter foreign antigen.

The Thymus: Site of T-Cell Maturation The **thymus** originates in the embryo as two lobes in the lower neck region that fuse into a triangular structure. Lymphocytes that originate in the bone marrow as naïve T lymphocytes migrate to the thymus to complete their maturation. Under the influence of thymic hormones, these cells develop specificity and are released into the circulation as mature T cells. The mature T cells subsequently migrate and settle in other secondary lymphoid organs, just as mature B cells do.

Secondary Lymphatic Organs One way to think of the secondary lymphatic organs is that they are where the action of immunity takes place. The lymph nodes, spleen, and various lymphoid tissues contain large numbers of T and B cells and are stationed throughout the body, ready to encounter antigen and become activated.

Lymph Nodes Lymph nodes are small, encapsulated, bean-shaped organs stationed, usually in clusters, along lymphatic channels and large blood vessels of the thoracic and abdominal cavities (see figure 12.3). Major aggregations of nodes occur in the loose connective tissue of the armpit (axillary nodes), groin (inguinal nodes), and neck (cervical nodes). Their job is to filter out materials in the lymph and to provide appropriate cells for immune reactions. In **figure 12.4,** the anatomy of a lymph node is illustrated. The outer rim of the lymph node is called the **cortex.** There are T lymphocytes in the **paracortical area,** and B lymphoctyes and macrophages

Medical Moment

Examining Lymph Nodes

Physicians routinely examine lymph nodes during a physical examination. Healthy lymph nodes feel well defined and slightly rubbery. They generally are less than 2 cm and are easily movable. The lymph nodes that can easily be felt when they are enlarged include the lymph nodes located behind the ears, along the neck, and in the armpits and the groin. Deeper lymph nodes, such as those in the chest, may be visible on an X ray or computed tomography (CT) scan.

Enlarged lymph nodes by themselves are not usually cause for concern, unless they are accompanied by other symptoms, such as fatigue, fever, weight loss, night sweats, or other constitutional symptoms. Blood tests, X rays, and biopsies may be used to determine the cause of swollen lymph nodes, particularly when other symptoms are present.

Q. A respiratory infection may lead to mild enlargement of the lymph nodes of the neck. What kind of condition(s) would you suspect if lymph nodes throughout the body were enlarged?

Answer in Appendix B.

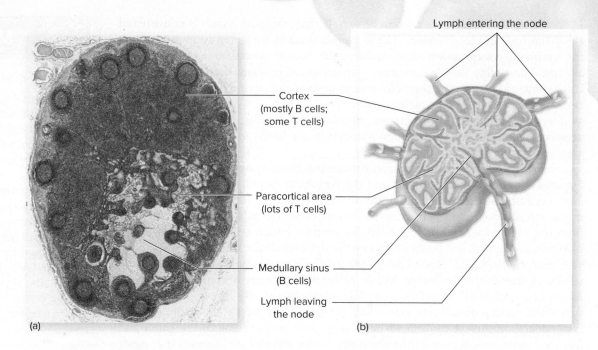

Figure 12.4 Diagram of a lymph node.   On the left is a micrograph of a lymph node that has been cut in half. On the right is a drawing of a lymph node.

Christine Eckel/McGraw-Hill Education (a); ©McGraw-Hill Education (b)

in the **medullary sinus.** Enlargement of the lymph nodes can provide physicians with important clues as to a patient's condition. Generalized lymph node enlargement may indicate the presence of a systemic illness, while enlargement of an individual lymph node may be evidence of a localized infection. Enlargement of lymph nodes reflects the replication of many lymphocyte clones during an adaptive immune response.

The Spleen   The spleen is a lymphoid organ in the upper left portion of the abdominal cavity. It is somewhat similar to a lymph node except that it serves as a filter for blood instead of lymph. While the spleen's primary function is to remove worn-out red blood cells from circulation, its most important immunologic function is the filtering of pathogens from the blood and their subsequent phagocytosis by macrophages in the spleen. Although adults whose spleens have been surgically removed can live a relatively normal life, children without spleens are severely immunocompromised. The spleen also acts as a storehouse of blood that can be released in the event of hemorrhage. It can hold up to 1 cup of blood. For this reason, injury to the spleen can result in profuse internal bleeding.

Associated Lymphoid Tissues   There are discrete bundles of lymphocytes at many sites on or just beneath the skin and mucosal surfaces all over the body. The tissues are often named for their location. For example, the **skin-associated lymphoid tissue** is called **SALT,** and the **mucosa-associated lymphoid tissue** is called **MALT.** The positioning of this diffuse system provides an effective first-strike potential against the constant influx of microbes and other foreign materials in food and air. In the pharynx, a ring of tissues called the **tonsils** provides an active source of lymphocytes. The breasts of pregnant and lactating women also become temporary sites of antibody-producing lymphoid tissues. The intestinal tract houses the best-developed collection of lymphoid tissue, called **gut-associated lymphoid tissue,** or **GALT.** Examples of GALT include the appendix, the lacteals (special lymphatic vessels stationed in each intestinal villus), and **Peyer's patches,** compact aggregations of lymphocytes in the ileum of the small intestine.

The Blood

The circulatory system consists of the heart, arteries, veins, and capillaries that circulate the blood. The lymphatic system includes lymphatic vessels and lymphatic organs (lymph nodes) that circulate lymph. As you will see, these two circulations parallel, interconnect with, and complement one another.

The substance that courses through the arteries, veins, and capillaries is **whole blood,** a liquid consisting of **blood cells** suspended in **plasma.** One can visualize these two components with the naked eye when a tube of unclotted blood is allowed to sit or is spun in a centrifuge. The cells' density causes them to settle into an opaque layer at the bottom of the tube, leaving the plasma, a clear, yellowish fluid, on top. Serum is essentially the same as plasma, except it is the clear fluid from clotted blood. Serum is often used in immune testing and therapy.

A Survey of Blood Cells   The production of blood cells is called **hematopoiesis** (hee″-mat-o-poy-ee′-sis). The primary precursor of new blood cells is a pool of undifferentiated cells called pluripotential **stem cells** in the bone marrow. "Pluripotent" means that the cells are able to become any type of blood cell that is needed. These cells are produced continually in the bone marrow, and subjected to a variety of growth factors and hormones that cause them to change over time. This process is called **differentiation.** The immature stem cells change their genetic expression, which in turn causes them to express different surface markers and respond to new signals. They become more specialized over time, eventually maturing either in the bone marrow or in other sites of the body. This is the process of blood cell development.

There are two primary lines of cells that arise from stem cells, and are still considered immature. These are (1) those that differentiate from a common myeloid cell, such as red blood cell precursors (erythroblasts) and platelet precursors (megakaryoblasts), and (2) those that differentiate from a common lymphoid precursor cell, such as precursors of many white blood cells (myeloblasts) and precursors of lymphocytes (lymphoblasts).

White blood cells **(leukocytes)** can generally be divided into two categories according to their staining patterns when viewed with a microscope: **granulocytes,** which have dark staining granules, and **agranulocytes,** which do not have granules and typically have large nuclei. The granules in granulocytes can be released to kill foreign cells or affect host tissues. **Figure 12.5** sorts out all these cells. They are vitally important to innate and adaptive immunity.

Cytokines: Critical for Cell Communication

Before we specifically address details of the first or second line of defense, we need to understand the role of some potent chemicals and proteins that influence defensive responses, and that immune cells use to communicate with one another.

Hundreds of small, active molecules are constantly being secreted to regulate, stimulate, suppress, and otherwise control the many aspects of cell development, inflammation, and immunity. These substances are the products of several types of cells, including monocytes, macrophages, lymphocytes, fibroblasts, mast cells, platelets, and the endothelial cells of blood vessels.

These cell products are generally called **cytokines.** There are dozens of cytokines that have been identified. Specific cytokines will be discussed as we cover various immune processes. **Table 12.1** provides four broad categories of cytokines based on their functions. One or two examples of each functional category are provided to demonstrate the breadth of their activities.

The Mononuclear Phagocyte System

The tissues of the body are permeated by a support network of connective tissue fibers, or a reticulum, that interconnects nearby cells and meshes with the massive connective tissue network surrounding all organs. The phagocytic cells enmeshed in this network are collectively called the mononuclear phagocyte system

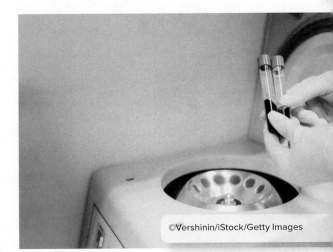

©Vershinin/iStock/Getty Images

Blood appears red because there are approximately 1,000× more red cells than "white" cells in this fluid.

©fStop/PunchStock

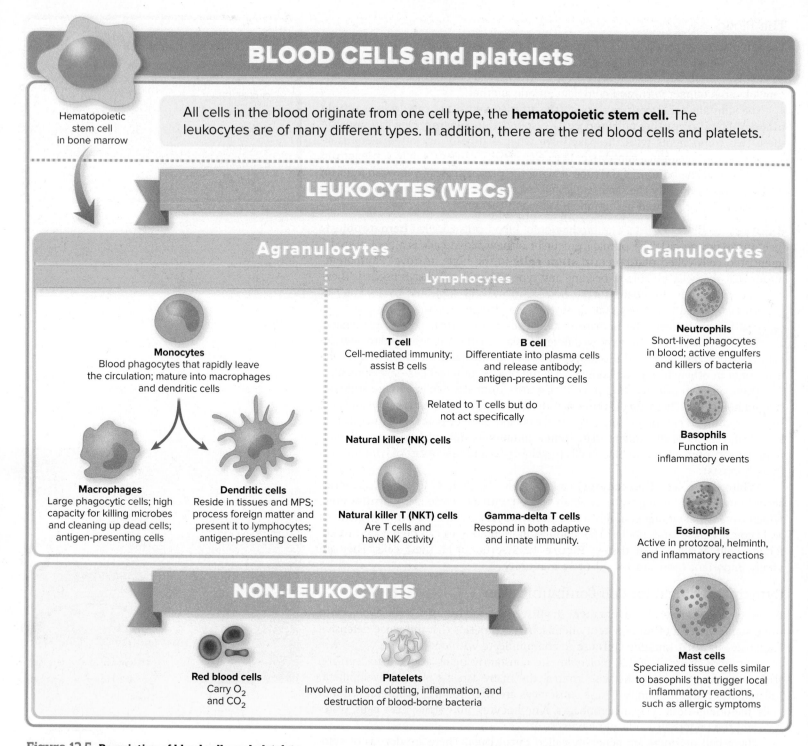

BLOOD CELLS and platelets

Hematopoietic stem cell in bone marrow

All cells in the blood originate from one cell type, the **hematopoietic stem cell.** The leukocytes are of many different types. In addition, there are the red blood cells and platelets.

LEUKOCYTES (WBCs)

Agranulocytes

Granulocytes

Lymphocytes

Monocytes
Blood phagocytes that rapidly leave the circulation; mature into macrophages and dendritic cells

Macrophages
Large phagocytic cells; high capacity for killing microbes and cleaning up dead cells; antigen-presenting cells

Dendritic cells
Reside in tissues and MPS; process foreign matter and present it to lymphocytes; antigen-presenting cells

T cell
Cell-mediated immunity; assist B cells

B cell
Differentiate into plasma cells and release antibody; antigen-presenting cells

Related to T cells but do not act specifically

Natural killer (NK) cells

Natural killer T (NKT) cells
Are T cells and have NK activity

Gamma-delta T cells
Respond in both adaptive and innate immunity.

Neutrophils
Short-lived phagocytes in blood; active engulfers and killers of bacteria

Basophils
Function in inflammatory events

Eosinophils
Active in protozoal, helminth, and inflammatory reactions

Mast cells
Specialized tissue cells similar to basophils that trigger local inflammatory reactions, such as allergic symptoms

NON-LEUKOCYTES

Red blood cells
Carry O_2 and CO_2

Platelets
Involved in blood clotting, inflammation, and destruction of blood-borne bacteria

Figure 12.5 Description of blood cells and platelets.

(MPS; **figure 12.6**). This system is intrinsic to the immune function because it provides a passageway within and between tissues and organs. The MPS is found in the thymus, where important white blood cells mature, and in the lymph nodes, tonsils, spleen, and lymphoid tissue in the mucosa of the gut and respiratory tract, where most of the MPS "action" takes place.

When differentiated macrophages and dendritic cells arrive in tissues that are near portals of entry or filtration organs, they often take up residence in these tissues permanently, waiting to attack foreign intruders **(figure 12.7)**. These tissues and specialized **histiocyte** cells include the liver (Kupffer cells), lungs (alveolar macrophages), skin (Langerhans cells), brain (microglia), and others.

Table 12.1 Cytokines

	Examples	Source	Target
Pro-inflammatory cytokines, which *encourage* adaptive and innate immune responses	Interleukin-1 (IL-1)	Macrophages, B cells, dendritic cells	B cells, T cells
	Tumor necrosis factor-β (TNF-β)	T cells	Phagocytes, tumor cells
Anti-inflammatory cytokines, which *discourage* adaptive and innate immune responses	Interleukin-10 (IL-10)	T cells	B cells, macrophages
Vasodilators and vasoconstrictors, which can change the diameter of blood vessels or vessel permeability	Serotonin	Platelets and intestinal cells	Cells in peripheral and central nervous system
	Histamine	Mast cells and basophils	Blood vessels, sensory nerves, neutrophils
Growth factor cytokines, which regulate lymphocyte growth or activation	Interleukin-7 (IL-7)	Bone marrow cells, epithelial cells	Stem cells (stimulates growth of B and T cells)
	Erythropoietin	Endothelial cells	Stem cells (stimulates production of red blood cells)

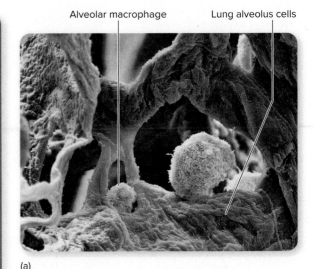

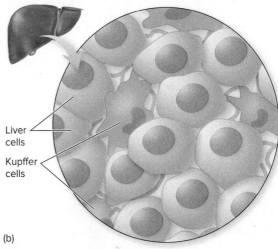

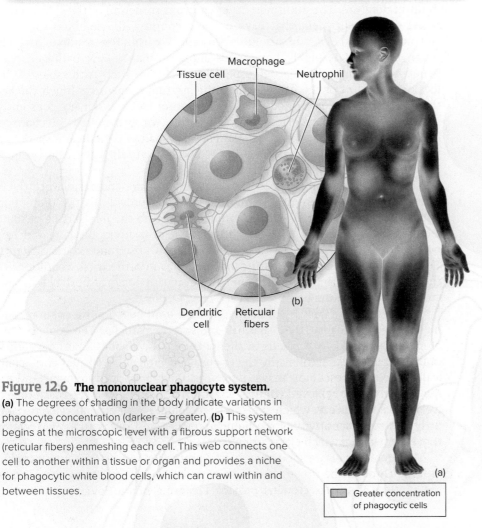

Figure 12.6 The mononuclear phagocyte system.
(a) The degrees of shading in the body indicate variations in phagocyte concentration (darker = greater). **(b)** This system begins at the microscopic level with a fibrous support network (reticular fibers) enmeshing each cell. This web connects one cell to another within a tissue or organ and provides a niche for phagocytic white blood cells, which can crawl within and between tissues.

Greater concentration of phagocytic cells

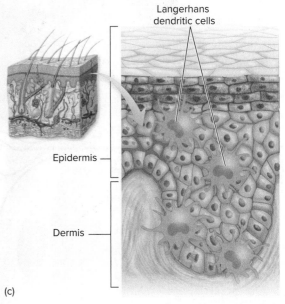

Figure 12.7 Macrophages.   **(a)** Scanning electron micrograph of an alveolar macrophage inside a lung. **(b)** Liver tissue with Kupffer cells. **(c)** Langerhans cells deep in the epidermis.

©Eye of Science/Science Source (a)

12.1 LEARNING OUTCOMES—Assess Your Progress

1. Summarize the three lines of host defenses.
2. Define "marker" and discuss its importance in the second and third lines of defense.
3. Name the body systems that participate in immunity.
4. Describe the structure and function of the lymphatic system and its connection with the circulatory system.
5. Name three kinds of blood cells that function in innate immunity.
6. Connect the mononuclear phagocyte system to innate immunity.
7. Describe how T and B lymphocytes are involved in adaptive immunity.
8. Summarize the importance of cytokines, and list one pro-inflammatory and one anti-inflammatory cytokine.

12.2 The First Line of Defense

The inborn defenses are the physical and chemical barriers that impede the entry of not only microbes but any foreign agent, whether living or not **(figure 12.8).**

Physical or Anatomical Barriers at the Body's Surface

The skin and mucous membranes of the respiratory and digestive tracts have several built-in defenses. The outermost layer (stratum corneum) of the skin is composed of epithelial cells that have become compacted, cemented together, and impregnated with an insoluble protein, keratin. The result is a thick, tough layer that is highly impervious and waterproof. Few pathogens can penetrate this unbroken barrier, especially in regions such as the soles of the feet or the palms of the hands, where the stratum corneum is much thicker than on other parts of the body. In addition, outer layers of skin are constantly sloughing off, taking associated microbes with them. Other cutaneous barriers include hair follicles and skin glands. The hair shaft is periodically shed, and the follicle cells are **desquamated** (des'-kwuh-mayt-ud). The flushing effect of sweat glands also helps remove microbes.

The mucous membranes of the digestive, urinary, and respiratory tracts and of the eye are moist and permeable. They do provide barrier protection but without a keratinized layer. The mucous coat on the free surface of some membranes impedes the entry and attachment of bacteria. Blinking and tear production (lacrimation) flush the eye's surface with tears and rid it of irritants. The constant flow of saliva helps carry microbes into the harsh conditions of the stomach. Vomiting and defecation also evacuate noxious substances or microorganisms from the body.

The respiratory tract is constantly guarded from infection by elaborate and highly effective adaptations. Nasal hair traps larger particles. The copious flow of mucus and fluids that occurs in allergy and colds provides a flushing action. In the respiratory tree (primarily the trachea and bronchi), a ciliated epithelium (called the ciliary escalator) moves foreign particles entrapped in mucus toward the pharynx to be removed **(figure 12.9).** Irritation of the nasal passage reflexively initiates a sneeze, which expels a large volume of air at high velocity. Similarly, the acute sensitivity of the bronchi, trachea, and larynx to foreign matter triggers coughing, which ejects irritants.

The genitourinary tract derives partial protection via the continuous trickle of urine through the ureters and from periodic bladder emptying that flushes the urethra. Vaginal secretions provide cleansing of the lower reproductive

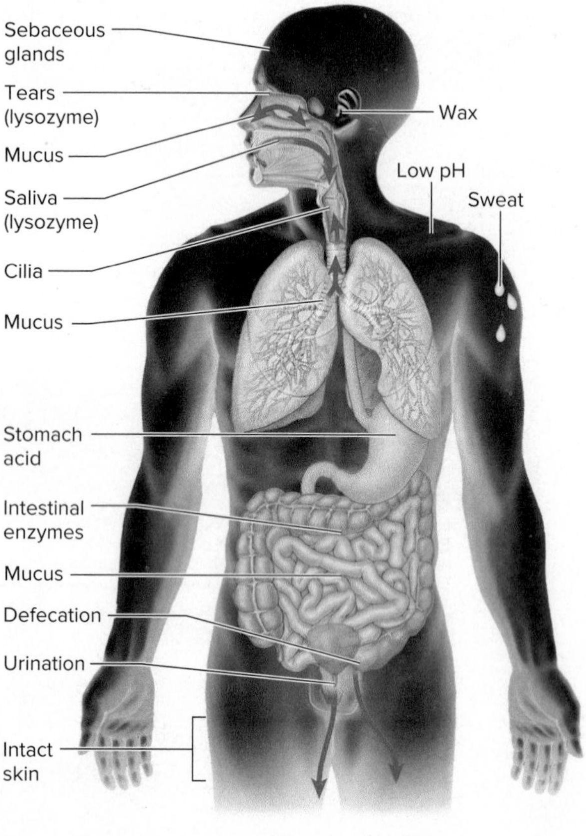

Sebaceous glands
Tears (lysozyme)
Mucus
Saliva (lysozyme)
Cilia
Mucus
Stomach acid
Intestinal enzymes
Mucus
Defecation
Urination
Intact skin
Wax
Low pH
Sweat

Figure 12.8 The primary physical and chemical defense barriers. The blue arrows indicate the direction of fluid flow.

tract in females. Women who are postmenopausal are sometimes prone to vaginal infections due to lack of vaginal secretions resulting from a decrease in estrogen production.

The human microbiome forms a type of structural barrier. Since humans have evolved with–and in response to–the constant presence of our microbiome, we can consider it as an integral part of our anatomy. Its presence can block the access of pathogens to epithelial surfaces and can create an unfavorable environment for pathogens by competing for limited nutrients or by altering the local pH.

New research stemming from microbiome research has continued to highlight the importance of the normal resident biota on the development of nonspecific defenses as well as innate and adaptive immunity. The presence of a robust commensal biota "trains" host defenses in such a way that commensals are kept in check and pathogens are eliminated. Evidence suggests that inflammatory bowel diseases, including Crohn's disease and ulcerative colitis, may well be a result of our overzealous attempts to free our environment of microbes and to overtreat ourselves with antibiotics. The result is an "ill-trained" gut defense system that responds inappropriately to commensal biota.

Nonspecific Chemical Defenses

The skin and mucous membranes offer a variety of chemical defenses. Sebaceous secretions exert an antimicrobial effect, and specialized glands of the eyelids lubricate the conjunctiva with an antimicrobial secretion. An additional defense in tears and saliva is **lysozyme,** an enzyme that hydrolyzes, or breaks the glycosidic bond between neighboring sugars in the peptidoglycan layer of bacterial cell walls. The high lactic acid and electrolyte concentrations of sweat and the skin's acidic pH and fatty acid content are also inhibitory to many microbes. Likewise, the hydrochloric acid in the stomach renders protection against many pathogens that are swallowed, and the intestine's digestive juices and bile are potentially destructive to microbes. Even semen contains an antimicrobial chemical that inhibits bacteria, and the vagina during reproductive years has a protective acidic pH maintained by normal biota.

The vital contribution of all types of barriers is clearly demonstrated in people who have lost them or never had them. Patients with severe skin damage due to burns are extremely susceptible to infections. Those with blockages in the salivary glands, tear ducts, intestine, and urinary tract are also at greater risk for infection. But as important as it is, the first line of defense alone is not sufficient to protect against infection. Because many pathogens find a way to get around the barriers by using their virulence factors (discussed in chapter 11), a whole new set of defenses–inflammation, phagocytosis, adaptive immune responses–are brought into play.

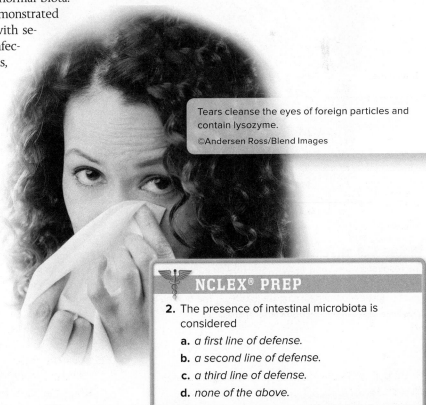

Tears cleanse the eyes of foreign particles and contain lysozyme.
©Andersen Ross/Blend Images

12.2 LEARNING OUTCOMES–Assess Your Progress

9. Identify the three components of the first line of defense.
10. Identify the four body systems that participate in the first line of defense.
11. Describe two examples of how the normal microbiota contribute to the first line of defense.

NCLEX® PREP

2. The presence of intestinal microbiota is considered
   a. *a first line of defense.*
   b. *a second line of defense.*
   c. *a third line of defense.*
   d. *none of the above.*

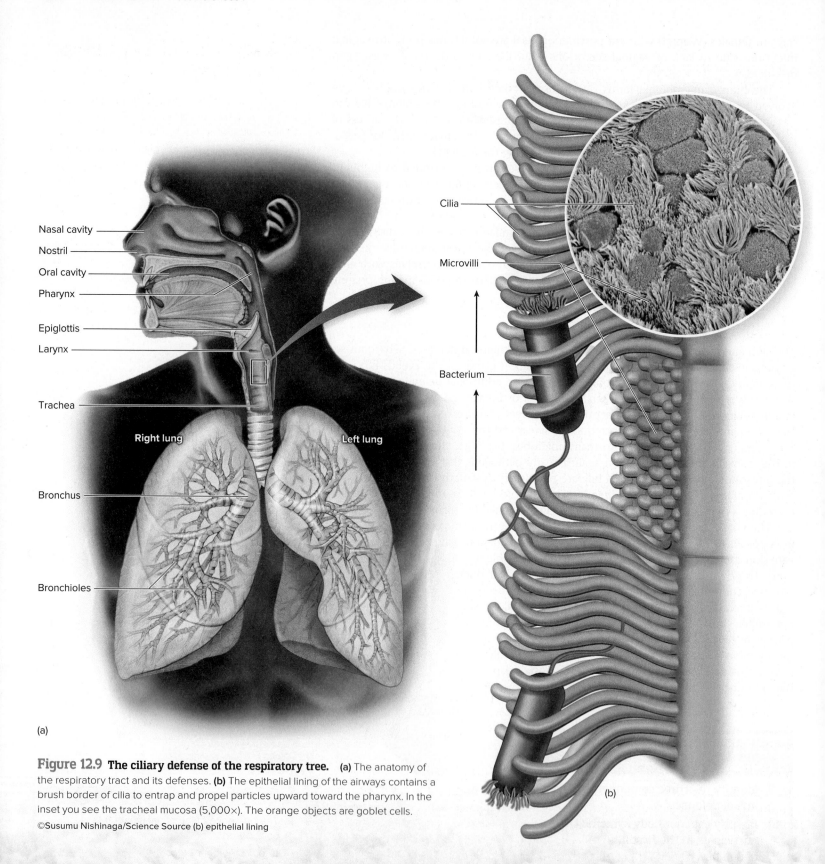

Nasal cavity

Nostril

Oral cavity

Pharynx

Epiglottis

Larynx

Trachea

Right lung

Left lung

Bronchus

Bronchioles

(a)

Cilia

Microvilli

Bacterium

(b)

Figure 12.9 **The ciliary defense of the respiratory tree.** **(a)** The anatomy of the respiratory tract and its defenses. **(b)** The epithelial lining of the airways contains a brush border of cilia to entrap and propel particles upward toward the pharynx. In the inset you see the tracheal mucosa (5,000×). The orange objects are goblet cells.

©Susumu Nishinaga/Science Source (b) epithelial lining

12.3 The Second Line of Defense

Now that we have introduced the principal anatomical and physiological framework of the immune system, we address four mechanisms that play important roles in host defenses: (1) phagocytosis, (2) inflammation, (3) fever, and (4) antimicrobial products. Because of the generalized nature of these defenses, they are considered innate in their effects, but they also support and interact with the adaptive immune responses described in chapter 13.

Phagocytosis: Cornerstone of Inflammation and Adaptive Immunity

By any standard, a phagocyte represents an impressive piece of living machinery, wandering through the tissues to seek, capture, and destroy a target. The general activities of phagocytes are as follows:

1. to survey the tissue compartments and discover microbes, particulate matter (dust, carbon particles, antigen-antibody complexes), and injured or dead cells;
2. to ingest and eliminate these materials; and
3. to read immunogenic information (antigens) from foreign matter.

It is generally accepted that all cells have some capacity to engulf materials, but professional phagocytes do it for a living. The three main types of phagocytes are neutrophils, macrophages, and dendritic cells.

Neutrophils

As previously stated, **neutrophils** are general-purpose phagocytes. Neutrophils have a limited ability to phagocytose due to their rapid death when exposed to their own toxic oxygen products that also kill engulfed bacteria. Dead neutrophils accumulate quickly near sites of injury or infection and are a primary component of pus. A common sign of bacterial infection is a high neutrophil count in the blood (neutrophilia), and neutrophils are also a primary component of pus.

Monocytes Lead to Macrophages and Dendritic Cells

After emigrating out of the bloodstream into the tissues, **monocytes** are transformed by various inflammatory mediators into **macrophages.** This process is marked by an increase in size and by enhanced development of lysosomes and other organelles. They live much longer than neutrophils and can consume a lot of material. Earlier we discussed specialized macrophages called histiocytes that live in a certain tissue and remain there during their life span. Remember that these are alveolar (lung) macrophages; the Kupffer cells in the liver; dendritic cells in the skin (see figure 12.7); and macrophages in the spleen, lymph nodes, bone marrow, kidney, bone, and brain. Other macrophages do not reside permanently in a particular tissue and drift nomadically throughout the MPS. Not only are macrophages dynamic scavengers, but they also process foreign substances and prepare them for reactions with B and T lymphocytes.

Mechanisms of Phagocytic Recognition, Engulfment, and Killing

The term *phagocyte* literally means "eating cell." But phagocytosis (the term for what phagocytes do) is more than just the physical process of engulfment because phagocytes also actively attack and dismantle foreign cells with a wide array of antimicrobial substances. Phagocytosis can occur as an isolated event performed by a few phagocytic cells responding to a minor irritant in their area or as part of the orchestrated events of inflammation described in the next section. The events in phagocytosis include chemotaxis, ingestion, phagolysosome formation, destruction, and excretion. **Table 12.2** provides the details of phagocytosis.

A Note About the Terms Innate and Adaptive

The terminology designating the second and third line of defenses has evolved. For a long time the second line of defense was called "nonspecific immunity," and the third line of defense was called "specific immunity." The terms referred to the binding of human defense cells to microbes. "Specific" bonds were known to operate in the third, adaptive, line of defenses, as you will see. But the second line of defense also requires specific binding between markers on the microbes and markers on the human cells. The binding is very specific on a molecular level, but can be carried out with a variety of invading microbes, even ones the human has not encountered before. To clear up confusion, that response (second line of defense) is usually called innate, since it does not require previous "training" of the human cells.

The third line of defense is fully activated by specific interactions between unique human markers and unique microbial markers. The best way to refer to it is as "adaptive immunity."

Table 12.2 Phagocytosis

1 Chemotaxis. Phagocytes migrate into a region of inflammation with a deliberate sense of direction, attracted by a gradient of stimulant products from the parasite and host tissue at the site of injury.

2 Adhesion. Phagocytes use their PRRs to recognize PAMPs on foreign cells. This receptor interaction causes the two to "stick" together.

3 4 Engulfment and Phagosome Formation. Once the phagocyte has made contact with its prey, it extends pseudopods that enclose the cells or particles in a pocket and internalize them in a vacuole called a phagosome. It also secretes more cytokines to further amplify the innate response.

5 Phagolysosome Formation and Killing. In a short time, lysosomes migrate to the scene of the phagosome and fuse with it to form a phagolysosome. Granules containing antimicrobial chemicals are released into the phagolysosome, forming a toxic brew designed to poison and then dismantle the ingested material.

6 Destruction. Two separate systems of destructive chemicals await the microbes in the phagolysosome. The oxygen-dependent system (known as the respiratory burst, or oxidative burst) involves several substances that were described in chapters 6 and 9. Myeloperoxidase, an enzyme found in granulocytes, forms halogen ions (OCl$^-$) that are strong oxidizing agents. Other products of oxygen metabolism such as hydrogen peroxide, the superoxide anion (O$_2^-$), activated or so-called singlet oxygen ($^{\bullet}$O$_2$), and the hydroxyl free radical (OH$^{\bullet}$) separately and together have formidable killing power. Other mechanisms that come into play are the liberation of lactic acid, lysozyme, and nitric oxide (NO), a powerful mediator that kills bacteria and inhibits viral replication. Cationic proteins that injure bacterial cytoplasmic membranes and a number of enzymes complete the job. These include those that can degrade DNA (DNase), RNA (RNase), and proteins (proteases).

7 Elimination. The small bits of undigestible debris are released from the macrophage by exocytosis.

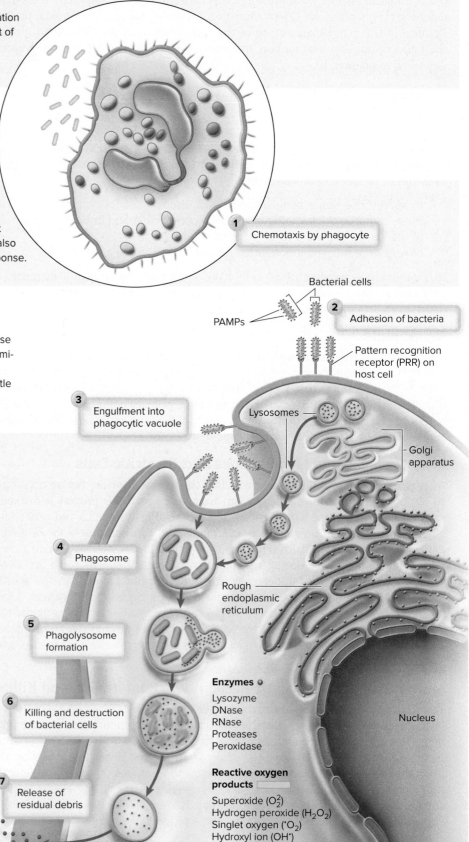

1 Chemotaxis by phagocyte

Bacterial cells

PAMPs

2 Adhesion of bacteria

Pattern recognition receptor (PRR) on host cell

3 Engulfment into phagocytic vacuole

Lysosomes

Golgi apparatus

4 Phagosome

Rough endoplasmic reticulum

5 Phagolysosome formation

Enzymes
Lysozyme
DNase
RNase
Proteases
Peroxidase

6 Killing and destruction of bacterial cells

Nucleus

Reactive oxygen products
Superoxide (O$_2^2$)
Hydrogen peroxide (H$_2$O$_2$)
Singlet oxygen ($^{\bullet}$O$_2$)
Hydroxyl ion (OH$^{\bullet}$)

7 Release of residual debris

Phagocytes and other defensive cells are able to recognize some microorganisms as foreign because of signal molecules that the microbes have on their surfaces. These pathogen-associated molecular patterns (PAMPs) are molecules shared by many microorganisms—but not present in mammals—and therefore serve as "red flags" for phagocytes and other cells of innate immunity.

Bacterial PAMPs include peptidoglycan and lipopolysaccharide. Double-stranded RNA, which is found only in some viruses, is also a PAMP. On the host side, phagocytes, dendritic cells, endothelial cells, and even lymphocytes possess pattern recognition receptors (PRRs) on their surfaces that recognize and bind PAMPs. The cells possess these PRRs all the time, whether or not they have encountered PAMPs before. Many phagocytic cells of the innate immune system contain PRRs inside their cytoplasm. These have a special name—**inflammasomes**—and they recognize microbial PAMPs as well as markers from damaged host cells once they have been phagocytosed. Recognition leads to the release of signals that initiate and regulate inflammation. (There are a lot of acronyms in immunology. Don't let them get away from you; keep up with them. If you know what all the acronyms stand for and what they do, you are halfway there in understanding host defenses!)

Figure 12.10 is an electron micrograph of a phagocyte engulfing bacteria.

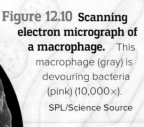

Figure 12.10 Scanning electron micrograph of a macrophage. This macrophage (gray) is devouring bacteria (pink) (10,000×).
SPL/Science Source

Bacteria
Macrophage

The Inflammatory Response: A Complex Reaction to Injury

At its most general level, the inflammatory response is a reaction to any traumatic event in the tissues. It is so commonplace that all of us manifest inflammation in some way every day. It appears in the nasty flare of a cat scratch, the blistering of a burn, the painful lesion of an infection, and the symptoms of allergy. When close to our external surfaces, it is readily identifiable by a classic series of signs and symptoms characterized succinctly by four Latin terms: *rubor, calor, tumor,* and *dolor*. Rubor (redness) is caused by increased circulation and vasodilation in the injured tissues; calor (warmth) is the heat given off by the increased flow of blood; tumor (swelling) is caused by increased fluid escaping into the tissues; and dolor (pain) is caused by the stimulation of nerve endings. A fifth symptom, loss of function, has been added to give a complete picture of the effects of inflammation.

It is becoming increasingly clear that some chronic diseases, such as cardiovascular disease, can be caused by chronic inflammation. While we speak of inflammation at a local site (such as a finger), inflammation can affect an entire system—such as blood vessels, lungs, skin, the joints, and so on.

Factors that can elicit inflammation include trauma from infection (the primary emphasis here), tissue injury or necrosis due to physical or chemical agents, and adaptive immune reactions. Although the details of inflammation are very complex, its chief functions can be summarized as follows:

1. to mobilize and attract immune components to the site of the injury,
2. to set in motion mechanisms to repair tissue damage and localize and clear away harmful substances, and
3. to destroy microbes and block their further invasion.

The inflammatory response is a powerful defensive reaction, a means for the body to maintain stability and restore itself after an injury. But it has the potential to actually *cause* tissue injury, destruction, and disease.

The Stages of Inflammation

The process leading to inflammation is a dynamic, predictable sequence of events that can be acute, lasting from a few minutes or hours, or chronic, lasting for days, weeks, or years. Once the initial injury has occurred, a chain reaction takes place at the site of damaged tissue, summoning beneficial cells and fluids into the injured area. As an example, we will look at an injury at the microscopic level and observe the flow of major events in **table 12.3.** Be sure to examine it before moving on.

Diapedesis and Chemotaxis   **Diapedesis** is the migration of WBCs out of blood vessels into tissues. It is made possible by several related characteristics of WBCs. For example, they are actively motile and readily change shape. This phenomenon is also assisted by the nature of the endothelial cells lining the venules. They contain complex adhesive receptors that capture the WBCs and help transport them from the venules into the extracellular spaces **(figure 12.11).**

Another factor in the migratory habits of these WBCs is chemotaxis, or the tendency of cells to migrate in response to a specific chemical stimulus given off at a site of injury or infection. This means that cells swarm from many compartments to the site of infection and remain there to perform general and specific immune functions. These basic properties are absolutely essential for the sort of intercommunication and deployment of cells required for most immune reactions.

The Benefits of Edema and Leaky Vessels   Both the secretion of fluids and the infiltration of neutrophils are physiologically beneficial activities. The influx of fluid dilutes toxic substances, and the fibrin clot can effectively trap microbes and prevent their further spread. The neutrophils that aggregate in the inflamed site are immediately involved in phagocytosing and destroying bacteria, dead tissues, and particulate matter. In some types of inflammation, accumulated phagocytes contribute to **pus,** a white, gooey mass of cells, liquefied cellular debris, and bacteria. Certain bacteria (streptococci, staphylococci, gonococci, and meningococci) are especially powerful attractants for neutrophils and are thus termed **pyogenic,** or pus-forming, bacteria.

Fever

An important systemic component of inflammation—and innate immunity in general—is fever, defined as an abnormally elevated body temperature. Although fever is very common in infection, it is also associated with certain allergies, cancers, and other organic illnesses. Fevers whose causes are unknown are called fevers of unknown origin, or FUO.

The body temperature is normally maintained by a control center in the hypothalamus region of the brain. This thermostat regulates the body's heat production and heat loss and sets the core temperature at around 37°C (98.6°F) with slight fluctuations during a daily cycle. Fever is initiated when circulating substances called **pyrogens** (py'-roh-jenz) reset the hypothalamic thermostat to a higher setting. This change signals the musculature to increase heat production and peripheral arterioles to decrease heat loss through vasoconstriction. Fevers range

The pain of a sore throat is caused by a combination of (1) damage inflicted by microbes, (2) the inflammatory response that can lead to pressure on surrounding nerves, and (3) the release of pain-inducing cytokines.

©Brand X Pictures/Punchstock

Table 12.3 Inflammation

1 Injury/Immediate Reactions

Following an injury, early changes occur in the vasculature (arterioles, capillaries, venules) in the vicinity of the damaged tissue. These changes are controlled by nervous stimulation, **chemical mediators,** and **cytokines** released by blood cells, tissue cells, and platelets in the injured area. Some mediators are **vasoactive**–that is, they affect the endothelial cells and smooth muscle cells of blood vessels–and others are **chemotactic factors,** also called **chemokines,** that affect white blood cells.

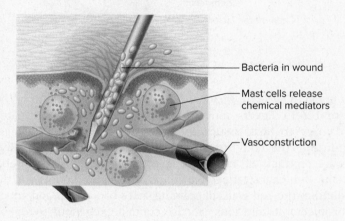

- Bacteria in wound
- Mast cells release chemical mediators
- Vasoconstriction

2 Vascular Reactions

In quick succession, the blood vessels in the vicinity dilate (widen), then constrict, and then dilate again. The wide-narrow-wide sequence is thought to be for the purpose of flushing irritants (such as bacteria) away from the area, then the narrowing is an attempt to stem blood leaving the blood vessels, followed by a long-term dilation to bring helpful blood components to the site. The overall effect of vasodilation is to increase the flow of blood into the area, which facilitates the influx of immune components and also causes redness and warmth.

Some substances cause the endothelial cells in the walls of postcapillary venules to contract and form gaps through which blood-borne components exude into the extracellular spaces. The fluid part that escapes is called the **exudate.**

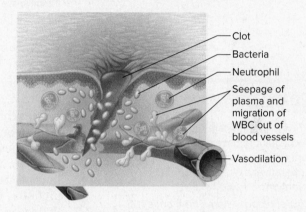

- Clot
- Bacteria
- Neutrophil
- Seepage of plasma and migration of WBC out of blood vessels
- Vasodilation

3 Edema and Pus Formation

Accumulation of this fluid in the tissues gives rise to local swelling and hardness, called edema. The fluid contains varying amounts of plasma proteins, such as globulins, albumin, the clotting protein fibrinogen, blood cells, and cellular debris. Depending on its content, the fluid may be clear (called **serous**), or it may contain red blood cells or pus. Pus is composed mainly of white blood cells and the debris generated by phagocytosis. In some types of edema, the fibrinogen is converted to fibrin threads that enmesh the injury site. Within an hour, multitudes of neutrophils responding chemotactically to special signaling molecules converge on the injured site.

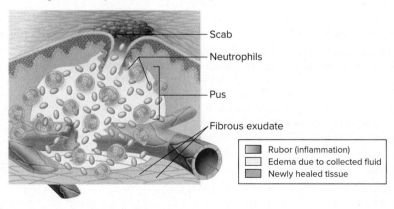

- Scab
- Neutrophils
- Pus
- Fibrous exudate

▨	Rubor (inflammation)
▢	Edema due to collected fluid
▨	Newly healed tissue

4 Resolution/Scar Formation

Repair is the last step and results either in complete resolution to healthy tissue or in formation of scar tissue, depending on the tissue type and the extent of the damage. Note here that macrophages are pictured leaving the blood vessels in a process called **diapedesis** (dye″-ah-puh-dee′-sis). Meanwhile, differentiated stem cells in the area begin to divide and repopulate the damaged site with new cells to replace those that were damaged.

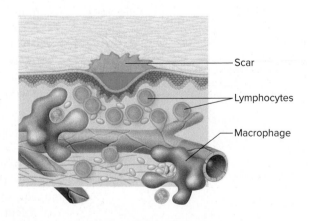

- Scar
- Lymphocytes
- Macrophage

Figure 12.11 Diapedesis and chemotaxis of leukocytes. **(a)** View of a venule depicts white blood cells squeezing themselves between spaces in the blood vessel wall via diapedesis. This process, shown in cross section, indicates how the pool of leukocytes adheres to the endothelial wall. From this site, they are poised to migrate out of the vessel into the tissue space. **(b)** This micrograph captures neutrophils in the process of diapedesis.

Courtesy Steve Kunkel (b)

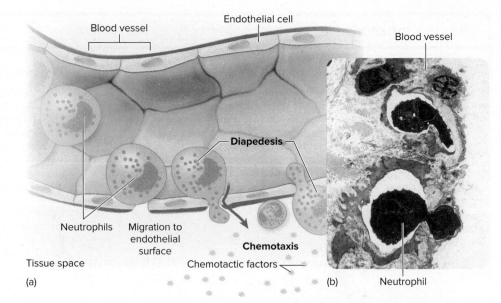

Blood vessel — Endothelial cell — Blood vessel

Diapedesis

Neutrophils — Migration to endothelial surface — Chemotaxis

Tissue space — Chemotactic factors

(a)   (b)   Neutrophil

in severity from low grade (37.7°C to 38.3°C, or 100°F to 101°F) to high (40.0°C to 41.1°C, or 104°F to 106°F). Fevers above 104°F to 105°F should be treated to prevent serious damage to host tissues. Pyrogens are described as **exogenous** (coming from outside the body) or **endogenous** (originating internally). Exogenous pyrogens are products of infectious agents such as viruses, bacteria, protozoa, and fungi. One well-characterized exogenous pyrogen is endotoxin, the lipopolysaccharide found in the cell walls of gram-negative bacteria. Blood, blood products, vaccines, or injectable solutions can also contain exogenous pyrogens. Endogenous pyrogens are released by monocytes, neutrophils, and macrophages during the process of phagocytosis and appear to be a natural part of the immune response. Two potent pyrogens released by macrophages are the cytokines interleukin-1 (IL-1) and tumor necrosis factor (TNF).

Benefits of Fever

Aside from its practical and medical importance as a sign of disease, increased body temperature has additional benefits:

- Fever inhibits multiplication of temperature-sensitive microorganisms such as the poliovirus, cold viruses, herpes zoster virus, and systemic and subcutaneous fungal pathogens.
- Fever interferes with the nutrition of bacteria by reducing the availability of iron.
- Fever increases metabolism and stimulates immune reactions and naturally protective physiological processes. It speeds up hematopoiesis, phagocytosis, and adaptive immune reactions and helps specific lymphocytes home in on sites of infection.

Thousands of years ago, people realized that fever was part of an innate protective response. Hippocrates offered the idea that it was the body's attempt to burn off a noxious agent. Sir Thomas Sydenham wrote in the seventh century: "Why, fever itself is Nature's instrument!" So widely held was the view that fever could be therapeutic that pyretotherapy (treating disease by inducing an intermittent fever) was once used to treat syphilis, gonorrhea, leishmaniasis (a protozoal infection), and cancer. This attitude fell out of favor when drugs for relieving fever (aspirin) first came into use in the early 1900s, and an adverse view of fever began to dominate.

Treatment of Fever

With this revised perspective on fever, whether to suppress it or not can be a difficult decision. Some advocates feel that a slight to moderate fever in an otherwise healthy person should be allowed to run its course, in light of its potential benefits and minimal side effects. Side effects of fever include tachycardia (rapid heart rate), tachypnea (elevated respiratory rate), and, in some individuals, a lowering of seizure threshold. Therefore, all medical experts do agree that high and prolonged fevers or fevers in patients with cardiovascular disease, head trauma, seizures, or respiratory ailments are risky and should be treated immediately with fever-reducing drugs.

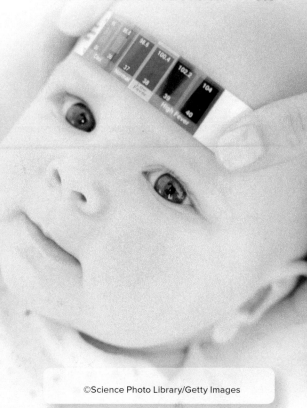

©Science Photo Library/Getty Images

Antimicrobial Products

In this section we will examine a broad group of products the host produces to battle microbial infections. We will look at four main categories: **interferons (IFN),** which are best known for targeting viruses, but can also target bacteria and even tumor cells; **complement,** which targets membranes of pathogens and also of pathogen-infected host cells; **antimicrobial peptides,** which can directly kill all manner of microbes; and **restriction factors,** which inhibit the multiplication of viruses in host cells.

Interferons

Interferons are small proteins produced naturally by certain white blood and tissue cells. Although the interferon system was originally thought to be directed exclusively against viruses, it is now known to be involved also in defenses against other microbes and in immune regulation and intercommunication. Three major types are *interferons alpha* and *beta*, which are produced by many cells, including lymphocytes, fibroblasts, and macrophages; and *interferon gamma*, a product of T cells.

Their biological activities are extensive. In all cases, interferons bind to cell surfaces and induce changes in genetic expression, but the exact results vary. In addition to antiviral effects discussed in the next section, all three IFNs can inhibit the expression of cancer genes and have tumor suppressor effects.

Characteristics of Interferons   When viruses, and sometimes other microbes or their component parts, bind to the receptors on a host cell, a signal is sent to the nucleus that directs the cell to synthesize interferons. After transcribing and translating the interferon genes, newly synthesized interferon molecules are rapidly secreted by the cell into the extracellular space, where they bind to other host cells. When interferons bind to the second cell, *that* cell produces proteins that inhibit viral multiplication. These proteins use a variety of mechanisms, such as degrading the viral RNA or preventing the translation of viral proteins **(figure 12.12).** Interferons can also induce the production of other proteins that combat infection, such as an enzyme that produces reactive oxygen chemicals that damage pathogens. Interferons are not microbe-specific, so their synthesis in response to one type of microbe will also protect against other types. Because these proteins are inhibitors of viruses, they have been used in medicine as a valuable treatment for a number of virus infections.

Complement

The immune system has another versatile system called complement that, like inflammation and phagocytosis, is brought into play at several levels. The complement system, named for its property of "complementing" immune reactions, consists of

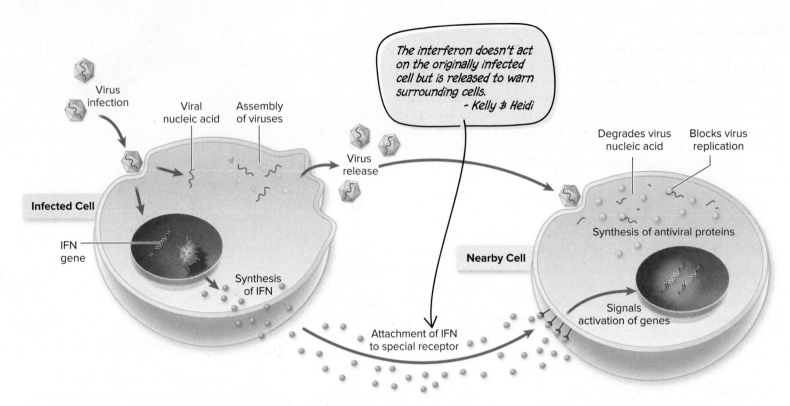

Figure 12.12 The antiviral activity of interferons. When a cell is infected by a virus (and sometimes other microbes), its nucleus is triggered to transcribe and translate the interferon (IFN) genes. Interferons diffuse out of the cell and bind to IFN receptors on nearby uninfected cells, where they induce production of proteins that eliminate genes from foreign organisms. Note that the original cell is not protected by IFNs and that IFNs do not prevent viruses (or other organisms) from invading the protected cells.

over 30 different blood proteins that work together to primarily destroy bacteria, but can have effects on viruses, other parasites, or nearby cells.

The concept of a cascade reaction is helpful in understanding how complement functions. A cascade reaction is a sequential physiological response like that of blood clotting, in which the first substance in a chemical series activates the next substance, which activates the next, and so on, until a desired end product is reached.

Four Stages of the Complement Cascade

In general, the complement system goes through four stages during its deployment:

- initiation,
- activation and cascade,
- polymerization, and
- membrane attack.

There are multiple ways in which the complement cascade can be initiated. When we study the third line of defense, we will learn about antibodies, which are host cell proteins that attach to the surfaces of pathogens. Complement action can be initiated by binding to the antibodies that are already bound to microbes. That scenario is called the **classical complement pathway.** Another important way that the complement cascade is initiated is called the **alternative complement pathway.** It does not require antibody to get started, but instead is initiated simply by the presence of foreign cell antigens. Because we have not yet studied antibodies, we will illustrate complement activity using the alternative pathway **(figure 12.13).**

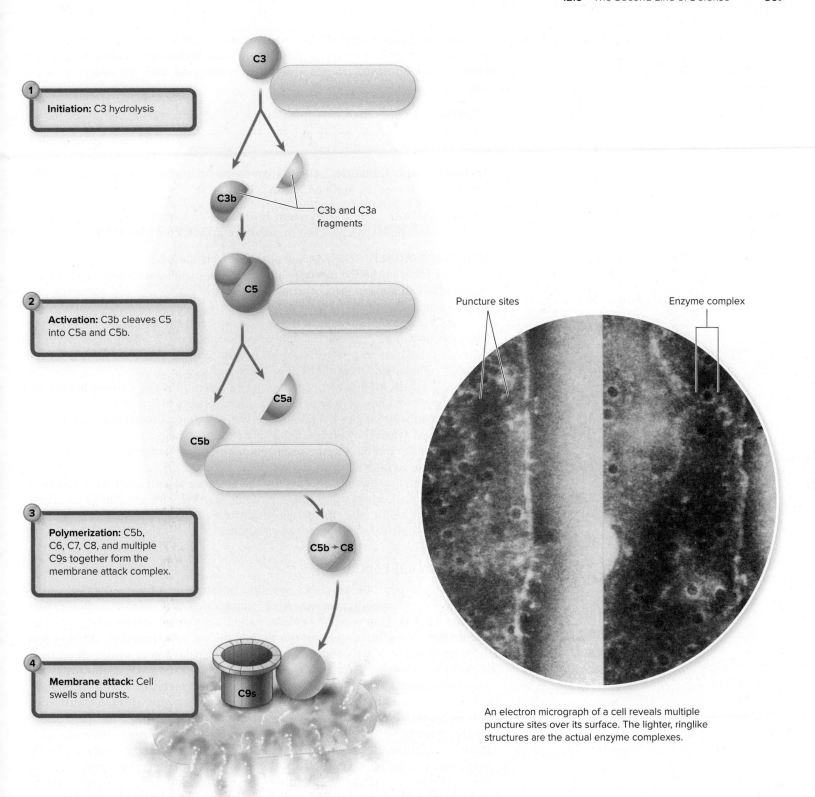

1 **Initiation:** C3 hydrolysis

C3

C3b

C3b and C3a fragments

2 **Activation:** C3b cleaves C5 into C5a and C5b.

C5

C5a

C5b

3 **Polymerization:** C5b, C6, C7, C8, and multiple C9s together form the membrane attack complex.

C5b → C8

4 **Membrane attack:** Cell swells and bursts.

C9s

Puncture sites

Enzyme complex

An electron micrograph of a cell reveals multiple puncture sites over its surface. The lighter, ringlike structures are the actual enzyme complexes.

Figure 12.13 Steps in the alternative complement pathway at a single site. All complement pathways function in a similar way, but the details differ. The alternative pathway is illustrated here.

Sucharit Bhakdi (micrograph)

The complement cascade can involve more than 30 different proteins. They are usually referred to as "C1," "C5," etc., and may be further divided, as in "C3b." Their numbered names do not necessarily correspond with the order in which they act.

Initiation In the alternative pathway, a C3 protein, either free or bound to a pathogen membrane, is hydrolyzed into two fragments, C3b and C3a (see figure 12.13).

Activation and Cascade The next steps involve further enzymatic action. The C3b protein cleaves the protein C5 into C5a and C5b.

Polymerization The C5b fragment is now free to form a complex with C6, C7, and C8. This complex is called the **membrane attack complex (MAC).**

Membrane Attack The MAC is positioned on the offending cell's membrane and forms pores in the membrane. This causes the membrane to lose its structural integrity, which leads to inappropriate flow of water and ions in and out of the cell, which eventually leads to the lysis of the cell.

The C3 molecule is readily available in tissues and the lymphatic system, often being left over from previous immune responses. Because it acts nonspecifically, it can react right away to new pathogens it may encounter. This also accounts for the fact that complement activation can go awry and lead to autoimmune dysfunction in hosts. The activation of complement by any means provides a strong linkage between the second line of defense and the third (adaptive) line of defense. It plays a role in regulating adaptive immunity as well. Binding of individual complement proteins, even when they do not result in a MAC, can help attract host antibodies. These antibodies can bind to the pathogen more easily, making them an easy target for phagocytes to identify them and eat them.

The classical pathway is activated by the presence of antibodies on cell surfaces and begins with "earlier" complement proteins, starting with one named C1. There are other ways besides the classical and alternative pathways to initiate complement activity, but these two are the most well-studied.

Antimicrobial Peptides

Antimicrobial peptides are short proteins, of between 12 and 50 amino acids, that have the capability of inserting themselves into bacterial membranes **(figure 12.14).** One type of peptide, called a defensin, inserts itself into cell membranes of bacteria, causing the membrane to fold around it. When a few defensin molecules gather in one area, they can make a pore in the membrane. If enough of these are formed, the cell could lyse. There are many other examples of antimicrobial peptides that have similar mechanisms for killing microbes, and some are even effective against eukaryotes or viruses. Because they will be effective against large groups of microorganisms, they are part of the innate immune system, but like other proteins in this section, they can have more far-reaching effects on other types of cells in the immune system as well. Many researchers are trying to turn these antimicrobial peptides into practical use as therapeutic drugs, while others are using computer programs to try to design synthetic peptides that will have as good an effect as these naturally occurring ones.

Host Restriction Factors

Some host cells and immune cells make molecules that can limit the ability of viruses to replicate once they are inside a host cell. These proteins and nucleic acids can bind to certain parts of the virus, prevent synthesis of new virus parts, prevent assembly of a new virus, or prevent virus release from host cells. These

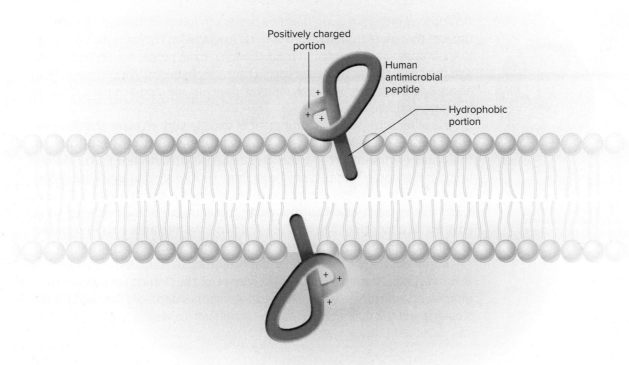

Figure 12.14 **Antimicrobial peptides.**   The cytoplasmic membrane is shown (white and yellow). The purple antimicrobial peptides insert themselves in the membrane and disrupt its integrity.

are collectively called *restriction factors*. One well-characterized set of restriction factors restricts the replication of retroviruses (such as HIV) by spontaneously changing cytosines to uracils and utilizing miRNAs that selectively target the viral DNA.

These apparently ancient innate defensive strategies are the subject of active research in the hopes that new antiviral medications can be developed, or that understanding them might aid in vaccine development.

To summarize, because it can be confusing, there are four main groups of antimicrobial products that we covered in this section:

- Interferon
- Complement
- Host restriction factors
- Antimicrobial peptides

12.3 LEARNING OUTCOMES—Assess Your Progress

12. List the four major categories of the second line of defense.

13. Outline the steps of phagocytosis.

14. Outline the steps of inflammation.

15. Discuss the mechanism of fever and how it helps defend the body.

16. Name four types of antimicrobial host-derived products.

17. Compose one good overview sentence about the purpose and the mode of action of the complement system, and another about the purpose and mode of action of interferon.

CASE FILE WRAP-UP

©Michael N. Paras/Pixtal/age fotostock
(health care worker); ©Al Telser/McGraw-
Hill Education (tissue)

Helicobacter pylori is a gram-negative bacterium found in the stomach. It is estimated that more than half of the world's population harbors this bacterium in their upper gastrointestinal tract. However, most people are asymptomatic and may never know that they carry the microorganism. Symptoms occur in about 20% of infected people. Symptoms may include stomach pain, bloating, belching, and nausea.

The stomach's acidic environment is usually very efficient at killing microorganisms that are swallowed. How does *H. pylori* survive and thrive in this harsh environment? *H. pylori* uses its flagella to burrow into the mucous lining of the stomach until it reaches a more neutral area. In a process known as chemotaxis, *H. pylori* can sense areas of acidity and can move to more hospitable areas. It also has the ability to neutralize acid in its immediate environment by using urease to break down urea into ammonia and carbon dioxide. The ammonia neutralizes harmful stomach acid.

An inflammatory response by the body tries to combat the invasion by *H. pylori.* A significant part of this response is for the stomach to produce even more acid. Unfortunately, the extra acid sometimes damages the lining of the stomach and the duodenum, causing ulcerations (ulcers).

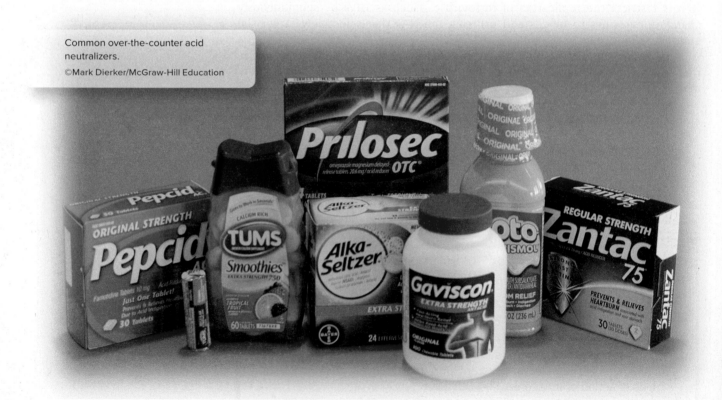

Common over-the-counter acid neutralizers.
©Mark Dierker/McGraw-Hill Education

Macrophages Shape Your Gut

In this chapter, you have seen just how fundamental macrophages are to the second line of defense. New data suggest that they are also fundamental to establishing a healthy gut microbiome. That seems a bit strange. How does it work?

According to scientists at the University of North Carolina at Chapel Hill, macrophages in the intestines are necessary for "creating" the correct gut microbiota. They did their studies in zebrafish. Zebrafish, like all fish, are animals, and they are frequently used for biological research on human traits.

They found that zebrafish with a mutation that caused them to produce many fewer macrophages in the intestine had a drastically different gut microbiota than those without the mutation. In the intestines that had a lot of macrophages (as seen in the photo to the right), in the intestinal epithelium, the bacteria in the gut were mainly commensal bacteria, and of varied types. In the mutants that had greatly decreased macrophages in the gut, the bacteria in the gut were dominated by rare or opportunistic bacteria, with less diversity than in the wild type animals. Many other microbiome studies have shown that when there is less diversity in the gut microbiome, negative health consequences can follow for the host.

So in the first paragraph, we said that it seemed a bit strange that macrophages are fundamental to a healthy microbiome. We said that because this chapter described macrophages as an early defense against microbes. In this study, we see that their presence is important for establishing the correct kind of microbes for health.

Source: https://www.cell.com/cell-reports/pdfExtended/S2211-1247(18)31455-4

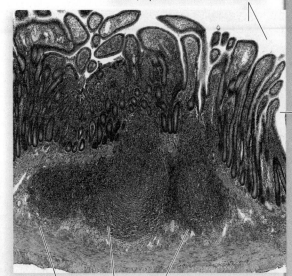

The lumen (the "inside tube") of the ileum, a portion of the small intestine

The villi of the small intestine

Peyer's patches, sites in the small intestine containing large numbers of macrophages and B cells and T cells

©Christine Eckel/McGraw-Hill Education

HOST DEFENSES PART I: OVERVIEW AND INNATE DEFENSES

12.1 DEFENSE MECHANISMS OF THE HOST: AN OVERVIEW

The interconnecting network of host protection against microbial invasion is organized into three lines of defense. The first line of defense consists of physical barriers, the microbiota barrier, and chemical barriers. The second line of defense consists of innate mechanisms. The third line of defense consists of adaptive responses to specific microorganisms (covered in chapter 13).

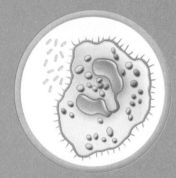

The first line of defense is made up of physical, chemical, and microbiome barriers. Mucous membranes and the skin are physical barriers. Sweat, lysozyme, fatty acids, and pH levels are typical chemical barriers.

THE FIRST LINE OF DEFENSE 12.2

12.3 THE SECOND LINE OF DEFENSE

Innate immune reactions are generalized responses to invasion, regardless of the source. Reactions include phagocytosis, inflammation, fever, and various antimicrobial products.

SmartGrid: From Knowledge to Critical Thinking

This *21 Question Grid* takes the topics from this chapter and arranges them with respect to the American Society for Microbiology's Undergraduate Curriculum guidelines—all six of the important "Concepts" as well as the important "Competency" of scientific literacy. Three questions are supplied about chapter content that refer to the Concept or Competency, in increasing levels of Bloom's Taxonomy for learning.

ASM Concept/ Competency	A. Bloom's Level 1, 2—Remember and Understand (Choose one.)	B. Bloom's Level 3, 4—Apply and Analyze	C. Bloom's Level 5, 6—Evaluate and Create
Evolution	1. A microorganism carries _____ markers and a B cell carries _____ markers. a. self, nonself b. nonself, self c. self, self d. nonself, nonself	2. Explain why the presence of lysozyme in human fluids such as tears and saliva is clear evidence of the coevolution of humans and microbes.	3. The pediatrician you work for has just recommended to the mother of a child with a cold and fever of 100°F *not* to treat the fever with children's Tylenol®. When the doctor leaves the room, the mother asks you to explain his recommendation. Use an evolutionary perspective to construct an answer for her.
Cell Structure and Function	4. Which of the following cells are lymphocytes? a. macrophages b. neutrophils c. red blood cells d. B cells	5. In what way is a phagocyte a tiny container of disinfectants?	6. Conduct research on the NK cells and NKT cells and describe distinguishing characteristics.
Metabolic Pathways	7. Cytokines are secreted by which cells? a. macrophages b. B cells c. T cells d. all of these	8. Explain the sequence of physiological events leading to fever.	9. Explain the physiological events that lead to each of the four cardinal signs of inflammation.
Information Flow and Genetics	10. The initial reaction to the presence of viruses in a human cell is the production of _____ by that cell. a. complement b. interferon c. antiviral protein d. fever	11. If the synthesis of antiviral proteins begins after exposure of the host cell to interferons, what kind of genes are the antiviral protein genes likely to be? a. constitutive b. repressed c. induced d. all of these	12. If scientists were going to recreate host cell restriction factors in the lab so that they could be used as antiviral drugs, how would they go about that?

ASM Concept/ Competency	A. Bloom's Level 1, 2—Remember and Understand (Choose one.)	B. Bloom's Level 3, 4—Apply and Analyze	C. Bloom's Level 5, 6—Evaluate and Create
Microbial Systems	13. An example of an exogenous pyrogen is a. interleukin-1. b. endotoxin. c. complement. d. interferon.	14. Why do you think that the intestines have one of the body's most well-developed sets of lymphoid tissues?	15. A less diverse gut microbiome has been linked to a reduction in healthy immune responses. Why might that be?
Impact of Microorganisms	16. The normal microbiota is part of the _____ line of defense. a. first b. second c. third d. None of these is correct.	17. People who smoke have higher rates of respiratory infections. Explain why this might be, invoking aspects of the first line of defense.	18. Recent studies suggest that even healthy blood contains a microbiome, most of which exists inside the white blood cells. Why are there microorganisms and/or their DNA there rather than in the red blood cells, and what might their role be?
Scientific Thinking	19. Figure 12.3 demonstrates that lymphatic fluid flows a. toward the heart. b. away from the heart. c. both toward and away from the heart. d. None of these is correct.	20. Suggest some reasons for so much redundancy and interactivity of immune responses.	21. There have been (controversial) attempts in the past to treat HIV+ people by infecting them with the malarial parasite. One of the reasons it was controversial is that it was ethically questionable for physicians to knowingly infect patients with a potentially serious organism (the malaria protozoan). Why did scientists think that a malaria infection might wipe out HIV in a patient's body?

Answers to the multiple-choice questions appear in Appendix A.

Visual Connections

This question connects content within and between chapters.

From chapter 11, figure 11.2. Relate specific events in inflammation to the symptoms of pneumonia pictured in this drawing.

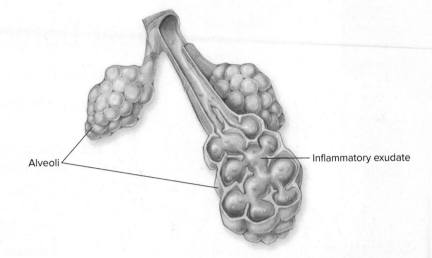

Alveoli

Inflammatory exudate

13

Host Defenses II
Adaptive Immunity and Immunization

IN THIS CHAPTER...

13.1 Adaptive Immunity: The Third and Final Line of Defense

1. Describe how the third line of defense is different from the other two.
2. Compare the terms *antigen*, *immunogen*, and *epitope*.
3. List the four stages of an adaptive immune response.
4. Discuss the role of cell markers in the immune response.
5. Describe the major histocompatibility complex in two sentences.

13.2 Stage I: The Development of Lymphocyte Diversity

6. Summarize the maturation process of both B cells and T cells.
7. Draw a diagram showing how lymphocytes are capable of responding to nearly any epitope imaginable.
8. Describe the structures of the B-cell receptor and the T-cell receptor.
9. Outline the processes of clonal deletion and clonal selection.

B-cell growth factors

Interleukins

13.3 Stage II: Presentation of Antigens

10. List characteristics of antigens that optimize their immunogenicity.
11. Describe how the immune system responds to alloantigens, superantigens, and allergens.
12. List the types of cells that can act as antigen-presenting cells.

13.4 Stages III and IV: T-Cell Response

13. Describe the main differences between T helper cells and T cytotoxic cells.
14. Explain how naive T cells become sensitized to an antigen.
15. Note the similarities and differences between gamma-delta T cells and the other T cells.

13.5 Stages III and IV: B-Cell Response

16. Diagram the steps in B-cell activation, including all types of cells produced.
17. Make a detailed drawing of an antibody molecule.
18. Explain the various end results of antibody binding to an antigen.
19. List the five types of antibodies and important facts about each.
20. Draw and label a graph—with time on the horizontal axis—that shows the development of the primary and secondary immune responses.

13.6 Adaptive Immunity and Vaccination

21. List and define the four different descriptors of adaptive immune states.
22. Discuss the qualities of an effective vaccine.
23. Name the two major categories of vaccines and then the subcategories under each.
24. Explain the principle of herd immunity and the risks that unfold when it is not maintained.

CASE FILE

When More Immune Response Is Not Better

I was an RN in the pediatric intensive care unit in spring 2020, in the middle of the COVID-19 pandemic. I cared for children with a wide range of conditions and diseases. One day, I was assigned to admit a patient from the emergency department. Trevor was a 15-year-old male who was previously healthy. He was tested for SARS-CoV-2 in the ED, and the test came back positive, even though he had not had COVID-19 symptoms. Trevor presented to my unit with fever, nausea, and an unusual skin rash that had begun 3 days ago.

When I assessed Trevor, I found him to be drowsy but responsive. He followed commands appropriately and answered questions. His heart rate was higher than normal and he had a fever. A generalized rash was present across his upper chest, neck, and back. He complained of abdominal pain and heart palpitations.

Because of Trevor's positive COVID-19 test, we maintained isolation precautions in the room. I gave him medication for his fever and monitored his vital signs closely. I administered intravenous fluids to counteract dehydration. I collected blood samples and sent them to the laboratory.

When the hospitalist arrived to evaluate Trevor, he immediately focused on his positive SARS-CoV-2 test. He asked him a series of questions, and checked the inflammatory markers, such as C-reactive protein, in his lab results.

Based on the laboratory values and Trevor's clinical presentation, a preliminary diagnosis of Kawasaki-like disease, COVID-19 related, was made. Intravenous steroids were ordered and the medical team discussed the treatment plan, focused upon suppressing Trevor's immune system.

- Why is the COVID-19 diagnosis relevant to this boy's condition?

Case File Wrap-Up appears at the end of the chapter.

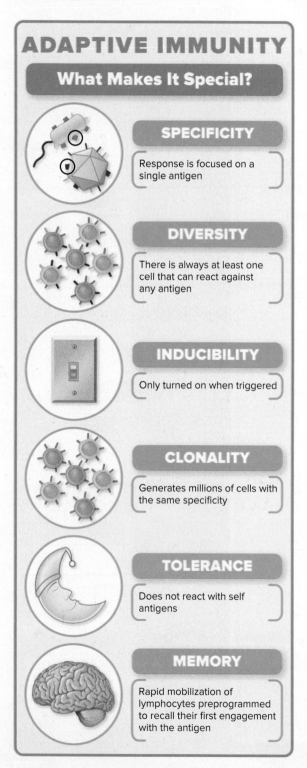

ADAPTIVE IMMUNITY

What Makes It Special?

SPECIFICITY

Response is focused on a single antigen

DIVERSITY

There is always at least one cell that can react against any antigen

INDUCIBILITY

Only turned on when triggered

CLONALITY

Generates millions of cells with the same specificity

TOLERANCE

Does not react with self antigens

MEMORY

Rapid mobilization of lymphocytes preprogrammed to recall their first engagement with the antigen

13.1 Adaptive Immunity: The Third and Final Line of Defense

When host barriers and innate defenses fail to control an infectious agent, a person with a normally functioning immune system has a mechanism to resist the pathogen–the third, specific line of immunity. This sort of immunity is not innate but adaptive. This means it is acquired only after an immunizing event such as an infection.

Adaptive immunity is the product of a dual system that we have previously mentioned–the B and T lymphocytes. During development, these lymphocytes undergo a selective process that prepares them for reacting only to one particular antigen or immunogen. During this time **immunocompetence,** the ability of the body to react with countless foreign substances, develops. An infant is born with the theoretical potential to react to an extraordinary array of different substances.

Antigens are molecules that can be seen and identified by the immune system. They may or may not provoke an immune response after being "sensed" by the immune system. If they do provoke a response they can be called **immunogens.** These molecules are usually protein or polysaccharide molecules on or inside all cells and viruses, including our own. (Environmental chemicals can also be antigens. These are covered in chapter 14.) In fact, any exposed or released protein or polysaccharide is potentially an antigen, even those on our own cells. For reasons we discuss later, our own antigens do not usually evoke a response from our own immune systems.

A lymphocyte's capacity to discriminate differences in molecular shape is so fine that it recognizes and responds to only a portion of the antigen molecule. This molecular fragment, called the **epitope,** is the primary signal that the molecule is foreign.

In chapter 12, we discussed pathogen-associated molecular patterns (PAMPs) that stimulate responses by phagocytic cells during an innate defense response. While PAMPs are molecules shared by many types of microbes that stimulate a innate response, antigens are highly individual and stimulate adaptive immunity.

To the left you will see six characteristics that make adaptive immunity unique. Two of the most important are **specificity** and **memory.** In general, the antibodies produced during an infection against the chickenpox virus will function against that virus and not against the measles virus. The property of memory refers to the rapid mobilization of lymphocytes that have been programmed to "recall" their first engagement with the invader and rush to the attack once again.

The elegance and complexity of immune function are largely due to lymphocytes working closely together with phagocytes. To simplify and clarify the network of immunologic development and interaction, we present it here as a series of stages, with each stage covered in a separate section (figure 13.1). The principal stages are as follows:

  I. Lymphocyte development and clonal deletion;
 II. Presentation of antigen and clonal selection;
III. The challenge of B and T lymphocytes by antigens;
IV. T-lymphocyte response: cell-mediated immunity; and B-lymphocyte response: the production and activities of antibodies.

We will give an overview here and spend the rest of the chapter filling in the details.

A Brief Overview of the Immune Response

Lymphocyte Development

Although all lymphocytes arise from the same basic stem cell type, at some point in development they diverge into two distinct types. Final development of B cells occurs in specialized bone marrow sites, and that of T cells occurs in the thymus. Both cell types subsequently migrate to separate areas in the lymphoid organs (for instance, nodes and spleen). B and T cells constantly recirculate through the circulatory system and lymphatics, migrating into and out of the lymphoid organs.

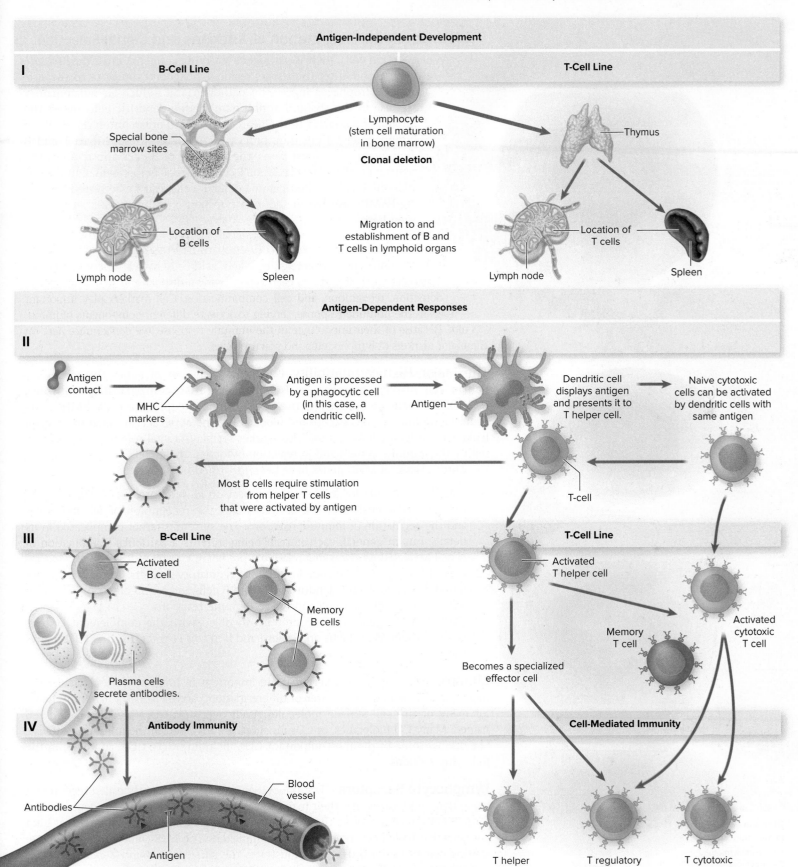

Antigen-Independent Development

I   **B-Cell Line**   **T-Cell Line**

Lymphocyte
(stem cell maturation
in bone marrow)
Clonal deletion

Special bone
marrow sites

Thymus

Migration to and
establishment of B and
T cells in lymphoid organs

Location of
B cells

Location of
T cells

Lymph node   Spleen   Lymph node   Spleen

Antigen-Dependent Responses

II

Antigen
contact

MHC
markers

Antigen is processed
by a phagocytic cell
(in this case, a
dendritic cell).

Antigen

Dendritic cell
displays antigen
and presents it to
T helper cell.

Naive cytotoxic
cells can be activated
by dendritic cells with
same antigen

Most B cells require stimulation
from helper T cells
that were activated by antigen

T-cell

III   **B-Cell Line**   **T-Cell Line**

Activated
B cell

Memory
B cells

Activated
T helper cell

Memory
T cell

Activated
cytotoxic
T cell

Plasma cells
secrete antibodies.

Becomes a specialized
effector cell

IV   **Antibody Immunity**   **Cell-Mediated Immunity**

Blood
vessel

Antibodies

Antigen

T helper   T regulatory   T cytotoxic

Figure 13.1 Overview of the stages of lymphocyte development and function. **I.** Development of B- and T-lymphocyte specificity and migration to lymphoid organs. **II.** Dendritic cell displays antigen and presents it to naive T helper cell. **III.** Lymphocyte activation, clonal expansion, and formation of memory B and T cells. **IV.** End result of lymphocyte activation. Left-hand side: antibody release; right-hand side: cell-mediated immunity.

Genes are the source of immunologic diversity.
©Comstock Images/PictureQuest

Entrance and Presentation of Antigens and Clonal Selection

When foreign cells, such as pathogens (carrying antigens), cross the first line of defense and enter the tissue, resident phagocytes migrate to the site. Tissue macrophages ingest the pathogen and induce an inflammatory response in the tissue if appropriate. Tissue dendritic cells ingest the antigen and migrate to the nearest lymphoid organ (often the draining lymph nodes). Here they process antigen and present it to T and B lymphocytes. In most cases, the response of B cells also requires the assistance of special classes of T cells called T helper cells. One special class of T cells, called gamma-delta T cells, can be activated quickly by PAMPs, as seen in the innate response, or by specific antigens as seen here.

The Role of Markers and Receptors in Presentation and Activation All cells—both foreign cells and "self" cells—have a variety of different markers on their surfaces, each type playing a distinct and significant role in detection, recognition, and cell communication. Cell markers play important roles in the immune response, serving to activate different components of immunity. Because of their importance in the immune response, we concentrate here on the major markers of lymphocytes and macrophages.

The Major Histocompatibility Complex One set of genes that codes for human cell markers or receptors is the **major histocompatibility complex (MHC).** This gene complex gives rise to a series of glycoproteins (called MHC molecules) found on all cells except red blood cells. The MHC is also called the human leukocyte antigen (HLA) system. This marker set plays a vital role in recognition of self by the immune system and in rejection of foreign tissue.

Three classes of MHC genes have been identified:

1. Class I genes code for markers that appear on all nucleated cells. They display unique characteristics of self and allow for the recognition of self molecules and the regulation of immune reactions. The system is rather complicated in its details, but in general, each human being inherits a particular combination of class I MHC (HLA) genes in a relatively predictable fashion.
2. Class II MHC genes also code for immune regulatory markers. These markers are found on macrophages, dendritic cells, and B cells and are involved in presenting antigens to T cells during cooperative immune reactions.
3. Class III MHC genes encode proteins involved with the complement system, among others. We'll focus on classes I and II in this chapter. **Figure 13.2** shows these two types of markers.

CD Molecules Other markers that are important in immunity are particular CD molecules. *CD* stands for "cluster of differentiation," and it is just a naming scheme for many of the cell surface molecules. Well over 400 CD molecules have been named. Many CD molecules, or CDs for short, are involved in the immune response. We discuss some of the most important CDs, including CD3, CD4, and CD8, in the following sections.

Lymphocyte Receptors Lymphocyte markers are frequently called receptors, a name that emphasizes that their major role is to "accept" or "grasp" antigens in some form. B cells have receptors that bind antigens, and T cells have receptors that bind antigens that have been processed and complexed with MHC molecules on the presenting cell surface. **Figure 13.3** illustrates the surfaces of B and T cells and their antigen receptors. There are potentially millions and even billions of unique types of antigens. The many sources of antigens include microorganisms, cancer cells, as well as an endless array of chemical compounds in the environment. We will soon see how T and B cells recognize so many different antigens.

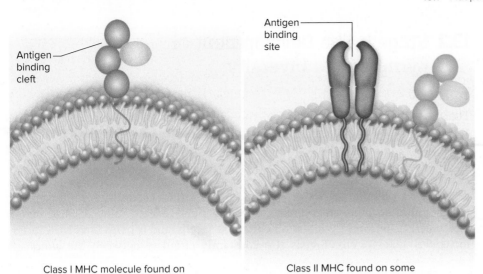

Antigen binding cleft

Antigen binding site

Class I MHC molecule found on all nucleated human cells.

Class II MHC found on some types of white blood cells (class I molecules here also, of course).

Figure 13.2 Classes I and II of molecules of the human major histocompatibility complex.

Challenging B and T Cells with Antigen

When challenged by antigen, both B cells and T cells *proliferate* and *differentiate*. The multiplication of a particular lymphocyte creates a clone, or group of genetically identical cells, some of which are memory cells that will ensure future reactiveness against that antigen. Because the B-cell and T-cell responses differ significantly from this point on in the sequence, they are summarized separately.

How T Cells Respond to Antigen

T-cell types and responses are extremely varied. When activated (sensitized) by antigen, a T cell gives rise to a variety of different cell types with different roles in the immune response. They generally fall into three categories:

1. helper T cells that activate macrophages, assist B-cell processes, and help activate cytotoxic T cells;
2. regulatory T cells that control the T-cell response by secreting anti-inflammatory cytokines or preventing proliferation; and
3. cytotoxic T cells that lead to the destruction of infected host cells and other "foreign" cells.

Although T cells secrete cytokines that help destroy pathogens and regulate immune responses, they do not produce antibodies.

How B Cells Respond to Antigen: Release of Antibodies

When a B cell is activated, or sensitized, by an antigen, it divides, giving rise to plasma cells, each with the same reactive profile. Plasma cells release antibodies into the tissue and blood. When these antibodies attach to the antigen for which they are specific, the antigen is marked for destruction or neutralization.

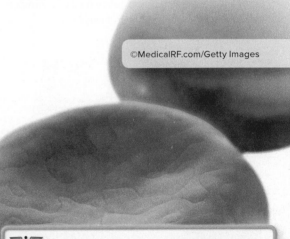

13.1 LEARNING OUTCOMES—Assess Your Progress

1. Describe how the third line of defense is different from the other two.
2. Compare the terms *antigen, immunogen,* and *epitope.*
3. List the four stages of an adaptive immune response.
4. Discuss the role of cell markers in the immune response.
5. Describe the major histocompatibility complex in two sentences.

NCLEX® PREP

1. The major histocompatibility complex (MHC) is a gene complex that gives rise to a series of glycoproteins (MHC molecules) found on all cells *except*
   a. *red blood cells.*   c. *plasma cells.*
   b. *white blood cells.*   d. *B cells.*

Medical Moment

The Thymus

The thymus is a small gland situated partly in the neck and partly in the thorax, behind the sternum. It weighs about 15 grams at birth, is largest at around puberty, and then atrophies slowly thereafter.

What is the main function of the thymus? Within the thymus, thymocytes (hematopoietic precursors from the bone marrow) mature into T cells. Once the T cells have become fully mature, they travel from the thymus and become the stock of circulating T cells responsible for many aspects of immunity. What happens when the thymus malfunctions, or when it is not working properly due to underlying disease? The following conditions may occur:

- **DiGeorge syndrome.** This rare genetic condition results in thymus deficiency. Patients with the syndrome may present with severe immunodeficiency.
- **Severe combined immunodeficiency (SCID) syndromes.** These are a group of genetic disorders resulting from defective hematopoietic precursor cells, causing a deficiency of both B lymphocytes and T lymphocytes.
- **Cancer.** Thymomas (tumors that originate from thymic epithelial cells) and thymic lymphomas (tumors originating from thymocytes) are the primary tumors affecting the thymus.

Q. Which of these three conditions would you consider to be most life-threatening, and why?

Answer in Appendix B.

13.2 Stage I: The Development of Lymphocyte Diversity

Specific Events in T-Cell Development

The maturation of most T cells and the development of their specific receptors are directed by the thymus and its hormones **(table 13.1).** Other T cells reach full maturity in the gastrointestinal tract. In addition to the antigen-specific T-cell receptor, all mature T lymphocytes express coreceptors called CD3. CD3 molecules surround the T-cell receptor and assist in binding. T cells also express either a CD4 or a CD8 coreceptor (see figure 13.3). CD4 is an accessory receptor protein mostly found on T helper cells that helps the T-cell receptor bind to MHC class II molecules. CD8 is mostly found on cytotoxic T cells, and it helps bind MHC class I molecules. Like B cells, T cells also constantly circulate between the lymphatic

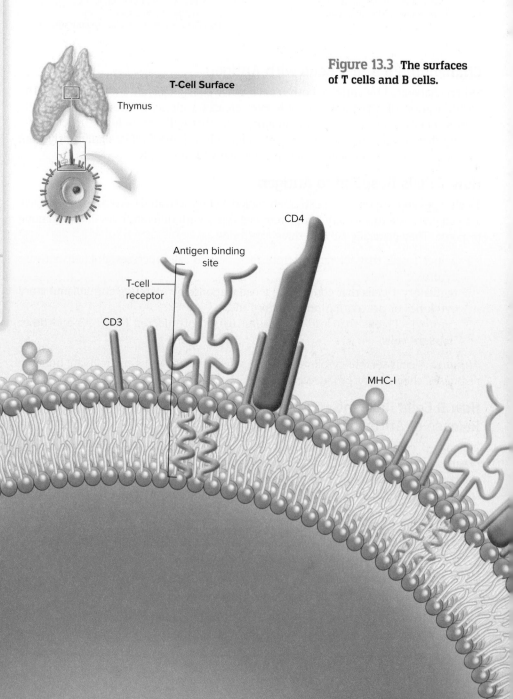

Figure 13.3 **The surfaces of T cells and B cells.**

and general circulatory systems, migrating to specific T-cell areas of the lymph nodes and spleen. It has been estimated that more than 10^9 T cells pass between the lymphatic and general circulations per day.

Specific Events in B-Cell Development

B cells develop in the bone marrow. As a result of gene modification and selection, hundreds of millions of distinct B cells develop. These naive lymphocytes circulate through the blood, "homing" to specific sites in the lymph nodes, spleen, and other lymphoid tissue, where they adhere to specific binding molecules. Here they will come into contact with antigens throughout life.

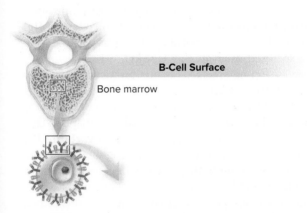

B-Cell Surface

Bone marrow

Table 13.1 Contrasting Properties of B Cells and T Cells

	B Cells	T Cells
Site of Maturation	Bone marrow	Thymus
Specific Surface Markers	Immunoglobulin	T-cell receptor Several CD molecules
Circulation in Blood	Low numbers	High numbers
Receptors for Antigen	B-cell receptor (immunoglobulin)	T-cell receptor
Distribution in Lymphatic Organs	Cortex (in follicles)	Paracortical sites (interior to the follicles)
Require Antigen Presented with MHC	No	Yes*
Product of Antigenic Stimulation	Plasma cells and memory cells	Several types of activated T cells and memory cells
General Functions	Production of antibodies to inactivate, neutralize, target antigens	Cells activated to help other immune cells; suppress or kill abnormal cells; mediate hypersensitivity; synthesize cytokines

*Gamma-delta T cells can be activated differently.

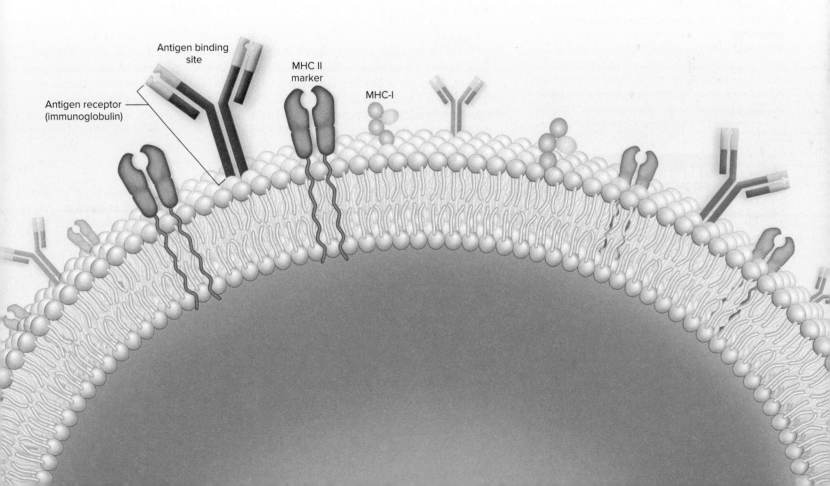

Antigen binding site

MHC II marker

MHC-I

Antigen receptor (immunoglobulin)

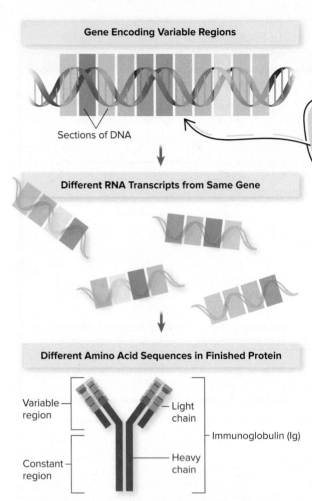

Gene Encoding Variable Regions

Sections of DNA

While the B cell is developing, segments of the DNA for the Ag receptors are rearranged extensively.
- Kelly & Heidi

Different RNA Transcripts from Same Gene

Different Amino Acid Sequences in Finished Protein

Variable region

Light chain

Immunoglobulin (Ig)

Constant region

Heavy chain

Figure 13.4 **The mechanism behind antibody variability.** The genes coding for the variable regions of antibody molecules have multiple different sections along their lengths. As a result of DNA recombination, very different RNA transcripts are created from the same original gene. When those transcripts are translated, the resulting protein will have extremely variable amino acid sequences—and therefore extremely variable shapes.

Medical Moment

Blood Exposure in Neonates

It is good medical practice to be very cautious about exposing neonates to blood products. Blood banks have developed practices to limit infant exposure to donor blood. Many centers dispense blood from a single donor in small amounts, or aliquots, for the duration of the patient's stay.

Even foreign proteins that are introduced for the baby's benefit, such as blood products, stimulate the immune system. Antigens present on red blood cells activate the immune response in the neonate. Immediately after harmless transfusion, lymphocyte counts increase.

(Continued)

Building Immunologic Diversity

By the time T and B cells reach the lymphoid tissues, each one is already equipped to respond to a unique antigen. This amazing specificity is generated by extensive rearrangements of more than 500 different gene segments that code for the antigen receptors—which are antibody molecules—on the T and B cells **(figure 13.4)**. In time, every possible recombination occurs, leading to a huge assortment of lymphocytes. It is estimated that each human produces antibodies with 10 trillion different specificities. While these DNA recombinations enable the recognition of a huge array of antigens (this is called the repertoire), keep in mind that a single naive lymphocyte can only recognize one antigen with its receptors.

The Specific B-Cell Receptor: An Immunoglobulin Molecule

In the case of B lymphocytes, the receptor genes that undergo the recombination described are genes coding for **immunoglobulin** (im″-yoo-noh-glahb′-yoo-lin) **(Ig)** synthesis. Immunoglobulins are large glycoprotein molecules that serve as the antigen receptors of B cells and, when secreted, as antibodies (see figure 13.3).

The immunoglobulin molecule is a conglomerate of four protein chains. Two of them are called light chains, and two of them are called heavy chains (see figure 13.4). The ends of the forks formed by the light and heavy chains contain pockets, called the **antigen binding sites.** These sites are highly variable in shape so that they can fit a wide range of antigens. This extreme versatility is due to **variable (V) regions** in antigen binding sites, where amino acid composition is highly varied from one clone of B lymphocytes to another, as a result of the gene rearrangements just discussed. The remainder of the light chains and heavy chains consist of constant (C) regions, whose amino acid content does not vary greatly from one antibody to another.

T-Cell Receptors

The T-cell receptor for antigen belongs to the same protein family as the B-cell receptor. It is similar to the B-cell receptor in being formed by genetic modification, having variable and constant regions, being inserted into the membrane, and having an antigen binding site formed from two parallel polypeptide chains (see figure 13.3). Unlike the B-cell receptor, the T-cell receptor is relatively small and is never secreted. Various other receptors and markers that are not antigen-specific are described in a later section.

Clonal Deletion and Selection

Table 13.2 illustrates the mechanism by which the exactly correct B or T cell is activated by any incoming antigen. This process is called **clonal selection.** After activation, the B or T cell multiplies rapidly in a process called clonal expansion. Two important features of clonal selection are that (1) lymphocyte specificity is preprogrammed, existing in the genetic makeup before an antigen has ever entered the tissues, and (2) each genetically distinct lymphocyte expresses only a single specificity and can react to that chemical epitope.

One potentially problematic outcome of random genetic assortment is the development of clones of lymphocytes able to react to *self*. This outcome can lead to severe damage if the immune system actually perceives self molecules as foreign and mounts a harmful response against the host's tissues. Any such clones are destroyed during development through **clonal deletion** (step 1 in the table). The removal of such potentially harmful clones is the basis of immune tolerance or tolerance to self. Because humans are exposed to many new antigenic substances during their lifetimes, T cells and B cells in the periphery of the body have mechanisms for *not* reacting to harmless antigens. Some

Table 13.2 Clonal Deletion and Clonal Selection of B and T Cells

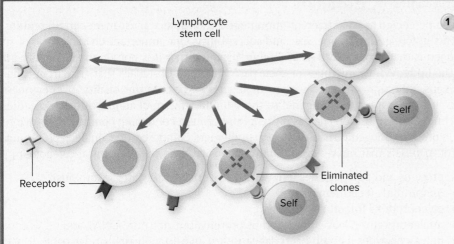

Lymphocyte stem cell

Receptors

Eliminated clones

Self

Self

1 Each genetically unique line of lymphocytes that arises from extensive recombinations of surface proteins is termed a **clone.** This stage of lymphocyte development does not require the actual presence of foreign antigens.

At the same time, any lymphocytes that develop a specificity for self molecules (and could be harmful) are eliminated from the pool of cells. This is called **clonal deletion** and leads to **immune tolerance.**

Repertoire of lymphocyte clones, each with unique receptor display

I

2 The specificity for a single epitope is programmed into the lymphocyte and is set for the life of a given cell. The end result is an enormous pool of mature but naive lymphocytes that are ready to further differentiate under the influence of certain immune stimuli.

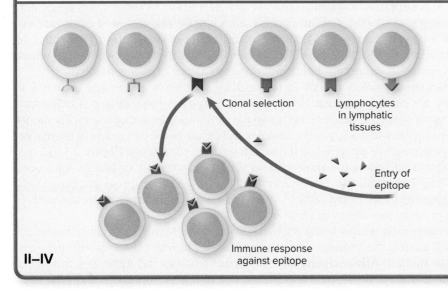

Clonal selection

Lymphocytes in lymphatic tissues

Entry of epitope

Immune response against epitope

II–IV

3 When any epitope enters the immune surveillance system, it encounters specific lymphocytes ready to recognize it. This stimulates activation of that clone, leading to genetic changes that cause it to differentiate into an effector cell. Mitotic divisions then expand it into a larger population of lymphocytes, all bearing the same specificity. This is clonal expansion.

diseases (autoimmunity) are thought to be caused by the loss of immune tolerance through the survival of certain "forbidden clones" or failure of these other systems (see chapter 14).

13.2 LEARNING OUTCOMES—Assess Your Progress

6. Summarize the maturation process of both B cells and T cells.
7. Draw a diagram showing how lymphocytes are capable of responding to nearly any epitope imaginable.
8. Describe the structures of the B-cell receptor and the T-cell receptor.
9. Outline the processes of clonal deletion and clonal selection.

Medical Moment
(*continued*)

Along with risk for blood transfusion complications, blood products may transmit infectious disease. Using consistent donor products in neonates limits these risks inherent to blood transfusion, while also reducing the patient's antigen exposure.

Q. What process is responsible for the increase in numbers of lymphocytes?

Answer in Appendix B.

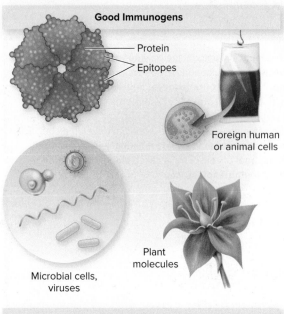

Good Immunogens

Protein

Epitopes

Foreign human
or animal cells

Plant
molecules

Microbial cells,
viruses

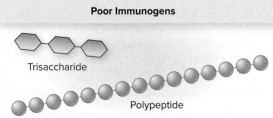

Poor Immunogens

Trisaccharide

Polypeptide

**Figure 13.5 A comparison of good immunogens
and poor immunogens.**   Top: Good immunogens are
large and complex. Bottom: Small molecules and linear
molecules are less likely to be good immunogens.

13.3 Stage II: Presentation of Antigens

Entrance of Antigens

To be perceived as an antigen or immunogen, a substance must meet certain requirements in foreignness, shape, size, and accessibility. One important characteristic of an antigen is that it be perceived as foreign, meaning that it is not a normal constituent of the body. Whole microbes or their parts, cells, or substances that arise from other humans, animals, plants, and various molecules all possess this quality of foreignness and thus are potentially antigenic to the immune system of an individual. Molecules of complex composition such as proteins and protein-containing compounds prove to be more immunogenic than repetitious polymers composed of a single type of unit. Most materials that serve as good immunogens fall into these chemical categories:

- proteins and polypeptides (enzymes, cell surface structures, exotoxins);
- lipoproteins (cell membranes);
- glycoproteins (blood cell markers);
- nucleoproteins (DNA complexed to proteins but not pure DNA); and
- polysaccharides (certain bacterial capsules) and lipopolysaccharides.

"Good" Immunogens

Antigens can be varying degrees of immunogenic. Characteristics of good immunogens (provoking a strong response) are (a) their chemical composition; (b) their context—meaning what types of cytokines are present; and (c) their size. We can generalize that large antigens are better than small antigens. However, large size alone is not sufficient for immunogenicity; glycogen, a polymer of glucose with a highly repetitious structure, has a molecular weight over 100,000 Daltons and is not normally antigenic, whereas insulin, a protein with a molecular weight of 6,000 Daltons, can be antigenic **(figure 13.5)**. Note that this aspect of size relates to the ability to stimulate immune "wakefulness." Remember that it is the smaller epitope to which the awakened immune cells bind, initiating the cascade of effects to follow.

Small foreign molecules that are too small by themselves to stimulate the immune response are termed **haptens.** However, if such an incomplete antigen is linked to a larger carrier molecule, the combined molecule can develop immunogenicity. The carrier group contributes to the size of the complex and enhances the proper spatial orientation of the determinative group, while the hapten serves as the epitope **(figure 13.6).**

Haptens include such molecules as drugs, metals, and ordinarily innocuous household, industrial, and environmental chemicals. Many haptens inappropriately develop antigenicity in the body by combining with large carrier molecules such as serum proteins (see allergy in chapter 14).

Because each human being is genetically and biochemically unique (except for identical twins), the proteins and other molecules of one person can be immunogenic to another. **Alloantigens** are cell surface markers and molecules that occur in some members of the same species but not in others. Alloantigens are the basis for an individual's blood group and major histocompatibility profile, and they are responsible for incompatibilities that can occur in blood transfusion or organ grafting.

Some bacterial toxins, which belong to a group of immunogens called **superantigens,** are potent stimuli for T cells. Their presence in an infection activates

Figure 13.6 Haptens.   Some small molecules are
poor immunogens but can be made more immunogenic
by combining them with a carrier.

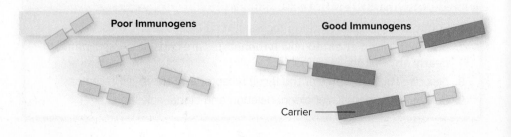

Poor Immunogens   **Good Immunogens**

Carrier

T cells at a rate 100 times greater than ordinary antigens. The result can be an overwhelming release of cytokines and cell death. Such diseases as toxic shock syndrome and certain autoimmune diseases are associated with this class of antigens.

Antigens that evoke allergic reactions, called **allergens,** are characterized in detail in chapter 14.

Antigen Processing and Presentation

In most immune reactions, the antigen must be formally presented to lymphocytes by cells called **antigen-presenting cells (APCs).** At least three different cells can serve as APCs: macrophages, B cells, and **dendritic** (den'-drih-tik) **cells.** Dendritic cells are the most potent and versatile of the APCs. Antigen-presenting cells grab the antigen-carrying microbe and ingest it. They degrade it and pass its antigens back out onto their membranes, complexed either with MHC-I or MHC-II markers. After processing is complete, the antigen is bound to the MHC receptor and moved to the surface of the APC so that it will be readily accessible to T lymphocytes during presentation. **Table 13.3** illustrates how antigen is presented to T helper cells. The activated T cell that is the product of this reaction is central to all of T-cell immunity and most of B-cell immunity.

Most antigens must be presented first to T cells, even though they will eventually activate both the T-cell and B-cell systems. However, a few antigens can trigger a response directly from B lymphocytes without the cooperation of APCs or T helper cells. These are called T-cell-independent antigens. They are usually simple molecules such

Table 13.3 How Antigen Is Presented to T Helper Cells

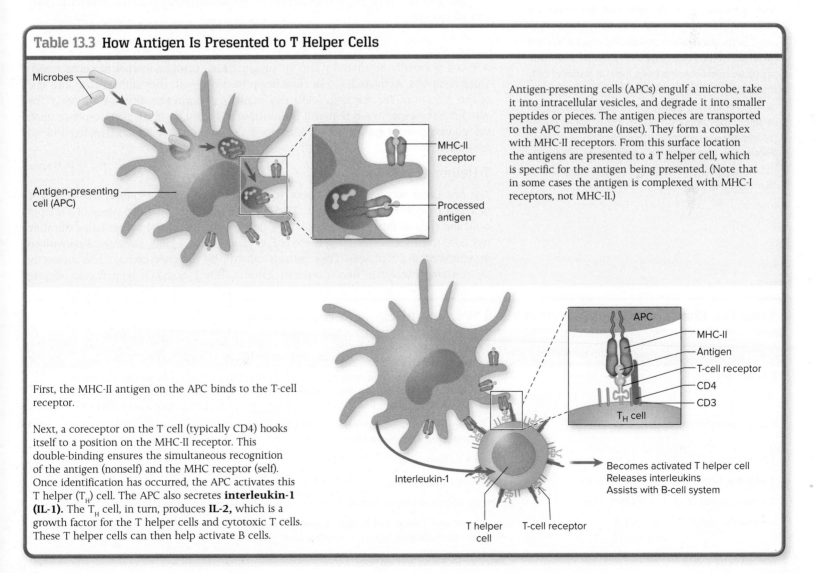

Antigen-presenting cells (APCs) engulf a microbe, take it into intracellular vesicles, and degrade it into smaller peptides or pieces. The antigen pieces are transported to the APC membrane (inset). They form a complex with MHC-II receptors. From this surface location the antigens are presented to a T helper cell, which is specific for the antigen being presented. (Note that in some cases the antigen is complexed with MHC-I receptors, not MHC-II.)

First, the MHC-II antigen on the APC binds to the T-cell receptor.

Next, a coreceptor on the T cell (typically CD4) hooks itself to a position on the MHC-II receptor. This double-binding ensures the simultaneous recognition of the antigen (nonself) and the MHC receptor (self). Once identification has occurred, the APC activates this T helper (T_H) cell. The APC also secretes **interleukin-1 (IL-1).** The T_H cell, in turn, produces **IL-2,** which is a growth factor for the T helper cells and cytotoxic T cells. These T helper cells can then help activate B cells.

A Note About Epitopes and Antigens

While up to now we have been calling the immunogenic substance "the antigen," it is more precisely termed the epitope. You could say, for instance, "the antigenic portion of the protein on a microbe is the epitope." You will also note that in practice, clinicians, and even other parts of this book, use the word "antigen" when the precise term is "epitope." You will know, however, that the part of the molecule that is actually recognized by the immune system is the epitope. This means, also, that every epitope can be recognized by B- and T-cell receptors that were formed during genetic reassortment. The particular tertiary structure and shape of this determinant must conform like a key to the receptor "lock" of the lymphocyte, which then responds to it. Certain amino acids accessible at the surface of proteins or protruding carbohydrate side chains are typical examples. Many foreign cells and molecules are very complex antigenically, with numerous component parts, each of which will elicit a separate and different lymphocyte response. Examples of these multiple, or *mosaic,* antigens include bacterial cells containing cell wall; membrane; and flagellar, capsular, and toxin antigens, as well as viruses. T-cell antigen receptors recognize these small pieces of antigens—epitopes—in combination with MHC molecules. And just to be complete, an antigen (or an epitope) that stimulates an immune response is termed *immunogenic.*

as carbohydrates with many repeating and invariable determinant groups. Examples include lipopolysaccharide from the cell wall of *Escherichia coli*, polysaccharide from the capsule of *Streptococcus pneumoniae,* and molecules from rabies and Epstein-Barr virus. Because so few antigens are of this type, most B-cell reactions require T helper cells.

13.3 LEARNING OUTCOMES—Assess Your Progress

10. List characteristics of antigens that optimize their immunogenicity.
11. Describe how the immune system responds to alloantigens and superantigens.
12. List the types of cells that can act as antigen-presenting cells.

13.4 Stages III and IV: T-Cell Response

T-cell reactions are among the most complex and diverse in the immune system and involve several subsets of T cells, whose particular actions are dictated by the APCs that activate them. We refer to T cells as "restricted"–that is, they require some type of MHC (self) recognition before they can be activated. T cells produce cytokines with a spectrum of biological effects. They also are influenced by cytokines secreted by other cells.

The end result of T-cell stimulation is the mobilization of other T cells, B cells, and phagocytes.

The Activation of T Cells and Their Differentiation into Subsets

A T cell is initially sensitized when an antigen/MHC complex comes in contact with T-cell receptors. Activated T cells then begin to divide, and they differentiate into one of the subsets of effector cells (cells that actually perform the ultimate action of the system) and memory cells that will be available to mount an immediate response upon subsequent contact **(table 13.4).** Memory T cells are some of the longest-lived blood cells known.

T Helper (T$_H$) Cells

The first thing you will notice about the T-cell response is the central role of a type of T cell called helper cells **(table 13.5).** There are many different types of T helper cells, and they all bear the CD4 marker. T helper cells are critical in regulating immune reactions to antigens, including those of B cells and other T cells. They are also involved in activating macrophages. They do this directly by receptor contact and indirectly by releasing cytokines like interferon gamma (IFNγ). Some T helper cells secrete

Table 13.4 Characteristics of Subsets of T-Cell Types in the Classic T-Cell Response

Types	Co-receptors on T cell	Functions/Important Features
T helper cell 1 (T$_H$1)	CD4	Activates the cell-mediated immunity pathway; secretes tumor necrosis factor and interferon gamma; also responsible for delayed hypersensitivity (allergy occurring several hours or days after contact); secretes IL-2
T helper cell 2 (T$_H$2)	CD4	Can activate macrophages to expel helminths or protozoans, phagocytose extracellular antigens; contributes to type 1 (allergic) hypersensitivity; can encourage tumor development
T helper cell 17 (T$_H$17)	CD4	Promotes inflammation
T follicular helper cell	CD4, CD40L	Drives B-cell proliferation, and aids B cells in antibody class switching
T regulatory cell (T$_{reg}$)	CD25, CD4	Controls adaptive immune response; prevents autoimmunity; can contribute to cancer progression
T cytotoxic cell (T$_C$)	CD8	Destroys a target foreign cell by lysis; important in destruction of complex microbes, cancer cells, virus-infected cells; graft rejection; requires MHC-I for function
Gamma-delta T cells	–	React in the innate and adaptive systems; responsive to lipid antigens

interleukin-2, which stimulates the primary growth and activation of many types of T cells, including cytotoxic T cells. Some T helper cells secrete interleukins-4, -5, and -6, which stimulate various activities of B cells. T helper cells are the most prevalent type of T cell in the blood and lymphoid organs, making up about 65% of this population. The severe depression of this class of T cells (with CD4 receptors) by HIV is what largely accounts for the pathology of AIDS.

The details of the T-cell reaction to antigen can seem complex. Spend some time with table 13.5 and the mechanism will start to be clear.

Cytotoxic T (T_C) Cells: Cells That Kill Other Cells

Cytotoxic T cells have one job: to destroy other cells. Target cells that T_C cells can destroy include the following:

- Virally infected cells. Cytotoxic cells recognize these because of telltale virus peptides expressed on their surface. Cytotoxic defenses are an essential protection against viruses. They also recognize and target cells carrying intracellular bacteria.
- Cancer cells. T cells constantly survey the tissues and immediately attack any abnormal cells they encounter **(figure 13.7).** The importance of this function is clearly demonstrated in the susceptibility of T-cell-deficient people to cancer (see chapter 14).
- Cells from other animals and humans. Cytotoxic cell-mediated immunity is the most important factor in graft rejection. In this instance, the T_C cells attack the foreign tissues that have been implanted into a recipient's body.

Keep in mind that there are always T helper cells and T cytotoxic cells in any given immune response.

Gamma-Delta T Cells

The subcategory of T cells called gamma-delta T cells is distinct from other T cells. They do have T-cell receptors that are rearranged to recognize a wide range of antigens, but they frequently respond to certain kinds of PAMPs on microorganisms the way white blood cells (WBCs) in the innate system do. This allows them to respond more quickly. However, they still produce memory cells when they are activated. For these reasons, they are considered a bridge between the innate and adaptive immune responses. They are particularly responsive to certain types of phospholipids and can recognize and react against tumor cells.

Additional Cells with Orders to Kill

Natural killer (NK) cells are a type of lymphocyte related to T cells that lack specificity for antigens. They circulate through the spleen, blood, and lungs, and are probably the first killer cells to attack cancer cells and virus-infected cells. They destroy such cells by similar mechanisms as T cells. They are not considered part of adaptive cell-mediated immunity, although their activities are acutely sensitive to cytokines such as interleukin-12 and interferon. Finally, there is a hybrid kind of cell that is part killer cell and part T cell, with T-cell receptors for antigen and the ability to release large amounts of cytokines very quickly, leading to cell death. These cells are called natural killer T cells, or NKT cells. Because they have T-cell receptors but respond very quickly, they are considered to be another important link between innate and adaptive immunity.

As you can see, the T-cell system is very complex. In summary, T cells differentiate into five different types of cells (and also memory cells), each of which contributes to the orchestrated immune response, under the influence of a multitude of cytokines.

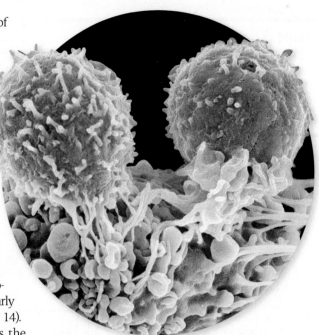

Figure 13.7 T cells attacking a cancer cell.
We've been showing drawings of the markers and receptors bringing immune cells and foreign cells together. That is how these T cells (pink) bound to this cancer cell.
Steve Gschmeissner/Science Source

Medical Moment

CAR-T Cells

A powerful new anti-cancer therapy was approved by the Food and Drug Administration in 2017 that has given new hope for curing some types of cancer. The therapy uses a patient's own T cells. The way it works is this: A patient's blood is drawn, and the T cells are separated from the other blood components. The T cells are infected with a virus that is engineered to carry a gene for the receptor that will bind to the cancerous cells. That gene, once delivered to the T cells, causes the receptor to be expressed on the surface of the T cell, and the T cells are then injected back into the patient. The new CAR-T cells (CAR stands for chimeric antigen receptor) multiply thousand-fold. They recognize, bind, and attack the patient's cancer cells. Currently CAR-T is approved to treat two kinds of cancers, large B cell lymphomas and acute lymphoblastic leukemia.

Q. In early trials of CAR-T, researchers discovered an unintended consequence of CAR-T therapy that threatened the lives of patients, even while their tumors were being destroyed. Can you think of what this complication might have been?

Answer in Appendix B.

13.4 LEARNING OUTCOMES—Assess Your Progress

13. Describe the main differences between T helper cells and T cytotoxic cells.
14. Explain how naive T cells become sensitized to an antigen.
15. Note the similarities and differences between gamma-delta T cells and the other T cells.

Table 13.5 T-Cell Activation

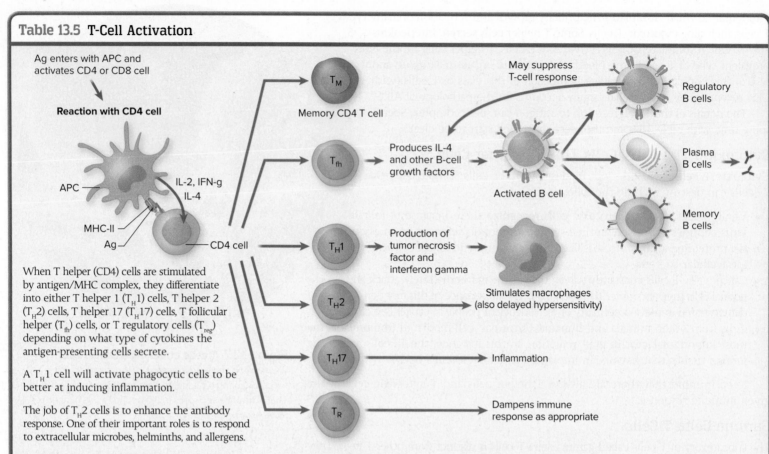

Ag enters with APC and activates CD4 or CD8 cell

Reaction with CD4 cell

APC — IL-2, IFN-g IL-4

MHC-II

Ag — CD4 cell

T_M — Memory CD4 T cell

May suppress T-cell response — Regulatory B cells

T_fh → Produces IL-4 and other B-cell growth factors → Activated B cell → Regulatory B cells / Plasma B cells / Memory B cells

T_H1 → Production of tumor necrosis factor and interferon gamma → Stimulates macrophages (also delayed hypersensitivity)

T_H2 → Stimulates macrophages (also delayed hypersensitivity)

T_H17 → Inflammation

T_R → Dampens immune response as appropriate

When T helper (CD4) cells are stimulated by antigen/MHC complex, they differentiate into either T helper 1 (T_H1) cells, T helper 2 (T_H2) cells, T helper 17 (T_H17) cells, T follicular helper (T_fh) cells, or T regulatory cells (T_reg) depending on what type of cytokines the antigen-presenting cells secrete.

A T_H1 cell will activate phagocytic cells to be better at inducing inflammation.

The job of T_H2 cells is to enhance the antibody response. One of their important roles is to respond to extracellular microbes, helminths, and allergens.

T_fh cells aid in B-cell differentiation.

T_H17 cells are so named because they secrete interleukin-17, which leads to the production of other cytokines that promote inflammation. Inflammation is useful, of course, but when excessive or inappropriate may lead to inflammatory diseases such as Crohn's disease or psoriasis. T_H17 may be critical to these conditions.

T_reg cells are also broadly in the T_H class, in that they also carry CD4 markers. But they are usually put in their own category. They act to control the inflammatory process, to prevent autoimmunity, and to make sure the immune response doesn't inappropriately target normal biota.

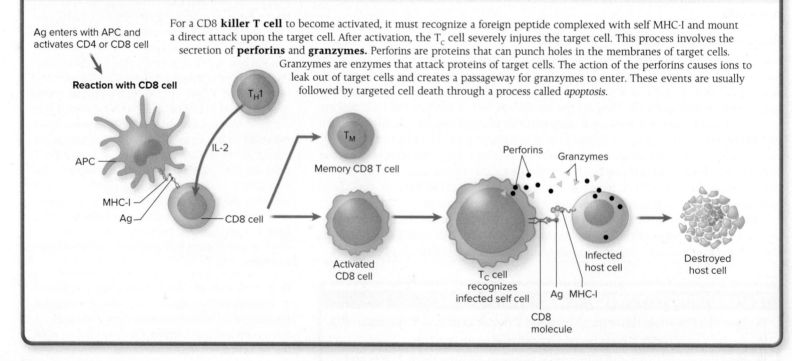

Ag enters with APC and activates CD4 or CD8 cell

Reaction with CD8 cell

For a CD8 **killer T cell** to become activated, it must recognize a foreign peptide complexed with self MHC-I and mount a direct attack upon the target cell. After activation, the T_C cell severely injures the target cell. This process involves the secretion of **perforins** and **granzymes.** Perforins are proteins that can punch holes in the membranes of target cells. Granzymes are enzymes that attack proteins of target cells. The action of the perforins causes ions to leak out of target cells and creates a passageway for granzymes to enter. These events are usually followed by targeted cell death through a process called *apoptosis.*

T_H1 — IL-2

APC

MHC-I

Ag — CD8 cell

T_M — Memory CD8 T cell

Activated CD8 cell

T_C cell recognizes infected self cell

Perforins   Granzymes

Infected host cell

CD8 molecule

Ag   MHC-I

Destroyed host cell

13.5 Stages III and IV: B-Cell Response

At the same time that the T-cell system is being activated by antigen, B cells are being stimulated as well. The immunologic activation of most B cells also requires a series of events. **Table 13.6** contains the details.

Products of B Lymphocytes: Antibody Structure and Functions

Earlier we saw how a basic immunoglobulin (Ig) molecule can have so many different variable regions. Let us view this structure once again, using one particular antibody type as a model. There are two functionally distinct segments called *fragments*. The two "arms" that bind antigen are termed *antigen binding fragments (Fabs)*, and the rest of the molecule is the crystallizable fragment (Fc). It is called that because it was the first to be crystallized in pure form. The basic immunoglobulin molecule is a composite of four polypeptide chains: a pair of identical heavy (H) chains and a pair of identical light (L) chains **(figure 13.8).** One light chain is bonded to one heavy chain, and the two heavy chains are bonded to one another with disulfide bonds, creating a symmetrical, Y-shaped arrangement.

The end of each Fab fragment (consisting of the variable regions of the heavy and light chains) folds into a groove that will accommodate one epitope. The presence of a special region at the site of attachment between the Fab and Fc fragments allows swiveling of the Fab fragments. In this way, they can change their angle to accommodate nearby antigen sites that vary slightly in distance and position. Figure 13.8 shows two views of antibody structure.

Don't Give Up!
Adaptive immunity is very complex. Don't be overwhelmed with all the illustrations and tables in this chapter. Seeing the processes with the use of drawings is better than just reading about them, but they are still intimidating. Even scientists studying immunology get overwhelmed. Look at the end results of each process and think about what is important. Ask your instructor what the key concepts are for T-cell activation, B-cell activation, and so on. Just don't ask them, "Will it be on the test?"!

©Ingram Publishing/SuperStock

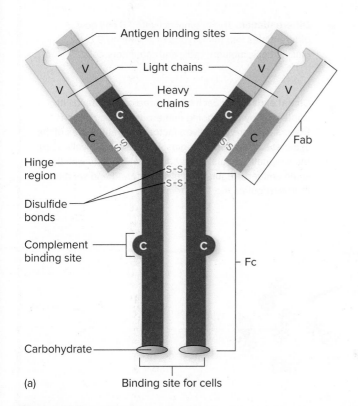

(a)

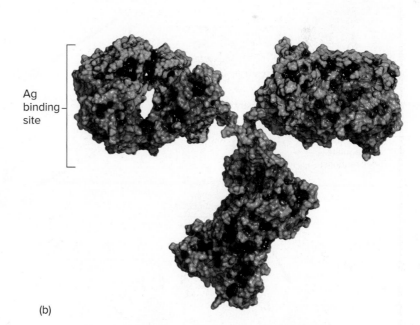

(b)

Figure 13.8 Antibody structure. **(a)** Diagrammatic view of IgG depicts the principal functional areas (Fabs and Fc) of the molecule. Each Fab contains a hypervariable region (V) and a constant, nonvariable region (C). **(b)** Realistic model of immunoglobulin shows the tertiary and quaternary structure achieved by additional intrachain and interchain bonds.

(b): Molekuul_be/Shutterstock

Table 13.6 B-Cell Activation

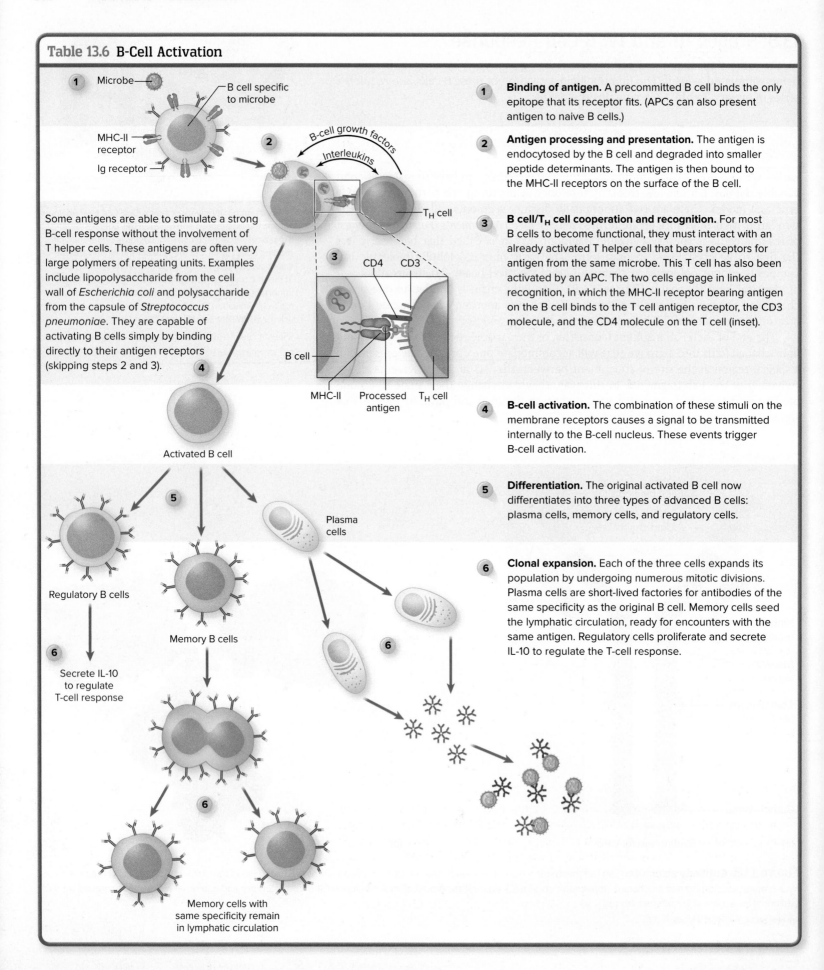

1 **Binding of antigen.** A precommitted B cell binds the only epitope that its receptor fits. (APCs can also present antigen to naive B cells.)

2 **Antigen processing and presentation.** The antigen is endocytosed by the B cell and degraded into smaller peptide determinants. The antigen is then bound to the MHC-II receptors on the surface of the B cell.

3 **B cell/T$_H$ cell cooperation and recognition.** For most B cells to become functional, they must interact with an already activated T helper cell that bears receptors for antigen from the same microbe. This T cell has also been activated by an APC. The two cells engage in linked recognition, in which the MHC-II receptor bearing antigen on the B cell binds to the T cell antigen receptor, the CD3 molecule, and the CD4 molecule on the T cell (inset).

4 **B-cell activation.** The combination of these stimuli on the membrane receptors causes a signal to be transmitted internally to the B-cell nucleus. These events trigger B-cell activation.

5 **Differentiation.** The original activated B cell now differentiates into three types of advanced B cells: plasma cells, memory cells, and regulatory cells.

6 **Clonal expansion.** Each of the three cells expands its population by undergoing numerous mitotic divisions. Plasma cells are short-lived factories for antibodies of the same specificity as the original B cell. Memory cells seed the lymphatic circulation, ready for encounters with the same antigen. Regulatory cells proliferate and secrete IL-10 to regulate the T-cell response.

Some antigens are able to stimulate a strong B-cell response without the involvement of T helper cells. These antigens are often very large polymers of repeating units. Examples include lipopolysaccharide from the cell wall of *Escherichia coli* and polysaccharide from the capsule of *Streptococcus pneumoniae*. They are capable of activating B cells simply by binding directly to their antigen receptors (skipping steps 2 and 3).

Labels in figure: Microbe; B cell specific to microbe; MHC-II receptor; Ig receptor; B-cell growth factors; Interleukins; T$_H$ cell; CD4; CD3; B cell; MHC-II; Processed antigen; T$_H$ cell; Activated B cell; Regulatory B cells; Secrete IL-10 to regulate T-cell response; Memory B cells; Plasma cells; Memory cells with same specificity remain in lymphatic circulation

Table 13.7 Summary of Antibody Functions

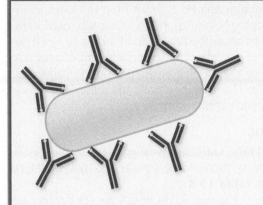

Antibodies coat the surface of a bacterium, preventing its normal function and reproduction in various ways.

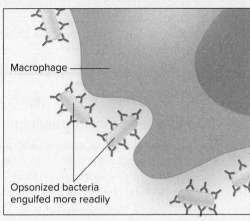

Antibodies called opsonins stimulate **opsonization** (ahp″-son-uh-zaz′-shun), a process that makes microbes more readily recognized by phagocytes, so that they can dispose of them. Opsonization has been likened to putting handles on a slippery object to provide phagocytes a better grip.

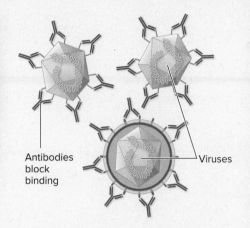

In **neutralization** reactions, antibodies fill the surface receptors on a virus or the active site on a microbial enzyme to prevent it from attaching normally.

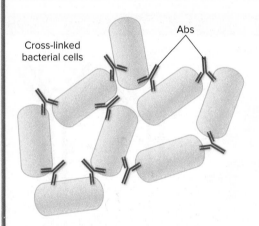

The capacity for antibodies to aggregate, or **agglutinate,** antigens is the consequence of their cross-linking cells or particles into large clumps. Agglutination renders microbes immobile and enhances their phagocytosis. This is a principle behind certain immune tests discussed in chapter 15.

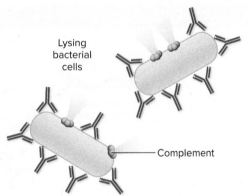

The interaction of an antibody with complement can result in the specific rupturing of cells and some viruses.

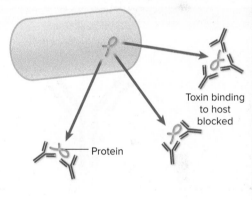

An **antitoxin** is a special type of antibody that neutralizes bacterial exotoxins.

Antibody-Antigen Interactions and the Function of the Fab

The site on the antibody where the epitope binds is composed of a *hypervariable region*, whose amino acid content can be extremely varied. The specificity of antigen binding sites for antigens is very similar to enzymes and substrates. Because the specificity of the two Fab sites is identical, an Ig molecule can bind epitope on the same cell or on two separate cells and thereby link them.

The principal activity of an antibody is to immobilize, call attention to, or neutralize the antigen for which it was formed. Follow along in **table 13.7** to see how each of these works.

Functions of the Fc Fragment

Although the Fab fragments bind antigen, the Fc fragment has a different binding function. In most classes of immunoglobulin, the Fc end can bind to receptors on the membranes of cells, such as macrophages, neutrophils, eosinophils, mast cells, basophils, and lymphocytes. The effect of an antibody's Fc end binding to a cell depends upon that cell's role. In the case of opsonization, the attachment of antibody to foreign cells and viruses is followed by the binding of the Fc end to phagocytes. The Fc end of the antibody of allergy (IgE) binds to basophils and mast cells, which causes the release of allergic mediators such as histamine.

The Classes of Immunoglobulins

Immunoglobulins exist as structural and functional classes called *isotypes*. The classes are differentiated with shorthand names (Ig, followed by a letter: IgG, IgA, IgM, IgD, IgE). Complete descriptions are found in **table 13.8**.

IgA is worth investigating a bit more. The two forms of IgA are (1) a monomer that circulates in small amounts in the blood and (2) a dimer that is a significant component of the mucous and serous secretions of the salivary glands, intestine, nasal membrane, breast, lung, and genitourinary tract. The dimer, called secretory IgA, is formed by two monomers held together by a J chain. To facilitate the transport of IgA

Table 13.8 Characteristics of the Immunoglobulin (Ig) Classes

	IgG	IgA	IgM	IgD	IgE
	Monomer	Dimer, Monomer	Pentamer	Monomer	Monomer
Number of Antigen Binding Sites	2	4, 2	10	2	2
Molecular Weight	150,000	170,000–385,000	900,000	180,000	200,000
Percentage of Total Antibody in Serum	80%	13%	6%	1%	0.002%
Average Half-Life in Serum (Days)	23	6	5	3	2.5
Crosses Placenta?	Yes	No	No	No	No
Fixes Complement?	Yes	No	Yes	No	No
Fc Binds to	Phagocytes				Mast cells and basophils
Biological Function	Monomer is produced by plasma cells in a primary response and by memory cells responding the second time to a given antigenic stimulus. It is the most prevalent antibody circulating throughout the tissue fluids and blood. It neutralizes toxins, opsonizes, fixes complement	Dimer is secretory antibody on mucous membranes; monomer is in small quantities in blood	Produced at first response to antigen. It can serve as B-cell receptor	Is the receptor on B cells and a triggering molecule for B-cell activation	Antibody of allergy and of worm infections. It also mediates anaphylaxis, asthma, etc.

across membranes, a secretory piece is later added. IgA coats the surface of mucous membranes and also is suspended in saliva, tears, colostrum, and mucus. It provides the most important adaptive local immunity to enteric, respiratory, and genitourinary pathogens. During lactation, the breast becomes a site for the proliferation of lymphocytes that produce IgA. The very earliest secretion of the breast, a thin, yellow milk called **colostrum,** is very high in IgA. These antibodies form a protective coating in the gastrointestinal tract of a nursing infant that guards against infection by a number of enteric pathogens (*Escherichia coli, Salmonella,* poliovirus, rotavirus). This is one of the reasons that new mothers are encouraged to nurse their newborns, at least for a few weeks, to take advantage of the immune properties of this first milk. Protection at this level is especially critical because an infant's own IgA and natural intestinal barriers are not yet well developed. As with immunity *in utero,* the necessary antibodies will be donated only if the mother herself has active immunity to the microbe through a prior infection or vaccination.

Monitoring Antibody Production over Time: Primary and Secondary Responses to Antigens

We can learn a great deal about how the immune system reacts to an antigen by studying the levels of antibodies in serum over time. This level is expressed quantitatively as the **titer** (ty'-tur), or concentration of antibodies. **Table 13.9** illustrates one of the most important features of adaptive immunity: memory.

Table 13.9 Primary and Secondary Response to Antigens

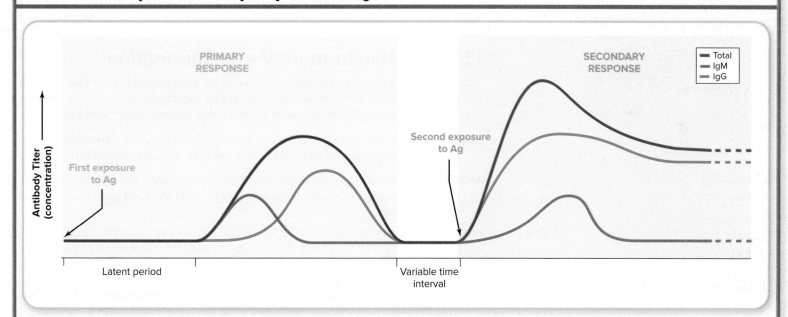

Upon the first exposure to an antigen, the system undergoes a **primary response.** The earliest part of this response, the *latent period,* is marked by a lack of antibodies for that antigen. Even so, much activity is occurring. During this time, the antigen is being concentrated in lymphoid tissue and is being processed by the correct clones of B lymphocytes. As plasma cells synthesize antibodies, the serum titer increases to a certain plateau and then tapers off to a low level over a few weeks or months. Early in the primary response, most of the antibodies are the IgM type, which is the first class to be secreted by plasma cells. Later, the class of the antibodies (but not their specificity) is switched to IgG or some other class (IgA or IgE).

After the initial response, there is no activity, but memory cells of the same specificity are seeded throughout the lymphatic system.

When the immune system is exposed again to the same immunogen within weeks, months, or even years, a **secondary response** occurs. The rate of antibody synthesis, the peak titer, and the length of antibody persistence are greatly increased over the primary response. The speed and intensity seen in this response are attributable to the memory B cells that were formed during the primary response. The secondary response is also called the **anamnestic response** (from the Greek word for "memory"). The advantage of this response is evident: It provides a quick and potent strike against subsequent exposures to infectious agents.

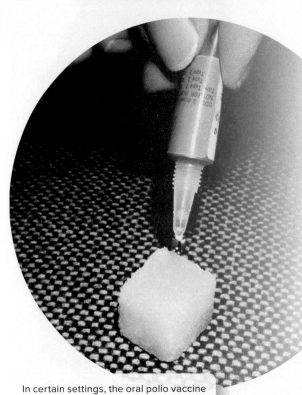

In certain settings, the oral polio vaccine is dispensed onto a sugar cube to make it palatable to children.

©Image Source/Getty Images

It is a well-accepted principle that memory B and T cells are only created from clones activated by a specific antigen. This provides a much quicker and more effective response on the second exposure and all exposures afterward. But researchers are now investigating a phenomenon that has been suspected for some time and confirmed in rigorous studies. It seems that exposure to a particular antigen can result in memory cells that will recognize antigens that are *chemically related* to it, even if those antigens have not been seen by the host. This might explain the well-known phenomenon, seen most clearly in developing countries, that vaccines against one disease can provide some protection against others. In Africa, for example, vaccinating against measles also cuts deaths from pneumonia, sepsis, and diarrhea by one-third.

This realization upsets the long-held view that memory only exists because of specific exposures, but it makes sense if we consider how activation of adaptive immunity occurs, via recognition of epitopes, small pieces of macromolecules on the surfaces of microbes. If other microbes share those chemical signatures (epitopes), memory cells will react against them as well.

13.5 LEARNING OUTCOMES—Assess Your Progress

16. Diagram the steps in B-cell activation, including all types of cells produced.
17. Make a detailed drawing of an antibody molecule.
18. Explain the various end results of antibody binding to an antigen.
19. List the five types of antibodies and important facts about each.
20. Draw and label a graph—with time on the horizontal axis—that shows the development of the primary and secondary immune responses.

13.6 Adaptive Immunity and Vaccination

Adaptive immunity in humans and other mammals is categorized using two different sets of criteria, which, when combined, result in four specific descriptors of the immune state. Immunity can either be natural or artificial. Also, it can be either active or passive.

Natural immunity encompasses any immunity that is acquired during the normal biological experiences of an individual rather than through medical intervention.

Artificial immunity is protection from infection obtained through medical procedures. This type of immunity is induced by immunization with vaccines and immune serum.

Active immunity occurs when an individual receives an immune stimulus (antigen) that activates the B and T cells, causing the body to produce immune substances such as antibodies. Active immunity is marked by several characteristics: (1) It creates a memory that renders the person ready for quick action upon reexposure to that same antigen; (2) it requires several days to develop; and (3) it lasts for a relatively long time, sometimes for life. Active immunity can be stimulated by natural or artificial means.

Passive immunity occurs when an individual receives immune substances (usually antibodies) that were produced actively in the body of another human or animal donor. The recipient is protected for a short time, even though he or she has not had prior exposure to the antigen. It is characterized by (1) lack of memory for the original antigen; (2) lack of production of new antibodies against that disease; (3) immediate onset of protection; and (4) short-term effectiveness because antibodies have a limited period of function and, ultimately, the recipient's body disposes of them. Passive immunity can also be natural or artificial in origin.

Table 13.10 illustrates the various possible combinations of adaptive immunities.

Table 13.10 The Four Types of Adaptive Immunity

Natural Immunity is acquired through the normal life experiences of a human and is not induced through medical means.

©Floresco Productions/Corbis

Active

After recovering from infectious disease, a person will generally be actively resistant to reinfection for a period that varies according to the disease. In the case of childhood viral infections such as measles, mumps, and rubella, this natural active stimulus provides nearly lifelong immunity. Other diseases result in a less extended immunity of a few months to years (such as pneumococcal pneumonia and shigellosis), and reinfection is possible. Even a subclinical infection can stimulate natural active immunity. This probably accounts for the fact that some people are immune to an infectious agent without ever having been noticeably infected with or vaccinated for it.

©Ingram Publishing/Superstock

Passive

Natural, passively acquired immunity occurs only as a result of the prenatal and postnatal mother-child relationship. During fetal life, IgG antibodies circulating in the maternal bloodstream are small enough to pass across the placenta. This natural mechanism provides an infant with a mixture of many maternal antibodies that can protect it for the first few critical months outside the womb, while its own immune system is gradually developing active immunity. Depending on the microbe, passive protection lasts anywhere from a few months to a year.

Another source of natural passive immunity comes to the baby by way of the mother's milk. Although the human infant acquires 99% of natural passive immunity *in utero* and only about 1% through nursing, the milk-borne antibodies provide a special type of intestinal protection that is not available from transplacental antibodies.

Artificial Immunity is that produced purposefully through medical procedures.

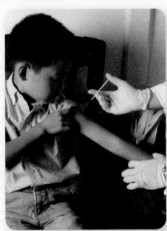

©Jill Braaten/McGraw-Hill Education

Active

Vaccination exposes a person to a specially prepared microbial (immunogenic) stimulus, which then triggers the immune system to produce antibodies and lymphocytes to protect the person upon future exposure to that microbe. As with natural active immunity, the degree and length of protection vary. Vaccines are also being developed for threats to human health that do not involve microbes at all (such as prostate cancer and heroin addiction).

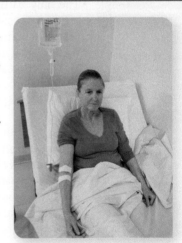

©Dave Moyer

Passive

Passive immunotherapy involves a preparation that contains specific antibodies against a particular infectious agent. Pooled human serum from donor blood (gamma globulin) and immune serum globulins containing high quantities of antibodies are frequently used.

COVID-19

In the spring of 2020, in the early days of the COVID-19 pandemic, the public became more aware of the option of passive immunotherapy. It was widely called "convalescent serum," referring to the fact that it uses serum, containing antibodies, from people who were ill with SARS-CoV-2, and had recovered (thus "convalescent"). This approach is useful when there is no vaccine yet available.

Immunization: A Lively History

The basic notion of immunization has existed for thousands of years. It probably stemmed from the observation that persons who had recovered from certain communicable diseases rarely got a second case. The earliest crude attempts involved bringing a susceptible person into contact with a diseased person or animal. The first recorded attempt at immunization occurred in sixth-century China. It consisted of drying and grinding up smallpox scabs and blowing them with a straw into the nostrils of vulnerable family members. By the 10th century, this practice had changed to the deliberate inoculation of dried pus from the smallpox pustules of one patient into the arm of a healthy person, a technique later called **variolation** (variola is the name

Figure 13.9 Ridicule of the new idea of vaccination. This cartoon appeared during Edward Jenner's advocacy for vaccination against smallpox.

Library of Congress Prints & Photographs Division [LC-USZC4-3147]

HOW A NEW VACCINE IS DEVELOPED, APPROVED, AND MONITORED

Before a new vaccine is ever given to people, extensive lab testing is done that can take several years. Once testing in people begins, it can take several more years before clinical studies are complete and the vaccine is licensed.

The Food and Drug Administration (FDA) sets rules for the three phases of clinical trials to ensure the safety of the volunteers. Researchers test vaccines with adults first.

PHASE I
20–100 healthy volunteers

Is this vaccine safe?
Does this vaccine seem to work?
Are there any serious side effects?
How is the size of the dose related to side effects?

PHASE II
Several hundred volunteers

What are the most common short-term side effects?
How are the volunteers' immune systems responding to the vaccine?

PHASE III
Hundreds or thousands of volunteers

How do people who get the vaccine and people who do not get the vaccine compare?
Is the vaccine safe?
Is the vaccine effective?
What are the most common side effects?

FDA licenses the vaccine only if:
• It's safe and effective • The benefits outweigh the risks

Vaccines are made in batches called lots.

Manufacturers must test all lots to make sure they are safe, pure, and potent. The lots can only be released once FDA reviews their safety and quality.

The FDA inspects manufacturing facilities regularly to ensure quality and safety.

---

for the smallpox virus). This method was used in parts of the Far East for centuries before it was brought to England in 1721.

Although the principles of the technique had some merit, unfortunately many recipients and their contacts died of smallpox. This outcome vividly demonstrates a cardinal rule for a workable vaccine: It must contain an antigen that will provide protection but not cause the disease.

Eventually, this human experimentation paved the way for the first really effective vaccine, developed by the English physician Edward Jenner in 1796. Jenner's work gave rise to the words **vaccine** and *vaccination* (from L., *vacca,* "cow"), which now apply to any immunity obtained by inoculation with selected antigens. Jenner was inspired by the case of a dairymaid who had been infected by a pustular infection called cowpox. This related virus afflicts cattle but causes a milder condition in humans. She explained that she and other milkmaids had remained free of smallpox. To test the effectiveness of this new vaccine, Jenner prepared material from human cowpox lesions and inoculated a young boy. When challenged 2 months later with an injection of crusts from a smallpox patient, the boy proved immune.

Jenner's discovery–that a less pathogenic agent could confer protection against a more pathogenic one–is especially remarkable in view of the fact that microscopy was still in its infancy and the nature of viruses was unknown. At first, the use of the vaccine was regarded with fear and skepticism **(figure 13.9).** When Jenner's method proved successful and word of its significance spread, it was eventually adopted in many other countries. In 1980, the World Health Organization declared that smallpox had been eradicated.

Passive Immunization

The only natural forms of passive immunization occur as (1) a fetus develops and encounters selected antibodies that are able to cross the placental barrier and (2) a newborn nurses and receives IgA in breast milk that is secreted at birth and for a short time afterward.

The first attempts at artificial passive immunization involved the transfusion of horse serum containing antitoxins to prevent tetanus and to treat patients exposed to diphtheria. Since then, antisera from animals have been replaced with products of human origin that function with various degrees of specificity. Intravenous immunoglobulin (IVIG), sometimes called *gamma globulin,* contains immunoglobulin extracted from the pooled blood of human donors. The method of processing IVIG concentrates the antibodies to increase potency and eliminates potential pathogens (such as hepatitis B and HIV). It is most useful in patients who have a diminished ability to mount their own immune response, or if the disease is so fast-acting that it could be fatal before the victim develops her own antibodies (tetanus, rabies).

A preparation called specific immune globulin (SIG) is derived from a more defined group of donors. Companies that prepare SIG obtain serum from patients who are convalescing and in a hyperimmune state after such infections as pertussis, tetanus, chickenpox, and hepatitis B. These globulins are preferable to IVIG because they contain higher titers of specific antibodies obtained from a smaller pool of patients. Although donated immunities only last a relatively short time, they act immediately and can help patients for whom no other useful medication or vaccine exists. This was demonstrated when COVID-19 patients were treated with SIG (convalescent sera) because there was no vaccine.

Artificial Active Immunity: Vaccination

The basic principle behind vaccination is to stimulate a primary response and a memory response that primes the immune system for future exposure to a virulent pathogen. If this pathogen enters the body, the immune response will be immediate, powerful, and sustained.

Vaccines have profoundly reduced the prevalence and impact of many infectious diseases that were once common and often deadly. Initially, the emphasis was on immunizing babies and children against formerly common childhood diseases, like measles, mumps, and rubella. Recent years have seen an additional push to immunize adolescents and adults against conditions such as human papillomavirus (HPV), *Streptococcus pneumoniae,* and shingles. Vaccines are also being developed for threats to human health that do not involve microbes at all.

In this section, we survey the principles of vaccine preparation and important considerations surrounding vaccine indication and safety. (Vaccines are also given specific consideration in later chapters on infectious diseases and organ systems.) Because there has been so much unfounded skepticism about vaccines, please take the time to read the infographics about how vaccines are developed and tested. In times of pandemics, such as the COVID-19 pandemic, vaccine production can be sped up, but quality control is still maintained.

Principles of Vaccine Preparation

In natural immunity, an infectious agent stimulates a relatively long-term protective response. In artificial active immunity, the objective is to obtain this same response with a modified version of the microbe or its components. Qualities of an ideal vaccine are as follows:

- It should protect against exposure to natural, wild forms of the pathogen.
- It should have a low level of adverse side effects or toxicity and not cause harm.
- It should stimulate both antibody (B-cell) response and cell-mediated (T-cell) response.
- It should have long-term, lasting effects (produce memory).
- It should not require numerous doses or boosters.
- It should be inexpensive, have a relatively long shelf life, and be easy to administer.

Vaccine preparations can be broadly categorized as either whole-organism or part-of-organism preparations. These categories also have subcategories:

1. Whole cells or viruses

   a. live, attenuated microbial cells or viruses
   b. killed cells or inactivated viruses

2. Part-of-organism preparations: antigen molecules derived from bacterial cells or viruses (subunits)

   a. subunits derived from cultures of cells or viruses
   b. subunits chemically synthesized to mimic natural molecules found on pathogens
   c. subunits manufactured via genetic engineering
   d. subunits conjugated with proteins (often from other microbes) to make them more immunogenic. These are called **conjugated vaccines.**

These categories are also shown in **table 13.11.**

Note: As of 2008, there were no vaccines used in the United States that consisted of killed whole bacteria. The last two available were those for cholera and plague.

Development of New Vaccines

Despite considerable successes, dozens of bacterial, viral, protozoal, and fungal diseases still remain without a functional vaccine. At the present time, no reliable vaccines are available for HIV/AIDS, various diarrheal diseases, respiratory diseases, and worm infections that affect over 200 million people per year worldwide. Worse than that, most existing vaccines are out of reach for much of the world's population.

DNA vaccines are one promising new approach to immunization. The technique in these formulations is very similar to gene therapy as described in chapter 8, except in this case, DNA is inserted into a plasmid vector and inoculated into a

HOW A VACCINE IS ADDED TO THE U.S. RECOMMENDED SCHEDULE

The Advisory Committee on Immunization Practices (ACIP) is a group of medical and public health experts. Members of the American Academy of Pediatrics and the American Academy of Family Physicians add their advice to the group. All available data about the vaccine from clinical trials and other studies are examined before a recommendation is made.

How safe is the vaccine when given at specific ages?
How well does the vaccine work at specific ages?
How serious is the disease the vaccine prevents?
How many children would get the disease the vaccine prevents if we didn't have the vaccine?

ACIP recommendations are not official until the CDC Director reviews and approves them.

New vaccine to protect against a disease is added to the immunization schedule.

HOW A VACCINE'S SAFETY CONTINUES TO BE MONITORED

FDA and CDC closely monitor vaccine safety now that hundreds of thousands of people are receiving it, looking for rare events that such a large group would reveal.

The purpose of monitoring is to watch for adverse events (possible side effects).

Monitoring a vaccine after it is licensed helps ensure that possible risks associated with the vaccine are identified.

VACCINE ADVERSE EVENT REPORTING SYSTEM (VAERS)

VAERS collects and analyzes reports of adverse events that happen after vaccination. Anyone can submit a report, including parents, patients, and health care professionals.

VACCINE SAFETY DATALINK

Network of health care organizations across the U.S.

Health care information available for population of over **9 million** people.

Scientists use VSD to conduct studies to evaluate the safety of vaccines and determine if side effects are actually associated with vaccination.

Adapted from the Centers for Disease Control and Prevention

Table 13.11 Types of Vaccines

Whole-Cell Vaccines

Whole cells or viruses are very effective immunogens because they are so large and complex. Depending on the vaccine, these are either killed or attenuated.

Killed vaccines (viruses are termed "inactivated" instead of "killed") are prepared by cultivating the desired strain or strains of a bacterium or virus and treating them with chemicals, radiation, heat, or some other agent that does not destroy antigenicity. The hepatitis A vaccine and three forms of the influenza vaccine contain inactivated viruses. Because the microbe does not multiply inside the host, killed vaccines often require a larger dose and more boosters to be effective.

Live attenuated vaccines contain live microbes whose virulence has been *attenuated*, or lessened/eliminated. This is usually achieved by modifying the growth conditions or manipulating microbial genes in a way that eliminates virulence factors. Vaccines for measles, mumps, polio (Sabin), and rubella contain live, nonvirulent viruses.

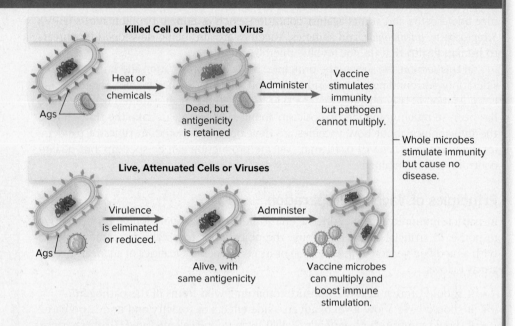

Killed Cell or Inactivated Virus

Ags — Heat or chemicals → Dead, but antigenicity is retained — Administer → Vaccine stimulates immunity but pathogen cannot multiply.

Live, Attenuated Cells or Viruses

Ags — Virulence is eliminated or reduced. → Alive, with same antigenicity — Administer → Vaccine microbes can multiply and boost immune stimulation.

Whole microbes stimulate immunity but cause no disease.

The advantages of live preparations are as follows:

1. Viable microorganisms can multiply and produce infection (but not disease) like the natural organism.
2. They confer long-lasting protection.
3. They usually require fewer doses and boosters than other types of vaccines.
4. They are particularly effective at inducing cell-mediated immunity.

Disadvantages of using live microbes in vaccines are that they require special storage facilities and could conceivably mutate back to become virulent again.

Subunit Vaccines (Parts of Organisms)

If the exact epitopes that stimulate immunity are known, it is possible to produce a vaccine based on a selected component of a microorganism. These vaccines for bacteria are called **subunit vaccines.** The antigens used in these vaccines may be taken from cultures of the microbes, produced by genetic engineering or synthesized chemically.

Examples of component antigens currently in use are the capsules of the pneumococcus and meningococcus, the protein surface antigen of anthrax, and the surface proteins of hepatitis B virus. A special type of vaccine is the **toxoid,** which consists of a purified bacterial exotoxin that has been chemically denatured. These vaccines cause humans to produce antitoxins that can neutralize the natural toxin. Toxoid vaccines provide protection against diseases such as diphtheria, tetanus, and pertussis.

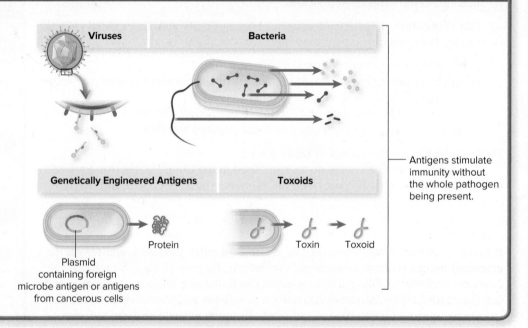

Viruses | **Bacteria**

Genetically Engineered Antigens | **Toxoids**

Plasmid containing foreign microbe antigen or antigens from cancerous cells → Protein

Toxin → Toxoid

Antigens stimulate immunity without the whole pathogen being present.

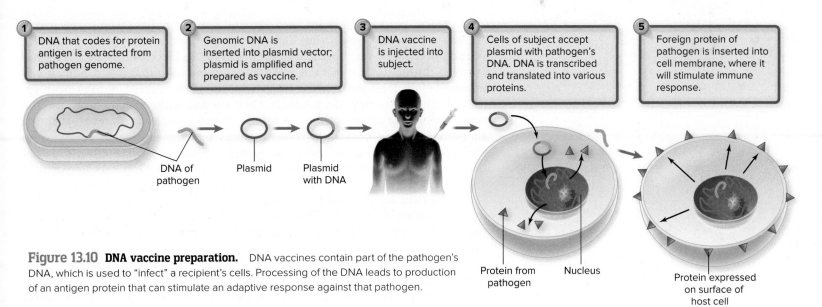

1 DNA that codes for protein antigen is extracted from pathogen genome.

2 Genomic DNA is inserted into plasmid vector; plasmid is amplified and prepared as vaccine.

3 DNA vaccine is injected into subject.

4 Cells of subject accept plasmid with pathogen's DNA. DNA is transcribed and translated into various proteins.

5 Foreign protein of pathogen is inserted into cell membrane, where it will stimulate immune response.

DNA of pathogen

Plasmid

Plasmid with DNA

Protein from pathogen

Nucleus

Protein expressed on surface of host cell

Figure 13.10 DNA vaccine preparation. DNA vaccines contain part of the pathogen's DNA, which is used to "infect" a recipient's cells. Processing of the DNA leads to production of an antigen protein that can stimulate an adaptive response against that pathogen.

recipient **(figure 13.10).** The expectation is that the human cells will take up some of the plasmids and express the foreign DNA in the form of proteins. Because these proteins are foreign, they will be recognized during immune surveillance and cause B and T cells to be sensitized and form memory cells. Currently there are more than 500 DNA vaccines in trial.

Vaccination strategies are now under intense investigation for the prevention and treatment of noninfectious diseases. One of the best examples is vaccination for Alzheimer's disease. Researchers are examining whether administering a peptide that is found in the brain plaques of Alzheimer's patients as a vaccine can lead to prevention of the disease. The principle is the same as with vaccines against microbial infection. If T and B memory cells that recognize these disease-inducing proteins can be created in a host via vaccination, once the proteins begin to form in the patient, the memory response eliminates them before they cause damage. One of these vaccines is in human trials now.

Immunotherapy as a treatment for cancer has been in use now for some years. In these situations, T cells or dendritic cells are removed from the patient. In the laboratory the cells are sensitized to known cancer antigens. These activated cells are then injected back into the patients, in the hopes that the actual tumor antigens match the ones used for activation in the lab, and the patient's own immune system will attack the tumor (see also the Medical Moment titled CAR-T about this topic in this chapter).

Route of Administration and Side Effects of Vaccines

Most vaccines are injected by subcutaneous, intramuscular, or intradermal routes. One type of the influenza vaccine comes in the form of a nasal spray. Oral (or nasal) vaccines are available for only a few diseases, but they have some distinct advantages. An oral or nasal dose of a vaccine can stimulate protection (IgA) on the mucous membrane of the portal of entry. Oral and nasal vaccines are also easier to give than injections, and are more readily accepted.

Some vaccines require the addition of a special binding substance, or **adjuvant** (ad'-joo-vunt). An adjuvant is any compound that enhances immunogenicity and prolongs antigen retention at the injection site. The adjuvant precipitates the antigen and holds it in the tissues so that it will be released gradually. Its gradual release presumably facilitates contact with antigen-presenting cells and lymphocytes. This helps involve the innate immune system as well.

NCLEX® PREP

3. Which characteristic is associated with passive artificial immunity?
   a. *long-term protection duration*
   b. *requires several days to develop protection*
   c. *immediate protection*
   d. *creation of memory in response to antigen exposure*

In recent years a significant "anti-vaccine" movement has taken shape among citizens of first-world countries. This is a science text, so we present important science facts about vaccines, and how they come to be recommended. Vaccines must go through many years of trials in experimental animals and human volunteers before they are licensed for general use. Even after they have been approved, like all therapeutic products, they are not without complications. The most common of these are local reactions at the injection site, fever, allergies, and other adverse outcomes. Some patients experience allergic reactions to the medium used to grow the vaccine organism (eggs or tissue culture) rather than to vaccine antigens. More serious reactions are extremely rare (fewer than 1 case in 220,000), but they can occur, and are intensively studied by scientists. All of the risks are detailed in materials provided to you by your doctor before the vaccination occurs. Vaccine companies have phased out certain preservatives, such as thimerosal, that are thought to cause allergies and other potential side effects.

Some people have attempted to link childhood vaccinations to later development of diabetes, asthma, and, most prominently, autism. However, the original 1998 scientific paper that suggested that vaccines might be responsible for autism was entirely discredited, and the principal author's medical license was revoked after authorities found the research and its claims fraudulent.

The World Health Organization named vaccine hesitancy as one of the top ten threats to global health in 2019. The biggest cause of vaccine hesitancy seems to be vaccine misinformation. Scientists at Quinnipiac University recently conducted a study with 17,229 participants, with almost half of them identifying as health care students or health care workers. They found an alarming amount of misconceptions, even among this population. For example, in the non–health care population, only 29% of respondents correctly agreed with this statement: *You cannot get the flu from the flu vaccine.* Among health care students or workers, only 49% agreed with that statement. (There is no risk of influenza from the vaccine.) In another example, 36% of all study participants agreed or were unsure of this statement: *There is evidence that the mercury in the measles vaccine can cause autism.* In fact, the measles vaccine has never contained mercury and has been shown beyond a doubt to not cause autism.

Outbreaks of measles, mumps, diphtheria, polio, typhoid fever, and whooping cough have popped up all over this country in college dormitories, in antivaccination religious communities, and in airplanes. These outbreaks are often attributed to a decrease in the level of herd immunity, a phenomenon in which a certain percentage of the population is vaccinated, which means that the microbe is unable to maintain its circulation through the population. Think about it—this means that getting vaccinated serves the common good as well as your individual good. At a time when most of the world's population is clamoring for vaccines, some in the developed world are refusing vaccination, essentially relying on others' willingness to be vaccinated to keep them and their children safe.

Some have speculated that vaccination has done too good of a job—at least in terms of being so effective for so long that many young parents have no memory of the prevaccination era and don't appreciate the much greater risk of not vaccinating compared to vaccinating. In the decade before measles vaccination began, 3 to 4 million cases occurred each year in the United States. Typically, 300 to 400 children died annually and 1,000 more were chronically disabled due to measles encephalitis. Put simply, childhood vaccines save the lives of 2.5 million children a year (worldwide), according to UNICEF. During the COVID-19 pandemic, many parents delayed well-visits to their medical providers, further decreasing vaccination rates.

Many people complain that they "got the flu" right after getting the vaccine. What they are actually experiencing is either fatigue from the immune system revving up to respond to the influenza antigens, or a cold that may have "snuck in" while the immune system was busy prepping your body for the much more dangerous influenza.

Kelly & Heidi

©FatCamera/Getty Images

Professionals involved in giving vaccinations must understand their inherent risks but also realize that the risks from the infectious disease almost always outweigh the chance of an adverse vaccine reaction.

13.6 LEARNING OUTCOMES—Assess Your Progress

21. List and define the four different descriptors of specific immune states.

22. Discuss the qualities of an effective vaccine.

23. Name the two major categories of vaccines and then the subcategories under each.

24. Explain the principle of herd immunity and the risks that unfold when it is not maintained.

CASE FILE WRAP-UP

Shutterstock/SamaraHeisz5

In mid-2020, it was just becoming apparent that some children and young adults who had been infected with SARS-CoV-2 experienced a debilitating condition that attacked their circulatory systems. Older "children"—adolescents—seemed to have more severe manifestations of this post-COVID condition. So 15-year-old Trevor had the rashes originating from blood vessel inflammation, but he also had heart complications.

When these patients started appearing in COVID-19 hotspots, clinicians quickly recognized the new syndrome as the immune system's inappropriate response to SARS-Cov-2. The symptoms are similar to those of Kawasaki disease, an inflammatory condition most frequently seen in young children. At the time of this RN's story, there was little else known about the condition, and the main treatments aimed to dampen the immune response and care for specific manifestation, such as the heart symptoms. At that time the condition was named multisystem inflammatory syndrome in children, or MIS-C.

Gut Bacteria Cause Blindness?

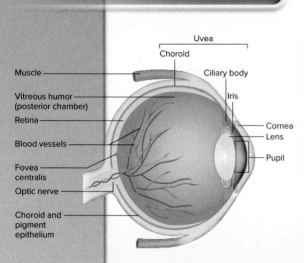

Muscle

Vitreous humor (posterior chamber)

Retina

Blood vessels

Fovea centralis

Optic nerve

Choroid and pigment epithelium

Uvea

Choroid

Ciliary body

Iris

Cornea

Lens

Pupil

There is a condition well known to ophthalmologists called *autoimmune uveitis (yoo-vee-eye'-tis)*. You can see from the illustration that the uvea is a central part of the eye. In autoimmune uveitis, T cells inappropriately attack retinal tissue in the uvea. This condition is responsible for up to 15% of severe vision problems and blindness in the developed world.

Scientists have struggled to understand this disease, for the following reason: The eye is a site of immunological privilege. This means that the blood vessels in the eye are less permeable and that there is no direct lymphatic vessel drainage. This prevents inflammatory products and immune cells from causing damage to the delicate eye tissue. But in autoimmune uveitis, T cells that recognize eye antigens cross through the blood vessel barrier and mount an immune response in the eye. So not only do the T cells breach the protected area, they do so after being sensitized to eye antigens that don't circulate in the lymphatic system. How does that happen?

A 2015 study conducted by researchers at the United States National Eye Institute, using a mouse model, found that T-cell activity in the mouse intestines increased just prior to the onset of uveitis. The intestines also contained increased levels of a cytokine released by T cells. The hypothesis is that T cells are activated by antigens from bacteria in the gut microbiome, and that they then mistakenly react with a similar protein in the eye.

This is an example of very interesting research—that is very preliminary. The study is suggestive of a cause, but is not proof. For example, researchers did not identify the exact substance that produces the cross-reaction with eye antigens. However, a 2017 study showed a similar pathway to age-related macular degeneration. In that study, particular members of the gut microbiomes were correlated with the occurrence of macular degeneration. Further studies will be required to move these associations from "correlated with the disease" to "causes the disease."

HOST DEFENSES PART II: ADAPTIVE IMMUNITY AND IMMUNIZATION

ADAPTIVE IMMUNITY: THE THIRD AND FINAL LINE OF DEFENSE

13.1

Adaptive immunity consists of four stages: I. Lymphocytes differentiate into B cells and T cells. II. Antigen-presenting cells detect invading pathogens and present these antigens to lymphocytes, which recognize the antigen and initiate the immune response. III. Clones of lymphocytes respond to the antigen. IV. Activated T cells participate directly in the response. Activated B cells release antibodies.

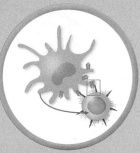

During development both B and T cells develop millions of genetically different clones, through reassortment of the DNA making up antigen receptor genes. Those that respond to self-antigens are mainly deleted (clonal deletion).

STAGE I: THE DEVELOPMENT OF LYMPHOCYTE DIVERSITY

13.2

STAGE II: PRESENTATION OF ANTIGENS

13.3

Antigen-presenting cells and macrophages engulf and process foreign antigen and bind the epitope to MHC class II molecules on their cell surface for presentation to CD4 T lymphocytes.

The three main classes of T cells are T helper cells, T regulatory cells, and T cytotoxic cells. Each subset of T cell produces a distinct set of cytokines that stimulate lymphocytes or destroy foreign or altered cells.

STAGES III AND IV: T-CELL RESPONSE

13.4

STAGE III AND IV: B-CELL RESPONSE

13.5

B cells produce five classes of antibody. They bind physically to identical epitopes that initiated their production. They immobilize or mark the antigen for destruction. The memory response ensures that the second exposure to an antigen is faster and more vigorous than the first.

Vaccination is an illustration of artificial active immunity, in which a piece of a microbe (a subunit) or an inactivated microbe is administered to a healthy subject to prime the adaptive immune response for a time when the subject is exposed to the actual microorganism naturally. Vaccines are being developed for a variety of noninfectious conditions as well.

ADAPTIVE IMMUNITY AND VACCINATION

13.6

SmartGrid: From Knowledge to Critical Thinking

This *21 Question Grid* takes the topics from this chapter and arranges them with respect to the American Society for Microbiology's Undergraduate Curriculum guidelines—all six of the important "Concepts" as well as the important "Competency" of scientific literacy. Three questions are supplied about chapter content that refer to the Concept or Competency, in increasing levels of Bloom's taxonomy for learning.

ASM Concept/ Competency	A. Bloom's Level 1, 2—Remember and Understand (Choose one.)	B. Bloom's Level 3, 4—Apply and Analyze	C. Bloom's Level 5, 6—Evaluate and Create
Evolution	1. A single bacterium has _____ epitope(s). a. a specific b. *multiple* c. *MHC* d. *clonal*	2. Would you suspect that the innate or the adaptive arms of the immune response evolved first in mammals? Explain your answer.	3. Provide an explanation to refute the following statement: Humans cannot develop specific immunity to a novel biological agent created in a laboratory.
Cell Structure and Function	4. The primary B-cell receptor is a. IgD. b. IgA. c. IgE. d. IgG.	5. Name three antigen-presenting cells, and what other functions each of them has.	6. Major histocompatibility molecules are critical for specific immunity. But they are named after another major function they have. How does histocompatibility work?
Metabolic Pathways	7. In humans, B cells mature in the ___ and T cells mature in the ____. a. GALT; liver b. bursa; thymus c. bone marrow; thymus d. lymph nodes; spleen	8. Explain how the memory response is the cornerstone of vaccination.	9. Conduct research on clonal deletion and write a paragraph that explains it to a typical high school biology student.
Information Flow and Genetics	10. Which of the following cells is capable of adaptively responding to a nearly infinite number of epitopes? a. B and T cells b. Plasma cells c. T cytotoxic cells d. All of these	11. Is antibody diversity generated at the DNA or RNA level? Explain.	12. In order for gene rearrangement of antigen receptors to be successful, the DNA sequence needs to remain in-frame after the rearrangement occurs. Explain what "in-frame" means and why it would be necessary in the generation of diversity in antigen receptors.
Microbial Systems	13. Some microbial products can activate B cells without the assistance of T cells. Which of the following can do this? a. Capsule of *S. pneumoniae* b. Lipopolysaccharide c. Some viral capsids d. All of these	14. Explain why it is useful that antibodies have two antigen-binding arms.	15. Using the details of T-cell activation, suggest a reason why they are so critical in recognizing and destroying virally infected host cells.

ASM Concept/Competency	A. Bloom's Level 1, 2—Remember and Understand (Choose one.)	B. Bloom's Level 3, 4—Apply and Analyze	C. Bloom's Level 5, 6—Evaluate and Create

Impact of Microorganisms	**16.** A vaccine that contains parts of viruses is called
	    a. acellular.
	    b. recombinant.
	    c. subunit.
	    d. attenuated.
		17. Explain how herd immunity works to protect the spread of infectious disease. Include two ways that people can be "immune" to infection.	**18.** Conduct research and discuss one current example illustrating how low herd immunity within a population has led to localized disease outbreaks in the United States.

Scientific Thinking	**19.** If you draw a blood sample from a patient to determine whether he or she has a herpes simplex infection, and the patient displays a large amount of IgG against the virus but low levels of IgM, what do you conclude?
	    a. The patient is newly infected.
	    b. The patient has had the infection for a while.
	    c. The patient is not infected.
	    d. It is impossible to draw conclusions.
		20. Chronic lymphocytic leukemia leads to the production of cancerous B cells, and treatment often involves bone marrow transplantation. Based upon your knowledge of lymphocyte development, explain how this procedure can lead to therapeutic effects.	**21.** Scientists have been developing a vaccine to prevent addiction to heroin and other opioids. Sketch out a scenario for how it might work.

Answers to the multiple-choice questions appear in Appendix A.

Visual Connections

This question connects content within and between chapters.

From table 13.9. In this figure describing primary and secondary responses to antigen, indicate where a vaccination might be most effective, and also indicate where natural infection would play a role.

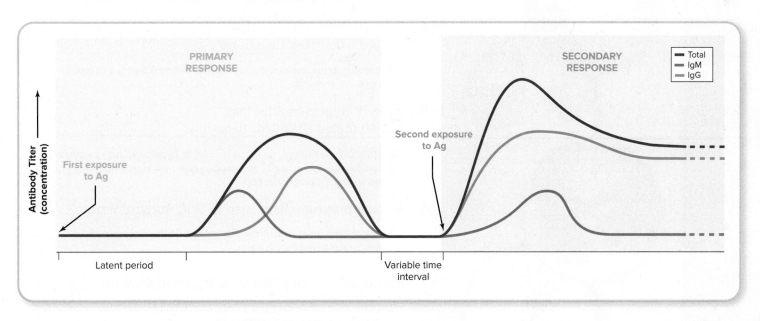

Chapter design elements: (Covid): CDC/ Alissa Eckert, MS; Dan Higgins, MAMS; (Note): McGraw-Hill Education; (NCLEX): Shutterstock/ Abert; (Doctor): David Gould/Getty Images

14

Disorders in Immunity

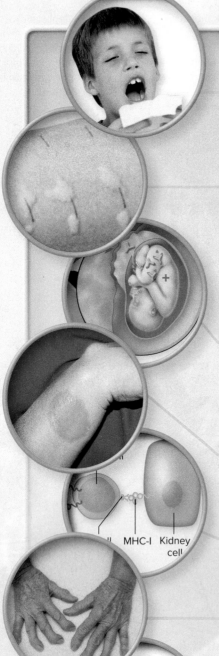

IN THIS CHAPTER...

14.1 The Immune Response: A Two-Sided Coin

1. Define *immunopathology*, and describe the two major categories of immune dysfunction.
2. Identify the four major categories of hypersensitivities, or overreactions to antigens.

14.2 Type I Allergic Reactions: Atopy and Anaphylaxis

3. Summarize genetic and environmental factors that influence allergy development.
4. Identify three conditions caused by IgE-mediated allergic reactions.
5. Identify the two clinical forms of anaphylaxis, explaining why one is more often fatal than the other.
6. List the three main ways to prevent or short-circuit type I allergic reactions.

14.3 Type II Hypersensitivities: Reactions That Lyse Foreign Cells

7. List the three immune components causing cell lysis in type II hypersensitivity reactions.
8. Explain the role of Rh factor in hemolytic disease development and how it is prevented in newborns.

14.4 Type III Hypersensitivities: Immune Complex Reactions

9. Identify commonalities and differences between type II and type III hypersensitivities.

14.5 Type IV Hypersensitivities: Cell-Mediated (Delayed) Reactions

10. Describe one example of a type IV delayed hypersensitivity reaction.
11. List four classes of grafts, and explain how host versus graft and graft versus host diseases develop.

14.6 An Inappropriate Response to Self: Autoimmunity

12. List at least three autoimmune diseases and the most important immunologic features in them.

14.7 Immunodeficiency Diseases: Hyposensitivities

13. Distinguish between primary and secondary immunodeficiencies, explaining how each develops.

CASE FILE

A Body Attacking Itself

A 57-year-old woman was admitted to the acute care unit where I was working as an RN. Her admitting diagnosis was rheumatoid arthritis (RA). Her past medical history was significant for gastric ulcers and insulin-dependent diabetes. She had been seen by her primary care physician in the clinic, who decided to admit her after examining her and listening to her history of symptoms.

This patient was unique in that her symptoms had begun quite suddenly 2 days prior, rather than insidiously as is usually the case in RA. She had awoken with moderately severe pain in her knees, ankles, hands, and feet. She also complained of extreme fatigue and stated that she felt as though she had the flu. She had a low-grade fever on admission. Her other vital signs were normal. Her hands, knees, and ankles were mildly swollen and were warm to the touch. She had difficulty moving due to joint pain and was having difficulty managing at home.

Blood work was ordered, including a complete blood count (CBC), chemistry panel, anticyclic citrullinated peptide (anti-CPP), erythrocyte sedimentation rate (ESR), rheumatoid factor (RF), and C-reactive protein. X rays were also ordered of the affected joints. Test results indicated that the patient was mildly anemic. The ESR was elevated and the patient was positive for both rheumatoid factor and anti-CPP. X rays showed soft tissue swelling around the affected joints, but little destruction of cartilage or damage to the bone was seen, which was not surprising considering the patient had had symptoms for such a short time.

Given that the patient had a history of gastrointestinal (GI) bleeding and had brittle diabetes, the physician started the patient on Enbrel (etanercept), a tumor necrosis factor (TNF) inhibitor. I questioned why the patient was not started on a steroid, as is usually the case. The physician explained that starting the patient on anti-inflammatory drugs might have precipitated another stomach ulcer, and steroids such as prednisone are known to affect blood glucose levels. For that reason, the physician believed the patient would do better on a TNF inhibitor such as Enbrel. Before starting the patient on the drug, the physician screened her for hepatitis and tuberculosis. All of her screening tests were negative, and within 2 weeks the patient was in remission and feeling great.

- How is tumor necrosis factor significant in rheumatoid arthritis?

- How do TNF inhibitors reduce inflammation?

Case File Wrap-Up appears at the end of the chapter.

14.1 The Immune Response: A Two-Sided Coin

Humans possess a powerful and necessary system of defense. However, by its very nature, it also carries the potential to cause injury and disease. In most instances, a defect in immune function results in irritating but nondebilitating symptoms such as those of hay fever and dermatitis. But abnormal immune functions are also actively involved in debilitating or life-threatening diseases such as asthma, anaphylaxis, diabetes, rheumatoid arthritis, and graft rejection.

The majority of our previous discussions of the immune response have centered around its many beneficial effects. In this chapter, we survey **immunopathology,** disease states caused by overreactivity or underreactivity of the immune response **(figure 14.1).** Overreactivity (also called hypersensitivity) takes the forms of allergy and autoimmunity. In these conditions, the tissues are innocent bystanders attacked by immune components that can't distinguish one's own tissues from foreign material. In immunodeficiency or **hyposensitivity diseases,** immune function is incompletely developed, is suppressed, or has been destroyed. We will start the chapter with a discussion of hypersensitivity, and then we'll discuss hyposensitivities.

Hypersensitivity: Four Types

Hypersensitivity reactions are classified into four major categories: type I ("common" allergy and anaphylaxis), type II (IgG- and IgM-mediated cell damage), type III (immune complex), and type IV (delayed hypersensitivity) **(table 14.1).** In general, types I, II, and III involve a B-cell–immunoglobulin response, and type IV involves a T-cell response. The antigens that elicit these reactions can be exogenous (originating from outside the body, for example, microbes, pollen grains, and foreign cells and proteins) or endogenous arising from self tissue (for example, autoimmunities).

Figure 14.1 Overview of disorders of the immune system. Just as the system of T cells and B cells provides necessary protection against infection and disease, the same system can cause serious and debilitating conditions by overreacting (hypersensitivity) or underreacting (hyposensitivity) to immune stimuli.
Courtesy Baylor College of Medicine, Public Affairs (David Vetter); Christopher Kerrigan/McGraw-Hill Education (taking medication); Pixtal/age fotostock (boy sneezing); Roc Canals Photography/ Getty Images (person getting blood transfusion); Dynamic Graphics/JupiterImages (woman wringing her hands); BW Folsom/ Shutterstock (a severe reaction to latex on the top of a male hand)

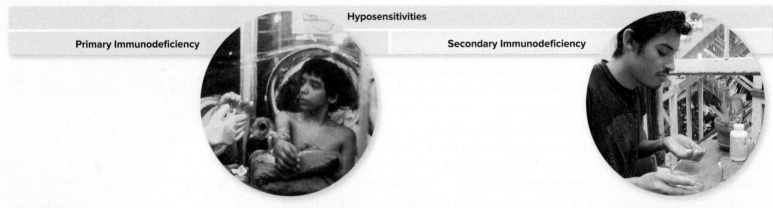

Hyposensitivities

| Primary Immunodeficiency | Secondary Immunodeficiency |

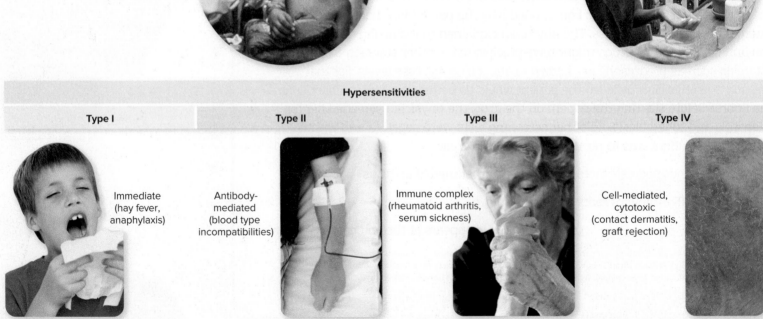

Hypersensitivities

| Type I | Type II | Type III | Type IV |

Immediate (hay fever, anaphylaxis)

Antibody-mediated (blood type incompatibilities)

Immune complex (rheumatoid arthritis, serum sickness)

Cell-mediated, cytotoxic (contact dermatitis, graft rejection)

Table 14.1 Hypersensitivity States

Type		Systems and Mechanisms Involved	Examples
I	Immediate hypersensitivity	IgE-mediated; involves mast cells, basophils, and allergic mediators	Anaphylaxis, allergies such as hay fever, asthma
II	Antibody-mediated	IgG, IgM antibodies plus complement act upon cells and cause cell lysis; includes some autoimmune diseases	Blood group incompatibility; pernicious anemia; myasthenia gravis
III	Immune complex-mediated	Antibody-mediated inflammation; circulating IgG complexes deposited in basement membranes of target organs; includes some autoimmune diseases	Systemic lupus erythematosus; rheumatoid arthritis; serum sickness; rheumatic fever
IV	T-cell-mediated	Delayed hypersensitivity and cytotoxic reactions in tissues; includes some autoimmune diseases	Infection reactions; contact dermatitis; graft rejection

One of the reasons allergies are easily mistaken for infections is that both involve tissue damage and can trigger the inflammatory response, as described in chapter 12. Many symptoms and signs of inflammation (redness, heat, skin eruptions, edema, and granuloma) are prominent features of allergies.

> ### 14.1 LEARNING OUTCOMES—Assess Your Progress
> 1. Define *immunopathology*, and identify the two major categories of immune dysfunction.
> 2. Identify the four major categories of hypersensitivities, or overreactions to antigens.

14.2 Type I Allergic Reactions: Atopy and Anaphylaxis

The term *allergy* refers to an exaggerated immune response that is manifested by inflammation. Allergic individuals are acutely sensitive to repeated contact with antigens, called *allergens*, which do not noticeably affect nonallergic individuals. All type I reactions share a similar physiological mechanism, are immediate in onset, and are associated with exposure to specific antigens. However, there are two levels of severity: **Atopy** is any chronic local allergy such as hay fever or asthma; **anaphylaxis** (an″-uh-fih-lax′-us) is a systemic, sometimes fatal reaction that involves airway obstruction and circulatory collapse.

Although the general effects of hypersensitivity are detrimental, we must be aware that it involves the very same types of immune reactions as those at work in protective immunities. Based upon this fact, *all* humans have the potential to develop allergies under particular circumstances.

Who Is Affected?

In the United States, nearly half of the population is affected by airborne allergens, such as dust, pollen, and mold. The incidence of some allergies in children has been increasing over the last 20 years **(figure 14.2)**. The majority of type I allergies are relatively mild, but certain forms such as asthma and some food allergies may require hospitalization and can cause death, especially in the youngest patients. In some individuals, atopic allergies last for a lifetime. Others "outgrow" them, and still others suddenly develop them later in life.

A predisposition to allergies seems to "run in families," that is, have a strong familial association. Be aware that the part that is hereditary is a *generalized susceptibility,* not the allergy to a specific substance. For example, a parent who is allergic to ragweed pollen can have a child who is allergic to cat hair. The prospect of a child's developing

Asthma is a form of type I hypersensitivity.
©Ian Hooton/Science Photo Library/Getty Images

Figure 14.2 Allergies in U.S. children ages 0 to 17 from 1997 to 2017. There have been statistically significant increases in both skin and food allergies over this time period.
Source: CDC

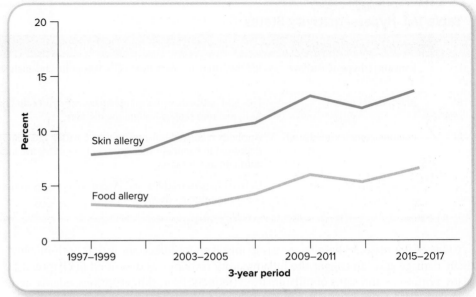

Source: CDC/NCHS, Health Data Interactive, National Health Interview Survey.

atopic allergy is at least 25% if one parent exhibits symptoms, and increases to nearly 50% if grandparents or siblings are also afflicted. The actual basis for atopy seems to be a genetic program that favors allergic antibody (IgE) production, increased reactivity of mast cells, and increased susceptibility of target tissue to allergic mediators.

The "hygiene hypothesis" provides one possible explanation for an environmental component to allergy development. This hypothesis suggests that the industrialized world has created a very hygienic environment, supplemented with antimicrobial products of all kinds and very well insulated homes, and that this has been bad for our immune systems. It seems that our immune systems need to be "trained" by interaction with microbes as we develop. In fact, children who grow up on farms have been found to have lower incidences of several types of allergies. Also, researchers have found that the combination of being delivered by cesarean section and a maternal history of allergy elevates the risk that a child will be allergic to foods by a factor of eight. As we saw in chapter 11, delivery by cesarean section keeps the baby from being exposed to vaginal and stool bacteria. Additional work has shown that babies need to be exposed to commensal bacteria in order for the IgA system to develop normally.

Another possible factor affecting allergy development appears to be related to breast feeding. Newborn babies that are breast fed exclusively for the first 4 months of life have a lower risk of asthma and eczema, especially if they have a family history of allergy. This is thought to come from the presence of cytokines and growth factors in human milk that act on the baby's gut mucosa to induce tolerance, rather than reactivity, to allergens. New information from the Human Microbiome Project reveals that nearly 600 species of bacteria can be transferred to infants through breast milk. Combined with data from other studies showing that a disruption of microbial populations in the gut may influence the development of asthma, it is clear that these organisms play an important role in the development of tolerance to foreign antigens.

Could allergies reflect some beneficial evolutionary adaptation? Most allergy sufferers would answer with a resounding "No!" Why would humans and other mammals evolve an allergic response that causes suffering, tissue damage, and even death? One possible explanation may be that the components involved in an allergic response exist to defend against helminthic worms and other multicellular human parasites. It is only relatively recently in our evolutionary history that developed countries have seen dramatically fewer infections with these parasites. One hypothesis is that the part of the immune system that fights helminthic worms is left idle in a population that has recently been "scrubbed" of these parasites, and goes awry.

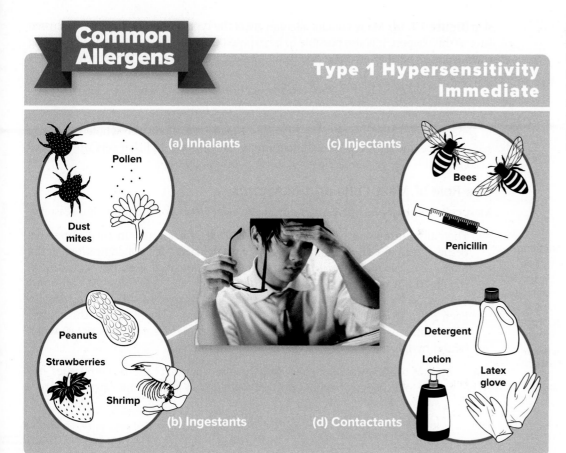

Common Allergens

Type 1 Hypersensitivity Immediate

(a) Inhalants
- Pollen
- Dust mites

(c) Injectants
- Bees
- Penicillin

(b) Ingestants
- Peanuts
- Strawberries
- Shrimp

(d) Contactants
- Detergent
- Lotion
- Latex glove

Figure 14.3 Common allergens, classified by portal of entry. **(a)** Common inhalants, or airborne environmental allergens, include pollen and insect parts. **(b)** Common ingestants, allergens that enter by mouth. **(c)** Common injectants, allergens that enter via the parenteral route. **(d)** Common contactants, allergens that enter through the skin. Rubberball/Alamy

The Nature of Allergens and Their Portals of Entry

As with all antigens, allergens have certain immunogenic characteristics. Proteins are more allergenic than carbohydrates, fats, or nucleic acids. Some allergens are haptens, nonprotein substances with a molecular weight of less than 1,000 that can form complexes with carrier molecules in the body (shown in figure 13.6). Organic and inorganic chemicals found in industrial and household products, cosmetics, food, and drugs are commonly this type of allergen.

Allergens typically enter through epithelial portals in the respiratory tract, gastrointestinal tract, and skin **(figure 14.3).** The mucosal surfaces of the gut and respiratory system present a thin, moist surface that is normally quite penetrable. The dry, tough keratin coating of skin is less permeable, but access still occurs through tiny breaks, glands, and hair follicles.

Airborne environmental allergens such as pollen, house dust, dander (shed skin scales), or fungal spores are termed *inhalants* **(figure 14.3*a*).** Each geographic region harbors a particular combination of airborne substances that varies with the season and humidity. Pollen is given off seasonally by trees and other flowering plants, while mold spores are released throughout the year. Airborne animal hair and dander, feathers, and the saliva of dogs and cats are common sources of allergens. The component of house dust that appears to account for most dust allergies is not soil or other debris but the decomposed bodies and feces of tiny mites that commonly live in this dust.

Allergens can be *ingestants, injectants,* or *contactants.* Ingestants often cause food allergies **(figure 14.3*b*).** Injectant allergies are triggered by drugs, vaccines, or hymenopteran (bee) venom **(figure 14.3*c*).** Contactants are allergens that enter through the

NCLEX® PREP

1. During a visit to his physician, a patient is asked about his allergies. The patient reports a generalized, raised rash following a dose of an intravenous antibiotic. This is an example of what type of reaction?
   a. *immediate hypersensitivity*
   b. *antibody-mediated hypersensitivity*
   c. *immune complex hypersensitivity*
   d. *cell-mediated hypersensitivity*
   e. *side effect*

skin (**figure 14.3d**). Many contact allergies are of the type IV (delayed) variety, discussed later in this chapter. It is also possible to be exposed to certain allergens, penicillin among them, during sexual intercourse due to the presence of allergens in the semen.

Mechanisms of Type I Allergy: Sensitization and Provocation

In general, type I allergies develop in stages. **Figure 14.4** tells the whole story. You see that it begins with the initial encounter with allergen, and that sets up the conditions for the allergy to manifest on subsequent encounters.

The Role of Mast Cells and Basophils

Mast cells and basophils play an important role in allergy for the following reasons:

1. Their ubiquitous location in tissues. Mast cells are located in the connective tissue of virtually all organs, but there are particularly high concentrations in the lungs, skin, gastrointestinal tract, and genitourinary tract. Basophils circulate in the blood but migrate readily into tissues.

2. Their capacity to bind IgE during sensitization (**figure 14.4a**) and *degranulate*. Each cell carries 30,000 to 100,000 cell receptors, which trigger the release of inflammatory cytokines from cytoplasmic granules (secretory vesicles) when bound by IgE that has been stimulated by the allergen.

The symptoms of allergy are not caused by the direct action of allergen on tissues, but rather by the physiological effects of mast-cell-derived allergic mediators on target

Figure 14.4 A schematic view of cellular reactions during the type I allergic response. **(a)** Sensitization (initial contact with sensitizing dose), 1–6. **(b)** Provocation (later contacts with provocative dose), 7–9. **(c)** The spectrum of reactions to inflammatory cytokines released by mast cells and the common symptoms they elicit in target tissues and organs.

Ingram Publishing (headache); robeo/iStock/Getty Images (hives); Colin Anderson/Brand X Pictures/Getty Images (stomach ache); Ian Hooton/Science Source (asthma inhaler use); StockBroker X/Stockbroker/Media Bakery (mature woman with flu or allergy symptoms)

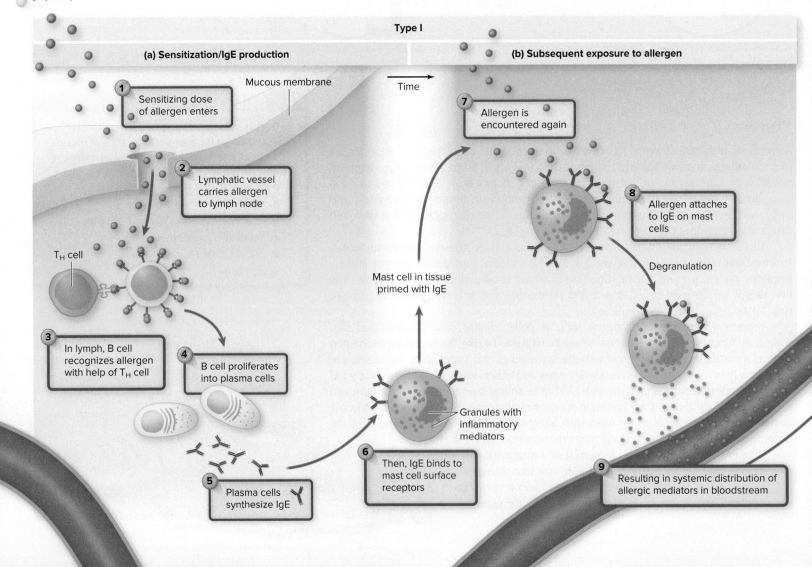

Type I

(a) Sensitization/IgE production

(b) Subsequent exposure to allergen

Mucous membrane

Time

1. Sensitizing dose of allergen enters

2. Lymphatic vessel carries allergen to lymph node

T_H cell

3. In lymph, B cell recognizes allergen with help of T_H cell

4. B cell proliferates into plasma cells

5. Plasma cells synthesize IgE

6. Then, IgE binds to mast cell surface receptors

Granules with inflammatory mediators

Mast cell in tissue primed with IgE

7. Allergen is encountered again

8. Allergen attaches to IgE on mast cells

Degranulation

9. Resulting in systemic distribution of allergic mediators in bloodstream

organs. In recent years researchers have also pinpointed a role for T-cell participation in allergic reactions to certain substances. This is not surprising since the orchestration of B-cell responses is usually influenced by T cells. So, as in most biological phenomena, it is more complicated than it appears on the surface. In this summary we will focus on the role of IgE.

Cytokines, Target Organs, and Allergic Symptoms

Numerous substances involved in mediating allergy have been identified. The principal chemical mediators produced by mast cells and basophils are histamine, serotonin, leukotriene, platelet-activating factor, prostaglandins, and bradykinin (**figure 14.4c**). These inflammatory cytokines, acting alone or in combination, account for the tremendous scope of allergic symptoms.

Histamine, the most profuse and fastest-acting allergic mediator, is a potent stimulator of secretory glands and smooth muscle. Histamine's actions on smooth muscle vary with location. It *constricts* the smooth muscle layers of the small bronchi and intestine, thereby causing labored breathing and increased intestinal motility. In contrast, histamine *relaxes* vascular smooth muscle and dilates arterioles and venules, resulting in *wheal-and-flare* reactions in the skin and pruritus (itching). Histamine can also stimulate eosinophils to release inflammatory cytokines, escalating the symptoms.

In allergic reactions, **bradykinin** causes prolonged smooth muscle contraction of the bronchioles, dilation of peripheral arterioles, increased capillary permeability, and increased mucus secretion. Although the exact role of **serotonin** in human allergy is uncertain, its effects appear to complement those of histamine and bradykinin.

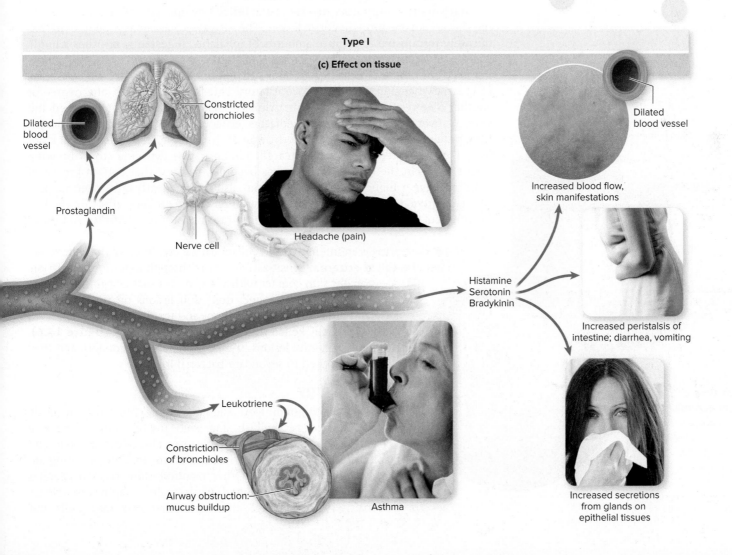

Type I

(c) Effect on tissue

Dilated blood vessel

Constricted bronchioles

Prostaglandin

Nerve cell

Headache (pain)

Dilated blood vessel

Increased blood flow, skin manifestations

Histamine
Serotonin
Bradykinin

Increased peristalsis of intestine; diarrhea, vomiting

Leukotriene

Constriction of bronchioles

Airway obstruction: mucus buildup

Asthma

Increased secretions from glands on epithelial tissues

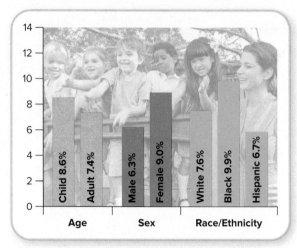

Source: National Health Interview Survey, National Center for Health Statistics, Centers for Disease Control and Prevention

Figure 14.5 Asthma prevalence in the United States.   Data are from 2015 and are displayed with respect to age, gender, and race/ethnicity.

Photo: goldenKB/Getty Images

Leukotriene (loo″-koh-try′-een) is known as the "slow-reacting substance of anaphylaxis" for its property of inducing gradual contraction of smooth muscle. This type of leukotriene is responsible for the prolonged bronchospasm, vascular permeability, and mucus secretion of the asthmatic individual. Other leukotrienes stimulate the activities of polymorphonuclear leukocytes, or granulocytes, which play a role in various immune functions (see chapter 12).

Prostaglandins are a group of powerful inflammatory agents. Normally, these substances regulate smooth muscle contraction (e.g., they stimulate uterine contractions during delivery). In allergic reactions, they are responsible for vasodilation, increased vascular permeability, increased sensitivity to pain, and bronchoconstriction. Nonsteroidal anti-inflammatory drugs (NSAIDs), such as aspirin and ibuprofen, work by preventing the actions of prostaglandins.

IgE- and Mast-Cell-Mediated Allergic Conditions

The mechanisms just described lead to the development of hay fever, allergic asthma, food allergy, drug allergy, and eczema. This section covers the main characteristics of these conditions.

Atopic Diseases

Hay fever is a generic term for allergic rhinitis, a seasonal reaction to inhaled plant pollen or molds, or a chronic, year-round reaction to a wide spectrum of airborne allergens or inhalants. The targets are typically respiratory membranes, and the symptoms include nasal congestion; sneezing; coughing; profuse mucus secretion; itchy, red, and teary eyes; and mild bronchoconstriction.

Asthma is a respiratory disease characterized by episodes of impaired breathing due to severe bronchoconstriction. The airways of asthmatic people are exquisitely responsive to minute amounts of inhalants, ingestants, or other stimuli, such as infectious agents. The symptoms of asthma range from occasional, annoying bouts of difficult breathing to fatal suffocation. Labored breathing, shortness of breath, wheezing, cough, and ventilatory **rales** are present to one degree or another. The respiratory tract of an asthmatic person is chronically inflamed and severely overreactive to allergic mediators, especially leukotrienes and serotonin from pulmonary mast cells. Upon activation of the allergic response, natural killer (NK) T cells are recruited and activated, adding to the cytokine storm brewing in the lungs. In the United States it has become apparent that certain groups of children have a higher rate of asthma, and also higher death rates from it, than others **(figure 14.5).** Since 2012 there has been an interagency effort by the U.S. government to reduce the excessively high rate of asthma in African-American children.

Atopic dermatitis is an intensely itchy inflammatory condition of the skin, sometimes also called **eczema.** Sensitization occurs through ingestion, inhalation, and, occasionally, skin contact with allergens. It usually begins in infancy with reddened, weeping, encrusted skin lesions on the face, scalp, neck, and inner surfaces of the limbs and trunk that may progress to a dry, scaly, thickened skin condition in adulthood **(figure 14.6).** The itchy, painful lesions cause considerable discomfort, and they are predisposed to secondary bacterial infections.

Food Allergy

The most common food allergens come from peanuts, fish, cow's milk, eggs, shellfish, and soybeans. Although the mode of entry is intestinal, food allergies can also affect the skin and respiratory tract. Gastrointestinal symptoms include vomiting, diarrhea, and abdominal pain. Other manifestations of food allergies include hives, rhinitis, asthma, and, occasionally, anaphylaxis. Classic food hypersensitivity involves IgE and degranulation of mast cells, but

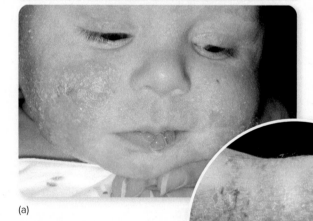

Type I—Immediate

(a)

(b)

Figure 14.6 Atopic dermatitis, or eczema.
(a) Vesicular, weepy, encrusted lesions are typical in afflicted infants. **(b)** In adulthood, lesions are more likely to be dry, scaly, and thickened.

Dr. P. Marazzi/Science Source (a); Biophoto Associates/Science Source (b) not all reactions involve this mechanism. (Do not confuse food allergy with food *intolerance*. Many people are lactose intolerant, for example, due to a deficiency in the enzyme that degrades the milk sugar.) Egg allergies, in particular, must be considered when vaccinating individuals, due to the presence of egg protein in many vaccine preparations.

Drug Allergy

Modern chemotherapy has been responsible for many medical advances. Unfortunately, it has also been hampered by the fact that drugs are foreign compounds capable of stimulating allergic reactions. In fact, allergy to drugs is one of the most common side effects of treatment (occurring in 5% to 10% of hospitalized patients). Depending on the allergen, route of entry, and individual sensitivities, virtually any tissue of the body can be affected, and reactions range from a rash **(figure 14.7)** to fatal anaphylaxis. Compounds implicated most often are antibiotics (penicillin), synthetic antimicrobials (sulfa drugs), aspirin, opiates, and contrast dye used in X rays. The actual allergen is not the intact drug itself but a hapten produced when the liver processes the drug.

Anaphylaxis: An Overpowering IgE-Mediated Allergic Reaction

The term *anaphylaxis*, or anaphylactic shock, refers to a swift reaction to allergens. Two clinical types of anaphylaxis are seen in humans. *Cutaneous anaphylaxis* is the wheal-and-flare inflammatory reaction to the local injection of allergen. *Systemic anaphylaxis*, on the other hand, is characterized by sudden respiratory and circulatory disruption that can be fatal within minutes. The allergen type and route of entry causing anaphylaxis vary, though bee stings and injection of antibiotics or serum are the most common causes. Bee venom is a complex material containing several allergens and enzymes that can create a sensitivity that can last for decades after exposure.

The underlying physiological events in systemic anaphylaxis parallel those of atopy, but the concentration of chemical mediators and the strength of the response are greatly amplified. The immune system of a sensitized person exposed to a provocative dose of allergen responds with a sudden, massive release of chemicals into the tissues and blood, which act rapidly on the target organs. Anaphylactic persons have been known to die within 15 minutes from complete airway blockage.

Diagnosis of Allergy

Because allergy mimics infection and other conditions, it is important to determine if a person is actually allergic and to identify the specific allergen or allergens. Allergy testing can be done on blood samples, or directly on the skin.

Blood Testing

The most widely used blood test is a radioallergosorbent test (RAST), which measures levels of IgE to specific allergens. A new test that can distinguish whether a patient has experienced an allergic attack measures elevated blood levels of tryptase, an enzyme released by mast cells that increases during an allergic response. Several types of specific *in vitro* tests can determine the allergic potential of a patient's blood sample. A differential blood cell count can reveal high levels of basophils and eosinophils, indicating allergy. The leukocyte histamine-release test measures the amount of histamine released from the patient's basophils when exposed to a specific allergen.

Skin Testing

A tried-and-true *in vivo* method to detect precise atopic or anaphylactic sensitivities is skin testing. With this technique, a patient's skin is injected, scratched, or pricked with a small amount of a pure allergen extract. There are hundreds of

Type I—Immediate

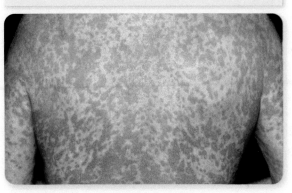

Figure 14.7 Drug allergy reaction. A typical rash that develops in an allergic reaction to an antibiotic.
Dr. P. Marazzi/Science Source

NCLEX® PREP

2. Anaphylaxis is characterized by which of the following clinical manifestations? Select all that apply.
   a. *circulatory disruption (tachycardia, low blood pressure)*
   b. *swelling of the lips, tongue, or throat*
   c. *loss of consciousness*
   d. *skin wheal and erythema*
   e. *itching*

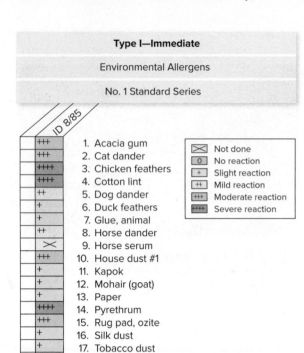

	Type I—Immediate	
	Environmental Allergens	
	No. 1 Standard Series	

ID 8/85

+++	1. Acacia gum	
+++	2. Cat dander	
++++	3. Chicken feathers	
++++	4. Cotton lint	
++	5. Dog dander	
+	6. Duck feathers	
+	7. Glue, animal	
++	8. Horse dander	
✕	9. Horse serum	
+++	10. House dust #1	
+	11. Kapok	
+	12. Mohair (goat)	
+	13. Paper	
++++	14. Pyrethrum	
+++	15. Rug pad, ozite	
+	16. Silk dust	
+	17. Tobacco dust	
+	18. Tragacanth gum	
+++++	19. Upholstery dust	
+++	20. Wool	

✕	Not done
0	No reaction
+	Slight reaction
++	Mild reaction
+++	Moderate reaction
++++	Severe reaction

No. 2 Airborne Particles	

ID 8/85

+++	1. Ant
+++++	2. Aphid
++++	3. Bee
++++	4. Housefly
✕	5. House mite
+++	6. Mosquito
++++	7. Moth
+++	8. Roach
++	9. Wasp
0	10. Yellow jacket

Airborne mold spores
++	11. *Alternaria*
+++	12. *Aspergillus*
++	13. *Cladosporium*
+++	14. *Hormodendrum*
0	15. *Penicillium*
+	16. *Phoma*
+++	17. *Rhizopus*
	18. ....

(b)

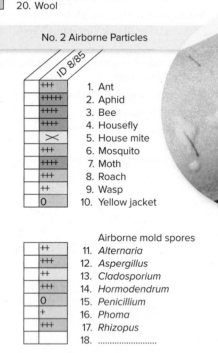

(a)

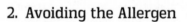

Figure 14.8 A method for conducting an allergy skin test.   The forearm (or back) is mapped and then injected with a selection of allergen extracts. The allergist must be very aware of the potential of anaphylactic attacks triggered by these injections. **(a)** Close-up of skin wheals showing a number of positive reactions (dark lines are measurer's marks). **(b)** An actual skin test record for some common environmental allergens [not related to **(a)**].
Southern Illinois University/Science Source (a)

these allergen extracts containing common airborne allergens and more unusual allergens (mule dander, theater dust, bird feathers). Unfortunately, skin tests for food allergies using food extracts are unreliable in most cases. In patients with numerous allergies, the allergist maps the skin on the inner aspect of the forearms or back and injects the allergens intradermally according to this predetermined pattern **(figure 14.8*a*)**. Approximately 20 minutes after antigenic challenge, each site is appraised for a wheal response indicative of histamine release. The diameter of the wheal is measured and rated on a scale of 0 (no reaction) to 4 (greater than 15 mm) **(figure 14.8*b*)**.

Treatment and Prevention of Allergy

In general, the methods of treating and preventing type I allergy involve the following **(figure 14.9)**:

1. taking drugs that block the action of lymphocytes, mast cells, or chemical mediators;
2. avoiding the allergen, although this may be very difficult in many instances; and
3. desensitization: controlled exposure to the antigen through ingestion, sublingual absorption, or injection to reset the allergic reaction.

1. Taking Drugs to Block Allergy

The aim of antiallergy medication is to block the progress of the allergic response somewhere along the route between IgE production and the appearance of symptoms. Oral anti-inflammatory drugs such as corticosteroids inhibit the activity of lymphocytes and thereby reduce the production of IgE, but they also have dangerous side effects and should not be taken for prolonged periods. Some drugs block the degranulation of mast cells and reduce the levels of inflammatory cytokines (Cromolyn). Asthma and rhinitis sufferers can find relief from a monoclonal antibody that binds to IgE in such a way that it is unable to dock with mast cells. Therefore no degranulation takes place (omalizumab [Xolair®]).

Widely used medications for preventing symptoms of atopic allergy are **antihistamines,** the active ingredients in most over-the-counter allergy-control drugs. Antihistamines interfere with histamine activity by binding to histamine receptors on target organs. Other drugs that relieve inflammatory symptoms are aspirin and acetaminophen, which reduce pain by interfering with prostaglandin, and theophylline, a bronchodilator that reverses spasms in the respiratory smooth muscles. Persons who suffer from anaphylactic attacks are urged to carry at all times injectable or aerosolized epinephrine (adrenaline) and an identification tag indicating their sensitivity. Epinephrine reverses constriction of the airways and slows the release of allergic mediators. Although epinephrine works quickly and well, it has a very short half-life. It is very common to require more than one dose in anaphylactic reactions. Injectable epinephrine buys the individual time to get to a hospital for continuing treatment.

2. Avoiding the Allergen

It is obvious that if you experience a negative health effect when exposed to a substance, then the first goal should be to avoid it. It is the most basic of all preventive measures and is practiced daily by millions of people.

3. Desensitization

In practice avoiding an allergen can be very difficult, as you may have noticed if you're ever on a plane and the flight crew announces that there will be no peanuts available because someone in the cabin has a peanut allergy. But the field of allergy medicine is changing rapidly. Desensitization, in the form of "allergy shots," has been used for decades. It involves injecting specific amounts of the allergen under the skin.

The allergen preparations contain pure, preserved suspensions of plant antigens, venoms, dust mites, dander, and molds (but so far, desensitization to foods has not proved very effective). The immunologic basis of this treatment is open to differences in interpretation. One hypothesis suggests that injected allergens stimulate the formation of allergen-specific IgG–**blocking antibodies**–that can remove allergen from the system before it can bind to IgE. It is also possible that allergen delivered in this fashion combines with the IgE itself and takes it from circulation.

Researchers are now looking at desensitization treatments that are delivered sublingually (under the tongue) and orally, in order to present the allergen through a mucosal surface, hoping to trigger IgA and IgG responses that would have a blocking effect on the allergic response.

Evidence also suggests that if children are allergic to fresh milk, having them consume it in small amounts in the form of baked goods (in which milk is an ingredient) can help desensitize them. As you know, heating proteins denatures them, changing their physical form. It is possible that the changed presentation of epitopes increases the IgG response–which blocks the antigen binding to the IgE on mast cells. This blocks degranulation.

Peanut allergies are one of the scariest for parents, as children can develop anaphylactic symptoms quickly from this allergy. Recent research has shown that "oral immunotherapy," a treatment in which children were given increasing amounts of peanut protein daily, can sometimes cure the peanut allergy in children. Again, this suggests that altering the way the antigen is presented (in this case, the dose; in the previous example, the physical form) can correct an inappropriate response.

There are also a great many clinical trials directed at more generalized parts of the immune system so those with multiple allergies could be helped with a single treatment. Probiotics have been shown to be helpful in some cases, and are still under investigation.

And if you remember, at the beginning of the chapter we inferred that the development of allergies might be the result of the sudden (in evolutionary time) disappearance of worm and protozoan pathogens in the developed world. The parts of immunity, such as IgE, that participate in allergy are the components that naturally react to helminths and larger microbes. So for several years now scientists (and some amateur allergy sufferers!) have experimented with using deliberate infections with *Trichuris suis*, a whipworm whose natural host is pigs. The worms can establish a brief colonization in humans without causing symptoms, during which time the immune system responds, and in some cases seems to reset itself so that it stops responding to the inappropriate antigen–the allergen. This approach is still experimental.

Figure 14.9 supplies a summary of ways to interfere with allergy symptoms.

14.2 LEARNING OUTCOMES—Assess Your Progress

3. Summarize genetic and environmental factors that influence allergy development.
4. Identify three conditions caused by IgE-mediated allergic reactions.
5. Identify the two clinical forms of anaphylaxis, explaining why one is more often fatal than the other.
6. List the three main ways to prevent or short-circuit type I allergic reactions.

① Blocking the Processes of Allergy

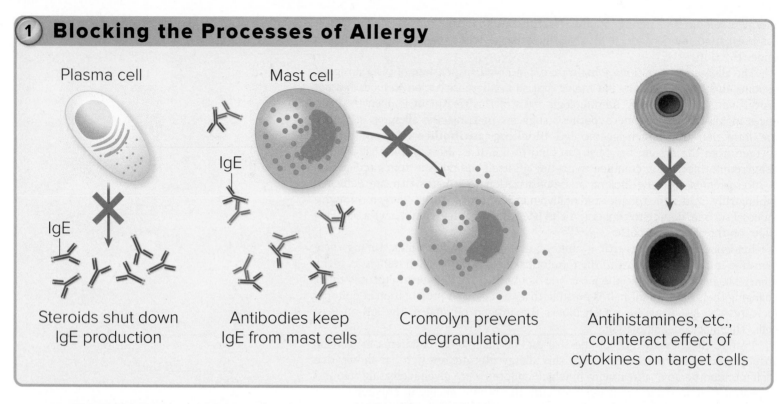

Plasma cell

Mast cell

IgE

IgE

Steroids shut down IgE production

Antibodies keep IgE from mast cells

Cromolyn prevents degranulation

Antihistamines, etc., counteract effect of cytokines on target cells

② Avoiding the Allergen

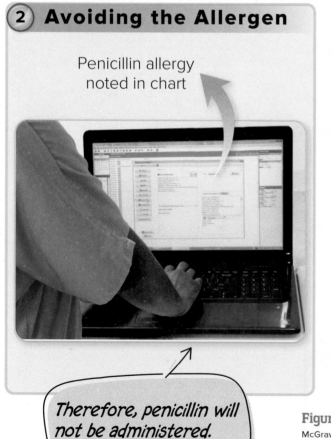

Penicillin allergy noted in chart

Therefore, penicillin will not be administered.
- Kelly & Heidi

③ Desensitization

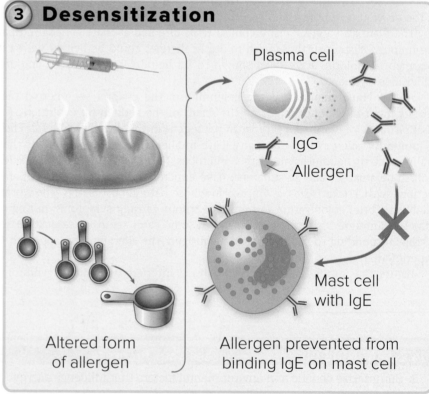

Plasma cell

IgG

Allergen

Mast cell with IgE

Altered form of allergen

Allergen prevented from binding IgE on mast cell

Figure 14.9 **Strategies for combating allergy attacks.**
McGraw-Hill Education

14.3 Type II Hypersensitivities: Reactions That Lyse Foreign Cells

The diseases termed type II hypersensitivities are a complex group of syndromes that involve complement-assisted destruction (lysis) of foreign cells by antibodies (IgG and IgM) directed against those cells' surface antigens. This category includes transfusion reactions and some types of autoimmunities (discussed in a later section). The cells targeted for destruction are often red blood cells, but other cells can be involved.

Chapters 12 and 13 described the functions of unique surface markers on cell membranes. Ordinarily, these molecules play essential roles in transport, recognition, and development, but they become medically important when the tissues of one person are placed into the body of another person. Blood transfusions and organ donations introduce **alloantigens** (molecules that differ in the same species) on donor cells that are recognized by the lymphocytes of the recipient. These reactions are not really immune dysfunctions the way that allergy and autoimmunity are. The immune system is in fact working normally, but it is not equipped to distinguish between the desirable foreign cells of a transplanted tissue and the undesirable ones of a microbe.

The Rh Factor and Its Clinical Importance

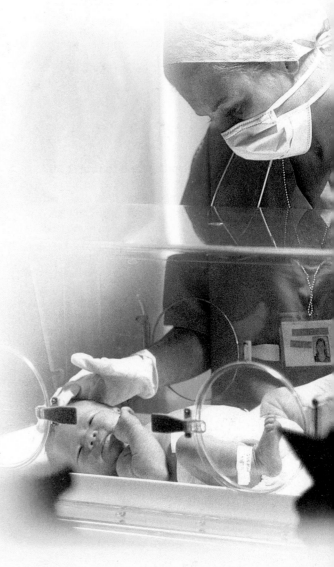

©Flying Colours Ltd/Getty Images

We are all aware of the need to match blood types when transfusing blood into a person. Those blood types are matched on the basis of the A, B, and O antigens on the surfaces of red blood cells, as well as another antigen, called the **Rh factor** (or D antigen). Because the Rh factor can cause disease in natural circumstances (a normal pregnancy), we will use it to illustrate type II hypersensitivities. This factor was first discovered in experiments exploring the genetic relationships among animals. Rabbits inoculated with the red blood cells (RBCs) of rhesus monkeys produced an antibody to an antigen on the rhesus blood cells. That antibody also reacted to a similar antigen on human blood cells. Further tests showed that the antigen that both monkeys and humans have on their blood cells (termed *Rh* for "rhesus") was present in about 85% of humans and absent in the other 15%. (These percentages vary in different ethnic groups.) The details of Rh inheritance are complicated, but in simplest terms, a person's Rh type results from a combination of two possible alleles—a dominant one that codes for the factor and a recessive one that does not. The "+" or "−" that appears after a blood type (i.e., O+) reflects the Rh status of the person. Unlike with the ABO antigens where antibodies are preformed against foreign blood antigens, the only ways one can develop antibodies against this factor are through exposure to a fetus's antigen while pregnant or through blood transfusion. Although the Rh factor should be matched for a transfusion to avoid this situation, it is acceptable to transfuse Rh− blood if the Rh type is not known.

Hemolytic Disease of the Newborn and Rh Incompatibility

The potential for placental sensitization occurs when a mother is Rh− and her unborn child is Rh+. The mother's immune system detects the foreign Rh factors on the fetal RBCs and is sensitized to them by producing antibodies and memory B cells. The first Rh+ child is usually not affected because the process begins so late in pregnancy that the child is born before maternal sensitization is completed. However, the mother's immune system has been strongly primed for a second contact with this factor in a subsequent pregnancy **(figure 14.10a).**

In the next pregnancy with an Rh+ fetus, fetal blood cells escape into the maternal circulation late in pregnancy and elicit a memory response. Maternal anti-Rh antibodies then cross the placenta into the fetal circulation, where they affix to fetal RBCs and cause complement-mediated lysis. The outcome is a potentially fatal **hemolytic disease of the newborn (HDN),** characterized by severe anemia and jaundice. It is also called *erythroblastosis fetalis* (eh-rith″-roh-blas-toh′-sis fee-tal′-is), reflecting the release of immature nucleated RBCs (erythroblasts) into the blood to compensate for destroyed RBCs.

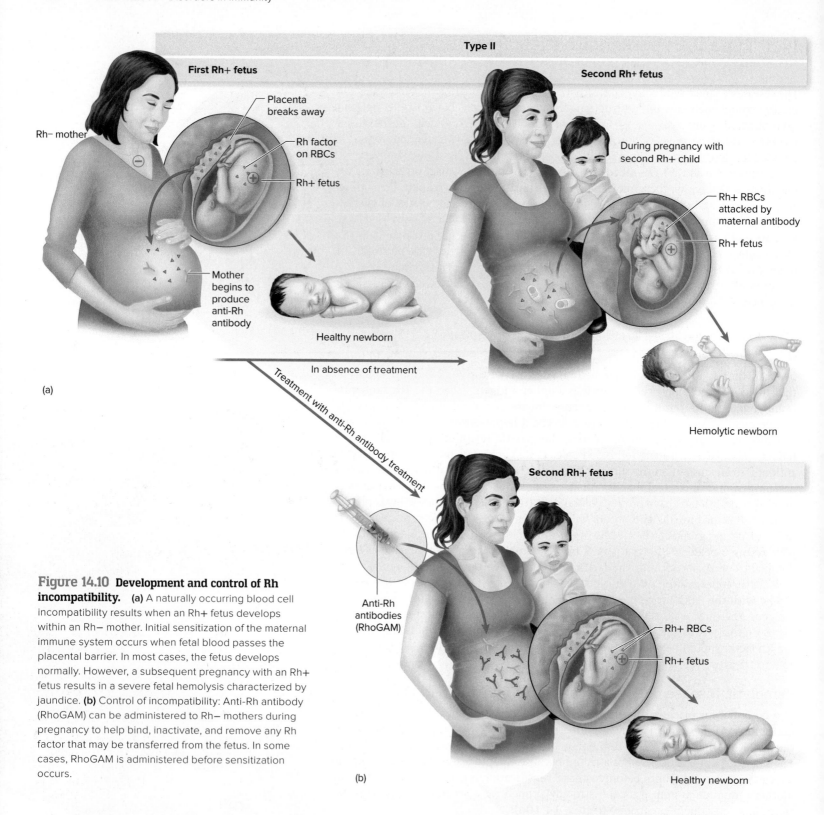

Figure 14.10 Development and control of Rh incompatibility. (a) A naturally occurring blood cell incompatibility results when an Rh+ fetus develops within an Rh− mother. Initial sensitization of the maternal immune system occurs when fetal blood passes the placental barrier. In most cases, the fetus develops normally. However, a subsequent pregnancy with an Rh+ fetus results in a severe fetal hemolysis characterized by jaundice. (b) Control of incompatibility: Anti-Rh antibody (RhoGAM) can be administered to Rh− mothers during pregnancy to help bind, inactivate, and remove any Rh factor that may be transferred from the fetus. In some cases, RhoGAM is administered before sensitization occurs.

Maternal-fetal incompatibilities are also possible in the ABO blood group, but adverse reactions occur less frequently than with Rh sensitization because the antibodies to these blood group antigens are IgM rather than IgG and are unable to cross the placenta in large numbers. In fact, the maternal-fetal relationship is a fascinating instance of foreign tissue not being rejected, despite the extensive potential for contact.

Preventing Hemolytic Disease of the Newborn

Once sensitization of the mother to Rh factor has occurred, all other Rh+ fetuses will be at risk for hemolytic disease of the newborn. Prevention requires a careful family history of an Rh− pregnant woman. If the father is also Rh−, the child will be Rh− and free of risk; but if the father is Rh+, there is a possibility that the fetus may be Rh+. In this case, the mother must be passively immunized with antiserum containing antibodies against the Rh factor. This antiserum is called RhoGAM. It is the immunoglobulin fraction of human anti-Rh serum, prepared from pooled human sera. This antiserum, injected at 28 to 32 weeks and again immediately after delivery, reacts with any fetal RBCs that have escaped into the maternal circulation, thereby preventing the sensitization of the mother's immune system to Rh factor **(figure 14.10b)**. Anti-Rh antibody must be given with each pregnancy that involves an Rh+ fetus.

14.3 LEARNING OUTCOMES—Assess Your Progress

7. List the three immune components causing cell lysis in type II hypersensitivity reactions.

8. Explain the role of Rh factor in hemolytic disease development and how it is prevented in newborns.

14.4 Type III Hypersensitivities: Immune Complex Reactions

Type III hypersensitivity involves the reaction of soluble antigen with antibody and the deposition of the resulting complexes in various tissues of the body. It is similar to type II because it involves the production of IgG and IgM antibodies after repeated exposure to antigens and the activation of complement. Type III differs from type II because the antigens that provoke it are not attached to the surface of a cell. The interaction of these antigens with antibodies produces free-floating complexes that can be deposited in the tissues, causing an **immune complex reaction,** or disease.

Mechanisms of Immune Complex Disease

After initial exposure to a large amount of antigen, the immune system produces large quantities of antibodies that circulate in the fluid compartments. When this antigen enters the system a second time, it reacts with the antibodies to form antigen-antibody complexes. These complexes recruit inflammatory components such as complement and neutrophils, which would ordinarily eliminate Ag-Ab complexes as part of the normal immune response. In an immune complex disease, however, there is an excess of antigen and small amounts of antibody. This causes cross linking of multiple antigens with the scarce antibodies, and the complexes can then deposit in the **basement membranes** of epithelial tissues and become inaccessible. In response to these events, neutrophils release lysosomal granules that digest tissues and cause a destructive inflammatory condition.

Types of Immune Complex Disease

In the early days of using antiserums as treatments for infectious diseases, hypersensitivity reactions to serum were common. In addition to anaphylaxis, two syndromes were identified, the **Arthus reaction** and **serum sickness,** associated with certain types of passive immunization (especially with animal serum).

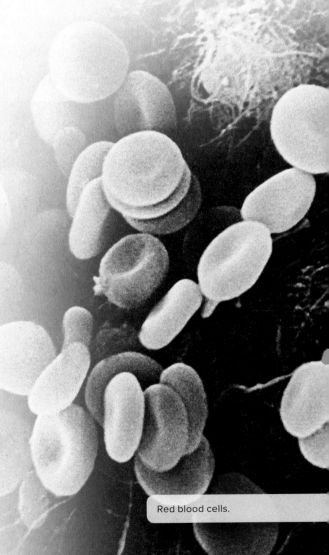
Red blood cells.

Type III

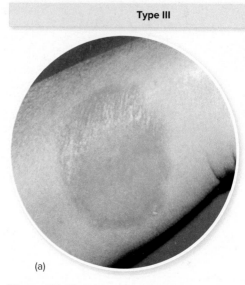

(a)

(b)

Figure 14.11 **Typical presentation of the Arthus reaction (a) and serum sickness (b), two immune complex diseases.**

Phototake (a); Courtesy Gary P. Wiliams, M.D. (b)

Medical Moment

Patch Testing

Patch testing is an alternative to skin testing for allergies. In patch testing, allergens are introduced onto a patch, which is then secured to the skin. Patch testing is useful for diagnosing delayed hypersensitivity reactions. It is also useful for determining whether a particular substance is causing contact dermatitis, or irritation of the skin caused by an allergic reaction. Patch testing can be used to detect allergy to hair dyes, fragrances, latex, resins, and metals such as silver.

The patches are worn on the skin (usually the back) for 48 hours, at which time the patches are removed and the skin is assessed for irritation. During patch testing, it is important to avoid bathing, showering, or activities that may result in heavy perspiration (i.e., working outdoors or exercising).

Q. Which cells of the immune system do you think are responsible for processing the antigen presented in a patch test?

Answer in Appendix B.

Serum sickness and the Arthus reaction are like anaphylaxis in that all three of them require sensitization, which leads to preformed antibodies. Characteristics that set serum sickness and the Arthus reaction apart from anaphylaxis are as follows:

1. they depend on IgG, IgM, or IgA (precipitating antibodies) rather than IgE;
2. they require large doses of antigen (not a minuscule dose as in anaphylaxis); and
3. their symptoms are delayed (a few hours to days).

The Arthus reaction is a *localized* dermal injury due to inflamed blood vessels in the vicinity of any injected antigen. Serum sickness, however, is a *systemic* injury initiated by antigen-antibody complexes that circulate in the blood and settle into membranes at various sites.

The Arthus Reaction

The Arthus reaction is usually an acute response to a second injection of drugs or vaccines (boosters) at the same site as the first injection. In a few hours, the area becomes red, hot to the touch, swollen, and very painful **(figure 14.11a)**. These symptoms are mainly due to the destruction of tissues in and around the blood vessels and the release of histamine from mast cells and basophils. Although the reaction is usually self-limiting and rapidly cleared, intravascular blood clotting can occasionally cause necrosis and loss of tissue.

Serum Sickness

Serum sickness was named for a condition that appeared in soldiers after repeated injections of horse serum to treat tetanus, though it is also caused by injections of animal hormones and drugs. The immune complexes enter the circulation; are carried throughout the body; and are eventually deposited in blood vessels of the kidney, heart, skin, and joints. The condition can become chronic, causing symptoms such as enlarged lymph nodes, rashes, painful joints, swelling, fever, and renal dysfunction **(figure 14.11b)**.

14.4 LEARNING OUTCOMES—Assess Your Progress

9. Identify commonalities and differences between type II and type III hypersensitivities.

14.5 Type IV Hypersensitivities: Cell-Mediated (Delayed) Reactions

The adverse immune responses we have covered so far are explained primarily by B-cell involvement and antibodies. Type IV hypersensitivity is different; it involves primarily the T-cell branch of the immune system. In general, type IV diseases result when T cells respond to antigens displayed on self tissues or transplanted foreign cells. Type IV immune dysfunction has traditionally been known as delayed hypersensitivity because the symptoms arise one to several days following the second contact with an antigen.

Infectious Allergy

A classic example of delayed-type hypersensitivity occurs when a person sensitized by tuberculosis infection is injected with an extract (tuberculin) of the bacterium *Mycobacterium tuberculosis*. The so-called tuberculin reaction is an acute skin inflammation at the injection site appearing within 24 to 48 hours. This has been the mainstay diagnostic tool for TB infection for decades, and still is. This form of hypersensitivity arises from time-consuming cellular events involving a specific class of T cells ($T_H 1$) and their release of cytokines that recruit various inflammatory cells such as macrophages, neutrophils, and eosinophils. The buildup of fluid and cells at the site gives rise to a red bump **(figure 14.12a)**.

Contact Dermatitis

The most common delayed allergic reaction, contact dermatitis, is caused by exposure to resins in poison ivy or poison oak, to simple haptens in household and personal articles (jewelry, cosmetics, elasticized undergarments), and to certain drugs. Like immediate atopic dermatitis, the reaction to these allergens requires a sensitizing dose followed by a provocative dose. The allergen first penetrates the outer skin layers, is processed by Langerhans cells (skin dendritic cells), and presented to T cells. When subsequent exposures attract lymphocytes and macrophages to this area, these cells release enzymes and inflammatory cytokines that damage the epidermis in the immediate vicinity. This response accounts for the intensely itchy papules and blisters that are the early symptoms **(figure 14.12b)**. As healing progresses, the epidermis is replaced by a thick, keratinized layer.

T Cells and Their Role in Organ Transplantation

Transplantation or grafting of organs and tissues is a common medical procedure. Although it is life-saving, this technique is plagued by the natural tendency of lymphocytes to seek out foreign antigens and mount a campaign to destroy them. The bulk of the damage that occurs in graft rejections can be attributed to cytotoxic T-cell action.

The Genetic and Biochemical Basis for Graft Rejection

In chapter 13, we learned that the genes and markers in major histocompatibility (MHC or HLA) classes I and II are extremely important in recognizing self and in regulating the immune response. Although the cells of each person can exhibit variability in the pattern of these cell surface molecules, the pattern is identical in different cells of the same person. Similarity is seen among related siblings and parents, but the more distant the relationship, the less likely that the MHC genes and markers will be alike. When donor tissue (a graft) displays surface molecules of a different MHC class, the T cells of the recipient (called the host) will recognize its foreignness and react against it.

T-Cell-Mediated Recognition of Foreign MHC Receptors

Host Rejection of Graft   When the cytotoxic T cells of a host recognize foreign class I MHC markers on the surface of grafted cells, they release interleukin-2 as part of a general immune mobilization **(figure 14.13)**. Antigen-specific helper and cytotoxic T cells bind to the grafted tissue and secrete lymphokines that begin the rejection process within 2 weeks of transplantation. Antibodies formed against the transplanted tissue contribute to the damage, resulting in the destruction of the vascular supply and death of the graft.

Graft Rejection of Host   Graft incompatibility is a two-way phenomenon. Some grafted tissues (especially bone marrow) contain an indigenous population

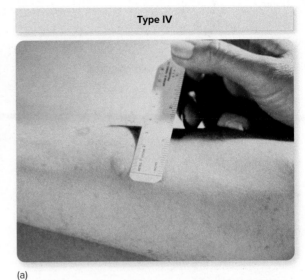

Type IV

(a)

(b)

Figure 14.12 Type IV delayed reactions.
(a) Positive tuberculin test. Intradermal injection of tuberculin extract in a person sensitized to tuberculosis yields a slightly raised red bump greater than 10 mm in diameter. **(b)** Contact dermatitis from poison oak, showing various stages of involvement: blister, scales, and thickened patches.
Source: CDC (a); carroteater/Shutterstock.com (b)

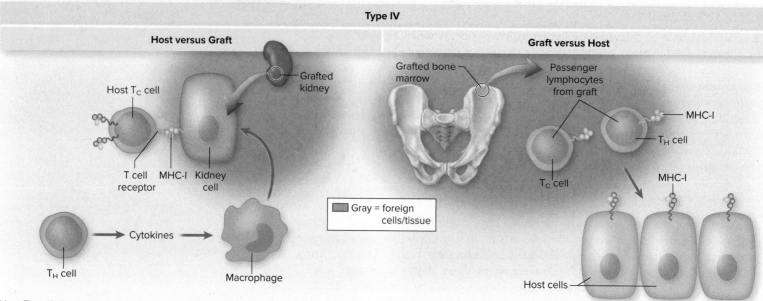

Type IV

Host versus Graft

Host T_C cell

Grafted
kidney

T cell
receptor

MHC-I

Kidney
cell

T_H cell

Cytokines

Macrophage

Gray = foreign
cells/tissue

Host T_C cells (and macrophages recruited by T_H cells to assist) attack grafted cells with foreign MHC-I markers.

Graft versus Host

Grafted bone
marrow

Passenger
lymphocytes
from graft

MHC-I

T_H cell

T_C cell

MHC-I

Host cells

Passenger lymphocytes from grafted tissue have donor MHC-I markers; attack recipient cells with different MHC-I specificity.

Figure 14.13 **Development of incompatible tissue graft reactions.**

called passenger lymphocytes (figure 14.13). This makes it quite possible for the graft to reject the host, causing **graft versus host disease (GVHD).** Because any host tissue bearing MHC markers foreign to the graft can be attacked, the effects of GVHD are widely systemic and toxic. A papular, peeling skin rash is the most common symptom, though other organs are also affected. GVHD typically occurs within 100 to 300 days of the graft.

Classes of Grafts

Grafts are generally classified according to the genetic relationship between the donor and the recipient **(figure 14.14).** Tissue transplanted from one site on an individual's body to another site on his or her body is known as an **autograft.** Typical examples are skin replacement in burn repair and the use of a vein to fashion a coronary artery bypass. In an **isograft,** tissue from an identical twin is used. Because isografts do not contain foreign antigens, they are not rejected. **Allografts,** the most common type of grafts, are exchanges between genetically different individuals belonging to the same species (two humans). A close genetic correlation is sought for most allograft transplants (see next section). A **xenograft** is a tissue exchange between individuals of different species.

Types of Transplants

Over 30,000 people receive transplants each year in the United States, while over 121,000 people are on a waiting list for one. Transplantation can be performed with every major organ, including parts of the brain and even faces. But it most often involves the kidney, liver, heart, skin, coronary artery, cornea, and bone marrow. Kidney transplants are by far the most common in the United States. The sources of organs and tissues are live donors, the recently deceased, and fetal tissues.

In the past decade, advancements in transplantation science have expanded the possibilities for treatment and survival. Fetal tissues have been used in the treatment of diabetes and Parkinson disease, while parents have successfully donated portions of their organs to help save their children suffering from the effects of cystic fibrosis or liver disease. Recent advances in stem cell technology have made it

Identical twins are the only people with exactly matching tissue antigens.
©Rubberball/Getty Images

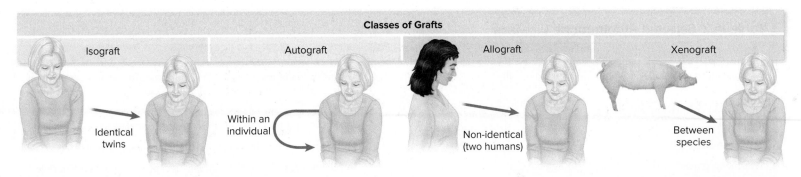

Figure 14.14 **Four classes of tissue grafts.**
iStockphoto/Getty Images

possible to isolate stem cells more efficiently from blood donors, and the use of umbilical cord blood cells has furthered progress in this area of science. Though many hurdles still exist, scientists are using genetic engineering technology to develop an ample supply of immunologically compatible, safe tissues for xenotransplantation.

Bone marrow transplantation is used in patients with immune deficiencies, aplastic anemia, leukemia and other cancers, and radiation damage. Before closely matched donor marrow can be infused, the patient is pretreated with chemotherapy and whole-body irradiation to destroy the person's own blood stem cells, preventing rejection of the new marrow cells. The donor marrow cells are then dripped intravenously into the circulatory system, and the new cells settle automatically into the appropriate bone marrow regions. However, because donor lymphoid cells can still cause GVHD, antirejection drugs may be necessary. Interestingly, after transplantation, a recipient's blood type may change to the blood type of the donor.

> **14.5 LEARNING OUTCOMES—Assess Your Progress**
>
> **10.** Describe one example of a type IV delayed hypersensitivity reaction.
> **11.** List four classes of grafts, and explain how host versus graft and graft versus host diseases develop.

14.6 An Inappropriate Response to Self: Autoimmunity

The immune diseases we have covered so far are all caused by foreign antigens. In the case of autoimmunity, individuals actually develop hypersensitivity to themselves. This pathologic process accounts for **autoimmune diseases,** in which **autoantibodies,** T cells, and, in some cases, both, mount an abnormal attack against self antigens. The scope of autoimmune disease includes *systemic,* involving several major organs, or *organ-specific* reactions, involving only one organ or tissue. There are more than 80 different autoimmune diseases, together affecting up to 23 million Americans. Many of these diseases result in catastrophic consequences. For example, people with rheumatoid arthritis have a 60% increased risk of death from cardiovascular disease. Some major diseases, their targets, and basic pathology are presented in **table 14.2.**

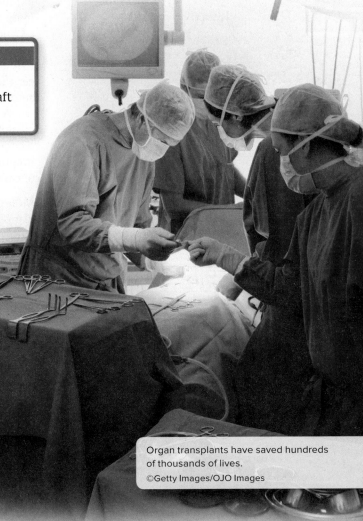

Organ transplants have saved hundreds of thousands of lives.
©Getty Images/OJO Images

Table 14.2 Selected Autoimmune Diseases

Disease	Target	Type of Hypersensitivity	Characteristics
Systemic lupus erythematosus (SLE)	Systemic	III	Inflammation of many organs; antibodies against red and white blood cells, platelets, clotting factors, nucleus DNA
Rheumatoid arthritis and ankylosing spondylitis	Systemic	II, III, and IV	Vasculitis; frequent target is joint lining; antibodies against other antibodies (rheumatoid factor), T-cell cytokine damage
Graves' disease	Thyroid	II	Antibodies against thyroid-stimulating hormone receptors
Myasthenia gravis	Muscle	II	Antibodies against the acetylcholine receptors on the nerve-muscle junction alter function
Type 1 diabetes	Pancreas	IV	T cells attack insulin-producing cells
Multiple sclerosis	Myelin	II and IV	T cells and antibodies sensitized to myelin sheath destroy neurons

Possible Causes of Autoimmune Disease

The root cause of autoimmune disease is still a frustrating mystery for the most part. We present some factors that influence—or may influence—the development of autoimmunity.

Genetics   Susceptibility can be influenced strongly by genetics and gender. Cases cluster in families, and even unaffected members tend to develop the autoantibodies for that disease. Studies show particular genes in the class I and II major histocompatibility complex coincide with certain autoimmune diseases. For example, autoimmune joint diseases such as rheumatoid arthritis and ankylosing spondylitis are more common in persons with the B-27 HLA type. With the expansion of genomic technology and the screening of whole genomes, many novel genes have recently been found to play a role in the pathway to autoimmunity. Some research suggests that autoimmune symptoms are a side effect of the body's immune response to cancer that is developing in the body (and often does not come to fruition, though the autoimmune disease remains).

A moderate, regulated amount of autoimmunity is probably required to dispose of old cells and cellular debris. Disease apparently arises when this regulatory or recognition apparatus goes awry. Sometimes the processes go awry due to genetic irregularities or inherent errors in the host's physiological processes. In a large subset of cases, however, microbes are behind the malfunctioning.

Molecular Mimicry   This is a process in which microbial antigens bearing molecular determinants similar to human cells induce the formation of antibodies that can cross-react with normal tissues. This is one possible explanation for the pathology of rheumatic fever. Similarly, T cells primed to react with streptococcal surface proteins also appear to react with keratin cells in the skin, causing them to proliferate. For this reason, psoriasis patients often report flare-ups after a strep throat infection.

Infection   Autoimmune disorders such as type 1 diabetes and multiple sclerosis are possibly triggered by *viral infection*. Viruses can noticeably alter cell receptors, thereby causing immune cells to attack the tissues bearing viral receptors.

The Gut Microbiome   Research has shown us the importance of the gut microbiome. We have also seen that the composition of the gut microbiome has changed over the last 50 years, due to the use of antibiotics, less exposure to the outdoors, and an altered, more artificial, diet. Many scientists connect the rise in autoimmune diseases over the same time period to this phenomenon. One direction these studies are going is the idea that a healthy microbiome is fundamentally important to training our immune system what to react against and what to tolerate. Drastic changes in the microbiome would be expected to disrupt this training process.

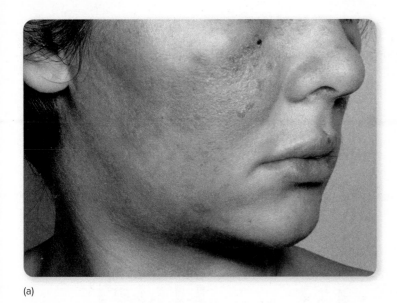

(a)

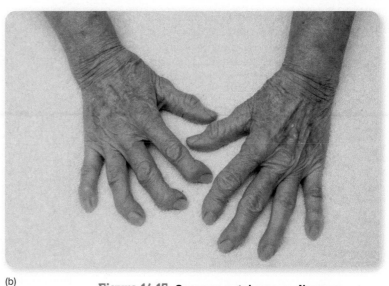

(b)

Figure 14.15 Common autoimmune diseases.
(a) Systemic lupus erythematosus. One symptom is a prominent rash across the bridge of the nose and on the cheeks. These papules and blotches can also occur on the chest and limbs. **(b)** Rheumatoid arthritis commonly targets the synovial membrane of joints. Over time, chronic inflammation causes thickening of this membrane, erosion of the articular cartilage, and fusion of the joint. These effects severely limit motion and can eventually swell and distort the joints.
ISM/Phototake (a); Aaron Roeth Photography (b)

Many researchers predict that altering the microbiome, or using chemicals from a healthy microbiome, will be a major treatment strategy for autoimmune diseases within 10 to 15 years.

Examples of Autoimmune Disease

Systemic Autoimmunities

One of the most severe chronic autoimmune diseases is systemic lupus erythematosus (SLE, or lupus). This name originated from the characteristic butterfly-shaped rash that drapes across the nose and cheeks **(figure 14.15a),** as ancient physicians thought the rash resembled a wolf bite on the face (*lupus* is Latin for "wolf"). Although the manifestations of the disease vary considerably, all SLE patients produce autoantibodies against a variety of targets, including organs and tissues or intracellular materials, such as the nucleoprotein of the nucleus and mitochondria. It is not known how such a generalized loss of self-tolerance arises, though viral infection and loss of normal immune response suppression are suspected.

Rheumatoid arthritis, another systemic autoimmune disease, causes progressive, debilitating damage to the joints and at times to the lungs, eyes, skin, and nervous system **(figure 14.15b).** In the joint form of the disease, autoantibodies form immune complexes that bind to the synovial membrane of the joints, which activates phagocytes to release cytokines. Chronic inflammation develops and leads to scar tissue and joint destruction. These cytokines (i.e., tumor necrosis factor [TNF]) can then trigger additional type IV delayed hypersensitivity responses. An IgM antibody, called rheumatoid factor (RF), is present in some, but not all, cases. Treatment has recently involved the targeting of TNF or TNF-mediated pathways, but new drugs, targeting other immune system components, are now appearing. This is a different disease than osteoarthritis, which is caused by the wearing down of cartilage between bones and is a more or less natural consequence of aging.

Autoimmunities of the Endocrine Glands

The underlying cause of **Graves' disease** is the attachment of autoantibodies to receptors on the thyroxin-secreting follicle cells of the thyroid gland. The abnormal stimulation of these cells causes the overproduction of this hormone and the symptoms of hyperthyroidism, which affect nearly every body system.

Type 1 diabetes is another condition that may be a result of autoimmunity. Insulin, secreted by the beta cells in the pancreas, regulates the utilization of glucose by cells. Molecular mimicry has been implicated in the sensitization of cytotoxic

Figure 14.16 Mechanism for involvement of autoantibodies in myasthenia gravis. Antibodies developed against receptors on the postsynaptic membrane block them so that acetylcholine cannot bind and muscle contraction is inhibited.

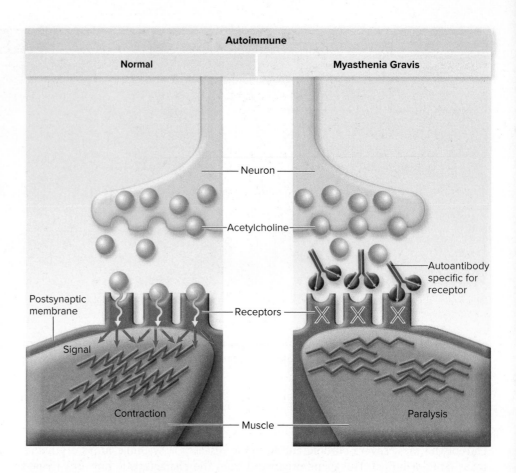

©Realistic Reflections

T cells in type 1 diabetes, which leads to the lysis of beta cells. The reduced amount of insulin underlies the symptoms of this disease.

Neuromuscular Autoimmunities

Myasthenia gravis is a syndrome caused by autoantibodies binding to the receptors for acetylcholine, a chemical required to transmit a nerve impulse across the synaptic junction to a muscle **(figure 14.16).** The first effects are usually felt in the muscles of the eyes and throat, but the syndrome eventually progresses to complete loss of skeletal muscle function and death. Current treatment usually includes immunosuppressive drugs and therapy to remove the auto-antibodies from the circulation.

Multiple sclerosis (MS) is a paralyzing neuromuscular disease associated with lesions in the insulating myelin sheath that surrounds neurons in the white matter of the central nervous system. T-cell and autoantibody-induced damage severely compromise the capacity of neurons to send impulses, resulting in muscular weakness and tremors, difficulties in speech and vision, and paralysis. Data suggest a possible association between infection with human herpesvirus 6 and the onset of disease. Immunosuppressants like cortisone and interferon beta alleviate symptoms, and the disease can be treated with monoclonal antibody therapy toward certain T-cell antigens.

14.6 LEARNING OUTCOMES—Assess Your Progress

12. List at least three autoimmune diseases and the most important immunologic features in them.

Table 14.3 General Categories of Immunodeficiency Diseases with Selected Examples

Primary Immune Deficiencies (Genetic)	Secondary Immune Deficiencies (Acquired)
B-cell defects (low levels of B cells and antibodies) Agammaglobulinemia (X-linked, non-sex-linked) Hypogammaglobulinemia Selective immunoglobulin deficiencies	**From natural causes** Infections (AIDS) or cancers Nutrition deficiencies Stress Pregnancy Aging
T-cell defects (lack of all classes of T cells) Thymic aplasia (DiGeorge syndrome)	**From immunosuppressive agents** Irradiation Severe burns Steroids Immunosuppressive drugs Removal of spleen
Combined B-cell and T-cell defects (usually caused by lack or abnormality of lymphoid stem cell) Severe combined immunodeficiency (SCID) disease Adenosine deaminase (ADA) deficiency	
Complement defects Lacking one of C components Hereditary angioedema associated with rheumatoid diseases	

14.7 Immunodeficiency Diseases: Hyposensitivities

Occasionally, errors occur in the development of the immune system, and a person is born with or develops weakened immune responses, called **immunodeficiencies.** The most obvious consequences of immunodeficiencies are recurrent, overwhelming infections, often with opportunistic microbes. Immunodeficiencies fall into two general categories: *primary diseases*, present at birth (congenital) and usually stemming from genetic errors, and *secondary diseases*, acquired after birth and caused by natural or artificial agents **(table 14.3).**

Primary Immunodeficiency Diseases

Primary deficiencies affect both adaptive defenses and innate ones such as phagocytosis. Consult **figure 14.17** to survey the places in the normal sequential development of lymphocytes, where defects can occur, and the possible consequences. In many cases, the deficiency is due to an inherited abnormality, though the exact nature of the abnormality is not known for a number of diseases. In some deficiencies, the lymphocyte in question is completely absent or is present at very low levels. In other cases, lymphocytes are present but do not function normally. Because the development of B cells and T cells diverges at some point, an individual can lack one or both cell lines. It must be emphasized, however, that some deficiencies affect other cell functions as well.

Clinical Deficiencies in B-Cell Development or Expression

Genetic deficiencies in B cells usually result in abnormal immunoglobulin expression. In some instances, only certain immunoglobulin classes are absent; in others, the levels of all types of immunoglobulin (Ig) are reduced.

The term **agammaglobulinemia** literally means the absence of gamma globulin, the component of serum that contains immunoglobulins. Because it is very rare for Ig to be completely absent, some physicians prefer the term **hypogammaglobulinemia.** The symptoms of recurrent, serious bacterial infections usually appear about 6 months after birth. The most common infection sites are the lungs, sinuses, meninges, and blood. Many Ig-deficient patients can have recurrent infections with viruses and protozoa, as well. The current treatment for this condition is passive immunotherapy with immune serum globulin.

Medical Moment

Hand Washing

The importance of hand washing cannot be overemphasized, particularly when caring for patients who have compromised immune systems. Such a simple step can be lifesaving for patients who are unable to fight off infection. Recognizing this fact, hospitals have made a conscious and concentrated effort to increase the number of hand washing stations available to increase compliance among health care workers.

Alcohol-based hand sanitizers have provided an additional option of hand antisepsis. Hand sanitizer should be applied liberally to the palm of one hand. The hands should then be rubbed together, dispersing the sanitizer over the surfaces of both hands, including the fingers, nails, and wrists. The longer you can leave the sanitizer on your hands (to evaporate), the more effective it will be. When hands are visibly soiled, clean running water and soap should be used.

Q. Using hand sanitizers is considered by many to be more effective than hand washing (except when hands are visibly soiled). Can you think of a reason why that might be?

Answer in Appendix B.

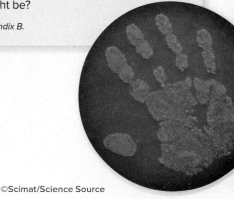

©Scimat/Science Source

Primary Immunodeficiency

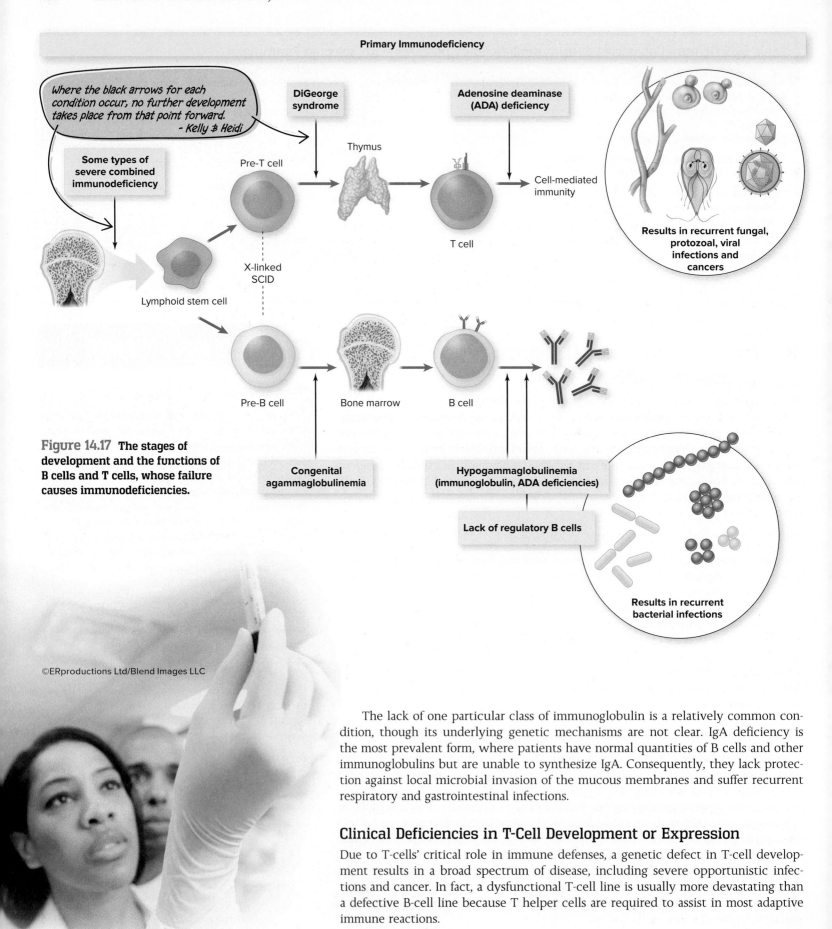

Where the black arrows for each condition occur, no further development takes place from that point forward.
- Kelly & Heidi

Some types of severe combined immunodeficiency

Lymphoid stem cell

Pre-T cell

DiGeorge syndrome

Thymus

T cell

X-linked SCID

Adenosine deaminase (ADA) deficiency

Cell-mediated immunity

Results in recurrent fungal, protozoal, viral infections and cancers

Pre-B cell

Bone marrow

B cell

Congenital agammaglobulinemia

Hypogammaglobulinemia (immunoglobulin, ADA deficiencies)

Lack of regulatory B cells

Results in recurrent bacterial infections

Figure 14.17 **The stages of development and the functions of B cells and T cells, whose failure causes immunodeficiencies.**

©ERproductions Ltd/Blend Images LLC

The lack of one particular class of immunoglobulin is a relatively common condition, though its underlying genetic mechanisms are not clear. IgA deficiency is the most prevalent form, where patients have normal quantities of B cells and other immunoglobulins but are unable to synthesize IgA. Consequently, they lack protection against local microbial invasion of the mucous membranes and suffer recurrent respiratory and gastrointestinal infections.

Clinical Deficiencies in T-Cell Development or Expression

Due to T-cells' critical role in immune defenses, a genetic defect in T-cell development results in a broad spectrum of disease, including severe opportunistic infections and cancer. In fact, a dysfunctional T-cell line is usually more devastating than a defective B-cell line because T helper cells are required to assist in most adaptive immune reactions.

Abnormal Development of the Thymus   The most severe of the T-cell deficiencies involves the congenital absence or immaturity of the thymus. Thymic aplasia, or **DiGeorge syndrome,** makes children highly susceptible to persistent infections by fungi, protozoa, and viruses. Vaccinations using live, attenuated microbes pose a danger, and common childhood infections such as chickenpox can be overwhelming and fatal in these children. Patients typically have reduced antibody levels as well.

Severe Combined Immunodeficiencies: Dysfunction in B and T Cells

Severe combined immunodeficiencies (SCIDs) are the most serious and potentially lethal forms of immunodeficiency disease because they involve dysfunction in both lymphocyte systems. Some SCIDs are due to the complete absence of the lymphocyte stem cell in the marrow; others are attributable to the dysfunction of B cells and T cells later in development. Infants with SCID usually manifest the T-cell deficiencies within days after birth by developing candidiasis, sepsis, pneumonia, or systemic viral infections.

In the two most common forms, Swiss-type agammaglobulinemia and thymic alymphoplasia, genetic defects in the development of the lymphoid cell line result in extremely low numbers of all lymphocyte types and poorly developed humoral and cellular immunity. A rare form of SCID, called **adenosine deaminase (ADA) deficiency,** is caused by an autosomal recessive defect in the metabolism of adenosine. In this case, lymphocytes develop but a metabolic product builds up abnormally and selectively destroys them. A small number of SCID cases are due to a developmental defect in receptors on B and T cells.

Because of their profound lack of specific adaptive immunities, SCID children require the most rigorous kinds of aseptic techniques to protect them from opportunistic infections. Aside from life in a sterile plastic bubble, exemplified by David Vetter, remembered as the "Boy in the Bubble" **(figure 14.18a),** the only serious option for their longtime survival is total replacement or correction of dysfunctional lymphoid cells. Some infants can benefit from fetal liver or stem cell grafts, though transplantation is complicated by graft versus host disease. A more lasting treatment for both X-linked and ADA types of SCID is gene therapy–insertion of normal genes to replace the defective genes (see chapter 8).

Secondary Immunodeficiency Diseases

Secondary acquired deficiencies in B cells and T cells are caused by one of four agents:

1. infection,
2. noninfectious metabolic disease,
3. chemotherapy, or
4. radiation.

One well-recognized infection-induced immunodeficiency is **AIDS** (see chapter 18). This syndrome is caused when several types of immune cells, including T helper cells, monocytes, macrophages, and antigen-presenting cells, are infected by the human immunodeficiency virus (HIV). The depletion of T helper cells and functional impairment of immune responses ultimately account for the cancers and opportunistic infections associated with this disease **(figure 14.18b).** Other infections that can deplete immunities are measles, leprosy, and malaria.

Cancers that target the bone marrow or lymphoid organs can be responsible for extreme malfunction of both humoral and cellular immunity. In leukemia, cancer cells outnumber normal cells, displacing them from bone marrow and blood. Plasma cell tumors produce large amounts of nonfunctional antibodies, while thymus tumors cause severe T-cell deficiencies.

Primary Immunodeficiency

(a)

Secondary Immunodeficiency

(b)

Figure 14.18 The two types of immuno-deficiency. (a) David Vetter, who lived from 1971 to his death in 1984 in a sterile bubble. (b) An AIDS patient.

Courtesy Baylor College of Medicine, Public Affairs (a); Christopher Kerrigan/McGraw-Hill Education (b)

©Comstock/Alamy

The ribbon for AIDS awareness.

An ironic outcome of lifesaving medical procedures is the possible suppression of a patient's immune system. Drugs that prevent graft rejection or decrease the symptoms of rheumatoid arthritis can likewise suppress beneficial immune responses, while radiation and anticancer drugs are damaging to the bone marrow and other body cells.

Some cancers are able to suppress the immune response, which allows the malignant cells to proliferate unchecked. Recent advances in cancer treatments sometimes involve reversing this suppression so that the immune system can do its job—destroying cells that have cancer markers (antigens) on them.

14.7 LEARNING OUTCOMES—Assess Your Progress

13. Distinguish between primary and secondary immunodeficiencies, explaining how each develops.

CASE FILE WRAP-UP

Rick Brady/McGraw-Hill Education (nurse); Steve Gschmeissner/Science Source (human blood cells, SEM)

Tumor necrosis factor (TNF) is found in higher amounts in the synovial fluid (the fluid that bathes the joints) of patients with rheumatoid arthritis (RA). TNF inhibitors such as Enbrel (etanercept) have been shown to reduce inflammation in RA patients.

Etanercept is a pharmaceutical that is composed of a genetically engineered receptor for human tumor necrosis factor, fused to the Fc portion of the immunoglobulin IgG. How does it work? The circulating TNF in the body binds to the drug instead of to the natural receptor in the body. This reduces the effective concentration of TNF acting in the body and reduces symptoms due to TNF.

The patient in this case had a past medical history of gastric ulcers and diabetes, which would make it risky for her to take NSAIDs (nonsteroidal anti-inflammatory drugs, such as ibuprofen) or many of the other drugs that are often the initial treatment for RA. Steroids may have been effective, but steroids can alter blood glucose. The patient was screened for hepatitis and tuberculosis because patients taking Enbrel may become more susceptible to infection. Patients who are taking etanercept must be cautioned to report signs of infection and stop taking the drug when they have an infection.

Asthma and the Airway—and Gut—Microbiome

Asthma is considered a form of allergy, an inappropriate antibody response to epitopes that should not provoke a reaction. It affects more than 25 million people in the United States, 7 million of them children under the age of 18. Scientists are not sure what causes it. They have found factors associated with it, but these can be as varied as poor living conditions, exposure to roaches, and even too *little* exposure to microbes.

Current research is showing that a disrupted microbiome in the airway is nearly always found in people who have asthma. This is surely not unexpected. But the gut microbiome also seems to be important. That gut microbiome is showing up everywhere! Scientists have been trying to tell us for years that the gut is almost like a second nervous system, and this asthma research bears that out.

Researchers looked at 319 babies in Canada and found a subgroup of 22 with a high risk of asthma. These babies were missing four types of bacteria that were present in healthy babies. The researchers concluded that the lack of these four bacterial groups put babies at a greatly increased risk for asthma. Importantly, they were able to determine the sequence of events because they sampled the babies before any evidence of disease. If they had sampled them at the same time they categorized them as having asthma or not, it would be impossible to determine if the lack of bacteria led to asthma, or if asthma somehow caused the absence of these bacteria. They have not yet teased out why the babies were missing those bacteria. They are investigating exposures such as cesarean birth, early antibiotic treatment, and breast-feeding.

Not only does the gut microbiome regulate and influence digestive functions, but it also is important in training the immune system to react appropriately throughout the body. Because the gut is the portal of entry for so many environmental microbes (through eating and drinking), it seems logical that it would be the "training center" for the body's defenses, saying: "Don't react to this; it is just a harmless bacterium coming in with this bite of carrot."

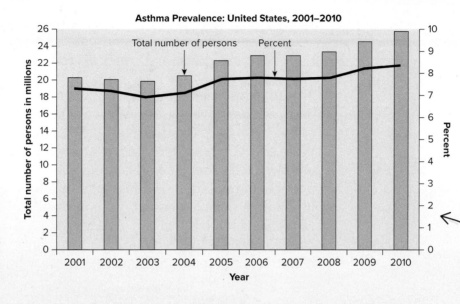

Asthma Prevalence: United States, 2001–2010

Total number of persons — Percent

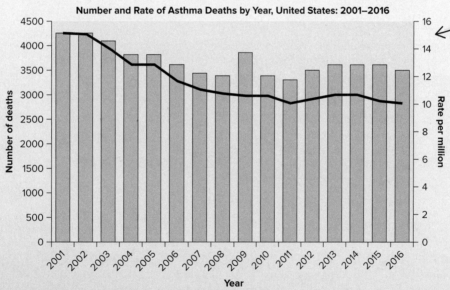

Number and Rate of Asthma Deaths by Year, United States: 2001–2016

Cases of asthma increased over this 10-year period, but deaths from asthma were decreasing over that period and beyond.
- Kelly & Heidi

435

Source: Shutterstock/Barbo

DISORDERS IN IMMUNITY

14.1 THE IMMUNE RESPONSE: A TWO-SIDED COIN

Immunopathology is the study of overactivity and underreactivity of the immune response. Overactivity is called hypersensitivity, of which there are 4 categories: type I (allergy and anaphylaxis), type II (complement, IgG- and IgM-mediated tissue destruction), type III (immune complex reactions), and type IV (delayed hypersensitivity reactions). Underactivity of the immune response can either be inherited or acquired.

Ian Hooton / Science Source

Avoiding the Allergen

Penicillin allergy noted in chart

McGraw-Hill Education

Antigens that trigger allergic reactions are called allergens. They either originate outside the body (exogenous) or are caused by the host's own tissues (endogenous). They can either be relatively mild (atopy) or life-threatening (anaphylaxis).

TYPE I ALLERGIC REACTIONS: ATOPY AND ANAPHYLAXIS 14.2

14.3 TYPE II HYPERSENSITIVITIES: REACTIONS THAT LYSE FOREIGN CELLS

These are complement-assisted reactions that occur when preformed antibodies (IgG or IgM) react with foreign cell-bound antigens, leading to complex formation, membrane attack, and lysis. Hemolytic disease of the newborn is a prime example of this reaction.

Image Source/Getty Images

Large quantities of antigen react with host antibody to form insoluble immune complexes that settle in tissue cell membranes, causing destructive inflammation. One main difference between type II and type III hypersensitivities is that type II involves soluble (not membrane-bound) antigens.

TYPE III HYPERSENSITIVITIES: IMMUNE COMPLEX REACTIONS 14.4

These occur when cytotoxic T cells attack either transplanted foreign cells or tissues or self antigens. Examples are rejection of transplanted tissue and the tuberculin reaction.

Rubberball/Getty Images

TYPE IV HYPERSENSITIVITIES: CELL-MEDIATED (DELAYED) REACTIONS 14.5

14.6 AN INAPPROPRIATE RESPONSE TO SELF: AUTOIMMUNITY

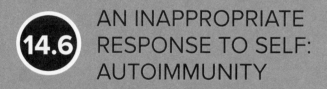

Autoimmunity reactions occur when host antibodies or host T cells mount an abnormal attack against self antigens. This can result in diseases such as rheumatoid arthritis, lupus, and multiple sclerosis.

Immunodeficiencies occur when the immune response is reduced or absent. Primary immunodeficiencies are genetically induced deficiencies of B cells, T cells, the thymus gland, or combinations of these. Secondary immune diseases are caused by infection (i.e., AIDS), other conditions including pregnancy, chemotherapy, and some malignancies.

Comstock/Alamy

IMMUNODEFICIENCY DISEASE: HYPOSENSITIVITIES 14.7

SmartGrid: From Knowledge to Critical Thinking

This *21 Question Grid* takes the topics from this chapter and arranges them with respect to the American Society for Microbiology's Undergraduate Curriculum guidelines—all six of the important "Concepts" as well as the important "Competency" of scientific literacy. Three questions are supplied about chapter content that refer to the Concept or Competency, in increasing levels of Bloom's taxonomy for learning.

ASM Concept/Competency	A. Bloom's Level 1, 2—Remember and Understand (Choose one.)	B. Bloom's Level 3, 4—Apply and Analyze	C. Bloom's Level 5, 6—Evaluate and Create

Evolution	1. Allergy and atopy might have evolved in human populations that have had a "sudden" decrease in
	  a. vaccinations.
	  b. exposure to vitamin D.
	  c. exposure to helminth infections.
	  d. intake of processed sugar.
		2. Can you think of a reason why humans have evolved the ability to mount an immune response to their own fetuses if they have a particular set of foreign antigens (as in the case of erythroblastosis fetalis)?	3. Do you suppose that if we continue to transplant allografts into human patients for thousands of years that the immune system will evolve to not react to those foreign antigens? Provide support for your argument.

Cell Structure and Function	4. Molecular mimicry is when _____ on host cells resemble _____ on pathogens, causing the host to mount an immune response to host tissues. (Same word for both blanks.)
	  a. antibodies
	  b. markers
	  c. antigens
	  d. two of the above
		5. What feature of antibodies makes them particularly suited for forming complexes with antigen, as in type III hypersensitivities?	6. An autoimmune disease called antiphospholipid syndrome manifests as thrombosis, or the formation of blood clots. It is characterized by the presence of antibodies directed against phospholipids. Connect the sign (elevated antibodies) to the symptom (thrombosis).

Metabolic Pathways	7. Which substance is most likely to be allergenic?
	  a. protein
	  b. fat
	  c. carbohydrate
	  d. nucleic acid
		8. One of the strategies for preventing type I hypersensitivity is to induce plasma cells to secrete IgG instead of IgE. Why would that block allergy symptoms?	9. Epi-pens containing epinephrine are devices used to instantly inject a person with epinephrine in the event of an anaphylactic exposure. What does epinephrine do to stop anaphylaxis? Predict what would happen if too much epinephrine was administered.

Information Flow and Genetics	10. What percent of the human population is positive for the Rh antigen?
	  a. 50%
	  b. 65%
	  c. 85%
	  d. 99%
		11. Steroids administered to stop an allergic response act in several ways. One of them is to shut down the synthesis of IgE by plasma cells. Identify some intracellular host macromolecules that may be bound (obstructed) by steroid molecules to result in this outcome.	12. A very small number of people have a condition called vibratory urticaria. In this condition, physical vibrations (bumpy bus rides, jackhammers, running) can cause skin rashes. Scientists have found a genetic mutation in a single gene for a protein that stabilizes the surface of immune cells. You take it from here: What would a mutation in that protein possibly lead to?

ASM Concept/ Competency	A. Bloom's Level 1, 2—Remember and Understand (Choose one.)	B. Bloom's Level 3, 4—Apply and Analyze	C. Bloom's Level 5, 6—Evaluate and Create

Microbial Systems	13. The hygiene hypothesis suggests that
	  a. there are still too many microorganisms in the environment.
	  b. we may need more contact with antimicrobials during our maturation.
	  c. there are not enough microbes on farms.
	  d. we may need more contact with microbes during our maturation.
		14. Probiotic supplements are sometimes promoted to alleviate the symptoms of hypersensitivity diseases such as rheumatoid arthritis and eczema. Trace the path to biological plausibility for that claim, if any.	15. Recent research shows that asthmatic patients have a much higher load of bacteria in their lower airways than do healthy subjects. How could that contribute to what we see in the pathology of asthma?

Impact of Microorganisms	16. Which statement is true of autoimmunity?
	  a. It involves misshapen antibodies.
	  b. It refers to "automatic immunity."
	  c. It often manifests as types II, III, and IV hypersensitivities.
	  d. It has an acute course and then usually resolves itself.
		17. It is often said that the normal microbiota of a human "trains" the immune system to react to foreign antigens and not react to self antigens. Use what you know about clonal deletion and clonal selection to suggest how this might happen.	18. Investigate the link between *Streptococcus pyogenes* infection and some types of obsessive-compulsive disorder. Identify similarities with some concepts in this chapter.

Scientific Thinking	19. Which disease would be most similar to AIDS in its pathology?
	  a. X-linked agammaglobulinemia
	  b. SCID
	  c. ADA deficiency
	  d. myasthenia gravis
		20. Do you think people with B-cell deficiencies can be successfully vaccinated against various microbes? Why or why not?	21. In the Microbiome box on page 435 this sentence appears: If they had sampled them at the same time they categorized them as having asthma or not, it would be impossible to determine if the lack of bacteria led to asthma, or if asthma somehow caused the absence of these bacteria. Explain why it would be impossible to determine.

Answers to the multiple-choice questions appear in Appendix A.

Visual Connections

This question connects content within and between chapters.

As you learned in **chapter 13,** B and T cells originate in tissues outside of the lymphatic system. With this in mind, provide at least one example of how an abnormality in these areas (i.e., in the bone marrow or thymus) can lead to immune deficiency.

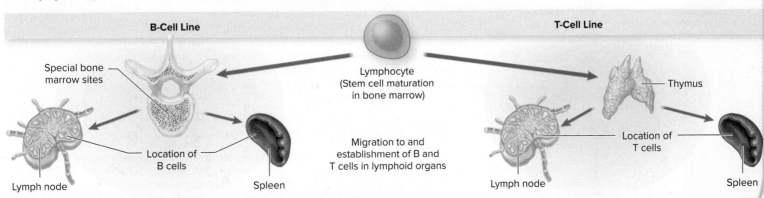

Chapter design elements: (Covid): CDC/Alissa Eckert, MS; Dan Higgins, MAMS; (Note): McGraw-Hill Education; (NCLEX): Shutterstock/ Abert; (Doctor): David Gould/Getty Images

15

Diagnosing Infections

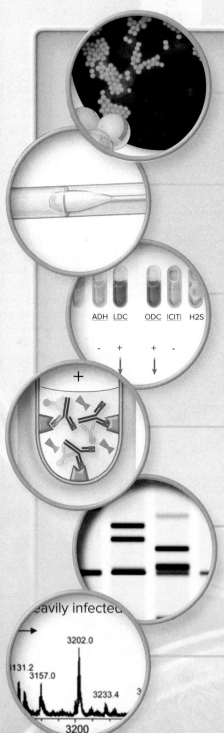

IN THIS CHAPTER...

15.1 What Is Causing This Condition?

1. List the three major categories of microbial identification techniques.
2. Provide a one-sentence description for each of these three categories.

15.2 First Steps: Specimen Collection

3. Identify factors that may affect the identification of an infectious agent from a patient sample.
4. Compare the types of tests performed on microbial isolates versus those performed on patients themselves.

15.3 Phenotypic Methods

5. List at least three different tests that fall in the direct identification category.
6. Explain the main principle behind biochemical testing, and identify an example of such tests.
7. Discuss two major drawbacks of phenotypic testing methods that require culturing the pathogen.

15.4 Immunologic Methods

8. Define the term *serology*, and explain the immunologic principle behind serological tests.
9. Identify two immunological diagnostic techniques that rely on a secondary antibody, and explain how they work.

15.5 Genotypic Methods

10. Explain why amplification techniques are useful for infectious disease diagnosis.
11. Name two examples of techniques that employ hybridization.
12. Explain how whole-genome sequencing can be used for diagnosis.
13. Identify the situations in which pulsed-field gel electrophoresis is most useful.

15.6 Additional Diagnostic Technologies

14. Describe the benefits of "lab on a chip" technologies for global public health.
15. Identify an advantage presented by mass spectrometry and also by imaging techniques as diagnostic tools.

CASE FILE

Tracing the Cause

When I was a lab tech student, my preceptor and I were asked to go to the emergency room to draw blood from a patient. The lab requisition filled out by the nurse caring for the patient stated the patient had "fever for 1 week with a decreased level of consciousness."

On reporting to the ER, we were directed to the patient, who was lying on a gurney in an exam room. The patient was a 78-year-old diabetic gentleman who appeared pale and thin. We introduced ourselves and told the patient that we were there to draw blood. The patient barely acknowledged our presence and seemed to have difficulty staying awake. The ER nurses had already collected a urine specimen from the patient's Foley catheter for urinalysis and culture. They handed us the specimens they had collected, which had been labeled correctly with the patient's name and the collection date and time.

We proceeded to draw blood according to the physician's order. We obtained blood for a complete blood count, electrolytes, blood glucose, liver function, and cardiac markers. We also obtained blood from two different sites for blood cultures. Once we had collected all of the samples, we notified the nursing staff that we were finished. We labeled all the collection vials with the necessary information and returned to the lab.

After running all of the tests that could be performed in-house, we sent copies of all of the reports to the ER. The patient's hemoglobin and hematocrit were slightly low. The white blood cell count was very high. The patient's potassium was high, and the sodium level was low. Liver function was normal, as were cardiac markers. The urinalysis showed ketones and protein in the urine, evidence of dehydration but not of infection. The patient's serum glucose was high but not dangerously so.

On the basis of these lab studies, the patient was rehydrated with intravenous fluids and his potassium and sodium levels were soon corrected. He was also given insulin to lower his blood glucose. The patient was started on broad-spectrum IV antibiotics pending the results of the urine and blood cultures. He continued to run a high fever and was only semiconscious following admission to the ICU.

The preliminary urine cultures failed to identify any bacteria in the patient's urine. When the blood culture results came back after 48 hours, they showed the presence of *Staphylococcus epidermidis* in the patient's blood. Sensitivity studies showed that the organism was sensitive to Penicillin G, rifampin, and vancomycin. It also revealed that the microorganism was resistant to the ceftriaxone that the patient was currently receiving. The patient was switched to Penicillin G every 6 hours intravenously, and he made a relatively rapid recovery soon thereafter, with his fever disappearing within 36 hours.

- Why are blood cultures collected from two different sites?
- Why was it important for the laboratory staff to notify the nursing staff after the samples had been collected?

Case File Wrap-Up appears at the end of the chapter.

Lisa Burgess/McGraw-Hill Education (Kirby-Bauer antibiotic testing); Image provided by AdvanDx, Inc. (*S. aureus* PNA FISH); Steven Puetzer/Getty Images (DNA sequencing gel, close-up full frame); McGraw-Hill Education (phlebotomist)

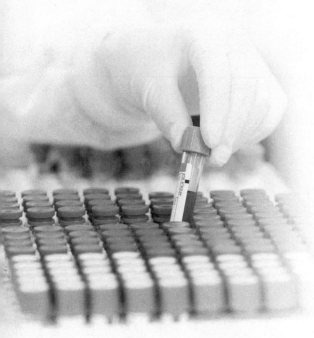

©Adam Gault/Getty Images

15.1 What Is Causing This Condition?

Chapters 16 through 21 cover the most clinically significant bacterial, fungal, parasitic, and viral diseases. Those chapters survey the most prevalent infectious conditions and the organisms that cause them. This chapter gets us started with an introduction to the how-to of diagnosing the infections.

For many students (and professionals), the most pressing topic in microbiology is *how to identify unknown bacteria* in patient specimens or in samples from nature. As seen in **table 15.1,** methods microbiologists use to identify bacteria to the level of genus and species fall into three main categories: *phenotypic,* which includes a consideration of morphology (microscopic and macroscopic) as well as bacterial physiology or biochemistry; *immunologic,* which entails serological analysis; and *genotypic* (or genetic) techniques. Data from a cross section of such tests can produce a unique

Table 15.1 Methods of Microbial Identification

Category	Description	Example
Phenotypic	Observation of microbe's microscopic and macroscopic morphology, physiology, and biochemical properties	
Immunologic	Analysis of microbe using antibodies, or of patients' antibodies using prepackaged antigens	Agglutinated mat Nonagglutinated pellet — Enlarged side view of wells
Genotypic	Analysis of microbe's DNA or RNA	

Arthur F. DiSalvo, M.D./CDC (phenotypic); Steven Puetzer/Getty Images (genotypic)

profile for any bacterium. Increasingly, genetic means of identification are being used as a sole resource for identifying bacteria. We are in the middle of a revolution in infectious disease diagnosis.

Sophisticated genetic techniques have become commonplace and affordable for hospitals and diagnostic labs to adopt on a wide-scale basis. Many of these techniques take less time than needed to actually culture microorganisms and therefore result in quicker results and better patient outcomes. There has also been an increase in what is called point-of-care diagnosis. These are tests that can be performed at the bedside, meaning that samples do not have to be "sent out" and patient care can continue without delay.

It should be noted that some other techniques can assist in diagnosis. For example, computerized tomography (CT) scans are often used to diagnose peritonsillar abscesses, after which the causative agent can be identified using the techniques described in this chapter. Magnetic resonance imaging (MRI) and positron emission tomography (PET) scans are also used to find areas of deep tissue infection.

Phenotypic Methods

You will remember that phenotype refers to the traits that an organism is expressing in the present. So, put simply, phenotypic methods of identifying microbes involve examining their appearance and their behavior. In this context, "behavior" is defined as what types of enzymatic activities it can carry out, what kind of physical conditions it thrives in, what antibiotics it is susceptible to, and the chemical composition of its walls and/or membranes.

©nicolas_/Getty Images

Immunologic Methods

As you learned in chapter 13, one immune response to antigens is the production of antibodies, which are designed to bind tightly to specific antigens. The nature of the antibody response is exploited for diagnostic purposes when a patient sample is tested for the presence of specific antibodies to a suspected pathogen (antigen). Alternatively, microbial antigens in the patient's tissues can be tested with antibodies "off the shelf." These immunological methods can be easier than trying to isolate the microbe itself. Laboratory kits based on this technique are available for immediate identification of a number of pathogens.

Genotypic Methods

Examining the genetic material itself (DNA and/or RNA) has revolutionized the identification and classification of bacteria. There are many advantages of genotypic methods over phenotypic methods, when they are available. The primary advantage is that culturing of the microorganisms is not always necessary. In recent decades, scientists have come to realize that there are many more microorganisms that we can't grow in the lab compared with those that we can. Numerous viable nonculturable (VNC) microbes are currently being identified in this manner, through studies such as the Human Microbiome Project. Another advantage is that genotypic methods are increasingly automated, producing rapid results that are often more precise than phenotypic methods.

Medical Moment

Should Be Obvious, But...

Most often the nurse is responsible for obtaining a specimen from a patient. Whenever there is any doubt on what or how to obtain a sample, nurses should consult with the medical laboratory staff on exactly what would be most useful. A recent article in a microbiology publication highlighted some best practices in communication between the floor staff and the lab staff, and included a funny anecdote about one such conversation.

A nurse called the lab about a child on her floor with a suspected enterovirus infection. This virus may be found in the respiratory tract, blood, cerebrospinal fluid, and the stool/rectum. The nurse was planning on taking a multisite swab, meaning one swab is used to sample more than one site, increasing the chances for finding microbes somewhere in the body. On this particular call, the nurse said she was going to swab the mouth and the rectum, and wanted to know if there was a preferred order in which she should do the swabbing? As the author of the article (the lab tech) states, "After a short silence, I suggested that she avoid placing the swab from the rectum into the patient's throat—and that it might be best to start from the top and work her way down." Source: Prinzi, A. 2019. *American Society for Microbiology.* Poop, Pus and Positive Results: Cultural Oddities from the Clinical Microbiology Lab. https://www.asm.org/Articles/2019/August/Poop,-Pus-and-Positive-Results-Cultural-Oddities-F.

Q. What can (nonpolio) enteroviruses cause in children?

Answer in Appendix B.

©Ingram/Media Bakery

15.2 First Steps: Specimen Collection

Regardless of the method of diagnosis, specimen collection is the common point that guides the health care decisions of every member of a clinical team. Indeed, the success of identification and treatment depends on how specimens are collected, handled, stored, and cultured. Specimens can be taken by a clinical laboratory scientist or medical technologist, nurse, physician, or even the patient. However, it is imperative that general aseptic procedures be used, including sterile sample containers and other tools to prevent contamination from the environment or the patient. **Figure 15.1** delineates the most common sampling sites and procedures.

In sites that normally contain resident microbiota, care should be taken to sample only the infected site and not the surrounding areas. Saliva is an especially undesirable contaminant because it contains millions of bacteria per milliliter, most of which are normal biota. Sputum, the mucus secretion that coats the lower respiratory surfaces, especially the lungs, is discharged by coughing or taken by a thin tube called a catheter to avoid contamination with saliva. In addition, throat and nasopharyngeal swabs should not touch the tongue, cheeks, or saliva. On occasion saliva samples are needed for dental diagnosis and are obtained by having the patient spit or drool into a container.

Urine is taken aseptically from the bladder with a catheter designed for that site. Another method, called a "clean catch," is taken by washing the external urethra and collecting the urine midstream. The latter method inevitably incorporates a few normal biota into the sample, but these can usually be differentiated from pathogens in an actual infection. Sometimes diagnostic techniques require first-voided "dirty catch" urine. The mucous lining of the urethra, vagina, or cervix can be sampled with a swab or applicator stick.

Depending on the nature of a skin lesion, skin can be swabbed or scraped with a scalpel to expose deeper layers. Wounds are sampled either by swabbing or by using a punch biopsy tool. Fluids such as blood, cerebrospinal fluid, and tissue fluids must be taken by sterile needle aspiration. Antisepsis of the puncture site is extremely important in these cases. Additional sources of specimens are the eye, ear canal, synovial fluid, nasal cavity (all by swab), and diseased tissue that has been surgically removed (biopsied).

After proper collection, the specimen is promptly transported to a lab and stored appropriately (usually refrigerated) if it must be held for a time. Nonsterile samples in particular, such as urine, feces, and sputum, are especially prone to deterioration at room temperature. Special swab and transport systems are designed to collect the specimen and maintain it in stable condition for several hours. These devices provide nonnutritive maintenance media (so microbes survive but do not grow), a buffering system, and an anaerobic environment to prevent destruction of oxygen-sensitive bacteria.

Knowing how to properly collect, transport, and store specimens is an important aspect of many health professionals' jobs. In addition, labeling and identifying specimens, as well as providing an accurate patient history, are crucial to obtaining timely and accurate results.

Overview of Laboratory Techniques

Analyzing the patient for signs of microbial infection (i.e., fever, wound exudate, mucus production, abnormal lesion) comes first. After that, specimens are collected

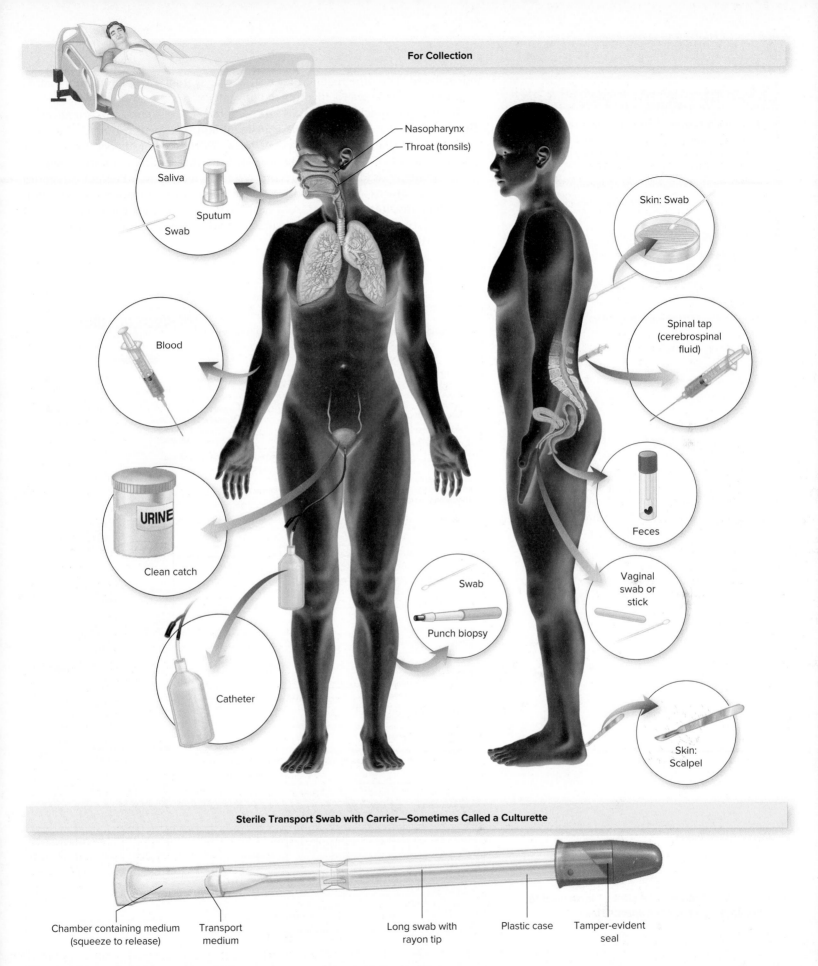

Saliva

Sputum

Swab

Nasopharynx

Throat (tonsils)

Skin: Swab

Blood

Spinal tap
(cerebrospinal
fluid)

URINE

Clean catch

Feces

Swab

Punch biopsy

Vaginal
swab or
stick

Catheter

Skin:
Scalpel

Sterile Transport Swab with Carrier—Sometimes Called a Culturette

Chamber containing medium
(squeeze to release)

Transport
medium

Long swab with
rayon tip

Plastic case

Tamper-evident
seal

Figure 15.1 Sampling sites and methods of collection for clinical laboratories.

NCLEX® PREP

1. An RN is training a new staff nurse about proper specimen collection. Consequences of improper specimen collection may include all of the following, except
   - **a.** *a delay in appropriate treatment.*
   - **b.** *exposing the patient to an unnecessary procedure.*
   - **c.** *having to repeat the test.*
   - **d.** *putting other patients at risk for incorrect treatment.*
   - **e.** *putting the patient at risk for incorrect treatment.*

and analyzed. The time required for testing ranges from a few minutes in a rapid strep test to several weeks in a tuberculosis infection.

Results of specimen analysis are entered in a summary patient chart **(figure 15.2)** that can be used in assessment and treatment regimens. The form pictured here is an "old school" paper form. We're showing it because it gives a more comprehensive view of the various tests that can be requested than the more commonly used electronic record system, which shows different categories on separate screens. You may notice that it compiles information on tests with which you are already familiar. As a health care provider, your understanding of and ability to communicate lab results are critical steps in the successful treatment of patients.

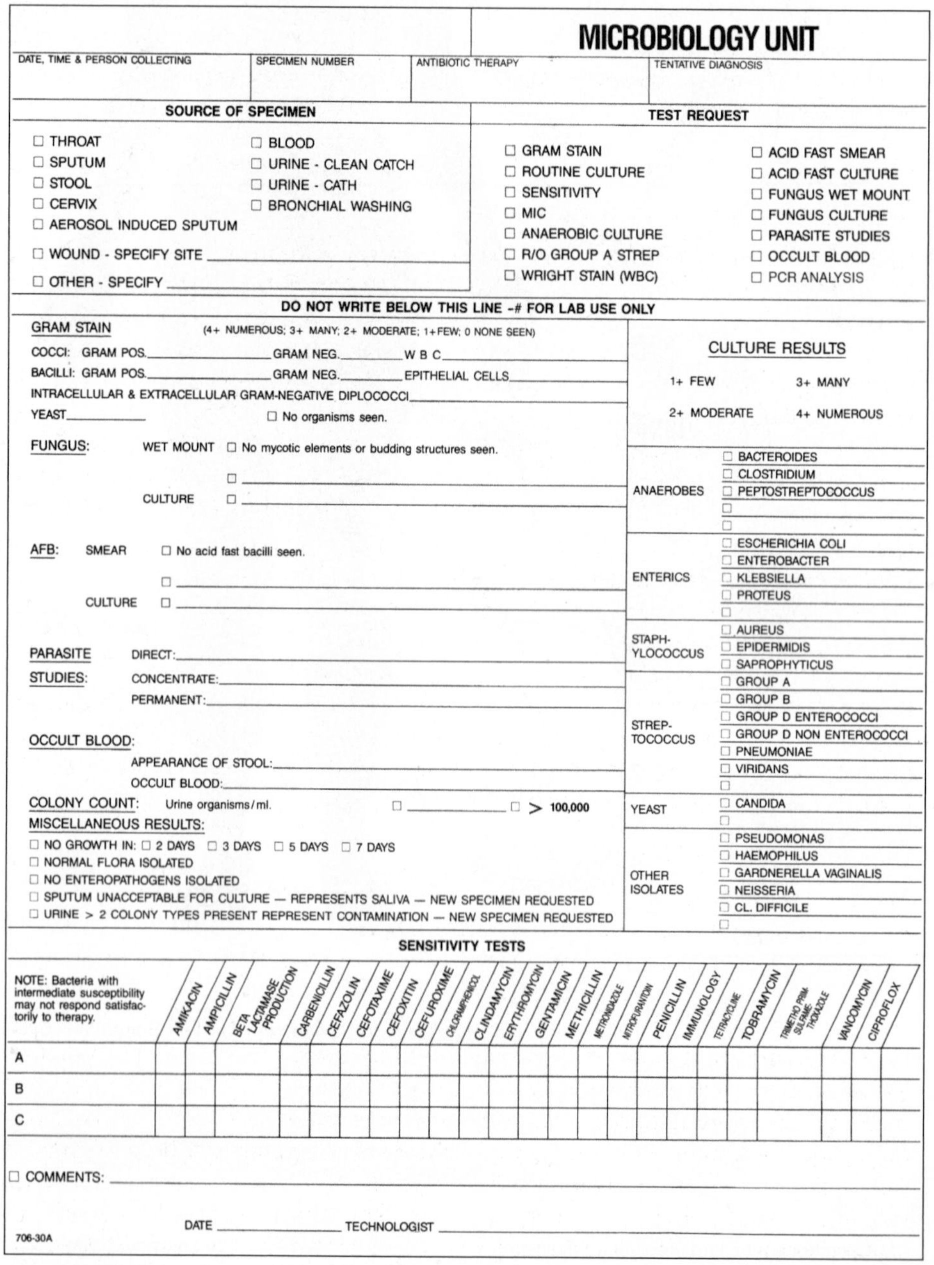

MICROBIOLOGY UNIT

DATE, TIME & PERSON COLLECTING | SPECIMEN NUMBER | ANTIBIOTIC THERAPY | TENTATIVE DIAGNOSIS

SOURCE OF SPECIMEN

☐ THROAT ☐ SPUTUM ☐ STOOL ☐ CERVIX ☐ AEROSOL INDUCED SPUTUM ☐ WOUND - SPECIFY SITE ☐ OTHER - SPECIFY
☐ BLOOD ☐ URINE - CLEAN CATCH ☐ URINE - CATH ☐ BRONCHIAL WASHING

TEST REQUEST

☐ GRAM STAIN ☐ ROUTINE CULTURE ☐ SENSITIVITY ☐ MIC ☐ ANAEROBIC CULTURE ☐ R/O GROUP A STREP ☐ WRIGHT STAIN (WBC)
☐ ACID FAST SMEAR ☐ ACID FAST CULTURE ☐ FUNGUS WET MOUNT ☐ FUNGUS CULTURE ☐ PARASITE STUDIES ☐ OCCULT BLOOD ☐ PCR ANALYSIS

DO NOT WRITE BELOW THIS LINE -# FOR LAB USE ONLY

GRAM STAIN (4+ NUMEROUS; 3+ MANY; 2+ MODERATE; 1+FEW; 0 NONE SEEN)
COCCI: GRAM POS. GRAM NEG. W B C
BACILLI: GRAM POS. GRAM NEG. EPITHELIAL CELLS
INTRACELLULAR & EXTRACELLULAR GRAM-NEGATIVE DIPLOCOCCI
YEAST ☐ No organisms seen.

FUNGUS: WET MOUNT ☐ No mycotic elements or budding structures seen. ☐ CULTURE ☐

AFB: SMEAR ☐ No acid fast bacilli seen. ☐ CULTURE ☐

PARASITE STUDIES: DIRECT: CONCENTRATE: PERMANENT:

OCCULT BLOOD: APPEARANCE OF STOOL: OCCULT BLOOD:

COLONY COUNT: Urine organisms/ml. ☐ ☐ > 100,000

MISCELLANEOUS RESULTS:
☐ NO GROWTH IN: ☐ 2 DAYS ☐ 3 DAYS ☐ 5 DAYS ☐ 7 DAYS
☐ NORMAL FLORA ISOLATED
☐ NO ENTEROPATHOGENS ISOLATED
☐ SPUTUM UNACCEPTABLE FOR CULTURE — REPRESENTS SALIVA — NEW SPECIMEN REQUESTED
☐ URINE > 2 COLONY TYPES PRESENT REPRESENT CONTAMINATION — NEW SPECIMEN REQUESTED

CULTURE RESULTS
1+ FEW 3+ MANY 2+ MODERATE 4+ NUMEROUS

ANAEROBES: ☐ BACTEROIDES ☐ CLOSTRIDIUM ☐ PEPTOSTREPTOCOCCUS ☐ ☐
ENTERICS: ☐ ESCHERICHIA COLI ☐ ENTEROBACTER ☐ KLEBSIELLA ☐ PROTEUS ☐
STAPHYLOCOCCUS: ☐ AUREUS ☐ EPIDERMIDIS ☐ SAPROPHYTICUS
STREPTOCOCCUS: ☐ GROUP A ☐ GROUP B ☐ GROUP D ENTEROCOCCI ☐ GROUP D NON ENTEROCOCCI ☐ PNEUMONIAE ☐ VIRIDANS ☐
YEAST: ☐ CANDIDA ☐
OTHER ISOLATES: ☐ PSEUDOMONAS ☐ HAEMOPHILUS ☐ GARDNERELLA VAGINALIS ☐ NEISSERIA ☐ CL. DIFFICILE ☐

SENSITIVITY TESTS

NOTE: Bacteria with intermediate susceptibility may not respond satisfactorily to therapy.	AMIKACIN	AMPICILLIN	BETA LACTAMASE PRODUCTION	CARBENICILLIN	CEFAZOLIN	CEFOTAXIME	CEFOXITIN	CEFUROXIME	CHLORAMPHENICOL	CLINDAMYCIN	ERYTHROMYCIN	GENTAMICIN	METHICILLIN	METRONIDAZOLE	NITROFURANTOIN	PENICILLIN	IMMUNOLOGY	TETRACYCLINE	TOBRAMYCIN	TRIMETHOPRIM SULFAMETHOXAZOLE	VANCOMYCIN	CIPROFLOX
A																						
B																						
C																						

☐ COMMENTS:

706-30A DATE TECHNOLOGIST

Figure 15.2 Example of a clinical form used to report data on a patient's specimens.

15.3 Phenotypic Methods

Immediate Direct Examination of Specimen

Direct microscopic observation of a fresh or stained specimen is one of the most rapid methods of determining presumptive and sometimes confirmatory microbial characteristics. The Gram stain and the acid-fast stain (see figure 2.18) are most often used for bacterial identification. As useful as these stains are, they can identify only a few organisms on their own. For that reason, we have a variety of other techniques.

Methods Requiring Growth

Selective and Differential Growth

Such a wide variety of media exist for microbial isolation that a certain amount of preselection must occur, based on the nature of the specimen. In cases in which the suspected pathogen is present in small numbers or is easily overgrown by normal biota, the specimen can be initially enriched with specialized media. Nonsterile specimens containing a diversity of bacterial species, such as urine and feces, are cultured on selective media to encourage the growth of only the suspected pathogen. For example, a large percentage of urinary tract infections are known to be caused by *Escherichia coli*, so selective media that will be sure to allow the growth of this common pathogen are chosen. Specimens are often inoculated into differential media to identify definitive characteristics, such as reactions in blood (blood agar) and fermentation patterns (mannitol salt and MacConkey agar).

A straightforward way to phenotypically identify specimens is to combine the results of tests such as Gram staining, growth on different media, and simple enzymatic tests. Usually a Gram stain is the first test. In most cases, the specimen will be either gram-negative or gram-positive. This can form the starting point for a **dichotomous key,** a graphic method that is essentially a flowchart leading to the identification of specimens **(figure 15.3).** The name "dichotomous" refers to the fact that at each branch of the flowchart there are two (di-) possible outcomes.

Biochemical Testing

The physiological reactions of bacteria to nutrients and other substrates provide excellent evidence of the types of enzyme systems present in a particular species. Knowing which enzymes an isolate has can often lead to its identity. Many of these tests are based on enzyme-mediated metabolic reactions that are visualized by a color change.

The microbe is cultured in a medium with a special substrate and then tested for a particular end product. Microbial expression of the enzyme is made visible by a colored dye; no coloration means it lacks the enzyme for utilizing the substrate in that particular way. The biochemical testing process has been fully automated by diagnostic companies. Even the growth and incubation step can be performed by the machines, and they can identify more than 2,500 different bacteria. With these machines, the sample is inoculated in some micro format, such as on a small card, inserted in a machine, and the whole process is performed by the machine. However,

Medical Moment

Qualitative versus Quantitative Diagnosis

Laboratory tests may be *qualitative* or *quantitative.* Qualitative tests are used to determine whether a substance is present in a sample, while quantitative tests measure the amounts of the substance that are present.

A pregnancy test is a perfect example of a test that may be qualitative or quantitative. Most home pregnancy tests are qualitative in that they test for the presence or absence of the hormone known as human chorionic gonadotropin (hCG), which will appear in the urine of pregnant women several days after conception.

A quantitative pregnancy test is determined from a blood sample. The hCG levels should double every 2 days in the first few weeks of pregnancy. When hCG levels do not rise as expected, there may be a problem with the pregnancy (such as miscarriage); extremely high levels may be associated with twins. Therefore, measuring the exact amount of hCG present in the bloodstream provides the ordering physician with much more information than a qualitative pregnancy test can, which only indicates whether or not a woman is pregnant.

Q. What technique in this chapter do most home pregnancy test kits utilize?

Answer in Appendix B.

Antibodies are used to "diagnose" noninfectious conditions as well, such as pregnancy.
©laflor/Getty Images

all over the world, and probably in your own microbiology lab course, the process is performed manually. Seeing it done manually makes it easier to visualize how it actually works **(figure 15.4).** Additionally, direct biochemical testing of patient samples now can be performed without incubation, which produces results within hours instead of days.

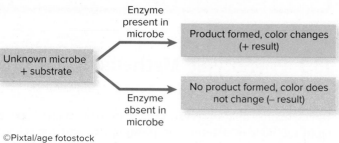

©Pixtal/age fotostock

Another type of phenotypic test is the MGIT system used to detect the growth of the slow-growing *Mycobacterium tuberculosis.* (MGIT stands for "mycobacterial growth indicator tube.") This system monitors oxygen levels in a tube that has been inoculated with a patient specimen (sputum, blood). The tube contains a medium that encourages the growth of the tuberculosis bacterium (which is excruciatingly slow-growing). There is a silicon chip at the bottom of the tube that is impregnated with a fluorescent substance that is sensitive to oxygen levels. When first inoculated, there will be a lot of free oxygen in the medium because even if the bacteria are present, there will not be many of them. But if the bacteria grow, they begin using the oxygen, and the decreased oxygen levels allow the silicon compound to fluoresce. In automated machines when a threshold level of fluorescence occurs, a visual and audible signal is given so a positive culture can be identified in the shortest possible time.

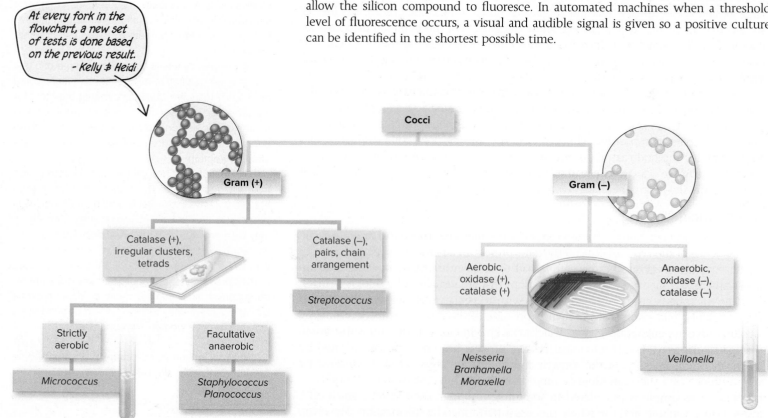

Figure 15.3 Flowchart to separate primary genera of various cocci. Identification scheme for cocci commonly involved in human diseases.

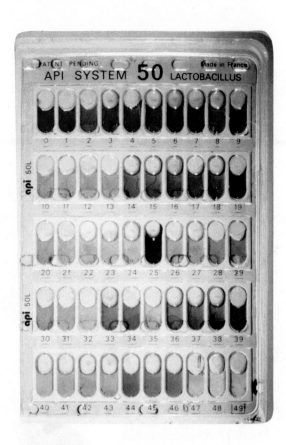

©Janice Haney Carr/CDC

Figure 15.4 Rapid biochemical test. Each of these 50 wells acts as a tiny test tube. They are pre-filled with liquid medium. In addition, a different enzymatic substrate, which will change color if it is acted upon, is present in each well. A single unknown bacterium is inoculated into each of the wells. If the bacterium has the enzyme that can metabolize the substrate in the well, the medium in the well will change color. The pattern of color changes is an easy way to detect which enzymes this bacterium has, and its identity can be deduced from that. This particular set of wells was inoculated with a bacterium that turned out to be *Clostridium ramosum*.

Antimicrobial Susceptibility Testing

As important as identifying the microbial pathogen is identifying what antimicrobial drugs it can be treated with. With the crisis-level rise in antimicrobial resistance, it is important not to waste valuable treatment time by simply guessing what drugs to use to treat an infection. Most of the automated phenotypic systems incorporate a panel of commonly used antimicrobials for the particular infection site, and simultaneously test susceptibility while identifying the pathogen. This is commonly accomplished using an adaptation of the tube-dilution method covered in chapter 12. Also, antimicrobial susceptibility tests can themselves provide the identity of some species, such as *Streptococcus, Clostridium,* and *Pseudomonas.*

Miscellaneous Tests

Phage typing relies on bacteriophages, viruses that attack bacteria in a very species-specific and strain-specific way. Such selection is useful in identifying some bacteria, primarily *Salmonella,* and is often used for tracing bacterial strains in epidemics. The technique of phage typing involves inoculating a lawn of bacterial cells onto agar, mapping it off into blocks, and applying a different phage to each sectioned area of growth **(figure 15.5).** Cleared areas corresponding to lysed cells indicate sensitivity to that phage, and a bacterial identification may be determined from this pattern.

The specificity of bacteriophages has been exploited by diagnostics manufacturers to create highly sensitive and rapid diagnostic tools. Many of these combine antibody reactions, the binding of bacteriophages, luminescence, and automation to identify bacteria, such as methicillin-resistant *Staphylococcus aureus* (MRSA) in as little as five hours, instead of the 18-24 hours required when cultures are performed.

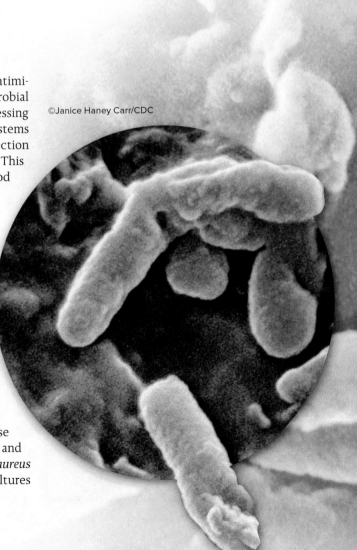

A liquid culture of the unknown bacterium is spread on an agar plate, and then each different phage preparation is "dotted" onto the agar in its respective square.
- Kelly & Heidi

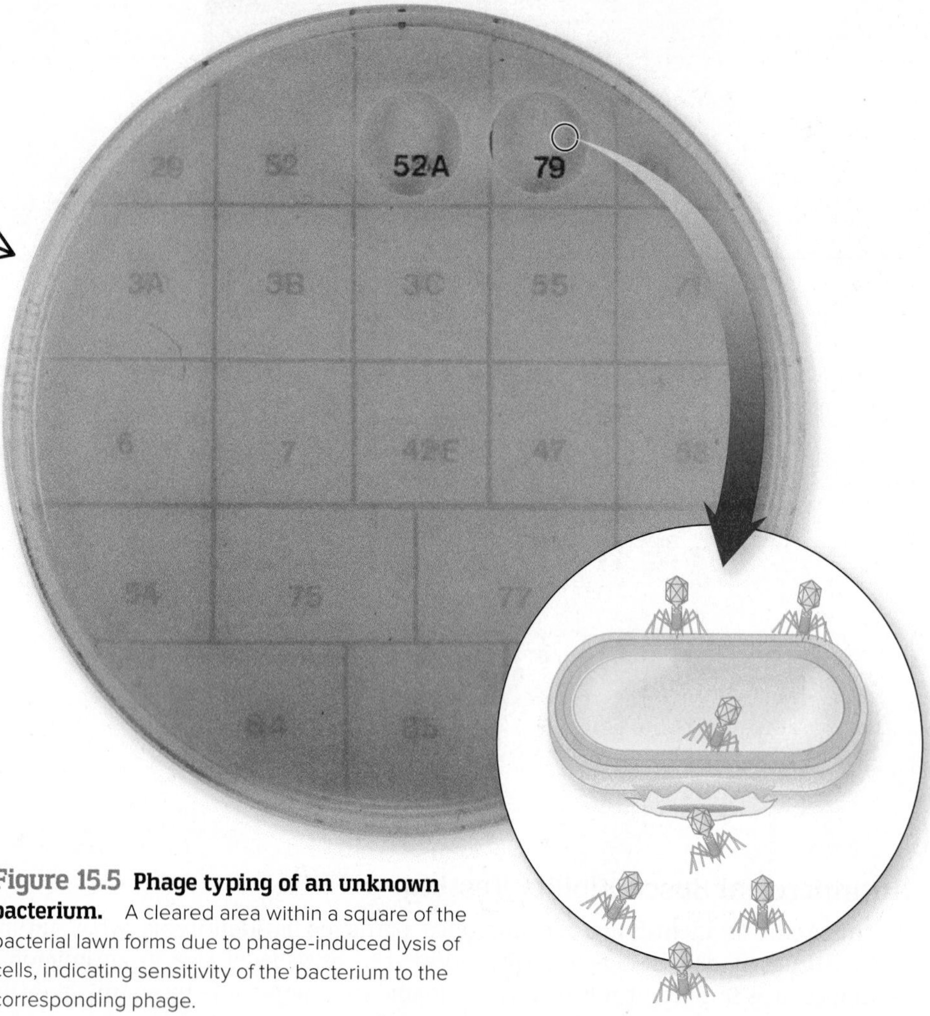

Figure 15.5 Phage typing of an unknown bacterium. A cleared area within a square of the bacterial lawn forms due to phage-induced lysis of cells, indicating sensitivity of the bacterium to the corresponding phage.
The University of Leeds

Susceptible animals are needed to cultivate bacteria such as *Mycobacterium leprae* and significant quantities of *Treponema pallidum*, whereas avian embryos and cell cultures can be used to grow host cell-dependent rickettsias, chlamydias, and viruses.

Determining Clinical Significance of Cultures

It is important to rapidly determine if an isolate from a specimen is clinically important or if it is merely a contaminant or normal biota. Although answering these questions may prove difficult, one can first focus on the number of microbes in a specimen. For example, a few colonies of *Escherichia coli* in a urine sample can simply indicate normal biota, whereas several hundred can mean active infection. In contrast, the presence of a single colony of a true pathogen, such as *Mycobacterium tuberculosis* in a sputum culture or an opportunist in sterile sites such as cerebrospinal fluid or blood, is highly suggestive of its role in disease. Furthermore, the repeated isolation of a relatively pure culture of any microorganism can mean it is an agent of disease.

As reliable as phenotypic methods can be, they have two major drawbacks: (1) When the microbe has to be cultured, it takes a minimum of 18 to 24 hours, and often longer; and (2) we are learning that many infectious conditions may be caused by nonculturable organisms, leaving open the possibility that the organism that we do culture is simply a bystander.

Table 15.2 summarizes phenotypic methods of diagnosis.

NCLEX® PREP

3. When determining the clinical significance of cultures,
   a. *the number of microbes is significant.*
   b. *the presence of a single colony of a true pathogen may indicate the presence of the disease if the culture comes from a site known to be sterile (i.e., cerebrospinal fluid).*
   c. *the repeated isolation of a relatively pure culture of any microorganism can mean it is an agent of disease, although this is not always the case.*
   d. *a range of tests may be needed to identify a pathogen.*
   e. *all of the above are true.*

Table 15.2

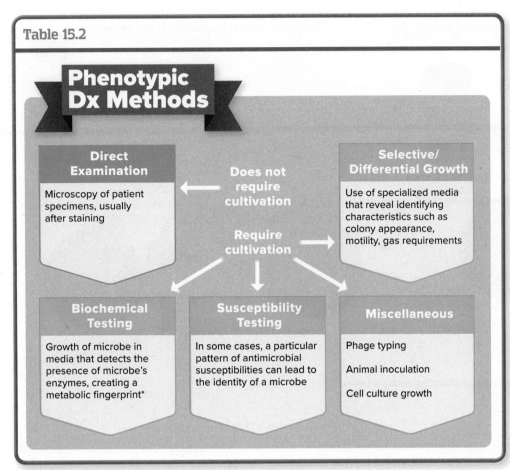

Phenotypic Dx Methods

Direct Examination

Microscopy of patient specimens, usually after staining

Does not require cultivation

Require cultivation

Selective/ Differential Growth

Use of specialized media that reveal identifying characteristics such as colony appearance, motility, gas requirements

Biochemical Testing

Growth of microbe in media that detects the presence of microbe's enzymes, creating a metabolic fingerprint*

Susceptibility Testing

In some cases, a particular pattern of antimicrobial susceptibilities can lead to the identity of a microbe

Miscellaneous

Phage typing

Animal inoculation

Cell culture growth

*Newer methods may not require cultivation.

15.3 LEARNING OUTCOMES—Assess Your Progress

5. List at least three different tests that fall in the direct identification category.

6. Explain the main principle behind biochemical testing, and identify an example of such tests.

7. Discuss two major drawbacks of phenotypic testing methods that require culturing the pathogen.

15.4 Immunologic Methods

The antibodies formed during an immune reaction are important in combating infection, but they also form the basis of a powerful diagnostic strategy. Characteristics of antibodies (such as their quantity or specificity) can reveal the history of a patient's contact with microorganisms or other antigens. This is the underlying basis of serological testing. **Serology** is the branch of immunology that traditionally deals with *in vitro* diagnostic testing of serum.

Serological testing is based on the principle that antibodies have extreme specificity for antigens, so when a particular antigen is exposed to its specific antibody, it will fit like a hand in a glove. The ability to visualize this interaction macroscopically or microscopically provides a powerful tool for detecting, identifying, and quantifying antibodies—or, on the other hand, antigens. One can detect or identify an unknown antibody using a known antigen, or one can use an antibody of known specificity to help detect or identify an unknown antigen **(figure 15.6).** Even though the strategy

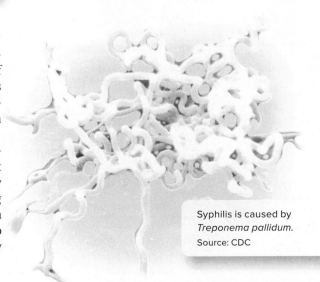

Syphilis is caused by *Treponema pallidum*.
Source: CDC

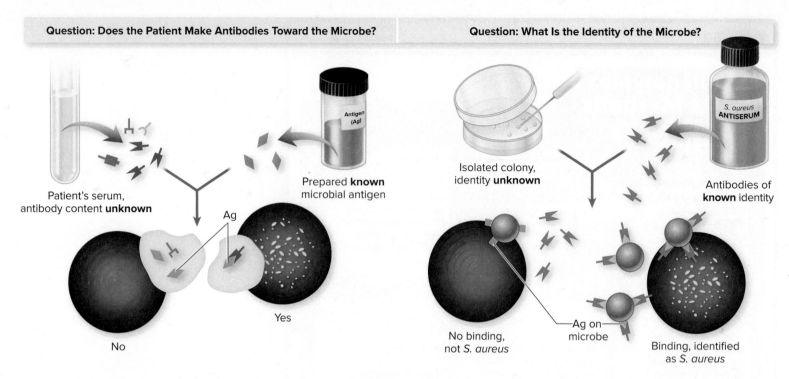

| Question: Does the Patient Make Antibodies Toward the Microbe? | Question: What Is the Identity of the Microbe? |

Patient's serum, antibody content **unknown**

Prepared **known** microbial antigen

Ag

No

Yes

Isolated colony, identity **unknown**

Antibodies of **known** identity

S. aureus ANTISERUM

No binding, not *S. aureus*

Ag on microbe

Binding, identified as *S. aureus*

Figure 15.6 **Basic principles of serological testing using antibodies and antigens.**

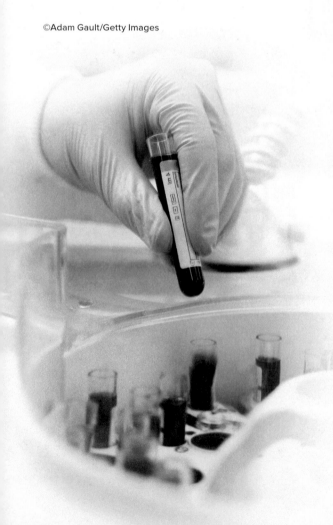

is called "serology" we now test all kinds of body samples, not just sera. These include urine, cerebrospinal fluid, whole tissues, and saliva. These and other immune tests help to determine the immunologic status of patients, confirm a suspected diagnosis, or screen individuals for disease. The tests are summarized in **table 15.3.**

Different Kinds of Antigen-Antibody Interactions

The molecular basis of immunologic testing is the binding of an antibody (Ab) to a specific site (epitope) on an antigen (Ag). The binding reaction is not visible to the naked eye nor in light microscopy. But if the antigens and antibody are suspended in a solution, under the right circumstances the presence of binding can be seen as clumps in the solution.

Agglutination and Precipitation Reactions

The essential differences between agglutination and precipitation as seen in table 15.3 are in size, solubility, and location of the antigen. In agglutination, the antigens are whole cells or organisms such as red blood cells, bacteria, or viruses displaying surface antigens **(figure 15.7).** In precipitation, the antigen examined is a soluble molecule. In both reactions, when antigen and antibody concentrations are optimal, one antigen is interlinked by several antibodies to form insoluble aggregates that settle out in solution. Agglutination is easily seen because it forms visible clumps of cells, such as in the *Weil-Felix reaction* used in diagnosing rickettsial infections. Agglutination is also used to determine blood compatibility.

These antigen-antibody binding reactions have been modified in many ways to make the reactions visible to the naked eye. One commonly used method is termed **immunochromatography,** often called the "lateral flow test." It can be found in drugstore pregnancy tests and rapid strep tests in the doctor's office. It usually consists of a plastic cartridge that contains some kind of porous material or polymer that directs fluid to flow in a particular direction. The fluid (the patient sample) will encounter antibodies along its route of flow. If the sample contains the correct antigen

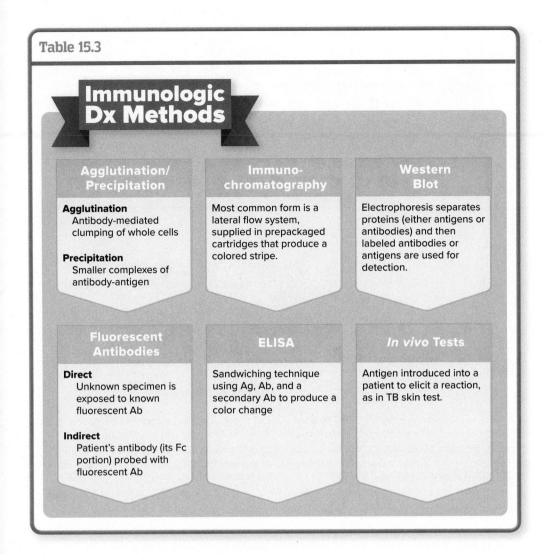

Table 15.3

Immunologic Dx Methods

Agglutination/ Precipitation	Immuno- chromatography	Western Blot
Agglutination Antibody-mediated clumping of whole cells **Precipitation** Smaller complexes of antibody-antigen	Most common form is a lateral flow system, supplied in prepackaged cartridges that produce a colored stripe.	Electrophoresis separates proteins (either antigens or antibodies) and then labeled antibodies or antigens are used for detection.
Fluorescent Antibodies	**ELISA**	***In vivo* Tests**
Direct Unknown specimen is exposed to known fluorescent Ab **Indirect** Patient's antibody (its Fc portion) probed with fluorescent Ab	Sandwiching technique using Ag, Ab, and a secondary Ab to produce a color change	Antigen introduced into a patient to elicit a reaction, as in TB skin test.

it will bind the antibodies and continue on to the next "station" in the cartridge. The next stage contains a third molecule that is impregnated on the paper in a stripe pattern. The third molecule binds the complexes and eventually this causes the stripe to change color. In this way, a microscopic antigen-antibody positive reaction is made visible to the naked eye **(figure 15.8).**

Antibody Titers

An antigen-antibody reaction in liquid is read as a titer, or the concentration of antibodies in a sample. Titer is determined by serially diluting patient serum into test tubes or wells of a microtiter plate, all containing equal amounts of bacterial cells (antigen). Titer is defined as the highest dilution of serum that still produces agglutination. In general, the more a serum sample can be diluted and still react with antigen, the greater the concentration of antibodies and thus its titer. Antibody titers are often used to diagnose autoimmune disorders such as rheumatoid arthritis and lupus, and also to determine past exposure to certain diseases such as rubella.

Serotyping

Serotyping is an antigen-antibody technique for identifying, classifying, and subgrouping certain bacteria into categories called serotypes. This method employs antisera against cell antigens such as the capsule, flagellum, and cell wall. Serotyping

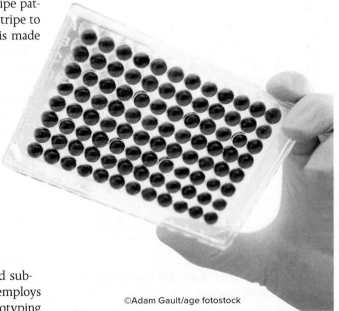

©Adam Gault/age fotostock

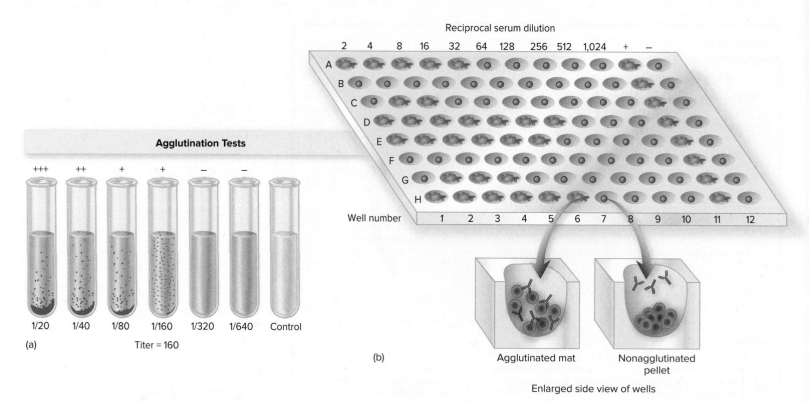

Figure 15.7 Agglutination tests. **(a)** Tube agglutination test for determining antibody titer. The same number of cells (antigen) is added to each tube, and the patient's serum is diluted in series. The titer in this example is 160 because there is no agglutination in the next tube in the dilution series (1/320). **(b)** A microtiter plate illustrating hemagglutination. Serial dilutions of the antibody are placed in wells 1–10. Positive controls (column 11) and negative controls (column 12) are included. Red blood cells (which display the antigen) are added to each well. If sufficient antibody is present to agglutinate the cells, they sink as a diffuse mat to the bottom of the well. If insufficient antibody is present, they form a tight pellet at the bottom.

is widely used in identifying *Salmonella* species and strains, and is the basis for differentiating the numerous pneumococcal and streptococcal serotypes.

The Western Blot Procedure

The **Western blot test** involves the separating of proteins by electrophoresis, and then using antibodies to detect the proteins **(figure 15.9).** First, a sample of proteins obtained from cells after lysing them is separated via electrical charge within a gel. The proteins embedded in the gel are then transferred and immobilized to a special filter. The filter is next incubated with antibody solutions, some of which have been labeled with radioactive, fluorescent, or luminescent molecules. After incubation, sites of specific antigen-antibody binding will appear as a pattern of bands that can be compared with known positive and negative controls. It is a highly specific and sensitive way to identify or verify the presence of microbial-specific antigens or antibodies in a patient sample.

Immunofluorescence Testing

The fundamental tool in immunofluorescence testing is a fluorescent antibody–a monoclonal antibody labeled by a fluorescent dye. Fluorescent antibodies (FAbs) can be used for diagnosis in two ways. In *direct testing*, an unknown test specimen or antigen is fixed to a slide and exposed to a FAb solution of known composition. If antibody-antigen complexes form, they will remain bound to the sample and will be visualized by fluorescence microscopy, thus indicating a positive result **(figure 15.10).** These tests are valuable for identifying and locating microbial antigens on cell surfaces or in tissues

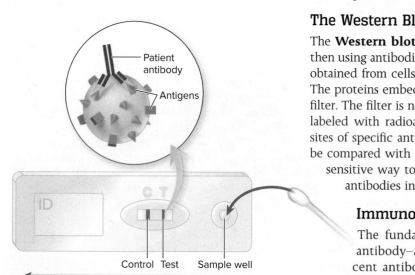

Figure 15.8 A lateral flow quick test relying on antigen-antibody binding. This is a positive test for *Neisseria gonorrhoeae.*

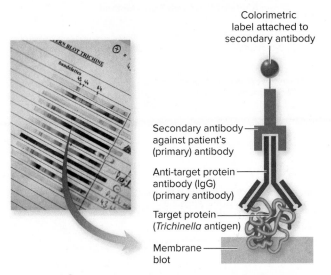

Colorimetric label attached to secondary antibody

Secondary antibody against patient's (primary) antibody

Anti-target protein antibody (IgG) (primary antibody)

Target protein (*Trichinella* antigen)

Membrane blot

Figure 15.9 **The Western blot procedure.** Major antigens from the nematode *Trichinella* were separated via electrophoresis and transferred to a filter. The filter was incubated with the sera of 10 separate patients (in the 10 rows). Sera contain primary antibodies. Secondary antibodies (blue in the drawing) and a colorimetric label were added to visualize the patient's bound antibody (maroon in the drawing). Complexes of *Trichinella* antigen and patient antibody are visualized as bands.

Phanie/Science Source (Western blot)

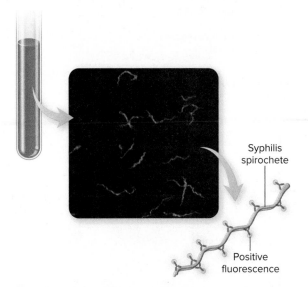

Syphilis spirochete

Positive fluorescence

Figure 15.10 **Direct fluorescence antigen test.** Photomicrograph of a direct fluorescence test for *Treponema pallidum*, the syphilis spirochete.

Source: CDC/Russell (micrograph)

and in identifying the causative agents of syphilis, gonorrhea, and meningitis, among others.

In contrast, FAbs used in *indirect testing* recognize the Fc region of antibodies in patient sera (remember that antibodies are antigenic themselves!). Known antigen (i.e., bacterial cells) is added to the test serum, and binding of the fluorescent antibody is visualized through fluorescence microscopy. Fluorescing aggregates or cells indicate that the FAbs have complexed with the microbe-specific antibodies in the test serum. This technique is frequently used to diagnose syphilis and various viral infections.

Enzyme-Linked Immunosorbent Assay (ELISA)

The **ELISA** test, also known as enzyme immunoassay (EIA), uses an enzyme-linked indicator antibody to visualize antigen-antibody reactions. This technique also relies on a solid support such as a plastic microtiter plate that can *adsorb* (attract on its surface) the reactants **(figure 15.11).**

The *indirect ELISA* test detects microbe-specific antibodies in patient sera. A known antigen is adsorbed to the surface of a well and mixed with unknown antibody. If an antibody-antigen complex forms, an added indicator antibody will bind, and subsequent development will produce a color change indicating a positive result. This is the common test used for antibody screening for HIV, various rickettsial species, hepatitis A and C, and *Helicobacter*. Because false positives can occur, a verification test (Western blot) may be necessary.

There are multiple ways to perform a direct ELISA. One of them is known as the sandwich test. In this setup, a known antibody is adsorbed to the bottom of a well and incubated with an unknown antigen. If an antibody-antigen complex forms, it will attract the indicator antibody, and color will develop in these wells, indicating a positive result. New detection systems utilize computer chips that sense minute changes in electrical current that occur when antibody-antigen complexes are formed.

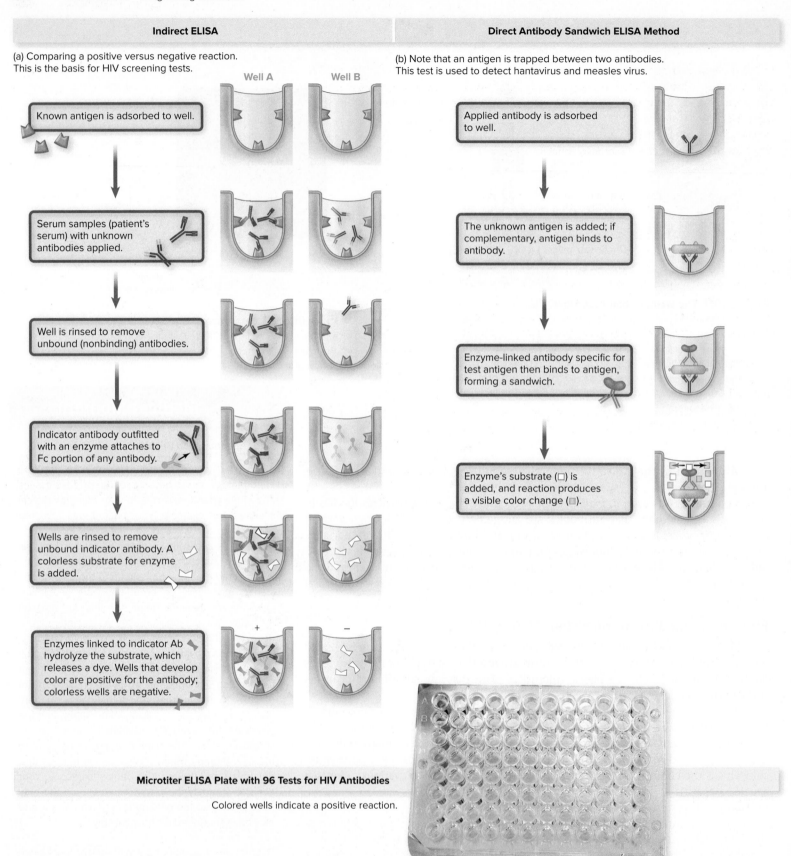

Indirect ELISA

(a) Comparing a positive versus negative reaction. This is the basis for HIV screening tests.

Well A   Well B

Known antigen is adsorbed to well.

Serum samples (patient's serum) with unknown antibodies applied.

Well is rinsed to remove unbound (nonbinding) antibodies.

Indicator antibody outfitted with an enzyme attaches to Fc portion of any antibody.

Wells are rinsed to remove unbound indicator antibody. A colorless substrate for enzyme is added.

+   −

Enzymes linked to indicator Ab hydrolyze the substrate, which releases a dye. Wells that develop color are positive for the antibody; colorless wells are negative.

Direct Antibody Sandwich ELISA Method

(b) Note that an antigen is trapped between two antibodies. This test is used to detect hantavirus and measles virus.

Applied antibody is adsorbed to well.

The unknown antigen is added; if complementary, antigen binds to antibody.

Enzyme-linked antibody specific for test antigen then binds to antigen, forming a sandwich.

Enzyme's substrate (□) is added, and reaction produces a visible color change (▪).

Microtiter ELISA Plate with 96 Tests for HIV Antibodies

Colored wells indicate a positive reaction.

Figure 15.11 Methods of ELISA testing. **(a)** Indirect ELISAs detect patient antibody. Here, both positive and negative tests are illustrated. **(b)** Direct sandwich ELISAs detect patient antigen (microbes). Here, only the positive reaction is illustrated.

Hank Morgan/Science Source

In Vivo Testing

In practice, *in vivo* tests employ principles similar to serological tests, except in this case an antigen is introduced into a patient to elicit some sort of visible reaction. The **tuberculin reaction,** where a small amount of purified protein derivative (PPD) from *Mycobacterium tuberculosis* is injected into the skin, is a classic example. The appearance of a red, raised, thickened lesion in 48 to 72 hours can indicate previous exposure to tuberculosis (shown in figure 14.12).

General Features of Immune Testing

The most effective diagnostic tests have a high degree of two characteristics: specificity and sensitivity. *Specificity* is the property of a test to focus on only a certain antibody or antigen and not to react with unrelated or distantly related ones. Put another way, a test with high specificity will have a low false-positive rate. *Sensitivity* refers to the detection of even minute quantities of antibodies or antigens in a specimen. A test with high sensitivity will have a low false-negative rate. Usually, a given test will be "better" at one or the other, and the trade-off must be considered when choosing a test. For example, in an epidemic of a deadly disease, such as Ebola, you might reason that you want to have a very high sensitivity to catch every case. People who test positive can then be screened using a different technique to determine if they actually have the disease. In other situations (for example, for that secondary screening test), you might want a higher specificity so that you don't have very many false positives. Other diagnostic methodologies besides serological are also characterized by a particular sensitivity and specificity.

15.4 LEARNING OUTCOMES—Assess Your Progress

8. Define the term *serology,* and explain the immunologic principle behind serological tests.
9. Identify two immunological diagnostic techniques that rely on a secondary antibody, and explain how they work.

15.5 Genotypic Methods

The sequence of nitrogenous bases within DNA or RNA is unique to every microorganism. Because of this and the many recent advances in genomic technology, nucleic-acid-based tests have become a mainstay of microbial identification. Genotypic methods are summarized in **table 15.4** and discussed below.

Polymerase Chain Reaction (PCR) and Other Nucleic Acid Amplification Techniques

Many nucleic acid tests use the polymerase chain reaction (PCR). PCR results in the production of numerous identical copies of DNA or RNA molecules within hours (see table 8.9). This method can amplify even minute quantities of nucleic acids present in a sample, which greatly improves the sensitivity of these tests. PCR amplification can be performed on genetic material from a wide variety of bacteria, viruses, protozoa, and fungi.

Adaptations or variations of the PCR method are constantly being made. Some of the most important ones being used in diagnosis of infections are described next. Often in procedure manuals you will see these lumped under the category of nucleic acid amplification tests, or NAATs.

Medical Moment

Detecting and Treating TB

In the phenotypic section earlier in this chapter, you saw an example of a test used to detect the slow-growing bacterium that causes tuberculosis, *Mycobacterium tuberculosis.* One excellent example of how genotypic methods have been used to improve on previous methodologies is a relatively new assay called the GeneXpert MTB/RIF test. It is an automated test using nucleic acid amplification (basically the PCR method). It accomplishes two things that are very difficult and very slow to do with *M. tuberculosis:* Identify the bacterium in a specimen and simultaneously assess its susceptibility to the frontline antibiotic, rifampin. This test is in a cartridge system and delivers answers to both those questions in 90 minutes, as opposed to weeks as was previously required. It is user-friendly and requires no special technical expertise, which makes it extremely useful in areas of the world where TB is endemic. It is a breakthrough that scientists have been working toward for decades.

Q. What are the consequences of taking several weeks to confirm a TB diagnosis, as opposed to 90 minutes?

Answer in Appendix B.

Table 15.4

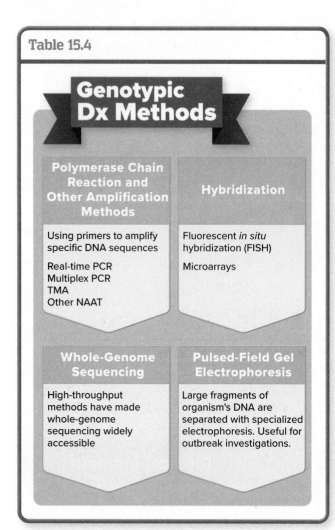

Genotypic Dx Methods

Polymerase Chain Reaction and Other Amplification Methods

Using primers to amplify specific DNA sequences

Real-time PCR
Multiplex PCR
TMA
Other NAAT

Hybridization

Fluorescent *in situ* hybridization (FISH)

Microarrays

Whole-Genome Sequencing

High-throughput methods have made whole-genome sequencing widely accessible

Pulsed-Field Gel Electrophoresis

Large fragments of organism's DNA are separated with specialized electrophoresis. Useful for outbreak investigations.

COVID-19

This is the CDC-supplied RT-PCR test for SARS-CoV-2. It detects viral RNA.

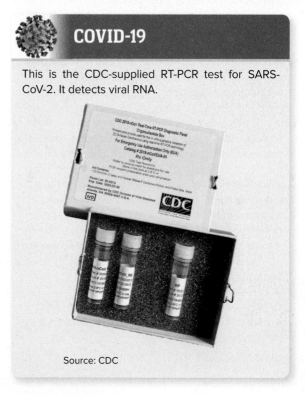

Source: CDC

Real-Time PCR

This technique is also known as "qPCR" because there is a quantitative aspect to it. It uses fluorescent labeling during the amplification procedure and the level of fluorescence is measured in real time as the reaction is running. It is fully automated and is faster than traditional PCR because analysis of the DNA after the reaction is finished is not necessary. Another advance is that qPCR assays often assess the antimicrobial susceptibilities at the same time they are identifying the organism. Note: Real-time PCR is often confused with reverse-transcriptase PCR (you can see why). Reverse-transcriptase PCR is rightly referred to as RT-PCR, and involves the creation of DNA out of RNA ("reverse" transcription).

Multiplex PCR

This is a type of diagnostic PCR that contains primers for multiple organisms instead of just a single primer. As with biochemical panels, a multiplex PCR will contain primers for multiple organisms in the differential diagnosis for the patient's symptoms. Most often, multiplex PCR is also real-time PCR.

Transcription-Mediated Amplification

This is an amplification method that, unlike PCR, does not require temperature changes. Also, it uses two enzymes, reverse transcriptase (to make DNA) and RNA polymerase (to transcribe RNA).

Hybridization: Probing for Identity

Hybridization is a technique that makes it possible to identify a microbe by analyzing segments of its genetic material. This requires small fragments of single-stranded DNA or RNA called **probes** that are known to be complementary to the specific sequences of nucleic acid isolated from a particular microbe. Base-pairing of the known probe to the nucleic acid can be observed, providing evidence of the microbe's identity. These tests have become more convenient and portable over the years. Probes are typically fluorescently labeled or attached to an enzyme that triggers a colorimetric change when hybridization occurs. This is based on the property of dyes such as fluorescein and rhodamine, which emit visible light in response to ultraviolet. This property of fluorescence has found numerous applications in diagnostic testing.

Fluorescent *in situ* hybridization (FISH) techniques involve the application of fluorescently labeled probes to intact cells within a patient specimen or an environmental sample **(figure 15.12)**. Microscopic analysis or automated processes are used to locate "glowing" cells and conclude the identity of a specific microbe. FISH is often used to confirm a diagnosis or to identify the microbial components within a biofilm. Due to the convenience and accuracy of new PCR methods, FISH methods are decreasing in use for diagnosis of infections, but continue to be used in cancer diagnoses, especially in personalized medicine, where a patient's DNA is examined for particular characteristics that can make a certain drug a better choice for them.

Microarrays

Microarrays designed for infectious disease diagnosis are "chips" (absorbent plates) that contain gene sequences from potentially thousands of different possible infectious agents, selected based on the syndrome being investigated (such as respiratory infection or meningitis symptoms). The arrays are selected based on a very large differential diagnosis; in other words, what possible microbes could cause disease in this syndrome? Arrays can be made to contain bacterial, viral, and fungal genes in a single test. In this scenario, patient samples (sputum, cerebrospinal fluid) or the nucleic acids isolated from them are incubated with labeled gene sequences on the microarray. Matching sequences hybridize to the chip, and the label (in most cases it

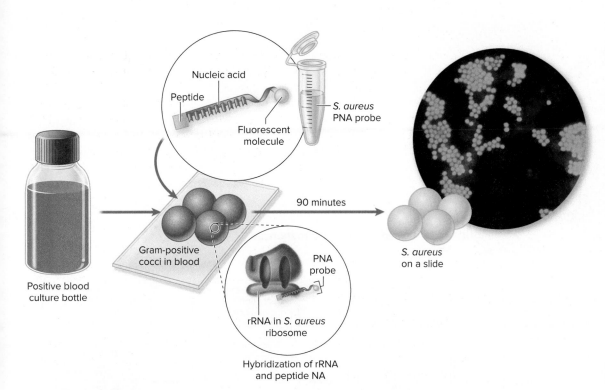

Figure 15.12 Peptide nucleic acid (PNA) FISH testing for *S. aureus*. Testing can identify blood-borne pathogens more quickly than other methods.

Image provided by AdvanDx, Inc. (*S. aureus*)

is fluorescence) is detected by a computer program, which provides the identity of the isolate or isolates **(figure 15.13).**

Whole-Genome Sequencing

The cost of whole-genome sequencing, in terms of time and money, is becoming so low that this technique is becoming commonplace in clinical and epidemiological laboratories around the world. It is particularly useful for rapid analysis of outbreaks and drug-resistant organisms. More importantly, a single genome can be scanned and analyzed multiple times in a process called "deep sequencing," which minimizes errors. Many scientists suspect that these types of sequencing will become so cheap and so routine that we will soon just "sequence everything" from a patient sample to find the one or more microbes causing symptoms.

Pulsed-Field Gel Electrophoresis: Microbial Fingerprints

In chapter 8, genetic fingerprinting was described as a method for analyzing short segments of DNA within a sample (see figure 8.17). Pulsed-field gel electrophoresis (PFGE) is a similar technique but it involves the separation of DNA fragments that are too large for conventional gel electrophoresis methods **(figure 15.14).** This is accomplished by slowly applying alternating voltage levels to the gel from three different directions, allowing even similarly sized DNA fragments to fully separate. Because the DNA is subjected to restriction enzymes, single changes in the DNA sequence (from mutations) will result in fragments of different sizes. PFGE is often used in acute outbreaks of foodborne and other infections. PulseNet is a program established by the CDC that uses PFGE to assist in the investigation of possible infectious disease outbreaks, such as those caused by foodborne pathogens or contact with reptiles or

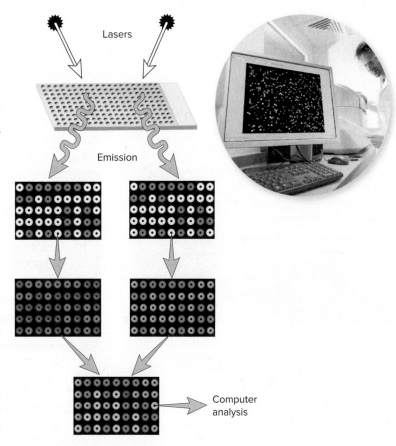

Figure 15.13 Microarrays. Digital view of a microarray.

Cultura RF/Getty Images (photo)

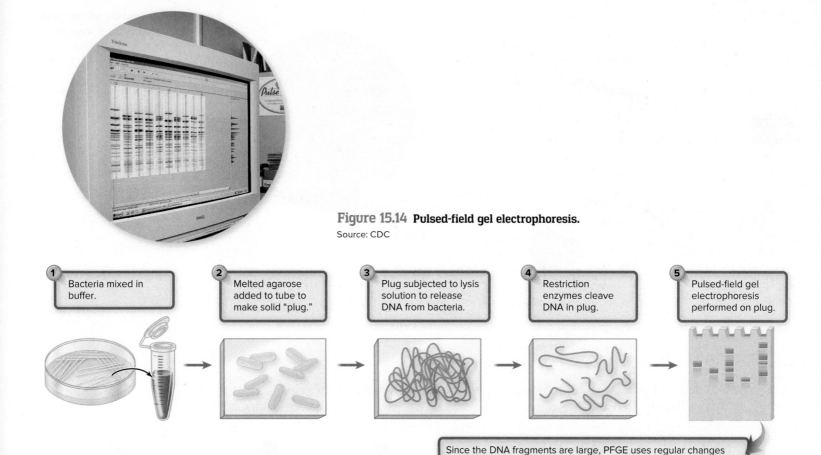

Figure 15.14 **Pulsed-field gel electrophoresis.**
Source: CDC

1 Bacteria mixed in buffer.

2 Melted agarose added to tube to make solid "plug."

3 Plug subjected to lysis solution to release DNA from bacteria.

4 Restriction enzymes cleave DNA in plug.

5 Pulsed-field gel electrophoresis performed on plug.

Since the DNA fragments are large, PFGE uses regular changes ("pulses") in the direction of the electrical field to tease them apart.

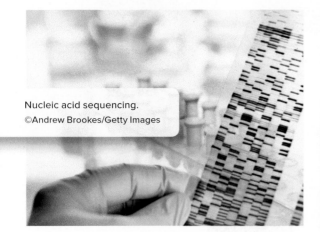

Nucleic acid sequencing.
©Andrew Brookes/Getty Images

backyard chickens. Scientists from public health facilities across the country are able to rapidly communicate and compare PFGE data from patient specimens and other samples, allowing identification of outbreaks to occur within hours versus days or weeks. The CDC reports that they are in the process of transitioning their PulseNet system to using whole-genome sequencing, but PFGE is still useful in some applications.

15.5 LEARNING OUTCOMES—Assess Your Progress

10. Explain why amplification techniques are useful for infectious disease diagnosis.

11. Name two examples of techniques that employ hybridization.

12. Explain how whole-genome sequencing can be used for diagnosis.

13. Identify the situations in which pulsed-field gel electrophoresis is most useful.

15.6 Additional Diagnostic Technologies

In addition to the traditional phenotypic and immunologic methods, you have seen in this chapter that highly automated processes for determining the genetic makeup of unknowns are being adopted at a rapid pace. In recent years breakthroughs in other scientific disciplines have also been brought into infectious disease diagnosis. We'll look at three of them: lab on a chip, mass spectrometry, and imaging technologies.

Lab on a Chip

Microarrays are contained on chips, as noted. Since the development of microarrays, many genetic tests have been miniaturized and placed on chips (integrated circuits) that are easy to use, requiring few supplies and little technical training. These have been facilitated by the use of microfluidics, using minuscule amounts of reagents and fluids to perform reactions that were previously done in large test tubes. This streamlines the testing and makes it easy to perform. DNA and RNA sequencing as well as PCR methods have been made available on "chips." While it is useful in clinical laboratories in developed countries, the biggest impact of labs on chips is predicted to be in developing countries, where diagnosis is often not possible due to lack of supplies, expertise, or even the refrigeration required to store an array of diagnostic reagents.

©laboratory/Alamy Stock Photo

Mass Spectrometry

Mass spectrometry has been utilized for years to determine the structure and composition of various chemical compounds and biological molecules. It is now being used, alone and in combination with other technologies such as PCR, to provide rapid and highly accurate microbial identification within just minutes. This technique, which is also often called MALDI-TOF, can be used to analyze a protein fingerprint from pure culture isolates or directly from samples isolated from patient specimens. It works by adding the patient sample to a metal plate and then striking it with a laser. This causes the sample to become ionized. The ions from the sample are guided into a machine that separates them and identifies them according to their mass-to-charge ratio. This technology has been applied to the identification of bacteria, viruses, and fungi, so far, and is becoming commonplace in many clinical and research laboratories due to its ability to produce rapid, precise, and cost-effective results compared to conventional phenotypic, genotypic, and immunologic methods. One distinct advantage of this technique is that clinical laboratories can construct databases of local strains of microorganisms by running profiles of known microbes. This provides for an infinitely customizable and expandable database.

Mass spectrometry technologies can also be used to simultaneously detect antibiotic susceptibilities–obviously, an important advantage **(figure 15.15).**

Imaging

For centuries imaging, in the form of X rays, has been used to assist in the diagnosis of tuberculosis. Now, in the 21st century, a variety of new imaging techniques

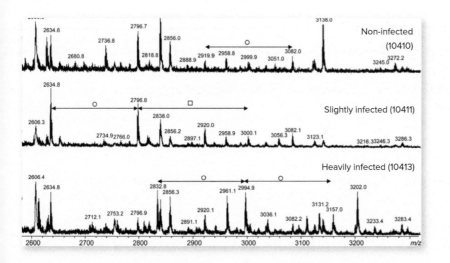

Figure 15.15 **Mass spectrometry of patient samples.**

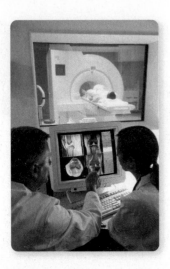

Figure 15.16 A patient undergoing MRI (magnetic resonance imaging) testing.
©Pixtal/AGE Fotostock

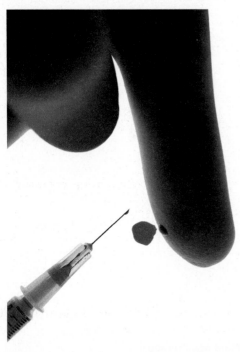

©Ingram Publishing

have been deployed to help localize infections. They offer a pair of eyes to inaccessible places, and can save the patient an invasive biopsy. Infections associated with hip implants, for example, may be difficult to access through blood samples. The bacteria may be growing in biofilms on the implanted materials, or they may be growing in an abscess deep in the hip joint. Magnetic resonance imaging, computerized tomography (CT) scans, and positron emission tomography (PET) scans have been increasingly employed to find areas of localized infection in deep tissue, which can later be biopsied to aspirate samples for culture. In the event no infection is found on the image, the patient has been spared an invasive procedure **(figure 15.16).**

Entirely New Diagnostic Strategies

Throughout the modern medical era, diagnosis of suspected microbial infections has focused on identifying the offending organism. But there is new thinking informing diagnosis and treatment. One strategy uses blood from a patient to check for the presence of seven genes that the host cells express in response to bacterial–but not viral–infection. If the seven genes are active, then the prescriber knows to prescribe an antibiotic. If they are not active, the symptoms are caused by virus(es) and therefore not going to be helped by antibiotics. This is an incredibly useful tool in a world where it is estimated that *the majority* of antibiotic prescriptions are incorrectly prescribed for conditions that are not caused by bacteria. Remember that many times in the United States (and most of the time in poorer countries) prescriptions are given without the benefit of diagnostic testing. Researchers at Stanford University, who developed this blood test, are working to make it more like *lab on a chip* technology, so that it can be quick and easy to use in all kinds of conditions.

15.6 LEARNING OUTCOMES—Assess Your Progress

14. Describe the benefits of "lab on a chip" technologies for global public health.

15. Identify an advantage presented by mass spectrometry and also by imaging techniques as diagnostic tools.

CASE FILE WRAP-UP

©McGraw-Hill Education (phlebotomist);
©Lisa Burgess/McGraw-Hill Education
(Kirby-Bauer antibiotic testing)

Increasing the number of blood cultures obtained increases the likelihood of isolating the offending organism and can be lifesaving. When two sets of blood cultures are obtained and a pathogen is identified from both cultures, it is highly unlikely that the organism cultured is a contaminant. Both aerobic and anaerobic blood cultures are obtained to provide the best chance of isolating the organism.

Blood cultures are indicated before beginning parenteral antibiotic therapy in patients with fever and leukocytosis (elevated white blood cell count). It is important to obtain blood cultures before IV antibiotics are started because the antibiotics may inhibit the growth of the microorganism before it can be isolated.

The patient in the opening Case File had a fever with a decreased level of consciousness, both signs of sepsis. Blood cultures are used to identify microorganisms that have invaded the bloodstream, causing life-threatening septicemia. It was important for the lab workers to notify the nursing staff as soon as the cultures were collected so that the nursing staff could administer the initial dose of IV antibiotics.

©Adam Gault/Getty Images

The Human Microbiome Project and Diagnosis of Noninfectious Disease

For decades, scientists have realized that culturing microbes for identification is not only time-consuming and inefficient but also incomplete. Many microbes are fastidious, requiring highly specialized growth conditions, and many others are considered "viable but nonculturable" (VNC). Some estimate that 99.9% of sequences found in the human body fall under the heading of VNC, and little is understood about the role that these elusive microbes play in the human microbiome and in disease.

Using techniques developed by J. Craig Venter in mapping the genome of microbes in the oceans, scientists have been able to map the genomes of VNC microorganisms in the Human Microbiome Project (HMP). Scientists at the Venter Institute developed multiple displacement amplification (MDA) technology. MDA is a non-PCR technique for copying fragments of DNA from a single cell until they reach the equivalent of the billions required for analysis. MDA is able to capture 90% of the genes from a single cell.

Using MDA, scientists are able to obtain information about susceptibility to antibiotics, signaling proteins used, and even how a VNC microbe lives and moves. All of this information is vital to the HMP. As you read in chapter 13, in the HMP, researchers sampled microbes from the mouth, nose, skin, intestine, and genitals of 242 volunteers (129 male and 113 female), resulting in over 5,000 samples from body sites. Using MDA, they collected about 3.5 terabases* of DNA sequences that have been entered into a gigantic comparative analysis system called the Integrated Microbial Genomes and Metagenomes for the Human Microbiome Project (IMG/M HMP).

Scientists point out that although there is only about a 0.01% difference between human genomes from one person to another, our microbiome genomes differ by as much as 50%. Researchers are beginning to use microbiome genome characteristics to predict who is at risk for type II diabetes, for example, and for leanness or obesity. While we should apply caution to "overinterpreting the microbiome," as some scientists say, there will clearly be important diagnostic tools coming from our knowledge of its genetics.

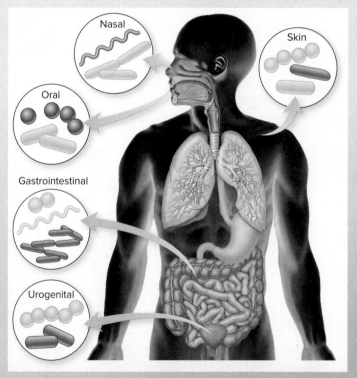

*A *terabase* is a genetic sequence of 10^{12} nucleic acid bases.

DIAGNOSING INFECTIONS

15.1 WHAT IS CAUSING THIS CONDITION?

Microbiologists use a variety of methods to diagnose infections. They range from "old-school" culturing and microscopy to cutting-edge genetic and imaging techniques.

Adam Gault/Getty Images

Saliva

Sputum

Swab

Many diagnostic strategies require a sample of the diseased tissue or fluid (such as blood or urine). The manner in which the sample—called the specimen—is collected, stored, and transported is extremely important. If these steps are mishandled, any diagnosis will be unreliable.

FIRST STEPS: SPECIMEN COLLECTION 15.2

15.3 PHENOTYPIC METHODS

Phenotypic reactions are those that can be observed either macroscopically or microscopically. These include culturing samples on media, biochemical testing of pure cultures, and examining under the microscope.

Also called serologic methods, immunologic methods use the principle of antibody-antigen specificity. Antibody-antigen reactions can be assessed in the lab, when their binding results in visible phenomena, or in the patient, when applying or injecting the antigen elicits a response.

IMMUNOLOGIC METHODS 15.4

15.5 GENOTYPIC METHODS

One of the most useful genotypic methods is amplifying DNA from the organism or the disease site. Also, DNA can be hybridized to known DNA sequences for identification. Whole-genome sequencing is increasingly affordable and useful.

Andrew Brookes/Getty Images

New diagnostic technologies are being created constantly. Mass spectrometry measures protein profiles of organisms or diseased sites. Imaging (MRIs, X rays, CT scans, etc.) are increasingly helpful ways to assess deep-tissue infections.

ADDITIONAL DIAGNOSTIC TECHNOLOGIES 15.6

SmartGrid: From Knowledge to Critical Thinking

This *21 Question Grid* takes the topics from this chapter and arranges them with respect to the American Society for Microbiology's Undergraduate Curriculum guidelines—all six of the important "Concepts" as well as the important "Competency" of scientific literacy. Three questions are supplied about chapter content that refer to the Concept or Competency, in increasing levels of Bloom's taxonomy for learning.

ASM Concept/ Competency	A. Bloom's Level 1, 2—Remember and Understand (Choose one.)	B. Bloom's Level 3, 4—Apply and Analyze	C. Bloom's Level 5, 6—Evaluate and Create
Evolution	1. When using pulsed-field gel electrophoresis, mutations in a microbe's genome will show up as a. a different pattern of bands. b. nonfluorescent bands. c. missing bands. d. all of these.	2. Explain why it is possible to identify some bacteria simply by knowing what antibiotic resistances they possess.	3. Serotyping identifies distinct members of the same species that are stably different from one another. Speculate about why these distinct serotypes are not different species instead.
Cell Structure and Function	4. Mass spectrometry identifies microbes via a. fluorescent antibodies. b. cell surface carbohydrates. c. protein fingerprints. d. DNA profiling.	5. Name some bacterial structures that might be useful to target when using fluorescent antibodies designed for diagnosis, and justify your choices.	6. You perform a lumbar puncture on a patient with meningitis symptoms and see that the spinal fluid is cloudy. However, PCR using bacterial primers comes up with nothing. What is a likely explanation?
Metabolic Pathways	7. Which category of diagnosis is represented by studying a microbe's utilization of nutrients? a. phenotypic b. genotypic c. immunologic d. none of these	8. Write a paragraph that explains the mycobacterial growth indicator tube test to a middle school science class.	9. You inoculated a biochemical test strip with a patient isolate in a rural clinic you work in and left it in the incubator overnight. When you came back in, you saw that there had been a power outage (with no backup power generator) and the temperature in the incubator was 25°C instead of 37°C. You examine the test strip and you see that the wells display growth. Do you trust the results or not? Why or why not?
Information Flow and Genetics	10. Which of the following diagnostic techniques is most likely to be affected by changes in growth conditions of the specimen? a. phenotypic b. immunologic c. genotypic	11. You perform a Kirby-Bauer disk diffusion test to determine antibiotic susceptibilities of a clinical isolate. There is a clear zone of several millimeters around the penicillin disk, but there are two to three distinct colonies in the middle of the clear halo. What is the probable explanation for those colonies?	12. Why might culture conditions affect the results of a mass spectrometry method of diagnosis?

ASM Concept/ Competency	A. Bloom's Level 1, 2—Remember and Understand (Choose one.)	B. Bloom's Level 3, 4—Apply and Analyze	C. Bloom's Level 5, 6—Evaluate and Create
Microbial Systems	13. Which of the following techniques is most likely to reveal that an infection is in biofilm form? a. ELISA b. whole-genome sequencing c. PFGE d. imaging	14. Why is it more important to use selective media when diagnosing a GI tract infection than when diagnosing a bloodstream infection?	15. What type of diagnostic method do you think would be most suited to determining if a bacterium causing a human infection was lysogenized by a bacteriophage that contains a virulence gene?
Impact of Microorganisms	16. T or F: Bacterial infection causes the expression of different human genes than does viral infection. a. T b. F	17. Many HIV tests look for patient antibody to the virus rather than the virus itself. Identify some advantages to this approach.	18. Nonhealing wounds on the surface of the body are often extremely difficult to manage, in part because the microbial cause of the lack of healing is often extremely difficult to identify. Create a list of reasons this might be the case.
Scientific Thinking	19. A test that results in a very large number of false positives probably has an unacceptable level of a. sensitivity. b. specificity.	20. When PCR is performed "by hand" (not with a prepackaged test kit), the lab environment has to be scrupulously clean. Explain why that might be.	21. What kind of a control would be important to run when performing a metabolic diagnostic test such as an API® strip?

Answers to the multiple-choice questions appear in Appendix A.

Visual Connections

This question connects content within and between chapters.

From chapter 14, figure 14.6b. Imagine that this patient is being seen by his or her physician for this unknown rash. What rapid phenotypic test could suggest that this condition is caused by a bacterium? Could a rapid immunoassay or fluorescent procedure be used to identify a specific viral cause? Explain your answer.

Biophoto Associates/Science Source

Chapter design elements: (Covid): CDC/Alissa Eckert, MS; Dan Higgins, MAMS; (Note): McGraw-Hill Education; (NCLEX): Shutterstock/ Abert); (Doctor): David Gould/Getty Images

16

Infectious Diseases Affecting the Skin and Eyes

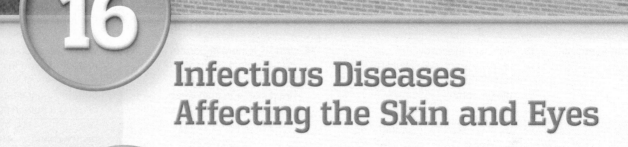

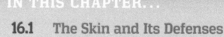

IN THIS CHAPTER...

16.1 The Skin and Its Defenses

1. Describe the important anatomical features of the skin.
2. List the natural defenses present in the skin.

16.2 Normal Biota of the Skin

3. List characteristics of the skin's normal microbiota.

16.3 Skin Diseases Caused by Microorganisms

4. Explain the important features of the "Highlight Disease" MRSA skin and soft-tissue infection.
5. List the possible causative agents, modes of transmission, virulence factors, diagnostic techniques, and prevention/treatment for the "Highlight Disease" maculopapular rash diseases.
6. Discuss important features of the other infectious skin diseases. These are impetigo, cellulitis, staphylococcal scalded skin syndrome, vesicular/pustular rash diseases, large pustular skin lesions, and cutaneous and superficial mycoses.
7. Discuss the relative dangers of rubella and measles (rubeola) viruses in different populations.

16.4 The Surface of the Eye and Its Defenses

8. Describe the important anatomical features of the eye.
9. List the natural defenses present in the eye.

16.5 Normal Biota of the Eye

10. List the types of normal biota presently known to occupy the eye.

16.6 Eye Diseases Caused by Microorganisms

11. List the possible causative agents, modes of transmission, virulence factors, diagnostic techniques, and prevention/treatment for the "Highlight Disease" conjunctivitis.
12. Discuss the most important causes of keratitis.

CASE FILE

A Rash of Symptoms

I was a newly graduated nurse working in the emergency room of a large urban children's hospital when Wyatt was brought in by his mother. Wyatt was a 6-year-old who, until a few days prior, had been very healthy. Wyatt's mother told the triage nurse that Wyatt was very sick and had a rash "all over." As was policy, Wyatt was given a mask to wear because he was coughing and was sent back out to the waiting room after having a brief history taken and a set of vital signs.

Wyatt had been triaged as urgent due to his cough, high fever, and elevated heart rate, so it wasn't long before I called Wyatt to an exam room to perform a more in-depth history and head-to-toe assessment. According to Wyatt's mother, Wyatt had initially complained of a sore throat. Within a day, he also complained of a headache and developed a high fever, which his mother treated appropriately with acetaminophen. He then developed a cough, which his mother assumed was part of a "bad cold." She was not overly concerned until Wyatt developed a rash, which started on his face and head and rapidly spread to his chest and back, and then to his arms and legs.

I had Wyatt's mother undress him and put him in an examination gown, and it was only then that I realized the extent of Wyatt's rash. It was winter and Wyatt had been covered in clothing from head to toe, having only his jacket removed when seen by the triage nurse. Dressed in a gown, I saw that Wyatt's mother was not exaggerating—Wyatt's rash covered almost all of his body. It was a maculopapular rash and did not look like any childhood rash that I had ever seen.

The physician on duty came in to see Wyatt and immediately asked Wyatt to open his mouth. She asked the mother about Wyatt's immunizations, and his mother admitted that she did not believe in immunizations and Wyatt had never been immunized. The physician then turned immediately to me and asked me to put Wyatt in an isolation room. Upon my return, she spoke to Wyatt's mother, and I will never forget what she said: "Well, it was a very bad decision to not have Wyatt immunized because now he has the measles. Not only does your son have the measles, but by bringing him here you exposed everyone in the waiting room to what can be a life-threatening illness." She then ordered a chest X ray and blood work to confirm the diagnosis and said that she would immediately notify the state health department.

- Why was a chest X ray performed?
- Why was the state health department contacted?

Case File Wrap-Up appears at the end of the chapter.

16.1 The Skin and Its Defenses

The organs under consideration in this chapter—the skin and eyes—form the boundary between the human and the environment. The skin, together with the hair, nails, and sweat and oil glands, forms the **integument.** The skin has a total surface area of 1.5 to 2 square meters. Its thickness varies from 1.5 millimeters at places such as the eyelids to 4 millimeters on the soles of the feet. Several distinct layers can be found in this thickness, and we summarize them here. Follow **figure 16.1** as you read.

The outermost portion of the skin is the epidermis, which is further subdivided into four or five distinct layers. On top is a thick layer of epithelial cells called the stratum corneum, about 25 cells thick. The cells in this layer are dead and have migrated from the deeper layers during the normal course of cell division. They are packed with a protein called **keratin,** which the cells produce in very large quantities. Cells emerge from the deepest levels of the epidermis. Because this process is continuous, the entire epidermis is replaced every 25 to 45 days. Keratin gives the cells their ability to withstand damage, abrasion, and water penetration. The surface of the skin is termed *keratinized* for this reason. Below the stratum corneum are three or four more layers of epithelial cells. The lowest layer, the stratum basale, or basal

Sweat is cooling and it has several antimicrobial properties.
©Rubberball/Getty Images

Figure 16.1 A cross section of skin.
Dennis Kunkel Microscopy/Science Source

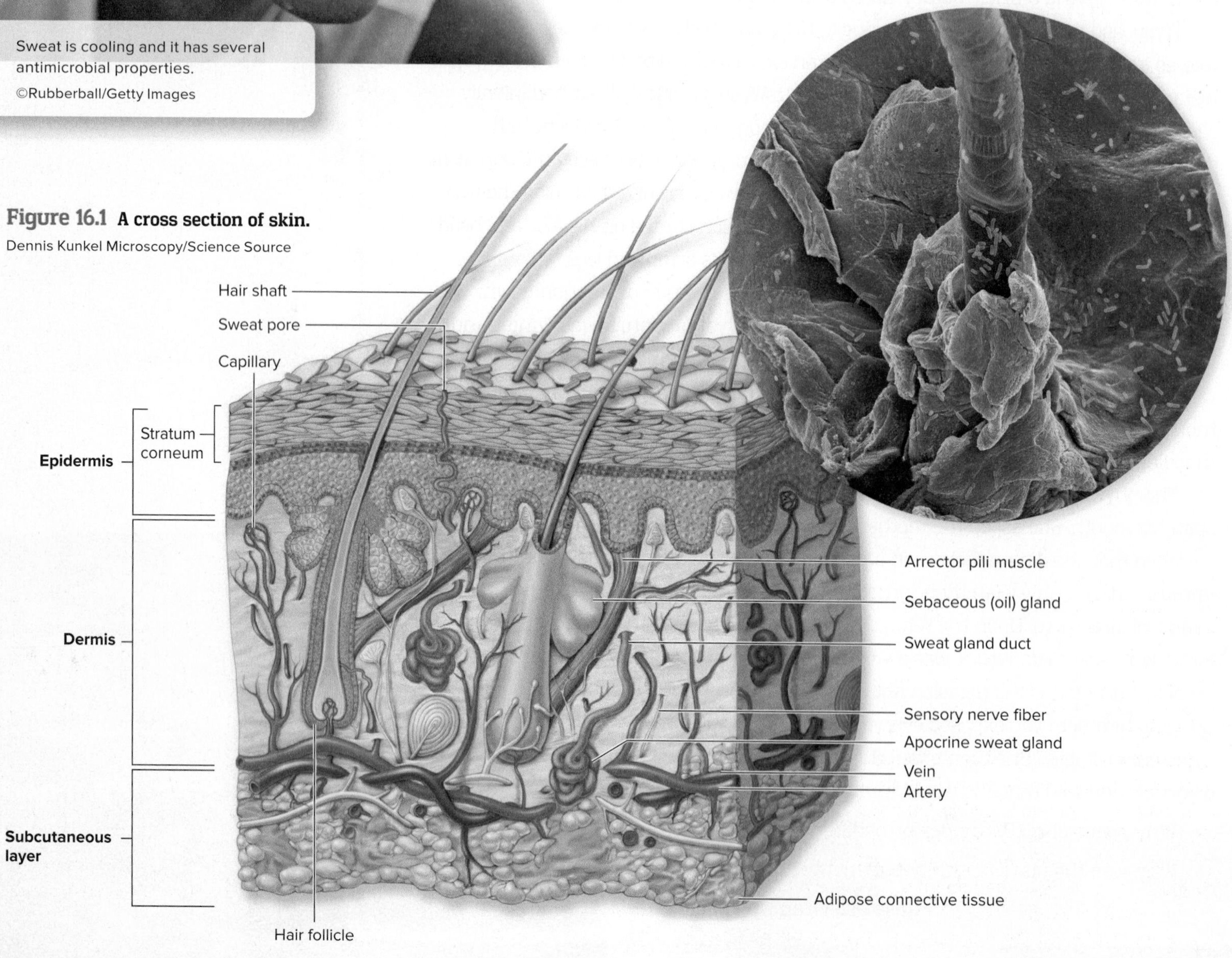

- Hair shaft
- Sweat pore
- Capillary
- Epidermis
  - Stratum corneum
- Dermis
- Subcutaneous layer
- Hair follicle
- Arrector pili muscle
- Sebaceous (oil) gland
- Sweat gland duct
- Sensory nerve fiber
- Apocrine sweat gland
- Vein
- Artery
- Adipose connective tissue

layer, is attached to the underlying dermis and is the source for all of the cells that make up the epidermis.

The dermis, underneath the epidermis, is composed of connective tissue instead of epithelium. This means that it is a rich matrix of fibroblast cells and fibers such as collagen, and it contains macrophages and mast cells. The dermis also harbors a dense network of nerves, blood vessels, and lymphatic vessels. Damage to the epidermis generally does not result in bleeding, whereas damage deep enough to penetrate the dermis results in broken blood vessels. Blister formation, the result of friction trauma or burns, represents a separation between the dermis and epidermis.

The "roots" of hairs are housed in follicles in the dermis. **Sebaceous** (oil) **glands** and scent glands are associated with the hair follicle. Separate sweat glands are also found in this tissue. All of these glands have openings on the surface of the skin, so they pass through the epidermis as well.

Millions of cells from the stratum corneum slough off every day, and attached microorganisms slough off with them. The skin is also brimming with antimicrobial substances. One of the most effective defenses of the skin is a class of molecules called antimicrobial peptides. These are positively charged chemicals that act by disrupting the negatively charged membranes of bacteria. There are many different types of these peptides, and they seem to be chiefly responsible for keeping the microbial count on skin relatively low.

The sebaceous glands' secretion, called **sebum,** has a low pH, which makes the skin inhospitable to many microorganisms. Sebum is oily due to its high concentration of lipids. The lipids can serve as nutrients for normal microbiota, but breakdown of the fatty acids contained in lipids leads to toxic by-products that inhibit the growth of microorganisms not adapted to the skin environment. This mechanism helps control the growth of potentially pathogenic bacteria. Sweat is also inhibitory to microorganisms because of both its low pH and its high salt concentration. Lysozyme is an enzyme found in sweat (and tears and saliva) that specifically breaks down peptidoglycan, which you learned in chapter 3 is a unique component of bacterial cell walls.

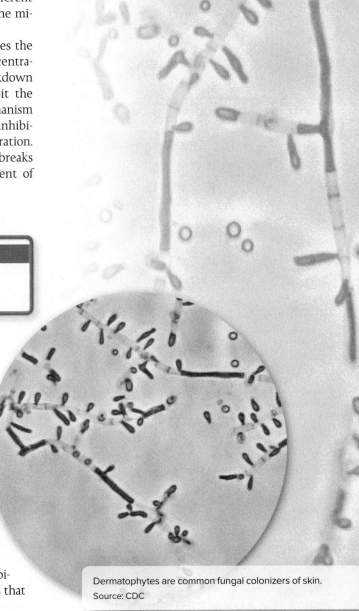

Dermatophytes are common fungal colonizers of skin.
Source: CDC

16.1 LEARNING OUTCOMES—Assess Your Progress

1. Describe the important anatomical features of the skin.
2. List the natural defenses present in the skin.

16.2 Normal Biota of the Skin

Microbes that do live on the skin surface as normal biota must be capable of living in the dry, salty conditions they find there. Microbes are relatively sparsely distributed over dry, flat areas of the body, such as on the back, but they can grow into dense populations in moist areas and skin folds, such as the underarm and groin areas. The normal microbiota also live in the protected environment of the hair follicles and glandular ducts.

Relying on cultivation-independent techniques such as 16S rRNA sequencing, recent microbiome studies have delivered fascinating information about our skin's biota. We are learning that hundreds of species of microbes, including some well-known pathogens, inhabit our epidermis, dermis, and subcutaneous skin layers. It is also common for different species to favor different areas of our bodies, and for different people to have different species. Last, in spite of this variation, it seems common for an individual's microbiota to remain relatively constant over time. In brief, these studies are showing us that the skin microbiota is far more diverse than we imagined.

Defenses and Normal Biota of the Skin

	Defenses	Normal Biota
Skin	Keratinized surface, sloughing, low pH, high salt, lysozyme, antimicrobial peptides	*Streptococcus, Staphylococcus, Corynebacterium, Propionibacterium, Pseudomonas, Lactobacillus;* yeasts such as *Candida*

16.2 LEARNING OUTCOMES—Assess Your Progress

3. List characteristics of the skin's normal microbiota.

16.3 Skin Diseases Caused by Microorganisms

HIGHLIGHT DISEASE

MRSA Skin and Soft-Tissue Infection

Methicillin-resistant *Staphylococcus aureus* (MRSA) is a common cause of skin lesions in non-hospitalized people. (The hospitalized population is more likely to acquire systemic, bloodstream infections from MRSA, addressed in chapter 18.) Even though the name mentions only "methicillin," these strains are usually resistant to multiple antibiotics.

Staphylococcus aureus is a gram-positive coccus that grows in clusters, like a bunch of grapes. It is nonmotile. Much of its destructiveness is due to its array of superantigens (see chapter 12). It can be highly virulent, but it also appears as "normal" biota on the skin of one-third of the population. Strains that are methicillin-resistant are also found on healthy people.

This species is considered the sturdiest of all non-endospore-forming pathogens, with well-developed capacities to withstand high salt (7.5% to 10%), extremes in pH, and high temperatures (up to 60°C for 60 minutes). *S. aureus* also remains viable after months of air drying and resists the effect of many disinfectants and antibiotics.

▶ Signs and Symptoms

MRSA infections of the skin tend to be raised, red, tender, localized lesions, often featuring pus and feeling hot to the touch **(figure 16.2).** They occur easily in breaks in the skin caused by injury, shaving, or even just abrasion. They may localize around a hair follicle. Fever is a common feature.

▶ Transmission and Epidemiology

MRSA is a common contaminant of all kinds of surfaces you touch daily, especially if the surfaces are not routinely sanitized. Gym equipment, airplane tray tables, electronic devices, razors, and so on, are all sources of indirect contact infection. Persons with active MRSA skin lesions should keep them covered in order to avoid direct contact transmission to others.

▶ Pathogenesis and Virulence Factors

All pathogenic *S. aureus* strains typically produce coagulase, an enzyme that coagulates plasma. Because 97% of all human isolates of *S. aureus* produce this enzyme, its presence is considered a highly diagnostic species characteristic.

Other enzymes expressed by *S. aureus* include hyaluronidase, which digests the intercellular "glue" (hyaluronic acid) that binds connective tissue in host tissues; staphylokinase, which digests blood clots; a nuclease that digests DNA (DNase); and lipases that help the bacteria colonize oily skin surfaces.

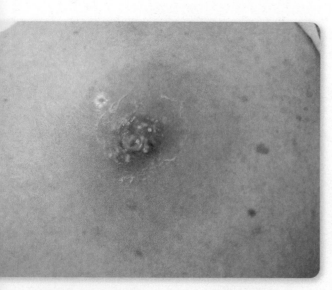

Figure 16.2 A typical MRSA skin lesion.
©Gregory Moran, M.D./CDC

▶ Culture and/or Diagnosis

Polymerase chain reaction (PCR) is routinely used to diagnose MRSA. Alternatively, cultivation on blood agar is a useful diagnostic technique **(figure 16.3).** For heavily contaminated specimens, selective media such as mannitol salt agar are used. The production of catalase, an enzyme that breaks down hydrogen peroxide accumulated during oxidative metabolism, can be used to differentiate the staphylococci, which produce it, from the streptococci, which do not.

One key technique for separating *S. aureus* from other species of *Staphylococcus* is the coagulase test **(figure 16.4).** By definition, any staph isolate that coagulates plasma is *S. aureus;* all others are coagulase-negative. You should be aware that an increasing number of dangerous skin infections are being caused by a gram-negative pathogen, *Vibrio vulnificus.* A Gram stain of the wound will point you in the right direction.

▶ Prevention and Treatment

Prevention is only possible with good hygiene. Treatment of these infections often starts with incision of the lesion and drainage of the pus. Antimicrobial treatment should include more than one antibiotic. Current recommendations in the United States are for the use of vancomycin, linezolid, or daptomycin. These recommendations, of course, will change based on antibiotic-resistance patterns.

Figure 16.3 *Staphylococcus aureus.* Blood agar plate growing *S. aureus*. Some strains show two zones of hemolysis, caused by two different hemolysins. The inner zone is clear, whereas the outer zone is fuzzy and appears only if the plate has been refrigerated after growth. Kathleen Talaro

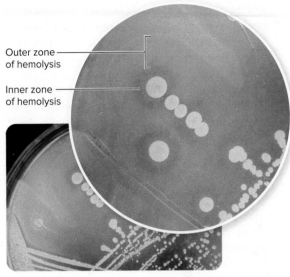

Outer zone of hemolysis

Inner zone of hemolysis

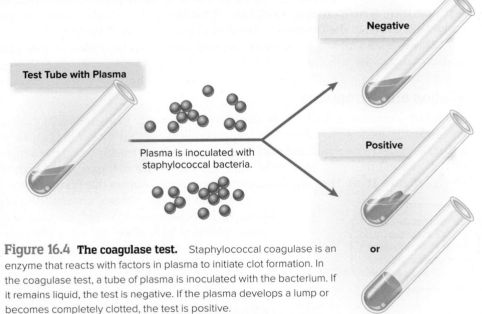

Negative

Test Tube with Plasma

Plasma is inoculated with staphylococcal bacteria.

Positive

or

Figure 16.4 **The coagulase test.** Staphylococcal coagulase is an enzyme that reacts with factors in plasma to initiate clot formation. In the coagulase test, a tube of plasma is inoculated with the bacterium. If it remains liquid, the test is negative. If the plasma develops a lump or becomes completely clotted, the test is positive.

Disease Table 16.1	MRSA Skin and Soft-Tissue Infection
Causative Organism(s)	Methicillin-resistant *Staphylococcus aureus* Ⓑ **G+**
Most Common Modes of Transmission	Direct contact, indirect contact
Virulence Factors	Coagulase, other enzymes, superantigens
Culture/Diagnosis	PCR, culture and Gram stain, coagulase and catalase tests, multitest systems
Prevention	Hygiene practices
Treatment	Vancomycin, linezolid, or daptomycin; in **Serious Threat** category in CDC Antibiotic Resistance Report
Epidemiology	Community-associated MRSA infections most common in children and young to middle-aged adults Incidence increasing in communities (decreasing in hospitals)

A Note About the Chapter Organization

In a clinical setting, patients present themselves to health care practitioners with a set of symptoms, and the health care team makes an "anatomical" diagnosis—such as a *generalized vesicular rash*. The anatomical diagnosis allows practitioners to narrow down the list of possible causes to microorganisms that are known to be capable of creating such a condition. Then the proper tests can be performed to arrive at an etiologic diagnosis (i.e., determining the exact microbial cause). So the order of events is as follows:

1. anatomical diagnosis,
2. differential diagnosis, and
3. etiologic diagnosis.

In this book, we organize diseases according to the anatomical diagnosis (which appears as a gold-colored heading). Then the agents in the differential diagnosis are each addressed, each of them appearing as turquoise headings. When we finish addressing each agent that could cause the condition, we sum them up in a Disease Table, whether there is only one possible cause or whether there are nine or ten.

In the Disease Tables you will also find a row featuring recommended treatment. Here we will identify the microbes that are on the CDC "Threat" list for their antibiotic resistance.

HIGHLIGHT DISEASE

Maculopapular Rash Diseases

There are a variety of microbes that can cause the type of skin eruptions classified as *maculopapular,* a term referring to flat to slightly raised colored bumps.

Measles

Every year hundreds of thousands of children in the developing world die from this disease (about 367 a day), even though an extremely effective vaccine has been available since 1963. Health campaigns all over the world seek to make the measles vaccine available to all and have been very effective. Ironically, it seems that more work and education need to be done in *developed* countries now. Many parents are opting to not have their children vaccinated, due to unfounded fears about the link between the vaccine and autism. We would do well to remember that before the vaccine was introduced, measles killed 6 million people worldwide each year. **Figure 16.5** depicts the sharp increase in measles cases in the United States in 2019. The year 2019 also saw a massive measles epidemic on the Pacific island of Samoa, population 200,000. In that year, there were more than 5,600 cases of measles, and 81 deaths. Another major outbreak occurred in the Democratic Republic of the Congo in 2019–2020.

Measles is also known as **rubeola.** Be very careful not to confuse it with the next maculopapular rash disease, rubella.

▶ Signs and Symptoms

The initial symptoms of measles are sore throat, dry cough, headache, conjunctivitis, lymphadenitis, and fever. In a short time, unusual oral lesions called *Koplik's spots* appear as a prelude to the characteristic red maculopapular **exanthem** (eg-zan′-thum) that erupts on the head and then progresses to the trunk and extremities until most of the body is covered **(figure 16.6).** The rash gradually coalesces into red patches that fade to brown.

In a small number of cases, children develop laryngitis, bronchopneumonia, and bacterial secondary infections such as ear and sinus infections. Occasionally (1 in 100 cases), measles progresses to pneumonia or encephalitis, resulting in various central nervous system (CNS) changes ranging from disorientation to coma. Permanent brain damage or epilepsy can result.

The most serious complication is **subacute sclerosing panencephalitis (SSPE),** a progressive neurological degeneration of the cerebral cortex, white matter, and brain stem. Once thought to be exceedingly rare (occurring in about 1 per million measles cases), a 2016 study found that the incidence is actually 1 in less than a thousand in children who get measles before being vaccinated. The pathogenesis of SSPE appears to involve a defective virus, one that has lost its ability to form a capsid and be released from an infected cell. Instead, it spreads unchecked through the brain by cell fusion, gradually destroying neurons and accessory cells and breaking down myelin. The disease causes profound intellectual and neurological impairment. The disease can occur years after the initial measles infection, and invariably leads to coma and death in a matter of months or years.

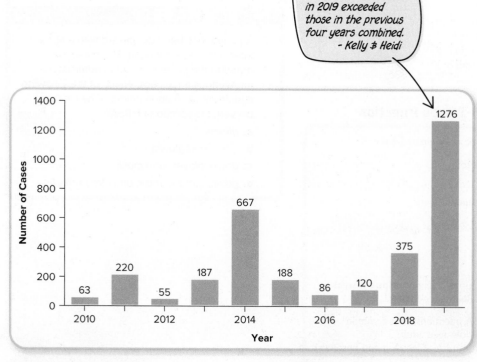

Figure 16.5 **Number of measles cases in the United States by year.**

▶ Pathogenesis and Virulence Factors

The virus implants in the respiratory mucosa and infects the tracheal and bronchial cells. From there, it travels to the lymphatic system, where it multiplies and then enters the bloodstream, a condition known as *viremia*. The bloodstream carries the virus to the skin and to various organs.

The measles virus induces the cell membranes of adjacent host cells to fuse into large syncytia (sin-sish'-uh), giant cells with many nuclei. These cells no longer perform their proper function. The virus seems proficient at disabling many aspects of the host immune response, especially cell-mediated immunity and delayed-type hypersensitivity. The virus also tends to erase or diminish the immune system's memory for other microbes, making its effects especially dangerous.

▶ Transmission and Epidemiology

Measles is one of the most contagious infectious diseases, transmitted principally by respiratory droplets. There is no reservoir other than humans, and a person is infectious during the periods of incubation, prodrome phase, and the skin rash but usually not during convalescence. In December 2014, a measles outbreak began in the Disney theme parks in California. Eventually 111 measles cases would be associated with that outbreak. Many of the people diagnosed with measles were unvaccinated: Some of them were babies too young to be vaccinated, but many were children and adults who chose not to be vaccinated. In July 2015, a woman in Washington State died of measles, the first measles death in the United States in 12 years.

▶ Culture and Diagnosis

The disease can be diagnosed on clinical presentation alone; but if further identification is required, an ELISA test is available that tests for patient IgM to measles antigen, indicating a current infection. A PCR test is also available.

▶ Prevention

The MMR vaccine (for measles, mumps, and rubella) contains live attenuated measles virus. A vaccine called ProQuad protects against measles, mumps, rubella, and varicella. Measles immunization is recommended for all healthy children at the age of 12 to 15 months, with a booster before the child enters school.

▶ Treatment

Treatment relies on reducing fever, suppressing cough, and replacing lost fluid. Complications require additional remedies to relieve neurological and respiratory symptoms and to sustain nutrient, electrolyte, and fluid levels. Vitamin A supplements are recommended by some physicians. They have been found effective in reducing the symptoms and decreasing the rate of complications.

Rubella

This disease is also known as German measles. Rubella is derived from the Latin for "little red," and that is a good way to remember it because it causes a relatively minor rash disease with few complications. Sometimes it is called the 3-day measles. The only exception to this mild course of events is when a fetus is exposed to the virus while in its mother's womb (*in utero*). Serious damage can occur, and for that reason women of childbearing years must be sure to be vaccinated before they plan to conceive.

▶ Signs and Symptoms

The two clinical forms of rubella are referred to as postnatal infection, which develops in children or adults, and **congenital** (prenatal) infection of the fetus, expressed in the newborn as various types of birth defects.

Postnatal Rubella   A rash of pink macules and papules first appears on the face and progresses down the trunk and toward the extremities, advancing and resolving in

Figure 16.6 **The rash of measles.**
Source: CDC

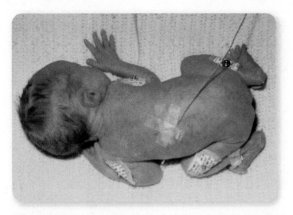

Figure 16.7 An infant born with congenital rubella can display a papular pink or purple rash.

James Stevenson/Science Source

A Note About Statistics in the Disease Tables

Each condition we study is summarized in a disease table. The last row of each table contains information about the epidemiology of the disease. The type of epidemiological information that is most relevant to a particular disease can vary. For example, it is vitally important to know the numbers of new cases (incidence) of some diseases. This is the case for measles in the United States (see Disease Table 16.2), as we track the resurgence of a disease once controlled by vaccination. For other conditions, such as fifth disease in the same table, it is more useful to know how many people have been affected by the time they reach a certain age. Prevalence, or the current number of people affected by the condition, is another common measure. And for some diseases the most informative statistic is how deadly they are (mortality rate). These tables contain the most useful information about each condition.

Source: CDC

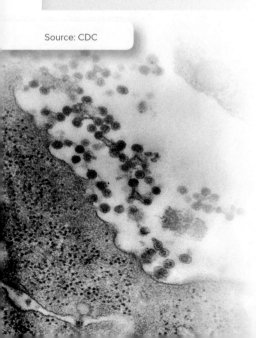

about 3 days. The rash is milder looking than the measles rash **(Disease Table 16.2).** Adult rubella is often characterized by joint inflammation and pain rather than a rash.

Congenital Rubella Rubella is a strongly **teratogenic** virus. Transmission of the rubella virus to a fetus *in utero* can result in a serious complication called **congenital rubella (figure 16.7).** The mother is able to transmit the virus even if she is asymptomatic. Infection in the first trimester is most likely to induce miscarriage or multiple permanent defects in the newborn. The most common of these is deafness and may be the only defect seen in some babies. Other babies may experience cardiac abnormalities, ocular lesions, deafness, and mental and physical retardation in varying combinations. Less drastic sequelae that usually resolve in time are anemia, hepatitis, pneumonia, carditis, and bone infection.

▶ Causative Agent

The rubella virus is a *Rubivirus,* in the family *Togaviridae.* The virus has the ability to stop mitosis, which is an important process in a rapidly developing embryo and fetus. It also induces apoptosis (programmed cell death) of normal tissue cells. This inappropriate cell death can do irreversible harm to organs it affects. And last, the virus damages vascular endothelium, leading to poor development of many organs.

▶ Transmission and Epidemiology

Rubella is a disease with worldwide distribution. Infection is initiated through contact with respiratory secretions and occasionally urine. The virus is shed during the prodromal phase and up to a week after the rash appears. Congenitally infected infants are contagious for a much longer period of time. Because the virus is only moderately communicable, close living conditions are required for its spread. Rubella and congenital rubella syndrome were declared eliminated from the United States in 2004, and from the Americas in 2015, but it is still common in other parts of the world.

▶ Culture and Diagnosis

Because it mimics other diseases, rubella should not be diagnosed on clinical grounds alone. IgM antibody to rubella virus can be detected early using an ELISA technique or a latex-agglutination card. Women in developed countries routinely undergo antibody testing at the beginning of pregnancy to determine their immune status.

▶ Prevention and Treatment

The attenuated rubella virus vaccine is usually given to children in the combined form (MMR or MMRV vaccination) at 12 to 15 months and a booster at 4 or 6 years of age. The vaccine for rubella can be administered on its own, without the measles and mumps components. Postnatal rubella is generally benign and requires only symptomatic treatment. No specific treatment is available for the congenital manifestations.

Fifth Disease

This disease, more precisely called *erythema infectiosum,* is so named because about 100 years ago it was the fifth of the diseases recognized by doctors to cause rashes in children. The first four were scarlet fever, measles, rubella, and another rash that was thought to be distinct but was probably not. Fifth disease is a very mild disease that often results in a characteristic "slapped-cheek" appearance because of a confluent reddish rash that begins on the face. Within 2 days, the rash spreads on the body but is most prominent on the arms, legs, and trunk. The rash of fifth disease may reoccur for several weeks and may be brought on by any activity that increases body heat (i.e., exercise, fever, sunlight, warm baths, and even high emotion).

The causative agent is parvovirus B19. You may have heard of "parvo" as a disease of dogs, but strains of this virus group infect humans. Fifth disease is usually diagnosed by the clinical presentation, but sometimes it is helpful to rule out rubella by testing for IgM against rubella.

This infection is very contagious. There is no vaccine and no treatment for this usually mild disease.

Disease Table 16.2 Maculopapular Rash Diseases

Disease	Measles (Rubeola)	Rubella	Fifth Disease	Roseola	Other Conditions to Consider
Causative Organism(s)	Measles virus Ⓥ	Rubella virus Ⓥ	Parvovirus B19 Ⓥ	Human herpesvirus 6 Ⓥ	Other conditions resulting in a rash that may look similar to these maculopapular conditions should be considered:
Most Common Modes of Transmission	Droplet contact	Droplet contact	Droplet contact, direct contact	Unknown	
Virulence Factors	Syncytium formation, ability to suppress CMI	In fetuses: inhibition of mitosis, induction of apoptosis, and damage to vascular endothelium	–	Ability to remain latent	Zika virus disease (covered in chapter 17)
Culture/Diagnosis	Clinical diagnosis; ELISA or PCR	Acute IgM, acute/convalescent IgG	Usually diagnosed clinically	Usually diagnosed clinically	Scarlet fever (chapter 19) Secondary syphilis (chapter 21)
Prevention	Live attenuated vaccine (MMR or MMRV)	Live attenuated vaccine (MMR or MMRV)	–	–	Rocky Mountain spotted fever (chapter 18)
Treatment	No antivirals; vitamin A, antibiotics for secondary bacterial infections	–	–	–	
Distinguishing Features of the Rashes	Starts on head, spreads to whole body, lasts over a week	Milder red rash, lasts approximately 3 days	"Slapped-face" rash first, spreads to limbs and trunk, tends to be confluent rather than distinct bumps	High fever precedes rash stage; rash not always present	
Epidemiological Features	Incidence increasing in North America; in developing countries incidence is 30 million cases/yr and 1 million deaths	Eliminated from the Americas; worldwide: 100,000 infants/yr born with congenital rubella syndrome	60% of population seropositive by age 20	Nearly 100% seropositive; 90% of disease cases occur before age of 2	
Appearance of Lesions	Source: CDC	Source: CDC	Dr. P. Marazzi/Science Source	Scott Camazine/Science Source	

Roseola

This disease is common in young children and babies. It is sometimes known as "sixth disease." It can result in a maculopapular rash, but a high percentage (up to 70%) of cases proceed without the rash stage. Children sick with this disease exhibit a high fever (up to 41°C, or 105°F) that comes on quickly and lasts for up to 3 days. Seizures may occur during this period, but other than that patients remain alert and do not act terribly ill. On the fourth day, the fever disappears, and it is at this point that a rash can appear, first on the chest and trunk and less prominently on the face and limbs. By the time the rash appears, the disease is almost over.

Roseola is caused by a human herpesvirus called HHV-6. Like all herpesviruses, it can remain latent in its host indefinitely after the disease has cleared. Very occasionally, the virus reactivates in childhood or adulthood, leading to mononucleosis-like or hepatitis-like symptoms. It is thought that 100% of the U.S. population is infected with this virus by adulthood. Some people experienced the disease roseola when they became infected, and some of them did not. The suggestion has been made that this virus causes other disease conditions later in life, such as multiple sclerosis. No vaccine and no treatment exist for roseola **Disease Table 16.2.**

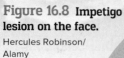

Figure 16.8 Impetigo lesion on the face.
Hercules Robinson/Alamy

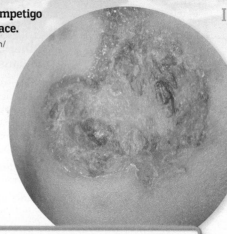

Impetigo

Impetigo is a superficial bacterial infection that causes the skin to flake or peel off **(figure 16.8)**. It is not a serious disease but is highly contagious, and children are the primary victims. Impetigo can be caused by either *Staphylococcus aureus* or *Streptococcus pyogenes*, and some cases are probably caused by a mixture of the two. It has been suggested that *S. pyogenes* begins most cases of the disease, and in some cases *S. aureus* later takes over and becomes the predominant bacterium cultured from lesions. Because *S. aureus* produces a bacteriocin (toxin) that can destroy *S. pyogenes*, it is possible that *S. pyogenes* is often missed in culture-based diagnosis.

Impetigo, whether it is caused by *S. pyogenes*, *S. aureus*, or both, is highly contagious and transmitted through direct contact but also via fomites and mechanical vector transmission. The peak incidence is in the summer and fall.

The only current prevention for impetigo is good hygiene. Vaccines are in development for both of the etiologic agents, but none are currently available.

▶ Signs and Symptoms

The "lesion" of impetigo looks variously like peeling skin, crusty and flaky scabs, or honey-colored crusts. Lesions are most often found around the mouth, face, and extremities, though they can occur anywhere on the skin, and occasionally in the vicinity of local trauma to the skin. It is very superficial and it itches. The symptomatology does not indicate whether the infection is caused by *Staphylococcus* or *Streptococcus*.

Impetigo Caused by *Staphylococcus aureus*

The most important virulence factors relevant to *S. aureus* impetigo are exotoxins called exfoliative toxins A and B, which are encoded by a phage that infects some *S. aureus* strains. At least one of the toxins attacks a protein that is very important for epithelial cell-to-cell binding in the outermost layer of the skin. Breaking up this protein leads to the characteristic blistering seen in the condition. The breakdown of skin architecture also facilitates the spread of the bacterium.

Impetigo Caused by *Streptococcus pyogenes*

Streptococcus pyogenes is thoroughly described in chapter 19. The important features are briefly summarized here, and the features pertinent to impetigo are listed in **Disease Table 16.3.**

S. pyogenes is a gram-positive coccus and is beta-hemolytic on blood agar. In addition to impetigo, it causes streptococcal pharyngitis (strep throat), scarlet fever, pneumonia, puerperal fever, necrotizing fasciitis, serious bloodstream infections, and poststreptococcal conditions such as rheumatic fever.

If the precise etiologic agent must be identified, there are well-established methods for identifying group A streptococci. Refer to chapter 19.

▶ Pathogenesis and Virulence Factors

Like *S. aureus*, this bacterium possesses a huge arsenal of enzymes and toxins. Some of these are listed in Disease Table 16.3.

Rarely, impetigo caused by *S. pyogenes* can be followed by acute poststreptococcal glomerulonephritis (see chapter 19). The strains that cause impetigo never cause rheumatic fever, however.

S. pyogenes is more often the cause of impetigo in newborns, and *S. aureus* is more often the cause of impetigo in older children, but both can cause infection in either age group.

NCLEX® PREP

3. A patient has been diagnosed with impetigo but has been noncompliant with medical treatment. What complication is associated with untreated impetigo?
   a. *future antibiotic allergies*
   b. *measles*
   c. *increased likelihood of roseola*
   d. *glomerulonephritis*
   e. *septicemia*

Medical Moment

Scabies

Scabies is a skin disease caused by infestation of the human itch mite (*Sarcoptes* sp.). It is typical for only 10 to 15 mites to invade initially. The mites burrow into the epidermis, where the female lives and lays eggs. The eggs hatch and larva mature in pouches underneath the skin's surface. The site of mite infestation develops a papular rash and can be extremely itchy. The burrowing sites may be visible as thin gray lines under the skin. Symptoms develop 2 to 6 weeks after infestation for those who have not been exposed to scabies previously.

Scabies infection is transmitted by prolonged skin contact or occasionally by fomites. It occurs most commonly in crowded areas, such as child care facilities, prisons, and nursing homes. Medication is available to kill the mites and their eggs. It is recommended that linens and clothing of infected persons and their household contacts be washed and dried in high heat to prevent transmission or reinfection.

Q. What does the fact that scabies can be transmitted by fomites suggest about the mite's need for a living host?

Answer in Appendix B.

Source: National Institute of Allergy and Infectious Diseases (NIAID)/NIH/USHHS

Disease Table 16.3	Impetigo	
Causative Organism(s)	*Staphylococcus aureus* Ⓑ G+	*Streptococcus pyogenes* Ⓑ G+
Most Common Modes of Transmission	Direct contact, indirect contact	Direct contact, indirect contact
Virulence Factors	Exfoliative toxin A, coagulase, other enzymes	Streptokinase, plasminogen-binding ability, hyaluronidase, M protein
Culture/Diagnosis	Routinely based on clinical signs; when necessary, culture and Gram stain, coagulase and catalase tests, multitest systems, PCR	Routinely based on clinical signs; when necessary, culture and Gram stain, coagulase and catalase tests, multitest systems, PCR
Prevention	Hygiene practices	Hygiene practices
Treatment	In uncomplicated cases, no treatment; topical mupirocin or retapamulin, oral dicloxacillin, cephalexin, or TMP-SMZ; (MRSA is in **Serious Threat** category in CDC Antibiotic Resistance Report)	In uncomplicated cases, no treatment; topical mupirocin or retapamulin; (erythromycin-resistant *Streptococcus pyogenes* is in **Concerning Threat** category in CDC Antibiotic Resistance Report)
Distinguishing Features	Seen more often in older children, adults	Seen more often in newborns
Epidemiological Features	Incidence approximately 2–5% of children in temperate climates; higher in tropical areas	

Disease Table 16.4	Cellulitis		
Causative Organism(s)	*Streptococcus pyogenes* Ⓑ G+	Methicillin-resistant *Staphylococcus aureus* Ⓑ G+	Other bacteria or fungi
Most Common Modes of Transmission	Parenteral implantation	Parenteral implantation	Parenteral implantation
Virulence Factors	Streptokinase, plasminogen-binding ability, hyaluronidase, M protein	Exfoliative toxin A, coagulase, other enzymes	–
Culture/Diagnosis	Based on clinical signs	Based on clinical signs	Based on clinical signs
Prevention	–	–	–
Treatment	Oral or IV antibiotic (penicillin); surgery sometimes necessary; (erythromycin-resistant *Streptococcus pyogenes* is in **Concerning Threat** category in CDC Antibiotic Resistance Report)	Oral or IV antibiotic (cefazolin, ceftriaxone, vancomycin, linezolid); surgery sometimes necessary; (MRSA is in **Serious Threat** category in CDC Antibiotic Resistance Report)	Aggressive treatment with oral or IV antibiotic; surgery sometimes necessary
Distinguishing Features	–	–	More common in immunocompromised
Epidemiological Features		Incidence highest among males 45-64	

Cellulitis

Cellulitis is a condition caused by a fast-spreading infection in the dermis and in the subcutaneous tissues below. It causes pain, tenderness, swelling, and warmth. Fever and swelling of the lymph nodes draining the area may also occur. Frequently, red lines leading away from the area are visible (a phenomenon called *lymphangitis*). This symptom is the result of microbes and inflammatory products being carried by the lymphatic system. Bacteremia could develop with this disease, but uncomplicated cellulitis has a good prognosis.

Cellulitis generally follows introduction of bacteria or fungi into the dermis, either through trauma or by subtle means (with no obvious break in the skin). The most common causes of the condition in healthy people are *Streptococcus pyogenes*, and occasionally *Staphylococcus aureus*, although almost any bacterium and some fungi can cause this condition in an immunocompromised patient. In infants, group B streptococci are a frequent cause (see chapter 21).

Medical Moment

Scrum Pox: Herpes Gladiatorum

Herpes gladiatorum is caused by herpes simplex virus type 1 (HSV-1). It often affects athletes such as wrestlers who come into close contact with each other (it is sometimes called "scrum pox" or "mat herpes" by rugby players or wrestlers). Outbreaks of the infection may occur in gyms or fitness centers.

The lesions occur in clusters of blisters that may or may not be painful. The lesions generally heal within a week to 10 days. Once a person has been infected with HSV-1, the person is infected for life and can pass the infection on to others. Periodic outbreaks can occur during times of stress, trauma, or sun exposure. During the first episode, the infected person may experience fever, sore throat, headache, and swollen glands. Some individuals can be infected with the virus but will fail to develop any skin lesions. These individuals carry the virus and can still pass the virus to others through skin-to-skin contact.

Q. What are the official names for the modes of transmission for herpes gladiatorum?

Answer in Appendix B.

Mild cellulitis responds well to oral antibiotics chosen to be effective against both *S. aureus* and *S. pyogenes*. Keep in mind that if it is caused by *S. aureus*, the bacterium is usually of the methicillin-resistant variety (MRSA). More involved infections and infections in immunocompromised people require intravenous antibiotics. If there are extensive areas of tissue damage, surgical debridement (duh-breed'-munt) is warranted **Disease Table 16.4.**

Staphylococcal Scalded Skin Syndrome (SSSS)

This syndrome is another **dermolytic** condition caused by *Staphylococcus aureus*. It affects mostly newborns and babies, although children and adults can experience the infection. Transmission may occur when caregivers carry the bacterium from one baby to another. Adults in the nursery can also directly transfer *S. aureus* because approximately 30% of adults are asymptomatic carriers. Carriers can harbor the bacteria in the nasopharynx, axilla, perineum, and even the vagina. (Fortunately, only about 5% of *S. aureus* strains are lysogenized by the type of phage that codes for the toxins responsible for this disease.)

Like impetigo, this is an exotoxin-mediated disease. The phage-encoded exfoliative toxins A and B are responsible for the damage. Unlike impetigo, the toxins enter the bloodstream from some focus of infection (the throat, the eye, or sometimes an impetigo infection) and then travel to the skin throughout the body. These toxins cause **bullous lesions,** which often appear first around the umbilical cord (in neonates) or in the diaper or axilla area. A split occurs in the epidermal tissue layers just above the stratum basale (see figure 16.1). Widespread desquamation of the skin follows, leading to the burned appearance referred to in the name **(figure 16.9).**

Once a tentative diagnosis of SSSS is made, immediate antibiotic therapy should be instituted **Disease Table 16.5.**

Vesicular or Pustular Rash Diseases

There are two diseases that present as generalized "rashes" over the body in which the individual lesions contain fluid. The lesions are often called *pox*, and the two diseases are chickenpox and smallpox **(Disease Table 16.6).** Chickenpox is very common and mostly benign, but even a single case of smallpox constitutes a public health emergency. Both are viral diseases.

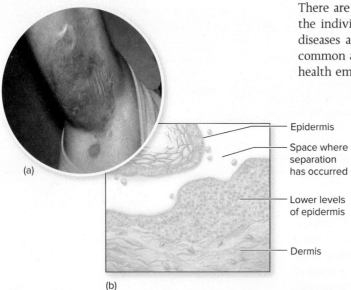

Figure 16.9 Staphylococcal scalded skin syndrome (SSSS) in the underarm area. **(a)** Exfoliative toxin produced in local infections causes blistering and peeling away of the outer layer of skin. **(b)** Drawing of a segment of skin affected with SSSS. The point of epidermal shedding, or desquamation, is in the epidermis. The lesions will heal well because the level of separation is so superficial.

DermPics/Science Source (a)

Disease Table 16.5	Scalded Skin Syndrome
Causative Organism(s)	*Staphylococcus aureus* **B** G+
Most Common Modes of Transmission	Direct contact, droplet contact
Virulence Factors	Exfoliative toxins A and B
Culture/Diagnosis	Histological sections; culture performed but false negatives common because toxins alone are sufficient for disease
Prevention	Eliminate carriers in contact with neonates
Treatment	Immediate systemic antibiotics (current recommendation is cloxacillin)
Distinguishing Features	Split in skin occurs *within* epidermis
Epidemiological Features	Mortality 1%-5% in children, 50%-60% in adults

Chickenpox

After an incubation period of 10 to 20 days, the first symptoms to appear are fever and an abundant rash that begins on the scalp, face, and trunk and radiates in sparse crops to the extremities. Skin lesions progress quickly from macules and papules to itchy vesicles filled with a clear fluid. In several days, they encrust and drop off, usually healing completely but sometimes leaving a tiny pit or scar. Lesions number from a few to hundreds and are more abundant when they are in adolescents and adults than in young children. **Figure 16.10** contains images of the chickenpox lesions in children and adults. The lesion distribution is *centripetal*, meaning that there are more in the center of the body and fewer on the extremities. (The distribution seen with smallpox is just the opposite.) The illness usually lasts 4 to 7 days. New lesions stop appearing after about 5 days. Patients are considered contagious until all of the lesions have crusted over.

Approximately 0.1% of chickenpox cases are followed by encephalopathy, or inflammation of the brain caused by the virus. It can be fatal, but in most cases recovery is complete.

▶ Shingles

After recuperation from chickenpox, the virus enters into the sensory nerve endings of cutaneous spinal nerve branches, especially those that serve the skin of the chest and head. From there, it becomes latent in the ganglia and may reemerge as **shingles** (also known as **herpes zoster**) with its characteristic asymmetrical distribution on the skin of the trunk or head **(figure 16.11).**

Shingles develops abruptly after reactivation by such stimuli as psychological stress, X-ray treatments, immunosuppressive and other drug therapy, surgery, or a

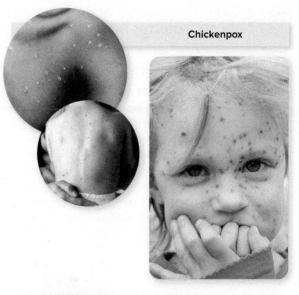
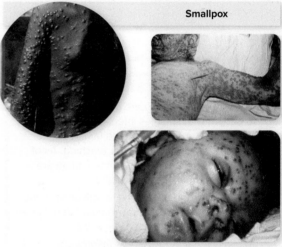

Figure 16.10 Images of chickenpox and smallpox.
David White/Alamy (chickenpox on face); Picture Partners/Alamy (chickenpox on back); Source: CDC (chickenpox on shoulder); Source: CDC/Dr. Robinson (smallpox on torso and arm); Source: CDC/Dr. John Noble, Jr. (smallpox on face); Everett Collection Historical/Alamy (smallpox on back)

The fact that the shingles lesions stop at the body's midline helps with diagnosis. This is because the virus moves down dermatomes on only one side of the spine.
- Kelly & Heidi

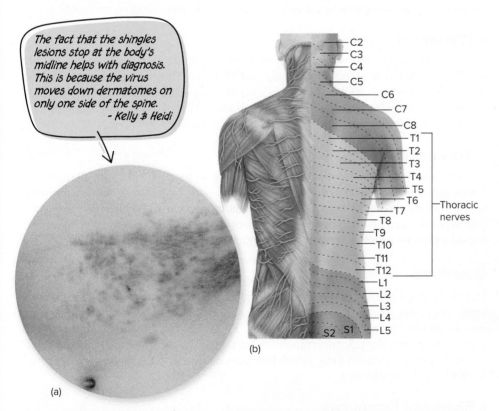

Figure 16.11 Varicella-zoster virus reemergence as shingles. **(a)** Clinical appearance of shingles lesions. **(b)** Dermatomes are areas of the skin served by a single ganglion. Shingles generally affects a single dermatome for this reason.
Franciscodiazpagador/Getty Images (a)

Note on Bioterror Agents

The Centers for Disease Control and Prevention maintains a list of the most dangerous infectious agents that would be most logically used for a bioterror attack. There are three categories: A, B, and C. Category A agents (the most dangerous) are those that are (1) easily disseminated, or transmissible person-to-person; (2) result in high mortality rates; (3) could incite panic and social disruption; and (4) require special or unique actions for preparedness. Category A agents will be pointed out in each of the disease tables in the next six chapters.

NCLEX® PREP

4. A nurse in an urgent care office is caring for a patient with shingles infection. Following education for the patient and family, the nurse evaluates the efficacy of her teaching. All of the following statements by the patient are true regarding shingles, except:
   a. *Shingles is caused by a reactivation of the varicella virus that may remain latent in the ganglia of nerve fibers for years.*
   b. *Shingles may cause eye inflammation and facial paralysis.*
   c. *Psychological stress can reactivate the virus that causes shingles.*
   d. *You can come down with shingles by being exposed to the fluid in shingles lesions.*

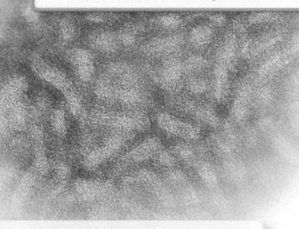

Vaccinia virus, the cause of cowpox, is used as a vaccine against smallpox, caused by the variola virus.
Source: CDC

developing malignancy. The virus is believed to migrate down the ganglion to the skin, where multiplication resumes and produces crops of tender, persistent vesicles. Inflammation of the ganglia and the pathways of nerves can cause pain and tenderness, known as postherpetic neuralgia, that can last for several months. Involvement of cranial nerves can lead to eye inflammation and ocular and facial paralysis.

▶ Causative Agent

Human herpesvirus 3 (HHV-3, also called **varicella** [var″-ih″sel′-ah]) causes chickenpox, as well as the condition called herpes zoster or shingles. The virus is sometimes referred to as the varicella-zoster virus (VZV). Like other herpesviruses, it is an enveloped DNA virus.

▶ Pathogenesis and Virulence Factors

HHV-3 enters the respiratory tract, attaches to respiratory mucosa, and then invades and enters the bloodstream. The viremia disseminates the virus to the skin, where the virus causes adjacent cells to fuse and eventually lyse, resulting in the characteristic lesions. The virus enters the sensory nerves at this site, traveling to the dorsal root ganglia.

The ability of HHV-3 to remain latent in ganglia is an important virulence factor because resting in this site protects it from attack by the immune system and provides a reservoir of virus for the reactivation condition of shingles.

▶ Transmission and Epidemiology

Humans are the only natural hosts for HHV-3. The virus is harbored in the respiratory tract but is communicable from both respiratory droplets and the fluid of active skin lesions. People can acquire a chickenpox infection by being exposed to the fluid of shingles lesions.

Infected persons are most infectious a day or two prior to the development of the rash. Chickenpox is so contagious that if you are exposed to it, you almost certainly will get it.

▶ Prevention

Live attenuated vaccine was licensed in 1995. There are two vaccines specifically for the prevention of shingles: Zostavax and Shingrix. They are intended for adults age 60 and over and are for the prevention of shingles. A large study has recently shown that vaccinated children have a much lower risk of developing shingles than do unvaccinated kids who contract chickenpox naturally.

▶ Treatment

Uncomplicated varicella is self-limiting and requires no therapy aside from alleviation of discomfort. Secondary bacterial infection is treated with topical or systemic antibiotics. Oral acyclovir or related antivirals should be administered within 24 hours of onset of the rash to people considered to be at risk for serious complications. Under no circumstances should aspirin be administered, as this may lead to Reye's syndrome.

Smallpox

Largely through the World Health Organization's comprehensive global efforts, naturally occurring smallpox is now a disease of the past. However, after the terrorist attacks on the United States on September 11, 2001, and the anthrax bioterrorism shortly thereafter, the U.S. government began taking the threat of smallpox bioterrorism very seriously. Vaccination, which had been discontinued, was once again offered to certain U.S. populations.

▶ Signs and Symptoms

Infection begins with fever and malaise, and later a rash begins in the pharynx, spreads to the face, and progresses to the extremities. Initially, the rash is *macular*, evolving in turn to *papular*, *vesicular*, and *pustular* before eventually crusting over, leaving nonpigmented sites pitted with scar tissue. There are two principal forms of smallpox: variola minor and variola major. Variola major is a highly virulent form that causes toxemia, shock, and intravascular coagulation. People who have survived any form of smallpox nearly always develop lifelong immunity.

It is vitally important for health care workers to be able to recognize the early signs of smallpox. The diagnosis of even a single suspected case must be treated as a health and law enforcement emergency. The symptoms of variola major progress as follows: After the prodrome period of high fever and malaise, a rash emerges, first in the mouth. Severe abdominal and back pain sometimes accompany this phase of the disease. A rash appears on the skin and spreads throughout the body within 24 hours. The rash typically occurs more on the extremities (a centrifugal distribution; see figure 16.10).

By the third or fourth day of the rash, the bumps become larger and fill with a thick opaque fluid. A major distinguishing feature of this disease is that the pustules are indented in the middle. Also, patients report that the lesions feel as if they contain a BB pellet. Within a few days, these pustules begin to scab over. After 2 weeks, most of the lesions will have crusted over. The patient remains contagious until the last scabs fall off because the crusts contain the virus. During the entire rash phase, the patient is very ill. The lesions occur at the dermal level, which is the reason that scars remain after the lesions are healed.

A patient with variola minor has a rash that is less dense and is generally less ill than someone with variola major.

Typical scar from a smallpox lesion
©SPL/Science Source

▶ Causative Agent

The causative agent of smallpox, the variola virus, is an orthopoxvirus, an enveloped DNA virus. Other members of this group are the monkeypox virus and the vaccinia virus from which smallpox vaccine is made. Variola is a hardy virus, surviving outside the host longer than most viruses.

▶ Transmission and Epidemiology

Smallpox is spread primarily through droplets, although fomites such as contaminated bedding and clothing can also spread it.

In the early 1970s, smallpox was endemic in 31 countries. Every year, 10 to 15 million people contracted the disease, and approximately 2 million people died from it. By 1977, after 11 years of intensive effort by the world health community, the last natural case occurred in Somalia. Currently, samples of stored smallpox virus exist in only two verified locations, at the CDC in Atlanta, Georgia, and at a state research center in Koltsovo, Russia.

▶ Prevention

In chapter 13, you read about Edward Jenner and his development of vaccinia virus to inoculate against smallpox. To this day, the vaccination for smallpox is based on the vaccinia virus. Immunizations were stopped in the United States in 1972. Since the terrorist events of 2001, some military branches are requiring that their personnel take the vaccination before deploying to certain parts of the world. Even though officially smallpox does not exist anymore except in labs in Russia and the United States, there are fears that other entities may have stockpiled the virus for use in biological warfare or terror attacks. That is why the United States maintains a stockpile of smallpox vaccine (and treatment) to use in the case of a smallpox attack.

Vaccination is also useful for postexposure prophylaxis, meaning that it can prevent or lessen the effects of the disease after you have already been exposed to it.

▶ Treatment

Two drugs have been approved by the Food and Drug Administration for smallpox. They are tecovirimat and cidofovir. The U.S. government also maintains a supply of these drugs to use in case of a public health emergency **Disease Table 16.6.**

Disease Table 16.6	Vesicular/Pustular Rash Diseases	
Disease	**Chickenpox**	**Smallpox**
Causative Organism(s)	Human herpesvirus 3 (varicella-zoster virus) Ⓥ	Variola virus Ⓥ
Most Common Modes of Transmission	Droplet contact, inhalation of aerosolized lesion fluid	Droplet contact, indirect contact
Virulence Factors	Ability to fuse cells, ability to remain latent in ganglia	Ability to dampen, avoid immune response
Culture/Diagnosis	Based largely on clinical appearance; PCR is available	Based largely on clinical appearance; if suspected refer to CDC
Prevention	Live attenuated vaccine; there is also vaccine to prevent reactivation of latent virus (shingles)	Live virus vaccine (vaccinia virus)
Treatment	None in uncomplicated cases; acyclovir for high risk	Tecovirimat, cidofovir
Distinguishing Features	No fever prodrome; lesions are superficial and in centripetal distribution (more in center of body)	Fever precedes rash, lesions are deep and in centrifugal distribution (more on extremities)
Epidemiological Features	Chickenpox: vaccine decreased hospital visits by 88%, ambulatory visits by 59%	Last natural case worldwide was in 1977; **Category A Bioterrorism Agent**
Appearance of Lesions	©Gabriel Blaj/Alamy	©Source: CDC/Dr. Charles Farmer, Jr.

Large Pustular Skin Lesions

Leishmaniasis

Two infections that result in large lesions (greater than a few millimeters across) deserve mention in this chapter on skin infections. The first is leishmaniasis, a zoonosis transmitted among various mammalian hosts by female sand flies. This infection can express itself in several different forms, depending on which species of the protozoan *Leishmania* is involved. Cutaneous leishmaniasis is a localized infection of the capillaries of the skin caused by *L. tropica*, found in Mediterranean, African, and Southeast Asian regions. A form of mucocutaneous leishmaniasis called espundia is caused by *L. brasiliensis*, endemic to parts of Central and South America. It affects both the skin and mucous membranes. Another form of this infection is systemic leishmaniasis.

Leishmania is transmitted to the mammalian host by the sand fly when it ingests the host's blood. The disease is endemic to equatorial regions that provide favorable conditions for the sand fly. At particular risk are travelers or immigrants who have never had contact with the protozoan and lack adaptive immunity.

There is no vaccine. Avoiding the sand fly is the only prevention.

Cutaneous Anthrax

This form of anthrax is the most common and least dangerous version of infection with *Bacillus anthracis*. (The spectrum of anthrax disease is discussed fully in chapter 18.) It is caused by endospores entering the skin through small cuts or

Disease Table 16.7	Large Pustular Skin Lesions	
Disease	**Leishmaniasis**	**Cutaneous Anthrax**
Causative Organism(s)	*Leishmania* spp. **P**	*Bacillus anthracis* **B G+**
Most Common Modes of Transmission	Biological vector (sand fly)	Direct contact with endospores
Virulence Factors	Multiplication within macrophages	Endospore formation; capsule, lethal factor, edema factor
Culture/Diagnosis	Usually microscopic visualization	Culture on blood agar; serology, PCR performed by CDC
Prevention	Avoiding sand fly	Avoid contact; vaccine available but not widely used
Treatment	Sodium stibogluconate, pentamidine	Ciprofloxacin, plus two additional antibiotics
Distinguishing Features	Mucocutaneous and systemic forms	Can be fatal
Epidemiological Features	Untreated visceral leishmaniasis mortality rate is 100%; 10% for cutaneous leishmaniasis	Untreated cutaneous anthrax mortality rate: 20%; treated mortality rate less than 1%; **Category A Bioterrorism Agent**
Appearance of Lesions	CDC	©Mediscan/Alamy Stock Photo

abrasions. Germination and growth of the pathogen in the skin are marked by the production of a papule that becomes increasingly necrotic and later ruptures to form a painless, black **eschar** (ess′-kar) **(Disease Table 16.7).** In the fall of 2001, 11 cases of cutaneous anthrax and 11 cases of inhalational anthrax occurred in the United States as a result of bioterrorism. Mail workers and others contracted the infection when endospores were sent through the mail. The infection can be naturally transmitted by contact with hides of infected animals (especially goats).

Left untreated, even the cutaneous form of anthrax is fatal approximately 20% of the time. A vaccine exists but is recommended only for high-risk persons and the military. **Disease Table 16.7.**

Cutaneous and Superficial Mycoses

Ringworm

A group of fungi that is collectively termed **dermatophytes** causes a variety of body surface conditions. These mycoses are strictly confined to the nonliving epidermal tissues (stratum corneum) and their derivatives (hair and nails). All these conditions have different names that begin with the word **tinea** (tin′-ee-ah), which derives from the erroneous belief that they were caused by worms. That misconception is also the reason these diseases are often called *ringworm*–ringworm of the scalp (tinea capitis), beard (tinea barbae), body (tinea corporis), groin (tinea cruris), foot (tinea pedis), and hand (tinea manuum). (Don't confuse these "tinea" terms with genus and species names. It is simply an old practice for naming the conditions.) Most of these conditions are caused by one of three different dermatophytes, which are discussed here.

The signs and symptoms of ringworm conditions are summarized in **table 16.1.**

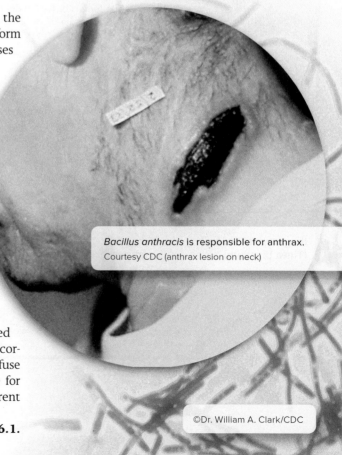

Bacillus anthracis is responsible for anthrax.
Courtesy CDC (anthrax lesion on neck)

©Dr. William A. Clark/CDC

Table 16.1 Signs and Symptoms of Cutaneous Mycoses

Ringworm of the Scalp (Tinea Capitis)	This mycosis results from the fungal invasion of the scalp and the hair of the head, eyebrows, and eyelashes.	
Ringworm of the Beard (Tinea Barbae)	This tinea, also called *barber's itch*, affects the chin and beard of adult males. Although once a common aftereffect of unhygienic barbering, it is now contracted mainly from animals.	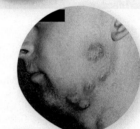
Ringworm of the Body (Tinea Corporis)	This extremely prevalent infection of humans can appear nearly anywhere on the body's glabrous (smooth and bare) skin.	
Ringworm of the Groin (Tinea Cruris)	Sometimes known as *jock itch*, crural ringworm occurs mainly in males on the groin, perianal skin, scrotum, and, occasionally, the penis. The fungus thrives under conditions of moisture and humidity created by sweating.	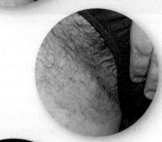
Ringworm of the Foot (Tinea Pedis)	Tinea pedis has more colorful names as well, including athlete's foot and jungle rot. Infections often begin with blisters between the toes that burst, crust over, and can spread to the rest of the foot and nails.	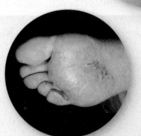
Ringworm of the Nail (Tinea Unguium)	Fingernails and toenails, being masses of keratin, are often sites for persistent fungus colonization. The first symptoms are usually superficial white patches in the nail bed. A more invasive form causes thickening, distortion, and darkening of the nail.	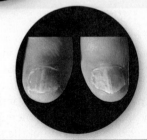

Source: CDC (tinea capitis); Source: CDC (tinea barbae); Biophoto Associates/Science Source (tinea corporis);
Dr. Harout Tanielian/Science Source (tinea cruris); Dr. P. Marazzi/Science Source (tinea pedis);
Dr Edwin P Ewing, Jr./CDC (tinea unguium)

▶ Causative Agents

There are about 39 species in the genera *Trichophyton*, *Microsporum*, and *Epidermophyton* that can cause the preceding conditions **(figure 16.12)**. The causative agent of a given type of ringworm varies from one geographic location to another and is not restricted to a particular genus and species.

Diagnosis of tinea of the scalp caused by some species is aided by use of a long-wave ultraviolet lamp that causes infected hairs to fluoresce. Samples of hair, skin scrapings, and nail debris treated with heated potassium hydroxide (KOH) show a thin, branching fungal mycelium if infection is present.

▶ Pathogenesis and Virulence Factors

The dermatophytes have the ability to invade and digest keratin, which is naturally abundant in the cells of the stratum corneum. They have also been found to suppress the ability of the immune system to respond to them, ensuring their long-term persistence. The fungi do not invade deeper epidermal layers.

▶ Transmission and Epidemiology

Transmission of the fungi that cause these diseases is via direct and indirect contact with other humans or with infected animals. Some of these fungi can be acquired from the soil. (Interestingly, cultures that walk barefoot most of the time have very low levels of athlete's foot, suggesting that it is the confined atmosphere of a shoe that encourages fungal growth.)

Therapy is usually a topical antifungal agent. Ointments containing tolnaftate, miconazole, itraconazole, terbinafine, or thiabendazole are applied regularly for several weeks. Some drugs work by speeding up loss of the outer skin layer.

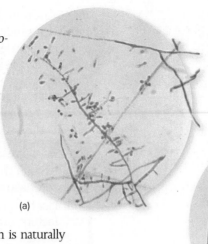

(a)

(b)

(c)

Figure 16.12 Examples of dermatophyte spores.
(a) Regular, numerous microconidia of *Trichophyton*. **(b)** Macroconidia of *Microsporum canis*, a cause of ringworm in cats, dogs, and humans. **(c)** Smooth-surfaced macroconidia in clusters characteristic of *Epidermophyton*.
Source: CDC/Dr. Lucille K. George (a–c)

Superficial Mycoses

Agents of **superficial mycoses** involve the outer epidermal surface and are ordinarily innocuous infections with cosmetic rather than outright disease-causing effects. Tinea versicolor is caused by the yeast genus *Malassezia*, a genus that has at least 10 species living on human skin. The yeast feeds on the high oil content of the skin glands. Even though these yeasts are very common (carried by nearly 100%

Disease Table 16.8	Cutaneous and Superficial Mycoses	
Disease	**Cutaneous Infections**	**Superficial Infections (Tinea Versicolor)**
Causative Organism(s)	*Trichophyton, Microsporum, Epidermophyton* Ⓕ	*Malassezia* species Ⓕ
Most Common Modes of Transmission	Direct and indirect contact, vehicle (soil)	Endogenous "normal biota"
Virulence Factors	Ability to degrade keratin, invoke hypersensitivity, avoidance of immune response	–
Culture/Diagnosis	Microscopic examination, KOH staining, culture	Usually clinical, KOH can be used
Prevention	Avoid contact	None
Treatment	Topical tolnaftate, itraconazole, terbinafine, miconazole, thiabendazole, oral terbinafine	Topical antifungals
Epidemiological Features	Among schoolchildren, 0%-19% prevalence, in humid climates up to 30%	Highest incidence among adolescents

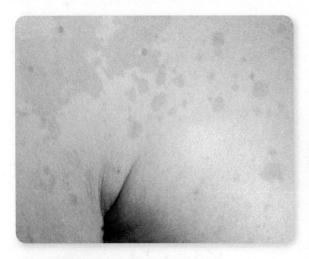

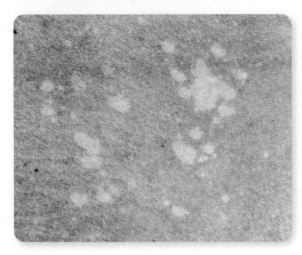

Figure 16.13 **Tinea versicolor.** Mottled, discolored skin pigmentation shown on two different skin tones. This is characteristic of superficial skin infection by *Malassezia furfur.*
Biophoto Associates/Science Source (light skin); 4FR/Getty Images (medium skin)

of humans tested), in some people its growth elicits mild, chronic scaling and interferes with production of pigment by melanocytes. The trunk, face, and limbs may take on a mottled appearance **(figure 16.13).** Other superficial skin conditions in which *Malassezia* is implicated are folliculitis, psoriasis, and seborrheic dermatitis (dandruff). A study in 2019 found that *Malassezia* species were found in the pancreas and could be associated with pancreatic cancer **Disease Table 16.8.**

16.3 LEARNING OUTCOMES—Assess Your Progress

4. Explain the important features of the "Highlight Disease" MRSA skin and soft-tissue infection.

5. List the possible causative agents, modes of transmission, virulence factors, diagnostic techniques, and prevention/treatment for the "Highlight Disease" maculopapular rash diseases.

6. Discuss important features of the other infectious skin diseases. These are impetigo, cellulitis, staphylococcal scalded skin syndrome, vesicular/pustular rash diseases, large pustular skin lesions, and cutaneous and superficial mycoses.

7. Discuss the relative dangers of rubella and measles (rubeola) viruses in different populations.

16.4 The Surface of the Eye and Its Defenses

The eye is a complex organ with many different tissue types, but for the purposes of this chapter we consider only its exposed surfaces, the *conjunctiva* and the *cornea* **(figure 16.14).** The **conjunctiva** is a very thin membranelike tissue that covers the eye (except for the cornea) and lines the eyelids. It secretes an oil- and mucus-containing fluid that lubricates and protects the eye surface. The **cornea** is the dome-shaped central portion of the eye lying over the iris (the colored part of the eye). It has five to six layers of epithelial cells that can regenerate quickly if they are superficially damaged. It has been called "the windshield of the eye."

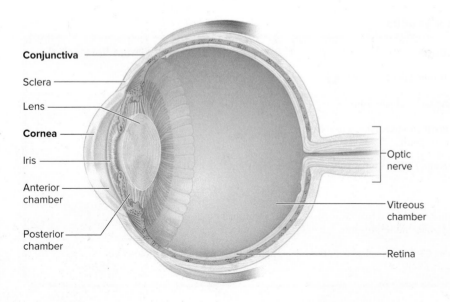

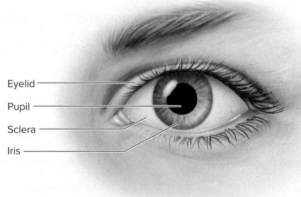

Figure 16.14 **The anatomy of the eye.**

The eye's best defense is the film of tears, which consists of an aqueous fluid, oil, and mucus. The tears are formed in the lacrimal gland at the outer and upper corner of each eye **(figure 16.15),** and they drain into the lacrimal duct at the inner corner. Tears contain sugars, lysozyme, and lactoferrin. These last two substances have antimicrobial properties. The mucous layer contains proteins and sugars and plays a protective role. And, of course, the flow of the tear film prevents the attachment of microorganisms to the eye surface.

Because the eye's primary function is vision, anything that hinders vision would be counterproductive. For that reason, inflammation does not occur in the eye as readily as it does elsewhere in the body. Flooding the eye with fluid containing a large number of light-diffracting objects, such as lymphocytes and phagocytes, in response to every irritant would mean almost constantly blurred vision. So even though the eyes are relatively vulnerable to infection (not being covered by keratinized epithelium), the evolution of the vertebrate eye has been toward reduced innate immunity and a corresponding reduction in inflammatory response. This characteristic is sometimes known as **immune privilege.**

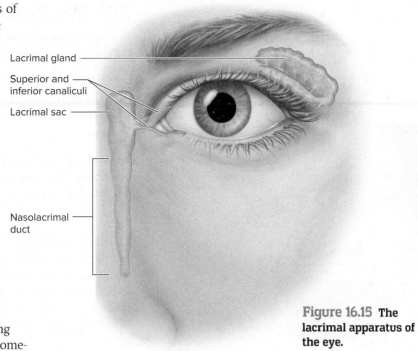

Lacrimal gland
Superior and inferior canaliculi
Lacrimal sac
Nasolacrimal duct

Figure 16.15 The lacrimal apparatus of the eye.

16.5 Normal Biota of the Eye

The eye was previously thought to be only sparsely populated by microbiota. 16S rRNA analysis of the healthy eye microbiome has revealed a more robust population, showing a lot of diversity in the bacteria found. In many cases *Corynebacterium* is the dominant genus. To a large extent, the eye microbiome resembles that of the skin.

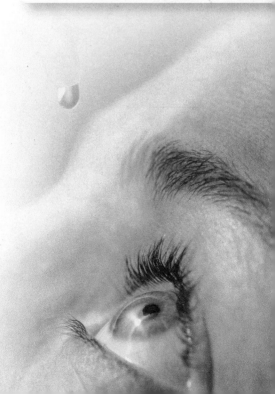

Defenses and Normal Biota of the Eyes		
	Defenses	**Normal Biota**
Eyes	Mucus in conjunctiva and in tears, lysozyme and lactoferrin in tears	*Corynebacterium* species and other skin colonizers

16.6 Eye Diseases Caused by Microorganisms

In this section, we cover the infectious agents that cause diseases of the surface structures of the eye–namely, the cornea and conjunctiva.

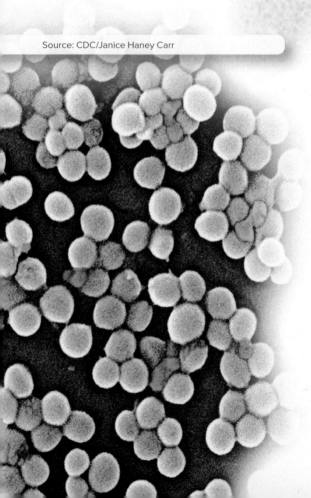

Figure 16.16 Neonatal conjunctivitis.
Medical-on-Line/Alamy

Source: CDC/Janice Haney Carr

> ## HIGHLIGHT DISEASE

Conjunctivitis

Infection of the conjunctiva is relatively common. It can be caused by specific microorganisms that have a predilection for eye tissues, by contaminants that proliferate due to the presence of a contact lens or an eye injury, or by accidental inoculation of the eye by a traumatic event.

▶ Signs and Symptoms

Just as there are many different causes of conjunctivitis, there are many different clinical presentations. Most bacterial infections produce a milky discharge, whereas viral infections tend to produce a clear watery exudate. It is typical for a patient to wake up in the morning with an eye "glued" shut by secretions that have accumulated and solidified through the night. Some conjunctivitis cases are caused by an allergic response, and these often produce copious amounts of clear fluid as well. The informal name for common conjunctivitis is pinkeye.

▶ Causative Agents and Their Transmission

Neonatal eye infection with *Neisseria gonorrhoeae* or *Chlamydia trachomatis* are usually transmitted vertically from a genital tract infection in the mother (discussed in chapter 21). Either one of these eye infections can lead to serious eye damage if not treated promptly **(figure 16.16).** Note that herpes simplex can also cause neonatal conjunctivitis, but it is often accompanied by generalized herpes infection (covered in chapter 21).

Bacterial conjunctivitis in other age groups is most commonly caused by *Staphylococcus epidermidis, Streptococcus pyogenes,* or *Streptococcus pneumoniae,* although *Haemophilus influenzae* and *Moraxella* species are also frequent causes. *N. gonorrhoeae* and *C. trachomatis* can also cause conjunctivitis in adults. These infections may result from autoinoculation from a genital infection or from sexual activity, although *N. gonorrhoeae* can be part of the normal biota in the respiratory tract. A wide variety of bacteria, fungi, and protozoa can contaminate contact lenses and lens cases and then be transferred to the eye, resulting in disease that may be very serious. Viral conjunctivitis is commonly caused by adenoviruses, although other viruses may be responsible. Both bacterial and viral conjunctivitis are transmissible by direct and even indirect contact and are usually highly contagious.

▶ Prevention and Treatment

Newborn children in the United States are administered antimicrobials in their eyes after delivery to prevent neonatal conjunctivitis from either *N. gonorrhoeae* or *C. trachomatis.* Treatment of those infections, if they are suspected, is started before lab results are available and usually is accomplished with erythromycin, both topical and oral. If *N. gonorrhoeae* is confirmed, oral therapy is usually switched to ceftriaxone. If antibacterial therapy is prescribed for other conjunctivitis cases, it should cover all possible bacterial pathogens. Ciprofloxacin eyedrops are a common choice. Erythromycin or gentamicin are also often used. Because conjunctivitis is usually diagnosed based on clinical signs, a physician may prescribe prophylactic antibiotics even if a viral cause is suspected. If symptoms don't begin improving within 48 hours, more extensive diagnosis may be performed. **Disease Table 16.9** lists the most common causes of conjunctivitis. Keep in mind that other microorganisms can also cause conjunctival infections.

Disease Table 16.9	Conjunctivitis		
Disease	**Neonatal Conjunctivitis**	**Bacterial Conjunctivitis**	**Viral Conjunctivitis**
Causative Organism(s)	*Chlamydia trachomatis* or *Neisseria gonorrhoeae* **B** **G⁻**	*Streptococcus pneumoniae, Staphylococcus epidermidis, Staphylococcus aureus, Haemophilus influenzae, Moraxella,* and also *Neisseria gonorrhoeae, Chlamydia trachomatis* **B**	Adenoviruses and others **V**
Most Common Modes of Transmission	Vertical	Direct, indirect contact	Direct, indirect contact
Virulence Factors	–	–	–
Culture/Diagnosis	Gram stain and culture	Clinical diagnosis	Clinical diagnosis
Prevention	Screen mothers, apply antibiotic ophthalmic solution to newborn eyes	Hygiene	Hygiene
Treatment	Topical and oral antibiotics (antibiotic-resistant *N. gonorrhoeae* is in **Urgent Threat** category in CDC Antibiotic Resistance Report)	Trimethoprim/polymyxin B for routine cases; azithromycin or levofloxacin for more serious cases (several of these bacteria appear in the CDC Antibiotic Resistance Report)	None, although antibiotics often given because type of infection not distinguished
Distinguishing Features	In babies <28 days old	Mucopurulent discharge	Serous (clear) discharge
Epidemiological Features	Less than 0.5% in developed world; higher incidence in developing world	More common in children	More common in adults

Keratitis

Keratitis is a more serious eye condition than conjunctivitis. Damage of deeper eye tissues occurs and can lead to complete corneal destruction. Any microorganism can cause this condition, especially after trauma to the eye, but this section focuses on two of the most common causes: miscellaneous bacteria (which cause 80% of the infectious keratitis cases) and herpes simplex virus. This virus can cause keratitis in the absence of predisposing trauma.

The usual cause of herpetic keratitis is a "misdirected" reactivation of (oral) herpes simplex virus type 1 (HSV-1). The virus, upon reactivation, travels into the ophthalmic rather than the mandibular branch of the trigeminal nerve. Infections with HSV-2 can also occur as a result of a sexual encounter with the virus or transfer of the virus from the genital to eye area. Blindness due to herpes is the leading infectious cause of blindness in the United States. Bacterial and fungal causes of keratitis are more common in developing countries.

The viral condition is treated with trifluridine or acyclovir or both.

Disease Table 16.10	Keratitis	
Causative Organism(s)	Herpes simplex virus **V**	Miscellaneous bacteria **B** or *Acanthamoeba* **P**
Most Common Modes of Transmission	Reactivation of latent virus, although primary infections can occur in the eye	Often traumatic introduction (parenteral)
Virulence Factors	Latency	Various
Culture/Diagnosis	Usually clinical diagnosis; viral culture or PCR if needed	Various
Prevention	–	–
Treatment	Topical trifluridine +/– oral acyclovir	Specific antimicrobials
Epidemiological Features	One-third worldwide population infected; in United States, annual incidence of 500,000	

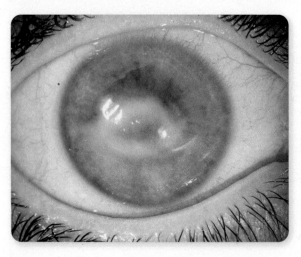

Figure 16.17 *Acanthamoeba* **infection of the eye.**
ISM/Phototake

In the last few years, another form of keratitis has been increasing in incidence. An amoeba called *Acanthamoeba* has been causing serious keratitis cases, especially in people who wear contact lenses. This free-living amoeba is everywhere—it lives in tap water, freshwater lakes, and the like. The infections are usually associated with less-than-rigorous contact lens hygiene, or previous trauma to the eye **(figure 16.17). Disease Table 16.10.**

16.6 LEARNING OUTCOMES—Assess Your Progress

11. List the possible causative agents, modes of transmission, virulence factors, diagnostic techniques, and prevention/treatment for the "Highlight Disease" conjunctivitis.
12. Discuss the most important causes of keratitis.

CASE FILE WRAP-UP

©Syda Productions/Shutterstock (health care worker); ©Dennis MacDonald/Alamy Images (hospital)

Wyatt, the patient in the chapter opening case file, was admitted to the hospital after his X ray confirmed a mild case of pneumonia, a complication of his measles. He was given IV antibiotics to treat his pneumonia and fluids to counter dehydration, as well as antipyretics to treat his fever and make him more comfortable. He was discharged home after 4 days and recovered completely.

The physician contacted the state health department, as must be done legally. State health department investigators contacted all of Wyatt's contacts to ensure they had not fallen ill. A few contacts were found who had never been vaccinated because their parents shared Wyatt's mother's erroneous view of vaccines. These children were quarantined at home to prevent the potential spread of the disease. Thankfully, all of the patients in the hospital waiting room were found to be immune to the disease and no one else became ill.

▶ Summing Up

Taxonomic Organization Microorganisms Causing Diseases of the Skin and Eyes

Microorganism	Pronunciation	Disease Table
Gram-positive bacteria		
Staphylococcus aureus	staf″-uh-lo-kok′-us are′-ee-us	MRSA skin and soft tissue infection, 16.1 Impetigo, 16.3 Cellulitis, 16.4 Scalded skin syndrome, 16.5
Streptococcus pyogenes	strep″-tuh-kok′-us pie′-ah″-gen-eez	Impetigo, 16.3 Cellulitis, 16.4 Maculopapular rash diseases, 16.2
Bacillus anthracis	buh-sill′-us an′-thray″-sus	Large pustular skin lesions, 16.7
Gram-negative bacteria		
Neisseria gonorrhoeae	nye-seer″-ee-uh′ gon′-uh-ree″-uh	Conjunctivitis, 16.9
Chlamydia trachomatis	kluh-mi″-dee-uh′ truh-koh′-muh-tis	Conjunctivitis, 16.9
DNA viruses		
Human herpesvirus 3	hew′-mun hur″-peez-vie′-russ	Vesicular/pustular rashes, 16.6
Variola virus	vayr′-ee-oh″-luh vie′-russ	Vesicular/pustular rashes, 16.6
Parvovirus B19	par″-voh-vie′-russ	Maculopapular rash diseases, 16.2
Human herpesvirus 6	hew′-mun hur″-peez-vie′-russ	Maculopapular rash diseases, 16.2
Herpes simplex virus	hur″-peez sim′-plex vie′-russ	Keratitis, 16.10
Adenovirus	a″-duh-no-vie′-russ	Conjuctivitis, 16.9
RNA viruses		
Measles virus	mee′-zulls vie′-russ	Maculopapular rash diseases, 16.2
Rubella virus	roo′-bell″-uh vie′-russ	Maculopapular rash diseases, 16.2
Fungi		
Trichophyton	try″-ko-fie′-tahn	Cutaneous and superficial mycoses, 16.8
Microsporum	my″-krow′-spoor′-um	Cutaneous and superficial mycoses, 16.8
Epidermophyton	ep′-uh-dur″-moh-fie′-tahn	Cutaneous and superficial mycoses, 16.8
Malassezia spp.	mal′-uh-see″-zee-uh	Cutaneous and superficial mycoses, 16.8
Protozoa		
Leishmania spp.	leesh-mayn″-ee-uh	Large pustular skin lesions, 16.7
Acanthamoeba	ay-kanth″-uh-mee′-buh	Keratitis, 16.10

Source: CDC/Cynthia Goldsmith

Do C-Section Babies Have a Different Microbiome Than Those Delivered Vaginally?

©Kevin Landwer-Johan/Getty Images

Think about your circle of female family and friends: How many of them have delivered babies via C-section? In some regions of the United States, the rates of C-section approach 50%. In some other countries (such as Brazil), it can be as high as 90%. But microbiome science suggests that it might not be the best start for babies—at least with respect to their gut microbiome. And we know how important the gut microbiome is. We have talked about how it can influence many aspects of our health, even outside our guts.

Scientists studied the gut microbiome of babies who were delivered vaginally versus that of those delivered via C-section. The vaginally born babies were colonized with mostly fecal bacteria from the mother. That might sound bad, but the bacteria help babies digest milk and are generally protective to gut health. On the other hand, the guts of babies delivered through C-section are generally colonized by the skin bacteria of their mother as well as hospital-acquired bacteria.

As research studies have continued into this phenomenon, we have found that it is possible that these differences fade fairly quickly and that the microbiomes become more similar within weeks.

It is frustrating when researchers can't tell us with precision which results are correct, but it is a feature of scientific research. Eventually, with a more time and more studies, the preponderance of evidence will point to a clearer answer.

Keratitis

Herpes simplex virus
Acanthamoeba
Various bacteria

Large Pustular Skin Lesions

Leishmania species
Bacillus anthracis

Staphylococcal Scalded Skin Syndrome

Staphylococcus aureus

Maculopapular Rash Diseases

Measles virus
Rubella virus
Parvovirus B19
Human herpesvirus 6

Impetigo

Staphylococcus aureus
Streptococcus pyogenes

Conjunctivitis

Neisseria gonorrhoeae
Chlamydia trachomatis
Various bacteria
Various viruses

Vesicular or Pustular Rash Disease

Human herpesvirus 3 (varicella)
Variola virus

Cellulitis

Streptococcus pyogenes
Staphylococcus aureus

Cutaneous and Superficial Mycoses

Trichophyton
Microsporum
Epidermophyton
Malassezia

MRSA Skin and Soft-Tissue Infection

Staphylococcus aureus

Helminths
Bacteria
Viruses
Protozoa
Fungi

**System Summary
Figure 16.18**

Deadliness and Communicability of Selected Diseases of the Skin and Eyes

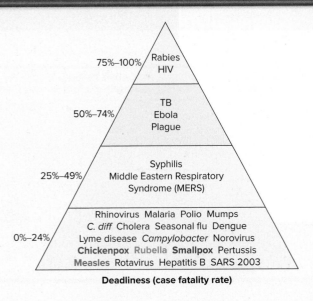

75%–100% | Rabies HIV

50%–74% | TB Ebola Plague

25%–49% | Syphilis Middle Eastern Respiratory Syndrome (MERS)

0%–24% | Rhinovirus Malaria Polio Mumps *C. diff* Cholera Seasonal flu Dengue Lyme disease *Campylobacter* Norovirus **Chickenpox** Rubella **Smallpox** Pertussis **Measles** Rotavirus Hepatitis B SARS 2003

Deadliness (case fatality rate)

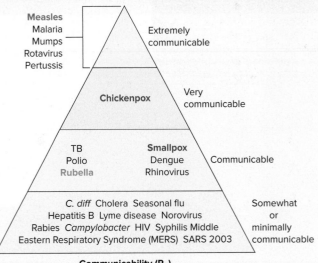

Measles Malaria Mumps Rotavirus Pertussis | Extremely communicable

Chickenpox | Very communicable

TB Polio Rubella | **Smallpox** Dengue Rhinovirus | Communicable

C. diff Cholera Seasonal flu Hepatitis B Lyme disease Norovirus Rabies *Campylobacter* HIV Syphilis Middle Eastern Respiratory Syndrome (MERS) SARS 2003 | Somewhat or minimally communicable

Communicability (R₀)

INFECTIOUS DISEASES AFFECTING THE SKIN AND EYES

The Skin and Its Defenses

The epidermal cells contain the protein keratin, which "waterproofs" the skin and protects it from microbial invasion. Other defenses include antimicrobial peptides, high salt, and lysozyme.

Normal Biota: Skin

The normal skin biota are far more diverse than previously understood from strictly culture techniques. The normal biota include hundreds of species of microorganisms, and differ from region to region on the body. They also differ between people.

Skin Diseases Caused by Microorganisms

- **MRSA Skin and Soft-Tissue Infection**
On the rise in the community environment.
- **Maculopapular Rash Diseases**
Measles, rubella, fifth disease, and roseola.
- **Impetigo**
Caused by *Streptococcus pyogenes* or *Staphylococcus aureus* or both.
- **Cellulitis**
On the rise in the community environment. Most often caused by *Streptococcus pyogenes* or *Staphylococcus aureus* or both.

- **Staphylococcus Scalded Skin Syndrome**
Similar to a systemic form of impetigo.
- **Vesicular or Pustular Rash Disease**
Chickenpox/shingles; smallpox.
- **Large Pustular Skin Lesions**
Leishmaniasis and cutaneous anthrax.
- **Cutaneous and Superficial Mycoses**
Ringworm (mycoses in nonliving epidermal tissues, hair, and nails). Superficial mycoses: involve only the outer epidermis.

The Surface of the Eye and Its Defenses

The flushing action of the tears, which contain lysozyme and lactoferrin, is the major protective feature of the eye.

Normal Biota: Eyes

The eye has similar microbes as the skin but in lower numbers. As with other body sites, cultivation-independent methods are showing that the eye is home to many more microbes than previously thought.

Eye Diseases Caused by Microorganisms

- **Conjunctivitis**
Infection of the conjunctiva. Can be caused by bacteria and viruses. Both are highly contagious.

- **Keratitis**
More serious than conjunctivitis; caused by herpes viruses, *Acanthamoeba*, and other microbes.

SmartGrid: From Knowledge to Critical Thinking

This *21 Question Grid* takes the topics from this chapter and arranges them with respect to the American Society for Microbiology's Undergraduate Curriculum guidelines—all six of the important "Concepts" as well as the important "Competency" of scientific literacy. Three questions are supplied about chapter content that refer to the Concept or Competency, in increasing levels of Bloom's taxonomy for learning.

ASM Concept/ Competency	A. Bloom's Level 1, 2—Remember and Understand (Choose one.)	B. Bloom's Level 3, 4—Apply and Analyze	C. Bloom's Level 5, 6—Evaluate and Create

Evolution	1. Which of the following infectious agents has evolved to maintain a persistent state in its host?
	  a. variola virus
	  b. herpes virus
	  c. vaccinia virus
	  d. *Staphylococcus aureus*
		2. Can you think of a plausible argument for why hospital cases of MRSA are declining and community cases of MRSA are increasing?	3. Smallpox has ravaged human populations for thousands of years. When it first came to the Americas in the 1500s, it had a high case-fatality rate. By the end of the 1800s it manifested as a much milder disease. Speculate on why that might be.

Cell Structure and Function	4. What is an antimicrobial enzyme found in sweat, tears, and saliva that can break down bacterial cell walls?
	  a. lysozyme
	  b. beta-lactamase
	  c. catalase
	  d. hyaluronidase
		5. What cell structure of *Bacillus anthracis* makes it capable of surviving in powdered form, and naturally in soil, for indefinite periods of time?	6. Both *Staphylococcus aureus* and *Streptococcus pyogenes* are "pyogenic." What does that mean, and what attributes of the two bacteria are likely to contribute to that phenomenon?

Metabolic Pathways	7. Which of the following organisms produces an enzyme that breaks down hydrogen peroxide?
	  a. *Streptococcus pyogenes*
	  b. *Staphylococcus aureus*
	  c. MRSA
	  d. two of these
		8. *Streptococcus pneumoniae*, which is a common inhabitant of healthy upper respiratory tracts, produces hydrogen peroxide to fight off microbial competitors. How is it possible then that *Staphylococcus aureus* is frequently a co-colonizer?	9. Some antifungal drugs inhibit an enzyme called squalene epoxidase. What does this enzyme do, and why is it an effective target for selective toxicity to the fungus and not the host?

Information Flow and Genetics	10. Which of these techniques has detected the larger number of normal microbiota on skin surfaces?
	  a. culturing
	  b. 16S rRNA sequencing
	  c. antibody probing
	  d. gel electrophoresis
		11. The first person to die of measles in the United States in 12 years was a woman in Washington State in July 2015. What kind of laboratory tests would you perform to determine whether the virus was the same that circulated at a Disney park in 2014?	12. Herpesviruses cause several of the diseases in this chapter. Considering some features all of these conditions have in common, hypothesize what the viral DNA does after initial invasion of a host cell.

ASM Concept/ Competency	A. Bloom's Level 1, 2–Remember and Understand (Choose one.)	B. Bloom's Level 3, 4–Apply and Analyze	C. Bloom's Level 5, 6–Evaluate and Create
Microbial Systems	13. Which of the following conditions is most likely to be a polymicrobial infection? a. measles b. rubella c. leishmaniasis d. impetigo	14. Describe how a virus infection of human cells can lead to syncytia.	15. Research the bacterium *Micavibrio aeruginosavorus*, which is often found as normal biota of the eye. Formulate a theory on how it keeps pathogens from harming the eye.
Impact of Microorganisms	16. *Staphylococcus aureus* is part of the differential diagnosis of which of the following diseases? a. impetigo b. maculopapular rash c. both of these d. neither of these	17. Compare and contrast the effects of the rubeola virus and the rubella virus on humans.	18. In a single (unvaccinated) family one winter, the 12-year-old and 4-year-old came down with chickenpox. The mother and father had experienced the disease as children and were unaffected this time. The 3-month-old baby did not become ill, either. However, 4 months later the baby exhibited a shingles outbreak. Using immunological phenomena, explain what was likely going on with the baby.
Scientific Thinking	19. Which steps of the diagnostic process are in order? a. differential diagnosis, anatomic diagnosis, etiologic diagnosis b. anatomic diagnosis, etiologic diagnosis, differential diagnosis c. anatomic diagnosis, differential diagnosis, etiologic diagnosis d. none of these	20. Explain why detecting a single case of smallpox would be considered a public health—and national security—emergency.	21. Construct a testable hypothesis for examining whether receiving the chickenpox vaccine results in fewer cases of shingles in the population.

Answers to the multiple-choice questions appear in Appendix A.

Visual Connections

This question connects content within and between chapters.

From chapter 11, figure 11.2. How does this figure help explain impetigo caused by *Staphylococcus aureus* or *Streptococcus pyogenes*?

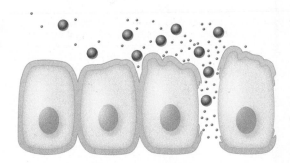

17

Infectious Diseases Affecting the Nervous System

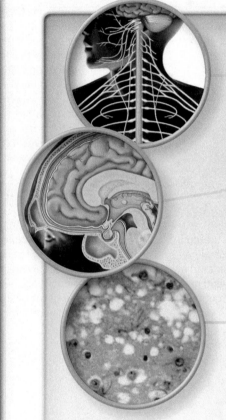

CASE FILE

Time Is of the Essence

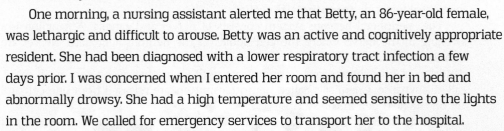

I was working in an assisted living facility. As an RN, I supervised the care of all the patients in the facility. My responsibilities included preparing and administering medications, performing daily patient assessments, and responding to changes in patient condition. I worked with a team of nursing assistants who helped residents with activities of daily living. Supervising a large number of patients was a challenge, but I could depend on my team to keep me alerted if any of our residents needed immediate attention.

One morning, a nursing assistant alerted me that Betty, an 86-year-old female, was lethargic and difficult to arouse. Betty was an active and cognitively appropriate resident. She had been diagnosed with a lower respiratory tract infection a few days prior. I was concerned when I entered her room and found her in bed and abnormally drowsy. She had a high temperature and seemed sensitive to the lights in the room. We called for emergency services to transport her to the hospital.

Once at the hospital, the medical team in the emergency department performed a thorough neurological assessment. Betty was confused and lethargic. She continued to exhibit photophobia and reported a headache. The provider decided to institute intravenous antibiotic therapy, as he was concerned for meningitis. Cefotaxime and vancomycin were ordered STAT. Labs were drawn, including a complete blood count (CBC) with differential, metabolic panel, and blood cultures. The provider performed a lumbar puncture.

Betty was admitted to an inpatient medical unit with presumed meningitis. The team planned to continue her antibiotic therapy and monitor her neurological status closely.

- What findings on a lumbar puncture are diagnostic of meningitis?

- How is medication dosing adjusted to cross the blood-brain barrier?

Case File Wrap-Up appears at the end of the chapter.

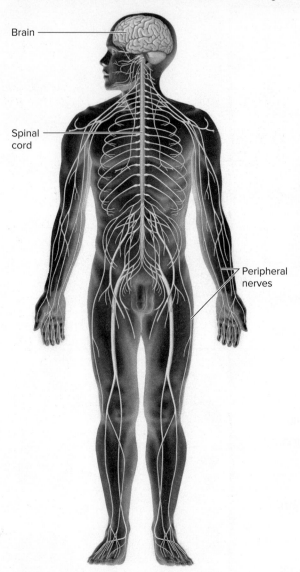

Brain

Spinal cord

Peripheral nerves

Figure 17.1 **The two component parts of the nervous system.** The central nervous system and the peripheral nerves.

17.1 The Nervous System and Its Defenses

The nervous system has two component parts: the central nervous system (CNS), consisting of the brain and spinal cord, and the peripheral nervous system (PNS), which contains the nerves that emanate from the brain and spinal cord to sense organs and to the periphery of the body **(figure 17.1).** The nervous system performs three important functions–sensory, integrative, and motor. The sensory function is fulfilled by sensory receptors at the ends of peripheral nerves. They generate nerve impulses that are transmitted to the central nervous system. There, the impulses are translated, or integrated, into sensation or thought, which in turn drives the motor function. The motor function necessarily involves structures outside of the nervous system, such as muscles and glands.

The brain and the spinal cord are dense structures made up of cells called **neurons.** They are both surrounded by bone. The brain is situated inside the skull, and the spinal cord lies within the spinal column **(figure 17.2),** which is composed of a stack of interconnected bones called vertebrae. The soft tissue of the brain and spinal cord is encased within a tough casing of three membranes called the **meninges.** The layers of these three membranes, from outermost to innermost position, are the dura mater, the arachnoid mater, and the pia mater. Between the arachnoid mater and pia mater is the subarachnoid space (i.e., the space under the arachnoid mater). The subarachnoid space is filled with a clear serumlike fluid called cerebrospinal fluid (CSF). The CSF provides nutrition to the CNS, while also providing a liquid cushion for the sensitive brain and spinal cord. The meninges are a common site of infection, and microorganisms can often be found in the CSF when meningeal infection **(meningitis)** occurs.

The peripheral nervous system consists of nerves and ganglia. A ganglion is a swelling in the nerve where the cell bodies of the neurons aggregate. Nerves are bundles of neuronal axons that receive and transmit nerve signals. The axons and dendrites of adjacent neurons communicate with each other over a very small space, called a synapse. Chemicals called neurotransmitters are released from one cell and act on the next cell in the synapse.

The defenses of the nervous system are mainly structural. The bony casings of the brain and spinal cord protect them from traumatic injury. The cushion of surrounding CSF also serves a protective function. The entire nervous system is served by the vascular system, but the interface between the blood vessels serving the brain and the brain itself is different from that of other areas of the body and provides a third structural protection. The cells that make up the walls of the blood vessels allow very few molecules to pass through. In other parts of the body, there is freer passage of ions, sugars, and other metabolites through the walls of blood vessels. The restricted permeability of blood vessels in the brain is called the **blood-brain barrier,** and it prohibits most microorganisms from passing into the central nervous system. The drawback of this phenomenon is that drugs and antibiotics are difficult to introduce into the CNS also.

The CNS is considered an "immunologically privileged" site. These sites are able to mount only a partial, or at least a different, immune response when exposed to immunologic challenge. The functions of the CNS are so vital for the life of an organism that even temporary damage that could potentially result from "normal" immune responses would be very detrimental. The uterus and parts of the eye are other immunologically privileged sites. Specialized cells in the central nervous system perform defensive functions. Microglia are a type of cell having phagocytic capabilities, and brain macrophages also exist in the CNS, although the activity of both of these types of cells is thought to be less than that of phagocytic cells elsewhere in the body.

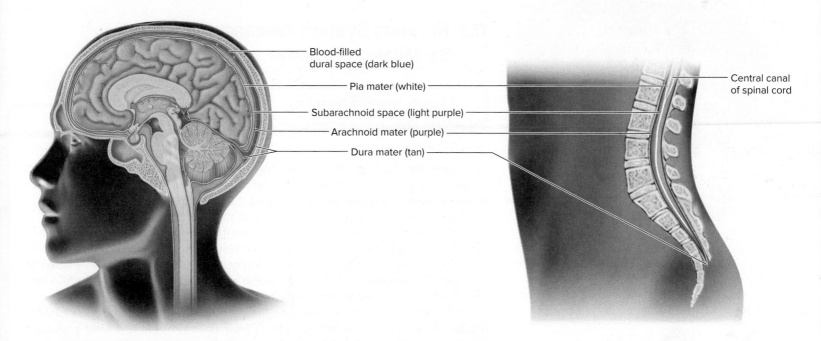

Blood-filled
dural space (dark blue)

Pia mater (white)

Subarachnoid space (light purple)

Arachnoid mater (purple)

Dura mater (tan)

Central canal
of spinal cord

Figure 17.2 **Detailed anatomy of the brain and spinal cord.**

17.1 LEARNING OUTCOMES—Assess Your Progress

1. Describe the important anatomical features of the nervous system.
2. List the natural defenses present in the nervous system.

17.2 Normal Biota of the Nervous System

It is still believed that there is no normal biota in either the CNS or PNS, and that finding microorganisms of any type in these tissues represents a deviation from the healthy state. Viruses such as herpes simplex live in a dormant state in the nervous system between episodes of acute disease, but they are not considered normal biota. Having said that, there is a lot of research suggesting that the gut microbiome influences the nervous system in many ways. In fact the very development of the brain, the blood-brain barrier, and proper construction of peripheral nerves are influenced by the microbiome in the developing gut.

Nervous System Defenses and Normal Biota		
	Defenses	**Normal Biota**
Nervous System	Bony structures, blood-brain barrier, microglial cells, and macrophages	None

---

17.2 LEARNING OUTCOMES—Assess Your Progress

3. Discuss the current state of knowledge of the normal microbiota of the nervous system.

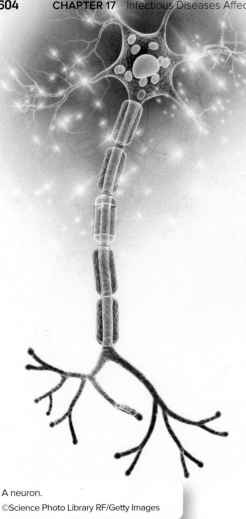

A neuron.
©Science Photo Library RF/Getty Images

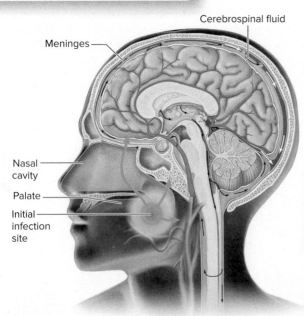

Figure 17.3 Dissemination of the meningococcus from a nasopharyngeal infection.   Bacteria spread to the roof of the nasal cavity, which borders a highly vascular area at the base of the brain. From this location, they can enter the blood, causing meningococcemia, and escape into the cerebrospinal fluid, leading to infection of the meninges.

17.3 Nervous System Diseases Caused by Microorganisms

Meningitis

Meningitis, an inflammation of the meninges, is an excellent example of an anatomical syndrome. Many different microorganisms can cause an infection of the meninges, and they produce a similar set of symptoms. Noninfectious causes of meningitis exist as well, but they are much less common than the infections listed here.

The more serious forms of acute meningitis are caused by bacteria, but it is thought that their entrance to the CNS is often facilitated by coinfection or previous infection with respiratory viruses. Meningitis in neonates is usually caused by different microorganisms than those causing the disease in children and adults, and therefore it is described separately in the following section.

Whenever meningitis is suspected, lumbar puncture (spinal tap) is performed to obtain CSF, which is then examined by Gram stain and/or culture. Most physicians will begin treatment with a broad-spectrum antibiotic immediately and shift treatment if necessary after a diagnosis has been confirmed.

▶ Signs and Symptoms

No matter the cause, meningitis results in these typical symptoms: severe headache, painful or stiff neck, fever, and nausea and vomiting. Early symptoms may be mistaken for flu symptoms. Photophobia (sensitivity to light) may also be noted. Skin rashes may be present in specific types of meningitis. There are usually an increased number of white blood cells in the CSF. Specific microorganisms may cause additional, and sometimes characteristic, symptoms, which are described in the individual sections that follow.

Like many other infectious diseases, meningitis can manifest as acute or chronic disease. Some microorganisms are more likely to cause acute meningitis, and others are more likely to cause chronic disease.

In a normal healthy person, it is very difficult for microorganisms to gain access to the nervous system. Those that are successful usually have specific virulence factors.

Neisseria meningitidis

Neisseria meningitidis appears as gram-negative diplococci (round cells occurring as joined pairs) and is commonly known as the meningococcus. It is often associated with epidemic forms of meningitis. This organism causes the most serious form of acute meningitis and accounts for 15% to 20% of all meningitis cases. Although 12 different strains with different capsular antigens exist, serotypes B, C, and Y are responsible for most cases of infection in the United States. In Africa other serotypes are prominent, and two serotypes (A and W) are associated with the Hajj, an annual pilgrimage made by Muslims to the city of Mecca in Saudi Arabia.

▶ Pathogenesis and Virulence Factors

The bacterium enters the body via the upper respiratory tract, moves into the blood, rapidly penetrates the meninges, and produces symptoms of meningitis. The most serious complications of meningococcal infection are due to meningococcemia **(figure 17.3)**, which can accompany meningitis but can also occur on its own. The pathogen releases endotoxin into the generalized circulation, which is a potent stimulus for certain white blood cells. Damage to the blood vessels caused by cytokines released by the white blood cells leads to vascular collapse, hemorrhage, and crops of lesions called **petechiae** (pee-tee′-kee-eye) on the trunk and appendages. A *petechia* (singular) is a small, 1- to 2-mm red or purple spot that may occur anywhere on the body **(figure 17.4).** In a small number of cases, meningococcemia becomes an overwhelming disease with a high mortality rate.

The disease has a sudden onset, marked by fever higher than 40°C or 104°F, sore throat, chills, delirium, severe widespread areas of bleeding under the skin, shock, and coma. Stomach pain is often an early symptom. Generalized intravascular clotting, cardiac failure, damage to the adrenal glands, and death can occur within a few hours. The bacterium has an IgA protease and a capsule, both of which help the microbe overcome the body's defenses.

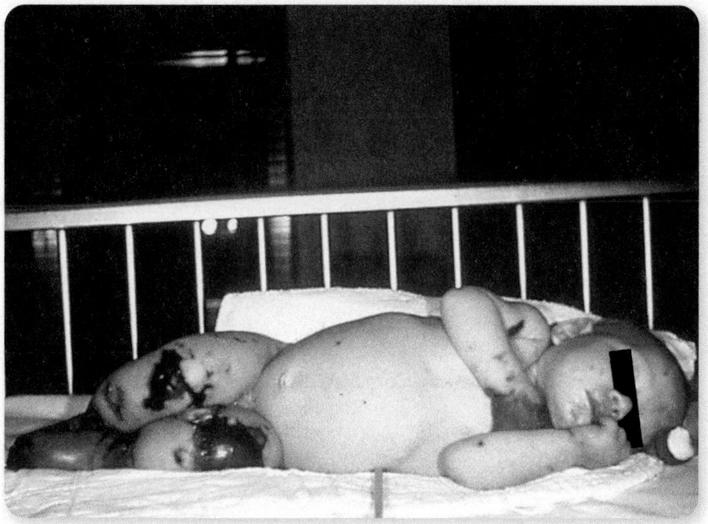

▶ Transmission and Epidemiology

Because meningococci do not survive long in the environment, these bacteria are usually acquired through close contact with secretions or droplets.

Meningococcal meningitis has a sporadic or epidemic incidence in late winter or early spring. The continuing reservoir of infection is humans who harbor the pathogen in the nasopharynx. The scene is set for transmission when carriers live in close quarters with nonimmune individuals, as might be expected in families, day care facilities, college dormitories, and military barracks. The highest risk groups are young children (6 to 36 months old) and older children and young adults (10 to 20 years old). Cases peak in January and February in the United States.

Figure 17.4 Vascular damage associated with meningococcal meningitis.
Mr Gust/CDC

▶ Culture and Diagnosis

Suspicion of bacterial meningitis constitutes a medical emergency, and differential diagnosis must be done with great haste and accuracy. It is most important to confirm (or rule out) meningococcal meningitis because it can be rapidly fatal. Treatment is usually begun with this bacterium in mind until it can be ruled out. Cerebrospinal fluid, blood, or nasopharyngeal samples are stained and observed directly for the typical gram-negative diplococci. Cultivation is the preferred method of diagnosis because it also enables a quick assessment of antimicrobial susceptibilities. Specific rapid tests are also available for detecting the capsular polysaccharide or the cells directly from specimens without culturing.

It is usually necessary to differentiate this species from normal *Neisseria* that also live in the human body and can be present in infectious fluids. Immediately after collection, specimens are streaked on Modified Thayer-Martin medium (MTM) or chocolate agar and incubated in a high CO_2 atmosphere. Presumptive identification of the genus is obtained by a Gram stain and oxidase testing on isolated colonies **(figure 17.5)**.

Figure 17.5 The oxidase test. A drop of oxidase reagent is placed on a suspected *Neisseria* or *Branhamella* colony. If the colony reacts with the chemical to produce a purple to black color, it is oxidase-positive; those that remain white to tan are oxidase-negative. Because several species of gram-negative rods are also oxidase-positive, this test is presumptive for these two genera only if a Gram stain has verified the presence of gram-negative cocci.
Kathleen Talaro

▶ Prevention and Treatment

The background infection rate in most populations is about 1%, so well-developed natural immunity to the meningococcus appears to be the rule. A sort of natural immunization occurs during the early years of life as one is exposed to the meningococcus and its close relatives. Because even treated meningococcemial disease has a mortality rate of up to 15%, it is vital that antibiotic therapy begin as soon as possible with one or more drugs. It is generally given in high doses intravenously. Patients may also require treatment for shock and intravascular clotting.

When family members, medical personnel, or children in day care or school have come in close contact with infected people, they should receive a vaccination. Preventive therapy with rifampin or tetracycline may also be warranted. In the United States, immunization begins at the age of 11 followed by a booster dose. Vaccines are also available for younger children and for adults over the age of 55 who are at high risk for infection. Routine immunization is with one of two meningococcal vaccines that protect against serotypes A, C, W, and Y. At about the time a booster is needed (16 years of age), the CDC recommends additionally the first dose of the new vaccine effective against serotype B, the serotype that has caused several outbreaks on college campuses in recent years.

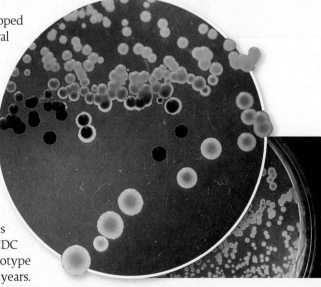

Disease Table 17.1	**Meningitis**		
Causative Organism(s)	*Neisseria meningitidis* Ⓑ **G⁻**	*Streptococcus pneumoniae* Ⓑ **G⁺**	*Haemophilus influenzae* Ⓑ **G⁻**
Most Common Modes of Transmission	Droplet contact	Droplet contact	Droplet contact
Virulence Factors	Capsule, endotoxin, IgA protease	Capsule, induction of apoptosis, hemolysin and hydrogen peroxide production	Capsule
Culture/Diagnosis	Gram stain/culture of CSF, blood, rapid antigenic tests, oxidase test	Gram stain/culture of CSF	Culture on chocolate agar
Prevention	Conjugated vaccine; ciprofloxacin, rifampin, or ceftriaxone used to protect contacts	Two vaccines: PCV13 and PPSV23	Hib (*H. influenzae* serotype b) vaccine, ciprofloxacin, rifampin, or ceftriaxone
Treatment	Penicillin, ceftriaxone, cefotaxime	Penicillin, vancomycin + ceftriaxone; in **Serious Threat** category in CDC Antibiotic Resistance Report	Ceftriaxone
Distinctive Features	Petechiae, meningococcemia, rapid decline	Serious, acute, most common meningitis in adults	Serious, acute, less common in United States since vaccine became available
Epidemiological Features	United States: 329 cases in 2018; meningitis "belt" in Africa: 1,000 cases per 100,000 annually	U.S. incidence before vaccine for children: 7.7 hospitalizations per 100,000. After vaccine for children: 2.6 per 100,000	Before Hib vaccine, 300,000–400,000 deaths worldwide per year

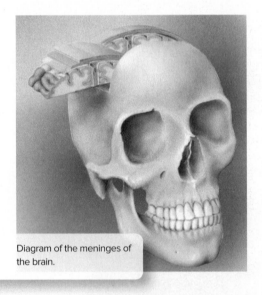

Diagram of the meninges of the brain.

Streptococcus pneumoniae

You will see in chapter 19 that *Streptococcus pneumoniae*, also referred to as the **pneumococcus,** causes the majority of bacterial pneumonias. Meningitis is also caused by this bacterium. Indeed, it is the most frequent cause of community-acquired meningitis and is also very severe. It does not cause the petechiae associated with meningococcal meningitis, and that difference is useful diagnostically. As many as 25% of pneumococcal meningitis patients will also have pneumococcal pneumonia. Pneumococcal meningitis is most likely to occur in patients with underlying susceptibility, such as alcoholic patients and patients with sickle-cell disease or those with absent or defective spleen function.

This bacterium is covered thoroughly in chapter 19 because it is a common cause of ear infections and pneumonia. It obviously has the potential to be highly pathogenic, while appearing as normal biota in many people. It can penetrate the respiratory mucosa; gain access to the bloodstream; and then, under certain conditions, enter the meninges.

Like the meningococcus, this bacterium has a polysaccharide capsule that protects it against phagocytosis. It also produces an alpha-hemolysin and hydrogen peroxide, both of which have been shown to induce damage in the CNS. It also appears capable of inducing brain cell apoptosis.

The bacterium is a small gram-positive flattened coccus that appears in end-to-end pairs. It has a distinctive appearance in a Gram stain of cerebrospinal fluid. Staining or culturing the nasopharynx is not useful because it is often normal biota there. Many strains of the bacterium are resistant to the first-line antibiotic, penicillin. In pneumococcal meningitis, initial treatment with vancomycin + ceftriaxone is recommended. If the isolate comes back as penicillin-sensitive (the cerebrospinal fluid must be cultured before beginning antibiotic treatment), then treatment can be switched.

Two vaccines are available for *S. pneumoniae*: a 13-valent conjugated vaccine (Prevnar), which is recommended as part of the childhood immunization schedule, and a 23-valent polysaccharide vaccine (Pneumovax 23), which is available for adults. Current recommendations for unvaccinated adults call for initial vaccination with Prevnar, followed by Pneumovax 6 to 12 months later.

Listeria monocytogenes **B** **G⁺**	Cryptococcus neoformans **F**	Coccidioides immitis **F**	Viruses **V**
Vehicle (food)	Vehicle (air, dust)	Vehicle (air, dust, soil)	Droplet contact
Intracellular growth	Capsule, melanin production	Granuloma (spherule) formation	Lytic infection of host cells
Cold enrichment, rapid methods	Negative staining, biochemical tests, DNA probes, cryptococcal antigen test	Identification of spherules, cultivation on Sabouraud's agar	Initially, absence of bacteria/fungi/protozoa, followed by viral culture or antigen tests
Cooking food, avoiding unpasteurized dairy products	–	Avoiding airborne endospores	–
Ampicillin, trimethoprim-sulfamethoxazole, gentamicin	Amphotericin B and fluconazole	Fluconazole or amphotericin B	Usually none (unless specific virus identified and specific antiviral exists)
Asymptomatic in healthy adults; meningitis in neonates, elderly, and immunocompromised	Acute or chronic, most common in AIDS patients	Almost exclusively in endemic regions	Generally milder than bacterial or fungal
Mortality can be as high as 33%	In United States, mainly a concern for HIV+ patients; 90% drop in incidence in the 1990s due to better management of AIDS; worldwide: 1 million new cases per year	Incidence increasing in recent years	In United States, 4 of 5 meningitis cases caused by viruses: 26,000–42,000 hospitalizations/year

Haemophilus influenzae

The meningitis caused by this bacterium is severe. Before the vaccine was introduced in 1988, it was a very common cause of severe meningitis and death. In the course of the last 13 years, meningitis caused by this bacterium is much less common in the United States, a situation that can always change if a lower percentage of people get the vaccine and herd immunity is compromised. Cases that do occur in the United States are now mostly caused by nonserotype B strains. Globally, it is still common and is an important cause of the disease in children under the age of 5.

Listeria monocytogenes

Listeria monocytogenes is a gram-positive bacterium that ranges in morphology from coccobacilli to long filaments in palisades formation **(figure 17.6)**. Cells do not produce capsules or endospores and have from one to four flagella. *Listeria* is not fastidious and is resistant to cold, heat, salt, pH extremes, and bile. It grows inside host cells and can move directly from an infected host cell to an adjacent healthy cell.

Listeriosis in healthy adults is often a mild or subclinical infection with nonspecific symptoms of fever, diarrhea, and sore throat. However, listeriosis in elderly or immuno-compromised patients, fetuses, and neonates (described later) usually affects the brain and meninges and results in septicemia. (*Septicemia* is a term that means the multiplication of bacteria in the bloodstream.) Some strains target the heart. The death rate in these populations is around 30%. Pregnant women are especially susceptible to infection, which can be transmitted to the fetus prenatally when the microbe crosses the placenta or postnatally through the birth canal. Intrauterine infections usually result in premature abortion and fetal death.

Apparently, the primary reservoir is soil and water, and animals, plants, and food are secondary sources of infection. Most cases of listeriosis are associated with ingesting contaminated dairy products, poultry, meat, and even hard-boiled eggs. Recent epidemics have spurred an in-depth investigation into the prevalence of *L. monocytogenes* in these sources. The pathogen has been isolated in 10% to 15% of ground beef

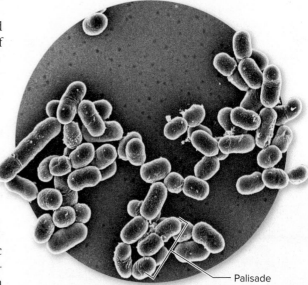

Figure 17.6 **Listeria monocytogenes.** The bacterium is generally rod shaped. In Gram stains, individual cells tend to stack up in structures called palisades. That arrangement, pointed out here, is more obvious on a Gram stain where many more bacteria are seen.
©Dr. Gary Gaugler/Science Source

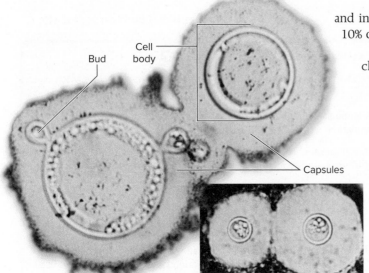

Figure 17.7 *Cryptococcus neoformans* from infected spinal fluid stained negatively with India ink. Halos around the large spherical yeast cells are thick capsules. Also note the buds forming on one cell. Encapsulation is a useful diagnostic sign for cryptococcosis, although the capsule is fragile and may not show up in some preparations (150×).

Gordon Love, M.D. VA, North CA Healthcare System, Martinez, CA

and in 25% to 30% of chicken and turkey carcasses and is also present in 5% to 10% of luncheon meats, hot dogs, and cheeses.

Diagnosing listeriosis is complicated by the difficulty in isolating it. The chances of isolation, however, can be improved by using a procedure called *cold enrichment,* in which the specimen is held at 4°C and periodically plated onto media, but this procedure can take 4 weeks. Rapid diagnostic kits using ELISA, immunofluorescence, and gene probe technology are now available for direct testing of dairy products and cultures. Antibiotic therapy should be started as soon as listeriosis is suspected. Ampicillin and trimethoprim-sulfamethoxazole are the first choices, followed by gentamicin. Prevention can be improved by adequate pasteurization temperatures and by proper washing, refrigeration, and cooking of foods that are suspected of being contaminated with animal manure or sewage. Pregnant women are cautioned by the U.S. Food and Drug Administration not to eat soft, unpasteurized cheeses.

Cryptococcus neoformans

The fungus *Cryptococcus neoformans* causes a more chronic form of meningitis with a more gradual onset of symptoms, although in AIDS patients the onset may be fast and the course of the disease more acute. It is sometimes classified as a meningoencephalitis (inflammation of both brain and meninges). Headache is the most common symptom, but nausea and neck stiffness are very common. This fungus is a widespread resident of human habitats. It has a spherical to ovoid shape, with small, constricted buds and a large capsule that is important in its pathogenesis **(figure 17.7)**.

The primary ecological niche of *C. neoformans* is the bird population. It is prevalent in urban areas where pigeons congregate, and it proliferates in the high-nitrogen environment of droppings that accumulate on pigeon roosts. Masses of dried yeast cells are readily scattered into the air and dust. Its role as an opportunist is supported by evidence that healthy humans have strong resistance to it and that obvious infection occurs primarily in debilitated patients.

Cryptococcal meningitis is a common occurrence among patients with AIDS. This meningitis is frequently fatal. Other conditions that predispose individuals to infection are steroid treatment, diabetes, and cancer. It is not considered communicable among humans.

▶ Prevention and Treatment

Systemic cryptococcosis requires immediate treatment with amphotericin B and fluconazole over a period of weeks or months. There is no prevention.

Coccidioides species

This fungus causes a condition that is often called "Valley Fever" in the U.S. Southwest. The morphology of *Coccidioides* is very distinctive. At 25°C, it forms a moist white to brown colony with abundant, branching, septate hyphae. These hyphae fragment into thick-walled, blocklike **arthroconidia** (arthrospores) at maturity **(figure 17.8a)**. On special media incubated at 37°C to 40°C, an arthrospore germinates into the parasitic phase, a small, spherical cell called a spherule **(figure 17.8b)** that can be found in infected tissues as well.

There are two species that cause this disease, found in different geographical places. *C. immitis* causes disease in California, and *C. posadasii* in northern Mexico, Central and South America, and the American Southwest, especially in Arizona. It has also recently been found in Washington State. Sixty percent of all U.S. infections occur in Arizona. **Figure 17.9** shows the incidence of this disease from 1998 to 2017 in the United States.

▶ Pathogenesis and Virulence Factors

This is a true systemic fungal infection of high virulence, as opposed to an opportunistic infection. It usually begins with pulmonary infection but can disseminate quickly

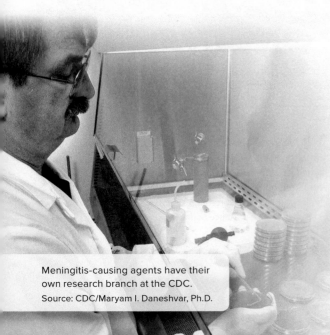

Meningitis-causing agents have their own research branch at the CDC.
Source: CDC/Maryam I. Daneshvar, Ph.D.

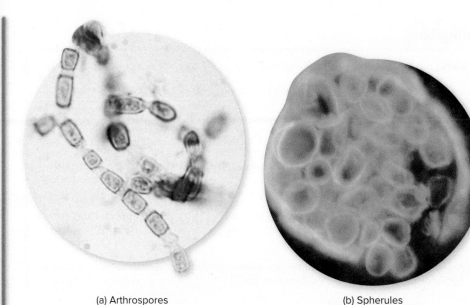

(a) Arthrospores

(b) Spherules containing endospores

Figure 17.8 **Two phases of *Coccidioides* infection.**
(a) Arthrospores are present in the environment and are inhaled. **(b)** In the lungs, the brain, or other tissues, arthrospores develop into spherules that are filled with endospores. Endospores are released and induce damage.
Source: Dr. Hardin/CDC (a); Source: CDC/Science Source (b)

 Note About Differential Diagnosis

Meningitis is a disease that illustrates how important it is for a clinician to think epidemiologically. When encountering a set of symptoms such as those of meningitis, a health care worker should be able to access a differential diagnosis in his or her mind. If the patient lives in Arizona, you definitely want to include *Coccidioides* in your differential diagnosis. Knowing the patient's vaccination history can help rule out some causative agents. And finally, if you look at our table of possible causative agents, you see that 80% of meningitis cases are caused by unidentified viruses and are mild. Here's where you see that simply knowing a list of possible causative agents is not enough. A little more information will tell you how likely each of them is. **Each Disease Table in this book serves to summarize the differential diagnosis.**

There's a saying among physicians: "When you hear hoofbeats, think horses, not zebras." This refers to the fact that you should consider the simplest possibility, which in the case of meningitis would be a virus, resulting in uncomplicated disease. But if the patient is very sick upon presentation, or has signs of petechiae on the skin, you err on the side of safety and treat for *Neisseria* until it is ruled out.

throughout the body. Coccidioidomycosis of the meninges is the most serious manifestation. All persons inhaling the arthrospores probably develop some degree of infection, but certain groups have a genetic susceptibility that gives rise to more serious disease. After the arthrospores are inhaled, they develop into spherules in the lungs. These spherules release scores of endospores into the lungs. (Unfortunately, *endospores* is the term used for this phase of the infection even though we have learned that endospores are *bacterial* structures.) At this point, the patient either experiences mild respiratory symptoms, which resolve themselves, or the endospores cause disseminated disease. Disseminated disease can include meningitis, osteomyelitis, and skin granulomas.

The highest incidence of coccidioidomycosis, estimated at 100,000 cases per year, occurs in the southwestern United States, although it also occurs in Washington State, Mexico, and parts of Central and South America. Especially concentrated reservoirs exist in the San Joaquin Valley of California and in southern Arizona. Outbreaks are usually associated with farming activity, archeological digs, construction, and mining. Climate change, which is drying out much of the southwestern and western United States, is predicted to increase the incidence and range of this disease in the western United States.

Viruses

It is estimated that four of five meningitis cases are caused by one of a wide variety of viruses. Because no bacteria, protozoa, nor fungi are found in the CSF in viral

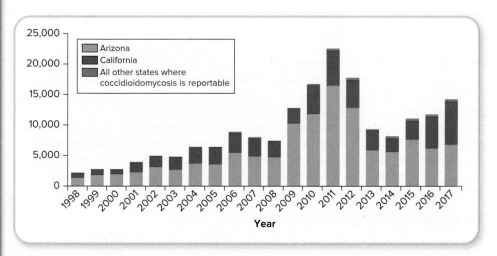

Figure 17.9 **Number of reported coccidioidomycosis cases in the United States, 1998–2017.**
Source: CDC

Disease Table 17.2	Neonatal and Infant Meningitis			
Causative Organism(s)	*Streptococcus agalactiae* **B** **G⁺**	*Escherichia coli*, strain K1 **B** **G⁻**	*Listeria monocytogenes* **B** **G⁺**	*Cronobacter sakazakii* **B** **G⁻**
Most Common Modes of Transmission	Vertical (during birth)	Vertical (during birth)	Vertical	Vehicle (baby formula)
Virulence Factors	Capsule	–	Intracellular growth	Ability to survive dry conditions
Culture/Diagnosis	Culture mother's genital tract on blood agar; CSF culture of neonate	CSF Gram stain/culture	Cold enrichment, rapid methods	Chromogenic differential agar, or rapid detection kits
Prevention	Culture and treatment of mother	–	Cooking food, avoiding unpasteurized dairy products	Safe preparation and use of, or avoidance of, powdered formula
Treatment	Ampicillin, penicillin G	Ceftazidime or cefepime +/− gentamicin	Ampicillin, trimethoprim-sulfamethoxazole	Begin with broad-spectrum drugs until susceptibilities determined
Distinctive Features	Most common; positive culture of mother confirms diagnosis	Suspected if infant is premature	–	–
Epidemiological Features	Before intrapartum antibiotics introduced in 1996: 1.8 cases per 1,000 live births After intrapartum antibiotics: 0.32 case per 1,000 live births	Estimated at 0.2-5 per 1,000 live births; 20% of pregnant women colonized	Mortality can be as high as 33%	Rare (a handful of documented cases in United States annually), but deadly

meningitis, the condition is often called *aseptic meningitis*. Aseptic meningitis may occasionally have noninfectious causes.

The majority of cases of viral meningitis occur in children, and 90% are caused by enteroviruses. A common cause of viral meningitis is initial infection with HSV-2, concurrent with a genital infection. But many other viruses also gain access to the central nervous system on occasion.

Viral meningitis is generally milder than bacterial or fungal meningitis, and it is usually resolved within 2 weeks. The mortality rate is less than 1%. Diagnosis begins with the failure to find bacteria, fungi, or protozoa in CSF and can be confirmed, depending on the virus, by viral culture or specific antigen tests. In most cases, no treatment is indicated.

The organisms in the differential diagnosis for acute meningitis are summarized in **Disease Table 17.1**.

Neonatal Meningitis

Meningitis in newborns is almost always a result of infection transmitted by the mother, either *in utero* or during passage through the birth canal. As more premature babies survive, the rates of neonatal meningitis increase because the condition is favored in patients with immature immune systems. In the United States, the two most common causes are *Streptococcus agalactiae* (known as group B streptococcus) and *Escherichia coli*. *Listeria monocytogenes* is also found frequently in neonates. It has already been covered here but is included in **Disease Table 17.2** as a reminder that it can cause neonatal cases as well. In the developing world, neonatal meningitis is more commonly caused by other organisms.

Streptococcus agalactiae

This species of *Streptococcus* belongs to group B of the streptococci. It colonizes 10% to 30% of female genital tracts and is the most frequent cause of neonatal meningitis (for details about this condition in women, see chapter 21). The treatment for neo-natal disease is ampicillin or penicillin G. Women are typically screened for the presence of these bacteria by means of a cervical and rectal swab at between 35 and 37 weeks of gestation. Women who are found to harbor the bacteria are offered intravenous antibiotics at the beginning of active labor and throughout labor until delivery is accomplished to avoid passing the bacteria to their infant during the birthing process. Penicillin is the drug of choice.

Escherichia coli

The K1 strain of *E. coli* is the second-most-common cause of neonatal meningitis. Most babies who suffer from this infection are premature, and their prognosis is poor. About 20% to 30% of them die, even with aggressive antibiotic treatment, and those who survive often have brain damage.

The bacterium is usually transmitted from the mother's birth canal. It causes no disease in the mother but can infect the vulnerable tissues of a neonate. It seems to have a predilection for the tissues of the central nervous system. Ceftazidime or cefepime +/− gentamicin is usually administered intravenously.

Cronobacter sakazakii

Cronobacter, formerly known as *Enterobacter sakazakii*, is found mainly in the environment but has been implicated in outbreaks of neonatal and infant meningitis transmitted through contaminated powdered infant formula. Although cases of *Cronobacter* meningitis are rare, mortality rates can reach 40%. The FDA and the CDC advise hospitals to use ready-to-feed or concentrated liquid formulas. They also recommend that home caregivers wash their hands and use clean feeding equipment when preparing formula, use fresh formula for each feeding, and discard any leftover formula **Disease Table 17.2.**

Zika Virus Disease

Starting in 2015 and continuing into 2016, an epidemic of babies born with abnormally small heads–a condition called microcephaly–became obvious in Brazil. The cause was quickly determined to be the Zika virus. Zika has been known to scientists since the mid-1900s, but the manifestation of hundreds of brain-damaged babies was new. Zika was first discovered in Uganda in Africa, where it seems to have circulated among human populations for some time. In the first half of the 20th century, the virus migrated to western Africa and then to Southeast Asia. In 2015 it was found to be circulating in Brazil, the first evidence of the virus in the Americas. In February 2016, the World Health Organization declared Zika an international health emergency. That emergency was ended in November of the same year. However, the virus continues to be a threat, and pregnant women in particular are cautioned not to travel to Zika-endemic areas. **Figure 17.10a** shows the distribution of Zika cases.

▶ Signs and Symptoms

When adults are infected with Zika, they can experience a range of symptoms, from none at all, to a skin rash, conjunctivitis, and muscle and joint pain. The virus also seems to trigger Guillain-Barré syndrome (GBS) in some adults. GBS is a neurological condition that can occur after infections with certain bacteria and viruses, and some-times after exposure to vaccines. Infection with the bacterium *Campylobacter jejuni*, a common foodborne bacterium, is one of the more common triggers of GBS. Zika infection has been found during this recent epidemic to lead to GBS. The syndrome is characterized by the immune system attacking peripheral nerves. The effects can be long-lasting, though usually they resolve completely. In a certain percentage of cases, respiratory muscles can be affected, and death may follow.

Top **5** things everyone needs to know about **ZIKA**

1 **Zika primarily spreads through infected mosquitoes. You can also get Zika through sex.**

Many areas in the United States have the type of mosquitoes that can spread Zika virus. There is no current local transmission of Zika virus in the continental United States, including Florida and Texas, which reported local transmission of Zika virus by mosquitoes for the first time in 2016–17.

2 **The best way to prevent Zika is to prevent mosquito bites.**

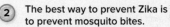

- Use insect repellent. It works!
- Wear long-sleeved shirts and long pants.
- When traveling, stay in places with air conditioning or window and door screens.
- Remove standing water around your home.

3 **Zika is linked to birth defects.**

Zika infection during pregnancy can cause a serious birth defect called microcephaly that is a sign of incomplete brain development. If you have a partner who lives in or has traveled to an area with risk of Zika, do not have sex, or use condoms every time you have sex during your pregnancy.

4 **Pregnant women should not travel to areas with risk of Zika.**

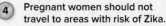

If you must travel to one of these areas, talk to your healthcare provider first and strictly follow steps to prevent mosquito bites during your trip.

5 **Returning travelers infected with Zika can spread the virus through mosquito bites.**

If you get infected with Zika and a mosquito bites you, you can pass the virus to the mosquito. The infected mosquito bites other people, who can get infected. Returning travelers should also use condoms or not have sex if they are concerned about passing it to their partners through sex.

Source: CDC

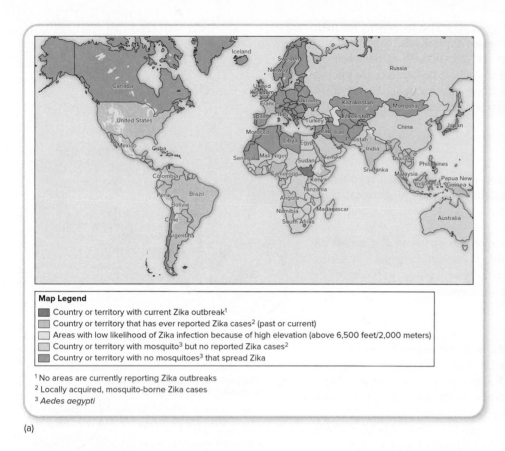

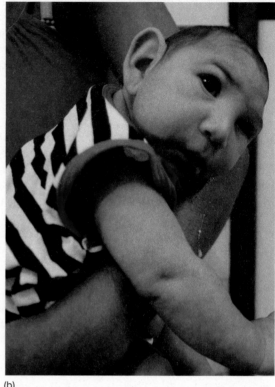

Map Legend

▪ Country or territory with current Zika outbreak[1]
▫ Country or territory that has ever reported Zika cases[2] (past or current)
▫ Areas with low likelihood of Zika infection because of high elevation (above 6,500 feet/2,000 meters)
▫ Country or territory with mosquito[3] but no reported Zika cases[2]
▪ Country or territory with no mosquitoes[3] that spread Zika

[1] No areas are currently reporting Zika outbreaks
[2] Locally acquired, mosquito-borne Zika cases
[3] *Aedes aegypti*

(a)   (b)

Figure 17.10 Zika Virus Disease.   (a) Map of Zika risk as of fall 2019. (b) A baby suffering from microcephaly due to Zika exposure in the womb.
©Brazil Photo Press/LatinContent Editorial/Getty Images (b)

Far more serious consequences affect babies who acquire Zika during gestation. In these cases it can cause congenital *Zika virus syndrome*. The most obvious manifestation is the characteristically small head, although it is not present in every case (**figure 17.10b**). Additional symptoms of the syndrome include vision problems, involuntary movements, seizures, and irritability. Symptoms of brain stem dysfunction such as swallowing problems are common as well.

▶ Causative Agent

Zika virus is an RNA virus in the Flaviviridae family. It is closely related to the viruses causing dengue fever, West Nile fever, and yellow fever.

▶ Transmission and Epidemiology

Scientists are still unraveling the pattern of Zika transmission and spread around the world. We do know that it is transmitted by the bite of the *Aedes* mosquito. The virus has also been transmitted via sexual intercourse, and the sharing of sex toys, with infected individuals. And of course, it can be transmitted vertically *in utero*. Its mosquito host, the genus *Aedes*, is spread throughout the world. Eighty percent of infections are asymptomatic. Between 5% and 10% of Zika-positive mothers have babies affected by the virus.

▶ Prevention and Treatment

Currently there is no vaccine for Zika, but the recent epidemics have spurred intense research in this area. Several vaccines are in development. At this point the only treatment is to provide supportive measures, for both adults and infants. Patients experiencing GBS should receive intensive physical therapy and may require mechanical ventilation **Disease Table 17.3**.

Disease Table 17.3	Zika Virus Infection
Causative Organism(s)	Zika virus
Most Common Modes of Transmission	Vertical, vector-borne, sexual contact, likely through blood transfusions (not yet confirmed)
Virulence Factors	–
Culture/Diagnosis	PCR testing
Prevention	Avoiding mosquitoes; vaccines in trials
Treatment	Supportive
Epidemiological Features	Originated in Africa but spreading throughout world; latest outbreak, leading to many microencephalies, started in 2015

Poliomyelitis

Poliomyelitis (poh″-lee-oh′-my″-eh′-ly′-tis) (polio) is an acute enteroviral infection of the spinal cord that can cause neuromuscular paralysis. Because it often affects small children, in the past it was called *infantile paralysis.* No civilization or culture has escaped the devastation of polio. The efforts of a WHO campaign have significantly reduced the global incidence of polio. It was the campaign's goal to eradicate all of the remaining wild polioviruses by 2000, and then by 2005. It didn't happen. Today an international partnership called the Global Polio Eradication Initiative (GPEI) is battling to eliminate polio. GPEI partners are the World Health Organization, Rotary International, the U.S. Centers for Disease Control and Prevention, the United Nations Children's Fund, and the Bill & Melinda Gates Foundation. It uses 20 million volunteers in 200 different countries and has vaccinated over 2.5 billion children since 1988.

▶ Signs and Symptoms

Most infections are contained as short-term, mild viremias. Some people develop mild nonspecific symptoms of fever, headache, nausea, sore throat, and myalgia. If the viremia persists, viruses can be carried to the central nervous system through its blood supply. The virus then spreads along specific pathways in the spinal cord and brain. The virus is **neurotropic,** that is, it attacks the nervous system. It infiltrates the motor neurons of the anterior horn of the spinal cord, although it can also attack spinal ganglia, cranial nerves, and motor nuclei.

In paralytic disease, invasion of motor neurons causes various degrees of flaccid (floppy) paralysis over a period of a few hours to several days. Depending on the level of damage to motor neurons, paralysis of the muscles of the legs, abdomen, back, intercostals, diaphragm, pectoral girdle, and bladder can result. In rare cases of **bulbar poliomyelitis,** the brain stem, medulla, or even cranial nerves are affected. This situation leads to loss of control of cardiorespiratory regulatory centers, requiring mechanical respirators. In time, the unused muscles begin to atrophy, growth is slowed, and severe deformities of the trunk and limbs develop. Common sites of deformities are the spine, shoulder, hips, knees, and feet. Because motor function but not sensation is compromised, the crippled limbs are often very painful.

In recent decades, a condition called post-polio syndrome (PPS) has been diagnosed in long-term survivors of childhood infection. PPS manifests as a progressive muscle deterioration that develops in about 25% to 50% of patients many years after their original polio attack.

▶ Causative Agent

The poliovirus is in the family Picornaviridae, genus *Enterovirus*–named for its small (*pico*) size and its RNA genome. It is nonenveloped and nonsegmented. The naked

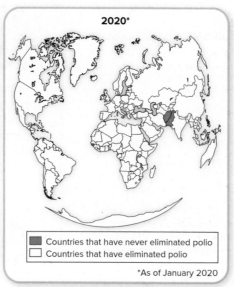

Figure 17.11 Progress in eliminating polio.
Source: CDC

Disease Table 17.4 — Poliomyelitis

Causative Organism(s)	Poliovirus **V**
Most Common Modes of Transmission	Fecal-oral, vehicle
Virulence Factors	Attachment mechanisms
Culture/Diagnosis	Viral culture, serology
Prevention	Live attenuated (OPV) (developing world) or inactivated vaccine (IPV) (developed world)
Treatment	None, palliative, supportive
Epidemiological Features	Eradicated from Western Hemisphere; still endemic in Pakistan and Afghanistan as of 2020. Vaccine strains caused 196 cases of paralytic polio in Africa.

capsid of the virus confers chemical stability and resistance to acid, bile, and detergents. This allows the virus to survive the gastric environment and other harsh conditions, which means it is transmitted easily.

▶ Pathogenesis and Virulence Factors

After being ingested, polioviruses adsorb to receptors of mucosal cells in the oropharynx and intestine. Here, they multiply in the mucosal epithelia and lymphoid tissue. Multiplication results in large numbers of viruses being shed into the throat and feces, and some of them leak into the blood. Depending on the number of viruses in the blood and their duration of stay there, an individual may exhibit no symptoms, mild nonspecific symptoms such as fever or short-term muscle pain, or devastating paralysis.

▶ Transmission and Epidemiology

Humans are the only known reservoir. The virus is passed within the population through food, water, hands, objects contaminated with feces, and mechanical vectors. Although the 20th century saw a very large rise in paralytic polio cases, it was also the century during which effective vaccines were developed. The infection was eliminated from the Western Hemisphere in the late 20th century. Because the virus cannot exist outside of a human, it is possible to totally eradicate the disease if we eliminate (i.e., vaccinate) all susceptible hosts. The decades of work performed by the Global Polio Eradication Initiative have made great progress **(figure 17.11),** but now that we are in the "endgame," progress has slowed. In the two countries in which natural infection still exists, Afghanistan and Pakistan, the number of cases quadrupled in 2019 vs. 2018.

Furthermore, during 2019, sixteen countries on the African continent—all of which had been declared polio-free—saw paralytic polio cases caused by a vaccine strain that reverted to its pathogenic state. Even with the massive efforts dedicated to polio eradication over at least 20 years, the virus is still hanging on.

▶ Prevention and Treatment

Treatment of polio rests largely on alleviating pain and suffering. During the acute phase, muscle spasm, headache, and associated discomfort can be alleviated by pain-relieving drugs. Respiratory failure may require artificial ventilation maintenance. Prompt physical therapy to diminish crippling deformities and to retrain muscles is recommended after the acute febrile phase subsides.

The mainstay of polio prevention is vaccination as early in life as possible, usually in four doses starting at about 2 months of age. Adult candidates for immunization are travelers and members of the armed forces. The two forms of vaccine currently in use are inactivated poliovirus vaccine (IPV), developed by Jonas Salk in 1954, and oral poliovirus vaccine (OPV), developed by Albert Sabin in the 1960s.

For many years, the oral vaccine was used in the United States because it is easily administered by mouth, but it is not free of medical complications. It contains an attenuated virus that can multiply in vaccinated people and be spread to others. The fact that it can be spread to others is the reason it is used in the developing world so that even those who don't receive the vaccine can receive protection. In very rare instances, the attenuated virus reverts to a virulent strain that causes disease rather than protects against it. This is what happened in Africa in 2019. For this reason, IPV, using killed virus, is the only vaccine used in the United States. **Disease Table 17.4.**

Meningoencephalitis

Two microorganisms cause a distinct disease called *meningoencephalitis* (disease in both the meninges and brain), and they are both amoebas. *Naegleria fowleri* and *Acanthamoeba* are accidental parasites that invade the body only under unusual circumstances.

Naegleria fowleri

Most cases of *Naegleria* infection reported worldwide occur in people who have been swimming in warm, natural bodies of freshwater. Infection can begin when amoebas are forced into human nasal passages as a result of swimming, diving, or other aquatic activities. Once the amoeba is inoculated into the favorable habitat of the nasal mucosa, it burrows in, multiplies, and uses the olfactory nerve to migrate into the brain and surrounding structures. The result is primary amoebic meningoencephalitis (PAM), a rapid, massive destruction of brain and spinal tissue that causes hemorrhage and coma and invariably ends in death within a week or so **(figure 17.12).** Note that this organism is very common—children often carry the amoeba as harmless biota, especially during the summer months, and the series of events leading to disease is exceedingly rare.

Unfortunately, *Naegleria* meningoencephalitis advances so rapidly that treatment usually proves futile. Studies have indicated that early therapy with amphotericin B, sulfadiazine, or tetracycline in some combination can be of some benefit. Because of the wide distribution of the amoeba and its hardiness, no general method of control exists. Public swimming pools and baths must be adequately chlorinated and checked periodically for the amoeba.

Acanthamoeba

This protozoan differs from *Naegleria* in its portal of entry. It invades broken skin, the conjunctiva, and occasionally the lungs and urogenital epithelia. It causes a meningoencephalitis somewhat similar to that of *Naegleria*. The course of infection is lengthier but nearly as deadly, with only a 2% to 3% survival rate. The disease is called granulomatous amoebic meningoencephalitis (GAM). At special risk for infection are people with traumatic eye injuries, contact lens wearers, and AIDS patients exposed to contaminated water. We discussed ocular infections in chapter 16. Cutaneous and CNS infections with this organism are occasional complications in AIDS.

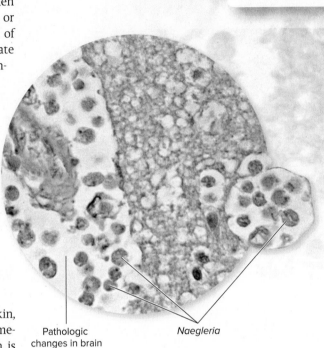

Source: CDC

Pathologic changes in brain

Naegleria

Figure 17.12 *Naegleria fowleri* **in the brain.**
The trophozoite form invades brain tissue, destroying it.
Source: CDC

Disease Table 17.5	Meningoencephalitis	
Disease	**Primary Amoebic Meningoencephalitis**	**Granulomatous Amoebic Meningoencephalitis**
Causative Organism(s)	*Naegleria fowleri* (P)	*Acanthamoeba* (P)
Most Common Modes of Transmission	Vehicle (exposure while swimming in water)	Direct contact
Virulence Factors	Invasiveness	Invasiveness
Culture/Diagnosis	Examination of CSF; brain imaging, biopsy	Examination of CSF; brain imaging, biopsy
Prevention	Limit warm freshwater or untreated tap water entering nasal passages	–
Treatment	Amphotericin B; mostly ineffective	Surgical excision of granulomas; ketoconazole may help
Epidemiological Features	United States: 0–7 cases a year; 97% case-fatality rate. Spreading to northern states as climate warms.	Predominantly occurs in immunocompromised patients

Pools are chlorinated to prevent *Naegleria* and other outbreaks.
©Lena Mirisola/Cultura Exclusive/Getty Images

Acute Encephalitis

Encephalitis (inflammation of the brain) can present as acute or **subacute.** It is always a serious condition, as the tissues of the brain are extremely sensitive to damage by inflammatory processes. Acute encephalitis is almost always caused by viral infection. One category of viral encephalitis is caused by viruses borne by insects (called *arboviruses,* which is short for *arthropod-borne viruses*), including West Nile virus. Alternatively, other viruses, such as the JC virus and members of the herpes family, are causative agents. Bacteria such as those covered under the meningitis section can also cause encephalitis, but the symptoms are usually more pronounced in the meninges than in the brain.

The signs and symptoms of encephalitis vary, but they may include behavior changes or confusion because of inflammation. Decreased consciousness and seizures frequently occur. Symptoms of meningitis are often also present. Few of these agents have specific treatments, but because swift initiation of acyclovir therapy can save the life if the patient is suffering from herpesvirus encephalitis, most physicians will begin empiric therapy with acyclovir in all seriously ill neonates and most other patients showing evidence of encephalitis. Treatment will, in any case, do no harm in patients who are infected with other agents.

Various Arthropod-Borne Viruses (Arboviruses)

Most arthropods that serve as infectious disease vectors feed on the blood of hosts. Infections show a peak incidence when the arthropod is actively feeding and reproducing, usually from late spring through early fall. Warm-blooded vertebrates also maintain the virus during the cold and dry seasons. Humans can serve as dead-end, accidental hosts, as in equine encephalitis, or they can be a maintenance reservoir, as in yellow fever (discussed in chapter 18).

Arboviral diseases have a great impact on humans. Although exact statistics are unavailable, it is believed that millions of people acquire infections each year, and thousands of them die. One common outcome of arboviral infection is an acute fever, often accompanied by rash. Viruses that primarily cause these symptoms are covered in chapter 18.

The arboviruses discussed in this chapter can cause encephalitis, and we consider them as a group because the symptoms and management are similar. The transmission and epidemiology of individual viruses are different, however, and are discussed for each virus in **table 17.1.**

▶ Pathogenesis and Virulence Factors

Arboviral encephalitis begins with an arthropod bite, the release of the virus into tissues, and its replication in nearby lymphatic tissues. Prolonged viremia establishes the virus in the brain, where inflammation can cause swelling and damage to the brain, nerves, and meninges. Symptoms are extremely variable and can include coma, convulsions, paralysis, tremor, loss of coordination, memory deficits, changes in speech and personality, and heart disorders. In some cases, survivors experience some degree of permanent brain damage. Young children and the elderly are most sensitive to injury by arboviral encephalitis. Having said that, most people who are infected will actually show no symptoms.

▶ Culture and Diagnosis

Except during epidemics, detecting arboviral infections can be difficult. The patient's history of travel to endemic areas or contact with vectors, along with serum analysis,

Public health officials regularly take samples of standing water in rain gutters, swimming pools, and elsewhere, looking for larval mosquitoes like those seen here.
Source: CDC

Table 17.1

Arbovirus Encephalitis

West Nile Virus

94%*

Emerged in US in 1999: By 2008 CDC reported that 1% of people in US had evidence of past or present infection, 70–80% who are infected have no symptoms. Birds are the amplifying hosts. Humans and other large mammals are dead-end hosts.

Number of cases in 2018: 2,647

Powassan Virus

0.7%

This virus is maintained in nature by ticks and groundhogs. Its geographic distribution is in the Northeast and the Great Lakes states.

21

La Crosse Virus

3%

La Crosse virus disease has only been reported in the eastern half of the US and in Texas. It is maintained by infection of small mammals, such as squirrels and chipmunks. Mosquitoes are the vector. Worst form of the disease occurs primarily in children under 16.

86

St. Louis Encephalitis Virus

0.3%

Most of those infected show no symptoms; severe neuroinvasive disease more common in adults. The virus is maintained in birds and transferred by mosquitoes to the dead-end host, humans.

8

Jamestown Canyon Virus

1%

This virus is maintained in nature by cycling between mosquitoes and deer or moose. The virus is found throughout North America but until recently was rarely seen in humans.

41

Eastern Equine Encephalitis Virus

0.2%

EEE is endemic to an area along the eastern coast of North America and Canada. The usual pattern is sporadic, but epidemics can occur in humans and horses. A vaccine exists for horses and is strongly recommended to eliminate the virus from its reservoir. In humans, the case fatality can be 70%.

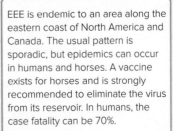

6

*Percent of all cases in United States as reported in 2018, the latest year for which data are available.

Source: ArboNET, Arboviral Diseases Branch, Centers for Disease Control and Prevention.

Disease Table 17.6	Acute Encephalitis		
Causative Organism(s)	Arboviruses (West Nile virus, La Crosse virus, Jamestown Canyon virus, St. Louis encephalitis virus, Powassan virus, eastern equine encephalitis virus) Ⓥ	Herpes simplex 1 or 2 Ⓥ	JC virus Ⓥ
Most Common Modes of Transmission	Vector (arthropod bites)	Vertical or reactivation of latent infection	? Ubiquitous
Virulence Factors	Attachment, fusion, invasion capabilities	–	–
Culture/Diagnosis	History, rapid serological tests, nucleic acid amplification tests	Clinical presentation, PCR, Ab tests, growth of virus in cell culture	PCR of cerebrospinal fluid
Prevention	Insect control	Maternal screening for HSV	None
Treatment	None	Acyclovir	Zidovudine or other antivirals
Distinctive Features	History of exposure to insect important	In infants, disseminated disease present; rare between 30 and 50 years	In severely immunocompromised, especially AIDS
Epidemiological Features	See incidence rates in table 17.1; 20% increase in U.S. incidence 2014–2018	HSV-1 more common cause of encephalitis; 2 cases per million per year	Affects 5% of adults with untreated AIDS

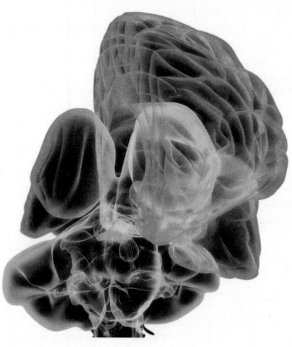

©MedicalRF.com

is highly supportive of a diagnosis. Rapid serological and nucleic acid amplification tests are available for some of the viruses.

► Treatment

No satisfactory treatment exists for any of the arboviral encephalitides (plural of *encephalitis*). As mentioned earlier, empiric acyclovir treatment may be begun in case the infection is actually caused by either herpes simplex virus or varicella zoster. Treatment of the other infections relies entirely on support measures to control fever, convulsions, dehydration, shock, and edema.

Most of the control safeguards for arbovirus disease are aimed at the arthropod vectors. Mosquito abatement by eliminating breeding sites and by broadcast-spreading insecticides has been highly effective in restricted urban settings. Birds play a role as reservoirs of the virus, but direct transmission between birds and humans does not occur.

Herpes Simplex Virus (HSV)

Herpes simplex type 1 and 2 viruses can cause encephalitis in newborns born to HSV-positive mothers. In this case, the virus is disseminated and the prognosis is poor. Older children and young adults (ages 5 to 30), as well as older adults (over 50 years old), are also susceptible to herpes simplex encephalitis caused most commonly by HSV-1. In these cases, the HSV encephalitis usually represents a reactivation of dormant HSV from the trigeminal ganglion.

It should be noted the varicella-zoster virus (see chapter 16) can also reactivate from the dormant state, and it is responsible for rare cases of encephalitis.

JC Virus

The **JC virus (JCV)** gets its name from the initials of the patient in whom it was first diagnosed as the cause of illness. Serological studies indicate that asymptomatic infection with this polyoma virus is commonplace. In patients with immune dysfunction, especially in those with AIDS, it can cause a condition called **progressive multifocal leukoencephalopathy** (loo″-koh-en-sef″-uh-lop′-uh-thee) **(PML).** This uncommon but generally fatal infection is a result of JC virus attack of accessory brain cells. The infection demyelinizes certain parts of the cerebrum. This virus should be considered when encephalitis symptoms are observed in AIDS patients. Recently, a few deaths from this condition have been prevented with high doses of zidovudine. **Disease Table 17.6.**

Subacute Encephalitis

When encephalitis symptoms take longer to show up, and when the symptoms are less striking, the condition is termed *subacute encephalitis*. The most common cause of subacute encephalitis is the protozoan *Toxoplasma*. Another form of subacute encephalitis can be caused by persistent measles virus as many as 7 to 15 years after the initial infection. A class of infectious agents known as prions can cause a condition called spongiform encephalopathy. Finally, the differential diagnosis of subacute encephalitis should consider a variety of infections with primary symptoms elsewhere in the body (see the last column of **Disease Table 17.7**).

Toxoplasma gondii

This protozoal infection in the fetus and in immunodeficient people, especially those with AIDS, is severe and often fatal. Although infection in otherwise healthy people is generally unnoticed, recent data suggest that it may have subtle but profound effects on their brain and the responses it controls. People with a history of *Toxoplasma* infection are often more likely to display thrill-seeking behaviors and seem to have slower reaction times. Researchers are zooming in on a possible association between schizophrenia and other mental illnesses, and a history of *Toxoplasma* infection.

▶ Signs and Symptoms

Most cases of toxoplasmosis are asymptomatic or marked by mild symptoms such as sore throat, lymph node enlargement, and low-grade fever. In patients whose immunity is suppressed by infection, cancer, or drugs, the outlook may be grim. A pregnant woman with toxoplasmosis has a 33% chance of transmitting the infection to her fetus. Congenital infection occurring in the first or second trimester is associated with stillbirth and severe abnormalities such as liver and spleen enlargement, liver failure, hydrocephalus, convulsions, and damage to the retina that can result in blindness.

▶ Pathogenesis and Virulence Factors

Toxoplasma is an obligate intracellular parasite, making its ability to invade host cells an important factor for virulence.

▶ Transmission and Epidemiology

T. gondii is a very successful parasite with so little host specificity that it can attack at least 200 species of birds and mammals. However, the parasite undergoes a sexual phase in the intestine of cats and is then released in feces, where it becomes an infective *oocyst* that survives in moist soil for several months. These

NCLEX® PREP

2. An infectious disease team is analyzing the epidemiology of vector-borne diseases, including those transmitted by arthropods. All of the following are arboviral diseases, except:
   a. *polio*
   b. *West Nile encephalitis.*
   c. *yellow fever.*
   d. *St. Louis encephalitis.*

Note About Neglected Parasitic Infections

You may have the impression that the United States has a low number of infections with protozoa and helminths, and that these are primarily "developing country" diseases. Unfortunately, that is not the case. The CDC has recently designated multiple diseases as "neglected parasitic infections (NPIs)" in the United States. Here are five of the most prevalent ones.

- Chagas disease (in chapter 18)—caused by *Trypanosome cruzi*, a protozoan
- Neurocysticercosis (chapter 20)—caused by the tapeworm *Taenia sodium*
- Toxocariasis (chapter 20)—caused by the roundworm *Toxocara*
- Toxoplasmosis (this chapter)—caused by the protozoan *Toxoplasma gondii*
- Trichomoniasis (chapter 21)—caused by the protozoan *Trichomonas vaginalis*

The numbers of people experiencing disease from these five organisms is in the millions—and that's just in the United States. We will point out these NPIs as we encounter them in the rest of this book.

NCLEX® PREP

3. For which disease processes are immunizations available as a method of prevention? Select all that apply.
   a. *Cryptococcus neoformans*
   b. *Listeria monocytogenes*
   c. *Haemophilus influenzae*
   d. *Streptococcus pneumoniae*
   e. *Neisseria meningitidis*

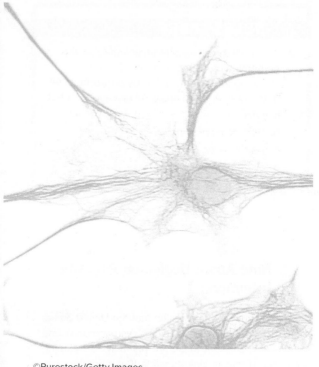

©Purestock/Getty Images

forms eventually enter an asexual cyst state in tissues, called a *pseudocyst*. Most of the time, the parasite does not cycle in cats alone and is spread by oocysts to intermediate hosts, including rodents and birds. The cycle returns to cats when they eat these infected prey animals. Cattle and sheep can also be infected.

Scientists have discovered that the protozoan crowds into a part of the rat brain that usually directs the rat to avoid the smell of cat urine (a natural defense against a domestic rat's major predator). When *Toxoplasma* infects rat brains, the rats lose their fear of cats. Infected rats are then easily eaten by cats, ensuring the continuing *Toxoplasma* life cycle. All other neurological functions in the rat are left intact.

Humans appear to be constantly exposed to the pathogen. The rate of prior infections, as detected through serological tests, can be as high as 90% in some populations. Many cases are caused by ingesting pseudocysts in undercooked contaminated meat, and other sources include contact with other mammals or even dirt and dust contaminated with oocysts. Fetuses may become infected when tachyzoites—a fast-reproducing form of the organism—cross the placenta.

In view of the fact that the oocysts are so widespread and resistant, hygiene is of paramount importance in controlling toxoplasmosis. Adequate cooking or freezing below −20°C destroys both oocysts and tissue cysts. Oocysts can also be avoided by washing the hands after handling cats or soil possibly contaminated with cat feces, especially sandboxes and litter boxes. Pregnant women should be especially attentive to these rules and should never clean the cat's litter box. *Toxoplasma* infection of the fetus can result in miscarriage, premature birth, or babies born with brain and/or vision problems.

Measles Virus: Subacute Sclerosing Panencephalitis

Subacute sclerosing panencephalitis is sometimes called a "slow virus infection." The symptoms appear years after an initial measles episode. SSPE seems to be caused by direct viral invasion of neural tissue. SSPE has no effective treatment and is always fatal. It is not clear what factors lead to persistence of the virus in some people. SSPE's important features are listed in **Disease Table 17.7.**

Prions

As you read in chapter 5, prions are *proteinaceous infectious particles* containing, apparently, no genetic material. They are known to cause diseases called **transmissible spongiform encephalopathies (TSEs),** neurodegenerative diseases with very long incubation periods (years) but rapid progressions once they begin. The human TSEs are **Creutzfeldt-Jakob disease (CJD),** Gerstmann-Strussler-Scheinker disease, and fatal familial insomnia. TSEs are also found in animals and include a disease called scrapie in sheep and goats, transmissible mink encephalopathy, and bovine spongiform encephalopathy (BSE). This last disease is commonly known as mad cow disease and was in the headlines in the 1990s due to its apparent link to a variant form of Creutzfeldt-Jakob human disease in Great Britain. In 2016 a new prion was discovered, one that is associated with a disease known as multiple system atrophy, a condition similar to Parkinson's disease. We consider CJD in this section.

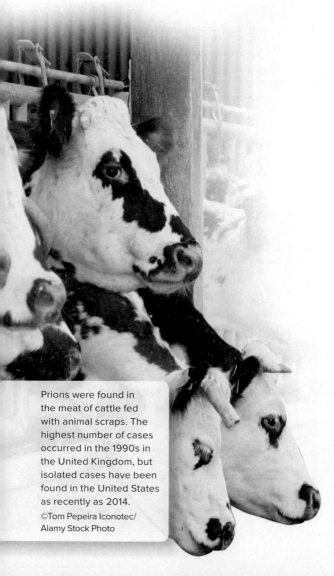

Prions were found in the meat of cattle fed with animal scraps. The highest number of cases occurred in the 1990s in the United Kingdom, but isolated cases have been found in the United States as recently as 2014.
©Tom Pepeira Iconotec/Alamy Stock Photo

▶ Signs and Symptoms of CJD

Symptoms of CJD include altered behavior, dementia, memory loss, impaired senses, delirium, and premature senility. Uncontrollable muscle contractions continue until death, which usually occurs within 1 year of diagnosis.

Disease Table 17.7	Subacute Encephalitis			
Causative Organism(s)	*Toxoplasma gondii* (P)	Subacute sclerosing panencephalitis	Prions	Other conditions to consider
Most Common Modes of Transmission	Vehicle (meat) or fecal-oral	Persistence of measles virus	CJD = direct/parenteral contact with infected tissue, or inherited vCJD = vehicle (meat, parenteral)	Other conditions may display subacute encephalitis symptoms:
Virulence Factors	Intracellular growth	Cell fusion, evasion of immune system	Avoidance of host immune response	Rickettsial diseases–Rocky Mountain spotted fever (chapter 18)
Culture/Diagnosis	Serological detection of IgM	EEGs, MRI, serology (Ab versus measles virus)	Biopsy, image of brain	Lyme disease (chapter 18)
Prevention	Personal hygiene, food hygiene	None	Avoiding infected meat or instruments; no prevention for inherited form	*Bartonella* or *Anaplasma* disease (chapter 18)
Treatment	Pyrimethamine and/or leucovorin and/or sulfadiazine	None	None	Tapeworm disease–*Taenia solium* (chapter 20)
Distinctive Features	Subacute, slower development of disease	History of measles	Long incubation period; fast progression once it begins	Syphilis–*Treponema pallidum* (chapter 21)
Epidemiological Features	15%–29% of U.S. population is seropositive; internationally, seroprevalence is up to 90%; disease occurs in 3%–15% of AIDS patients; designated a "neglected parasitic infection" in the United States by the CDC	Occurs in 1 in 609 persons who had measles	CJD: 1 case per year per million worldwide; seen in older adults vCJD: 98% cases originated in United Kingdom	

▶ Causative Agent of CJD

The transmissible agent in CJD is a prion. In some forms of the disease, it involved the transformation of a normal host protein (called PrP), a protein that is supposed to function to help the brain develop normally. Once this happens, the abnormal PrP itself becomes catalytic and able to spontaneously convert other normal human PrP proteins into the abnormal form. This becomes a self-propagating chain reaction that creates a massive accumulation of altered PrP, leading to plaques, spongiform damage (i.e., holes in the brain) **(figure 17.13),** and severe loss of brain function.

Using the term *transmissible agent* may be a bit misleading, however, as some cases of CJD arise through genetic mutation of the *PrP* gene, which can be a heritable trait. So it seems that although one can acquire a defective PrP protein via transmission, one can also have an altered *PrP* gene passed on through heredity.

Prions are incredibly hardy "pathogens." They are highly resistant to chemicals, radiation, and heat. They can even withstand prolonged autoclaving.

▶ Transmission and Epidemiology

In the late 1990s, it became apparent that humans were contracting a variant form of CJD (vCJD) after ingesting meat from cattle that had been afflicted by bovine spongiform encephalopathy. Presumably, meat products had been contaminated with fluid or tissues infected with the prion. Cases of this disease were centered around Great Britain, where many cows were found to have BSE. The median age at death of

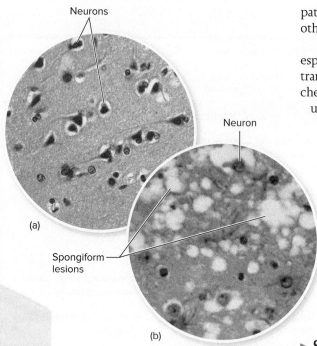

Neurons

(a)

Neuron

Spongiform
lesions

(b)

Figure 17.13 The microscopic effects of spongiform encephalopathy. **(a)** Normal cerebral cortex section, showing neurons and glial cells. **(b)** Sectioned cortex in CJD patient shows numerous round holes, producing a "spongy" appearance. This destroys brain architecture and causes massive loss of neurons and glial cells.
Michael Abbey/Science Source (a); Source: CDC/Dr. Al Jenny (b)

Bats are one of the vectors for rabies.
©Hugh Lansdown/Shutterstock

patients with vCJD is 28 years. In contrast, the median age at death of patients with other forms of CJD is 68 years.

Health care professionals should be aware of the possibility of CJD in patients, especially when surgical procedures are performed, as cases have been reported of transmission of CJD via contaminated surgical instruments. Due to the heat and chemical resistance of prions, normal disinfection and sterilization procedures are usually not sufficient to eliminate them from instruments and surfaces. The latest CDC guidelines for handling of CJD patients in a health care environment should be consulted. There is no known treatment for CJD, and mortality appears to be 100%; but there is active research into treatments for prion diseases **Disease Table 17.7.**

Rabies

Rabies is a slow, progressive zoonotic disease characterized by a fatal encephalitis. It is so distinctive in its pathogenesis and its symptoms that we discuss it separately from the other encephalitides. It is distributed nearly worldwide, except for perhaps a few countries that have remained rabies-free by practicing rigorous animal control.

▸ Signs and Symptoms

The average incubation period of rabies is 2 weeks to even years, depending on the wound site, its severity, and the inoculation dose. The incubation period is shorter in facial, scalp, or neck wounds because of closer proximity to the brain. The prodromal phase begins with fever, nausea, vomiting, headache, fatigue, and other nonspecific symptoms.

Until recently, humans were never known to survive rabies. But a handful of patients have recovered in recent years after receiving intensive, long-term treatment.

▸ Pathogenesis and Virulence Factors

Infection with rabies virus typically begins when an infected animal's saliva enters a puncture site. The virus occasionally is inhaled or inoculated through the membranes of the eye. The rabies virus remains up to a week at the trauma site, where it multiplies. The virus then gradually enters nerve endings and advances toward the ganglia, spinal cord, and brain. Viral multiplication throughout the brain is eventually followed by migration to such diverse sites as the eye, heart, skin, and oral cavity. The infection cycle is completed when the virus replicates in the salivary gland and is shed into the saliva. Untreated rabies proceeds through several distinct stages that almost inevitably end in death, unless post-exposure vaccination is performed before symptoms begin.

Scientists have discovered that virulence is associated with an envelope glycoprotein that seems to give the virus its ability to spread in the CNS and to invade certain types of neural cells.

▸ Transmission and Epidemiology

The primary reservoirs of the virus are wild mammals such as canines, skunks, raccoons, badgers, cats, and bats that can spread the infection to domestic dogs and cats. Both wild and domestic mammals can spread the disease to humans through bites, scratches, and inhalation of droplets. The annual worldwide total for human rabies is estimated to be about 35,000 to 50,000 cases, but only a tiny number of these cases occur in the United States. The majority of these are transmitted to humans from bats. Most U.S. cases of rabies occur in wild animals (about 6,000 to 7,000 cases per year), while dog rabies has declined **(figure 17.14).** Globally, though, over 90% of human rabies cases are still acquired from dogs.

The epidemiology of animal rabies in the United States varies. The most common wild animal reservoir hosts are raccoons, bats, and skunks. Regional differences in the dominant reservoir also occur. Skunks are the most common carriers of rabies in California, while raccoons are the predominant carriers in the East, and foxes dominate in Texas.

Diagnosis requires multiple tests. Reverse transcription PCR is used with saliva samples but must be accompanied by detection of antibodies to the virus in serum or spinal fluid. Skin biopsies are also used.

▶ Prevention and Treatment

A bite from a wild or stray animal demands assessment of the animal, meticulous care of the wound, and a specific treatment regimen. A wild mammal, especially a skunk, raccoon, fox, or coyote, that bites without provocation is presumed to be rabid, and therapy for rabies is immediately begun. Rabies has been transmitted to humans in the absence of a bite; aerosols of bat saliva are thought to be capable of transmitting the virus. For that reason, people who have found a bat in their house can be encouraged to undergo the postexposure prophylaxis regimen.

Rabies is one of the few infectious diseases for which a combination of passive and active postexposure immunization is indicated (and successful). Initially the wound is infused with human rabies immune globulin (HRIG) to impede the spread of the virus, and globulin is also injected intramuscularly to provide immediate systemic protection. A full course of vaccination is started simultaneously. The routine postexposure vaccination entails intramuscular or intradermal injection on days 0, 3, 7, and 14, sometimes with two additional boosters. High-risk groups such as veterinarians, animal handlers, laboratory personnel, and travelers

Disease Table 17.8	Rabies
Causative Organism(s)	Rabies virus Ⓥ
Most Common Modes of Transmission	Parenteral (bite trauma), droplet contact
Virulence Factors	Envelope glycoprotein
Culture/Diagnosis	Direct fluorescent antigen testing
Prevention	Inactivated vaccine
Treatment	Postexposure passive and active immunization; induced coma and ventilator support if symptoms have begun
Epidemiological Features	United States: 1-5 cases per year Worldwide: 35,000-55,000 cases annually

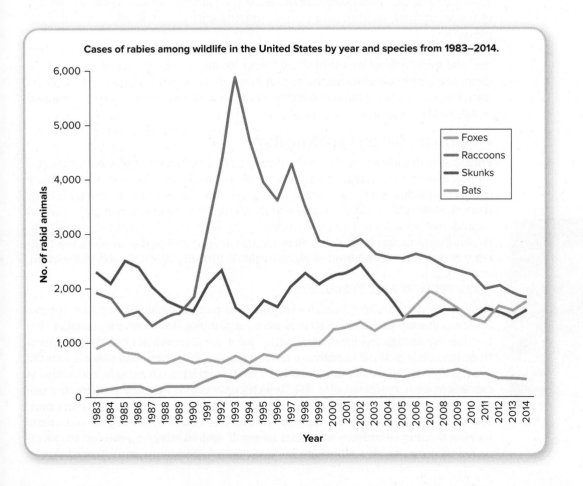

Figure 17.14 Most common carriers of rabies in the United States.

Source: CDC

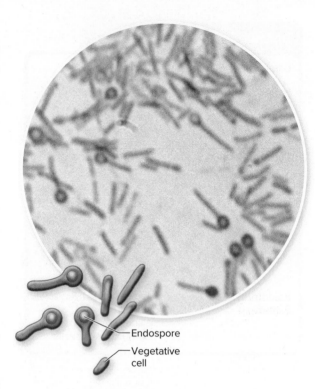

Figure 17.15 *Clostridium tetani.* Its typical tennis racket morphology is created by terminal endospores that swell the end of the cell (170×).
©Biophoto Associates/Science Source

should receive three doses to protect against possible exposure. A DNA vaccine for rabies is in development. **Disease Table 17.8.**

Tetanus

Tetanus is a neuromuscular disease whose alternate name, lockjaw, refers to an early effect of the disease on the jaw muscle. The etiologic agent, *Clostridium tetani,* is a common resident of cultivated soil and the gastrointestinal tracts of animals. It is a gram-positive, spore-forming rod. The endospores it produces often cause the vegetative cell to swell **(figure 17.15).** The endospores are produced only under anaerobic conditions.

▶ Signs and Symptoms

C. tetani releases a powerful exotoxin that is a neurotoxin, **tetanospasmin,** that binds to target sites on peripheral motor neurons, in the spinal cord and brain, and in the sympathetic nervous system. The toxin acts by blocking the inhibition of muscle contraction. Without inhibition of contraction, the muscles contract uncontrollably, resulting in spastic paralysis. The first symptoms are clenching of the jaw, followed in succession by extreme arching of the back, flexion of the arms, and extension of the legs **(figure 17.16).** Lockjaw confers the bizarre appearance of *risus sardonicus* (sardonic grin), which looks eerily as though the person is smiling **(figure 17.17).** Death most often occurs due to paralysis of the respiratory muscles and respiratory arrest.

▶ Pathogenesis and Virulence Factors

The mere presence of endospores in a wound is not sufficient to initiate infection because the bacterium is unable to invade damaged tissues readily. It is also a strict anaerobe, and the endospores cannot become established unless tissues at the site of the wound are necrotic and poorly supplied with blood, conditions that favor germination.

As the vegetative cells grow, the tetanospasmin toxin is released into the infection site. The toxin spreads to nearby motor nerve endings in the injured tissue, binds to them, and travels via axons to the ventral horns of the spinal cord (figure 17.17). The toxin blocks the release of neurotransmitters. Only a small amount of toxin is required to initiate the symptoms.

▶ Transmission and Epidemiology

Endospores usually enter the body through accidental puncture wounds, burns, umbilical stumps, frostbite, and crushed body parts. The incidence of tetanus is low in North America. Most cases occur among geriatric patients and intravenous drug abusers. Historically, however, the worldwide incidence of maternal and neonatal tetanus has been high, causing 60,000 deaths each year. In response, the World Health Organization (WHO) has made dramatic global progress in reducing mortality through the promotion of more hygienic birthing practices and vaccination.

▶ Prevention and Treatment

A patient with a clinical appearance suggestive of tetanus should immediately receive antitoxin therapy with human tetanus immune globulin (TIG) and also penicillin G.

The recommended vaccination series for 1- to 3-month-old babies consists of three injections of DTaP (diphtheria, tetanus, and acellular pertussis) given 2 months apart, followed by booster doses about 1 and 4 years later. Alternately, they may be vaccinated with a vaccine called "DT," which protects only against diphtheria and tetanus. Children thus immunized probably have protection for 10 years. At that point, and every 10 years thereafter, they should receive a dose of TD, tetanus-diphtheria vaccine. Additional protection against neonatal tetanus may be achieved by vaccinating pregnant women, whose antibodies will be passed to the fetus. Toxoid should

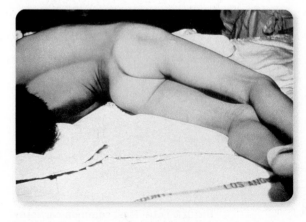

Figure 17.16 Late-stage tetanus.
Image courtesy of CDC

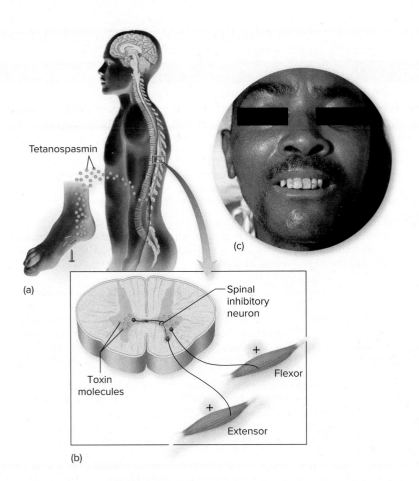

(a)

(b)

(c)

Disease Table 17.9	Tetanus
Causative Organism(s)	*Clostridium tetani* Ⓑ G+
Most Common Modes of Transmission	Parenteral, direct contact
Virulence Factors	Tetanospasm exotoxin
Culture/Diagnosis	Symptomatic
Prevention	Tetanus toxoid immunization
Treatment	Combination of passive antitoxin and tetanus toxoid active immunization, metronidazole; sedation
Epidemiological Features	United States: Approximately 30 cases/year; worldwide, +/− 35,000 newborn deaths annually in developing countries

Figure 17.17 **The events in tetanus.** **(a)** After traumatic injury, bacteria infecting the local tissues secrete tetanospasmin, which is absorbed by the peripheral axons and is carried to the target neurons in the spinal column. **(b)** In the spinal cord, the toxin attaches to the junctions of regulatory neurons that inhibit inappropriate contraction. Released from inhibition, the muscles, even opposing members of a muscle group, receive constant stimuli and contract uncontrollably. **(c)** Muscles contract spasmodically, without regard to regulatory mechanisms or conscious control. Note the clenched jaw, called *risus sardonicus*.
Source: CDC, AFIP, C Farmer (c)

also be given to injured persons who have never been immunized, have not completed the series, or whose last booster was received more than 10 years previously. The vaccine can be given simultaneously with passive TIG immunization to achieve immediate as well as long-term protection. **Disease Table 17.9.**

Botulism

Botulism is an **intoxication** (i.e., caused by an exotoxin) associated with eating poorly preserved foods, although it can also occur as a true infection. Until recent times, it was relatively common and frequently fatal, but modern techniques of food preservation and medical treatment have reduced both its incidence and its fatality rate.

▶ Signs and Symptoms

There are three major forms of botulism, distinguished by their means of transmission and the population they affect. **Table 17.2** summarizes these. The symptoms are largely the same in all three forms, however. From the circulatory system, an exotoxin called the **botulinum toxin** travels to its principal site of action, the neuromuscular junctions of skeletal muscles. The effect of botulinum is to prevent the release of the neurotransmitter, acetylcholine, that initiates the signal for muscle contraction. This results in the

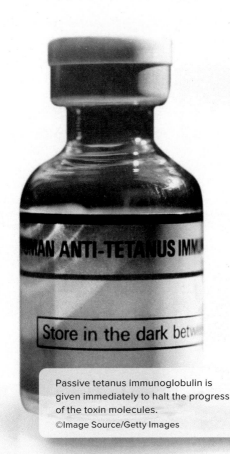

Passive tetanus immunoglobulin is given immediately to halt the progress of the toxin molecules.
©Image Source/Getty Images

Table 17.2 Three Types of Botulism

		Transmission and Epidemiology	Culture and Diagnosis
Infant Botulism 66%*	Infection followed by intoxication	This is currently the most common type of botulism in the United States, with approximately 140 cases reported annually. The exact food source is not always known, although raw honey has been implicated in some cases, and the endospores are common in dust and soil. Apparently, the immature state of the neonatal intestine and microbial biota allows the endospores to gain a foothold, germinate, and give off neurotoxin. As in adults, babies exhibit flaccid paralysis, usually manifested as a weak sucking response, generalized loss of tone (the "floppy-baby syndrome"), and respiratory complications. Although adults can also ingest botulinum endospores in contaminated vegetables and other foods, the adult intestinal tract normally inhibits this sort of infection.	Finding the toxin or the organism in the feces confirms the diagnosis.
Wound Botulism 14%	Infection followed by intoxication	Perhaps three or four cases of wound botulism occur each year in the United States. In this form of the disease, endospores enter a wound or puncture, much as in tetanus, but the symptoms are similar to those of foodborne botulism. Cases of this form of botulism have increased several-fold in the last several years due to intravenous drug use.	The toxin should be demonstrated in the serum, or the organism should be grown from the wound.

Foodborne Botulism
 	
10%
	Pure intoxication	Many botulism outbreaks occur in home-processed foods, including canned vegetables, smoked meats, and cheese spreads.
		     Several factors in food processing can lead to botulism. Endospores can be present on the vegetables or meat at the time of gathering and are difficult to remove completely. When contaminated food is put in jars and steamed in a pressure cooker that does not reach reliable pressure and temperature, some endospores survive (botulinum endospores are highly heat resistant). At the same time, the pressure is sufficient to evacuate the air and create anaerobic conditions. Storage of the jars at room temperature favors endospore germination and vegetative growth, and one of the products of the cell's metabolism is botulinum, the most potent microbial toxin known.
		     Bacterial growth may not be evident in the appearance of the jar or can or in the food's taste or texture, and only minute amounts of toxin may be present. Botulism is never transmitted from person to person.
			Some laboratories attempt to identify the toxin in the offending food. Alternatively, if multiple patients present with the same symptoms after ingesting the same food, a presumptive diagnosis can be made. The cultivation of *C. botulinum* in feces is considered confirmation of the diagnosis because the carrier rate is very low.

*Percentage of botulism cases in United States that are of this type, in 2017. Percentages do not add up to 100 due to cases with undetermined causes.

opposite effect of tetanus. The type of paralysis is called flaccid paralysis, where muscle contraction is inhibited **(figure 17.18).** The usual time before onset of symptoms is 12 to 72 hours, depending on the size of the dose. Neuromuscular symptoms first affect the muscles of the head and include double vision, difficulty in swallowing, and dizziness, but there is no sensory or mental lapse. Later symptoms are descending muscular paralysis and respiratory compromise. In the past, death resulted from respiratory arrest, but mechanical respirators have reduced the fatality rate to about 10%.

You are probably familiar with the use of the botulinum toxin, or "Botox," to relieve facial wrinkles. This treatment exploits the action of the exotoxin to inhibit muscle contractions, leading to smoother surface skin. Botox treatment (in developed countries) is currently being used for several medical conditions that have nothing to do with vanity. People suffering from migraine, excessive sweating, urinary incontinence, spasmocity associated with multiple sclerosis, and even tennis elbow are experiencing relief from Botox injections.

▶ Causative Agent

Clostridium botulinum, like *Clostridium tetani*, is an endospore-forming anaerobe that does its damage through the release of an exotoxin. *C. botulinum* commonly inhabits soil and water and occasionally the intestinal tract of animals. It is distributed worldwide but occurs most often in the Northern Hemisphere. The species has seven distinctly different types (designated A, B, C, D, E, F, and G) that vary in distribution among animals, regions of the world, and types of exotoxin. Human disease is usually associated with types A, B, E, and F, and animal disease with types A, B, C, D, and E.

Both *C. tetani* and *C. botulinum* produce neurotoxins; but tetanospasmin, the toxin made by *C. tetani*, results in spastic paralysis (uncontrolled muscle contraction).

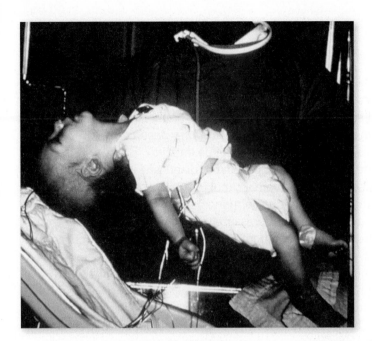

Figure 17.18 Infant botulism. The child is 6 weeks old and displays the flaccid paralysis characteristic of botulism.
Source: CDC

Disease Table 17.10	Botulism
Causative Organism(s)	*Clostridium botulinum* **B** G⁺
Most Common Modes of Transmission	Vehicle (foodborne toxin, airborne organism); direct contact (wound); parenteral (injection)
Virulence Factors	Botulinum exotoxin
Culture/Diagnosis	Culture of organism; demonstration of toxin
Prevention	Food hygiene; toxoid immunization available for laboratory professionals
Treatment	Antitoxin, penicillin G for wound botulism, supportive care
Epidemiological Features	United States: 66% of botulism is infant botulism; wound botulism increasing with an increase in injecting drug use. **Category A bioterror agent;** 1 gm of aerosolized botulinum toxin estimated to be able to kill 1.5 million people

In contrast, botulinum, the *C. botulinum* neurotoxin, results in flaccid paralysis, a loss of ability to contract the muscles.

▶ Culture and Diagnosis

Diagnostic standards are slightly different for the three different presentations of botulism. A suspected case of botulism should trigger a phone call to the state health department or the CDC before proceeding with diagnosis or treatment.

▶ Prevention and Treatment

The CDC maintains a supply of antitoxin, which, when administered soon after diagnosis, can prevent the worst outcomes of the disease. Patients are also managed with respiratory and cardiac support systems. In all cases, hospitalization is required and recovery takes weeks. There is an overall 5% mortality rate.

17.3 LEARNING OUTCOMES—Assess Your Progress

4. List the possible causative agents, modes of transmission, virulence factors, diagnostic techniques, and prevention/treatment for the "Highlight Disease" meningitis.

5. Identify the most common and also the most deadly of the multiple possible causes of meningitis.

6. Explain the difference between the oral polio vaccine and the inactivated polio vaccine, and under which circumstances each is appropriate.

7. Discuss important features of the diseases most directly involving the brain. These are Zika virus disease, meningoencephalitis, encephalitis, and subacute encephalitis.

8. Identify which encephalitis-causing viruses you should be aware of in your geographic area.

9. Name three causes of subacute encephalitis and discuss the epidemiology of each.

10. Discuss important features of the disease rabies.

11. Compare and contrast the diseases tetanus and botulism.

CASE FILE WRAP-UP

©Exactostock/SuperStock (health care worker);
©Steve Gschmeissner/Science Photo Library/
Alamy Stock Photo (blood cells)

Meningitis is an infection of the meninges, which surround the brain and spinal cord. In Betty's case it is possible that the organism causing her respiratory illness migrated from the nasal cavity to the brain. The most common form of meningitis in adults is caused by *Streptococcus pneumoniae,* the same organism that causes most cases of bacterial pneumonia.

Lumbar puncture is an important part of the diagnostic workup for meningitis. A needle is inserted into the spinal canal to obtain a sample of cerebral spinal fluid (CSF). During the procedure, a manometer is used to measure the pressure in the CNS system. Elevated pressure is indicative of swelling in the brain. Lab results from a lumbar puncture can help clarify the pathogen—viral, fungal, or bacterial. For a patient with bacterial meningitis, one would expect low glucose, high protein, and high white blood cell count in the CSF.

Due to the high risk of complications and mortality from meningitis, antibiotics should be administered as soon as possible, regardless of whether the CSF sample has been collected. Prompt and well-monitored treatment is essential to prevent the neurological complications of meningitis, including hearing loss, intellectual deficits, partial paralysis, and seizures.

Monitoring serum drug levels is an essential part of care. The metabolic panel gives an overview of organ function, particularly the kidney and liver, both of which play an important role in clearing medications from the bloodstream. If organ dysfunction causes accumulation of an antibiotic, toxicity may occur. In the case of Betty's meningitis, monitoring medication levels was important for another reason. The blood-brain barrier is a protective barrier that prohibits the introduction of any substances to the central nervous system. In order for the antibiotics to be effective in treating the infection, this barrier must be overcome. In patients with meningitis, higher doses and/or more frequent administration of antibiotics is necessary to make sure they get to the CNS.

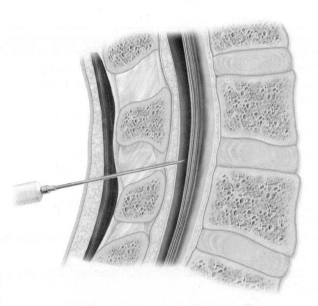

Diagram of the needle placement for lumbar puncture.

▶ Summing Up

Taxonomic Organization Microorganisms Causing Disease in the Nervous System		
Microorganism	**Pronunciation**	**Disease Table**
Gram-positive endospore-forming bacteria		
Clostridium botulinum	klos-trid″-ee-um bot′-yew-lyn″-um	Botulism, 17.10
Clostridium tetani	klos-trid″-ee-um tet′-a-nie	Tetanus, 17.9
Gram-positive bacteria		
Streptococcus agalactiae	strep″-tuh-kok′-us ay-ga-lact′-tee-ay	Neonatal and infant meningitis, 17.2
Streptococcus pneumoniae	strep″-tuh-kok′-us nu-mo′-nee-ay	Meningitis, 17.1
Listeria monocytogenes	lis-teer′-ee-uh mon′-oh-sy-toj″-eh-nees	Meningitis, 17.1
		Neonatal and infant meningitis, 17.2
Gram-negative bacteria		
Cronobacter sakazakii	krow″-no-bak′-tur sock″-uh-zock′-ee	Neonatal and infant meningitis, 17.2
Escherichia coli	esh′-shur-eesh″-ee-uh col′-eye	Neonatal and infant meningitis, 17.2
Haemophilus influenzae	huh-mah′-fuh-luss in′-floo-en″-zay	Meningitis, 17.1
Neisseria meningitidis	nye-seer″-ee-uh′ men′-in-jit″-ih-dus	Meningitis, 17.1
DNA viruses		
Herpes simplex virus type 1 and 2	hur′-peez sim′-plex vie′-russ	Acute encephalitis, 17.6
JC virus	jay′-cee″ vie′-russ	Acute encephalitis, 17.6
RNA viruses		
Arboviruses	ar′-bow-vie′-russ-suz	Acute encephalitis, 17.6
West Nile virus, La Crosse virus, Jamestown Canyon virus, St. Louis encephalitis virus, Powassan virus, eastern equine encephalitis virus	west nyl vie′-russ, luh-kross vie′-russ, jeymz′-toun kan′-yun vie′-russ, saynt lew′iss en-cef′-uh-ly′-tuss vie′-russ, poe-was′-uhn vie′-russ, ee′-stern ee′-cwine en-sef-uh-ly′-tuss vie′-russ	
Poliovirus	poh′-lee-oh-vie′-russ	Poliomyelitis, 17.4
Rabies virus	ray′-bees vie′-russ	Rabies, 17.8
Zika virus	zee′kuh vie′-rus	Zika virus infection, 17.3
Fungi		
Cryptococcus neoformans	crip-tuh-kok′-us nee′--oh-for″-mans	Meningitis, 17.1
Coccidioides	cox-sid″-ee-oid′-ees	Meningitis, 17.1
Prions		
Creutzfeldt-Jakob prion	croytz′-felt yaw′-cob pree′-on	Subacute encephalitis, 17.7
Protozoa		
Acanthamoeba	ay-kanth″-uh-mee′-buh	Meningoencephalitis, 17.5
Naegleria fowleri	nay-glar′-ee-uh fow-lahr′-ee	Meningoencephalitis, 17.5
Toxoplasma gondii	tox′-oh-plas″-mah gon′-dee-eye	Subacute encephalitis, 17.7

The Gut and the Brain

Have you ever heard the term "gut-brain axis"? For many years, it has been recognized that there is an important and comprehensive connection between the gastrointestinal tract and the brain. They are connected through hormonal, endocrine, and neuronal mechanisms, so that one affects the other. This connection is so important that the gut is sometimes called "the second brain." We know this instinctively because our gut reacts when we think certain thoughts, such as "I have to give a class presentation in 5 minutes." Situations and thoughts that make us extremely uneasy or happy have a noticeable effect on our digestive system.

Since the early 2000s we have realized that there is another huge influence on our central nervous system that comes from the gut: our gut microbiota. It may seem incredible, but the composition of our gut microbiota has been shown to be closely correlated with the following characteristics of our brain biology:

- The way our brain develops *in utero*. The gut microbiome appears to influence the number of neurons created during embryonic development and the number of neurons that are disposed of as part of the normal process of brain development before birth.
- The relative activity of microglia—the resident phagocytic cells in the brain, which account for 10% to 15% of all brain cells. With a disrupted (or absent) microbiota, these cells have less immune responsiveness.
- The process of myelination of nerves. Myelination occurs during fetal development and afterwards. It lays down a fatty layer (myelin) around nerves to enhance signal conduction. The process is notably disrupted when the gut microbiota is not present or is significantly altered. This can have a significant effect on nerve function, both in the central and in the peripheral nervous system.
- The creation of a sound blood-brain barrier (BBB). With a disrupted gut microbiota, the BBB is leaky and doesn't function the way it does in animals with healthy microbiota.

What is the upshot of these phenomena? Recent research is finding many connections between healthy neural functioning of organisms and their microbiome—and the findings described here suggest the underlying mechanistic reasons for this. We have mentioned elsewhere that there are correlative studies linking the gut microbiome and such disorders as depression, schizophrenia, and even autism spectrum disorder. Clearly, this is a rich area for exploration.

Note About Correlation vs. Causation This area of research is a good example to illustrate a very important distinction in science. When we say that it is a rich area of research, it is "rich" in two senses. There will certainly be groundbreaking findings, of the type where we can say definitively that "an *x* type of microbiome leads to (or causes) *y* neurological outcome." These will indeed lead to better management of neurological disorders via the microbiome. There will also certainly be "red herrings"—studies that find an association between the microbiome and certain brain conditions that turn out to be unsubstantiated. This is the way scientific research proceeds, since early studies usually only establish *correlation* (that two phenomenon were seen together), and it takes more in-depth studies to establish *causation* (that one factor actually resulted in the other). Unfortunately, news outlets don't often appreciate the distinction in these two types of findings, so you will frequently see media reports of new studies that merely found a correlation, breathlessly proclaiming that "Factor A causes Factor B."

Encephalitis

Arboviruses
Herpes simplex virus type 1 or 2
JC virus

Subacute Encephalitis

Toxoplasma gondii
Prions

Rabies

Rabies virus

Tetanus

Clostridium tetani

Zika Virus Disease

Zika virus

- Bacteria
- Viruses
- Protozoa
- Fungi
- Prions

Creutzfeldt-Jakob Disease

Prion

Meningoencephalitis

Naegleria fowleri
Acanthamoeba

Meningitis

Neisseria meningitidis
Streptococcus pneumoniae
Haemophilus influenzae
Listeria monocytogenes
Cryptococcus neoformans
Coccidioides species
Various viruses

Neonatal Meningitis

Streptococcus agalactiae
Escherichia coli
Listeria monocytogenes
Cronobacter sakazakii

Polio

Poliovirus

Botulism

Clostridium botulinum

System Summary
Figure 17.19

Deadliness and Communicability of Selected Diseases of the Nervous System

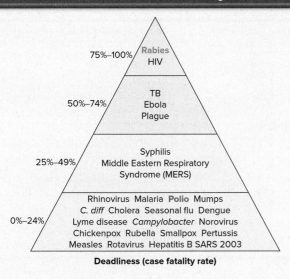

75%–100%   Rabies HIV

50%–74%   TB Ebola Plague

25%–49%   Syphilis Middle Eastern Respiratory Syndrome (MERS)

0%–24%   Rhinovirus Malaria **Polio** Mumps
C. diff Cholera Seasonal flu Dengue
Lyme disease *Campylobacter* Norovirus
Chickenpox Rubella Smallpox Pertussis
Measles Rotavirus Hepatitis B SARS 2003

Deadliness (case fatality rate)

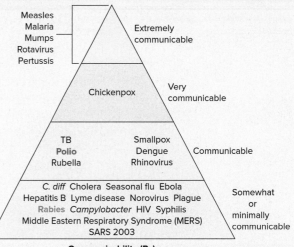

Measles
Malaria
Mumps
Rotavirus
Pertussis   — Extremely communicable

Chickenpox   Very communicable

TB **Polio** Rubella   Smallpox Dengue Rhinovirus   Communicable

C. diff Cholera Seasonal flu Ebola
Hepatitis B Lyme disease Norovirus Plague
Rabies *Campylobacter* HIV Syphilis
Middle Eastern Respiratory Syndrome (MERS)
SARS 2003   Somewhat or minimally communicable

Communicability (R_0)

INFECTIOUS DISEASES AFFECTING THE NERVOUS SYSTEM

Anatomy & Defenses: Nervous System

The nervous system consists of the central nervous system (the brain and spinal cord) and the peripheral nervous system (nerves and ganglia). The soft tissue of the brain and spinal cord is encased within the tough casing of three membranes called the meninges. The nervous system is protected by the blood-brain barrier, which limits the passage of substances from the bloodstream to the brain and spinal cord.

Normal Biota: Nervous System

At the present time, we believe there is no normal biota in either the central nervous system or the peripheral nervous system.

Nervous System Diseases Caused by Microorganisms

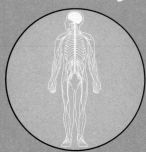

- **Meningitis**
The more serious forms are caused by bacteria, though the most common forms are caused by viruses. Fungi and protozoa are less frequent causes. The most serious bacterial causes are *Neisseria meningitidis* and *Streptococcus pneumoniae*.

- **Neonatal Meningitis**
Usually transmitted vertically. Group B streptococcus is the most common cause.

- **Zika Virus Disease**
In neonates can cause microcephaly; in adults symptoms may be unnoticed or mild or may cause the onset of Guillain-Barré syndrome.

- **Poliomyelitis**
Acute enterovirus infection of spinal cord; can cause neuromuscular paralysis. After a massive effort, most of the world is now polio-free, but it is proving difficult to finish the eradication.

- **Meningoencephalitis**
Caused mainly by two amoebas, *Naegleria fowler* and *Acanthamoeba*.

- **Acute Encephalitis**
Usually caused by viral infection. Arboviruses carried by arthropods are often responsible. Majority of infections in U.S. caused by West Nile virus. Depending on geographical location, can also be caused by LaCrosse virus, Jamestown Canyon virus, St. Louis encephalitis virus, Powassan, or eastern equine encephalitis. Nonarboviral causes: Herpes simplex virus, JC virus.

- **Subacute Encephalitis**
Symptoms take longer to manifest than in acute encephalitis. Causes quite often are the protozoan *Toxoplasma gondii*, prions, and a delayed consequence of measles infection called subacute sclerosing panencephalitis.

- **Rabies**
Slow, progressive zoonotic disease characterized by fatal encephalitis; caused by rabies virus.

- **Tetanus**
Neuromuscular disease caused by *Clostridium tetani* neurotoxin called tetanospasmin. Leads to muscle stiffness that can lead to death.

- **Botulism**
Neuromuscular disease caused by *Clostridium botulinum* exotoxin. There are three main forms: infant botulism, foodborne botulism, and wound botulism. Leads to potentially fatal flaccid paralysis.

SmartGrid: From Knowledge to Critical Thinking

This *21 Question Grid* takes the topics from this chapter and arranges them with respect to the American Society for Microbiology's Undergraduate Curriculum guidelines—all six of the important "Concepts" as well as the important "Competency" of scientific literacy. Three questions are supplied about chapter content that refer to the Concept or Competency, in increasing levels of Bloom's taxonomy for learning.

ASM Concept/ Competency	A. Bloom's Level 1, 2—Remember and Understand (Choose one.)	B. Bloom's Level 3, 4—Apply and Analyze	C. Bloom's Level 5, 6—Evaluate and Create
Evolution	1. Which pathogen has evolved to make its rodent host less avoidant of cats? a. *Cryptococcus neoformans* b. *Neisseria meningitidis* c. rabies virus d. *Toxoplasma gondii*	2. *Neisseria meningitidis* occurs as noninvasive normal biota in the nares (noses) of approximately 35% of the population. Those that have developed the ability to be invasive, entering the nervous system, are said to be at an "evolutionary dead end." Speculate on why the invasive strains are characterized with that label.	3. Speculate on why the poliovirus has changed over evolutionary time to cause severe disease less often in humans, while a disease such as rabies has not become less virulent for humans over time.
Cell Structure and Function	4. What cellular structure do several of the organisms that cause meningitis share? a. capsule b. pilus c. fimbria d. endospore	5. Why is it only necessary to include the exotoxin (in toxoid form) in vaccines for tetanus and botulism?	6. Justify the use of only a Gram stain in diagnosing some types of meningitis.
Metabolic Pathways	7. Which of the following organisms is anaerobic? a. poliovirus b. *Cryptococcus* c. *Clostridium* d. *Coccidioides*	8. When cultivating specimens for *N. meningitidis*, nonselective chocolate agar is often used. But when culturing for another *Neisseria* species, *N. gonorrhoeae*, a different medium, Modified Thayer-Martin (MTM) medium, which contains multiple antibiotics, is preferred. Speculate on why this is.	9. What do you expect to happen to the geographical distribution of *Coccidioides* cases over the next 10 years? Justify your answer.
Information Flow and Genetics	10. Which disease is caused by an infectious agent that carries no nucleic acid? a. CJD b. rabies c. polio d. meningitis	11. Why would PCR be the best method for identifying microbes in the cerebrospinal fluid if antibiotic treatment was begun before the fluid was sampled?	12. Scientists have analyzed the "transcriptome" (the RNA transcripts) of mouse brains that have been infected with *Toxoplasma*. In what ways might that be informative?

ASM Concept/ Competency	A. Bloom's Level 1, 2—Remember and Understand (Choose one.)	B. Bloom's Level 3, 4—Apply and Analyze	C. Bloom's Level 5, 6—Evaluate and Create

Microbial Systems	**13.** The normal gut microbiota in adults, but not infants, inhibit the growth of which pathogen?
	  a. *Neisseria meningitidis*
	  b. *Clostridium botulinum*
	  c. *Clostridium tetani*
	  d. *Naegleria fowleri*
		14. Botulinum toxin has been proposed as a treatment for the symptoms of tetanus. Discuss why this might be useful.	**15.** Recent research shows that when scientists infect *Aedes aegypti* with a bacterium called *Wolbachia,* they become less likely to spread Zika and other viruses to humans. Speculate on what *Wolbachia* may be doing in the mosquito.

Impact of Microorganisms	**16.** Subacute encephalitis can be caused by
	  a. *Toxoplasma gondii.*
	  b. *Streptococcus agalactiae.*
	  c. *Naegleria fowleri.*
	  d. *Haemophilus influenzae.*
		17. "Accidental herd immunity" conferred by the oral polio vaccine in the developing world has aided the elimination of polio from many countries. Explain what accidental herd immunity is and how it works.	**18.** Conduct research about the multistage meningitis outbreak linked to steroid injections in 2012. Speculate as to why this "nonpathogenic" fungal species caused such severe disease and death.

Scientific Thinking	**19.** Mosquito eradication could change the epidemiology of
	  a. polio.
	  b. Zika.
	  c. West Nile.
	  d. two of these.
		20. The paralytic form of poliomyelitis began to occur in epidemics in North America and Europe in the early part of the 1900s. Many scientists believe this increase in incidence and in disease manifestations was due to significant improvements in sanitation that occurred around that time. Use what you know about how the virus is transmitted to justify this seemingly puzzling outcome. (Hint: Maternal antibodies might be involved.)	**21.** In The Microbiome feature, the link between the gut microbiome and brain health is discussed. The feature also mentions that many of the published studies show correlation, and not causation. Using laboratory mice as your subjects, design an experiment that could show that prolonged antibiotic exposure disrupts the gut microbiome and *causes* altered brain development.

Answers to the multiple-choice questions appear in Appendix A.

Visual Connections

This question connects content within and between chapters.

From chapter 2, figure 2.19a. Without looking back to the figure in chapter 2, speculate on which meningitis-causing organism you are seeing here. How could your presumptive diagnosis be confirmed?

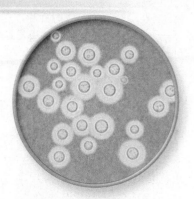

Source: CDC/Dr. Leanor Haley

18

Infectious Diseases Affecting the Cardiovascular and Lymphatic Systems

IN THIS CHAPTER...

18.1 The Cardiovascular and Lymphatic Systems and Their Defenses

1. Describe the important anatomical features of the cardiovascular and lymphatic systems.
2. List the natural defenses present in the cardiovascular and lymphatic systems.

18.2 Normal Biota of the Cardiovascular and Lymphatic Systems

3. Explain the "what" and the "why" of the normal biota of the cardiovascular and lymphatic systems.

18.3 Cardiovascular and Lymphatic System Diseases Caused by Microorganisms

4. List the possible causative agents, modes of transmission, virulence factors, diagnostic techniques, and prevention/treatment for the "Highlight Diseases" malaria and HIV.
5. Discuss the epidemiology of malaria.
6. Describe the epidemiology of HIV infection in the developing world.
7. Discuss the important features of infectious cardiovascular conditions that have more than one possible cause. These are the two forms of endocarditis, septicemia, hemorrhagic fever diseases, and nonhemorrhagic fever diseases.
8. Identify factors that distinguish hemorrhagic and nonhemorrhagic fever diseases.
9. Outline the series of events that may lead to septicemia and how it should be prevented and treated.
10. Discuss the important features of infectious cardiovascular diseases that have only one possible cause. These are plague, tularemia, Lyme disease, infectious mononucleosis, Chagas disease, and anthrax.
11. Describe what makes anthrax a good agent for bioterrorism, and list the important presenting signs to look for in patients.

CASE FILE

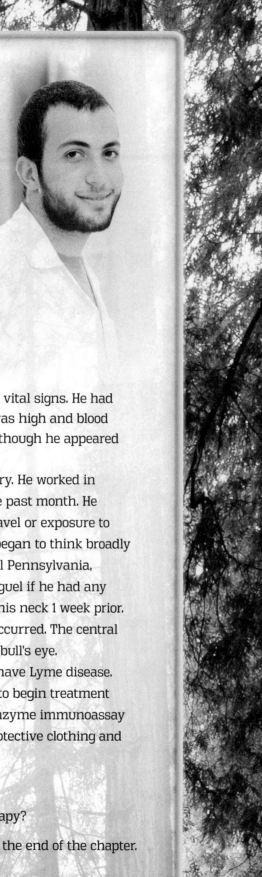

Bitten by a Tick

Miguel, a 28-year-old male, presented to the emergency department where I was working as a nurse practitioner. Miguel reported that for the past 5 days, he had been experiencing a persistent headache and fatigue. He had an intermittent fever with chills. Miguel sought medical attention with his primary care provider and was sent to the emergency department with concern for a serious or systemic illness.

I brought Miguel into a treatment room and checked his vital signs. He had a fever and reported a persistent headache. His heart rate was high and blood pressure was normal. He was able to answer my questions, though he appeared lethargic.

I gathered more information about Miguel's health history. He worked in construction and had been working on a new job site for the past month. He had no significant past medical history. He denied recent travel or exposure to infectious disease. With his unremarkable health history, I began to think broadly about what could have caused his symptoms. Living in rural Pennsylvania, exposure to ticks was common in forested areas. I asked Miguel if he had any recent insect bites. He recalled finding a tick on the back of his neck 1 week prior. Upon my exam, I found a red area where the tick bite had occurred. The central red area was surrounded by another red ring, resembling a bull's eye.

Based on his symptoms, I suspected that Miguel might have Lyme disease. I ordered a 3-week course of antibiotics and instructed him to begin treatment immediately. To confirm the diagnosis, I ordered a serum enzyme immunoassay test. I also counseled Miguel about the use of appropriate protective clothing and DEET insect repellant to prevent tick bites at his job site.

- Why is it important to treat Lyme disease early?

- Will Miguel be cured of the disease after antibiotic therapy?

Case File Wrap-Up appears at the end of the chapter.

18.1 The Cardiovascular and Lymphatic Systems and Their Defenses

The Cardiovascular System

The *cardiovascular system* is the pipeline of the body. It is composed of the blood vessels, which carry blood to and from all regions of the body, and the heart, which pumps the blood. This system moves the blood in a closed circuit, and it is therefore known as the *circulatory system*. The cardiovascular system provides tissues with oxygen and nutrients and carries away carbon dioxide and waste products, delivering them to the appropriate organs for removal. A closely related but largely separate system, the *lymphatic system*, is a major source of immune cells and fluids, and it serves as a one-way passage, returning fluid from the tissues to the cardiovascular system. **Figure 18.1** shows you how the two systems work together. You first saw this figure in chapter 12.

©Purestock/Getty Images

Figure 18.1 **The anatomy of the cardiovascular and lymphatic systems.**

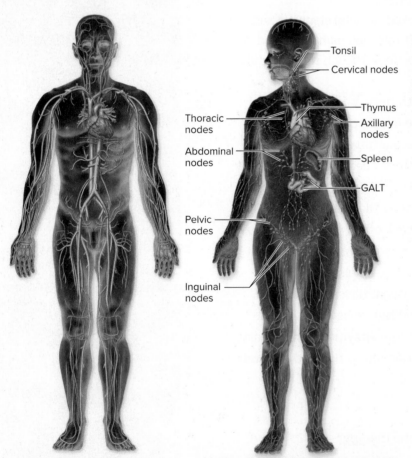

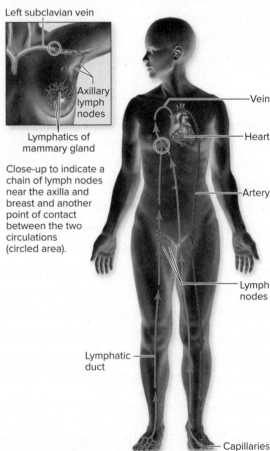

The Circulatory System

Body compartments are screened by circulating WBCs in the cardiovascular system.

The Lymphatic System

The lymphatic system consists of a branching network of vessels that extend into most body areas. Note the higher density of lymphatic vessels in the "dead-end" areas of the hands, feet, and breast, which are frequent contact points for infections. Other lymphatic organs include the lymph nodes, spleen, gut-associated lymphoid tissue (GALT), the thymus, and the tonsils.

The Lymphatic and Circulatory Systems

Comparison of the generalized circulation of the lymphatic system and the blood. Although the lymphatic vessels parallel the regular circulation, they transport in only one direction unlike the cyclic pattern of blood. Direct connection between the two circulations occurs at points near the heart where large lymph ducts empty their fluid into veins (circled area).

The heart is a fist-size muscular organ that pumps blood through the body. It is divided into two halves, each of which is divided into an upper and lower chamber **(figure 18.2).** The upper chambers are called atria (singular, *atrium*), and the lower are ventricles. The entire organ is encased in a fibrous covering, the pericardium, which is an occasional site of infection. The actual wall of the heart has three layers: From outer to inner, they are the epicardium, the myocardium, and the endocardium. The endocardium also covers the valves of the heart, and it is a relatively common target of microbial infection.

The blood vessels consist of *arteries, veins,* and *capillaries.* Arteries carry oxygenated blood away from the heart under relatively high pressure. They branch into smaller vessels called arterioles. Veins actually begin as smaller venules in the periphery of the body and coalesce into veins. The smallest blood vessels, the capillaries, connect arterioles to venules. Both arteries and veins have walls made of three layers of tissue. The innermost layer is composed of a smooth epithelium called endothelium. Its smooth surface encourages the smooth flow of cells and platelets through the system. The next layer is composed of connective tissue and muscle fibers. The outside layer is a thin layer of connective tissue. Capillaries, the smallest vessels, have walls made of only one layer of endothelium.

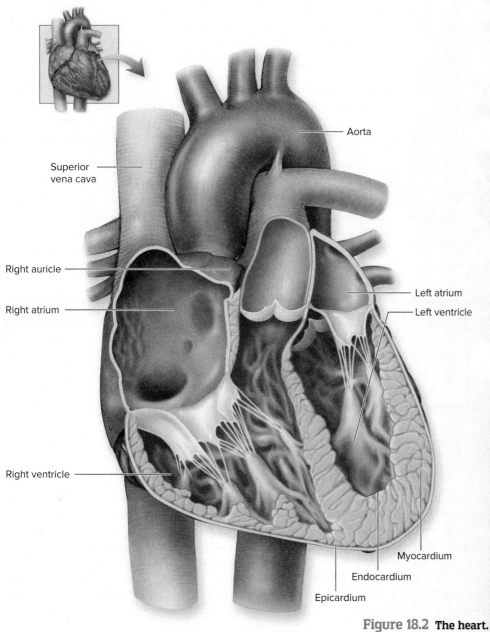

Aorta

Superior
vena cava

Right auricle

Right atrium

Left atrium

Left ventricle

Right ventricle

Myocardium

Endocardium

Epicardium

Conceptual image of blood
cells in an artery.
Purestock/Getty Images

Figure 18.2 The heart.

The Lymphatic System

The lymphatic system consists mainly of the lymph vessels, which roughly parallel the blood vessels; lymph nodes, which cluster at body sites such as the groin, neck, armpit, and intestines; and the spleen. It serves to collect fluid that has left the blood vessels and entered tissues, filter it of impurities and infectious agents, and return it to the blood.

Defenses of the Cardiovascular and Lymphatic Systems

The cardiovascular system is highly protected from microbial infection. Microbes that successfully invade the system, however, gain access to every part of the body, and every system may potentially be affected. For this reason, bloodstream infections are called **systemic** infections.

There are multiple modes of defense against infection in the bloodstream. The blood is full of leukocytes, with approximately 5,000 to 10,000 white blood cells per microliter of blood. The various types of white blood cells include the lymphocytes, responsible for adaptive immunity, and the phagocytes, which are so critical to innate as well as adaptive immune responses. Only a few microbes can survive in the blood with so many defensive elements. That said, a handful of infectious agents have nonetheless evolved sophisticated mechanisms for avoiding blood-borne defenses.

Medical conditions involving the blood often have the suffix *-emia*. For instance, viruses that cause meningitis can travel to the nervous system via the bloodstream. Their presence in the blood is called **viremia.** When fungi are in the blood, the condition is termed **fungemia,** and bacterial presence is called **bacteremia,** a general term denoting only their *presence*. Bacteria frequently are introduced into the bloodstream during the course of daily living. Brushing your teeth or tearing a hangnail can introduce bacteria from the mouth or skin into the bloodstream. This situation is usually temporary. But when bacteria flourish and grow in the bloodstream, the condition is termed **septicemia.** Septicemia (also called sepsis) can very quickly lead to cascading immune responses, resulting in decreased systemic blood pressure, which can lead to **septic shock,** a life-threatening condition.

18.1 LEARNING OUTCOMES—Assess Your Progress

1. Describe the important anatomical features of the cardiovascular and lymphatic systems.
2. List the natural defenses present in the cardiovascular and lymphatic systems.

18.2 Normal Biota of the Cardiovascular and Lymphatic Systems

Like the nervous system, the cardiovascular and lymphatic systems are "closed" systems with no normal access to the external environment. Therefore, it has long been believed that they possess no normal biota. Thus, in the healthy state, no microorganisms *colonize* either the lymphatic or cardiovascular system. Of course, this is biology, and it is never quite that simple. Recent studies from the Human Microbiome Project have suggested that the bloodstream is not completely sterile, even during periods of apparent health. There is evidence that the blood cells–especially white blood cells–do contain bacteria of various types. It is tempting to speculate that these low-level microbial "infections" may contribute to diseases for which no etiology has previously been found or for conditions currently thought to be noninfectious.

Cardiovascular and Lymphatic System Defenses and Normal Biota

	Defenses	Normal Biota
Cardiovascular System	Blood-borne components of innate and adaptive immunity–including phagocytosis, adaptive immunity	Sparse, mostly in white blood cells
Lymphatic System	Numerous immune defenses reside here.	Unclear

18.2 LEARNING OUTCOMES—Assess Your Progress

3. Explain the "what" and the "why" of the normal biota of the cardiovascular and lymphatic systems.

18.3 Cardiovascular and Lymphatic System Diseases Caused by Microorganisms

Categorizing cardiovascular and lymphatic infections according to clinical presentation is somewhat difficult because most of these conditions are systemic, with effects on multiple organ systems. We start with two extremely important conditions, malaria and HIV.

HIGHLIGHT DISEASE

Malaria

From prehistoric time until the present, malaria has been one of the greatest afflictions, in the same rank as bubonic plague, influenza, and tuberculosis. Even now, as the dominant protozoal disease, it threatens 40% of the world's population every year. The origin of the name is from the Italian words *mal*, "bad," and *aria*, "air."

▶ Signs and Symptoms

After a 10- to 16-day incubation period, the first symptoms are malaise, fatigue, vague aches, and nausea with or without diarrhea, followed by bouts of chills, fever, and sweating. These symptoms occur at 48- or 72-hour intervals, as a result of the synchronized rupturing of red blood cells. The interval, length, and regularity of symptoms reflect the type of malaria. Patients with falciparum malaria, the most virulent type, often display persistent fever, cough, and weakness for weeks without relief. Complications of malaria are hemolytic anemia from lysed blood cells and organ enlargement and rupture due to cellular debris that accumulates in the spleen, liver, and kidneys. One of the most serious complications of falciparum malaria is termed *cerebral malaria*. In this condition, small blood vessels in the brain become obstructed due to the increased ability of red blood cells (RBCs) to adhere to vessel walls (a condition called *cytoadherence* induced by the protozoan). The resulting decrease in oxygen in brain tissue can result in coma and death. In general, malaria has the highest death rate in the acute phase, especially in children. Certain kinds of malaria (those caused by *Plasmodium vivax* and *P. ovale*) are subject to relapses because some infected liver cells harbor dormant protozoans for up to 5 years.

Note on the Global Fund to Fight AIDS, Tuberculosis, and Malaria

In 2002 a unique partnership between governments, the private sector, civil society, and people affected by AIDS, tuberculosis, or malaria was formed. Its aim is to raise money and distribute it to programs that work in countries where the epidemics are the worst, and to save lives. The Bill and Melinda Gates Foundation is a major contributor to the Global Fund. And it has been very successful. There are one-third fewer deaths due to those "Big 3" diseases in countries where the Global Fund is active. They report that 17 million lives have been saved since 2002, at the rate of about 2 million a year. This chapter covers malaria and AIDS, and tuberculosis will be addressed in chapter 19. When COVID-19 hit, the Global Fund announced it was providing one billion dollars to help countries fight COVID-19, and to shore up their efforts fighting HIV, TB, and malaria during the COVID-19 pandemic.

Mosquitos transmit the malaria protozoan.
©Renaud Visage/Digital Vision/ Getty Images

▶ **Causative Agent**

Plasmodium species are protozoa in the sporozoan group. The genus *Plasmodium* contains over 200 species, but only five are known to commonly infect humans: *P. malariae, P. vivax, P. knowlesi, P. ovale,* and *P. falciparum.* The five species show variations in the pattern and severity of disease. For instance, *P. falciparum* is responsible for the vast majority of deaths.

Table 18.1 Life Cycle of the Malarial Parasite

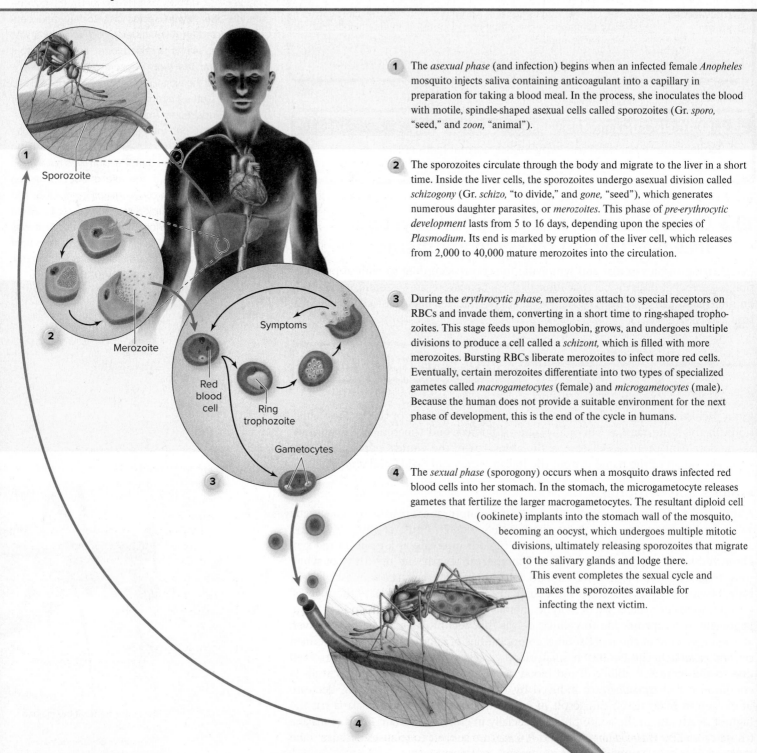

1. The *asexual phase* (and infection) begins when an infected female *Anopheles* mosquito injects saliva containing anticoagulant into a capillary in preparation for taking a blood meal. In the process, she inoculates the blood with motile, spindle-shaped asexual cells called sporozoites (Gr. *sporo,* "seed," and *zoon,* "animal").

2. The sporozoites circulate through the body and migrate to the liver in a short time. Inside the liver cells, the sporozoites undergo asexual division called *schizogony* (Gr. *schizo,* "to divide," and *gone,* "seed"), which generates numerous daughter parasites, or *merozoites.* This phase of *pre-erythrocytic development* lasts from 5 to 16 days, depending upon the species of *Plasmodium.* Its end is marked by eruption of the liver cell, which releases from 2,000 to 40,000 mature merozoites into the circulation.

3. During the *erythrocytic phase,* merozoites attach to special receptors on RBCs and invade them, converting in a short time to ring-shaped trophozoites. This stage feeds upon hemoglobin, grows, and undergoes multiple divisions to produce a cell called a *schizont,* which is filled with more merozoites. Bursting RBCs liberate merozoites to infect more red cells. Eventually, certain merozoites differentiate into two types of specialized gametes called *macrogametocytes* (female) and *microgametocytes* (male). Because the human does not provide a suitable environment for the next phase of development, this is the end of the cycle in humans.

4. The *sexual phase* (sporogony) occurs when a mosquito draws infected red blood cells into her stomach. In the stomach, the microgametocyte releases gametes that fertilize the larger macrogametocytes. The resultant diploid cell (ookinete) implants into the stomach wall of the mosquito, becoming an oocyst, which undergoes multiple mitotic divisions, ultimately releasing sporozoites that migrate to the salivary glands and lodge there. This event completes the sexual cycle and makes the sporozoites available for infecting the next victim.

Development of the malarial parasite is divided into two distinct phases: the asexual phase, carried out in the human, and the sexual phase, carried out in the mosquito. **Table 18.1** lists the steps of the malarial life cycle.

Figure 18.3 illustrates the ring trophozoite stage in a malarial infection.

▶ Pathogenesis and Virulence Factors

The invasion of the merozoites into RBCs leads to the release of fever-inducing chemicals into the bloodstream. Chills and fevers often occur in a cyclic pattern. *Plasmodium* also metabolizes glucose at a very high rate, leading to hypoglycemia in the human host. The damage to RBCs results in anemia. The accumulation of malarial products in the liver and the immune stimulation in the spleen can lead to enlargement of these organs. Individual protozoa within a host can express distinctly different surface antigens, making it difficult for the host immune system to battle.

▶ Transmission and Epidemiology

All forms of malaria are spread primarily by the female *Anopheles* mosquito. Although malaria was once distributed throughout most of the world, the control of mosquitoes in temperate areas has successfully restricted it mostly to a belt extending around the equator **(figure 18.4)**. Despite this achievement, approximately 200 million new cases are still reported each year, about 90% of them in Africa. The most frequent victims are children and young adults, of whom around 500,000 die annually. A particular form of the malarial protozoan causes damage to the placenta in pregnant women, leading to excess mortality among fetuses and newborns. The total case rate in the United States is about 1,000 to 2,000 new cases a year. While most of these are found in people who acquired it in a known endemic area, locally transmitted infections are on the rise.

▶ Culture and Diagnosis

Malaria can be diagnosed definitively by the discovery of a typical stage of *Plasmodium* in stained blood smears (see figure 18.3). Newer Ag-specific tests have been developed, but the smears are still considered the gold standard. Other indications are knowledge of the patient's residence or travel in endemic areas and symptoms such as recurring chills, fever, and sweating.

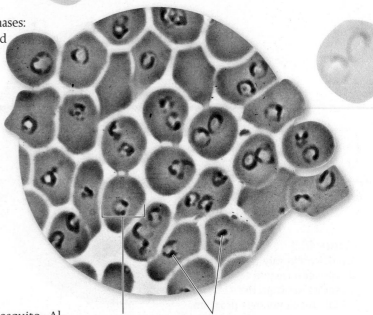

Red blood cell   Ring trophozoites

Figure 18.3 The ring trophozoite stage in a *Plasmodium falciparum* infection.   A smear of peripheral blood shows ring forms in red blood cells. Some RBCs have multiple trophozoites.
Stephen B. Aley, Ph.D., University of Texas at El Paso

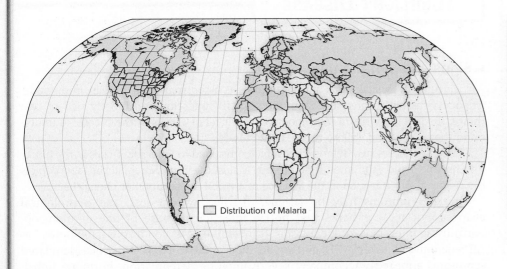

Figure 18.4 The malaria belt. Yellow zones outline the major regions that harbor malaria.   The malaria belt corresponds to a band around the equator.

□ Distribution of Malaria

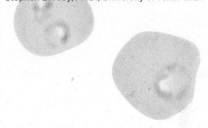

NCLEX® PREP

2. After returning from international travel, a 4-year-old female is diagnosed with falciparum malaria. All of the following complications may be associated with this disease, except
   a. *splenic rupture.*
   b. *acute kidney injury.*
   c. *altered mental status.*
   d. *peptic ulcer.*
   e. *anemia.*

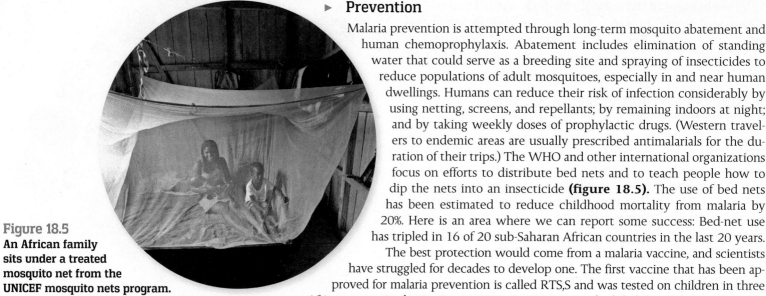

Figure 18.5
An African family sits under a treated mosquito net from the UNICEF mosquito nets program.

Anthony Asael/Art in All of Us/Corbis News/Getty Images

Disease Table 18.1	Malaria
Causative Organism(s)	*Plasmodium falciparum, P. vivax, P. ovale, P. malariae, P. knowlesi* **P**
Most Common Modes of Transmission	Biological vector (mosquito), vertical
Virulence Factors	Multiple life stages; multiple antigenic types; ability to scavenge glucose, GPI, cytoadherence
Culture/Diagnosis	Blood smear; serological methods, PCR in cases where protozoa not detectable in blood
Prevention	Mosquito control; use of bed nets; vaccine for children in endemic areas now in use; prophylactic antiprotozoal agents
Treatment	Artemisinin combination therapy, or chloroquine; consult WHO
Epidemiological Features	United States: cases are generally in travelers or immigrants; internationally, 200 million cases in "malaria belt"; 0.5 million deaths per year; more deadly in children

HIV budding out of an infected immune cell.

Source: CDC/Cynthia Goldsmith

▶ Prevention

Malaria prevention is attempted through long-term mosquito abatement and human chemoprophylaxis. Abatement includes elimination of standing water that could serve as a breeding site and spraying of insecticides to reduce populations of adult mosquitoes, especially in and near human dwellings. Humans can reduce their risk of infection considerably by using netting, screens, and repellants; by remaining indoors at night; and by taking weekly doses of prophylactic drugs. (Western travelers to endemic areas are usually prescribed antimalarials for the duration of their trips.) The WHO and other international organizations focus on efforts to distribute bed nets and to teach people how to dip the nets into an insecticide **(figure 18.5)**. The use of bed nets has been estimated to reduce childhood mortality from malaria by 20%. Here is an area where we can report some success: Bed-net use has tripled in 16 of 20 sub-Saharan African countries in the last 20 years.

The best protection would come from a malaria vaccine, and scientists have struggled for decades to develop one. The first vaccine that has been approved for malaria prevention is called RTS,S and was tested on children in three African countries beginning in 2019. In 2021 it is scheduled for widespread use on that continent. Many scientists are excited about new research in which the CRISPR gene drive technique has been used to engineer a gene into mosquitoes that makes them resistant to carrying malaria. This technique causes the inserted gene to be passed on to its offspring, until all of its progeny quickly become resistant to infection. This could eliminate malaria transmission altogether.

▶ Treatment

Quinine has long been a mainstay of malaria treatment. Chloroquine, the least toxic type, is used in nonresistant forms of the disease. The malarial protozoan has developed resistance to nearly every drug used for its treatment. A treatment called artemisinin combination therapy (ACT) is also recommended as a first-line treatment. Artemisinin is a plant-derived compound from the wormwood tree, discovered in 1972 by a Chinese scientist. She was later awarded the Nobel Prize for this life-saving find. **Disease Table 18.1.**

HIGHLIGHT DISEASE

HIV Infection and AIDS

▶ Signs and Symptoms

A spectrum of clinical signs and symptoms is associated with human immunodeficiency virus (HIV) infection. Symptoms in HIV infection are directly tied to two things: the level of virus in the blood and the level of T cells in the blood. To understand the progression, follow **table 18.2** closely.

The table shows two different lines that correspond to virus and T cells in the blood. Another line depicts the amount of antibody against the virus. Note that the table depicts the course of HIV infection in the absence of medical intervention or chemotherapy.

Initial symptoms may be fatigue, diarrhea, weight loss, and neurological changes, but most patients first notice infection because of one or more opportunistic infections or neoplasms (cancers). These conditions are known as AIDS-defining illnesses (ADIs) and are detailed in **table 18.3**. Other disease-related symptoms appear to accompany severe immune deregulation, hormone imbalances, and metabolic disturbances. Pronounced wasting of body mass is a consequence of weight loss, diarrhea, and poor nutrient absorption.

Table 18.2 The Progression of HIV in the Absence of Treatment

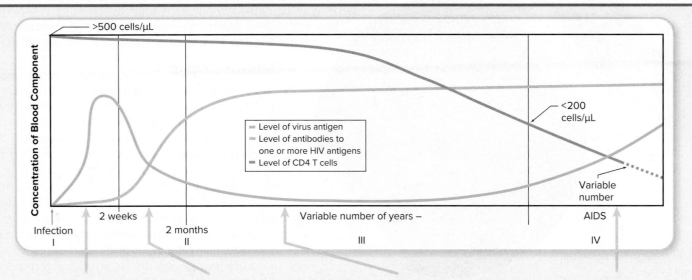

Initial infection is characterized by vague, mononucleosis-like symptoms that soon disappear. This phase corresponds to the initial high levels of virus (the green line above). Antibodies are not yet abundant.

In the second phase, virus numbers in blood drop dramatically and antibody levels become detectable. CD4 T cells begin to decrease in number.

A long period of mostly asymptomatic infection ensues. During this time, which can last from 2 to 15 years, lymphadenopathy may be the prominent symptom. During the mid- to late-asymptomatic period, the number of T cells in the blood is steadily decreasing. Once the T-cell level reaches a (low) threshold, the symptoms of AIDS ensue.

Once T cells drop below 200 cells/μL, AIDS results. Note that even though antibody levels remain high, virus levels in the blood begin to rise.

Table 18.3 AIDS-Defining Illnesses

Skin and/or Mucous Membranes (Includes Eyes)	Nervous System	Cardiovascular and Lymphatic System or Multiple Organ Systems	Respiratory Tract	Gastrointestinal Tract	Genitourinary and/or Reproductive Tract
Cytomegalovirus retinitis (with loss of vision)	Cryptococcosis, extrapulmonary	Coccidioidomycosis, disseminated or extrapulmonary	Candidiasis of trachea, bronchi, or lungs	Candidiasis of esophagus, GI tract	Invasive cervical carcinoma (HPV)
Herpes simplex chronic ulcers (>1 month duration)	HIV encephalopathy	Cytomegalovirus (other than liver, spleen, nodes)	Herpes simplex bronchitis or pneumonitis	Herpes simplex chronic ulcers (>1 month duration) or esophagitis	Herpes simplex chronic ulcers (>1 month duration)
Kaposi's sarcoma	Lymphoma, primarily in brain	Histoplasmosis	*Mycobacterium avium* complex	Isosporiasis, intestinal	
	Progressive multifocal leukoencephalopathy	Burkitt's lymphoma	Tuberculosis (*Mycobacterium tuberculosis*)	Cryptosporidiosis, chronic intestinal (>1 month duration)	
	Toxoplasmosis of the brain	Immunoblastic lymphoma	*Pneumocystis jiroveci* pneumonia		
		Mycobacterium kansasii, disseminated or extrapulmonary	Pneumonia, recurrent		
		Mycobacterium tuberculosis, disseminated or extrapulmonary	Cryptococcosis		
		Salmonella septicemia, recurrent			
		Wasting syndrome			

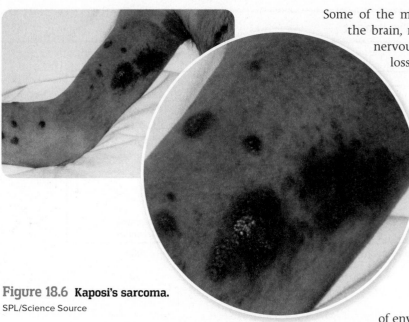

Figure 18.6 **Kaposi's sarcoma.**
SPL/Science Source

Some of the most virulent complications are neurological. Lesions occur in the brain, meninges, spinal column, and peripheral nerves. Patients with nervous system involvement show some degree of persistent memory loss, withdrawal, spasticity, sensory loss, and progressive AIDS dementia. **Figure 18.6** depicts a common ADI, Kaposi's sarcoma.

▶ Causative Agent

HIV is a retrovirus in the genus *Lentivirus*. Many retroviruses have the potential to cause cancer and produce dire, often fatal diseases and are capable of altering the host's DNA in profound ways. They are named "retroviruses" because they reverse the usual order of transcription. They contain an unusual enzyme called reverse transcriptase (RT) that catalyzes the replication of double-stranded DNA from single-stranded RNA. Not only can this retroviral DNA be incorporated into the host genome as a provirus that can be passed on to progeny cells, but some retroviruses also transform cells (make them malignant) and regulate certain host genes.

HIV and other retroviruses display structural features typical of enveloped RNA viruses **(figure 18.7a)**. The outermost component is a lipid envelope with transmembrane glycoprotein spikes that mediate viral adsorption to the host cell. HIV can only infect host cells that present the required receptors, which is a combination receptor consisting of the CD4 marker plus a coreceptor called CCR-5. The virus uses these receptors to gain entrance to several types of leukocytes and tissue cells **(figure 18.7b)**.

▶ Pathogenesis and Virulence Factors

HIV enters a mucous membrane or the skin and travels to dendritic cells beneath the epithelium. In the dendritic cell, the virus grows and is shed from the cell without killing it. The virus is amplified by macrophages in the skin, lymph organs, bone marrow, and blood. One of the great ironies of HIV is that it infects and destroys many of the very cells needed to combat it, including the helper (T4 or CD4) class of lymphocytes, monocytes, macrophages, and even B lymphocytes. The virus is adapted to docking onto its host cell's surface receptors. It then induces viral fusion with the cell membrane and creates syncytia.

Table 18.4 illustrates the life cycle of the virus.

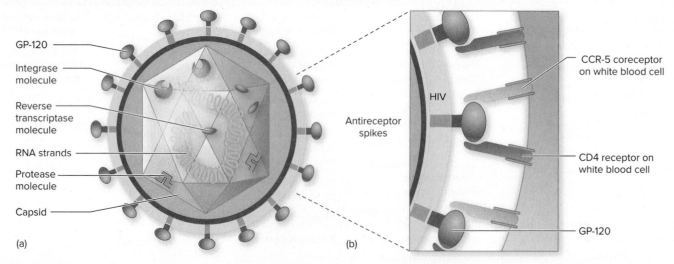

GP-120

Integrase molecule

Reverse transcriptase molecule

RNA strands

Protease molecule

Capsid

(a)

Antireceptor spikes

HIV

CCR-5 coreceptor on white blood cell

CD4 receptor on white blood cell

GP-120

(b)

Figure 18.7 **The general structure of HIV.** **(a)** The virus consists of glycoprotein (GP) spikes in the envelope, two identical RNA strands, and several molecules of reverse transcriptase, protease, and integrase encased in a protein capsid. **(b)** The snug attachment of HIV glycoprotein molecules to their specific receptors on a human cell membrane. These receptors are CD4 and a coreceptor called CCR-5 (fusin) that permit docking with the host cell and fusion with the cell membrane.

Table 18.4 The Multiplication Cycle of HIV

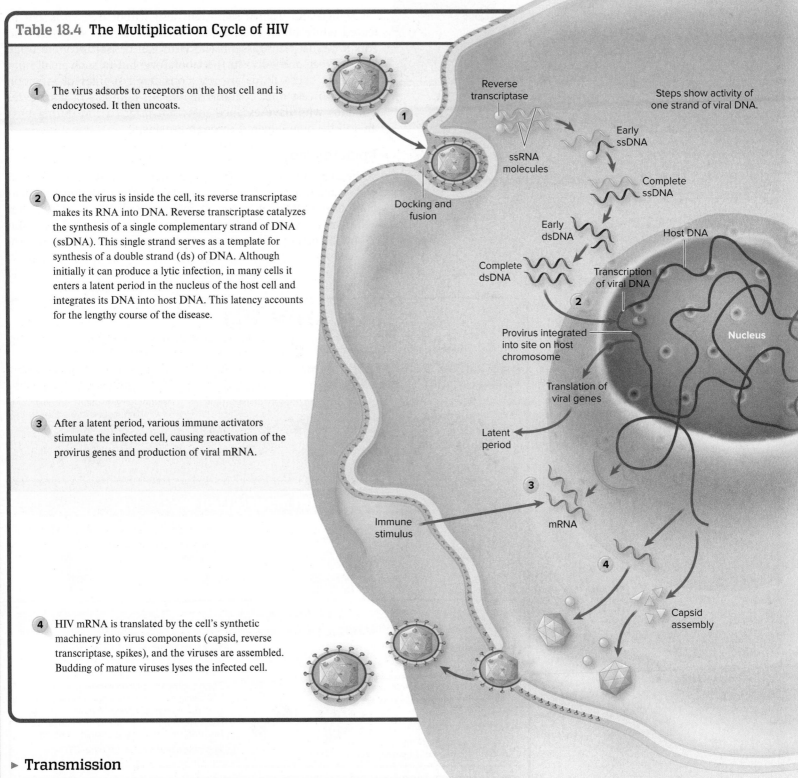

1 The virus adsorbs to receptors on the host cell and is endocytosed. It then uncoats.

2 Once the virus is inside the cell, its reverse transcriptase makes its RNA into DNA. Reverse transcriptase catalyzes the synthesis of a single complementary strand of DNA (ssDNA). This single strand serves as a template for synthesis of a double strand (ds) of DNA. Although initially it can produce a lytic infection, in many cells it enters a latent period in the nucleus of the host cell and integrates its DNA into host DNA. This latency accounts for the lengthy course of the disease.

3 After a latent period, various immune activators stimulate the infected cell, causing reactivation of the provirus genes and production of viral mRNA.

4 HIV mRNA is translated by the cell's synthetic machinery into virus components (capsid, reverse transcriptase, spikes), and the viruses are assembled. Budding of mature viruses lyses the infected cell.

Reverse transcriptase

Steps show activity of one strand of viral DNA.

ssRNA molecules

Early ssDNA

Complete ssDNA

Docking and fusion

Early dsDNA

Host DNA

Complete dsDNA

Transcription of viral DNA

Nucleus

Provirus integrated into site on host chromosome

Translation of viral genes

Latent period

Immune stimulus

mRNA

Capsid assembly

▶ Transmission

HIV transmission occurs mainly through two forms of contact: sexual intercourse and transfer of blood or blood products **(figure 18.8).** Babies can also be infected before or during birth, as well as through breast feeding. The mode of transmission is similar to that of hepatitis B virus, except that the AIDS virus does not survive for as long outside the host and it is far more sensitive to heat and disinfectants. Additionally, HIV is not transmitted through saliva, as hepatitis B can be. Health care workers should be aware that fluids they may come in contact with during childbirth or invasive procedures can also transmit the virus. These are amniotic fluid, synovial fluid, and spinal fluid.

One of your authors (Kelly Cowan) attended this rally. Remember, in these days, every HIV diagnosis was a death sentence. One of my friends was memorialized with a panel in this quilt.
- Kelly & Heidi

The AIDS quilt was launched in the late 1980s. Each panel commemorates a single victim of the disease.
©Carol M. Highsmith/Library of Congress

Semen and vaginal secretions also harbor free virus and infected white blood cells; for this reason, they are significant factors in sexual transmission. The virus can be isolated from urine, tears, sweat, and saliva in the laboratory–but in such small numbers that these fluids are not considered sources of infection. Because breast milk contains significant numbers of leukocytes, neonates who have escaped infection prior to and during birth can still become infected through nursing.

▶ **Epidemiology**

Since the beginning of the AIDS epidemic in the early 1980s, approximately 35 million people have died worldwide. The best global estimate of the number of individuals currently infected with HIV is 37 .9 million (2018), with over 1.2 million in the United States. There is good news: Between 2005 and 2016, there was a significant decrease in new infections in the

HIV 101

Without treatment, HIV (human immunodeficiency virus) can make a person very sick and even cause death. Learning the basics about HIV can keep you healthy and prevent transmission.

HIV CAN BE TRANSMITTED BY

Sexual contact	Sharing needles to inject drugs	Mother to baby during pregnancy, birth, or breastfeeding

HIV IS **NOT** TRANSMITTED BY

Air or water	Saliva, sweat, tears, or closed-mouth kissing	Insects or pets	Sharing toilets, food, or drinks

PROTECT YOURSELF FROM HIV

- Get tested at least once or more often if you are at risk.
- Use condoms the right way every time you have anal or vaginal sex.
- Choose activities with little to no risk like oral sex.
- Limit your number of sex partners.
- Don't inject drugs, or if you do, don't share needles or works.

- If you are at very high risk for HIV, ask your health care provider if pre-exposure prophylaxis (PrEP) is right for you.
- If you think you've been exposed to HIV within the last 3 days, ask a health care provider about postexposure prophylaxis (PEP) right away. PEP can prevent HIV, but it must be started within 72 hours.
- Get tested and treated for other STDs.

KEEP YOURSELF HEALTHY AND PROTECT OTHERS IF YOU ARE LIVING WITH HIV

- Find HIV care. It can keep you healthy and greatly reduce your chance of transmitting HIV.
- Take your medicines the right way every day.
- Stay in HIV care.

- Tell your sex or drug-using partners that you are living with HIV. Use condoms the right way every time you have sex, and talk to your partners about PrEP.
- Get tested and treated for other STDs.

Figure 18.8 HIV transmission. Adapted from the Centers for Disease Control and Prevention.
Source: CDC

United States. However, between 2000 and 2016 the number of annual new infections increased in 25-35 year olds. Since then, no more decreases have been seen, probably due to the difficulty of reaching those most likely to be infected in the South, in rural areas, and among African Americans and Latinos.

Treatment of HIV-infected mothers with an anti-HIV drug has dramatically decreased the rate of maternal-to-infant transmission of HIV during pregnancy. Current treatment regimens result in a transmission rate of approximately 11%, with some studies of multidrug regimens claiming rates as low as 5%. Evidence suggests that giving mothers protease inhibitors can reduce the transmission rate to around 1%. (Untreated mothers pass the virus to their babies at the rate of 33%.) In 2015, Cuba became the first country to completely eliminate mother-to-child transmission of the virus, through comprehensive testing and treatment of mothers. Since then, a dozen other countries (but not the United States) have reached this milestone.

▶ Culture and Diagnosis

A person is diagnosed as having HIV infection, as opposed to having AIDS, if he or she has tested positive for the human immunodeficiency virus. The U.S. Preventive Services Task Force recommends that all people between the ages of 13 and 64 be tested for HIV. People outside of that age group who are at high risk, as well as pregnant women, should also be tested. It is estimated that more than 154,000 people in the United States are currently infected but are unaware of it. This population is responsible for about 30% of the transmission of the virus.

Most initial testing is based on detection of antibodies specific to the virus in serum or other fluids **(Figure 18.9).** Newer tests detect antibodies to the virus and viral antigens simultaneously, and are also widely used.

False-negative results can occur when testing is performed too early in the infection. It takes a few weeks for antibodies to be detectable in body fluids, and the presence of viral antigen is also unpredictable in the early weeks (see table 18.2). Persons who test negative but feel they may have been exposed should be tested a second time 3 to 6 months later.

Positive tests always require follow-up with a more specific test. An antibody differentiation immunoassay determines whether the infection is with HIV-1 or HIV-2. Uncertain results are followed by an HIV nucleic acid amplification test (NAAT).

In the United States, people are diagnosed with Stage 3 HIV infection (AIDS), as opposed to HIV, if they meet the following criteria: (1) They are positive for the virus *and* (2) they fulfill one of these additional criteria:

- They have a CD4 (helper T cell) count of fewer than 200 cells per microliter of blood.
- Their CD4 cells account for fewer than 14% of all lymphocytes.
- They experience one or more of a CDC-provided list of AIDS-defining illnesses (ADIs).

▶ Prevention

Avoidance of sexual contact with infected persons is a cornerstone of HIV prevention. A sexually active person should consider every partner to be infected unless proven otherwise. Barrier protection (condoms) should be used when having sex with anyone whose HIV status is not known with certainty to be negative. Although avoiding intravenous drugs is an important preventive measure, many drug addicts do not, or cannot, choose this option. In such cases, risk can be decreased by not sharing syringes or needles. Preexposure prophylaxis, called PrEP, is currently recommended for people who are at risk for becoming infected; for example, if their partner is positive and they are negative. It consists of a two-drug combination marketed as Truvada. It is designed to be used in combination with condoms, so that each is the fallback for the possible failure of the other. The risk of being infected through sex, if you are taking PrEP every day, is decreased by 99%. It is around 74% effective in reducing risk due to injecting drug use. A second drug called Descovy has also been approved for PrEP but is not approved for receptive vaginal sex.

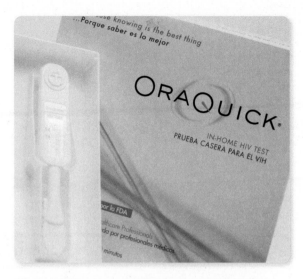

Figure 18.9 At-home HIV tests help increase awareness and promote early detection.
Kristoffer Tripplaar/Alamy Stock Photo

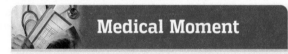

Medical Moment

Postexposure Prophylaxis

Health care workers are at risk of acquiring infectious disease from exposure to body fluids. The first line of defense is proper use of gloves and other personal protective equipment. Needles are designed with safety mechanisms to protect the user from inadvertent sticks. Despite these safety measures, exposure to patients' blood still occurs.

When a health care worker is exposed to body fluids, such as by puncture from a contaminated needle, they should seek medical attention as soon as possible. It is recommended that testing be run on the exposure source's blood to determine HIV status. Though the rate of transmission from occupational exposure is low, health care workers should be offered treatment with postexposure prophylaxis (PEP).

PEP is a 4-week antiretroviral regimen. It is effective in preventing HIV infection, though it does not offer complete protection. PEP must be started within 72 hours of exposure, and the sooner treatment begins, the greater the likelihood that it will work. These medications also have some unpleasant side effects so it is important to discuss compliance with the drug regimen. Following PEP treatment, the health care worker should undergo HIV testing for 4 to 6 months after the exposure to monitor for infection.

Q. Health care workers are also vaccinated against what blood-borne virus that can be transmitted by needle sticks?

Answer in Appendix B.

Source: cdc.gov

▶ Treatment

There is no cure for HIV. However, new drugs and new insights into their use have greatly improved the quality of life for some people living with HIV and AIDS. Unfortunately, much of the world is too poor to afford or properly use the latest medications. Two great challenges are (1) to make more widely available the drugs we do have and (2) to develop drugs and delivery mechanisms that are more affordable and practical.

Clear-cut guidelines exist for treating people who test HIV-positive. These guidelines are updated regularly. Treatment is referred to as antiretroviral therapy, or ART. Initially it is generally a combination of three drugs from two different classes **(table 18.5).** It

Table 18.5 Mechanisms of Action of Anti-HIV Drugs

The first effective drugs developed were the synthetic nucleoside analogs (reverse transcriptase inhibitors) such as azidothymidine (AZT). They interrupt the HIV multiplication cycle by mimicking the structure of actual nucleosides and being added to viral DNA by reverse transcriptase. Because these drugs lack all of the correct binding sites for further DNA synthesis, viral replication is terminated.

Also seen here is one of the latest additions to the arsenal, enfuvirtide (Fuzeon), a drug classified as a fusion inhibitor. It prevents the virus from fusing with the membrane of target cells, thereby stopping infection altogether.

Another important class of drugs is the protease inhibitors, which block the action of the HIV enzyme (protease) involved in the final assembly and maturation of the virus. Examples of these drugs include indinavir (Crixivan), ritonavir (Norvir), and amprenavir (Agenerase).

A class of drugs called integrase inhibitors prevent integration of viral sequences into host DNA. This stops virus multiplication.

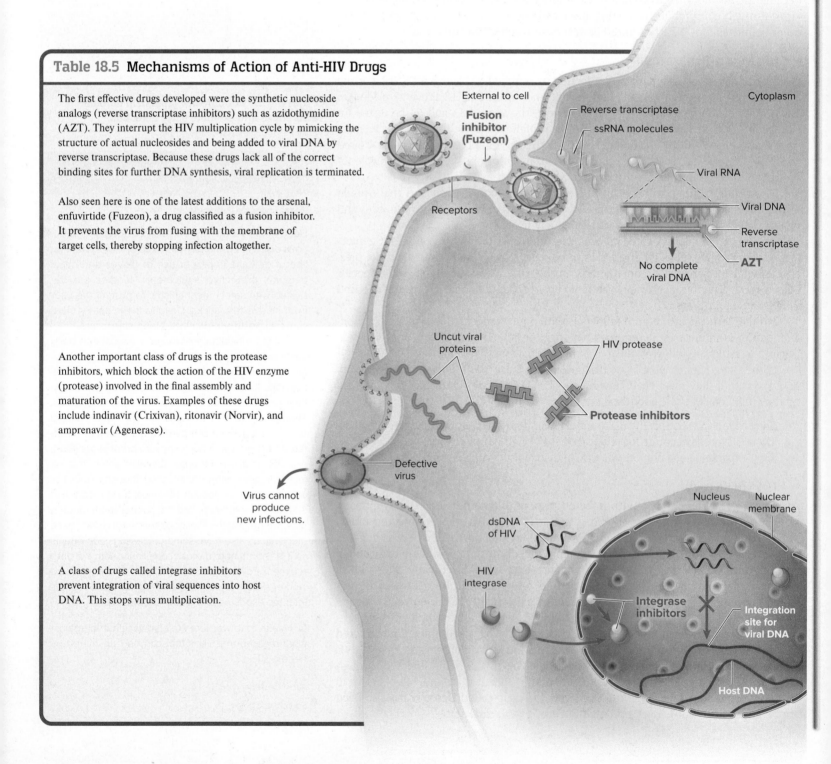

is adjusted when needed as a patient's response is monitored. The newer recommendations call for treatment to begin immediately after HIV diagnosis. In addition to antiviral chemotherapy, AIDS patients should receive a wide array of drugs to prevent or treat a variety of opportunistic infections and other ADIs such as wasting disease. **Disease Table 18.2.**

Endocarditis

Endocarditis is an inflammation of the endocardium, or inner lining of the heart. Most of the time, endocarditis refers to an infection of the valves of the heart, often the mitral or aortic valve **(figure 18.10).** Two variations of infectious endocarditis have been described: acute and subacute. Each has distinct groups of possible causative agents, both primarily bacterial. On rare occasions, infections may be caused by fungi and perhaps viruses. Physical trauma can also cause such inflammation.

Patients with prosthetic valves can acquire acute endocarditis if bacteria are introduced during the surgical procedure, typically with high rates of morbidity and mortality. Alternatively, the prosthetic valves can serve as infection sites for the subacute form of endocarditis long after the surgical procedure.

Because the symptoms and the diagnostic procedures are similar for both forms of endocarditis, they are discussed first; then the specific aspects of acute and subacute endocarditis are addressed.

▸ Signs and Symptoms

The signs and symptoms are similar for both types of endocarditis, except that in the subacute condition they develop more slowly and are less pronounced than with the acute disease. Symptoms include fever, fatigue, joint pain, edema (swelling of feet, legs, and abdomen), weakness, anemia, abnormal heartbeat, and sometimes symptoms similar to myocardial infarction (heart attack), including shortness of breath or chills. Abdominal or side pain is sometimes reported. The patient may look very ill and may have petechiae (small red-to-purple discolorations) over the upper half of the body and under the fingernails (splinter hemorrhages). Red, painless skin spots on the palms and soles (Janeway lesions) and small painful nodes on the pads of fingers and toes (Osler's nodes) may also be apparent on examination. In subacute cases, an enlarged spleen may have developed over time. Cases of extremely long duration (years) can lead to clubbed fingers and toes (a condition in which the ends of the fingers and toes are enlarged) due to lack of oxygen in the blood.

Acute Endocarditis

Acute endocarditis is most often the result of an overwhelming bloodstream challenge with bacteria. Some of these bacteria seem to have the ability to colonize normal heart valves. Accumulations of bacteria on the valves (vegetations) hamper their function and can lead directly to cardiac malfunction and death. Alternatively, pieces of the bacterial vegetation can break off and create emboli (blockages) in vital organs. The bacterial biofilms can also provide a constant source of blood-borne bacteria, with the accompanying systemic inflammatory response and shock.

▸ Causative Agents

Forty percent of acute cases of endocarditis are caused by *Staphylococcus aureus.* Other agents that cause it are *Streptococcus pyogenes, Streptococcus pneumoniae, Enterococcus,* and *Pseudomonas aeruginosa,* as well as a host of other bacteria. Physicians report that *Staphylococcus aureus* is the most dangerous cause of endocarditis.

Disease Table 18.2	HIV Infection and AIDS
Causative Organism(s)	Human immunodeficiency virus 1 or 2 **V**
Most Common Modes of Transmission	Direct contact (sexual), parenteral (blood-borne), vertical (perinatal and via breast milk)
Virulence Factors	Attachment, syncytia formation, reverse transcriptase, high mutation rate
Culture/ Diagnosis	Immunoassay to detect antibodies as well as HIV antigen
Prevention	Avoidance of contact with infected sex partner, contaminated blood, breast milk; pre-exposure prophylaxis (PrEP) for high-risk individuals
Treatment	Multiple simultaneous antiretroviral drugs
Epidemiological Features	Global infections: 37.9 million, 1.7 million new infections in 2018. U.S. infections: approximately 1.2 million. Global deaths to date: 35 million.

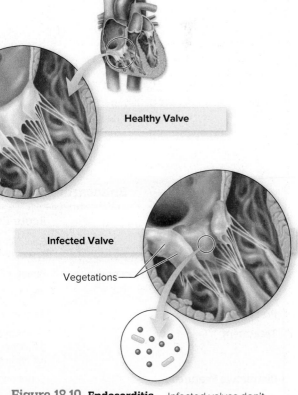

Figure 18.10 **Endocarditis.** Infected valves don't work properly.

Healthy Valve

Infected Valve

Vegetations

▶ Transmission and Epidemiology

The most common route of transmission for acute endocarditis is parenteral—that is, via direct entry into the body. Intravenous or subcutaneous drug users have been a growing risk group for the condition. Traumatic injuries and surgical procedures can also introduce the large number of bacteria required for the acute form of endocarditis. The heroin epidemic currently sweeping the United States has led to a large increase in the incidence of acute endocarditis, usually caused by *Staphylococcus aureus*. In many hospitals, on any given day, the majority of the infectious disease physician's list of consults is endocarditis secondary to injecting drug use. These patients often also suffer from infections of the vertebrae and epidural abscesses, caused by bacteria from the bloodstream forming abscesses along the spine.

Subacute Endocarditis

Subacute forms of this condition are almost always preceded by some form of damage to the heart valves or by congenital malformation. Irregularities in the valves encourage the attachment of bacteria, which then form biofilms and impede normal function, as well as provide an ongoing source of bacteria to the bloodstream. People who have suffered rheumatic fever and the accompanying damage to heart valves are particularly susceptible to this condition (see chapter 19 for a complete discussion of rheumatic fever).

▶ Causative Agents

Most commonly, subacute endocarditis is caused by bacteria of low pathogenicity, often originating in the oral cavity. Alpha-hemolytic streptococci, such as *Streptococcus sanguis*, *S. oralis*, and *S. mutans*, are most often responsible, although normal biota from the skin and other bacteria can also colonize abnormal valves and lead to this condition.

▶ Transmission and Epidemiology

Minor disruptions in the skin or mucous membranes, such as those induced by overly vigorous toothbrushing, dental procedures, or relatively minor cuts and lacerations, can introduce bacteria into the bloodstream and lead to valve colonization. The bacteria are not, therefore, transmitted from other people or from the environment.

▶ Prevention

Prophylactic antibiotics are prescribed for patients before surgery in order to prevent this condition. The same practice was used for decades in advance of routine dental procedures but is no longer used unless there is a high risk for endocarditis. **Disease Table 18.3.**

Septicemia (or Sepsis)

Septicemia occurs when organisms are actively multiplying in the blood. Many different bacteria (and a few fungi) can cause this condition. Patients suffering from these infections are sometimes described as "septic."

Disease Table 18.3	Endocarditis	
Disease	**Acute Endocarditis** Ⓑ	**Subacute Endocarditis** Ⓑ
Causative Organism(s)	*Staphylococcus aureus, Streptococcus pyogenes, S. pneumoniae, Enterococcus, Pseudomonas,* others	Alpha-hemolytic streptococci, others
Most Common Modes of Transmission	Parenteral	Endogenous transfer of normal biota to bloodstream
Culture/Diagnosis	Blood culture	Blood culture
Prevention	Aseptic surgery, injections	Prophylactic antibiotics before invasive procedures
Treatment	Vancomycin surgery may be necessary. Be aware that several of these bacteria are on CDC "Threat" list for antibiotic resistance.	Surgery may be necessary.
Distinctive Features	Acute onset, high fatality rate	Slower onset
Epidemiological Features	Greatly increased incidence due to opioid epidemic	—

▶ Signs and Symptoms

Fever is a prominent feature of septicemia. The patient appears very ill and may have an altered mental state, shaking chills, and gastrointestinal symptoms. Often an increased breathing rate is exhibited, accompanied by respiratory alkalosis (increased tissue pH due to breathing disorder). Low blood pressure is a hallmark of this condition and is caused by the inflammatory response to infectious agents in the bloodstream, which leads to a loss of fluid from the vasculature. This condition is the most dangerous feature of the disease, often culminating in death.

▶ Causative Agents

Most are caused by bacteria, and they are approximately evenly divided between gram-positives and gram-negatives. MRSA is a very common cause. Perhaps 10% are caused by fungal infections. An important emerging cause of sepsis is the yeast *Candida auris*. This microbe was just discovered in 2009 but has quickly become a great concern to public health specialists. It is resistant to several classes of antimicrobials. Polymicrobial bloodstream infections increasingly are being identified in which more than one microorganism is causing the infection.

▶ Pathogenesis and Virulence Factors

Gram-negative bacteria multiplying in the blood release large amounts of endotoxin into the bloodstream, stimulating a massive inflammatory response mediated by a host of cytokines. This response invariably leads to a drastic drop in blood pressure, a condition called **endotoxic shock.** Gram-positive bacteria can instigate a similar cascade of events when fragments of their cell walls are released into the blood.

▶ Transmission and Epidemiology

In many cases, septicemias can be traced to parenteral introduction of the microorganisms via intravenous lines or surgical procedures. Other infections may arise from serious urinary tract infections or from renal, prostatic, pancreatic, or gallbladder abscesses. Patients with underlying spleen malfunction may be predisposed to multiplication of microbes in the bloodstream. Meningitis, osteomyelitis (bone infections), and pneumonia can all lead to sepsis. Hospitalization for sepsis has more than doubled in recent years. More alarming is the fact there is a 20% to 50% mortality rate, reflecting the high risks of this disease even when treatment is available. One in three patients who die in a hospital die from sepsis.

▶ Culture and Diagnosis

Because the infection is in the bloodstream, a blood culture is the obvious route to diagnosis. Increasingly, deep sequencing techniques are being used for faster and more appropriate treatment. A blood marker called procalcitonin can be found in a blood test and be an early indicator of sepsis. Empiric therapy should be started immediately before culture and susceptibility results are available. Also, different agents are more likely to be the causative agents in specific patient populations, such as neonates, injecting drug users, splenectomized patients, and recent international travelers.

▶ Prevention and Treatment

Empiric therapy, which is begun immediately after blood cultures are taken, often begins with a broad-spectrum antibiotic. Once the organism is identified and its antibiotic susceptibility is known, treatment can be adjusted accordingly, as data show that rapid diagnosis and treatment are of paramount importance. Increasingly, deep sequencing techniques are being used for faster and more appropriate treatment. **Disease Table 18.4.**

Plague

Although pandemics of plague have occurred since antiquity, the first one that was reliably chronicled killed an estimated 100 million people in the sixth century AD.

Disease Table 18.4	Septicemia
Causative Organism(s)	Bacteria or fungi, often MRSA, increasingly *Candida auris*
Most Common Modes of Transmission	Parenteral, endogenous transfer
Virulence Factors	Cell wall or membrane components
Culture/Diagnosis	Blood culture, deep sequencing
Prevention	–
Treatment	Broad-spectrum antibiotic until identification and susceptibilities tested
Epidemiological Features	United States: 1.7 million cases/year; 270,000 deaths per year

Prairie dogs in the southwest United States sometimes harbor *Yersinia pestis*.
©Chase Swift/Getty Images

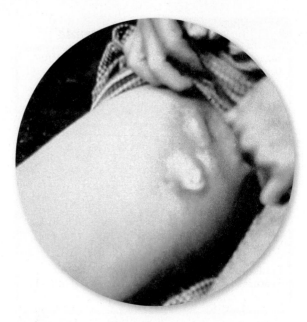

Figure 18.11 A classic inguinal bubo of bubonic plague. This hard nodule is very painful and can rupture onto the surface.
Source: CDC

Figure 18.12 *Yersinia pestis.* Note the more darkly stained poles of the bacterium (purple), lending it a "safety pin" appearance.
Science Source

▶ Signs and Symptoms

Three possible manifestations of infection occur with the bacterium causing plague. **Pneumonic plague** is a respiratory disease, described in chapter 19. In **bubonic plague,** the bacterium, which is injected by the bite of a flea, enters the lymph and is filtered by a local lymph node. Infection causes inflammation and necrosis of the node, resulting in a swollen lesion called a **bubo,** usually in the groin or axilla **(figure 18.11).** The incubation period lasts 2 to 8 days, ending abruptly with the onset of fever, chills, headache, nausea, weakness, and tenderness of the bubo. Mortality rates, even with treatment, can reach up to 15%.

These cases often progress to massive bacterial growth in the blood termed **septicemic plague.** The presence of the bacteria in the blood results in disseminated intravascular coagulation, subcutaneous hemorrhage, and **purpura** that may degenerate into necrosis and gangrene. Mortality rates, once the disease has progressed to this point, are 30% to 50% with treatment and 100% without treatment. Because of the visible darkening of the skin, the plague has often been called the "Black Death."

▶ Causative Agent

The cause of this dreadful disease is a tiny gram-negative rod, *Yersinia pestis*, a member of the family Enterobacteriaceae. *Y. pestis* displays unusual bipolar staining that makes it look like a safety pin **(figure 18.12).**

▶ Pathogenesis and Virulence Factors

The number of bacteria required to initiate a plague infection is small—perhaps only 3 to 50 cells.

▶ Transmission and Epidemiology

The plague bacterium resides in over 200 species of mammalian hosts. Some of these, such as mice and voles, serve as long-term endemic reservoirs, which are not affected by the disease. Other species, including rats and rabbits, are amplifying reservoirs, which get sick and tend to be closely connected to outbreaks of plague in humans.

The principal agents in the transmission of the plague bacterium are fleas. After a flea ingests a blood meal from an infected animal, the bacteria multiply in its gut. The bacterium promotes its spread by causing coagulation and blockage of the flea's esophagus. Being unable to feed properly, the ravenous flea jumps from animal to animal in a futile attempt to get nourishment. Regurgitated infectious material then is inoculated into the bite wound.

The distribution of plague is extensive. Although the incidence of disease has been reduced in the developed world, it has actually been increasing in Africa and other parts of the world. There was a rather large outbreak of plague in 2017 in Madagascar,

Disease Table 18.5	Plague
Causative Organism(s)	*Yersinia pestis* **B** G⁻
Most Common Modes of Transmission	Vector, biological; also droplet contact (pneumonic) and direct contact with body fluids
Virulence Factors	Capsule, plasminogen activator
Culture/Diagnosis	Rapid genomic methods
Prevention	Flea and/or animal control; vaccine available for high-risk individuals
Treatment	Gentamicin or doxycycline
Epidemiological Features	United States: endemic in all western and southwestern states; internationally, 95% of human cases occur in Africa, including Madagascar; **Category A Bioterrorism Agent**

an island nation off the eastern coast of Africa. In the United States, sporadic cases (usually 10 to 20 per year) occur as a result of contact with wild and domestic animals. This disease is considered endemic in U.S. western and southwestern states. Persons most at risk for developing plague are veterinarians and people living and working near woodlands and forests. Dogs and cats can be infected with the plague, often from contact with infected wild animals such as prairie dogs. **Disease Table 18.5.**

Tularemia

▶ Signs and Symptoms

After an incubation period ranging from a few days to 3 weeks, acute symptoms of headache, backache, fever, chills, coughing, and weakness appear. Further clinical manifestations are tied to the portal of entry. They include ulcerative skin lesions, swollen lymph glands, conjunctival inflammation, sore throat, intestinal disruption, and pulmonary involvement. Unfortunately, clinicians can misinterpret these signs and symptoms, delaying effective treatment. The death rate in the most serious forms of disease is 30%, but proper treatment with gentamicin or streptomycin reduces mortality to almost zero.

▶ Causative Agent

The causative agent of tularemia is a facultative intracellular gram-negative bacterium called *Francisella tularensis*. It is a zoonotic disease of assorted mammals and is endemic to the Northern Hemisphere. Because it has been associated with outbreaks of disease in wild rabbits, it is sometimes called rabbit fever. It is currently listed as a Category A bioterrorism agent, along with anthrax, plague, and others.

▶ Transmission and Epidemiology

Although rabbits and rodents (muskrats and ground squirrels) are the chief reservoirs, other wild animals (skunks, beavers, foxes, opossums) and some domestic animals are implicated as well. The chief route of transmission in the past had been through the activity of skinning rabbits, but with the decline of rabbit hunting, transmission via tick bites is more common. Ticks are the most frequent arthropod vector, followed by biting flies, mites, and mosquitoes. It can also be transmitted from aerosols.

With an estimated infective dose of between 10 and 50 organisms, *F. tularensis* is often considered one of the most infectious of all bacteria. Cases of tularemia have appeared in people who have accidentally run over small dead animals while lawn mowing, presumably from inhaling aerosolized bacteria. In 2009, two different people in Alaska acquired tularemia after wresting infected rabbits from their dogs' mouths.

▶ Prevention

Because the intracellular persistence of *F. tularensis* can lead to relapses, antimicrobial therapy must not be discontinued prematurely. Postexposure prophylaxis with doxycycline or ciprofloxacin can prevent the disease in lab workers or others who may have been exposed. Laboratory workers and other occupationally exposed personnel must wear gloves, masks, and eyewear. **Disease Table 18.6.**

Lyme Disease

▶ Signs and Symptoms

Lyme disease is slow-acting, and it often evolves into a slowly progressive syndrome that mimics neuromuscular and rheumatoid conditions. An early symptom in some cases is a rash at the site of a tick bite. The lesion, called *erythema migrans,* can look something like a bull's eye, with a raised erythematous (reddish) ring that gradually spreads outward and a pale central region **(figure 18.13).** Until recently, this lesion was thought to be the most common presentation of erythema migrans. But it only has this appearance in about 10% of cases. It can also be flat and scaly with no clear areas, or it can be pustular. It can mimic the appearance of ringworm. Other early symptoms

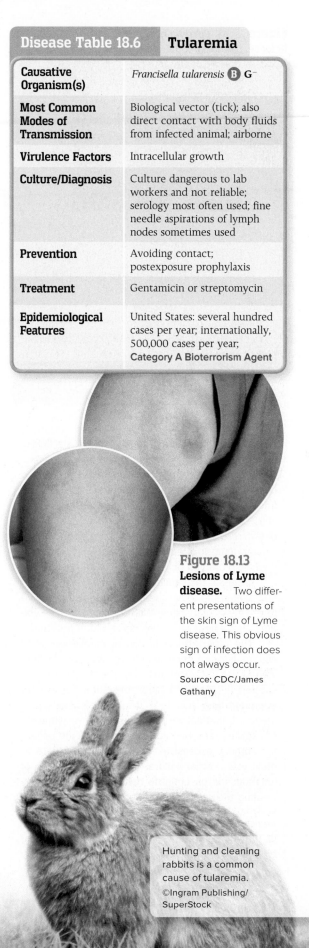

Disease Table 18.6	Tularemia
Causative Organism(s)	*Francisella tularensis* **B** G⁻
Most Common Modes of Transmission	Biological vector (tick); also direct contact with body fluids from infected animal; airborne
Virulence Factors	Intracellular growth
Culture/Diagnosis	Culture dangerous to lab workers and not reliable; serology most often used; fine needle aspirations of lymph nodes sometimes used
Prevention	Avoiding contact; postexposure prophylaxis
Treatment	Gentamicin or streptomycin
Epidemiological Features	United States: several hundred cases per year; internationally, 500,000 cases per year; **Category A Bioterrorism Agent**

Figure 18.13
Lesions of Lyme disease. Two different presentations of the skin sign of Lyme disease. This obvious sign of infection does not always occur. Source: CDC/James Gathany

Hunting and cleaning rabbits is a common cause of tularemia. ©Ingram Publishing/SuperStock

Figure 18.14 *Borrelia* **has 3 to 10 loose, irregular coils.**   Each individual Borrelia cell measures from 10–30 microns, on average. Source: CDC/Janice Haney Carr

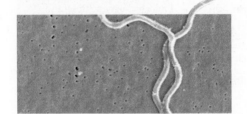

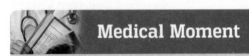
Medical Moment

Where Does the Fluid Go?

A characteristic feature of sepsis is the loss of fluid in the intravascular space, also known as "third spacing." Where is this third space? In homeostasis, fluid constantly moves between the intracellular and extracellular spaces. Intracellular fluid is found within cells. Extracellular fluid makes up approximately one-third of the body's fluid volume and includes plasma (fluid in the blood vessels) and interstitial fluid (fluid bathing body tissues and organs).

In sepsis, the capillaries become leaky, causing fluid volume to leave the blood vessels and move to the interstitial space. With depleted plasma volume, the patient experiences low blood pressure.

The rapid administration of intravenous fluid volume is often necessary to resuscitate a patient with sepsis. Medications may be started to help support the patient's blood pressure, but without adequate fluid volume in the intravascular space, these are ineffective. While treating the symptoms of capillary leak with fluid may stabilize the patient temporarily, excessive fluid loss may progress to multisystem organ failure. Timely administration of antibiotics is essential to halt this progression by treating the organism that is causing the capillary leak.

Q. What is the term for low blood pressure caused by loss of fluids in the vessels, when it is caused by a microorganism?

Answer in Appendix B.

---

are fever, headache, stiff neck, and dizziness. If not treated or if treated too late, the disease can advance to the second stage, during which cardiac and neurological symptoms, such as facial palsy, can develop. After several weeks or months, the third stage may occur. This involves a crippling arthritis; some people acquire chronic neurological complications that are severely disabling.

▶ Causative Agent

Borrelia burgdorferi is the cause of Lyme disease. These bacteria are unusual spirochetes in that they are large, ranging from 0.2 to 0.5 μm in width and from 10 to 20 μm in length, and contain 3 to 10 irregularly spaced and loose coils **(figure 18.14).**

▶ Pathogenesis and Virulence Factors

The bacterium is very good at evading the immune system. It changes its surface antigens while it is in the tick and again after it has been transmitted to a mammalian host. It provokes a strong humoral and cellular immune response, but this response is largely ineffective, perhaps because of the bacterium's ability to switch its antigens. Indeed, it is possible that the immune response contributes to the pathology of the infection. Closely related *Borrelia burgdorferi*–like strains can also cause the disease.

▶ Transmission and Epidemiology

B. burgdorferi is transmitted primarily by hard ticks of the genus *Ixodes*. In the northeastern part of the United States, *Ixodes scapularis* (the black-legged deer tick, **figure 18.15**) passes through a complex 2-year cycle that involves two principal hosts. In California, the transmission cycle involves *Ixodes pacificus*, another black-legged tick, and the dusky-footed woodrat as reservoir.

The greatest concentrations of Lyme disease are found in areas having large populations of both the intermediate and definitive hosts **(figure 18.16).**

▶ Culture and Diagnosis

Culture of the organism is not useful. Diagnosis in the early stages, if the rash is present, is usually accomplished based on symptoms and history of possible exposure to ticks because the organism is not easily detectable at this stage. Acute and convalescent sera may be helpful. In late Lyme disease, ELISAs and/or Western blots can be used to detect antibodies to the organism in blood.

It is important to consider coinfection with *Anaplasma* or *Babesia* (see Nonhemorrhagic Fever Diseases), because these organisms are transmitted by the same kind of tick that transmits Lyme disease, and the tick may be infected with two or more different microbes. Lingering cases of Lyme disease may occasionally be due to failure to consider these microbes.

▶ Prevention and Treatment

Anyone involved in outdoor activities should wear protective clothing, boots, leggings, and insect repellent containing DEET. Individuals exposed to the outdoors should routinely inspect their bodies for ticks and remove ticks gently without crushing, preferably with forceps or fingers protected with gloves, because it is possible to become infected by tick feces or fluids.

Early, prolonged (2 weeks) treatment with doxycycline or amoxicillin is effective. Unfortunately, 10% to 20% of treated patients develop a syndrome called posttreatment Lyme disease syndrome (PTLDS). This is a condition lasting weeks or months characterized by fatigue and muscle and joint pain. Previously PTLDS was often treated with prolonged antibiotic therapy; the National Institute of Allergy and Infectious Disease, after rigorous study, no longer supports that treatment. Evidence suggests that some form of the spirochete–even if just its peptidoglycan wall–persists and that these forms elicit antibodies. Either the spirochete or the antibodies may be responsible for the lingering symptoms. **Disease Table 18.7.**

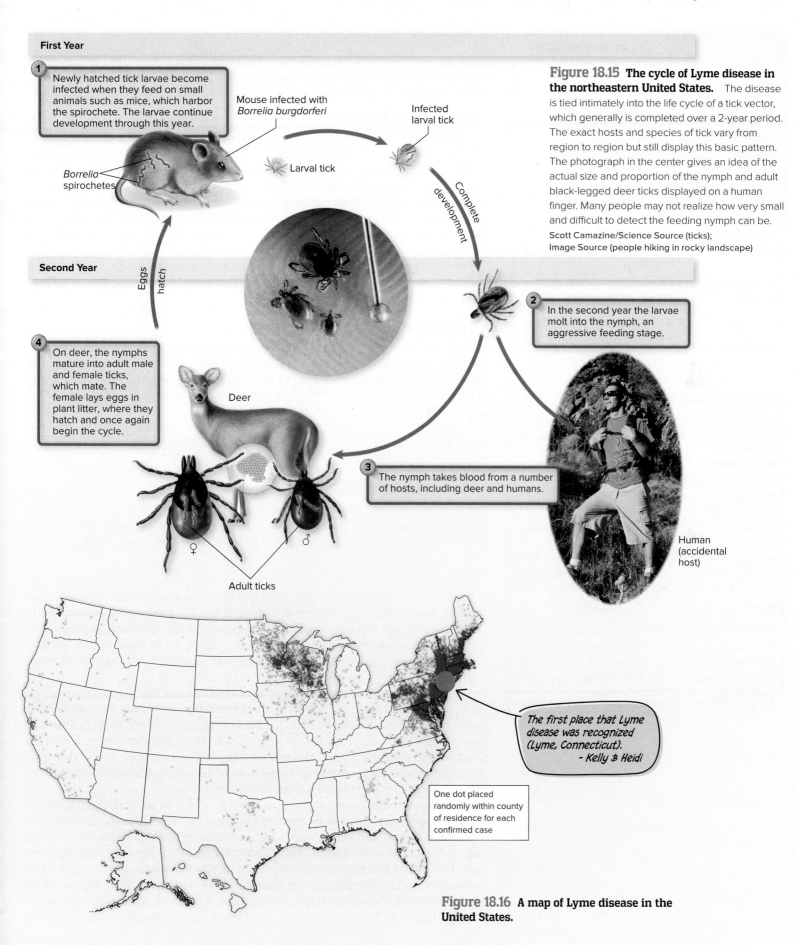

First Year

1 Newly hatched tick larvae become infected when they feed on small animals such as mice, which harbor the spirochete. The larvae continue development through this year.

Borrelia spirochetes

Mouse infected with *Borrelia burgdorferi*

Larval tick

Infected larval tick

Complete development

Second Year

Eggs hatch

4 On deer, the nymphs mature into adult male and female ticks, which mate. The female lays eggs in plant litter, where they hatch and once again begin the cycle.

Deer

Adult ticks

♀   ♂

2 In the second year the larvae molt into the nymph, an aggressive feeding stage.

3 The nymph takes blood from a number of hosts, including deer and humans.

Human (accidental host)

Figure 18.15 The cycle of Lyme disease in the northeastern United States. The disease is tied intimately into the life cycle of a tick vector, which generally is completed over a 2-year period. The exact hosts and species of tick vary from region to region but still display this basic pattern. The photograph in the center gives an idea of the actual size and proportion of the nymph and adult black-legged deer ticks displayed on a human finger. Many people may not realize how very small and difficult to detect the feeding nymph can be.
Scott Camazine/Science Source (ticks);
Image Source (people hiking in rocky landscape)

The first place that Lyme disease was recognized (Lyme, Connecticut).
- Kelly & Heidi

One dot placed randomly within county of residence for each confirmed case

Figure 18.16 A map of Lyme disease in the United States.

Disease Table 18.7	Lyme Disease
Causative Organism(s)	*Borrelia burgdorferi* and closely related species **B** G⁻
Most Common Modes of Transmission	Biological vector (tick)
Virulence Factors	Antigenic shifting, adhesins
Culture/ Diagnosis	ELISA for Ab, Western blot; acute and convalescent sera testing
Prevention	Tick avoidance
Treatment	Doxycycline and/or amoxicillin (2 weeks)
Epidemiological Features	Endemic in North America, Europe, and Asia; approximately 300,000 new cases/year in the United States; only 10% diagnosed/reported

Disease Table 18.8	Infectious Mononucleosis
Causative Organism(s)	Epstein-Barr virus (EBV) **V**
Most Common Modes of Transmission	Direct, indirect contact; parenteral
Virulence Factors	Latency, ability to incorporate into host DNA
Culture/ Diagnosis	Differential blood count, monospot test for heterophile antibody, specific ELISA
Prevention	–
Treatment	Supportive
Distinctive Features	Lifelong persistence
Epidemiological Features	United States: 500 cases per 100,000 per year

Infectious Mononucleosis

This lymphatic system disease, which is often simply called "mono" or the "kissing disease," can be caused by a number of bacteria or viruses, but the vast majority of cases are caused by the **Epstein-Barr virus (EBV),** a member of the herpesvirus family.

▶ Signs and Symptoms

The symptoms of mononucleosis are sore throat, high fever, and cervical lymphadenopathy, which develop after a long incubation period (30 to 50 days). Many patients also have a gray-white exudate in the throat, a skin rash, and enlarged spleen and liver. A notable sign of mononucleosis is sudden leukocytosis (an increase in number of white blood cells), consisting initially of infected B cells and later T cells. Fatigue is a hallmark of the disease. Patients remain fatigued for a period of weeks. During that time, they are advised not to engage in strenuous activity due to the possibility of injuring their enlarged spleen (or liver).

Eventually, the strong, cell-mediated immune response is decisive in controlling the infection and preventing complications. But after recovery, people usually remain chronically infected with EBV.

▶ Transmission and Epidemiology

More than 90% of the world's population is infected with EBV. In general, the virus causes no noticeable symptoms, but the time of life when the virus is first encountered seems to matter. In the case of EBV, infection during the teen years results in disease between 30 and 77% of the time, whereas infection before or after this period may be entirely asymptomatic.

Direct oral contact and contamination with saliva are the principal modes of transmission, although transfer through blood transfusions, sexual contact, and organ transplants is possible.

▶ Prevention and Treatment

The usual treatments for infectious mononucleosis are directed at symptomatic relief of fever and sore throat. Hospitalization is rarely needed. Occasionally, rupture of the spleen necessitates immediate surgery to remove it **Disease Table 18.8**.

Hemorrhagic Fever Diseases

A number of agents that infect the blood and lymphatics cause extreme fevers, some of which are accompanied by internal hemorrhaging. The presence of the virus in the bloodstream causes capillary fragility and disrupts the blood-clotting system, which leads to various degrees of pathology, including death. All of these viruses are RNA enveloped viruses. Their geographic distribution is dependent on where their natural hosts live.

The fast-spreading Ebola epidemic of 2014 demonstrates that diseases that may get minimal amounts of space in a textbook can suddenly become worldwide threats. While sporadic infections are still occurring, the Ebola epidemic has been contained through the often-heroic efforts of medical professionals in Africa and many from around the world to help fight the disease. This was the biggest Ebola outbreak ever. **Table 18.6** shows the numbers of Ebola outbreaks and cases in the last decade. See also the Medical Moment about Ebola for specifics about the epidemic.

The diseases yellow fever, chikungunya, and dengue fever are all notable for the fact that they are spread by the *Aedes* genus of mosquito. All three diseases are common in South America and Africa; and dengue fever has exploded into the Americas in recent years. One of the diseases discussed in chapter 17 is Zika virus disease. Zika is also transmitted by the *Aedes* mosquito. When we discussed that disease, we mentioned that Zika virus spread rapidly through the Americas, and continues to do so. **Figure 18.17** depicts the likely habitat of one common *Aedes* species, *Aedes aegypti*. The vector *A. aegypti* used to be everywhere in the United States, bringing with it multiple

Table 18.6 Ebola Virus Outbreaks by Species and Size

Country	Town	Cases	Deaths	Species	Year
Dem. Rep. of Congo, Uganda	Multiple	Ongoing	Ongoing	Zaire ebolavirus	2018–2020
Dem. Rep. of Congo	Bikoro	54	33	Zaire ebolavirus	2018
Dem. Rep. of Congo	Likati	8	4	Zaire ebolavirus	2017
Dem. Rep. of Congo	Multiple. Equateur province	66	49	Zaire ebolavirus	2014
Guinea, Sierra Leone, and Liberia	Multiple	28,652	11,325	Zaire ebolavirus	2014–2016
Uganda	Luwero District	6	3	Sudan ebolavirus	2012
Dem. Rep. of Congo	Isiro Health Zone	36	13	Bundibugyo ebolavirus	2012
Uganda	Kibaale District	11	4	Sudan ebolavirus	2012

Source: www.cdc.gov/vhf/ebola/history/distribution-map.html

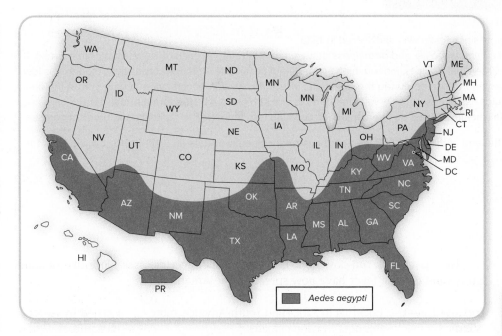

Figure 18.17 **Estimated range in the United States of *Aedes aegypti*.**
Source: CDC

diseases, notably malaria, but it was largely eliminated in temperate regions in the 1800s. Now it is back, as temperatures have allowed it into more northern regions. Changing temperatures inevitably lead to new disease patterns. In the period between 2004 and 2016, diseases spread by mosquitos, fleas, and ticks have increased threefold in the United States.

Currently in several places in the world, trials are being conducted in which male *A. aegypti* mosquitoes infected with the bacterium *Wolbachia* are being released in areas with heavy dengue burden. When these males mate with females, the resultant eggs are not delivered. Then the mosquito population is greatly reduced or eliminated. This is viewed as a much better approach than killing mosquitoes with insecticide, a technique we have relied on for decades. Mosquitoes, like bacteria, acquire resistance to chemicals that are initially lethal to them.

NCLEX® PREP

4. A patient presents to his primary care provider with a tick bite, surrounded by an erythematic rash. He has been experiencing fever, lethargy, and headache. The nurse provides education for the patient regarding Lyme disease. Which of the following statements by the patient demonstrates a proper understanding of the teaching?

a. *"It is important that I take my antiviral medications as prescribed."*

b. *"My family is at risk of contracting Lyme disease from me."*

c. *"If the disease progresses, I may develop a facial droop."*

d. *"Lyme disease is preventable with a vaccination."*

Medical Moment

Ebola Epidemic

Prior to 2014, sporadic small-scale outbreaks of Ebola had occurred from time to time in rural areas in Africa. Because the population density was low in those places, when health care workers descended into these areas to care for patients and prevent the spread of the disease, the outbreaks remained contained. In 2014 that all changed. The virus exploded in crowded urban areas of Liberia, Guinea, and Sierra Leone. In all, more than 28,000 people were infected, and more than 11,000 of them died. There was another outbreak in the Democratic Republic of Congo in

©Dr. Heidi Soeters/CDC

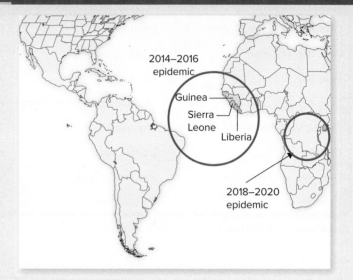

2018–2020, but the outbreak in 2014 remains the largest since Ebola was discovered in 1976.

Ebola is an excellent example of the concepts of *communicability* and *deadliness* that we are illustrating in each disease chapter in this text. If you look at that figure at the end of this chapter, you will see that Ebola is, indeed, high on the deadliness scale. Perhaps surprisingly, though, it is low on the communicability scale, despite how easily it seemed to spread during the epidemic. Most transmission occurs when a patient is in the final stages of the disease and hemorrhaging virus-filled fluids heavily. So caregivers at the bedside are at highest risk. Preparing bodies for burial is also a risky activity, which is why one of the first things public health officials do in an epidemic situation is to manage the handling and burial of corpses.

Q. Look at the "deadliness" of Ebola in the feature at the end of this chapter. Based on the largest epidemic to date (in 2014–2016), do you think the case fatality rate of Ebola might be revised?

Answer in Appendix B.

Here is a quick run-down on the hemorrhagic fever diseases. Note that chikungunya and dengue fever have now spread to the Americas.

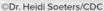

©Sarah2/Shutterstock

Yellow Fever Virus:	Endemic in Africa and South America; more frequent in rainy climates.	Carried by *Aedes* mosquitoes.
Dengue Fever:	Endemic in southeast Asia and Africa; epidemics have occurred in South America, Central America, and the Caribbean.	Carried by *Aedes* mosquitoes.
Chikungunya:	Endemic in Africa; arrived in Central America in 2013 and in U.S. and Europe in 2014; went from zero to over 1.7 million cases in the Americas in 3 years.	Carried by *Aedes* mosquitoes.
Ebola and Marburg Fevers:	Endemic to Africa; capillary fragility is extreme and patients can bleed from their orifices and mucous membranes.	Bats thought to be natural reservoir of Ebola.
Lassa Fever:	Endemic to West Africa. Asymptomatic in 80% of cases. In others, severe symptoms develop.	Reservoir of virus is multimammate rat.

These diseases are summarized in **Disease Table 18.9.**

Disease Table 18.9	Hemorrhagic Fevers				
Disease	**Yellow Fever**	**Dengue Fever**	**Chikungunya**	**Ebola and/or Marburg**	**Lassa Fever**
Causative Organism(s)	Yellow fever virus Ⓥ	Dengue fever virus Ⓥ	Chikungunya virus Ⓥ	Ebola virus, Marburg virus Ⓥ	Lassa fever virus Ⓥ
Most Common Modes of Transmission	Biological vector (*Aedes* mosquito)	Biological vector (*Aedes* mosquito)	Biological vector (*Aedes* mosquito)	Direct contact, body fluids	Droplet contact (aerosolized rodent excretions), direct contact with infected fluids
Virulence Factors	Disruption of clotting factors	Disruption of clotting factors	Disruption of clotting factors	Disruption of clotting factors	Disruption of clotting factors
Culture/Diagnosis	ELISA, PCR	Rise in IgM titers	PCR	PCR, viral culture (conducted at CDC)	ELISA
Prevention	Live attenuated vaccine available	New vaccine available but has special requirements. Consult CDC or WHO.	Vaccine in development	New Ebola vaccine suitable for epidemic situations	Avoiding rats, safe food storage
Treatment	Supportive	Supportive	Supportive	Two new drugs developed during the ongoing outbreak in the Democratic Republic of the Congo	Ribavirin
Distinctive Features	Accompanied by jaundice	"Breakbone fever"–so named due to severe pain in some forms	Arthritic symptoms	Massive hemorrhage; rash sometimes present	Chest pain, deafness as long-term sequelae
Epidemiological Features	United States: only sporadic cases in travelers; internationally, 200,000 cases annually, 30,000 deaths; 90% of cases in Africa	United States: only sporadic cases, small outbreaks. Most cases in Puerto Rico, the U.S. Virgin Islands, Samoa, and Guam; internationally, 50–300 million people infected every year and tens of thousands of deaths occur, mostly among children; **Category A Bioterrorism Agent**	First local transmission in United States in 2014; has exploded in the Americas since its arrival with an estimated 1.7 million suspected cases	United States: only imported infections; internationally, sporadic outbreaks in Africa; major Ebola outbreak 2014–2016; **Category A Bioterrorism Agent**	United States: no reported cases; internationally, estimated 100,000–300,000 cases annually in West Africa; **Category A Bioterrorism Agent**

Nonhemorrhagic Fever Diseases

In this section, we examine some infectious diseases that result in a syndrome characterized by high fever but without the capillary fragility that leads to hemorrhagic symptoms. All of the diseases in this section, except for Babesiosis, are caused by bacteria.

Brucellosis

This common disease goes by several other names, including Malta fever, undulant fever, and Bang's disease. Brucellosis often causes severe outbreaks of placental infections in livestock, which result in devastating economic impacts. The potential economic impact is one reason the CDC lists it as a possible bioterrorism agent.

▶ Signs and Symptoms

The *Brucella* bacteria responsible for this disease live in phagocytic cells. These cells carry the bacteria into the bloodstream, creating focal lesions in the liver, spleen, bone marrow, and kidney. The cardinal manifestation of human brucellosis is a fluctuating pattern of fever, which is the origin of the name *undulant fever*.

▶ Causative Agent

The bacterial genus *Brucella* contains tiny, aerobic, gram-negative coccobacilli. Several species can cause this disease in humans: *B. melitensis*, *B. abortus*, and *B. suis*.

Even though a principal manifestation of the disease in animals is an infection of the placenta and fetus, human placentas do not become infected.

▶ Pathogenesis and Virulence Factors

Brucella enters through damaged skin or via mucous membranes of the digestive tract, conjunctiva, and respiratory tract. From there it is taken up by phagocytic cells. Because it is able to avoid destruction in the phagocytes, the bacterium is transported easily through the bloodstream and to various organs, such as the liver, kidney, breast tissue, or joints. Scientists suspect that the up-and-down nature of the fever is related to unusual properties of the bacterial lipopolysaccharide.

Q Fever

The name of this disease arose from the frustration created by not being able to identify its cause. The Q stands for "query." Its cause, a bacterium called *Coxiella burnetii*, was finally identified in the mid-1900s. The clinical manifestations of acute Q fever are abrupt onset of fever, chills, head and muscle ache, and, occasionally, a rash. The disease is sometimes complicated by pneumonitis (30% of cases), hepatitis, and endocarditis. About a quarter of the cases are chronic rather than acute and result in vascular damage and endocarditis-like symptoms.

C. burnetii is a very small pleomorphic (variously shaped) gram-negative bacterium, and for a time it was considered a rickettsia, in the same genus as the bacterium that causes Rocky Mountain spotted fever (below). *C. burnetii* is apparently harbored by a wide assortment of vertebrates and arthropods, especially ticks, which play an essential role in transmission between wild and domestic animals. Ticks do not transmit the disease to humans, however. Humans acquire infection largely by means of environmental contamination and airborne spread. Birth products, such as placentas, of infected domestic animals contain large numbers of bacteria. Other sources of infectious material include urine, feces, milk, and airborne particles from infected animals. The primary portals of entry are the lungs, skin, conjunctiva, and gastrointestinal tract.

People at highest risk are farm workers, meat cutters, veterinarians, laboratory technicians, and consumers of raw milk products.

Cat-Scratch Disease

This disease is one of a group of diseases caused by different species of the small gram-negative rod *Bartonella*. *Bartonella* species are considered to be emerging pathogens. They are fastidious but not obligate intracellular parasites, so they will grow on blood agar. Two other species of *Bartonella* cause diseases known as trench fever and Carrión's disease, but cat-scratch fever is most common.

B. henselae is the agent of cat-scratch disease (CSD), an infection connected with being clawed or bitten by a cat. It is transmitted among cats by fleas. The pathogen is present in over 40% of cats, especially kittens. There are approximately 25,000 cases per year in the United States, 80% of them in children 2 to 14 years old. The symptoms start after 1 to 2 weeks, with a cluster of small papules at the site of inoculation. One to three weeks after infection, lymph nodes become very enlarged and tender. Most infections remain localized and resolve in a few weeks, but drugs such as azithromycin, erythromycin, and rifampin can be effective therapies. The disease can be prevented by flea control and by thorough antiseptic cleansing of a cat bite or scratch.

Ehrlichiosis

Ehrlichia is a small intracellular bacterium with a strict parasitic existence and association with ticks (*Ixodes* species). The species of tick varies with the geographic location in the United States and Europe. Its distribution can be seen in **figure 18.18.** The signs and symptoms include an acute febrile state resulting in headache, muscle pain, and rigors. Most patients recover rapidly with no lasting effects, but about 5% of older, chronically ill patients can die.

Rapid diagnosis is done through PCR tests and indirect fluorescent antibody tests. It can be critical to differentiate or detect coinfection with Lyme disease *Borrelia*, which is carried by the same tick. Doxycycline will clear up most infections within 7 to 10 days.

Anaplasmosis

Anaplasma is another small intracellular bacterium. It shares lifestyle characteristics with *Ehrlichia* and causes nearly identical clinical manifestations. But the two bacteria have important differences in geographic distributions (see figure 18.18) and are carried by two different species of ticks. Treatment of anaplasmosis is also with doxycycline.

Babesiosis

Babesia is a protozoan that infects red blood cells. It produces similar symptoms to *Ehrlichia* and *Anaplasma*. It is also carried by ticks. It is found in the upper Great Lakes region as well as in the northeast United States. It is often diagnosed via a blood smear; the protozoan is visible inside red blood cells **(figure 18.19).**

Because it is a protozoan, treatment is different than for *Ehrlichia* and *Anaplasma*. Combined therapy of either atovaquone (an antiprotozoal) + azithromycin, or clindamycin + quinine (another antiprotozoal) is recommended.

Spotted Fever Rickettsioses

This is a category of diseases of which Rocky Mountain Spotted Fever (RMSF) is the most well-known. It is caused by the bacterium *Rickettsia rickettsii*. In recent years it has become clear that other *Rickettsia* species cause similar diseases. The various *Rickettsia* species are not distinguishable with lab tests, so the spectrum of diseases has been renamed spotted fever rickettsiosis (SFR). Interestingly, while RMSF is named for the region in which it was first detected in the United States–the Rocky Mountains of Montana and Idaho–the disease occurs infrequently in the western United States. The majority of cases are concentrated in the Southeast and Eastern Seaboard regions. It also occurs in Canada and Central and South America. Infections occur most frequently in the spring and summer, when the tick vector is most active.

RMSF is caused by a bacterium called *Rickettsia rickettsii* transmitted by hard ticks such as the wood tick, the American dog tick, and the Lone Star tick. The dog tick is probably most responsible for transmission to humans because it is the major vector in the southeastern United States.

After 2 to 4 days of incubation, the first symptoms are sustained fever, chills, headache, and muscular pain. A distinctive spotted rash usually comes on within 2 to 4 days after the prodrome **(figure 18.20),** which usually appears first on the wrists, forearms, and ankles, before spreading. Early lesions are slightly mottled like

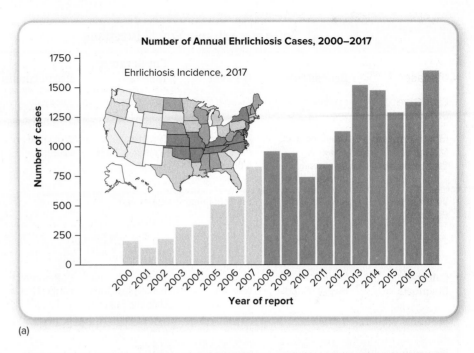

(a)

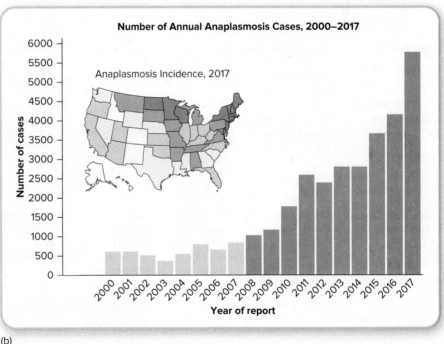

(b)

Figure 18.18 Number of annual ehrlichiosis and anaplasmosis cases in the United States from 2000–2017. In the maps, the darker a state is colored, the more cases it has.
Source: CDC

Disease Table 18.10 **Nonhemorrhagic Fever Diseases**

Disease	Brucellosis	Q Fever	Cat-Scratch Disease	Ehrlichiosis	Anaplasmosis	Babesiosis	Spotted Fever Rickettsiosis
Causative Organism(s)	*Brucella* species **B** **G⁻**	*Coxiella burnetii* **B** **G⁻**	*Bartonella henselae* **B** **G⁻**	*Ehrlichia* species **B** **G⁻**	*Anaplasma* species **B** **G⁻**	*Babesia* species **P**	*Rickettsia* species **B** **G⁻**
Most Common Modes of Transmission	Direct contact, airborne, parenteral (needlesticks)	Airborne, direct contact, foodborne	Parenteral (cat scratch or bite)	Biological vector (tick)	Biological vector (tick)	Biological vector (tick)	Biological vector (tick)
Virulence Factors	Intracellular growth; avoidance of destruction by phagocytes	Endospore-like structure	Endotoxin	–	–	–	Induces apoptosis in cells lining blood vessels
Culture/ Diagnosis	Gram stain of biopsy material	Serological tests for antibody	Biopsy of lymph nodes plus Gram staining; ELISA (performed by CDC)	PCR, indirect antibody test	PCR, indirect antibody	Blood smear	Fluorescent antibody, PCR
Prevention	Animal control, pasteurization of milk	Vaccine for high-risk population	Clean wound sites, control fleas	Avoid ticks	Avoid ticks	Avoid ticks	Avoid ticks
Treatment	Doxycycline plus gentamicin or streptomycin	Doxycycline	Azithromycin	Doxycycline	Doxycycline	Combination therapy with antibacterial + antiprotozoal	Doxycycline
Distinctive Features	Undulating fever, muscle aches	Airborne route of transmission, variable disease presentation	History of cat bite or scratch; fever not always present	Southeast, south central United States	Upper Midwest and northeastern United States	Northeastern and upper midwestern United States	Rocky Mountain Spotted Fever is most severe of the rickettsioses
Epidemiological Features	United States: fewer than 100 cases per year; internationally, 500,000 cases per year	One-third of cases occur in four states: Colorado, California, Texas, and Illinois	United States: estimated incidence is 9.3 cases per 100,000; internationally, seroprevalence from 0.6% to 37% depending on cat population	Great increase in incidence since mid-1990s	Great increase in incidence since mid-1990s	–	Tenfold increase since 2000

measles, but later ones can change shape to look like other types of rashes. In the most severe untreated cases, the enlarged lesions merge and can become necrotic, predisposing the patient to gangrene of the toes or fingertips.

Although the spots are the most obvious symptom of the disease, the most serious manifestations are cardiovascular disruption, including hypotension, thrombosis, and hemorrhage. Conditions of restlessness, delirium, convulsions, tremor, and coma are signs of the often overwhelming effects on the central nervous system. Fatalities occur in an average of 20% of untreated cases and 5% to 10% of treated cases. **Disease Table 18.10.**

Chagas Disease

Chagas disease is sometimes called "the American trypanosomiasis." The causative agent is the flagellated protozoan *Trypanosoma cruzi*. A different trypanosome,

T. brucei, causes sleeping sickness on the African continent. Chagas disease has been called "the new AIDS of the Americas" because it has a long incubation time and is very difficult to cure.

▶ Signs and Symptoms

Once the trypanosomes are transmitted by a group of insects called the triatomines **(figure 18.21),** they multiply in muscle and blood cells. From time to time, the blood cells rupture and large numbers of trypanosomes are released into the bloodstream. The disease manifestations are divided into acute and chronic phases. Soon after infection, the acute phase begins. Symptoms are relatively nondescript and range from mild to severe fever, nausea, and fatigue. A swelling called a "chagoma" at the site of the bug bite may be present. If the bug bite is close to the eyes, a distinct condition called Romana's sign, swelling of the eyelids, may appear. The acute phase lasts for weeks or months, after which the condition becomes chronic, which is virtually asymptomatic for a period of years or indefinitely. Eventually, the trypanosomes are found in numerous sites around the body; in later years, this may lead to inflammation and disruption of function in organs such as the heart, the brain, and the intestinal tract.

▶ Transmission and Epidemiology

Estimates put the prevalence of this disease at 8 million people, 300,000 of whom live in the United States. Chagas disease is one of the five "Neglected Parasitic Infections" in the United States as designated by the CDC (see Medical Moment in chapter 17). Most U.S. cases were acquired in an endemic area by travelers or other people who have since immigrated to this country.

As already noted, the disease is transmitted through the bite of a triatomine bug. The trypanosome can also be transmitted vertically because it crosses the placenta, and via blood transfusion with infected blood. Argentina, Brazil, and Mexico are the top three countries in terms of numbers of cases. Recently, the United States began screening all donated blood for this disease.

▶ Prevention

No vaccine exists for Chagas disease. In endemic areas, pesticides and improved building materials in houses are used to minimize the presence of the bug.

▶ Treatment

Treatment is most successful if begun during the acute phase. However, it is often not accomplished because the acute phase of the disease is not necessarily suggestive of Chagas. Drugs for treatment are only available through the CDC. During the chronic phase of the disease, symptomatic treatment of cardiac and other problems may be indicated. **Disease Table 18.11.**

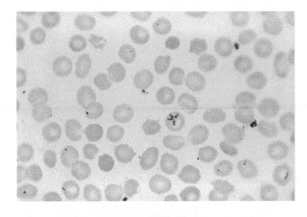

Figure 18.19 *Babesia* **cells inside red blood cells.**
Source: CDC

Figure 18.20 The rash in RMSF. This case occurred in a child several days after the onset of fever.
Source: CDC (hand); Source: CDC/Andrew J. Brooks (tick)

The body of this bug is about 2 centimeters long. For comparison, the tick shown in figure 18.20 has a body length of about 3 millimeters.
- Kelly & Heidi

Approximate size of a tick

Figure 18.21 A representative triatomine bug, the carrier of *T. cruzi.*
schlyx/Shutterstock

Disease Table 18.11	Chagas Disease
Causative Organism	*Trypanosoma cruzi* Ⓟ
Most Common Modes of Transmission	Biological vector (triatomine bug), vertical
Virulence Factors	Antioxidant enzymes, co-opting host antigens; induces autoimmunity
Culture/Diagnosis	Blood smear in acute phase; serological methods in later stages
Prevention	Insect control
Treatment	Consult CDC
Epidemiological Features	Endemic in Central and South America; 300,000 cases present in United States

Anthrax

Anthrax causes disease in the lungs and in the skin, and is addressed elsewhere in this book. We discuss anthrax in this chapter because it multiplies in large numbers in the blood and because septicemic anthrax is a possible outcome of all forms of anthrax.

For centuries, anthrax has been known as a zoonotic disease of herbivorous livestock (sheep, cattle, and goats). It has an important place in the history of medical microbiology because Robert Koch used anthrax as a model for developing his postulates in 1877, and later, Louis Pasteur used the disease to prove the usefulness of vaccination.

▶ Signs and Symptoms

As just noted, anthrax infection can exhibit its primary symptoms in various locations of the body: on the skin (cutaneous anthrax), in the lungs (pulmonary anthrax), in the gastrointestinal tract (acquired through ingestion of contaminated foods), and in the central nervous system (anthrax meningitis). The cutaneous and pulmonary forms of the disease are the most common. In all of these forms, the anthrax bacterium gains access to the bloodstream, and death, if it occurs, is usually a result of an overwhelming septicemia. Pulmonary anthrax—and the accompanying pulmonary edema and hemorrhagic lung symptoms—can sometimes be the primary cause of death, although it is difficult to separate the effects of septicemia from the effects of pulmonary infection.

In addition to symptoms specific to the site of infection, septicemic anthrax results in headache, fever, and malaise. Bleeding in the intestine and from mucous membranes and orifices may occur in late stages of septicemia.

▶ Causative Agent

Bacillus anthracis is a gram-positive endospore-forming rod that is among the largest of all bacterial pathogens. It is composed of block-shaped, angular rods 3 to 5 μm long and 1 to 1.2 μm wide. Central endospores develop under all growth conditions except in the living body of the host **(figure 18.22)**. Because the primary habitat of many *Bacillus* species, including *B. anthracis*, is the soil, endospores are continuously dispersed by means of dust into water and onto the bodies of plants and animals.

▶ Pathogenesis and Virulence Factors

The main virulence factors of *B. anthracis* are its polypeptide capsule and what is referred to as a "tripartite" toxin—a protein complex composed of three separate exotoxins. The end result of exotoxin action is massive inflammation and initiation of shock. Clinicians have noticed over time that some people experience noticeably lighter symptoms than others when infected with *B. anthracis*. Researchers have discovered that some people have a gene that codes for a surface protein that makes it more difficult for the bacterium's toxins to enter the cell. This could provide a way to predict who would be most affected in a large outbreak, as well as novel treatment strategies.

Additional virulence factors for *B. anthracis* include hemolysins and other enzymes that damage host membranes.

▶ Transmission and Epidemiology

The anthrax bacillus undergoes its cycle of vegetative growth and sporulation in the soil. Animals become infected while grazing on grass contaminated with endospores. When the pathogen is returned to the soil in animal excrement or carcasses, it can sporulate and become a long-term reservoir of infection for the animal population. The majority of natural anthrax cases are reported in livestock from Africa, Asia, and the Middle East. Most recent (natural) cases in the United States have occurred in textile workers handling imported animal hair or hide or products made from them.

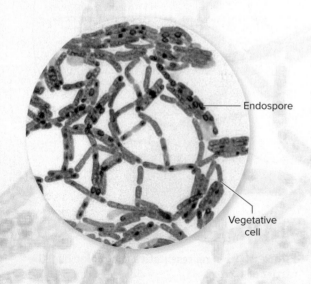

Figure 18.22 *Bacillus anthracis.* Note the centrally placed endospores and streptobacillus arrangement (600×).
Larry Stauffer/Oregon State Public/CDC

Endospore

Vegetative cell

Because of effective control procedures, the number of cases in the United States is extremely low (fewer than 10 per year).

Twenty years ago, envelopes containing freeze-dried endospores of *B. anthracis* were sent through the mail to two United States senators and several media outlets. It was an act of bioterrorism. During that attack, 22 people acquired anthrax and 5 people died.

▶ Culture and Diagnosis

Diagnosis requires a high index of suspicion. This means that anthrax must be present as a possibility in the clinician's mind or it is likely not to be diagnosed because it is such a rare disease in the developed world and because, in all of its manifestations, it can mimic other infections that are not so rare. First-level (presumptive) diagnosis begins with culturing the bacterium on blood agar and performing a Gram stain. Ultimately, samples should be handled by the Centers for Disease Control and Prevention, which will perform confirmatory tests, usually involving direct fluorescent antibody testing and phage typing tests.

▶ Prevention and Treatment

Humans should be vaccinated with the purified toxoid vaccine if they have occupational contact with livestock or products such as hides and bone. Military personnel may also receive the vaccine in certain circumstances. Effective vaccination requires five inoculations given over the space of a year, with annual boosters. The cumbersome nature of the vaccination has spurred research and development of more manageable vaccines. Persons who are suspected of being exposed to the bacterium are given prophylactic antibiotics, which seem to be effective at preventing disease even after exposure.

The recommended treatment for anthrax is usually doxycycline or ciprofloxacin. However, there is still debate about the best way to treat anthrax exposure because antibiotic usage can sometimes worsen symptoms by releasing large amounts of toxin into the bloodstream. Treatment of human cases is conducted in consultation with the CDC. **Disease Table 18.12.**

Disease Table 18.12	Anthrax
Causative Organism(s)	*Bacillus anthracis* **B** G⁺
Most Common Modes of Transmission	Vehicle (air, soil), indirect contact (animal hides), vehicle (food)
Virulence Factors	Triple exotoxin
Culture/Diagnosis	Culture, direct fluorescent antibody tests
Prevention	Vaccine for high-risk population
Treatment	In consultation with the CDC
Epidemiological Features	Internationally, 2,000–20,000 cases annually, most cutaneous; **Category A Bioterrorism Agent**

18.3 LEARNING OUTCOMES—Assess Your Progress

4. List the possible causative agents, modes of transmission, virulence factors, diagnostic techniques, and prevention/treatment for the "Highlight Diseases" malaria and HIV.

5. Discuss the epidemiology of malaria.

6. Describe the epidemiology of HIV infection in the developing world.

7. Discuss the important features of infectious cardiovascular conditions that have more than one possible cause. These are the two forms of endocarditis, septicemia, hemorrhagic fever diseases, and nonhemorrhagic fever diseases.

8. Identify factors that distinguish hemorrhagic and nonhemorrhagic fever diseases.

9. Outline the series of events that may lead to septicemia and how it should be prevented and treated.

10. Discuss the important features of infectious cardiovascular diseases that have only one possible cause. These are plague, tularemia, Lyme disease, infectious mononucleosis, Chagas disease, and anthrax.

11. Describe what makes anthrax a good agent for bioterrorism, and list the important presenting signs to look for in patients.

CASE FILE WRAP-UP

©ZouZou//Shutterstock (health care worker)

Source: Photographs in the Carol M. Highsmith Archive, Library of Congress, Prints and Photographs Division [LC-DIG-highsm-11900] (forest)

If Lyme disease is not promptly treated, neurological and cardiovascular complications can occur. As the infection spreads through the body, cranial nerves and the spinal cord may be affected. The patient may experience acute neurological changes, such as facial palsy, nerve pain, and sleep disturbances. The disease may also inhibit conduction of electrical activity in the heart.

The late stages of untreated Lyme disease involve chronic symptoms. Motor and sensory deficits lead to limited mobility and neuropathy. Cognitive abilities, memory, and personality may be affected. Patients may develop swollen joints and debilitating arthritis.

If patients are treated promptly with antibiotics, full recovery from Lyme disease is likely. In a small percentage of treated individuals, posttreatment Lyme disease syndrome (PTLDS) may occur. Fatigue, musculoskeletal pain, and cognitive problems persist for 6 months or longer after treatment. Damage to tissues and the immune system may be to blame for these lingering symptoms. Unfortunately long-term antibiotic therapy and corticosteroids have not shown to be beneficial. It is a lengthy recovery, but most patients with PTLDS eventually return to their original state of health.

▶ Summing Up

Taxonomic Organization Summing up Microorganisms Causing Disease in the Cardiovascular and Lymphatic Systems

Microorganism	Pronunciation	Disease Table
Gram-positive endospore-forming bacteria		
Bacillus anthracis	buh-sill′-us an-thray′-sus	Anthrax, 18.12
Gram-positive bacteria		
Staphylococcus aureus	staf″-uh-lo-kok′-us are′-ee-us	Endocarditis, 18.3; septicemia, 18.4
Streptococcus pyogenes	strep″-tuh-kok′-us pie′-ah″-gen-eez	Endocarditis, 18.3
Streptococcus pneumoniae	strep″-tuh-kok′-us nu-mo′-nee-ay	Endocarditis, 18.3
Gram-negative bacteria		
Yersinia pestis	yur-sin′-ee-uh pes′-tiss	Plague, 18.5
Francisella tularensis	fran-si′-sell″-uh tew′-luh-ren″-sis	Tularemia, 18.6
Borrelia burgdorferi	bor-rill′-ee-ah berg-dorf′-fur-eye	Lyme disease, 18.7
Brucella abortus, B. suis	bru-sell′-uh uh-bort′-us, bru-sell′-uh soo′-is	Nonhemorrhagic fever diseases, 18.10
Coxiella burnetii	cox-ee-ell′-uh bur-net′-tee-eye	Nonhemorrhagic fever diseases, 18.10
Bartonella henselae	bar-ton-nell′-uh hen′-sell-ay	Nonhemorrhagic fever diseases, 18.10
Ehrlichia spp.	air′-lick″-ee-uh	Nonhemorrhagic fever diseases, 18.10
Anaplasma spp.	an′-uh-plaz″-muh	Nonhemorrhagic fever diseases, 18.10
Enterococcus species	en″-teh-ro-kok′-us	Endocarditis, 18.3
Pseudomonas species	soo″-duh-moh′-nus	Endocarditis, 18.3
Rickettsia spp.	ri-ket′-see-uh	Nonhemorrhagic fever diseases, 18.10
Candida auris	kan′-did-ah aw′-ris	Septicemia, 18.4
DNA viruses		
Epstein-Barr virus	ep′-steen bar″ vie′-russ	Infectious mononucleosis, 18.8
RNA viruses		
Yellow fever virus	yel′-loh fee′-ver vie′-russ	Hemorrhagic fever, 18.9
Dengue fever virus	den′-gay fee′-ver vie′-russ	Hemorrhagic fever, 18.9
Ebola and Marburg viruses	ee-bowl′-uh and mar′-berg vie′-russ-suz′	Hemorrhagic fever, 18.9
Lassa fever virus	lass′-sah fee′-ver vie′-russ	Hemorrhagic fever, 18.9
Chikungunya virus	chick-un-goon′-yah vie′-russ	Hemorrhagic fever, 18.9
Retroviruses		
Human immunodeficiency virus 1 and 2	hew′-mun im′-muh-noh-dee-fish″-shun-see vie′-russ	HIV infection and AIDS, 18.2
Protozoa		
Plasmodium falciparum, P. vivax, P. knowlesi, P. ovale, P. malariae	plas-moh′-dee-um fal-sip′-uh-rum, plas-moh′-dee-um vy′-vax, plas-moh′-dee-um nohles′-eye, plas-moh′-dee-um oh-val′-ee, plas-moh′-dee-um ma-lair′-ee-ay	Malaria, 18.1
Babesia species	buh-bee′-see-uh	Nonhemorrhagic fever diseases, 18.10
Trypanosoma cruzi	tri-pan″-uh-sohm′-uh krewz′-ee-eye	Chagas disease, 18.11

The Microbiome and Heart Failure

In 2019 a research review was published in the journal *Cardiac Failure Review*. It painted a picture of how a disrupted gut microbiome could worsen heart disease. They were specifically interested in the worsening of patients who were already in some degree of heart failure.

Many people are living with heart failure, also known as congestive heart failure, to varying degrees. These people have what is called *decreased output,* which means that there is less overall blood flow, including intestinal blood flow, which leads to less oxygen, which can predispose the gut to the growth of more pathogenic bacteria.

At the same time, the venous system is also not efficient and causes buildup of fluids in the bowel, leading to buildup in bacterial overgrowth in the mucus layer there. Because of the swelling and permeability, bacteria in the gut can escape to the bloodstream, and, if gram-negative, lead to more circulating lipopolysaccharide (LPS; endotoxin). As you know, LPS causes systemic inflammation.

So far we have less oxygen in the gut due to decreased heart output (favored by pathogenic bacteria). We also have more leakage in the gut, leading to bacteria escaping into the circulation. Next, the increased bacterial growth in the gut results in increase in a substance called TMAO (trimethylamine oxide), a metabolite of bacteria. Its role is not entirely clear, but its presence has been correlated with negative outcomes in heart failure patients.

All of these circumstances lead to damage to the heart. The usual mechanisms of scar tissue apply here—increasing inflammation and macrophage and cytokine activation, leading to repair (scarring) that ends up hindering the function of the heart muscle.

Interestingly, the authors hypothesize (but do not test) the idea that consuming less red meat in the diet can be beneficial in this series of events because gut bacteria produce TMAO from two components of red meat, choline and lecithin. Here is an important example of the difference between research findings (the events detailed in the above paragraphs) and speculation that comes from the findings (the idea of dietary effects). The research detailed here thus leads to new questions to be researched.

Source: 2019. https://www.ncbi.nlm.nih.gov/pmc/articles/PMC6545994/

Shutterstock /Magic mine

Infectious Diseases Affecting
The Cardiovascular and Lymphatic Systems

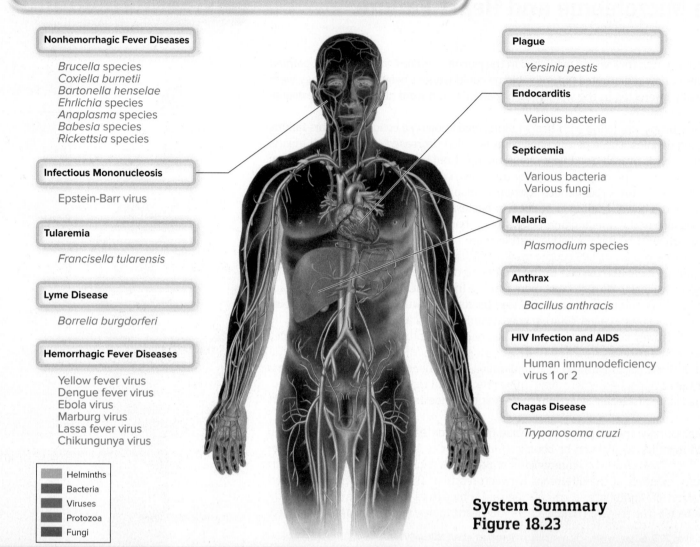

Nonhemorrhagic Fever Diseases

Brucella species
Coxiella burnetii
Bartonella henselae
Ehrlichia species
Anaplasma species
Babesia species
Rickettsia species

Infectious Mononucleosis

Epstein-Barr virus

Tularemia

Francisella tularensis

Lyme Disease

Borrelia burgdorferi

Hemorrhagic Fever Diseases

Yellow fever virus
Dengue fever virus
Ebola virus
Marburg virus
Lassa fever virus
Chikungunya virus

Plague

Yersinia pestis

Endocarditis

Various bacteria

Septicemia

Various bacteria
Various fungi

Malaria

Plasmodium species

Anthrax

Bacillus anthracis

HIV Infection and AIDS

Human immunodeficiency
virus 1 or 2

Chagas Disease

Trypanosoma cruzi

Helminths
Bacteria
Viruses
Protozoa
Fungi

**System Summary
Figure 18.23**

Deadliness and Communicability of Selected Diseases of the Cardiovascular and Lymphatic Systems

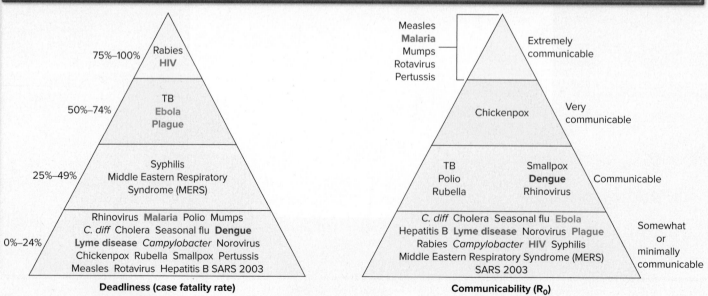

75%–100% — Rabies **HIV**

50%–74% — TB Ebola Plague

25%–49% — Syphilis Middle Eastern Respiratory Syndrome (MERS)

0%–24% — Rhinovirus **Malaria** Polio Mumps *C. diff* Cholera Seasonal flu **Dengue** **Lyme disease** *Campylobacter* Norovirus Chickenpox Rubella Smallpox Pertussis Measles Rotavirus Hepatitis B SARS 2003

Deadliness (case fatality rate)

Measles **Malaria** Mumps Rotavirus Pertussis — Extremely communicable

Chickenpox — Very communicable

TB Polio Rubella Smallpox **Dengue** Rhinovirus — Communicable

C. diff Cholera Seasonal flu Ebola Hepatitis B **Lyme disease** Norovirus **Plague** Rabies *Campylobacter* **HIV** Syphilis Middle Eastern Respiratory Syndrome (MERS) SARS 2003 — Somewhat or minimally communicable

Communicability (R₀)

INFECTIOUS DISEASES AFFECTING THE CARDIOVASCULAR AND LYMPHATIC SYSTEMS

The Cardiovascular and Lymphatic Systems and their Defenses

The cardiovascular system is composed of blood vessels and the heart. The lymphatic system is a one-way passage, returning fluid from the tissues to the cardiovascular system. These systems are somewhat protected from microbial infection, compared to the other body systems.

Normal Biota: Cardiovascular and Lymphatic Systems

Recent research suggest that blood cells, especially white cells, can contain normal microbiota, or at least their component parts and/or DNA.

Cardiovascular and Lymphatic Diseases Caused by Microorganisms

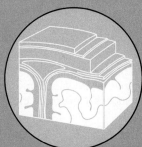

- **Malaria**
Usually caused by one of four different species of the protozoan *Plasmodium*. Transmitted by *Anopheles* mosquitoes.
- **HIV Infection and AIDS**
HIV is a retrovirus, meaning it creates DNA from RNA. The disease has been transformed from a death sentence to a chronic disease, for those who have access to appropriate treatment.
- **Endocarditis**
Inflammation of the endocardium, usually due to infection of the valves of the heart. Acute endocarditis involves a very large dose of bacteria being introduced into the bloodstream. Subacute endocarditis is usually preceded by some form of damage to the heart valves.
- **Septicemia**
Microbes actively multiplying in the blood.
- **Plague**
The plague is an ancient disease, which can manifest in three main ways: pneumonic plague, bubonic plague, and septicemic plague. Caused by the bacterium *Yersinia pestis*.
- **Tularemia**
Causative agent is the intracellular bacterium *Francisella tularensis*. Category A bioterrorism agent.

- **Lyme Disease**
Caused by *Borrelia burdorferii* and related species. Transmitted by ticks. Can have long-term effects.
- **Infectious Mononucleosis**
Vast majority of cases are caused by Epstein-Barr virus.
- **Hemorrhagic Fever Diseases**
Extreme fevers and internal bleeding. Common diseases are yellow fever, dengue fever, Ebola and Marbug diseases, and Lassa fever. Caused by RNA enveloped viruses.
- **Nonhemorrhagic Fever Diseases**
High fever without the capillary damage seen in hemorrhagic diseases. Diseases in this category include brucellosis, Q fever, cat-scratch disease, ehrlichiosis, anaplasmosis, babesiosis, and the spotted fever rickettsioses.
- **Chagas Disease**
Trypanosoma cruz, transmitted by insects in South and Central America. Increasingly frequent in the United States.
- **Anthrax**
Bacillus anthracis can produce symptoms in the lungs, the skin, and the central nervous system.

SmartGrid: From Knowledge to Critical Thinking

This *21 Question Grid* takes the topics from this chapter and arranges them with respect to the American Society for Microbiology's Undergraduate Curriculum guidelines—all six of the important "Concepts" as well as the important "Competency" of scientific literacy. Three questions are supplied about chapter content that refer to the Concept or Competency, in increasing levels of Bloom's taxonomy for learning.

ASM Concept/ Competency	A. Bloom's Level 1, 2—Remember and Understand (Choose one.)	B. Bloom's Level 3, 4—Apply and Analyze	C. Bloom's Level 5, 6—Evaluate and Create
Evolution	1. Which of the following microbes have evolved an intracellular lifestyle? a. *Bacillus anthracis* b. *Coxiella burnetii* c. MRSA d. two of these	2. Why do you think that malarial infection is more often fatal in children than in adults in areas where it is endemic?	3. In chapter 17 you were asked to speculate on why a disease such as polio has become less virulent over time, while rabies has not. Malaria has also not become less virulent over time. There is evidence that HIV is becoming less virulent in humans. Is there a pattern you can discern in microbes that don't become less virulent over time?
Cell Structure and Function	4. Which of the following is a G⁺ bacterium? a. *Staphylococcus aureus* b. *Borrelia burgdorferi* c. *Coxiella burnetii* d. *Trypanosoma cruzi*	5. What characteristic(s) of *Bacillus anthracis* make it a good candidate for bioterror?	6. Construct an immunological argument about why Lyme disease might continue to cause symptoms after the bacterium has been cleared.
Metabolic Pathways	7. Which of the following diseases is characterized by the formation of a biofilm? a. plague b. HIV c. endocarditis d. Chagas disease	8. Name some ways in which you think intracellular bacteria such as *Coxiella* and *Brucella* might resemble viruses.	9. Argue for the bloodstream being an advantageous environment for microbial pathogens. Now argue the opposite.
Information Flow and Genetics	10. Which of the following diseases is caused by a retrovirus? a. Lassa fever b. Ebola c. anthrax d. HIV	11. Discuss whether human genetics plays a role in HIV infection, providing at least one example to illustrate your position.	12. Use a unique characteristic of HIV to provide an explanation of why it is so easy for HIV to integrate into its host cell's DNA and linger there for months/years.
Microbial Systems	13. The bite of a tick can cause a. ehrlichiosis. b. Lyme disease. c. anaplasmosis. d. all of these.	14. Discuss the merits and problems inherent in eliminating all mosquitoes in an effort to battle diseases such as yellow fever, dengue fever, and chikungunya.	15. This text frequently discusses the increase in infectious diseases in humans caused by the warming climate. Can you describe a hypothetical microbial pathogen that would cause less disease in a warming climate?

ASM Concept/ Competency	A. Bloom's Level 1, 2—Remember and Understand (Choose one.)	B. Bloom's Level 3, 4—Apply and Analyze	C. Bloom's Level 5, 6—Evaluate and Create
Impact of Microorganisms	16. Normal biota found in the oral cavity are most likely to cause a. acute endocarditis. b. subacute endocarditis. c. malaria. d. tularemia.	17. In the Middle Ages, during a massive plague epidemic, one of the control measures instituted was the quarantine of infected people. Why was this not successful?	18. For decades public health officials have been trying to eliminate malaria. There are some ecologists and anthropologists who suggest that eliminating malaria will lead to overpopulation and famine in "the malaria belt." Defend one of these viewpoints.
Scientific Thinking	19. Lyme disease is most likely to occur in a. North Dakota. b. Connecticut. c. Oklahoma. d. Arkansas.	20. Explain why the incidence of HIV infection in the United States has declined, while the prevalence of HIV⁺ persons has increased.	21. When female mosquitoes mate with males infected with the bacterium *Wolbachia*, they no longer produce viable eggs. This is being tested as a control strategy for dengue fever and other mosquito-borne diseases. What questions would be important to answer before making this a widespread practice?

Answers to the multiple-choice questions appear in Appendix A.

Visual Connections

This question connects content within and between chapters.

1. **From chapter 12, figure 12.11.** Imagine that the WBCs shown in this illustration are unable to control the microorganisms in the blood. Could the change that has occurred in the vessel wall help the organism spread to other locations? If so, how?

2. If the organisms are able to survive phagocytosis, how could that impact the progress of this disease? Explain your answer.

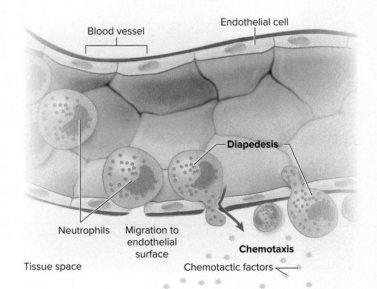

Chapter design elements: (Covid): CDC/Alissa Eckert, MS; Dan Higgins, MAMS; (Note): McGraw-Hill Education; (NCLEX): Shutterstock/ Abert; (Doctor): David Gould/Getty Images

19

Infectious Diseases Affecting the Respiratory Systems

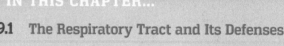

CASE FILE

Very Sick, Very Fast

I was a respiratory therapist student completing a rotation in pediatrics when I became involved in caring for Michael, a 6-month-old who was admitted to the pediatric unit with a respiratory infection.

Michael was a healthy infant who was born at term (39 weeks' gestation). He was current with all of his immunizations. Michael was the youngest of three children, with two older siblings who were school age. Michael attended day care on a part-time basis when his mother worked.

Michael and both his siblings had become ill during recent days with cold symptoms. At first, Michael had a runny nose and a slight cough. He was fussy and irritable and his mother stated he had been sleeping more than usual. Michael was breast-fed. However, when he became ill, he fussed at the breast and fed less often, even though his mother offered him the breast frequently to encourage fluid intake and provide comfort. Having three children, Michael's mother was well versed in caring for a sick child and was not prone to panic over a common cold.

However, on the third day of his illness, Michael took an alarming turn for the worse. Although he acted as though he was hungry, he would stop feeding almost as soon as he started. Michael had copious nasal discharge, and his mother rightly tried to clear his nostrils using a bulb syringe, but Michael still fed poorly. He had only had one slightly wet diaper over the course of the night and had not urinated at all in the morning. He was running a high fever of 39.6°C (103.3°F). His eyes appeared sunken in his face and there were no tears when he cried, both signs of dehydration. He was coughing frequently. Most concerning, Michael's respiratory rate was 48 breaths/minute. His nail beds were cyanosed (blue) and he was very pale in color. His oxygen saturation was 89% on room air.

Michael was seen in the emergency room and was immediately admitted to the pediatric unit by the physician on call, who was very concerned. My preceptor and I were called to assess Michael on the unit. When we arrived, Michael had a chest X ray and blood work done. He was lying in a crib with a nasal cannula supplying humidified oxygen taped to his face. His color was still very pale but his nail beds were pink. His oxygen saturation had increased to 95% with oxygen flow at 5 L/minute. I listened to Michael's chest and heard wheezes throughout Michael's lungs. After I assessed Michael, my preceptor asked me what I thought might be going on with Michael. Hedging my bets, I replied that all I could tell for certain was that Michael had a respiratory tract illness and a virus was the likely culprit. My preceptor promptly replied, "You're right that Michael has a viral illness of the respiratory tract, but which one?" When I confessed that I did not know, my preceptor informed me that Michael's symptoms were consistent with respiratory syncytial virus (RSV). Michael's tests would confirm that.

- How is RSV spread?
- How might Michael have contracted the respiratory syncytial virus?

Case File Wrap-Up appears at the end of the chapter.

19.1 The Respiratory Tract and Its Defenses

The respiratory tract is the most common place for infectious agents to gain access to the body. We breathe 24 hours a day, and anything in the air we breathe passes at least temporarily into this organ system.

The structure of the system is illustrated in **figure 19.1a.** Most clinicians divide the system into two parts, the *upper* and *lower respiratory tracts.* The upper respiratory tract includes the mouth, the nose, the nasal cavity and sinuses above it, the throat or pharynx, and the epiglottis and larynx. The lower respiratory tract begins with the trachea, which feeds into the bronchi and bronchioles in the lungs. Attached to the bronchioles are small balloonlike structures called alveoli, which inflate and deflate with inhalation and exhalation. These are the sites of oxygen exchange in the lungs.

Several anatomical features of the respiratory system protect it from infection. As described in chapter 12, nasal hair serves to trap particles. Cilia **(figure 19.1b)** on the epithelium of the trachea and bronchi (the ciliary escalator) propel particles upward and out of the respiratory tract. Mucus on the surface of the mucous membranes lining the respiratory tract is a natural trap for invading microorganisms. Once the microorganisms are trapped, involuntary responses such as coughing, sneezing, and swallowing can move them out of sensitive areas. These are first-line defenses.

The second and third lines of defense also help protect the respiratory tract. Complement action, antimicrobial peptides, and increased levels of cytokines all help battle pathogens in the lungs. Macrophages inhabit the alveoli of the lungs and the clusters of lymphoid tissue (tonsils) in the throat. Secretory IgA against specific pathogens can be found in the mucus secretions as well.

Medical Moment

Epiglottitis

The epiglottis is a small piece of cartilage that partially covers the larynx. Its job is to help prevent the inhalation of food and fluids into your lungs. Epiglottitis occurs when an infection or an injury causes the epiglottis to swell, which may result in the inability to draw air into the lungs.

Symptoms of epiglottitis may include fever, an extremely sore throat, a muffled or hoarse voice, stridor (a high-pitched sound that occurs during inhalation), and difficulty breathing and swallowing. Children may drool due to their inability to swallow oral secretions and may appear very anxious.

Epiglottitis is a medical emergency and may cause complete blockage of the airway. The most common infectious cause of epiglottitis historically was *Haemophilus influenzae* b. Because a vaccine for that bacterium is now in common use, the condition is less common, with less than 20,000 cases a year in the United States.

Q. Name two other serious infectious conditions that have decreased since the introduction of the Hib vaccine.

Answer in Appendix B.

19.1 LEARNING OUTCOMES—Assess Your Progress

1. Draw or describe the anatomical features of the respiratory tract.
2. List the natural defenses present in the respiratory tract.

19.2 Normal Biota of the Respiratory Tract

The latest research shows that a healthy upper respiratory system harbors thousands of commensal microorganisms and that even the lungs have a normal, if limited, biota. Part of this normal biota can cause serious disease, especially in immunocompromised individuals. These include *Streptococcus pyogenes, Haemophilus influenzae, Streptococcus pneumoniae, Neisseria meningitidis,* and *Staphylococcus aureus.* Yeasts, especially *Candida albicans,* also colonize the mucosal surfaces of the mouth in the upper respiratory tract. Other fungi can be found throughout the respiratory tract. The composition of the lung microbiome differs in patients suffering from lung disorders such as chronic obstructive pulmonary disease (COPD), asthma, and cystic fibrosis. Smokers and nonsmokers also appear to have different microbiota.

Respiratory Tract Defenses and Normal Biota

	Defenses	Normal Biota
Upper Respiratory Tract	Nasal hair, ciliary escalator, mucus, involuntary responses such as coughing and sneezing, secretory IgA	Many species of bacteria and fungi, including *Neisseria meningitidis, Staphylococcus, Streptococcus, Aspergillus,* and other fungi
Lower Respiratory Tract	Mucus, alveolar macrophages, secretory IgA	Still unclear; low levels of colonization by multiple species probable

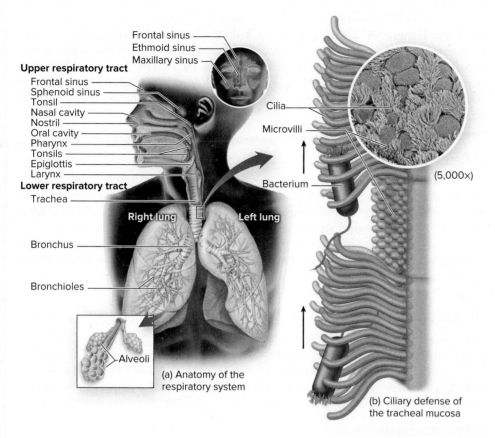

Figure 19.1 The respiratory tract. **(a)** Important structures in the upper and lower respiratory tract. The four pairs of sinuses are pictured in the inset. **(b)** Ciliary defense of the respiratory tract.

©Susumu Nishinaga/Science Source (micrograph)

In the respiratory system, as in some other organ systems, the normal biota performs the important function of microbial antagonism (see chapter 11). This reduces the chances of pathogens establishing themselves in the same area by competing with them for resources and space. To illustrate this point, *Lactobacillus sakei*, a known member of the sinus microbiome, can suppress the pathogenic potential of another normal biota organism, *Corynebacterium tuberculostearicum*, reducing the incidence of sinus infection.

19.2 LEARNING OUTCOMES—Assess Your Progress

3. List the types of normal biota presently known to occupy the respiratory tract.

19.3 Upper Respiratory Tract Diseases Caused by Microorganisms

HIGHLIGHT DISEASE

Pharyngitis

▶ Signs and Symptoms

The name says it all–this is an inflammation of the throat, which the host experiences as pain and swelling. The severity of pain can range from moderate to severe, depending on the causative agent. Viral sore throats are generally mild and sometimes lead to hoarseness. Sore throats caused by bacteria are generally more painful than those caused by viruses, and they are more likely to be accompanied by fever, headache, and nausea.

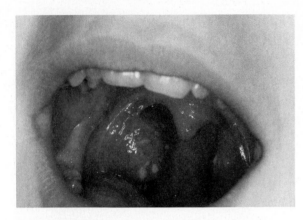

Figure 19.2 The appearance of the throat in pharyngitis and tonsillitis. The pharynx and tonsils become bright red and suppurative. Whitish pus nodules may also appear on the tonsils.

©Pulse Picture Library/CMP Images/PhotoTake

Clinical signs of a sore throat are reddened mucosa, swollen tonsils, and sometimes white packets of inflammatory products visible on the walls of the throat, especially in streptococcal disease **(figure 19.2).** The mucous membranes may be swollen, affecting speech and swallowing. Often pharyngitis results in foul-smelling breath. The incubation period for most sore throats is generally 2 to 5 days.

▶ Causative Agents

The same viruses causing the common cold are those most often causing a sore throat, which can also accompany other diseases, such as infectious mononucleosis (described in chapter 18). Pharyngitis may simply be the result of mechanical irritation from prolonged shouting or from drainage of an infected sinus cavity. The most serious cases of pharyngitis are caused by *Streptococcus pyogenes*, a group A streptococcus, and *Fusobacterium necrophorum*, an anaerobic gram-negative bacterium.

Streptococcus pyogenes

S. pyogenes is a gram-positive coccus that grows in chains. It does not form endospores, is nonmotile, and forms capsules and slime layers. *S. pyogenes* is a facultative anaerobe that ferments a variety of sugars. It does not produce catalase, but it does have a peroxidase system for inactivating hydrogen peroxide, which allows its survival in the presence of oxygen.

▶ Pathogenesis

Untreated streptococcal throat infections can occasionally result in serious complications, either right away or days to weeks after the throat symptoms subside. The post-streptococcal conditions can be caused by the presence of an extra toxin (causing scarlet fever) or by the deposition of antigen-antibody complexes in the body (causing glomerulonephritis). Another cause of post-streptococcal complications is immune system attack of self tissues triggered by streptococcal "superantigens" that cause immune activation to similar or even unrelated (human) proteins. This is responsible for rheumatic fever and for some types of obsessive-compulsive disorder (brain tissue is targeted) and psoriasis (skin is targeted). More details about scarlet fever and rheumatic fever are presented here.

Scarlet Fever   Scarlet fever is the result of infection with an *S. pyogenes* strain that is itself infected with a bacteriophage. This lysogenic virus gives the streptococcus the ability to produce erythrogenic toxin, described in the section on virulence. Scarlet fever is characterized by a sandpaper-like rash, most often on the neck, chest, elbows, and inner surfaces of the thighs. High fever accompanies the rash. It most often affects school-age children and was a source of great suffering in the United States in the early part of the 20th century. In epidemic form, the disease can have a fatality rate of up to 95%. Although the disease had all but disappeared in the last 100 years, there has been a resurgence of cases in various parts of the world.

Rheumatic Fever   Rheumatic fever is thought to be due to an immunologic cross-reaction between the streptococcal M protein and heart muscle. This means that the lymphocyte clones activated by the M protein also react with an epitope on the heart muscle. It tends to occur approximately 3 weeks after pharyngitis has subsided. It can

---

NCLEX® PREP

1. What information should be included in health promotion classes for parents of school-age children with regard to the treatment of sore throats?

   a. *As long as the child does not have a fever, there is no need to seek medical treatment.*

   b. *Provide fluids as tolerated and keep the child well hydrated.*

   c. *Parents should take their child to a health care provider so that a rapid throat culture can be obtained.*

   d. *If the child can swallow with minimal pain, all that is needed is increased fluid.*

result in permanent damage to heart valves. Other symptoms include arthritis in multiple joints and the appearance of nodules over bony surfaces just under the skin.

► Virulence Factors

The virulence of *S. pyogenes* is a result of two main phenomena: the ability of its surface antigens to mimic host proteins and its possession of superantigens. Superantigens were described in chapter 12.

Streptococci display numerous surface antigens **(figure 19.3)**. Specialized polysaccharides on the surface of the cell wall help to protect the bacterium from being dissolved by the lysozyme of the host. Lipoteichoic acid (LTA) contributes to the adherence of *S. pyogenes* to epithelial cells in the pharynx. A spiky surface projection called *M protein* contributes to virulence by resisting phagocytosis and possibly by contributing to adherence. A capsule made of hyaluronic acid (HA) is formed by most *S. pyogenes* strains. It probably contributes to the bacterium's adhesiveness.

Extracellular Toxins Group A streptococci owe some of their virulence to the effects of hemolysins called **streptolysins.** The two types are streptolysin O (SLO) and streptolysin S (SLS). Both types cause beta-hemolysis of sheep blood agar (see "Culture and Diagnosis"). Both hemolysins rapidly injure many cells and tissues, including leukocytes and liver and heart muscle (in other forms of streptococcal disease).

A key toxin in the development of scarlet fever is **erythrogenic** (eh-rith′-roh-jen′-ik) **toxin.** This toxin is responsible for the bright red rash typical of this disease, and it also induces fever by acting upon the temperature regulatory center in the brain. Only lysogenic strains of *S. pyogenes* that contain genes from a temperate bacteriophage can synthesize this toxin. (For a review of the concept of lysogeny, see chapter 5.)

► Transmission and Epidemiology

Physicians estimate that 30% of sore throats may be caused by *S. pyogenes,* adding up to several million cases each year. Most transmission of *S. pyogenes* is via respiratory droplets or direct contact with mucus secretions. Humans are the only significant reservoir of *S. pyogenes.*

More than 80 serotypes of *S. pyogenes* exist, and that means that people can experience multiple infections throughout their lives because immunity is serotype-specific.

► Culture and Diagnosis

The failure to recognize group A streptococcal infections can have devastating effects. Rapid cultivation and diagnostic techniques to ensure proper treatment and prevention measures are essential. Several different rapid diagnostic test kits are used in clinics and doctors' offices to detect group A streptococci from pharyngeal swab samples **(figure 19.4a)**.

If a culture is needed, it is generally taken at the same time as the rapid swab and is plated on sheep blood agar. *S. pyogenes* displays a beta-hemolytic pattern due to its streptolysins (and hemolysins) **(figure 19.4b)**. If the pharyngitis is caused by a virus, the blood agar dish will show a variety of colony types, representing the normal bacterial biota. Active infection with *S. pyogenes* will yield a plate with a majority of beta-hemolytic colonies. Newer

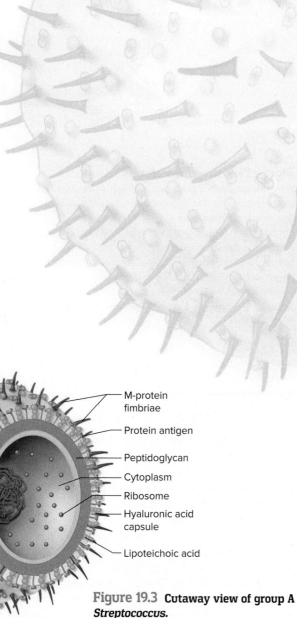

Figure 19.3 **Cutaway view of group A** ***Streptococcus.***

- M-protein fimbriae
- Protein antigen
- Peptidoglycan
- Cytoplasm
- Ribosome
- Hyaluronic acid capsule
- Lipoteichoic acid

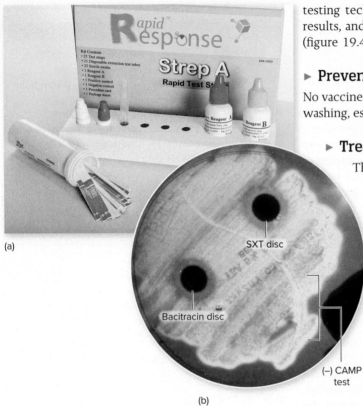

(a)

SXT disc

Bacitracin disc

(–) CAMP test

(b)

testing techniques are performed at the point of care (POC), take 15 minutes for results, and don't require confirmation with culture. A positive bacitracin disc test (figure 19.4b) provides additional evidence for group A.

▶ Prevention

No vaccine exists for group A streptococci. Infection can be prevented by good hand washing, especially after coughing and sneezing and before preparing foods or eating.

▶ Treatment

The antibiotic of choice for *S. pyogenes* is penicillin. In patients with penicillin allergies, a first-generation cephalosporin, such as cephalexin, is prescribed. Although most sore throats caused by *S. pyogenes* can resolve on their own, they should be treated with antibiotics because serious sequelae are a distinct possibility.

Fusobacterium necrophorum

In the last two decades, more and more cases of severe sore throats caused by this bacterium have been identified. Some studies suggest it is as common a cause of pharyngitis as *S. pyogenes* in adolescents and young adults. It can manifest as a severe sore throat, and then it can progress to a condition called Lemierre's syndrome. In that situation, the bacterium can invade the jugular vein in the neck and spread to the whole circulatory system, sometimes resulting in death. **Disease Table 19.1**.

Figure 19.4 Streptococcal tests. **(a)** A rapid, immunologic test for diagnosis of group A infections. **(b)** Bacitracin disc test. With very few exceptions, only *Streptococcus pyogenes* is sensitive to a minute concentration (0.02 µg) of bacitracin. Any zone of inhibition around the B disc is interpreted as a presumptive indication of this species. (Note: Group A streptococci are negative for sulfamethoxazole-trimethoprim [SXT] sensitivity and the CAMP test.)
©PHOTO RF/Science Source (a); ©Van Bucher/SPL/Science Source (b)

Disease Table 19.1	Pharyngitis		
Causative Organism(s)	*Streptococcus pyogenes* Ⓑ G⁺	*Fusobacterium necrophorum* Ⓑ G⁻	Viruses Ⓥ
Most Common Modes of Transmission	Droplet or direct contact	Usually endogenous	All forms of contact
Virulence Factors	LTA, M protein, hyaluronic acid capsule, SLS and SLO, superantigens, induction of autoimmunity	Invasiveness, endotoxin	–
Culture/Diagnosis	Beta-hemolytic on blood agar, sensitive to bacitracin, rapid antigen tests	Culture anaerobically, CT scan for abscess(es)	Goal is to rule out *S. pyogenes*, further diagnosis usually not performed
Prevention	Hygiene practices	?	Hygiene practices
Treatment	Penicillin, cephalexin in penicillin-allergic	Penicillin	Symptom relief only
Distinctive Features	Generally more severe than viral pharyngitis	Can lead Lemierre's syndrome	Hoarseness frequently accompanies viral pharyngitis
Epidemiological Features	United States: 20%–30% of all cases of pharyngitis	Causes up to 15% of acute pharyngitis in teens/young adults	Ubiquitous; responsible for 40%–60% of all pharyngitis

The Common Cold

Everyone is familiar with the symptoms of a common cold: sneezing, scratchy throat, and runny nose, which usually begin 2 or 3 days after infection. An uncomplicated cold generally is not accompanied by fever, although children can experience low fevers (less than 102°F). The incubation period is usually 2 to 5 days. People with asthma and other underlying respiratory conditions, such as chronic obstructive pulmonary disease (COPD), often suffer more severe symptoms triggered by the common cold.

The common cold is caused by one of over 200 different kinds of viruses. The most common type of virus leading to rhinitis is the group called rhinoviruses, of which there are more than 150 serotypes. Coronaviruses and adenoviruses are also frequent causes. Also, the respiratory syncytial virus (RSV) causes colds in most people it infects, but in some, especially infants and children, they can lead to more serious respiratory tract symptoms (discussed later in the chapter). **Disease Table 19.2.**

Sinusitis

Commonly called a *sinus infection*, this inflammatory condition of any of the four pairs of sinuses in the skull (see figure 19.1a) can actually be caused by allergy (most common), infections, or simply structural problems such as narrow passageways or a deviated nasal septum. The infectious agents that may be responsible for the condition commonly include a variety of viruses or bacteria and, less commonly, fungi. **Disease Table 19.3.**

©ballyscanlon/
Getty Images

Disease Table 19.2	The Common Cold
Causative Organism(s)	Approximately 200 viruses Ⓥ
Most Common Modes of Transmission	Indirect contact, droplet contact
Virulence Factors	Attachment proteins; most symptoms induced by host response
Culture/ Diagnosis	Not necessary
Prevention	Hygiene practices
Treatment	For symptoms only
Epidemiological Features	Highest incidence among preschool and elementary schoolchildren with average of three to eight colds per year; adults and adolescents: two to four colds per year

Disease Table 19.3	Sinusitis		
Causative Organism(s)	Viruses Ⓥ	Various bacteria, often mixed Ⓑ infection	Various fungi Ⓕ
Most Common Modes of Transmission	Direct contact, indirect contact	Endogenous (opportunism)	Introduction by trauma or opportunistic overgrowth
Virulence Factors	–	–	–
Culture/Diagnosis	Culture not usually performed; diagnosis based on clinical presentation	Culture not usually performed; diagnosis based on clinical presentation; occasionally X rays or other imaging technique used	Same
Prevention	Hygiene	–	–
Treatment	None	Recommendation is for no antibiotics unless it remains unresolved for some weeks	Physical removal of fungus; in severe cases, antifungals used
Distinctive Features	Viral and bacterial much more common than fungal	Viral and bacterial much more common than fungal	Suspect in immunocompromised patients
Epidemiological Features	Commonly follows the common cold	United States: affects 1 of 7 adults; between 12 and 30 million diagnoses per year	Fungal sinusitis varies with geography; in United States, more common in SE and SW; internationally: more common in India, North Africa, Middle East

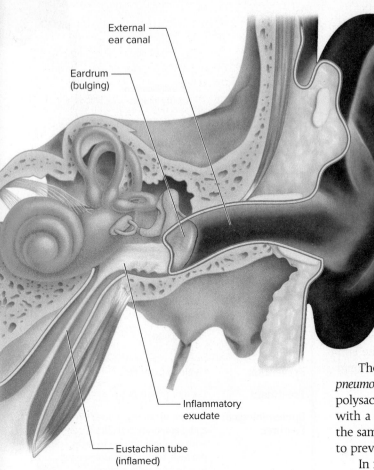

External
ear canal

Eardrum
(bulging)

Inflammatory
exudate

Eustachian tube
(inflamed)

Figure 19.5 An infected middle ear.

Acute Otitis Media (Ear Infection)

Viral infections of the upper respiratory tract lead to inflammation of the eustachian tubes and the buildup of fluid in the middle ear, which can lead to bacterial multiplication in those fluids. Although the middle ear normally has no biota, bacteria can migrate along the eustachian tube from the upper respiratory tract **(figure 19.5).** When bacteria encounter mucus and fluid buildup in the middle ear, they multiply rapidly. Their presence increases the inflammatory response, leading to pus production and continued fluid secretion. This fluid is referred to as *effusion*.

Another condition, known as chronic otitis media, occurs when fluid remains in the middle ear for indefinite periods of time. Until recently, physicians considered it to be the result of a noninfectious immune reaction because they could not culture bacteria from the site and because antibiotics were not effective. New data suggest that this form of otitis media is caused by a mixed biofilm of bacteria that is attached to the mucosa of the middle ear. Biofilm bacteria generally are less susceptible to antibiotics (as discussed in chapter 3), and their presence in biofilm form would explain the difficulty in culturing them from ear fluids. Scientists now believe that the majority of acute and chronic otitis media cases are mixed infections with viruses and bacteria acting together.

The single most common bacterium seen in acute otitis media is *Streptococcus pneumoniae*. The vaccine (Prevnar) is a conjugated vaccine (see chapter 13). It contains polysaccharide capsular material from 13 different strains of the bacterium complexed with a chemical that makes it more antigenic. It is distinct from another vaccine for the same bacterium (Pneumovax), which is primarily targeted to the older population to prevent pneumococcal pneumonia.

In recent years, a troubling new cause of otitis media has emerged. *Candida auris* was first recognized in a Japanese patient in 2009, and since then has appeared in over 30 countries, including the United States. It is difficult to diagnose and is also resistant to multiple antifungal drugs. It can spread from a localized ear infection to a bloodstream infection, in which case it can be fatal.

The current treatment recommendation for uncomplicated acute otitis media with a fever below 104°F is "watchful waiting" for 72 hours to allow the body to clear the infection, avoiding the use of antibiotics. In patients less than six months

Disease Table 19.4	Otitis Media		
Causative Organism(s)	*Streptococcus pneumoniae* **B** **G+**	*Candida auris* **F**	Other bacteria/viruses
Most Common Modes of Transmission	Endogenous (may follow upper respiratory tract infection by *S. pneumoniae* or other microorganisms)	Not known	Endogenous
Virulence Factors	Capsule, hemolysin	Adherence, biofilm formation, enzymes	–
Culture/Diagnosis	Usually relies on clinical symptoms and failure to resolve within 72 hours	MALDI-TOF or PCR; CDC will identify if requested	Usually relies on clinical symptoms and failure to resolve within 72 hours
Prevention	Pneumococcal conjugate vaccine	–	None
Treatment	Wait for resolution; if needed, amoxicillin (high rates of resistance) or amoxicillin + clavulanate or cefuroxime; in **Serious Threat** category in CDC Antibiotic Resistance Report	Consult with CDC; in **Urgent Threat** category in CDC Antibiotic Resistance Report	Wait for resolution; if needed, a broad-spectrum antibiotic (azithromycin) may be used in absence of etiologic diagnosis
Distinctive Features	–		–
Epidemiological Features	United States: 70% of children experience at least one case before age 2; developing world: chronic otitis media results in significant hearing loss in 100s of millions and death in approx. 30,000 per year (in absence of treatment)		

old, broad-spectrum antibiotics should be considered. Children who experience frequent recurrences of ear infections sometimes have small tubes placed through the tympanic membranes into their middle ears to provide a channel for the fluids to drain out of the ear. **Disease Table 19.4**.

19.3 LEARNING OUTCOMES—Assess Your Progress

4. List the possible causative agents, modes of transmission, virulence factors, diagnostic techniques, and prevention/treatment for the "Highlight Disease" pharyngitis.

5. Identify the most dangerous causes of pharyngitis and discuss possible sequelae of the infection.

6. Discuss important features of the other infectious diseases of the upper respiratory tract. These are the common cold, sinusitis, and acute otitis media.

19.4 Lower Respiratory Tract Diseases Caused by Microorganisms

HIGHLIGHT DISEASE

Pneumonia

Pneumonia is a classic example of a disease characterized by an *anatomical diagnosis*. It is defined as an inflammatory condition of the lung in which fluid fills the alveoli. The set of symptoms that we call pneumonia can be caused by a wide variety of different microorganisms. In a sense, the microorganisms need only to have appropriate characteristics to allow them to circumvent the host's defenses and to penetrate and survive in the lower respiratory tract. In particular, the microorganisms must avoid being phagocytosed by alveolar macrophages, or at least avoid being killed once inside the macrophage. Bacteria, fungi, and a wide variety of viruses can cause pneumonias, and there is a lot of variation in the virulence of different pathogenic agents. Pneumonia can be deadly, and across the globe, more children under the age of 5 die from pneumonia than any other infectious disease.

Even though all pneumonias have similar kinds of symptoms, physicians often distinguish between two forms of pneumonia, characterized by different modes of transmission and pathogenic agents. *Community-acquired pneumonia (CAP)* is experienced by persons in the general population. *Healthcare-associated pneumonia (HCAP)* develops in individuals receiving treatment at health care facilities, including hospitals.

▶ Causative Agents of Community-Acquired Pneumonia

SARS-CoV-2 was first recognized as causing a deadly pneumonia. Although it causes damage to many parts of the body, we consider it here. In non-pandemic times, *Streptococcus pneumoniae* accounts for up to 40% of community-acquired bacterial pneumonia cases. It causes more lethal pneumonia cases than any other microorganism. *Legionella* is a less common but also serious cause of the disease. A number of bacteria cause a milder form of pneumonia that is often referred to as "walking pneumonia." Two of these are *Mycoplasma pneumoniae* and *Chlamydophila pneumoniae* (formerly known as *Chlamydia pneumoniae*). *Histoplasma capsulatum* is a fungus that infects many people but causes a pneumonia-like disease in relatively few. The hantavirus, which emerged in 1993, causes a very serious pneumonia. (Be aware that *Bacillus anthracis*, covered in chapters 18 and 20, can also cause a pneumonic form of anthrax.) Also, recall that many, if not most, deaths stemming from influenza are caused by pneumonia, either from the original virus or from secondary infection.

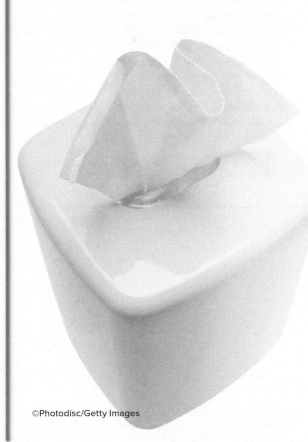
©Photodisc/Getty Images

Note for Covid

In 2002, a virus in the Coronavirus family, previously known to cause only coldlike symptoms, burst onto the world stage as it started to cause pneumonias and death in Hong Kong. That SARS epidemic ended nearly as quickly as it started, thanks to a massive global public health effort, and the fact that this SARS virus was not transmissible until the host developed symptoms. Since 2004, new cases of SARS have not been detected anywhere on the planet. In 2013, another new coronavirus emerged from Saudi Arabia. It was named MERS (Middle East Respiratory Syndrome). It had a 30% fatality rate. Thankfully, it was not highly transmissible human-to-human. In 2019 another novel coronavirus emerged from Wuhan, China. That was SARS-CoV-2, resulting in the COVID-19 pandemic.

COVID-19

COVID-19 information in this chapter is current as of July 2020. You are in the position of knowing how the progress of the pandemic continued after this. Since it was a brand new microbe for humans, everything about it was unknown when it hit in late 2019. The methods of epidemiology and public health are critical in a time like this and new facts coming from new data emerge every day. No doubt you have seen "facts" declared one day, only to be reversed in later days. This is the nature of emerging infections, and indeed, of science. So you have a front-row seat to the process of science, for better or worse. In the case of COVID-19, there is a lot of "worse," since it is causing so much disease and suffering.

SARS-CoV-2   On December 31, 2019, a cluster of pneumonia cases of unknown etiology was reported in Wuhan, China. By January 9, 2020, the Chinese CDC identified the cause as a novel coronavirus, which was later officially named SARS-CoV-2. The disease it causes is called COVID-19. As of early September 2020, the virus has caused over 26 million cases worldwide, and killed over 870,000. In the United States, as of Sept 4, 2020, 6.2 million confirmed cases and 187,618 deaths have occurred. Both of these numbers were still increasing.

▶ Signs and Symptoms

COVID-19 may cause no symptoms at all, or it may lead to death quickly. Although it was first recognized for its pneumonia manifestations, it can damage blood vessels all over the body, leading to damage to the heart, kidneys, brain, and other organs. Blood clots are a serious complicating factor. Those with the worst respiratory symptoms require mechanical ventilation. Patients who have been hospitalized as well as those with milder symptoms report post-COVID symptoms that reflect the systemic nature of this virus, lasting for weeks to months. Children and young adults also are at risk for a condition called MIS-C (multi-system inflammatory syndrome in children), an aggressive inflammatory disease that follows COVID-19 in some young patients.

▶ Causative Agent

SARS-CoV-2 is a coronavirus. Generally speaking, coronaviruses are mild. Four different varieties of coronaviruses circulate constantly among humans, causing the common cold. This one probably originated in bats, and may also have circulated in a mammal, such as a pangolin. (Animals are commonly infected by their own varieties of coronaviruses.) It is an RNA virus with spikes (glycoproteins) which allow it to attach to host cells. These spikes are visible under the electron microscope and give it its name *corona*, for crown.

▶ Pathogenesis and Virulence Factors

The virus, like many viruses when they initially cross into humans, is highly pathogenic. Although it was first identified as a pneumonia-causing virus, it also has systemic effects. It triggers widespread activation of bradykinins, which cause damage to lungs and other tissues. Early studies suggest that the virus has proteins that block the production of interferon, crippling an important part of the innate immune response.

▶ Transmission and Epidemiology

SARS-CoV-2 is transmitted through droplet and airborne contact. Other routes of transmission are not yet well established. One factor making the pandemic difficult to control is that it is transmissible even in the absence of symptoms, so "well" people circulating between public spaces and home are reservoirs and carriers.

▶ Culture and Diagnosis

Scientists acted quickly and learned to grow the virus in cell culture. But testing for the virus was not terribly successful in the beginning. Original tests used RT-PCR of nasopharyngeal swab material but there was a high rate of false-negatives. Blood tests were developed to detect IgG to the virus, but these were also plagued with low accuracy.

▶ Prevention and Treatment

No specific treatment was available as of mid-2020, though a drug called remdesivir shows some success. Vaccine development started immediately but will take time to test and deploy. The best preventions in lieu of a vaccine are physical distancing and the use of face masks. Many governments worldwide imposed quarantines of people who tested positive and also urged or mandated shutdowns of businesses, physical distancing, and the use of face-masks **(figure 19.6)**. Nothing like this had been seen in the United States since the 1918 influenza pandemic.

Figure 19.6 Sign in a pharmacy during the COVID-19 pandemic.
©Martin berry/Alamy Stock Photo

Streptococcus pneumoniae   This bacterium, which is often simply called the pneumococcus, is a small gram-positive flattened coccus that often appears in pairs, lined up end to end. It is alpha-hemolytic on blood agar. Factors that favor the ability of the pneumococcus to cause disease are old age, the season (rate of infection is highest in the winter), underlying viral respiratory disease, diabetes, and chronic abuse of alcohol or narcotics. Healthy people commonly inhale this and other microorganisms into the respiratory tract without serious consequences because of the host defenses present there.

Because the pneumococcus is such a frequent cause of pneumonia in older adults, this population is encouraged to seek immunization with the older pneumococcal polysaccharide vaccine PPSV23 or with PCV13.

Legionella pneumophila   *Legionella* is a weakly gram-negative bacterium that has a range of shapes, from coccus to filaments. Several species or subtypes have been characterized, but *L. pneumophila* ("lung-loving") is the one most frequently isolated from infections.

Legionella's ability to survive and persist in natural habitats has been something of a mystery, yet it appears to be widely distributed in aqueous habitats as diverse as tap water, cooling towers, spas, ponds, and other freshwaters. It is resistant to chlorine. The bacterium can live in close association with free-living amoebas. It is released during aerosol formation and can be carried for long distances. Cases have been traced to supermarket vegetable sprayers, hotel fountains, air-conditioning vents, and even misting towers at Disney parks.

Atypical Pneumonias   Pneumonias caused by *Mycoplasma* (as well as those caused by *Chlamydophila* and some other microorganisms) are often called atypical pneumonia—atypical in the sense that the symptoms do not resemble those of pneumococcal or other severe pneumonias. *Mycoplasma* pneumonia is transmitted by aerosol droplets among people confined in close living quarters, especially families, students, and the military. Lack of acute illness in most patients has given rise to the name "walking pneumonia."

Hantavirus   In 1993, hantavirus suddenly burst into the American consciousness. A cluster of unusual cases of severe lung edema among healthy young adults arose in the Four Corners area of New Mexico. Most of the patients died within a few days. They were later found to have been infected with hantavirus, an agent that had previously only been known to cause severe kidney disease and hemorrhagic fevers in other parts of the world. The new condition was named hantavirus pulmonary syndrome (HPS). Since 1993, the disease has occurred sporadically, but it has a mortality rate of at least 33%.

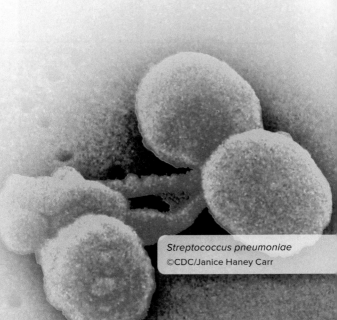

Streptococcus pneumoniae
©CDC/Janice Haney Carr

Figure 19.7 **Signature tents in Yosemite National Park.** These tent types harbored hantavirus in the 2012 Yosemite outbreak.
©Jebb Harris/ZUMA Press/Newscom

Very soon after the initial cases in 1993, it became clear that the virus was associated with the presence of mice in close proximity to the victims. Investigators eventually determined that the virus, an enveloped virus of the Bunyaviridae family, is transmitted via airborne dust contaminated with the urine, feces, or saliva of infected rodents. Deer mice and other rodents can carry the virus with few apparent symptoms. Small outbreaks of the disease are usually correlated with increases in the local rodent population. In 2012, hantavirus caused a localized outbreak among visitors to Yosemite National Park. Visitors staying in "signature tent cabins" were exposed to mouse droppings containing the virus **(figure 19.7)**. Three people died, and at least 10 were sickened. The signature tent cabins were of a new double-walled design that was meant to be safer, but the space between the walls turned out to be a perfect nesting space for the mice.

Histoplasma capsulatum This organism is widely distributed on all continents except Australia. Its highest rates of incidence occur in the eastern and central regions of the United States, especially in the Ohio Valley. This fungus appears to grow most abundantly in moist soils high in nitrogen content, especially those supplemented by bird and bat droppings **(figure 19.8)**.

A useful tool for determining the distribution of *H. capsulatum* is to inject a fungal extract into the skin and monitor for allergic reactions (much like the TB skin test). Application of this test has verified the extremely widespread distribution of the fungus. In high-prevalence areas such as southern Ohio, Illinois, Missouri, Kentucky, Tennessee, Michigan, Georgia, and Arkansas, 80% to 90% of the population show signs of prior infection.

Pneumocystis (carinii) jirovecii Although the fungus *Pneumocystis jirovecii* (formerly called *P. carinii*) was discovered in 1909, it remained relatively obscure until it was suddenly propelled into clinical prominence as the agent of *Pneumocystis* pneumonia. PCP is one of the most frequent opportunistic infections in AIDS patients, most of whom will develop one or more episodes during their lifetimes. Cancer patients and others with extreme immunosuppression are also at risk for this disease. There is some debate about how *Pneumocystis* is acquired, but inhalation of spores is probably common, and healthy people may even harbor it as "normal biota" in their lungs.

Figure 19.8 **Sign in wooded area in Kentucky.**
The sign is covered in bird droppings. Up to 90% of the population in the Ohio Valley show evidence of past infection with *Histoplasma*.
©Tom Volk, TomVolkFungi.net

Disease Table 19.5	Community-Acquired Pneumonia							
Causative Organism(s)	SARS-CoV-2 **V**	*Streptococcus pneumoniae* **B** **G⁺**	*Legionella* species **B** **G⁻**	*Mycoplasma pneumoniae* **B**	Hantavirus **V**	*Histoplasma capsulatum* **F**	*Pneumocystis jirovecii* **F**	Other respiratory viruses **V**
Most Common Modes of Transmission	Droplet contact, airborne transmission, possibly contact with contaminated fomites	Droplet contact or endogenous transfer	Vehicle (water droplets)	Droplet contact	Vehicle–airborne virus emitted from rodents	Vehicle–inhalation of contaminated soil	Droplet contact	Droplet contact or endogenous transfer
Virulence Factors	Asymptomatic transmission, shutting down interferon production, production of bradykinins	Capsule	–	Adhesins	Ability to induce inflammatory response	Survival in phagocytes	–	–
Culture/ Diagnosis	Diagnostic test is RT-PCR; other tests in development	Gram stain often diagnostic, alpha-hemolytic on blood agar; nucleic acid amplification test	Requires selective charcoal yeast extract agar; PCR available	PCR; rule out other etiologic agents	Serology (IgM), PCR identification of antigen in tissue	Rapid antigen tests, microscopy	PCR	Failure to find bacteria or fungi
Prevention	Avoidance	Pneumococcal polysaccharide vaccine (PPSV23) or the conjugate vaccine PCV13	–	No vaccine, no permanent immunity	Avoid mouse habitats and droppings	Avoid contaminated soil/bat, bird droppings	Antifungals given to AIDS patients to prevent this	Hygiene
Treatment	Remdesivir	Cefotaxime, ceftriaxone, with or without vancomycin; in **Serious Threat** category in CDC Antibiotic Resistance Report	Levofloxacin, azithromycin, moxifloxacin	Doxycycline	Supportive	Amphotericin B; itraconazole	Trimethoprim-sulfamethoxazole	None
Distinctive Features		Patient usually severely ill	Mild pneumonias in healthy people; can be severe in elderly or immuno-compromised	Usually mild; "walking pneumonia"	Rapid onset; high mortality rate	Many infections asymptomatic	Majority occur in AIDS patients	Usually mild
Epidemiological Features	Worldwide, sudden emergence in late 2019	5% of CAP cases; in 2009 influenza epidemic, 29% of fatalities were co-infected with this bacterium	United States: on the rise; 6,000 to 8,000 cases annually; internationally: 2 million cases per year	20%-40% of CAP cases	United States: 10-40 cases per year; fatality rate 25%-75%	In United States, 250,000 infected per year; only 5%-10% have symptoms	Almost exclusively in severely immuno-compromised patients.	30% of CAP cases

Traditional antifungal drugs are ineffective against *Pneumocystis* pneumonia because the chemical makeup of the organism's cell wall differs from that of most fungi.

Respiratory Viruses Viruses are very common causes of community-acquired pneumonia. They are viruses that are either residents in the upper respiratory tract or acquired through our daily activities. Viral pneumonias are generally mild. Recently a very common cause of respiratory infections, including pneumonia, was identified. Human metapneumovirus can now be tested for via RT-PCR or immunoassay. **Disease Table 19.5.**

©Travel Ink/Getty Images

Disease Table 19.6	Healthcare-Associated Pneumonia
Causative Organism(s)	Gram-negative and gram-positive bacteria from upper respiratory tract or stomach; environmental contamination of ventilator
Most Common Modes of Transmission	Endogenous (aspiration)
Virulence Factors	–
Culture/Diagnosis	Culture of lung fluids
Prevention	Elevating patient's head, preoperative education, care of respiratory equipment
Treatment	Broad-spectrum antibiotics
Epidemiological Features	United States: 300,000 cases per year; occurs in 0.5%-1.0% of admitted patients; mortality rate in United States and internationally is 20%-50%

NCLEX® PREP

4. A nurse is developing a health promotion class on identification and prevention of tuberculosis (TB). Which clients would the nurse include in TB testing? Select all that apply.
   a. *client who is HIV-positive*
   b. *male, age 38, who works as a prison guard*
   c. *nursing student with no comorbid conditions*
   d. *migrant farm worker with limited health access who presents to the clinic for information on diabetes*
   e. *female, age 26, who recently traveled abroad visiting endemic areas*

Healthcare-Associated Pneumonia

Up to 1% of hospitalized or institutionalized people experience the complication of pneumonia. Together with surgical site infections, it is the most common healthcare-associated infection. It is most commonly associated with mechanical ventilation, via an endotracheal or tracheostomy tube. This is sometimes labeled "ventilator associated pneumonia," or VAP. The mortality rate is quite high–between 30% and 50%. The most frequent cause of all forms of healthcare-associated pneumonia today are MRSA strains of *Staphylococcus aureus*, as well as nonresistant strains. MRSA is on the CDC's list as a Serious Threat. After MRSA, gram-negative bacteria are most common. These include *Klebsiella pneumoniae, Enterobacter, E. coli, Pseudomonas aeruginosa*, and *Acinetobacter*. Some *Enterobacter* strains are highly antibiotic-resistant. When they are resistant to a last-line antibiotic (carbapenem), they are designated CRE and are in the Urgent Threat category from the CDC. Likewise, *Acinetobacter* is likely to be multidrug-resistant. In those cases it is in the Serious Threat category. Further complicating matters, many healthcare-associated pneumonias appear to be polymicrobial in origin–meaning that there are multiple microorganisms multiplying in the alveolar spaces.

Culture of sputum or of tracheal swabs is not very useful in diagnosing healthcare-associated pneumonia because the condition is usually caused by normal biota. Cultures of fluids obtained through endotracheal tubes or from bronchoalveolar lavage provide better information but are fairly intrusive. It is also important to remember that if the patient has already received antibiotics, culture results will be affected.

Because most healthcare-associated pneumonias are caused by microorganisms aspirated from the upper respiratory tract, measures that discourage the transfer of microbes into the lungs are very useful for preventing the condition. Elevating patients' heads to a 30- to 45-degree angle helps reduce aspiration of secretions. Good preoperative education of patients about the importance of deep breathing and frequent coughing can reduce postoperative infection rates. Proper care of mechanical ventilation and respiratory therapy equipment is essential as well.

Studies have shown that delaying antibiotic treatment of suspected healthcare-associated pneumonia leads to a greater likelihood of death. Even in this era of conservative antibiotic use, empiric therapy should be started as soon as healthcare-associated pneumonia is suspected, using multiple antibiotics that cover both gram-negative and gram-positive organisms. **Disease Table 19.6.**

Influenza

The "flu" is a very important disease to study for several reasons. First of all, everyone is familiar with the cyclical increase of influenza infections occurring during the winter months in the United States. Second, many conditions are erroneously termed the "flu," while in fact only diseases caused by influenza viruses are actually the flu. Third, the way that influenza viruses evolve provides an excellent illustration of the way other viruses can, and do, change to cause more serious diseases than they did previously.

Influenzas that occur every year are called "seasonal" flus. Often these are the only flus that circulate each year. Occasionally another flu strain appears, one that is new and may cause worldwide pandemics. In some years, such as in 2009, both of these flus were issues. They may have different symptoms, affect different age groups, and have separate vaccine protocols.

▶ Signs and Symptoms

Influenza begins in the upper respiratory tract but in serious cases may also affect the lower respiratory tract. There is a 1- to 4-day incubation period, after which symptoms begin very quickly. These include headache, chills, dry cough, body aches, fever, stuffy nose, and sore throat. Even the sum of all these symptoms can't describe how a person actually feels: lousy. The flu is known to "knock you off your feet." Extreme fatigue can last for a few days or even a few weeks. An infection with influenza can leave patients vulnerable to secondary infections, often bacterial. Influenza infection occasionally leads to a pneumonia that can cause rapid death, even in young healthy adults.

▶ Causative Agent

All influenza is caused by one of three influenza viruses: A, B, or C. They belong to the family Orthomyxoviridae. They are spherical particles with an average diameter of 80 to 120 nanometers. Each virus has a lipoprotein envelope that is studded with glycoprotein spikes acquired during viral maturation **(figure 19.9)**. Also note that the envelope contains proteins that form a channel for ions into the virus. The two glycoproteins that make up the spikes of the envelope and contribute to virulence are called hemagglutinin (H) and neuraminidase (N). The name hemagglutinin is derived from this glycoprotein's agglutinating action on red blood cells, which is the basis for viral assays used to identify the viruses. Hemagglutinin contributes to infectivity by binding to host cell receptors of the respiratory mucosa, a process that facilitates viral penetration. Neuraminidase breaks down the protective mucous coating of the respiratory tract, assists in viral budding and release, keeps viruses from sticking together, and participates in host cell fusion.

The ssRNA genome of the influenza virus is known for its extreme variability. It is subject to constant genetic changes that alter the structure of its envelope glycoproteins. This constant mutation of the glycoproteins is called **antigenic drift**–the antigens gradually change their amino acid composition, resulting in decreased ability of host memory cells to recognize them. Antigenic drift is the reason that a new vaccine is required for each year. This vaccine is called the seasonal vaccine.

An even more serious phenomenon is known as **antigenic shift.** The genome of the virus consists of just 10 genes, encoded on 8 separate RNA strands. Antigenic shift is the swapping out of one of those genes or strands with a gene or strand from a different influenza virus. For example, we know that certain influenza viruses infect both humans and swine. Other influenza viruses infect birds and swine. All of these viruses have 10 genes coding for the same important influenza proteins (including H and N)–but the actual sequence of the genes is different in the different types of viruses. Second, when the two viruses just described infect a single swine host, with both virus types infecting the same host cell, the viral packaging step can accidentally produce a human influenza virus that contains seven human influenza virus RNA strands plus a single duck influenza virus RNA strand **(figure 19.10).** When that virus infects a human, no immunologic recognition of the protein that came from the duck virus occurs. Influenza A viruses are named according to the different types of H and N spikes they display on their surfaces. In 2009, a flu called H1N1 caused a limited pandemic. It reappeared in 2014. In 2013, the main circulating virus was H3N2. This strain–along with H5N1 and H7N9–is called bird flu because it originates in poultry.

Some people express frustration that public health officials seem to raise the alarm about new flu threats that don't pan out. Now that you have learned how easily a new flu strain can become lethal to a new species (i.e., humans), and knowing that when that happened in 1918 the flu killed 20-50 million people, you

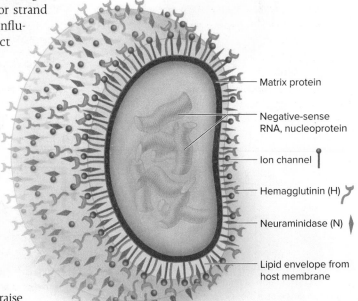

Matrix protein

Negative-sense RNA, nucleoprotein

Ion channel

Hemagglutinin (H)

Neuraminidase (N)

Lipid envelope from host membrane

Figure 19.9 Schematic drawing of influenza virus.

can appreciate the science—and the necessary precaution—behind these warnings. The warnings are usually accompanied by extra measures of preparation, such as additional vaccine, antivirals, and diagnostic kits being made available. "Better safe than sorry" is the thinking. Influenza B viruses are not divided into subtypes because they are thought to undergo only antigenic drift and not antigenic shift. Influenza C viruses are thought to cause only minor respiratory disease and are probably not involved in epidemics.

▶ Pathogenesis and Virulence Factors

The influenza virus binds primarily to ciliated cells of the respiratory mucosa. Infection causes the rapid shedding of these cells along with a load of viruses. Stripping the respiratory epithelium to the basal layer eliminates protective ciliary clearance. Sometimes the immune system responds too aggressively in a phenomenon called a "cytokine storm," and this leads to more severe irritation and inflammation in the lungs. The illness is further aggravated by fever, headache, and the other symptoms just described. The viruses tend to remain in the respiratory tract rather than spread to the bloodstream. As the normal ciliated epithelium is restored in a week or two, the symptoms subside.

▶ Transmission and Epidemiology

Inhalation of virus-laden aerosols and droplets is the major route of influenza infection, although fomites can play a secondary role. Transmission is greatly facilitated by crowding and poor ventilation in classrooms, barracks, nursing homes,

Influenza viruses (orange) budding from a host cell (83,300×)
©Science Photo Library/Getty Images

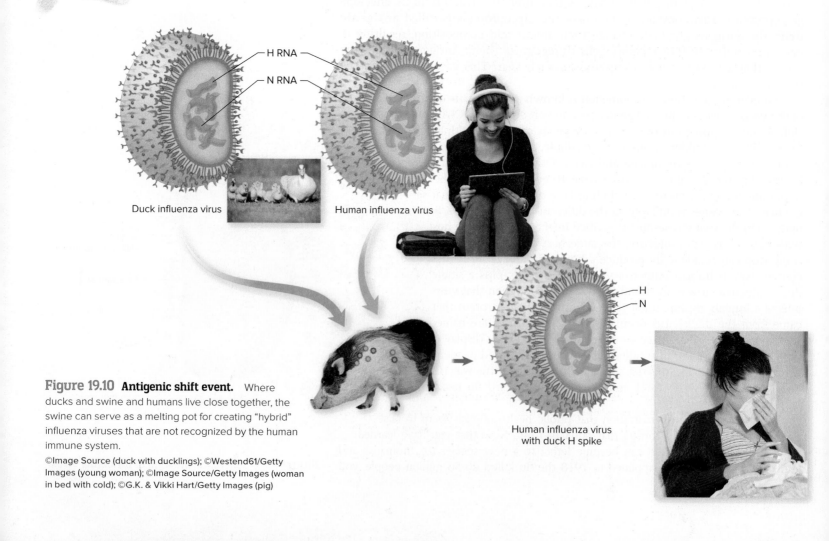

Figure 19.10 Antigenic shift event. Where ducks and swine and humans live close together, the swine can serve as a melting pot for creating "hybrid" influenza viruses that are not recognized by the human immune system.
©Image Source (duck with ducklings); ©Westend61/Getty Images (young woman); ©Image Source/Getty Images (woman in bed with cold); ©G.K. & Vikki Hart/Getty Images (pig)

dormitories, and military installations in the late fall and winter. The drier air of winter facilitates the spread of the virus, as the moist particles expelled by sneezes and coughs become dry very quickly, helping the virus remain airborne for longer periods of time. In addition, the dry cold air makes respiratory tract mucous membranes more brittle, with microscopic cracks that facilitate invasion by viruses.

Influenza is highly contagious and affects people of all ages. Annually, deaths from seasonal influenza and its complications (usually pneumonia) occur mainly among the very young and the very old. Numbers in the United States range from 12,000 to 60,000, depending on the year. The number of deaths each year depends on several factors, including the relative virulence of the virus circulating that season, the percentage of the population that received vaccination, and how well the vaccine strains matched the actual circulating strains **(figure 19.11).**

The most recent year in which a pandemic strain circulated was during the 2009-2010 flu season. In that season, a new H1N1 virus appeared in the United States in April. It was called a quadruple reassortment virus, meaning that its genes came from four different influenza virus sources (including swine flu, avian flu, and human flu). The World Health Organization declared the emergence of this virus a "Public Health Emergency of International Concern." In hindsight, the pandemic was kept in check, but it could have gone differently.

▶ Culture and Diagnosis

There are a wide variety of culture-based and non-culture-based methods to diagnose the infection. RT-PCR is by far the preferred method for diagnosis. In 2009, officials did not often test for H1N1 but tested for influenza A or B virus, assuming if it was A then

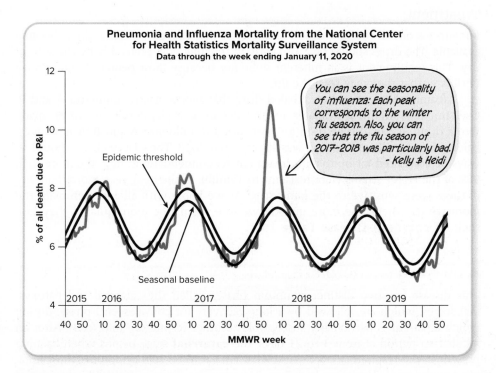

Figure 19.11 Epidemic curve for deaths from influenza/pneumonia.   The horizontal axis is marked in weeks of the year. Clearly influenza peaks in mid-winter. The bottom black line in the graph indicates the seasonal baseline number of cases calculated from years of data. The top black line is the threshold beyond which influenza is declared an epidemic. We are using data from before COVID-19, as pneumonia caused by SARS-CoV-2 has been difficult to differentiate from influenza deaths.
Source: CDC

it was H1N1 because the circulating seasonal virus was influenza B. When specimens were tested, 100% of the influenza A isolates were in fact the H1N1.

▶ Prevention

Preventing influenza infections and epidemics is one of the top priorities for public health officials. There are currently several types of vaccines. Some are produced the traditional way–in chicken eggs. Some are made in animal cell culture, and some are produced through recombinant methods.

Vaccines can be administered intramuscularly (injected) or intradermally. There is a nasal spray vaccine as well. Some of the vaccines consist of inactivated (killed) viruses; others contain live attenuated virus. Those over the age of 65 are urged to get the high-dose vaccine to improve their immunity to the virus.

Each year the CDC chooses three or four strains to put in the vaccine, depending on what type of influenza is circulating in the population. This is called the seasonal vaccine. You should get a new seasonal vaccine each year, due to the phenomenon of antigenic drift. If a pandemic strain suddenly appears on the world stage, public health officials will create an additional vaccine, targeted to the pandemic strain. Pandemic strains emerge from antigenic shift. The recommendations for type of vaccine change every year. The CDC website should be consulted for the latest recommendations.

One of the most promising new vaccine prospects is a vaccine that would protect against *all* flu viruses and not need to be given every year. This vaccine, still experimental, targets epitopes that most flu viruses share in common. This discovery has the possibility of revolutionizing influenza prevention.

Treatment

Influenza is one of the first viral diseases for which effective antiviral drugs became available. The drugs should be taken early in the infection, preferably by the second day. This requirement is an inherent difficulty because most people do not realize until later that they may have the flu.

Zanamivir (Relenza) is an inhaled drug that works against influenza A and B. Oseltamivir (Tamiflu) is available in capsules or as a powdered mix to be made into a drink. It can also be used for prevention of influenza A and B. A new drug, baloxavir (trade name Xofluza), was approved in 2018. The CDC website should always be consulted for up-to-date treatment recommendations. In 2008, more than 98% of the H1N1 strains were resistant to Tamiflu. In the next year, more than 98% of them were sensitive to the same drug. As we know with all antimicrobials, the more we use them, the more quickly we lose them (the more quickly they lose their effectiveness). **Disease Table 19.7.**

Whooping Cough (Pertussis)

This disease has two distinct symptom phases called the catarrhal and paroxysmal stages, which are followed by a long recovery (or convalescent) phase, during which a patient is particularly susceptible to other respiratory infections. After an incubation period of from 3 to 21 days, the **catarrhal** stage begins when bacteria present in the respiratory tract cause what appear to be cold symptoms, most notably a runny nose. This stage lasts 1 to 2 weeks. The disease worsens in the second **(paroxysmal)** stage, which is characterized by severe and uncontrollable coughing (a *paroxysm* can be thought of as a convulsive attack). The common name for the disease comes from the whooping sound a patient makes as he or she tries to grab a breath between uncontrollable bouts of coughing. As in any disease, the convalescent phase is the time when numbers of bacteria are decreasing and no longer cause ongoing symptoms.

Disease Table 19.7	Influenza
Causative Organism(s)	Influenza A, B, and C viruses Ⓥ
Most Common Modes of Transmission	Droplet contact, direct contact, or indirect contact
Virulence Factors	Glycoprotein spikes, overall ability to change genetically, ability to slow down immune system
Culture/Diagnosis	Gold standard is RT-PCR
Prevention	Annual vaccination with one of several types of vaccines
Treatment	Zofluza, Tamiflu, or Relenza
Epidemiological Features	For seasonal flu, deaths vary from year to year. United States: range from 12,000–60,000; internationally: range from 250,000–500,000

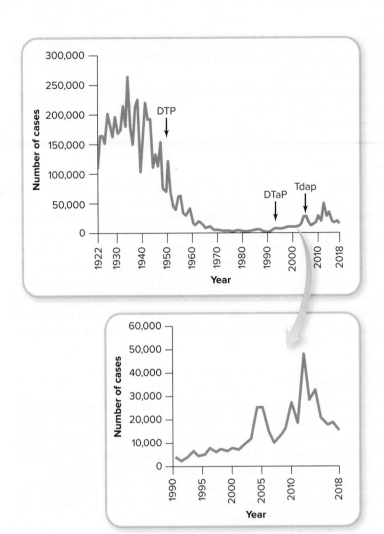

Figure 19.12 Number of cases of pertussis in the United States from 1922 to 2018. The inset graph magnifies the last 25 years.
Source: CDC

Disease Table 19.8	Pertussis (Whooping Cough)
Causative Organism(s)	*Bordetella pertussis* **B** G⁻
Most Common Modes of Transmission	Droplet contact
Virulence Factors	Fimbrial hemagglutinin (adhesion), pertussis toxin and tracheal cytotoxin, endotoxin
Culture/ Diagnosis	PCR or grown on B-G, charcoal, or potato-glycerol agar; diagnosis can be made on symptoms
Prevention	Acellular vaccine (DTaP), azithromycin, sulfamethoxazole for contacts
Treatment	Azithromycin or erythromycin; in **Watch List** category in CDC Antibiotic Resistance Report
Epidemiological Features	Resurging in United States and other countries; internationally: similar pattern of gradual increase

Pertussis was all but wiped out by the 1970s, after the introduction of the first vaccine in the 1950s **(figure 19.12)**. This progress was reversed in 2012-2013, but the rates are coming back down. All of the deaths from pertussis in 2013 were in babies younger than 6 months, who had not yet been fully vaccinated and caught the disease from adults and caregivers who were not effectively vaccinated. Researchers are working on a vaccine for babies, and also a vaccine that could be given to an expectant mother to protect her newborn from pertussis. **Disease Table 19.8.**

Respiratory Syncytial Virus Infection

As its name indicates, respiratory syncytial virus (RSV) infects the respiratory tract and produces giant multinucleated cells (syncytia). Outbreaks of droplet-spread RSV disease occur regularly throughout the world, with peak incidence in the winter and early spring. Children 6 months of age or younger, as well as premature babies, are especially susceptible to serious disease caused by this virus. RSV is the most prevalent cause of respiratory infection in the newborn age group, and nearly all children have experienced it by age 2. An estimated 100,000 children are hospitalized with RSV infection each year in the United States. Infection in older children and adults usually manifests as a cold. Older adults, particularly those with chronic diseases, are also susceptible to severe manifestations of RSV.

Disease Table 19.9	RSV Disease
Causative Organism(s)	Respiratory syncytial virus (RSV) Ⓥ
Most Common Modes of Transmission	Droplet and indirect contact
Virulence Factors	Syncytia formation
Culture/ Diagnosis	RT-PCR preferred
Prevention	Passive antibody (humanized monoclonal) in high-risk children
Treatment	Ribavirin in severe cases
Epidemiological Features	United States: general population, less than 1% mortality rates, 3% to 5% mortality in premature infants or those with congenital heart defects; internationally: 7 times higher fatality rate in children in developing countries

The virus is highly contagious and is transmitted through droplet contact but also through fomite contamination. The afflicted child is conspicuously ill, with signs typical of pneumonia and bronchitis.

There is no RSV vaccine available yet, but an effective passive antibody preparation, called Palivizumab, is used as prevention in high-risk children and babies born prematurely. **Disease Table 19.9.**

Tuberculosis

Mummies from the Stone Age, ancient Egypt, and Peru provide evidence that tuberculosis (TB) is an ancient human disease. In fact, historically it has been such a prevalent cause of death that it was called "Captain of the Men of Death" and "White Plague." After the discovery of streptomycin in 1943, the rates of tuberculosis in the developed world declined rapidly. But since the mid-1980s, it has reemerged as a serious threat. Worldwide, 1.7 billion people are currently infected. That is about 25% of the world's population! The cause of tuberculosis is primarily the bacterial species *Mycobacterium tuberculosis*, informally called the tubercle bacillus. In this discussion, we will first address the general aspects of the infection and then turn to its most troubling form today, drug-resistant tuberculosis.

▶ Signs and Symptoms

A clear-cut distinction can be made between infection with the TB bacterium and the disease it causes. In general, humans are rather easily infected with the bacterium but are resistant to the disease. Estimates project that only about 5% to 10% of infected people actually develop a clinical case of tuberculosis. The majority (85%) of TB cases are contained in the lungs, even though disseminated TB bacteria can result in tuberculosis in any organ of the body. Clinical tuberculosis is divided into primary tuberculosis, secondary (reactivation or reinfection) tuberculosis, and disseminated or extrapulmonary tuberculosis.

Primary Tuberculosis The minimum infectious dose for lung infection is low, around 10 bacterial cells. Alveolar macrophages phagocytose these cells, but they are not killed and continue to multiply inside the macrophages. This period of hidden infection is asymptomatic or is accompanied by mild fever. Some bacteria escape from the lungs into the blood and lymphatics. After 3 to 4 weeks, the immune system mounts a complex, cell-mediated assault against the bacteria. The large influx of mononuclear cells into the lungs plays a part in the formation of specific infection sites called **tubercles.** Tubercles are granulomas that consist of a central core containing TB bacteria in enlarged macrophages and an outer wall made of fibroblasts, lymphocytes, and macrophages **(figure 19.13).** Although this response further checks spread of infection and helps prevent the disease, it also carries a potential for damage. Frequently, as neutrophils come on the scene and release their enzymes, the centers of tubercles break down into necrotic **caseous** (kay′-see-us) **lesions** that gradually heal by *calcification*–normal lung tissue is replaced by calcium deposits. The response of T cells to *M. tuberculosis* proteins also causes a cell-mediated immune response evident in the skin test called the **tuberculin reaction,** a valuable diagnostic and epidemiological tool.

TB infection outside of the lungs is more common in immunosuppressed patients and young children. Organs most commonly involved in **extrapulmonary TB** are the regional lymph nodes, kidneys, long bones, genital tract, brain, and meninges. Because of the debilitation of the patient and the high load of TB bacteria, these complications are often fatal.

Secondary (Reactivation) Tuberculosis Although the majority of adequately treated TB patients recover more or less completely from the primary episode of infection, live bacteria can remain dormant and become reactivated weeks, months, or years later, especially in people with weakened immunity. In chronic tuberculosis, tubercles filled with masses of bacteria expand, cause cavities in the lungs, and drain into the bronchial tubes and upper respiratory tract. The patient gradually experiences more severe symptoms, including violent coughing, greenish or bloody sputum, low-grade fever, anorexia, weight loss, extreme fatigue, night sweats, and chest pain. It is the gradual wasting of the body that accounts for an older name for tuberculosis–*consumption.* Untreated secondary disease has nearly a 60% mortality rate.

▶ Causative Agents

M. tuberculosis, the cause of tuberculosis in most patients, is an acid-fast rod, long and thin. It is a strict aerobe, and technically speaking, there is still debate about whether it is a gram-positive or a gram-negative organism. It is rarely called gram anything, however, because its acid-fast nature is much more relevant in a clinical setting. It grows very slowly. With a generation time of 15 to 20 hours, a period of up to 6 weeks is required for colonies to appear in culture. (Note: The prefix *Myco-* might make you think of fungi, but this is a bacterium. The prefix in the name came from the mistaken impression that colonies growing on agar resembled fungal colonies. Be sure to differentiate this bacterium from *Mycoplasma–* they are not related.)

Robert Koch identified that *M. tuberculosis* often forms serpentine cords while growing, and he called the unknown substance causing this style of growth *cord factor.* Cord factor appears to be associated with virulent strains, and it is a lipid component of the mycobacterial cell wall. All mycobacterial species have walls that have a very high content of complex lipids, including mycolic acid and waxes. This chemical characteristic makes them relatively impermeable to stains and difficult to decolorize (acid-fast) once they are stained. The lipid wall of the bacterium also influences its virulence and makes it resistant to drying and disinfectants.

In recent decades, tuberculosis-like conditions caused by *Mycobacterium avium* and related mycobacterial species (sometimes referred to as the *M. avium* complex, or MAC) have been found in AIDS patients and other immunocompromised people. In this section, we consider only *M. tuberculosis.*

▶ Pathogenesis and Virulence Factors

The course of the infection–and all of its possible variations–was described under "Signs and Symptoms." Important characteristics of the bacterium that contribute to its virulence are its waxy surface (contributing to both its survival in the environment and its survival within macrophages) and its ability to stimulate a strong cell-mediated immune response that contributes to the pathology of the disease.

▶ Transmission and Epidemiology

Mycobacterium tuberculosis is transmitted almost exclusively by fine droplets of respiratory mucus suspended in the air. The TB bacterium is highly resistant and can survive for 8 months in fine aerosol particles. Although larger particles become trapped in mucus and are expelled, tinier ones can be inhaled into the bronchioles and alveoli. This effect is especially pronounced among people sharing small closed rooms with limited access to sunlight and fresh air.

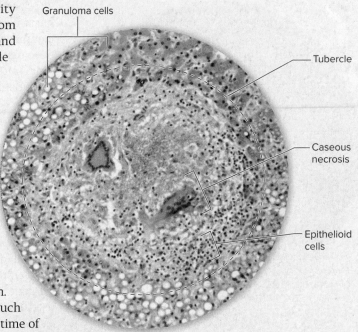

Granuloma cells

Tubercle

Caseous necrosis

Epithelioid cells

Figure 19.13   Tubercle formation.   Photomicrograph of a tubercle. The massive granuloma infiltrate has obliterated the alveoli and set up a dense collar of fibroblasts, lymphocytes (granuloma cells), and epithelioid cells. The core of this tubercle is a caseous (cheesy) material containing the bacilli.
©Steve Gschmeissner/Science Source

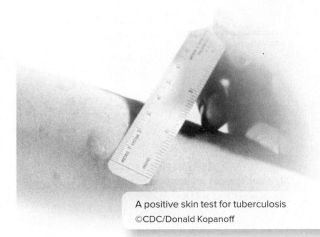

A positive skin test for tuberculosis
©CDC/Donald Kopanoff

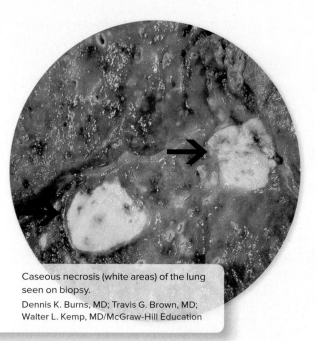

Caseous necrosis (white areas) of the lung seen on biopsy.

Dennis K. Burns, MD; Travis G. Brown, MD; Walter L. Kemp, MD/McGraw-Hill Education

Factors that significantly affect people's susceptibility to tuberculosis are inadequate nutrition, debilitation of the immune system, poor access to medical care, lung damage, and their own genetics. Put simply, TB is an infection of poverty. People in developing countries are often infected as infants and harbor the microbe for many years until the disease is reactivated in young adulthood.

TB infection rates have been falling in the United States since a high in the early 1990s. Some populations are at higher risk of infection or of developing life-threatening forms of the disease. For instance, about 60% of cases in the United States are among foreign-born persons. People who work or live in certain communities such as nursing homes, hospitals, or jails are also among those at greater risk. This is important to know as a health care provider so you can be alert for TB in certain populations.

▶ Culture and Diagnosis

You are probably familiar with several methods of detecting tuberculosis in humans. Clinical diagnosis of tuberculosis relies on three techniques: (1) tuberculin testing, (2) interferon-gamma release assays (IGRAs) performed on a blood sample, and (3) gene amplification and antimicrobial susceptibility testing. Acid-fast staining is used as a supplement to these techniques.

Tuberculin Sensitivity and Testing   Because infection with the TB bacterium can lead to delayed hypersensitivity to the proteins of the microbe, testing for hypersensitivity has been an important way to screen populations for tuberculosis infection and disease. Although there are newer methods available, the most widely used test is still the tuberculin skin test, called the **Mantoux test.** It involves local injection of purified protein derivative (PPD), a standardized solution taken from culture fluids of *M. tuberculosis.* The injection is done intradermally into the forearm to produce an immediate small bleb. After 48 and 72 hours, the site is observed for a red wheal called an **induration,** which is measured and interpreted as positive or negative according to size.

Blood Testing   Because the main immune response to *Mycobacterium* comes from T cells, the blood test for tuberculosis looks for T-cell activities in the form of cytokine release. The assay is called interferon gamma release assay (IGRA). It uses whole blood and can be completed within 24 hours.

Gene Amplification and Antibiotic Susceptibility Testing   Trade name Cepheid Xpert MTB/RIF accomplishes identification and susceptibility to rifampin simultaneously. It uses respiratory specimens.

Acid-Fast Staining   The diagnosis of tuberculosis in people with positive skin tests or X rays can be backed up by acid-fast staining of sputum or other specimens. Several variations on the acid-fast stain are currently in use. The Ziehl-Neelsen stain produces bright red acid-fast bacilli (AFB) against a blue background **(figure 19.14).** Fluorescence staining shows luminescent yellow-green bacteria against a dark background.

▶ Prevention

Preventing TB in the United States is accomplished by limiting exposure to infectious airborne particles. Extensive precautions, such as isolation in negative-pressure rooms, are used in health care settings when a person with active TB is identified. Vaccine is

M. tuberculosis

Figure 19.14 Ziehl-Neelsen staining of *Mycobacterium tuberculosis* **in sputum.**
©Dr. George P. Kubica/CDC

generally not used in the United States, although an attenuated vaccine called BCG is used in many countries. BCG stands for Bacille Calmette-Guerin, named for two French scientists who created the vaccine in the early 1900s. It is a live strain of a bovine tuberculosis bacterium that has been made avirulent by many passages through artificial media. Remember that persons vaccinated with BCG may respond positively to a tuberculin skin test.

In the past, prevention in the context of tuberculosis referred to preventing a person with latent TB from experiencing reactivation. This strategy is more accurately referred to as treatment of latent infection and is considered in the next section, "Treatment."

▶ Treatment

Treatment of latent TB infection is effective in preventing full-blown disease in persons who have positive skin tests and who are at risk for reactivated TB. Treatment of latent TB is with three drugs: isoniazid, rifampin, and rifapentine. The rifampin and rifapentine are taken for 4 and 3 months, respectively, and the isoniazid is continued for 9 months.

Treatment of active TB infection occurs in two phases. In the first, four drugs—rifampin, isoniazid, ethambutol, and pyrazinamide—are used for 2 months. The second phase uses only rifampin and isoniazid and lasts either 4 or 7 months, decided on a case-by-case basis.

One of the biggest problems with TB therapy is noncompliance on the part of the patient. It is very difficult, even under the best of circumstances, to keep to a regimen of multiple antibiotics daily for months—and most TB patients are living under conditions that are far from the best of circumstances. Failure to adhere to the antibiotic regimen leads to antibiotic resistance in the slow-growing microorganism; in fact, many *M. tuberculosis* isolates are now found to be **MDR-TB,** or multidrug-resistant TB.

Multidrug-Resistant Tuberculosis (MDR-TB)

Multidrug-resistant tuberculosis is defined as being resistant to at least isoniazid and rifampin. It requires treatment of 18 to 24 months with four to six drugs. It is particularly common in people who have been previously treated for tuberculosis. In 2017, approximately 500,000 new cases of MDR-TB were diagnosed. Of these, 45% were in India, China, and the Russian Federation.

People with MDR-TB are generally sicker and have higher mortality rates than those infected with non-MDR-TB. This is true even when they are being treated, as the multiple-drug combination has severe side effects. Even with treatment, about half of these patients die.

Extensively Drug-Resistant Tuberculosis (XDR-TB)

MDR-TB strains with resistance to two additional drugs are called **XDR-TB.** These strains have been reported in patients in 117 countries. Worldwide, it is estimated that 9% of the MDR-TB cases also qualify as XDR-TB. Patients with XDR-TB have few treatment options and their mortality rate is estimated to be about 70% within months of diagnosis. **Disease Table 19.10.**

Medical Moment

Tuberculosis—A Lesson in Inadequate Treatment

Tuberculosis is such a slow-growing bacterium that to be effective, antibiotic treatment has to continue uninterrupted for months, even in an uncomplicated case. Because of the demographics of TB infection (i.e., the countries hardest hit, the socioeconomic status of individuals who become infected), sustained treatment has always been very difficult to achieve. That is how the first MDR-TB strains arose. When a patient stops taking antibiotics prematurely, the bacteria that have not yet been killed are the ones that are least sensitive to the antibiotics. Those bacteria can then spread to other people, and if some of those people are also noncompliant with the rigorous treatment schedule, the bacterium can become even more resistant. Repeat this pattern hundreds of thousands of times and you can see why we have MDR-TB and XDR-TB.

Q. In addition to patient noncompliance with the lengthy antibiotic regimen, can you think of other factors that might have led to the development of resistant strains?

Answer in Appendix B.

Disease Table 19.10	Tuberculosis	
Causative Organism(s)	*Mycobacterium tuberculosis* Ⓑ	MDR-TB and XDR-TB Ⓑ
Most Common Modes of Transmission	Vehicle (airborne)	
Virulence Factors	Lipids in wall, ability to stimulate strong cell-mediated immunity (CMI)	
Culture/Diagnosis	Rapid methods; initial tests are skin testing, chest X ray	
Prevention	Avoiding airborne *M. tuberculosis*; BCG vaccine in other countries	
Treatment	Isoniazid, rifampin, and pyrazinamide + ethambutol or streptomycin for varying lengths of time (always lengthy)	Multiple-drug regimen, which may include bedaquiline; in **Serious Threat** category in CDC Antibiotic Resistance Report
Distinctive Features	Remains airborne for long periods; extremely slow-growing, which has implications for diagnosis and treatment.	Much higher fatality rate over shorter duration
Epidemiological Features	United States: approx. 10,000 cases/year, 16% of cases whites, 84% ethnic minorities; internationally: 1.5 million deaths in 2018	United States: fewer than 100/year; worldwide: approximately 500,000 new cases MDR-TB in 2017

19.4 LEARNING OUTCOMES—Assess Your Progress

7. List the possible causative agents, modes of transmission, virulence factors, diagnostic techniques, and prevention/treatment for the "Highlight Disease" community-acquired pneumonia.

8. Discuss important features of the other infectious diseases of the upper and lower respiratory tracts. These are healthcare-associated pneumonia, influenza, pertussis, RSV disease, and tuberculosis.

9. Compare and contrast *antigenic drift* and *antigenic shift* in influenza viruses.

10. Discuss the problems associated with MDR-TB and XDR-TB.

CASE FILE WRAP-UP

©Purestock/SuperStock (health care worker);
©MedicalRF.com (RSV)

Respiratory syncytial virus (RSV) is prevalent during the winter months and is a common cause of lower respiratory tract infection during infancy and childhood. Older children and adults generally experience mild disease, but infants ages 6 months and younger may develop bronchiolitis (inflammation of the smaller airways) or pneumonia as a consequence of the infection. Premature infants are particularly at risk and may develop life-threatening disease. There is no vaccine for RSV, but premature infants and/or infants born with cardiac or lung disease may be given a series of injections of the monoclonal antibody palivizumab (Synagis), which has proven to be moderately effective at preventing the disease.

RSV spreads easily by droplet contact. It can also survive for hours on inanimate objects and surfaces such as tables. Michael, the infant in the case file, may have contracted RSV at his day care, or he may have been exposed to the virus from his school-aged siblings. By 2 to 3 years of age, most children in the United States will have encountered RSV. Unfortunately, there is no vaccine for RSV, but researchers are getting closer to developing one.

▶ Summing Up

Taxonomic Organization Microorganisms Causing Disease in the Respiratory Tract

Microorganism	Pronunciation	Disease Table
Gram-positive bacteria		
Streptococcus pneumoniae	strep′-tuh-kok″-us nu-moh′-nee-ay	Otitis media, 19.4
		Community-acquired pneumonia, 19.5
Streptococcus pyogenes	strep′-tuh-kok″-us pie-ah′-gen-eez	Pharyngitis, 19.1
Gram-negative bacteria		
Candida auris	can′-di-duh or′-is	Otitis media, 19.4
Bordetella pertussis	bor′-duh-tell′-uh per-tuss′-is	Pertussis whooping cough, 19.8
Fusobacterium necrophorum	few′-soh-bak-teer″-ee-um nek′-row-for″-um	Pharyngitis, 19.1
Legionella spp.	lee″-juhn-el′-uh	Community-acquired pneumonia, 19.5
Pseudomonas aeruginosa	sue′-doe-moe″-ness air-roo′-gi-no″-sa	Healthcare-associated pneumonia, 19.6
Acinetobacter baumannii	a′-ci-nay′-toe-bak″-ter bow-mah″-nee-i	Healthcare-associated pneumonia, 19.6
Other bacteria		
Mycobacterium tuberculosis*	my″-co-back-teer′-ee-em tuh-ber′-cue-loh-sis	Tuberculosis, 19.10
Mycoplasma pneumoniae	my″-co-plazz′-muh nu-moh′-nee-ay	Community-acquired pneumonia, 19.5
RNA viruses		
SARS-CoV-2	sars-coh-vee-two	Community-acquired pneumonia, 19.5
Respiratory syncytial virus	ress″-pur-uh-tor′-ee sin-sish′-ull vie′-russ	RSV disease, 19.9
Influenza virus A, B, and C	in″-floo-en′-zuh vie′-russ	Influenza, 19.7
Hantavirus	haun″-tuh-vie′-russ	Community-acquired pneumonia, 19.5
Fungi		
Pneumocystis jirovecii	new-moh-siss′-tiss yee″-row-vet′-zee	Community-acquired pneumonia, 19.5
Histoplasma capsulatum	hiss″-toe-plazz′-muh cap″-sue-lah′-tum	Community-acquired pneumonia, 19.5

*There is some debate about the gram status of the genus *Mycobacterium*; it is generally not considered gram-positive or gram-negative.

Gut Bacteria Involved in Deadly Lung Disease?

Acute respiratory distress syndrome (ARDS) is a dangerous condition with sudden onset that affects more than 200,000 people a year in the United States. It has a case fatality rate of nearly 50%. Recently a group of researchers from the University of Michigan found a new clue about the pathological process of this disease.

They studied 68 patients with the condition and compared their lung microbiome to the lung microbiome of healthy volunteers. They found microbes in the ill patients' lungs that are generally only found in the GI tract. This led them to hypothesize that treatments for the severe condition were missing an important component. They characterize ARDS as a "vicious cycle," in which critically ill patients develop systemic inflammation, which can lead to tight junctions in the intestines becoming leaky, allowing gut bacteria into the bloodstream. In the bloodstream, they would gain access possibly to the lungs. This, of course, will increase inflammation and the whole system becomes amped up, resulting in injury to lung tissue and other inflammatory effects.

They point out that even before the importance of the microbiome was appreciated, critical care teams recognized that the gut microbiome influenced the outcome of a critical illness. In some European countries, patients in intensive care units—even for noninfectious reasons—are treated with antibiotics, a practice believed to decrease organ system failure. This technique is called selective decontamination of the digestive tract. The new findings from the Michigan team might provide a reason that the practice works.

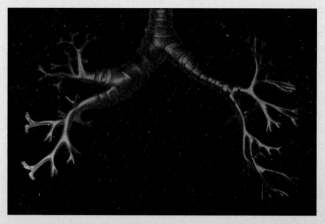

©MedicalRF.com

Infectious Diseases Affecting
The Respiratory System

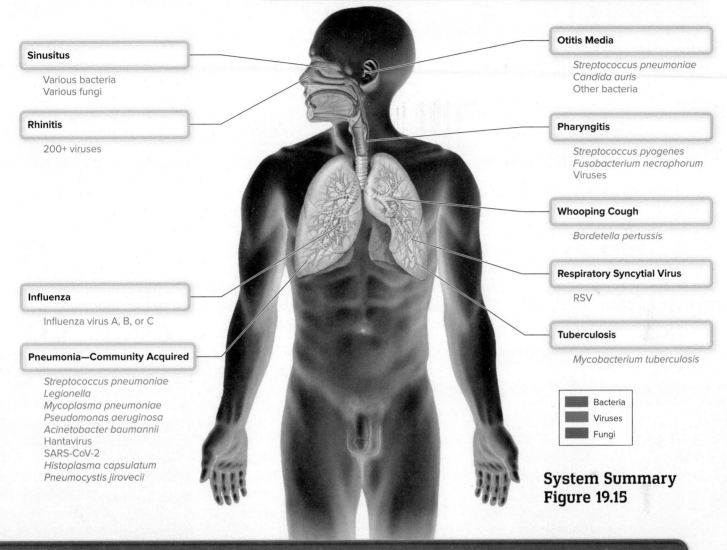

Sinusitus

Various bacteria
Various fungi

Rhinitis

200+ viruses

Influenza

Influenza virus A, B, or C

Pneumonia—Community Acquired

Streptococcus pneumoniae
Legionella
Mycoplasma pneumoniae
Pseudomonas aeruginosa
Acinetobacter baumannii
Hantavirus
SARS-CoV-2
Histoplasma capsulatum
Pneumocystis jirovecii

Otitis Media

Streptococcus pneumoniae
Candida auris
Other bacteria

Pharyngitis

Streptococcus pyogenes
Fusobacterium necrophorum
Viruses

Whooping Cough

Bordetella pertussis

Respiratory Syncytial Virus

RSV

Tuberculosis

Mycobacterium tuberculosis

Bacteria
Viruses
Fungi

**System Summary
Figure 19.15**

Deadliness and Communicability of Selected Diseases of the Respiratory System

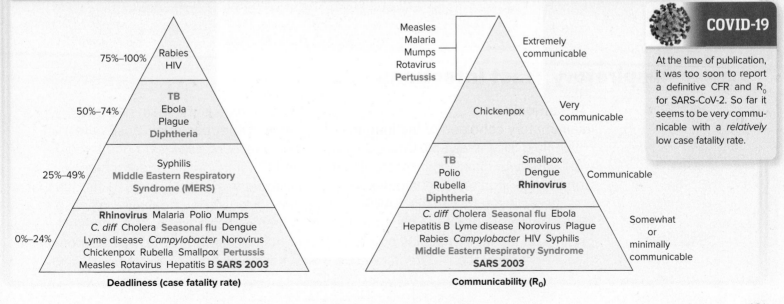

Deadliness (case fatality rate)

75%–100% — Rabies, HIV

50%–74% — TB, Ebola, Plague, **Diphtheria**

25%–49% — Syphilis, **Middle Eastern Respiratory Syndrome (MERS)**

0%–24% — **Rhinovirus** Malaria Polio Mumps *C. diff* Cholera **Seasonal flu** Dengue Lyme disease *Campylobacter* Norovirus Chickenpox Rubella Smallpox **Pertussis** Measles Rotavirus Hepatitis B **SARS 2003**

Communicability (R_0)

Measles, Malaria, Mumps, Rotavirus, **Pertussis** — Extremely communicable

Chickenpox — Very communicable

TB, Polio, Rubella, **Diphtheria** / Smallpox, Dengue, **Rhinovirus** — Communicable

C. diff Cholera **Seasonal flu** Ebola Hepatitis B Lyme disease Norovirus Plague Rabies *Campylobacter* HIV Syphilis **Middle Eastern Respiratory Syndrome SARS 2003** — Somewhat or minimally communicable

COVID-19

At the time of publication, it was too soon to report a definitive CFR and R_0 for SARS-CoV-2. So far it seems to be very communicable with a *relatively* low case fatality rate.

INFECTIOUS DISEASES AFFECTING THE RESPIRATORY SYSTEM

Anatomy & Defenses: Respiratory Tract

The respiratory tract is divided into upper and lower sections. It has many natural defenses: the ciliary escalator, mucus on the surface of mucous membrane, alveolar macrophages, and secretory IgA. Involuntary responses such as coughing and sneezing also serve a defensive function.

Normal Biota: Respiratory Tract

Normal biota include a wide variety of gram-positive and gram-negative bacteria, viruses, and fungi. Many of the colonizing microbes can cause disease in other people or other circumstances. The lower respiratory tract, previously believed to have no normal biota, has now been shown to be colonized, even in health.

Upper Respiratory Tract Infections

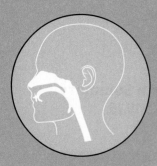

• **Pharyngitis**
Viruses cause the majority of throat infections. The bacteria *Streptococcus pyogenes* and *Fusobacterium necrophorum* cause more serious forms of the condition.

• **The Common Cold**
Caused by one of over 200 different kinds of virus, most commonly the rhinovirus, followed by coronaviruses. Respiratory syncytial virus (RSV) causes colds in the most people, but in young babies it can be very serious.

• **Sinusitis**
Inflammatory condition of the sinuses in the skull, most commonly caused by allergies or infections by a variety of viruses or bacteria and, less commonly, fungi.

• **Acute Otitis Media**
Most common cause is *Streptococcus pneumoniae*, though multiple organisms are usually present in infections.

Lower Respiratory Tract Infections

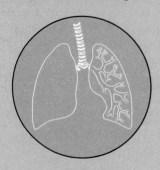

• **Pneumonia**
Inflammatory condition of the lung in which fluid fills the alveoli. Caused by a wide variety of different microorganisms. Community-acquired pneumonias are often caused by viruses. SARS-CoV-2 emerged in 2019 to cause a pandemic of pneumonia plus other disease manifestations. *Streptococcus pneumoniae* is the main bacterial cause. Other bacterial causes are *Legionella, Mycoplasma pneumoniae,* and *Chlamydophila pneumoniae.* *Histoplasma capsulatum* is a fungus that can cause a pneumonia-like disease. A hantavirus causes a pneumonia-like condition named hantavirus pulmonary syndrome (HPS).

• Health-Associated Pneumonia
Many cases are ventilator-associated. HAP is commonly caused by *S. aureus*, often MRSA strains. Furthermore, many healthcare-associated pneumonias appear to be polymicrobial in origin.

• Influenza
The ssRNA genome of the influenza virus is subject to constant genetic changes that alter the structure of its envelope glycoproteins. Antigenic drift refers to constant mutations of these glycoproteins. Antigenic shift is less frequent, but can lead to pandemics. It involves the eight separate RNA strands involved in the swapping out of one of those genes or strands with a gene or strand from a different influenza virus.

• Whooping Cough
Causative agent *Bordetella pertussis* releases multiple *exotoxins*, including *pertussis toxin* and *tracheal cytotoxin*, that damage ciliated respiratory epithelial cells and cripple other components of the host defenses.

• Respiratory Syncytial Virus Infection
RSV infects the respiratory tract and produces giant multinucleated cells (syncytia).

• Tuberculosis
Cause is primarily the bacterial species *Mycobacterium tuberculosis*. Multidrug-resistant tuberculosis (MDR-TB) and extensively drug-resistant tuberculosis (XDR-TB) are growing problems and have higher fatality rates.

SmartGrid: From Knowledge to Critical Thinking

This *21 Question Grid* takes the topics from this chapter and arranges them with respect to the American Society for Microbiology's Undergraduate Curriculum guidelines—all six of the important "Concepts" as well as the important "Competency" of scientific literacy. Three questions are supplied about chapter content that refer to the Concept or Competency, in increasing levels of Bloom's taxonomy for learning.

ASM Concept/ Competency	A. Bloom's Level 1, 2—Remember and Understand (Choose one.)	B. Bloom's Level 3, 4—Apply and Analyze	C. Bloom's Level 5, 6—Evaluate and Create
Evolution	1. Which of the following organisms is most problematic with respect to its resistance to antibiotics? a. *Streptococcus pyogenes* b. *Haemophilus influenzae* c. *Mycobacterium tuberculosis* d. *Bordetella pertussis*	2. Explain why an antigenic shift event in the influenza virus can lead to a catastrophic pandemic.	3. It has recently been established that the lower respiratory tract—the lungs—has a microbiome. Use what you know about human evolution to explain why this is not surprising, and how it is probably highly advantageous to health.
Cell Structure and Function	4. Which of these microbes has an unusual, waxy, wall structure that contributes to its virulence? a. *Streptococcus pyogenes* b. *Haemophilus influenzae* c. *Mycobacterium tuberculosis* d. *Bordetella pertussis*	5. What cell structure of *Streptococcus pneumoniae* makes it such a successful and versatile pathogen, in your estimation? Defend your answer.	6. Scientists have worked for years to create a vaccine for the common cold. One effort attempts to create a vaccine that is targeted to the host epithelial cell instead of viral antigens. What is the logic behind this approach? What antigens would you include in such a vaccine?
Metabolic Pathways	7. The beta-hemolysis of blood agar observed with *Streptococcus pyogenes* is due to the presence of a. streptolysin. b. M protein. c. hyaluronic acid. d. capsule.	8. Would you expect the microbiome in the lung to be aerobes, strict anaerobes, or facultative anaerobes? Explain your answer.	9. In some diseases and outbreaks of respiratory diseases caused by certain strains of hantavirus and influenza, healthy young adults suffer the most pathology and mortality. This is an unusual pattern; it is more common for the younger and older vulnerable patients to suffer the worst effects of infectious diseases. Speculate on what is happening when the most robust suffer the most pathology.
Information Flow and Genetics	10. Which of the following pathogens uses the product of a bacteriophage to cause disease? a. *Streptococcus pneumoniae* b. *Mycobacterium tuberculosis* c. *Bordetella pertussis* d. *Streptococcus pyogenes*	11. Explain why we are advised to get a flu vaccine every year. Then explain why there may be years when we are advised to get two flu vaccines.	12. SARS-CoV-2 was shown to be transmissible even when the carrier has no symptoms. How did this characteristic make it difficult to contain the pandemic, particularly in comparison to the 2003 SARS pandemic?

ASM Concept/Competency	A. Bloom's Level 1, 2—Remember and Understand (Choose one.)	B. Bloom's Level 3, 4—Apply and Analyze	C. Bloom's Level 5, 6—Evaluate and Create
Microbial Systems	13. Which of the following infections often has/have a polymicrobial cause? a. otitis media b. healthcare-associated pneumonia c. sinusitis d. all of these	14. Outline how a progression from influenza infection to pneumonia to death might occur, focusing on the microbes responsible and the symptoms that ensue.	15. Explain what difficulties scientists might encounter in trying to develop a vaccine for sinusitis.
Impact of Microorganisms	16. The vast majority of pneumonias caused by this organism occur in AIDS patients. a. hantavirus b. *Histoplasma capsulatum* c. *Pneumocystis jirovecii* d. *Mycoplasma pneumoniae*	17. A distraught mother recently posted on Facebook that although she had not been sick since the birth of her child 3 months ago, she was responsible for her baby being in the ICU with whooping cough. Explain this (accurate) assessment.	18. Conducting your own research, describe the total impact on human population from the two diseases malaria and tuberculosis.
Scientific Thinking	19. "Watchful waiting" is the recommended response to which infection? a. otitis media b. tuberculosis c. COVID-19 d. healthcare-associated pneumonia	20. What effect can you imagine the higher specificity vs. sensitivity in the RT-PCR test for SARS-CoV-2 has on the public perception of how prevalent COVID-19 is?	21. Explain everything we can learn from figure 19.9, the influenza surveillance graph. Be sure to explain the meaning of the black lines as well as the red.

Answers to the multiple-choice questions appear in Appendix A.

Visual Connections

This question connects content within and between chapters.

From chapter 2, figure 2.18b. Although there are many different organisms present in the respiratory tract, an acid-fast stain of sputum like the one shown here along with patient symptoms can establish a presumptive diagnosis of tuberculosis. Explain why.

©Richard J. Green/Science Source

Acid-fast stain
Reddish-purple cells are acid-fast.
Blue cells are non-acid-fast.

20

Infectious Diseases Affecting the Gastrointestinal Tract

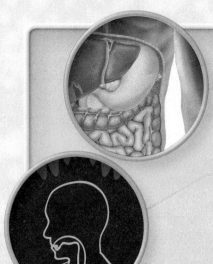

IN THIS CHAPTER...

20.1 The Gastrointestinal Tract and Its Defenses

1. Draw or describe the anatomical features of the gastrointestinal tract.
2. List the natural defenses present in the gastrointestinal tract.

20.2 Normal Biota of the Gastrointestinal Tract

3. List the types of normal biota presently known to occupy the gastrointestinal tract.
4. Describe how our view has changed of normal biota present in the stomach.

20.3 Gastrointestinal Tract Diseases Caused by Microorganisms (Nonhelminthic)

5. List the possible causative agents, modes of transmission, virulence factors, diagnostic techniques, and prevention/treatment for the "Highlight Disease" acute diarrhea.
6. Discuss important features of food poisoning and chronic diarrhea.
7. Discuss important features of the two categories of oral conditions: dental caries and periodontitis.
8. Identify the most important features of mumps, gastritis, and gastric ulcers.
9. Differentiate among the main types of hepatitis, and discuss the causative agents, mode of transmission, diagnostic techniques, prevention, and treatment of each.

20.4 Gastrointestinal Tract Diseases Caused by Helminths

10. Describe some distinguishing characteristics and commonalities seen in helminthic infections.
11. List four helminths that cause primarily intestinal symptoms, and identify which life cycle they follow and one unique fact about each one.
12. Explain reasons for and the consequences of the following statement: "Cysticercosis is underdiagnosed in the United States."

CASE FILE

"Blood and Guts"

I was working as an LPN in a rural facility when Peter came into the emergency room one evening. My supervising RN and I realized almost immediately that Peter was very ill, and we worked together quickly to obtain Peter's history and vital signs.

Peter stated that he had been ill since the morning, when he woke up earlier than normal with diarrhea. He stated that he had been well the night before when he went to bed and had awoken in the morning with severe abdominal cramping. The diarrhea started almost immediately after the cramps began. Peter estimated that he had had approximately 20 bouts of watery, foul-smelling diarrhea throughout the day. He also stated that there was blood present in his stool.

I took Peter's vital signs. He had a fever of 38.4°C (101°F). He was tachycardic with a heart rate of 118 beats per minute. His blood pressure was low at 100/50 mmHg. He was pale and diaphoretic (sweaty) and complained of chills and a headache. He also complained of intermittent severe cramping. His abdomen was tender in all four quadrants on palpation and his bowel sounds were hyperactive on auscultation. His mucous membranes were dry and he complained of thirst. His skin turgor was poor.

My supervising RN called the physician to report Peter's symptoms and vital signs. The doctor stated that he would be coming in right away and to notify the lab. He asked us to start an intravenous on Peter.

We inserted a large-bore IV and started an infusion of normal saline at 250 mL/hour to rehydrate Peter. The physician and lab technician arrived almost simultaneously, and blood was drawn by the lab tech while the physician examined Peter and asked about his symptoms. Peter was asked to provide a stool sample for culture and an ova and parasites test (O&P test), which he was able to provide in a short time. The physician asked if Peter's wife and daughter were ill and if they had eaten the same food as Peter had in the last few days.

Peter's blood work came back. His white blood cell count was elevated and his potassium level was low. Peter was admitted to the hospital for rehydration and was started on broad-spectrum intravenous antibiotics for full coverage of potential pathogens while we awaited the stool culture results.

Peter's stool cultures came back, revealing that Peter had *Shigella*, which was sensitive to sulfamethoxazole/trimethoprim and ciprofloxacin. Peter had an allergy to ciprofloxacin, so he was started on sulfamethoxazole/trimethoprim in oral form twice a day and IV fluids were continued. After 5 days in the hospital, Peter had recovered enough to go home. He continued the antibiotic therapy for another week and eventually fully recovered. The source of the infection was never determined.

- How is *Shigella* transmitted?

- How can *Shigella* be prevented?

Case File Wrap-Up appears at the end of the chapter.

©Kate Mitchell/Corbis

20.1 The Gastrointestinal Tract and Its Defenses

The gastrointestinal (GI) tract can be thought of as a long tube, extending from mouth to anus. It is a very sophisticated delivery system for nutrients, composed of eight main sections and augmented by four accessory organs. The eight sections are the mouth, pharynx, esophagus, stomach, small intestine, large intestine, rectum, and anus. Along the way, the salivary glands, liver, gallbladder, and pancreas add digestive fluids and enzymes to assist in digesting and processing the food we take in **(figure 20.1).** The GI tract is often called the *digestive tract* or the *enteric tract.*

The GI tract has a very heavy load of microorganisms, and it encounters millions of new ones every day. Because of this, defenses against infection are extremely important. All intestinal surfaces are coated with a layer of mucus, which confers mechanical protection. Secretory IgA can also be found on most intestinal surfaces. The muscular walls of the GI tract keep food (and microorganisms) moving through the system through the action of peristalsis. Various fluids in the GI tract have antimicrobial properties. Saliva contains the antimicrobial proteins lysozyme and lactoferrin. The stomach fluid is antimicrobial by virtue of its extremely high acidity. Bile is also antimicrobial.

The entire system is outfitted with cells of the immune system, collectively called gut-associated lymphoid tissue (GALT). The tonsils and adenoids in the oral cavity and pharynx, small areas of lymphoid tissue in the esophagus, Peyer's patches in the small intestine, and the appendix are all packets of lymphoid tissue consisting of T and B cells as well as cells of innate immunity. One of their jobs is to produce IgA, but they perform a variety of other immune functions.

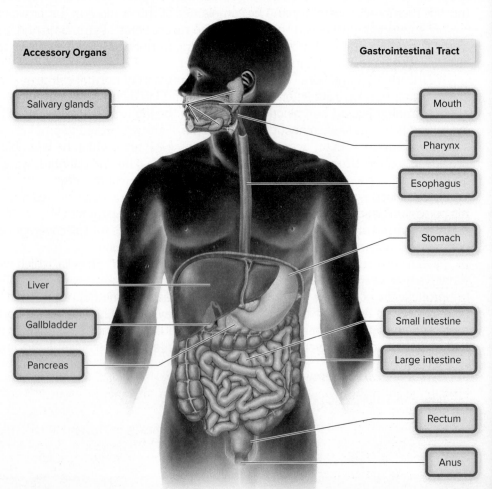

Figure 20.1 Major organs of the digestive system.

A huge population of commensal organisms lives in this system, especially in the large intestine. They avoid immune destruction through various mechanisms, including cloaking themselves with host sugars they find on the intestinal walls.

> ### 20.1 LEARNING OUTCOMES—Assess Your Progress
> 1. Draw or describe the anatomical features of the gastrointestinal tract.
> 2. List the natural defenses present in the gastrointestinal tract.

20.2 Normal Biota of the Gastrointestinal Tract

As just mentioned, the GI tract is home to a large variety of normal biota. The oral cavity alone is populated by more than 600 known species of microorganisms, including *Actinomyces, Lactobacillus, Neisseria, Prevotella, Streptococcus, Treponema,* and *Veillonella* species. Fungi such as *Candida albicans* are also numerous. A few protozoa (*Trichomonas tenax, Entamoeba gingivalis*) also call the mouth "home." Bacteria live on the teeth as well as the soft structures in the mouth. Numerous species of normal biota bacteria live on the teeth in a synergistic community called dental plaque, which is a type of biofilm (see chapter 3). Bacteria are held in the biofilm by specific recognition molecules. Alpha-hemolytic streptococci are generally the first colonizers of the tooth surface after it has been cleaned. The streptococci attach specifically to proteins in the **pellicle,** a mucinous glycoprotein covering on the tooth. Then other species attach specifically to proteins or sugars on the surface of the streptococci, and so on.

The pharynx (throat) contains a variety of microorganisms, many of which were described in chapter 19. Although the stomach was previously thought to be sterile due to its very low pH, researchers have found the molecular signatures of 128 different species of microorganisms in the stomach. Though many of these are likely to be "just passing through," a few, including *Bacillus, Clostridium, Staphylococcus,* and *Streptococcus,* are permanent residents. The large intestine has always been known to be a haven for billions of microorganisms (10^{11} per gram of contents), including the bacteria *Bacteroides, Bifidobacterium, Clostridium, Enterobacter, Escherichia, Fusobacterium, Lactobacillus, Peptostreptococcus, Staphylococcus,* and *Streptococcus;* the fungus *Candida;* and several protozoa as well. Researchers have also found archaea species there.

You recall from chapter 17 that it is now clear the gut microbiome influences the development of the nervous system, including portions of the brain, and can have effects on behavior and neurological wellness. Within the gut itself, the normal biota in the gut provide a protective function, but they perform other jobs as well. Some of them help with digestion. Some provide nutrients that we can't produce ourselves. *E. coli,* for instance, synthesizes vitamin K. Its mere presence in the large intestine seems to be important for the proper formation of epithelial cell structure. And the normal biota in the gut plays an important role in "teaching" our immune system to react to microbial antigens. As you have seen throughout this book, we now know that the composition of the gut microbiome has far-reaching consequences on many aspects of human health. The coming years will bring us a more precise understanding, and the possibility of "healing" the microbiome, or manipulating it in ways that improve various disease conditions.

> ### NCLEX® PREP
> 1. Gut-associated lymphoid tissue (GALT) includes
>   a. *Peyer's patches in the small intestine.*
>   b. *the appendix.*
>   c. *tonsils.*
>   d. *adenoids.*
>   e. *all of the above.*
>   f. *none of the above.*

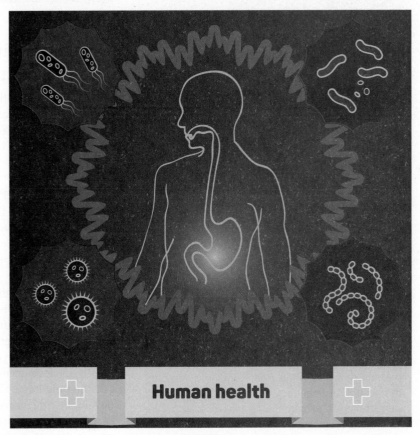

©Anna Smirnova/Alamy Stock Vector

Gastrointestinal Tract Defenses and Normal Biota

	Defenses	Normal Biota
Oral Cavity	Saliva, slgA, lysozyme, tonsils, adenoids	Hundreds of gram-positive and gram-negative bacteria, protozoa, and fungi
Rest of GI Tract	GALT, lymphoid tissue, Peyer's patches, appendix, slgA, normal biota	Thousands of microbes of all kinds, aerobic and anaerobic

Chickens and their eggs can be colonized by *Salmonella*.
©Pixtal/age fotostock

The accessory organs (salivary glands, gallbladder, liver, and pancreas) are free of a natural microbiome but can be exposed to microbes when normal barriers in the gut are disrupted by a condition broadly called *dysbiosis*. This simply refers to an unhealthy mix of gut microbes in the intestinal tract. It can result in leakage of bacteria or their metabolic products into internal organs. Research has shown that gut dysbiosis can play a role in various diseases in the accessory organs.

20.2 LEARNING OUTCOMES—Assess Your Progress

3. List the types of normal biota presently known to occupy the gastrointestinal tract.
4. Describe how our view has changed of normal biota present in the stomach.

20.3 Gastrointestinal Tract Diseases Caused by Microorganisms (Nonhelminthic)

In this section, we first address microbes that cause diarrhea of various types. Then we discuss oral diseases (dental caries and periodontitis), stomach conditions (gastritis and gastric ulcers), and the spectrum of hepatitis infections.

HIGHLIGHT DISEASE

Acute Diarrhea

Diarrhea—usually defined as three or more loose stools in a 24-hour period—needs little explanation. In recent years, on average, citizens of the United States experienced 1.2 to 1.9 cases of diarrhea per person per year, and among children that number is twice as high. In developing countries, children may experience more than 10 episodes of diarrhea a year. In fact, more than 700,000 infants a year die from diarrhea, mostly in developing countries. In developing countries, the high mortality rate is not the only issue. Children who survive dozens of bouts with diarrhea during their developmental years are likely to have permanent physical and cognitive effects. Diarrheal illnesses are often accompanied by fever, abdominal pain and/or cramping, nausea, vomiting, and dehydration.

In the United States, up to a third of all acute diarrhea is transmitted by contaminated food. Accurate numbers are hard to come by, but the CDC estimates that 48 million people are sickened each year by a foodborne illness, and 3,000 of them die. Dozens of microbes are capable of causing foodborne illness. The most common culprits in the United States are *Salmonella*, norovirus, *Campylobacter*, *E. coli* STEC strains (STEC stands for Shiga-toxin-producing *E. coli*), *Staphylococcus aureus*, and *Clostridium perfringens*. **Figure 20.2** highlights the work done by CDC to investigate dozens of foodborne outbreaks every year.

Although most diarrhea episodes are self-limiting and therefore do not require treatment, others can have devastating effects. In most diarrheal illnesses, antimicrobial treatment is contraindicated (inadvisable), but some, such as shigellosis, call for quick treatment with antibiotics. For public health reasons, it is important to know which agents are causing diarrhea in the community, but in many cases identification of the agent is not performed.

Medical Moment

Dehydration

Dehydration is a common symptom experienced by individuals suffering from conditions affecting the gastrointestinal tract. Recognizing and treating dehydration is a high-priority task of health care workers.

Dehydration can be thought of as an imbalance of fluids resulting from excessive fluid loss (as in vomiting and diarrhea) or inadequate fluid intake (due to nausea or loss of appetite).

Health care workers classify dehydration as mild, moderate, or severe—based on the estimated amount of weight lost. This is often expressed in percentages. Capillary refill, skin turgor, and breathing are considered to be the most accurate signs in estimating degree of dehydration in adults.

Q. Dehydration is most dangerous for babies and young children. Why do you think that is?

Answer in Appendix B.

FOODBORNE DISEASE OUTBREAKS

The 3 types of data used to link illnesses to contaminated foods and solve outbreaks

Public health and regulatory officials gather 3 types of data during an investigation:
EPIDEMIOLOGIC | TRACEBACK | FOOD & ENVIRONMENTAL TESTING

EPIDEMIOLOGIC *data*

| Patterns in where and when people got sick, and past outbreaks caused by the same microbe | Interviews with sick people to look for foods or other exposures occurring more often than expected | Discovery of clusters of unrelated sick people who ate at the same restaurant, shopped at the same grocery store, or attended the same event |

TRACEBACK *data*

| A common point of contamination in the distribution chain from farm to fork, identified by reviewing records collected from restaurants or stores where sick people ate or shopped | Inspections in food production facilities, on farms, and in restaurants that identify food safety risks |

FOOD & ENVIRONMENTAL TESTING *data*

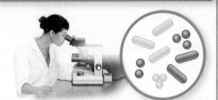

| The microbe that caused illness is found in a food item collected from a sick person's home, from a retail location, or in the food production environment | The same DNA fingerprint linking microbes found in foods or production environments to microbes is found in sick people |

Figure 20.2 The types of data used to investigate foodborne disease outbreaks.
Source: CDC

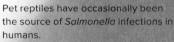

Pet reptiles have occasionally been the source of *Salmonella* infections in humans.
©Ryan McVay/Getty Images

In this section, we describe acute diarrhea with infectious agents as the cause. In the sections following this one, we discuss acute diarrhea and vomiting caused by toxins, commonly known as food poisoning, and chronic diarrhea and its causes.

Salmonella

A decade ago, one of every three chickens destined for human consumption was contaminated with *Salmonella,* but the rate is now about 10%. Other poultry, such as ducks and turkeys, are also affected. Eggs are infected as well because the bacteria may actually enter the egg while the shell is being formed in the chicken. *Salmonella* is a very large genus of bacteria but with a complicated nomenclature. The disease

A screen capture from the CDC *Salmonella* site, outlining sources of *Salmonella* infections in late 2019 and early 2020.

Source: Centers for Disease Control and Prevention. Salmonella. https://www.cdc.gov/salmonella/outbreaks-active.html

condition in which we are interested is caused by *Salmonella enterica* subspecies *enterica*. This subspecies is further subdivided into serogroups that have formal names such as "*Salmonella enterica* subspecies *enterica* serovar Typhi." In clinical situations, this latter organism is usually just called "*Salmonella typhi.*"

Salmonellae are motile; they ferment glucose with acid and sometimes gas. Most, too, produce hydrogen sulfide (H_2S) but not urease. They grow readily on most laboratory media and can survive outside the host in inhospitable environments such as freshwater and freezing temperatures. These pathogens are resistant to chemicals such as bile and dyes, which are the basis for isolation on selective media.

▶ Signs and Symptoms

The genus *Salmonella* causes a variety of illnesses in the GI tract and beyond. Until fairly recently, its most severe manifestation was typhoid fever. Since the mid-1900s, a milder disease usually called salmonellosis has been much more common. Sometimes the condition is also called enteric fever or gastroenteritis. Whereas typhoid fever is caused by the Typhi serotype, gastroenteritises are generally caused by the serotypes known as Typhimurium, Enteritidis, Heidelberg, Newport, and Javiana. Most of these strains come from animals, unlike the Typhi serotype, which infects humans exclusively. *Salmonella* bacteria are normal intestinal biota in cattle, poultry, rodents, and reptiles and each has been a documented source of infection and disease in humans.

Salmonellosis can be relatively severe, with an elevated body temperature and septicemia as more prominent features than GI tract disturbance. But it can also be fairly mild, with gastroenteritis—vomiting, diarrhea, and mucosal irritation—as its major feature. Blood can appear in the stool. In otherwise healthy adults, symptoms spontaneously subside after 2 to 5 days. Death is infrequent except in debilitated persons.

▶ Pathogenesis and Virulence Factors

Salmonella serotypes vary in their virulence due to genetic differences in their ability to adhere to the gut mucosa, and also to evade the immune system. Endotoxin, a lipopolysaccharide component of the outer membrane of gram-negative bacteria, is an important virulence factor for *Salmonella*.

▶ Transmission and Epidemiology

The CDC estimates that *Salmonella* causes approximately 1.2 million infections a year in the United States. An important factor to consider in all diarrheal pathogens is how many organisms must be ingested to cause disease (their ID_{50}). It varies widely. *Salmonella* has a high ID_{50}, meaning a lot of organisms have to be ingested in order for disease to result. Animal products such as meat and milk can be readily contaminated with *Salmonella* during slaughter, collection, and processing.

Most cases are traceable to a common food source such as milk or eggs. Some cases may be due to poor sanitation. In one outbreak, about 60 people became infected after visiting the Komodo dragon exhibit at the Denver zoo. They picked up the infection by handling the rails and fence of the dragon's cage.

▶ Prevention and Treatment

The only prevention for salmonellosis is avoiding contact with the bacterium. Uncomplicated cases of salmonellosis are treated with fluid and electrolyte replacement. If the patient has underlying immunocompromise or if the disease is severe, ciprofloxacin is recommended.

Shigella

The *Shigella* bacteria are gram-negative rods, nonmotile and non-endospore-forming **(figure 20.3)**. They do not produce urease or hydrogen sulfide. Knowing this helps in their identification. They are primarily human parasites, though they can infect apes. All produce a similar disease that can vary in intensity. These bacteria resemble some types of pathogenic *E. coli* very closely.

▶ Signs and Symptoms

The symptoms of shigellosis include frequent, watery stools, as well as fever, and often intense abdominal pain. Nausea and vomiting are common. Stools often contain obvious blood, and even more often are found to have occult (not visible to the naked eye) blood. Diarrhea containing blood is also called **dysentery.** Mucus from the GI tract will also be present in the stools.

▶ Pathogenesis and Virulence Factors

Shigellosis is different from many GI tract infections in that *Shigella* invades the villus cells of the large intestine rather than the small intestine. In addition, it is not as invasive as *Salmonella* and does not perforate the intestine or invade the blood. It enters the intestinal mucosa by means of special cells in Peyer's patches. Once in the mucosa, *Shigella* instigates an inflammatory response that causes extensive tissue destruction. The release of endotoxin causes fever. **Enterotoxin,** an exotoxin that affects the enteric (or GI) tract, damages the mucosa and villi. Local areas of erosion give rise to bleeding and heavy secretion of mucus **(figure 20.4).** *Shigella dysenteriae* (and some of the other species) produces a heat-labile exotoxin called **Shiga toxin,** which seems to be responsible for the more serious damage to the intestine as well as any systemic effects, including injury to nerve cells.

▶ Transmission and Epidemiology

In addition to the usual oral route, shigellosis is also acquired through direct person-to-person contact, largely because of the small infectious dose required (from 10 to 200 bacteria). The disease is mostly associated with lax sanitation, malnutrition, and crowding. It is also spread epidemically in day care centers, prisons, mental institutions, nursing homes, and military camps. *Shigella* can establish a chronic carrier condition in some people that lasts several months.

▶ Prevention and Treatment

The only prevention of this and most other diarrheal diseases is good hygiene and avoiding contact with infected persons. In uncomplicated cases, no antibiotics are indicated. If patient is immunocompromised, ciprofloxacin is recommended.

Shiga-Toxin-Producing *E. coli* (STEC)

Dozens of different strains of *E. coli* exist, most of which cause no disease at all. A handful of them cause various degrees of intestinal symptoms, as described in this and the following section. Some of them cause urinary tract infections (see chapter 21). *E. coli* O157:H7 was the first representative of this group in the news, having caused an outbreak originating in a chain of Jack-in-the-Box restaurants in the Pacific Northwest. Now there are other "numbered" *E. coli* strains that produce Shiga toxins; all of them are dangerous pathogens. Collectively we call them Shiga-toxin-producing *E. coli.*

▶ Signs and Symptoms

Shiga-toxin-producing *E. coli* is the agent of a spectrum of conditions, ranging from mild gastroenteritis with fever to bloody diarrhea. A minority of patients will develop

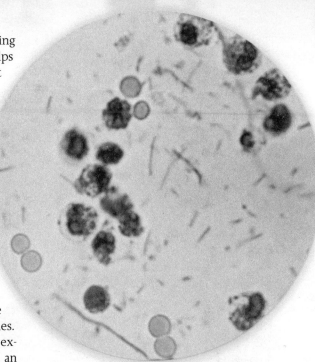

Figure 20.3 Fecal smear from patient with ***Shigella* infection.** The bacteria (small red rods), red blood cells (light red discs), and white blood cells (the larger cells with intracellular granules) are all evident.
Source: CDC

 Note About "Enterotoxin"

You have already learned about two broad classes of toxins: exotoxins and endotoxins. (Do you remember the definitions?) In this chapter you learn the term *enterotoxin*. It sounds similar to the other two terms. But enterotoxin is a descriptive term for a particular type of exotoxin: one that targets the enteric—or the gastrointestinal—tract. It might be worth making a note to yourself!

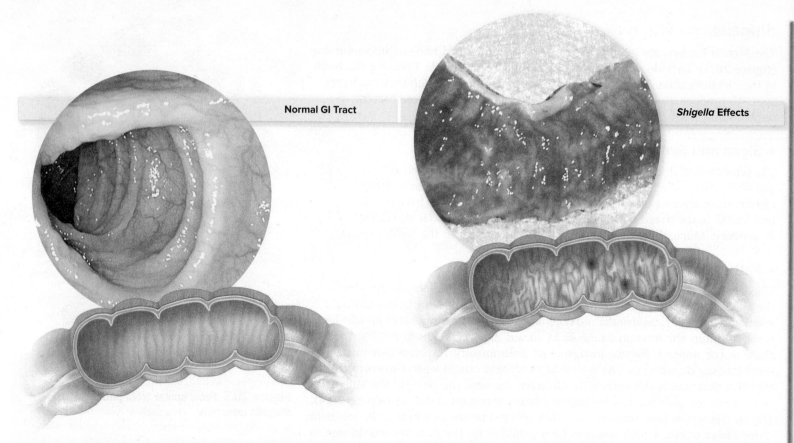

Normal GI Tract

***Shigella* Effects**

Figure 20.4 **The appearance of the large intestinal mucosa in *Shigella* dysentery.**   Note the patches of blood and mucus, the erosion of the lining, and the absence of perforation.

Gastrolab/Science Source (normal); Source: CDC (diseased)

hemolytic uremic syndrome (HUS), a severe hemolytic anemia that can cause kidney damage and failure. Neurological symptoms such as blindness, seizure, and stroke (and long-term debilitation) are also possible.

▶ Pathogenesis and Virulence Factors

These *E. coli* owe much of their virulence to Shiga toxins (so named because they are identical to the Shiga exotoxin secreted by virulent *Shigella* species). Shiga toxin genes are present on prophage genes donated by bacteriophage in *E. coli* but are on the chromosome of *Shigella dysenteriae,* suggesting that the *E. coli* acquired the virulence factor through phage-mediated transfer (transduction). As described for *Shigella,* the Shiga toxin interrupts protein synthesis in its target cells. It seems to be responsible especially for the systemic effects of this infection.

Another important virulence determinant for STEC is the ability to efface (rub out or destroy) enterocytes, which are gut epithelial cells. The net effect is a lesion in the gut (effacement), usually in the large intestine. The microvilli are lost from the gut epithelium, and the lesions produce bloody diarrhea.

▶ Transmission and Epidemiology

The most common mode of transmission for STEC is the ingestion of contaminated foodstuffs.

In 2015 the Chipotle chain of restaurants was the source of two distinct *E. coli* O26 outbreaks in the United States. In 2016 there was a multistate outbreak of two STEC strains, *E. coli* O12 and *E. coli* O26, which were traced to flour manufactured by General Mills in Kansas City, Missouri.

▶ Culture and Diagnosis

Infection with this type of *E. coli* should be confirmed with stool culture and also tested for Shiga toxin. Positive specimens should be sent to state or local health departments for detailed characterization and outbreak monitoring.

▶ Prevention and Treatment

The best prevention for this disease is good food hygiene. The Shiga toxin is heat-labile, and the *E. coli* is killed by heat as well.

No vaccine exists for *E. coli* O157:H7 or other STEC strains. However, some countries vaccinate cattle against *E. coli* O157:H7 as a means to protect human populations.

Antibiotics may be contraindicated for this infection, as they may increase the pathology by releasing more toxin, leading to HUS. Supportive therapy, including plasma transfusions to dilute toxin in the blood, is a good option.

Other *E. coli*

At least five other categories of *E. coli* can cause diarrheal diseases. Scientists call these **enterotoxigenic** *E. coli* (ETEC), **enteroinvasive** *E. coli* (EIEC), **enteropathogenic** *E. coli* (EPEC), diffusely adherent *E. coli* (DAEC), and **enteroaggregative** *E. coli* (EAEC). In clinical practice, most physicians are interested in differentiating Shiga-toxin-producing *E. coli* (STEC) from all the others. In **Disease Table 20.1**, the non-Shiga-toxin-producing *E. coli* are grouped together in one column.

Campylobacter

Although you may never have heard of *Campylobacter*, it is one of the most common bacterial causes of diarrhea in the United States. The symptoms of campylobacteriosis are frequent watery stools, fever, vomiting, headaches, and abdominal pain. The symptoms may last longer than most acute diarrheal episodes, sometimes extending beyond 2 weeks. They may subside and then recur over a period of weeks.

Campylobacter jejuni is the most common cause, although there are other pathogenic *Campylobacter* species. Campylobacters are slender, curved, or spiral gram-negative bacteria propelled by polar flagella at one or both poles, often appearing in S-shaped or gull-winged pairs. These bacteria tend to be microaerophilic inhabitants of the intestinal tract, genitourinary tract, and oral cavity of humans and animals. Transmission of this pathogen takes place via the ingestion of contaminated beverages and food, especially water, milk, meat, and chicken. In late 2019 and early 2020, an outbreak occurred in at least 30 states in which people were infected by contact with healthy-looking puppies they had purchased or worked with as pet store employees.

Once ingested, *C. jejuni* cells reach the mucosa at the last segment of the small intestine (ileum) near its junction with the colon. They adhere, burrow through the mucus, and multiply. Symptoms begin after an incubation period of 1 to 7 days. The mechanisms of pathology appear to involve a heat-labile enterotoxin that stimulates a secretory diarrhea like that of cholera. In a small number of cases, infection with this bacterium can lead to a serious neuromuscular paralysis called Guillain-Barré syndrome.

Guillain-Barré syndrome (GBS) (pronounced gee″-luhn-buh-ray′) is the leading cause of acute paralysis in the United States since the eradication of polio. The good news is that many patients recover completely from this paralysis. The condition is still mysterious in many ways, but it seems to be an autoimmune reaction that can be brought on by infection with viruses and bacteria, by vaccination in rare cases, and even by surgery. The single most common precipitating event for the onset of GBS is *Campylobacter* infection. Twenty to

Hand washing is a very effective way to reduce transmission of fecal bacteria.
©BananaStock/PunchStock

forty percent of GBS cases are preceded by infection with *Campylobacter*. The reasons for this are not clear. (Note that even though 20% to 40% of GBS cases are preceded by *Campylobacter* infection, only about 1 in 1,000 cases of *Campylobacter* infection results in GBS.)

Resolution of *Campylobacter* infection occurs in most instances with simple, nonspecific rehydration and electrolyte balance therapy. In more severely affected patients, it may be necessary to administer azithromycin. Antibiotic resistance is increasing in these bacteria, in large part due to the use of fluoroquinolones in the poultry industry. Because vaccines are yet to be developed, prevention depends on rigid sanitary control of water and milk supplies and care in food preparation.

Clostridioides difficile

Clostridioides difficile is a gram-positive endospore-forming rod found as normal biota in the intestine. It was once considered relatively harmless but now is known to cause a condition called pseudomembranous colitis, also called antibiotic-associated colitis. In the earliest cases, this infection was precipitated by therapy with broad-spectrum antibiotics such as ampicillin, clindamycin, or cephalosporins. Now it is well-established in health care facilities and is an important healthcare-associated infection. Although *C. difficile* is relatively noninvasive, it is able to superinfect the large intestine when drugs have disrupted the normal biota. It produces two enterotoxins, toxins A and B, that cause areas of necrosis in the wall of the intestine. More severe cases exhibit abdominal cramps, fever, and leukocytosis. The colon is inflamed and gradually sloughs off loose, membranelike patches called pseudomembranes consisting of fibrin and cells **(figure 20.5).** If the condition is not stopped, perforation of the cecum and death can result.

The epidemiology of *C. diff* is changing. In the most comprehensive study conducted to date in the United States in 2011, an estimated 453,000 people acquired *C. diff*. Of these U.S. cases, 66% were healthcare-associated, even though 24% of infections did not occur until after the patient had gone home. If a patient is receiving clindamycin, ceftriaxone, or a fluoroquinolone for a different infection and displays *C. diff* symptoms, the first step is to withdraw the offending antibiotic. In mild *C. diff* infections, metronidazole should be administered. If it is more severe, the drug of choice is vancomycin. It can be very difficult to eradicate, and stubborn infections can significantly degrade a patient's quality of life. Fecal implants, or stool transplantation from healthy donors, is the treatment with the most success in these patients.

A key point to remember is that *C. diff* releases endospores, which contaminate the environment. Hospitalized patients must be put in isolation conditions, and constant attention to disinfection and infection control is required.

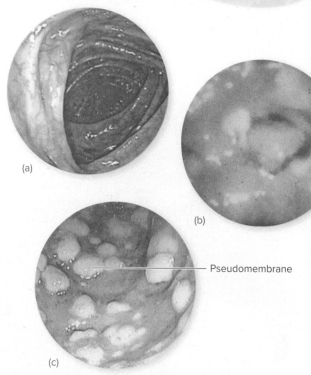

(a)

(b)

Pseudomembrane

(c)

Figure 20.5 *Clostridioides difficile* **colitis.**
(a) Normal colon. **(b)** A mild form of colitis with diffuse, inflammatory patches. **(c)** Heavy yellow plaques, or pseudomembranes, typical of more severe cases. Photographs were made by a sigmoidoscope, an instrument capable of photographing the interior of the colon.

(a) David Musher/Science Source; (b) Courtesy Fred Pittman; (c) David M. Martin, M.D./Science Source

Vibrio cholerae

Cholera has been a devastating disease for centuries. It is not an exaggeration to say that the disease has shaped a good deal of human history in Asia and Latin America, where it has been endemic. These days we have come to expect outbreaks of cholera to occur after natural disasters, war, or large refugee movements, especially in underdeveloped parts of the world.

These bacteria are rods with a single polar flagellum. They belong to the family *Vibrionaceae*. A freshly isolated specimen of *Vibrio cholerae* contains quick, darting cells that slightly resemble a comma. *Vibrio* shares many characteristics with members of the *Enterobacteriaceae* family. Vibrios are fermentative and grow on ordinary or selective media containing bile at 37°C. They possess unique O and H antigens and membrane receptor antigens that provide some basis for classifying members of the family. There are two major types, called *classic* and *El Tor*.

▶ Signs and Symptoms

After an incubation period of a few hours to a few days, symptoms begin abruptly with vomiting, followed by copious watery feces called secretory diarrhea. The intestinal contents are lost very quickly, leaving only secreted fluids. This voided fluid contains flecks of mucus—hence, the description "rice-water stool." (See the photo.) Fluid losses of nearly 1 liter per hour have been reported in severe cases, and an untreated patient can lose up to 50% of body weight during the course of this disease. The diarrhea causes loss of blood volume, acidosis from bicarbonate loss, and potassium depletion, which manifest in muscle cramps, severe thirst, flaccid skin, sunken eyes, and, in young children, coma and convulsions. Secondary circulatory consequences can include hypotension, tachycardia, cyanosis, and collapse from shock within 18 to 24 hours. If cholera is left untreated, death can occur in less than 48 hours, and the mortality rate is between 55% and 70%.

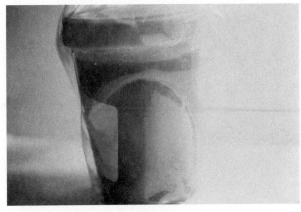

Sample of stool from a cholera patient. The mucus flecks are settling in the bottom of the container.
©CDC (rice-water)

▶ Pathogenesis and Virulence Factors

V. cholerae has a relatively high infectious dose (10^8 cells). At the junction of the duodenum and jejunum, the vibrios penetrate the mucous barrier using their flagella, adhere to the microvilli of the epithelial cells, and multiply there. The bacteria never enter the host cells or invade the mucosa. The virulence of *V. cholerae* is due to an enterotoxin called cholera toxin (CT), which disrupts the normal physiology of intestinal cells. Under the influence of this system, the cells shed large amounts of electrolytes into the intestine, an event accompanied by profuse water loss.

▶ Transmission and Epidemiology

The pattern of cholera transmission and the onset of epidemics are greatly influenced by the season of the year and the climate. Cold, acidic, dry environments inhibit the migration and survival of *Vibrio*, whereas warm, monsoon, alkaline, and saline conditions favor them. The bacteria survive in water sources for long periods of time. Recent outbreaks in several parts of the world have been traced to giant cargo ships that pick up ballast water in one port and empty it in another elsewhere in the world. Cholera ranks among the top seven causes of morbidity and mortality, affecting several million people in endemic regions of Asia and Africa.

In nonendemic areas such as the United States, the microbe is spread by water and food contaminated by asymptomatic carriers, but it is relatively uncommon.

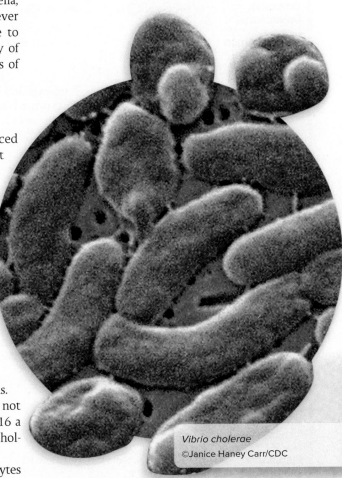

Vibrio cholerae
©Janice Haney Carr/CDC

▶ Prevention and Treatment

Effective prevention is contingent on proper sewage treatment and water purification. Vaccines are available for travelers and people living in endemic regions. There are two vaccines approved by the World Health Organization that are not available in the United States; their effectiveness is between 60% and 67%. In 2016 a new vaccine was approved for use in the United States for people traveling to cholera-endemic areas.

The key to cholera therapy is prompt replacement of water and electrolytes because their loss accounts for the severe morbidity and mortality. This therapy can be accomplished by various rehydration techniques that replace the lost fluid and electrolytes. One of these, oral rehydration therapy (ORT), is incredibly simple and astonishingly effective. Until the 1970s, the treatment, if one could access it, was rehydration through an IV drip. This treatment usually required traveling to the nearest clinic, often miles or days away. Most affected children received no treatment at all, and 3 million of them died every year. Then scientists tested a simple sugar-salt solution that patients could drink.

The relatively simple solution, developed by the World Health Organization (WHO), consists of a mixture of the electrolytes sodium chloride, sodium bicarbonate, potassium chloride, and glucose or sucrose dissolved in water. When administered early in amounts ranging from 100 to 400 milliliters per hour, the solution can restore patients in 4 hours, often bringing them literally back from the brink of death. This therapy has several advantages, especially for countries with few resources. It does not require medical facilities, high-technology equipment, or complex medication protocols. It also eliminates the need for clean needles, which is a pressing issue in many parts of the world.

In the United States and other developed countries antibiotics, such as doxycycline, are given orally along with rehydration. These can shorten periods of both diarrhea and bacterial excretion.

Disease Table 20.1   Acute Diarrhea

Bacterial Causes

Causative Organism(s)	*Salmonella* B G⁻	*Shigella* B G⁻	Shiga-toxin-producing *E. coli* (STEC) B G⁻	Other *E. coli* (non-Shiga-toxin-producing) B G⁻	*Campylobacter* B G⁻
Most Common Modes of Transmission	Vehicle (food, beverage), fecal-oral	Fecal-oral, direct contact	Vehicle (food, beverage), fecal-oral	Vehicle, fecal-oral	Vehicle (food, water), fecal-oral
Virulence Factors	Adhesins, endotoxin	Endotoxin, enterotoxin, Shiga toxins in some strains	Shiga toxins; proteins for attachment, secretion, effacement	Various: proteins for attachment, secretion, effacement; heat-labile and/or heat-stable exotoxins; invasiveness	Adhesins, exotoxin, induction of autoimmunity
Culture/Diagnosis	Stool culture, not usually necessary	Stool culture, antigen testing for Shiga toxin	Stool culture, antigen testing for Shiga toxin; involve state or local health department	Stool culture not usually necessary in absence of blood, fever	Stool culture, not usually necessary; dark-field microscopy
Prevention	Food hygiene and personal hygiene	Food hygiene and personal hygiene	Avoid live *E. coli* (cook meat, wash vegetables, avoid raw flour)	Food hygiene and personal hygiene	Food hygiene and personal hygiene
Treatment	Rehydration; no antibiotic for uncomplicated disease; in **Serious Threat** category in CDC Antibiotic Resistance Report	Rehydration unless there is indication for treatment; in that case ciprofloxacin; in **Serious Threat** category in CDC Antibiotic Resistance Report	Antibiotics may be contraindicated, supportive measures	Rehydration, antimotility agent	Rehydration, azithromycin in severe cases (antibiotic resistance rising); in **Serious Threat** category in CDC Antibiotic Resistance Report
Fever Present?	Usually	Often	Often	Sometimes	Usually
Blood in Stool?	Sometimes	Often	Usually	Sometimes	No
Distinctive Features	Often associated with chickens, reptiles	Very low ID$_{50}$	Hemolytic uremic syndrome	EIEC, ETEC, EPEC, EAEC	Guillain-Barré syndrome
Epidemiological Features	United States: +/− 1.2 million cases per year; 20% of all cases require hospitalization; death rate of 0.6%	United States: estimated 450,000 cases per year; internationally: 165 million cases per year	Internationally: causes HUS in 10% of patients; 25% of HUS patients suffer neurological complications; 50% have chronic renal sequelae	—	United States: 2.4 million cases per year; internationally: 400 million cases per year

Non-cholera *Vibrio* Species

In the United States, infection with non-cholera *Vibrio* species is more likely. These infections are called vibrioses, as opposed to cholera. Two species are most prominent, *V. vulnificus* and *V. parahaemolyticus*. Infection can be from exposure to seawater but more often is associated with eating contaminated shellfish. It is a relatively rare infection but very much on the increase. Scientists suspect that the increase is due to three factors: (1) the increased demand for raw oysters; (2) increased awareness, meaning that more people are diagnosed; and (3) climate change causing a wider habitat for these bacteria in bodies of water. In people who are immunocompromised, the infections can be fatal, especially in the case of *V. vulnificus*. These infections, along with those by *V. cholerae*, are nationally notifiable diseases.

			Nonbacterial Causes		
Clostridioides difficile **B G+**	*Vibrio cholerae* **B G⁻**	Non-cholera *Vibrio* species **B G⁻**	*Cryptosporidium* **P**	Rotavirus **V**	Norovirus **V**
Endogenous (normal biota)	Vehicle (water and some foods), fecal-oral	Vehicle (food or natural bodies of water)	Vehicle (water, food), fecal-oral	Fecal-oral, vehicle, fomite	Fecal-oral, vehicle
Enterotoxins A and B	Cholera toxin (CT)	–	Intracellular growth	–	–
Stool culture, PCR, ELISA demonstration of toxins in stool	Clinical diagnosis, microscopic techniques, serological detection of antitoxin	Culture of stool or blood	Fluorescence microscopy	Usually not performed	Rapid antigen test
–	Water hygiene	Avoiding raw shellfish	Water treatment, proper food handling	Oral live virus vaccine	Hygiene
Withdrawal of antibiotic; vancomycin; in **Urgent Threat** category in CDC Antibiotic Resistance Threat Report	Rehydration; in severe cases, doxycycline	Doxycycline	None; nitazoxanide used sometimes	Rehydration	Rehydration
Sometimes	No	Yes	Often	Often	Sometimes
Not usually; mucus prominent	No	No	Not usually	No	No
Antibiotic-associated diarrhea	Rice-water stools	Sepsis can follow	Resistant to chlorine disinfection	Severe in babies	–
United States: 3 million cases per year	Global estimate: 21,000–143,000 deaths annually	Cause 90% of seafood-related deaths in United States	United States: estimated 748,000 cases per year; 30% seropositive	United States: incidence drastically declined since vaccine was introduced; internationally: 125 million cases of infantile diarrhea annually	United States: most common cause of diarrhea, second-most common cause of foodborne illness hospitalization

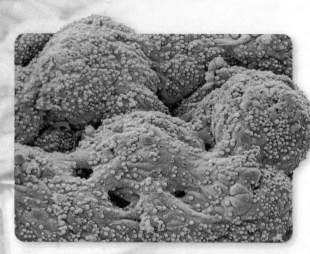

Figure 20.6 Scanning electron micrograph of *Cryptosporidium* **(green) attached to the intestinal epithelium.**
Moredun Animal Health Ltd/Science Source

NCLEX® PREP

2. A nurse is preparing to assess a patient admitted for persistent diarrhea. It was reported that the patient has experienced a 6% weight loss in the past 24 hours. Which of the following clinical manifestations may be expected? Select all that apply.

   a. *increased respiratory rate*

   b. *hypertension*

   c. *headache*

   d. *extreme thirst*

   e. *sustained tachycardia*

©Erica Simone Leeds

Figure 20.7 Norovirus transmission in food.   The pie chart shows where foodborne transmission of the virus occurs. Adapted from the Centers for Disease Control and Prevention

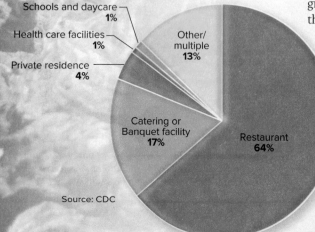

Schools and daycare 1%
Health care facilities 1%
Private residence 4%
Other/multiple 13%
Catering or Banquet facility 17%
Restaurant 64%

Source: CDC

Cryptosporidium

Cryptosporidium is an intestinal protozoan of the apicomplexan type (see chapter 4) that infects a variety of mammals, birds, and reptiles. For many years, cryptosporidiosis was considered an intestinal ailment exclusive to calves, pigs, chickens, and other poultry, but it is clearly a zoonosis as well. The organism's life cycle includes a hardy intestinal oocyst as well as a tissue phase. Humans accidentally ingest the oocysts with water or food that has been contaminated by feces from infected animals. The oocyst "excysts" once it reaches the intestines and releases sporozoites that attach to the epithelium of the small intestine **(figure 20.6).** The organism penetrates the intestinal cells and lives intracellularly in them. It undergoes asexual and sexual reproduction in these cells and produces more oocysts, which are released into the gut lumen, excreted from the host, and after a short time become infective again. The oocysts are highly infectious and extremely resistant to treatment with chlorine and other disinfectants.

The prominent symptoms mimic other types of gastroenteritis, with headache, sweating, vomiting, severe abdominal cramps, and diarrhea. AIDS patients may experience chronic persistent cryptosporidial diarrhea that can be used as a criterion to help diagnose AIDS. The agent can be detected in fecal samples or in biopsies using ELISA or acid-fast staining. Stool cultures should be performed to rule out other (bacterial) causes of infection.

Half of the outbreaks of diarrhea associated with swimming pools are caused by *Cryptosporidium*. Because chlorination is not entirely successful in eradicating the cysts, most treatment plants use filtration to remove them, but even this method can fail.

Treatment is not usually required for otherwise healthy patients. Antidiarrheal agents (antimotility drugs) may be used. Although no curative antimicrobial agent exists for *Cryptosporidium*, physicians will often try nitazoxanide, which can be effective against protozoa in immunocompetent patients.

Rotavirus

Rotavirus is a member of the *Reovirus* group, which consists of an unusual double-stranded RNA genome with both an inner and an outer capsid. Globally, rotavirus is the primary viral cause of morbidity and mortality resulting from diarrhea, accounting for nearly 50% of all cases. Before the rotavirus vaccine was introduced in 2006, 1 million cases of rotavirus infection occurred in the United States every year, leading to 70,000 hospitalizations. The virus gets its name from its physical appearance, which is said to resemble a spoked wheel.

The virus is transmitted by the fecal-oral route, including through contaminated food, water, and fomites. For this reason, disease is most prevalent in areas of the world with poor sanitation.

The effects of infection vary with the age, nutritional state, general health, and living conditions of the patient. Babies from 6 to 24 months of age lacking maternal antibodies have the greatest risk for fatal disease. These children present symptoms of watery diarrhea, fever, vomiting, dehydration, and shock. The intestinal mucosa can be damaged in a way that chronically compromises nutrition, and long-term or repeated infections can retard growth. Newborns seem to be protected by maternal antibodies. Adults can also acquire this infection, but it is generally mild and self-limiting.

Norovirus

Noroviruses are the second most common cause of hospitalizations from foodborne diseases in the United States. One particular setting that is frequently affected by norovirus outbreaks is the cruise ship industry. Overall, norovirus causes about five times as much foodborne illness as *Salmonella*, the most common cause of hospitalization from foodborne disease. It's just that many fewer norovirus cases require hospitalization.

Transmission is fecal-oral or via contamination of food and water **(figure 20.7).** Viruses generally cause a profuse, watery diarrhea lasting 3 to 5 days. Vomiting may accompany the disease, especially in the early phases. Mild fever is often seen.

Treatment of these infections always focuses on rehydration. **Disease Table 20.1.**

Food Poisoning

If a patient presents with severe nausea and frequent vomiting accompanied by diarrhea, and reports that companions with whom he or she shared a recent meal (within the last 1 to 6 hours) are suffering the same fate, food poisoning should be suspected. **Food poisoning** refers to symptoms in the gut that are caused by a preformed toxin of some sort. In many cases, the toxin comes from *Staphylococcus aureus*. In others, the source of the toxin is *Bacillus cereus* or *Clostridium perfringens*. The toxin occasionally comes from nonmicrobial sources such as fish, shellfish, or mushrooms. In any case, if the symptoms are violent and the incubation period is very short, this condition, which is an *intoxication* (the effects of a toxin) rather than an *infection*, should be considered.

Staphylococcus aureus Exotoxin

This illness is associated with eating foods such as custards, sauces, cream pastries, processed meats, chicken salad, or ham that have been contaminated by handling and then left unrefrigerated for a few hours. Because of the high salt tolerance of *S. aureus*, even foods containing salt as a preservative are not exempt. The toxins produced by the multiplying bacteria do not noticeably alter the food's taste or smell. The exotoxin (which is an enterotoxin) is heat-stable; inactivation requires 100°C for at least 30 minutes. For that reason, heating the food after toxin production may not prevent disease. The ingested toxin acts upon the gastrointestinal epithelium and stimulates nerves, with acute symptoms of cramping, nausea, vomiting, and diarrhea. Recovery is also rapid, usually within 24 hours. Most often, the *S. aureus* comes from a food handler's skin or nose.

This condition is almost always self-limiting, and antibiotics are definitely not warranted.

Bacillus cereus Exotoxin

Bacillus cereus is a sporulating gram-positive bacterium that is naturally present in soil. As a result, it is a common resident on vegetables and other products in close contact with soil. It produces two exotoxins, one of which causes a diarrheal-type disease, the other of which causes an **emetic** (ee-met'-ik) or vomiting disease. The type of disease that takes place is influenced by the type of food that is contaminated by the bacterium. The emetic form is most frequently linked to rice and pasta, especially when it has been cooked and kept warm for long periods of time. These conditions are apparently ideal for the expression of the low-molecular-weight, heat-stable exotoxin having an emetic effect. The diarrheal form of the disease is usually associated with cooked meats or vegetables that are held at a warm temperature for long periods of time. These conditions apparently favor the production of the high-molecular-weight, heat-labile exotoxin. The symptom in these cases is a watery, profuse diarrhea that lasts only for about 24 hours.

In both cases, the only prevention is the proper handling of food.

Clostridium perfringens Exotoxin

Another sporulating gram-positive bacterium that causes intestinal symptoms is *Clostridium perfringens*. Endospores from *C. perfringens* can contaminate many kinds of foods. Those most frequently implicated in disease are animal flesh (meat, fish) and vegetables such as beans that have not been cooked thoroughly enough to destroy endospores. When these foods are cooled, endospores germinate, and the germinated cells multiply, especially if the food is left unrefrigerated. If the food is eaten without adequate reheating, live *C. perfringens* cells enter the small intestine and release exotoxin. The toxin, acting upon epithelial cells, initiates acute abdominal pain, diarrhea, and nausea in 8 to 16 hours. Recovery is rapid, and deaths are extremely rare.

C. perfringens also causes an enterocolitis infection similar to that caused by *C. difficile*. This infectious type of diarrhea is acquired from contaminated food, or it may be transmissible by inanimate objects. **Disease Table 20.2.**

Disease Table 20.2	Food Poisoning		
Causative Organism(s)	*Staphylococcus aureus* exotoxin **B** **G+**	*Bacillus cereus* **B** **G+**	*Clostridium perfringens* **B** **G+**
Most Common Modes of Transmission	Vehicle (food)	Vehicle (food)	Vehicle (food)
Virulence Factors	Heat-stable exotoxin	Heat-stable toxin, heat-labile toxin	Heat-labile toxin
Culture/Diagnosis	Usually based on epidemiological evidence	Microscopic analysis of food or stool	Detection of toxin in stool
Prevention	Proper food handling	Proper food handling	Proper food handling
Treatment	Supportive	Supportive	Supportive
Fever Present	Not usually	Not usually	Not usually
Blood in Stool	No	No	No
Distinctive Features	Suspect in foods with high salt or sugar content	Two forms: emetic and diarrheal	Acute abdominal pain
Epidemiological Features	United States: estimated 240,000 cases per year	United States: estimated 63,000 cases per year	United States: estimated 966,000 cases per year

Chronic Diarrhea

Chronic diarrhea is defined as lasting longer than 14 days. It can have infectious causes or can reflect noninfectious conditions. Most of us are familiar with diseases that cause long-term bowel symptoms, such as irritable bowel syndrome and ulcerative colitis, neither of which is directly caused by a microorganism as far as we know. Increasing evidence suggests that a chronically disrupted intestinal biota (from long-term use of antibiotics, for example) can predispose people to these conditions.

People suffering from AIDS frequently suffer from chronic diarrhea. These patients have diarrhea caused by a variety of opportunistic microorganisms, including *Cryptosporidium, Mycobacterium avium,* and so forth. A patient's HIV status should be considered if he or she presents with chronic diarrhea.

Here we examine a few of the microbes that can be responsible for chronic diarrhea in otherwise healthy people.

Enteroaggregative *E. coli* (EAEC)

In the section on acute diarrhea, you read about the various categories of *E. coli* that can cause disease in the gut. One type, the enteroaggregative *E. coli* (EAEC), is particularly associated with chronic disease, especially in children. This bacterium is distinguished by its ability to adhere to human cells in aggregates rather than as single cells **(figure 20.8)**. Its presence stimulates secretion of large amounts of mucus in the gut, which may be part of its role in causing chronic diarrhea.

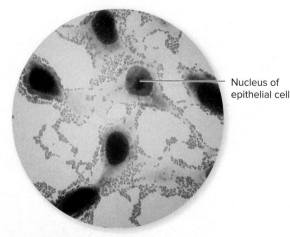

Nucleus of epithelial cell

Figure 20.8 Enteroaggregative *E. coli* adhering to epithelial cells.

Iruka Okeke

Cyclospora

The first case of human disease diagnosed as being caused by the protozoan *Cyclospora cayetanensis* was in 1979. But since 1990, the disease has become very common. Its mode of transmission is fecal-oral, and most cases have been associated with consumption of fresh produce and water presumably contaminated with feces. This disease occurs worldwide, and although primarily of human origin, it is not spread directly from person to person. This is due to its life cycle. When it is shed in the feces of infected people, it is not in its infective stage but rather in a noninfective oocyst stage. It takes days or weeks for the oocyst to sporulate and become infective to humans. Outbreaks have been traced to imported raspberries, salad made with fresh greens, and drinking water. In 2019, a multistate outbreak of *Cyclospora* was traced to fresh basil.

Giardia

Giardia duodenalis (also known as *Giardia intestinalis* and *Giardia lamblia*) is a pathogenic flagellated protozoan first observed by Antonie van Leeuwenhoek in his own feces. In

Cyst

Nuclei

Trophozoite

Ventral View

Nucleus

Ventral depression

Lateral View

(a)

(b)

Figure 20.9 *Giardia* **trophozoite.**
(a) Schematic drawing. **(b)** Scanning electron micrograph of intestinal surface, revealing (on the left) the lesion left behind by adhesive disc of a *Giardia* that has detached.
(b) Source: CDC/Dr. Stan Erlandsen

The trophozoite here is lying on its back and is revealing its adhesive disc.
- Kelly & Heidi

fact, it is the most common flagellate isolated in clinical specimens. Observed straight on, the trophozoite has a unique symmetrical heart shape with organelles positioned in such a way that it resembles a face **(figure 20.9).** Four pairs of flagella emerge from the ventral surface, which is concave and acts like a suction cup for attachment to a substrate. *Giardia* cysts are small and compact and contain four nuclei.

Typical symptoms include diarrhea of long duration, abdominal pain, and flatulence. Stools have a greasy, foul-smelling quality to them. Fever is usually not present.

▶ Transmission and Epidemiology of Giardiasis

Giardiasis has a complex epidemiological pattern. The protozoan has been isolated from the intestines of beavers, cattle, coyotes, cats, and human carriers, but the precise reservoir is unclear at this time. Although both trophozoites and cysts escape in the stool, the cysts play a greater role in transmission. Unlike other pathogenic flagellates, *Giardia* cysts can survive for 2 months in the environment. Cysts are usually ingested with water and food or swallowed after close contact with infected people or contaminated objects. Infection can occur with a dose of only 10 to 100 cysts.

▶ Prevention and Treatment

There is a vaccine against *Giardia* that can be given to animals, including dogs. No human vaccine is available. Avoiding drinking from freshwater sources is the major preventive measure that can be taken.

Treatment is with tinidazole or nitazoxanide.

Entamoeba

Amoebas are widely distributed in aqueous habitats and are frequent parasites of animals, but only a small number of them have the necessary virulence to invade tissues and cause serious pathology. One of the most significant pathogenic amoebas is *Entamoeba histolytica* (en″-tah-mee′-bah his″-toh-lit′-ih-kuh). The relatively simple life cycle of this parasite alternates between a large trophozoite that is motile by means of pseudopods and a smaller, compact, nonmotile cyst **(figure 20.10a–c).**

©Barry Barker/McGraw-Hill Education

The trophozoite lacks most of the organelles of other eukaryotes, and it has a large single nucleus that contains a prominent nucleolus called a *karyosome*. The mature cyst is encased in a thin yet tough wall and contains four nuclei as well as distinctive cigar-shaped bodies called *chromatoidal bodies,* which are actually dense clusters of ribosomes.

▶ Signs and Symptoms

Clinical amoebiasis exists in intestinal and extraintestinal forms. The initial targets of intestinal amoebiasis are the cecum, appendix, colon, and rectum. The amoeba secretes enzymes that dissolve tissues, and it actively penetrates deeper layers of the mucosa, leaving erosive ulcerations **(figure 20.10*d*).** This phase is marked by dysentery (bloody, mucus-filled stools), abdominal pain, fever, diarrhea, and weight loss. The most life-threatening manifestations of intestinal infection are hemorrhage, perforation, appendicitis, and tumorlike growths called amoebomas. Lesions in the mucosa of the colon have a characteristic flasklike shape.

Extraintestinal infection occurs when amoebas invade the viscera of the peritoneal cavity. The most common site of invasion is the liver. Here, abscesses containing necrotic tissue and trophozoites develop and cause amoebic hepatitis. Another rarer complication is pulmonary amoebiasis. Other infrequent targets of infection are the spleen, adrenals, kidney, skin, and brain. Severe forms of the disease result in about a 10% fatality rate.

▶ Transmission and Epidemiology of Amoebiasis

Entamoeba is harbored by chronic carriers whose intestines favor the encystment stage of the life cycle. Cyst formation cannot occur in active dysentery because the feces are so rapidly flushed from the body; but after recuperation, cysts are continuously shed in feces.

Infection is usually acquired by ingesting food or drink contaminated with cysts released by an asymptomatic carrier. The amoeba is thought to be carried in the intestines of one-tenth of the world's population, and it kills up to 100,000 people a year. Occurrence is highest in tropical regions (Africa, Asia, and Latin America), where "night soil" (human excrement) or untreated sewage is used to fertilize crops, and sanitation of water and food can be substandard. Although the prevalence of the disease is lower in the United States, as many as 10 million people could harbor the agent.

▶ Prevention and Treatment

Prevention of the disease relies on purification of water because no vaccine currently exists. Because regular chlorination of water supplies does not kill cysts, more rigorous methods such as boiling or iodine are required.

Figure 20.10 *Entamoeba histolytica.*
(a) A trophozoite containing a single nucleus, a karyosome, and red blood cells. **(b)** A mature cyst with four nuclei and two blocky chromatoidals. **(c)** Stages in excystment. Divisions in the cyst create four separate cells, or metacysts, that differentiate into trophozoites and are released. **(d)** Intestinal amoebiasis and dysentery of the cecum. Red patches are sites of amoebic damage to the intestinal mucosa. **(e)** Trophozoite of *Entamoeba histolytica* (green). Note the fringe of very fine pseudopods it uses to invade and feed on tissue.
(d) Source: CDC; (e) Eye of Science/Science Source

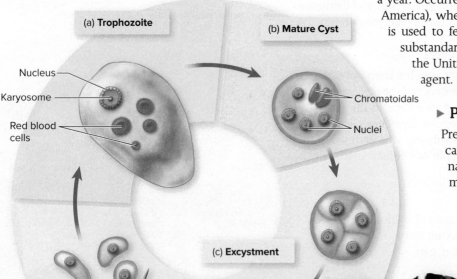

(a) **Trophozoite**
Nucleus
Karyosome
Red blood cells

(b) **Mature Cyst**
Chromatoidals
Nuclei

(c) **Excystment**

(d) **Erosion of the Intestine**
Ulcerations

(e)

Disease Table 20.3	Chronic Diarrhea			
Causative Organism(s)	Enteroaggregative *E. coli* (EAEC) **B** **G⁻**	*Cyclospora cayetanensis* **P**	*Giardia duodenalis* **P**	*Entamoeba histolytica* **P**
Most Common Modes of Transmission	Vehicle (food, water), fecal-oral	Fecal-oral, vehicle	Vehicle, fecal-oral, direct and indirect contact	Vehicle, fecal-oral
Virulence Factors	?	Invasiveness	Attachment to intestines alters mucosa	Lytic enzymes, induction of apoptosis, invasiveness
Culture/Diagnosis	Difficult to distinguish from other *E. coli*	Stool examination, PCR	Stool examination, ELISA	PCR, stool examination, ELISA, serology
Prevention	?	Washing, cooking food, personal hygiene	Water hygiene, personal hygiene	Water hygiene, personal hygiene
Treatment	None, or ciprofloxacin	TMP-SMZ	Tinidazole, nitazoxanide	Metronidazole or paramomycin
Fever Present	No	Usually	Not usually	Yes
Blood in Stool	Sometimes, mucus also	No	No, mucus present (greasy and malodorous)	Yes
Distinctive Features	Chronic in the malnourished	–	Frequently occurs in backpackers, campers	–
Epidemiological Features	Developing countries: 87% of chronic diarrhea in children >2 years old	United States: estimated 16,000 cases per year; internationally: endemic in 27 countries, mostly tropical	United States: estimated 1.2 million cases per year; internationally: prevalence rates from 2% to 5% in industrialized world; 40-50 million cases per year	Internationally: 40,000-100,000 deaths annually

Effective treatment usually involves the use of metronidazole (Flagyl) or paromomycin. Other drugs are given to relieve diarrhea and cramps, while lost fluid and electrolytes are replaced by oral or intravenous therapy. Infection with *E. histolytica* provokes antibody formation against several antigens, but permanent immunity is unlikely and reinfection can occur. **Disease Table 20.3.**

Tooth and Gum Infections

It is difficult to pinpoint exactly when the "normal biota biofilm" described for the oral environment becomes a "pathogenic biofilm." If left undisturbed, the biofilm structure eventually contains anaerobic bacteria that can damage the soft tissues and bones (referred to as the periodontium) surrounding the teeth. Also, the introduction of carbohydrates to the oral cavity can result in breakdown of hard tooth structure due to the production of acid by certain oral streptococci in the biofilm. These two separate circumstances are discussed here.

Dental Caries (Tooth Decay)

Dental caries is the most common infectious disease of human beings. The process involves the dissolution of solid tooth surface due to the metabolic action of bacteria. (**Figure 20.11** depicts the structure of a tooth.) The symptoms are often not noticeable but range from minor disruption in the outer (enamel) surface of the tooth to complete destruction of the enamel and then destruction of deeper

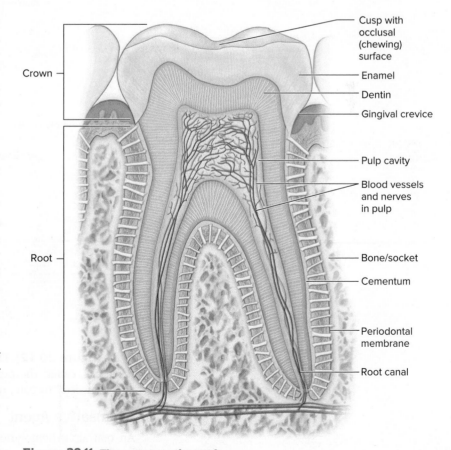

Figure 20.11 The anatomy of a tooth.

Labels: Cusp with occlusal (chewing) surface; Enamel; Dentin; Gingival crevice; Pulp cavity; Blood vessels and nerves in pulp; Bone/socket; Cementum; Periodontal membrane; Root canal; Crown; Root

Figure 20.12 Stages in plaque development and cariogenesis. **(a)** A microscopic view of pellicle and plaque formation, acidification, and destruction of tooth enamel. **(b)** Progress and degrees of cariogenesis.

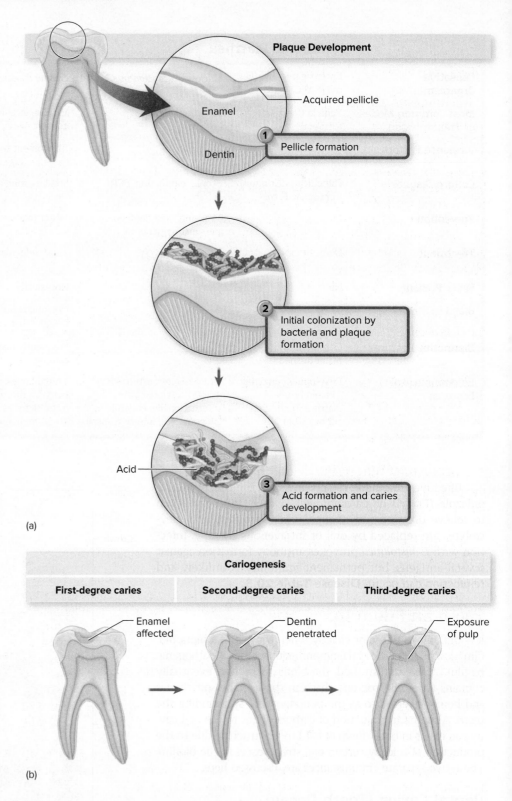

Plaque Development

Acquired pellicle

Enamel

Dentin

1 Pellicle formation

2 Initial colonization by bacteria and plaque formation

Acid

3 Acid formation and caries development

(a)

Cariogenesis

First-degree caries	Second-degree caries	Third-degree caries
Enamel affected	Dentin penetrated	Exposure of pulp

(b)

layers **(figure 20.12)**. Deeper lesions can result in infection to the soft tissue inside the tooth, called the pulp, which contains blood vessels and nerves. These deeper infections lead to pain, referred to as a "toothache."

▶ Causative Agent

An oral alpha-hemolytic streptococcus, *Streptococcus mutans*, was long thought to be the primary cause of dental caries. More recent techniques analyzing the oral microbiome in caries points to a polymicrobial consortium of bacteria that contributes to the destructive acid production. Note that in the absence of dietary carbohydrates, bacteria do not cause decay.

▶ Pathogenesis and Virulence Factors

In the presence of sucrose and, to a lesser extent, other carbohydrates, *S. mutans* and other streptococci produce sticky polymers of glucose called fructans and glucans. These adhesives help bind them to the smooth enamel surfaces and contribute to the sticky bulk of the plaque biofilm **(figure 20.13).** If mature plaque is not removed from sites that readily trap food, it can result in a carious lesion. This is due to the action of the streptococci and other bacteria that produce acid as they ferment the carbohydrates. If the acid is immediately flushed from the plaque and diluted in the mouth, it has little effect. However, in the denser regions of plaque, the acid can accumulate in direct contact with the enamel surface and lower the pH to below 5, which is acidic enough to begin to dissolve (decalcify) the calcium phosphate of the enamel in that spot. This initial lesion can remain localized in the enamel and can be repaired with various inert materials (fillings). Once the deterioration has reached the level of the dentin, tooth destruction speeds up and the tooth can be rapidly destroyed.

▶ Transmission and Epidemiology

The bacteria that cause dental caries are transmitted to babies and children by their close contacts, especially the mother or closest caregiver. There is evidence for transfer of oral bacteria between children in day care centers, as well.

▶ Culture and Diagnosis

Dental professionals diagnose caries based on the tooth condition. Culture of the lesion is not routinely performed.

▶ Prevention and Treatment

The best way to prevent dental caries is through dietary restriction of sucrose and other refined carbohydrates. Regular brushing and flossing to remove plaque are also important. Most municipal communities in the United States add trace amounts of fluoride to their drinking water because fluoride, when incorporated into the tooth structure, can increase tooth (as well as bone) hardness. Fluoride can also encourage the remineralization of teeth that have begun the demineralization process. The CDC estimates that the rate of tooth decay is decreased by 25% by the addition of fluoride to drinking water. Nearly 75% of people in the United States using public drinking water supplies are using fluoridated water. The World Health Organization recommends the use of fluoride in many different forms for all countries. Fluoride is also added to toothpastes and mouth rinses and can be applied in gel form.

Treatment of a carious lesion involves removal of the affected part of the tooth (or the whole tooth in the case of advanced caries), followed by restoration of the tooth structure with an artificial material. **Disease Table 20.4.**

Periodontitis

The initial stage of periodontal disease is **gingivitis,** the signs of which are swelling, loss of normal contour, patches of redness, and increased bleeding of the gums (gingiva). Spaces or pockets of varying depth also develop between the tooth and the gingiva. If this condition persists, a more serious disease called periodontitis results. The deeper involvement increases the size of the pockets and can cause bone

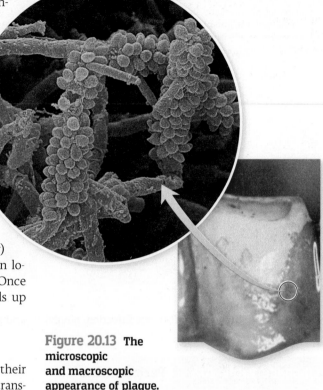

Figure 20.13 The microscopic and macroscopic appearance of plaque.
Disclosing tablets containing vegetable dye stain heavy plaque accumulations at the junction of the tooth and gingiva and other parts of the tooth. The blown-up image is a scanning electron micrograph of the plaque biofilm with long filamentous forms and "corn cobs" that are mixed bacterial aggregates.
David Scharf/Science Source (micrograph); BSIP SA/Alamy Stock Photo (tooth)

Disease Table 20.4	Dental Caries
Causative Organism(s)	A polymicrobial mixture of acid-producing bacteria
Most Common Modes of Transmission	Direct contact
Virulence Factors	Adhesion, acid production
Culture/Diagnosis	–
Prevention	Oral hygiene, fluoride supplementation
Treatment	Removal of diseased tooth material
Epidemiological Features	Globally, 60%-90% prevalence in school-age children

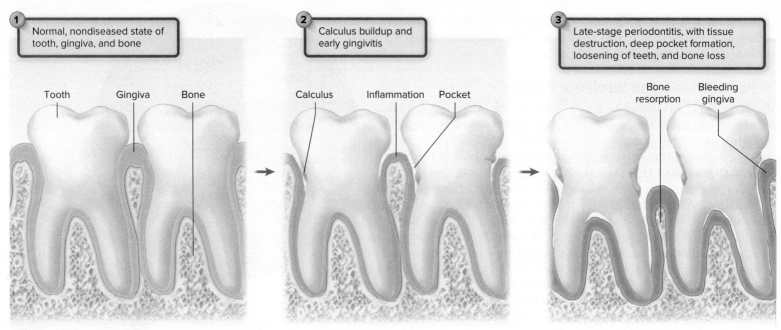

Figure 20.14 **Stages in soft-tissue infection, gingivitis, and periodontitis.**

Disease Table 20.5	Periodontitis
Causative Organism(s)	Polymicrobial community including some or all of *Tannerella forsythia, Aggregatibacter actinomycetemcomitans, Porphyromonas gingivalis*, others
Most Common Modes of Transmission	–
Virulence Factors	Induction of inflammation, enzymatic destruction of tissues
Culture/ Diagnosis	–
Prevention	Oral hygiene
Treatment	Removal of plaque and calculus, gum reconstruction, possibly anti-inflammatory treatments
Epidemiological Features	United States: smokers = 11%, nonsmokers = 2%; internationally: 10%-15% of adults

NCLEX® PREP

3. A client has been diagnosed with Guillain-Barré syndrome (GBS). Which observation found in the client's past medical history is relevant?
   a. *congestive heart failure*
   b. Campylobacter *infection*
   c. *dehydration*
   d. *diabetes mellitus*

resorption severe enough to loosen the tooth in its socket. If the condition is allowed to progress, the tooth can be lost **(figure 20.14).**

▶ **Causative Agents**

Data from the Human Microbiome Project reveal that the composition of the microbial *community*, rather than single organisms, relates directly to risk of dental caries or periodontitis. When the polymicrobial biofilms consist of the right combination of bacteria, such as the anaerobes *Tannerella forsythia* (formerly *Bacteroides forsythus*), *Aggregatibacter actinomycetemcomitans, Porphyromonas gingivalis*, and perhaps *Fusobacterium* and spirochete species, the periodontal destruction process begins. The most common predisposing condition occurs when the plaque becomes mineralized (calcified) with calcium phosphate crystals. This process produces a hard, porous substance called **calculus** above and below the gingival margin (edge) that can induce varying degrees of periodontal damage **(figure 20.15).** The presence of calculus leads to a series of inflammatory events that characterize the disease.

Recent research has also shown that some of these microbes can escape into the bloodstream and increase the risk of coronary diseases, autoimmune diseases, and even meningitis.

Most periodontal disease is treated by removal of calculus and plaque and maintenance of good oral hygiene. Often, surgery to reduce the depth of periodontal pockets is required. Antibiotic therapy, either systemic or applied in periodontal packings, may also be utilized. **Disease Table 20.5.**

Mumps

The word *mumps* is Old English for "lump" or "bump." The symptoms of this viral disease are so distinctive that Hippocrates clearly characterized it in the fifth century BC as a self-limited, mildly epidemic illness associated with painful swelling at the angle of the jaw **(figure 20.16).**

▶ **Signs and Symptoms**

After an average incubation period of 2 to 3 weeks, symptoms of fever, nasal discharge, muscle pain, and malaise develop. These may be followed by inflammation of the salivary glands (especially the parotids), producing the classic gopherlike swelling of the cheeks on one or both sides. Swelling of the gland is called parotitis, and it can cause considerable discomfort. Viral multiplication in salivary glands is followed by invasion of other organs, especially the testes, ovaries, thyroid gland, pancreas, meninges, heart, and kidney. Despite the invasion of multiple organs, the prognosis of most infections is complete, uncomplicated recovery with permanent immunity.

Complications in Mumps In 20% to 30% of young adult males, mumps infection localizes in the epididymis and testis, usually on one side only. The resultant syndrome of orchitis and epididymitis may be rather painful, but no permanent damage usually occurs.

▶ Transmission and Epidemiology of Mumps Virus

Humans are the exclusive natural hosts for the mumps virus. It is communicated primarily through salivary and respiratory secretions. Most cases occur in children under the age of 15, and as many as 40% are subclinical. Because lasting immunity follows any form of mumps infection, no long-term carrier reservoir exists in the population. The incidence of mumps in the United States spiked in 2006, and in 2016 and 2017. In each of these years, there were more than 6,000 reported cases. Mumps is most likely to break out in settings where people gather closely, such as in college dorms, sporting clubs, etc.

▶ Prevention and Treatment

The general pathology of mumps is mild enough that symptomatic treatment to relieve fever, dehydration, and pain is usually adequate. Vaccine recommendations call for a dose of MMR at 12 to 15 months and a second dose at 4 to 6 years. Health care workers and college students who haven't already had both doses are advised to do so. In outbreak situations, the CDC may recommend an additional MMR dose for at-risk people. **Disease Table 20.6.**

Gastritis and Gastric Ulcers

The curved cells of *Helicobacter* were first detected by J. Robin Warren in 1979 in stomach biopsies from ulcer patients. He and a partner, Barry J. Marshall, isolated the microbe in culture and even served as guinea pigs by swallowing a large inoculum to prove that it would cause gastric ulcers. Warren and Marshall won the Nobel Prize in Medicine in 2005 for their discovery.

▶ Signs and Symptoms

Gastritis is experienced as sharp or burning pain emanating from the abdomen. Gastric or peptic ulcers are actual lesions in the mucosa of the stomach (gastric ulcers) or in the uppermost portion of the small intestine (duodenal ulcers). Severe ulcers can be accompanied by bloody stools, vomiting, or both. The symptoms are often worse at night, after eating, or under conditions of psychological stress.

The sixth most common cancer in the world is stomach cancer, and ample evidence suggests that long-term infection with *Helicobacter pylori* is a major contributing factor.

▶ Causative Agent

Helicobacter pylori is a curved gram-negative rod, closely related to *Campylobacter,* which we studied earlier in this chapter.

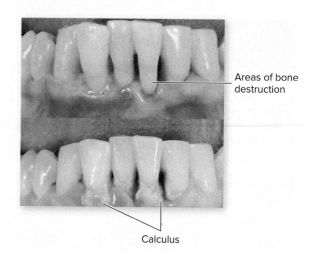

Areas of bone destruction

Calculus

Figure 20.15 The effect of calculus. The bottom picture shows the lower front teeth before cleaning. Note the deposits of calculus on the root surfaces. After removal of the calculus, the extent of bone destruction and gum recession, caused by the calculus, can be seen.
danielzgombic/Getty Images

Figure 20.16 The external appearance of a swollen parotid gland in mumps (parotitis).
Source: CDC

Disease Table 20.6	Mumps
Causative Organism(s)	Mumps virus (genus *Paramyxovirus*) 🅥
Most Common Modes of Transmission	Droplet contact
Virulence Factors	Spike-induced syncytium formation
Culture/Diagnosis	Clinical, fluorescent Ag tests, ELISA for Ab
Prevention	MMR live attenuated vaccine
Treatment	Supportive
Epidemiological Features	United States: fluctuates between a few hundred cases a year and several thousand; internationally: epidemic peaks every 2–5 years

Disease Table 20.7	Gastritis and Gastric Ulcers
Causative Organism(s)	*Helicobacter pylori* Ⓑ G⁻
Most Common Modes of Transmission	Direct contact, fecal oral
Virulence Factors	Adhesins, urease
Culture/Diagnosis	Endoscopy, urea breath test, stool antigen test
Prevention	None
Treatment	Acid suppression and antibiotics
Epidemiological Features	United States: infection (not disease) rates at 35% of adults; internationally: infection rates at 50%

Medical Moment

Assessing Jaundice

Jaundice is caused by elevated levels of bilirubin in the blood. How is jaundice recognized in a patient? Jaundice progresses caudally from face to trunk and extremities, so you should begin by looking at the patient's face. In some patients, jaundice will be easily recognized because the patient's sclerae (the white portions of the eye) will be yellow—sometimes startlingly so! The skin may also have a yellowish discoloration. In dark-skinned individuals, especially older adults, the sclerae may not be reliable indicators as the sclerae may have a yellowish tint even in patients without jaundice. You can assess the palms and soles of the feet in patients with dark skin. Blanching the skin (applying pressure to the skin with your fingertip) also helps to reveal the underlying skin color. The mucous membranes of affected individuals will also reveal a yellow discoloration.

Q. Why are the effects of jaundice so widespread in the body?

Answer in Appendix B.

▶ Pathogenesis and Virulence Factors

Once the bacterium passes into the gastrointestinal tract, it bores through the outermost mucous layer that lines the stomach epithelial tissue. Then it attaches to specific binding sites on the cells and entrenches itself. It produces an enzyme, urease, that breaks down urea in its vicinity to form carbon dioxide and ammonia. In this way it can neutralize the highly acidic environment of the stomach.

Before the bacterium was discovered, spicy foods, high-sugar diets (which increase acid levels in the stomach), and psychological stress were considered to be the cause of gastritis and ulcers. Now it appears that these factors merely aggravate the underlying infection.

▶ Transmission and Epidemiology

The bacterium has probably been in human stomachs for tens of thousands of years. However, there is evidence that the percentage of people with *Helicobacter* in their stomachs is decreasing with each generation. Scientists suspect that this is due to our more sanitary environment and, of course, the use of antibiotics. While you might think this is a good thing, in light of the fact that it can cause ulcers and stomach cancer, there is evidence to suggest that its absence leads to higher incidences of acid reflux and even asthma. It should not surprise us by now that a single microbe can have both positive and negative consequences, depending on the circumstances. It should also not surprise us that obliterating a long-standing member of our microbiome in a short period of (evolutionary) time could lead to health imbalances and consequences.

H. pylori is probably transmitted from person to person by the oral-oral or fecal-oral route. It seems to be acquired early in life and carried asymptomatically unless its activities begin to damage the digestive mucosa. Because other animals are also susceptible to *H. pylori* and even develop chronic gastritis, it has been proposed that the bacterium is a zoonosis transmitted from an animal reservoir. The bacterium has also been found in water sources, suggesting that perhaps proper sanitation may reduce transmission.

▶ Prevention and Treatment

The only preventive approaches available currently are those that diminish some of the aggravating factors just mentioned. Many over-the-counter remedies offer symptom relief; most of them act to neutralize stomach acid. The best treatment is a course of antibiotics augmented by acid suppressors. *Helicobacter* is becoming resistant to a number of antibiotics. It is best to find out what types of antibiotics the patient has taken previously before deciding on a treatment. **Disease Table 20.7.**

Hepatitis

When certain viruses infect the liver, they cause **hepatitis,** an inflammatory disease marked by necrosis of hepatocytes (liver cells) and an inflammatory response that swells and disrupts the liver architecture. This pathologic change interferes with the liver's excretion of bile pigments such as bilirubin into the intestine. When bilirubin, a greenish-yellow pigment, accumulates in the blood and tissues, it causes **jaundice,** a yellow tinge in the skin and eyes. The condition can be caused by a variety of different viruses, including cytomegalovirus, Epstein-Barr virus, and a virus known as GBV-C. The others are all called "hepatitis viruses" but only because they all can cause this inflammatory condition in the liver. They are quite different from one another. While there are some recently discovered hepatitis viruses, they are not yet well characterized so we will cover the five that are well understood, named hepatitis A-E.

Note that noninfectious conditions can also cause inflammation and disease in the liver, including some autoimmune conditions, drugs, and alcohol overuse.

Hepatitis A and E Viruses

Hepatitis A virus (HAV) and hepatitis E virus (HEV) are both single-stranded non-enveloped RNA viruses. These two viruses are considered together because they are both transmitted through the fecal-oral route, and both cause relatively minor, self-limited hepatitis. The important exception to this is when HEV infects pregnant women, in whom it causes a 10% to 30% fatality rate. It can also be dangerous to people who have received an organ transplant and are on immunosuppressive therapy.

▶ Signs and Symptoms

Most infections by these viruses are either subclinical or accompanied by vague, flulike symptoms. In more overt cases, the presenting symptoms may include jaundice and swollen liver. The viruses are not oncogenic (cancer causing), and in most everyone besides pregnant women and transplant patients, complete uncomplicated recovery results.

▶ Transmission and Epidemiology

In general, the disease is associated with deficient personal hygiene and lack of public health measures. In countries with inadequate sewage control, most outbreaks are associated with fecally contaminated water and food. Most infections result from close institutional contact, unhygienic food handling, eating infected shellfish, sexual transmission, or travel to other countries.

Hepatitis A occasionally can be spread by blood or blood products, but this is the exception rather than the rule. In developing countries, children are the most common victims because exposure to the virus tends to occur early in life, whereas in North America and Europe, more cases appear in adults. Because the virus is not carried chronically, the principal reservoirs are asymptomatic, short-term carriers (often children) or people with clinical disease.

▶ Prevention and Treatment

Prevention of hepatitis A is based primarily on immunization. An inactivated viral vaccine (Havrix) has been in use since the mid-1990s. Short-term protection can be conferred by passive immune globulin. When people are exposed to infection via contact with an infected person or with a restaurant that was a source of a known outbreak, passive immunoglobulin or the vaccine can be administered. Administering a combined hepatitis A/hepatitis B vaccine, called Twinrix, is recommended for people who may be at risk for both diseases, such as people with chronic liver dysfunction, intravenous drug users, and anyone engaging in anal-oral intercourse. Travelers to areas with high rates of both diseases should obtain vaccine coverage as well. Hepatitis E has no vaccine.

No specific medicine is available for hepatitis A or E once the symptoms begin. Drinking lots of fluids and avoiding liver irritants such as aspirin or alcohol will speed recovery.

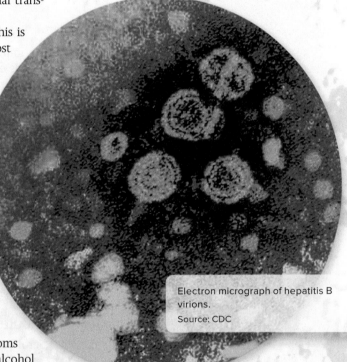

Electron micrograph of hepatitis B virions.
Source: CDC

Hepatitis B Virus (and Hepatitis D)

Hepatitis B virus (HBV) is an enveloped DNA virus in the family Hepadnaviridae. Intact viruses are often called Dane particles. The genome is partly double-stranded and partly single-stranded. **Hepatitis D** is an enveloped RNA virus. It is actually a *subvirus satellite* of HBV; that is, it can propagate only in the presence of HBV.

▶ Signs and Symptoms

In addition to the direct damage to liver cells just outlined, the spectrum of hepatitis disease may include fever, chills, malaise, anorexia, abdominal discomfort, diarrhea, and nausea. Rashes may appear and arthritis may occur. Hepatitis B infection can be very serious, even life-threatening. A small number of patients develop glomerulonephritis and arterial inflammation. Complete liver regeneration and restored function occur in most patients; however, a small number of patients develop chronic liver disease in the form of necrosis or cirrhosis (permanent liver scarring and loss of tissue). In some cases, chronic HBV infection can lead to liver cancer.

> ⚕ **NCLEX® PREP**
>
> **4.** A patient with hepatitis is admitted to an inpatient medical unit. The nurse educates her about the clinical manifestations that accompany the disease. All of the following findings may be associated with jaundice, except:
>
> **a.** *yellow sclerae.*
>
> **b.** *pruritus.*
>
> **c.** *clay-colored stools.*
>
> **d.** *decreased bilirubin levels.*
>
> **e.** *dark-colored urine.*

Patients who become infected as children have significantly higher risks of long-term infection and disease. In fact, 90% of neonates infected at birth develop chronic infection, as do 30% of children infected between the ages of 1 and 5, but only 6% of persons infected after the age of 5. This finding is one of the major justifications for the routine vaccination of children. Also, infection becomes chronic more often in men than in women. The mortality rate is 15% to 25% for people with chronic infection.

When patients are infected with both HBV and hepatitis D virus (HDV), the disease can become more severe and is more likely to progress to permanent liver damage.

▶ Pathogenesis and Virulence Factors

The hepatitis B virus enters the body through a break in the skin or mucous membrane or by injection into the bloodstream. Eventually, it reaches the liver cells (hepatocytes) where it multiplies and releases viruses into the blood during an incubation period of 4 to 24 weeks (7 weeks average). Surprisingly, the majority of those infected exhibit few overt symptoms and eventually develop an immunity to HBV, but some people experience the symptoms described earlier.

▶ Transmission and Epidemiology

An important factor in the transmission pattern of hepatitis B virus is that it multiplies exclusively in the liver, which continuously seeds the blood with viruses. Electron microscopic studies have revealed up to 10^7 virions per milliliter of infected blood. Even a minute amount of blood (a *millionth* of a milliliter) can transmit infection. The abundance of circulating virions is so high and the minimal dose so low that such simple practices as sharing a toothbrush or a razor can transmit the infection. HBV can be transmitted via semen and vaginal secretions. Vertical transmission is possible, and it predisposes the child to development of the carrier state and increased risk of liver cancer. It is sometimes known as *serum hepatitis*.

This virus is one of the major infectious concerns for health care workers. Needlesticks can easily transmit the virus, and therefore most workers are required to have the full series of HBV vaccinations. Unlike the more notorious HIV, HBV remains infective for days in dried blood, for months when stored in serum at room temperature, and for decades if frozen. Although it is not inactivated after 4 hours of exposure to 60°C, boiling for the same period can destroy it. Disinfectants containing chlorine, iodine, and glutaraldehyde show potent anti-hepatitis B activity.

▶ Culture and Diagnosis

Serological tests can detect either virus antigen or antibodies. Radioimmunoassay and ELISA testing are used to detect the important surface antigen (S antigen) of HBV very early in infection. Antibody tests are most valuable in patients who are negative for the antigen.

▶ Prevention and Treatment

The primary prevention for HBV infection is vaccination. The vaccines are recombinant, containing the pure surface antigen cloned in yeast cells. Vaccination is a must for medical and dental workers and students, patients receiving multiple transfusions, immunodeficient persons, and cancer patients. The vaccine is also now strongly recommended for all newborns as part of a routine immunization schedule. A vaccine called Pediarix contains protection against hepatitis B, diphtheria, tetanus, pertussis, and polio. It is recommended for children between 6 weeks and seven years. As just mentioned, a combined vaccine for HAV/HBV may be appropriate for certain people.

Passive immunization with hepatitis B immune globulin (HBIG) gives significant immediate protection to people who have been exposed to the virus through needle puncture, broken blood containers, or skin and mucosal contact with blood. Another group for whom passive immunization is highly recommended is neonates born to infected mothers.

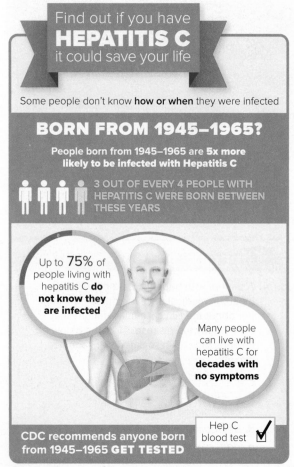

Find out if you have
HEPATITIS C
it could save your life

Some people don't know **how or when** they were infected

BORN FROM 1945–1965?

People born from 1945–1965 are **5x more likely to be infected with Hepatitis C**

3 OUT OF EVERY 4 PEOPLE WITH HEPATITIS C WERE BORN BETWEEN THESE YEARS

Up to **75%** of people living with hepatitis C **do not know they are infected**

Many people can live with hepatitis C for **decades with no symptoms**

CDC recommends anyone born from 1945–1965 GET TESTED

Hep C blood test ✓

Source: CDC

Disease Table 20.8	Hepatitis		
Causative Organism(s)	Hepatitis A or E virus Ⓥ	Hepatitis B virus Ⓥ	Hepatitis C virus Ⓥ
Most Common Modes of Transmission	Fecal-oral, vehicle	Parenteral (blood contact), direct contact (especially sexual), vertical	Parenteral (blood contact), vertical
Virulence Factors	–	Latency	Core protein suppresses immune function?
Culture/Diagnosis	IgM serology	Serology (ELISA, radioimmunoassay)	Serology, also PCR
Prevention	Hepatitis A vaccine or combined HAV/HBV vaccine	HBV recombinant vaccine	–
Treatment	HAV: hepatitis A vaccine or immune globulin; HEV: immune globulin	Interferon, tenofovir, or entecavir	Sofosbuvir/ledipasvir combination drug (DAA)
Incubation Period	2-4 weeks	1-6 months	2-8 weeks
Epidemiological Features	Hepatitis A, United States: 20,000 cases annually and 40% of adults show evidence of prior infection; internationally: 1.4 million cases per year; hepatitis E, internationally: 20 million infections per year; 60% in East and Southeast Asia	United States: 19,000 new cases per year; 800,000 to 2.2 million have chronic infection; internationally: 240 million	United States: estimated 17,000 new cases per year; 3-4 million with chronic HCV; internationally: 150 million chronically infected

Mild cases of hepatitis B may be managed by symptomatic treatment and supportive care. Severe or chronic infection can be controlled with recombinant human interferon, tenofovir, or entevir. Each of these can help to slow virus multiplication and prevent liver damage in many but not all patients. None of the drugs is considered curative.

©Christopher Kerrigan/McGraw-Hill Education

Hepatitis C Virus

Hepatitis C virus (HCV) is sometimes referred to as the "silent epidemic" because more than 2.4 million Americans are infected with the virus, but it takes many years to cause noticeable symptoms. Liver failure from hepatitis C is one of the most common reasons for liver transplants in this country. Hepatitis C is an RNA virus in the Flaviviridae family. It used to be known as "non-A non-B" virus. It is usually diagnosed with a blood test for antibodies to the virus.

▶ Signs and Symptoms

People have widely varying experiences with this infection. It shares many characteristics of hepatitis B disease, but it is much more likely to become chronic. Of those infected, 75% to 85% will remain infected indefinitely. (In contrast, only about 6% of persons who acquire hepatitis B after the age of 5 will be chronically infected.) With HCV infection, it is possible to have severe symptoms without permanent liver damage, but it is more common to have chronic liver disease even if there are no overt symptoms. Cancer may also result from chronic hepatitis C virus (HCV) infection. Worldwide, HBV infection is the most common cause of liver cancer, but in the United States it is more likely to be caused by HCV. The CDC recommends that all baby boomers (those born between 1945 and 1965) be tested for HCV.

▶ Transmission and Epidemiology

This virus is acquired in similar ways to HBV. It is more commonly transmitted through blood contact (both "sanctioned," such as in blood transfusions, and "unsanctioned," such as needle sharing by injecting drug users) than through transfer of other body fluids. Vertical transmission is also possible.

Anyone with a history of exposure to blood products or organs before 1992 (when effective screening became available) is at higher risk for this infection, as is anyone with a history of injecting drug use.

▶ Prevention and Treatment

There is currently no vaccine for hepatitis C. In 2013 a two-drug regimen became available that has had excellent results, including complete cure. The treatment is generally known as Direct Acting Agents (DAA). **Disease Table 20.8.**

20.3 LEARNING OUTCOMES—Assess Your Progress

5. List the possible causative agents, modes of transmission, virulence factors, diagnostic techniques, and prevention/treatment for the "Highlight Disease" acute diarrhea.

6. Discuss important features of the conditions food poisoning and chronic diarrhea.

7. Discuss important features of the two categories of oral conditions: dental caries and periodontitis.

8. Identify the most important features of mumps, gastritis, and gastric ulcers.

9. Differentiate among the main types of hepatitis, and discuss the causative agents, mode of transmission, diagnostic techniques, prevention, and treatment of each.

Figure 20.17 Four basic helminth life and transmission cycles.

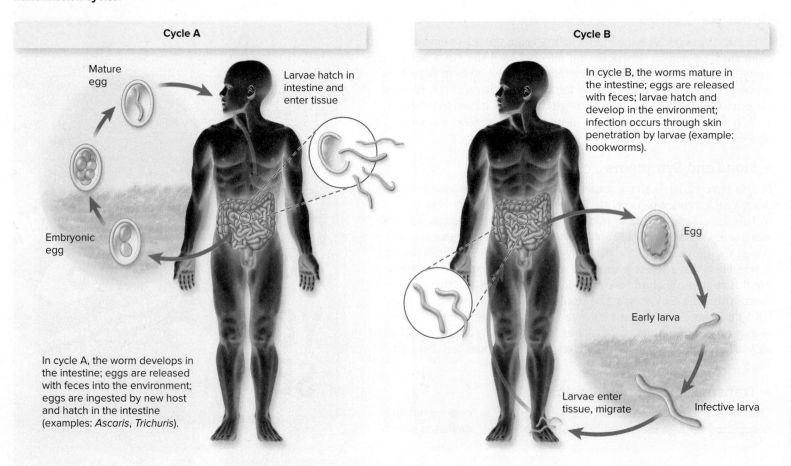

Cycle A

Mature egg

Larvae hatch in intestine and enter tissue

Embryonic egg

In cycle A, the worm develops in the intestine; eggs are released with feces into the environment; eggs are ingested by new host and hatch in the intestine (examples: *Ascaris, Trichuris*).

Cycle B

In cycle B, the worms mature in the intestine; eggs are released with feces; larvae hatch and develop in the environment; infection occurs through skin penetration by larvae (example: hookworms).

Egg

Early larva

Infective larva

Larvae enter tissue, migrate

20.4 Gastrointestinal Tract Diseases Caused by Helminths

Helminths that parasitize humans are amazingly diverse, ranging from barely visible roundworms (0.3 mm) to huge tapeworms (25 m long). In the introduction to these organisms in chapter 4, we grouped them into three categories: nematodes (roundworms), trematodes (flukes), and cestodes (tapeworms), and we discussed basic characteristics of each group. You may want to review those sections before continuing. In this section, we examine the intestinal diseases caused by helminths. Although they can cause symptoms that may be mistaken for some of the diseases discussed elsewhere in this chapter, helminthic diseases are usually accompanied by an additional set of symptoms that arise from the host response to helminths. Worm infection usually provokes an increase in granular leukocytes called eosinophils, which have a specialized capacity to destroy multicellular parasites. This increase, termed **eosinophilia,** is a hallmark of helminthic infection and is detectable in blood counts. If the following symptoms occur coupled with eosinophilia, helminthic infection should be suspected.

Helminthic infections may be acquired through the fecal-oral route or through penetration of the skin, but most of these organisms spend part of their lives in the intestinal tract. **Figure 20.17** depicts the four different types of life cycles of the helminths. While the worms are in the intestines, they can produce a gamut of intestinal symptoms. Some of them also produce symptoms outside of the intestines.

Clinical Considerations

We start with diagnosis, pathogenesis and prevention, and treatment of the helminths as a group in the next subsections. We'll then highlight one of the most common forms of helminthic disease, and finish with summaries of the others.

Trained sushi chefs are skilled at detecting helminth infection in the raw fish they prepare.
©George Doyle/stockbyte/Getty Images

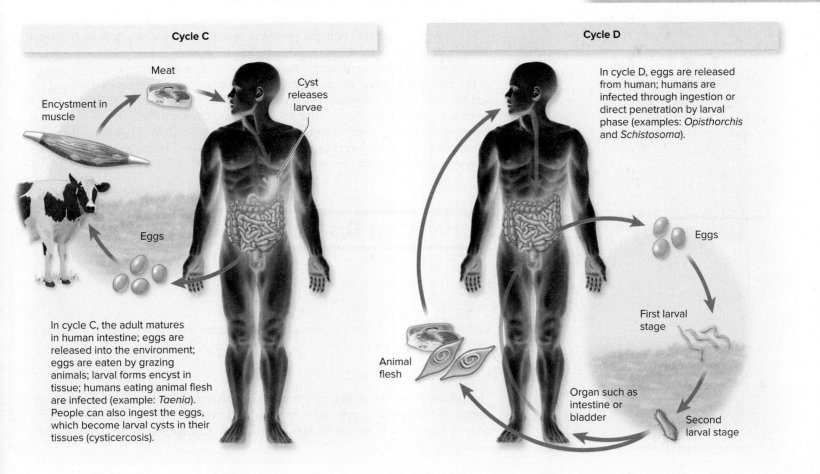

Cycle C

Meat

Encystment in muscle

Cyst releases larvae

Eggs

In cycle C, the adult matures in human intestine; eggs are released into the environment; eggs are eaten by grazing animals; larval forms encyst in tissue; humans eating animal flesh are infected (example: *Taenia*). People can also ingest the eggs, which become larval cysts in their tissues (cysticercosis).

Cycle D

In cycle D, eggs are released from human; humans are infected through ingestion or direct penetration by larval phase (examples: *Opisthorchis* and *Schistosoma*).

Eggs

First larval stage

Animal flesh

Organ such as intestine or bladder

Second larval stage

▶ Pathogenesis and Virulence Factors in General

Helminths have numerous adaptations that allow them to survive in their hosts. They have specialized mouthparts for attaching to tissues and for feeding, enzymes with which they liquefy and penetrate tissues, and a cuticle or other covering to protect them from host defenses. In addition, their organ systems are usually reduced to the essentials: getting food and processing it, moving, and reproducing. The damage they cause in the host is very often the result of the host's response to the presence of the invader.

Many helminths have more than one host during their lifetimes. If this is the case, the host in which the *adult* worm is found is called the definitive host (usually a vertebrate). Sometimes the actual definitive host is not the host usually used by the parasite but an accidental bystander. Humans often become the accidental definitive hosts for helminths whose normal definitive host is a cow, pig, or fish. *Larval* stages of helminths are found in intermediate hosts. Humans can serve as intermediate hosts, too. Helminths may require no intermediate host at all or may need one or more intermediate hosts to complete their entire life cycle.

▶ Diagnosis in General

Diagnosis of almost all helminthic infections follows a similar series of steps. The most common diagnostic method is the "O&P test," which is microscopic examination of a stool sample to look for ova or parasites. A differential blood count showing increased eosinophils and serological tests indicating sensitivity to helminthic antigens all provide indirect evidence of worm infection. A history of travel to the tropics or immigration from those regions is also helpful, even if it occurred years ago, because some flukes and nematodes persist for decades. The worms are sufficiently distinct in morphology that positive identification can be based on any stage, including eggs. That said, not all of these diseases result in eggs or larval stages that can easily be found in stool.

▶ Prevention and Treatment in General

There are no vaccines for the helminthic infections described here. In populations in which the infections are common, prophylactic treatment twice a year with antihelminthic drugs has been shown to keep people healthy.

Although several useful antihelminthic medications exist, the cellular physiology of the eukaryotic parasites resembles that of humans, and drugs toxic to them can also be toxic to us. Some antihelminthic drugs suppress a metabolic process that is more important to the worm than to the human. Others inhibit the worm's movement and prevent it from maintaining its position in a certain organ. Note that some helminths have developed resistance to the drugs used to treat them. In some cases, surgery may be necessary to remove worms or larvae.

HIGHLIGHT DISEASE

Helminthic Infections: Intestinal Distress as the Primary Symptom

Both tapeworms and roundworms can infect the intestinal tract in such a way as to cause primary symptoms there.

Enterobius vermicularis

This nematode is often called the pinworm, or seatworm. It is the most common worm disease of children in temperate zones. The transmission of this roundworm is of the cycle A type. Freshly deposited eggs have a sticky coating that causes them to lodge beneath the fingernails and to adhere to fomites. Upon drying, the eggs become airborne and settle in house dust. Eggs are ingested from contaminated food or drink and from self-inoculation from one's own fingers. Eggs hatch in the small intestine and release larvae that migrate to the large intestine. There the larvae mature into adult worms and mate.

The hallmark symptom of this condition is pronounced anal itching when the mature female emerges from the anus and lays eggs. Although infection is not fatal, and most cases are asymptomatic, the afflicted child can suffer from disrupted sleep and sometimes nausea, abdominal discomfort, and diarrhea. A simple rapid test can be performed by pressing a piece of transparent adhesive tape against the anal skin and then applying it to a slide for microscopic examination. When one member of the family is diagnosed, the entire family should be tested and/or treated because it is likely that multiple members are infected. Eggs are not present in stool, so standard "O&P" is not useful.

Other Helminths Responsible for Intestinal Distress

Trichuris trichiura, the whipworm, follows the cycle A lifestyle. Trichuriasis has its highest incidence in areas of the tropics and subtropics that have poor sanitation. Symptoms of this infection may include localized hemorrhage of the bowel caused by worms burrowing and piercing intestinal mucosa. This can also provide a portal of entry for secondary bacterial infection. Heavier infections can cause dysentery, loss of muscle tone, and rectal prolapse, which can prove fatal in children.

The tapeworm *Diphyllobothrium latum* has an intermediate host in fish. It follows cycle C. It is common in the Great Lakes, Alaska, and Canada. Mammals, including humans, act as definitive hosts. It develops in the intestine and can cause long-term symptoms. It can be transmitted in raw food such as sushi and sashimi made from salmon. It is the largest human tapeworm known, growing up to 30 feet.

Hymenolepis species are small tapeworms and are the most common human tapeworm infections in the world. They follow cycle C. There are two species: *Hymenolepis nana*, known as the dwarf tapeworm because it is only 15 to 40 mm in length, and *H. diminuta*, the rat tapeworm, which is usually 20 to 60 cm in length as an adult.

Disease Table 20.9	Intestinal Distress			
Causative Organism(s)	*Enterobius vermicularis* (pinworm) Ⓗ	*Trichuris trichiura* (whipworm) Ⓗ	*Diphyllobothrium latum* (fish tapeworm) Ⓗ	*Hymenolepis nana* and *H. diminuta* Ⓗ
Most Common Modes of Transmission	Cycle A: vehicle (food, water), fomites, self-inoculation	Cycle A: vehicle (soil), fecal-oral	Cycle C: vehicle (seafood)	Cycle C: vehicle (ingesting insects), fecal-oral
Virulence Factors	–	Burrowing and invasiveness	Vitamin B$_{12}$ usage	–
Culture/Diagnosis	Adhesive tape + microscopy	Blood count, serology, egg or worm detection	Blood count, serology, egg or worm detection	Blood count, serology, egg or worm detection
Prevention	Hygiene	Hygiene, sanitation	Cook meat	Hygienic environment
Treatment	Mebendazole (consider for whole family)	Mebendazole	Praziquantel	Praziquantel
Distinctive Features	Common in United States	Humans sole host	Large tapeworm; anemia	Most common tapeworm infection
Epidemiological Features	Most common nematode infection in United States; 40 million cases per year	United States: prevalence approx. 0.1%; internationally: prevalence as high as 80% in Southeast Asia, Africa, the Caribbean, and Central and South America	Estimated 20 million infections worldwide	United States: prevalence approximately 0.4%; internationally: the single most prevalent tapeworm infection

Helminthic Infections: Intestinal Distress Accompanied by Migratory Symptoms

A diverse group of helminths enter the body as larvae or eggs, mature to the worm stage in the intestine, and then migrate into the circulatory and lymphatic systems, after which they travel to the heart and lungs, migrate up the respiratory tree to the throat, and are swallowed. This journey puts the mature worms in the intestinal tract, where they then take up residence. All of these conditions, in addition to causing symptoms in the digestive tract, may induce inflammatory reactions along their migratory routes, resulting in eosinophilia and, during their lung stage, pneumonia. Examples of this type of infection appear in **Disease Table 20.10**.

Cysticercosis

Taenia solium is a tapeworm. Adult worms are usually around 5 meters long and have a scolex with hooklets and suckers to attach to the intestine **(figure 20.18)**. This helminth follows cycle C in figure 20.17, in which humans are infected by eating animal flesh that contains the worm eggs, or even the worms themselves.

Disease caused by *T. solium* (the pig tapeworm) is distributed worldwide but is mainly concentrated in areas where humans live in close proximity with pigs or eat undercooked pork. The cycle starts in the pigs, when the eggs hatch in the small intestine and the released larvae migrate throughout the organs. Ultimately, they encyst in the pig's muscles, becoming *cysticerci*, young tapeworms that are the infective stage for humans. When humans ingest a live cysticercus in pork, the coat is digested and the organism is flushed into the intestine, where it firmly attaches by the scolex and develops into an adult tapeworm. Infection with *T. solium* can take another form when humans ingest the tapeworm eggs rather than cysticerci. Although humans are not the usual intermediate hosts, the eggs can still hatch in the intestine, releasing tapeworm larvae that migrate to all tissues. They form bladderlike sacs throughout the body that can cause serious damage. This condition is called **cysticercosis (figure 20.19)**, one of the five neglected parasitic infections (NPIs) in the United States. It is estimated that tens of thousands of Latinos living in the United States are affected

Disease Table 20.10	Intestinal Distress Plus Migratory Symptoms		
Causative Organism(s)	*Toxocara* species 🄷	*Ascaris lumbricoides* (intestinal roundworm) 🄷	*Necator americanus* and *Ancylostoma duodenale* (hookworms) 🄷
Most Common Modes of Transmission	Cycle A: dog or cat feces	Cycle A: vehicle (soil/fecal-oral), fomites, self-inoculation	Cycle B: vehicle (soil), fomite
Virulence Factors	—	Induction of hypersensitivity, adult worm migration, abdominal obstruction	Induction of hypersensitivity, adult worm migration, abdominal obstruction
Culture/Diagnosis	Blood count, serology, egg or worm detection	Blood count, serology, egg or worm detection	Blood count, serology, egg or worm detection
Prevention	Hygiene	Hygiene	Sanitation
Treatment	Albendazole	Albendazole	Albendazole
Distinctive Features	Can cause migration symptoms or blindness	Most cases mild, unnoticed	Penetrates skin, serious intestinal symptoms
Epidemiological Features	Nearly 100% of newborn puppies in United States infected; 14% of people in United States have been infected; designated a "neglected parasitic infection" in the United States by the CDC	Internationally: up to 25% prevalence, 80,000–100,000 deaths per year	United States: widespread in Southeast until early 1900s; internationally: 800 million infected

Disease Table 20.11	Cysticercosis
Causative Organism(s)	*Taenia solium* (pork tapeworm) **H**
Most Common Modes of Transmission	Cycle C: vehicle (pork), fecal-oral
Virulence Factors	--
Culture/Diagnosis	Blood count, serology, egg or worm detection
Prevention	Cook meat, avoid pig feces
Treatment	Praziquantel
Distinctive Features	Ingesting larvae embedded in pork leads to intestinal tapeworms; ingesting eggs (fecal-oral route) causes cysticercosis, larval cysts embedded in tissue of new host
Epidemiological Features	United States: a common cause of seizures; designated a "neglected parasitic infection" in the United States by the CDC; internationally: very common in Latin America and Asia

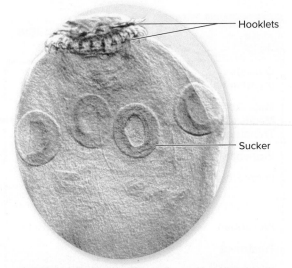

(a) Tapeworm scolex showing sucker and hooklets.

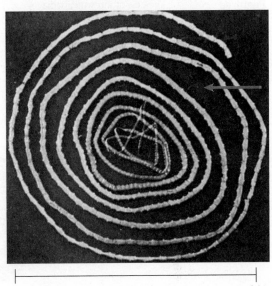

24 in

(b) Adult tapeworm. The arrow points to the (barely visible) scolex; the remainder of the tape, called the strobila, has a total length of 5 meters.

Figure 20.18 Tapeworm characteristics.

(a) Eric Grave/SPL/Getty Images; (b) Dr. Dickson D. Despommier

by cysticercosis, but it is not often recognized because American physicians may not know to look for it. A particularly nasty form of this condition is neurocysticercosis, in which the larvae encyst in the brain. It is estimated to be responsible for 10% of seizures requiring emergency room visits in some U.S. cities. **Disease Table 20.11.**

Schistosomiasis: Liver Disease

When liver swelling or malfunction is accompanied by eosinophilia, **schistosomiasis** should be suspected. Schistosomiasis has afflicted humans for thousands of years. The disease described here is caused by the blood flukes *Schistosoma mansoni* or *S. japonicum*, species that are morphologically and geographically distinct but share similar life cycles, transmission methods, and general disease manifestations. It is one of the few infectious agents that can invade intact skin.

Another species called *Schistosoma haematobium* causes disease in the bladder.

▶ Signs and Symptoms

The most severe consequences associated with chronic infection are hepatomegaly, liver disease, and splenomegaly. Occasionally, eggs from the worms are carried into the central nervous system and heart, and create a severe granulomatous response. Adult flukes can live for many years and, by eluding the immune defenses, cause a chronic affliction.

▶ Causative Agent

Schistosomes are trematodes, or flukes (see chapter 4), but they are more cylindrical than flat **(figure 20.20)**. They are often called blood flukes. Humans are the definitive hosts for the blood fluke, and snails are the intermediate host.

▶ Pathogenesis and Virulence Factors

This parasite is clever indeed. Once inside the host, it coats its outer surface with proteins from the host's bloodstream, basically "cloaking" itself from the host defense system. This coat reduces its surface antigenicity and allows it to remain in the host indefinitely.

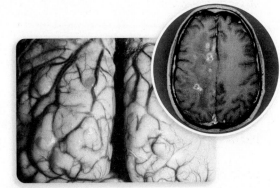

Figure 20.19 Cysticerci in the brain caused by *Taenia solium*.

PR Bouree/age fotostock (brain); Science Photo Library/Getty Images (scan)

Disease Table 20.12	Schistosomiasis Liver Disease
Causative Organism(s)	*Schistosoma mansoni,* *S. japonicum* Ⓗ
Most Common Modes of Transmission	Cycle D: vehicle (contaminated water)
Virulence Factors	Antigenic "cloaking"
Culture/Diagnosis	Identification of eggs in feces, scarring of intestines detected by endoscopy
Prevention	Avoiding contaminated vehicles
Treatment	Praziquantel
Distinctive Features	Penetrates skin, lodges in blood vessels of intestine, damages liver
Epidemiological Features	Internationally: 230 million new infections per year by these and the urinary schistosome

▶ Transmission and Epidemiology

The life cycle of the schistosome is of the "D" type, and is very complex (see figure 20.20). The cycle begins when infected humans release eggs into irrigated fields or ponds, either by deliberate fertilization with excreta or by defecating or urinating directly into the water.

The disease is endemic to 74 countries located in Africa, South America, the Middle East, and the Far East. Schistosomiasis (including the urinary tract form) is the second most prominent parasitic disease after malaria, probably affecting 200 million people at any one time worldwide.

20.4 LEARNING OUTCOMES—Assess Your Progress

10. Describe some distinguishing characteristics and commonalities seen in helminthic infections.

11. List four helminths that cause primarily intestinal symptoms, and identify which life cycle they follow and one unique fact about each one.

12. Explain reasons for and the consequences of the following statement: "Cysticercosis is underdiagnosed in the United States."

Figure 20.20 Stages in the life cycle of *Schistosoma.*
Courtesy Harvey Blankespoor (cercaria); Dr. Shirley Maddison/CDC (mating); ©Sinclair Stammers/Science Source (miracidium)

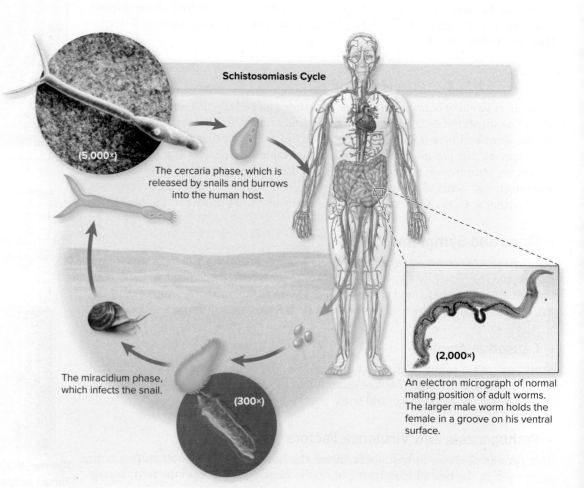

Schistosomiasis Cycle

(5,000×)

The cercaria phase, which is released by snails and burrows into the human host.

(2,000×)

The miracidium phase, which infects the snail.

(300×)

An electron micrograph of normal mating position of adult worms. The larger male worm holds the female in a groove on his ventral surface.

CASE FILE WRAP-UP

©Purestock/SuperStock (health care worker);
Source: CDC (Shigella)

Shigellosis is the disease caused by a group of bacteria called *Shigella*. Symptoms of *Shigella* include fever, abdominal cramps, and diarrhea that may be bloody. Some people with shigellosis are asymptomatic. Illness requiring hospitalization is uncommon—most people recover without treatment within a week. However, one form of *Shigella* (*Shigella dysenteriae* type 1) causes deadly epidemics in developing countries.

Shigellosis is spread through direct contact with someone who is carrying the bacteria. Poor hygiene is often a factor. *Shigella* may also be found in contaminated food and water. Proper food handling can help prevent shigellosis, as can frequent hand washing, especially following a trip to the bathroom. Peter, the patient diagnosed with *Shigella* in the case file at the beginning of this chapter, may have come in contact with someone who was infected and did not wash his or her hands after a trip to the bathroom; or he may have been exposed to contaminated food, although no one else in his family fell ill.

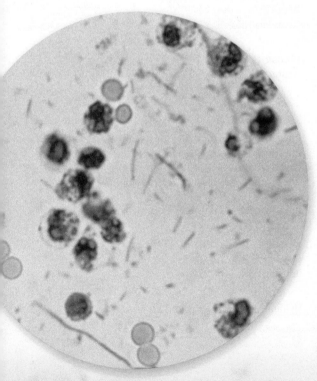

Source: CDC

▶ Summing Up

Taxonomic Organization Microorganisms Causing Disease in the GI Tract

Microorganism	Pronunciation	Disease Table
Gram-positive endospore-forming bacteria		
Clostridioides difficile	klos-trid″-ee-oy′-deez dif″-i-sil	Acute diarrhea, 20.1
Clostridium perfringens	klos-trid″-ee-um purr-frinj′-unz	Food poisoning, 20.2
Bacillus cereus	buh-sill′-us seer′-ee-uhs	Food poisoning, 20.2
Gram-positive bacteria		
Streptococcus spp.	strep″-tuh-kok′-us	Dental caries, 20.4
Staphylococcus aureus	staf″-uh-lo-kok′-us are′-ee-us	Food poisoning, 20.2
Gram-negative bacteria		
Campylobacter jejuni	cam″-pi-lo-bac′-ter juh-june′-ee	Acute diarrhea, 20.1
Helicobacter pylori	heel″-i-coe-back′-tur pie-lor′-ee	Gastritis and gastric ulcers, 20.7
Shiga-toxin-producing Escherichia coli	esh′-shur-eesh″-ee-uh-col′-eye	Acute diarrhea 20.1
Other E. coli		Acute diarrhea, 20.1
		Chronic diarrhea, 20.3
Salmonella	sal′-muh-nel″-luh	Acute diarrhea, 20.1
Shigella	shi-gel′-luh	Acute diarrhea, 20.1
Vibrio cholerae	vib′-ree-oh col′-er-ee	Acute diarrhea, 20.1
Non-cholera Vibrio	vib′-ree′-oh	Acute diarrhea, 20.1
Tannerella forsythia, Aggregatibacter actinomycetemcomitans, Porphyromonas gingivalis, Treponema vincentii, Prevotella intermedia, Fusobacterium	tan-er-rel′-ah for-sye′-thee-ah, ag-gruh-ga″-ti-bac′-tur ack-tin′-oh-my-see″-tem-cah′-mi-tans, por-fuhr′-oh-moan″-as jin-ji-vall′-is, trep′-oh-nee-ma vin-cen′-tee-eye, prev′-oh-tell″-ah in-ter-meed′-ee-ah, few′-zo-bac-teer′-ee′-um	Periodontitis 20.5
DNA viruses		
Hepatitis B virus	hep-uh-tie′-tis B vie′-russ	Hepatitis, 20.8
RNA viruses		
Hepatitis A virus	hep-uh-tie′-tis A vie′-russ	Hepatitis, 20.8
Hepatitis C virus	hep-uh-tie′-tis C vie′-russ	Hepatitis, 20.8
Hepatitis E virus	hep-uh-tie′-tis E vie′-russ	Hepatitis, 20.8
Mumps virus	mumps vie′-russ	Mumps, 20.6
Rotavirus	ro′-ta-vie′-russ	Acute diarrhea, 20.1
Norovirus	no″-row-vie′-russ	Acute diarrhea, 20.1
Protozoa		
Entamoeba histolytica	en″-tah-mee′-ba his″-toh-lit′-ih-kuh	Chronic diarrhea, 20.3
Cryptosporidium	crip′-toe-spor-id″-ee-um	Acute diarrhea, 20.1
Cyclospora	Sie″-clo-spor′-ah	Chronic diarrhea, 20.3
Giardia duodenalis	jee-ar′-dee-ah dwa′-duh-nal″-us	Chronic diarrhea, 20.3
Helminths—nematodes		
Ascaris lumbricoides	a-scare′-is lum′-bri-coi″-dees	Intestinal distress plus migratory symptoms, 20.10
Enterobius vermicularis	en′-ter-oh″-bee-us ver-mick″-u-lar′-is	Intestinal distress, 20.9
Trichuris trichiura	tri-cur′-is trick-ee-ur′-ah	Intestinal distress, 20.9
Necator americanus and Ancylostoma duodenale	neh-cay′-ter a-mer′-i-can″-us, an′-sy-lo-sto″-mah dew-ah′-den-al″-ee	Intestinal distress plus migratory symptoms, 20.10
Toxocara spp.	tox′-uh-kair″-uh	Intestinal distress plus migratory symptoms, 20.10
Helminths—cestodes		
Hymenolepis	hie′-men-oh″-lep-is	Intestinal distress, 20.9
Taenia solium	te′-ne-ah so′-lee-um	Cysticercosis, 20.11
Diphyllobothrium latum	dif′-oh-lo-both″-ree-um lah′-tum	Intestinal distress, 20.9
Helminths—trematodes		
Schistosoma mansoni and S. japonicum	shis′-toh-so-ma man-sohn″-ee, ja-pawn′-i-cum	Helminthic liver disease, 20.12

Crohn's Disease and the Gut Microbiome

Throughout this book, you have read about the importance of the gut microbiome and its possible influence on mood, body weight, autoimmune diseases, and many other things. Let us examine for a moment the role the gut microbiome might play in a mostly gut-focused disease: Crohn's disease.

Physicians consider Crohn's to be idiopathic, meaning they are not sure what causes it. Genetic analysis of Crohn's sufferers' DNA compared to that of healthy subjects shows some difference in some Crohn's patients, but not all. It would not be unreasonable to expect the microbiome to have some influence on gut health.

The important question is, if there *is* a difference in microbiota between Crohn's patients and healthy people, is that what causes the Crohn's? Or does Crohn's pathology lead to a different microbiome?

Many studies have been conducted in an effort to characterize the microbiome in these two circumstances (Crohn's and health). So far, what has been found is that there are fewer different types of microbes in Crohn's patients than in healthy guts. This decrease in diversity is seen in many types of deviations from health, but in very few of them has it been found to *precede* the condition. So the question remains. And all we can say is that there is an *association*—a term that does not imply which causes which.

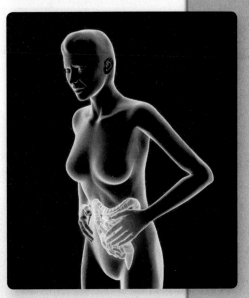

©Roger Harris/Science Photo Library
RF/Science Source

Infectious Diseases Affecting
The Gastrointestinal Tract

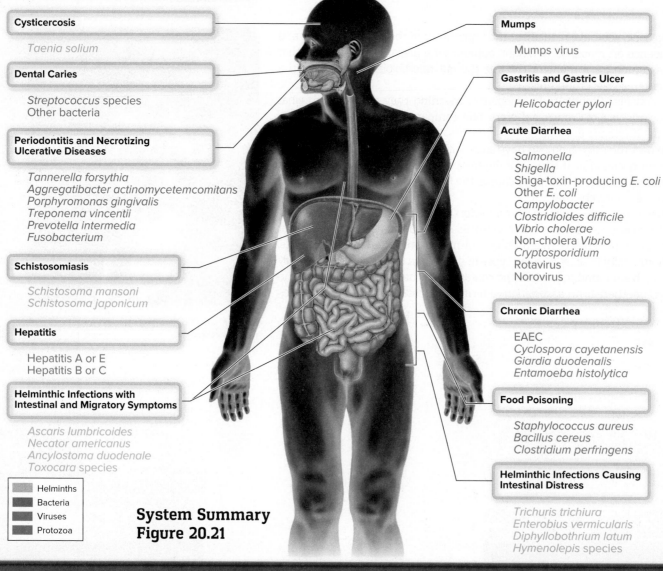

Cysticercosis

Taenia solium

Dental Caries

Streptococcus species
Other bacteria

Periodontitis and Necrotizing Ulcerative Diseases

Tannerella forsythia
Aggregatibacter actinomycetemcomitans
Porphyromonas gingivalis
Treponema vincentii
Prevotella intermedia
Fusobacterium

Schistosomiasis

Schistosoma mansoni
Schistosoma japonicum

Hepatitis

Hepatitis A or E
Hepatitis B or C

Helminthic Infections with Intestinal and Migratory Symptoms

Ascaris lumbricoides
Necator americanus
Ancylostoma duodenale
Toxocara species

Mumps

Mumps virus

Gastritis and Gastric Ulcer

Helicobacter pylori

Acute Diarrhea

Salmonella
Shigella
Shiga-toxin-producing *E. coli*
Other *E. coli*
Campylobacter
Clostridioides difficile
Vibrio cholerae
Non-cholera *Vibrio*
Cryptosporidium
Rotavirus
Norovirus

Chronic Diarrhea

EAEC
Cyclospora cayetanensis
Giardia duodenalis
Entamoeba histolytica

Food Poisoning

Staphylococcus aureus
Bacillus cereus
Clostridium perfringens

Helminthic Infections Causing Intestinal Distress

Trichuris trichiura
Enterobius vermicularis
Diphyllobothrium latum
Hymenolepis species

Legend:
Helminths
Bacteria
Viruses
Protozoa

System Summary Figure 20.21

Deadliness and Communicability of Selected Diseases of the Gastrointestinal Tract

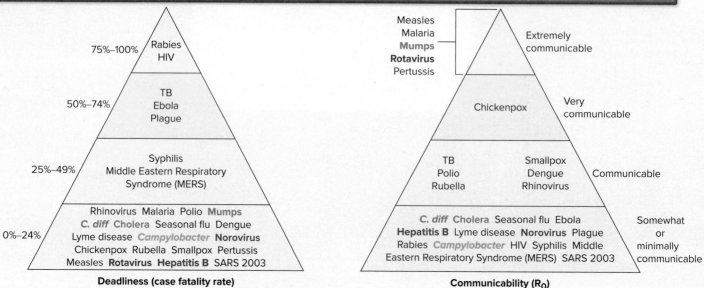

Deadliness (case fatality rate)

75%–100%
Rabies
HIV

50%–74%
TB
Ebola
Plague

25%–49%
Syphilis
Middle Eastern Respiratory Syndrome (MERS)

0%–24%
Rhinovirus Malaria Polio **Mumps**
***C. diff* Cholera** Seasonal flu Dengue
Lyme disease *Campylobacter* **Norovirus**
Chickenpox Rubella Smallpox Pertussis
Measles **Rotavirus Hepatitis B** SARS 2003

Communicability (R_0)

Measles
Malaria
Mumps
Rotavirus
Pertussis
— Extremely communicable

Chickenpox
— Very communicable

TB   Smallpox
Polio   Dengue
Rubella   Rhinovirus
— Communicable

***C. diff* Cholera** Seasonal flu Ebola
Hepatitis B Lyme disease **Norovirus** Plague
Rabies *Campylobacter* HIV Syphilis Middle
Eastern Respiratory Syndrome (MERS) SARS 2003
— Somewhat or minimally communicable

INFECTIOUS DISEASES AFFECTING THE GASTROINTESTINAL TRACT

Anatomy & Defenses: Gastrointestinal Tract

The GI tract has a heavy load of microorganisms, and it encounters millions of new ones every day. There are significant mechanical, chemical, and antimicrobial defenses to combat invading microbes.

Normal Biota: Gastrointestinal Tract

Microbes abound in all sections of the GI tract. The composition of the intestinal microbiota has profound effects on many aspects of human health.

Gastrointestinal Diseases Caused by Microorganisms (Nonhelminthic)

- **Acute Diarrhea**
Up to one-third of these cases are caused by foodborne infections. The most frequent causes are *Salmonella*, *Shigella*, Shiga-toxin-producing *E. coli* (STEC), other *E. coli*, *Campylobacter*, *Clostridioides difficile*, *Vibrio cholerae*, *Cryptosporidium*, rotavirus, and norovirus.

- **Food Poisoning**
Symptoms in the gut caused by a preformed bacterial toxin. Caused most often by exotoxins from *Staphyloccus aureus*, *Bacillus cereus*, and *Clostridium perfringens*.

- **Chronic Diarrhea**
Caused most often by enteroaggregative *E. coli* (EAEC), or the protozoa *Cyclospora cayetanensis* or *Entamoeba histolytica*.

- **Tooth and Gum Infections**
Dental caries is a polymicrobial condition in which acid-producing bacteria damage the hard surface of teeth.

- **Periodontitis**
Inflammation and destruction of the supporting tissues of the teeth, caused by a mostly anaerobic consortium of bacteria.

- **Mumps**
Caused by an enveloped, single-stranded RNA virus (mumps virus) from the genus *Paramyxovirus*.

- **Gastritis and Gastric Ulcers**
Helicobacter pylori, a curved gram-negative rod, is the causative agent.

- **Hepatitis**
Inflammatory disease marked by damage to hepatocytes and a mononuclear response that swells and disrupts the liver, causing jaundice. Caused by a variety of different viruses: Hepatitis A or E viruses are RNA viruses spread via the fecal-oral or foodborne route, and cause a relatively mild disease. Hepatitis B is a DNA virus transmitted by blood and other bodily fluids and can result in long-term illness. Major concern for health care workers. Hepatitis C is an RNA virus that causes serious disease and is more likely to become chronic than even hepatitis B.

Gastrointestinal Diseases Caused by Microorganisms (Helminthic)

- **Helminthic Infections**
Intestinal Distress as the Primary Symptom The pinworm, *Enterobius vermicularis,* is the most common helminthic disease of children in temperate zones.

- **Helminthic Infections**
Intestinal Distress Accompanied by Migratory Symptoms *Toxocara* is categorized as a neglected parasitic infection by the CDC. *Ascaris lumbricoides* can obstruct the GI tract or leave the GI tract and migrate throughout the body. The hookworms *Necator americanus* and *Anclyostoma duodenale* penetrate intact skin and travel through the bloodstream to the lungs and then to the GI tract.

- **Cysticercosis**
Larval tapeworms embed in brain and other tissues, often causing seizures. Also a neglected parasitic infection.

- **Schistosomiasis**
Liver Disease Schistosomiasis in the intestines is caused by blood flukes *Schistosoma mansoni* and *S. japonicum*. Symptoms include fever, chills, diarrhea, and liver and spleen disease.

SmartGrid: From Knowledge to Critical Thinking

This *21 Question Grid* takes the topics from this chapter and arranges them with respect to the American Society for Microbiology's Undergraduate Curriculum guidelines—all six of the important "Concepts" as well as the important "Competency" of scientific literacy. Three questions are supplied about chapter content that refer to the Concept or Competency, in increasing levels of Bloom's taxonomy for learning.

ASM Concept/ Competency	A. Bloom's Level 1, 2—Remember and Understand (Choose one.)	B. Bloom's Level 3, 4—Apply and Analyze	C. Bloom's Level 5, 6—Evaluate and Create
Evolution	1. Which of the following causes of acute diarrhea is of the most concern to the CDC due to its antibiotic resistance characteristics? a. *Salmonella* b. *Clostridioides difficile* c. *Cryptosporidium* d. *Shigella*	2. Healthy teeth are important for good nutrition and overall human health. Yet despite the fact that humans and their oral microbiome have coevolved over thousands of years, there was a vast increase in the incidence of human dental caries with the advent of Western civilization. What factor(s) are likely responsible for that?	3. Considering that *Helicobacter* has probably been colonizing human stomachs for thousands of years, would you suspect that its primary role is as a pathogen or as normal biota? Support your answer.
Cell Structure and Function	4. This endospore-forming bacterium contaminates meat and vegetables, and in other situations causes gas gangrene. a. *Bacillus cereus* b. *Clostridium perfringens* c. *Shigella* d. *Staphylococcus aureus*	5. Why is heating food contaminated with *Staphylococcus aureus* no guarantee that potential food poisoning will be prevented?	6. Several pathogens in this chapter have the ability to form endospores or cysts. Create an argument, using what you know about transmission mechanisms, for why these organisms are particularly suited for GI tract disease.
Metabolic Pathways	7. Which of the following bacteria produces an enzyme that breaks down urea? a. *Helicobacter pylori* b. *Salmonella* c. *Shigella* d. Shiga-toxin-producing *E. coli*	8. You probably think of the oral cavity as an aerobic environment. Considering the anatomy of the teeth and gums, where would you expect the anaerobes that colonize the mouth to survive?	9. *Vibrio cholerae* can survive in a wide range of aquatic conditions, from brackish water to seawater, and in habitats that have scarce nutrients. It readily forms biofilms on submerged surfaces, including other organisms such as phytoplankton. How would living in a biofilm protect *Vibrio* in a low-nutrient environment?
Information Flow and Genetics	10. Which of these organisms has an unusual double-stranded RNA genome? a. norovirus b. *Schistosoma* c. hepatitis C d. rotavirus	11. Why is it thought that the Shiga toxin in Shiga-toxin-producing *E. coli* originated in *Shigella*, and not the other way around? Also, explain how this characteristic was transferred between the two species.	12. Conduct research on how the hepatitis B vaccine is manufactured, and write a paragraph responding to a patient who is unhappy about receiving a recombinant vaccine.
Microbial Systems	13. The normal biota of the GI tract is most diverse in the a. pharynx. b. stomach. c. small intestine. d. large intestine.	14. The normal biota of the GI tract seems to include a lot of disease-causing organisms. How can that be? That is, if they are "normal," why are they also potential pathogens?	15. The *Salmonella* strain causing typhoid fever is transmitted human-to-human. The *Salmonella* strains causing gastroenteritis generally are transmitted from animals. Sketch a scenario for why gastroenteritis is now the more common outcome of *Salmonella* infection, and not typhoid fever.

ASM Concept/Competency	A. Bloom's Level 1, 2—Remember and Understand (Choose one.)	B. Bloom's Level 3, 4—Apply and Analyze	C. Bloom's Level 5, 6—Evaluate and Create

Impact of Microorganisms	**16.** Which of these microorganisms is associated with Guillain-Barré syndrome?
	  a. *E. coli*
	  b. *Salmonella*
	  c. *Campylobacter*
	  d. *Shigella*
		17. Describe the populations that are most at risk for hepatitis C, and why.	**18.** Use what you know about inflammation to explain how periodontal organisms might be responsible for diseases elsewhere in the body, such as the heart.

Scientific Thinking	**19.** You are examining a patient who is a recent immigrant from Southeast Asia. She is complaining of diarrhea and vomiting. Which of these would you add to your differential diagnosis in light of her history?
	  a. *Vibrio cholerae*
	  b. *Salmonella*
	  c. *Shigella*
	  d. *Campylobacter*
		20. An outbreak of cholera occurred in Haiti following a devastating earthquake in 2010. It is still ongoing. Based on your knowledge of the bacterium involved, discuss the factors that probably favored this outbreak.	**21.** Consider the function of the liver. Then speculate on why so many different types of viruses can cause pathology there.

Answers to the multiple-choice questions appear in Appendix A.

Visual Connections

This question connects content within and between chapters.

From chapter 11, figure 11.2a. Imagine for a minute that the organism in this illustration is a Shiga-toxin-producing *E. coli*. What would be one reason not to treat a patient having this infection with powerful antibiotics?

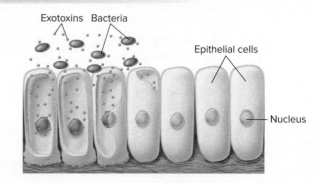

Chapter design elements: (Covid): CDC/Alissa Eckert, MS; Dan Higgins, MAMS; (Note): McGraw-Hill Education; (NCLEX): Shutterstock/Abert; (Doctor): David Gould/Getty Images

21

Infectious Diseases Affecting the Genitourinary System

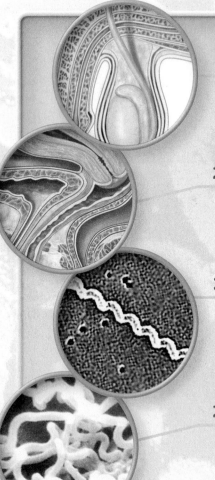

IN THIS CHAPTER...

21.1 The Genitourinary Tract and Its Defenses

1. Draw or describe the anatomical features of the genitourinary tracts of both sexes.
2. List the natural defenses present in the genitourinary tracts.

21.2 Normal Biota of the Genitourinary Tract

3. List the types of normal biota presently known to occupy the genitourinary tracts of both sexes.

21.3 Urinary Tract Diseases Caused by Microorganisms

4. List the possible causative agents, modes of transmission, virulence factors, diagnostic techniques, and prevention/treatment for the "Highlight Disease" urinary tract infection.
5. Discuss important features of leptospirosis.

21.4 Reproductive Tract Diseases Caused by Microorganisms

6. List the possible causative agents, modes of transmission, virulence factors, and prevention/treatment for the "Highlight Diseases" gonorrhea and *Chlamydia* infection.
7. Distinguish between vaginitis and vaginosis.
8. Discuss prostatitis.
9. Name three diseases that result in genital ulcers, and discuss their important features.
10. Differentiate between the two diseases causing warts in the reproductive tract.
11. List at least three reasons the HPV vaccine is so strongly recommended for young adolescents.
12. Identify the most important risk group for group B *Streptococcus* infection and explain why that group is important.

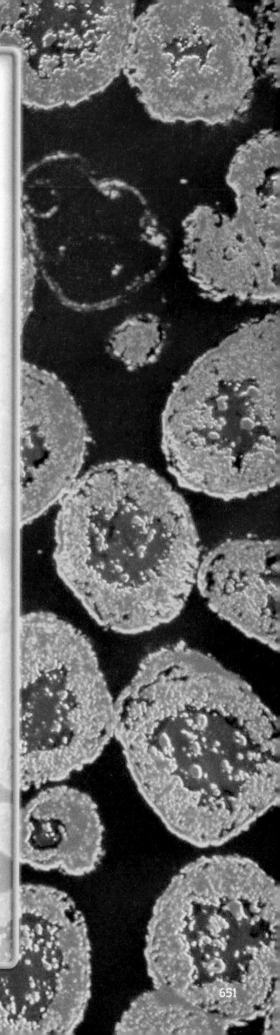

CASE FILE

It's All in the Walk

One of the most disturbing and gut-wrenching cases I was ever involved in concerned a 10-year-old girl who was brought to the emergency room by her mother, who was afraid her daughter had appendicitis. What her daughter turned out to have was far more troubling.

The patient had developed a high fever (39.4°C [103°F]) and was complaining of severe lower abdominal pain. Upon questioning, the patient also reported painful urination. As I led her into an examining room, I noticed that she walked taking very small steps. She did not lift her feet off the floor and seemed reluctant or unable to straighten up fully. Her odd gait raised a red flag in my mind.

After performing a quick abdominal examination and obtaining vital signs, I found the physician on call and told her about the patient's symptoms. I also described how the patient moved with an odd shuffling gait. Although I could not recall what I had learned about the symptom, I knew that it was a diagnostic feature. The physician asked me how old the patient was and asked if I had taken the opportunity to ask the patient whether she was sexually active. I had not done so due to the patient's age. To my surprise, the physician asked me to prepare for an internal (vaginal) exam in another room while she examined the patient and spoke to her mother.

- What condition did the physician suspect?

- Is the condition reportable? If so, to whom would you report it?

Case File Wrap-Up appears at the end of the chapter.

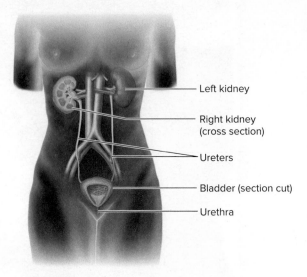

Figure 21.1 **The urinary system.**

21.1 The Genitourinary Tract and Its Defenses

As suggested by the name, the structures considered in this chapter are really two distinct organ systems. The *urinary tract* has the job of removing substances from the blood, regulating certain body processes, and forming urine and transporting it out of the body. The *genital system* has reproduction as its major function. It is also called the *reproductive system*.

The urinary tract includes the kidneys, ureters, bladder, and the urethra **(figure 21.1)**. The kidneys remove metabolic wastes from the blood, acting as a sophisticated filtration system. Ureters are tubular organs extending from each kidney to the bladder. The bladder is a collapsible organ that stores urine and empties it into the urethra, which is the conduit of urine to the exterior of the body. In males, the urethra is also the terminal organ of the reproductive tract, but in females the urethra is separate from the vagina, which is the outermost organ of the reproductive tract.

The most obvious defensive mechanism in the urinary tract is the flushing action of the urine flowing out of the system. The flow of urine also encourages the desquamation (shedding) of the epithelial cells lining the urinary tract. For example, each time a person urinates, he or she loses hundreds of thousands of epithelial cells! Any microorganisms attached to them are also shed, of course. Probably the most common microbial threat to the urinary tract is the group of microorganisms that comprise the normal biota in the gastrointestinal tract because the two organ systems are in close proximity. But the cells of the epithelial lining of the urinary tract have different chemicals on their surfaces than do those lining the GI tract. For that reason, most (but not all) bacteria that are adapted to adhere to the chemical structures in the GI tract cannot gain a foothold in the urinary tract.

Urine, in addition to being acidic, also contains two antibacterial proteins, lysozyme and lactoferrin. You may recall that lysozyme is an enzyme that breaks down peptidoglycan. Lactoferrin is an iron-binding protein that inhibits bacterial growth. Finally, secretory IgA specific for previously encountered microorganisms can be found in the urine.

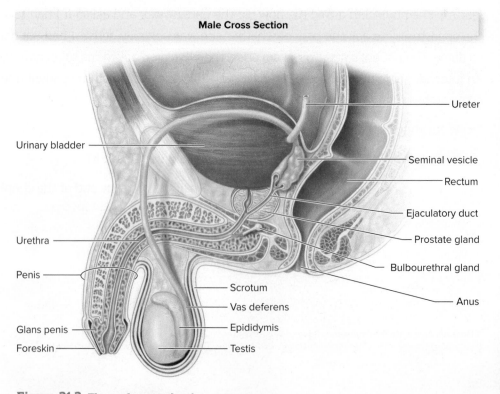

Male Cross Section

Figure 21.2 **The male reproductive system.**

The male reproductive system produces, maintains, and transports sperm cells and is the source of male sex hormones. It consists of the *testes*, which produce sperm cells and hormones, and the *epididymides*, which are coiled tubes leading out of the testes. Each epididymis terminates in a *vas deferens*, which combines with the seminal vesicle and terminates in the ejaculatory duct **(figure 21.2)**. The contents of the ejaculatory duct empty into the urethra during ejaculation. The *prostate gland* is a walnut-shaped structure at the base of the urethra. It also contributes to the released fluid (semen). The external organs are the scrotum, containing the testes, and the *penis*, a cylindrical organ that houses the urethra. As for its innate defenses, the male reproductive system also benefits from the flushing action of the urine, which helps move microorganisms out of the system.

The female reproductive system consists of the *uterus*, the *fallopian tubes* (also called uterine tubes), *ovaries*, and *vagina* **(figure 21.3)**. One very important tissue of the female reproductive tract is the *cervix*, which is the lower one-third of the uterus and the part that connects to the vagina. The cervix serves as the opening to the uterus. The cervix is a common site of infection in the female reproductive tract.

The natural defenses of the female reproductive tract vary over the lifetime of the woman. The vagina is lined with mucous membranes and, thus, has the protective covering of secreted mucus. During childhood and after menopause, this mucus is the major innate defense of this system. Secretory IgA antibodies specific for any previously encountered infections will also be present on these surfaces. During a woman's reproductive years, a major portion of the defense is provided by changes in the pH of the vagina brought about by the release of estrogen. This hormone stimulates the vaginal mucosa to secrete glycogen, which certain bacteria can ferment into acid, lowering the pH of the vagina to between 4.2 and 5.0. Before puberty, a girl produces little estrogen and little glycogen and has a vaginal pH of about 7. The change in pH beginning in adolescence results in a vastly different normal biota in the vagina, described later. The biota of women in their childbearing years is thought to prevent the establishment and invasion of microbes that might have the potential to harm a developing fetus.

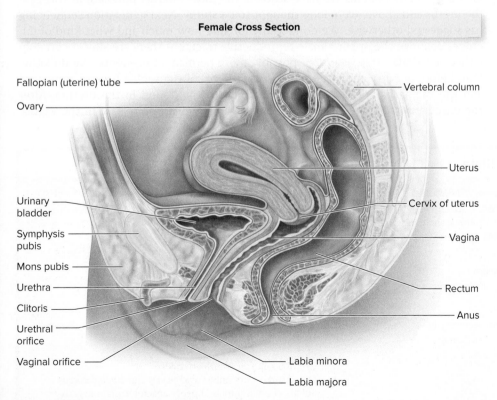

Female Cross Section

Fallopian (uterine) tube

Ovary

Vertebral column

Uterus

Urinary bladder

Cervix of uterus

Symphysis pubis

Vagina

Mons pubis

Urethra

Rectum

Clitoris

Anus

Urethral orifice

Vaginal orifice

Labia minora

Labia majora

Figure 21.3 The female reproductive system.

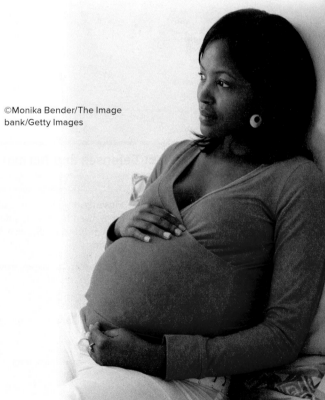

Normal biota in the female genital tract is different in women of childbearing age than in younger or older women.
©Ariel Skelley/Blend Images

21.2 Normal Biota of the Genitourinary Tract

As with all other organ systems, recent research is showing that the urinary system harbors a more diverse microbiota than we used to think. The lower urethra has a well-established microbiota, while the upper urinary tract appears to have fewer types and lower abundance. The exact microbial composition varies between men and women—and also among individuals. Because the urethra in women is so short (about 3.5 cm long) and is in such close proximity to the anus, it can act as a pipeline for bacteria from the GI tract to the bladder, resulting in urinary tract infections. The outer surface of the penis is colonized by *Pseudomonas* and *Staphylococcus* species—aerobic bacteria. In an uncircumcised penis, the area under the foreskin is colonized by anaerobic gram-negatives.

Normal Biota of the Male Genital Tract

Because the terminal "tube" of the male genital tract is the urethra, the normal biota of the male genital tract (i.e., in the urethra) is comprised of the same residents just described. However, after sexual activity begins, microbes associated with sexually transmitted infections (STIs) can sometimes become long-term residents. Circumcision has a large effect on the "normal" biota of the penis. The uncircumcised penis hosts a larger, more diverse population of microbes, and more anaerobic species.

Normal Biota of the Female Genital Tract

Like other body systems we have studied, the most internal portions, in this case the uterus and above, were long thought to be sterile. Next-generation sequencing has suggested otherwise. We don't know for sure how much and what kind of microbes colonize the upper female reproductive tract, but there are almost certainly either occasional "trespassers" or possibly more permanent residents. We do know that, before puberty and after menopause, the pH of the vagina is close to neutral and the vagina harbors a biota that is similar to that found in the urethra. After the onset of puberty, estrogen production leads to glycogen release in the vagina,

Genitourinary Tract Defenses and Normal Biota

	Defenses	**Normal Biota**
Urinary Tract (Both Sexes)	Flushing action of urine; specific attachment sites not recognized by most nonnormal biota; shedding of urinary tract epithelial cells, secretory IgA, lysozyme, and lactoferrin in urine	Nonhemolytic *Streptococcus*, *Staphylococcus*, *Corynebacterium*, *Lactobacillus*, *Prevotella*, *Veillonella*, *Gardnerella*
Female Genital Tract (Childhood and Postmenopause)	Mucus secretions, secretory IgA	Same as for urinary tract
Female Genital Tract (Childbearing Years)	Acidic pH, mucus secretions, secretory IgA	Variable, but often *Lactobacillus* predominates; also *Prevotella*, *Sneathia*, *Streptococcus*, and *Candida albicans*
Male Genital Tract	Same as for urinary tract	Urethra: same as for urinary tract Outer surface of penis: *Pseudomonas* and *Staphylococcus* Sulcus of uncircumcised penis: anaerobic gram-negatives

resulting in an acidic pH. *Lactobacillus* species thrive in the acidic environment and contribute to it, converting sugars to acid. Their predominance in the vagina, combined with the acidic environment, discourages the growth of many microorganisms. The estrogen-glycogen effect continues, with minor disruptions, throughout the childbearing years until menopause, when the biota gradually returns to a mixed population similar to that of prepuberty. Note that the fungus *Candida albicans* is also present at low levels in the healthy female reproductive tract. When the normal vaginal microbiota is altered and is unable to keep *Candida albicans* in check, an overgrowth of the fungus may occur, resulting in a symptomatic yeast infection.

21.2 LEARNING OUTCOMES—Assess Your Progress

3. List the types of normal biota presently known to occupy the genitourinary tracts of both sexes.

21.3 Urinary Tract Diseases Caused by Microorganisms

We consider two types of diseases in this section. **Urinary tract infections (UTIs)** result from invasion of the urinary system by bacteria or other microorganisms. Leptospirosis, by contrast, is a spirochete-caused disease transmitted by contact of broken skin or mucous membranes with contaminated animal urine.

©Nancy R. Cohen/Photodisc/Getty Images

HIGHLIGHT DISEASE

Urinary Tract Infections (UTIs)

Even though the flushing action of urine helps to keep infections to a minimum in the urinary tract, urine itself is a good growth medium for many microorganisms. When urine flow is reduced, or bacteria are accidentally introduced into the bladder, an infection of that organ (known as *cystitis*) can occur. Occasionally, the infection can also affect the kidneys, in which case it is called *pyelonephritis*. If an infection is limited to the urethra, it is called *urethritis*.

▶ Signs and Symptoms

Cystitis is a disease of sudden onset. Symptoms include pain, frequent urges to urinate even when the bladder is empty, and burning pain accompanying urination (called *dysuria*). The urine can be cloudy due to the presence of bacteria and white blood cells. It may have an orange tinge from the presence of red blood cells (*hematuria*). Low-grade fever and nausea are frequently present. If back pain is present and fever is high, it is an indication that the kidneys may also be involved (pyelonephritis). Pyelonephritis is a serious infection that can result in permanent damage to the kidneys if improperly or inadequately treated. If only the bladder is involved, the condition is sometimes called acute uncomplicated UTI.

▶ Causative Agents

In the same way we saw in the discussion of pneumonia, it is important to distinguish between UTIs that are acquired in health care facilities and those acquired outside of the health care setting. When they occur in health care facilities, they are almost always a result of catheterization and are therefore called *catheter-associated UTIs* (*CA-UTIs*). (Be careful! The abbreviation "CA" in other infections often refers to "community-acquired"–just the opposite of what is meant here! We will spell out "community" in referring to non-healthcare-associated UTIs.)

⚕ NCLEX® PREP

1. In providing discharge information to a client who is being treated for a urinary tract infection with phenazopyridine (Pyridium), which information should the nurse include? Select all that apply.

a. *Increase fluid intake during the course of treatment.*

b. *Urine may turn orange and cause "staining" to clothing.*

c. *Continue the medication for the full 7-day course.*

d. *Take an analgesic to treat burning and pain with urination.*

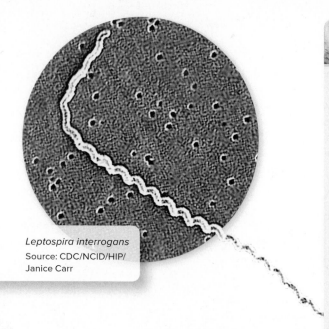

Leptospira interrogans
Source: CDC/NCID/HIP/
Janice Carr

Medical Moment

Cranberry versus UTI

It has long been claimed that cranberries help prevent, or even cure, urinary tract infections (UTIs). Cranberry supplementation is available in the form of fresh fruit, dried fruit, juice, and tablets.

Could the cranberry have a role in preventing a UTI? While results have been promising in animal studies, those in humans have been less conclusive. We do know that cranberry products can prevent the adhesion of uropathogens to the epithelium of the genitourinary tract, at least in the laboratory. However, the variation in composition and dosage of cranberry products has posed challenges in extrapolating a lot of the data to humans. In some populations, using certain forms of cranberry, the data for reduction in occurrence of UTIs appear promising. For young and middle-aged women, cranberry may be effective in preventing recurrent UTIs. Data do not support effectiveness in older women, particularly those in nursing homes.

There is no evidence that cranberry products treat an active UTI. Antibiotic therapy is the recommended form of treatment. Additionally, cranberry interacts with many medications so their use should be discussed with a pharmacist or provider.

Q. Finding an effective preventive agent could be seen as a particularly pressing need for this particular infection. Why is that?

Answer in Appendix B.

In 95% of UTIs, the cause is bacteria that are normal biota in the gastrointestinal tract. *Escherichia coli* is by far the most common of these, accounting for approximately 80% of community-acquired urinary tract infections. *Staphylococcus saprophyticus* and other members of the bacterial family that contains *E. coli*, Enterobacteriaceae, especially *Klebsiella pneumoniae*, and *Proteus mirablis* are also common culprits. These last two are only referenced in **Disease Table 21.1** following the discussion of *E. coli*.

The *E. coli* species that cause UTIs are ones that exist as normal biota in the gastrointestinal tract. They are not the ones that cause diarrhea and other digestive tract diseases.

▶ Transmission and Epidemiology

Community-acquired UTIs are nearly always "transmitted" *not* from one person to another but from one organ system to another, namely from the GI tract to the

Disease Table 21.1	**Urinary Tract Infections**		
Causative Organism(s)	*Escherichia coli* **B** G⁻	*Staphylococcus saprophyticus* **B** G+	Other Enterobacteriaceae (*Klebsiella pneumoniae, Proteus mirablis*) **B** G⁻
Most Common Modes of Transmission	Opportunism: transfer from GI tract (community-acquired) or environment or GI tract (via catheter)		
Virulence Factors	Adhesins, motility	–	–
Culture/Diagnosis	Usually culture-based; antimicrobial susceptibilities always checked		
Prevention	Hygiene practices; in case of CA-UTIs, limit catheter usage		
Treatment	Usually nitrofurantoin; members of family Enterobacteriaceae in **Urgent Threat** category in CDC Antibiotic Resistance Threat Report	Usually nitrofurantoin	Based on susceptibility testing; members of family Enterobacteriaceae in **Urgent Threat** category in CDC Antibiotic Resistance Threat Report
Epidemiological Features	Causes 90% of community UTIs and 50%–70% of CA-UTIs	Causes small percentage of community UTIs and even lower percentage of CA-UTIs	Frequent cause of CA-UTIs

urinary system. They are much more common in women than in men because of the nearness of the female urethral opening to the anus (see figure 21.3). Many women experience what have been referred to as "recurrent urinary tract infections," although it is now known that some *E. coli* can invade the deeper tissue of the urinary tract and therefore avoid being destroyed by antibiotics. They can emerge later to cause symptoms again. It is not clear how many "recurrent" infections are actually infections that reactivate in this way.

The National Healthcare Safety Network is now recommending minimizing the use of urinary catheters as much as possible to limit the incidence of these infections.

▶ Treatment

A drug called nitrofurantoin (Macrobid) is most often used for UTIs of various etiologies. Often another nonantibiotic drug called phenazopyridine (Pyridium) is administered simultaneously. This drug relieves the very uncomfortable symptoms of burning and urgency. However, some physicians are reluctant to administer this medication for fear that it may mask worsening symptoms. When Pyridium is used, it should be used only for a maximum of 2 days. Pyridium is an azo dye and causes the urine to turn a dark orange to red color. It may also color contact lenses. A large percentage of *E. coli* strains are resistant to penicillin derivatives, so these should be avoided. Importantly, some representatives of the family Enterobacteriaceae (three of which are considered here) have become resistant to carbapenem, and are designated "CRE" for carbapenem-resistant Enterobacteriaceae. These are in the Urgent Threat category for antibiotic resistance. They usually remain susceptible to very few antibiotics, and are most common in health care facilities. **Disease Table 21.1.**

Leptospirosis

©Corbis

This infection is a zoonosis associated with wild animals and domesticated animals. It can affect the kidneys, liver, brain, and eyes. It is considered in this section because it can have its major effects on the kidneys and because its presence in animal urinary tracts causes it to be shed into the environment through animal urine.

▶ Signs and Symptoms

Leptospirosis has two phases. During the early, or leptospiremic, phase, the pathogen appears in the blood and cerebrospinal fluid. Symptoms are sudden high fever, chills, headache, muscle aches, conjunctivitis, and vomiting. During the second phase, the blood infection is cleared by natural defenses. This period is marked by milder fever, headache due to leptospiral meningitis, and, in rare cases, *Weil's syndrome*. This is a cluster of symptoms characterized by kidney invasion, hepatic disease, jaundice, anemia, and neurological disturbances. Long-term disability and even death can result from damage to the kidneys and liver, but they occur primarily with the most virulent strains and in elderly persons.

▶ Causative Agent

Leptospires are typical spirochete bacteria marked by tight, regular, individual coils with a bend or hook at one or both ends. *Leptospira interrogans* (lep″-toh-spy′-ruh in-ter′-ruh-ganz) is the species that causes leptospirosis in humans and animals. There are nearly 200 different serotypes of this species distributed among various animal groups, which accounts for extreme variations in the disease manifestations in humans.

▶ Transmission and Epidemiology

Infection occurs almost entirely through contact of skin abrasions or mucous membranes with animal urine or some environmental source containing urine. In 1998, dozens of athletes competing in the swimming phase of a triathlon in Illinois contracted leptospirosis from the water. It is a common pathogen in areas of Latin America and Asia.

Disease Table 21.2	Leptospirosis
Causative Organism(s)	*Leptospira interrogans* Ⓑ G⁻
Most Common Modes of Transmission	Vehicle: contaminated soil or water
Virulence Factors	Adhesins, invasion proteins
Culture/Diagnosis	In United States, CDC will culture specimens
Prevention	Avoiding contaminated vehicles
Treatment	Doxycycline, penicillin G, or ceftriaxone
Epidemiological Features	United States: 100–200 cases per year, half in Hawaii; internationally: 80% of people in tropical areas are seropositive

▶ Treatment

Early treatment with doxycycline, penicillin G, or ceftriaxone rapidly reduces symptoms and shortens the course of disease, but delayed therapy is less effective. Other spirochete diseases, such as syphilis (described later), exhibit this same pattern of being susceptible to antibiotics early in the infection but less so later on. **Disease Table 21.2.**

21.3 LEARNING OUTCOMES—Assess Your Progress

4. List the possible causative agents, modes of transmission, virulence factors, diagnostic techniques, and prevention/treatment for the "Highlight Disease" urinary tract infection.
5. Discuss important features of leptospirosis.

21.4 Reproductive Tract Diseases Caused by Microorganisms

Not all reproductive tract diseases are sexually transmitted, though many are. Vaginitis/vaginosis may or may not be; prostatitis probably is not.

It was very difficult to choose a "highlight" disease for this section, though we were sure we wanted it to be a sexually transmitted disease (now more appropriately called *sexually transmitted infections,* due to the fact that so many of them are silent, while still being serious). The WHO reports that there are 1 million new sexually transmitted infections *every day* worldwide. In the United States, the CDC reports that chlamydia, gonorrhea, and syphilis are at their highest rates ever, after decades of progress **(figure 21.4).** The CDC blames these increases on budget cuts to state and local STD clinics. In a recent year, 20 STD clinics closed altogether. They point out that this harms lower-income individuals disproportionately. Syphilis, in particular, causes devastating effects, including miscarriages, stillbirths, blindness, and strokes. At one point, syphilis was close to being eliminated in the United States; now it is increasing at an alarming rate, showing nearly a 70% increase between 2014 and 2018. The discharge diseases *Chlamydia* and gonorrhea are responsible for unprecedented numbers of infertility cases. Herpes infections are incurable and therefore simply increase in their prevalence over time. Americans have become more familiar with the dangers of HPV infection since the introduction of the HPV vaccine, so, in a split decision, we have decided to highlight the dangers of curable–but usually undetected–discharge diseases.

A Note About STI Statistics

It is difficult to compare the incidence of different STIs to one another, for several reasons. The first is that many, many infections are "silent," and therefore infected people don't access the health care system and don't get counted. Of course, we know that many silent infections are actually causing damage that won't be noticed for years—and when it is, the original causative organism is almost never sought out. The second reason is that only some STIs are officially reportable to the CDC, and states' regulations vary. *Chlamydia* infection and gonorrhea are nationally reportable, for example, but herpes and HPV infections are not. In each section, we present estimates of the prevalence and/or incidence of the diseases wherever possible.

HIGHLIGHT DISEASE

Discharge Diseases with Major Manifestation in the Genitourinary Tract

Discharge diseases are those in which the infectious agent causes an increase in fluid discharge in the male and female reproductive tracts. Examples are trichomoniasis, gonorrhea, and *Chlamydia* infection. The causative agents are transferred to new hosts when the fluids in which they live contact the mucosal surfaces of the receiving partner.

Gonorrhea

Gonorrhea has been known as a sexually transmitted disease since ancient times. It was named by the Greek physician Claudius Galen, who thought that it was caused by an excess flow of semen.

▶ Signs and Symptoms

In the male, infection of the urethra elicits urethritis, painful urination, and a yellowish discharge, although a relatively large number of cases are asymptomatic. In most cases, infection is limited to the distal urogenital tract, but it can occasionally spread from the urethra to the prostate gland and epididymis (refer to figure 21.2). Scar tissue formed in the spermatic ducts during healing of an invasive infection can render a man infertile.

In the female, it is likely that both the urinary and genital tracts will be infected during sexual intercourse. A mucopurulent (containing mucus and pus) or bloody vaginal discharge occurs in a minority of the cases, along with painful urination if the urethra is affected. Major complications occur when the infection ascends from the vagina and cervix to higher reproductive structures such as the uterus and fallopian tubes **(figure 21.5)**. One disease resulting from this progression is **salpingitis** (sal″-pin-jy′-tis). This inflammation of the fallopian tubes may be isolated, or it may also include inflammation of other parts of the upper reproductive tract, termed *pelvic inflammatory disease (PID)*. It is not unusual for the microbe that initiates PID to become involved in mixed infections with anaerobic bacteria. The buildup of scar tissue from PID can block the fallopian tubes, causing sterility or ectopic pregnancies.

Serious consequences of gonorrhea can occur outside of the reproductive tract. In a small number of cases, the gonococcus enters the bloodstream and is disseminated to the joints and skin. Involvement of the wrist and ankle can lead to chronic arthritis and a painful, sporadic, papular rash on the limbs. Rare complications of gonococcal bacteremia are meningitis and endocarditis.

Children born to gonococcus carriers are also in danger of being infected as they pass through the birth canal. Because of the potential harm to the fetus, physicians usually screen pregnant mothers for its presence. Gonococcal eye infections are very serious and often result in keratitis, ophthalmia neonatorum, and even blindness **(figure 21.6)**. A universal precaution to prevent such complications is the use of antibiotic eyedrops or ointments (usually erythromycin) for newborn babies. The pathogen may also infect the pharynx and respiratory tract of neonates. Finding gonorrhea in children other than

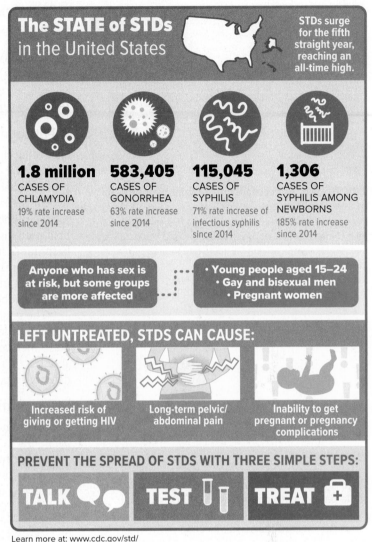

Learn more at: www.cdc.gov/std/

Figure 21.4 STI statistics from the CDC.
Centers for Disease Control and Prevention. Sexually Transmitted Disease Surveillance 2018. https://www.cdc.gov/std/stats18/infographic.htm

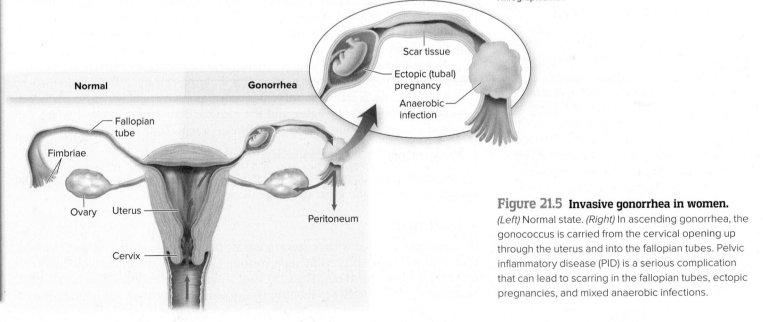

Figure 21.5 Invasive gonorrhea in women.
(Left) Normal state. *(Right)* In ascending gonorrhea, the gonococcus is carried from the cervical opening up through the uterus and into the fallopian tubes. Pelvic inflammatory disease (PID) is a serious complication that can lead to scarring in the fallopian tubes, ectopic pregnancies, and mixed anaerobic infections.

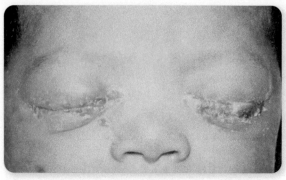

Figure 21.6 Gonococcal ophthalmia neonatorum in a week-old infant. The infection is marked by intense inflammation and edema; if allowed to progress, it causes damage that can lead to blindness. Fortunately, this infection is completely preventable and treatable.
Source: CDC/J. Pledger

Figure 21.7 Gram stain of urethral pus from a male patient with gonorrhea (1,000×). Note the intracellular (phagocytosed) gram-negative diplococci in polymorphonuclear leukocytes (neutrophils).
Dr. Norman Jacobs/CDC

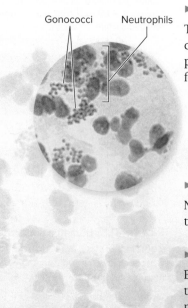

Gonococci   Neutrophils

neonates is strong evidence of sexual abuse by infected adults, and it calls for child welfare consultation along with thorough bacteriologic analysis.

▶ Causative Agent

Neisseria gonorrhoeae is a pyogenic (pus-forming) gram-negative diplococcus. It appears as pairs of kidney- or coffee bean–shaped bacteria, with their flat sides touching **(figure 21.7).**

▶ Pathogenesis and Virulence Factors

Successful attachment is key to the organism's ability to cause disease. Gonococci use specific chemical groups on the tips of fimbriae to anchor themselves to mucosal epithelial cells. Once the bacterium attaches, it invades the cells and multiplies within the basement membrane.

The fimbriae may also play a role in slowing down effective immunity. The fimbrial proteins are controlled by genes that can be turned on or off, depending on the bacterium's situation. This phenotypic change is called phase variation. In addition, the genes can rearrange themselves to put together fimbriae of different configurations. This antigenic variation confuses the body's immune system. Antibodies that previously recognized fimbrial proteins may not recognize them once they are rearranged.

The gonococcus also possesses an enzyme called IgA protease, which can cleave IgA molecules stationed for protection on mucosal surfaces. In addition, pieces of its outer membrane are shed during growth. These "blebs," containing endotoxin, probably play a role in pathogenesis because they can stimulate portions of the innate defense response, resulting in localized damage.

▶ Transmission and Epidemiology

Except for neonatal infections, the gonococcus is spread through some form of sexual contact.

Gonorrhea is a strictly human infection that occurs worldwide and ranks among the most common sexually transmitted diseases. Infections were declining for years after 2006, and were at historic lows, but have been climbing again, with approximately 580,000 new infections in 2018.

It is important to consider the reservoir of asymptomatic males and females when discussing the transmission of the infection. Because approximately 10% of infected males and 50% of infected females experience no symptoms, it is often spread unknowingly.

▶ Culture and Diagnosis

The best method for diagnosis is a PCR test of secretions. A Gram stain of male secretions usually will yield visible gonococci inside polymorphonuclear cells, but this procedure is not considered sensitive enough to rule out infection if no bacteria are found (see figure 21.7).

N. gonorrhoeae grows best in an atmosphere containing increased CO_2. Because *Neisseria* is so fragile, it is best to inoculate it onto media directly from the patient rather than using a transport tube. Gonococci produce catalase, enzymes for fermenting various carbohydrates, and the enzyme cytochrome oxidase, which can be used for identification as well. Gonorrhea is a reportable disease.

▶ Prevention

No vaccine is yet available for gonorrhea. Using condoms is an effective way to avoid transmission of this and other discharge diseases.

▶ Treatment

Because those infected with *N. gonorrhoeae* are frequently coinfected with *Chlamydia,* treatment recommendations include treating for that bacterium as well, unless its presence has been ruled out. The CDC runs a program called the Gonococcal Isolate Surveillance Project (GISP) to monitor the growing prevalence of antibiotic-resistant strains of *N. gonorrhoeae.* Every month in 28 local STD clinics around the country, *N. gonorrhoeae* isolates from the first 25 males diagnosed with the infection are sent to regional testing labs, their antibiotic sensitivities are determined, and the data are provided to the GISP program at the CDC. The CDC has this bacterium in its Urgent Threat category for antibiotic resistance because it has developed resistance to nearly every drug that could be used for treatment. The recommended treatment is two antibiotics simultaneously to try to slow the growth of antibiotic resistance.

Chlamydia

Genital *Chlamydia trachomatis* infection is the most common reportable infectious disease in the United States. Annually, more than 1.5 million cases are reported, but the actual infection rate may be five to seven times that number. **Figure 21.8** demonstrates how the infection rate has been increasing steadily. The vast majority of cases are asymptomatic. When we consider the serious consequences that may follow *Chlamydia* infection, those facts are very disturbing.

▶ Signs and Symptoms

In males who experience symptoms of *Chlamydia* infection, the bacterium causes an inflammation of the urethra. The symptoms mimic gonorrhea—namely, discharge and painful urination. Untreated infections may lead to epididymitis. Females who experience symptoms have cervicitis, a discharge, and often salpingitis. Pelvic inflammatory disease is a frequent sequela of female chlamydial infection. A woman is even more likely to experience PID as a result of a *Chlamydia* infection than as a result of gonorrhea. Up to 75% of *Chlamydia* infections are asymptomatic, which puts women at risk for developing PID because they don't seek treatment for initial infections. The PID itself may be acute and painful, or it may be relatively asymptomatic, allowing damage to the upper reproductive tract to continue unchecked.

Certain strains of *C. trachomatis* can invade the lymphatic tissues, resulting in another condition called lymphogranuloma venereum. This condition is accompanied by headache, fever, and muscle aches. The lymph nodes near the lesion begin to fill with granuloma cells and become enlarged and tender. These "nodes" can cause long-term lymphatic obstruction that leads to chronic, deforming edema of the genitalia

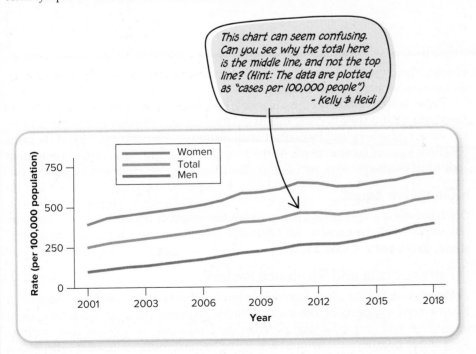

This chart can seem confusing. Can you see why the total here is the middle line, and not the top line? (Hint: The data are plotted as "cases per 100,000 people")
- Kelly & Heidi

Source: CDC

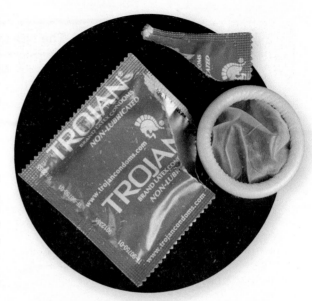

©Christopher Kerrigan/McGraw-Hill Education

📝 A Note About Pelvic Inflammatory Disease (PID) and Infertility

The National Center for Health Statistics estimates that more than 6 million women in the United States have impaired fertility. There are many different reasons for infertility, but the leading cause is pelvic inflammatory disease, or PID. PID is caused by infection of the upper reproductive structures in women—namely, the uterus, fallopian tubes, and ovaries. These organs have no normal biota, and when bacteria from the vagina are transported higher in the tract, they start a chain of inflammatory events that may or may not be noticeable to the patient. The inflammation can be acute, resulting in pain, abnormal vaginal discharge, fever, and nausea, or it can be chronic, with less noticeable symptoms. In acute cases, women usually seek care; in some ways, these can be considered the lucky ones. If the inflammation is curbed at an early stage by using antibiotics to kill the bacteria, chances are better that the long-term sequelae of PID can be avoided. *Chlamydia* infection is the leading cause of PID, followed closely by *N. gonorrhoeae* infection. But other bacteria, perhaps also including normal biota of the reproductive tract, can also cause PID if they are traumatically introduced into the uterus.

Figure 21.8 Reported rates of *Chlamydia* in the United States from 2000–2018.

Table 21.1 The Life Cycle of *Chlamydia*

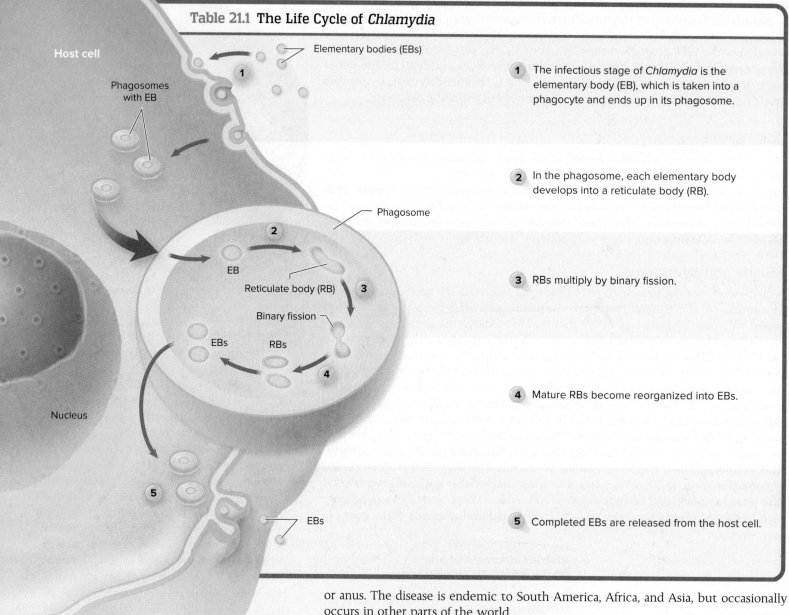

1. The infectious stage of *Chlamydia* is the elementary body (EB), which is taken into a phagocyte and ends up in its phagosome.

2. In the phagosome, each elementary body develops into a reticulate body (RB).

3. RBs multiply by binary fission.

4. Mature RBs become reorganized into EBs.

5. Completed EBs are released from the host cell.

or anus. The disease is endemic to South America, Africa, and Asia, but occasionally occurs in other parts of the world.

Babies born to mothers with *Chlamydia* infections can develop eye infections and also pneumonia if they become infected during passage through the birth canal. Infant conjunctivitis caused by contact with maternal *Chlamydia* infection is the most prevalent form of conjunctivitis in the United States. Antibiotic drops or ointment applied to newborns' eyes are used to eliminate both *Chlamydia* and *N. gonorrhoeae*.

▶ Causative Agent

C. trachomatis is a very small gram-negative bacterium. It lives inside host cells as an obligate intracellular parasite. All *Chlamydia* species alternate between two distinct stages, illustrated in **table 21.1.**

▶ Pathogenesis and Virulence Factors

Chlamydia's ability to grow intracellularly contributes to its virulence because it escapes certain aspects of the host's immune response. Also, the bacterium has a unique cell wall that apparently prevents the phagosome from fusing with the lysosome inside phagocytes. The presence of the bacteria inside cells causes the release of cytokines that provoke intense inflammation. This defensive response leads to most of the actual tissue damage in *Chlamydia* infection. Of course, the last step of inflammation is repair, which often results in scarring. This can have disastrous effects on a narrow-diameter structure like the fallopian tube.

▶ Transmission and Epidemiology

The reservoir of pathogenic strains of *C. trachomatis* is the human body. The microbe shows an astoundingly broad distribution within the population, and incidence is rising. Adolescent women are more likely than older women to harbor the bacterium because it prefers to infect cells that are particularly prevalent on the adolescent cervix. It is transmitted through sexual contact and also vertically. Fifty percent of (untreated) babies born to infected mothers will acquire chlamydial conjunctivitis or pneumonia.

▶ Culture and Diagnosis

Infection with this microorganism is usually detected initially using a rapid technique such as PCR or ELISA. A urine test is available, which has definite advantages for widespread screening, but it is slightly less accurate for females than males.

▶ Prevention

Avoiding contact with infected tissues and secretions through abstinence or barrier protection (condoms) is the only means of prevention.

▶ Treatment

The CDC recommends annual screening of young women for this often asymptomatic infection. It is also recommended that older women with some risk factor (new sexual partner, for instance) also be screened. Treatment is usually with doxycycline and/or azithromycin. Coinfection with gonorrhea should be assumed and treated similarly. Many patients become reinfected soon after treatment; therefore, the recommendation is that patients be rechecked for *Chlamydia* infection 3 to 4 months after treatment. Treatment of all sexual partners of the patient is also recommended to prevent reinfection. Repeated infections with *Chlamydia* increase the likelihood of PID and other serious sequelae.

Disease Table 21.3	Genital Discharge Diseases (in Addition to Vaginitis/Vaginosis)	
	Gonorrhea Ⓑ G⁻	**Chlamydia** Ⓑ G⁻
Causative Organism(s)	*Neisseria gonorrhoeae*	*Chlamydia trachomatis*
Most Common Modes of Transmission	Direct contact (STI), also vertical	Direct contact (STI), vertical
Virulence Factors	Fimbrial adhesins, antigenic variation, IgA protease, membrane blebs/endotoxin	Intracellular growth resulting in avoiding immune system and cytokine release, unusual cell wall preventing phagolysosome fusion
Culture/Diagnosis	Gram stain in males, rapid tests (PCR, ELISA) for females, culture on Thayer-Martin agar	PCR or ELISA, can be followed by cell culture
Prevention	Avoid contact; condom use	Avoid contact; condom use
Treatment	Coinfection with gonorrhea and *C. trachomatis* should be assumed; treat with doxycycline and azithromycin. Be on alert for multidrug-resistant *N. gonorrhoeae*, which is in **Urgent Threat** category in CDC Antibiotic Resistance Report.	
Distinctive Features	Rare complications including arthritis, meningitis, endocarditis	More commonly asymptomatic than gonorrhea
Effects on Fetus	Eye infections, blindness	Eye infections, pneumonia
Epidemiological Features	United States: rates increasing since 2014; 2018 incidence = 583,000; internationally: 26 million cases	United States: increasing since 2000; 2018 incidence = 1.8 million; internationally: eye infection (trachoma) has 90% prevalence rate in developing world

Vaginitis and Vaginosis

▶ Signs and Symptoms

Vaginitis, an inflammation of the vagina, is a condition characterized by some degree of vaginal itching, depending on the etiologic agent. Symptoms may also include burning and sometimes a discharge, which may take different forms as well. Vaginosis is similar but does not include significant inflammation.

▶ Causative Agents

While a variety of bacteria and even protozoa can cause vaginitis, the most well-known agent is the fungus *Candida albicans*. The vaginal condition caused by this fungus is known as a *yeast infection*.

Candida albicans

C. albicans is a dimorphic fungus that is normal biota in the majority of humans, living in low numbers on many mucosal surfaces such as the mouth, gastrointestinal tract, vagina, and so on. The vaginal condition it causes is also called vulvovaginal candidiasis. The yeast is easily detectable on a wet prep or a Gram stain of material obtained during a pelvic exam **(figure 21.9).** The presence of pseudohyphae in the smear is a clear indication that the yeast is growing rapidly and causing a yeast infection.

In otherwise healthy people, the fungus is not invasive and limits itself to this surface infection. Please note, however, that *Candida* infections of the bloodstream do occur and they have high mortality rates. They do not normally stem from vaginal infections with the fungus, however, but are seen most frequently in hospitalized patients. AIDS patients are also at risk of developing systemic *Candida* infections.

▶ Transmission and Epidemiology

Vaginal infections with this organism are nearly always opportunistic. Disruptions of the normal bacterial biota or even minor damage to the mucosal epithelium in the vagina can lead to overgrowth by this fungus. Disruptions may be mechanical, such as trauma to the vagina, or they may be chemical, as when broad-spectrum antibiotics taken for some other purpose temporarily diminish the vaginal bacterial population. Diabetics and pregnant women are also predisposed to vaginal yeast overgrowths. Some women are prone to this condition during menstruation.

It is possible to transmit this yeast through sexual contact, especially if a woman is experiencing an overgrowth of it. The recipient's immune system may well subdue the yeast so that it acts as normal biota in them. But the yeast may be passed back to the original partner during further sexual contact after treatment. Women with HIV infection experience frequently recurring yeast infections. Also, a small percentage of women with no underlying immune disease experience chronic or recurrent vaginal infection with *Candida* for reasons that are not clear.

▶ Prevention and Treatment

No vaccine is available for *C. albicans*. Topical and oral azole drugs are used to treat vaginal candidiasis, and many of them are now available over the counter. Many women experience this condition multiple times in their lives, but if infections recur frequently or fail to resolve, it is important to see a physician.

Gardnerella Species

The bacterium *Gardnerella* is associated with a particularly common condition in women in their childbearing years. This condition is usually called *vaginosis* rather than *vaginitis* because it doesn't appear to induce inflammation in the vagina. It is also known as BV, or bacterial vaginosis. Despite the absence of an inflammatory response, a vaginal discharge is associated with the condition, which often has a fishy odor. Itching is also common, but it is also true that many women have this condition with no noticeable symptoms.

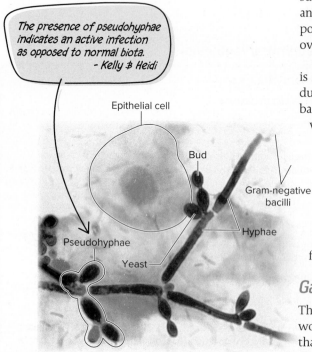

The presence of pseudohyphae indicates an active infection as opposed to normal biota.
- Kelly & Heidi

Epithelial cell
Bud
Gram-negative bacilli
Hyphae
Pseudohyphae
Yeast

Figure 21.9 Gram stain of *Candida albicans* in a vaginal smear.
Danny L. Wiedbrauk

Disease Table 21.4	Vaginitis/Vaginosis		
Causative Organism(s)	*Candida albicans* **F**	Mixed infection, usually including *Gardnerella* **B**	*Trichomonas vaginalis* **P**
Most Common Modes of Transmission	Opportunism	Opportunism or STI?	Direct contact (STI)
Virulence Factors	–	–	–
Culture/Diagnosis	Wet prep or Gram stain	Visual exam of vagina, or clue cells seen in Pap smear or other smear	Protozoa seen on Pap smear or Gram stain; culture is gold standard
Prevention	–	–	Barrier use during intercourse
Treatment	Topical or oral azole drugs, some over-the-counter drugs	Metronidazole or clindamycin	Metronidazole, tinidazole
Distinctive Features	White curdlike discharge	Discharge may have fishy smell	Discharge may be greenish
Epidemiological Features	United States: causes 20% of all vaginitis cases; 75% of women report at least one case during lifetime	United States: 30% prevalence rate; internationally: prevalence rates vary from 20% to 50%	United States: 3.7 million people infected; is one of CDC's "neglected parasitic infections"

Vaginosis is most likely a result of a shift from a predominance of "good" bacteria (lactobacilli) in the vagina to a predominance of "bad" bacteria, and one of those is *Gardnerella vaginalis*. This genus of bacteria is a facultative anaerobe and grampositive, although in a Gram stain it usually appears gram-negative. (Some texts refer to it as *gram-variable* for this reason.) Probably a mixed infection leads to the condition, however. Anaerobic streptococci and other bacteria, particularly a genus known as *Mobiluncus*, that are normally found in low numbers in a healthy vagina can also often be found in high numbers in this condition. The fishy odor comes from the metabolic by-products of anaerobic metabolism by these bacteria.

▶ Pathogenesis and Virulence Factors

The mechanism of damage in this disease is not well understood, but some of the outcomes are. Besides the symptoms just mentioned, vaginosis can lead to complications such as pelvic inflammatory disease, infertility, and, more rarely, ectopic pregnancies. Babies born to some mothers with vaginosis have low birth weights.

▶ Transmission and Epidemiology

Thirty percent of women between the ages of 15 and 44 in the United States are estimated to have bacterial vaginosis. Rates are noticeably greater in non-White women: White women have a prevalence of 23%; African Americans, 51%; and Mexican Americans, 32%. This mixed infection is not necessarily considered to be sexually transmitted, although women who have never had sex are less likely to develop the condition. It is very common in sexually active women. It may be that the condition is associated with sex but not transmitted by it. This situation could occur if the act of penetration or the presence of semen (or saliva) causes changes in the vaginal epithelium or in the vaginal biota. We do not know exactly what causes the increased numbers of *Gardnerella* and other normally rare biota. The low pH typical of the vagina is usually higher in vaginosis. It is not clear whether this causes or is caused by the change in bacterial biota.

▶ Culture and Diagnosis

The condition can be diagnosed by a variety of methods. Sometimes a simple stain of vaginal secretions is used to examine sloughed vaginal epithelial cells. In vaginosis, some cells will appear to be nearly covered with adherent bacteria. In normal times, vaginal epithelial cells are sparsely covered with bacteria. These cells are called clue cells and are a helpful diagnostic indicator **(figure 21.10)**. They can also be found on Pap smears, but sometimes genomic analyses are needed.

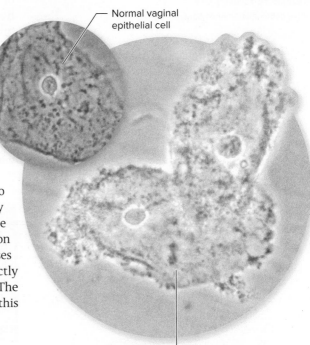

Normal vaginal epithelial cell

Vaginal epithelial cell with numerous bacteria (clue cell)

Figure 21.10 Clue cell in bacterial vaginosis.
These epithelial cells came from a pelvic exam. The cells in the large circle have an abundance of bacteria attached to them.
Source: CDC/M. Rein

A single *Trichomonas* cell.
©Eye of Science/Science Source

▶ Prevention and Treatment

Women who find the condition uncomfortable or who are planning on becoming pregnant should be treated. Women who use intrauterine devices (IUDs) for contraception should also be treated because IUDs can provide a passageway for the bacteria to gain access to the upper reproductive tract. The usual treatment is oral or topical metronidazole or clindamycin.

Trichomonas vaginalis

Trichomonads are small, pear-shaped protozoa with four anterior flagella and an undulating membrane. *Trichomonas vaginalis* seems to cause asymptomatic infections. Trichomonads are considered asymptomatic infectious agents rather than normal biota because of evidence that some people experience long-term negative effects. Even though *Trichomonas* is a protozoan, it has no cyst form, and it does not survive long outside of the host.

Many cases are asymptomatic, and men seldom have symptoms. Women often have vaginitis symptoms, which can include a white to green frothy discharge. Chronic infection can make a person more susceptible to other infections, including HIV. Also, women who become infected during pregnancy are predisposed to premature labor and low-birth-weight infants. Chronic infection may also lead to infertility.

Because *Trichomonas* is common biota in so many people, it is easily transmitted through sexual contact. It has been called the most common nonviral sexually transmitted infection. It does not appear to undergo opportunistic shifts within its host (i.e., to become symptomatic under certain conditions), but rather, the protozoan causes symptoms when transmitted to a noncarrier. **Disease Table 21.4.**

Prostatitis

Prostatitis is an inflammation of the prostate gland (see figure 21.2). It can be acute or chronic. Acute prostatitis is virtually always caused by bacterial infection. The bacteria are usually normal biota from the intestinal tract or may have caused a previous urinary tract infection. Chronic prostatitis is also often caused by bacteria. Researchers have found that chronic prostatitis, often unresponsive to antibiotic treatment, can be caused by mixed biofilms of bacteria in the prostate. Some forms of chronic prostatitis have no known microbial cause, though many infectious disease specialists feel that one or more bacteria are involved, but they are simply not culturable with current techniques.

Symptoms may include pain in the groin and lower back, frequent urge to urinate, difficulty in urinating, blood in the urine, and painful ejaculation. Treatment is with ciprofloxacin or levofloxacin.

Disease Table 21.5	Prostatitis
Causative Organism(s)	GI tract biota
Most Common Modes of Transmission	Endogenous transfer from GI tract; otherwise unknown
Virulence Factors	Various
Culture/Diagnosis	Digital rectal exam to examine prostate; culture of urine or semen
Prevention	None
Treatment	Antibiotics, muscle relaxers, alpha blockers
Distinctive Features	Pain in genital area and/or back, difficulty urinating
Epidemiological Features	United States: 50% of men experience during lifetime

Genital Ulcer Diseases

Three common infectious conditions can result in lesions (ulcers) on the genitals: syphilis, chancroid, and genital herpes. In this section, we consider each of these. One very important fact to remember about the ulcer diseases is that having one of them increases the chances of infection with HIV because of the open lesions.

Syphilis

Untreated syphilis is marked by distinct clinical stages designated as *primary, secondary,* and *tertiary syphilis.* The disease also has latent periods of varying duration during which it is quiescent. The spirochete appears in the lesions and blood during the primary and secondary stages and, therefore, is transmissible at these times. During the early latency period between secondary and tertiary syphilis, it is also transmissible. Syphilis is largely nontransmissible during the "late latent" and tertiary stages.

▶ Primary Syphilis

The earliest indication of syphilis infection is the appearance of a hard **chancre** (shang′-ker) at the site of entry of the pathogen. Because these ulcers tend to be painless, they may escape notice, especially when they are on internal surfaces. The chancre heals spontaneously without scarring in 3 to 6 weeks, but the healing is deceptive because the spirochete has escaped into the circulation and is entering a period of tremendous activity.

▶ Secondary Syphilis

About 3 weeks to 6 months after the chancre heals, the secondary stage appears. By then, many systems of the body have been invaded, and the signs and symptoms are more profuse and intense. Initial symptoms are fever, headache, and sore throat, followed by lymphadenopathy and a peculiar red or brown rash that breaks out on all skin surfaces, including the palms of the hands and the soles of the feet. A person's hair often falls out. Like the chancre, the lesions contain viable spirochetes and disappear spontaneously in a few weeks. The major complications of this stage, occurring in the bones, hair follicles, joints, liver, eyes, and brain, can linger for months and years.

▶ Latency and Tertiary Syphilis

After resolution of secondary syphilis, about 30% of infections enter a highly varied latent period that can last for 20 years or longer. During latency, although antibodies to the bacterium are readily detected, the bacterium itself is not. The final stage of the disease, tertiary syphilis, is relatively rare today because of widespread use of antibiotics. But it is so damaging that it is important to recognize. By the time a patient reaches this phase, numerous pathologic complications occur in susceptible tissues and organs. Cardiovascular syphilis results from damage to the small arteries in the aortic wall. As the fibers in the wall weaken, the aorta is subject to distension and fatal rupture. The same pathologic process can damage the aortic valves, resulting in heart failure.

In one form of tertiary syphilis, painful swollen syphilitic tumors called **gummas** (goo-mahz′) develop in tissues such as the liver, skin, bone, and cartilage **(figure 21.11).** Gummas are usually benign and only occasionally lead to death, but they can impair function. Neurosyphilis can involve any part of the nervous system, but it shows particular affinity for the blood vessels in the brain, cranial nerves, and dorsal roots of the spinal cord. The diverse results include severe headaches, convulsions, atrophy of the optic nerve, blindness, dementia, and a sign called the Argyll-Robertson pupil—a condition caused by adhesions along the inner edge of the iris that fix the pupil's position into a small irregular circle.

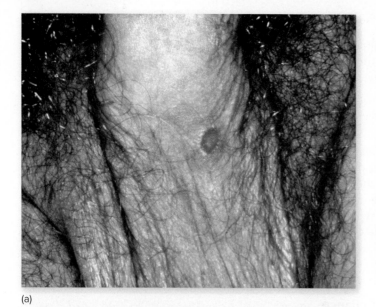

(a)

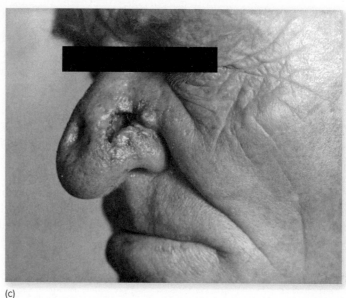

(c)

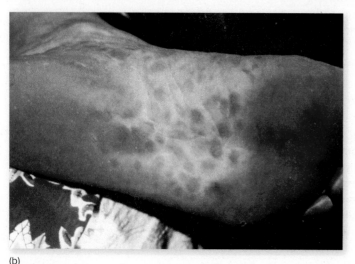

(b)

Figure 21.11 **Primary, secondary, and tertiary syphilis.** **(a)** The lesion (chancre) of primary syphilis. **(b)** Secondary syphilis produces a generalized rash that appears also on the palms of the hands and soles of the feet. **(c)** Tertiary syphilis patient with gummas on his nose. Sources: CDC/Dr. N.J. Fiumara, Dr. Gavin Hart (a); CDC (b); CDC/ Susan Lindsley (c)

▶ Congenital Syphilis

The syphilis bacterium can pass from a pregnant woman's circulation into the placenta and can be carried throughout the fetal tissues. The pathogen inhibits fetal growth and disrupts critical periods of development with varied consequences, ranging from mild to the extremes of spontaneous miscarriage or stillbirth. Infants often demonstrate such signs as profuse nasal discharge **(figure 21.12a),** skin eruptions, bone deformation, and nervous system abnormalities. The late form gives rise to an unusual assortment of problems in the bones, eyes, inner ear, and joints, and causes the formation of Hutchinson's teeth **(figure 21.12b).**

▶ Causative Agent

Treponema pallidum, a spirochete, is a thin, regularly coiled cell with a gram-negative cell wall. It is a strict parasite with complex growth requirements that necessitate cultivating it in living host cells.

▶ Pathogenesis and Virulence Factors

Brought into direct contact with mucous membranes or abraded skin, *T. pallidum* binds avidly by its hooked tip to the epithelium **(figure 21.13).** At the binding site, the spirochete multiplies and penetrates the capillaries nearby. Within a short time, it moves into the circulation, and the body is literally transformed into a large receptacle for incubating the pathogen. Virtually any tissue is a potential target.

▶ Transmission and Epidemiology

Humans are evidently the sole natural hosts and source of *T. pallidum.* The bacterium is extremely fastidious and sensitive and cannot survive for long outside the host, being rapidly destroyed by heat, drying, disinfectants, soap, high oxygen tension, and pH changes. It survives a few minutes to hours when protected by body secretions and about 36 hours in stored blood. The risk of infection from an infected sexual partner is 12% to 30% per encounter.

For centuries, syphilis was a common and devastating disease in the United States, so much so that major medical centers had "Departments of Syphilology." Its effect on social life was enormous. This effect diminished quickly when antibiotics were discovered. But since 2003, the rates have been increasing again in the United States, especially since 2011 **(figure 21.14)**. And syphilis continues to be a serious problem worldwide, especially in Africa and Asia. As mentioned previously, persons with syphilis often suffer concurrent infections with other STIs. Coinfection with the AIDS virus can be an especially deadly combination with a rapidly fatal course.

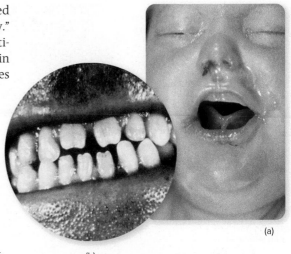

(a)

(b)

Figure 21.12 Congenital syphilis. (a) An early sign is snuffles, a profuse nasal discharge that obstructs breathing. **(b)** A common characteristic of late congenital syphilis is notched, barrel-shaped incisors (Hutchinson's teeth).
Source: CDC/Dr. Norman Cole (a); Robert Sumpter/CDC (b)

▶ Culture and Diagnosis

Syphilis can be detected in patients most rapidly by using dark-field microscopy of a suspected lesion (figure 21.13). A single negative test is not enough to exclude syphilis because the patient may have removed the organisms by washing, so follow-up tests are recommended.

Very commonly, blood tests are used for this diagnosis. The best test is that which specifically reacts with treponemal antigens. There is also a test called the rapid plasmin reagin (RPR), which is a long-standing test used for syphilis that is still useful. It tests for antigens that are not from the bacterium but that appear in the host during infection by the bacterium. Unfortunately, these antigens can appear in some other conditions. So the RPR is almost always coupled with an immunoassay specific for treponemal antigens.

▶ Prevention

The core of an effective prevention program depends on detection and treatment of the sexual contacts of syphilitic patients. Public health departments and physicians are charged with the task of questioning patients and tracing their contacts. All individuals identified as being at risk, even if they show no signs of infection, are given immediate prophylactic penicillin in a single long-acting dose.

The barrier effect of a condom provides excellent protection during the primary phase. Protective immunity apparently does arise in humans, allowing the prospect of an effective immunization program in the future, although no vaccine exists currently.

▶ Treatment

Throughout most of history, the treatment for syphilis was a dose of mercury or even a "mercurial rub" applied to external lesions. In 1906, Paul Ehrlich discovered that a derivative of arsenic called salvarsan could be very effective. The fact that toxic compounds like mercury and arsenic were used to treat syphilis gives some indication of how dreaded the disease was and to what lengths people would go to rid themselves of it.

Current recommendations are for ciprofloxacin or levofloxacin.

Chancroid

This ulcerative disease usually begins as a soft papule, or bump, at the point of contact. It develops into a "soft chancre" (in contrast to the hard syphilis chancre), which is very painful in men, but may be unnoticed in women **(Disease Table 21.6)**. Inguinal lymph nodes can become very swollen and tender.

Chancroid is caused by a pleomorphic gram-negative rod called *Haemophilus ducreyi*. Recent research indicates that a hemolysin (exotoxin) is important in the pathogenesis of chancroid disease. It is very common in the tropics and subtropics and also appears in the United States. Chancroid is transmitted exclusively through direct contact, especially sexually. This disease is associated with sex workers and poor hygiene. Uncircumcised men seem to be more commonly infected than those who have been circumcised. People may carry this bacterium asymptomatically.

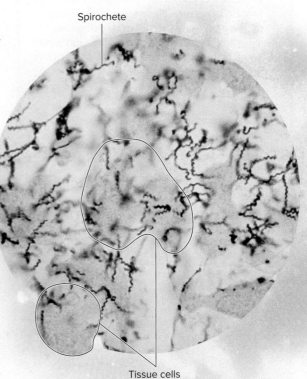

Spirochete

Tissue cells

Figure 21.13 *Treponema pallidum* from a syphilitic chancre, viewed with dark-field illumination. Its tight spirals are highlighted next to human cells and tissue debris.
Source: CDC

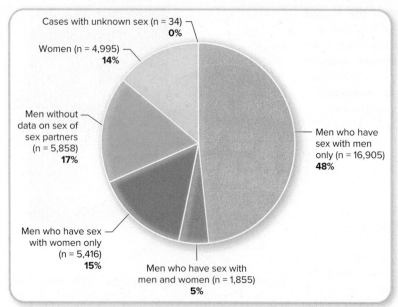

Figure 21.14 Primary and secondary syphilis in 2018. Who is experiencing primary and/or secondary syphilis?

Source: CDC

NCLEX® PREP

3. A neonate is brought in to the pediatrician's office by her parents for a 1-month checkup. Findings indicate inflammation and edema of the left eye. Parents deny any traumatic injury to the left eye. Which information would be most important to ascertain in order to develop a plan of care?

 a. *delivery history of the neonate—vaginal or C-section*

 b. *documentation of vitamin K injection postdelivery*

 c. *Apgar score following delivery*

 d. *maternal history of STIs*

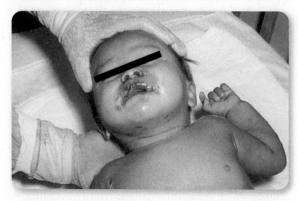

Figure 21.15 Neonatal herpes simplex. Babies can be born with the lesions or develop them 1 to 2 weeks after birth.

Source: CDC/J.D. Millar

Genital Herpes

Genital herpes is much more common than most people think. It is caused by herpes simplex viruses (HSVs). Two types of HSV have been identified, HSV-1 and HSV-2.

▶ Signs and Symptoms

Genital herpes infection has multiple presentations. After initial infection, a person may notice no symptoms. Alternatively, herpes could cause the appearance of single or multiple vesicles on the genitalia, breasts, perineum, thigh, and buttocks. The vesicles are small and are filled with a clear fluid (see **Disease Table 21.6**). They can be intensely painful to the touch. The appearance of lesions the first time you get them can be accompanied by malaise, anorexia, fever, and bilateral swelling and tenderness in the groin. Occasionally central nervous system symptoms such as meningitis or encephalitis can develop. So we see that initial infection can either be completely asymptomatic or be serious enough to require hospitalization.

After recovery from initial infection, a person may have recurrent episodes of lesions. They are generally less severe than the original symptoms, although the whole gamut of possible severity is seen here as well. Some people never have recurrent lesions. Others have nearly constant outbreaks with little recovery time between them. On average, the number of recurrences is four or five a year. Their frequency tends to decrease over the course of years.

In most cases, patients remain asymptomatic or experience recurrent "surface" infections indefinitely. Very rarely, complications can occur. Every year, one or two persons per million with chronic herpes infections develop encephalitis. The virus disseminates along nerve pathways to the brain (although it can also infect the spinal cord). The effects on the central nervous system begin with headache and stiff neck and can progress to mental disturbances and coma.

▶ Herpes of the Newborn

Although HSV infections in healthy adults are annoying and unpleasant, only rarely are they life-threatening. However, in the neonate and the fetus **(figure 21.15)**, HSV infections are very destructive and can be fatal. Most cases occur when infants are contaminated by the mother's reproductive tract immediately before or during birth, but they have also been traced to hand transmission from the mother's lesions to the baby.

Pregnant women with a history of recurrent infections must be monitored for any signs of viral shedding, especially in the last 4 weeks of pregnancy. If no evidence of recurrence is seen, vaginal birth is indicated, but any evidence of an outbreak at the time of delivery necessitates a cesarean section.

▶ Causative Agent

Both HSV-1 and HSV-2 can cause genital herpes if the virus contacts the genital epithelium, although HSV-1 is thought of as a virus that infects the oral mucosa, resulting in "cold sores" or "fever blisters" **(figure 21.16),** and HSV-2 is thought of as the genital virus. In reality, either virus can infect either region, depending on the type of contact.

▶ Pathogenesis and Virulence Factors

Herpesviruses have a strong tendency to become latent. During latency, some type of signal causes most of the HSV genome not to be transcribed. This allows the virus to be maintained within cells of the nervous system between episodes. Recent research has found that microRNAs are in part responsible for the latency of HSV-1. It is further suggested that in some peripheral cells, viral replication takes place at a constant, slow rate, resulting in constant low-level shedding of the virus without lesion production.

Reactivation of the virus can be triggered by a variety of stimuli, including stress, UV radiation (sunlight), injury, menstruation, or another microbial infection. At that point, the virus begins manufacturing large numbers of entire virions, which cause new lesions on the surface of the body served by the neuron, usually in the same site as previous lesions.

▶ Transmission and Epidemiology

It might surprise you to know that 50% to 80% of adults in the United States are seropositive for HSV-1, and 20% to 40% are positive for HSV-2. Worldwide, the WHO estimates that 4 billion people are infected. That's 4 billion of about 7.7 billion total people. Because these viruses are relatively sensitive to the environment, transmission is primarily through direct exposure to secretions containing the virus. People with active lesions are the most significant source of infection, but studies indicate that genital herpes can be transmitted even when no lesions are present (due to the constant shedding just discussed).

Earlier in this chapter, you read that *Chlamydia* infection is the most common *reported* infectious disease in the United States. Elsewhere you might hear that gonorrhea is one of the most common reportable *STIs* in the United States. Both statements are true. It is also true that genital herpes is much more common than either of these diseases. Herpes, however, is not a nationally *reportable* disease.

Research has shown that 50% to 90% of people who are infected don't even know it, either because they have rare symptoms that they fail to recognize or because they have no symptoms at all.

▶ Culture and Diagnosis

These two viruses are sometimes diagnosed based on the characteristic lesions alone. PCR tests are available to test for these viruses directly from lesions. Alternatively, antibody to either of the viruses can be detected from blood samples.

▶ Prevention

No vaccine is currently licensed for HSV, but more than one is being tested in clinical trials, meaning that vaccines may become available very soon. In the meantime, avoiding contact with infected body surfaces is the only way to avoid HSV. Condoms provide good protection when they actually cover the site where the lesion is, but lesions can occur outside of the area covered by a condom.

Mothers with cold sores should be careful in handling their newborns; they should never kiss their infants on the mouth.

▶ Treatment

Several agents are available for treatment. These agents often result in reduced viral shedding and a decrease in the frequency of lesion occurrence. They are not curative. Acyclovir and its derivatives (famciclovir or valacyclovir) are very effective. Topical formulations can be applied directly to lesions, and pills are available as well. Sometimes medicines are prescribed on an ongoing basis to decrease the frequency of recurrences, and sometimes they are prescribed to be taken at the beginning of a recurrence to shorten it. **Disease Table 21.6.**

Wart Diseases

In this section, we describe two viral STIs that cause wartlike growths. The more serious disease is caused by the *human papillomavirus* (HPV). The other condition, called *molluscum contagiosum,* apparently has no serious effects.

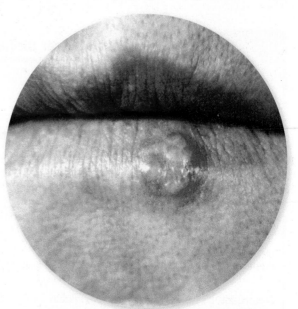

Figure 21.16 Oral herpes infection. Tender itchy papules erupt around the mouth and progress to vesicles that burst, drain, and scab over. These sores and fluid are highly infectious and should not be touched.
Source: CDC

Medical Moment

Crabs

Pediculosis pubis, commonly known as crabs, is caused by infestation of the pubic hair by *Phthirus pubis,* a tiny (1 mm or less) insect that can multiply rapidly. The condition can be spread from person to person through sexual contact and can therefore be considered a sexually transmitted condition. The eyebrows, eyelashes, chest hair, scalp hair, and facial hair may also be affected.

Although these tiny creatures do not pose much of a risk to human health, infestation with crabs may lead to a higher risk for other STIs. It is estimated that approximately 3 million people are infected yearly in the United States.

Symptoms include intense itching of the affected areas. Some people do not experience itching but while they are bathing may observe the insects, as the lice are visible to the naked eye.

Treatment is applied topically in the form of creams, shampoos, or lotions. Manual removal of adult insects and their eggs is recommended using a fine-toothed metal comb. Clothing and bedding should be laundered in hot water to prevent reinfestation.

Q. Why do you think a case of crabs can make you more susceptible to other STIs?

Answer in Appendix B.

Disease Table 21.6	**Genital Ulcer Diseases**		
	Syphilis	**Chancroid**	**Herpes**
Causative Organism(s)	*Treponema pallidum* Ⓑ G⁻	*Haemophilus ducreyi* Ⓑ G⁻	Herpes simplex viruses 1 and 2 Ⓥ
Most Common Modes of Transmission	Direct contact and vertical	Direct contact (vertical transmission not documented)	Direct contact, vertical
Virulence Factors	Lipoproteins	Rule out other ulcer diseases	Latency
Culture/ Diagnosis	Direct tests (immunofluorescence, dark-field microscopy), blood tests for treponemal and nontreponemal antibodies	Culture from lesion	Clinical presentation, PCR, antibody tests, growth of virus in cell culture
Prevention	Antibiotic treatment of all possible contacts, avoiding contact	Avoiding contact	Avoiding contact, antivirals can reduce recurrences
Treatment	Ciprofloxacin, levofloxacin	Ceftriaxone or azithromycin	Acyclovir and derivatives
Distinctive Features	Three stages of disease plus latent period, possibly fatal	No systemic effects	Ranges from asymptomatic to frequent recurrences
Effects on Fetus	Congenital syphilis	None	Blindness, disseminated herpes infection
Appearance of Lesions			Vesicles
Epidemiological Features	United States: estimated 115,000 new cases per year; internationally: estimated 12 million new infections per year	United States: no more than a handful per year; internationally: estimated 7 million cases annually	United States: 50%-80% prevalence of seropositivity for HSV-1; 20%-40% for HSV-2; worldwide: more than half are seropositive for one of them

Clinical Photography, Central Manchester University Hospitals NHS Foundation Trust, UK/Science Source (Syphilis); Dr. M.A. Ansary/Science Source (Chancroid); Dr. P. Marazzi/Science Source (Herpes)

A Note About HIV and Hepatitis B and C

This chapter is about diseases *whose major (presenting) symptoms occur in the genitourinary tract.* But some sexually transmitted diseases do not have their major symptoms in this system. HIV and hepatitis B and C can all be transmitted in several ways, one of them being through sexual contact. HIV is considered in chapter 18 because its major symptoms occur in the cardiovascular and lymphatic systems. Because the major disease manifestations of hepatitis B and C occur in the gastrointestinal tract, these diseases are discussed in chapter 20. Anyone diagnosed with any sexually transmitted disease should also be tested for HIV.

Human Papillomavirus Infection

These viruses are the causative agents of genital warts. But an individual can be infected with these viruses without having any warts, while still risking serious consequences.

▶ Signs and Symptoms

Symptoms, if present, may manifest as warts—outgrowths of tissue on the genitals **(Disease Table 21.7).** In females, these growths can occur on the vulva and in and around the vagina. In males, the warts can occur in or on the penis and the scrotum. In both sexes, the warts can appear in or on the anus and even on the skin around the groin, such as the area between the thigh and the pelvis. The warts themselves range from tiny, flat, inconspicuous bumps to extensively branching, cauliflower-like masses called **condyloma acuminata.** The warts are unsightly and can be obstructive, but they don't generally lead to more serious symptoms.

Other types of HPV can lead to more subtle symptoms. Certain types of the virus infect cells on the female cervix. This infection may be "silent," or it may lead to abnormal cell changes and malignancies of the cervix. Approximately 4,300 women die each year in the United States from cervical cancer, and the vast majority of these are caused by HPV.

Males can also get cancer from infection with these viruses. The sites most often affected are the penis and the anus. These cases are less common than cervical cancer. Mouth and throat cancers in both sexes are also caused by HPV infection and are thought to be a consequence of oral sex.

▶ Causative Agent

The human papillomaviruses are a group of nonenveloped DNA viruses belonging to the Papovaviridae family. There are more than 100 different types of HPV. Some types are specific for the mucous membranes; others invade the skin. Some of these viruses are the cause of plantar warts, which often occur on the soles of the feet. Other HPVs cause the common or "seed" warts and flat warts. In this chapter, we are concerned only with the HPVs that colonize the genital tract.

Among the HPVs that infect the genital tract, some are more likely to cause the appearance of warts. Others that have a preference for growing on the cervix can lead to cancerous changes. Five types in particular, HPV-16, -18, -31, -33, and -35, are closely associated with development of cervical and anal cancer. Other types put you at higher risk for vulvar or penile cancer.

▶ Pathogenesis and Virulence Factors

Scientists are working hard to understand how viruses cause the growths we know as warts and also how some of them can cause cancer. The major virulence factor for cancer-causing HPVs is their **oncogenes,** which code for proteins that interfere with normal host cell function, resulting in uncontrolled growth.

▶ Transmission and Epidemiology

The CDC states, "Nearly all sexually active adults will get HPV at some time in their lives." It is estimated that 14% of female college students become infected with this incurable condition each year.

The mode of transmission is direct contact. Autoinoculation is also possible– meaning that the virus can be spread to other parts of the body by touching warts. Indirect transmission occurs but is more common for nongenital warts caused by HPV.

▶ Culture and Diagnosis

PCR-based screening tests can be used to test samples from a pelvic exam for the presence of dangerous HPV types. These tests are now recommended for women over the age of 30. The Pap smear is still the single best screening procedure available for cervical cancer. It will detect nearly all precancerous cells if conducted on the recommended schedule.

▶ Prevention

Infection with these viruses is prevented the same way other sexually transmitted infections are prevented–by avoiding direct, unprotected contact, but also through vaccination. The vaccines prevent infection by up to nine types of HPV and are recommended in both girls and boys as young as age 9. There has been an unusual amount of controversy about this vaccine, which is no doubt due to the fact that it protects against a disease transmitted by sex, and parents don't like to think about their kids having sex. It has been repeatedly shown to be safe, and it has been shown to provide nearly 100% protection against cervical cancer as well as warts. **Table 21.2** provides guidelines for who should get the vaccine.

Preventing infection prevents the cancers that may have resulted from previous infection by an HPV. It is important to note that even though women now have access to the vaccines, cervical cancer can still result from HPV types not included in the vaccines. The good news is that cervical cancer is slow in developing, so that even if a woman is infected with a malignant HPV type, regular screening of the cervix can detect abnormal changes early. The standardized screen for cervical cell changes is the Pap smear. Precancerous changes show up very early, and the development process can be stopped by removal of the affected tissue. Women should have their first Pap smear by age 21 or within 3 years of their first sexual

Table 21.2 **Who Should Get the HPV Vaccine**	
Ideal	11- to 12-year-olds of both sexes should get two doses, 6 to 12 months apart
"Catch-Up"	Females who missed this through the age of 26 should get three doses Males who missed it through the age of 21 should get three doses
Others for "Catch-Up"	Men who identify as gay or bisexual through the age of 26 Transgender people through the age of 26 Immunocompromised people through the age of 26

©Image Point Fr/Shutterstock (woman); Science Photo Library/Getty Images (virus)

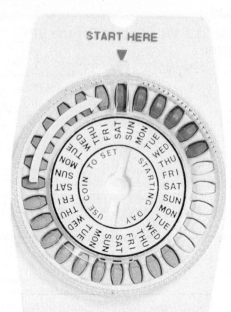

The introduction of oral contraceptives revolutionized many aspects of society. Unfortunately, they do not protect against STIs.
©Comstock/Alamy Stock Photo

activity, whichever comes first. New Pap smear technologies have been developed. Depending on which one your physician uses, it is now possible that you need to be screened only once every 2 or 3 years. But you should base your screening practices on the sound advice of a physician.

▶ Treatment

Infection with any HPV is incurable, though, unlike herpesviruses, it may resolve itself over a period of months to years. Genital warts can be removed through a variety of methods, but that does not eliminate the virus.

Molluscum Contagiosum

An unclassified virus in the family Poxviridae can cause a condition called molluscum contagiosum. This disease can take the form of skin lesions, and it can also be transmitted sexually. The wartlike growths that result from this infection can be found on the mucous membranes or the skin of the genital area (see **Disease Table 21.7**). Few problems are associated with these growths beyond the warts themselves. We present them here mostly for purposes of differentiating them from HPV.

The virus causing these growths can also be transmitted through fomites such as clothing or towels and through autoinoculation.

Disease Table 21.7	Wart Diseases	
	HPV	**Molluscum Contagiosum**
Causative Organism(s)	Human papillomaviruses Ⓥ	Poxvirus, sometimes called the molluscum contagiosum virus (MCV) Ⓥ
Most Common Modes of Transmission	Direct contact (STI), also autoinoculation, indirect contact	Direct contact (STI), also autoinoculation, indirect contact
Virulence Factors	Oncogenes (in the case of malignant types of HPV)	–
Culture/Diagnosis	PCR tests for certain HPV types, clinical diagnosis, Pap smear	Clinical diagnosis, also histology, PCR
Prevention	Vaccines available; avoid direct contact; prevent cancer by screening cervix	Avoid direct contact
Treatment	Warts or precancerous tissue can be removed; virus not treatable	Warts can be removed; virus not treatable
Distinguishing Features	Infection may or may not result in warts; infection may result in malignancy	Wartlike growths are only known consequence of infection
Effects on Fetus	May cause laryngeal warts	–
Appearance of Lesions		
Epidemiological Features	United States: eventually all sexually active adults will likely have one or more types; 13,800 new cases of HPV-associated cervical cancer annually	United States: affects 2%-10% of children annually

BSIP/age fotostock (HPV); Dr. Jack Jerjian/Phototake (Molluscum Contagiosum); Moredun Animal Health Ltd/Science Source (*Neisseria gonorrhoeae*, background)

Group B *Streptococcus* "Colonization"—Neonatal Disease

Ten to forty percent of women in the United States are colonized, asymptomatically, by a beta-hemolytic *Streptococcus* in Lancefield group B. Nonpregnant women experience no ill effects from this colonization. But colonization of pregnant women with this organism is associated with preterm delivery. Additionally, about half of their infants become colonized by the bacterium during passage through the birth canal or by ascension of the bacteria through ruptured membranes. For those reasons, this colonization is considered a reproductive tract disease.

A small percentage of infected infants experience life-threatening bloodstream infections, meningitis, or pneumonia. If they recover from these acute conditions, they may have permanent disabilities, such as developmental disabilities, hearing loss, or impaired vision. In some cases, the mothers also experience disease, such as amniotic infection or subsequent stillbirths. Although group B *Streptococcus* infections have declined sharply in the United States, they remain a major threat to infant morbidity and mortality worldwide.

The CDC has reiterated recommendations that all pregnant women be screened for group B *Streptococcus* colonization at 35 to 37 weeks of pregnancy. Recommendations for earlier testing are sometimes warranted because colonization has been associated with preterm birth. Women positive for the bacterium should be treated with penicillin or ampicillin, unless the bacterium is found to be resistant to these and unless allergy to penicillin is present, in which case erythromycin may be used.

Disease Table 21.8	Group B *Streptococcus* Colonization
Causative Organism(s)	Group B *Streptococcus* B G+
Most Common Modes of Transmission	Vertical
Virulence Factors	–
Culture/Diagnosis	Culture of mother's genital tract
Prevention/Treatment	Treat mother with penicillin/ampicillin; watch for clindamycin-resistant strains as they are in **Concerning Threat** category on CDC Antibiotic Resistance Report
Epidemiological Features	United States: vaginal carriage rates 15%-45%; neonatal sepsis due to this occurs in 1.8-3.2 per 1,000 live births; internationally: vaginal carriage rates 12%-27%

21.4 LEARNING OUTCOMES—Assess Your Progress

6. List the possible causative agents, modes of transmission, virulence factors, and prevention/treatment for the "Highlight Diseases" gonorrhea and *Chlamydia* infection.
7. Distinguish between vaginitis and vaginosis.
8. Discuss prostatitis.
9. Name three diseases that result in genital ulcers, and discuss their important features.
10. Differentiate between the two diseases causing warts in the reproductive tract.
11. List at least three reasons the HPV vaccine is so strongly recommended for young adolescents.
12. Identify the most important risk group for group B *Streptococcus* infection and explain why that group is important.

CASE FILE WRAP-UP

Transmission electron micrograph of
Neisseria gonorrhoeae.

©Syda Productions/Shutterstock (health care
worker); ©Moredun Animal Health Ltd/Science
Source (*Neisseria gonorrhoeae*)

The young patient featured in the opening case study was given a pelvic exam after consent was obtained from the patient's mother. The patient was found to have a purulent and foul-smelling vaginal discharge, in addition to fever and lower abdominal pain. Cultures of the vaginal discharge and the cervix were obtained. Blood work and an ultrasound were also ordered. The blood work revealed an elevated white blood cell count and erythrocyte sedimentation rate (ESR), while the ultrasound revealed an abscess near the left ovary. The patient was admitted with a diagnosis of pelvic inflammatory disease (PID) for IV antibiotics and possible surgical drainage of the abscess. Cultures eventually yielded the specific causative agent, *Neisseria gonorrhoeae*. The "PID shuffle" is a term used to describe the typical gait of a patient with PID, in which the feet are advanced in a shuffling manner to avoid jarring the pelvic organs, which can result in severe pain.

The patient was only 10 years of age, which should raise suspicion of child abuse. Not only should the case be reported to the appropriate state health authority, but it should also be reported to social services or the police in order to determine whether the patient was a victim of abuse.

Save the World with the Vaginal Microbiome

A few years ago, a scientist named Gregor Reid made a provocative statement to a meeting of the American Society for Microbiology. He said, "To not place a huge focus on the human vaginal microbiome is like putting human survival at risk." Here is his reasoning: A healthy vaginal microbiome is essential to successful pregnancy and delivery. Disturbances in the microbiome change the pH of the vagina, inhibit fertility, cause spontaneous abortions, and induce early-term delivery. Currently, 30% of American women have abnormal vaginal microbiomes; the rates are as high as 60% in inner-city populations. This condition is known as bacterial vaginosis. Many women who have it have no symptoms at all, and it only causes problems when women try to get pregnant or during pregnancy or delivery.

The causes of the condition are poorly understood, but douching, smoking, obesity, and stress have all been associated with it. Dr. Reid and others reserve a particular disdain for the use of douche. Douching has been linked to preterm birth, an elevated risk of acquiring HIV, ectopic pregnancies, cervical cancer, and endometriosis. Some physicians feel that there is a direct link between this practice and the unusually high rates of preterm delivery and infant mortality in many urban areas.

The vaginal microbiome as the cradle of civilization? Gregor Reid thinks that if the vaginal microbiome were to suddenly shift across the human population, it would not be unreasonable to expect that the human race to go extinct. This is obviously a worst-case scenario, and an unlikely event, but it does put a spotlight on an overlooked aspect of women's and children's health.

©Arthimedes/Shutterstock

▶ Summing Up

Taxonomic Organization Microorganisms Causing Disease in the Genitourinary Tract

Microorganism	Pronunciation	Disease Table
Gram-positive bacteria		
Staphylococcus saprophyticus	staf″-uh-lo-kok′-us sap′-pro-fit″-uh-cus	Urinary tract infection, 21.1
Gardnerella (note: stains gram-negative)	gard′-ner-el″-uh	Vaginitis/vaginosis, 21.4
Group B *Streptococcus*	groop bee strep′-tuh-kok″-us	Group B *Streptococcus* colonization, 21.8
Gram-negative bacteria		
Escherichia coli	esh′-shur-eesh″-ee-uh col′-eye	Urinary tract infection, 21.1
Other *Enterobacteriaceae*	en″-tur-oh-bak-tier′-ee-ay-cee	Urinary tract infection, 21.1
Leptospira interrogans (spirochete)	lep″-toh-spy′-ruh in-ter′-ruh-ganz	Leptospirosis, 21.2
Neisseria gonorrhoeae	nye-seer″-ee-uh′ gon′-uh-ree″-uh	Genital discharge diseases (in addition to vaginitis/vaginosis), 21.3
Chlamydia trachomatis	kluh-mi″-dee-uh′ truh-koh′-muh-tis	Genital discharge diseases (in addition to vaginitis/vaginosis), 21.3
Treponema pallidum (spirochete)	trep′-oh-nee″-ma pal′-uh-dum	Genital ulcer disease, 21.6
Haemophilus ducreyi	huh-mah′-fuh-luss doo-cray′-ee-eye	Genital ulcer disease, 21.6
DNA viruses		
Herpes simplex virus 1 and 2	hur′-peez sim′-plex vie′-russ	Genital ulcer disease, 21.6
Human papillomavirus	hew′-mun pap′-uh-loh″-muh-vie′-russ	Wart diseases, 21.7
Poxvirus	pox′-vie′-russ	Wart diseases, 21.7
Fungi		
Candida albicans	can′-duh-duh al″-buh-cans′	Vaginitis/vaginosis, 21.4
Protozoa		
Trichomonas vaginalis	trick″-uh-moh′-nus vaj′-ih-nal″-us	Vaginitis/vaginosis, 21.4

Infectious Diseases Affecting
The Female Genitourinary System

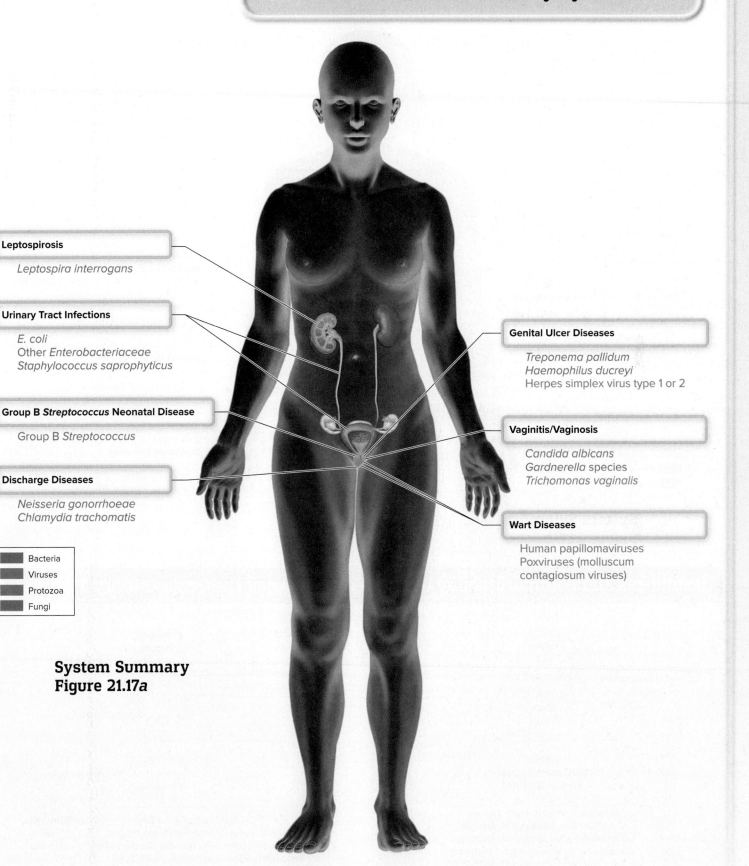

Leptospirosis

Leptospira interrogans

Urinary Tract Infections

E. coli
Other *Enterobacteriaceae*
Staphylococcus saprophyticus

Group B *Streptococcus* Neonatal Disease

Group B *Streptococcus*

Discharge Diseases

Neisseria gonorrhoeae
Chlamydia trachomatis

Genital Ulcer Diseases

Treponema pallidum
Haemophilus ducreyi
Herpes simplex virus type 1 or 2

Vaginitis/Vaginosis

Candida albicans
Gardnerella species
Trichomonas vaginalis

Wart Diseases

Human papillomaviruses
Poxviruses (molluscum
contagiosum viruses)

Bacteria
Viruses
Protozoa
Fungi

**System Summary
Figure 21.17*a***

Infectious Diseases Affecting
The Male Genitourinary System

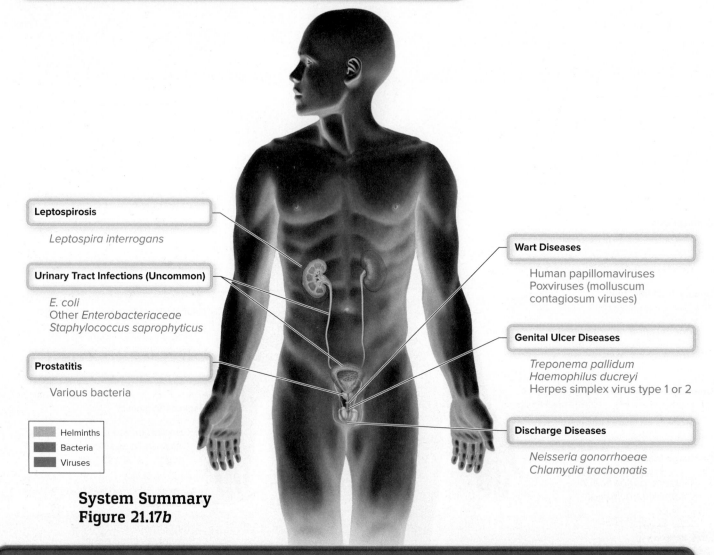

Leptospirosis

Leptospira interrogans

Urinary Tract Infections (Uncommon)

E. coli
Other *Enterobacteriaceae*
Staphylococcus saprophyticus

Prostatitis

Various bacteria

Wart Diseases

Human papillomaviruses
Poxviruses (molluscum
contagiosum viruses)

Genital Ulcer Diseases

Treponema pallidum
Haemophilus ducreyi
Herpes simplex virus type 1 or 2

Discharge Diseases

Neisseria gonorrhoeae
Chlamydia trachomatis

Helminths
Bacteria
Viruses

System Summary
Figure 21.17b

Deadliness and Communicability of Selected Diseases of the Genitourinary Tract

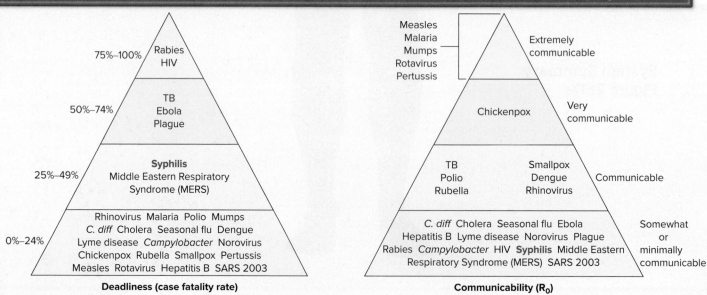

Deadliness (case fatality rate)

75%–100% — Rabies / HIV

50%–74% — TB / Ebola / Plague

25%–49% — **Syphilis** / Middle Eastern Respiratory Syndrome (MERS)

0%–24% — Rhinovirus Malaria Polio Mumps *C. diff* Cholera Seasonal flu Dengue Lyme disease *Campylobacter* Norovirus Chickenpox Rubella Smallpox Pertussis Measles Rotavirus Hepatitis B SARS 2003

Communicability (R₀)

Measles / Malaria / Mumps / Rotavirus / Pertussis — Extremely communicable

Chickenpox — Very communicable

TB Polio Rubella / Smallpox Dengue Rhinovirus — Communicable

C. diff Cholera Seasonal flu Ebola Hepatitis B Lyme disease Norovirus Plague Rabies *Campylobacter* HIV **Syphilis** Middle Eastern Respiratory Syndrome (MERS) SARS 2003 — Somewhat or minimally communicable

INFECTIOUS DISEASES AFFECTING THE GENITOURINARY TRACT

Anatomy & Defenses: Genitourinary Tract

The urinary system allows excretion of fluid and wastes from the body. It has mechanical and chemical defense mechanisms. The reproductive tract is protected by normal mucosal defenses and specialized features such as low pH.

Normal Biota: Genitourinary Tract

The most abundant normal biota occurs in the distal parts of the urinary and reproductive tracts. Normal biota in the female reproductive tract changes over the course of her lifetime.

Urinary Tract Diseases Caused By Microorganisms

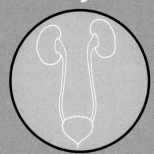

Urinary Tract Infections (UTIs)
Can occur in the bladder (cystitis), the kidneys (pyelonephritis), and the urethra (urethritis). Most common causes are *Escherichia coli,* other members of the family Enterobacteriaceae, and also *Staphylococcus saprophyticus.*

Community-acquired UTIs are most often transmitted from the GI tract to the urinary system.

Leptospirosis
Zoonosis associated with wild animals that affects the kidneys, liver, brain, and eyes. Caused by *Leptospira interrogans.*

Reproductive Tract Diseases Caused By Microorganisms

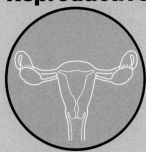

Discharge Diseases with Major Manifestation in the Genitourinary Tract
• Gonorrhea
Gonorrhea can elicit urethritis in males, but many cases are asymptomatic. In females, both the urinary and genital tracts may be infected during sexual intercourse. One outcome is salpingitis, which can lead to PID. Causative agent, *Neisseria gonorrhoeae,* is a gram-negative diplococcus.
• *Chlamydia*
Genital *Chlamydia* infection is the most common reportable infectious disease in the United States. In males: an inflammation of the urethra. In females: cervicitis, discharge, salpingitis, and frequently PID. Often asymptomatic.

Vaginitis and Vaginosis
• Vaginitis
Most commonly caused by *Candida albicans.* Nearly always an opportunistic infection.
• Vaginosis
Gardnerella is associated with vaginosis; has a discharge but no inflammation. Could lead to complications such as pelvic inflammatory disease (PID).
• *Trichomonas vaginalis*
Causes asymptomatic to mild infections in females and males. Considered a "neglected parasitic disease" by the CDC.
Prostatitis
Inflammation of the prostate; can be acute or chronic. Not all cases have microbial cause, but most do.

Reproductive Tract Diseases Caused By Microorganisms (cont'd)

Genital Ulcer Diseases

• Syphilis

Caused by spirochete *Treponema pallidum*. Three distinct clinical stages: primary, secondary, and tertiary syphilis. Congenital syphilis can lead to severe disease.

• Chancroid

Caused by *Haemophilus ducreyi*.

• Genital Herpes

Caused by HSV-1 or HSV-2. Symptoms range from none to serious encephalitis.

Wart Diseases

• Human papillomavirus

Causes genital warts and also cervical and other cancers. Vaccine for cancerous types is available.

• Molluscum contagiosum

Caused by a virus in the family Poxviridae. Can take the form of wartlike growths on genitals. Not serious.

Group B *Streptococcus* "Colonization"—Neonatal Disease

Asymptomatic colonization of women by a beta-hemolytic *Streptoccus* in Lancefield group B is very common. It can cause preterm delivery and infections in newborns.

SmartGrid: From Knowledge to Critical Thinking

This *21 Question Grid* takes the topics from this chapter and arranges them with respect to the American Society for Microbiology's Undergraduate Curriculum guidelines—all six of the important "Concepts" as well as the important "Competency" of scientific literacy. Three questions are supplied about chapter content that refer to the Concept or Competency, in increasing levels of Bloom's taxonomy for learning.

ASM Concept/ Competency	A. Bloom's Level 1, 2—Remember and Understand (Choose one.)	B. Bloom's Level 3, 4—Apply and Analyze	C. Bloom's Level 5, 6—Evaluate and Create
Evolution	1. Which of the following is on the CDC's "Urgent Threat" list because of its potential resistance to antibiotics? a. *Staphylococcus saprophyticus* b. *Trichomonas vaginalis* c. Herpesvirus d. Enterobacteriaceae	2. Some of the *E. coli* strains that are normal biota in the GI tract have evolved adhesive structures that can attach to the urinary tract and cause disease there. What sequence of molecular events (in the bacteria) do you suppose led to this?	3. Many STIs are asymptomatic, or remain latent for long periods of time. What evolutionary advantage do you think that gives them?
Cell Structure and Function	4. Which of the following microbes is described as "dimorphic"? a. *Trichomonas vaginalis* b. *Gardnerella* c. *Candida albicans* d. *E. coli*	5. Three microorganisms in this chapter can be identified via a Gram stain. Name them and explain why this is possible for each of them.	6. The drug Flagyl (metronidazole) has the unusual characteristic of being effective against some bacteria as well as some protozoa. What cellular structure(s) would it be most likely to target?
Metabolic Pathways	7. Which of the following bacteria is known to produce substances that raise the local pH? a. *Gardnerella* b. *E. coli* c. *Lactobacillus* d. *Leptospira*	8. What advantage does the particular life cycle of *Chlamydia* give to the pathogen?	9. Infection with *Gardnerella* seems not to induce inflammation in the vagina. Speculate on characteristics of the infection or the organism that could result in lack of inflammation.
Information Flow and Genetics	10. Which of the following bacteria behaves most similarly to viruses? a. *E. coli* b. *Chlamydia* c. *Neisseria* d. *Haemophilus ducreyi*	11. Using your knowledge of viral oncogenesis, outline the steps that the human papillomavirus might take from infection to cervical cancer.	12. The bacterium *Treponema palladium* is known to express an unusually low density of proteins in its outer membrane. Speculate on why that might contribute to its ability for long-term infection of the host.
Microbial Systems	13. Which part of the urinary tract has the most abundant and diverse microbiome? a. lower urethra b. upper urethra c. bladder d. kidneys	14. Why do many of the infections discussed in this chapter make the host more vulnerable to HIV infection? Provide detail in your answer.	15. What are some of the stimuli that can trigger reactivation of a latent herpesvirus infection? Speculate on why.

ASM Concept/ Competency	A. Bloom's Level 1, 2—Remember and Understand (Choose one.)	B. Bloom's Level 3, 4—Apply and Analyze	C. Bloom's Level 5, 6—Evaluate and Create
Impact of Microorganisms	16. Bacterial vaginosis is commonly associated with which organism? a. *Candida albicans* b. *Gardnerella* c. *Trichomonas* d. any of these	17. A 38-year-old male living in Hawaii tried to rescue some of the family's belongings as the basement flooded with water from a nearby flooded stream. Within a few days, he developed flulike symptoms, which cleared in a few days. Several weeks later he developed a painful headache and jaundice. The physician requested a urinalysis. What disease do you suspect and why?	18. Use science to explain this statement: Forensic microbiology can help identify sexual abuse of minors.
Scientific Thinking	19. Which of the following STIs is curable with antimicrobials? a. HPV b. herpes c. *Chlamydia* d. two of these	20. Why do you suppose a urine screening test for *Chlamydia* is more accurate for males than for females?	21. How would a laboratory technologist identify a case of vaginosis versus a case of vaginitis from a vaginal swab specimen?

Answers to the multiple-choice questions appear in Appendix A.

Visual Connections

This question connects content within and between chapters.

From chapters 18 and 21, figure 18.21 and figure 21.11*b*.
Compare these two rashes. What kind of information would help you determine the diagnosis in both cases?

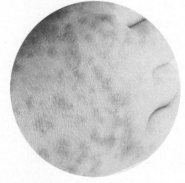

Source: CDC

Source: CDC

22

One Health

The Interconnected Health of the Environment, Humans, and Other Animals

Contributions by Ronald M. Atlas, University of Louisville

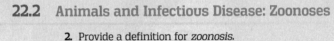
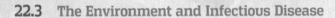

CASE FILE

Viral Zoonoses

In January 2020, my nursing supervisor informed me of a mandatory training that was offered on two different days to accommodate our schedules. Because the novel coronavirus had just emerged from China, I figured we would be learning about it. But when I showed up in the staff auditorium, the title slide said "SARS AND MERS: Lessons Learned."

SARS stands for severe acute respiratory syndrome and MERS is the Middle East respiratory syndrome. When the speaker from the state health department stepped up, the first thing she said was, "Both SARS and MERS are case studies for understanding any new coronavirus that pops up—including SARS-CoV-2, the cause of COVID-19 disease, as well as the next one that will inevitably emerge."

The original SARS popped up in China in 2002. It quickly spread to more than two dozen countries and was finally contained in July 2003, after being declared a public health emergency. There have been no cases since. The Middle East respiratory virus popped up in 2012 in Saudi Arabia. It is caused by a novel coronavirus that was an infection of camels but jumped species to be able to infect humans who had contact with infected camels. It is still causing sporadic cases in 2020.

The speaker explained that these cross-species infections become worrisome when the virus then becomes able to be transmitted human-to-human. This opens up the possibility of a pandemic because humans are naive to the virus and have no adaptive immunity.

Two important features of any emerging respiratory virus, according to our speaker, are (a) how transmissible between humans it is and (b) how deadly it is (the case-fatality rate). You were introduced to this concept in the disease chapters, where we catalogued both things—deadliness and communicability—for many microbes.

MERS turned out to be only moderately transmissible (1 infected person theoretically infected 0.7 other people). This is why it did not ultimately lead to a pandemic. SARS-CoV-1 was more highly transmissible (1 person infecting 2.5 others), but unlike SARS-CoV-2, it was generally only transmissible when the carrier was symptomatic. Strict temperature screenings and quarantines in China and other affected countries led to its control. In terms of deadliness, MERS had a case-fatality rate of 35%. That meant that more than a third of people with the symptoms died. SARS-CoV-1 had a case fatality rate of around 10%. At this writing, it is too soon to know these percentages for COVID-19.

- What is the difference between isolation and quarantine in an epidemic situation?

- What are infectious diseases of humans that originate in animals called?

Case File Wrap-Up appears at the end of the chapter.

©BambooSIL/SuperStock

22.1 One Health

We live in a complex and ever-changing world in which human health, animal health, and environmental health are tightly connected. So much today is global in nature—global travel, global economy, global movement of livestock, global climate change. It is no wonder, then, that human health, animal health, and environmental health are inextricably interrelated—and that we are beginning to appreciate that the health of all life on earth is connected.

The concept that the health of humans, animals, and the environment should be viewed in a holistic (all-inclusive) way is called *One Health*. This is not a new idea for microbiologists, who have always needed to consider all the causes that lead to the spread of infectious diseases and the ability to control them. Pioneering microbiologists like Louis Pasteur and Robert Koch recognized the interconnection between animal and human health. They carried out research on diseases such as anthrax and rabies, which affected both humans and other animals. Medical and veterinary educators of that era also freely crossed the boundaries of human and animal health.

The contemporary revival of the *One Health* concept has expanded beyond the human–animal health interface to encompass the health and sustainability of all the world's ecosystems, including plant life. Microorganisms circulate among human hosts, animal hosts, and environmental reservoirs. Disruption of the environment can lead to transmission of pathogens to animals and humans. Evolution of new microbial traits can occur in response to changes in the environment. In addition, reservoirs of pathogens and virulence traits can persist in the environment, poised to enter humans and other animals at an opportune time.

Global Mixing Bowl

One way to think about *One Health* is to picture three overlapping spheres representing humans, animals, and the environment **(figure 22.1)**. A change in any one of these spheres impacts the others and with that the microbes that each contains. The mixing of microbes in different animal hosts and under different environmental conditions can foster the evolution of new and potentially deadly pathogens. Human activities in particular can promote the emergence of infectious diseases, for example, through ecological disturbances and movement of animals. This occurs frequently for some microbes. Consider the example of influenza viruses, in which the mixing of different strains of influenza viruses in birds, swine, and humans results in the evolution of new recombinant strains with the potential to spread globally every year. And of course COVID-19 jumped into humans for the first time in 2019 and caused a global pandemic.

We use the term "mixing bowl" because within the last few decades, which is no time at all in evolutionary terms, humans have created an incredible array of unique opportunities and challenges for microbes. For example, we have the ability to fly between continents in the space of a few hours (with our microbes). We have the ability to transport bacteria from one ocean to another in the bilge water of tanker ships. We use antimicrobial drugs that put selective pressures on microbes to an unprecedented extent. The result has been the dawn of a new era of emerging and reemerging diseases **(table 22.1)**.

The patterns of newly emergent infectious diseases, such as West Nile fever (in New York in 1999) and Zika virus disease (in the 2000s), reflect contemporary demographic and environmental changes—more people living closer together and in greater contact with wild and domesticated animals, and habitat destruction. Another issue is the multipurpose use of resources—such as using the same water for waste disposal, irrigation, and human consumption. These changes have been

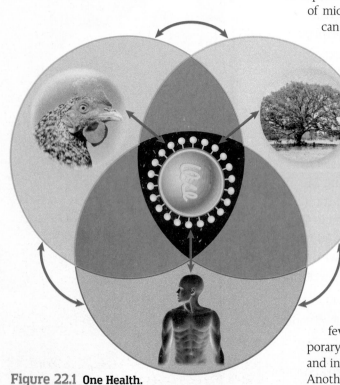

Figure 22.1 One Health.
Pixtal/age fotostock (chicken); OGphoto/Getty Images (tree)

Table 22.1 Emerging and Reemerging Infectious Diseases—2020*

Babesiosis	Protozoan tick-borne disease
Bartonella henselae	Bacterial cat scratch disease
Candida auris	Fungal disease causing local and bloodstream infections, multiply drug-resistant
Chikungunya and Mayaro	Two arbovirus diseases that have recently moved into the Americas
Clostridioides difficile	Endospore-forming bacterium causing severe GI tract infection
SARS-CoV-2	Acute respiratory distress, damage to other organ systems
Ebola virus disease	Hemorrhagic fever; large epidemic in 2014-2015
Ehrlichiosis	Tick-borne bacterial disease
Elizabethkingia	Common environmental bacterium causing HAIs
Lassa fever	Viral hemorrhagic fever, confined to Africa currently
Leishmaniasis	Protozoan disease of skin and tissue
Marburg virus	Viral hemorrhagic fever, similar to Ebola
Measles	Threatens to reemerge due to lower vaccination rates
MERS	Middle East respiratory syndrome
Mumps	Threatens to reemerge, partially due to lower vaccination rates
Nipah virus	Encephalitis and respiratory illness; Asia and Africa
SARS	Sudden acute respiratory syndrome, emerged suddenly in 2003
Shiga-toxin-producing *E. coli*	Foodborne and water-borne infection
Zika virus	Mosquito-borne illness affecting fetuses most severely

*List compiled from CDC, WHO, and other sources. Note that many of these are mosquito- and tick-borne, and are creeping ever farther north as their vectors move with the warming climate.

Source: U.S. Fish & Wildlife Service/Wyman Meinzer

driving repeated pathogen spillover from wildlife and the spread of newly evolved pathogens in dense human populations. Over the past four decades, the rate of infectious disease emergence has increased in both humans and animals. Just as troubling are reemerging diseases. These are infections that have been minimized previously through vaccinations, vector controls, or sometimes no human intervention at all, yet have resurged again—or threaten to resurge—in the human population. You will note that table 22.1 contains both measles and mumps for this reason.

22.1 LEARNING OUTCOMES—Assess Your Progress

1. Draw your own representation or logo depicting how animal health, human health, and environmental conditions interact.

22.2 Animals and Infectious Disease: Zoonoses

The term *zoonosis* describes the transfer of disease-causing microorganisms from animals to humans. Worldwide, diseases that humans acquire from animals result in millions of cases of illness and perhaps a million deaths each year **(figure 22.2)**. Today, however, we also recognize that microbes can transfer back *and* forth between animals and humans. If you are trekking with gorillas in Rwanda, you may want to worry more about the danger of transmitting disease-causing microbes from yourself to the gorillas than about your own safety.

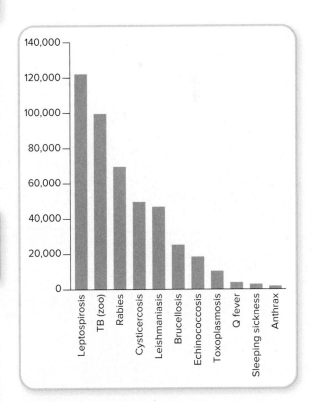

Figure 22.2 Annual number of human deaths worldwide caused by zoonoses.

Source: International Livestock Research Institute

©William Leaman/Alamy Stock Photo

Over the last three decades, approximately 75% of newly emerging human diseases have been zoonotic, and many have come from or through wildlife. Given that there are 50,000 known vertebrate species and assuming that each has 20 endemic viruses (which is likely to be an underestimate; bats alone harbor 20,000), there probably are more than 1 million vertebrate viruses. Only 2,000 or so viruses have been described, so 99.8% of vertebrate viruses remain to be discovered. These calculations begin to give you an appreciation of the enormous potential for future zoonotic diseases to emerge.

Once introduced into human populations, some zoonotic agents can then spread from human to human—in some cases causing global pandemics. For example, the HIV/AIDS epidemic, which has resulted in more than 38 million human cases worldwide, had its origin as a chimpanzee retrovirus that jumped species and then adapted itself to human-to-human transmission. Such epidemics are extremely difficult to control and can have high rates of morbidity and mortality.

A Classic Example: Lyme Disease

In 1976, a number of children living near Lyme, Connecticut, began developing arthritis-like symptoms. Because arthritis is not an infectious disease, the cluster of children with these symptoms was unusual, suggesting an underlying microbial cause. Soon this disease would be recognized as a bacterial infection caused by the spirochete *Borrelia burgdorferi* and given the name *Lyme disease*. It is now the most commonly reported arthropod-borne illness in the United States **(figure 22.3)** and Europe and is also found in Asia. The life cycle of the bacterium, depicted in figure 18.15, involves ticks, small mammals, and large mammals. The reasons for the emergence of Lyme disease reflect changing demographics and human behavior, including efforts to restore disrupted environments.

As suburban areas of the northeastern United States became heavily populated, and growing numbers of housing developments backed up to woodland areas, conditions became ideal for *B. burgdorferi*-infected ticks to come into frequent contact with humans.

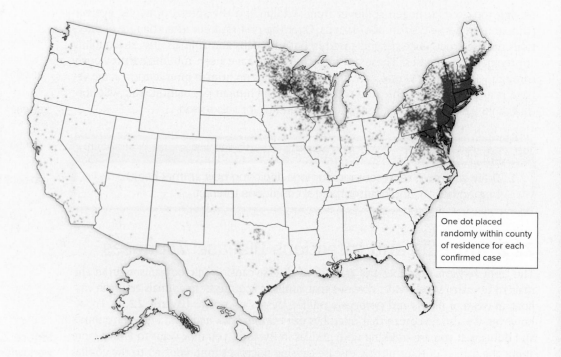

One dot placed randomly within county of residence for each confirmed case

Figure 22.3 **A map of Lyme disease in the United States.**

During the past 40 years, the infection has continued to spread in the northeastern United States. Climate change (covered later in this chapter) is also influencing the epidemiology of Lyme disease. Scientists mapped the places where temperatures have increased in North America between 1971 and 2020, and found a corresponding increase in the tick population in those areas. Warmer conditions mean that larval ticks are more likely to survive to maturity. This is another reason that Lyme disease–and a host of other tick-borne diseases–are on the rise in temperate climates.

A New Example—COVID-19

On the very last day of 2019, December 31, China confirmed the existence of dozens of cases of a virus brand new to humans, later identified as SARS-CoV-2, which causes a syndrome that was soon named COVID-19.

It seems to have originally been a bat virus and then cycled in intermediate mammalian hosts before jumping to humans in 2019. As you know, when humans encounter a microbe new to them, they have no adaptive immunity, and the "collision" between the two species can be deadly to the humans. Within weeks, tens of thousands of people had died of the infection, and millions had been infected. Unprecedented measures were taken to slow the spread of the disease. In the United States, many schools were closed beginning in March. Churches canceled services. Restaurants and bars were closed. Disney parks closed for the first time since the terrorist attacks of September 11, 2001. These efforts were part of a strategy called social distancing: decreasing the density of people in the same space to limit transmission.

Important epidemiological concepts were illustrated by the COVID-19 event. At this writing (2020), neither the communicability coefficient, R_0, nor the case fatality rate can be definitively calculated. However, one factor pushing the high transmissibility of the virus is the fact that it can be transmitted from someone who is asymptomatic, or presymptomatic. This greatly complicates screening using fever-checks, one of the factors that helped bring the SARS pandemic in 2003 to a halt. When the pandemic was first recognized in early 2020, health officials began to speak about a concept called "flattening the curve." This is illustrated in **figure 22.4.** With the exponential growth of the number of infections, as illustrated in the first curve, the number of severe cases would quickly overwhelm hospitals, the number of health care providers, and equipment such as ventilators that can save lives. In such a situation, people needing emergency care for other conditions, such as heart attacks, accidents, etc., will also experience difficulty accessing care. The blue curve in figure 22.4 is how the epidemic curve can be shifted so that the number of cases does not peak at a place that exceeds the health care capacity (the black dotted line). This is known as *flattening the curve.*

The worldwide response to the COVID-19 pandemic will be studied for years to come. Different countries took different measures. South Korea and New Zealand were viewed as success stories in keeping case counts very low. At the same time, countries such as Italy, Brazil, and the United States saw a quick and dramatic increase in cases and deaths.

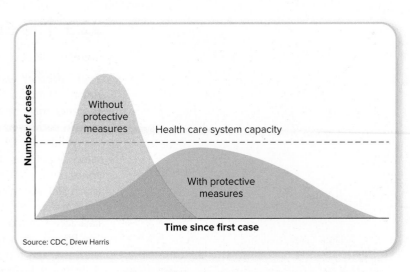

Source: CDC, Drew Harris

Figure 22.4 Representation of how "flattening the curve" works. Start with the dotted line—which represents the capacity for hospitals to accommodate patients in acute respiratory distress as in severe cases of COVID-19. The first epidemic curve (orange) is a hypothetical curve corresponding to rapid transmission and spread throughout a population. The second curve (blue) depicts a flattening of the curve that could occur if social distancing and other mitigating measures are taken.

22.2 LEARNING OUTCOMES—Assess Your Progress

2. Provide a definition for *zoonosis.*

3. Discuss the effects of deforestation and reforestation on the eventual emergence of Lyme disease in the northeastern United States.

4. Explain the concepts of social distancing and "flattening the curve."

22.3 The Environment and Infectious Disease

Clean Water and Infectious Disease

Every drop of water we drink contains thousands of microbes. This includes the nearly 30 million gallons of bottled water consumed in the United States daily. Fortunately, most municipal and bottled water consumed in the United States is free of human pathogens–but this is not the case everywhere.

While the United States and many other nations take great care to ensure the safety of potable (drinkable) water, this is not the case worldwide–particularly in rural communities of developing countries. In fact, a large percentage of the world's population lack access to safe drinking water. According to the World Health Organization (WHO), more than 1 billion people lack access to clean water, and 2.6 billion people lack access to basic sanitation **(figure 22.5).**

Poor water sanitation and a lack of safe drinking water take a greater human toll than war and terrorism. Waterborne diarrheal disease alone accounts for 4.1% of the daily global disease burden, and more than 3.4 million deaths each year are attributable to unsafe drinking water–primarily among children in developing countries. According to a U.N. report, 4,000 children die each day as a result of infectious diseases acquired by drinking contaminated water. The poorest and most vulnerable members of society bear the greatest burden of waterborne and diarrheal disease.

Figure 22.5 Much of the world's population uses the same source of water for many different purposes, including bathing, relieving themselve, and drinking.
robertharding/Alamy Stock Photo

Drinking Water Quality–Microbiological Safety of Potable Water

One of the reasons potable water is generally safe in the United States and many developed nations is that great care is taken to disinfect the water and test it for bacterial content before it is consumed. Water is generally disinfected by filtration or by treatment with chlorine to ensure that it is free of potential pathogens. A typical water purification process is shown in **figure 22.6.** When there is a breach in the water delivery system and the quality of the water cannot be ensured, a boil-water advisory is issued; that is, the public is told to boil the water to kill most potential pathogens before drinking the water.

Routine bacterial monitoring of potable waters involves testing for **indicator bacteria,** that is, microorganisms normally found in mammalian gastrointestinal tracts whose presence would indicate likely contamination with fecal matter **(figure 22.7).** Testing of the actual pathogens of concern (protozoa like *Giardia* and *Cryptosporidium;* bacteria like *Salmonella, Shigella, Vibrio,* and *Campylobacter;* and viruses like hepatitis A and Norwalk) would be difficult and may not provide the necessary margin of safety given the large volumes of water that are distributed. The intestinal bacteria most useful in the routine monitoring of microbial pollution are gram-negative rods called **coliforms** (*E. coli* and similar bacteria), which are gram-negative bacteria that ferment lactose and produce gas. These coliform bacteria, as well as fecal streptococci, which also are useful indicator bacteria, survive in natural waters but do not multiply there. For that reason, finding them in high numbers indicates recent fecal contamination and the possible presence of intestinal pathogens. When significant numbers of coliform bacteria or fecal streptococci are detected, the water is not considered safe to drink.

Although total coliform counts are useful indicators for ensuring water safety, they can be misleading. In many circumstances, it is important to differentiate between coliforms that can be naturally found in soils and uncontaminated waters (*Enterobacter, Klebsiella, Citrobacter*) and **fecal coliforms** (*E. coli*) that live mainly in human and animal intestines. Although most *E. coli* strains are not pathogenic, they almost always come from a mammal's intestinal tract, so their presence in a sample is a clear indicator of fecal contamination.

Sometimes the public and the media confuse *coliform* with *E. coli.* This can cause unwarranted concerns. For example, in recent years, there was a minor panic when media outlets reported that iced tea from restaurants in Cincinnati, Ohio, contained significant numbers of "fecal coliforms." One headline read, "Iced Tea Worse Than

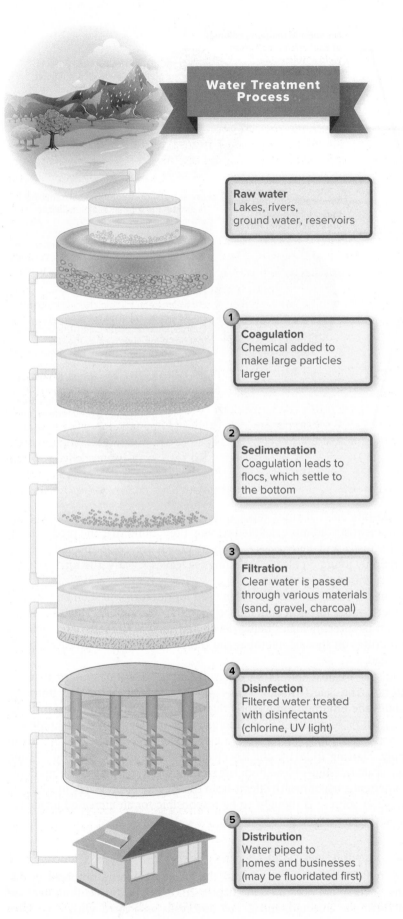

Water Treatment Process

Raw water
Lakes, rivers, ground water, reservoirs

1
Coagulation
Chemical added to make large particles larger

2
Sedimentation
Coagulation leads to flocs, which settle to the bottom

3
Filtration
Clear water is passed through various materials (sand, gravel, charcoal)

4
Disinfection
Filtered water treated with disinfectants (chlorine, UV light)

5
Distribution
Water piped to homes and businesses (may be fluoridated first)

Medical Moment

Plastic Bottles for Clean Water

Every week around the world, 30,000 people die from lack of clean water. Ninety percent of these are children under 5 years old. Clean water—taken for granted in the developed world—is a resource more precious than gold on the rest of the planet. Even though we take it for granted, the processes and infrastructure used to deliver it to us are complex and expensive. How can we export those to other settings? Maybe we don't have to. Solar water disinfection is a method of safely disinfecting drinking water by simply placing contaminated water in a transparent plastic bottle and leaving it in the sun for 6 hours. Ultraviolet light kills bacteria and parasites and inactivates viruses, making the water safe. This technique has been used all over the world in impoverished nations where citizens have no access to clean drinking water, and it has proven to be an effective way of preventing diarrheal disease.

©Khalil Senosi/AP Images

Q. Would you suspect that the water treated this way becomes sterile?

Answer in Appendix B.

Figure 22.6 A common pathway for water purification.

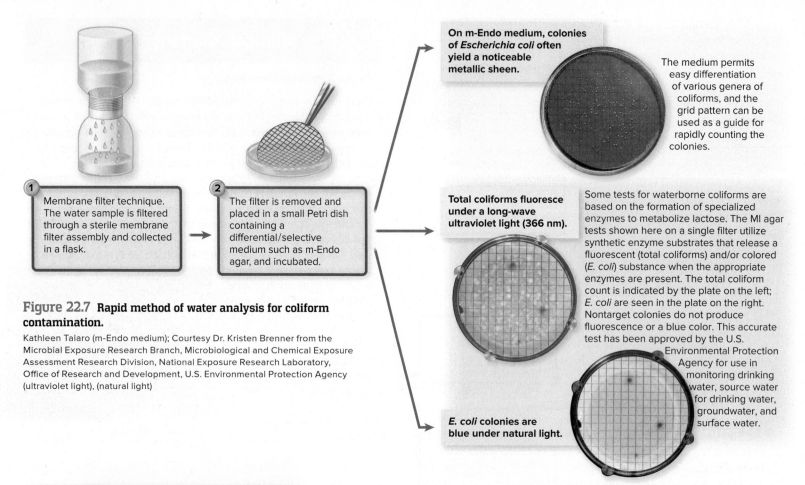

On m-Endo medium, colonies of *Escherichia coli* often yield a noticeable metallic sheen.

The medium permits easy differentiation of various genera of coliforms, and the grid pattern can be used as a guide for rapidly counting the colonies.

1 Membrane filter technique. The water sample is filtered through a sterile membrane filter assembly and collected in a flask.

2 The filter is removed and placed in a small Petri dish containing a differential/selective medium such as m-Endo agar, and incubated.

Total coliforms fluoresce under a long-wave ultraviolet light (366 nm).

Some tests for waterborne coliforms are based on the formation of specialized enzymes to metabolize lactose. The MI agar tests shown here on a single filter utilize synthetic enzyme substrates that release a fluorescent (total coliforms) and/or colored (*E. coli*) substance when the appropriate enzymes are present. The total coliform count is indicated by the plate on the left; *E. coli* are seen in the plate on the right. Nontarget colonies do not produce fluorescence or a blue color. This accurate test has been approved by the U.S. Environmental Protection Agency for use in monitoring drinking water, source water for drinking water, groundwater, and surface water.

E. coli colonies are blue under natural light.

Figure 22.7 **Rapid method of water analysis for coliform contamination.**

Kathleen Talaro (m-Endo medium); Courtesy Dr. Kristen Brenner from the Microbial Exposure Research Branch, Microbiological and Chemical Exposure Assessment Research Division, National Exposure Research Laboratory, Office of Research and Development, U.S. Environmental Protection Agency (ultraviolet light), (natural light)

Medical Moment

Cryptosporidium in Your Tap Water?

For those who are immunocompromised, even potable water may carry risk. *Cryptosporidium,* as you recall from chapter 20, is an example of a zoonotic disease. The protozoan infects humans when they consume water that has been contaminated by infected animal feces. The protozoa represent a risk for those who are neutropenic, such as patients on immunosuppressant drugs (such as cancer patients or posttransplant patients), with disorders of immunity, or with AIDS.

Because municipalities have different standards for water sanitation, *Cryptosporidium* may be present in drinking water in some regions. For the general population, this poses a minimal risk. Immunocompromised persons should consult with their public health department as further precautions may be necessary to prevent cryptosporidiosis.

Q. What further precautions can vulnerable people take if their municipal water supply does not eliminate *Cryptosporidium?*

Answer in Appendix B.

River Water." Despite such alarmist reports, no one became sick from drinking the iced tea. When scientists did more detailed testing, they found that the predominant species were *Klebsiella* and *Enterobacter,* both of which are coliforms and both of which are commonly found on plants, including tea leaves. Clearly, the total coliform count was misleading in this situation.

No doubt the coliform test that has been used for decades will soon be replaced with more direct testing for fecal contamination, either through specific testing for *E. coli* or through new technologies being developed such as **biosensors,** machines that detect structures on specific microbes and transmit electrical or physical signals that are easily read.

Microorganisms are just one of several substances in water sources that are closely monitored. The Safe Drinking Water Act is a federal law dating from the 1970s that guides the Environmental Protection Agency and state and local jurisdictions on water quality. In addition to microorganisms, the levels of several different chemicals is monitored. Flint, Michigan, is a city in the news because of a crisis in its water supply. Starting in 2014, officials changed the source of the city's water, leading to large-scale leaching of lead from the city's aging water pipes. City residents continue to experience major health effects from the elevated lead levels. In addition, the switch in water sources is thought to be responsible for an increased incidence of *Legionella* cases in the city.

Microbial Contamination of the Food Chain

You have only to look at the news to find the latest foodborne outbreak of disease, and many of these are described in chapter 20. *E. coli* outbreaks from beef are caused when cattle are infected, and *E. coli* outbreaks associated with vegetables are generally triggered when crops are watered with unclean water containing fecal matter or, possibly, when infected animals or birds have access to the crops. Many helminth diseases are acquired through a contaminated environment (organisms burrowing through skin or being swallowed) or from ingesting infected food. As has been stated already, the concept of *One Health* is an old one indeed; but in many parts of the research and regulatory communities, the last few decades have seen increasing compartmentalization. Now the connectedness is becoming appreciated once again.

While pathogens can easily be transmitted from water sources, the food we eat can also carry harmful microorganisms. Hepatitis A virus is a pathogen that can be transmitted via water or food contaminated with human fecal matter. Hepatitis A virus is widespread throughout the world. Transmission rates are very high in areas where there is no sewage treatment. Hepatitis A viral infections have an especially high incidence in developing countries and rural areas. In rural areas of South Africa, the seroprevalence is 100%. This means that all of the people in those areas have been infected with hepatitis A, although many are asymptomatic and show no signs of disease even though they are carriers of the hepatitis A virus.

Hepatitis A infections also occur in developed countries. Approximately 25,000 cases of hepatitis A viral infections are reported in the United States every year–but because many cases are asymptomatic or not reported, the actual annual incidence of hepatitis A infections is estimated to be more than 260,000. In 2016 an epidemic of hepatitis A was declared in the the U.S. and is still ongoing in 2020. The hospitalization rate is higher (57%) than previously seen with this infection.

Global Climate Change and Biology

Over the past 40 years, the concentrations of carbon dioxide and methane in the atmosphere have increased. They are now at their highest levels since the first humans set foot on earth **(figure 22.8).** As a result of the increased concentrations of carbon dioxide and methane, the climate is changing. Because these compounds absorb long-wavelength energy coming from the earth's surface and reflect some of it back to earth, they cause a rise in temperatures–so things get warmer, much like in a greenhouse.

Climate change can be detected as changes in weather–more rainfall and generally higher average temperatures over the year with more variable weather patterns. Most scientists argue that human activities are the root cause and that if we do not change course, there will be devastating impacts on human, animal, and environmental health. Because microbes assist with carbon metabolism in the soil and in animal guts, they have a huge impact on climate.

We hear quite frequently about rising sea levels and the threat of devastating storms if we do not halt the atmospheric buildup of greenhouse gases **(figure 22.9).** We should also understand that climate change may increase the global burden of disease. Drought and lack of water get worse in some places, while in other areas increased rainfall likely will lead to increases in some

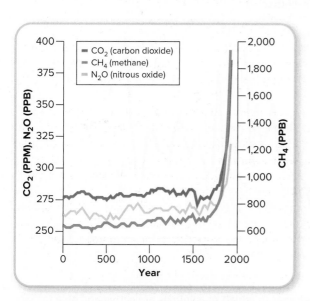

Figure 22.8 **The increase in greenhouse gas concentrations over the past 2,000 years.** PPM = parts per million; PPB = parts per billion.

CLIMATE CHANGE & HEALTH

WEATHER EXTREMES

Heavy rains, floods, hurricanes, wildfires, etc., endanger health as well as destroy property and livelihoods. In 2019, deadly wildfires, fueled by record-breaking temperatures and drought, raged in California and Australia. Historically, developing countries are hardest hit by major weather events.

RECORD HIGH AND LOW TEMPERATURES

Short-term fluctuations in temperature can affect health—causing heat stress or extreme cold—and lead to increased death rates from heart and respiratory disease.

ALLERGEN LEVELS

Pollen and other airborne allergen levels are higher in extreme heat. These can trigger asthma, which affects hundreds of millions of people worldwide. Ongoing temperature increases are expected to increase this burden.

RISING SEA LEVELS

Rising sea levels increase the risk of coastal flooding, and could cause population displacement, which puts people at increased risk for epidemic disease. More than half of the world's population now lives within 60 kilometers (about 37 miles) of shorelines.

MORE VARIABLE RAINFALL PATTERNS

Fresh water can become more scarce. Globally, water scarcity affects 4 of 10 people. Lack of water can compromise hygiene and health and lead to diarrhea, and death. Alternatively, excess rain can lead to deadly flooding.

WATER- AND MOSQUITO-BORNE DISEASES

Climate-sensitive diseases are among the largest global killers. Diarrhea, malaria, and malnutrition alone cause millions of deaths annually, and are expected to be affected by climate changes.

Source: WHO

pathogens and vectors of infectious disease. *Vibrio cholerae,* for example, thrives in warm waters. There are predictions that the global incidence of cholera will increase as a result of global warming. Mosquito populations likely also will increase in areas of increased rainfall and with them the global incidence of infectious diseases like yellow fever and malaria. Mathematical models show that a 2- to 3-degree rise in temperature could increase the number of people affected by malaria by several hundred million. A recent report by the group Climate Central revealed that mosquitoes known to carry several serious diseases were a threat to humans in San Francisco for 47 more days in 2017 than they were in 1970 **(figure 22.9)**. The diseases include Zika virus disease (chapter 17), dengue fever (chapter 18), and Chikungunya disease (chapter 18).

The United States saw an elegant example of how weather patterns can lead to a sudden outbreak of disease in 1993, when hantavirus suddenly began killing healthy young adults in the Four Corners area of the American Southwest. The virus had never before been known to cause the severe, overwhelming respiratory distress and death seen in that outbreak (previously, it was known as a kidney pathogen). After the outbreak was contained and research conducted into the causes of the outbreak, the weather phenomenon El Niño, a warming trend that originates over the Pacific Ocean, looked like the ultimate culprit. Here's how: In the year preceding this outbreak, El Niño caused wetter winter conditions in the Southwest. This led to a great increase in the pine nut population, a favorite delicacy of deer mice. By the spring and summer of 1993, there was a very large deer mouse population. Deer mice are the reservoir for hantavirus, and the people who became infected all had exposure to large amounts of deer mouse feces and/ or urine. When these substances dried out and were aerosolized, they delivered lethal doses of the virus to the lungs of humans. Note the interconnectedness of the environment (weather patterns), plant world (pine nuts), animal world (deer mice), and humans in this event.

Climate change will undeniably have far-reaching effects, where the change in temperature of even a few degrees can start a cascade of events that can easily lead to unusual occurrences of infectious disease. When we discussed the Zika virus earlier in this book, we pointed out that a warming climate has carried its mosquito vector to places that were previously too cool for it. We can expect future movements of vectors and the diseases they carry to result in increased disease outbreaks if climate change continues on its current course.

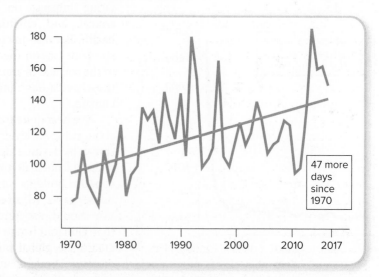

Figure 22.9 Number of days per year in which mosquitoes are active and able to transmit disease. Data are for the city of San Francisco.

Climate Central. U.S. Faces a Rise in Mosquito "Disease Danger Days." https://www.climatecentral.org/news/us-faces-a-rise-in-mosquito-disease-danger-days-21903

22.4 Microbes to the Rescue

While the presence of pathogens in the environment is clearly a threat to human and animal health, non-disease-causing microbes are of extreme importance in *saving us* from the wastes and pollutants that we produce. In New York City alone, wastewater treatment plants, with the help of microbes, treat over 1 billion gallons of sewage a day, removing an average of 65% of the organic matter. Urban solid waste in the United States amounts to roughly 150 million tons per year. Microbes also help us by cleaning up our pollutants, such as industrial chemicals and oil spills, both when we specifically "ask them to" and when we do not.

Liquid Waste Treatment—Sewage

Each year, trillions of gallons of human waste are released into the environment. Low levels of human wastes can be accommodated because natural waters have self-purification capacities. The naturally occurring nonpathogenic microbes in those ecosystems can degrade the wastes. However, in areas of high population densities, the wastes can be too great. When waters are overwhelmed by concentrated inputs of organic matter (including microbes), they exhibit a high demand for oxygen. Exhaustion of the dissolved oxygen occurs, and the water becomes putrid and septic. Pathogens also spread. In areas of high population densities, therefore, it is generally necessary to treat the waste to reduce risk. Several different sewage treatment processes are used, depending upon the population density and the environment into which the wastewater effluents can be released.

Many rural and suburban areas with relatively low population densities rely on septic tanks. These are containers into which the sewage flows. The solid material settles and is subject to microbial decomposition. The liquid, with its greatly reduced organic content, is allowed to overflow and is distributed through a series of perforated pipes into the surrounding soil. As long as the houses are far enough apart, the reduction of organic matter is sufficient to reduce the concentrations of organic matter being released to levels that can be accommodated without causing environmental harm. Such systems, however, are inadequate to handle the wastes from densely populated communities.

For some municipalities, the answer is to pump the sewage away—sometimes offshore into the ocean away from beaches and shellfish beds and sometimes to agricultural fields away from the city. Mexico City, for example, only treats a small percentage of the wastewater it collects, and only has a maximum capacity for treating 25% of the sewage generated by the 25 million inhabitants of the city. The remaining untreated sewage is pumped to agricultural areas—largely alfalfa but also barley, corn, and wheat fields—that collectively cover an area the size of the state of Rhode Island.

Most other major cities—from New York City to Johannesburg, South Africa—have extensive sewage treatment plants. The border cities of Tijuana in Mexico and San Diego in the United States have developed an international agreement for jointly treating their wastewater. Sewage from Tijuana is piped across the border to a treatment plant on the U.S. side of the border. This plant also treats the wastewater for San Diego.

The sewage that is piped to wastewater treatment plants is subjected to physical, biological (microbial), and sometimes chemical processes. The main aim of wastewater

Source: NPS photo by John Good

©Andrew Penner/Getty Images

(a)

(b)

Figure 22.10 **Treatment of sewage and wastewater.** **(a)** Digester tanks used in the primary phase of treatment; each tank can process several million gallons of raw sewage a day. **(b)** View inside the secondary reactor shows the large stirring paddle that mixes the sludge to aerate it to encourage microbial decomposition.
Courtesy Sanitation Districts of Los Angeles County

treatment is to reduce the organic matter content, that is, to lessen the **biological oxygen demand,** or requirement for oxygen, when the wastewater is eventually released into the environment. Normally, the numbers of pathogens also decline during the treatment processes so that the water released from the plant is safer.

The processes involved in wastewater treatment are divided into (1) the primary phase **(figure 22.10a),** which involves physical separation of solid materials, largely through settling; (2) a secondary phase in which the liquid portion of the waste is subjected to microbial decomposition that is largely aerobic **(figure 22.10b);** and (3) an optional tertiary phase, which may be chemical, physical, or biological, to remove additional inorganic substances like ammonia, nitrate, and phosphate and to eliminate pathogens.

The *trickling filter system* is a relatively simple and inexpensive secondary treatment system. Liquid waste is sprayed over a porous bed of rocks and allowed to flow downward for collection or directly into the groundwater. The rocks are covered by a biofilm of microorganisms. As the wastewater slowly percolates downward, the microorganisms in the biofilms aerobically degrade the organic compounds in the wastewater, reducing the biological oxygen demand. These trickling filters, though, have limited capacities and are not suitable for very high volumes of wastewater.

When high volumes of wastewater require treatment, the *activated sludge process* is used in the secondary stage. In this process, the liquid waste is placed into large tanks. Air is forced into the tanks and the wastewater is rigorously mixed to ensure aerobic conditions. An active microbial community develops that degrades the organic compounds in the wastewater. A sludge forms that is allowed to settle. Most of the sludge is removed for further treatment, but a portion of the sludge containing high numbers of microbes is reintroduced into the aerated treatment tank along with the next batch of wastewater to be treated. This component is called activated sludge because the microbes are already adapted to degrading the organic compounds in the wastewater.

The activated sludge process generally reduces the biological oxygen demand by 85% to 90%. It also greatly reduces the numbers of pathogens, which are largely outcompeted by the nonpathogens and tend to settle in the sludge. The concentrations of *Salmonella, Shigella,* enteroviruses, and other pathogens are generally 90% to 99% lower in the effluent water that is discharged from the plant than in the water that comes into the treatment facility. The water from the secondary treatment process often is released into nearby water bodies.

Sometimes, prior to release, the water from the secondary treatment tanks is subjected to tertiary treatment. Tertiary treatment removes nutrients that could support algal blooms and is important if the wastewater is going to be released into a pristine lake. The tertiary phase may also involve disinfection of the wastewater–for example, by filtration or chlorine treatment–to eliminate pathogens.

The sludge from primary settling tanks and activated sludge treatment tanks can be treated in **anaerobic digestors.** Here, anaerobic bacteria and archaea further decompose the solid wastes, producing a stable solid material and methane that can be collected and used as a fuel. The gas produced from the anaerobic digestor treatment of the sludge can be used to heat the treatment facility or can be purified and sold. In this way, waste treatment and energy generation can be coupled to improve environmental health.

Solid Waste Treatment—Composting and Landfills

Urban solid waste (what most of us know as "garbage") production in the United States amounts to roughly 150 million tons per year. Much of this material is glass, metal, and plastic that is not subject to microbial degradation. The rest is decomposable organic compounds, such as food scraps and paper.

Today, we attempt to remove the nonbiodegradable material from the solid waste and recycle it. This includes paper that is relatively difficult for microbes to decompose. In some cases, the organic wastes are dumped offshore or simply discarded on land, but excessive dumping can cause serious problems. Most municipalities in the United States, therefore, use a combination of sanitary landfills and composting to treat solid wastes.

Composting is an aerobic process in which air and inorganic nutrients support the growth of diverse aerobic microbial communities of bacteria and fungi that are able to decompose the organic wastes. The wastes are generally arranged in a heap or pile that can be managed to ensure adequate aeration and moisture retention **(figure 22.11).** Sometimes, homeowners compost leaves and kitchen and garden wastes to get rid of them. Initially, the decomposition process is carried out by mesophilic microorganisms; but as the decomposition proceeds, heat builds up and there is a shift to thermophiles. The end result of the decomposition of the organic wastes is a material that can be added to soils as a soil conditioner or fertilizer.

The alternative to composting or decomposition in an anaerobic digestor is to send the wastes to a landfill. Generally, organic and inorganic solid wastes are deposited together on land that has minimal real estate value. Because exposed waste can cause esthetic and odor issues and also attract insects and rodents, each day's waste is covered with a layer of soil–creating a sanitary landfill **(figure 22.12).** Eventually, the waste and soil layers build to form a hill. As the buried waste is decomposed, largely by anaerobic microorganisms, the land mass subsides. After 30 to 50 years, the land often is stable and suitable for other uses like housing development. One of the campuses of your author Kelly Cowan's university is built on top of a landfill!

Biodegradation and Bioremediation—Oil and Chemical Spills

Microorganisms have an extraordinary capacity to degrade and to transform chemicals that enter our environment through either natural processes or human activities. Without these microbial activities, all life would cease on earth within a few weeks. Yet, human exploitation of fossil fuels that sometimes results in major oil spills and the release into the environment of xenobiotics–novel, chemically synthesized compounds that do not naturally occur–is proving challenging to microbial **biodegradation** and the sustainability of environmental quality.

©PhotoAlto

Figure 22.11 A backyard composting bin.
Wave Royalty Free/Alamy Stock Photo

Figure 22.12 **A typical landfill.**
Comstock Images/Jupiterimages

Some xenobiotic compounds are totally resistant to microbial attack. When these compounds enter the environment, they accumulate and can cause environmental harm. In some cases, it has been possible to redesign these compounds so that they are biodegradable. Laundry detergents, for example, now have linear alkyl benzyl sulfonates, which are readily biodegradable, instead of the previously used branched (nonlinear) alkyl benzyl sulfonates. It was discovered that these were recalcitrant and accumulated in the environment. The story of laundry detergent design is significant because it was one of the first instances in which a xenobiotic chemical was redesigned to make it biodegradable. Today, we also have biodegradable plastics and various other substances that have been designed to be readily degraded by microorganisms.

In contrast to recalcitrant xenobiotic compounds for which microorganisms have not had sufficient time to evolve the capacity to degrade, microorganisms have long been exposed to petroleum hydrocarbons that routinely seep into the environment. Over 10 million tons of oil pollutants enter the world's oceans each year from natural seepages, accidental spillages, and the disposal of oily wastes. Microorganisms that naturally occur in the oceans have the ability to degrade most of the compounds in petroleum, which is why the oceans are not covered with a layer of oil. However, as we periodically witness, the sudden release of large amounts of oil from a supertanker accident or well blowout can overwhelm the microbial biodegradative capacity.

In the case of the 1989 *Exxon Valdez* spill, far more oil washed up on the shorelines of Prince William Sound, Alaska, than the microbes could quickly biodegrade. In particular, there was a lack of sufficient inorganic nutrients to support the microbial growth needed to consume the hydrocarbons in the oil rapidly. To overcome this limitation, inorganic nitrogen- and phosphate-containing fertilizers were added to stimulate the growth of the naturally occurring oil-degrading microbes—a process called *bioremediation*. No microbes were added—they were already there and just needed the added fertilizer to allow them to grow faster.

©Vanessa Vick/Science Source

Likewise, within 2 to 6 days, microbes in the Gulf of Mexico were estimated to biodegrade half of the dispersed oil released from the BP *Deepwater Horizon* well explosion in 2010 **(figure 22.13).** Even in the cold, deep waters of the Gulf, microbes consumed over 90% of the oil within a month of its release. Molecular analyses showed that diverse microbes, particularly *Oceanospirillum* and *Colwellia*, were responsible for the rapid biodegradation of this oil. Nature has an amazing self-cleaning capacity—one that is fully dependent upon microorganisms.

Counteracting Climate Change?

Some scientists think that it may be possible to manage microbial activities in soils and oceans to combat global warming. Microbes are responsible for the production and consumption of the major greenhouse gases, including carbon dioxide and methane. Some microbes produce carbon dioxide, whereas others, like cyanobacteria and algae, consume it. Some microbes, like the archaea in the guts of herbivorous animals, produce large amounts of methane. In fact, according to the U.S. Environmental Protection Agency, cows release more methane into the atmosphere than automobiles. So you see, some microorganisms contribute to the production of greenhouse gases, whereas others tend to reduce atmospheric concentrations of carbon dioxide and methane and thereby lessen the impact of human activities.

It is unclear whether scientists will be able to—and also whether they should—manipulate microbes to help counteract climate change. It is also unclear whether the balance of microbial production and consumption of carbon dioxide and methane can be controlled. What does seem certain is that if the concentrations of greenhouse gases continue to increase and climate changes continue, human, animal, and environmental health will also change—and the ultimate outcome may well depend upon the microbes.

Figure 22.13 **The 2010 Gulf oil spill.**
Chief Petty Officer John Kepsimelis/U.S. Coast Guard

22.4 LEARNING OUTCOMES—Assess Your Progress

7. Define *biological oxygen demand.*

8. Outline the three phases in wastewater treatment.

9. Provide an example of a xenobiotic that has been made biodegradable.

CASE FILE WRAP-UP

©ASDF_MEDIA/Shutterstock (health care workers);
©Bob Riha, Jr./Archive Photos/Getty Images (stadium)

Although patients sick with MERS were subject to isolation precautions, there were no wide-scale quarantines instituted. In the SARS-CoV-1 outbreak of 2003, Chinese authorities used both isolation and quarantine measures, which ultimately led to the disappearance of the infection. Isolation is used for those who are already infected and/or ill. Quarantine is used for non-ill contacts of an ill person or even populations of people who are very likely to have had contact with someone carrying the virus.

MERS originated in camels, SARS-CoV-1 originated in bats (and was maintained in civet cats), and the 2019 virus is thought (as of this writing) to have originated in bats, as well. All of these, then, are zoonoses, or human infections originating in animals.

Thanks to the Sponge, and Its Microbiome, for Letting Us Breathe

©Carson Ganci/Design Pics

This is our last microbiome story in the book, but it is one of the first microbiome stories in the history of human evolution. As you know, early days on our planet were anaerobic; the development of complex animals and humans could not have happened before the atmosphere contained sufficient oxygen. There is evidence that a very specific microbiome—the microbiome of deep sea sponges—kick-started the accumulation of oxygen on earth.

Recently scientists made a discovery that explained at least partly the immense influx of oxygen onto the planet about 750 million years ago that enabled the Cambrian explosion. The Cambrian explosion refers to the rapid appearance of many types of organisms and animals that we see currently on our planet, with credit going to the sudden appearance of large amounts of oxygen, mainly in the oceans covering the earth.

But where did the oxygen come from? The research suggested that it came from sponges—those sea creatures that seem to have great capacities for pumping and filtering water. But how did that lead to the introduction of all that oxygen? Because sponges live on very low levels of oxygen, and do all that pumping and filtering, they seem to have been producing oxygen in their habitats in the deep sea. But from what?

The oxygen was being produced by photosynthetic microorganisms in the vicinity of the sponges. The active pumping of the oxygen by the sponges created large deposits of phosphate (because phosphorus—abundant in the ocean—combines with oxygen to form the phosphate deposits). Scientists deduced that the large amounts of phosphate that started appearing in the ocean depths around this time are evidence for oxygen-producing bacteria—in the form of the microbiome associated with sponges. So it is quite possible that sea sponges, and their microbes, created the conditions that made oxygenic life on earth possible.

This chapter is titled "One Health." The contribution of animals and environmental factors to the existence of humans is beautifully illustrated with this research.

ONE HEALTH: THE INTERCONNECTED HEALTH OF THE ENVIRONMENT, HUMANS, AND OTHER ANIMALS

Source: Shutterstock/Barbo

22.1 ONE HEALTH

The health of the environment, the health of animals, and the health of humans are intertwined. Scientists call this concept *One Health*.

©Pixtal/age fotostock (chicken);
©OGphoto/Getty Images (tree)

Human infectious diseases that originated in animals are called zoonoses (singular, *zoonosis*). Human diseases can also be transmitted to animals. Lyme disease is a classic example of a disease whose transmission is affected by plant life and the environment, and the movement of humans.

©William Leaman/Alamy Stock Photo

ANIMALS AND INFECTIOUS DISEASE: ZOONOSES 22.2

22.3 THE ENVIRONMENT AND INFECTIOUS DISEASE

Drinking water free of pathogens is fundamental to life. Most of the world's population struggles to obtain clean water. Climate change results in changes in our weather patterns and the environment and plants and animals that are affected by it. This leads to changes in the transmission of diseases to humans.

©Chief Petty Officer John Kepsimelis/
U.S. Coast Guard

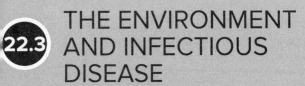

Humans have learned how to take advantage of microbial processes to help us maintain a safe environment and clean up pollution that we create. Sewage treatment and solid waste treatment are assisted in multiple ways by microbes.

MICROBES TO THE RESCUE 22.4

SmartGrid: From Knowledge to Critical Thinking

This *21 Question Grid* takes the topics from this chapter and arranges them with respect to the American Society for Microbiology's Undergraduate Curriculum guidelines—all six of the important "Concepts" as well as the important "Competency" of scientific literacy. Three questions are supplied about chapter content that refer to the Concept or Competency, in increasing levels of Bloom's taxonomy for learning.

ASM Concept/ Competency	A. Bloom's Level 1, 2—Remember and Understand (Choose one.)	B. Bloom's Level 3, 4—Apply and Analyze	C. Bloom's Level 5, 6—Evaluate and Create
Evolution	1. Which of the following is/ are emerging or reemerging diseases for which vaccines are widely available? a. Zika virus b. measles c. Ebola d. *C. diff*	2. In Malaysia, a form of malaria that was previously only seen in macaque monkeys is being seen in humans because deforestation activities bring the monkeys and humans closer together. Diagram a chain of events that could lead to human infection, including plant life, mosquitoes, monkeys, and humans.	3. Develop an argument about how a sexually transmitted microbe might benefit from evolving toward lower virulence while a mosquito-borne virus would not.
Cell Structure and Function	4. Which of the following emerging pathogens is a spirochete? a. West Nile virus b. *E. coli* c. *Borrelia burgdorferi* d. hantavirus	5. Why do you think microbes are so adept at breaking down sewage and wastewater?	6. Are most zoonotic viruses likely to be enveloped or nonenveloped? Provide justification for your answer.
Metabolic Pathways	7. Contaminated water has a. lower biological oxygen demand than pure water. b. higher biological oxygen demand than pure water. c. the same biological oxygen demand as pure water. d. no demand for oxygen.	8. Why do microbes exist that are capable of cleaning up oil spills in the environment?	9. The text states, "...cows release more methane into the atmosphere than automobiles." Would it be a good idea to dose cows with probiotics containing methane-consuming bacteria? Why or why not?
Information Flow and Genetics	10. From what bacterium did the Shiga-toxin-producing *E. coli* acquire its plasmid-encoded toxin? a. *Salmonella* b. *Shigella* c. *Borrelia* d. *Shigaria*	11. Until the 1993 outbreak, hantaviruses were only known to cause hemorrhagic fevers accompanied by kidney damage. Genetic changes may have made it capable of causing lung disease. What aspect of its genetic makeup would suggest that it mutates easily?	12. With what you have learned about increasingly affordable high-throughput DNA sequencing for diagnosing infectious diseases, speculate on the pros and cons of using that technology to replace fecal coliform screening.

ASM Concept/ Competency	A. Bloom's Level 1, 2—Remember and Understand (Choose one.)	B. Bloom's Level 3, 4—Apply and Analyze	C. Bloom's Level 5, 6—Evaluate and Create
Microbial Systems	13. Which of the following refers to a process in which microbial activity is *discouraged?* a. composting b. trickling filter system c. bioremediation d. none of these	14. Explain how a warming climate is likely to impact the epidemiology of mosquito-borne diseases.	15. Conduct research to find out where the microorganisms used in wastewater treatment come from.
Impact of Microorganisms	16. In which way can infectious agents pass between humans and other animals? a. from animals to humans b. from humans to animals c. Both a and b are possible. d. Neither a nor b is possible.	17. Create an infographic or visual representation that describes the similarities and differences between *E. coli* and other coliforms.	18. Zika virus infection in adults may be asymptomatic or result in a skin rash, but it is devastating to fetuses. Suggest reasons for this and identify another virus known for the same characteristics.
Scientific Thinking	19. Use figure 22.2 to identify which of the following zoonoses leads to the most human deaths annually. a. Q fever b. cysticercosis c. leishmaniasis d. toxoplasmosis	20. Defend or refute this statement with information from this chapter: The environment would be a safer place for human health if all microbes could be eliminated from it.	21. Research the Coalition for Epidemic Preparedness Innovations, and write a paragraph describing the experiences that form the group's motivation for creating vaccines for emerging diseases.

Answers to the multiple-choice questions appear in Appendix A.

Visual Connections

This question connects content within and between chapters.

From chapter 18, **figure 18.15.** Use this illustration to explain how the transmission of Lyme disease can be increased by changes in the weather.

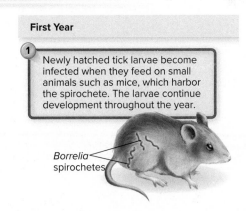

First Year

1 Newly hatched tick larvae become infected when they feed on small animals such as mice, which harbor the spirochete. The larvae continue development throughout the year.

Borrelia spirochetes

Answers to Multiple-Choice Questions in SmartGrid

Chapter 1
1. d
4. b
7. d
10. a
13. d
16. c
19. c

Chapter 2
1. d
4. a
7. d
10. a
13. b
16. d (a & b)
19. d

Chapter 3
1. d
4. b
7. c
10. b
13. d
16. d
19. b

Chapter 4
1. c
4. b
7. c
10. c
13. a
16. b
19. d

Chapter 5
1. b
4. a
7. a
10. c
13. a
16. a
19. b

Chapter 6
1. d
4. a
7. a
10. c
13. b
16. a
19. b

Chapter 7
1. b
4. a
7. b
10. d (a & c)
13. d
16. c
19. d

Chapter 8
1. a
4. d
7. b
10. b
13. a
16. d
19. c

Chapter 9
1. a
4. c
7. c
10. c
13. d
16. a
19. d

Chapter 10
1. d
4. b
7. b
10. b
13. d
16. b
19. b

Chapter 11
1. d
4. d
7. d
10. b
13. d
16. b
19. b

Chapter 12
1. b
4. d
7. d
10. b
13. b
16. a
19. a

Chapter 13
1. b
4. a
7. c
10. a
13. d
16. c
19. b

Chapter 14
1. c
4. b
7. a
10. c
13. d
16. c
19. b

Chapter 15
1. a
4. c
7. a
10. a
13. d
16. a
19. a

Chapter 16
1. b
4. a
7. d (b & c)
10. b
13. d
16. a
19. c

Chapter 17
1. d
4. a
7. c
10. a
13. b
16. a
19. d (b & c)

Chapter 18
1. b
4. a
7. c
10. d
13. d
16. b
19. b

Chapter 19
1. c
4. c
7. a
10. d
13. d
16. c
19. a

Chapter 20
1. b
4. b
7. a
10. d
13. d
16. c
19. a

Chapter 21
1. d
4. c
7. a
10. b
13. a
16. b
19. c

Chapter 22
1. b
4. c
7. b
10. b
13. d
16. c
19. b

B

Answers to the Medical Moments and NCLEX® Questions

Chapter 1
Medical Moments

Medications from Microbes

You might reason that it would be advantageous for one microbe to limit the number of competitors in its environment, and might therefore have evolved the ability to secrete chemicals that damage other microbes. This is, in fact, true!

Delivering Essential Nutrients

The paragraph indicates that the nutrition is delivered intravenously, which bypasses the digestive tract altogether. The gastrointestinal tract is often called the enteric tract. From the context, you could deduce that "parenteral" means, at the very least, "not via the digestive tract." In fact, it means "through direct penetration of the outer tissue (skin)."

NCLEX® Prep

1. a, b, d, e   2. a

Chapter 2
Medical Moments

The Making of the Flu Vaccine

Yes, they can. You will learn later that viruses have molecules on their surfaces that match up with molecules on the surface of only certain cells of certain species. Because influenza can infect the embryonic cells in chickens (birds), you would predict that they can also infect adult birds. This is in fact true. You have no doubt heard of "bird flu," a type of influenza virus that can infect humans, as well as birds.

Gram-Positive versus Gram-Negative Bacteria

Because medical doctors are more interested in the effects of the molecule on a patient, they are likely to call it "endotoxin." Researchers are more likely to identify it as a structural component of the bacterial cell, and identify it as "lipopolysaccharide."

NCLEX® Prep

1. b   2. a, b, c, d   3. c

Chapter 3
Medical Moments

Healthcare-Associated Infections

Surgery compromises important barriers and can allow microbes from the environment, or from internal sites in the body, to contaminate the surgical site. Indeed, surgical-site infections are one of the main HAIs.

Collecting Sputum

Sputum would contain large quantities of normal biota from the throat and mouth. Cultures of sputum would contain these isolates as well as possible *Mycobacterium* colonies. This is one of the reasons acid-fast staining is so useful. Other bacteria will not resist the acid wash and therefore will not be stained by the main stain.

NCLEX® Prep

1. b

Chapter 4
Medical Moments

Vaginal Candidiasis

Diarrhea can have many different causes, from infection to chronic autoimmune dysfunction. Limited postantibiotic diarrhea is often caused by the temporary disruption of the normal gut biota, through the same "bystander" effects we described for vaginal yeast infections. Usually the effect is short-lived, but if the diarrhea is severe and long-lasting, other conditions should be considered.

Opportunistic Fungal Infection

Fungi are eukaryotic cells, just like human cells. It is difficult to cause harm in the invading eukaryote without causing harm to the eukaryotic human cells. Bacteria are prokaryotes.

Neglected Parasitic Infections

Malaria

NCLEX® Prep

1. a   2. c   3. a

Chapter 5
Medical Moments

Why Antibiotics Are Ineffective Against Viruses

Vaccinations were developed to prevent viral infections because there was no effective treatment of them. Eventually vaccines against bacterial diseases were also developed.

Differentiating Between Bacterial and Viral Infections

No, a light microscope would not be useful to diagnose a viral infection because viruses are too small to be seen with light microscopes. However, light microscopy can reveal the absence of bacteria, suggesting a viral cause. It can also reveal the presence of white blood cells, which can be evidence of an infection of some type.

NCLEX® Prep

1. d, e   2. a   3. b, c, e   4. c   5. b

Chapter 6
Medical Moments

Osmosis and IV Fluids

When discussing the uses of IV fluids of varying tonicity, the semipermeable "membranes" are actually the blood vessel walls, acting to exclude larger solutes but not water.

MRSA: PCR over Culture

Patients with MRSA should preferably be housed in private rooms. Their rooms should be prioritized for daily cleaning. If possible, stethoscopes and other frequently used equipment should be dedicated to that room. The CDC has a comprehensive list of precautions and when to use them: www.cdc.gov/mrsa/healthcare/clinicians/precautions.html

NCLEX® Prep

1. d   2. e   3. a   4. b, c, e

Chapter 7
Medical Moments

Facultative Anaerobes

Listeria is psychrotrophic, meaning it thrives in cooler temperatures, including in refrigerated foods.

Amino Acids: Essential, Nonessential, and Conditionally Essential Amino Acids

Tyrosine is produced by mammalian enzymes from the starting material phenylalanine.

NCLEX® Prep

1. b, c   2. a   3. b

Chapter 8
Medical Moments

Gene Mutations

The redundancy of the genetic code means that fewer cases of these diseases occur than if the redundancy did not exist.

Mutations Caused by Life-Saving Radiation

Because DNA mutations caused by radiation therapy are induced, they are not able to be passed on through the genetic code. A notable exception is mutations in germ cells (eggs and sperm), which may be inherited by future generations.

NCLEX® Prep

1. a, b, c, d   2. c   3. d   4. a, b

Chapter 9
Medical Moments

The Use of Alcohol-Based Hand Cleansers

An enveloped virus is more likely to be inactivated by alcohol because alcohol's major action is to dissolve membrane lipids.

Zap VAP: The Role of Chlorhexidine

Chlorhexidine targets bacterial cell walls and membranes, as well as proteins.

NCLEX® Prep

1. a   2. b   3. c

Chapter 10
Medical Moments

Why Do Antibiotics Cause Diarrhea?

Technically, it is neither. It is a side effect caused by the disruption of the gut microbiome and not a result of the drug harming human cells directly (toxicity), or of an inappropriate immune response to it (allergy).

NCLEX® Prep

1. b   2. c   3. c   4. d   5. a

Chapter 11

Medical Moments

When the Portal of Entry Is Compromised

Smoking damages the cilia in the upper respiratory tract, which help clear out infectious agents before they can establish themselves.

Eye on Careers: Infection Control Practitioner

Two of the most frequent cases of health care worker exposure to infection are accidental needlesticks and blood splashes.

Typhoid Mary

The most obvious benefit is bringing the epidemic to the attention of health officials. It is also possible then to work backwards to determine who has already been affected, and, of course, work forward to be on the alert for new cases.

NCLEX® Prep

1. b, d, e   2. d   3. d   4. b, c, e   5. a   6. b

Chapter 12

Medical Moments

Examining Lymph Nodes

This could indicate a systemic condition, infectious or noninfectious. Examples would include sepsis, a blood cancer, or an autoimmune disease.

NCLEX® Prep

1. c   2. a   3. a, b, c, e   4. a

Chapter 13

Medical Moments

The Thymus

Severe combined immunodeficiency is probably the most serious of these conditions, due to the fact that both the B- and T-cell systems are absent. Patients have essentially no specific immune response. They have to be completely sequestered from the environment and other humans. Bone marrow transplantation has helped some of these patients, and gene therapy has had mixed success.

Blood Exposure in Neonates

Clonal expansion after antigen encounter causes B cells and T cells to proliferate.

NCLEX® Prep

1. a   2. b   3. c

Chapter 14

Medical Moments

Hand Washing

There may be several reasons for that. Hand sanitizers contain either alcohol or povidone-iodine or some other antimicrobial chemical, while soaps may only contain the surfactant molecule. Probably most important is the fact that hand sanitizers are left on the hands to evaporate, instead of being rinsed off, allowing more time for the ingredients to act.

Patch Testing

The dendritic cells, just under the surface of the skin, will process and present the antigen to other cells of the immune system.

NCLEX® Prep

1. a   2. a, b, d   3. b   4. d

Chapter 15

Medical Moments

Should Be Obvious, But . . .

A wide variety of illnesses, such as hand, foot, and mouth disease; acute flaccid myelitis; and meningitis.

Qualitative versus Quantitative Diagnosis

The lateral-flow test, which is an immunochromatographic method.

Understanding Lab Results

It was not necessary for you to know which antibiotic should be used for which bacterium; this information is likely to change over time anyway. What is important is that you read the lab results carefully and noticed that the antibiotic he was being treated with did not inhibit the bacterium he is infected with in lab tests. Your thoroughness got him on the correct drug as much as half a day earlier than would have otherwise happened.

Detecting and Treating TB

The infected person can continue to spread TB in the time it takes to definitively diagnose the disease. Also, the infected person will experience more pathology from the disease before being treated with an effective antibiotic.

NCLEX® Prep

1. d   2. e   3. e

Chapter 16

Medical Moments

Scabies

It suggests that its need for a living host is not absolute. In fact, it can live on inanimate objects for 24 to 48 hours.

Scrum Pox: Herpes Gladiatorum

The modes described in the Medical Moment are *direct contact* and *indirect contact* (fomites).

NCLEX® Prep

1. c   2. b   3. d   4. d

Chapter 17

Medical Moments

Acute Flaccid Myelitis

Botulism is characterized by flaccid paralysis.

NCLEX® Prep

1. a   2. a   3. c, d, e

Chapter 18

Medical Moments

Postexposure Prophylaxis

Hepatitis B

Where Does the Fluid Go?

Septic shock

Ebola Epidemic

Because this is the largest number of Ebola patients seen to date, the data could indeed inform a new calculation of the case fatality rate. In this epidemic, the CFR was closer to 39%. Other factors are also considered when designating a CFR, and they are rarely calculated using only a single epidemic.

NCLEX® Prep

1. b   2. d   3. a, b, e   4. c   5. a

Chapter 19

Medical Moments

Epiglottitis

Meningitis, otitis media, and pneumonia can be caused by *Haemophilus influenzae* b (Hib), and the incidence of these diseases caused by Hib has fallen drastically.

Tuberculosis—A Lesson in Inadequate Treatment

In poor countries, sometimes the correct antibiotics are not available. Even when they are, supplies can be unreliable, and treatment may be interrupted for even the most compliant patients. Also, especially before there was much awareness that resistant strains were being created, physicians incorrectly treated tuberculosis sequentially with single antibiotics, encouraging the development of resistance.

NCLEX® Prep

1. c   2. d   3. b   4. a, b, d, e

Chapter 20

Medical Moments

Dehydration

Dehydration harms bodies by depriving vital organs of the correct tonicity of fluids. Smaller bodies have smaller margins for loss of water (increasing hypertonicity disproportionately). Organ damage, and death, occur more quickly.

Assessing Jaundice

Because jaundice is the visible effect of elevated bilirubin levels in the bloodstream, and the blood infiltrates every organ, including the skin and eyes, every organ will have symptoms of the bilirubin elevation.

NCLEX® Prep

1. e   2. a, b, c, d, e   3. b   4. d   5. f

Chapter 21

Medical Moments

Cranberry versus UTI

Because the main causes of UTIs can be resistant to many antibiotics, and may eventually become resistant to most if not all antibiotics.

Crabs

Crabs lay their eggs at the base of pubic hairs, which can cause itching. Scratching the affected areas can damage the skin and break down the barriers to other microbes entering.

NCLEX® Prep

1. a, b, c   2. c   3. a, d

Chapter 22

Medical Moments

Plastic Bottles for Clean Water

No. UV light has to come into direct contact with microbes in order to kill them. The water is contained in plastic, which reduces the UV's effectiveness. But the sunlight treatment does reduce the microbial load to levels that make the water safer.

Cryptosporidium in Your Tap Water?

Water should be boiled or filtered after it comes out of the tap. Ice should be made from water treated this way, as well. Alternatively, bottled water can be substituted for tap water. Raw fruits and vegetables should only be washed with treated water before eating.

Glossary

A

abiogenesis The belief in spontaneous generation as a source of life.

abscess An inflamed, fibrous lesion enclosing a core of pus.

acid-fast stain A solution containing carbol fuchsin, which, when bound to lipids in the envelopes of *Mycobacterium* species, cannot be removed with an acid wash.

acidophilic Thriving in low-pH environments.

actin A globular protein that forms microfilaments and provides structural support in cells.

active immunity Immunity acquired through direct stimulation of the immune system by antigen.

active transport Nutrient transport method that requires carrier proteins in the membranes of the living cells and the expenditure of energy.

acute Characterized by rapid onset and short duration.

adenine (A) One of the nitrogen bases found in DNA and RNA with a purine form.

adenosine deaminase (ADA) deficiency An immunodeficiency disorder and one type of SCID that is caused by an inborn error in the metabolism of adenine. The accumulation of adenine destroys both B and T lymphocytes.

adenosine triphosphate (ATP) A nucleotide that is the primary source of energy to cells.

adhesion The process by which microbes gain a more stable foothold at the portal of entry; often involves a specific interaction between the molecules on the microbial surface and the receptors on the host cell.

adjuvant In immunology, a chemical vehicle that enhances antigenicity, presumably by prolonging antigen retention at the injection site.

adsorption A process of adhering one molecule onto the surface of another molecule.

aerobic respiration Respiration in which the final electron acceptor in the electron transport chain is oxygen (O_2).

agammaglobulinemia Also called *hypogammaglobulinemia*. The absence of or severely reduced levels of antibodies in serum.

agar A polysaccharide found in seaweed and commonly used to prepare solid culture media.

agglutination The aggregation by antibodies of suspended cells or similar-size particles (agglutinogens) into clumps that settle.

agranulocyte One form of leukocyte (white blood cell) having globular, nonlobed nuclei and lacking prominent cytoplasmic granules.

AIDS Acquired immunodeficiency syndrome. The complex of signs and symptoms characteristic of the late phase of human immunodeficiency virus (HIV) infection.

akaryote A designation for bacteria and archaea.

algae Photosynthetic, plantlike organisms that generally lack the complex structure of plants; they may be single-celled or multicellular and inhabit diverse habitats such as marine and freshwater environments, glaciers, and hot springs.

allergen A substance that provokes an allergic response.

allergy The altered, usually exaggerated, immune response to an allergen. Also called *hypersensitivity*.

alloantigen An antigen that is present in some but not all members of the same species.

allograft Relatively compatible tissue exchange between nonidentical members of the same species. Also called *homograft*.

alternative complement pathway Complement cascade initiated by spontaneous breakdown of a blood protein called C3 in the presence of microbes.

amino acids The building blocks of protein. Amino acids exist in 20 naturally occurring forms that impart different characteristics to the various proteins they compose.

amphibolism Pertaining to the metabolic pathways that serve multiple functions in the breakdown, synthesis, and conversion of metabolites.

amphitrichous Having a single flagellum or a tuft of flagella at opposite poles of a microbial cell.

anabolism The energy-consuming process of incorporating nutrients into protoplasm through biosynthesis.

anaerobic digesters Closed chambers used in a microbial process that converts organic sludge from waste treatment plants into useful fuels such as methane and hydrogen gases. Also called *bioreactors*.

anaerobic respiration Respiration in which the final electron acceptor in the electron transport chain is an inorganic molecule containing sulfate, nitrate, nitrite, carbonate, and so on.

anamnestic response In immunology, an augmented response or memory related to a prior stimulation of the immune system by antigen. It boosts the levels of immune substances.

anaphylaxis The unusual or exaggerated allergic reaction to antigen that leads to severe respiratory and cardiac complications.

antagonism Relationship in which microorganisms compete for survival in a common environment by taking actions that inhibit or destroy another organism.

antibody A large protein molecule evoked in response to an antigen that interacts specifically with that antigen.

anticodon The trinucleotide sequence of transfer RNA that is complementary to the trinucleotide sequence of messenger RNA (the codon).

antigen Any cell, particle, or chemical that induces a specific immune response by B cells or T cells and can stimulate resistance to an infection or a toxin. See **immunogen.**

antigen binding site Specific region at the ends of the antibody molecule that recognize specific antigens. These sites have numerous shapes to fit a wide variety of antigens.

antigenic drift Minor antigenic changes in the influenza A virus due to mutations in the spikes' genes.

antigenic shift Major changes in the influenza A virus due to recombination of viral strains from two different host species.

antigen-presenting cells (APCs) Cells of the immune system that digest foreign cells and particles and place pieces of them on their own surfaces in such a way that other cells of the immune system recognize them.

antihistamine A drug that counters the action of histamine and is useful in allergy treatment.

antimicrobial peptides Short protein molecules found in epithelial cells; have the ability to kill bacteria.

antiparallel A description of the two strands of DNA, which are parallel to each other, but the orientation of the deoxyribose and phosphate groups run in the opposite directions, with the 5′ carbon at the top of the leading strand and the 3′ carbon at the top of the lagging strand.

antiseptic A growth-inhibiting agent used on tissues to prevent infection.

antitoxin Globulin fraction of serum that neutralizes a specific toxin. Also refers to the specific antitoxin antibody itself.

apoenzyme The protein part of an enzyme, as opposed to the nonprotein or inorganic cofactors.

appendages Accessory structures that sprout from the surface of bacteria. They can be divided into two major groups: those that provide motility and those that enable adhesion.

aqueous Referring to solutions in which water is used as the solvent.

archaea Prokaryotic single-celled organisms of primitive origin that have unusual anatomy, physiology, and genetics and live in harsh habitats; when capitalized **(Archaea),** the term refers to one of the three domains of living organisms as proposed by Woese.

arthroconidia Reproductive body of *Coccidioides immitis*; also *arthrospore*.

Arthus reaction An immune complex phenomenon that develops after repeat injection. This localized inflammation results from aggregates of antigen and antibody that bind, complement, and attract neutrophils.

artificial immunity Immunity that is induced as a medical intervention, either by exposing an individual to an antigen or administering immune substances to him or her.

asepsis A condition free of viable pathogenic microorganisms.

aseptic technique Method of handling microbial cultures, patient specimens, and other sources of microbes in a way that prevents infection of the handler and others who may be exposed.

assay medium Microbiological medium used to test the effects of specific treatments to bacteria, such as antibiotic or disinfectant treatment.

assembly (viral) The step in viral multiplication in which capsids and genetic material are packaged into virions.

asthma A type of chronic allergy in which the airways become constricted and produce excess mucus in reaction to allergens, exercise, stress, or cold temperatures.

asymptomatic An infection that produces no noticeable symptoms even though the microbe is active in the host tissue.

atopy Allergic reaction classified as type I, with a strong familial relationship; caused by allergens such as pollen, insect venom, food, and dander; involves IgE antibody; includes symptoms of hay fever, asthma, and skin rash.

ATP synthase A unique enzyme located in the mitochondrial cristae and chloroplast grana that harnesses the flux of hydrogen ions to the synthesis of ATP.

autoantibody An "anti-self" antibody having an affinity for tissue antigens of the subject in which it is formed.

autoclave A sterilization chamber that allows the use of steam under pressure to sterilize materials. The most common temperature/pressure combination for an autoclave is 121°C and 15 psi.

autograft Tissue or organ surgically transplanted to another site on the same subject.

autoimmune disease The pathologic condition arising from the production of antibodies against autoantigens. Example: rheumatoid arthritis. Also called *autoimmunity*.

autotroph A microorganism that requires only inorganic nutrients and whose sole source of carbon is carbon dioxide.

B

bacillus Bacterial cell shape that is cylindrical (longer than it is wide).

back-mutation A mutation that counteracts an earlier mutation, resulting in the restoration of the original DNA sequence.

bacteremia The presence of viable bacteria in circulating blood.

Bacteria When capitalized, can refer to one of the three domains of living organisms proposed by Woese containing all nonarchaea prokaryotes.

bacteria (singular, *bacterium*) Category of prokaryotes with peptidoglycan in their cell walls and circular chromosome(s). This group of small cells is widely distributed in the earth's habitats.

bacterial chromosome A circular body in bacteria that contains the primary genetic material. Also called *nucleoid*.

bactericide An agent that kills bacteria.

bacteriophage A virus that specifically infects bacteria.

bacteristatic Any process or agent that inhibits bacterial growth.

barophile A microorganism that thrives under high (usually hydrostatic) pressure.

basement membrane A thin layer (1-6 μm) of protein and polysaccharide found at the base of epithelial tissues.

beta oxidation The degradation of long-chain fatty acids. Two-carbon fragments are formed as a result of enzymatic attack directed against the second or beta carbon of the hydrocarbon chain. Aided by coenzyme A, the fragments enter the Krebs cycle and are processed for ATP synthesis.

binary fission The formation of two new cells of approximately equal size as the result of parent cell division.

binomial system Scientific method of assigning names to organisms that employs two names to identify every organism–genus name plus species name.

biodegradation The breaking down of materials through the action of microbes or insects.

biofilm A complex association that arises from a mixture of microorganisms growing together on the surface of a habitat.

biogenesis Belief that living things can only arise from others of the same kind.

biological oxygen demand Requirement for oxygen by microbes in a niche; used to measure drinking water quality.

bioremediation Decomposition of harmful chemicals by microbes or consortia of microbes.

biosensor A device used to detect microbes or trace amounts of compounds through PCR, genome techniques, or electrochemical signaling.

biota Beneficial or harmless resident bacteria commonly found on and/or in the human body.

biotechnology The intentional use by humans of living organisms or their products to accomplish a goal related to health or the environment.

blocking antibody The IgG class of immunoglobulins that competes with IgE antibody for allergens, thus blocking the degranulation of basophils and mast cells.

blood cells Cellular components of the blood consisting of red blood cells, primarily responsible for the transport of oxygen and carbon dioxide, and white blood cells, primarily responsible for host defense and immune reactions.

blood-brain barrier Decreased permeability of the walls of blood vessels in the brain, restricting access to that compartment.

botulinum toxin *Clostridium botulinum* toxin. Ingestion of this potent exotoxin leads to flaccid paralysis.

bradykinin An active polypeptide that is a potent vasodilator released from IgE-coated mast cells during anaphylaxis.

broad-spectrum Denotes drugs that have an effect on a wide variety of microorganisms.

bubo The swelling of one or more lymph nodes due to inflammation.

bubonic plague The form of plague in which bacterial growth is primarily restricted to the lymph and is characterized by the appearance of a swollen lymph node referred to as a *bubo*.

bulbar poliomyelitis Complication of polio infection in which the brain stem, medulla, or cranial nerves are affected. Leads to loss of respiratory control and paralysis of the trunk and limbs.

bullous Consisting of fluid-filled blisters.

C

calculus Dental deposit formed when plaque becomes mineralized with calcium and phosphate crystals. Also called *tartar*.

capsid The protein covering of a virus's nucleic acid core. Capsids exhibit symmetry due to the regular arrangement of subunits called *capsomers*. See **icosahedron.**

capsomere A subunit of the virus capsid shaped as a triangle or disc.

capsular staining Any staining method that highlights the outermost polysaccharide and/or protein structure on a bacterial, fungal, or protozoal cell.

capsule   In bacteria, the loose, gel-like covering or slime made chiefly of polysaccharides. This layer is protective and can be associated with virulence.

carbohydrate   A compound containing primarily carbon, hydrogen, and oxygen in a 1:2:1 ratio.

carbohydrate fermentation medium   A growth medium that contains sugars that are converted to acids through fermentation. Usually contains a pH indicator to detect acid protection.

carrier   A person who harbors infections and inconspicuously spreads them to others. Also, a chemical agent that can accept an atom, chemical radical, or subatomic particle from one compound and pass it on to another.

caseous lesion   Necrotic area of lung tubercle superficially resembling cheese. Typical of tuberculosis.

catabolism   The chemical breakdown of complex compounds into simpler units to be used in cell metabolism.

catalyst   A substance that alters the rate of a reaction without being consumed or permanently changed by it. In cells, enzymes are catalysts.

catarrhal   A term referring to the secretion of mucus or fluids; term for the first stage of pertussis.

cell   An individual membrane-bound living entity; the smallest unit capable of an independent existence.

cell wall   In bacteria, a rigid structure made of peptidoglycan that lies just outside the cytoplasmic membrane; eukaryotes also have a cell wall but may be composed of a variety of materials.

cellulose   A long, fibrous polymer composed of β-glucose; one of the most common substances on earth.

cestode   The common name for tapeworms that parasitize humans and domestic animals.

chancre   The primary sore of syphilis that forms at the site of penetration by *Treponema pallidum*. It begins as a hard, dull red, painless papule that erodes from the center.

chemical mediators   Small molecules that are released during inflammation and specific immune reactions that allow communication between the cells of the immune system and facilitate surveillance, recognition, and attack.

chemoautotroph   An organism that relies upon inorganic chemicals for its energy and carbon dioxide for its carbon. Also called a *chemolithotroph*.

chemoheterotroph   Microorganisms that derive their nutritional needs from organic compounds.

chemokine   Chemical mediators (cytokines) that stimulate the movement and migration of white blood cells.

chemostat   A growth chamber with an outflow that is equal to the continuous inflow of nutrient media. This steady-state growth device is used to study such events as cell division, mutation rates, and enzyme regulation.

chemotactic factors   Chemical mediators that stimulate the movement of white blood cells. See **chemokine.**

chemotaxis   The tendency of organisms to move in response to a chemical gradient (toward an attractant or to avoid adverse stimuli).

chemotroph   Organism that oxidizes compounds to feed on nutrients.

chitin   A polysaccharide similar to cellulose in chemical structure. This polymer makes up the horny substance of the exoskeletons of arthropods and certain fungi.

chloroplast   An organelle containing chlorophyll that is found in photosynthetic eukaryotes.

cholesterol   Best-known member of a group of lipids called *steroids*. Cholesterol is commonly found in cell membranes and animal hormones.

chromatin   The genetic material of the nucleus. Chromatin is made up of nucleic acid and stains readily with certain dyes.

chromosome   The tightly coiled bodies in cells that are the primary sites of genes.

class   In the levels of classification, the division of organisms that follows phylum.

classical complement pathway   Pathway of complement activation initiated by a specific antigen-antibody interaction.

clonal deletion   The selective elimination of lymphocytes that would recognize self markers.

clonal selection   A conceptual explanation for the development of lymphocyte specificity and variety during immune maturation.

clone   A colony of cells (or group of organisms) derived from a single cell (or single organism) by asexual reproduction. All units share identical characteristics. Also used as a verb to refer to the process of producing a genetically identical population of cells or genes.

cloning host   An organism such as a bacterium or a yeast that receives and replicates a foreign piece of DNA inserted during a genetic engineering experiment.

coccobacillus   An elongated coccus; a short, thick, oval-shaped bacterial rod.

coccus   A spherical-shaped bacterial cell.

codon   A specific sequence of three nucleotides in mRNA (or the sense strand of DNA) that constitutes the genetic code for a particular amino acid.

coenzyme   A complex organic molecule, several of which are derived from vitamins (e.g., nicotinamide, riboflavin). A coenzyme operates in conjunction with an enzyme. Coenzymes serve as transient carriers of specific atoms or functional groups during metabolic reactions.

cofactor   An enzyme accessory. It can be organic, such as coenzymes, or inorganic, such as Fe^{2+}, Mn^{2+}, or Zn^{2+} ions.

coliform   A collective term that includes normal enteric bacteria that are gram-negative and lactose-fermenting.

colonize   The act of taking up long-term residence; as in microbes establishing a steady relationship with a host.

colony   A macroscopic cluster of cells appearing on a solid medium, each arising from the multiplication of a single cell.

colostrum   The clear yellow early product of breast milk that is very high in secretory antibodies. Provides passive intestinal protection.

commensalism   An unequal relationship in which one species derives benefit without harming the other.

common-source epidemic   An outbreak of disease in which all affected individuals were exposed to a single source of the pathogen, even if they were exposed at different times.

communicable infection   Capable of being transmitted from one individual to another.

competent   Referring to bacterial cells that are capable of absorbing free DNA in their environment either naturally or through induction by exposure to chemicals or electrical currents.

competitive inhibition   Control process that relies on the ability of metabolic analogs to control microbial growth by successfully competing with a necessary enzyme to halt the growth of bacterial cells.

complement   In immunology, serum protein components that act in a definite sequence when set in motion either by an antigen-antibody complex or by factors of the alternative (properdin) pathway.

complementary DNA (cDNA)   DNA created by using reverse transcriptase to synthesize DNA from RNA templates.

condyloma acuminata   Extensive, branched masses of genital warts caused by infection with human papillomavirus.

congenital   Transmission of an infection from mother to fetus.

congenital rubella   Transmission of the rubella virus to a fetus *in utero*. Injury to the fetus is generally much more serious than it is to the mother.

conidia   Asexual fungal spores shed as free units from the tips of fertile hyphae.

conidiospore   A type of asexual spore in fungi; not enclosed in a sac.

conjugated vaccines   Subunit vaccines combined with carrier proteins, often from other microbes, to make them more immunogenic.

conjugation   In bacteria, the contact between donor and recipient cells associated with the transfer of genetic material such as plasmids.

Can involve special (sex) pili. Also a form of sexual recombination in ciliated protozoa.

conjunctiva   The thin fluid-secreting tissue that covers the eye and lines the eyelid.

constitutive enzyme   An enzyme present in bacterial cells in constant amounts, regardless of the presence of substrate. Enzymes of the central catabolic pathways are typical examples.

contagious   Communicable; transmissible by direct contact with infected people and their fresh secretions or excretions.

contaminant   An impurity; any undesirable material or organism.

continuation   The phase in a microbial infection in which either the organism lingers after symptoms subside or the symptoms continue after the organism is no longer detectable.

convalescence   Recovery; the period between the end of a disease and the complete restoration of health in a patient.

cornea   The transparent, dome-shaped tissue covering the iris, pupil, and anterior chamber of the eye composed of five to six layers of quickly regenerating epithelial cells.

cortex   The outer rim of a lymph node.

Creutzfeldt-Jakob disease (CJD)   A spongiform encephalopathy caused by infection with a prion. The disease is marked by dementia, impaired senses, and uncontrollable muscle contractions.

CRISPR   Clustered regularly interspaced short palindromic repeats found in bacteria and archaea used to combat invasion by foreign DNA. Used by scientists to direct DNA manipulation.

crista   The infolded inner membrane of a mitochondrion that is the site of the respiratory chain and oxidative phosphorylation.

culture   The visible accumulation of microorganisms in or on a nutrient medium. Also, the propagation of microorganisms with various media.

cyst   The resistant, dormant but infectious form of protozoa. Can be important in spread of infectious agents such as *Entamoeba histolytica* and *Giardia lamblia*.

cysteine   A nonessential amino acid that is related to the essential amino acid cystine.

cysticercosis   Condition caused by *Taenia solium* tapeworm. Larval cysts embed in brain, muscle, and other tissue, causing seizures and other symptoms.

cystine   An amino acid, $HOOC-CH(NH_2)-CH_2-S-S-CH_2-CH(NH_2)COOH$. An oxidation product of two cysteine molecules in which the OSH (sulfhydryl) groups form a disulfide union. Also called *dicysteine*.

cytochrome   A group of heme protein compounds whose chief role is in electron and/or hydrogen transport occurring in the last phase of aerobic respiration.

cytokine   Regulatory chemical released by cells of the immune system that serves as signal between different cells.

cytopathic effect   The degenerative changes in cells associated with virus infection. Examples: the formation of multinucleate giant cells (Negri bodies), the prominent cytoplasmic inclusions of nerve cells infected by rabies virus.

cytoplasm   Dense fluid encased by the cytoplasmic membrane; the site of many of the cell's biochemical and synthetic activities.

cytosine (C)   One of the nitrogen bases found in DNA and RNA, with a pyrimidine form.

cytotoxicity   The property of being able to damage or kill cells.

D

deamination   The removal of an amino group from an amino acid.

death phase   End of the cell growth due to lack of nutrition, depletion of environment, and accumulation of wastes. Population of cells begins to die.

definitive host   The organism in which a parasite develops into its adult or sexually mature stage. Also called the *final host*.

denaturation   The loss of normal characteristics resulting from some molecular alteration. Usually in reference to the action of heat or chemicals on proteins whose function depends upon an unaltered tertiary structure.

dendritic cell   A large, antigen-processing cell characterized by long, branchlike extensions of the cell membrane.

denitrification   The end of the nitrogen cycle when nitrogen compounds are returned to the reservoir in the air.

deoxyribose   A 5-carbon sugar that is an important component of DNA.

dermatophytes   A group of fungi that cause infections of the skin and other integument components. They survive by metabolizing keratin.

dermolytic   Capable of damaging the skin.

desiccation   To dry thoroughly. To preserve by drying.

desquamate   To shed the cuticle in scales; to peel off the outer layer of a surface.

diapedesis   The migration of intact blood cells between endothelial cells of a blood vessel such as a venule.

dichotomous key   A flowchart at which there are two choices at each fork in the chart; used for microbial identification.

differential medium   A single substrate that discriminates between groups of microorganisms on the basis of differences in their appearance due to different chemical reactions.

differential stain   A technique that utilizes two dyes to distinguish between different microbial groups or cell parts by color reaction.

differentiation   Making one object distinguishable from another; also, the process of biological cells changing from one state to a more mature state.

diffusion   The dispersal of molecules, ions, or microscopic particles propelled down a concentration gradient by spontaneous random motion to achieve a uniform distribution.

DiGeorge syndrome   A birth defect usually caused by a missing or incomplete thymus that results in abnormally low or absent T cells and other developmental abnormalities.

dimorphic   In mycology, the tendency of some pathogens to alter their growth form from mold to yeast in response to rising temperature.

direct or total cell count   1. Counting total numbers of individual cells being viewed with magnification. 2. Counting isolated colonies of organisms growing on a plate of media as a way to determine population size.

disaccharide   A sugar containing two monosaccharides. Example: sucrose (fructose + glucose).

disease   Any deviation from health, as when the effects of microbial infection damage or disrupt tissues and organs.

division   In the levels of classification, an alternate term for *phylum*.

DNA polymerase   Enzyme responsible for the replication of DNA. Several versions of the enzyme exist, each completing a unique portion of the replication process.

DNA profiling   A pattern of restriction enzyme fragments that is unique for an individual organism.

DNA sequencing   Determining the exact order of nucleotides in a fragment of DNA. Most commonly done using the Sanger dideoxy sequencing method.

DNA vaccine   A newer vaccine preparation based on inserting DNA from pathogens into host cells to encourage them to express the foreign protein and stimulate immunity.

domain   In the levels of classification, the broadest general category to which an organism is assigned. Members of a domain share only one or a few general characteristics.

doubling time   Time required for a complete fission cycle–from parent cell to two new daughter cells. Also called *generation time*.

drug resistance   An adaptive response in which microorganisms begin to tolerate an amount of drug that would ordinarily be inhibitory.

dysentery   Diarrheal illness in which stools contain blood and/or mucus.

E

ectoplasm   Viscous outer layer of cytoplasm in some eukaryotic cells.

eczema   An acute or chronic allergy of the skin associated with itching and burning sensations. Typically, red, edematous, vesicular lesions erupt, leaving the skin scaly and sometimes hyperpigmented.

edema   The accumulation of excess fluid in cells, tissues, or serous cavities. Also called *swelling*.

ELISA   Abbreviation for **e**nzyme-**l**inked **i**mmuno**s**orbent **a**ssay, a very sensitive serological test used to detect antibodies in diseases such as AIDS.

emetic   Inducing to vomit.

endemic disease   A native disease that prevails continuously in a geographic region.

endergonic reaction   A chemical reaction that occurs with the absorption and storage of surrounding energy. See **exergonic reaction.**

endocytosis   The process whereby solid and liquid materials are taken into the cell through membrane invagination and engulfment into a vesicle.

endogenous   Originating or produced within an organism or one of its parts.

endoplasmic reticulum (ER)   An intracellular network of flattened sacs or tubules with or without ribosomes on their surfaces.

endospore   A small, dormant, resistant derivative of a bacterial cell that germinates under favorable growth conditions into a vegetative cell. The bacterial genera *Bacillus* and *Clostridium* are typical endospore formers.

endosymbiosis   The theory that eukaryotic cells came to be when primordial cell types engulfed bacteria or other primordial cell types that later became organelles inside the more complex cells.

endotoxic shock   A massive drop in blood pressure caused by the release of endotoxin from gram-negative bacteria multiplying in the bloodstream.

enriched medium   A nutrient medium supplemented with blood, serum, or some growth factor to promote the multiplication of fastidious microorganisms.

enteroaggregative   The term used to describe certain types of intestinal bacteria that tend to stick to each other in large clumps.

enteroinvasive   Predisposed to invade the intestinal tissues.

enteropathogenic   Pathogenic to the alimentary canal.

enterotoxigenic   Having the capacity to produce toxins that act on the intestinal tract.

enterotoxin   A bacterial toxin that specifically targets intestinal mucous membrane cells. Enterotoxigenic strains of *Escherichia coli* and *Staphylococcus aureus* are typical sources.

enumeration medium   Microbiological medium that does not encourage growth and allows for the counting of microbes in food, water, or environmental samples.

enveloped virus   A virus whose nucleocapsid is enclosed by a membrane derived in part from the host cell. It usually contains exposed glycoprotein spikes specific for the virus.

enzyme   A protein biocatalyst that facilitates metabolic reactions.

enzyme induction   One of the controls on enzyme synthesis. This occurs when enzymes appear only when suitable substrates are present.

enzyme repression   The inhibition of enzyme synthesis by the end product of a catabolic pathway.

eosinophilia   An increase in eosinophil concentration in the bloodstream, often in response to helminthic infection.

epidemic   A sudden and simultaneous outbreak or increase in the number of cases of disease in a community.

epidemiology   The study of the factors affecting the prevalence and spread of disease within a community.

epigenetic   Refers to "above the gene," specifically additions to the DNA that do not change the gene but can change its expression.

epimutation   Heritable change in gene expression that does not involve change in the gene itself but chemical additions to it.

epitope   The precise molecular group of an antigen that defines its specificity and triggers the immune response.

Epstein-Barr virus (EBV)   Herpesvirus linked to infectious mononucleosis, Burkitt's lymphoma, and nasopharyngeal carcinoma.

erythrogenic toxin   An exotoxin produced by lysogenized group A strains of β-hemolytic streptococci that is responsible for the severe fever and rash of scarlet fever in the nonimmune individual. Also called a *pyrogenic toxin*.

eschar   A dark, sloughing scab that is the lesion of anthrax and certain rickettsioses.

essential nutrient   Any ingredient such as a certain amino acid, fatty acid, vitamin, or mineral that cannot be formed by an organism and must be supplied in the diet. A growth factor.

etiologic agent   The microbial cause of disease; the pathogen.

Eukarya   One of the three domains (sometimes called superkingdoms) of living organisms, as proposed by Woese; contains all eukaryotes.

eukaryotic cell   A cell that differs from a prokaryotic cell chiefly by having a nuclear membrane (a well-defined nucleus), membrane-bound subcellular organelles, and mitotic cell division.

evolution   Scientific principle that states that living things change gradually through hundreds of millions of years, and these changes are expressed in structural and functional adaptations in each organism. Evolution presumes that those traits that favor survival are preserved and passed on to following generations, and those traits that do not favor survival are lost.

exanthem   An eruption or rash of the skin.

exergonic reaction   A chemical reaction associated with the release of energy to the surroundings. See **endergonic reaction.**

exoenzyme   An extracellular enzyme chiefly for hydrolysis of nutrient macromolecules that are otherwise impervious to the cell membrane. It functions in saprobic decomposition of organic debris and can be a factor in invasiveness of pathogens.

exogenous   Originating outside the body.

exon   A stretch of eukaryotic DNA coding for a corresponding portion of mRNA that is translated into peptides. Intervening stretches of DNA that are not expressed are called *introns*. During transcription, exons are separated from introns and are spliced together into a continuous mRNA transcript.

exotoxin   A toxin (usually protein) that is secreted and acts upon a specific cellular target. Examples: botulin, tetanospasmin, diphtheria toxin, and erythrogenic toxin.

exponential   Pertaining to the use of exponents, numbers that are typically written as a superscript to indicate how many times a factor is to be multiplied. Exponents are used in scientific notation to render large, cumbersome numbers into small workable quantities.

exponential growth phase   The period of maximum growth rate in a growth curve. Cell population increases logarithmically.

extrapulmonary tuberculosis   A condition in which tuberculosis bacilli have spread to organs other than the lungs.

exudate   Fluid that escapes cells into the extracellular spaces during the inflammatory response.

F

facultative   Pertaining to the capacity of microbes to adapt or adjust to variations; not obligate. Example: The presence of oxygen is not obligatory for a facultative anaerobe to grow. See **obligate.**

family   In the levels of classification, a midlevel division of organisms that groups more closely related organisms than previous levels. An order is divided into families.

fastidious   Requiring special nutritional or environmental conditions for growth; said of bacteria.

fecal coliforms   Any species of gram-negative, lactose-positive bacteria (primarily *Escherichia coli*) that live primarily in the intestinal tract and not the environment. Finding evidence of these bacteria in a water or food sample is substantial evidence of fecal contamination and potential for infection. See **coliform.**

fermentation   The extraction of energy through anaerobic degradation of substrates into simpler, reduced metabolites. In large industrial processes, fermentation can mean any use of microbial metabolism to manufacture organic chemicals or other products.

fertility (F) factor   Donor plasmid that allows synthesis of a pilus in bacterial conjugation. Presence of the factor is indicated by F⁺, and lack of the factor is indicated by F⁻.

filament   A helical structure composed of proteins that is part of bacterial flagella.

fimbria   A short, numerous-surface appendage on some bacteria that provides adhesion but not locomotion.

Firmicutes   Taxonomic category of bacteria that have gram-positive cell envelopes.

flagellar staining   A staining method that highlights the flagellum of a bacterium.

flagellum   A structure that is used to propel the organism through a fluid environment.

fluorescence   The property possessed by certain minerals and dyes to emit visible light when excited by ultraviolet radiation. A fluorescent dye combined with specific antibody provides a sensitive test for the presence of antigen.

food poisoning   Symptoms in the intestines (which may include vomiting) induced by preformed exotoxin from bacteria.

frameshift mutation   An insertion or deletion mutation that changes the codon reading frame from the point of the mutation to the final codon. Almost always leads to a nonfunctional protein.

fructose   One of the carbohydrates commonly referred to as sugars. Fructose is commonly fruit sugars.

fungemia   The condition of fungi multiplying in the bloodstream.

fungi (singular, **fungus**)   Macroscopic and microscopic heterotrophic eukaryotic organisms that can be uni- or multicellular.

G

gastritis   Inflammation or infection of the stomach.

gel electrophoresis   A laboratory technique for separating DNA fragments according to length by employing electricity to force the DNA through a gel-like matrix typically made of agarose. Smaller DNA fragments move more quickly through the gel, thereby moving farther than larger fragments during the same period of time.

gene   A site on a chromosome that provides information for a certain cell function. A specific segment of DNA that contains the necessary code to make a protein or RNA molecule.

gene drive   A genetic engineering technique that increases the density of a particular gene in a population.

gene probe   Short strand of single-stranded nucleic acid that hybridizes specifically with complementary stretches of nucleotides on test samples and thereby serves as a tagging and identification device.

gene therapy   The introduction of normal functional genes into people with genetic diseases such as sickle-cell anemia and cystic fibrosis. This is usually accomplished by a virus vector.

general-purpose media   Laboratory agar or broth that allow most microorganisms to grow, without differentiating among them.

generation time   Time required for a complete fission cycle–from parent cell to two new daughter cells. Also called *doubling time*.

genetic engineering   A field involving deliberate alterations (recombinations) of the genomes of microbes, plants, and animals through special technological processes.

genetics   The science of heredity.

genome   The complete set of chromosomes and genes in an organism.

genomic libraries   Collections of DNA fragments representing the entire genome of an organism inserted into plasmids and stored in vectors such as bacteria or yeasts.

genomics   The systematic study of an organism's genes and their functions.

genotype   The genetic makeup of an organism. The genotype is ultimately responsible for an organism's phenotype, or expressed characteristics.

genus   In the levels of classification, the second-most-specific level. A family is divided into several genera.

germ theory of disease   A theory first originating in the 1800s that proposed that microorganisms can be the cause of diseases. The concept is actually so well established in the present time that it is considered a fact.

germicide   An agent lethal to non-endospore-forming pathogens.

gingivitis   Inflammation of the gum tissue in contact with the roots of the teeth.

glucose   One of the carbohydrates commonly referred to as sugars. Glucose is characterized by its 6-carbon structure.

glycerol   A 3-carbon alcohol, with three OH groups that serve as binding sites.

glycocalyx   A filamentous network of carbohydrate-rich molecules that coats cells.

glycolysis   The energy-yielding breakdown (fermentation) of glucose to pyruvic or lactic acid. It is often called *anaerobic glycolysis* because no molecular oxygen is consumed in the degradation.

Golgi apparatus   An organelle of eukaryotes that participates in packaging and secretion of molecules.

Gracilicutes   Taxonomic category of bacteria that have gram-negative envelopes.

graft versus host disease (GVHD)   A condition associated with a bone marrow transplant in which T cells in the transplanted tissue mount an immune response against the recipient's (host) normal tissues.

Gram stain   A differential stain for bacteria useful in identification and taxonomy. Gram-positive organisms appear purple from crystal violet mordant retention, whereas gram-negative organisms appear red after loss of crystal violet and absorbance of the safranin counterstain.

gram-negative   A category of bacterial cells that describes bacteria with an outer membrane, a cytoplasmic membrane, and a thin cell wall.

gram-positive   A category of bacterial cells that describes bacteria with a thick cell wall and no outer membrane.

granulocyte   A mature leukocyte that contains noticeable granules in a Wright stain. Examples: neutrophils, eosinophils, and basophils.

granuloma   A solid mass or nodule of inflammatory tissue containing modified macrophages and lymphocytes. Usually a chronic pathologic process of diseases such as tuberculosis or syphilis.

granzymes   Enzymes secreted by cytotoxic T cells that damage proteins of target cells.

Graves' disease   A malfunction of the thyroid gland in which autoantibodies directed at thyroid cells stimulate an overproduction of thyroid hormone (hyperthyroidism).

growth curve   A graphical representation of the change in population size over time. This graph has four periods known as lag phase, exponential or log phase, stationary phase, and death phase.

growth factor   An organic compound such as a vitamin or amino acid that must be provided in the diet to facilitate growth. An essential nutrient.

guanine (G)   One of the nitrogen bases found in DNA and RNA in the purine form.

guide RNA   A sequence of RNA used to target specific gene sequences during use of the CRISPR system.

Guillain-Barré syndrome   A neurological complication of infection or vaccination.

gumma   A nodular, infectious granuloma characteristic of tertiary syphilis.

gut-associated lymphoid tissue (GALT)   A collection of lymphoid tissue in the gastrointestinal tract that includes the appendix, the lacteals, and Peyer's patches.

H

halophile   A microbe whose growth is either stimulated by salt or requires a high concentration of salt for growth.

hapten   An incomplete or partial antigen. Although it constitutes the determinative group and can bind antigen, hapten cannot stimulate a full immune response without being carried by a larger protein molecule.

hay fever   A form of atopic allergy marked by seasonal acute inflammation of the conjunctiva and mucous membranes of the respiratory passages. Symptoms are irritative itching and rhinitis.

healthcare-associated infection (HAI)   Formerly referred to as *nosocomial infection*, any infection acquired as a direct result of a patient's presence in a hospital or health care setting.

helical   Having a spiral or coiled shape.

helical capsid   Protein covering of viruses made of rod-shaped capsomeres that form a coiled nucleic acid/capsid structure.

helminth   A term that designates all parasitic worms.

hematopoiesis   The process by which the various types of blood cells are formed, such as in the bone marrow.

hematopoietic stem cell   The cell from which blood cells of all types originate.

hemolysin   Any biological agent that is capable of destroying red blood cells and causing the release of hemoglobin. Many bacterial pathogens produce exotoxins that act as hemolysins.

hemolytic disease of the newborn (HDN)   Incompatible Rh factor between mother and fetus causes maternal antibodies to attack the fetus and trigger complement-mediated lysis in the fetus.

hemolytic uremic syndrome (HUS)   Severe hemolytic anemia leading to kidney damage or failure; can accompany *E. coli* O157:H7 intestinal infection.

hemolyze   When red blood cells burst and release hemoglobin pigment.

hepatitis   Inflammation and necrosis of the liver, often the result of viral infection.

hepatitis D   The delta agent; a defective RNA virus that cannot reproduce on its own unless a cell is also infected with the hepatitis B virus.

heredity   Genetic inheritance.

hermaphroditic   Containing the sex organs for both male and female in one individual.

herpes zoster   A recurrent infection caused by latent chickenpox virus. Its manifestation on the skin tends to correspond to dermatomes and to occur in patches that "girdle" the trunk. Also called *shingles*.

heterotroph   An organism that relies upon organic compounds for its carbon and energy needs.

hexose   A 6-carbon sugar such as glucose and fructose.

histamine   A cytokine released when mast cells and basophils release their granules. An important mediator of allergy, its effects include smooth muscle contraction, increased vascular permeability, and increased mucus secretion.

histiocyte   Another term for *macrophage*.

histone   Proteins associated with eukaryotic DNA. These simple proteins serve as winding spools to compact and condense the chromosomes.

holoenzyme   An enzyme complete with its apoenzyme and cofactors.

horizontal gene transfer   Transmission of genetic material from one cell to another through nonreproductive mechanisms, that is, from one organism to another living in the same habitat.

host range   The limitation imposed by the characteristics of the host cell on the type of virus that can successfully invade it.

human immunodeficiency virus (HIV)   A retrovirus that causes acquired immunodeficiency syndrome (AIDS).

human leukocyte antigen (HLA)   One name for the human version of the major histocompatibility complex.

human microbiome   The complete complement of microorganisms that live in or on humans.

Human Microbiome Project (HMP)   A project of the National Institutes of Health to identify microbial inhabitants of the human body and their role in health and disease; uses metagenomic techniques instead of culturing.

hybridization   A process that matches complementary strands of nucleic acid (DNA-DNA, RNA-DNA, RNA-RNA). Used for locating specific sites or types of nucleic acids.

hypertonic   Having a greater osmotic pressure than a reference solution.

hyphae   The tubular threads that make up filamentous fungi (molds). This web of branched and intertwining fibers is called a *mycelium*.

hypogammaglobulinemia   An inborn disease in which the gamma globulin (antibody) fraction of serum is greatly reduced. The condition is associated with a high susceptibility to pyogenic infections.

hyposensitivity diseases   Diseases in which there is a diminished or lack of immune reaction to pathogens due to incomplete immune system development, immune suppression, or destruction of the immune system.

hypotonic   Having a lower osmotic pressure than a reference solution.

I

icosahedron   A regular geometric figure having 20 surfaces that meet to form 12 corners. Some virions have capsids that resemble icosahedral crystals.

immune complex reaction   Type III hypersensitivity of the immune system. It is characterized by the reaction of soluble antigen with antibody and the deposition of the resulting complexes in basement membranes of epithelial tissue.

immune privilege   The restriction or reduction of immune response in certain areas of the body that reduces the potential damage to tissues that a normal inflammatory response could cause.

immune tolerance   Tolerance to self; the inability of one's immune system to react to self proteins or antigens.

immunochromatography   Use of devices with antigens or antibodies embedded in matrices that detect the presence of antigens or antibodies in patient samples.

immunocompetence   The ability of the body to recognize and react with multiple foreign substances.

immunodeficiency   Immune function is incompletely developed, suppressed, or destroyed.

immunogen   Any substance that induces a state of sensitivity or resistance after processing by the immune system of the body.

immunoglobulin (Ig)   The chemical class of proteins to which antibodies belong.

immunology   The study of the system of body defenses that protect against infection.

immunopathology   The study of disease states associated with overreactivity or underreactivity of the immune response.

incidence   In epidemiology, the number of new cases of a disease occurring during a period.

incubate   To isolate a sample culture in a temperature-controlled environment to encourage growth.

incubation period   The period from the initial contact with an infectious agent to the appearance of the first symptoms.

index case   The first case of a disease identified in an outbreak or epidemic.

indicator bacteria   In water analysis, any easily cultured bacteria that may be found in the intestine and can be used as an index of fecal contamination. The category includes coliforms and enterococci. Discovery of these bacteria in a sample means that pathogens may also be present.

induced mutation   Any alteration in DNA that occurs as a consequence of exposure to chemical or physical mutagens.

induction   The process whereby a bacteriophage in the prophage state is activated, begins replication, and enters the lytic cycle.

induration   Area of hardened, reddened tissue associated with the tuberculin test.

infection   The entry, establishment, and multiplication of pathogenic organisms within a host.

infectious disease   The state of damage or toxicity in the body caused by an infectious agent.

inflammasome   A large protein in phagocytic cells that contains pattern recognition receptors (PRRs) to help these cells initiate the inflammatory response.

inflammation   A natural, nonspecific response to tissue injury that protects the host from further damage. It stimulates immune reactivity and blocks the spread of an infectious agent.

inoculation The implantation of microorganisms into or upon culture media.

integument The outer surfaces of the body: skin, hair, nails, sweat glands, and oil glands.

interferon (IFN) Natural human chemical that inhibits viral replication; used therapeutically to combat viral infections and cancer.

interleukins A class of chemicals released from host cells that have potent effects on immunity.

intermediate filament Proteinaceous fibers in eukaryotic cells that help provide support to the cells and their organelles.

intoxication Poisoning that results from the introduction of a toxin into body tissues through ingestion or injection.

intron The segments on split genes of eukaryotes that do not code for polypeptide. They can have regulatory functions. See **exon.**

irradiation The application of radiant energy for diagnosis, therapy, disinfection, or sterilization.

isograft Transplanted tissue from one monozygotic twin to the other; transplants between highly inbred animals that are genetically identical.

isolation The separation of microbial cells by serial dilution or mechanical dispersion on solid media to create discrete colonies.

isotonic Two solutions having the same osmotic pressure such that, when separated by a semipermeable membrane, there is no net movement of solvent in either direction.

J

jaundice The yellowish pigmentation of skin, mucous membranes, sclera, deeper tissues, and excretions due to abnormal deposition of bile pigments. Jaundice is associated with liver infection, as with hepatitis B virus and leptospirosis.

JC virus (JCV) Causes a form of encephalitis (progressive multifocal leukoencephalopathy), especially in AIDS patients.

K

keratin Protein produced by outermost skin cells provides protection from trauma and moisture loss.

killed or inactivated vaccine A whole cell or intact virus preparation in which the microbes are dead or preserved and cannot multiply but are still capable of conferring immunity.

killer T cells A T lymphocyte programmed to directly affix cells and kill them. See **cytotoxicity.**

kingdom In the levels of classification, the second division from more general to more specific. Each domain is divided into kingdoms.

Koch's postulates A procedure to establish the specific cause of disease. In all cases of infection: (1) The agent must be found; (2) inoculations of a pure culture must reproduce the same disease in animals; (3) the agent must again be present in the experimental animal; and (4) a pure culture must again be obtained.

Krebs cycle or **tricarboxylic acid (TCA) cycle** The second pathway of the three pathways that complete the process of primary catabolism. Also called the *citric acid cycle.*

L

L form A stage in the lives of some bacteria in which they have no peptidoglycan.

labile In chemistry, molecules or compounds that are chemically unstable in the presence of environmental changes.

lactose One of the carbohydrates commonly referred to as sugars. Lactose is commonly found in milk.

lactose (*lac*) operon Control system that manages the regulation of lactose metabolism. It is composed of three DNA segments, including a regulator, a control locus, and a structural locus.

lag phase The early phase of population growth during which no signs of growth occur.

lagging strand The newly forming 5′ DNA strand that is discontinuously replicated in segments (Okazaki fragments).

latency The state of being inactive. Example: a latent virus or latent infection.

leading strand The newly forming 3′ DNA strand that is replicated in a continuous fashion without segments.

lesion A wound, injury, or some other pathologic change in tissues.

leukocidin A heat-labile substance formed by some pyogenic cocci that impairs and sometimes lyses leukocytes.

leukocytes White blood cells. The primary infection-fighting blood cells.

leukocytosis An abnormally large number of leukocytes in the blood, which can be indicative of acute infection.

leukopenia A lower-than-normal leukocyte count in the blood that can be indicative of blood infection or disease.

leukotriene An unsaturated fatty acid derivative of arachidonic acid. Leukotriene functions in chemotactic activity, smooth muscle contractility, mucus secretion, and capillary permeability.

ligase An enzyme required to seal the sticky ends of DNA pieces after splicing.

lipase A fat-splitting enzyme. Example: Triacylglycerol lipase separates the fatty acid chains from the glycerol backbone of triglycerides.

lipid A term used to describe a variety of substances that are not soluble in polar solvents, such as water, but will dissolve in nonpolar solvents such as benzene and chloroform. Lipids include triglycerides, phospholipids, steroids, and waxes.

lipopolysaccharide A molecular complex of lipid and carbohydrate found in the bacterial cell wall. The lipopolysaccharide (LPS) of gram-negative bacteria is an endotoxin with generalized pathologic effects such as fever.

lipoteichoic acid Anionic polymers containing glycerol that are anchored in the cytoplasmic membranes of gram-positive bacteria.

liquid [media] Growth-supporting substance in fluid form.

lithoautotroph Bacteria that rely on inorganic minerals to supply their nutritional needs. Sometimes referred to as *chemoautotrophs.*

live attenuated vaccines Vaccines composed of living organisms that have been weakened and cannot cause disease.

logarithmic or log phase Maximum rate of cell division during which growth is geometric in its rate of increase. Also called *exponential growth phase.*

lophotrichous Describing bacteria having a tuft of flagella at one or both poles.

lymphadenitis Inflammation of one or more lymph nodes. Also called *lymphadenopathy.*

lymphatic system A system of vessels and organs that serve as sites for development of immune cells and immune reactions. It includes the spleen, thymus, lymph nodes, and gut-associated lymphoid tissue (GALT).

lyophilization A method for preserving microorganisms (and other substances) by freezing and then drying them directly from the frozen state.

lyse To burst.

lysis The physical rupture or deterioration of a cell.

lysogenic conversion A bacterium acquires a new genetic trait due to the presence of genetic material from an infecting phage.

lysogeny The indefinite persistence of bacteriophage DNA in a host without bringing about the production of virions.

lysosome A cytoplasmic organelle containing lysozyme and other hydrolytic enzymes.

lysozyme An enzyme found in sweat, tears, and saliva that breaks down bacterial peptidoglycan.

M

macromolecules Large, molecular compounds assembled from smaller subunits, most notably biochemicals.

macronutrient A chemical substance required in large quantities (phosphate, for example).

macrophage A white blood cell derived from a monocyte that leaves the circulation and enters tissues. These cells are important in nonspecific phagocytosis and in regulating, stimulating, and cleaning up after immune responses.

major histocompatibility complex (MHC) A set of genes in mammals that produces molecules on surfaces of cells that differentiate among different individuals in the species. See **HLA.**

maltose One of the carbohydrates referred to as sugars. A fermentable sugar formed from starch.

Mantoux test An intradermal screening test for tuberculin hypersensitivity. A red, firm patch of skin at the injection site greater than 10 mm in diameter after 48 hours is a positive result that indicates current or prior exposure to the TB bacillus.

marker Any trait or factor of a cell, virus, or molecule that makes it distinct and recognizable; example: a genetic marker.

maximum temperature The highest temperature at which an organism will grow.

MDR-TB Multidrug-resistant tuberculosis.

medium (plural, *media*) A nutrient used to grow organisms outside of their natural habitats.

medullary sinus Anatomical portion of a lymph node where B cells reside.

membrane attack complex A group of proteins that insert themselves into bacterial or infected cell membranes as a result of complement activation.

memory (immunologic memory) The capacity of the immune system to recognize and act against an antigen upon second and subsequent encounters.

Mendosicutes Taxonomic category of bacteria that have unusual cell walls; archaea.

meninges The tough tri-layer membrane covering the brain and spinal cord. Consists of the dura mater, arachnoid mater, and pia mater.

meningitis An inflammation of the membranes (meninges) that surround and protect the brain. It is often caused by bacteria such as *Neisseria meningitidis* (the meningococcus) and *Haemophilus influenzae.*

mesophile Microorganisms that grow at intermediate temperatures.

messenger RNA (mRNA) A single-stranded transcript that is a copy of the DNA template that corresponds to a gene.

metabolism A general term for the totality of chemical and physical processes occurring in a cell.

metabolomics The study of the complete complement of small chemicals present in a cell at any given time.

metagenomics Analysis of total DNA from an environmental or patient sample, without culturing any microorganisms.

microbial antagonism Relationship in which microorganisms compete for survival in a common environment by taking actions that inhibit or destroy another organism.

microbicides Chemicals that kill microorganisms.

microbiology A specialized area of biology that deals with living things ordinarily too small to be seen without magnification, including bacteria, archaea, fungi, protozoa, and viruses.

microbistatic The quality of inhibiting the growth of microbes.

micronutrient A chemical substance required in small quantities (trace metals, for example).

microorganism A living thing ordinarily too small to be seen without magnification; an organism of microscopic size.

microscopic Invisible to the naked eye.

microtubules Long hollow tubes in eukaryotic cells; maintain the shape of the cell and transport substances from one part of cell to another; involved in separating chromosomes in mitosis.

minimum inhibitory concentration (MIC) The smallest concentration of drug needed to visibly control microbial growth.

minimum temperature The lowest temperature at which an organism will grow.

missense mutation A mutation in which a change in the DNA sequence results in a different amino acid being incorporated into a protein, with varying results.

mitochondrion A double-membrane organelle of eukaryotes that is the main site for aerobic respiration.

monocyte A large mononuclear leukocyte normally found in the lymph nodes, spleen, bone marrow, and loose connective tissue. This type of cell makes up 3% to 7% of circulating leukocytes.

monomer A simple molecule that can be linked by chemical bonds to form larger molecules.

mononuclear phagocyte system (MPS) A collection of monocytes and macrophages scattered throughout the extracellular spaces that function to engulf and degrade foreign molecules.

monosaccharide A simple sugar such as glucose that is a basic building block for more complex carbohydrates.

monotrichous Describing a microorganism that bears a single flagellum.

morbidity rate The number of persons afflicted with an illness under question or with illness in general, expressed as a numerator, with the denominator being some unit of population (as in $x/100,000$).

mortality rate The number of persons who have died as the result of a particular cause or due to all causes, expressed as a numerator, with the denominator being some unit of population (as in $x/100,000$).

motility Self-propulsion.

mucosa-associated lymphoid tissue (MALT) Patches of lymphatic tissue containing B and T cells that underlie the surface of most mucosal surfaces in the body.

mutagen Any agent that induces genetic mutation; examples: certain chemical substances, ultraviolet light, radioactivity.

mutant strain A subspecies of microorganism that has undergone a mutation, causing expression of a trait that differs from other members of that species.

mutualism Organisms living in an obligatory but mutually beneficial relationship.

mycelium The filamentous mass that makes up a mold. Composed of hyphae.

N

nanotubes Extensions of bacterial membranes that are channels for nutrient or energy exchange.

narrow-spectrum Denotes drugs that are selective and limited in their effects. For example, they inhibit either gram-negative or gram-positive bacteria, but not both.

natural immunity Any immunity that arises naturally in an organism via previous experience with the antigen.

negative stain A staining technique that renders the background opaque or colored and leaves the object unstained so that it is outlined as a colorless area.

nematode A common name for helminths called roundworms.

neurons Cells that make up the tissues of the brain and spinal cord that receive and transmit signals to and from the peripheral nervous system and central nervous system.

neurotropic Having an affinity for the nervous system. Most likely to affect the spinal cord.

neutralization The process of combining an acid and a base until they reach a balanced proportion, with a pH value close to 7.

neutrophil A mature granulocyte present in peripheral circulation, exhibiting a multilobular nucleus and numerous cytoplasmic granules that retain a neutral stain. The neutrophil is an active phagocytic cell in bacterial infection.

nitrogen base A ringed compound of which pyrimidines and purines are types.

nitrogenous base A nitrogen-containing molecule found in DNA and RNA that provides the basis for the genetic code. Adenine, guanine, and cytosine are found in both DNA and RNA, while thymine is found exclusively in DNA and uracil is found exclusively in RNA.

noncommunicable   An infectious disease that does not arrive through transmission of an infectious agent from host to host.

noncompetitive inhibition   Form of enzyme inhibition that involves binding of a regulatory molecule to a site other than the active site.

nonself   Molecules recognized by the immune system as containing foreign markers, indicating a need for immune response.

nonsense mutation   A mutation that changes an amino-acid-producing codon into a stop codon, leading to premature termination of a protein.

nosocomial infection   An infection not present upon admission to a hospital but incurred while being treated there.

nucleic acid   A polymeric strand of nucleotides; exists in two forms: ribonucleic acid (RNA) and deoxyribonucleic acid (DNA).

nucleocapsid   In viruses, the close physical combination of the nucleic acid with its protective covering.

nucleoid   The basophilic nuclear region or nuclear body that contains the bacterial chromosome.

nucleolus   A granular mass containing RNA that is contained within the nucleus of a eukaryotic cell.

nucleotide   The basic structural unit of DNA and RNA; each nucleotide consists of a phosphate, a sugar (ribose in RNA, deoxyribose in DNA), and a nitrogenous base such as adenine, guanine, cytosine, thymine (DNA only), or uracil (RNA only).

O

obligate   A description that means that whatever word comes after it is a required state for that organism.

Okazaki fragment   In replication of DNA, a segment formed on the lagging strand in which biosynthesis is conducted in a discontinuous manner dictated by the $5' \rightarrow 3'$ DNA polymerase orientation.

oncogene   A naturally occurring type of gene that when activated can transform a normal cell into a cancer cell.

oncovirus   Mammalian virus capable of causing malignant tumors.

operator   In an operon sequence, the DNA segment where transcription of structural genes is initiated.

operon   A genetic operational unit that regulates metabolism by controlling mRNA production. In sequence, the unit consists of a regulatory gene, inducer or repressor control sites, and structural genes.

opportunistic   In infection, ordinarily nonpathogenic or weakly pathogenic microbes that cause disease primarily in an immunologically compromised host.

opsonization   The process of stimulating phagocytosis by affixing molecules (opsonins such as antibodies and complement) to the surfaces of foreign cells or particles.

optimum temperature   The temperature at which a species shows the most rapid growth rate.

order   In the levels of classification, the division of organisms that follows class. Increasing similarity may be noticed among organisms assigned to the same order.

organelle   A small component of eukaryotic cells that is bounded by a membrane and specialized in function.

osmophile   A microorganism that thrives in a medium having high osmotic pressure.

osmosis   The diffusion of water across a selectively permeable membrane in the direction of lower water concentration.

outer membrane (OM)   An additional membrane possessed by gram-negative bacteria; a lipid bilayer containing specialized proteins and polysaccharides. It lies outside of the cell wall.

oxidation   In chemical reactions, the loss of electrons by one reactant.

oxidation-reduction   Redox reactions in which paired sets of molecules participate in electron transfers.

oxidative phosphorylation   The synthesis of ATP using energy given off during the electron transport phase of respiration.

P

paleontology   The study of the history of life on earth as seen through fossils.

palindrome   A word, verse, number, or sentence that reads the same forward or backward. Palindromes of nitrogen bases in DNA have genetic significance as transposable elements, as regulatory protein targets, and in DNA splicing.

palisades   The characteristic arrangement of *Corynebacterium* cells resembling a row of fence posts and created by snapping.

pandemic   A disease afflicting an increased proportion of the population over a wide geographic area (often worldwide).

paracortical area   A layer of the lymph node internal to the cortex; houses T cells.

parasite   An organism that lives on or within another organism (the host) from which it obtains nutrients and enjoys protection. The parasite produces some degree of harm in the host.

parasitism   A relationship between two organisms in which the host is harmed in some way while the colonizer benefits.

paroxysmal   Events characterized by sharp spasms or convulsions; sudden onset of a symptom such as fever and chills.

passive immunity   Specific resistance that is acquired indirectly by donation of preformed immune substances (antibodies) produced in the body of another individual.

pasteurization   Heat treatment of perishable fluids such as milk, fruit juices, or wine to destroy heat-sensitive vegetative cells, followed by rapid chilling to inhibit growth of survivors and germination of spores. It prevents infection and spoilage.

pathogen   Any agent (usually a virus, bacterium, fungus, protozoan, or helminth) that causes disease.

pathogen-associated molecular patterns (PAMPs)   Molecules on the surfaces of many types of microbes that are not present on host cells that mark the microbes as foreign.

pathogenicity   The capacity of microbes to cause disease.

pattern recognition receptors (PRRs)   Molecules on the surface of host defense cells that recognize pathogen-associated molecular patterns on microbes.

pellicle   A membranous cover; a thin skin, film, or scum on a liquid surface; a thin film of salivary glycoproteins that forms over newly cleaned tooth enamel when exposed to saliva.

penetration (viral)   The step in viral multiplication in which virus enters the host cell.

pentose   A monosaccharide with five carbon atoms per molecule; examples: arabinose, ribose, xylose.

peptide   Molecule composed of short chains of amino acids, such as a dipeptide (two amino acids), a tripeptide (three), and a tetrapeptide (four).

peptide bond   The covalent union between two amino acids that forms between the amine group of one and the carboxyl group of the other. The basic bond of proteins.

peptidoglycan (PG)   A network of polysaccharide chains cross-linked by short peptides that forms the rigid part of bacterial cell walls. Gram-negative bacteria have a smaller amount of this rigid structure than do gram-positive bacteria.

perforin   Proteins released by cytotoxic T cells that produce pores in target cells.

peritrichous   In bacterial morphology, having flagella distributed over the entire cell.

petechiae   Minute hemorrhagic spots in the skin that range from pinpoint- to pinhead-size.

Peyer's patches   Oblong lymphoid aggregates of the gut located chiefly in the wall of the terminal and small intestine. Along with the tonsils and appendix, Peyer's patches make up the gut-associated lymphoid tissue that responds to local invasion by infectious agents.

pH The symbol for the negative logarithm of the H ion concentration; p (power) or $[H^+]_{10}$. A system for rating acidity and alkalinity.

phagocyte A class of white blood cells capable of engulfing other cells and particles.

phagocytosis A type of endocytosis in which the cell membrane actively engulfs large particles or cells into vesicles.

phase variation The process of bacteria turning on or off a group of genes that changes its phenotype in a heritable manner.

phenotype The observable characteristics of an organism produced by the interaction between its genetic potential (genotype) and the environment.

phosphate An acidic salt containing phosphorus and oxygen that is an essential inorganic component of DNA, RNA, and ATP.

phosphorylation Process in which inorganic phosphate is added to a compound.

photoautotroph An organism that utilizes light for its energy and carbon dioxide chiefly for its carbon needs.

photosynthesis A process occurring in plants, algae, and some bacteria that traps the sun's energy and converts it to ATP in the cell. This energy is used to fix CO_2 into organic compounds.

phototrophs Microbes that use photosynthesis to feed.

phylum In the levels of classification, the third level of classification from general to more specific. Each kingdom is divided into numerous phyla. Sometimes referred to as a *division*.

pili (singular, *pilus*) Long, tubular structures made of pilin protein produced by gram-negative bacteria and used for conjugation.

pinocytosis The engulfment, or endocytosis, of liquids by extensions of the cell membrane.

plaque In virus propagation methods, the clear zone of lysed cells in tissue culture or chick embryo membrane that corresponds to the area containing viruses. In dental application, the filamentous mass of microbes that adheres tenaciously to the tooth and predisposes to caries, calculus, or inflammation.

plasma The carrier fluid element of blood.

plasmids Extra chromosomal genetic units characterized by several features. A plasmid is a double-stranded DNA that is smaller than and replicates independently of the cell chromosome; it bears genes that are not essential for cell growth; it can bear genes that code for adaptive traits; and it is transmissible to other bacteria.

pleomorphism Normal variability of cell shapes in a single species.

pneumococcus Common name for the bacterium *Streptococcus pneumoniae*.

pneumonia An inflammation of the lung leading to accumulation of fluid and respiratory compromise.

pneumonic plague The acute, frequently fatal form of pneumonia caused by *Yersinia pestis*.

point mutation A change that involves the loss, substitution, or addition of one or a few nucleotides.

point-source epidemic An outbreak of disease in which all affected individuals were exposed to a single source of the pathogen at a single point in time.

polymer A macromolecule made up of a chain of repeating units; examples: starch, protein, DNA.

polymerase chain reaction (PCR) A technique that amplifies segments of DNA for testing. Using denaturation, primers, and heat-resistant DNA polymerase, the number can be increased several-million-fold.

polymicrobial Involving multiple distinct microorganisms.

polypeptide A relatively large chain of amino acids linked by peptide bonds.

polyribosomal complex An assembly line for mass production of proteins composed of a chain of ribosomes involved in mRNA transcription.

polysaccharide A carbohydrate that can be hydrolyzed into a number of monosaccharides; examples: cellulose, starch, glycogen.

portal of entry Route of entry for an infectious agent; typically a cutaneous or membranous route.

portal of exit Route through which a pathogen departs from the host organism.

positive stain A method for coloring microbial specimens that involves a chemical that sticks to the specimen to give it color.

posttranslational Referring to modifications to the protein structure that occur after protein synthesis is complete, including removal of formyl methionine, further folding of the protein, addition of functional groups, or addition of the protein to a quaternary structure.

prebiotics Nutrients used to stimulate the growth of favorable biota in the intestine.

prevalence The total number of cases of a disease in a certain area and time period.

primary lymphatic organ Sites where B and T lymphocytes are generated and become mature. In the human, the red bone marrow and the thymus are primary lymphatic organs.

primary response The first response of the immune system when exposed to an antigen.

primary structure Initial protein organization described by type, number, and order of amino acids in the chain. The primary structure varies extensively from protein to protein.

primers Synthetic oligonucleotides of known sequence that serve as landmarks to indicate where DNA amplification will begin.

prion A concocted word to denote "proteinaceous infectious agent"; a cytopathic protein associated with the slow-virus spongiform encephalopathies of humans and animals.

probes Small fragments of single-stranded DNA (RNA) that are known to be complementary to the specific sequence of DNA being studied.

probiotics Preparations of live microbes used as a preventive or therapeutic measure to displace or compete with potential pathogens.

prodromal stage A short period of mild symptoms occurring at the end of the period of incubation. It indicates the onset of disease.

progressive multifocal leukoencephalopathy (PML) An uncommon, fatal complication of infection with JC virus (polyoma virus).

prokaryotic cells Small cells lacking special structures such as a nucleus and organelles. All **prokaryotes** are microorganisms.

promoter Part of an operon sequence. The DNA segment that is recognized by RNA polymerase as the starting site for transcription.

promoter region The site composed of a short signaling DNA sequence that RNA polymerase recognizes and binds to commence transcription.

propagated epidemic An outbreak of disease in which the causative agent is passed from affected persons to new persons over the course of time.

prophage A lysogenized bacteriophage; a phage that is latently incorporated into the host chromosome instead of undergoing viral replication and lysis.

prostaglandin A hormonelike substance that regulates many body functions. Prostaglandin comes from a family of organic acids containing 5-carbon rings that are essential to the human diet.

protease Enzymes that act on proteins, breaking them down into component parts.

protein Predominant organic molecule in cells, formed by long chains of amino acids.

proteomics The study of an organism's complement of proteins (its *proteome*) and functions mediated by the proteins.

protozoa A group of single-celled, eukaryotic organisms.

provirus The genome of a virus when it is integrated into a host cell's DNA.

pseudohypha A chain of easily separated, spherical to sausage-shaped yeast cells partitioned by constrictions rather than by septa.

psychrophile A microorganism that thrives at low temperature (0°C–20°C), with a temperature optimum of 0°C–15°C.

pure culture A container growing a single species of microbe whose identity is known.

purine A nitrogen base that is an important encoding component of DNA and RNA. The two most common purines are adenine and guanine.

purpura Purple-colored spots or blotches on the skin.

pus The viscous, opaque, usually yellowish matter formed by an inflammatory infection. It consists of serum exudate, tissue debris, leukocytes, and microorganisms.

pyogenic   Pertains to pus formers, especially the pyogenic cocci: pneumococci, streptococci, staphylococci, and neisseriae.

pyrimidine   Nitrogen bases that help form the genetic code on DNA and RNA. Uracil, thymine, and cytosine are the most important pyrimidines.

pyrogen   A substance that causes a rise in body temperature. It can come from pyrogenic microorganisms or from polymorphonuclear leukocytes (endogenous pyrogens).

Q

quaternary structure   Most complex protein structure, characterized by the formation of large, multiunit proteins by more than one of the polypeptides. This structure is typical of antibodies and some enzymes that act in cell synthesis.

quinine   A substance derived from cinchona trees that was used as an antimalarial treatment; has been replaced by synthetic derivatives.

quorum sensing   The ability of bacteria to regulate their gene expression in response to sensing bacterial density.

R

rabies   The only rhabdovirus that infects humans. Zoonotic disease characterized by fatal meningoencephalitis.

radiation   Electromagnetic waves or rays, such as those of light given off from an energy source.

rales   Sounds in the lung, ranging from clicking to rattling; indicate respiratory illness.

real image   An image formed at the focal plane of a convex lens. In the compound light microscope, it is the image created by the objective lens.

recombinant   An organism that contains genes that originated in another organism, whether through deliberate laboratory manipulation or natural processes.

recombinant DNA technology   A technology, also known as *genetic engineering*, that deliberately modifies the genetic structure of an organism to create novel products, microbes, animals, plants, and viruses.

recombination   A type of genetic transfer in which DNA from one organism is donated to another.

reducing agent   An atom or a compound that can donate electrons in a chemical reaction.

reducing medium   A growth medium that absorbs oxygen and allows anaerobic bacteria to grow.

redundancy   The property of the genetic code that allows an amino acid to be specified by several different codons.

refractive index   The measurement of the degree of light that is bent, or refracted, as it passes between two substances such as air, water, or glass.

regulated enzymes   Enzymes whose extent of transcription or translation is influenced by changes in the environment.

regulator   DNA segment that codes for a protein capable of repressing an operon.

release   The final step in the multiplication cycle of viruses in which the assembled virus particle exits the host cell and moves on to infect another cell.

reportable disease   Those diseases that must be reported to health authorities by law.

repressor   The protein product of a repressor gene that combines with the operator and arrests the transcription and translation of structural genes.

reproduction rate   The average number of susceptible persons one infected host will spread an infection to; describes communicability of an infectious disease.

reservoir   In disease communication, the natural host or habitat of a pathogen.

resistance (R) factor   Plasmids, typically shared among bacteria by conjugation, that provide resistance to the effects of antibiotics.

resolving power   The capacity of a microscope lens system to accurately distinguish between two separate entities that lie close to each other. Also called *resolution*.

respiratory syncytial virus (RSV)   An RNA virus that infects the respiratory tract. RSV is the most prevalent cause of respiratory infection in newborns.

restriction endonuclease   An enzyme present naturally in cells that cleaves specific locations on DNA. It is an important means of inactivating viral genomes, and it is also used to splice genes in genetic engineering.

restriction factors   Host cell molecules that inhibit viral replication.

restriction fragment length polymorphisms (RFLPs)   Variations in the lengths of DNA fragments produced when a specific restriction endonuclease acts on different DNA sequences.

restriction fragments   Short pieces of DNA produced when DNA is exposed to restriction endonucleases.

reverse transcriptase (RT)   The enzyme possessed by retroviruses that carries out the reversion of RNA to DNA–a form of reverse transcription.

Rh factor   An isoantigen that can trigger hemolytic disease in newborns due to incompatibility between maternal and infant blood factors.

ribose   A 5-carbon monosaccharide found in RNA.

ribosomal RNA (rRNA)   A single-stranded transcript that is a copy of part of the DNA template.

ribosome   A bilobed macromolecular complex of ribonucleoprotein that coordinates the codons of mRNA with tRNA anticodons and, in so doing, constitutes the peptide assembly site.

RNA polymerase   Enzyme process that translates the code of DNA to RNA.

rolling circle   An intermediate stage in viral replication of circular DNA into linear DNA.

rough endoplasmic reticulum (RER)   Microscopic series of tunnels that originates in the outer membrane of the nuclear envelope and is used in transport and storage. Large numbers of ribosomes, partly attached to the membrane, give the rough appearance.

rubeola (red measles)   Acute disease caused by infection with *Morbillivirus*.

S

S layer   Single layer of thousands of copies of a single type of protein linked together on the surface of a bacterial cell that is produced when the cell is in a hostile environment.

saccharide   Scientific term for sugar. Refers to a simple carbohydrate with a sweet taste.

salpingitis   Inflammation of the fallopian tubes.

saprobe   A microbe that decomposes organic remains from dead organisms. Also known as a *saprophyte* or *saprotroph*.

sarcina   A cubical packet of 8, 16, or more cells; the cellular arrangement of the genus *Sarcina* in the family Micrococcaceae.

schistosomiasis   Infection by blood fluke, often as a result of contact with contaminated water in rivers and streams. Symptoms appear in liver, spleen, or urinary system depending on species of *Schistosoma*. Infection may be chronic.

sebaceous glands   The sebum- (oily, fatty) secreting glands of the skin.

sebum   Low pH, oil-based secretion of the sebaceous glands.

secondary lymphatic organs   Tissue locations where T and B lymphocytes perform their actions. Examples are lymph nodes, MALT, SALT.

secondary response   The rapid rise in antibody titer following a repeat exposure to an antigen that has been recognized from a previous exposure. This response is brought about by memory cells produced as a result of the primary exposure.

secondary structure   Protein structure that occurs when the functional groups on the outer surface of the molecule interact by forming hydrogen bonds. These bonds cause the amino acid chain to either twist, forming a helix, or pleat into an accordion pattern called a *β-pleated sheet*.

selective media   Nutrient media designed to favor the growth of certain microbes and to inhibit undesirable competitors.

selectively toxic   Property of an antimicrobial agent to be highly toxic against its target microbe while being far less toxic to other cells, particularly those of the host organism.

self   Natural markers of the body that are recognized by the immune system.

semiconservative replication   In DNA replication, the synthesis of paired daughter strands, each retaining a parent strand template.

semisolid [media]   Nutrient media with a firmness midway between that of a broth (a liquid medium) and an ordinary solid medium; motility media.

sepsis   The state of putrefaction; the presence of pathogenic organisms or their toxins in tissue or blood.

septic shock   Blood infection resulting in a pathological state of low blood pressure accompanied by a reduced amount of blood circulating to vital organs. Endotoxins of all gram-negative bacteria can cause shock, but most clinical cases are due to gram-negative enteric rods.

septicemia   Systemic infection associated with microorganisms multiplying in circulating blood.

septicemic plague   A form of infection with *Yersinia pestis* occurring mainly in the bloodstream and leading to high mortality rates.

septum (plural, *septa*)   A partition or cellular cross wall, as in certain fungal hyphae.

sequela   A morbid complication that follows a disease.

sequence maps   Finding patterns found in the patterns among DNA sequences.

serology   The branch of immunology that deals with *in vitro* diagnostic testing of serum.

serotonin   A vasoconstrictor that inhibits gastric secretion and stimulates smooth muscle.

serotyping   The subdivision of a species or subspecies into an immunologic type, based upon antigenic characteristics.

serous   Referring to serum, the clear fluid that escapes cells during the inflammatory response.

serum sickness   A type of immune complex disease in which immune complexes enter circulation, are carried throughout the body, and are deposited in the blood vessels of the kidney, heart, skin, and joints. The condition may become chronic.

severe combined immunodeficiency (SCID)   A collection of syndromes occurring in newborns caused by a genetic defect that knocks out both B- and T-cell types of immunity. There are several versions of this disease.

Shiga toxin   Heat-labile exotoxin released by some *Shigella* species and by *E. coli* O157:H7; responsible for worst symptoms of these infections.

shingles   Lesions produced by reactivated human herpesvirus 3 (chickenpox) infection; also known as herpes zoster.

sign   Any abnormality uncovered upon physical diagnosis that indicates the presence of disease. A *sign* is an objective assessment of disease, as opposed to a *symptom*, which is the subjective assessment perceived by the patient.

silent mutation   A mutation that, because of the degeneracy of the genetic code, results in a nucleotide change in both the DNA and mRNA but not the resultant amino acid and, thus, not the protein.

simple stain   Type of positive staining technique that uses a single dye to add color to cells so that they are easier to see. This technique tends to color all cells the same color.

single nucleotide polymorphism (SNP)   A mutation of a single nucleotide in DNA causing differences among individuals carrying that mutation from those not carrying it.

skin-associated lymphoid tissue (SALT)   Patches of lymphatic tissue containing B and T cells that underlie the surface of many skin surfaces in the body.

slime layer   A diffuse, unorganized layer of polysaccharides and/or proteins on the outside of some bacteria.

smooth endoplasmic reticulum (SER)   A microscopic series of tunnels lacking ribosomes that functions in the nutrient processing function of a cell.

species   In the levels of classification, the most specific level of organization.

specificity   In immunity, the concept that some parts of the immune system only react with antigens that originally activated them.

spirillum   A type of bacterial cell with a rigid spiral shape and external flagella.

spirochete   A coiled, spiral-shaped bacterium that has endoflagella and flexes as it moves.

spliceosome   A molecule composed of RNA and protein that removes introns from eukaryotic mRNA before it is translated by forming a loop in the intron, cutting it from the mRNA, and joining exons together.

spontaneous generation   Early belief that living things arose from vital forces present in nonliving, or decomposing, matter.

spontaneous mutation   A mutation in DNA caused by random mistakes in replication and not known to be influenced by any mutagenic agent. These mutations give rise to an organism's natural, or background, rate of mutation.

sporadic   Description of a disease that exhibits new cases at irregular intervals in unpredictable geographic locales.

sporangiospore   A form of asexual spore in fungi; enclosed in a sac.

sporangium   A fungal cell in which asexual spores are formed by multiple cell cleavage.

spore   A differentiated, specialized cell form that can be used for dissemination, for survival in times of adverse conditions, and/or for reproduction. Spores are usually unicellular and may develop into gametes or vegetative organisms.

sporozoite   One of many minute elongated bodies generated by multiple division of the oocyst. It is the infectious form of the malarial parasite that is harbored in the salivary gland of the mosquito and inoculated into the victim during feeding.

sporulation   The process of spore formation.

stationary growth phase   Survival mode in which cells either stop growing or grow very slowly.

stem cells   Pluripotent, undifferentiated cells.

sterile   Completely free of all life forms, including spores and viruses.

streptolysin   A hemolysin produced by streptococci.

structural gene   A gene that codes for the amino acid sequence (peptide structure) of a protein.

subacute   Indicates an intermediate status between acute and chronic disease.

subacute sclerosing panencephalitis (SSPE)   A complication of measles infection in which progressive neurological degeneration of the cerebral cortex invariably leads to coma and death.

subclinical   A period of inapparent manifestations that occurs before symptoms and signs of disease appear.

substrate   The specific molecule upon which an enzyme acts.

subunit vaccine   A vaccine preparation that contains only antigenic fragments such as surface receptors from the microbe. Usually in reference to virus vaccines.

sucrose   One of the carbohydrates commonly referred to as sugars. Common table or cane sugar.

superantigens   Bacterial toxins that are potent stimuli for T cells and can be a factor in diseases such as toxic shock.

superficial mycosis   A fungal infection located in hair, nails, and the epidermis of the skin.

superinfection   An infection occurring during antimicrobial therapy that is caused by an overgrowth of drug-resistant microorganisms.

symbiosis   An intimate association between individuals from two species; used as a synonym for *mutualism*.

symptom   The subjective evidence of infection and disease as perceived by the patient.

syncytium (plural, *syncytia*)   A multinucleated protoplasmic mass formed by consolidation of individual cells.

syndrome   The collection of signs and symptoms that, taken together, paint a portrait of the disease.

synergism   The coordinated or correlated action by two or more drugs or microbes that results in a heightened response or greater activity.

synthesis (viral)   The step in viral multiplication in which viral genetic material and proteins are made through replication and transcription/translation.

systemic   Occurring throughout the body; said of infections that invade many compartments and organs via the circulation.

T

taxa   Taxonomic categories.

taxonomy   The formal system for organizing, classifying, and naming living things.

teichoic acid   Anionic polymers containing glycerol that appear in the walls of gram-positive bacteria.

temperate phage   A bacteriophage that enters into a less virulent state by becoming incorporated into the host genome as a prophage instead of in the vegetative or lytic form that eventually destroys the cell.

template strand   The strand in a double-stranded DNA molecule that is used as a model to synthesize a complementary strand of DNA or RNA during replication or transcription.

Tenericutes   Taxonomic category of bacteria that lack cell walls.

teratogenic   Causing abnormal fetal development.

tertiary structure   Protein structure that results from additional bonds forming between functional groups in a secondary structure, creating a three-dimensional mass.

tetanospasmin   The neurotoxin of *Clostridium tetani*, the agent of tetanus. Its chief action is directed upon the inhibitory synapses of the anterior horn motor neurons.

tetrads   Groups of four.

theory of evolution   The evidence cited to explain how evolution occurs.

therapeutic index   The ratio of the toxic dose to the effective therapeutic dose that is used to assess the safety and reliability of the drug.

thermal death point (TDP)   The lowest temperature that achieves sterilization in a given quantity of broth culture upon a 10-minute exposure. Examples: 55°C for *Escherichia coli*, 60°C for *Mycobacterium tuberculosis*, and 120°C for endospores.

thermal death time (TDT)   The least time required to kill all cells of a culture at a specified temperature.

thermophile   A microorganism that thrives at a temperature of 50°C or higher.

thymine (T)   One of the nitrogen bases found in DNA but not in RNA. Thymine is in a pyrimidine form.

thymus   Butterfly-shaped organ near the tip of the sternum that is the site of T-cell maturation.

tincture   A medicinal substance dissolved in an alcoholic solvent.

tinea   Ringworm; a fungal infection of the hair, skin, or nails.

titer   In immunochemistry, a measure of antibody level in a patient, determined by agglutination methods.

tonsils   A ring of lymphoid tissue in the pharynx that acts as a repository for lymphocytes.

topoisomerases   Enzymes that can add or remove DNA twists and thus regulate the degree of supercoiling.

toxin   A specific chemical product of microbes, plants, and some animals that is poisonous to other organisms.

toxoid   A toxin that has been rendered nontoxic but is still capable of eliciting the formation of protective antitoxin antibodies; used in vaccines.

trace elements   Micronutrients (zinc, nickel, and manganese) that occur in small amounts and are involved in enzyme function and maintenance of protein structure.

transcription   Messenger RNA (mRNA) synthesis; the process by which a strand of RNA is produced against a DNA template.

transduction   The transfer of genetic material from one bacterium to another by means of a bacteriophage vector.

transfection   The introduction of DNA into eukaryotic cells from the environment by exposing cells to chemicals or electrical currents; similar to **transformation.**

transfer RNA (tRNA)   A form of RNA that serves to connect amino acids with another type of RNA (mRNA) during the process of translation.

transformation   In microbial genetics, the transfer of genetic material contained in "naked" DNA fragments from a donor cell to a competent recipient cell.

translation   Protein synthesis; the process of decoding the messenger RNA code into a polypeptide.

transmissible spongiform encephalopathies (TSEs)   Diseases caused by proteinaceous infectious particles (also known as *prions*).

transport medium   Microbiological medium that is used to transport specimens.

transposon   A DNA segment with an insertion sequence at each end, enabling it to migrate to another plasmid, to the bacterial chromosome, or to a bacteriophage.

trematode   A category of helminth; also known as *flatworm* or *fluke*.

triglyceride   A type of lipid composed of a glycerol molecule bound to three fatty acids.

trophozoite   A vegetative protozoan (feeding form) as opposed to a resting (cyst) form.

true pathogen   A microbe capable of causing infection and disease in healthy persons with normal immune defenses.

tubercle   In tuberculosis, the granulomatous well-defined lung lesion that can serve as a focus for latent infection.

tuberculin reaction   A diagnostic test in which PPD, or purified protein derivative (of *M. tuberculosis*), is injected superficially under the skin and the area of reaction measured; also called the *Mantoux test*.

turbid   Cloudy appearance of nutrient solution in a test tube due to growth of microbe population.

U

ubiquitous   Present everywhere at the same time.

ultraviolet (UV) radiation   Radiating energy with wavelengths between 100 and 400 nm.

uncoating   The process of removal of the viral coat and release of the viral genome by its newly invaded host cell.

uracil (U)   One of the nitrogen bases in RNA but not in DNA. Uracil is in a pyrimidine form.

urinary tract infection (UTI)   Invasion and infection of the urethra and bladder by bacterial residents, most often *E. coli*.

V

vaccine   Originally used in reference to inoculation with the cowpox or vaccinia virus to protect against smallpox. In general, the term now pertains to injection of whole microbes (killed or attenuated), toxoids, or parts of microbes as a prevention or cure for disease.

vacuoles   In the cell, membrane-bounded sacs containing fluids or solid particles to be digested, excreted, or stored.

variable (V) region   The antigen-binding fragment of an immunoglobulin molecule, consisting of a combination of heavy and light chains whose molecular conformation is specific for the antigen.

varicella   Informal name for virus responsible for chickenpox as well as shingles; also known as *human herpesvirus 3* (HHV-3).

variolation   A hazardous, outmoded process of deliberately introducing smallpox material scraped from a victim into the nonimmune subject in the hope of inducing resistance.

vasoactive   Referring to chemical mediators involved in the immune response that act on endothelial cells or the smooth muscle of blood vessels, causing them to either restrict or relax.

vector   An animal that transmits infectious agents from one host to another, usually a biting or piercing arthropod like the tick, mosquito, or fly. Infectious agents can be conveyed mechanically by simple contact or biologically, whereby the parasite develops in the vector. A genetic element such as a plasmid or a bacteriophage used to introduce genetic material into a cloning host during recombinant DNA experiments.

vegetative In describing microbial developmental stages, a metabolically active feeding and dividing form, as opposed to a dormant, seemingly inert, nondividing form. Examples: a bacterial cell versus its spore; a protozoal trophozoite versus its cyst.

viable nonculturable (VNC) A description of a state in which bacteria are alive but are not metabolizing at an appreciable rate and will not grow when inoculated onto laboratory medium.

vibrio A curved, rod-shaped bacterial cell.

viremia The presence of viruses in the bloodstream.

virion An elementary virus particle in its complete morphological and thus infectious form. A virion consists of the nucleic acid core surrounded by a capsid, which can be enclosed in an envelope.

viroid An infectious agent that, unlike a virion, lacks a capsid and consists of a closed circular RNA molecule. Although known viroids are all plant pathogens, it is conceivable that animal versions exist.

virome The complete array of viruses in a given ecosystem, such as the human body.

virtual image In optics, an image formed by diverging light rays; in the compound light microscope, the second, magnified visual impression formed by the ocular from the real image formed by the objective.

virulence In infection, the relative capacity of a pathogen to invade and harm host cells.

virulence factors A microbe's structures or capabilities that allow it to establish itself in a host and cause damage.

virus Microscopic, acellular agent composed of nucleic acid surrounded by a protein coat.

vitamins A component of coenzymes critical to nutrition and the metabolic function of coenzyme complexes.

W

Western blot test A procedure for separating and identifying antigen or antibody mixtures by two-dimensional electrophoresis in polyacrylamide gel, followed by immune labeling.

whole blood A liquid connective tissue consisting of blood cells suspended in plasma.

wild type The natural, nonmutated form of a genetic trait.

wobble A characteristic of amino acid codons in which the third base of a codon can be altered without changing the code for the amino acid.

X

XDR-TB Extensively drug-resistant tuberculosis (worse than multidrug-resistant tuberculosis).

xenograft The transfer of a tissue or an organ from an animal of one species to a recipient of another species.

Z

zoonosis An infectious disease indigenous to animals that humans can acquire through direct or indirect contact with infected animals.

Index

Note: Page numbers followed by t and f refer to tables and figures, respectively. Page numbers in **boldface** refer to boxed material, and page numbers in *italics* refer to definitions or introductory discussions.